Digital Logic Circuits

for BE | BTech—EEE Branch of All Universities and Autonomous Colleges in Tamil Nadu

with VHDL Program

Other CBS Titles by *A Nagoor Kani*

- A Simplified Text in Electrical Machine Design, 3/e
- Advanced Control Theory, 3/e
- Circuit Analysis, 2/e (*Anna University/ECE*)
- Control Systems, 5/e
- Control Systems (*Anna University/EEE*)
- Control Systems Engineering (*Anna University/ECE*)
- Control Systems Engineering, 2/e
- Circuit Theory, 2/e (*Anna University/EEE*)
- Design of Electrical Apparatus (*Anna University/EEE*)
- Digital Signal Processing (*Anna University/EEE | E&I | ICE*)
- Digital Systems Design *(ECE)*
- Microprocessors and Microcontrollers (*Anna University/ECE*)
- Microprocessors and Microcontrollers (*Anna University/EEE*)
- Microprocessors and Microcontrollers, 3/e
- Power System Analysis
- Signals and Systems–Simplified, 2/e (*for Anna University*)
- 8085 Microprocessor and its Applications, 4/e
- 8086 Microprocessor and its Applications, 3/e
- Discrete Time Signal Processing (*Anna University/ECE*)
- Electric Circuits, 2/e
- Signals and Systems, 2/e

Digital Logic Circuits

for BE | BTech—EEE Branch of All Universities and Autonomous Colleges in Tamil Nadu

with VHDL Program

A Nagoor Kani
Founder
RBA Educational Group
Chennai

CBS Publishers & Distributors Pvt Ltd
New Delhi • Bengaluru • Chennai • Kochi • Kolkata • Lucknow • Mumbai
Hyderabad • Jharkhand • Nagpur • Patna • Pune • Uttarakhand

Disclaimer

Science and technology are constantly changing fields. New research and experience broaden the scope of information and knowledge. The author has tried his best in giving information available to him while preparing the material for this book. Although all efforts have been made to ensure optimum accuracy of the material, yet it is quite possible some errors might have been left uncorrected. The publisher, the printer and the author will not be held responsible for any inadvertent errors or inaccuracies.

ISBN: 978-93-5466-998-9

First Edition: 2024

Published by Satish Kumar Jain and produced by Varun Jain for
CBS Publishers & Distributors Pvt Ltd
4819/XI Prahlad Street, 24 Ansari Road, Daryaganj, New Delhi 110 002, India
Ph: 011–23289259, 23266861 Website: www.cbspd.com e-mail: delhi@cbspd.com

Corporate Office: 204 FIE, Industrial Area, Patparganj, Delhi 110 092, India
Ph: 011–49344934 Fax: 011–49344935 e-mail: publishing@cbspd.com; publicity@cbspd.com

Branches

- **Bengaluru:** Seema House 2975, 17th Cross, K.R. Road, Banasankari 2nd Stage, Bengaluru 560 070, Karnataka, India
  Ph: +91-80-26771678/79 Fax: +91-80-26771680 e-mail: bangalore@cbspd.com
- **Chennai:** 7, Subbaraya Street, Shenoy Nagar, Chennai 600 030, Tamil Nadu, India
  Ph: +91-44-26680620, 26681266 Fax: +91-44-42032115 e-mail: chennai@cbspd.com
- **Kochi:** 42/1325, 1326, Power House Road, Opposite KSEB, Power House, Ernakulum 682018, Kochi, Kerala, India
  Ph: +91-484-4059061–67 Fax: +91-484-4059065 e-mail: kochi@cbspd.com
- **Kolkata:** 147, Hind Ceramics Compound, 1st Floor, Nilgunj Road, Belghoria, Kolkata 700056, West Bengal, India
  Ph: +91-33-25330055/56 e-mail: kolkata@cbspd.com
- **Lucknow:** Basement, Khushuma Complex, 7 Meerabai Marg (behind Jawahar Bhawan), Lucknow 226001, UP, India
  Ph: +91-522-4000032 e-mail: tiwari.lucknow@cbspd.com
- **Mumbai:** PWD Shed, Gala No. 25/26, Ramchandra Bhatt Marg, Next JJ Hospital Gate No. 2, Opp. Union Bank of India, Noorbaug, Mumbai 400009, Maharashtra, India
  Ph: +91-22-66661880/89 e-mail: mumbai@cbspd.com

Representatives

- **Hyderabad** 0-9885175004
- **Jharkhand** 0-9811541605
- **Nagpur** 0-8692091830
- **Patna** 0-9334159340
- **Pune** 0-9664372571
- **Uttarakhand** 0-9716462459

Printed at: Mudrak, Noida, UP, India.

to

Mr Satish K Jain

CMD

CBS Publishers & Distributors

PREFACE

The main objective of this book is to explore the basic concepts of digital logic circuits in a simple and easy-to-understand manner. This text on digital logic circuits has been crafted for latest syllabus of BE|BTech—EEE branch students requirements of various universities and autonomous colleges in Tamil Nadu. Considering the difficulty in logical analysis needed in this subject, more emphasis has been given on the problem-solving methodology. Considerable effort has been made to elucidate the logical explanations in a step-by-step manner. Exercise problems with various difficulty levels are given in the text to help students get an intuitive grasp on the subject.

This book with its lucid writing style and germane pedagogical features will prove to be a master text for engineering students.

Salient features

- Proof of important concepts and theorems are clearly highlighted by shaded boxes
- Wherever required, problems are solved in multiple methods
- Additional explanations for solutions and proofs are provided in separate boxes
- Different types of fonts are used for text, proof, solved problems
- Keywords are highlighted by bold, italic fonts
- Easy, concise and accurate study material
- Extremely precise edition where concepts are reinforced by pedagogy
- Ample figures and examples to enhance students' understanding
- Practice through MCQs
- VHDL programs which are useful for laboratory practice and other projects

This book is organized with 5 Chapters.

Chapter 1 starts with fundamental concepts of digital electronics such as number systems, binary codes, error detection and correction codes, Boolean operators, theorems and functions and basic concepts of logic gates. Boolean function minimization using theorems, K-maps and Quine-McCluskey methods are also discussed in this chapter.

The characteristics of the devices of various logic families are also included in **Chapter 1**. The development of logic gates in various logic families are explained in detail in this chapter.

Chapter 2 is devoted to combinational logic circuit analysis and design. The analysis and design of adders, subtractors, decoders, encoders, multiplexers and de-multiplexers are presented with clear examples. Parity generator and checker and the combinational circuit design using standard ICs are also included in this chapter.

Chapter 3 starts with detailed working of various types of flip-flops used to build synchronous circuits. The characteristics of flip-flops and the conversion of flip-flops from one type to other types are clearly explained in this chapter. The analysis and design of Mealy and Moore models of synchronous logic circuits are presented with step-by-step explanations. The basic concepts of state and state variables are explained clearly. The analysis and design of synchronous circuits like counters, sequence detectors and registers are explained with appropriate examples. The sequential circuit design using standard ICs are also included in this chapter.

The analysis and design of asynchronous logic circuits is discussed in **Chapter 4**. The races and hazards are discussed in details and various methods for eliminating races and hazards are discussed with examples. State reduction using implication table and merger diagram and race-free state assignment are clearly explained with appropriate examples. Design of fundamental mode asynchronous sequential circuits using latches are presented with step-by-step explanations. The system design using programmable ICs like PROM, PLA and PAL are explained with appropriate examples. A short note on advanced programmable ICs CPLD and FPGA are also included in this chapter.

Chapter 5 is dedicated to VHDL programming. A brief introduction to VHDL language is presented for better understanding of the readers. VHDL programs for combinational and sequential circuit design are presented in this chapter and test bench is also included for some of the VHDL programs. A brief introduction about RTL design methodology is included with programming examples.

All VHDL statement/program examples in this book are simulated and verified in license free Xilinx VHDL tool Vivado 2019.2, web edition, which can be directly downloaded from the following link.

https://www.xilinx.com/support/download/index.html/content/xilinx/en/downloadNav/vivado-design-tools/archive.html

File name: *Vivado HLx 2019.2: All OS installer Single-File Download (TAR/GZIP-26.55GB).*

I have taken special care in presenting the concepts of digital logic circuits in simple manner supported by carefully chosen solved problems and hope that the teaching and student community will welcome the book. The readers can feel free to convey their criticism and suggestions to nagoorkani65@yahoo.com, for further improvement of the book.

A Nagoor Kani

ACKNOWLEDGEMENTS

I express my heartfelt thanks to my wife Mrs C Gnanaparanjothi Nagoor Kani and my sons N Bharath Raj Alias Chandrakani Allaudeen and N Vikram Raj for the support, encouragement and cooperation they have extended to me throughout my career. I thank Ms T A Benazir, Manager, RBA Group, and all my office-staff for their cooperation in carrying out my day-to-day activities.

It is my pleasure to acknowledge the contributions of our technical editors, Ms S Saranya and Ms P Kanimozhi and Ms K G Sathyapriya for editing, proof-reading and type-setting of the manuscript and preparing the layout of the book.

My sincere thanks to all reviewers Dr R Swarnalatha, Associate Professor, BITS, Pilani Dubai Campus, Mrs K Sivagami, Research scholar, IIITDM, Kancheepuram, Mrs S Chitra Lakshmi and Mrs S G Vijayakumari, Assistant professor, SRMIST, for their valuable suggestions and comments which helped me to explore the subject to a greater depth.

I am also grateful to Mr Satish K Jain, CMD, CBS Publishers & Distributors, for his keen interest in publishing this work in CBS banner. My sincere thanks to all team members of CBS Publishers & Distributors, for their concern and care in publishing this work.

Finally, a special note of appreciation is due to my sisters, brothers, relatives, friends, students and the entire teaching community for their overwhelming support and encouragement to my writing.

A Nagoor Kani

CONTENTS

CHAPTER 2: COMBINATIONAL CIRCUITS 2.1–2.134

CHAPTER 4: Asynchronous Sequential Circuits and Programmability Logic Devices 4.1–4.182

CHAPTER 5: VHDL 5.1–5.130

CHAPTER 1

NUMBER SYSTEMS AND DIGITAL LOGIC FAMILIES

1.1 Number System

There are four types of popular number systems.

1. Decimal
2. Binary
3. Octal
4. Hexadecimal

Every number system has a radix, r. The radix, r is also called ***base***. The ***radix*** or ***base*** is the number of unique symbols used to represent the number system. Hence the number systems are also called ***radix number systems*** or ***based number systems***.

Table 1.1: Radix/Base and Symbols of Number Systems

Number Systems	Radix/Base	Symbols
Decimal	10	0, 1, 2, 3, 4, 5, 6, 7, 8, 9
Binary	2	0, 1
Octal	8	0, 1, 2, 3, 4, 5, 6, 7
Hexadecimal	16	0, 1, 2, 3, 4, 5, 6, 7, 8, 9, A, B, C, D, E, F

Note: $A_{16} = 10_{10}$; $B_{16} = 11_{10}$; $C_{16} = 12_{10}$; $D_{16} = 13_{10}$; $E_{16} = 14_{10}$; $F_{16} = 15_{10}$

In order to differentiate different number system the base or radix is marked as suffix as shown in following examples.

Examples of decimal number: 1024.765_{10} , 79645.23_{10}

Examples of binary number: 1011.101_{2} , 11010.11_{2}

Examples of octal number: 7421.342_{8} , 12467.57_{8}

Examples of hexadecimal number: $2A6B.7C4_{16}$, $1296B.CA_{16}$

1.1.1 Number Representation

A number in general is represented in the following format.

$$\underbrace{d_{n-1}\ .....d_4\ d_3\ d_2\ d_1\ d_0}_{\text{n digit integer part}} . \underbrace{d_{-1}\ d_{-2}\ d_{-3}\ .....\ d_{-m}}_{\text{m digit fraction part}}$$

Here, d represent the digit in a number. The dot is the seperation between integer part and fraction part.

Each symbol in the number is called ***digit***. Each digit has a ***weight*** which is given by the power of radix.

Table 1.2: Weights of Digits in Number Systems

Number Systems	Weights of Digits .... d_4 d_3 d_2 d_1 d_0 . d_{-1} d_{-2} d_{-3} ....
Decimal	.... 10^4 10^3 10^2 10^1 10^0 . 10^{-1} 10^{-2} 10^{-3} ....
Binary	.... 2^4 2^3 2^2 2^1 2^0 . 2^{-1} 2^{-2} 2^{-3} ....
Octal	.... 8^4 8^3 8^2 8^1 8^0 . 8^{-1} 8^{-2} 8^{-3} ....
Hexadecimal	.... 16^4 16^3 16^2 16^1 16^0 . 16^{-1} 16^{-2} 16^{-3} ....

The following example illustrates the relation between digits and weights of number system.

$$786.35_{10} = 7 \times 10^2 + 8 \times 10^1 + 6 \times 10^0 + 3 \times 10^{-1} + 5 \times 10^{-2} = 786.35_{10}$$

$$572.24_8 = 5 \times 8^2 + 7 \times 8^1 + 2 \times 8^0 + 2 \times 8^{-1} + 4 \times 8^{-2} = 378.3125_{10}$$

$$110.11_2 = 1 \times 2^2 + 1 \times 2^1 + 0 \times 2^0 + 1 \times 2^{-1} + 1 \times 2^{-2} = 6.75_{10}$$

$$A79.B2_{16} = A \times 16^2 + 7 \times 16^1 + 9 \times 16^0 + B \times 16^{-1} + 2 \times 16^{-2}$$

$$= \underbrace{10_{10} \times 16^2 + 7 \times 16 + 9 \times 1}_{2681_{10}} + \underbrace{11_{10} \times 16^{-1} + 2 \times 16^{-2}}_{.6953125_{10}} = 2681.6953125_{10}$$

> ***Note:*** *Use calculator to compute the above calculations. The result has 11 digits and so in 10-digit calculator the calculation has to be performed separately for integer part and fraction part.*

The above examples also shows the method to convert any number to decimal number.

1.1.2 Decimal Number System

Decimal number system is the common and widely used number system. The invention of digital systems leads to usage of binary number system. The octal and hexadecimal are useful for short-hand representation of binary numbers. Since decimal number system is the standard number system, the value of a number in any other number system is verified/cross-checked by converting to an equivalent decimal number.

The ***decimal number system*** is a radix number system in which the radix r is 10. Since the radix is 10, the number system uses 10 symbols which are chosen as 0, 1, 2, 3, 4, 5, 6, 7, 8 and 9.

A decimal number in general is represented in the following format.

$$\underbrace{d_{n-1} \ldots d_4\, d_3\, d_2\, d_1\, d_0}_{\text{n digit integer part}} \underset{\substack{\uparrow \\ \text{Decimal point}}}{.} \underbrace{d_{-1}\, d_{-2}\, d_{-3} \ldots d_{-m}}_{\text{m digit fraction part}}$$

In the above representation, d represents the digit in the decimal number. The suffix in each digit represent power in the weight of the digit.

The dot in decimal number is called ***decimal point*** and it separates the integer part and fraction part.

Each digit of a decimal number has a weight which is given by the power of radix, 10. The digits and corresponding weights of a decimal number are shown in Table 1.3.

Table 1.3: Digits and Weights of Decimal Number

Digits	d_{n-1}	....	d_4	d_3	d_2	d_1	d_0	.	d_{-1}	d_{-2}	d_{-3}	....	d_{-m}
Weights	10^{n-1}	....	10^4	10^3	10^2	10^1	10^0	.	10^{-1}	10^{-2}	10^{-3}	....	10^{-m}

The value of a decimal number can be expressed as a sum of products of decimal digits and its corresponding weights as shown below:

$$d_{n-1} \times 10^{n-1} + \ldots + d_4 \times 10^4 + d_3 \times 10^3 + d_2 \times 10^2 + d_1 \times 10^1 + d_0 \times 10^0 + d_{-1} \times 10^{-1} + d_{-2} \times 10^{-2} + d_{-3} \times 10^{-3} + \ldots + d_{-m} \times 10^{-m}$$

Example: $1932.25_{10} = 1 \times 10^3 + 9 \times 10^2 + 3 \times 10^1 + 2 \times 10^0 + 2 \times 10^{-1} + 5 \times 10^{-2}$

1.1.3 Binary Number System

The ***binary number system*** is a radix number system in which the radix r is 2. Since the radix is 2, the number system uses only two symbols 0 and 1.

A binary number in general is represented in the following format.

$$\underbrace{d_{n-1} \ldots d_4\, d_3\, d_2\, d_1\, d_0}_{\text{n digit integer}} \underset{\substack{\uparrow \\ \text{Binary point}}}{.} \underbrace{d_{-1}\, d_{-2}\, d_{-3} \ldots d_{-m}}_{\text{m digit fraction}}$$

In the above representation, d represents the digit in the binary number. The suffix in each digit represent power in the weight of the digit.

The dot in binary number is called ***binary point*** and it separates the integer part and fraction part.

Each digit of a binary number has a weight which is given by the power of radix, 2. The digits and corresponding weights of a binary number are shown in Table 1.4.

Table 1.4: Digits and Weights of Binary Number

Digits	d_{n-1}	....	d_4	d_3	d_2	d_1	d_0	.	d_{-1}	d_{-2}	d_{-3}	....	d_{-m}
Weights	2^{n-1}	....	2^4	2^3	2^2	2^1	2^0	.	2^{-1}	2^{-2}	2^{-3}	....	2^{-m}

The value of a binary number can be expressed as a sum of products of binary digits and its corresponding weights as shown ahead:

$$d_{n-1} \times 2^{n-1} + ... + d_4 \times 2^4 + d_3 \times 2^3 + d_2 \times 2^2 + d_1 \times 2^1 + d_0 \times 2^0 + d_{-1} \times 2^{-1} + d_{-2} \times 2^{-2} + d_{-3} \times 2^{-3} + ... + d_{-m} \times 2^{-m}$$

Example: $1011.01_2 = 1 \times 2^3 + 0 \times 2^2 + 1 \times 2^1 + 1 \times 2^0 + 0 \times 2^{-1} + 1 \times 2^{-2}$

Table 1.5: Decimal Numbers and Equivalent Binary Numbers

Decimal Number	Binary Number	Decimal Number	Binary Number	Decimal Number	Binary Number	Decimal Number	Binary Number
0	0	11	1011	21	10101	31	11111
1	1	12	1100	22	10110	32	100000
2	10	13	1101	23	10111	33	100001
3	11	14	1110	24	11000	34	100010
4	100	15	1111	25	11001	35	100011
5	101	16	10000	26	11010	36	100100
6	110	17	10001	27	11011	37	100101
7	111	18	10010	28	11100	38	100110
8	1000	19	10011	29	11101	39	100111
9	1001	20	10100	30	11110	40	101000
10	1010						

Decimal to Binary Conversion

Integer part: Divide the integer part of decimal number by 2 and list the quotient and remainder as shown below. Repeat division until quotient is less than 2. The remainders are arranged in the order shown below to get the binary equivalent of decimal integer part.

Fraction Part: Multiply the fraction part of decimal number by 2 as shown below and record the overflow. Repeat multiplication of fraction part alone till the result is zero. The overflows are arranged in the order shown below to get the binary equivalent of decimal fraction.

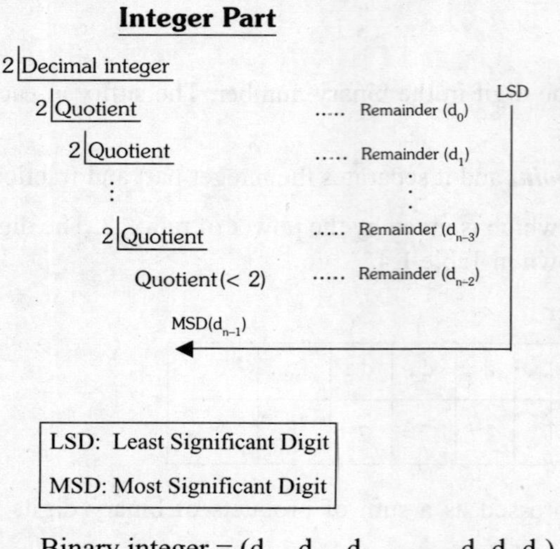

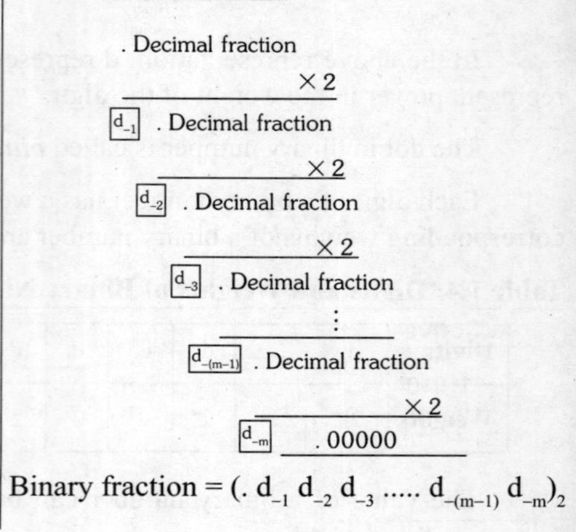

> ***Note:*** *In case of recurring fraction we will not get zero result in repeated multiplication and so stop multiplication when the required number of digits are obtained.*

Binary to Decimal Conversion

For binary to decimal conversion multiply each binary digit by its weight and sum up the product.

Consider the binary number,

$$.........d_3\ d_2\ d_1\ d_0\ .\ d_{-1}\ d_{-2}\ d_{-3}\ ....$$

The decimal number can be obtained from following calculation.

$$\text{Binary number} = ...d_3\ d_2\ d_1\ d_0\ .\ d_{-1}\ d_{-2}\ d_{-3}.....$$

$$= .....d_3 \times 2^3 + d_2 \times 2^2 + d_1 \times 2^1 + d_0 \times 2^0 + d_{-1} \times 2^{-1} + d_{-2} \times 2^{-2} + d_{-3} \times 2^{-3} + ...$$

$$= \text{Equivalent decimal number}$$

Example 1.1

Convert 26.375_{10} to binary and check the result by reverse conversion.

Solution

Integer Part

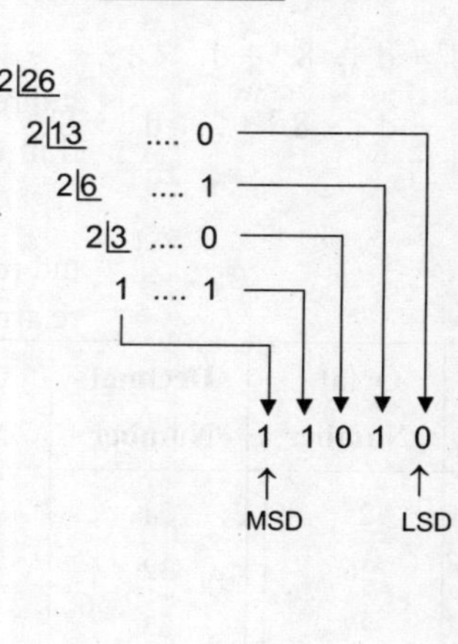

Fraction Part

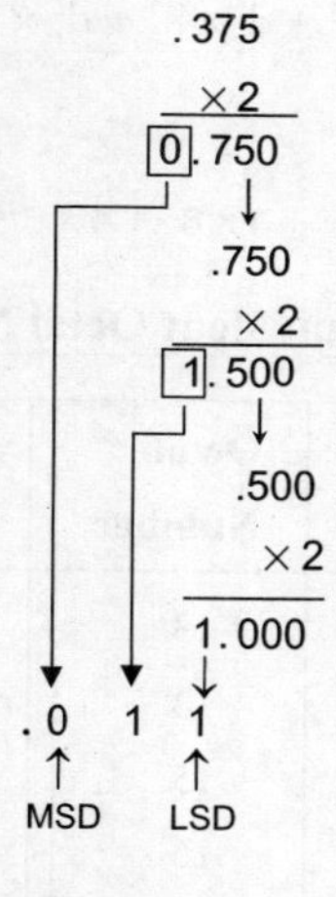

$26_{10} = 11010_2$

$.375_{10} = .011_2$

$\therefore\ 26.375_{10} = 11010.011_2$

Reverse Conversion

$$11010.011_2 = 1\times2^4 + 1\times2^3 + 0\times2^2 + 1\times2^1 + 0\times2^0 + 0\times2^{-1} + 1\times2^{-2} + 1\times2^{-3}$$

$$= 16 + 8 + 0 + 2 + 0 + 0 + .25 + .125$$

$$= 26.375_{10}$$

1.1.4 Octal Number System

The *octal number system* is a radix number system in which the radix r is 8. Since the radix is 8, the number system uses 8 symbols which are chosen as 0, 1, 2, 3, 4, 5, 6 and 7.

An octal number in general is represented in the following format.

$$\underbrace{d_{n-1} \ldots\ldots d_4\, d_3\, d_2\, d_1\, d_0}_{\text{n digit integer part}} \, . \, \underbrace{d_{-1}\, d_{-2}\, d_{-3} \ldots\ldots d_{-m}}_{\text{m digit fraction part}}$$

In the above representation, d represents the digit in the octal number. The suffix in each digit represent power in the weight of the digit.

Each digit of an octal number has a weight which is given by the power of radix, 8. The digits and corresponding weights of an octal number are shown in Table 1.6.

Table 1.6: Digits and Weights of Octal Number

Digits	d_{n-1}	....	d_4	d_3	d_2	d_1	d_0	.	d_{-1}	d_{-2}	d_{-3}	....	d_{-m}
Weights	8^{n-1}	....	8^4	8^3	8^2	8^1	8^0	.	8^{-1}	8^{-2}	8^{-3}	....	8^{-m}

The value of an octal number can be expressed as a sum of products of octal digits and its corresponding weights as shown below:

$$d_{n-1} \times 8^{n-1} + \ldots + d_4 \times 8^4 + d_3 \times 8^3 + d_2 \times 8^2 + d_1 \times 8^1 + d_0 \times 8^0 + d_{-1} \times 8^{-1} + d_{-2} \times 8^{-2} + d_{-3} \times 8^{-3} + \ldots + d_{-m} \times 8^{-m}$$

Example: $246.34_8 = 2 \times 8^2 + 4 \times 8^1 + 6 \times 8^0 + 3 \times 8^{-1} + 4 \times 8^{-2}$

Table 1.7: Decimal Numbers and Equivalent Octal Numbers

Decimal Number	Octal Number	Decimal Number	Octal Number	Decimal Number	Octal Number	Decimal Number	Octal Number
0	0	11	13	21	25	31	37
1	1	12	14	22	26	32	40
2	2	13	15	23	27	33	41
3	3	14	16	24	30	34	42
4	4	15	17	25	31	35	43
5	5	16	20	26	32	36	44
6	6	17	21	27	33	37	45
7	7	18	22	28	34	38	46
8	10	19	23	29	35	39	47
9	11	20	24	30	36	40	50
10	12						

Decimal to Octal Conversion

Integer part: Divide the integer part of decimal number by 8 and list the quotient and remainder as shown below. Repeat division until quotient is less than 8. The remainders are arranged in the order shown below to get the octal equivalent of decimal integer part.

Fraction Part: Multiply the fraction part of decimal number by 8 as shown below and record the overflow. Repeat multiplication of fraction part alone till the result is zero. The overflow are arranged in the order shown below to get the octal equivalent of decimal fraction.

> ***Note:*** *In case of recurring fraction we will not get zero result in repeated multiplication and so stop multiplication when the required number of digits are obtained.*

Integer Part

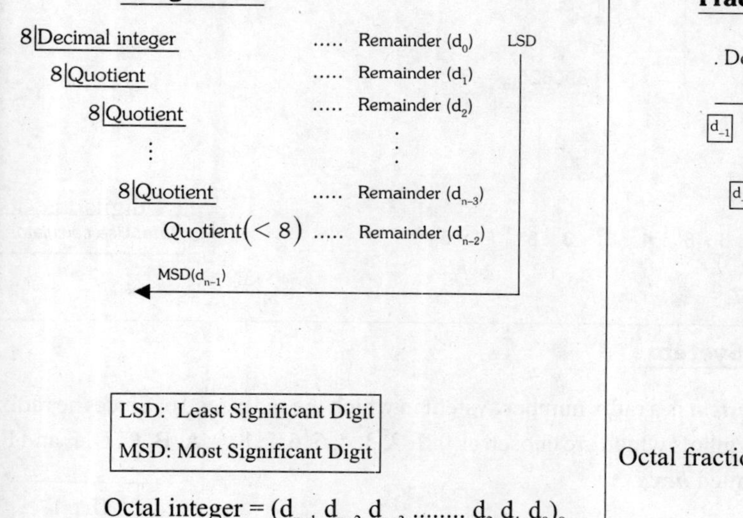

Octal integer = $(d_{n-1}\ d_{n-2}\ d_{n-3}\ ......\ d_2\ d_1\ d_0)_8$

Fraction Part

. Decimal fraction
× 8
$\boxed{d_{-1}}$. Decimal fraction
× 8
$\boxed{d_{-2}}$. Decimal fraction
× 8
$\boxed{d_{-3}}$. Decimal fraction
⋮
$\boxed{d_{-(m-1)}}$. Decimal fraction
× 8
$\boxed{d_{-m}}$. 00000

Octal fraction = $(.\ d_{-1}\ d_{-2}\ d_{-3}\ .....\ d_{-(m-1)}\ d_{-m})_8$

Octal to Decimal Conversion

For octal to decimal conversion multiply each octal digit by its weight and sum up the product.

Consider the octal number,

$.........d_3\ d_2\ d_1\ d_0\ .\ d_{-1}\ d_{-2}\ d_{-3}\ ....$

The decimal number can be obtained from following the calculation.

$$\text{Octal number} = ...d_3\ d_2\ d_1\ d_0\ .\ d_{-1}\ d_{-2}\ d_{-3}......$$

$$= .....d_3 \times 8^3 + d_2 \times 8^2 + d_1 \times 8^1 + d_0 \times 8^0 + d_{-1} \times 8^{-1} + d_{-2} \times 8^{-2} + d_{-3} \times 8^{-3} + ...$$

$$= \text{Equivalent decimal number}$$

Example 1.2

Convert 1012.390625_{10} to octal and check the result by reverse conversion.

Solution

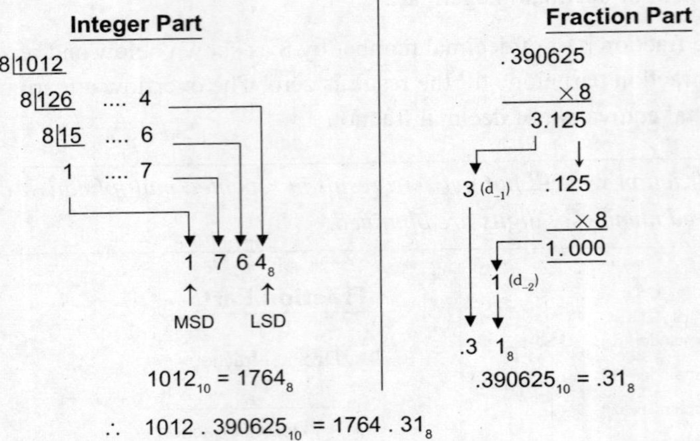

$1012_{10} = 1764_8$ $\qquad$ $.390625_{10} = .31_8$

$\therefore \quad 1012.390625_{10} = 1764.31_8$

Reverse Conversion

$$1764.31_8 = 1\times8^3 + 7\times8^2 + 6\times8^1 + 4\times8^0 + 3\times8^{-1} + 1\times8^{-2}$$

$$= 1012.390625_{10}$$

Note: *Use calculator.*

1.1.5 Hexadecimal Number System

The ***hexadecimal number system*** is a radix number system in which the radix r is 16. Since the radix is 16, the number system uses 16 symbols which are chosen as 0, 1, 2, 3, 4, 5, 6, 7, 8, 9, A, B, C, D, E and F. The hexadecimal in short can be called ***hexa***.

A hexadecimal number in general is represented in the following format.

$$\underbrace{d_{n-1} \ldots d_4\, d_3\, d_2\, d_1\, d_0}_{\text{n digit integer part}} \;.\; \underbrace{d_{-1}\, d_{-2}\, d_{-3} \ldots d_{-m}}_{\text{m digit fraction part}}$$

In the above representation, d repesents the digit in the hexa number. The suffix in each digit represent power in the weight of the digit.

Each digit of a hexa number has a weight which is given by the power of radix,16. The digits and corresponding weights of a hexa number are shown in Table 1.8.

Table 1.8: Digits and Weights of Hexadecimal Number

Digits	d_{n-1}	....	d_4	d_3	d_2	d_1	d_0	.	d_{-1}	d_{-2}	d_{-3}	....	d_{-m}
Weights	16^{n-1}	....	16^4	16^3	16^2	16^1	16^0	.	16^{-1}	16^{-2}	16^{-3}	....	16^{-m}

The value of a hexadecimal number can be expressed as a sum of products of hexa digits and its corresponding weights as shown below:

$$d_{n-1} \times 16^{n-1} + ... + d_4 \times 16^4 + d_3 \times 16^3 + d_2 \times 16^2 + d_1 \times 16^1 + d_0 \times 16^0 + d_{-1} \times 16^{-1} + d_{-2} \times 16^{-2} + d_{-3} \times 16^{-3} + ... + d_{-m} \times 16^{-m}$$

Example: $AC2.7B = A \times 16^2 + C \times 16^1 + 2 \times 16^0 + 7 \times 16^{-1} + B \times 16^{-2}$

Table 1.9: Decimal Number and Equivalent Hexadecimal Number

Decimal Number	Hexadecimal Number	Decimal Number	Hexadecimal Number	Decimal Number	Hexadecimal Number	Decimal Number	Hexadecimal Number
0	0	11	B	21	15	31	1F
1	1	12	C	22	16	32	20
2	2	13	D	23	17	33	21
3	3	14	E	24	18	34	22
4	4	15	F	25	19	35	23
5	5	16	10	26	1A	36	24
6	6	17	11	27	1B	37	25
7	7	18	12	28	1C	38	26
8	8	19	13	29	1D	39	27
9	9	20	14	30	1E	40	28
10	A						

Decimal to Hexadecimal Conversion

Integer part: Divide the integer part of decimal number by 16 and list the quotient and remainder as shown below. Repeat division until quotient is less than 16. The remainders are arranged in the order shown below to get the hexadecimal equivalent of decimal integer part.

Fraction Part: Multiply the fraction part of decimal number by 16 as shown below and record the overflow. Repeat multiplication of fraction part alone till the result is zero. The overflow are arranged in the order shown below to get the hexadecimal equivalent of decimal fraction.

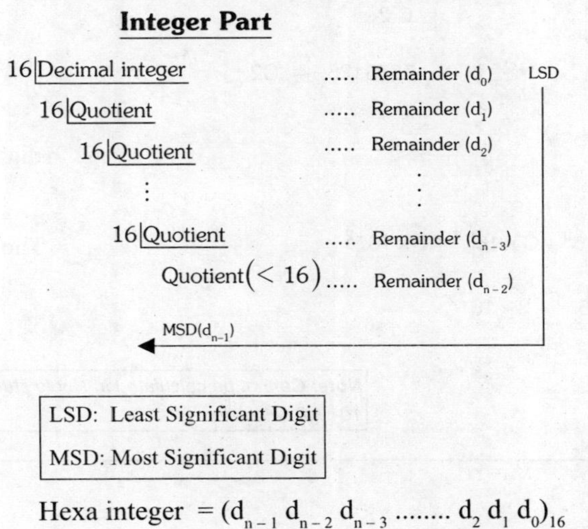

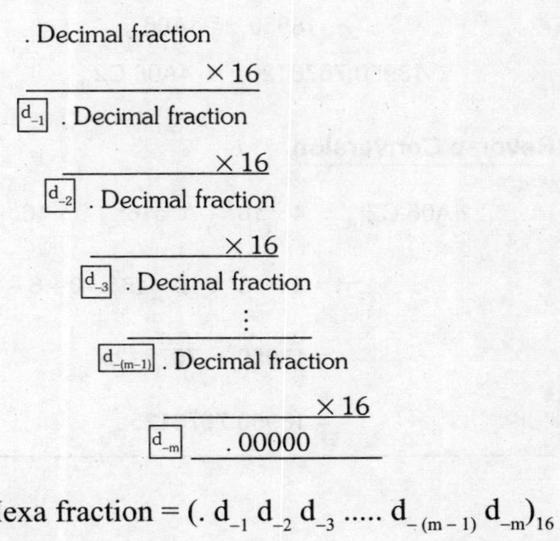

> ***Note:*** *In case of recurring fraction we will not get zero result in repeated multiplication and so stop multiplication when the required number of digits are obtained.*

Hexadecimal to Decimal Conversion

For hexadecimal to decimal conversion multiply each hexa digit by its weight and sum up the product.

Consider the hexadecimal number,

$$\ldots\ldots d_3\, d_2\, d_1\, d_0\, .\, d_{-1}\, d_{-2}\, d_{-3}\, \ldots$$

Now the decimal number can be obtained from the following calculation.

$$\text{Hexadecimal number} = \ldots d_3\, d_2\, d_1\, d_0\, .\, d_{-1}\, d_{-2}\, d_{-3}$$

$$= \ldots d_3 \times 16^3 + d_2 \times 16^2 + d_1 \times 16^1 + d_0 \times 16^0 + d_{-1} \times 16^{-1} + d_{-2} \times 16^{-2} + d_{-3} \times 16^{-3}$$

$$= \text{Equivalent decimal number}$$

Example 1.3

Convert 18950.7578125_{10} to hexadecimal number and cross-check result by reverse conversion.

Solution

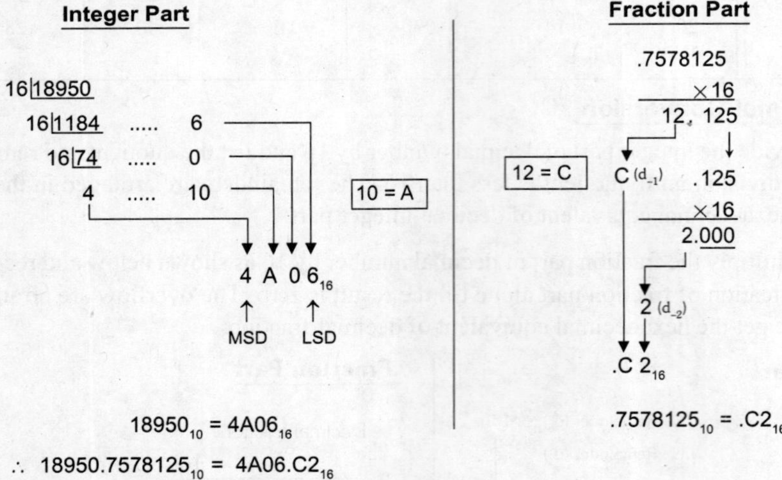

$18950_{10} = 4A06_{16}$ $\quad$ $.7578125_{10} = .C2_{16}$

$\therefore\ 18950.7578125_{10} = 4A06.C2_{16}$

Reverse Conversion

$$4A06.C2_{16} = 4 \times 16^3 + A \times 16^2 + 0 \times 16^1 + 6 \times 16^0 + C \times 16^{-1} + 2 \times 16^{-2}$$

$$= 4 \times 16^3 + 10 \times 16^2 + 0 + 6 + 12 \times 16^{-1} + 2 \times 16^{-2}$$

$$= 18950 + .7578125$$

$$= 18950.7578125_{10}$$

> ***Note:*** *Cannot be calculated in single step by 10-digit calculator.*

1.1.6 Relation between Binary, Octal and Hexadecimal Number Systems

The radix of octal is 8, the radix of hexa is 16.

Here, $8 = 2^3$ and $16 = 2^4$.

Therefore, a hexa digit can be represented by 4 digit binary and an octal digit by 3 digit binary. The octal and hexa decimal numbers are short hand form of binary.

An octal number can be directly converted to binary and vice-versa.

Similarly, a hexa number can be directly converted to binary and vice-versa.

Table 1.10: Binary Equivalent of Octal Digit

Octal	Binary
0	000
1	001
2	010
3	011
4	100
5	101
6	110
7	111

Table 1.11: Binary Equivalent of Hexa Digit

Hexa	Binary	Hexa	Binary
0	0000	8	1000
1	0001	9	1001
2	0010	A	1010
3	0011	B	1011
4	0100	C	1100
5	0101	D	1101
6	0110	E	1110
7	0111	F	1111

Example 1.4

Convert 726.35_8 to binary and cross-check binary and octal by converting to decimal.

Solution

Note: *Each octal digit is represented by three digit binary.*

$$726.35_8 = 111\ 010\ 110.011\ 101_2$$
$$= 111010110.011101_2$$

Cross-Check

$$726.35_8 = 7\times8^2+2\times8^1+6\times8^0+3\times8^{-1}+5\times8^{-2} = 470.453125_{10}$$

$$111010110.011101_2 = 1\times2^8+1\times2^7+1\times2^6+0\times2^5+1\times2^4+0\times2^3+1\times2^2+1\times2^1+0\times2^0$$
$$+0\times2^{-1}+1\times2^{-2}+1\times2^{-3}+1\times2^{-4}+0\times2^{-5}+1\times2^{-6}$$
$$= 2^8+2^7+2^6+2^4+2^2+2+2^{-2}+2^{-3}+2^{-4}+2^{-6}$$
$$= 470.453125_{10}$$

Example 1.5

Convert $A52.42_{16}$ to binary and cross-check binary and hexa by converting to decimal.

Solution

Note: *Each hexa digit is represented by four digit binary.*

$$A52.42_{16} = 1010\ 0101\ 0010.0100\ 0010_2$$
$$= 101001010010.0100001_2$$

Cross-Check

$$A52.42_{16} = A\times16^2+5\times16^1+2\times16^0+4\times16^{-1}+2\times16^{-2}$$

$$= 10\times16^2+5\times16+2+4\times16^{-1}+2\times16^{-2}$$

$$= 2642+0.2578125 = 2642.2578125_{10}$$

$$101001010010.0100001_2 = 1\times2^{11}+0\times2^{10}+1\times2^9+0\times2^8+0\times2^7+1\times2^6+0\times2^5+1\times2^4$$
$$+0\times2^3+0\times2^2+1\times2^1+0\times2^0+0\times2^{-1}+1\times2^{-2}+0\times2^{-3}$$
$$+0\times2^{-4}+0\times2^{-5}+0\times2^{-6}+1\times2^{-7}$$

$$= 2^{11}+2^9+2^6+2^4+2+2^{-2}+2^{-7}$$

$$= 2642+0.2578125 = 2642.2578125_{10}$$

***Note:** When using 10-digit calculator, calculate integer part and fraction part separately.*

Example 1.6

Convert 10101001010.10101101_2 to octal and hexa number and cross-check binary, octal and hexa by converting to decimal.

Solution

Separate the binary into group of three digits to find octal number. Represent each three digit binary by an equivalent octal digit.

$$10101001010.10101101_2 = 10\ 101\ 001\ 010.101\ 011\ 01_2$$

$$= 010\ 101\ 001\ 010.101\ 011\ 010_2$$

Append with leading and trailing zero.

$$= 2\quad 5\quad 1\quad 2\ .\ 5\quad 3\quad 2_8$$

$$= 2512.532_8$$

Separate the given binary into group of four digit binary to find hexa number. Represent each four digit binary by an equivalent hexa digit.

$$10101001010.10101101_2 = 101\ 0100\ 1010.1010\ 1101_2$$

$$= 0101\ 0100\ 1010.1010\ 1101_2$$

Append with leading zero.

$$= 5\quad 4\quad A\ .\ A\quad D_{16}$$

$$= 54A.AD_{16}$$

Cross-Check

$$10101001010.10101101_2 = 1\times2^{10}+1\times2^8+1\times2^6+1\times2^3+1\times2^1+1\times2^{-1}$$
$$+1\times2^{-3}+1\times2^{-5}+1\times2^{-6}+1\times2^{-8}$$

$$= 1354.67578125_{10}$$

***Note:** Weights of zeros are not shown.*

$$2512.532_8 = 2\times8^3+5\times8^2+1\times8^1+2\times8^0+5\times8^{-1}+3\times8^{-2}+2\times8^{-3}$$

$$= 1354.67578125_{10}$$

$$54A.AD_{16} = 5\times16^2+4\times16^1+A\times16^0+A\times16^{-1}+D\times16^{-2}$$

$$= 5\times16^2+4\times16+10+10\times16^{-1}+13\times16^{-2}$$

$$= 1354.67578125_{10}$$

Example 1.7

Convert 732.51_8 to binary and hexa and cross-check binary and hexa by converting to decimal.

Solution

$732.51_8 = 111\ 011\ 010.101\ 001_2$ — Represent each octal digit by three digit binary.

$= 111011010.101001_2$

$732.51_8 = 111\ 011\ 010.101\ 001_2$ — Rearrange the binary representation into group of four digits.

$= 1\ 1101\ 1010.1010\ 01_2$ — Append with leading and trailing zeros.

$= 0001\ 1101\ 1010.1010\ 0100_2$

$= \ 1 \quad D \quad A \ . \ A \quad 4_{16}$ — Represent each group of four digit binary by an equivalent hexa digit.

$= 1DA.A4_{16}$

Cross-Check

$$732.51_8 = 7\times8^2+3\times8^1+2\times8^0+5\times8^{-1}+1\times8^{-2}$$

$$= 474.640625_{10}$$

$$111011010.101001_2 = 1\times2^8+1\times2^7+1\times2^6+1\times2^4+1\times2^3+1\times2^1+1\times2^{-1}+1\times2^{-3}+1\times2^{-6}$$

$$= 474.640625_{10}$$

Note: *Weights of zeros are not shown.*

$$1DA.A4_{16} = 1\times16^2+D\times16^1+A\times16^0+A\times16^{-1}+4\times16^{-2}$$

$$= 1\times16^2+13\times16+10+10\times16^{-1}+4\times16^{-2}$$

$$= 474.640625_{10}$$

1.1.7 Unsigned and Signed Binary Number Systems

The binary number system can be unsigned binary or signed binary number system. The unsigned binary has only positive numbers. The ***signed binary*** has positive and negative numbers. There are three types of signed binary number system. They are,

1. Sign-magnitude Form
2. One's Complement Form
3. Two's Complement Form

1.1.8 Unsigned Binary

The general format of an unsigned binary number is given below:

$$d_{n-1} \ldots\ldots d_4\, d_3\, d_2\, d_1\, d_0$$

Each binary digit has a weight. The weight of digit, d_k is r^k. Here r = 2. Therefore weight is 2^k.

In a n-bit unsigned binary number the digit d_0 has the weight $2^0 = 1$, which is the lowest weight. Hence the digit d_0 is called ***LSD (Least Significant Digit)***.

In a n-bit unsigned binary number the digit d_{n-1} has the weight 2^{n-1}, which is the largest weight. Hence the digit d_{n-1} is called ***MSD (Most Significant Digit)***.

Range of Unsigned Binary Number

With n-bit binary it is possible to generate 2^n binary numbers.

When $n = 1$, $2^n = 2^1 = 2$, 1-bit binary numbers: 0, 1

When $n = 2$, $2^n = 2^2 = 4$, 2-bit binary numbers: 00, 01, 10, 11

When $n = 3$, $2^n = 2^3 = 8$, 3-bit binary numbers: 000, 001, 010, 011, 100, 101, 110, 111

and so on.

Therefore, the range of decimal number that can be represented by n-bit unsigned binary is 0 to 2^n-1.

The binary equivalent of decimal numbers 0 to 31 are listed in Table 1.12.

Sometime we have to write the binary number in a required size of 4-bit or 6-bit or 8-bit, etc. In order to express the unsigned binary to required size, append required number of zeros before MSD. The 8-bit representation of the binary numbers are also shown in Table 1.12.

Table 1.12: Decimal Number and its Equivalent Unsigned Binary Number

Decimal Number	Unsigned Binary Number	8-bit Unsigned Binary Number	Decimal Number	Unsigned Binary Number	8-bit Unsigned Binary Number
0	0	0000 0000	16	10000	0001 0000
1	1	0000 0001	17	10001	0001 0001
2	10	0000 0010	18	10010	0001 0010
3	11	0000 0011	19	10011	0001 0011
4	100	0000 0100	20	10100	0001 0100
5	101	0000 0101	21	10101	0001 0101
6	110	0000 0110	22	10110	0001 0110
7	111	0000 0111	23	10111	0001 0111
8	1000	0000 1000	24	11000	0001 1000
9	1001	0000 1001	25	11001	0001 1001
10	1010	0000 1010	26	11010	0001 1010
11	1011	0000 1011	27	11011	0001 1011
12	1100	0000 1100	28	11100	0001 1100
13	1101	0000 1101	29	11101	0001 1101
14	1110	0000 1110	30	11110	0001 1110
15	1111	0000 1111	31	11111	0001 1111

1.1.9 Sign-Magnitude Form

In n-bit binary representation of sign-magnitude form of binary, 1-bit is reserved for sign and n–1 bits are used to represent the magnitude.

For positive number the sign bit is "0" and for negative number the sign bit is "1". The magnitude bits are same for a given number with opposite signs.

In order to represent the given decimal number in n-bit sign-magnitude form, first represent the magnitude as n – 1 bits unsigned number. Then put "0" in MSD for positive number and put "1" in MSD position for negative number. The drawback in sign-magnitude representation is that there will be two zeros + 0 and – 0. (Refer Table 1.13).

An example of representing $+5_{10}$ and -5_{10} in sign-magnitude form is shown below:

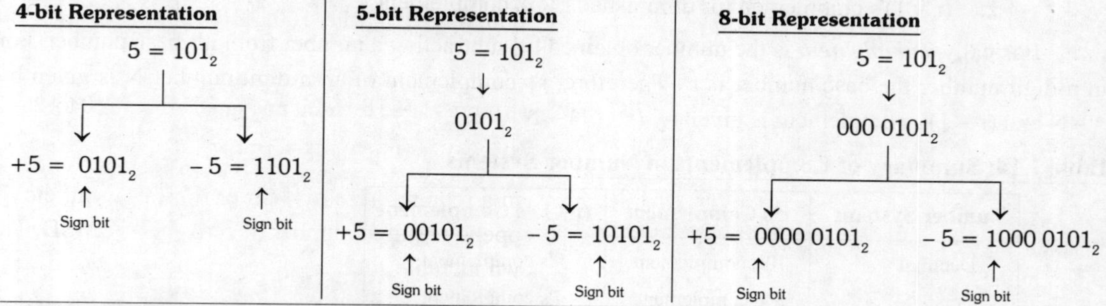

Range of Sign-Magnitude Binary Number

The range of decimal numbers that can be represented in n-bit sign-magnitude form of binary number is,

$$-(2^{n-1}-1) \quad \text{to} \quad +(2^{n-1}-1)$$

The range of decimal numbers that can be represented in sign-magnitude form of binary for n = 4 and 5 are shown below:

n = 4

Range = $-(2^{4-1}-1)$ to $+(2^{4-1}-1)$

= -7_{10} to $+7_{10}$

n = 5

Range = $-(2^{5-1}-1)$ to $+(2^{5-1}-1)$

= -15_{10} to $+15_{10}$

Table 1.13: Decimal and Sign-Magnitude form of Binary Number

Decimal Number	4-bit Sign-magnitude Binary Number	Standard 8-bit Sign-magnitude Binary Number
+0	0000	0000 0000
+1	0001	0000 0001
+2	0010	0000 0010
+3	0011	0000 0011
+4	0100	0000 0100
+5	0101	0000 0101
+6	0110	0000 0110
+7	0111	0000 0111
–0	1000	1000 0000
–1	1001	1000 0001
–2	1010	1000 0010
–3	1011	1000 0011
–4	1100	1000 0100
–5	1101	1000 0101
–6	1110	1000 0110
–7	1111	1000 0111

1.1.10 Complement of Number Systems

Complement of positive numbers are used to represent negative numbers in digital system.

In radix number system, two types of complement are used.

1. r's complement or radix complement.
2. (r – 1)'s complement or diminished radix complement.

Basically, ***complement*** is the number obtained by subtracting a number from it's base number. For an n-digit number the base number is r^n. Therefore, r's complement of an n-digit number N is given by $r^n - N$ and (r – 1)'s complement is given by $(r^n - 1) - N$.

Table 1.14: Summary of Complements in Number Systems

Number Systems	r's Complement	(r – 1)'s Complement
Decimal	10's complement	9's complement
Octal	8's complement	7's complement
Hexa	16's complement	15's complement
Binary	2's complement	1's complement

Importance of Complement

The basic mathematical operation is addition. Subtraction is addition of positive and negative number.

Example: 75 – 42 = 72 + (– 42)

Multiplication is successive addition.

Example: 6 × 3 = 6 + 6 + 6 = 3 + 3 + 3 + 3 + 3 + 3

Division is successive subtraction. Subtraction is again addition of positive and negative number.

Example: 10 ÷ 3 ⇒ 10 – 3 = 7 – 3 = 4 – 3 = 1 (Remainder)
Quotient = 1 + 1 + 1 = 3 (Count of number of subtraction)

Therefore in digital system all mathematical operations are realized in terms of addition.

Therefore, negative number is represented in complement form in digital system, so that every arithmetic operation is performed by addition.

Note: *The ALU (Arithmetic Logical Unit) of a microprocesser is basically an adder capable of performing arithmetic and logical operations in terms of addition.*

1.1.11 One's Complement

In one's complement form of binary number representation, the positive numbers will be same as that of unsigned binary. The negative of a n-digit positive number N is given by $(2^n - 1) - N$. This is same as bit-by-bit complement of the positive number which is changing every 1 to 0 and 0 to 1 of the positive number. Therefore for positive numbers the MSD is "0" and for negative numbers MSD is "1". In order to express the one's complement number in the required size, sign extension technique is employed. In this technique the upper most bit is extended to the required size.

The drawback in one's complement form is that there will be two zeros +0 and – 0. (Refer Table 1.15).

An example of representing $+5_{10}$ and -5_{10} in one's complement form is shown below:

$5_{10} = 101_2$

⇓ Append sign bit

$0101_2 = +5$

Sign bit (leftmost bit)

⇓ Complement

$1010_2 = -5$ (One's complement form)

Sign bit (leftmost bit)

$\therefore\ +5_{10} = 0101_2$ (4-bit representation) $= 0000\ 0101_2$ (sign extended to 8-bit)

$-5_{10} = 1010_2$ (4-bit representation) $= 1111\ 1010_2$ (sign extended to 8-bit)

Range of One's Complement Binary Number

The range of decimal numbers that can be represented in n-bit one's complement form of binary number is,

$$-(2^{n-1}-1) \quad \text{to} \quad +(2^{n-1}-1)$$

The range of decimal numbers that can be represented in one's complement form of binary for n = 4 and 5 are shown below:

n = 4

Range $= -(2^{4-1}-1)$ to $+(2^{4-1}-1)$

$= -7_{10}$ to $+7_{10}$

n = 5

Range $= -(2^{5-1}-1)$ to $+(2^{5-1}-1)$

$= -15_{10}$ to $+15_{10}$

Table 1.15: Decimal and One's Complement form of Binary Number

Decimal Number	4-bit One's Complement Binary Number	Sign-extended One's Complement 8-bit Binary Number
+0	0000	0000 0000
+1	0001	0000 0001
+2	0010	0000 0010
+3	0011	0000 0011
+4	0100	0000 0100
+5	0101	0000 0101
+6	0110	0000 0110
+7	0111	0000 0111
–7	1000	1111 1000
–6	1001	1111 1001
–5	1010	1111 1010
–4	1011	1111 1011
–3	1100	1111 1100
–2	1101	1111 1101
–1	1110	1111 1110
–0	1111	1111 1111

One's Complement of Binary Fraction

For one's complement of binary fraction numbers every bit is complemented like integer number. If one's complement is made twice we will get original number.

Example:

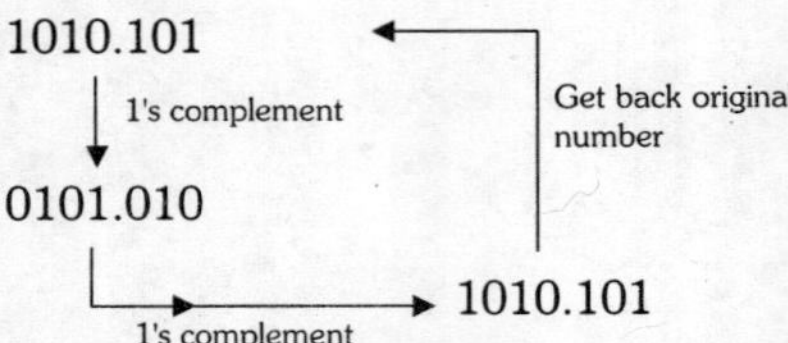

1.1.12 Two's Complement

In two's complement form of binary number representation, the positive numbers will be same as that of unsigned binary. The negative of n-digit positive number N is given by $2^n - N$ which is same as bit-by-bit complementing the positive number and then adding one. Here also the MSD is "0" for positive number and "1" for negative number. The advantage in two's complement is that there is only one zero. (Refer Table 1.16). Due to unique representation for zero, two's complement binary number system is practically followed in all digital systems.

In two's complement representation the sign-extention technique is followed to represent the number in required size.

An example of representing $+5_{10}$ and -5_{10} in two's complement form is shown below:

$5_{10} = 101_2$

⇓ Append sign bit

$0101_2 = +5_{10}$

Sign bit (MSB)

⇓ Complement

1010

⇓ Add 1

$1011_2 = -5_{10}$ (Two's complement form)

Sign bit (MSB)

$\therefore +5_{10} = 0101_2$ (4-bit representation) $=$ $0000\ 0101_2$ (sign extended to 8-bit)

$-5_{10} = 1011_2$ (4-bit representation) $=$ $1111\ 1011_2$ (sign extended to 8-bit)

Range of Two's Complement Binary Number

The range of decimal numbers that can be represented in n-bit two's complement form of binary numbers is,

$$-(2^{n-1}) \quad \text{to} \quad +(2^{n-1}-1)$$

The range of decimal numbers that can be represented in two's complement form of binary for n = 4 and 5 are shown ahead:

n = 4

Range = $-(2^{4-1})$ to $+(2^{4-1}-1)$

$= -8_{10}$ to $+7_{10}$

n = 5

Range = $-(2^{5-1})$ to $+(2^{5-1}-1)$

$= -16_{10}$ to $+15_{10}$

Table 1.16: Decimal and Two's Complement form of Binary Number

Decimal Number	4-bit Two's Complement Binary Number	Sign-extended Two's Complement 8-bit Binary Number
+0	0000	0000 0000
+1	0001	0000 0001
+2	0010	0000 0010
+3	0011	0000 0011
+4	0100	0000 0100
+5	0101	0000 0101
+6	0110	0000 0110
+7	0111	0000 0111
–8	1000	1111 1000
–7	1001	1111 1001
–6	1010	1111 1010
–5	1011	1111 1011
–4	1100	1111 1100
–3	1101	1111 1101
–2	1110	1111 1110
–1	1111	1111 1111

wo's Complement of Binary Fraction

For two's complement of binary fraction number take one's complement of every bit and then add to right most digit. If two's complement is made twice we will get original number.

xample:

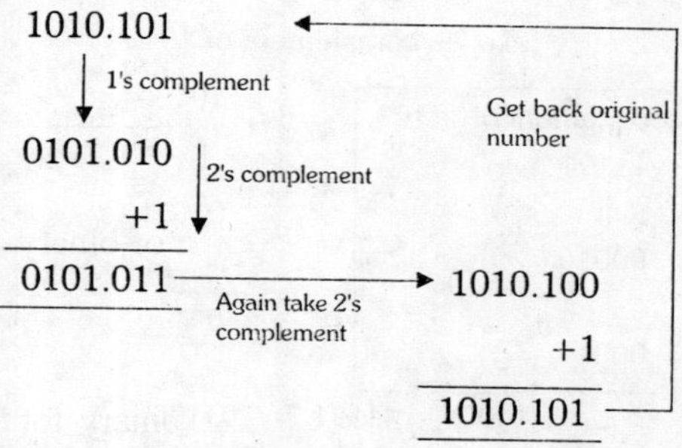

1.1.13 One's and Two's Complement Subtraction

In one's and two's complement subtraction the subtrahend, is represented in complement form and added to minuend.

In complement subtraction,

If carry = 1 ; Result is positive, discard carry in two's complement subtraction but add carry to sum in one's complement subtraction.

If no carry ; Result is negative, take respective complement of sum and put "–" sign.

Example 1.8

Perform the following subtraction by one's and two's complement subtraction.

a) $11101_2 - 10001_2$ b) $1001_2 - 11101_2$

Solution

a) $11101_2 - 10001_2$

One's complement subtraction

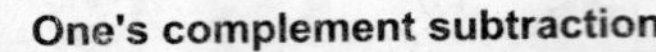

Minuend: 11101_2 ⇒ 11101

Subtrahend: -10001_2 ⇒(1's complement) + 01110

[1]01011 → Add carry 1 → 01100

1's complement of 10001: 10001 Comp ⇓ 01110

$\therefore 11101_2 - 10001_2 = 01100_2$

Two's complement subtraction

Minuend: 11101_2 ⇒ 11101

Subtrahend: -10001_2 ⇒(2's complement) + 01111

[1] 01100 ↑ Discard carry

2's complement of 11101: 10001 Comp ⇓ 01110 +1 = 01111

$\therefore 11101_2 - 10001_2 = 01100_2$

b) $10001_2 - 11101_2$

One's complement subtraction

Minuend: 10001_2 ⇒ 10001

Subtrahend: -11101_2 ⇒(1's complement) + 00010

10011 ↑ No carry

∴ Take 1's complement of 10011

10011 Comp ⇓ 01100

1's complement of 11101: 11101 Comp ⇓ 00010

$\therefore 10001_2 - 11101_2 = -01100_2$

Two's complement subtraction

Minuend: 10001_2 ⇒ 10001

Subtrahend: -11101_2 ⇒(2's complement) + 00011

10100 ↑ No carry

∴ Take 2's complement of 10100

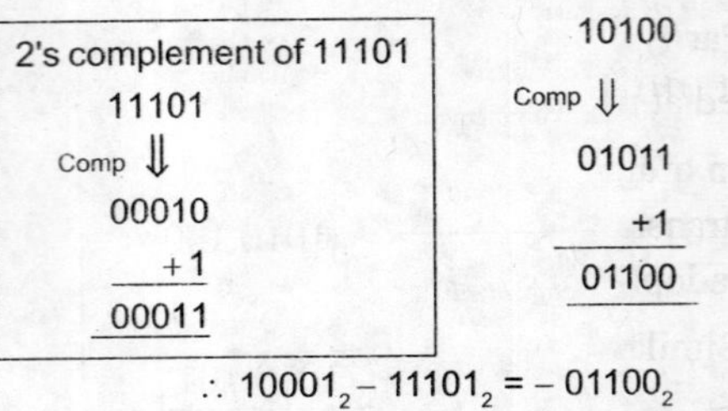

$\therefore 10001_2 - 11101_2 = -01100_2$

Example 1.9

Perform one's complement and two's complement subtraction of fractional numbers 1101.101_2 and 1000.001_2 and compare the results with actual subtraction.

Solution

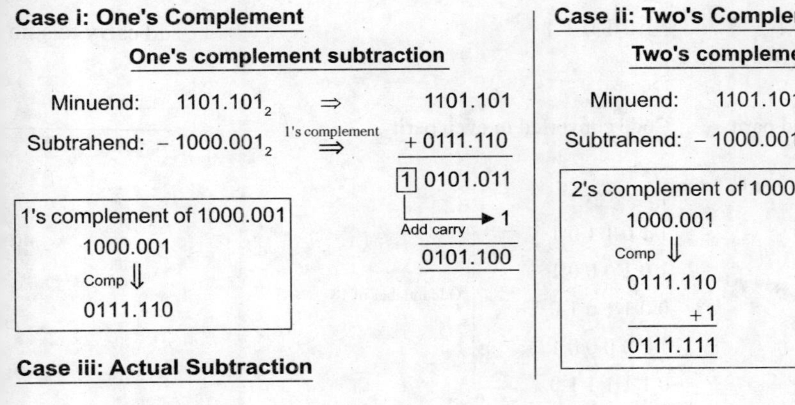

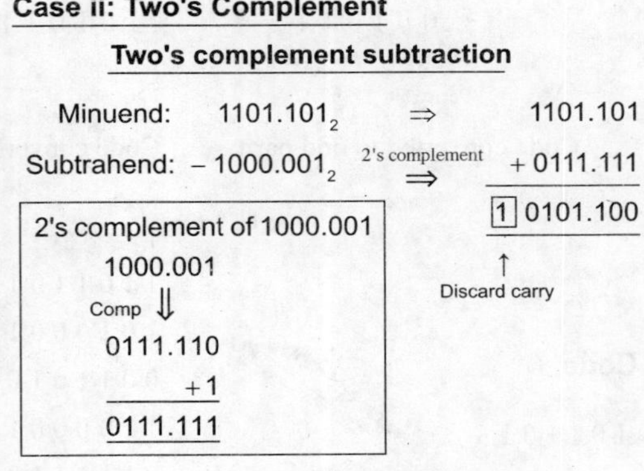

Case iii: Actual Subtraction

```
  1101.101
– 1000.001
  0101.100
```

Note: The results are same in direct subtraction, one's and two's complement subtraction.

1.2 Error Detection and Correction

In digital systems binary information is transmitted from one system to another system or one device to another device. While transmitting the binary information or codes, there may be errors. Different techniques have been developed to detect and correct errors in transmission of digital information or codes.

Parity is popular method of error detection and correction by sending a request to retransmit the data.

Hamming code is another popular error deduction and correction technique to deduct and correct 1-bit error in the received data.

1.2.1 Parity

Parity refer to number of 1's in an n-bit digital code. When the number of 1's in an n-bit code is even, then the code is said to have even parity. When the number of 1's in an n-bit code is odd, then the code is said to have odd parity.

Parity of an n-bit code can be modified by appending an extra bit called ***parity bit***. The parity bit is appended at most significant bit position.

In a digital transmission system, the information to be transmitted will be coded as n-bit binary. During transmisssion it is usual practice to transmit every n-bit code with same parity. When the transmission system adopt even parity then every n-bit code is converted to even parity by inserting a parity bit.

Similarly, when odd parity is adopted every n-bit code is converted to odd parity.

Converting the coded binary to even parity or odd parity is a popular and simple technique to detect errors in transmission of digital codes.

Example:

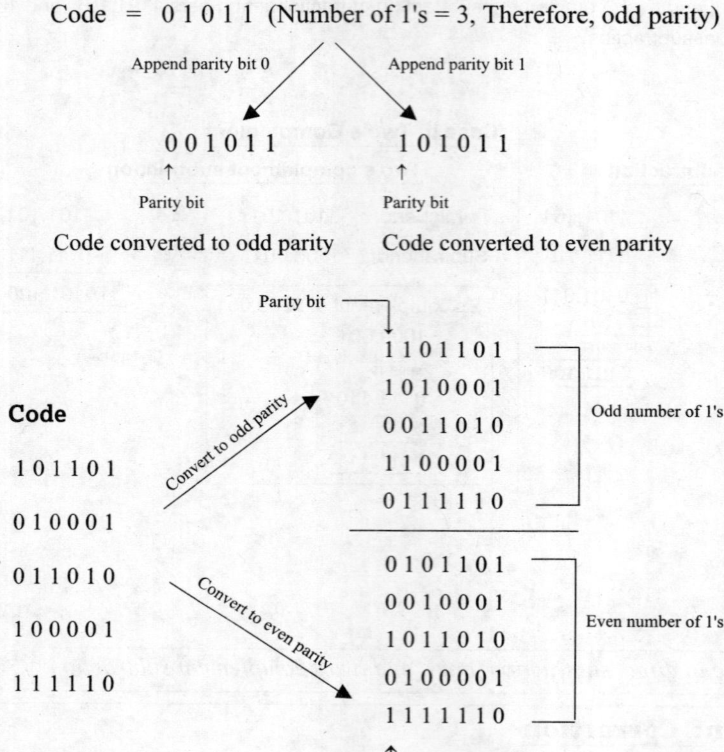

1.2.2 Hamming Code

(AU, Apr/May'19, 6 Marks)

Hamming code is developed by **Richard W Hamming** of Bell Telephone Laboratories. The Hamming code is error detecting and correcting code. It can accuratly detect 1-bit error.

The binary information to be coded is called data. In Hamming coding technique additional bits called Hamming or parity bits are inserted at various positions in between data bits. The combined data and Hamming bits is called Hamming code. In a particular transmission system, the position of data and Hamming bits are fixed and known to transmitter and receiver.

Number of Hamming or Parity Bits

In order to code n-bit binary data k number of Hamming or parity bits are needed. Here, n and k should satisfy the condition,

$$2^k \geq n + k + 1 \quad (\text{or } 2^k - k - 1 \geq n)$$

There will be multiple values of k that satisfy the above condition, but minimum value of k is chosen to restrict the size of coded data.

Let us decide the number of Hamming or parity bits for 7-bit binary data.

$\therefore$ n = 7

Let, $k = 3$

$\therefore\ 2^k = 2^3 = 8$

Now, $n + k + 1 = 7 + 3 + 1 = 11$

When $k = 3$ the condition, $2^k \geq n + k + 1$, is not satisfied.

Let, $k = 4$

$\therefore\ 2^k = 2^4 = 16$

Now, $n + k + 1 = 7 + 4 + 1 = 12$

When $k = 4$ the condition, $2^k \geq n + k + 1$, is satisfied.

Therefore, a minimum of 4 Hamming or parity bits are required to code 7-bit binary data.

Let us decide the number of Hamming or parity bits for 12-bit binary data.

$\therefore\ n = 12$

Let, $k = 4$

$\therefore\ 2^k = 2^4 = 16$

Now, $n + k + 1 = 12 + 4 + 1 = 17$

When $k = 4$ the condition, $2^k \geq n + k + 1$, is not satisfied.

Let, $k = 5$

$\therefore\ 2^k = 2^5 = 32$

Now, $n + k + 1 = 12 + 5 + 1 = 18$

When $k = 5$ the condition, $2^k \geq n + k + 1$, is satisfied.

Therefore, a minimum of 5 Hamming or parity bits are required to code 12-bit binary data.

The range of data bits that satisfy the condition $2^k \geq n + k + 1$ for $k = 3, 4, 5, 6$ and 7 are listed in Table 1.17.

Table 1.17: Range of Data Bits for k Parity Bits

Number of Parity Bits k	Range of Data Bits n
3	2 to 4
4	5 to 11
5	12 to 26
6	27 to 57
7	58 to 120

Combining the Data and Hamming or Parity Bits

In order to code n-bit binary data, $(d_n, d_{n-1}, \ldots\ldots d_3, d_2, d_1)$ determine k number of Hamming or parity bits $(P_1, P_2, \ldots. P_{k-1}, P_k)$ that satisfy the condition $2^k \geq n + k + 1$.

Let the bits of coded data is denoted as $b_1, b_2, \ldots. b_{n+k-1}, b_{n+k}$. In the coded data the binary bit positions that corresponds to binary weights are reserved for parity bits. Therefore, $b_1, b_2, b_4, b_8, b_{16}, \ldots..$ are reserved for parity bits.

Consider a 5-bit data to be coded by 4 Hamming bits,

$\therefore$ n = 5, k = 4, $2^k = 2^4 = 16$

$\therefore$ n + k + 1 = 5 + 4 + 1 = 10

Here the condition, $2^k \geq n + k + 1$, is satisfied.

Let, Binary data = $d_5\ d_4\ d_3\ d_2\ d_1$

Parity bits = $P_1\ P_2\ P_3\ P_4$

The size of coded data is 5 + 4 = 9 bits

Let, coded data = $b_1\ b_2\ b_3\ b_4\ b_5\ b_6\ b_7\ b_8\ b_9$

Here, the binary bit positions that corresponds to binary weight are reserved for parity bits. Therefore, parity bits are inserted in b_1, b_2, b_4 and b_8 and data bits are inserted in other bit positions as shown below:

Coded data : $b_1\ b_2\ b_3\ b_4\ b_5\ b_6\ b_7\ b_8\ b_9$

$P_1\ P_2\ d_5\ P_3\ d_4\ d_3\ d_2\ P_4\ d_1$

The format of 6, 7, 8 and 9 bits coded data are shown below:

6-bit Data

n = 6, k = 4, n + k = 10

The size of coded data is 10-bits.

Data bits : $d_6\ d_5\ d_4\ d_3\ d_2\ d_1$

Parity bits : $P_1\ P_2\ P_3\ P_4$

Coded data bits : $b_1\ b_2\ b_3\ b_4\ b_5\ b_6\ b_7\ b_8\ b_9\ b_{10}$

Hamming code : $P_1\ P_2\ d_6\ P_3\ d_5\ d_4\ d_3\ P_4\ d_2\ d_1$

7-bit Data

n = 7, k = 4, n + k = 11

The size of coded data is 11-bits.

Data bits : $d_7\ d_6\ d_5\ d_4\ d_3\ d_2\ d_1$

Parity bits : $P_1\ P_2\ P_3\ P_4$

Coded data bits : $b_1\ b_2\ b_3\ b_4\ b_5\ b_6\ b_7\ b_8\ b_9\ b_{10}\ b_{11}$

Hamming code : $P_1\ P_2\ d_7\ P_3\ d_6\ d_5\ d_4\ P_4\ d_3\ d_2\ d_1$

8-bit Data

n = 8, k = 4, n + k = 12

The size of coded data is 12-bits.

Data bits : $d_8\ d_7\ d_6\ d_5\ d_4\ d_3\ d_2\ d_1$

Parity bits : $P_1\ P_2\ P_3\ P_4$

Coded data bits : b_1 b_2 b_3 b_4 b_5 b_6 b_7 b_8 b_9 b_{10} b_{11} b_{12}

Hamming code : P_1 P_2 d_8 P_3 d_7 d_6 d_5 P_4 d_4 d_3 d_2 d_1

9-bit Data

$n = 9, k = 4, n + k = 13$

The size of coded data is 13-bits.

Data bits : d_9 d_8 d_7 d_6 d_5 d_4 d_3 d_2 d_1

Parity bits : P_1 P_2 P_3 P_4

Coded data bits : b_1 b_2 b_3 b_4 b_5 b_6 b_7 b_8 b_9 b_{10} b_{11} b_{12} b_{13}

Hamming code : P_1 P_2 d_9 P_3 d_8 d_7 d_6 P_4 d_5 d_4 d_3 d_2 d_1

Generation of Parity Bits

Consider the binary values of decimal numbers 1 to 13 shown in Table 1.18.

Table 1.18: Binary Value of Decimal Number

Decimal Number	4-bit Binary Values to be Considered for				
	9-bit Code	10-bit Code	11-bit Code	12-bit Code	13-bit Code
	$w_8w_4w_2w_1$	$w_8w_4w_2w_1$	$w_8w_4w_2w_1$	$w_8w_4w_2w_1$	$w_8w_4w_2w_1$
1	0 0 0 1	0 0 0 1	0 0 0 1	0 0 0 1	0 0 0 1
2	0 0 1 0	0 0 1 0	0 0 1 0	0 0 1 0	0 0 1 0
3	0 0 1 1	0 0 1 1	0 0 1 1	0 0 1 1	0 0 1 1
4	0 1 0 0	0 1 0 0	0 1 0 0	0 1 0 0	0 1 0 0
5	0 1 0 1	0 1 0 1	0 1 0 1	0 1 0 1	0 1 0 1
6	0 1 1 0	0 1 1 0	0 1 1 0	0 1 1 0	0 1 1 0
7	0 1 1 1	0 1 1 1	0 1 1 1	0 1 1 1	0 1 1 1
8	1 0 0 0	1 0 0 0	1 0 0 0	1 0 0 0	1 0 0 0
9	1 0 0 1	1 0 0 1	1 0 0 1	1 0 0 1	1 0 0 1
10		1 0 1 0	1 0 1 0	1 0 1 0	1 0 1 0
11			1 0 1 1	1 0 1 1	1 0 1 1
12				1 1 0 0	1 1 0 0
13					1 1 0 1

9-bit Code (b_1 b_2 b_3 b_4 b_5 b_6 b_7 b_8 b_9)

In Table 1.18, the following observations are made.

Excluding the number 1, the decimal numbers 3, 5, 7, 9 has 1 in binary weight position w_1.

$\therefore P_1 = b_3 \oplus b_5 \oplus b_7 \oplus b_9$

Excluding the numbers 1, 2, the decimal numbers 3, 6, 7 has 1 in binary weight position w_2.

$\therefore P_2 = b_3 \oplus b_6 \oplus b_7$

Excluding the numbers 1, 2, 3, 4, the decimal numbers 5, 6, 7 has 1 in binary weight position w_4.

$\therefore P_3 = b_5 \oplus b_6 \oplus b_7$

Excluding the numbers 1, 2, 3, 4, 5, 6, 7, 8, the decimal numbers 9 has 1 in binary weight position w_8.

$\therefore P_4 = b_9$

By following the procedure for 9-bit code the parity bits of 9, 10, 11, 12 and 13 bits Hamming codes can be obtained using the following equations.

10-bit Code (b_1 b_2 b_3 b_4 b_5 b_6 b_7 b_8 b_9 b_{10})

$P_1 = b_3 \oplus b_5 \oplus b_7 \oplus b_9$

$P_2 = b_3 \oplus b_6 \oplus b_7 \oplus b_{10}$

$P_3 = b_5 \oplus b_6 \oplus b_7$

$P_4 = b_9 \oplus b_{10}$

11-bit Code (b_1 b_2 b_3 b_4 b_5 b_6 b_7 b_8 b_9 b_{10} b_{11})

$P_1 = b_3 \oplus b_5 \oplus b_7 \oplus b_9 \oplus b_{11}$

$P_2 = b_3 \oplus b_6 \oplus b_7 \oplus b_{10} \oplus b_{11}$

$P_3 = b_5 \oplus b_6 \oplus b_7$

$P_4 = b_9 \oplus b_{10} \oplus b_{11}$

12-bit Code (b_1 b_2 b_3 b_4 b_5 b_6 b_7 b_8 b_9 b_{10} b_{11} b_{12})

$P_1 = b_3 \oplus b_5 \oplus b_7 \oplus b_9 \oplus b_{11}$

$P_2 = b_3 \oplus b_6 \oplus b_7 \oplus b_{10} \oplus b_{11}$

$P_3 = b_5 \oplus b_6 \oplus b_7 \oplus b_{12}$

$P_4 = b_9 \oplus b_{10} \oplus b_{11} \oplus b_{12}$

13-bit Code (b_1 b_2 b_3 b_4 b_5 b_6 b_7 b_8 b_9 b_{10} b_{11} b_{12} b_{13})

$P_1 = b_3 \oplus b_5 \oplus b_7 \oplus b_9 \oplus b_{11} \oplus b_{13}$

$P_2 = b_3 \oplus b_6 \oplus b_7 \oplus b_{10} \oplus b_{11}$

$P_3 = b_5 \oplus b_6 \oplus b_7 \oplus b_{12} \oplus b_{13}$

$P_4 = b_9 \oplus b_{10} \oplus b_{11} \oplus b_{12} \oplus b_{13}$

Error Detection and Correction

In order to detect error in transmitted data check bits are generated. The number of check bits will be same as number of parity bits. When there is no error in received data all check bits will be zero. When there is error the check bits will give the binary value of bit position that has error.

Let, check bits = C_4 C_3 C_2 C_1

For example,

If C_4 C_3 C_2 C_1 = 0000, then no error in received code.

If C_4 C_3 C_2 C_1 = 0011, then error in b_3 bit in the received code.

If C_4 C_3 C_2 C_1 = 0111, then error in b_7 bit in the received code.

The check bits are generated by XOR of the parity bit position along with the bits that are used to generate parity bits.

Check Bits for 9-bit Code ($b_1\ b_2\ b_3\ b_4\ b_5\ b_6\ b_7\ b_8\ b_9$)

$$C_1 = b_1 \oplus b_3 \oplus b_5 \oplus b_7 \oplus b_9$$
$$C_2 = b_2 \oplus b_3 \oplus b_6 \oplus b_7$$
$$C_3 = b_4 \oplus b_5 \oplus b_6 \oplus b_7$$
$$C_4 = b_8 \oplus b_9$$

Check Bits for 10-bit Code ($b_1\ b_2\ b_3\ b_4\ b_5\ b_6\ b_7\ b_8\ b_9\ b_{10}$)

$$C_1 = b_1 \oplus b_3 \oplus b_5 \oplus b_7 \oplus b_9$$
$$C_2 = b_2 \oplus b_3 \oplus b_6 \oplus b_7 \oplus b_{10}$$
$$C_3 = b_4 \oplus b_5 \oplus b_6 \oplus b_7$$
$$C_4 = b_8 \oplus b_9 \oplus b_{10}$$

Check Bits for 11-bit Code ($b_1\ b_2\ b_3\ b_4\ b_5\ b_6\ b_7\ b_8\ b_9\ b_{10}\ b_{11}$)

$$C_1 = b_1 \oplus b_3 \oplus b_5 \oplus b_7 \oplus b_9 \oplus b_{11}$$
$$C_2 = b_2 \oplus b_3 \oplus b_6 \oplus b_7 \oplus b_{10} \oplus b_{11}$$
$$C_3 = b_4 \oplus b_5 \oplus b_6 \oplus b_7$$
$$C_4 = b_8 \oplus b_9 \oplus b_{10} \oplus b_{11}$$

Check Bits for 12-bit Code ($b_1\ b_2\ b_3\ b_4\ b_5\ b_6\ b_7\ b_8\ b_9\ b_{10}\ b_{11}\ b_{12}$)

$$C_1 = b_1 \oplus b_3 \oplus b_5 \oplus b_7 \oplus b_9 \oplus b_{11}$$
$$C_2 = b_2 \oplus b_3 \oplus b_6 \oplus b_7 \oplus b_{10} \oplus b_{11}$$
$$C_3 = b_4 \oplus b_5 \oplus b_6 \oplus b_7 \oplus b_{12}$$
$$C_4 = b_8 \oplus b_9 \oplus b_{10} \oplus b_{11} \oplus b_{12}$$

Check Bits for 13-bit Code ($b_1\ b_2\ b_3\ b_4\ b_5\ b_6\ b_7\ b_8\ b_9\ b_{10}\ b_{11}\ b_{12}\ b_{13}$)

$$C_1 = b_1 \oplus b_3 \oplus b_5 \oplus b_7 \oplus b_9 \oplus b_{11} \oplus b_{13}$$
$$C_2 = b_2 \oplus b_3 \oplus b_6 \oplus b_7 \oplus b_{10} \oplus b_{11}$$
$$C_3 = b_4 \oplus b_5 \oplus b_6 \oplus b_7 \oplus b_{12} \oplus b_{13}$$
$$C_4 = b_8 \oplus b_9 \oplus b_{10} \oplus b_{11} \oplus b_{12} \oplus b_{13}$$

Example 1.10

(AU, Apr/May'19, 2 Marks)

Generate the Hamming code for an 8-bit data string 10101101. Introduce error in a data bit and prove that Hamming code detect the error.

Solution

For 8 data bits, 4 parity bits are required.

$\therefore$ n = 8, k = 4

Here, $2^k = 2^4 = 16$

n + k + 1 = 8 + 4 + 1 = 13

The condition, $2^k \geq n + k + 1$ is satisfied.

The size of Hamming code is n + k = 8 + 4 = 12 bits.

Let, Hamming code = $b_1 b_2 b_3 b_4 b_5 b_6 b_7 b_8 b_9 b_{10} b_{11} b_{12}$

In Hamming code, the bit positions 1, 2, 4 and 8 are reserved for parity bits and remaining bits are used for data bits. The data and parity bits are combined to form Hamming code as shown below:

Hamming code : $b_1\ b_2\ b_3\ b_4\ b_5\ b_6\ b_7\ b_8\ b_9\ b_{10}\ b_{11}\ b_{12}$

$P_1\ P_2\ d_8\ P_3\ d_7\ d_6\ d_5\ P_4\ d_4\ d_3\ d_2\ d_1$

⇓ Insert data bits

$P_1\ P_2\ 1\ P_3\ 0\ 1\ 0\ P_4\ 1\ 1\ 0\ 1$

Parity bits are calculated as follows.

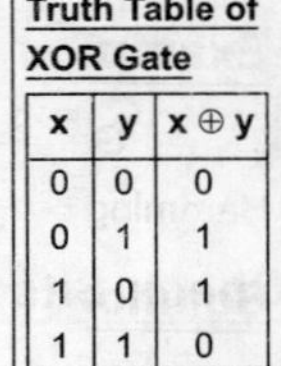

Truth Table of XOR Gate

x	y	x ⊕ y
0	0	0
0	1	1
1	0	1
1	1	0

$$P_1 = b_3 \oplus b_5 \oplus b_7 \oplus b_9 \oplus b_{11} = 1 \oplus 0 \oplus 0 \oplus 1 \oplus 0 = 0$$

$$P_2 = b_3 \oplus b_6 \oplus b_7 \oplus b_{10} \oplus b_{11} = 1 \oplus 1 \oplus 0 \oplus 1 \oplus 0 = 1$$

$$P_3 = b_5 \oplus b_6 \oplus b_7 \oplus b_{12} = 0 \oplus 1 \oplus 0 \oplus 1 = 0$$

$$P_4 = b_9 \oplus b_{10} \oplus b_{11} \oplus b_{12} = 1 \oplus 1 \oplus 0 \oplus 1 = 1$$

$$\therefore P_1 P_2 P_3 P_4 = 0\,1\,0\,1$$

Substitute the parity bits in Hamming code.

Hamming code : $P_1\ P_2\ 1\ P_3\ 0\ 1\ 0\ P_4\ 1\ 1\ 0\ 1$

⇓ Insert parity bits

0 1 1 0 0 1 0 1 1 1 0 1

Case i: Received Code has No Error

Received Hamming code : $b_1\ b_2\ b_3\ b_4\ b_5\ b_6\ b_7\ b_8\ b_9\ b_{10}\ b_{11}\ b_{12}$

0 1 1 0 0 1 0 1 1 1 0 1

The check bits are generated as follows:

$$C_1 = b_1 \oplus b_3 \oplus b_5 \oplus b_7 \oplus b_9 \oplus b_{11} = 0 \oplus 1 \oplus 0 \oplus 0 \oplus 1 \oplus 0 = 0$$

$$C_2 = b_2 \oplus b_3 \oplus b_6 \oplus b_7 \oplus b_{10} \oplus b_{11} = 1 \oplus 1 \oplus 1 \oplus 0 \oplus 1 \oplus 0 = 0$$

$$C_3 = b_4 \oplus b_5 \oplus b_6 \oplus b_7 \oplus b_{12} = 0 \oplus 0 \oplus 1 \oplus 0 \oplus 1 = 0$$

$$C_4 = b_8 \oplus b_9 \oplus b_{10} \oplus b_{11} \oplus b_{12} = 1 \oplus 1 \oplus 1 \oplus 0 \oplus 1 = 0$$

$$\therefore C_4 C_3 C_2 C_1 = 0\,0\,0\,0$$

Here all the check bits are zeros and hence no error in received Hamming code.

Case ii: Introduce Error in b_5 and Check for Error

Received Hamming code : $b_1\ b_2\ b_3\ b_4\ b_5\ b_6\ b_7\ b_8\ b_9\ b_{10}\ b_{11}\ b_{12}$

0 1 1 0 1 1 0 1 1 1 0 1

The check bits are generated as follows:

$$C_1 = b_1 \oplus b_3 \oplus b_5 \oplus b_7 \oplus b_9 \oplus b_{11} = 0 \oplus 1 \oplus 1 \oplus 0 \oplus 1 \oplus 0 = 1$$

$$C_2 = b_2 \oplus b_3 \oplus b_6 \oplus b_7 \oplus b_{10} \oplus b_{11} = 1 \oplus 1 \oplus 1 \oplus 0 \oplus 1 \oplus 0 = 0$$

$$C_3 = b_4 \oplus b_5 \oplus b_6 \oplus b_7 \oplus b_{12} = 0 \oplus 1 \oplus 1 \oplus 0 \oplus 1 = 1$$

$$C_4 = b_8 \oplus b_9 \oplus b_{10} \oplus b_{11} \oplus b_{12} = 1 \oplus 1 \oplus 1 \oplus 0 \oplus 1 = 0$$

$\therefore C_4\,C_3\,C_2\,C_1 = 0\,1\,0\,1$

$\Downarrow$

5 (In received Hamming code, error in bit b_5)

Now, complement the bit b_5 to get the correct code.

	b_1	b_2	b_3	b_4	b_5	b_6	b_7	b_8	b_9	b_{10}	b_{11}	b_{12}
Received code =	0	1	1	0	1	1	0	1	1	1	0	1

$\downarrow$ Complement b_5

Corrected code =	0	1	1	0	0	1	0	1	1	1	0	1

Example 1.11

Generate the Hamming code for an 9-bit data string 100101001. Introduce error in a data bit and prove that Hamming code detect the error.

Solution

For 9 data bits, 4 parity bits are required.

$\therefore n = 9,\ k = 4$

Here, $2^k = 2^4 = 16$

$n + k + 1 = 9 + 4 + 1 = 14$

The condition, $2^k \geq n + k + 1$ is satisfied.

The size of Hamming code is $n + k = 9 + 4 = 13$ bits.

Let, Hamming code = $b_1\,b_2\,b_3\,b_4\,b_5\,b_6\,b_7\,b_8\,b_9\,b_{10}\,b_{11}\,b_{12}\,b_{13}$

In Hamming code, the bit positions 1, 2, 4 and 8 are reserved for parity bits and remaining bits are used for data bits. The data and parity bits are combined to form Hamming code as shown below:

Hamming code :	b_1	b_2	b_3	b_4	b_5	b_6	b_7	b_8	b_9	b_{10}	b_{11}	b_{12}	b_{13}
	P_1	P_2	d_9	P_3	d_8	d_7	d_6	P_4	d_5	d_4	d_3	d_2	d_1

$\Downarrow$ Insert data bits

$P_1\ P_2\ 1\ P_3\ 0\ 0\ 1\ P_4\ 0\ 1\ 0\ 0\ 1$

Parity bits are calculated as follows.

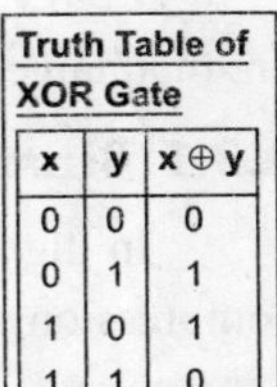

Truth Table of XOR Gate

x	y	x ⊕ y
0	0	0
0	1	1
1	0	1
1	1	0

$P_1 = b_3 \oplus b_5 \oplus b_7 \oplus b_9 \oplus b_{11} \oplus b_{13} = 1 \oplus 0 \oplus 1 \oplus 0 \oplus 0 \oplus 1 = 1$

$P_2 = b_3 \oplus b_6 \oplus b_7 \oplus b_{10} \oplus b_{11} = 1 \oplus 0 \oplus 1 \oplus 1 \oplus 0 = 1$

$P_3 = b_5 \oplus b_6 \oplus b_7 \oplus b_{12} \oplus b_{13} = 0 \oplus 0 \oplus 1 \oplus 0 \oplus 1 = 0$

$P_4 = b_9 \oplus b_{10} \oplus b_{11} \oplus b_{12} \oplus b_{13} = 0 \oplus 1 \oplus 0 \oplus 0 \oplus 1 = 0$

$\therefore P_1P_2P_3P_4 = 1\,1\,0\,0$

Substitute the parity bits in Hamming code.

Hamming code : $P_1\ P_2\ 1\ P_3\ 0\ 0\ 1\ P_4\ 0\ 1\ 0\ 0\ 1$

$\Downarrow$ Insert parity bits

1 1 1 0 0 0 1 0 0 1 0 0 1

Case i: Received Code has No Error

Received Hamming code :	b_1	b_2	b_3	b_4	b_5	b_6	b_7	b_8	b_9	b_{10}	b_{11}	b_{12}	b_{13}
	1	1	1	0	0	0	1	0	0	1	0	0	1

The check bits are generated as follows:

$$C_1 = b_1 \oplus b_3 \oplus b_5 \oplus b_7 \oplus b_9 \oplus b_{11} \oplus b_{13} = 1 \oplus 1 \oplus 0 \oplus 1 \oplus 0 \oplus 0 \oplus 1 = 0$$

$$C_2 = b_2 \oplus b_3 \oplus b_6 \oplus b_7 \oplus b_{10} \oplus b_{11} = 1 \oplus 1 \oplus 0 \oplus 1 \oplus 1 \oplus 0 = 0$$

$$C_3 = b_4 \oplus b_5 \oplus b_6 \oplus b_7 \oplus b_{12} \oplus b_{13} = 0 \oplus 0 \oplus 0 \oplus 1 \oplus 0 \oplus 1 = 0$$

$$C_4 = b_8 \oplus b_9 \oplus b_{10} \oplus b_{11} \oplus b_{12} \oplus b_{13} = 0 \oplus 0 \oplus 1 \oplus 0 \oplus 0 \oplus 1 = 0$$

$$\therefore C_4\,C_3\,C_2\,C_1 = 0\,0\,0\,0$$

Here all the check bits are zeros and hence no error in received Hamming code.

Case ii: Introduce Error in b_6 and Check for Error

Received Hamming code : b_1 b_2 b_3 b_4 b_5 b_6 b_7 b_8 b_9 b_{10} b_{11} b_{12} b_{13}

1 1 1 0 0 1 1 0 0 1 0 0 1

The check bits are generated as follows:

$$C_1 = b_1 \oplus b_3 \oplus b_5 \oplus b_7 \oplus b_9 \oplus b_{11} \oplus b_{13} = 1 \oplus 1 \oplus 0 \oplus 1 \oplus 0 \oplus 0 \oplus 1 = 0$$

$$C_2 = b_2 \oplus b_3 \oplus b_6 \oplus b_7 \oplus b_{10} \oplus b_{11} = 1 \oplus 1 \oplus 1 \oplus 1 \oplus 1 \oplus 0 = 1$$

$$C_3 = b_4 \oplus b_5 \oplus b_6 \oplus b_7 \oplus b_{12} \oplus b_{13} = 0 \oplus 0 \oplus 1 \oplus 1 \oplus 0 \oplus 1 = 1$$

$$C_4 = b_8 \oplus b_9 \oplus b_{10} \oplus b_{11} \oplus b_{12} \oplus b_{13} = 0 \oplus 0 \oplus 1 \oplus 0 \oplus 0 \oplus 1 = 0$$

$$\therefore C_4\,C_3\,C_2\,C_1 = 0\,1\,1\,0$$

⇓

6 (In received Hamming code, error in bit b_6)

Now, complement the bit b_6 to get the correct code.

b_1 b_2 b_3 b_4 b_5 b_6 b_7 b_8 b_9 b_{10} b_{11} b_{12} b_{13}

Received code = 1 1 1 0 0 1 1 0 0 1 0 0 1

↓ Complement b_6

Corrected code = 1 1 1 0 0 0 1 0 0 1 0 0 1

1.3 Codes and Code Conversions

Codes are used to represent the information in different forms so that secrecy of the information are maintained. Also coded form of information is easy for storage and transmission in digital systems.

1.3.1 Binary Codes

In digital systems the informations can be coded in binary for various reasons like security, data compression, transfer of information from human to machine (ASCII codes), etc.

Binary codes are also used to code numbers as well as alphabets.

A ***binary code*** is a group of n-bit binary used to represent the set of elements of an information. Each binary code of the group is assigned to a member/element of a set that is being coded.

The size n, of binary code depends on number of codes needed to assign a unique code to each member/element of a set of information. With n-bit binary code it is possible to generate 2^n distinct binary codes.

When $n = 1$, $2^1 = 2$, Two codes are possible

$n = 2$, $2^2 = 4$, Four codes are possible

n = 3, $2^3 = 8$, Eight codes are possible

n = 4, $2^4 = 16$, Sixteen codes are possible and so on.

Some of the binary codes are weighted codes. In weighted codes, each digit has a weight and on adding the weights we will get the decimal value of the codes.

Table 1.19: Binary Codes for n = 1 to 5

Size of Binary Code (n)	Number of Codes 2^n	Binary Codes
1	$2^1 = 2$	0
		1
2	$2^2 = 4$	00
		01
		10
		11
3	$2^3 = 8$	000
		001
		010
		011
		100
		101
		110
		111
4	$2^4 = 16$	0000
		0001
		0010
		0011
		0100
		0101
		0110
		0111
		1000
		1001
		1010
		1011
		1100
		1101
		1110
		1111

Size of Binary Code (n)	Number of Codes 2^n	Binary Codes
5	$2^5 = 32$	00000
		00001
		00010
		00011
		00100
		00101
		00110
		00111
		01000
		01001
		01010
		01011
		01100
		01101
		01110
		01111
		10000
		10001
		10010
		10011
		10100
		10101
		10110
		10111
		11000
		11001
		11010
		11011
		11100
		11101
		11110
		11111

1.3.2 Types of BCD Codes

The different types of BCD codes are,

- 8421 BCD (Generally called BCD)
- 2421 code
- 5421 code
- 8, 4, −2, −1 code
- Excess-3 code

8421, 2421, 5421 and 8, 4, −2, −1 are ***weighted codes***. In these codes on adding the weights of 1's in the code we get the value of decimal symbol.

Example: 5421 code: $8 \Rightarrow 1011 \Rightarrow 5+0+2+1=8$

$7 \Rightarrow 0111 \Rightarrow 0+4+2+1=7$

Example: 2421 code: $5 \Rightarrow 1011 \Rightarrow 2+0+2+1=5$

$6 \Rightarrow 1100 \Rightarrow 2+4+0+0=6$

Example: 8, 4, −2, −1: $4 \Rightarrow 0100 \Rightarrow 0+4+0+0=4$

$9 \Rightarrow 1111 \Rightarrow 8+4+(-2)+(-1)=9$

$7 \Rightarrow 1001 \Rightarrow 8+0+0+(-1)=7$

The 2421 and Excess-3 are self complementing code.

In ***self complementing code***, the complement of the code will give the code for 9's complement representation of decimal digit.

Some examples of self complementing code are given below:

$5 \xrightarrow[\text{BCD}]{2421} 1011 \xrightarrow{\text{Complement}} 0100 = 4$

(4 in 2421 BCD)

9's complement of 5
$$\begin{array}{r} 9 \\ -5 \\ \hline 4 \end{array}$$

$37 \xrightarrow[\text{BCD}]{2421} 0011\,1101 \xrightarrow{\text{Complement}} 1100\,0010 = 62$

(62 in 2421 code)

9's complement of 37
$$\begin{array}{r} 99 \\ -37 \\ \hline 62 \end{array}$$

$6 \xrightarrow{\text{Excess-3}} 1001 \xrightarrow{\text{Complement}} 0110 = 3$

(3 in Excess-3)

9's complement of 6
$$\begin{array}{r} 9 \\ -6 \\ \hline 3 \end{array}$$

$452 \xrightarrow{\text{Excess-3}} 0111 \quad 1000 \quad 0101$

↓ Complement

$1000 \quad 0111 \quad 1010 = 547$

(547 in Excess-3)

9's complement of 452
$$\begin{array}{r} 999 \\ -452 \\ \hline 547 \end{array}$$

Table 1.20: Summary of Various BCD Codes for Decimal Digit 0 to 9

Decimal Digit	8421 BCD	2421 BCD	5421 BCD	8 4 −2 −1 BCD	Excess-3 BCD
0	0000	0000	0000	0000	0011
1	0001	0001	0001	0111	0100
2	0010	0010	0010	0110	0101
3	0011	0011	0011	0101	0110
4	0100	0100	0100	0100	0111
5	0101	1011	1000	1011	1000
6	0110	1100	1001	1010	1001
7	0111	1101	1010	1001	1010
8	1000	1110	1011	1000	1011
9	1001	1111	1100	1111	1100

***Note:** In 4-bit binary code $2^4 = 16$ binary codes are possible. But in decimal representation using 4-bit codes, only 10 codes are used and 6 codes are not used. The unused codes are illegal codes in that respective codes.*

1.3.3 8421 BCD

The ***8421 BCD*** code is a binary coded decimal that uses 4-digit binary to represent each decimal digit. BCD is a weighted code with weight as 8, 4, 2 and 1 for the 4-digits of binary. In general, the 8421 BCD is simply called ***BCD***.

In BCD the 10 symbols of decimal number system is represented by a unique 4-bit code which is binary value of the decimal symbol itself.

In order to convert a decimal digit to BCD first convert the decimal digit to binary and then append with leading zeros to form 4-digit binary. The BCD for decimal digits 0 to 9 are listed in Table 1.20.

The weight of the four digits of BCD code are 8, 4, 2 and 1 and they are same as weight of binary number system.

$$8 \quad 4 \quad 2 \quad 1 \Rightarrow 2^3 \; 2^2 \; 2^1 \; 2^0$$

The BCD code can be converted to decimal using binary to decimal conversion technique, in which add the weights of code wherever 1 is available as shown below:

$0000 = 0 + 0 + 0 + 0 = 0$

$0001 = 0 + 0 + 0 + 1 = 1$

$0010 = 0 + 0 + 2 + 0 = 2$

$0011 = 0 + 0 + 2 + 1 = 3$

$0100 = 0 + 4 + 0 + 0 = 4$

$0101 = 0 + 4 + 0 + 1 = 5$

$0110 = 0 + 4 + 2 + 0 = 6$

$0111 = 0 + 4 + 2 + 1 = 7$

$1000 = 8 + 0 + 0 + 0 = 8$

$1001 = 8 + 0 + 0 + 1 = 9$

In BCD representation of multi digit decimal number each digit of decimal number is represented by 4-digit BCD. Some examples of BCD representation of multi digit decimal numbers are shown in Table 1.21.

Table 1.21: Examples of BCD Representation of Multi Digit Decimal Number

Decimal Number	BCD			
23			0010	0011
45			0100	0101
376		0011	0111	0110
498		0100	1001	1000
1746	0001	0111	0100	0110
8532	1000	0101	0011	0010

The advantage of BCD is that the digital system can directly perform mathematical operations on decimal number in BCD without converting them to binary number system.

BCD Addition

BCD addition is performed by adding the BCD representation directly and then result is corrected to BCD.

In BCD addition, in order to correct the result, a correction 6_{10} $(0110)_2$ is added to BCD sum of digits whenever sum of BCD digits exceeds 9 or if there is a carry to next addition.

Example 1.12

Perform BCD addition of following decimal numbers.

i) $6_{10} + 3_{10}$ **ii)** $7_{10} + 8_{10}$ **iii)** $42_{10} + 35_{10}$ **iv)** $72_{10} + 19_{10}$

v) $84_{10} + 63_{10}$ **vi)** $327_{10} + 915_{10}$ **vii)** $4831_{10} + 2974_{10}$

Solution

i) $6_{10} + 3_{10}$

		BCD addition		
6	Convert to BCD ⟹	0110		
+ 3		+ 0011		
9		1001	Convert to decimal ⟶	9_{10}

ii) $7_{10} + 8_{10}$

BCD addition

```
   7     Convert to BCD       0111
 + 8   ================>   + 1000
 ---                        -----
  15                         1111
                           + 0110  (Add 6)      Sum exceed 9 and so correction 6 is added
                           ------
                           1 0101
                             ⇓ Represent in BCD
                         0001  0101  --Convert to decimal-->  15₁₀
```

iii) $42_{10} + 35_{10}$

BCD addition

```
  42     Convert to BCD       0100   0010
+ 35   ================>   +  0011   0101
----                       -------------
  77                          0111   0111  --Convert to decimal-->  77₁₀
```

iv) $72_{10} + 19_{10}$

BCD addition

```
  72     Convert to BCD       0111   0010
+ 19   ================>   +  0001   1001
----                       -------------
  91                          1000   1011      Sum of first digits exceed 9 and so
                                 + 0110 (Add 6)  correction 6 is added to first digit.
                           -------------
                              1001   0001  --Convert to decimal-->  91₁₀
```

v) $84_{10} + 63_{10}$

BCD addition

```
  84     Convert to BCD       1000   0100
+ 63   ================>   +  0110   0011
----                       -------------
 147                          1110   0111
                           +  0110          (Add 6)
                           -------------
                            1 0100   0111      Sum of second digits exceed 9 and so
                                               correction 6 is added to second digit.
                              ⇓ Represent in BCD
                         0001 0100   0111  --Convert to decimal-->  147₁₀
```

vi) $327_{10} + 915_{10}$

BCD addition

```
  327                      0011   0010   0111
           Convert to BCD
+ 915      ------------->  1001   0001   0101
 1242                      1100   0011   1100
                          +0110         +0110  (Add 6)
                         1 0010   0100   0010
                         ⇓ Represent in BCD
                      0001 0010   0100   0010   --Convert to decimal-->  1242₁₀
```

Sum exceed 9 in addition of first and third digits and so correction 6 is added to first and third digits.

---

vii) $4831_{10} + 2974_{10}$

BCD addition

```
  4831                     0100   1000   0011   0001
           Convert to BCD
+ 2974     ------------->  0010   1001   0111   0100
  7805                     0111   0001   1010   0101
                                 +0110  +0110  (Add 6)
                           0111   1000   0000   0101  --Convert to decimal-->  7805₁₀
```

Sum of second digits exceed 9 and so correction 6 is added to second digit.

Sum of third digits generate a carry to next digit and so correction 6 is added to third digit.

1.3.4 2421 Code

The ***2421 code*** is a binary coded decimal that uses 4-digit binary to represent each decimal digit. It is a weighted code with weight as 2, 4, 2 and 1 for the 4-digits of binary.

The 2421 code for decimal digits 0 to 9 are listed in Table 1.20.

The decimal digit can be converted to 2421 code by realizing the decimal digit as sum of weights and then replacing the weight by 1 as shown below:

$1_{10} \xrightarrow[\text{of 2421 to form 1}]{\text{Take weight}} 0+0+0+1 \xrightarrow[\text{in the place weight are considered}]{\text{Put 1}} 0001$

$2_{10} \xrightarrow[\text{of 2421 to form 2}]{\text{Take weight}} 0+0+2+0 \xrightarrow[\text{in the place weight are considered}]{\text{Put 1}} 0010$

$3_{10} \xrightarrow[\text{of 2421 to form 3}]{\text{Take weight}} 0+0+2+1 \xrightarrow[\text{in the place weight are considered}]{\text{Put 1}} 0011$

$4_{10} \xrightarrow[\text{of 2421 to form 4}]{\text{Take weight}} 0+4+0+0 \xrightarrow[\text{in the place weight are considered}]{\text{Put 1}} 0100$

$5_{10} \xrightarrow[\text{of 2421 to form 5}]{\text{Take weight}} 2+0+2+1 \xrightarrow[\text{in the place weight are considered}]{\text{Put 1}} 1011$

$6_{10} \xrightarrow[\text{of 2421 to form 6}]{\text{Take weight}} 2+4+0+0 \xrightarrow[\text{in the place weight are considered}]{\text{Put 1}} 1100$

$7_{10} \xrightarrow[\text{of 2421 to form 7}]{\text{Take weight}} 2+4+0+1 \xrightarrow[\text{in the place weight are considered}]{\text{Put 1}} 1101$

$$8_{10} \xrightarrow[\text{of 2421 to form 8}]{\text{Take weight}} 2+4+2+0 \xrightarrow[\text{in the place weight are considered}]{\text{Put 1}} 1110$$

$$9_{10} \xrightarrow[\text{of 2421 to form 9}]{\text{Take weight}} 2+4+2+1 \xrightarrow[\text{in the place weight are considered}]{\text{Put 1}} 1111$$

The 2421 code can be converted to decimal by adding the weights wherever 1 is available in the code as shown below:

$0000 = 0 + 0 + 0 + 0 = 0$

$0001 = 0 + 0 + 0 + 1 = 1$

$0010 = 0 + 0 + 2 + 0 = 2$

$0011 = 0 + 0 + 2 + 1 = 3$

$0100 = 0 + 4 + 0 + 0 = 4$

$1011 = 2 + 0 + 2 + 1 = 5$

$1100 = 2 + 4 + 0 + 0 = 6$

$1101 = 2 + 4 + 0 + 1 = 7$

$1110 = 2 + 4 + 2 + 0 = 8$

$1111 = 2 + 4 + 2 + 1 = 9$

In 2421 code representation of multi digit decimal number, each digit of decimal number is represented by 4-digit 2421 code. Some examples of 2421 code representation of multi digit decimal numbers are shown in Table 1.22.

Table 1.22: Examples of 2421 and 5421 Code Representation of Multi Digit Decimal Number

Decimal Number	2421 Code				5421 Code			
23			0010	0011			0010	0011
45			0100	1011			0100	1000
376		0011	1101	1100		0011	1010	1001
498		0100	1111	1110		0100	1100	1011
1746	0001	1101	0100	1100	0001	1010	0100	1001
8532	1110	1011	0011	0010	1011	1000	0011	0010

Example 1.13

Convert 5_{10} to 2421 code and verify by back converting to decimal.

Solution

The 2421 code can be obtained expressing 5_{10} as sum of 2421 weight and then inserting 1 in the place of weight that are used to form the decimal value.

$$5_{10} \xrightarrow[\text{of 2421 to form 5}]{\text{Take weight}} 2+0+2+1 \xrightarrow[\text{in the place weight are considered}]{\text{Put 1}} 1011$$

$\therefore$ 2421 code for 5_{10} = 1011

To convert 2421 code to decimal add the weight wherever 1 is present in the code.

$$1011 \xrightarrow[\text{of weights where 1 is available}]{\text{Take sum}} 2+0+2+1 = 5_{10}$$

∴ Decimal value of 2421 code 1011 = 5_{10}

1.3.5 5421 Code

The ***5421 code*** is a binary coded decimal that uses 4-digit binary to represent each decimal digit. It is a weighted code with weight as 5, 4, 2 and 1 for the 4-digits of binary.

The 5421 code for decimal digits 0 to 9 are listed in Table 1.20.

The decimal digit can be converted to 5421 code by realizing the decimal digit as sum of weights and then replacing the weight by 1 as shown below:

$$1_{10} \xrightarrow[\text{of 5421 to form 1}]{\text{Take weight}} 0+0+0+1 \xrightarrow[\text{in the place weight are considered}]{\text{Put 1}} 0001$$

$$2_{10} \xrightarrow[\text{of 5421 to form 2}]{\text{Take weight}} 0+0+2+0 \xrightarrow[\text{in the place weight are considered}]{\text{Put 1}} 0010$$

$$3_{10} \xrightarrow[\text{of 5421 to form 3}]{\text{Take weight}} 0+0+2+1 \xrightarrow[\text{in the place weight are considered}]{\text{Put 1}} 0011$$

$$4_{10} \xrightarrow[\text{of 5421 to form 4}]{\text{Take weight}} 0+4+0+0 \xrightarrow[\text{in the place weight are considered}]{\text{Put 1}} 0100$$

$$5_{10} \xrightarrow[\text{of 5421 to form 5}]{\text{Take weight}} 5+0+0+0 \xrightarrow[\text{in the place weight are considered}]{\text{Put 1}} 1000$$

$$6_{10} \xrightarrow[\text{of 5421 to form 6}]{\text{Take weight}} 5+0+0+1 \xrightarrow[\text{in the place weight are considered}]{\text{Put 1}} 1001$$

$$7_{10} \xrightarrow[\text{of 5421 to form 7}]{\text{Take weight}} 5+0+2+0 \xrightarrow[\text{in the place weight are considered}]{\text{Put 1}} 1010$$

$$8_{10} \xrightarrow[\text{of 5421 to form 8}]{\text{Take weight}} 5+0+2+1 \xrightarrow[\text{in the place weight are considered}]{\text{Put 1}} 1011$$

$$9_{10} \xrightarrow[\text{of 5421 to form 9}]{\text{Take weight}} 5+4+0+0 \xrightarrow[\text{in the place weight are considered}]{\text{Put 1}} 1100$$

The 5421 code can be converted to decimal by adding the weights wherever 1 is available in the code as shown below:

$0000 = 0+0+0+0 = 0$

$0001 = 0+0+0+1 = 1$

$0010 = 0+0+2+0 = 2$

$0011 = 0+0+2+1 = 3$

$0100 = 0+4+0+0 = 4$

$1000 = 5+0+0+0 = 5$

$1001 = 5 + 0 + 0 + 1 = 6$

$1010 = 5 + 0 + 2 + 0 = 7$

$1011 = 5 + 0 + 2 + 1 = 8$

$1100 = 5 + 4 + 0 + 0 = 9$

In 5421 code representation of multi digit decimal number, each digit of decimal number is represented by 4-digit 5421 code. Some examples of 5421 code representation of multi digit decimal numbers are shown in Table 1.22.

Example 1.14

Convert 6_{10} to 5421 code and verify by back converting to decimal.

Solution

The 5421 code can be obtained by expressing 6_{10} as sum of 5421 weights and then inserting 1 in the place of weights that are used to form decimal value.

$$6_{10} \xrightarrow[\text{of 5421 to form 6}]{\text{Take weight}} 1 + 0 + 0 + 1 \xrightarrow[\text{in the place weight are considered}]{\text{Put 1}} 1001$$

$\therefore$ 5421 code for 6_{10} = 1001

To convert the 5421 code to decimal add the weights wherever 1 is present in the code.

$$1001 \xrightarrow[\text{of weights where 1 is available}]{\text{Take sum}} 5 + 0 + 0 + 1 = 6_{10}$$

$\therefore$ Decimal value of 5421 code 1001 = 6_{10}

1.3.6 8, 4, −2, −1 Code

The ***8, 4, −2, −1 code*** is a binary coded decimal that uses 4-digit binary to represent each decimal digit. It is a weighted code with weight as 8, 4, −2 and −1 for the 4-digits of binary.

The 8, 4, −2, −1 code for decimal digits 0 to 9 are listed in Table 1.20.

The decimal digit can be converted to 8, 4, −2, −1 code by realizing the decimal digit as sum of weights and then replacing the weight by 1 as shown below:

$$1_{10} \xrightarrow[\text{of 8, 4, – 2, – 1 to form 1}]{\text{Take weight}} 0 + 4 - 2 - 1 \xrightarrow[\text{in the place weight are considered}]{\text{Put 1}} 0111$$

$$2_{10} \xrightarrow[\text{of 8, 4, – 2, – 1 to form 2}]{\text{Take weight}} 0 + 4 - 2 + 0 \xrightarrow[\text{in the place weight are considered}]{\text{Put 1}} 0110$$

$$3_{10} \xrightarrow[\text{of 8, 4, – 2, – 1 to form 3}]{\text{Take weight}} 0 + 4 + 0 - 1 \xrightarrow[\text{in the place weight are considered}]{\text{Put 1}} 0101$$

$$4_{10} \xrightarrow[\text{of 8, 4, – 2, – 1 to form 4}]{\text{Take weight}} 0 + 4 + 0 + 0 \xrightarrow[\text{in the place weight are considered}]{\text{Put 1}} 0100$$

$$5_{10} \xrightarrow[\text{of 8, 4, – 2, – 1 to form 5}]{\text{Take weight}} 8 + 0 - 2 - 1 \xrightarrow[\text{in the place weight are considered}]{\text{Put 1}} 1011$$

$$6_{10} \xrightarrow[\text{of 8, 4, – 2, – 1 to form 6}]{\text{Take weight}} 8 + 0 - 2 + 0 \xrightarrow[\text{in the place weight are considered}]{\text{Put 1}} 1010$$

$$7_{10} \xrightarrow[\text{of 8, 4, – 2, – 1 to form 7}]{\text{Take weight}} 8 + 0 + 0 - 1 \xrightarrow[\text{in the place weight are considered}]{\text{Put 1}} 1001$$

$$8_{10} \xrightarrow[\text{of 8, 4, – 2, – 1 to form 8}]{\text{Take weight}} 8 + 0 + 0 + 0 \xrightarrow[\text{in the place weight are considered}]{\text{Put 1}} 1000$$

$$9_{10} \xrightarrow[\text{of 8, 4, – 2, – 1 to form 9}]{\text{Take weight}} 8 + 4 - 2 - 1 \xrightarrow[\text{in the place weight are considered}]{\text{Put 1}} 1111$$

The 8, 4, −2, −1 code can be converted to decimal by adding the weights wherever 1 is available in the code as shown below:

$$0000 = 0 + 0 + 0 + 0 = 0$$

$$0111 = 0 + 4 - 2 - 1 = 1$$

$$0110 = 0 + 4 - 2 + 0 = 2$$

$$0101 = 0 + 4 + 0 - 1 = 3$$

$$0100 = 0 + 4 + 0 + 0 = 4$$

$$1011 = 8 + 0 - 2 - 1 = 5$$

$$1010 = 8 + 0 - 2 + 0 = 6$$

$$1001 = 8 + 0 + 0 - 1 = 7$$

$$1000 = 8 + 0 + 0 + 0 = 8$$

$$1111 = 8 + 4 - 2 - 1 = 9$$

In 8, 4, −2, −1 code representation of multi digit decimal number, each digit of decimal number is represented by 4-digit 8, 4, −2, −1 code. Some examples of 8, 4, −2, −1 code representation of multi digit decimal numbers are shown in Table 1.23.

Table 1.23: Examples of 8, 4, −2, −1 and Excess-3 Code Representation of Multi Digit Decimal Number

Decimal Number	8, 4, −2, −1 Code	Excess-3 Code
23	0110 0101	0101 0110
45	0100 1011	0111 1000
376	0101 1001 1010	0110 1010 1001
498	0100 1111 1000	0111 1100 1011
1746	0111 1001 0100 1010	0100 1010 0111 1001
8532	1000 1011 0101 0110	1011 1000 0110 0101

Example 1.15

Convert 3_{10} to 8, 4, −2, −1 code and verify by back converting to decimal.

Solution

The 8, 4, −2, −1 code can be obtained by expressing 3_{10} as sum of 8, 4, −2, −1 weights and then inserting 1 in the place of weights that are used to form decimal value.

$$3_{10} \xrightarrow[\text{of 8, 4, – 2, – 1 to form 3}]{\text{Take weight}} 0 + 4 + 0 - 1 \xrightarrow[\text{in the place weight are considered}]{\text{Put 1}} 0101$$

∴ 8, 4, –2, –1 code for 3_{10} = 0101

To convert the 8, 4, –2, –1 code to decimal add the weights wherever 1 is present in the code.

$$0101 \xrightarrow[\text{of weights where 1 is available}]{\text{Take sum}} 0+4+0-1 = 3_{10}$$

∴ Decimal value of 8, 4, –2, –1 code 0101 = 3_{10}

1.3.7 Excess-3 Code

The ***excess-3 code*** is a binary coded decimal that uses 4-digit binary to represent each decimal digit. The excess-3 is non-weighted code. The excess-3 code for decimal digits 0 to 9 are listed in Table 1.20.

Excess-3 code for a decimal digit is obtained by adding 4-bit binary value of 3_{10} (0011_2) to 8421 BCD of the decimal digit as shown below:

$0 \Rightarrow 0000 + 0011 \Rightarrow 0011$

$1 \Rightarrow 0001 + 0011 \Rightarrow 0100$

$2 \Rightarrow 0010 + 0011 \Rightarrow 0101$

$3 \Rightarrow 0011 + 0011 \Rightarrow 0110$

$4 \Rightarrow 0100 + 0011 \Rightarrow 0111$

$5 \Rightarrow 0101 + 0011 \Rightarrow 1000$

$6 \Rightarrow 0110 + 0011 \Rightarrow 1001$

$7 \Rightarrow 0111 + 0011 \Rightarrow 1010$

$8 \Rightarrow 1000 + 0011 \Rightarrow 1011$

$9 \Rightarrow 1001 + 0011 \Rightarrow 1100$

The excess-3 code can be converted to decimal by subtracting 3_{10} (011_2) to get BCD and then converting BCD to decimal using binary to decimal convertion technique.

In excess-3 code representation of multi digit decimal number, each digit of decimal number is represented by 4-digit excess-3 code. Some examples of excess-3 code representation of multi digit decimal numbers are shown in Table 1.23.

Example 1.16

Convert 8_{10} to excess-3 code and verify by back converting to decimal.

Solution

The excess-3 code can be obtained by adding 011_2 to binary value of the decimal digit.

$$8_{10} \xrightarrow{\text{Convert to binary}} 1000_2 \xrightarrow{\text{Add 011}} 1011$$

∴ Excess-3 code for 8_{10} = 1011

To convert excess-3 code to decimal, subtract 011_2 to get binary value and then convert binary to decimal.

$$1011 \xrightarrow{\text{Subtract } 011_2} 1000_2 \xrightarrow[\text{to decimal}]{\text{Convert}} 8_{10}$$

∴ Decimal value of excess-3 code 1011 = 8_{10}

Example 1.17

Represent the decimal number 2569 in **(a)** BCD **(b)** Excess-3 code **(c)** 2421 code **(d)** 5421 code **(e)** 8, 4, – 2, – 1 code

Solution

a) Convert to BCD

$2_{10} \xrightarrow[\text{to 4-bit bianry}]{\text{Convert}} 0010_2$

$5_{10} \xrightarrow[\text{to 4-bit bianry}]{\text{Convert}} 0101_2$

$6_{10} \xrightarrow[\text{to 4-bit bianry}]{\text{Convert}} 0110_2$

$9_{10} \xrightarrow[\text{to 4-bit bianry}]{\text{Convert}} 1001_2$

$\therefore$ BCD of 2569_{10} = 0010 0101 0110 1001

b) Convert to Excess-3

Add 011_2 to binary value of every decimal digit to get excess-3 code.

$2_{10} = 0010_2$	$5_{10} = 0101_2$	$6_{10} = 0110_2$	$9_{10} = 1001_2$
+011	+011	+011	+011
0101	1000	1001	1100

$\therefore$ Excess-3 code for 2569_{10} = 0101 1000 1001 1100

c) Convert to 2421 code

$2_{10} \xrightarrow[\text{of 2421 to form 2}]{\text{Take weight}} 0+0+2+0 \xrightarrow[\text{in the place weight are considered}]{\text{Put 1}} 0010$

$5_{10} \xrightarrow[\text{of 2421 to form 5}]{\text{Take weight}} 2+0+2+1 \xrightarrow[\text{in the place weight are considered}]{\text{Put 1}} 1011$

$6_{10} \xrightarrow[\text{of 2421 to form 6}]{\text{Take weight}} 2+4+0+0 \xrightarrow[\text{in the place weight are considered}]{\text{Put 1}} 1100$

$9_{10} \xrightarrow[\text{of 2421 to form 9}]{\text{Take weight}} 2+4+2+1 \xrightarrow[\text{in the place weight are considered}]{\text{Put 1}} 1111$

$\therefore$ 2421 code for 2569_{10} = 0010 1011 1100 1111

d) Convert to 5421 code

$2_{10} \xrightarrow[\text{of 5421 to form 2}]{\text{Take weight}} 0+0+2+0 \xrightarrow[\text{in the place weight are considered}]{\text{Put 1}} 0010$

$5_{10} \xrightarrow[\text{of 5421 to form 5}]{\text{Take weight}} 5+0+0+0 \xrightarrow[\text{in the place weight are considered}]{\text{Put 1}} 1000$

$6_{10} \xrightarrow[\text{of 5421 to form 6}]{\text{Take weight}} 5+0+0+1 \xrightarrow[\text{in the place weight are considered}]{\text{Put 1}} 1001$

$9_{10} \xrightarrow[\text{of 5421 to form 9}]{\text{Take weight}} 5+4+0+0 \xrightarrow[\text{in the place weight are considered}]{\text{Put 1}} 1100$

$\therefore$ 5421 code for 2569_{10} = 0010 1000 1001 1100

e) Convert to 8, 4, –2, –1 code

$$2_{10} \xrightarrow[\text{of 8, 4, – 2, – 1 to form 2}]{\text{Take weight}} 0+4-2-0 \xrightarrow[\text{in the place weight are considered}]{\text{Put 1}} 0110$$

$$5_{10} \xrightarrow[\text{of 8, 4, – 2, – 1 to form 5}]{\text{Take weight}} 8+0-2-1 \xrightarrow[\text{in the place weight are considered}]{\text{Put 1}} 1011$$

$$6_{10} \xrightarrow[\text{of 8, 4, – 2, – 1 to form 6}]{\text{Take weight}} 8+0-2+0 \xrightarrow[\text{in the place weight are considered}]{\text{Put 1}} 1010$$

$$9_{10} \xrightarrow[\text{of 8, 4, – 2, – 1 to form 9}]{\text{Take weight}} 8+4-2-1 \xrightarrow[\text{in the place weight are considered}]{\text{Put 1}} 1111$$

∴ 8, 4, –2, –1 code for 2569_{10} = 0110 1011 1010 1111

1.3.8 Gray Code

In ***gray code*** only one bit changes from either "0" to "1" or "1" to "0" between two consecutive codes.

The 1-bit gray code can be used to represent decimal numbers 0 and 1.

The 2-bit gray code can be used to represent decimal numbers 0 to 3.

The 3-bit gray code can be used to represent decimal numbers 0 to 7.

The 4-bit gray code can be used to represent decimal numbers 0 to 15_{10}.

Table 1.24: Decimal and Equivalent Gray Code

Decimal Number	1-bit Gray Code	2-bit Gray Code	3-bit Gray Code	4-bit Gray Code
0	0	00	000	0000
1	1	01	001	0001
2		11	011	0011
3		10	010	0010
4			110	0110
5			111	0111
6			101	0101
7			100	0100
8				1100
9				1101
10				1111
11				1110
12				1010
13				1011
14				1001
15				1000

The gray code is constructed as follows:

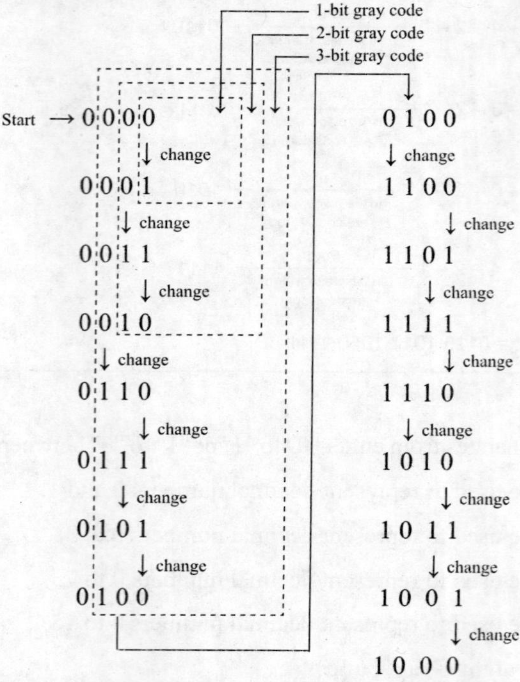

Gray code can be formed by another method as shown below:

Step-1: For 2^n gray code first form 2^{n-1} gray codes.

Step-2: Then write the gray codes of step-1 in reverse order as next set of 2^{n-1} gray codes.

Step-3: Append the gray codes of step-1 with leading "0".

Step-4: Append the gray codes of step-2 with leading "1".

We can start with 1-bit gray code which is 0 and 1 then proceed to higher size as shown below:

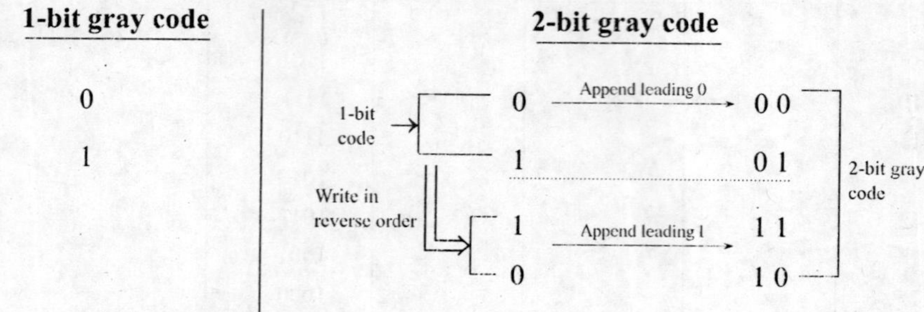

3-bit gray code

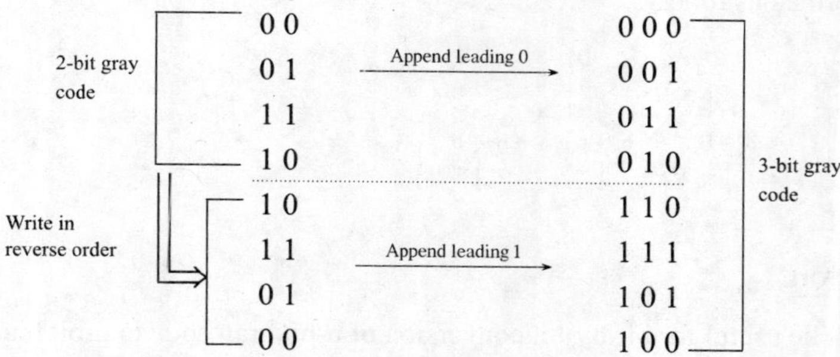

4-bit gray code

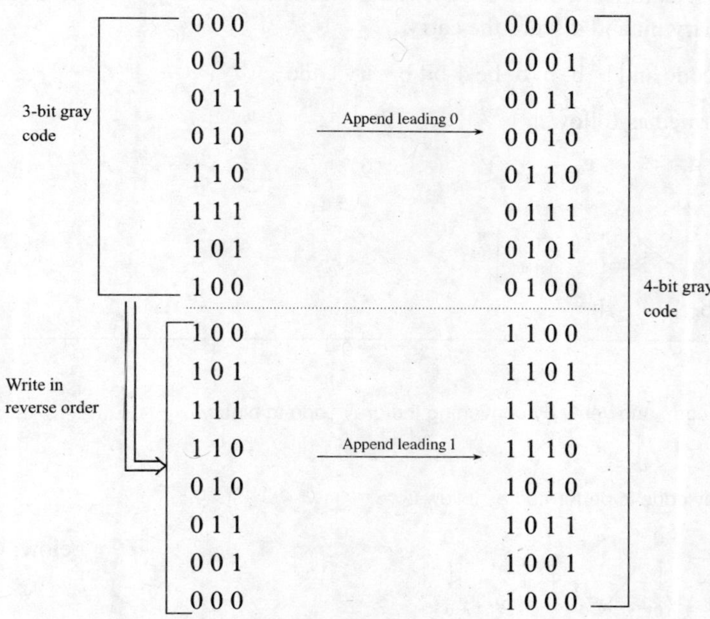

Similarly we can construct 5-bit, 6-bit, etc, gray codes.

The two methods discussed above will be useful to find the full range of n-bit gray code. But will not be useful to convert a given single binary code to gray code. The following procedure can be followed to convert a given single binary code to gray code and vice-versa.

Binary to Gray Code Conversion

The following procedure will be useful for bit-by-bit conversion of n-bit binary to n-bit gray code.

Step-1: The most significant bit of binary and gray code are same. Hence retain the most significant bit of binary as most significant bit of gray code.

Step-2: Starting from most significant bit of binary, add two consecutive bits of binary and record the sum as bit of gray code and discard the carry in every addition.

Let, $b_3\ b_2\ b_1\ b_0$ be 4-bit binary code and $g_3\ g_2\ g_1\ g_0$ be 4-bit gray code

Now, the conversion is performed as follows:

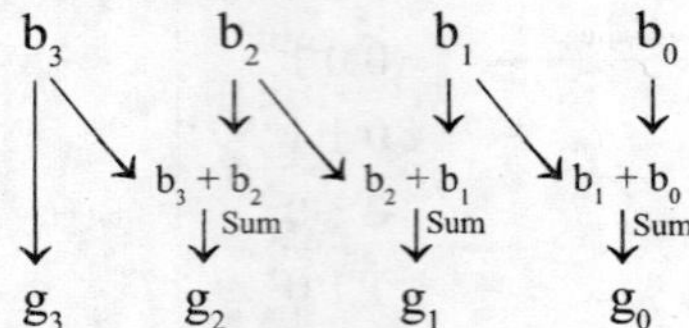

Gray Code to Binary Conversion

The following procedure will be useful for bit-by-bit conversion of n-bit gray code to n-bit binary.

Step-1: The most significant bit of gray code and binary are same. Hence retain the most significant bit of gray code as most significant bit of binary.

Step-2: Starting from most significant bit of binary, add the generated binary bit to next bit of gray code and record the sum as binary bit and discard the carry.

Let, $g_3\ g_2\ g_1\ g_0$ be 4-bit gray code and $b_3\ b_2\ b_1\ b_0$ be 4-bit binary code

Now, the conversion is performed as follows:

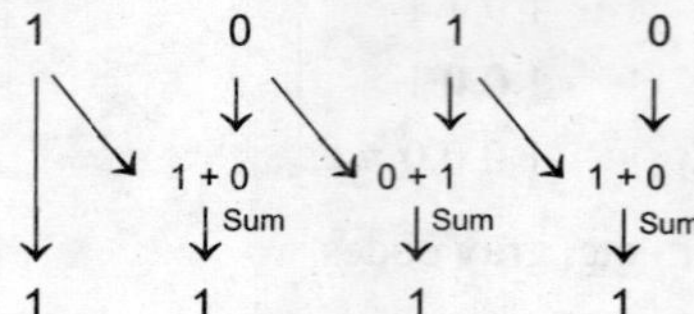

Example 1.18

Convert the binary 1010 to gray code and verify by converting the gray code to binary.

Solution

Conversion of binary code to gray code is performed as follows:

1 0 1 0

1 + 0 0 + 1 1 + 0

Sum Sum Sum

1 1 1 1

∴ Gray code for binary 1010 = 1111

Conversion of gray code to binary is performed as follows:

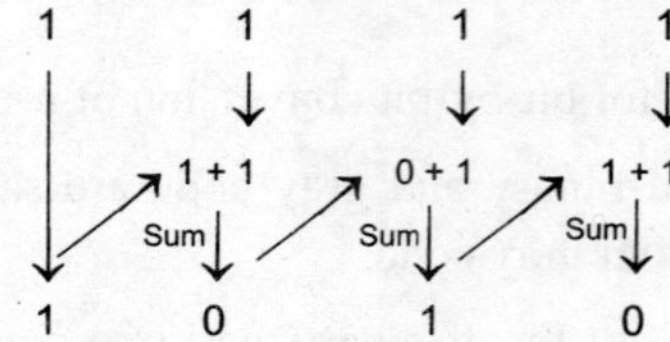

∴ Binary for gray code 1111 = 1010

Application for Gray Code

In rotating machines the angular position of shaft and hence speed of rotating shaft are popularly measured using encoding disk shown in Fig. 1.1.

The disk is mounted at one end of shaft of rotating machine.

In this encoding disk 16 angular positions of rotating shaft are encoded in 4-bit gray code.

Since gray code changes only in one bit position when the rotating shaft move from one angular position to other, the errors if any in encoded binary can be easily detected.

The encoding disk is divided into 16-angular sectors and 4-circular regions to mark black and white pattern. The black and white pattern are created such that 4-bit gray code is generated when sensed using four light sensors. Here white pattern generate "0" and black pattern generate "1", it can be vice-versa also.

For more precision the angular sectors can be divided in to 32 angular sectors so that 5-bit gray code is generated and 64 angular sectors so that 6-bit gray code generated and so on .

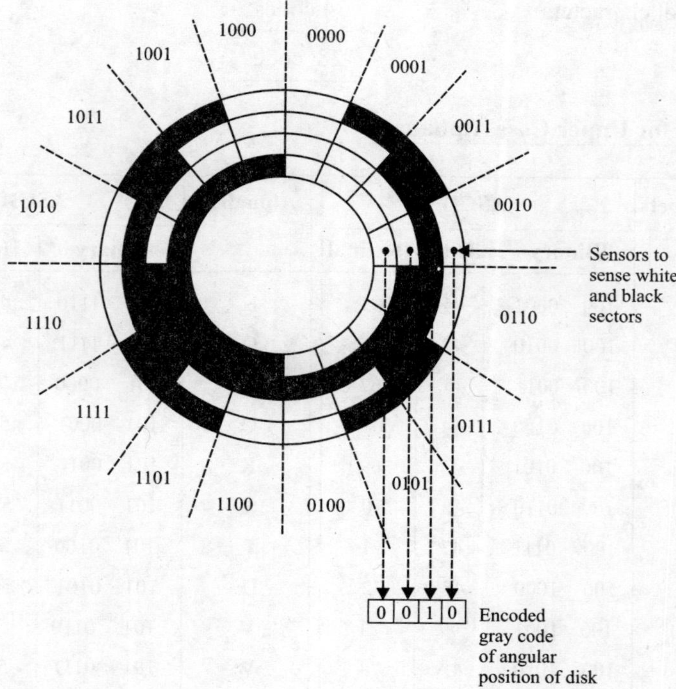

Fig. 1.1: Gray code encoding disk.

1.3.9 Alphanumeric Code (ASCII)

ASCII: American Standard Code for Information Interchange

ASCII was developed in the early stage of digital systems in order to standardize the coding system for human and machine interface.

ASCII is 7-bit binary code that includes most of the symbols in early type-writing machine.

ASCII includes codes for upper and lower case english alphabets and numerals (0 to 9).

With 7-bit binary code, $2^7 = 128$ ASCII codes are possible. They are assigned as follows:

Upper case alphabets,	A to Z	=	26 codes
Lower case alphabets,	a to z	=	26 codes
Numerals,	0 to 9	=	10 codes
Special printable characters		=	32 codes
Control characters		=	34 codes
			128 codes

Table 1.25: ASCII for Upper Case Alphabets

Alphabets	ASCII			Alphabets	ASCII		
	Binary	Hexa	Decimal		Binary	Hexa	Decimal
A	100 0001	41	65	N	100 1110	4E	78
B	100 0010	42	66	O	100 1111	4F	79
C	100 0011	43	67	P	101 0000	50	80
D	100 0100	44	68	Q	101 0001	51	81
E	100 0101	45	69	R	101 0010	52	82
F	100 0110	46	70	S	101 0011	53	83
G	100 0111	47	71	T	101 0100	54	84
H	100 1000	48	72	U	101 0101	55	85
I	100 1001	49	73	V	101 0110	56	86
J	100 1010	4A	74	W	101 0111	57	87
K	100 1011	4B	75	X	101 1000	58	88
L	100 1100	4C	76	Y	101 1001	59	89
M	100 1101	4D	77	Z	101 1010	5A	90

Table 1.26: ASCII for Lower Case Alphabets

Alphabet	ASCII			Alphabet	ASCII		
	Binary	Hexa	Decimal		Binary	Hexa	Decimal
a	110 0001	61	97	n	110 1110	6E	110
b	110 0010	62	98	o	110 1111	6F	111
c	110 0011	63	99	p	111 0000	70	112
d	110 0100	64	100	q	111 0001	71	113
e	110 0101	65	101	r	111 0010	72	114
f	110 0110	66	102	s	111 0011	73	115
g	110 0111	67	103	t	111 0100	74	116
h	110 1000	68	104	u	111 0101	75	117
i	110 1001	69	105	v	111 0110	76	118
j	110 1010	6A	106	w	111 0111	77	119
k	110 1011	6B	107	x	111 1000	78	120
l	110 1100	6C	108	y	111 1001	79	121
m	110 1101	6D	109	z	111 1010	7A	122

Table 1.27: ASCII for Special Printable Characters

Special Characters	ASCII			Special Characters	ASCII		
	Binary	Hexa	Decimal		Binary	Hexa	Decimal
!	010 0001	21	33	;	011 1011	3B	59
“	010 0010	22	34	<	011 1100	3C	60
#	010 0011	23	35	=	011 1101	3D	61
$	010 0100	24	36	>	011 1110	3E	62
%	010 0101	25	37	?	011 1111	3F	63
&	010 0110	26	38	@	100 0000	40	64
‘	010 0111	27	39	[	101 1011	5B	91
(	010 1000	28	40	\	101 1100	5C	92
)	010 1001	29	41	]	101 1101	5D	93
*	010 1010	2A	42	^	101 1110	5E	94
+	010 1011	2B	43	_	101 1111	5F	95
,	010 1100	2C	44	`	110 0000	60	96
–	010 1101	2D	45	{	111 1011	7B	123
.	010 1110	2E	46	|	111 1100	7C	124
/	010 1111	2F	47	}	111 1101	7D	125
:	011 1010	3A	58	~	111 1110	7E	126

Table 1.28: ASCII for Control Characters

Control Characters	ASCII			Control Characters	ASCII		
	Binary	Hexa	Decimal		Binary	Hexa	Decimal
NUL	000 0000	00	0	DC1	001 0001	11	17
SOH	000 0001	01	1	DC2	001 0010	12	18
STX	000 0010	02	2	DC3	001 0011	13	19
ETX	000 0011	03	3	DC4	001 0100	14	20
EOT	000 0100	04	4	NAK	001 0101	15	21
ENQ	000 0101	05	5	SYN	001 0110	16	22
ACK	000 0110	06	6	ETB	001 0111	17	23
BEL	000 0111	07	7	CAN	001 1000	18	24
BS	000 1000	08	8	EM	001 1001	19	25
HT	000 1001	09	9	SUB	001 1010	1A	26
LF	000 1010	0A	10	ESC	001 1011	1B	27
VT	000 1011	0B	11	FS	001 1100	1C	28
FF	000 1100	0C	12	GS	001 1101	1D	29
CR	000 1101	0D	13	RS	001 1110	1E	30
SO	000 1110	0E	14	US	001 1111	1F	31
SI	000 1111	0F	15	SP	010 0000	20	32
DLE	001 0000	10	16	DEL	111 1111	7F	127

Table 1.29: ASCII for Numerals

Numeral	ASCII			Numeral	ASCII		
	Binary	Hexa	Decimal		Binary	Hexa	Decimal
0	011 0000	30	48	5	011 0101	35	53
1	011 0001	31	49	6	011 0110	36	54
2	011 0010	32	50	7	011 0111	37	55
3	011 0110	33	51	8	011 1000	38	56
4	011 0100	34	52	9	011 1001	39	57

Table 1.30: Name/Meaning of ASCII Control Characters

Symbol	Name/Meaning	Symbol	Name/Meaning
NUL	Null	DC2	Device Control 2
SOH	Start of Heading	DC3	Device Control 3
STX	Start of Text	DC4	Device Control 4
ETX	End of Text	NAK	Negative Acknowledge
EOT	End of Transmission	SYN	Synchronous Idle
ENQ	Enquiry	ETB	End-of-Transmission Block
ACK	Acknowledge	CAN	Cancel
BEL	Bell	EM	End of Medium
BS	Backspace	SUB	Substitute
HT	Horizondal Tab	ESC	Escape
LF	Line Feed	FS	File Separator
VT	Vertical Tab	GS	Group Separator
FF	Form Feed	RS	Record Separator
CR	Carriage Return	US	Unit Separator
SO	Shift Out	SP	Space
SI	Shift In	DEL	Delete
DLE	Data-Link Escape		
DC1	Device Control 1		

1.4 Boolean Algebra

George Boole developed Boolean algebra in **1854**.

Boolean algebra is an algebraic structure that includes a set of elements consisting of

- Binary operators, " + " and " . "
- Binary variables, x, y, z .....
- Binary elements, 0 and 1
- Boolean postulates and theorems

Let x and y be two boolean variables that can take all possible combination of binary value. The rules for binary operators " + " and " . " are listed in Table 1.31 using the two binary variables x and y.

Table 1.31: Rules for Binary Operators " + " and " . "

x	y	x . y	x + y	x	x'
0	0	0	0	0	1
0	1	0	1	1	0
1	0	0	1		
1	1	1	1		

Duality

The *duality* principle of Boolean algebra states that every algebraic expression deducible from the postulates of Boolean algebra remains valid if the operator and identity element are interchanged. Here, the operators are " + " and " . " and the identity elements are " 0 " and " 1 ".

Table 1.32: Basic Example of Duality

x	y	x + y
0	0	0
0	1	1
1	0	1
1	1	1

Change 0 to 1 and 1 to 0

1	1	1
1	0	0
0	1	0
0	0	0

Change "+" to "•"

x	y	x . y
0	0	0
0	1	0
1	0	0
1	1	1

Operator Precedence while Evaluating Boolean Expression

A statement written with Boolean variables, constants and operators is called ***Boolean expression***.

Examples of Boolean expression: $x + y . z'$

$(x . y) + z' + (x + y . z)'$

In order to evaluate Boolean expression the operator precedence is as follows:

Step-1: Evaluate expression with in paranthesis, i.e, within ( ).

Step-2: Evaluate complement

Step-3: Evaluate " . "

Step-4: Evaluate " + "

Operator precedence in evaluation of Boolean expressions are shown in the following example.

Example 1.19

Demonstrate the operator precedence in the evaluation of following Boolean expression.

a) $x + y . z'$ **b)** $(x.y) + z' + (x + y . z)'$

Solution

a) $x + y . z'$

$x + y . z'$
↓
z' First stage
↓
$y . z'$ Second stage
↓
$x + y . z'$ Third stage

b) $(x.y) + z' + (x + y . z)'$

$(x . y) + z' + (x + y . z)'$
↓ ↓ ↓
$(x . y)$ z' $(y . z)$ First stage
↓
$x + (y . z)$ Second stage
↓
$(x + y . z)'$ Third stage
↓
$(x . y) + z' + (x + y . z)'$ Fourth stage

1.4.1 Postulates of Boolean Algebra

Postulates of Boolean algebra are developed by **E. V. Huntington** in **1904.**

Boolean structure have to satisfy the following postulates.

1. a) The structure is closed with the respect to operator " + "

b) The structure is closed with respect to operator " . "

2. a) The element 0 is an identity element with respect to operator " + "

$\therefore$ $x + 0 = x$

$0 + x = x$

where x is a Boolean variable that can take a value either 0 or 1.

b) The element 1 is an identity element with respect to operator " . "

$\therefore$ $x \cdot 1 = x$

$1 \cdot x = x$

where x is Boolean variable that can take a value either 0 or 1.

3. a) The structure is commutative with respect to operator " + "

$\therefore$ $x + y = y + x$

where x and y are two Boolean variables.

b) The structure is commutative with respect to operator " ."

$\therefore$ $x \cdot y = y \cdot x$

4. a) The operator " . " is distributive over operator " + "

$\therefore$ $x \cdot (y + z) = (x \cdot y) + (x \cdot z)$

b) The operator " + " is distributive over operator " . "

$\therefore$ $x + (y \cdot z) = (x + y) \cdot (x + z)$

5. a) For every variable x there exists an element x' called complement of x

$\therefore$ $x + x' = 1$ and $x \cdot x' = 0$

6. There exists at least two variables x and y such that, $x \neq y$.

Verification of Postulates

1. From the Table 1.31 it is obvious that results of "+" and "." operations are either 0 or 1 and so structure is closed with respect to operators " + " and " . "

2. $x + 0 = x$, when $x = 0$, $x + 0 = 0 + 0 = 0 = x$

$x = 1$, $x + 0 = 1 + 0 = 1 = x$

$0 + x = x$, when $x = 0$, $0 + x = 0 + 0 = 0 = x$

$x = 1$, $0 + x = 0 + 1 = 1 = x$

3. $x \cdot 1 = x$, when $x = 0$, $x \cdot 1 = 0 \cdot 1 = 0 = x$

$x = 1$, $x \cdot 1 = 1 \cdot 1 = 1 = x$

$1 . x = x$, when $x = 0$, $1 . x = 1 . 0 = 0 = x$

$x = 1$, $1 . x = 1 . 1 = 1 = x$

4. Let x, y, z be three Boolean variables that takes all possible combinations of binary value. Construct a truth table for all possible combinations of x, y and z as shown in Table 1.33 to prove,

$x . (y + z) = (x . y) + (x . z)$

Table 1.33: Verification of x . (y + z) = (x . y) + (x . z)

x	y	z	y + z	x . (y + z)	x . y	x . z	(x . y) + (x . z)
0	0	0	0	0	0	0	0
0	0	1	1	0	0	0	0
0	1	0	1	0	0	0	0
0	1	1	1	0	0	0	0
1	0	0	0	0	0	0	0
1	0	1	1	1	0	1	1
1	1	0	1	1	1	0	1
1	1	1	1	1	1	1	1

Similarly, Table 1.34 is constructed to prove,

$x + (y . z) = (x + y) . (x + z)$

Table 1.34: Verification of x + (y . z) = (x + y) . (x + z)

x	y	z	y . z	x + (y . z)	(x + y)	(x + z)	(x + y) . (x + z)
0	0	0	0	0	0	0	0
0	0	1	0	0	0	1	0
0	1	0	0	0	1	0	0
0	1	1	1	1	1	1	1
1	0	0	0	1	1	1	1
1	0	1	0	1	1	1	1
1	1	0	0	1	1	1	1
1	1	1	1	1	1	1	1

5. $x + x' = 1$

when $x = 0$, $x' = 1$, $\therefore\ x + x' = 0 + 1 = 1$

when $x = 1$, $x' = 0$, $\therefore\ x + x' = 1 + 0 = 1$

$x . x' = 0$

when $x = 0$, $x' = 1$, $\therefore\ x . x' = 0 . 1 = 0$

when $x = 1$, $x' = 0$, $\therefore\ x . x' = 1 . 0 = 0$

6. Boolean algebra has two elements 1 and 0, where $1 \neq 0$.

$\therefore$ If x and y are two variables, then there is a possibility that

$x = 0$, $y = 1$, $\therefore\ x \neq y$

and $x = 1$, $y = 0$, $\therefore\ x \neq y$

1.4.2 Boolean Theorems

Theorem 1: $x + x = x$

$x . x = x$

Theorem 2: $x + 1 = 1$

$x . 0 = 0$

Theorem 3: (Involution) $(x')' = x$

Theorem 4: (Associative) $x + (y + z) = (x + y) + z$

$x . (y . z) = (x . y) . z$

Theorem 5: (DeMorgan's Law or DeMorgan's Theorem) $(x + y)' = x' . y'$

$(x . y)' = x' + y'$

Theorem 6: (Absorption) $x + (x . y) = x$

$x . (x + y) = x$

(y is absorbed)

Proof of Theorems

Theorem 1:

$x + x = x$

when, $x = 0$, $x + x = 0 + 0 = 0 = x$

when, $x = 1$, $x + x = 1 + 1 = 1 = x$

$x . x = x$

when, $x = 0$, $x . x = 0 . 0 = 0 = x$

when, $x = 1$, $x . x = 1 . 1 = 1 = x$

Theorem 2:

$x + 1 = 1$

when, $x = 0$, $x + 1 = 0 + 1 = 1$

when, $x = 1$, $x + 1 = 1 + 1 = 1$

$x . 0 = 0$

when, $x = 0$, $x . 0 = 0 . 0 = 0$

when, $x = 1$, $x . 0 = 1 . 0 = 0$

Theorem 3:

$(x')' = x$

when, $x = 0$, $(x')' = (0')' = 1' = 0 = x$

when, $x = 1$, $(x')' = (1')' = 0' = 1 = x$

Theorem 4:

$x + (y + z) = (x + y) + z$

Construct a truth table for all possible combinations of x, y and z as shown in Table 1.35 to prove,

$x + (y + z) = (x + y) + z$

$x . (y . z) = (x . y) . z$

Construct a truth table for all possible combinations of x, y and z as shown in Table 1.36 to prove,

$x . (y . z) = (x . y) . z$

Table 1.35: Verification of x + (y + z) = (x + y) + z

x	y	z	y + z	x + (y + z)	x + y	(x + y) + z
0	0	0	0	0	0	0
0	0	1	1	1	0	1
0	1	0	1	1	1	1
0	1	1	1	1	1	1
1	0	0	0	1	1	1
1	0	1	1	1	1	1
1	1	0	1	1	1	1
1	1	1	1	1	1	1

Table 1.36: Verification of x . (y . z) = (x . y) . z

x	y	z	y . z	x . (y . z)	x . y	(x . y) . z
0	0	0	0	0	0	0
0	0	1	0	0	0	0
0	1	0	0	0	0	0
0	1	1	1	0	0	0
1	0	0	0	0	0	0
1	0	1	0	0	0	0
1	1	0	0	0	1	0
1	1	1	1	1	1	1

Theorem 5:

$(x + y)' = x' . y'$

Construct a truth table for all possible combinations of x and y as shown in Table 1.37 to prove,

$(x + y)' = x' . y'$

$(x . y)' = x' + y'$

Construct a truth table for all possible combinations of x and y as shown in Table 1.38 to prove,

$(x . y)' = x' + y'$

Table 1.37: Verification of (x + y)' = x' . y'

x	y	x + y	(x + y)'	x'	y'	x' . y'
0	0	0	1	1	1	1
0	1	1	0	1	0	0
1	0	1	0	0	1	0
1	1	1	0	0	0	0

Table 1.38: Verification of (x . y)' = x' + y'

x	y	x . y	(x . y)'	x'	y'	x' + y'
0	0	0	1	1	1	1
0	1	0	1	1	0	1
1	0	0	1	0	1	1
1	1	1	0	0	0	0

Theorem 6:

$x + (x \cdot y) = x$

Construct a truth table for all possible combinations of x and y as shown in Table 1.39 to prove,

$x + (x \cdot y) = x$

Table 1.39: Verification of x + (x . y) = x

x	y	x . y	x + (x . y)
0	0	0	0
0	1	0	0
1	0	0	1
1	1	1	1

$x \cdot (x + y) = x$

Construct a truth table for all possible combinations of x and y as shown in Table 1.40 to prove,

$x \cdot (x + y) = x$

Table 1.40: Verification of x . (x + y) = x

x	y	x + y	x . (x + y)
0	0	0	0
0	1	1	0
1	0	1	1
1	1	1	1

1.4.3 Binary Logic

Binary logic consists of binary variables and a set of logical operations. The binary variables are denoted by letters of alphabet, A, B, C, ..., x, y, z, ...

Each binary variable can take only two possible values 0 or 1. In positive logic, 0 is **low** and 1 is **high**. The basic logical operations are AND, OR and NOT.

AND operation is same as " **.** " operation. In positive logic, the logical AND of two or more variables will be 1 if and only if the value of all the variables is 1.

OR operation is same as " + " operation. In positive logic, the logical OR of two or more variables will be 1 if the value of any one of the variables is 1.

NOT operation is same as complement operation.

The results or outputs of logical operations of two or more variables for all possible combinations of the variables can be listed in a table called ***truth table***. The truth tables of basic logical operations are shown in Table 1.41.

Table 1.41: Truth Tables of AND, OR and NOT Operation of Two Variables

AND operation

x	y	x . y
0	0	0
0	1	0
1	0	0
1	1	1

OR operation

x	y	x + y
0	0	0
0	1	1
1	0	1
1	1	1

NOT operation

x	x'
0	1
1	0

Positive and Negative Logic

Binary constants are 1 and 0. Physically in a digital circuit they represent two voltage levels.

The two voltage levels are called **high** and **low.**

The voltage level of **high** and **low** depends on technology used to fabricate the gates.

In TTL logic (Transistor-Transistor Logic) **high** is + 5 V and **low** is 0 V.

In CMOS (Complementry Symmetry MOSFET) **high** is +3.3 *V* and **low** is 0 *V*.

If, 1 represent **high** voltage and 0 represent **low** voltage then logic system is called ***positive logic***.

If, 1 represent **low** voltage and 0 represent **high** voltage then the logic system is called ***negative logic***.

Table 1.42: Positive and Negative Logic Levels

Logic Type (or Logic System)	**Technology**	
	TTL	**CMOS**
Positive Logic	1 = High = +5 *V* 0 = Low = 0 *V*	1 = High = + 3.3 *V* 0 = Low = 0 *V*
Negative Logic	0 = High = + 5 *V* 1 = Low = 0 *V*	0 = High = + 3.3 *V* 1 = Low = 0 *V*

1.4.4 Logic Gates

Logic gates are electronic devices or circuits that perform logical operations on one or more input logical variables and produce a binary output. The basic logic gates are AND, OR and NOT gates.

The logic gates have standard symbols as shown in Figs. 1.2 to 1.6.

AND Gate

AND gate is an electronic device that performs logical AND operation of two or more variables.

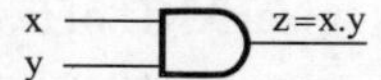

Fig. 1.2: Two input AND gate.

Fig. 1.3: Three input AND gate.

OR Gate

OR gate is an electronic device that performs logical OR operation of two or more variables.

Fig. 1.4: Two input OR gate.

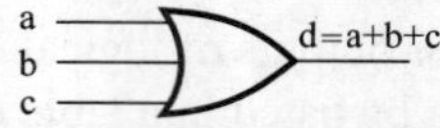

Fig. 1.5: Three input OR gate.

NOT Gate (or inverter)

NOT gate is an electronic device that performs complement operation of a Boolean variable.

x —▷o— x′

Fig. 1.6: NOT gate.

> ***Note:*** *Theoretically, AND and OR gates can have any number of inputs.*

Other Logical Operations

With n variables it is possible to form 2^{2n} Boolean functions.

When, n = 2, $2^{2n} = 2^{2 \times 2} = 2^4 = 16$

Therefore, with two variables we can form 16 Boolean functions.

All the 16 possible function of two variables are listed in Table 1.43.

The output of 16 functions are 16 binary combinations of 4-bit binary.

Let, F_0, F_1, F_2, ..., F_{15} be 16 possible Boolean functions for two variables x, y.

Table 1.43: Two-Variable Boolean Functions

Functions	x	y	F	Functions	x	y	F
Null operation	0	0	0	NOR operation	0	0	1
$F = F_0 = 0$	0	1	0	$F = F_8 = (x + y)'$	0	1	0
	1	0	0		1	0	0
	1	1	0		1	1	0
AND operation	0	0	0	Exclusive NOR	0	0	1
$F = F_1 = x \cdot y$	0	1	0	(or equivalence)	0	1	0
	1	0	0	$F = F_9 = (x \oplus y)'$	1	0	0
	1	1	1	$= x \cdot y + x' \cdot y'$	1	1	1
Inhibition	0	0	0	Complement of y	0	0	1
$F = F_2 = x \cdot y'$	0	1	0	$F = F_{10} = y'$	0	1	0
(x true, but y false)	1	0	1		1	0	1
	1	1	0		1	1	0
Transfer	0	0	0	Implication	0	0	1
$F = F_3 = x$	0	1	0	$F = F_{11} = x + y'$	0	1	0
(Transfer x)	1	0	1	(If $y = 1$, $F = x$	1	0	1
	1	1	1	If $y = 0$, $F = y'$)	1	1	1
Inhibition	0	0	0	Complement of x	0	0	1
$F = F_4 = x' \cdot y$	0	1	1	$F = F_{12} = x'$	0	1	1
(y true, but x false)	1	0	0		1	0	0
	1	1	0		1	1	0
Transfer	0	0	0	Implication	0	0	1
$F = F_5 = y$	0	1	1	$F = F_{13} = x' + y$	0	1	1
(Transfer y)	1	0	0	(If $x = 1$, $F = y$	1	0	0
	1	1	1	If $x = 0$, $F = x'$)	1	1	1
XOR operation	0	0	0	NAND operation	0	0	1
$F = F_6 = x \oplus y$	0	1	1	$F = F_{14} = (x \cdot y)'$	0	1	1
$= x \cdot y' + x' \cdot y$	1	0	1		1	0	1
	1	1	0		1	1	0
OR operation	0	0	0	Identity	0	0	1
$F = F_7 = x + y$	0	1	1	$F = F_{15} = 1$	0	1	1
	1	0	1		1	0	1
	1	1	1		1	1	1

Summary of Logic Gates

The AND, OR and NOT are basic gates.

Using these basic gates some more useful logical operations can be defined.

They are NAND, NOR, Exclusive-OR and Exclusive-NOR.

The summary of popular logic gates used in digital electronics are listed in Table 1.44.

Table 1.44: Summary of Popular Logic Gates

Gate	Symbol	Truth Table
AND Gate	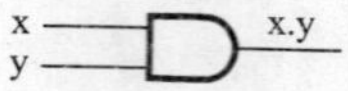 x, y → x.y	x y | x . y: 0 0 | 0; 0 1 | 0; 1 0 | 0; 1 1 | 1
OR Gate	 x, y → x+y	x y | x + y: 0 0 | 0; 0 1 | 1; 1 0 | 1; 1 1 | 1
Inverter Gate (or NOT gate)	 x → x′	x | x′: 0 | 1; 1 | 0
Buffer	 x → x	x: 0; 1
NAND Gate	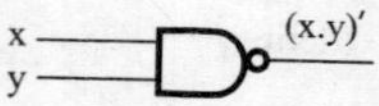 x, y → (x.y)′	x y | x.y | (x.y)′: 0 0 | 0 | 1; 0 1 | 0 | 1; 1 0 | 0 | 1; 1 1 | 1 | 0

Table 1.44: Continued...

<table>
<tr><th>Gate</th><th>Symbol</th><th>Truth Table</th></tr>
<tr><td>NOR Gate</td><td> 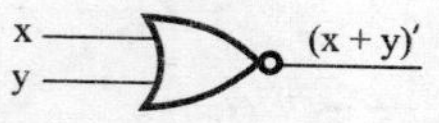x, y → (x + y)′</td><td>

x	y	x + y	(x + y)′
0	0	0	1
0	1	1	0
1	0	1	0
1	1	1	0

</td></tr>
<tr><td>Exclusive-OR Gate (XOR)</td><td>x, y → x.y′ + x′.y = x⊕y</td><td>

x	y	x . y′	x′ . y	x ⊕ y
0	0	0	0	0
0	1	0	1	1
1	0	1	0	1
1	1	0	0	0

</td></tr>
<tr><td>Exclusive-NOR Gate (or equivalence)</td><td>x, y → x.y + x′.y′ = (x⊕y)′</td><td>

x	y	x⊕y	(x⊕y)′
0	0	0	1
0	1	1	0
1	0	1	0
1	1	0	1

</td></tr>
</table>

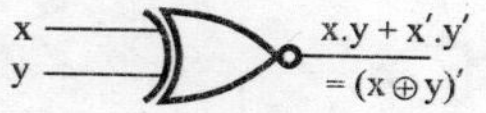

***Note:** Bubble at output of gate indicates NOT operation.*

1.4.5 Universal Gates

NAND and NOR gates are called ***universal gates***, because any Boolean function can be realized only using NAND gates or only using NOR gates.

The NAND gate is a combination of AND followed by NOT gate. The NOR gate is a combination of OR followed by NOT gate.

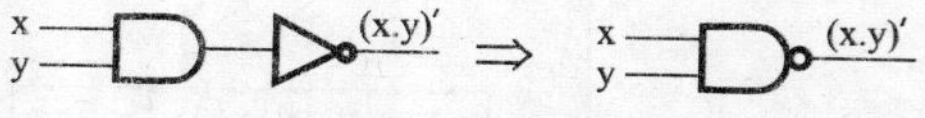

x, y → (x.y)′ ⇒ x, y → (x.y)′

***Fig. 1.7:** NAND gate.*

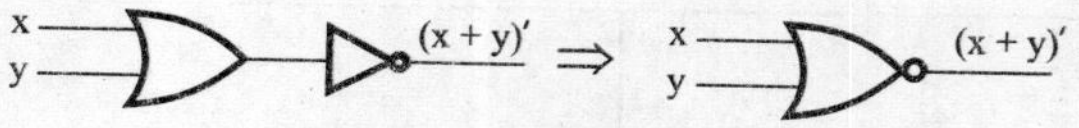

x, y → (x + y)′ ⇒ x, y → (x + y)′

***Fig. 1.8:** NOR gate.*

Table 1.45: Truth Table of NAND Gate

x	y	(x . y)′
0	0	1
0	1	1
1	0	1
1	1	0

Table 1.46: Truth Table of NOR Gate

x	y	(x + y)′
0	0	1
0	1	0
1	0	0
1	1	0

The realization of basic gates using only NAND gates are shown in Table 1.47.

The realization of basic gates using only NOR gates are shown in Table 1.48.

Table 1.47: Realization of Basic Logic Gates using NAND Gates

<table>
<tr><th>Gate</th><th>Logic Circuit using NAND</th><th>Truth Table</th></tr>
<tr><td>Inverter/NOT gate using NAND</td><td>x
(x.x)′=x′</td><td><table><tr><th>x</th><th>x′</th></tr><tr><td>0</td><td>1</td></tr><tr><td>1</td><td>0</td></tr></table></td></tr>
<tr><td>AND using NAND</td><td>x
y
(x.y)′
x.y</td><td><table><tr><th>x</th><th>y</th><th>x . y</th></tr><tr><td>0</td><td>0</td><td>0</td></tr><tr><td>0</td><td>1</td><td>0</td></tr><tr><td>1</td><td>0</td><td>0</td></tr><tr><td>1</td><td>1</td><td>1</td></tr></table></td></tr>
<tr><td>OR using NAND</td><td>x
x′
(x′.y′)′ = x + y
y
y′</td><td><table><tr><th>x</th><th>y</th><th>x + y</th></tr><tr><td>0</td><td>0</td><td>0</td></tr><tr><td>0</td><td>1</td><td>1</td></tr><tr><td>1</td><td>0</td><td>1</td></tr><tr><td>1</td><td>1</td><td>1</td></tr></table></td></tr>
<tr><td>NOR using NAND</td><td>x
x′
(x′.y′)′
x′.y′=(x+y)′
y
y′</td><td><table><tr><th>x</th><th>y</th><th>x + y</th><th>(x + y)′</th></tr><tr><td>0</td><td>0</td><td>0</td><td>1</td></tr><tr><td>0</td><td>1</td><td>1</td><td>0</td></tr><tr><td>1</td><td>0</td><td>1</td><td>0</td></tr><tr><td>1</td><td>1</td><td>1</td><td>0</td></tr></table></td></tr>
<tr><td>XOR using NAND</td><td>x
x′
(x′.y)′
((x′.y)′.(x.y′)′)′
= x′.y + x.y′
= x ⊕ y
y
y′
(x.y′)′</td><td><table><tr><th>x</th><th>y</th><th>x . y′</th><th>x′. y</th><th>x ⊕ y</th></tr><tr><td>0</td><td>0</td><td>0</td><td>0</td><td>0</td></tr><tr><td>0</td><td>1</td><td>0</td><td>1</td><td>1</td></tr><tr><td>1</td><td>0</td><td>1</td><td>0</td><td>1</td></tr><tr><td>1</td><td>1</td><td>0</td><td>0</td><td>0</td></tr></table>
$x \oplus y = x y' + x' y$</td></tr>
</table>

Table 1.48: Realization of Basic Logic Gates using NOR Gates

Gate	Logic Circuit using NOR	Truth Table
Inverter/NOT gate using NOR	x, x′, (x + x)′ = x′	**x** | **x′** 0 | 1 1 | 0
OR using NOR	x, y, (x+y)′, x+y	**x** **y** | **x + y** 0 0 | 0 0 1 | 1 1 0 | 1 1 1 | 1
AND using NOR	x, x′, y, y′, (x′+y′)′ = x.y	**x** **y** | **x . y** 0 0 | 0 0 1 | 0 1 0 | 0 1 1 | 1
NAND using NOR	x, x′, y, y′, (x′+y′)′, x′+y′=(x.y)′	**x** | **y** | **x.y** | **(x.y)′** 0 | 0 | 0 | 1 0 | 1 | 0 | 1 1 | 0 | 0 | 1 1 | 1 | 1 | 0
XOR using NOR	x, x′, y, y′, (x′ + y)′ = xy′, (x + y′)′ = x′y, (xy′ + x′y)′, xy′ + x′y = x⊕y	**x** **y** | **x . y′** | **x′ . y** | **x ⊕ y** 0 0 | 0 | 0 | 0 0 1 | 0 | 1 | 1 1 0 | 1 | 0 | 1 1 1 | 0 | 0 | 0 $x \oplus y = xy' + x'y$

Commercially Available Logic Gates

Logic gate devices are available in the form of ICs manufactured using semiconductors. Usually 4 to 6 logic gates will be packed in a single IC package and will have a standard number. Some of the logic gate devices are,

7400 : 2-input NAND gate, pack of 4 gates

7402 : 2-input NOR gate, pack of 4 gates

7404 : Inverter gate, pack of 6 gates

7408 : 2-input AND gate, pack of 4 gates

7432 : 2-input OR gate, pack of 4 gates

7486 : 2-input XOR gate, pack of 4 gates

The pin diagram of the above ICs are shown in Fig. 1.9.

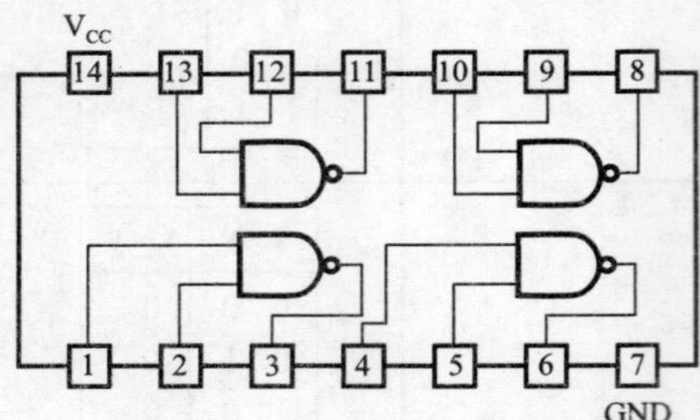

Fig. a: 2-input NAND gate 7400.

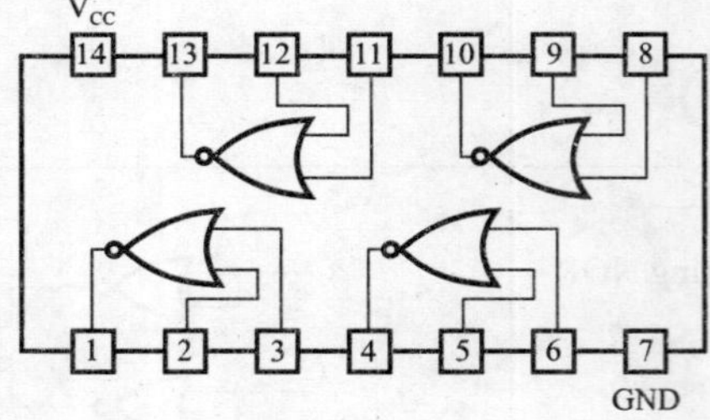

Fig. b: 2-input NOR gate 7402.

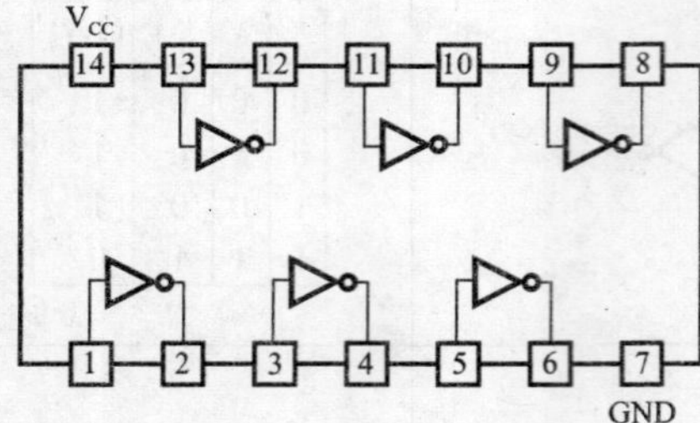

Fig. c: Inverters 7404.

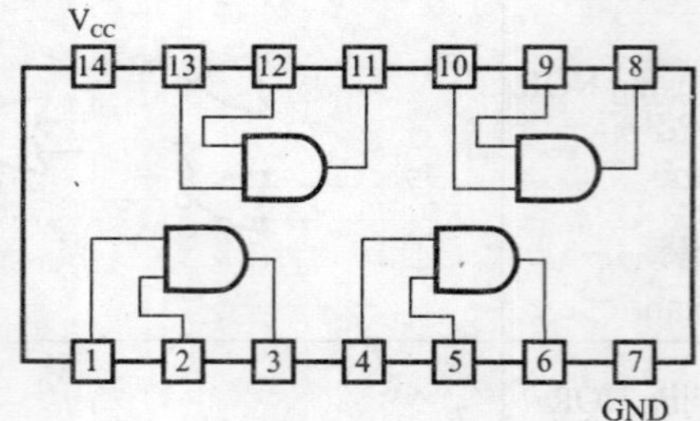

Fig. d: 2-input AND gate 7408.

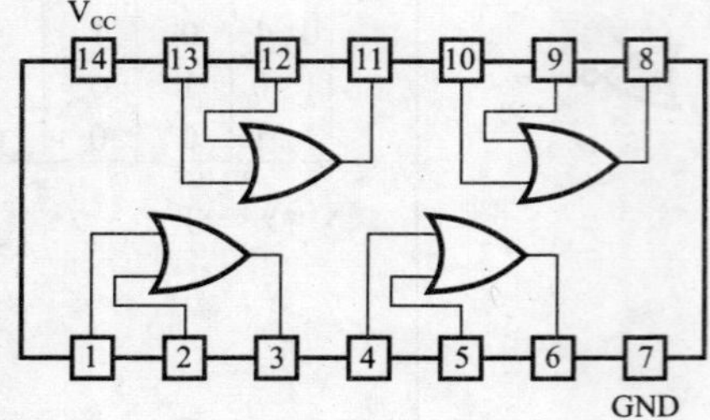

Fig. e: 2-input OR gate 7432.

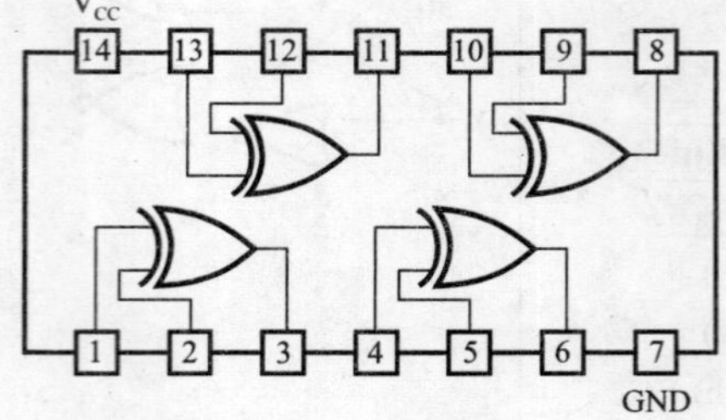

Fig. f: 2-input XOR gate 7486.

***Fig. 1.9:** Digital gates in IC packages with identification numbers and pin assignments.*

Extension of Inputs of Logic Gates

The AND and OR logic gates theoretically, can have any number of inputs. But commercially available logic devices will have only few inputs. When larger number of inputs are required, the logic gate can be cascaded as shown in Figs. 1.10(a) and (b).

Since NAND and NOR logic do not satisfy associative property, they cannot be cascaded like AND and OR gates. But multi-input NAND gate can be realized by cascaded AND gate followed by NOT gate as shown in Fig. 1.10(c). Similarly, multi-input NOR gate can be realized by cascaded OR gate followed by NOT gate as shown in Fig. 1.10(d).

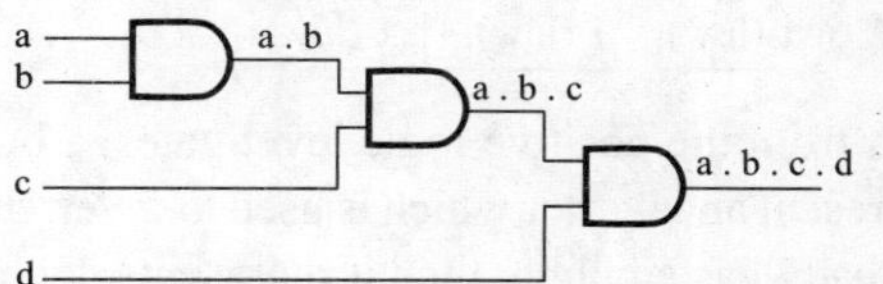

***Fig. a:** 4-input AND gate using 2-input AND gates.*

a
b
a + b
a + b + c
c
a + b + c + d
d

***Fig. b:** 4-input OR gate using 2-input OR gates.*

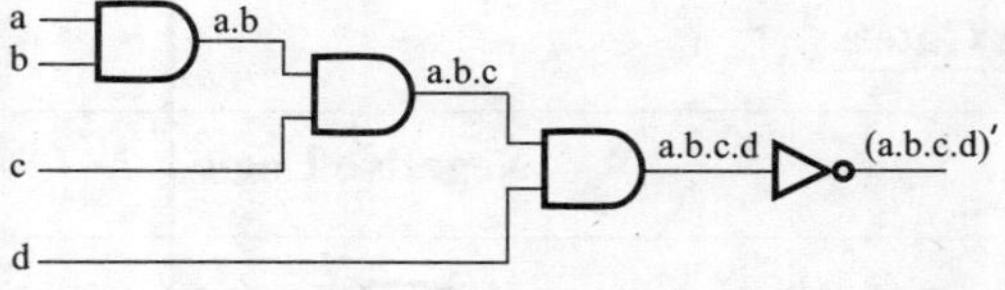

***Fig. c:** 4-input NAND gate using 2-input AND gates.*

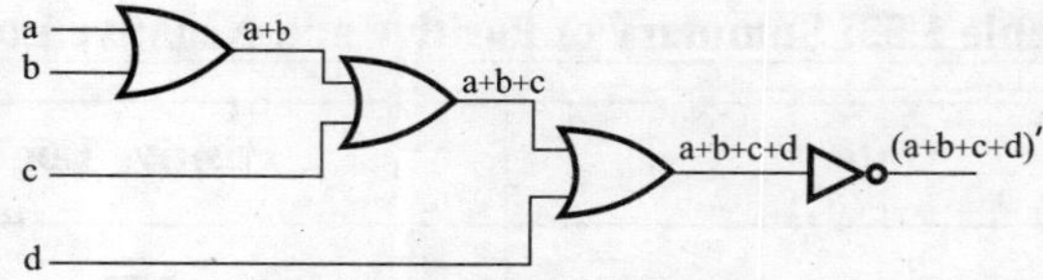

***Fig. d:** 4-input NOR gate using 2-input OR gates.*

***Fig. 1.10:** Multi-input logic gates.*

1.4.6 Positive and Negative Logic Gates

In positive logic, 1 is logic **high** and 0 is logic **low**. In negative logic, 0 is logic **high** and 1 is logic **low**. The logic gates discussed in Section 1.4.4 are positive logic gates which means that the logic levels of inputs and outputs are positive logic levels.

Logic gates can also be designed to work with negative logic levels. The AND and OR operation in positive and negative logic are given below:

In this book, only positive logic gates are used for analysis and design of digital logic circuit.

AND operation: AND operation of two or more variables is **high** only if all the variables are **high**.

Table 1.49: Truth Table of Positive Logic AND Gate

Inputs		Output
a	**b**	**a . b**
0 (low)	0 (low)	0 (low)
0 (low)	1 (high)	0 (low)
1 (high)	0 (low)	0 (low)
1 (high)	1 (high)	1 (high)

Table 1.50: Truth Table of Negative Logic AND Gate

Inputs		Output
a	**b**	**a . b**
0 (high)	0 (high)	0 (high)
0 (high)	1 (low)	1 (low)
1 (low)	0 (high)	1 (low)
1 (low)	1 (low)	1 (low)

OR operation: OR operation of two or more variables is **high** if one of the variable is **high**.

Table 1.51: Truth Table of Positive Logic OR Gate

Inputs		Output
a	b	a + b
0 (low)	0 (low)	0 (low)
0 (low)	1 (high)	1 (high)
1 (high)	0 (low)	1 (high)
1 (high)	1 (high)	1 (high)

Table 1.52: Truth Table of Negative Logic OR Gate

Inputs		Output
a	b	a . b
0 (high)	0 (high)	0 (high)
0 (high)	1 (low)	0 (high)
1 (low)	0 (high)	0 (high)
1 (low)	1 (low)	1 (low)

In order to differentiate negative logic level gates from the positive logic level gates, a bubble is added at every input and output. The bubble basically represent an inverter which is used to invert the logic levels. When a bubble is added to the output end that already has a bubble then it represents double time inversion and so that output end will not have a bubble in negative logic representation. The conversion of positive logic level gates to negative logic level gates are listed in Table 1.53.

Table 1.53: Summary of Positive and Negative Logic Gates

Gate	Positive Logic	Negative Logic
AND Gate	x, y; $z = x.y$	x, y; $z = (x'.y')'$
OR Gate	x, y; $z = x+y$	x, y; $z = (x'+y')'$
Inverter Gate	x; x'	x; x'
Buffer	x; x	x; x
NAND Gate	x, y; $z = (x.y)'$	x, y; $z = x'.y'$
NOR Gate	x, y; $z = (x+y)'$	x, y; $z = x'+y'$
XOR gate	x, y; $z = x \oplus y$	x, y; $z = x' \oplus y'$

Using the output equations of logic gates listed in Table 1.53, the truth table of positive and negative logic gates are obtained as shown in Tables 1.54 to 1.57.

Table 1.54: Truth Table of AND Gate

a) Positive Logic

Inputs		Output
x	y	z = x . y
0	0	0
0	1	0
1	0	0
1	1	1

b) Negative Logic

Inputs		Complement of Inputs		Output
x	y	x′	y′	z = (x′ . y′)′
0	0	1	1	0
0	1	1	0	1
1	0	0	1	1
1	1	0	0	1

Table 1.55: Truth Table of OR Gate

a) Positive Logic

Inputs		Output
x	y	z = x + y
0	0	0
0	1	1
1	0	1
1	1	1

b) Negative Logic

Inputs		Complement of Inputs		Output
x	y	x′	y′	z = (x′ + y′)′
0	0	1	1	0
0	1	1	0	0
1	0	0	1	0
1	1	0	0	1

In negative logic AND gate,

$z = (x'.y')' \Rightarrow z = x + y$ Using DeMorgan's theorem

Therefore, we can say that the negative logic AND gate is same as positive logic OR gate.

In negative logic OR gate,

$z = (x' + y')' \Rightarrow z = x.y$ Using DeMorgan's theorem

Therefore, we can say that the negative logic OR gate is same as positive logic AND gate.

Table 1.56: Truth Table of NAND Gate

a) Positive Logic

Inputs		Output
x	y	z = (x . y)′
0	0	1
0	1	1
1	0	1
1	1	0

b) Negative Logic

Inputs		Complement of Inputs		Output
x	y	x′	y′	z = x′ . y′
0	0	1	1	1
0	1	1	0	0
1	0	0	1	0
1	1	0	0	0

Table 1.57: Truth Table of NOR Gate

a) Positive Logic

Inputs		Output
x	y	z = (x + y)'
0	0	1
0	1	0
1	0	0
1	1	0

b) Negative Logic

Inputs		Complement of Inputs		Output
x	y	x'	y'	z = x' + y'
0	0	1	1	1
0	1	1	0	1
1	0	0	1	1
1	1	0	0	0

In negative logic NAND gate,

$$z = x' . y' \Rightarrow z = (x + y)'$$ Using DeMorgan's theorem

Therefore, we can say that the negative logic NAND gate is same as positive logic NOR gate.

In negative logic NOR gate,

$$z = x' + y' \Rightarrow z = (x . y)'$$ Using DeMorgan's theorem

Therefore, we can say that the negative logic NOR gate is same as positive logic NAND gate.

1.5 Switching Function (Boolean Function)

A ***Boolean function*** is described by a Boolean expression which consists of binary variables, binary constants 0 and 1 and logical operators, AND, OR and NOT. Binary variables are denoted by either lower case (a, b, c, ... x, y, z) or upper case (A, B, C, ... X, Y, Z) alphabets.

Some examples of Boolean functions are given below:

$$F = x . y + z'$$

$$F_1 = (x . z) + y'$$

$$F_2 = (x . y . z) + (y . z')$$

Alternatively,

$$F(x, y, z) = x . y + z'$$

$$F_1(x, y, z) = x . z + y'$$

$$F_2(x, y, z) = (x . y . z) + (y . z')$$

A Boolean function can be evaluated for all possible combinations of binary values of the variables of the function. The variables of a function are also known as ***input variables*** or ***inputs***.

Let, n = Number of variables in a Boolean function

2^n = Number of combinations of binary values of n variables.

Truth table for a Boolean function can be constructed with 2^n combinations of n input variables.

Note: *To simplify Boolean expression dot or AND operation is represented without operator.*

$x . y \Rightarrow xy$ $\quad$ $y . z \Rightarrow yz$ $\quad$ $x . y . z \Rightarrow xyz$

Example 1.20

Construct the truth table for the following functions.

a) $F_1 = xy + xy' + y'z$ b) $F_2 = bc + a'c'$

Solution

a) $F_1 = xy + xy' + y'z$

Table 1: Truth Table of Function, F_1

Input Variables and Complement				Product Terms			Function Output
x	y	z	y'	x y	x y'	y'z	F_1
0	0	0	1	0	0	0	0
0	0	1	1	0	0	1	1
0	1	0	0	0	0	0	0
0	1	1	0	0	0	0	0
1	0	0	1	0	1	0	1
1	0	1	1	0	1	1	1
1	1	0	0	1	0	0	1
1	1	1	0	1	0	0	1

b) $F_2 = bc + a'c'$

Table 1: Truth Table of Function, F_2

Input Variables and Complements					Product Terms		Function Output
a	b	c	a'	c'	bc	a'c'	F_2
0	0	0	1	1	0	1	1
0	0	1	1	0	0	0	0
0	1	0	1	1	0	1	1
0	1	1	1	0	1	0	1
1	0	0	0	1	0	0	0
1	0	1	0	0	0	0	0
1	1	0	0	1	0	0	0
1	1	1	0	0	1	0	1

1.5.1 Implementation of Boolean Functions by Logic Gates

A Boolean function basically involves product of variables, sum of variables and complement of variables. Therefore, the straight forward implementation of Boolean function can be made by using basic gates AND, OR and NOT.

The complement of a variable is obtained using NOT gate. The product term is realized using AND gate. The sum term is realized using OR gate.

Example 1.21

Implement the following functions using basic logic gates.

a) $F_1 = x\,y + x\,y' + y'z$ b) $F_2 = bc + a'c'$

Solution

a) $F_1 = x\,y + x\,y' + y'z$

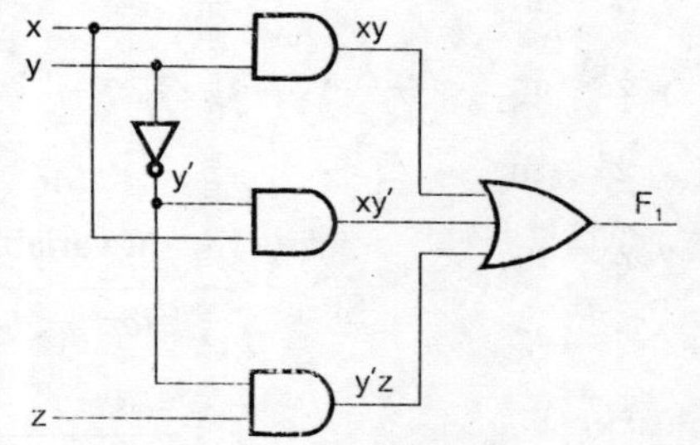

Fig. 1: Logic circuit for function, F_1.

b) $F_2 = bc + a'c'$

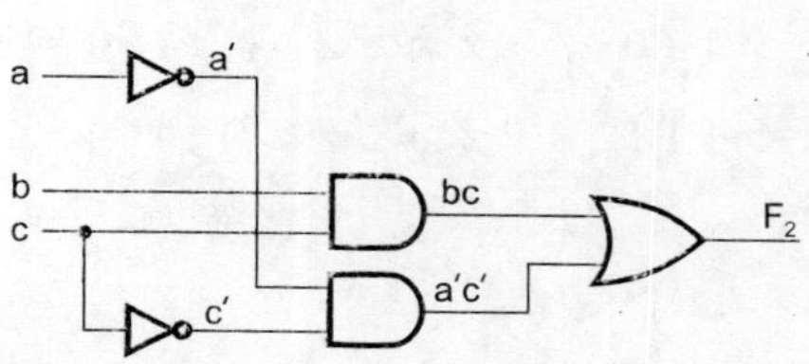

Fig. 2: Logic circuit for function, F_2.

1.5.2 Minterms

Minterms are 2^n possible combinations of AND terms (or product terms) with n variables such that the logical AND of all the variables is 1.

The minterms can be formed with two, three, four, ... variables. The 2^n minterms are denoted as $m_0, m_1, m_2, m_3 \ldots\ldots, m_q$ where $q = 2^n - 1$.

The 2^n combination of AND terms, for n = 2 and 3 are shown below. The AND terms are formed from 2^n combinations of n-bit binary. In AND terms a literal is primed if its value is 0 and unprimed if its value is 1. So that the AND of all literals is always 1.

Let, n = 2

Let x and y be the two variables to denote the two bits of binary.

When n = 2, $2 = 2^2 = 4$

Therefore, four combinations of AND terms with two variables are possible as shown below:

0 0		$x'.y'$	$(0'.0' = 1)$		$x'y' = m_0$
0 1	$\Rightarrow$	$x'.y$	$(0'.1 = 1)$	$\Rightarrow$	$x'y = m_1$
1 0		$x.y'$	$(1.0' = 1)$		$xy' = m_2$
1 1		$x.y$	$(1.1 = 1)$		$xy = m_3$

Let, n = 3

Let x, y and z be the three variables to denote the three bits of binary.

When n = 3, $2^n = 2^3 = 8$

Therefore, eight combinations of AND terms with three variables are possible as shown below:

0 0 0		$x'.y'.z'$	$(0'.0'.0' = 1)$		$x'y'z' = m_0$
0 0 1	$\Rightarrow$	$x'.y'.z$	$(0'.0'.1 = 1)$	$\Rightarrow$	$x'y'z = m_1$
0 1 0		$x'.y.z'$	$(0'.1.0' = 1)$		$x'yz' = m_2$
0 1 1		$x'.y.z$	$(0'.1.1 = 1)$		$x'yz = m_3$
1 0 0		$x.y'.z'$	$(1.0'.0' = 1)$		$xy'z' = m_4$
1 0 1	$\Rightarrow$	$x.y'.z$	$(1.0'.1 = 1)$	$\Rightarrow$	$xy'z = m_5$
1 1 0		$x.y.z'$	$(1.1.0' = 1)$		$xyz' = m_6$
1 1 1		$x.y.z$	$(1.1.1 = 1)$		$xyz = m_7$

Table 1.58: Two-Variable Minterms

x	y	Minterm	Notation
0	0	x'. y'	m_0
0	1	x'. y	m_1
1	0	x . y'	m_2
1	1	x . y	m_3

Table 1.59: Three-Variable Minterms

x	y	z	Minterm	Notation
0	0	0	x'. y'. z'	m_0
0	0	1	x'. y'. z	m_1
0	1	0	x'. y. z'	m_2
0	1	1	x'. y. z	m_3
1	0	0	x . y'. z'	m_4
1	0	1	x . y'. z	m_5
1	1	0	x . y . z'	m_6
1	1	1	x . y . z	m_7

Table 1.60: Four-Variable Minterms

a	b	c	d	Minterm	Notation
0	0	0	0	a'. b'. c'. d'	m_0
0	0	0	1	a'. b'. c'. d	m_1
0	0	1	0	a'. b'. c . d'	m_2
0	0	1	1	a'. b'. c . d	m_3
0	1	0	0	a'. b . c'. d'	m_4
0	1	0	1	a'. b . c'. d	m_5
0	1	1	0	a'. b . c . d'	m_6
0	1	1	1	a'. b . c . d	m_7
1	0	0	0	a . b'. c'. d'	m_8
1	0	0	1	a . b'. c'. d	m_9
1	0	1	0	a . b'. c . d'	m_{10}
1	0	1	1	a . b'. c . d	m_{11}
1	1	0	0	a . b . c'. d'	m_{12}
1	1	0	1	a . b . c'. d	m_{13}
1	1	1	0	a . b . c . d'	m_{14}
1	1	1	1	a . b . c . d	m_{15}

1.5.3 Maxterms

Maxterms are 2^n possible combinations of OR terms (or sum terms) with n variables such that the logical OR of all the variables is 0.

The maxterms can be formed with two, three, four, ... variables.

The 2^n maxterms are denoted as $M_0, M_1, M_2, M_3, ..., M_q$ where $q = 2^n - 1$.

The 2^n combinations of OR terms, for n = 2 and 3 are shown below. The OR terms are formed from 2^n combinations of n-bit binary. In OR terms a literal is primed if its value is 1, and unprimed if its value is 0 so that the OR of all literals is always 0.

Let, n = 2

Let x and y be the two variables to denote two bits of binary.

When n = 2, $2^n = 2^2 = 4$

Therefore, four combinations of OR terms with two variables are possible as shown below:

$$\begin{array}{llll} 0\ 0 & x+y & (0+0=0) & x+y = M_0 \\ 0\ 1 & x+y' & (0+1'=0) & x+y' = M_1 \\ 1\ 0 \Rightarrow & x'+y & (1'+0=0) \Rightarrow & x'+y = M_2 \\ 1\ 1 & x'+y' & (1'+1'=0) & x'+y' = M_3 \end{array}$$

Let, n = 3

Let x, y and z be the three variables to denote three bits of binary.

When n = 3, $2^n = 2^3 = 8$

Therefore, eight combinations of OR terms with three variables are possible as shown below:

$$\begin{array}{lllll} 0\,0\,0 & & x+y+z & (0+0+0=0) & & x+y+z = M_0 \\ 0\,0\,1 & & x+y+z' & (0+0+1'=0) & & x+y+z' = M_1 \\ 0\,1\,0 & & x+y'+z & (0+1'+0=0) & & x+y'+z = M_2 \\ 0\,1\,1 & \Rightarrow & x+y'+z' & (0+1'+1'=0) & \Rightarrow & x+y'+z' = M_3 \\ 1\,0\,0 & & x'+y+z & (1'+0+0=0) & & x'+y+z = M_4 \\ 1\,0\,1 & & x'+y+z' & (1'+0+1'=0) & & x'+y+z' = M_5 \\ 1\,1\,0 & & x'+y'+z & (1'+1'+0=0) & & x'+y'+z = M_6 \\ 1\,1\,1 & & x'+y'+z' & (1'+1'+1'=0) & & x'+y'+z' = M_7 \end{array}$$

Table 1.61: Two-Variable Maxterms

x	y	Maxterm	Notation
0	0	x + y	M_0
0	1	x + y'	M_1
1	0	x' + y	M_2
1	1	x' + y'	M_3

Table 1.62: Three-Variable Maxterms

x	y	z	Maxterm	Notation
0	0	0	x + y + z	M_0
0	0	1	x + y + z'	M_1
0	1	0	x + y' + z	M_2
0	1	1	x + y' + z'	M_3
1	0	0	x' + y + z	M_4
1	0	1	x' + y + z'	M_5
1	1	0	x' + y' + z	M_6
1	1	1	x' + y' + z'	M_7

Table 1.63: Four-Variable Maxterms

a	b	c	d	Maxterm	Notation
0	0	0	0	a + b + c + d	M_0
0	0	0	1	a + b + c + d'	M_1
0	0	1	0	a + b + c' + d	M_2
0	0	1	1	a + b + c' + d'	M_3
0	1	0	0	a + b' + c + d	M_4
0	1	0	1	a + b' + c + d'	M_5
0	1	1	0	a + b' + c' + d	M_6
0	1	1	1	a + b' + c' + d'	M_7
1	0	0	0	a' + b + c + d	M_8
1	0	0	1	a' + b + c + d'	M_9
1	0	1	0	a' + b + c' + d	M_{10}
1	0	1	1	a' + b + c' + d'	M_{11}
1	1	0	0	a' + b' + c + d	M_{12}
1	1	0	1	a' + b' + c + d'	M_{13}
1	1	1	0	a' + b' + c' + d	M_{14}
1	1	1	1	a' + b' + c' + d'	M_{15}

1.5.4 Standard Forms

There are three types of standard forms of expressing a Boolean function. They are,

1. Sum-of-products (SOP) form
2. Product-of-sums (POS) form
3. Canonical form

The ***canonical form*** is also expressed in SOP or POS form. The SOP or POS form is said to be canonical only if all the variables are present in every term of the SOP or POS form.

In SOP or POS form each term of the Boolean expression need not have all the variables of the function. Hence SOP and POS forms are also simplified forms of Boolean functions.

1.5.5 Sum-of-Products and Product-of-Sums Simplification

Sum-of-products (SOP) can be realized by AND operation of literals followed by OR operation of output of AND. Here, AND operation of literals are the products and OR operation of output of AND is the sum of product terms. ***Product-of-sums (POS)*** can be realized by OR operation of literals followed by AND operation of output of OR. Here, OR operation of literals are the sums and AND operation of output of OR is the product of sum terms. The realizations in SOP and POS forms are called ***two-level realization of standard forms***.

The two-level realization of SOP form can be obtained by using AND gates to implement product terms followed by OR gate to get sum of product terms. The two-level realization of POS form can be obtained by using OR gates to implement sum terms followed by AND gate to get product of sum terms.

Since NAND and NOR gates are universal gates, the two-level realization of SOP form of a Boolean function can be obtained using only NAND gates. Similarly, two-level realization of POS form of a Boolean function can be obtained using only NOR gates.

It can be observed that complement of sum-of-products will be product-of-sums and vice-versa

i.e, Product-of-sums $\xleftrightarrow{\text{Complement}}$ Sum-of-products

> ***Note:*** *This concept holds good for complement of canonical forms also.*

Example 1.22

Draw the two-level realization of the following functions.

a) $F_1 = y' + x\,y + x'y\,z'$ b) $F_2 = x\,(y' + z)\,(x' + y + z')$

Solution

a) $F_1 = y' + x\,y + x'y\,z'$ - Sum-of-products

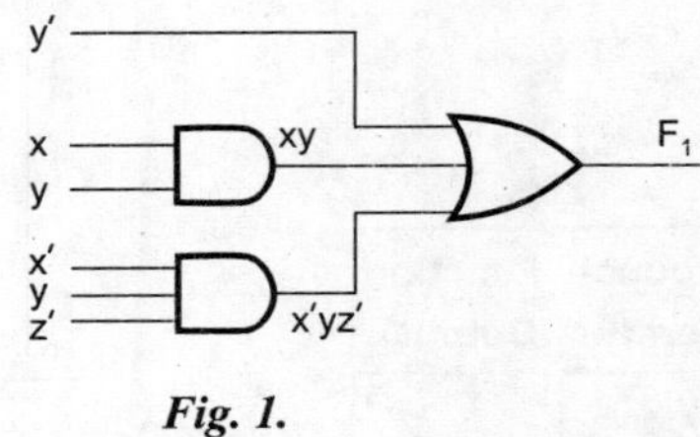

Fig. 1.

b) $F_2 = x\,(y' + z)\,(x' + y + z')$ - Product-of-sums

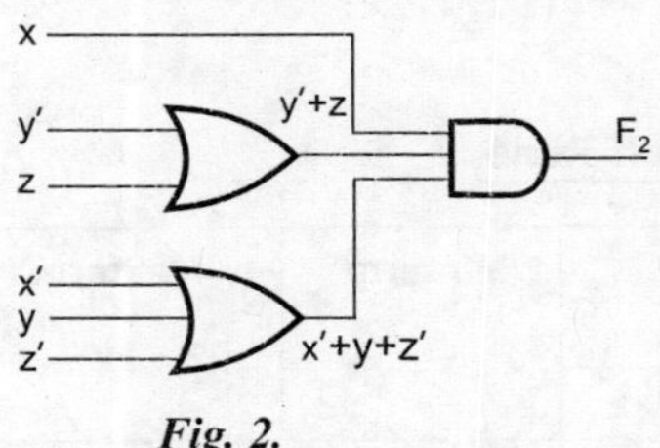

Fig. 2.

Example 1.23

Determine the complement of the following functions and show that SOP becomes POS and vice-versa.

a) $F_1 = y' + x\,y + x'y\,z$ b) $F_2 = x\,(y' + z)\,(x' + y + z')$

Solution

a) $F_1 = y' + x\,y + x'y\,z$ - Sum-of-products

$\Downarrow$ Complement

$F_1' = (y' + x\,y + x'y\,z)'$

$= (y')' \,.\, (x\,y)' \,.\, (x'y\,z)'$

$= y \,.\, (x' + y') \,.\, (x + y' + z')$ - Product-of-sums

> Using DeMorgan's theorem
> $(a + b)' = a' \,.\, b'$
> $(a \,.\, b)' = a' + b'$

b) $F_2 = x (y' + z) (x' + y + z')$ - Product-of-sums

⇓ Complement

Using DeMorgan's theorem
$(a + b)' = a' . b'$
$(a . b)' = a' + b'$

$$F_2' = (x (y' + z) (x' + y + z'))'$$

$$= x' + (y' + z)' + (x' + y + z')'$$

$$= x' + (y . z') + (x . y' . z)$$ - Sum-of-products

1.5.6 Canonical Form

A Boolean function can be expressed algebraically from its truth table by forming a minterm for each combination of the variables that produces a 1 in the output and then taking the OR of all the minterms.

Similarly, a Boolean function can be expressed algebraically from its truth table by forming a maxterm for each combination of the variable that produces a 0 in the output and then taking the AND of all the maxterm. A Boolean function expressed as a sum of minterms or product of maxterms are said to be in canonical form.

The given function can be expressed in canonical form without using truth table. For sum of minterms insert sum of missing literal and its complement with "." operation and expand. Similarly for product of maxterms insert product of missing literal and its complement with "+" operation and expand.

1.5.7 Implementation of Boolean Expressions using Universal Gates

The NAND and NOR gates are called ***universal gates***. When a Boolean function is expressed in SOP form, it can be realized using only NAND gates. When a Boolean function is expressed in POS form, it can be realized using only NOR gates.

Example 1.24

Express the function $F = x + y'z$, in sum of minterms and product of maxterms and verify the result by simplification using Boolean theorems and postulates.

Solution

Table 1: Truth Table

Input Variables			Minterm		Maxterm		Complement of y	Product Term	Function Output
x	y	z					y'	y'z	F
0	0	0	m_0	x'y'z'	M_0	x + y + z	1	0	0
0	0	1	m_1	x'y'z	M_1	x + y + z'	1	1	1
0	1	0	m_2	x'y z'	M_2	x + y' + z	0	0	0
0	1	1	m_3	x'y z	M_3	x + y' + z'	0	0	0
1	0	0	m_4	x y'z'	M_4	x' + y + z	1	0	1
1	0	1	m_5	x y'z	M_5	x' + y + z'	1	1	1
1	1	0	m_6	x y z'	M_6	x' + y' + z	0	0	1
1	1	1	m_7	x y z	M_7	x' + y' + z'	0	0	1

A function can be expressed as a sum of minterms for which the function output is 1. Here, F is 1, when m_1, m_4, m_5, m_6 and m_7 are inputs.

$$\therefore\ F = m_1 + m_4 + m_5 + m_6 + m_7$$
$$= \Sigma m(1, 4, 5, 6, 7)$$
$$= x'y'z + xy'z' + xy'z + xyz' + xyz$$

A function can be expressed as a product of maxterms for which the function output is 0. Here, F is 0, when M_0, M_2 and M_3 are inputs

$$F = M_0\, M_2\, M_3$$
$$= \Pi\, M(0, 2, 3)$$
$$= (x + y + z)\,(x + y' + z)\,(x + y' + z')$$

Proof:

Case i:

$$F = m_1 + m_4 + m_5 + m_6 + m_7$$
$$= x'y'z + xy'z' + xy'z + xyz' + xyz$$ $a + a' = 1$
$$= x'y'z + xy'(z' + z) + xy(z' + z)$$ $a . 1 = a$
$$= x'y'z + xy' . 1 + xy . 1$$
$$= x'y'z' + xy' + xy$$
$$= x'y'z + x . (y' + y)$$
$$= x'y'z + x . 1$$
$$= x'y'z + x$$
$$= (x' + x) . (y'z + x)$$
$$= 1 . y'z + x$$
$$= x + y'z$$

Case ii:

$$F = M_0\, M_2\, M_3$$
$$= (x + y + z)\,(x + y' + z)\,(x + y' + z')$$
$$= (x + z + y)\,\ (x + z + y')\,\ (x + y' + z')$$
$$= (x + z + (y . y'))\,(x + y' + z')$$ $a . a' = 0$
$$= (x + z + 0)\,(x + y' + z')$$ $a + 0 = a$
$$= x + (z . (y' + z'))$$
$$= x + z . y' + zz'$$
$$= x + y' . z + 0$$
$$= x + y'z$$

Example 1.25

Express the following Boolean function in canonical form and implement the Boolean functions using either only NAND or NOR gates.

a) $F_1 = xy + y'z$ **b)** $F_2 = (x + y) . (y + z)$ **c)** $F_3 = x + y'z$

Solution

a) $F_1 = x\,y + y'z$

Case i: SOP

Missing literal in first term is z and second term is x.

$a + a' = 1$

$a \,.\, 1 = a$

$$F_1 = x\,y + y'z$$

$$= x\,y\,.\,(z + z') + (x + x')\,.\,y'z$$

$$= x\,y\,z + x\,y\,z' + x\,y'z + x'y'z$$

$$(m_7)\quad (m_6)\quad (m_5)\quad (m_1)$$

$$= x'y'z + x\,y'z + x\,y\,z' + x\,y\,z$$

$$= m_1 + m_5 + m_6 + m_7$$

The SOP form of F_1 is realized using only NAND gates as shown in Fig. 1.

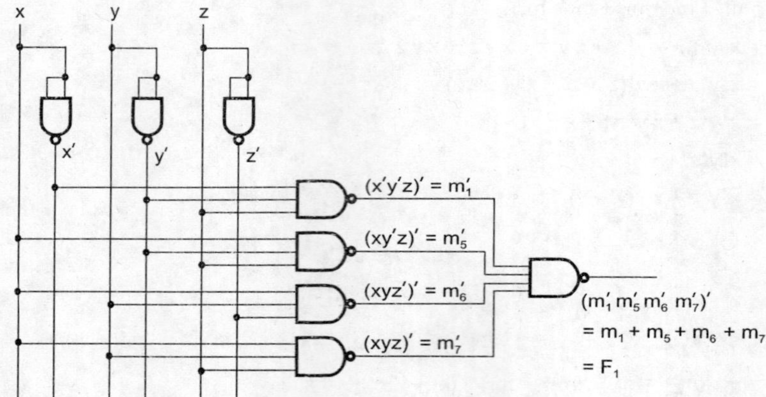

Fig. 1: Logic circuit of F_1 using only NAND gates.

Case ii: POS

$a \,.\, a' = 0$

$a + 0 = a$

Missing literal in first term is z, second term is y and third term is x.

$$F_1 = x\,y + y'z$$

$$= (x + y'z)\,.\,(y + y'z)$$

$$= (x + y')\,.\,(x + z)\,.\,(y + y')\,.\,(y + z)$$

$$= (x + y' + (z\,.\,z'))\,.\,(x + z + (y\,.\,y'))\,((x\,.\,x') + y + z)$$

$$= (x + y' + z)\,.\,(x + y' + z')\,.\,(x + z + y)\,.\,(x + z + y')\,.\,(x + y + z)\,.\,(x' + y + z)$$

$$= (x + y' + z)\,.\,(x + y' + z')\,.\,(x + y + z)\,.\,(x + y' + z)\,.\,(x + y + z)\,.\,(x' + y + z)$$

$$(M_2)\qquad (M_3)\qquad (M_0)\qquad (M_2)\qquad (M_0)\qquad (M_4)$$

$$= (x + y + z)\,(x + y' + z)\,(x + y' + z')\,(x' + y + z)$$

$$= M_0\,M_2\,M_3\,M_4$$

Repeated maxterms are considered only one time

The POS form of F_1 is realized using only NOR gates as shown in Fig. 2.

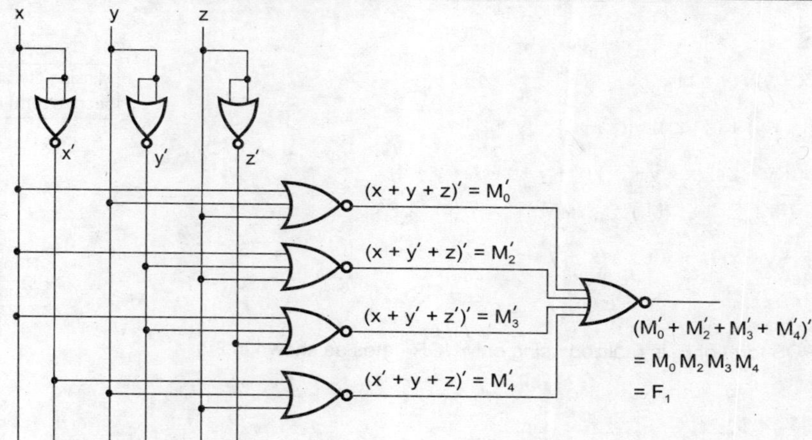

***Fig. 2:** Logic circuit of F_1 using only NOR gates.*

b) $F_2 = (x + y) . (y + z)$

Case i: SOP

$F_2 = (x + y) . (y + z) = x . (y + z) + y . (y + z)$

$= x . y + x . z + y . y + y . z = xy + xz + y + yz$ [y . y = y]

$= x y (z + z') + x z (y + y') + (x + x') y + (x + x') y z$ [Insert missing literals]

$= xyz + xyz' + xzy + xzy' + xy + x'y + xyz + x'yz$ [a + a' = 1] [a . 1 = a]

$= xyz + xyz' + xzy + xzy' + xy (z + z') + x'y (z + z') + xyz + x'yz$

$= xyz + xyz' + xyz + xy'z + xyz + xyz' + x'yz + x'yz' + xyz + x'yz$

$(m_7)\ (m_6)\ (m_7)\ (m_5)\ (m_7)\ (m_6)\ (m_3)\ (m_2)\ (m_7)\ (m_3)$

$= x'yz' + x'yz + xy'z + xyz' + xyz$ [Repeated minterms are considered only one time]

$= m_2 + m_3 + m_5 + m_6 + m_7$

The SOP form of F_2 is realized using only NAND gates as shown in Fig. 3.

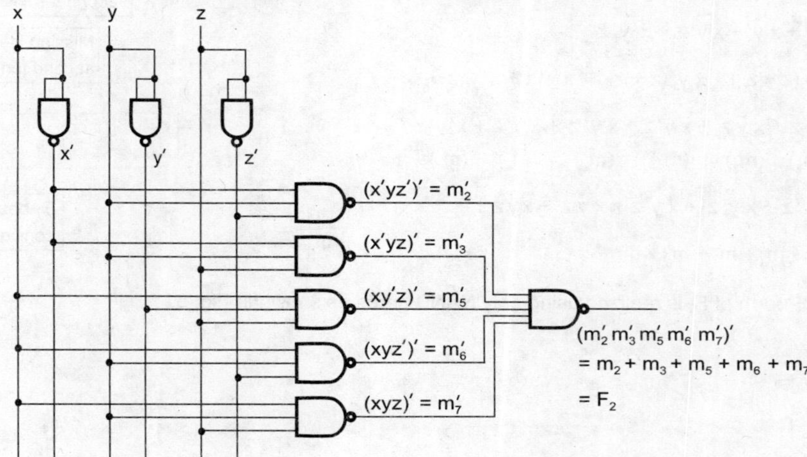

***Fig. 3:** Logic circuit of F_2 using only NAND gates.*

Case ii: POS

$F_2 = (x + y) . (y + z)$

$= ((x + y) + (z . z')) . ((x . x') + (y + z))$

$= (x + y + z) . (x + y + z') (x + y + z) (x' + y + z)$

$(M_0) \quad (M_1) \quad (M_0) \quad (M_4)$

$= (x + y + z) (x + y + z') (x' + y + z)$

$= M_0 M_1 M_4$

Missing literal in first term is z and second term is x.

a . a' = 0

a + 0 = a

Repeated maxterms are considered only one time

The POS form of F_2 is realized using only NOR gates as shown in Fig. 4.

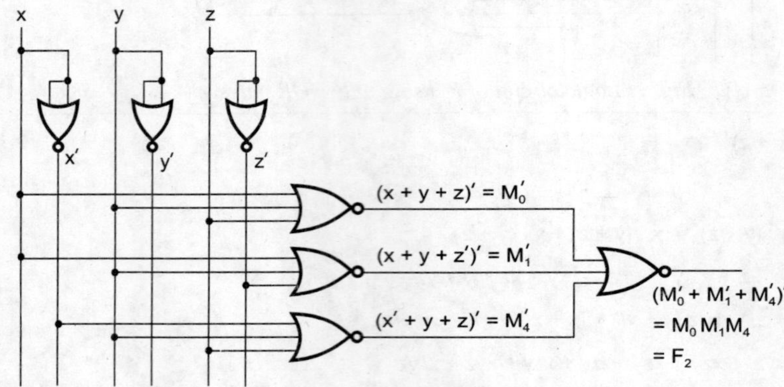

***Fig. 4:** Logic circuit of F_2 using only NOR gates.*

c) $F_3 = x + y'z$

Case i: SOP

$F_3 = x + y'z$

$= x . (y + y') + (x + x') y'z$

$= x y + x y' + x y'z + x'y'z$

$= x y (z + z') + x y'(z + z') + x y'z + x'y'z$

$= x y z + x y z' + x y'z + x y'z' + x y'z + x'y'z$

$(m_7) \quad (m_6) \quad (m_5) \quad (m_4) \quad (m_5) \quad (m_1)$

$= x'y'z + x y'z' + x y'z + x y z' + x y z$

$= m_1 + m_4 + m_5 + m_6 + m_7$

a . 1 = a

a + a' = 1

Missing literal in first term is y and second term is x.

Missing literal in first term and second term is z.

Repeated minterms are considered only one time

The SOP form of F_3 is realized using only NAND gates as shown in Fig. 5.

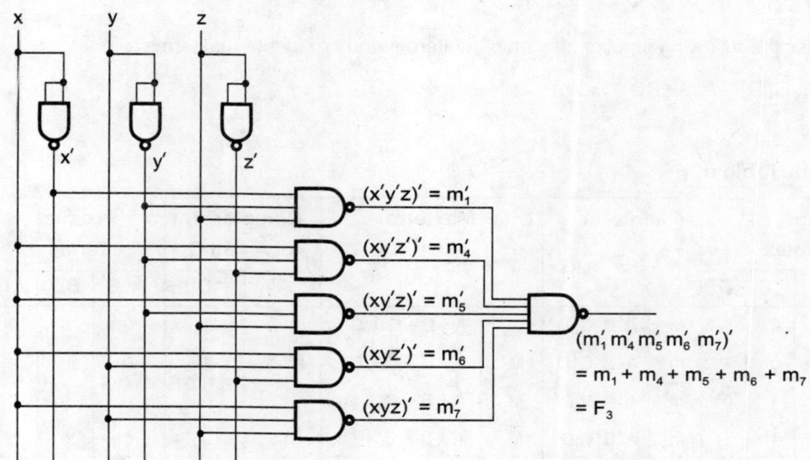

***Fig. 5:** Logic circuit of F_3 using only NAND gates.*

Case ii: POS

$F_3 = x + y'\,z$

$= (x + y') . (x + z)$

$= ((x + y') + (z . z')) . ((x + z) + (y . y'))$

$= (x + y' + z)\,(x + y' + z')\,(x + y + z)\,(x + y' + z)$

$\quad (M_2) \qquad (M_3) \qquad (M_0) \qquad (M_2)$

$= (x + y + z)\,(x + y' + z)\,(x + y' + z')$

$= M_0\,M_2\,M_3$

a . 1 = a
a + a' = 1

Missing literal in first term is z and second term is y.

Repeated maxterms are considered only one time

The POS form of F_3 is realized using only NOR gates as shown in Fig. 6.

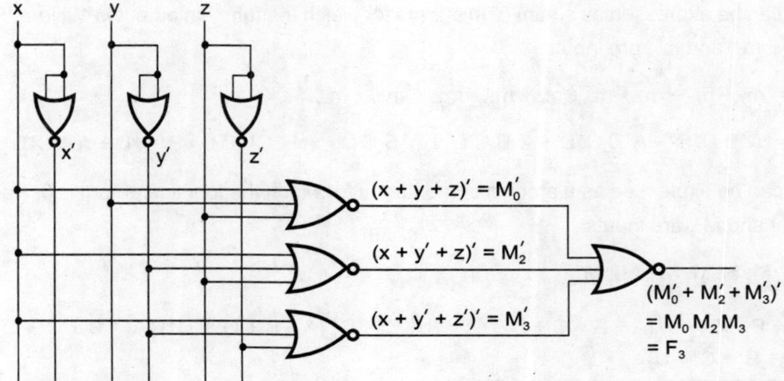

***Fig. 6:** Logic circuit of F_3 using only NOR gates.*

Example 1.26

Express the following function in sum of minterms and product of maxterms

F(A, B, C, D) = A′B + BD + AC′

Solution

Table 1: Truth Table of F

Input Variables				Minterm		Maxterm		Complement of A and C		Product Terms			Function Output
A	B	C	D					A′	C′	A′B	BD	AC′	F
0	0	0	0	m_0	A′B′C′D′	M_0	A + B + C + D	1	1	0	0	0	0
0	0	0	1	m_1	A′B′C′D	M_1	A + B + C + D′	1	1	0	0	0	0
0	0	1	0	m_2	A′B′C D′	M_2	A + B + C′ + D	1	0	0	0	0	0
0	0	1	1	m_3	A′B′C D	M_3	A + B + C′ + D′	1	0	0	0	0	0
0	1	0	0	m_4	A′B C′D′	M_4	A + B′ + C + D	1	1	1	0	0	1
0	1	0	1	m_5	A′B C′D	M_5	A + B′ + C + D′	1	1	1	1	0	1
0	1	1	0	m_6	A′B C D′	M_6	A + B′ + C′ + D	1	0	1	0	0	1
0	1	1	1	m_7	A′B C D	M_7	A′ + B′ + C′ + D′	1	0	1	1	0	1
1	0	0	0	m_8	AB′C′D′	M_8	A′ + B + C + D	0	1	0	0	1	1
1	0	0	1	m_9	AB′C′D	M_9	A + B′ + C′ + D	0	1	0	0	1	1
1	0	1	0	m_{10}	AB′C D′	M_{10}	A′ + B + C′ + D	0	0	0	0	0	0
1	0	1	1	m_{11}	AB′C D	M_{11}	A′ + B + C′ + D′	0	0	0	0	0	0
1	1	0	0	m_{12}	A B C′D′	M_{12}	A′ + B′ + C + D	0	1	0	0	1	1
1	1	0	1	m_{13}	A B C′D	M_{13}	A′ + B′ + C + D′	0	1	0	1	1	1
1	1	1	0	m_{14}	A B C D′	M_{14}	A′ + B′ + C′ + D	0	0	0	0	0	0
1	1	1	1	m_{15}	A B C D	M_{15}	A′ + B′ + C′ + D′	0	0	0	1	0	1

A function can be expressed as a sum of minterms for which the function output is 1. Here, F is 1, when m_4, m_5, m_6, m_7, m_8, m_9, m_{12}, m_{13} and m_{15} are inputs.

$$\therefore\ F = m_4 + m_5 + m_6 + m_7 + m_8 + m_9 + m_{12} + m_{13} + m_{15}$$

$$= A'BC'D' + A'BC'D + A'BCD' + A'BCD + AB'C'D' + AB'C'D + ABC'D' + ABC'D + ABCD$$

A function can be expressed as a product of maxterms for which the function output is 0. Here, F is 0, when M_0, M_1, M_2, M_3, M_{10}, M_{11} and M_{14} are inputs

$$F = M_0\, M_1\, M_2\, M_3\, M_{10}\, M_{11}\, M_{14}$$

$$= (A + B + C + D)\,(A + B + C + D')\,(A + B + C' + D)\,(A + B + C' + D')\,(A' + B + C' + D)\,(A' + B + C' + D')\,(A' + B' + C' + D)$$

Example 1.27

Express the Boolean function F =A + B′C in a sum of minterms.

Solution

Table 1: Truth Table of F = A + B′C

Input Variables				Minterm		Complement of B	Product Term	Function Output
A	B	C	D			B′	B′C	F
0	0	0	0	m_0	A′B′C′D′	1	0	0
0	0	0	1	m_1	A′B′C′D	1	0	0
0	0	1	0	m_2	A′B′C D′	1	1	1
0	0	1	1	m_3	A′B′C D	1	1	1
0	1	0	0	m_4	A′B C′D′	0	0	0
0	1	0	1	m_5	A′B C′D	0	0	0
0	1	1	0	m_6	A′B C D′	0	0	0
0	1	1	1	m_7	A′B C D	0	0	0
1	0	0	0	m_8	AB′C′D′	1	0	1
1	0	0	1	m_9	AB′C′D	1	0	1
1	0	1	0	m_{10}	AB′C D′	1	1	1
1	0	1	1	m_{11}	AB′C D	1	1	1
1	1	0	0	m_{12}	AB C′D′	0	0	1
1	1	0	1	m_{13}	AB C′D	0	0	1
1	1	1	0	m_{14}	AB C D′	0	0	1
1	1	1	1	m_{15}	AB C D	0	0	1

A function can be expressed as a sum of minterms for which the function output is 1. Here, F is 1, when m_2, m_3, m_8, m_9, m_{10}, m_{11}, m_{12}, m_{13}, m_{14} and m_{15} are inputs

$$\therefore\ F = m_2 + m_3 + m_8 + m_9 + m_{10} + m_{11} + m_{12} + m_{13} + m_{14} + m_{15}$$

$$= A'B'CD' + A'B'CD + AB'C'D' + AB'C'D + AB'CD' + AB'CD + ABC'D' + ABC'D + ABCD' + ABCD$$

1.5.8 Simplification of Boolean Expressions or Functions

Each variable with in a term of a Boolean expression is called a ***literal***. Example of 7 literals and 10 literals Boolean expression are shown below:

a′ b + c d + a b d
↓ ↓ ↓ ↓ ↓ ↓ ↓
(1)(2) (3)(4) (5)(6)(7) → 7 literals

a b c + b′c + a c′ + a b′ c
(1)(2)(3) (4)(5) (6)(7) (8)(9)(10) → 10 literals

Reducing the number of literals in a Boolean expression or function will simplify the implementation of function by logic gates with minimum number of gates.

The Boolean expression can be reduced or simplified using postulates and theorems of Boolean algebra. There is no standard procedure for simplification using theorems.

Alternatively, Boolean functions can be simplified using following two methods.

- Karnaugh map method
- Quine-McCluskey method or Table method

These two methods have standard procedure for simplification of Boolean functions.

Example 1.28

Reduce the number of literals in following expressions.

a) $xy + xy'$ **b)** $xy + x(wz + wz')$ **c)** $(A + B)'(A' + B')'$

Solution

a) $xy + xy'$

$$xy + xy' = x.(y + y') \qquad \boxed{y + y' = 1}$$
$$= x.1 \qquad \boxed{x.1 = x}$$
$$= x$$

The given expression $xy + xy'$ is 4 literal expression which is reduced to single literal.

b) $xy + x(wz + wz')$

$$xy + x(wz + wz') = xy + x(w(z + z'))$$
$$= xy + x(w.1) \qquad \boxed{z + z' = 1}$$
$$= xy + xw \qquad \boxed{w.1 = w}$$
$$= x.(y + w)$$

The given expression $xy + x(wz + wz')$ is 7 literals expression which is reduced to three literals.

c) $(A + B)'(A' + B')'$

$$(A + B)'(A' + B')' = (A + B)'.((A')'.(B')') \qquad \boxed{\text{Using DeMorgan's law } (x + y)' = x'.y'}$$
$$= (A'.B').(A.B)$$
$$= A'.B'.A.B \qquad \boxed{x.x' = 0}$$
$$= A'.A.B.B' = 0.0 = 0$$

The given expression is a 4 literal expression which is reduced to null expression.

Complement of a Function

If, F is a function then F′ is complement of F. The complement of a function can be directly evaluated using Boolean theorems and postulates. Alternatively, the complement can also be obtained by using duality. In duality replace "+" by "." and "." by "+" and variables by its complement.

Example 1.29

Determine the complement of the following Boolean functions.

a) $F_1 = x\,y + x'z$ b) $F_2 = w\,x + y\,z$ c) $F_3 = x\,y' + x'\,y$

Solution

a) $F_1 = x\,y + x'z$

Case i: Direct Evaluation of Complement

$$F_1 = x\,y + x'z$$

⇓ Complement

$$F_1' = (x\,y + x'z)'$$
$$= (x\,y)' . (x'z)'$$
$$= (x' + y') . ((x')' + z')$$
$$= (x' + y') . (x + z')$$

Using DeMorgan's theorem
$(a . b)' = a' + b'$
$(a + b)' = a' . b'$

Case ii: Complement using Duality

$$x\ .\ y\ +\ x'.\ z$$

↓ ↓ ↓ ↓ ↓ ↓ ↓ Replace by dual elements

$$x' +\ y'.\ x\ +\ z'$$

⇓

$$(x' + y') . (x + z') = F_1'$$

b) $F_2 = w\,x + y\,z$

Case i: Direct Evaluation of Complement

$$F_2 = w\,x + y\,z$$

⇓ Complement

$$F_2' = (w\,x + y\,z)'$$
$$= (w\,x)' . (y\,z)' = (w' + x') . (y' + z')$$

Using DeMorgan's theorem
$(a + b)' = a' . b'$
$(a . b)' = a' + b'$

Case ii: Complement using Duality

$$w\ .\ x\ +\ y\ .\ z$$

↓ ↓ ↓ ↓ ↓ ↓ ↓ Replace by dual elements

$$w' +\ x'.\ y' + z'$$

⇓

$$(w' + x') . (y' + z') = F_2'$$

c) $F_3 = x\,y' + x'\,y$

Case i: Direct Evaluation of Complement

$$F_3 = x\,y' + x'\,y$$

⇓ Complement

$$F_3' = (x\,y' + x'\,y)'$$
$$= (x\,y')' . (x'\,y)' = (x' + y) . (x + y')$$

Using DeMorgan's theorem
$(a + b)' = a' . b'$
$(a . b)' = a' + b'$

Case ii: Complement using Duality

x . y' + x'. y

↓ ↓ ↓ ↓ ↓ ↓ ↓ Replace by dual elements

x'+ y . x + y'

⇓

$(x'+y).(x+y') = F_3'$

1.6 Minimization using K-map

K-map is a pictorial form of truth table and used to simplify Boolean functions. Simplification of Boolean function using K-map is simple and straight forward. A K-map is a diagram made up of squares with each square representing one minterm. Alternatively, maxterm can be used to construct K-map in which each square represent a maxterm.

When minterms are considered for simplification the resultant Boolean function will be in sum-of-products form. When maxterms are considered for simplification the resultant Boolean function will be in product-of-sums form.

Note: *The simplified Boolean function using K-map is not unique. Sometimes there may be multiple solutions.*

For n variable Boolean function the K-map will have 2^n squares. Each square represent a minterm or maxterm. The literals of minterm are split and arranged as rows and columns. While arranging the literal/minterms only one change is allowed if we move from one row to next row or from one column to next column.

1.6.1 Realization of Sum-of-Products Form using K-map

1. Construct the truth table of Boolean function.
2. Draw K-map with 2^n squares where n is number of variables in Boolean function.
3. Enter the value of function or output in the squares as either 1 or 0.
4. The squares with entry 1 are the minterms that represent the function.
5. Each square with 1 is an ***implicant***.
6. Combine adjacent 1's to form ***prime implicants***.
7. A prime implicant with single 1 represent n literal product term.
8. A prime implicant with two 1's represent $n-1$ literal product term.
9. A prime implicant with four 1's represent $n-2$ literal product term and so on.
10. The literals represented by a prime implicant are the literals that present in all the squares of the prime implicant. (or the literals common to all the squares of the prime implicant).
11. While forming prime implicants any number of over lapping is allowed in horizontal and vertical directions. But the diagonal elements cannot be combined to form prime implicants.
12. While forming prime implicants the K-map can be folded in any direction to get adjacent 1's.
13. The simplified Boolean function is sum of all the product terms of prime implicant.
14. If all minterms are 1's then all squares of K-map will be filled by 1 and function value is 1.

Table 1.64: Relationship between Number of Adjacent Squares and Literals in the Product Term

Number of Adjacent Squares in Prime Implicants	Number of Literals in a Product Term in an n-variable Map			
	n = 2	n = 3	n = 4	n = 5
1	2	3	4	5
2	1	2	3	4
4	0	1	2	3
8		0	1	2
16			0	1
32				0

1.6.2 Two-Variable K-map

Let, n = Number of variables in Boolean function

x, y = Two variables of Boolean function

Therefore, n = 2, for two-variable K-map

Here, $2^n = 2^2 = 4$

Therefore, two-variable, K-map will have 4 squares. The two-variable K-map and arrangements of literals and minterms are shown in Fig. 1.11.

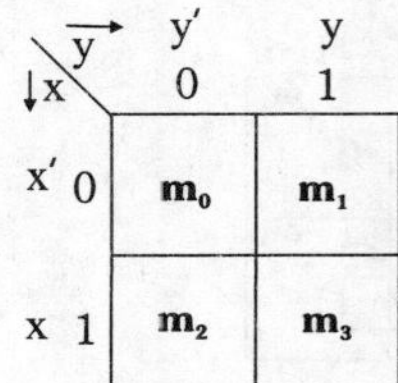

***Fig. a:** Arrangement of literals and minterms in 2-variable K-map.*

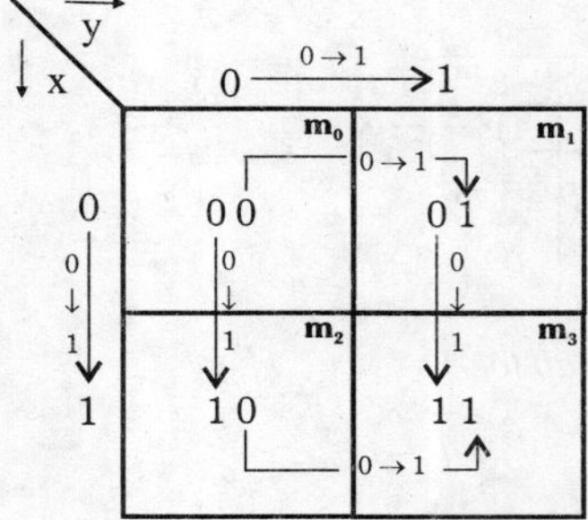

***Fig. b:** Change in bit values of literals and minterms in 2-variable K-map.*

***Fig. 1.11:** Two-variable K-map.*

Some examples of two-variable K-map with single prime implicant formed using single 1's is shown in Figs. 1.12 and 1.13. Since the prime implicants shown in Figs. 1.12 and 1.13 are formed using single 1's in two-variable K-map, each prime implicant represents a product term of two literals.

The K-maps in Fig. 1.12 have only one prime implicant and so the Boolean function, F is given by prime implicant of the K-map. The K-maps in Fig. 1.13 has two prime implicants and so the Boolean function, F is given by sum of the two prime implicants of the K-map.

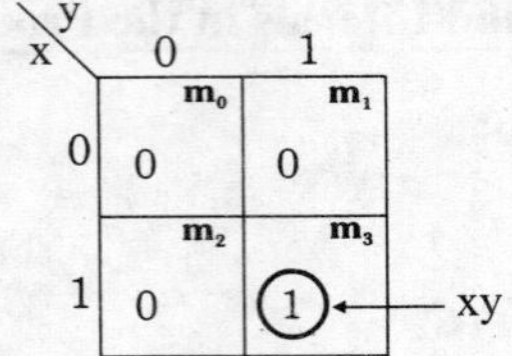

Fig. a: K-map for, F = xy.

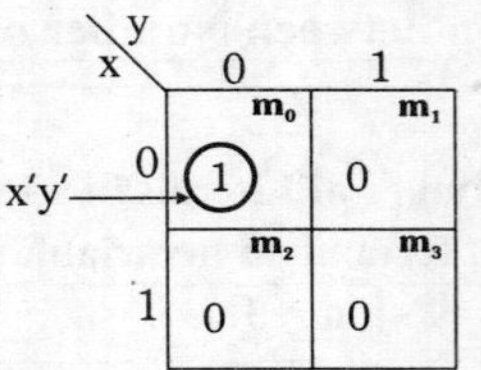

Fig. b: K-map for, F = x′y′.

Fig. 1.12: Two-variable K-map with single prime implicant formed using single 1's.

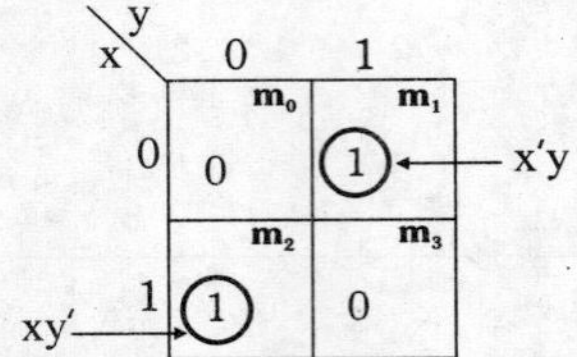

Fig. a: K-map for, F = x′y + xy′.

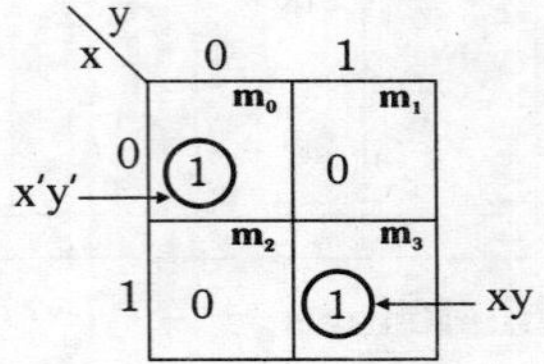

Fig. b: K-map for, F = x′y′ + xy.

Fig. 1.13: Two-variable K-map with two prime implicants formed using single 1's.

Some examples of two-variable K-map with single prime implicant formed using two adjacent 1's is shown in Fig. 1.14. Since the prime implicants shown in Fig. 1.14 are formed using two adjacent 1's in two-variable K-map, each prime implicant represents a single literal. Since the K-map have only one prime implicant the Boolean function, F is given by prime implicant of the K-map.

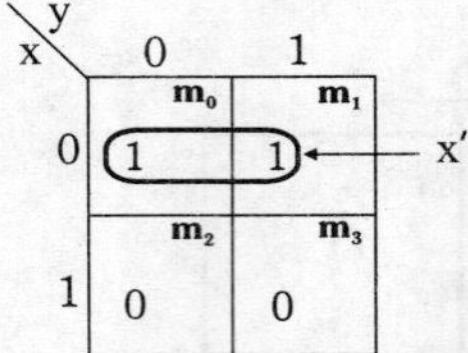

Fig. a: K-map for, F = x′.

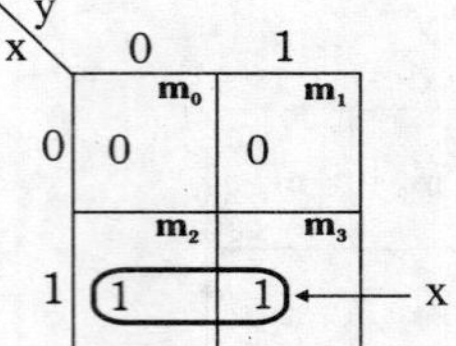

Fig. b: K-map for, F = x.

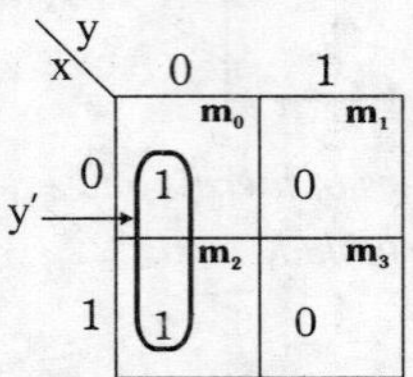

Fig. c: K-map for, F = y′.

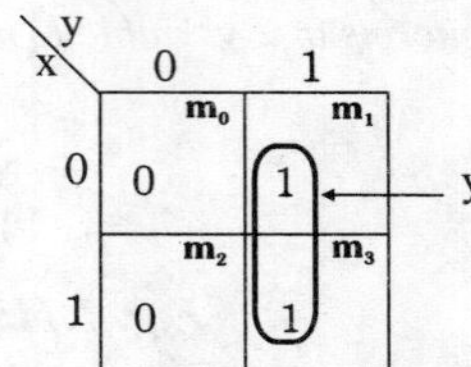

Fig. d: K-map for, F = y.

Fig. 1.14: Two-variable K-map with prime implicant formed using two adjacent 1's.

Some examples of two-variable K-map with two overlapping prime implicants formed using two adjacent 1's is shown in Fig. 1.15. Since the prime implicants shown in Fig. 1.15 are formed using two adjacent 1's in two-variable K-map, each prime implicant represents a single literal. Since the K-map has two prime implicant the Boolean function, F is given by sum of two prime implicants of the K-map.

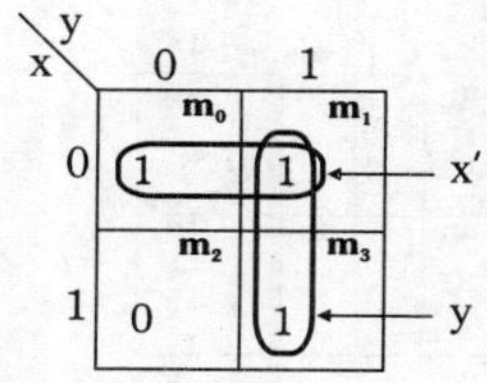

Fig. a: K-map for, F = x′ + y.

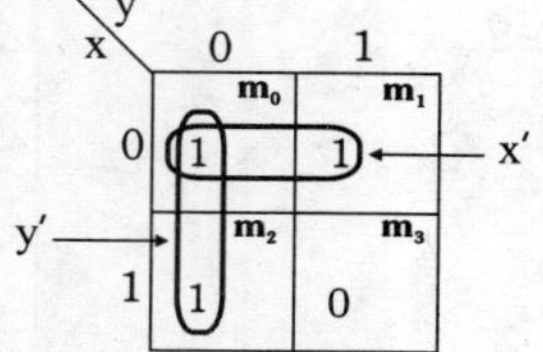

Fig. b: K-map for, F = x′ + y′.

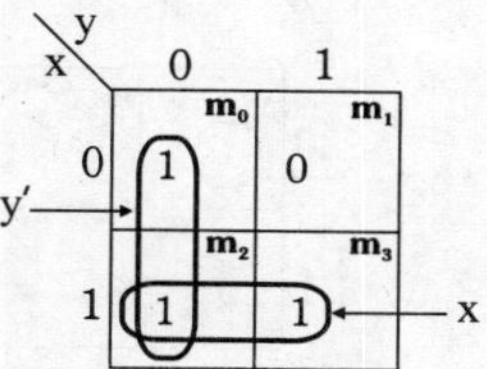

Fig. c: K-map for, F = x + y′.

Fig. 1.15: Two-variable K-map with two overlapping prime implicants formed using two adjacent 1's.

When all the entries of K-map (or all minterms) are 1's then the Boolean function value is 1 (F = 1) as shown in Fig. 1.16.

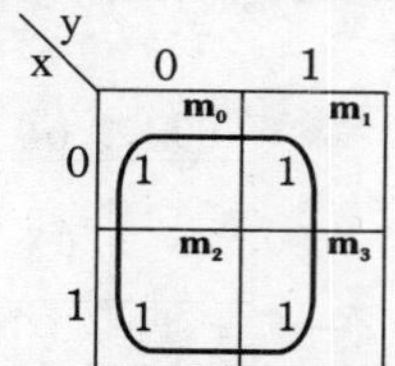

Fig. 1.16: Two-variable K-map with all minterms equal to 1 and F=1.

1.6.3 Three-Variable K-map

Let, n = Number of variables in Boolean function

x, y, z = Three variables of Boolean function

Therefore, n = 3, for three-variable K-map

Here, $2^n = 2^3 = 8$

Therefore, three-variable, K-map will have 8 squares. The three-variable K-map and arrangements of literals and minterms are shown in Fig. 1.17.

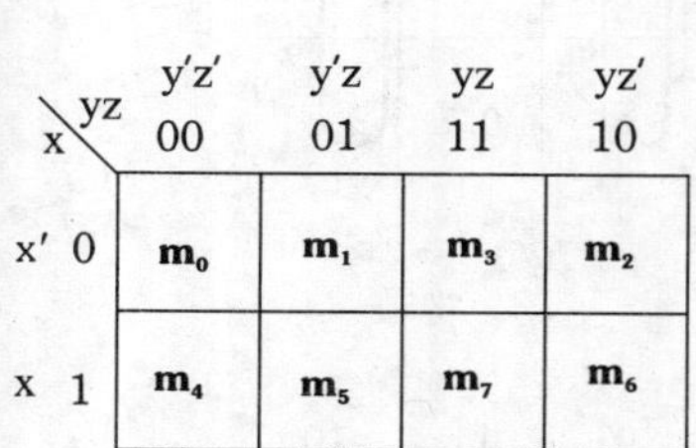

Fig. a: Arrangement of literals and minterms in 3-variable K-map.

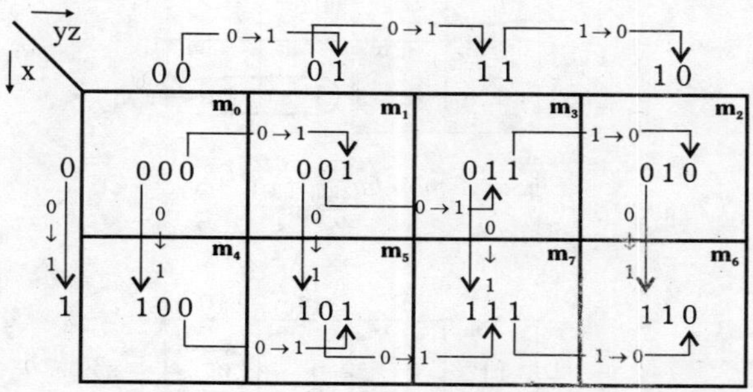

Fig. b: Change in bit values of literals and minterms in 3-variable K-map.

Fig. 1.17: Three-variable K-map.

Some examples of three-variable K-map with prime implicant formed using single 1's are shown in Figs. 1.18 and 1.19. Since the prime implicants shown in Figs. 1.18 and 1.19 are formed using single 1's in three-variable K-map, each prime implicant represents a product term of three literals.

The K-maps in Fig. 1.18 have only one prime implicant and so the Boolean function, F is given by prime implicant of the K-map. The K-maps in Fig. 1.19 has two prime implicant and so the Boolean function, F is given by sum of the two prime implicants of the K-map.

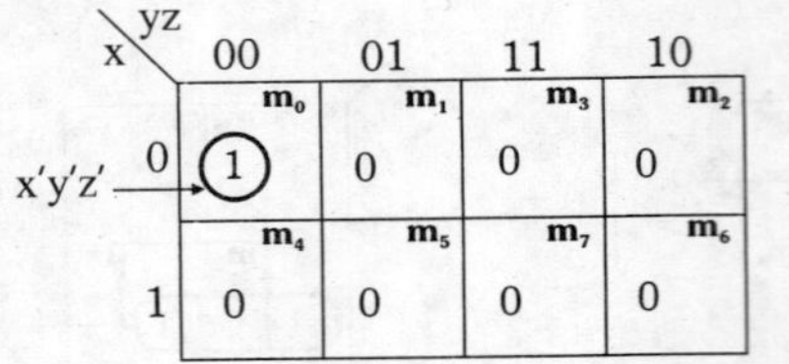

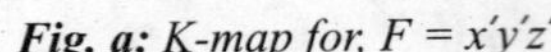

Fig. a: K-map for, F = x'y'z'.

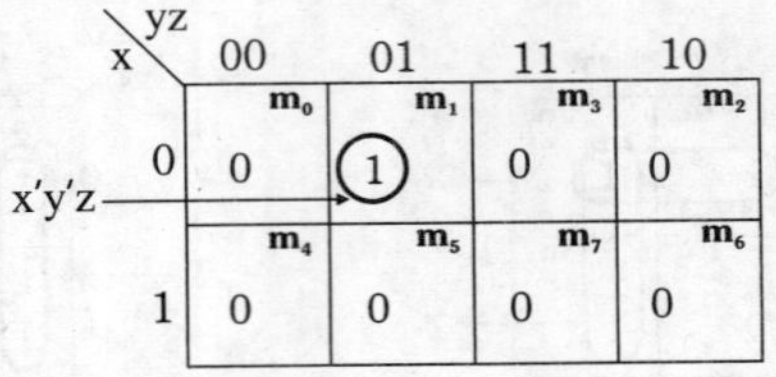

Fig. b: K-map for, F = x'y'z.

***Fig. 1.18:** Three-variable K-map with prime implicant formed using single 1's.*

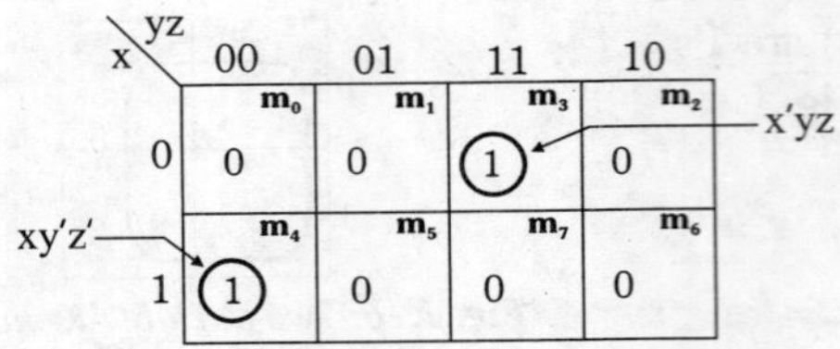

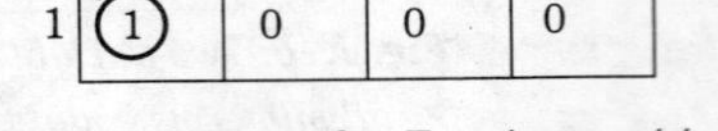

Fig. a: K-map for, F = x'yz + xy'z'.

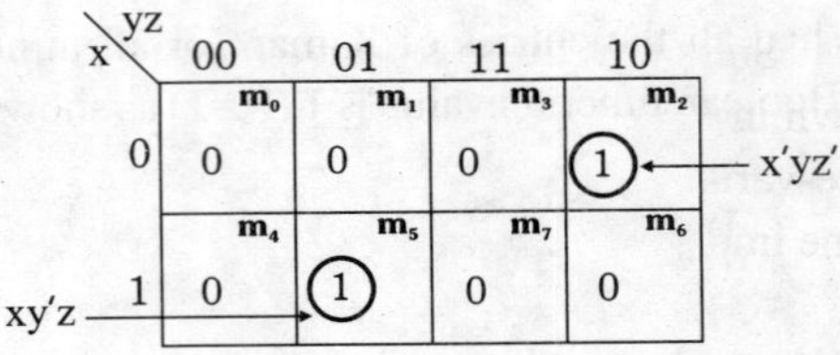

Fig. b: K-map for, F = x'yz' + xy'z.

***Fig. 1.19:** Three-variable K-map with two prime implicants formed using single 1's.*

Some examples of three-variable K-map with two prime implicants formed using two adjacent 1's are shown in Fig. 1.20. Since the prime implicants shown in Fig. 1.20 are formed using two adjacent 1's in three -variable K-map, each prime implicant represents a product term of two literals. Since the K-map has two prime implicants the Boolean function, F is given by sum of the two prime implicants of the K-map.

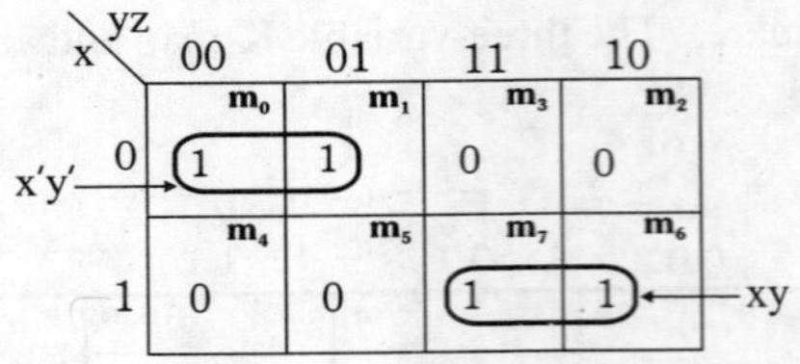

Fig. a: K-map for, F = x'y' + xy.

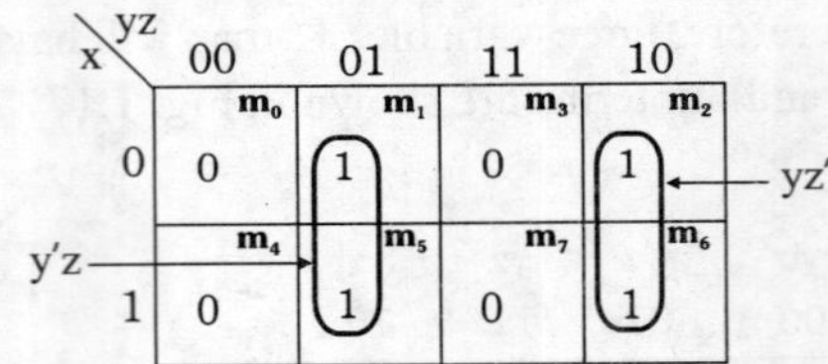

Fig. b: K-map for, F = y'z + yz'.

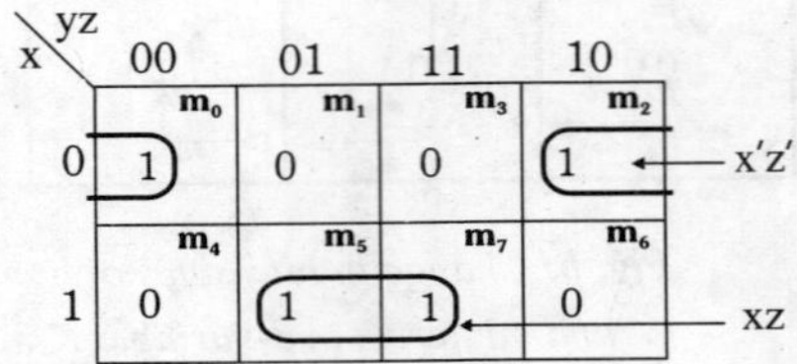

Fig. c: K-map for, F = x'z' + xz.

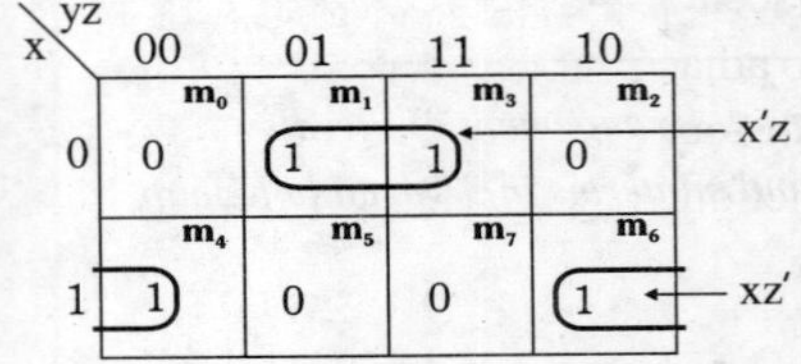

Fig. d: K-map for, F = x'z + xz'.

***Fig. 1.20:** Three-variable K-map with two prime implicants formed using two adjacent 1's.*

Some examples of three-variable K-map with two overlapping prime implicants formed using two adjacent 1's and a prime implicant with single 1 are shown in Fig. 1.21. The prime implicants formed using two adjacent 1's in three-variable K-map, represents a product term of two literals and the prime implicant with single 1 represents a product term of three literal. Since the K-map has three prime implicant the Boolean function, F is given by sum of three prime implicants of the K-map.

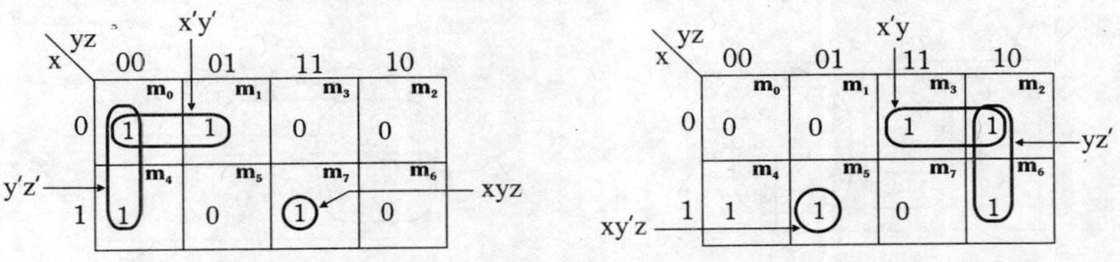

Fig. a: *K-map for, F = xyz + x'y' + y'z'.*

Fig. b: *K-map for, F = xy'z + yz' + x'y.*

Fig. 1.21: *Three-variable K-map with prime implicants formed using single 1's and two overlapping two adjacent 1's.*

Some examples of three-variable K-map with prime implicant formed using four adjacent 1's are shown in Fig. 1.22. Since the prime implicants shown in Fig. 1.22 are formed using four adjacent 1's in three-variable K-map, each prime implicant represents a single literal. Since the K-map have only one prime implicant the Boolean function, F is given by prime implicant of the K-map.

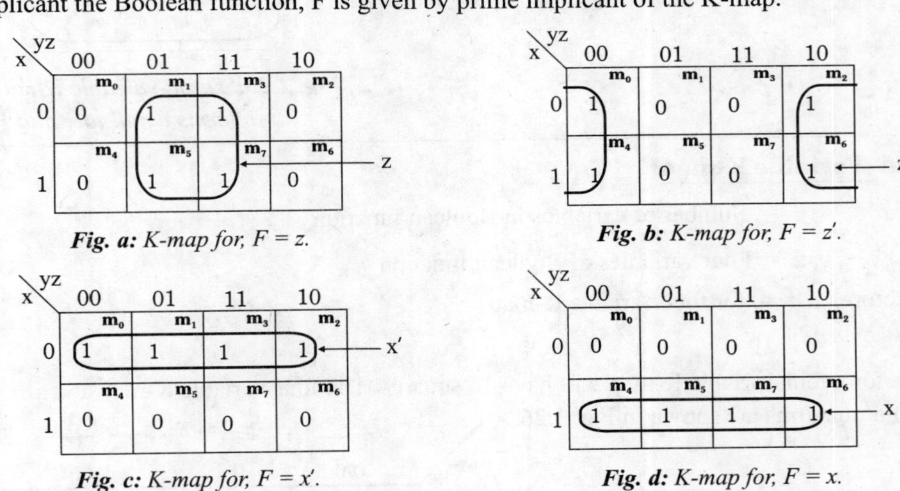

Fig. a: *K-map for, F = z.*

Fig. b: *K-map for, F = z'.*

Fig. c: *K-map for, F = x'.*

Fig. d: *K-map for, F = x.*

Fig. 1.22: *Three-variable K-map with prime implicant formed using four adjacent 1's.*

Some examples of three-variable K-map with two overlapping prime implicants formed using four adjacent 1's and two adjacent 1's are shown in Figs. 1.23 and 1.24. The prime implicants formed using two adjacent 1's in three-variable K-map, represents a product term of two literals and the prime implicant with four adjacent 1's represents a single literal. Since the K-map has two prime implicant the Boolean function, F is given by sum of two prime implicants of the K-map.

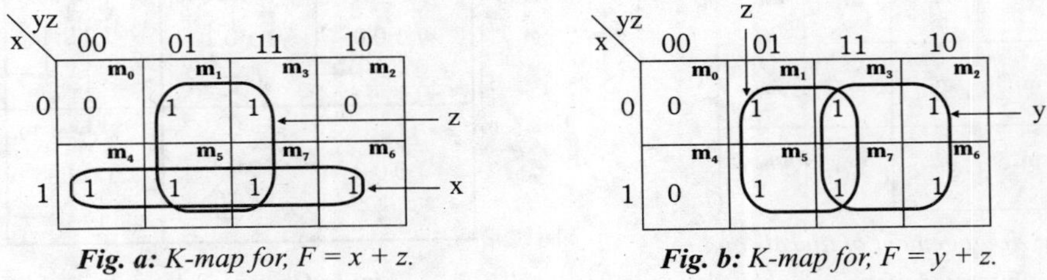

Fig. a: *K-map for, F = x + z.*

Fig. b: *K-map for, F = y + z.*

Fig. 1.23: *Three-variable K-map with two overlapping prime implicants formed using four adjacent 1's.*

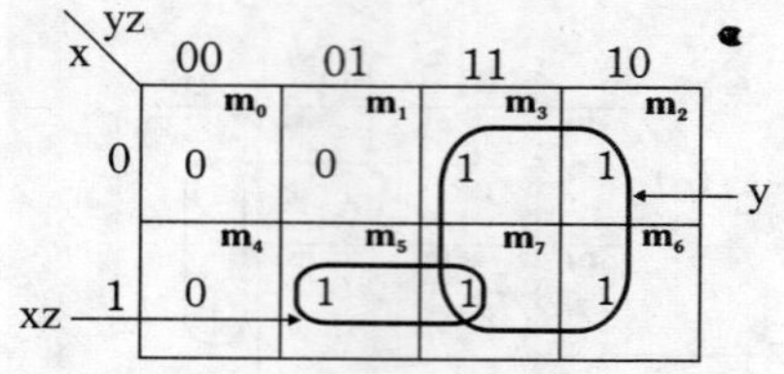

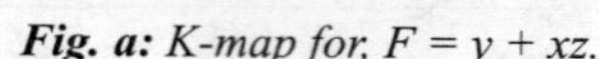
Fig. a: *K-map for, F = y + xz.*

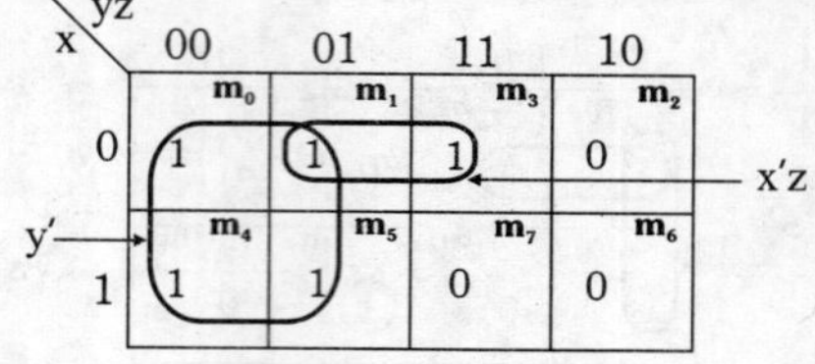

Fig. b: *K-map for, F = y′ + x′z.*

Fig. 1.24: *Three-variable K-map with overlapping prime implicants formed using two adjacent 1's and four adjacent 1's.*

When all the entries of K-map (or all minterms) are 1's then the Boolean function value is 1 (F = 1) as shown in Fig. 1.25.

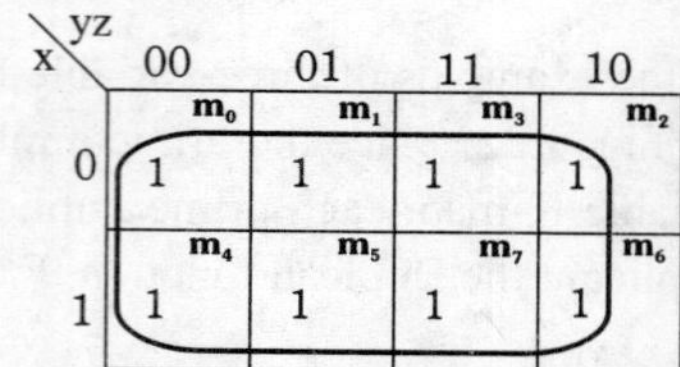

Fig. 1.25: *Three-variable K-map with all minterms equal to 1 and F=1.*

1.6.4 Four-Variable K-map

Let, n = Number of variables in Boolean function

w, x, y, z = Four variables of Boolean function

Therefore, n = 4, for four-variable K-map

Here, $2^n = 2^4 = 16$

Therefore, four-variable K-map will have 16 squares. The four-variable K-map and arrangements of literals and minterms are shown in Fig. 1.26.

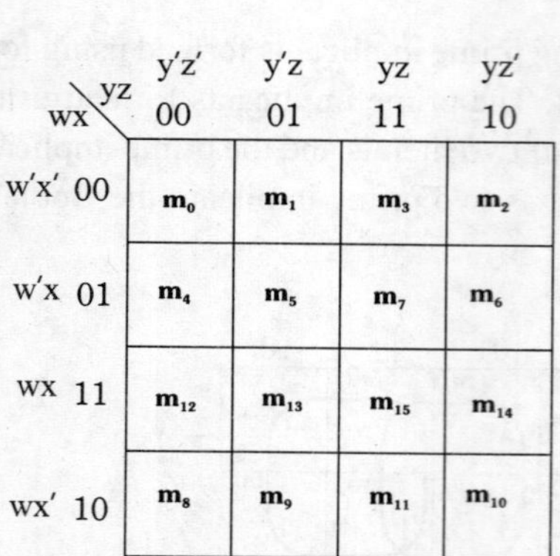

Fig. a: *Arrangement of literals and minterms in four-variable K-map.*

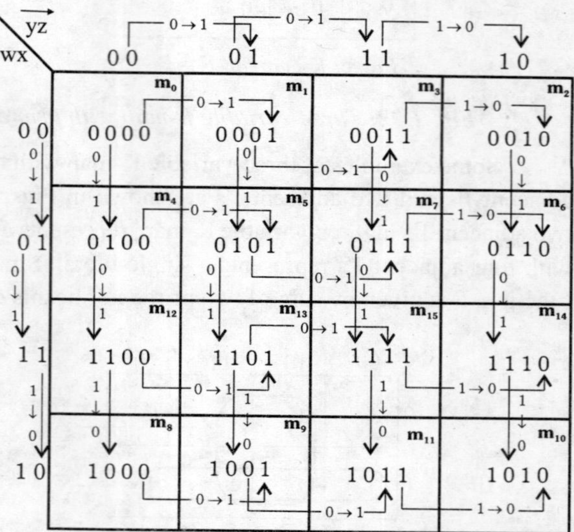

Fig. b: *Change in bit values of literals and minterms in four-variable K-map.*

Fig. 1.26: *Four-variable K-map.*

Some examples of four-variable K-map with prime implicants formed using four adjacent 1's, two adjacent 1's and single 1's are shown in Fig. 1.27. In four-variable K-map, the prime implicants formed using four adjacent 1's represents a product term of two literals, the prime implicants with two adjacent 1's represents a product term of three literals and the prime implicant with single 1 represents a product term of four literals. Boolean function, F is given by sum of prime implicants of the K-map.

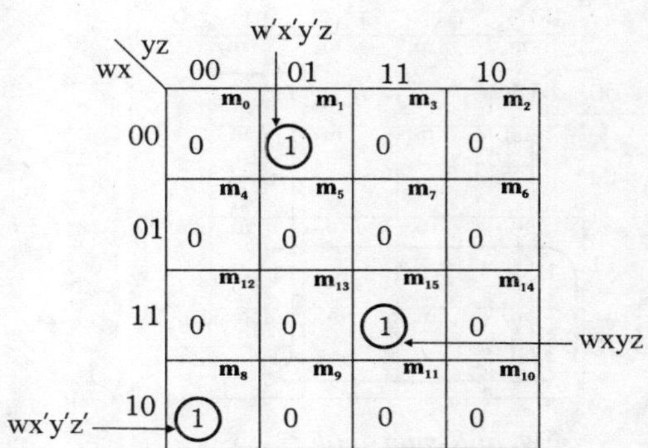

Fig. a: *K-map for,* $F = w'x'y'z + wxyz + wx'y'z'$.

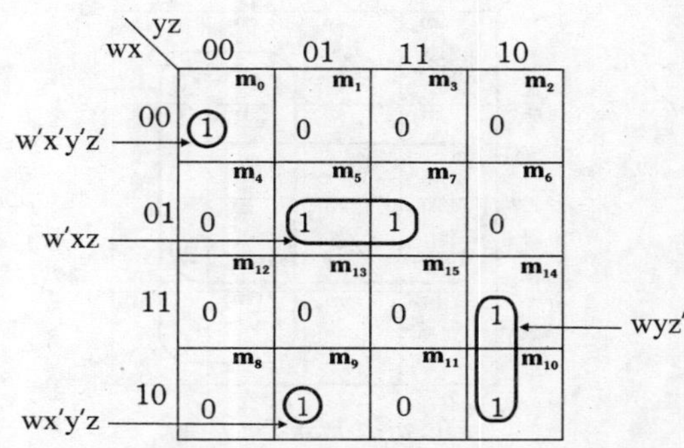

Fig. b: *K-map for,* $F = w'x'y'z' + wx'y'z + w'xz + wyz'$.

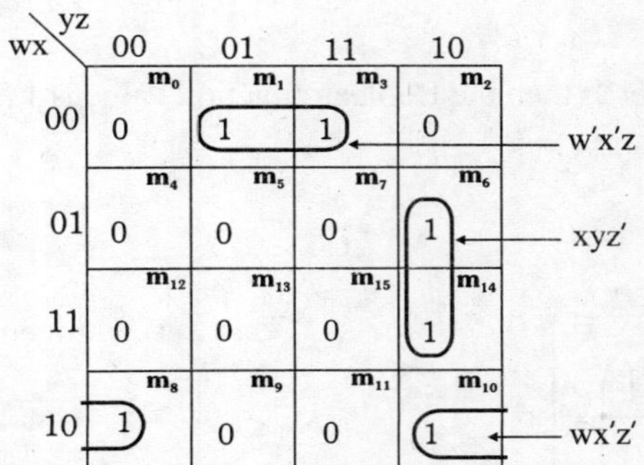

Fig. c: *K-map for,* $F = w'x'z + xyz' + wx'z'$.

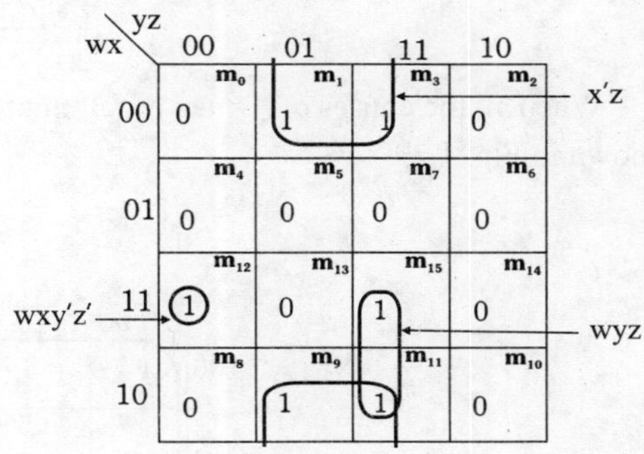

Fig. d: *K-map for,* $F = wxy'z' + wyz + x'z$.

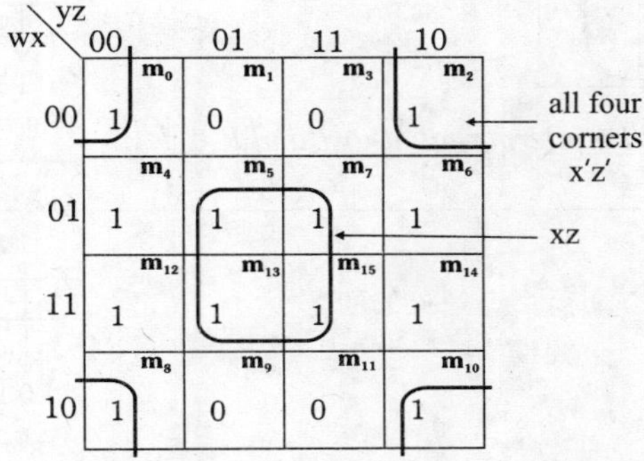

Fig. e: *K-map for,* $F = xz + x'z'$.

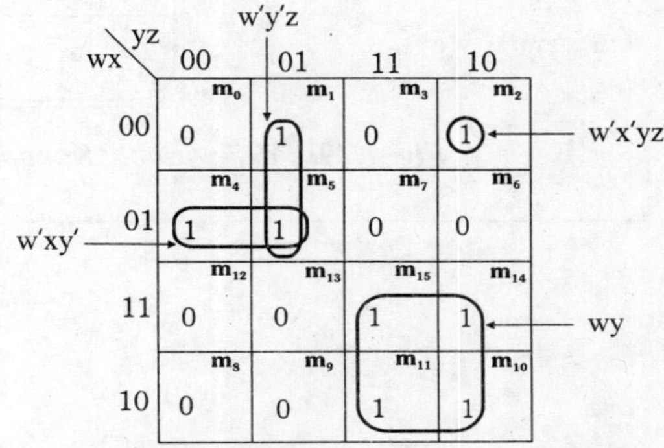

Fig. f: *K-map for,* $F = w'x'yz' + w'y'z + w'xy' + wy$.

Fig. 1.27: *Four-variable K-maps.*

Some examples of four-variable K-map with two overlapping prime implicants formed using eight adjacent 1's are shown in Fig. 1.28. Since the prime implicants shown in Fig. 1.28 are formed using eight adjacent 1's in four-variable K-map, each prime implicant represents a single literal. Since the K-map has two prime implicant the Boolean function, F is given by sum of two prime implicants of the K-map.

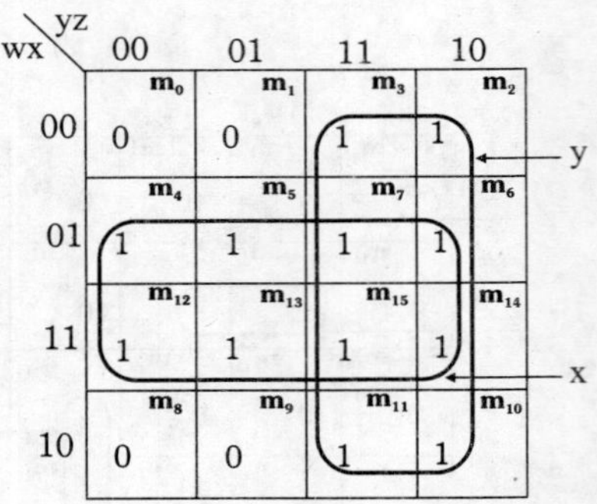

Fig. a: *K-map for, F = x + y.*

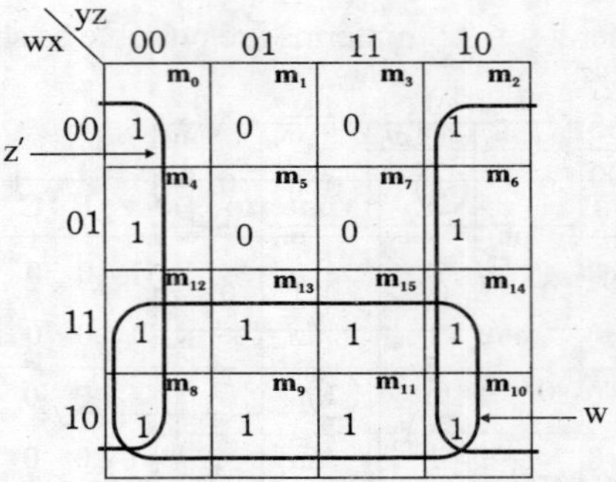

Fig. b: *K-map for, F = w + z′.*

Fig. 1.28: *Four-variable K-map with overlapping prime implicants formed using eight adjacent 1's.*

When all the entries of K-map (or all minterms) are 1's then the Boolean function value is 1 (F = 1) as shown in Fig. 1.29.

wx \ yz	00	01	11	10
00	m_0: 1	m_1: 1	m_3: 1	m_2: 1
01	m_4: 1	m_5: 1	m_7: 1	m_6: 1
11	m_{12}: 1	m_{13}: 1	m_{15}: 1	m_{14}: 1
10	m_8: 1	m_9: 1	m_{11}: 1	m_{10}: 1

Fig. 1.29: *Four-variable K-map with all minterms equal to 1 and F = 1.*

1.6.5 Five-Variable K-map

Let A, B, C, D and E be the five-variables.

Now, n = 5 and $2^5 = 32$

Therefore, 32 minterms are possible as shown in Table 1.65.

Table 1.65: Truth Table

A	B	C	D	E	Minterm	A	B	C	D	E	Minterm
0	0	0	0	0	m_0	1	0	0	0	0	m_{16}
0	0	0	0	1	m_1	1	0	0	0	1	m_{17}
0	0	0	1	0	m_2	1	0	0	1	0	m_{18}
0	0	0	1	1	m_3	1	0	0	1	1	m_{19}
0	0	1	0	0	m_4	1	0	1	0	0	m_{20}
0	0	1	0	1	m_5	1	0	1	0	1	m_{21}
0	0	1	1	0	m_6	1	0	1	1	0	m_{22}
0	0	1	1	1	m_7	1	0	1	1	1	m_{23}
0	0	1	0	0	m_8	1	1	0	0	0	m_{24}
0	1	0	0	1	m_9	1	1	0	0	1	m_{25}
0	1	0	1	0	m_{10}	1	1	0	1	0	m_{26}
0	1	0	1	1	m_{11}	1	1	0	1	1	m_{27}
0	1	1	0	0	m_{12}	1	1	1	0	0	m_{28}
0	1	1	0	1	m_{13}	1	1	1	0	1	m_{29}
0	1	1	1	0	m_{14}	1	1	1	1	0	m_{30}
0	1	1	1	1	m_{15}	1	1	1	1	1	m_{31}

The 32 minterms can be arranged in 2 numbers of 4 variable K-map, one for A = 0 and other for A = 1 as shown in Figs. 1.30 and 1.31. These two K-maps can be placed one over the other as shown in Fig. 1.32. While forming prime implicants any two square that lie one over the other are considered adjacent.

Apart from this, the guidelines for forming prime implicants of 4-variable K-map can be applied individually to Figs. 1.30 and 1.31 for the adjacent squares that do not lie one over the other in Fig. 1.32. Some examples of formation of prime implicants in 5-variable K-map are shown in Figs. 1.33 and 1.34.

A = 0
DE
BC
00 01 11 10
00 m0 m1 m3 m2
01 m4 m5 m7 m6
11 m12 m13 m15 m14
10 m8 m9 m11 m10

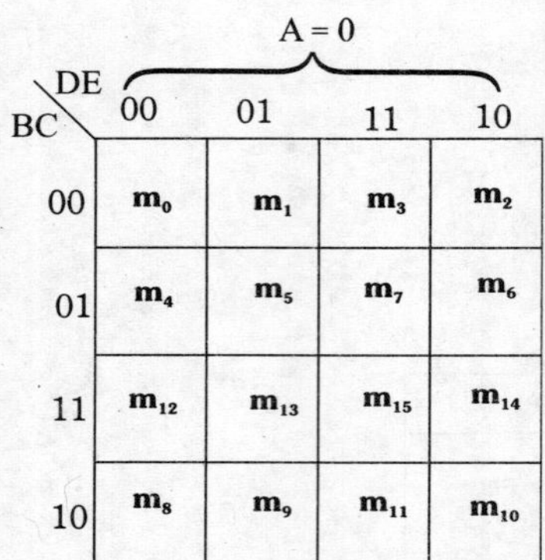

Fig. 1.30: 4-variable K-map for A = 0.

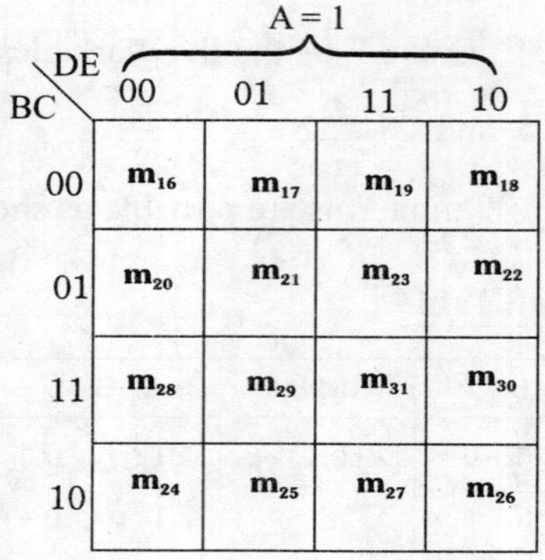

A = 1
DE
BC
00 01 11 10
00 m16 m17 m19 m18
01 m20 m21 m23 m22
11 m28 m29 m31 m30
10 m24 m25 m27 m26

Fig. 1.31: 4-variable K-map for A = 1.

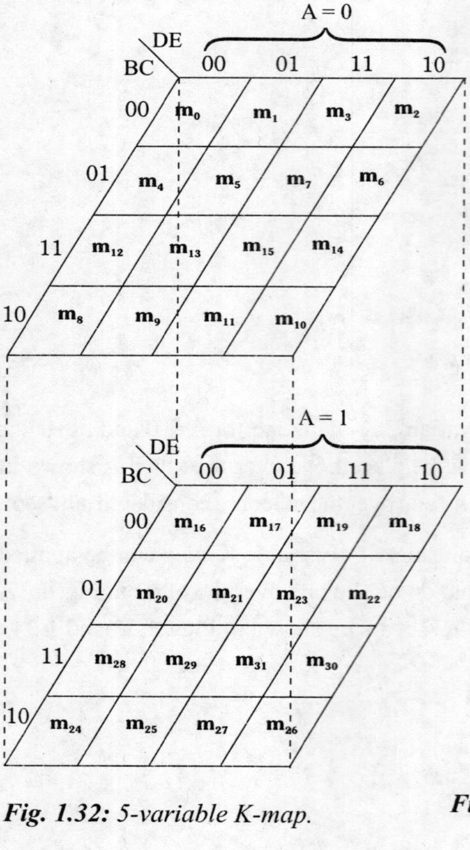

A = 0
DE
BC
00 01 11 10
00 m0 m1 m3 m2
01 m4 m5 m7 m6
11 m12 m13 m15 m14
10 m8 m9 m11 m10
A = 1
DE
BC
00 01 11 10
00 m16 m17 m19 m18
01 m20 m21 m23 m22
11 m28 m29 m31 m30
10 m24 m25 m27 m26

Fig. 1.32: 5-variable K-map.

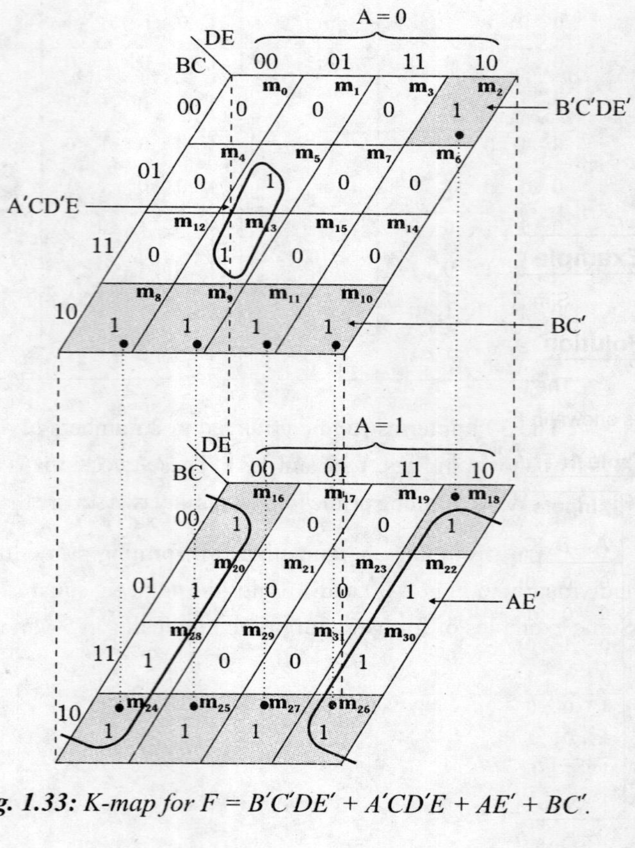

A = 0
DE
BC
00 01 11 10
00 m0 0 m1 0 m3 0 m2 1
01 m4 0 m5 1 m7 0 m6 0
11 m12 0 m13 1 m15 0 m14 0
10 m8 1 m9 1 m11 1 m10 1
B′C′DE′
A′CD′E
BC′
A = 1
DE
BC
00 01 11 10
00 m16 1 m17 0 m19 0 m18 1
01 m20 1 m21 0 m23 0 m22 1
11 m28 1 m29 0 m31 0 m30 1
10 m24 1 m25 1 m27 1 m26 1
AE′

Fig. 1.33: K-map for F = B′C′DE′ + A′CD′E + AE′ + BC′.

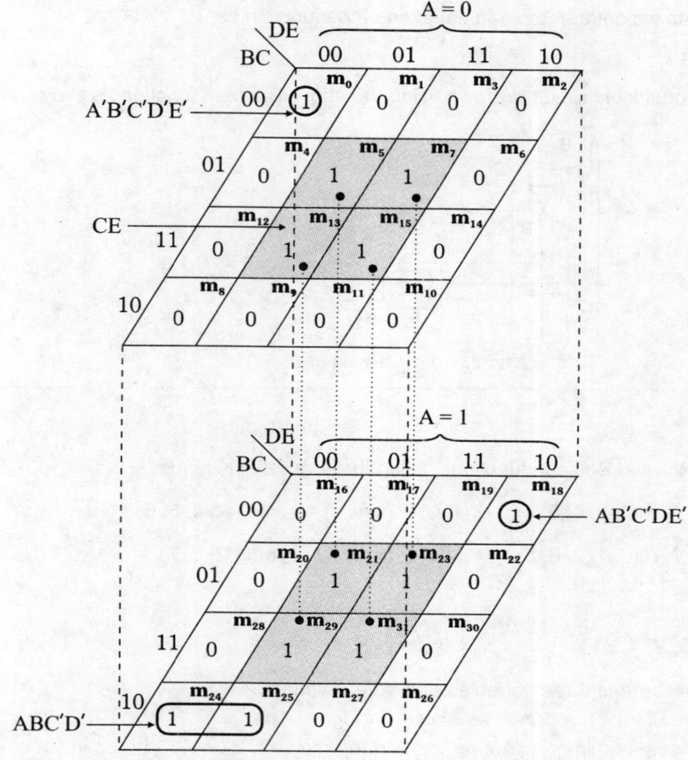

Fig. 1.34: K-map for F = A′B′C′D′E′ + CE + AB′C′DE′ + ABC′D′.

Example 1.30

(AU, Apr/May'19, 7 Marks)

Simplify and implement the logic function F(A, B, C) = Σ m(0, 1, 4, 5, 7) using logic gates.

Solution

The truth table of logic function is shown in Table 1 and the three variable K-map is constructed using Table 1 as shown in Fig. 1. The function output, F is 1 for minterms m_0, m_1, m_4, m_5 and m_7.

Table 1: Truth Table

Inputs			Minterm	Output
A	B	C		F
0	0	0	m_0	1
0	0	1	m_1	1
0	1	0	m_2	0
0	1	1	m_3	0
1	0	0	m_4	1
1	0	1	m_5	1
1	1	0	m_6	0
1	1	1	m_7	1

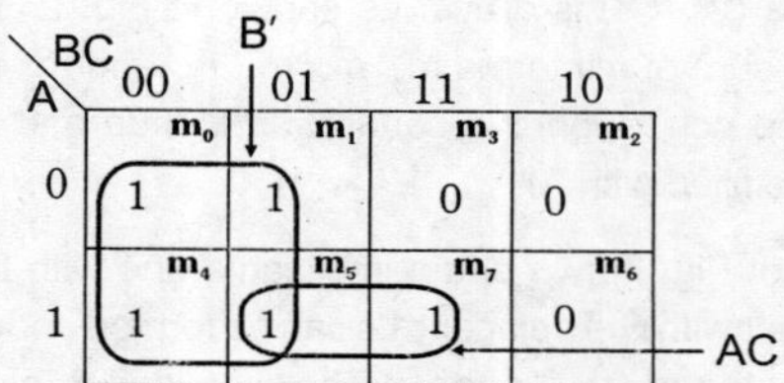

Fig. 1: K-map for F.

From the K-map we get the Boolean equation of the function as,

F = AC + B′

The combinational logic circuit is drawn using the above Boolean function as shown in Fig. 2.

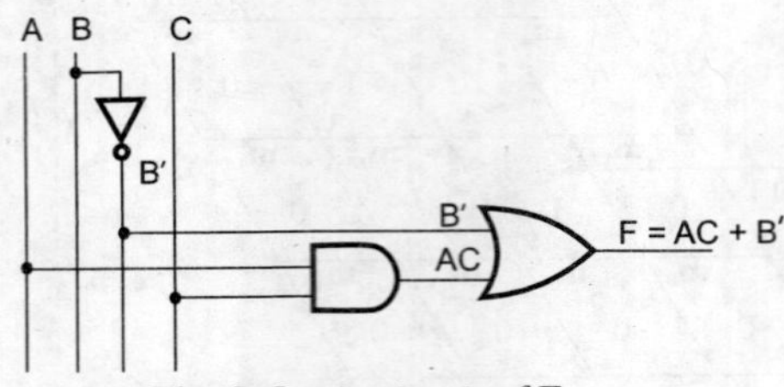

Fig. 2: Logic circuit of F.

Example 1.31

Simplify the following Boolean functions, using three variable K-maps:

a) F (x, y, z) = $\sum$ m(0, 2, 4, 5) **b)** F (x, y, z) = $\sum$ m(0, 2, 4, 5, 6)

c) F (x, y, z) = $\sum$ m(0, 1, 2, 3, 5) **d)** F (x, y, z) = $\sum$ m(1, 2, 3, 7)

Solution

a) F (x, y, z) = $\sum$ m(0, 2, 4, 5)

The given function has three variables and so a 3-variable K-map with 8 squares (2^3 = 8) is drawn as shown in Fig. 1. The function is defined as sum of minterms m_0, m_2, m_4 and m_5 and so a "1" is filled in the corresponding squares in K-map and the remaining squares are filled with "0".

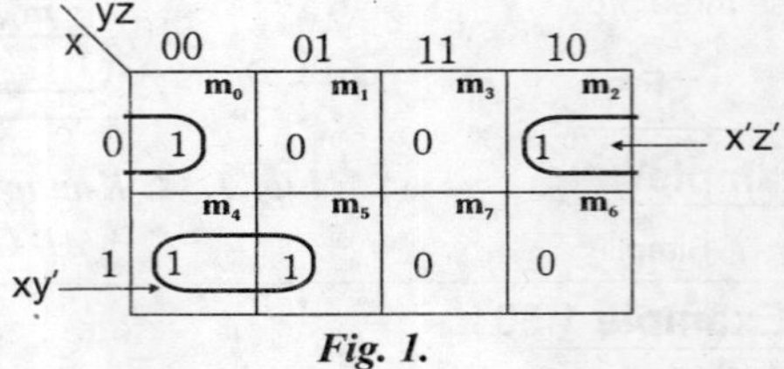

Fig. 1.

In the K-map of Fig. 1, two prime implicants each with two adjacent 1's can be formed and the corresponding product terms are xy′ and x′z′. The simplified Boolean function is given by sum of these two product terms.

∴ F = x y′ + x′z′

b) F (x, y, z) = $\sum$ m(0, 2, 4, 5, 6)

The given function has three variables and so a 3-variable K-map with 8 squares (2^3 = 8) is drawn as shown in Fig. 2. The function is defined as sum of minterms m_0, m_2, m_4, m_5 and m_6 and so a "1" is filled in the corresponding squares in K-map and the remaining squares are filled with "0".

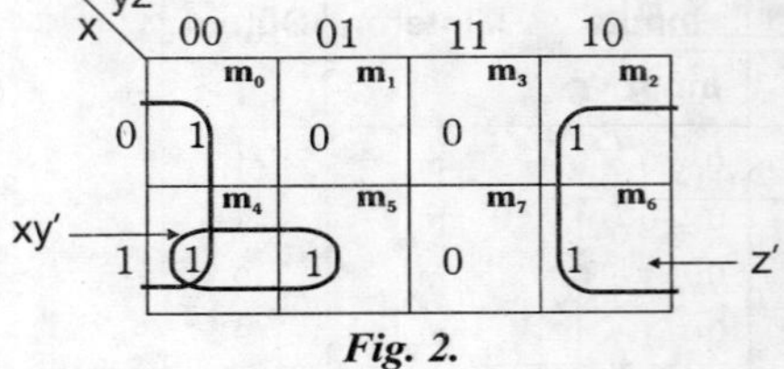

Fig. 2.

In the K-map of Fig. 2, two prime implicants one with four adjacent 1's and another with two adjacent 1's can be formed and the corresponding product terms are z′ and xy′. The simplified Boolean function is given by sum of these two product terms.

∴ F = z′ + x y′

c) F (x, y, z) = $\sum$ m(0, 1, 2, 3, 5)

The given function has three variables and so a 3-variable K-map with 8 squares ($2^3 = 8$) is drawn as shown in Fig. 3. The function is defined as sum of minterms m_0, m_1, m_2, m_3 and m_5 and so a "1" is filled in the corresponding squares in K-map and the remaining squares are filled with "0".

x \ yz	00	01	11	10
0	m_0: 1	m_1: 1	m_3: 1	m_2: 1
1	m_4: 0	m_5: 1	m_7: 0	m_6: 0

x′ (row 0 group), y′z (column 01 group)

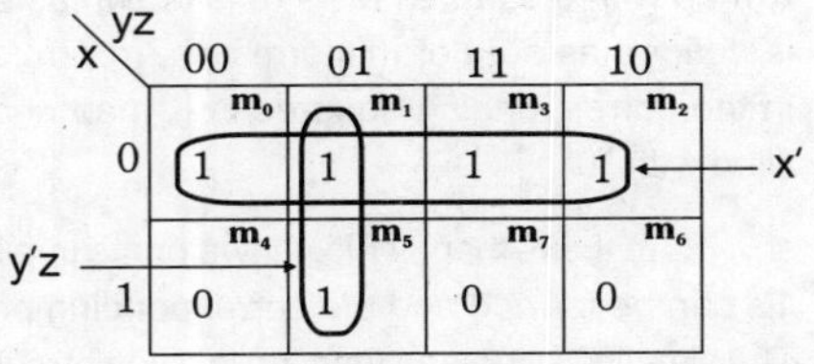

Fig. 3.

In the K-map of Fig. 3, two prime implicants one with four adjacent 1's and another with two adjacent 1's can be formed and the corresponding product terms are x′ and y′z. The simplified Boolean function is given by sum of these two product terms.

$\therefore$ F = x′ + y′z

d) F (x, y, z) = $\sum$ m(1, 2, 3, 7)

The given function has three variables and so a 3-variable K-map with 8 squares ($2^3 = 8$) is drawn as shown in Fig. 4. The function is defined as sum of minterms m_1, m_2, m_3 and m_7 and so a "1" is filled in the corresponding squares in K-map and the remaining squares are filled with "0".

x \ yz	00	01	11	10
0	m_0: 0	m_1: 1	m_3: 1	m_2: 1
1	m_4: 0	m_5: 0	m_7: 1	m_6: 0

x′z, x′y, yz

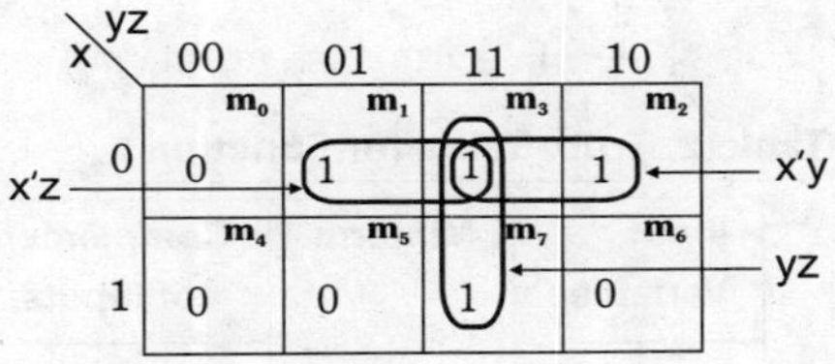

Fig. 4.

In the K-map of Fig. 4, three prime implicants each with two adjacent 1's can be formed and the corresponding product terms are x′z, x′y and yz. The simplified Boolean function is given by sum of these three product terms.

$\therefore$ F = x′z + x′y + yz

Example 1.32

Simplify the following Boolean function, using three-variable K-maps:

a) F_1 = xy + x′y′z′ + x′yz′ **b)** F_2 (x, y, z) = x′y + yz′ + y′z′

Solution

a) F_1 = xy + x′y′z′ + x′yz′

The truth table of the given function is formed as shown in Table 1 and the minterms for which function output is "1" are determined and using these minterms the function F_1 can be expressed as,

$F_1 = \Sigma (m_0, m_2, m_6, m_7)$

Table 1: Truth Table for Functions F_1

Input Variables						Minterm	Product Terms			Function Output
x	y	z	x′	y′	z′		xy	x′y′z′	x′yz′	F_1
0	0	0	1	1	1	m_0	0	1	0	1
0	0	1	1	1	0	m_1	0	0	0	0
0	1	0	1	0	1	m_2	0	0	1	1
0	1	1	1	0	0	m_3	0	0	0	0
1	0	0	0	1	1	m_4	0	0	0	0
1	0	1	0	1	0	m_5	0	0	0	0
1	1	0	0	0	1	m_6	1	0	0	1
1	1	1	0	0	0	m_7	1	0	0	1

The given function has three variables and so a 3-variable K-map with 8 squares ($2^3 = 8$) is drawn as shown in Fig. 1. The function is defined as sum of minterms m_0, m_2, m_6 and m_7 and so a "1" is filled in the corresponding squares in K-map and the remaining squares are filled with "0".

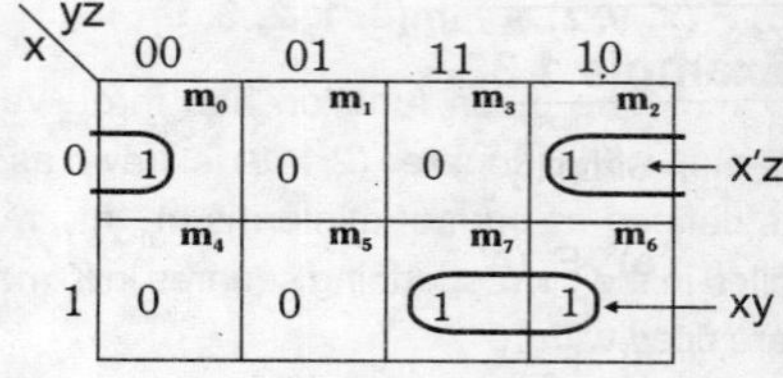

Fig. 1: K-map for function F_1.

In the K-map of Fig. 1, two prime implicants each with two adjacent 1's can be formed and the corresponding product terms are xy and x'z'. The simplified Boolean function is given by sum of these two product terms.

$\therefore F_1 = xy + x'z'$

---

b) $F_2 (x, y, z) = x'y + yz' + y'z'$

The truth table of the given function is formed as shown in Table 2 and the minterms for which function output is "1" are determined and using these minterms the function F_2 can be expressed as,

$$F_2 = \Sigma (m_0, m_2, m_3, m_4, m_6)$$

Table 2: Truth Table for Function F_2

Input Variables			Minterm	Complement of Inputs			Product Terms			Function Output
x	y	z		x'	y'	z'	x'y	yz'	y'z'	F_2
0	0	0	m_0	1	1	1	0	0	1	1
0	0	1	m_1	1	1	0	0	0	0	0
0	1	0	m_2	1	0	1	1	1	0	1
0	1	1	m_3	1	0	0	1	0	0	1
1	0	0	m_4	0	1	1	0	0	1	1
1	0	1	m_5	0	1	0	0	0	0	0
1	1	0	m_6	0	0	1	0	1	0	1
1	1	1	m_7	0	0	0	0	0	0	0

The given function has three variables and so a 3-variable K-map with 8 squares ($2^3 = 8$) is drawn as shown in Fig. 2. The function is defined as sum of minterms m_0, m_2, m_3, m_4 and m_6 and so a "1" is filled in the corresponding squares in K-map and the remaining squares are filled with "0".

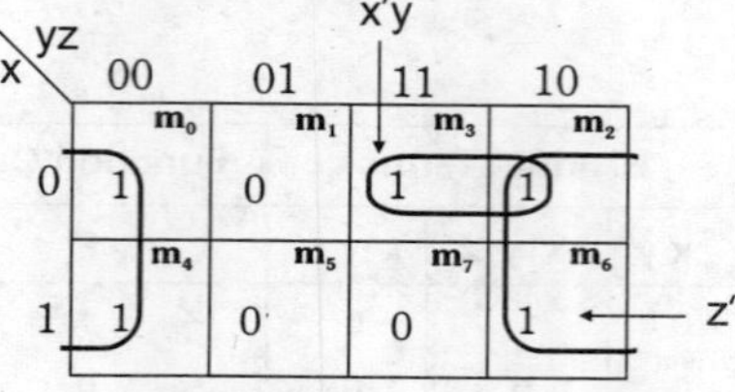

Fig. 2: K-map for function F_2.

In the K-map of Fig. 2, two prime implicants one with four adjacent 1's and another with two adjacent 1's can be formed and the corresponding product terms are z' and x'y. The simplified Boolean function is given by sum of these two product terms.

$\therefore F_2 = z' + x'y$

Example 1.33

Simplify the following Boolean functions, using Karnaugh maps.

a) F_1 (A, B, C, D) = Σ m(3, 7, 11, 13, 14, 15) b) F_2(w, x, y, z) = Σ m(2, 3, 12, 13, 14, 15)

c) F_3 (A, B, C, D) = Σ m(0, 2, 4, 5, 8, 10, 14, 15)

Solution

a) F_1 (A, B, C, D) = Σ m(3, 7, 11, 13, 14, 15)

The given function has four variables and so a 4-variable K-map with 16 squares (2^4 = 16) is drawn as shown in Fig. 1. The function is defined as sum of minterms m_3, m_7, m_{11}, m_{13}, m_{14} and m_{15} and so a "1" is filled in the corresponding squares in K-map and the remaining squares are filled with "0".

In the K-map of Fig. 1, three prime implicants one with four adjacent 1's and the other two with two adjacent 1's can be formed and the corresponding product terms are CD, ABC and ABD. The simplified Boolean function is given by sum of these three product terms.

$\therefore$ F_1 = CD + ABC + ABD

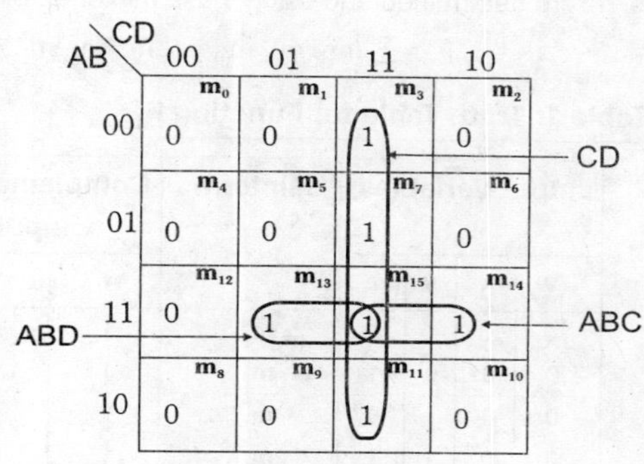

Fig. 1: K-map for F_1.

b) F_2 (w, x, y, z) = Σ m(2, 3, 12, 13, 14, 15)

The given function has four variables and so a 4-variable K-map with 16 squares (2^4 = 16) is drawn as shown in Fig. 2. The function is defined as sum of minterms m_2, m_3, m_{12}, m_{13}, m_{14} and m_{15} and so a "1" is filled in the corresponding squares in K-map and the remaining squares are filled with "0".

In the K-map of Fig. 2, two prime implicants one with four adjacent 1's and the other with two adjacent 1's can be formed and the corresponding product terms are wx and w'x'y. The simplified Boolean function is given by sum of these two product terms.

$\therefore$ F_2 = wx + w'x'y

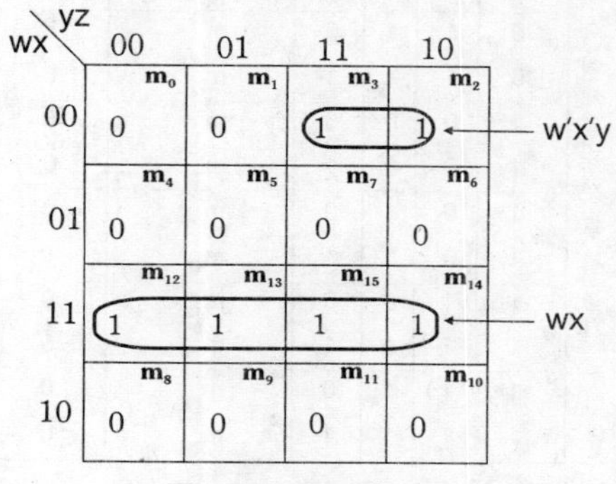

Fig. 2: K-map for F_2.

c) F_3 (A, B, C, D) = Σ m(0, 2, 4, 5, 8, 10, 14, 15)

The given function has four variables and so a 4-variable K-map with 16 squares (2^4 = 16) is drawn as shown in Fig. 3. The function is defined as sum of minterms m_0, m_2, m_4, m_5, m_8, m_{10}, m_{14} and m_{15} and so a "1" is filled in the corresponding squares in K-map and the remaining squares are filled with "0".

In the K-map of Fig. 3, three prime implicants one with four adjacent 1's and the other two with two adjacent 1's can be formed and the corresponding product terms are B'D', A'BC', and ABC. The simplified Boolean function is given by sum of these three product terms.

$\therefore$ F_3 = B'D' + A'BC' + ABC

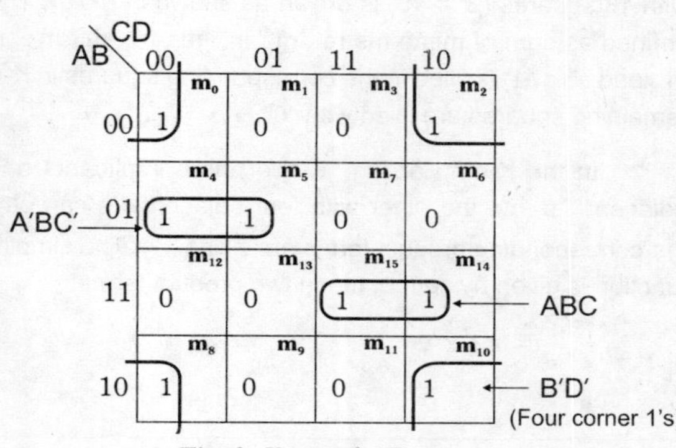

Fig. 3: K-map for F_3.

Example 1.34

Simplify the following Boolean expressions, using four-variable maps:

a) $F_1 = w'z + xz + x'y + wx'z$ b) $F_2 = AB'C + B'C'D' + BCD + ACD' + A'B'C + A'BC'D$

Solution

a) $F_1 = w'z + xz + x'y + wx'z$

The truth table of the given function is formed as shown in Table 1 and the minterms for which function output is "1" are determined and using these minterms the function F_1 can be expressed as,

$$F_1 = \sum (m_1, m_2, m_3, m_5, m_7, m_9, m_{10}, m_{11}, m_{13}, m_{15})$$

Table 1: Truth Table for Function F_1

Input Variables				Minterm	Complement of Inputs		Product Terms				Function Output
w	x	y	z		w'	x'	w'z	xz	x'y	wx'z	F_1
0	0	0	0	m_0	1	1	0	0	0	0	0
0	0	0	1	m_1	1	1	1	0	0	0	1
0	0	1	0	m_2	1	1	0	0	1	0	1
0	0	1	1	m_3	1	1	1	0	1	0	1
0	1	0	0	m_4	1	0	0	0	0	0	0
0	1	0	1	m_5	1	0	1	1	0	0	1
0	1	1	0	m_6	1	0	0	0	0	0	0
0	1	1	1	m_7	1	0	1	1	0	0	1
1	0	0	0	m_8	0	1	0	0	0	0	0
1	0	0	1	m_9	0	1	0	0	0	1	1
1	0	1	0	m_{10}	0	1	0	0	1	0	1
1	0	1	1	m_{11}	0	1	0	0	1	1	1
1	1	0	0	m_{12}	0	0	0	0	0	0	0
1	1	0	1	m_{13}	0	0	0	1	0	0	1
1	1	1	0	m_{14}	0	0	0	0	0	0	0
1	1	1	1	m_{15}	0	0	0	1	0	0	1

The given function has four variables and so a 4-variable K-map with 16 squares ($2^4 = 16$) is drawn as shown in Fig. 1. The function is defined as sum of minterms m_1, m_2, m_3, m_5, m_7, m_9, m_{10}, m_{11}, m_{13} and m_{15} and so a "1" is filled in the corresponding squares in K-map and the remaining squares are filled with "0".

In the K-map of Fig. 1, two prime implicants one with eight adjacent 1's and the other with four adjacent 1's can be formed and the corresponding product terms are z and x'y. The simplified Boolean function is given by sum of these two product terms.

$\therefore F_1 = z + x'y$

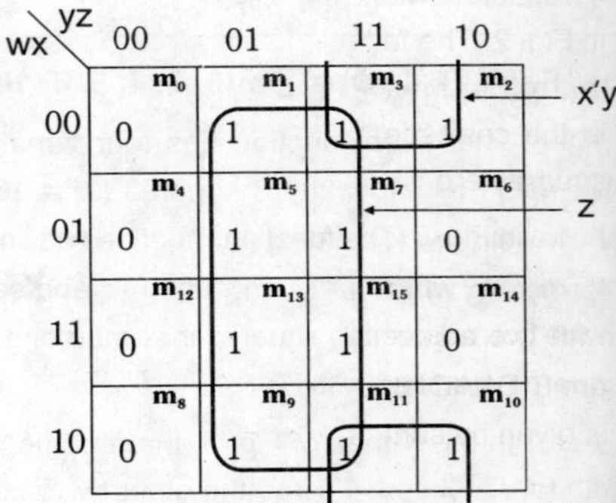

Fig. 1: K-map for function F_1.

b) $F_2 = AB'C + B'C'D' + BCD + ACD' + A'B'C + A'BC'D$

The truth table of the given function is formed as shown in Table 2 and the minterms for which function output is "1" are determined and using these minterms the function F_2 can be expressed as,

$$F_2 = \sum (m_0, m_2, m_3, m_5, m_7, m_8, m_{10}, m_{11}, m_{14}, m_{15})$$

Table 2: Truth Table for Function F_2

Input Variables A B C D	Minterm	Complement of Inputs A'	B'	C'	D'	Product Terms AB'C	B'C'D'	BCD	ACD'	A'B'C	A'BC'D	Function Output F_2
0 0 0 0	m_0	1	1	1	1	0	1	0	0	0	0	1
0 0 0 1	m_1	1	1	1	0	0	0	0	0	0	0	0
0 0 1 0	m_2	1	1	0	1	0	0	0	0	1	0	1
0 0 1 1	m_3	1	1	0	0	0	0	0	0	1	0	1
0 1 0 0	m_4	1	0	1	1	0	0	0	0	0	0	0
0 1 0 1	m_5	1	0	1	0	0	0	0	0	0	1	1
0 1 1 0	m_6	1	0	0	1	0	0	0	0	0	0	0
0 1 1 1	m_7	1	0	0	0	0	0	1	0	0	0	1
1 0 0 0	m_8	0	1	1	1	0	1	0	0	0	0	1
1 0 0 1	m_9	0	1	1	0	0	0	0	0	0	0	0
1 0 1 0	m_{10}	0	1	0	1	1	0	0	1	0	0	1
1 0 1 1	m_{11}	0	1	0	0	1	0	0	0	0	0	1
1 1 0 0	m_{12}	0	0	1	1	0	0	0	0	0	0	0
1 1 0 1	m_{13}	0	0	1	0	0	0	0	0	0	0	0
1 1 1 0	m_{14}	0	0	0	1	0	0	0	1	0	0	1
1 1 1 1	m_{15}	0	0	0	0	0	0	1	0	0	0	1

The given function has four variables and so a 4-variable K-map with 16 squares ($2^4 = 16$) is drawn as shown in Fig. 2. The function is defined as sum of minterms m_0, m_2, m_3, m_5, m_7, m_8, m_{10}, m_{11}, m_{14} and m_{15} and so a "1" is filled in the corresponding squares in K-map and the remaining squares are filled with "0".

In the K-map of Fig. 2, four prime implicants can be formed in which three with four adjacent 1's and the other with two adjacent 1's and the corresponding product terms are B'D', AC, B'C and A'BD. The simplified Boolean function is given by sum of these four product terms.

$$\therefore F_2 = B'D' + AC + B'C + A'BD$$

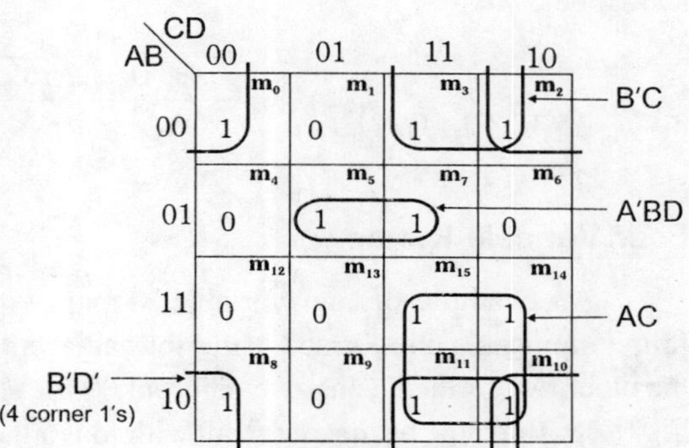

Fig. 2: K-map for function F_2.

1.6.6 Realization of Product-of-Sums Form using K-map

In truth table the function output value with "0" represent maxterms. Therefore in K-map if squares with 0's are considered for forming prime implicants and the complement of the resultant sum-of-products form will give product-of-sums form.

Two-Variable K-map

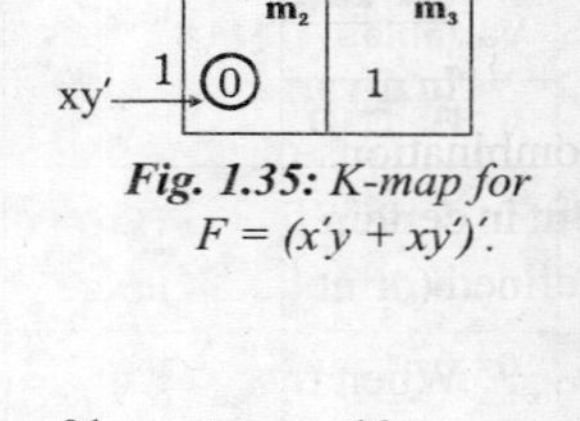

Fig. 1.35: K-map for F = (x'y + xy')'.

An example of two-variable K-map with two prime implicants formed using single 0's are shown in Fig. 1.35. Here, each prime implicant represents a product term of two literals. Now, the POS form of Boolean function is given by complement of sum of these two product terms as shown below:

$$F = (x'y + xy')' = (x'y)' . (xy')'$$

$$= (x + y') . (x' + y) \quad \text{Using DeMorgan's Theorem}$$

Three-Variable K-map

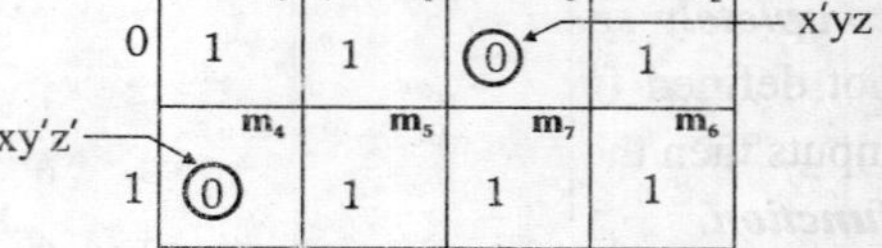

Fig. 1.36: K-map for F = (x'yz + xy'z')'.

An example of three-variable K-map with two prime implicants formed using single 0's are shown in Fig. 1.36. Here, each prime implicant represents a product term of three literals. Now, the POS form of Boolean function is given by complement of sum of these two product terms as shown below:

$$F = (x'yz + xy'z')'$$

$$= (x'yz)' . (xy'z')' \quad \text{Using DeMorgan's Theorem}$$

$$= (x + y' + z') . (x' + y + z)$$

An example of three-variable K-map with two prime implicants formed using two adjacent 0's are shown in Fig. 1.37. Here, each prime implicant represents a product term of two literals. Now, the POS form of Boolean function is given by complement of sum of these two product terms as shown below:

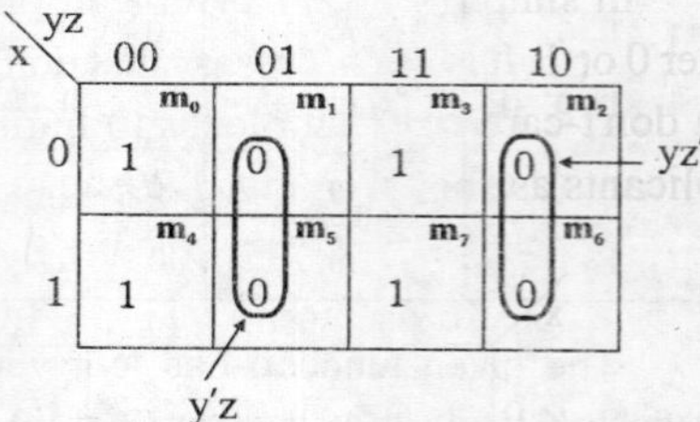

Fig. 1.37: K-map for F = (yz' + y'z)'.

$$F = (yz' + y'z)' \quad \text{Using DeMorgan's Theorem}$$

$$= (yz')' . (y'z)'$$

$$= (y' + z) . (y + z')$$

Four-Variable K-map

An example of four-variable K-map with three prime implicants one formed using four adjacent 0's and the other two formed using two adjacent 0's are shown in Fig. 1.38. Here, the prime implicant with four adjacent 0's represent a product term of two literals and each prime implicant with two adjacent 0's represent a product term of three literals. Now, the POS form of Boolean function is given by complement of sum of these three product terms as shown ahead:

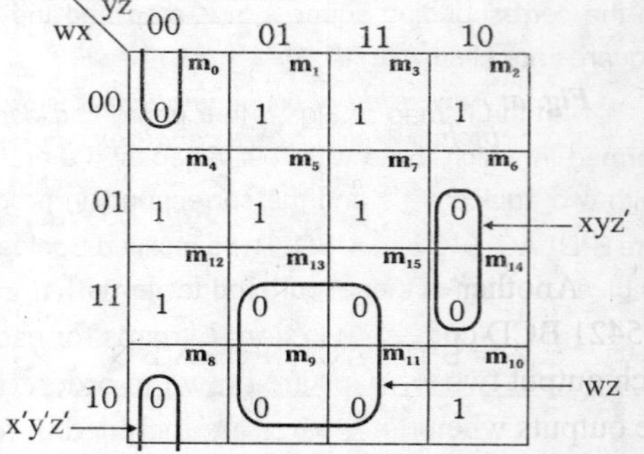

Fig. 1.38: K-map for F = (wz + x'y'z' + xyz')'.

$$F = (wz + x'y'z' + x\,y\,z')'$$

$$= (wz)' \,.\, (x'y'z')' \,.\, (x\,y\,z')' \quad \text{Using DeMorgan's Theorem}$$

$$= (w' + z') \,.\, (x + y + z) \,.\, (x' + y' + z)$$

1.6.7 Completely and Incompletely Specified Functions (K-map with Don't-Care Conditions)

In n-variable Boolean function there are 2^n possible combinations of inputs and so there will be 2^n outputs. But in certain functions all the 2^n possible outputs are not defined (or not used).

When the outputs are defined for all the possible 2^n combinations of inputs then the function is called ***completely specified function***. When the outputs are not defined for some of the possible combinations of inputs then the function is called ***incompletely specified function***.

Consider the truth table shown in Table 1.66 and the corresponding K-map shown in Fig. 1.39. In Table 1.60 the undefined function outputs are called ***don't-care conditions*** and denoted by ×.

Table 1.66: Truth Table with Don't-cares

Inputs x y z	Minterm	Output F
0 0 0	m_0	0
0 0 1	m_1	1
0 1 0	m_2	×
0 1 1	m_3	×
1 0 0	m_4	0
1 0 1	m_5	1
1 1 0	m_6	1
1 1 1	m_7	0

In simplification of Boolean functions using K-maps, don't-care conditions can be considered as either 0 or 1. It is not necessary that don't-care conditions to be included in prime implicants, but if inclusion of a don't-care condition leads to minimization of a literal then, don't-care can be included in the prime implicants as shown in Fig. 1.39.

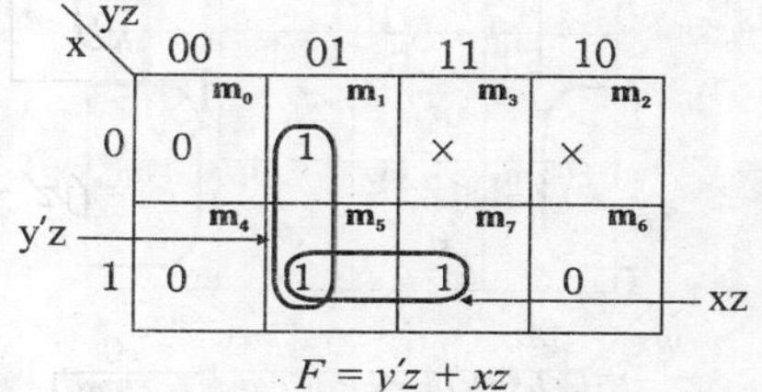

$F = y'z + xz$

Fig. a: *Formation of prime implicants without including don't-care condition.*

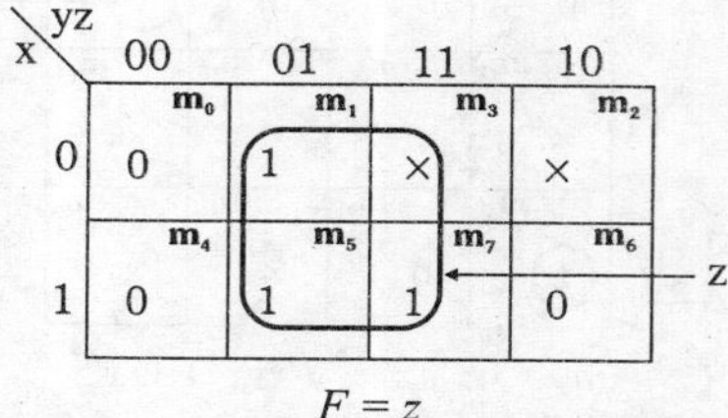

$F = z$

Fig. b: *Formation of prime implicants including don't-care condition.*

Fig. 1.39: *K-map with don't-care conditions.*

Another example of truth table with don't-cares is shown in Table 1.67 which is truth table of binary to 5421 BCD code conversion. K-maps for each output a_1, a_2, a_4 and a_5 are shown in Figs. 1.40 to 1.43. For each output two K-maps are drawn in order to show that the literals are reduced in Boolean expression of the outputs when don't-cares are included in formation of prime implicants.

Table 1.67: Truth Table of BCD

Decimal	Minterm	Input Binary w x y z	Output 5421 BCD a_5 a_4 a_2 a_1
0	m_0	0 0 0 0	0 0 0 0
1	m_1	0 0 0 1	0 0 0 1
2	m_2	0 0 1 0	0 0 1 0
3	m_3	0 0 1 1	0 0 1 1
4	m_4	0 1 0 0	0 1 0 0
5	m_5	0 1 0 1	1 0 0 0
6	m_6	0 1 1 0	1 0 0 1
7	m_7	0 1 1 1	1 0 1 0
8	m_8	1 0 0 0	1 0 1 1
9	m_9	1 0 0 1	1 1 0 0
-	m_{10}	1 0 1 0	× × × ×
-	m_{11}	1 0 1 1	× × × ×
-	m_{12}	1 1 0 0	× × × ×
-	m_{13}	1 1 0 1	× × × ×
-	m_{14}	1 1 1 0	× × × ×
-	m_{15}	1 1 1 1	× × × ×

(6 don't-care conditions: m_{10} to m_{15})

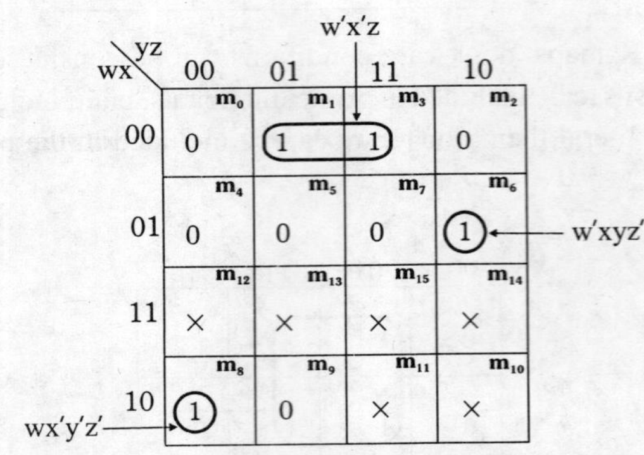

$a_1 = wx'y'z' + w'xyz' + w'x'z$

Fig. a: Prime implicants formed without including don't-cares.

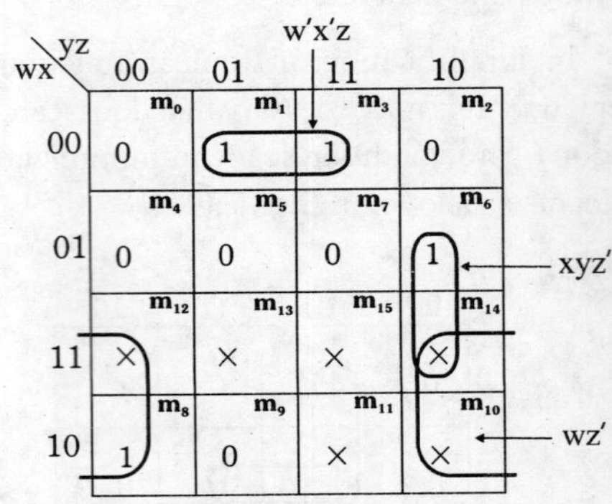

$a_1 = wz' + xyz' + w'x'z$

Fig. b: Prime implicants formed by including don't-cares.

***Fig. 1.40:** K-map for a_1.*

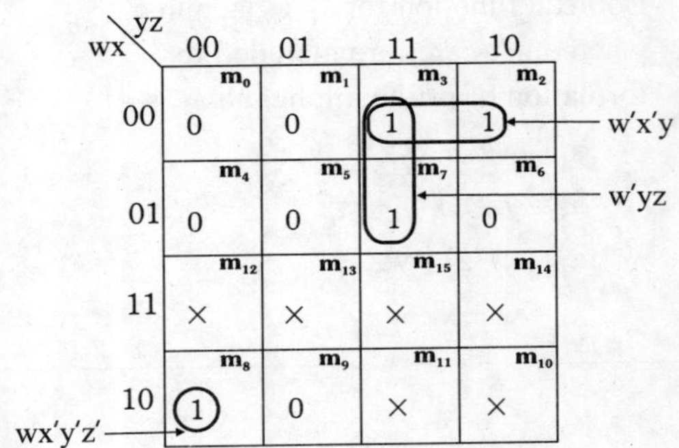

$a_2 = wx'y'z' + w'yz + w'x'y$

***Fig. a:** Prime implicants formed without including don't-cares.*

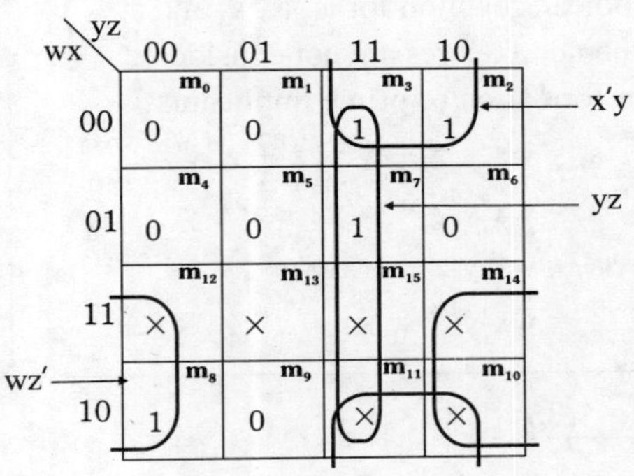

$a_2 = yz + x'y + wz'$

***Fig. b:** Prime implicants formed by including don't-cares.*

Fig. 1.41: *K-map for a_2.*

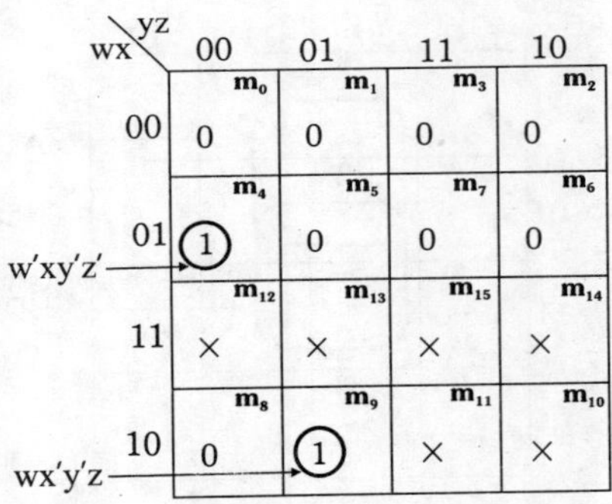

$a_4 = w'xy'z' + wx'y'z$

***Fig. a:** Prime implicants formed without including don't-cares.*

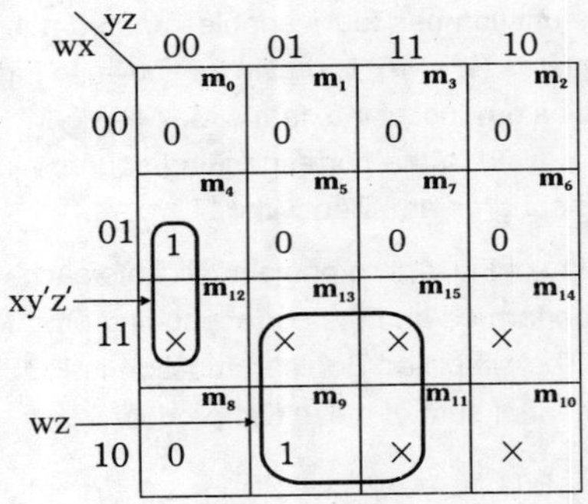

$a_4 = xy'z' + wz$

***Fig. b:** Prime implicants formed by including don't-cares.*

Fig. 1.42: *K-map for a_4.*

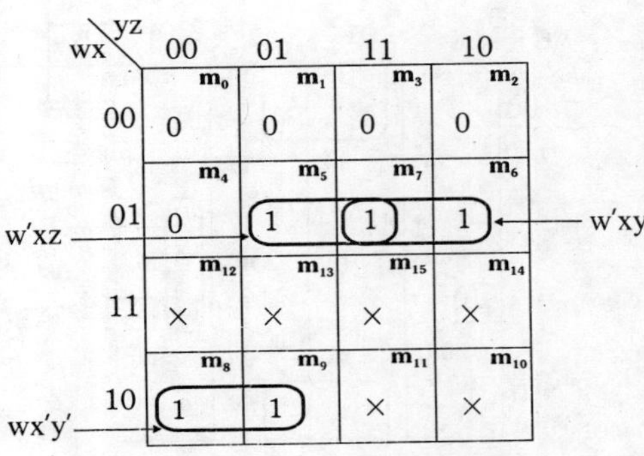

$a_5 = wx'y' + w'xz + w'xy$

***Fig. a:** Prime implicants formed without including don't-cares.*

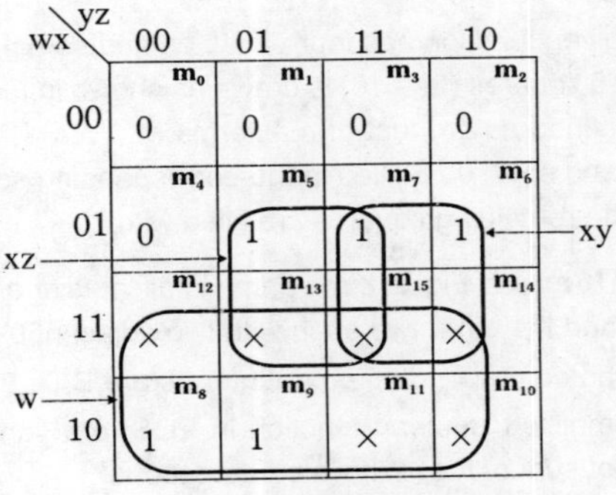

$a_5 = w + xz + xy$

***Fig. b:** Prime implicants formed by including don't-cares.*

Fig. 1.43: *K-map for a_5.*

Boolean function for a_1, a_2, a_4 and a_5 when don't-cares are not considered for formation of prime implicants	Boolean function for a_1, a_2, a_4 and a_5 when don't-cares are included for formation of prime implicants
$a_1 = wx'y'z' + w'xyz' + w'x'z$	$a_1 = wz' + xyz' + w'x'z$
$a_2 = wx'y'z' + w'yz + w'x'y$	$a_2 = yz + x'y + wz'$
$a_4 = w'xy'z' + wx'y'z$	$a_4 = xy'z' + wz$
$a_5 = wx'y' + w'xz + w'xy$	$a_5 = w + xz + xy$

Example 1.35

Simplify the following Boolean functions.

a) F_1 (A, B, C, D) = $\prod$ M(1, 3, 5, 7, 13, 15) b) F_2 (A, B, C, D) = $\prod$ M(1, 3, 6, 9, 11, 12, 14)

Solution

a) F_1 (A, B, C, D) = $\prod$ M(1, 3, 5, 7, 13, 15)

The given function has four variables and so a 4-variable K-map with 16 squares ($2^4 = 16$) is drawn as shown in Fig. 1. The function is defined as product of maxterms M_1, M_3, M_5, M_7, M_{13} and M_{15} and so a "0" is filled in the corresponding squares in K-map and the remaining squares are filled with "1".

In the K-map of Fig. 1, two prime implicants each with four adjacent 0's can be formed and the corresponding product terms are A'D and BD. The simplified Boolean function in POS form is given by complement of sum of these two product terms.

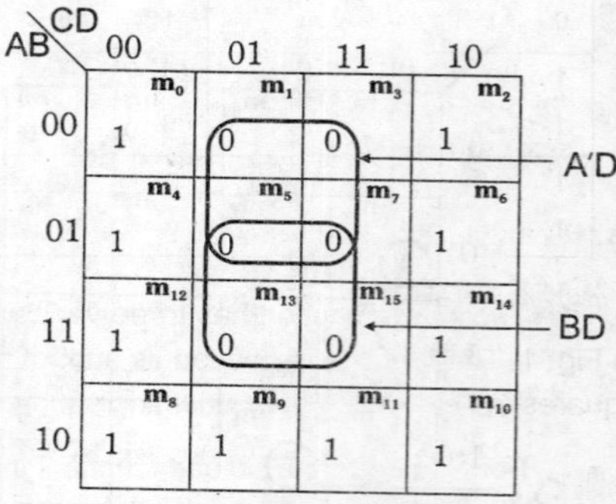

Fig. 1: K-map for F_1.

F_1 = (A'D + B D)'

= (A'D)' . (B D)' Using DeMorgan's Theorem

= (A + D') . (B' + D')

b) F_2 (A, B, C, D) = $\prod$ M(1, 3, 6, 9, 11, 12, 14)

The given function has four variables and so a 4-variable K-map with 16 squares ($2^4 = 16$) is drawn as shown in Fig. 2. The function is defined as product of maxterms M_1, M_3, M_6, M_9, M_{11}, M_{12} and M_{14} and so a "0" is filled in the corresponding squares in K-map and the remaining squares are filled with "1".

In the K-map of Fig. 2, three prime implicants one with four adjacent 0's and the other two each with two adjacent 0's can be formed and the corresponding product terms are B'D, BCD' and ABD'. The simplified Boolean function in POS form is given by complement of sum of these three product terms.

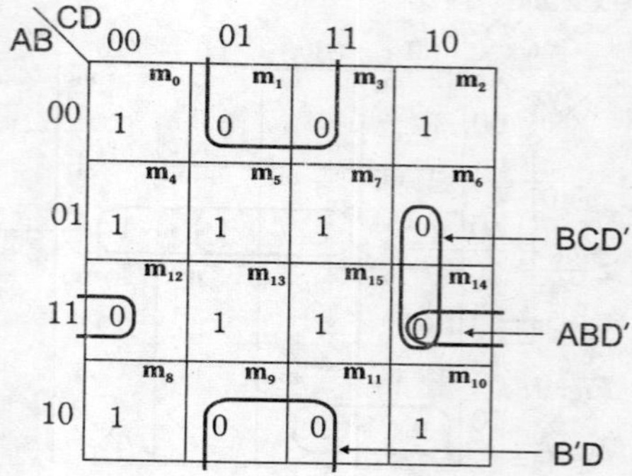

Fig. 2: K-map for F_2.

F_2 = (B'D + B C D' + A B D')'

= (B'D)' . (B C D')' . (A B D')' Using DeMorgan's Theorem

= (B + D') . (B' + C' + D) . (A' + B' + D)

Example 1.36

Simplify the following function to sum-of-products and product-of-sums.

F = x' z' + y' z' + y z' + x y

Solution

The truth table of the given function is formed as shown in Table 1 and the minterms for which function output is "1" are determined and using these minterms the function F can be expressed as,

$$F = \sum (m_0, m_2, m_4, m_6, m_7)$$

Table 1: Truth Table

Input Variables			Minterm	Complement of Inputs			Product Terms				Function Output
x	y	z		x'	y'	z'	x'z'	y'z'	y z'	x y	F
0	0	0	m_0	1	1	1	1	1	0	0	1
0	0	1	m_1	1	1	0	0	0	0	0	0
0	1	0	m_2	1	0	1	1	0	1	0	1
0	1	1	m_3	1	0	0	0	0	0	0	0
1	0	0	m_4	0	1	1	0	1	0	0	1
1	0	1	m_5	0	1	0	0	0	0	0	0
1	1	0	m_6	0	0	1	0	0	1	1	1
1	1	1	m_7	0	0	0	0	0	0	1	1

The given function has three variables and so a 3-variable K-map with 8 squares ($2^3 = 8$) is drawn as shown in Fig. 1. The function is defined as sum of minterms m_0, m_2, m_4, m_6 and m_7 and so a "1" is filled in the corresponding squares in K-map and the remaining squares are filled with "0".

For SOP form of Boolean function, the prime implicants are formed by considering the squares filled with 1's. In the K-map of Fig. 1, two prime implicants one with four adjacent 1's and the other with two adjacent 1's can be formed and the corresponding product terms are z' and xy. The simplified Boolean function in SOP form is given by sum of these two product terms.

For POS form of Boolean function, the prime implicants are formed by considering the squares filled with 0's. In the K-map of Fig. 2, two prime implicants each with two adjacent 0's can be formed and the corresponding product terms are x'z and y'z. The simplified Boolean function in POS form is given by complement of sum of these two product terms.

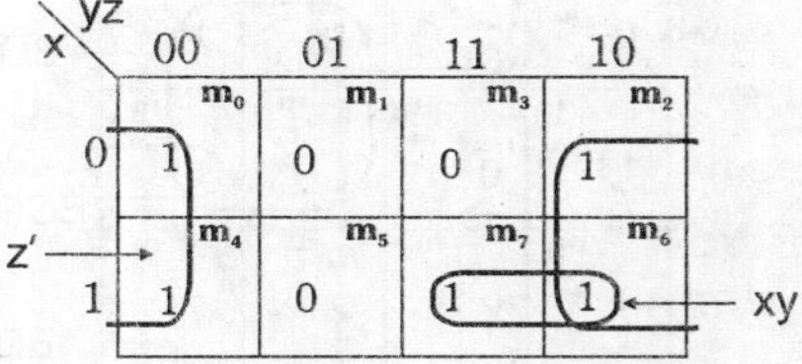

Fig. 1: K-map for SOP (Sum-of-Products).

F = z' + x y

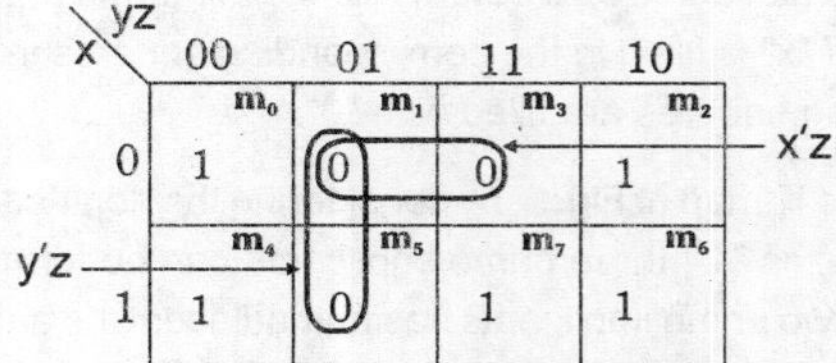

Fig. 2: K-map for POS (Product-of-Sums).

F = (x' z + y' z)'

= (x' z)' . (y' z)' — Using DeMorgan's Theorem

= (x + z') . (y + z')

Example 1.37

Simplify the following Boolean functions, together with the don't-care conditions and then express the simplified function in sum-of-products.

a) F_1 (A, B, C, D) = Σ m(0, 6, 8, 13, 14)
d_1 (A, B, C, D) = Σ d(2, 4, 10)

b) F_2 (A, B, C, D) = Σ m(5, 6, 7, 12, 14, 15)
d_2 (A, B, C, D) = Σ d(1, 3, 9, 11)

Solution

a) F_1 (A, B, C, D) = Σ m(0, 6, 8, 13, 14)

d_2 (A, B, C, D) = Σ d(2, 4, 10)

The given function has four variables and so a 4-variable K-map with 16 squares (2^4 = 16) is drawn as shown in Fig. 1. The function is defined as sum of minterms m_0, m_6, m_8, m_{13} and m_{14} and so a "1" is filled in the corresponding squares in K-map. The function also has don't-care outputs for the minterms m_2, m_4 and m_{10} and so a "x" is filled in the corresponding squares in K-map and the remaining squares are filled with "0".

In the K-map of Fig. 1, by considering the don't-care condition in m_2 and m_{10} as "1", three prime implicants can be formed in which two has four adjacent 1's and one have single 1 and the corresponding product terms are B'D', CD' and ABC'D. The simplified Boolean function is given by sum of these three product terms.

$\therefore$ F_1 = B'D' + CD' + ABC'D

AB \ CD	00	01	11	10
00	m_0: 1	m_1: 0	m_3: 0	m_2: ×
01	m_4: ×	m_5: 0	m_7: 0	m_6: 1
11	m_{12}: 0	m_{13}: 1	m_{15}: 0	m_{14}: 1
10	m_8: 1	m_9: 0	m_{11}: 0	m_{10}: ×

ABC'D; CD'; B'D' (4 corner 1's)

Fig. 1: K-map for F_1.

> ***Note:** Don't-care in m_2 and m_{10} are considered as 1.*

b) F_2 (A, B, C, D) = Σ m(5, 6, 7, 12, 14, 15)

d_2 (A, B, C, D) = Σ d(1, 3, 9, 11)

The given function has four variables and so a 4-variable K-map with 16 squares (2^4 = 16) is drawn as shown in Fig. 2. The function is defined as sum of minterms m_5, m_6, m_7, m_{12}, m_{14} and m_{15} and so a "1" is filled in the corresponding squares in K-map. The function also has don't-care outputs for the minterms m_1, m_3, m_9 and m_{11} and so a "x" is filled in the corresponding squares in K-map and the remaining squares are filled with "0".

In the K-map of Fig. 2, by considering the don't-care condition in m_1 and m_3 as "1", three prime implicants can be formed in which each of the two prime implicants has four adjacent 1's and one prime implicant have two adjacent 1's and the corresponding product terms are A'D, BC and ABD'. The simplified Boolean function is given by sum of these three product terms.

$\therefore$ F_2 = A'D + BC + ABD'

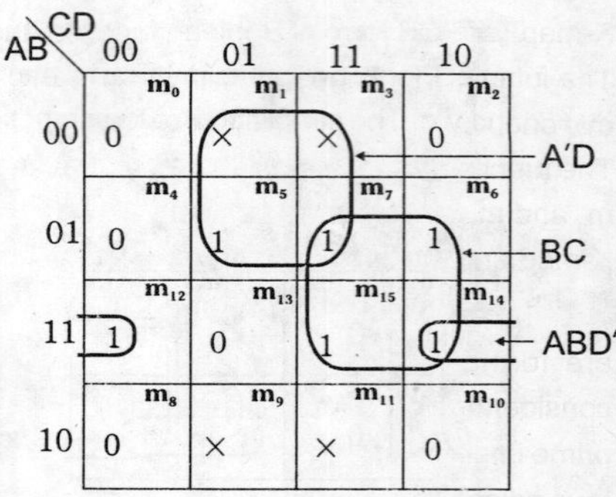

Fig. 2: K-map for F_2.

> ***Note:** Don't-care in m_1 and m_3 are considered as 1.*

Example 1.38

Simplify the following Boolean function with don't care in POS form.

a) F_1 (A, B, C, D) = $\sum$ m(0, 2, 6, 12, 13, 14) **b)** F_2 (A, B, C, D) = $\sum$ m(5, 6, 12, 14, 15)

d_1 (A, B, C, D) = $\sum$ d(1, 4, 10) d_2 (A, B, C, D) = $\sum$ d(1, 3, 9, 11)

Solution

a) F_1 (A, B, C, D) = $\sum$ m(0, 2, 6, 12, 13, 14)

d_1 (A, B, C, D) = $\sum$ d(1, 4, 10)

The given function has four variables and so a 4-variable K-map with 16 squares (2^4 = 16) is drawn as shown in Fig. 1. The function is defined as sum of minterms m_0, m_2, m_6, m_{12}, m_{13} and m_{14} and so a "1" is filled in the corresponding squares in K-map. The function also has don't-care outputs for the minterms m_1, m_4 and m_{10} and so a "x" is filled in the corresponding squares in K-map and the remaining squares are filled with "0".

For POS form of Boolean function, the prime implicants are formed by considering 0's. In the K-map of Fig. 1, by considering the don't-care condition in m_1 and m_{10} as "0", three prime implicants can be formed in which all three has four adjacent 0's and the corresponding product terms are A'D, AB'and CD. The simplified Boolean function in POS form is given by complement of sum of these three product terms.

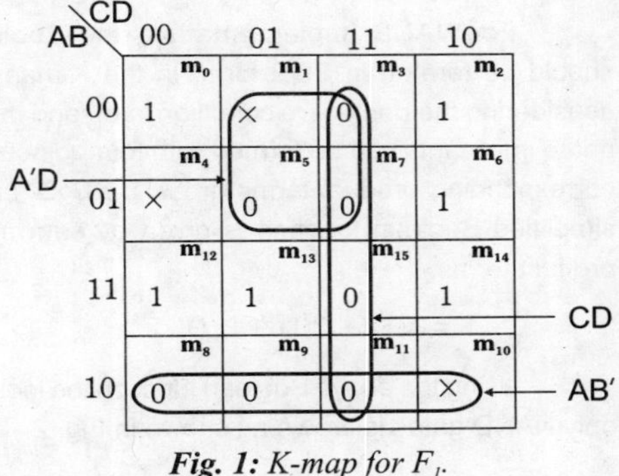

Fig. 1: K-map for F_1.

F_1 = (A'D + AB' + CD)'

= (A'D)' . (AB')' . (CD)'

= (A + D') . (A' + B) . (C' + D') Using DeMorgan's Theorem

---

b) F_2 (A, B, C, D) = $\sum$ m(5, 6, 12, 14, 15)

d_2 (A, B, C, D) = $\sum$ d(1, 3, 9, 11)

The given function has four variables and so a 4-variable K-map with 16 squares (2^4 = 16) is drawn as shown in Fig. 2. The function is defined as sum of minterms m_5, m_6, m_{12}, m_{14} and m_{15} and so a "1" is filled in the corresponding squares in K-map. The function also has don't-care outputs for the minterms m_1, m_3, m_9 and m_{11} and so a "x" is filled in the corresponding squares in K-map and the remaining squares are filled with "0".

For POS form of Boolean function, the prime implicants are formed by considering 0's. In the K-map of Fig. 2, by considering the don't-care condition in m_3 and m_9 as "0", four prime implicants can be formed in which one prime implicant has four adjacent 0's and each of the other three prime implicants has two adjacent 0's and the corresponding product terms are B', A'C'D', AC'D and A'CD. The simplified Boolean function in POS form is given by complement of sum of these four product terms.

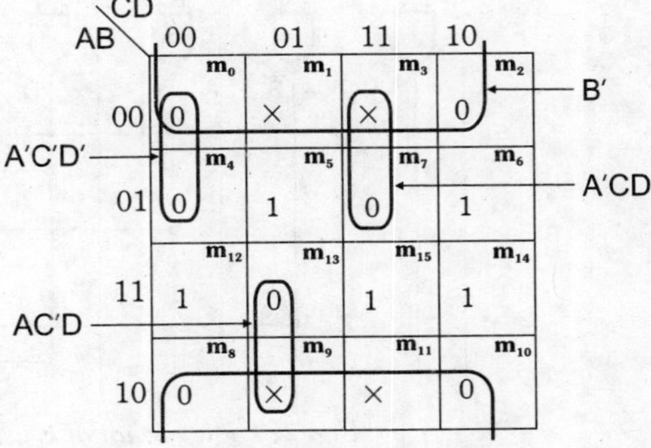

Fig. 2: K-map for F_2.

F_2 = (B' + A'C'D' + AC'D + A'CD)'

= B . (A'C'D')' . (AC'D)' . (A'CD)' Using DeMorgan's Theorem

= B . (A + C + D) . (A' + C + D') . (A + C' + D')

Example 1.39

(AU, Apr/May'19, 15 Marks)

Use Karnaugh Map method to simplify the following Boolean function

$F(A, B, C, D) = \sum m(2, 4, 6, 10, 12) + \sum d(0, 8, 9, 13)$

a) Implement the Boolean function, F using only NAND gates.

b) Implement the Boolean function, F using only NOR gates.

Solution

a) To Implement the Boolean function, F using only NAND gates

The given function has four variables and so a 4-variable K-map with 16 squares ($2^4 = 16$) is drawn as shown in Fig. 1. The function is defined as sum of minterms m_2, m_4, m_6, m_{10} and m_{12} and so a "1" is filled in the corresponding squares in K-map. The function also has don't-care outputs for the minterms m_0, m_8, m_9 and m_{13} and so a "x" is filled in the corresponding squares in K-map and the remaining squares are filled with "0".

For NAND implementation, the Boolean function should be formed in SOP form. In the K-map of Fig. 1, by considering the don't-care condition in m_0 and m_8 as "1", three prime implicants can be formed with four adjacent 1's and the corresponding product terms are A'D', C'D' and B'D'. The simplified Boolean function is given by sum of these three product terms.

$\therefore\ F = A'D' + C'D' + B'D'$

Using the above Boolean function the logic circuit using only NAND gates is drawn as shown in Fig. 2.

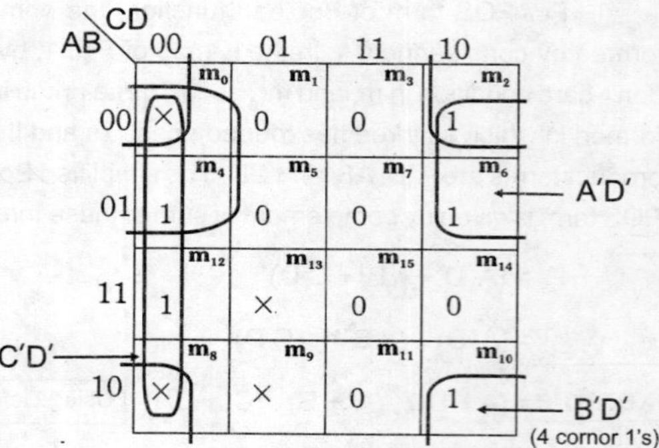

Fig. 1: K-map for F in SOP form.

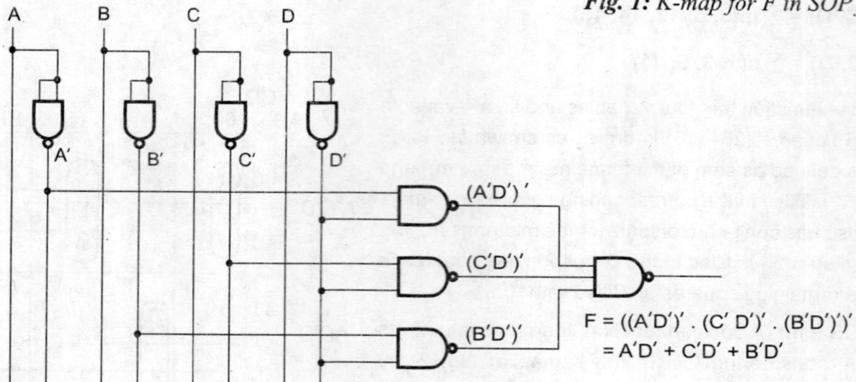

Fig. 2: Logic circuit of F using only NAND gates.

b) To Implement the Boolean function, F using only NOR gates

For NOR implementation, the Boolean function should be formed in POS form. For POS form of Boolean function, the prime implicants are formed by considering 0's. In the K-map of Fig. 3, by considering the don't-care condition in m_9 and m_{13} as "0" two prime implicants can be formed in which one prime implicant has eight adjacent 0's and one prime implicants has two adjacent 0's and the corresponding product terms are D and ABC. The simplified Boolean function in POS form is given by complement of sum of these two product terms.

$F = (D + ABC)'$

$= D' . (ABC)'$

$= D' . (A' + B' + C')$

Using the above POS form of Boolean function, the logic circuit using only NOR gates is drawn as shown in Fig. 4.

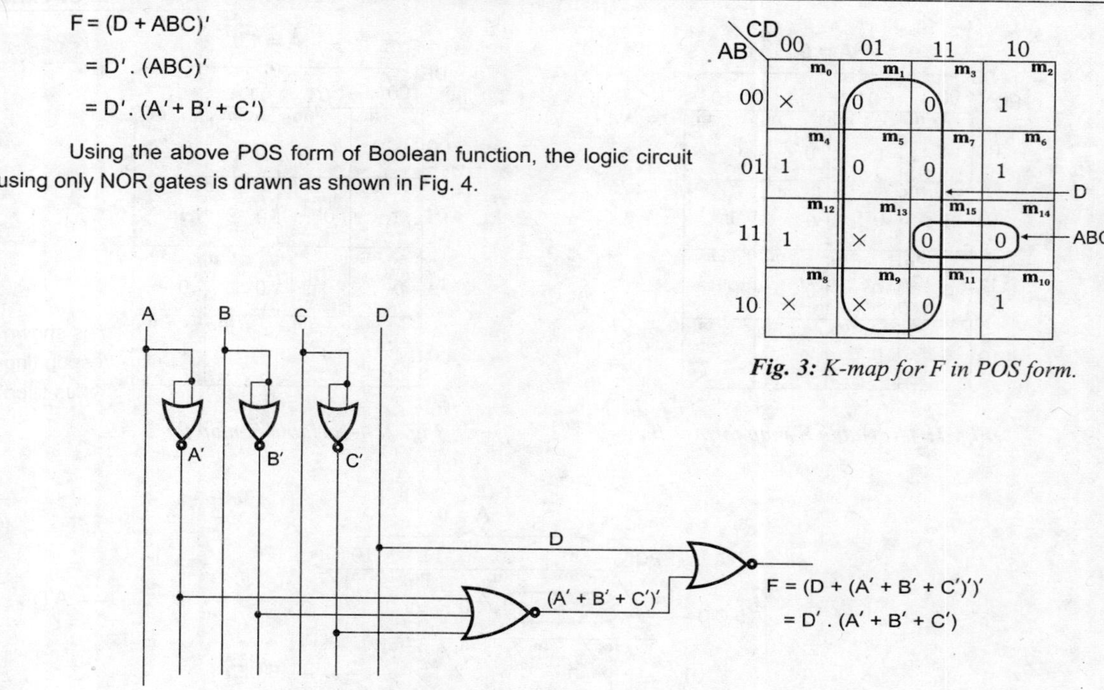

Fig. 3: K-map for F in POS form.

Fig. 4: Logic circuit of F using only NOR gates.

Example 1.40

Find the minimum sum of products (MSOP) representation for,

$F(A, B, C, D, E) = \sum m(1, 4, 6, 10, 20, 22, 24, 26) + \sum d(0, 11, 16, 27)$

using K-map. Draw the logical circuit of the minimal expression using ony NAND gates.

Solution

The given function has five variables and so two 4-variable K-map each with 16 squares ($2^4 = 16$) are drawn as shown in Figs. 1 and 2. The K-map of Fig. 1 is drawn for minterms m_0 to m_{15} in which A = 0. The K-map of Fig. 2 is drawn for minterms m_{16} to m_{31} in which A = 1. The function is defined as sum of minterms m_1, m_4, m_6, m_{10} m_{20}, m_{22}, m_{24} and m_{26} and so a "1" is filled in the corresponding squares in K-maps. The function also has don't-care outputs for the minterms m_0, m_{11}, m_{16} and m_{27} and so a "x" is filled in the corresponding squares in K-maps and the remaining squares are filled with "0".

For NAND realization the Boolean function should be formed in SOP form. In order to find adjacent squares the K-maps of Figs. 1 and 2 are placed one over the other as shown in Fig. 3. In the K-map of Fig. 3, by considering the don't-care condition in m_0, m_{11} and m_{27} as "1", four prime implicants can be formed in which two has four adjacent 1's and the other two has two adjacent 1's and the corresponding product terms are B′CE′, BC′D, A′B′C′D′ and ABC′E′. The simplified Boolean function is given by sum of these four product terms.

$\therefore \quad F = B'CE' + BC'D + A'B'C'D' + ABC'E'$

Using the above Boolean function the logic circuit using only NAND gate is drawn as shown in Fig. 4.

A = 0

BC \ DE	00	01	11	10
00	m_0 ×	m_1 1	m_3 0	m_2 0
01	m_4 1	m_5 0	m_7 0	m_6 1
11	m_{12} 0	m_{13} 0	m_{15} 0	m_{14} 0
10	m_8 0	m_9 0	m_{11} ×	m_{10} 1

***Fig. 1:** 4-variable K-map for A = 0.*

A = 1

BC \ DE	00	01	11	10
00	m_{16} ×	m_{17} 0	m_{19} 0	m_{18} 0
01	m_{20} 1	m_{21} 0	m_{23} 0	m_{22} 1
11	m_{28} 0	m_{29} 0	m_{31} 0	m_{30} 0
0	m_{24} 1	m_{25} 0	m_{27} ×	m_{26} 1

***Fig. 2:** 4-variable K-map for A = 1.*

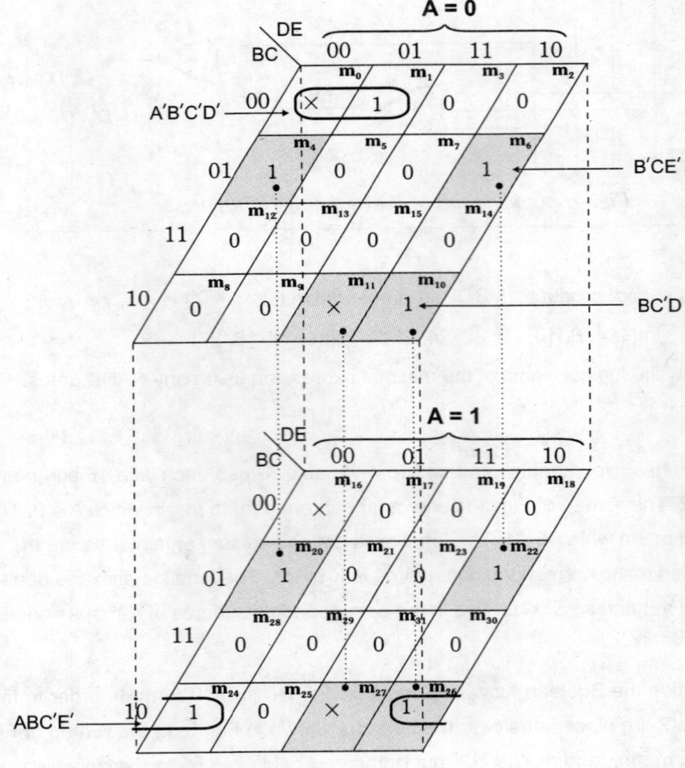

***Fig. 3:** 5-variable K-map.*

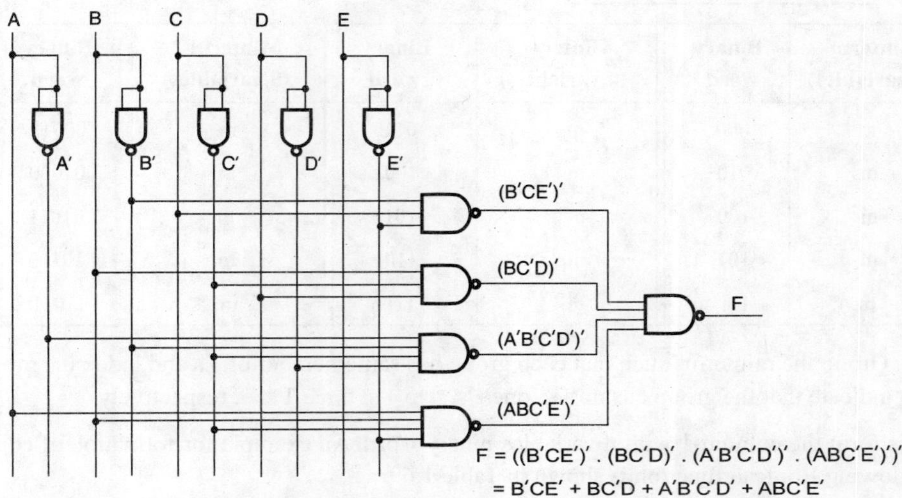

Fig. 4: Logic circuit of function, F using only NAND gates.

1.7 Minimization using Quine-McCluskey (QM) Method (Table Method of Minimization)

Quine-McCluskey method is a table based method to determine prime implicants and essential prime implicants of a Boolean function and hence to find simplified Boolean expression of a function.

Quine-McCluskey method has two phases:

1. Expansion phase
2. Covering phase.

The ***expansion phase*** is the first phase in which minterms are obtained from truth table and implicants are identified after eliminating unwanted/reduntant literals.

The ***covering phase*** is the second phase in which the essential prime implicants are selected from the prime implicants of first phase such that all minterms are covered with minimal number of prime implicants.

Procedure:

1. Construct truth table of given function and determine the minterms. If don't-cares are available then consider all the don't-cares as minterms.

(If minterms and don't-cares are given directly then no need to construct truth table).

2. Determine the binary word corresponding to minterms. The binary word is the binary value of the decimal number used to represent the minterm. An example for formation of binary word is shown in Table 1.68.

Table 1.68: Formation of Binary Word

Minterm (3 variables)	Binary word	Minterm (4 variables)	Binary word	Minterm (5 variables)	Binary word
m_0	000	m_6	0110	m_3	00011
m_2	010	m_9	1001	m_8	01000
m_4	100	m_{10}	1010	m_{11}	01011
m_5	101	m_{12}	1100	m_{20}	10100
m_7	111	m_{15}	1111	m_{27}	11011

3. Group the minterms such that each group has same number of 1's and index the group as 0, 1, 2, 3 ..... to indicate that the group has no 1's, one 1's, two 1's, three 1's,.... respectively.

This form the column-1 with n variables binary word. An example for formation of column-1 for the the following Boolean function is shown in Table 1.69.

$$F(A, B, C, D) = \sum m\ (0, 2, 4, 6, 8, 14, 15)$$

Table 1.69: Formation of Column-1 Grouping Minterms with Same Number of 1's

Minterm and binary word of function, F		Column-1		
			Minterm	Binary word
m_0	0000	Index 0	0	0000
m_2	0010			
m_4	0100	Index 1	2	0010
m_6	0110		4	0100
m_8	1000		8	1000
m_{14}	1110			
m_{15}	1111	Index 2	6	0110
		Index 3	14	1110
		Index 4	15	1111

4. Next, form column-2 with one variable/literal minimized. In order to reduce one variable, identify a change in any one digit position in the two binary words in consecutive group.

Starting from index 0, each binary word of an index is compared with a binary word of next index, if a change in any bit position from 0 to 1 or 1 to 0 is encountered, then that literal (or variable) can be removed and put an underscore ("_") in that position and write the reduced binary word in column-2.

The column-2 is formed after all possible comparisons. Whenever the comparison of a pair of minterms lead to reduction of a literal, put a √ mark (tick mark) in the corresponding binary words of column-1 to indicate that they are considered for reduction. The formation of column-2 using the column-1 in Table 1.69 is shown in Table 1.70.

Table 1.70: Formation of Column-2 after Comparison of Minterm Pairs in Column-1

Column-1			Column-2	
	Minterm	**Binary word**	**Minterm**	**Binary word**
Index 0	0	0000 ✓	0, 2	00_0
			0, 4	0_00
Index 1	2	0010 ✓	0, 8	_000
	4	0100 ✓		
	8	1000 ✓	2, 6	0_10
			4, 6	01_0
Index 2	6	0110 ✓		
			6, 14	_110
Index 3	14	1110 ✓		
			14, 15	111_
Index 4	15	1111 ✓		

***Note:** When we compare minterms 8 and 6 there are three changes and so the minterm pairs 8 and 6 could not participate in reduction process shown in formation of column-2 in Table 1.68.*

Column-1 has 4 variables binary word which reduced to 3 variables binary word in column-2 and the binary word of column-2 is combination of two minterms.

5. Similarly form column-3 to get combination of 4 minterms and 2 variables binary word.

6. The reduction can be continued till 1 variable binary word or until no more reduction is possible.

7. After all possible reductions the unticked binary words are the prime implicants of the Boolean function. Mark the unticked binary words of all columns as P_0, P_1, P_2, P_3 ...... etc.

8. Then choose the ***essential prime implicants*** from P_0, P_1, P_2, P_3 ...... such that all minterms of function are covered with minimum number of prime implicants.

Example 1.41

Simplify the following Boolean function using table method.

F (A, B, C, D) = Σ m(0, 2, 3, 6, 7, 8, 9, 10, 13)

Solution

The minterms and corresponding binary words of the Boolean function are listed in Table 1. The given Boolean function is 4-variable function and so 4-bit binary words are used to represent minterms in Table 1.

Table 1: Minterms and Binary Words

Inputs A B C D	Minterm	Binary Word
0 0 0 0	m_0	0000
0 0 1 0	m_2	0010
0 0 1 1	m_3	0011
0 1 1 0	m_6	0110
0 1 1 1	m_7	0111
1 0 0 0	m_8	1000
1 0 0 1	m_9	1001
1 0 1 0	m_{10}	1010
1 1 0 1	m_{13}	1101

The minterms with no 1's are grouped as index 0, the minterms with one 1's are grouped as index 1, the minterms with two 1's are grouped as index 2 and the minterms with three 1's are grouped as index 3 and listed in column-1 of Table 2. The minterms of column-1 are 4-variable minterms.

The column-2 of Table 2 is formed by reducing one variable of binary words in column-1. In order to reduce a variable, starting from index 0, compare a binary word of an index with a binary word of next index and if a single change, either 1 to 0, or 0 to 1, is identified then remove that variable and form reduced binary word with an underscore in the place of removed variable. Also, a tick mark is made against the minterms of column-1 that are involved in comparison and reduction of a variable. The comparisons are performed with all possible pairs. The reduced binary words along with corresponding pair of minterms are listed in column-2. The minterms of column-2 are three variable minterms.

The column-3 of Table 2 is formed by reducing one variable of binary words in column-2. The reduced words are determined by following the procedure similar to that for reducing a literal of binary words of column-2. The minterms of column-3 are two variable minterms.

In this problem further reduction is not possible and so comparison is stopped with column-2. Now, the unticked binary words in columns-2 and 3 of Table 2 are the prime implicants. There are four unticked binary words in Table 2. Let us denote the four binary words as prime implicants P_0, P_1, P_2 and P_3. The reduced binary words along with minterms and prime implicants are listed in Table 3. Repeating words in column-3, can be considered once.

Table 2: Formation of Columns by Reducing a Variable of Binary Words

Column-1			Column-2 (First Reduction)		Column-3 (Second Reduction)	
Minterm		Binary word	Minterm	Binary word	Minterm	Binary word
Index 0	0	0000 √	0, 2	00_0 √	0, 2, 8, 10	_0_0 ← Unticked binary words
			0, 8	_000 √	~~0, 8, 2, 10~~	~~_0_0~~
Index 1	2	0010 √				
	8	1000 √	2, 3	001_ √	2, 3, 6, 7	0_1_ ← Unticked binary words
			2, 6	0_10 √	~~2, 6, 3, 7~~	~~0_1_~~
	3	0011 √	2, 10	_010 √		
	6	0110 √	8, 9	100_ ← Unticked binary words		
Index 2	9	1001 √	8, 10	10_0 √		
	10	1010 √				
			3, 7	0_11 √		
	7	0111 √	6, 7	011_ √		
Index 3	13	1101 √	9, 13	1_01 ← Unticked binary words		

Table 3: Reduced Binary Words and Prime Implicants

Minterm	Reduced Binary Word	Prime Implicants
8, 9	100_	P_0 = AB'C'(D is removed)
9, 13	1_01	P_1 = AC'D (B is removed)
0, 2, 8, 10	_0_0	P_2 = B'D' (A and C are removed)
2, 3, 6, 7	0_1_	P_3 = A'C (B and D are removed)

The minterms of the function are 0, 2, 3, 6, 7, 8, 9, 10 and 13. Form Table 4 by taking prime implicants P_0, P_1, P_2 and P_3 on rows and minterms of function on columns.

Table 4: Selection of Essential Prime Implicants

Prime implicants	Minterm								
	0	2	3	6	7	8	9	10	13
P_0 8, 9						×	×		
P_1 9, 13							×		⊗
P_2 0, 2, 8, 10	⊗	×				×		⊗	
P_3 2, 3, 6, 7		×	⊗	⊗	⊗				

The following observations are made from Table 4.

1. The minterms 0 and 10 are covered only in P_2 and so P_2 is essential.
2. The minterms 3, 6 and 7 are covered only in P_3 and so P_3 is essential.
3. The minterms 13 is covered only in P_1 and so P_1 is essential.
4. The prime implicants P_0 has two minterms 8 and 9. And in this 8 is already covered in P_2 and 9 is already covered in P_1 and so P_0 is not essential.
5. Finally the reduced set of prime implicants are P_1, P_2 and P_3.

The function F is given by sum of the prime implicants P_1, P_2 and P_3.

$\therefore F = P_1 + P_2 + P_3$

$= AC'D + B'D' + A'C$

Example 1.42

Find all the prime implicants for the following Boolean function and determine the essential prime implicants and minimum SOP form of the function.

$F(w, x, y, z) = \Sigma\, m(0, 2, 4, 5, 6, 7, 8, 10, 13, 15)$

Solution

The minterms and corresponding binary words of the Boolean function are listed in Table 1. The given Boolean function is 4-variable function and so 4-bit binary words are used to represent minterms in Table 1.

Table 1: Minterms and Binary Words

Inputs w x y z	Minterm	Binary Word
0 0 0 0	m_0	0000
0 0 1 0	m_2	0010
0 1 0 0	m_4	0100
0 1 0 1	m_5	0101
0 1 1 0	m_6	0110
0 1 1 1	m_7	0111
1 0 0 0	m_8	1000
1 0 1 0	m_{10}	1010
1 1 0 1	m_{13}	1101
1 1 1 1	m_{15}	1111

The minterms with no 1's are grouped as index 0, the minterms with one 1's are grouped as index 1, the minterms with two 1's is grouped as index 2, the minterms with three 1's are grouped as index 3 and the minterms with four 1's are grouped as index 4 and listed in column-1 of Table 2. The minterms of column-1 are 4-variable minterms.

The column-2 of Table 2 is formed by reducing one variable of binary words in column-1. In order to reduce a variable, starting from index 0, compare a binary word of an index with a binary word of next index and if a single change, either 1 to 0, or 0 to 1, is identified then remove that variable and form reduced binary word with an underscore in the place of removed variable. Also, a tick mark is made against the minterms of column-1 that are involved in comparison and reduction of a variable. The comparisons are performed with all possible pairs. The reduced binary words along with corresponding pair of minterms are listed in column-2. The minterms of column-2 are three variable minterms.

The column-3 of Table 2 is formed by reducing one variable of binary words in column-2. The reduced words are determined by following the procedure similar to that for reducing a literal of binary words of column-2. The minterms of column-3 are two variable minterms.

In this problem further reduction is not possible and so comparison is stopped with column-2. Now, the unticked binary words in column-3 of Table 2 are the prime implicants. There are eight unticked binary words in Table 2 and in this each word is repeating twice and so four reduced binary words without repetition are selected as prime implicants. Let us denote the four binary words as prime implicants P_0, P_1, P_2 and P_3. The reduced binary words along with minterms and prime implicants are listed in Table 3.

Table 2: Formation of Columns by Reducing a Variable of Binary Words

	Column-1		Column-2 (First Reduction)		Column-3 (Second Reduction)	
	Minterm	**Binary word**	**Minterm**	**Binary word**	**Minterm**	**Binary word**
Index 0	0	0000 √	0, 2	00_0 √	0, 2, 4, 6	0__0
Index 1	2	0010 √	0, 4	0_00 √	0, 2, 8, 10	_0_0
	4	0100 √	0, 8	_000 √	~~0, 4, 2, 6~~	~~0__0~~
	8	1000 √	2, 6	0_10 √	~~0, 8, 2, 10~~	~~_0_0~~
Index 2	5	0101 √	2, 10	_010 √	4, 5, 6, 7	01__
	6	0110 √	4, 5	010_ √	~~4, 6, 5, 7~~	~~01__~~
	10	1010 √	4, 6	01_0 √	5, 7, 13, 15	_1_1
Index 3	7	0111 √	8, 10	10_0 √	~~5, 13, 7, 15~~	~~_1_1~~
	13	1101 √	5, 7	01_1 √	Unticked binary words	
Index 4	15	1111 √	5, 13	_101 √		
			6, 7	011_ √		
			7, 15	_111 √		
			13, 15	11_1 √		

Table 3: Reduced Binary words and Prime Implicants

Minterm	Reduced Binary Word	Prime Implicants
0, 2, 4, 6	0_ _0	P_0 = w'z' (x and y are removed)
0, 2, 8, 10	_0_0	P_1 = x'z' (w and y are removed)
4, 5, 6, 7	01_ _	P_2 = w'x (y and z are removed)
5, 7, 13, 15	_1_1	P_3 = x z (w and y are removed)

Table 4: Selection of Essential Prime Implicants

Prime implicants	Minterms									
	0	2	4	5	6	7	8	10	13	15
P_0 0, 2, 4, 6	$\times$	$\times$	$\times$		$\times$					
P_1 0, 2, 8, 10	$\times$	$\times$					$\otimes$	$\otimes$		
P_2 4, 5, 6, 7			$\times$	$\times$	$\times$	$\times$				
P_3 5, 7, 13, 15				$\times$		$\times$			$\otimes$	$\otimes$

The minterms of the function are 0, 2, 4, 5, 6, 7, 8, 10, 13 and 15. Form Table 4 by taking prime implicants P_0, P_1, P_2 and P_3 on rows and minterms of function on columns. The following observations are made from Table 4.

1. Minterms 8 and 10 are covered only in P_1 and so P_1 is an essential prime implicant.
2. Minterms 13 and 15 are covered only in P_3 and so P_3 is an essential prime implicant.
3. Apart from P_1 and P_3 if we select P_2 all minterms of the function are covered.

The minimum SOP form of the function, F is given by sum of prime implicants P_1, P_2 and P_3.

$$\therefore F = P_1 + P_2 + P_3 = x'z' + w'x + xz$$

Example 1.43

Determine the minimum SOP form of the following Boolean function using Quine-McCulskey method

$F(A, B, C, D) = \sum m(6, 7, 8, 9) + \sum d(10, 11, 12, 13, 14, 15)$

Solution

The minterms and corresponding binary words of the Boolean function are listed in Table 1. Here the don't-cares are also considered as minterms. The given Boolean function is 4-variable function and so 4-bit binary words are used to represent minterms in Table 1.

Table 1: Minterms and Binary Words

Inputs A B C D	Minterm	Binary Word
0 1 1 0	m_6	0110
0 1 1 1	m_7	0111
1 0 0 0	m_8	1000
1 0 0 1	m_9	1001
1 0 1 0	m_{10}	1010
1 0 1 1	m_{11}	1011
1 1 0 0	m_{12}	1100
1 1 0 1	m_{13}	1101
1 1 1 0	m_{14}	1110
1 1 1 1	m_{15}	1111

The minterms with no 1's are grouped as index 0, the minterms with one 1's are grouped as index 1, the minterms with two 1's are grouped as index 2, the minterms with three 1's are grouped as index 3 and the minterms with four 1's are grouped as index 4 and listed in column-1 of Table 2. The minterms of column-1 are 4-variable minterms.

The column-2 of Table 2 is formed by reducing one variable of binary words in column-1. In order to reduce a variable, starting from index 0, compare a binary word of an index with a binary word of next index and if a single change, either 1 to 0, or 0 to 1, is identified then remove that variable and form reduced binary word with an underscore in the place of removed variable. Also, a tick mark is made against the minterms of column-1 that are involved in comparison and reduction of a variable. The comparisons are performed with all possible pairs. The reduced binary words along with corresponding pair of minterms are listed in column-2. The minterms of column-2 are three variable minterms.

The column-3 of Table 2 is formed by reducing one variable of binary words in column-2. The reduced words are determined by following the procedure similar to that for reducing a literal of binary words of column-2. The minterms of column-3 are two variable minterms.

The column-4 of Table 2 is formed by reducing one variable of binary words in column-3. The reduced words are determined by following the procedure similar to that for reducing a literal of binary words of column-2. The repeated reduced binary words of column-3 are ignored for comparison. The minterms of column-4 are one variable minterms.

In this problem further reduction is not possible and so comparison is stopped with column-4. Now, the unticked binary words in columns-3 and 4 of Table 2 are the prime implicants. There is one unticked binary word each in column-3 and 4 of Table 2. Let us denote the two binary words as prime implicants P_0 and P_1. The reduced binary words along with minterms and prime implicants are listed in Table 3.

Table 2: Formation of Columns by Reducing a Variable of Binary Words

Column-1		Column-2 (First Reduction)		Column-3 (Second Reduction)		Column-4 (Third Reduction)	
Minterm	**Binary word**	**Minterm**	**Binary word**	**Minterm**	**Binary word**		
Index 0 0	-						
Index 1 8	1000 √	8, 9	100_ √	8, 9, 10, 11	10__ √	8, 9, 10, 11 12, 13, 14, 15	1___
		8, 10	10_0 √	8, 9, 12, 13	1_0_ √		
Index 2 6	0110 √	8, 12	1_00 √	~~8, 10, 9, 11~~	~~10__~~	~~8, 9, 12, 13~~ ~~10, 11, 14, 15~~	~~1___~~
9	1001 √			8, 10, 12, 14	1__0 √		
10	1010 √	6, 7	011_ √	~~8, 12, 9, 13~~	~~1_0_~~	~~8, 10, 12, 14~~ ~~9, 11, 13, 15~~	~~1___~~
12	1100 √	6, 14	_110 √	~~8, 12, 10, 14~~	~~1__0~~		
		9, 11	10_1 √	6, 7, 14, 15	_11_		
Index 3 7	0111 √	9, 13	1_01 √	~~6, 14, 7, 15~~	~~_11_~~		
11	1011 √	10, 11	101_ √	9, 11, 13, 15	1__1 √	Unticked binary words	
13	1101 √	10, 14	1_10 √	~~9, 13, 11, 15~~	~~1__1~~		
14	1110 √	12, 13	110_ √	10, 11, 14, 15	1_1_ √		
		12, 14	11_0 √	~~10, 14, 11, 15~~	~~1_1_~~		
Index 4 15	1111 √			12, 13, 14, 15	11__ √		
		7, 15	_111 √	~~12, 14, 13, 15~~	~~11__~~		
		11, 15	1_11 √				
		13, 15	11_1 √				
		14, 15	111_ √				

Table 3: Reduced Binary words and Prime Implicants

Minterm	Reduced Binary Word	Prime Implicants
6, 7, 14, 15	_11_	P_0 = BC (A and D are removed)
8, 9, 10, 11 12, 13, 14, 15	1___	P_1 = A (B, C and D are removed)

From Table 3 it is observed that both the prime implicants P_0 and P_1 are essential so that all the minterms of the function are included.

The function F is given by sum of the prime implicants P_0 and P_1.

$$\therefore F = BC + A$$

1.8 Digital Logic Families

Digital ICs (Integrated Circuits) are manufactured using different semiconductor technologies. The collection of different logical devices that perform different logic functions but manufactured using same semiconductor technology is called a ***logic family***. The logic devices of a family will have similar input, output and internal circuit characteristics.

The digital ICs of the same logic family can be connected to implement a logic circuit. Whereas, the digital ICs of different logic families cannot be connected to implement a logic circuit, because they may not have compatible characteristics.

Some of the popular digital logic families are,

- RTL (Resistor-transistor logic)
- DTL (Diode-transistor logic)
- TTL (Transistor-transistor logic)
- ECL (Emitter-coupled logic)
- CMOS (Complementary metal-oxide semiconductor)

The RTL, DTL, TTL and ECL digital logic family of ICs are constructed using bipolar NPN transistors as fundamental switching device. The CMOS digital logic family of devices are constructed using unipolar MOSFET (Metal Oxide Semiconductor Field Effect Transistor) as fundamental switching device.

The important characteristics of digital logic family of integrated circuits (ICs) that are used to compare the performance of various family of devices are,

- Logic levels
- Propagation delay
- Power dissipation
- Fan-out and Fan-in
- Noise margin

1.8.1 Evolution of Logic Family

The first logic family developed was Resistor-transistor logic (RTL). The RTL (Resistor-Transistor Logic) family of devices uses resistors to receive inputs and use transistor as switching device and so the name Resistor-Transistor Logic. The primary advantage of RTL technology was that it involved a minimum number of transistors. The disadvantage of RTL is its high power dissipation during conduction state.

Diode-Transistor Logic (DTL) is a class of digital circuits built from diodes, resistors and bipolar junction transistors (BJT). In DTL (Diode-Transistor Logic) family of devices the resistances of RTL family are replaced by diodes. Therefore, DTL family of devices uses diodes to receive inputs and use transistor as switching device and so the name diode-transistor logic.

The TTL logic family has been developed to overcome the disadvantages in RTL and DTL family. In TTL (Transistor-Transistor Logic) family of devices the diodes of DTL family are replaced by transistors.

The p-n junction of a diode has considerable capacitance, so changing the logic level of an input connected to a diode, as in DTL, requires considerable time and energy. In TTL, the base-emitter junction is used to receive inputs instead of diodes in DTL. Therefore, TTL family of devices uses transistors to receive inputs as well as switching device and so the name transistor-transistor logic.

In RTL, DTL and TTL families the transistor switches are driven from saturation to cut-off and vice-versa, during switching operation. This switching action takes larger propagation time and power dissipation.

The need for fast switching leads to development of ECL (Emitter-Coupled Logic) family which works in active region with a small difference in two logic voltage levels which is decided by the currents. Hence, ECL is non-saturated logic family of devices and has low power dissipation and very low propagation delay. The, other bipolar logic families are based on voltage mode logic, ECL utilizes current mode logic. The invention of MOSFET leads to development very low power CMOS switches. The modern trend is using various versions of CMOS family of devices with low power and high-speed switching.

1.9 Operation and Characteristics of Digital Logic Families

The basic devices in the logic families are BJT and CMOS transistor and they are operated in saturation and cut-off region. In saturation the transistor is considered as ON and in cut-off the transistor is considered as OFF. The ON-OFF characteristics of BJT and CMOS transistors are discussed in sections 1.9.6 and 1.9.7.

The performance of logic family of devices are measured in terms of logic levels, propagation delay, power dissipation, noise margin, fan-in and fan-out and they are also called performance characteristics.

1.9.1 Logic Levels

The digital ICs operates on binary constants 1 and 0, which are physically two different voltages in a digital circuit or device. The two different voltage levels are called ***logic levels***. In positive logic, the binary constant 1 is **high** and 0 is **low**. Different logic families will have different values of voltages for logic levels which is decided by the fundamental device used in the logic families.

In general, the logic levels will have a range of voltages and also associated with currents. The **high** and **low** logic levels will have a minimum and maximum value of voltage and current. The definition of minimum and maximum voltage and current levels of digital logic devices are given below:

$\mathbf{V_{IH\_min}}$: It is the minimum value for a **high** logic level input voltage at an input of a logical device to recognize a logical 1 or **high**.

$\mathbf{V_{IL\_max}}$: It is the maximum value for a **low** logic level input voltage at an input of a logical device to recognize a logical 0 or **low**.

$\mathbf{V_{OH\_min}}$: It is the minimum value for a **high** logic level output voltage at an output of a logical device to recognize a logical 1 or **high** at a standard load condition.

$\mathbf{V_{OH\_max}}$: It is the maximum value for a **high** logic level output voltage at an output of a logical device to recognize a logical 0 or **low** at a standard load condition.

$\mathbf{I_{IH}}$: It is the current that flows into an input when a **high** logic level voltage is applied to that input.

$\mathbf{I_{IL}}$: It is the current that flows into an input when a **low** logic level voltage is applied to that input.

$\mathbf{I_{OH}}$: It is the current that flows from an output when a **high** logic level voltage exists at output at a standard load condition.

I_{OL}: It is the current that flows from the output when a **low** logic level voltage exists at output at a standard load condition.

The currents and voltages during a **high** state and a **low** state are shown in Fig. 1.44.

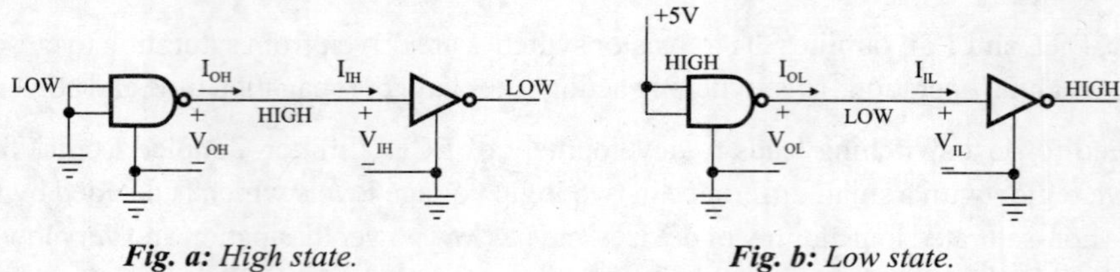

Fig. a: High state. *Fig. b: Low state.*

Fig. 1.44: Currents and voltages in the high and low states.

1.9.2 Propagation Delay

In digital circuits or devices, the logic levels change from **high** to **low** or vice-versa. The logic voltage level changes always experience a delay for a change in input to be available as a change in the output. The propagation delay is defined as the time required for a change in input to be available in the output. Two delay times are defined in order to estimate the propagation delay with reference to an inverter or NOT gate.

t_{PLH}: Delay time for a **low** at input to realize as a **high** in output.

t_{PHL}: Delay time for a **high** at input to realize as a **low** in output.

Figure 1.45 shows the two delay times of an inverter. The delay times are measured between the 50 percent points on the input and output transitions. In general, t_{PHL} and t_{PLH} will not be same and so the propagation delay is defined as the average of the delay times, t_{PHL} and t_{PLH}.

$$\text{Propagation delay, } t_{pd} = \frac{t_{PHL} + t_{PLH}}{2}$$

The propagation delay time will be in the order of nano seconds.

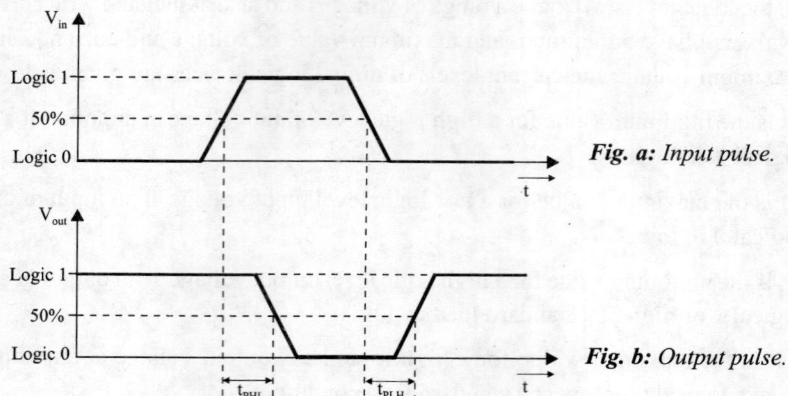

Fig. 1.45: Propagation delay for a change in logic levels in an inverter.

1.9.3 Power Requirement

Every Integrated circuit (IC) require an electrical power supply connection at the prescribed voltage level. Hence, every IC will consume a definite power during its working time.

The amount of power that an IC consumes is given by the product of current (I_{CC}) drawn by the IC, and the voltage level (V_{cc}) of power supply connected to that IC.

Power = $I_{CC}\ V_{CC}$

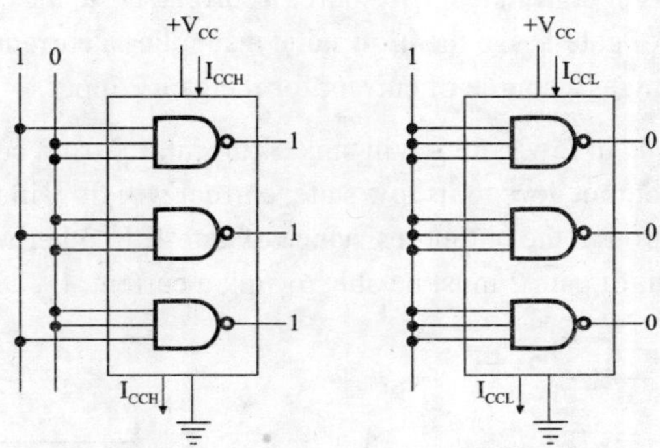

Fig. a: High state. *Fig. b: Low state.*

Fig. 1.46: High and low state currents in a NAND gate.

In digital ICs the current drawn from the power supply will vary and depends on the logic states of the devices in the chip. For example, consider a NAND gate IC shown in Fig. 1.46(a) when all the gate outputs are **high**. Now, the current drawn from the power supply is called I_{CCH}. Likewise, consider a NAND gate IC shown in Fig. 1.46(b) when all the gate outputs are **low**. Now, the current drawn from the power supply is called I_{CCL}.

In general, I_{CCH} and I_{CCL} will not be same and so an average current can be estimated as shown below:

$$\text{Average current, } I_{CC\_av} = \frac{I_{CCH} + I_{CCL}}{2}$$

This average current can be used to estimate the power consumed by the IC.

$$\text{Average Power, } P_{av} = V_{CC}\ I_{CC\_av}$$

A common means for measuring and comparing the overall performance of an IC family is the speed-power product, which is obtained by multiplying the propagation delay by the power dissipation.

For example, suppose an IC family has an average propagation delay of 2 ns and an average power dissipation of 4 *milli-Watt.*

Now, Speed-power product = 2 ns × 5 mW = 2 × 10^{-9} × 5 × 10^{-3} *Watt-second*

= 10 × 10^{-12} *Watt-second* or *Joules* = 10 *pico-Joules (pJ)*

1.9.4 Fan-out and Fan-in

In general, a logic-circuit output is connected to several other logic inputs. The fan-out is defined as the maximum number of standard load that an output can drive reliably. A standard load is usually defined as the amount of current needed by an input of another gate in the same logic family. Therefore, fan-out is the maximum number of inputs of other logic devices that can be connected to the output of a gate, and is expressed by a number.

When the output of a gate is in **high** state it will source a current, I_{OH} to the connected inputs as shown in Fig. 1.47(a). When the output of gate 1 is in the **high** state, it supplies a current I_{IH} to the input of gate 2. Thus, the output of gate 1 is acting as a source of current for the gate 2 input.

When the output of a gate is in **low** state it will sink a current, I_{OL} from connected inputs as shown in Fig. 1.47(b). When the gate 1 output goes to its **low** state, current will flow in the direction shown from the input circuit of gate 2 back through the output resistance of gate 1. In other words, in the **low** state the circuit output that drives the input of gate 2 must be able to sink a current, I_{IL}, coming from that input.

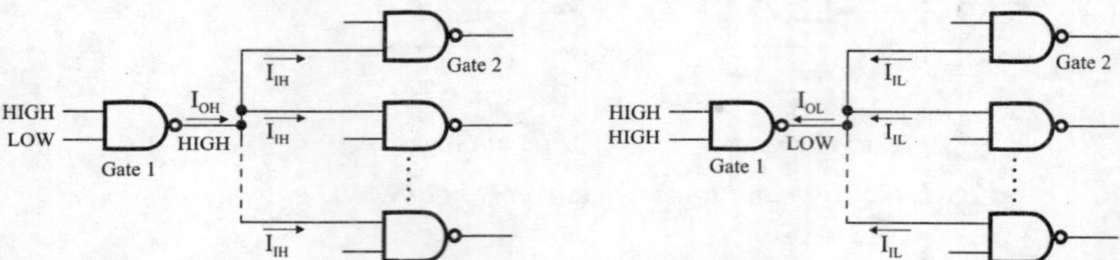

Fig. a: Current sourcing in high state. *Fig. b: Current sinking in low state.*

Fig. 1.47: Current sourcing and sinking.

The fan-out of a gate is calculated from the ratio I_{OH}/I_{IH} or I_{OL}/I_{IL}. The lowest among the two ratio is chosen as fan-out.

$$\text{Fan-out (Low)} = \frac{I_{OL}}{I_{IL}}$$

$$\text{Fan-out (High)} = \frac{I_{OH}}{I_{IH}}$$

Fan-in is the number of inputs provided (or available) in a logic gate and it is decided during manufacturing itself which in turn depends on technology used for manufacturing the IC. Larger number of inputs will increase the propagation delay and so number of inputs are normally limited to 2 or 3.

1.9.5 Noise Margin

Stray electric and magnetic fields can induce voltages on the connecting wires between logic circuits. These unwanted, spurious signals are called noise and can sometimes cause the voltage at the input of a logic circuit to drop below V_{IH_min} or rise above V_{IL_max}, which could produce unpredictable operation. The noise immunity of a logic circuit refers to the circuit's ability to tolerate noise voltages without causing unwanted changes in the output voltage. A quantitative measure of noise immunity in terms of voltage is called ***noise margin***.

Figure 1.48 shows the range of voltages that can occur at a logic-circuit output and in this range, any voltages greater than V_{OH_min} are considered as logic 1, and any voltages lower than V_{OL_max} are considered as logic 0. Voltages in the disallowed range should not appear at a logic circuit output under normal conditions.

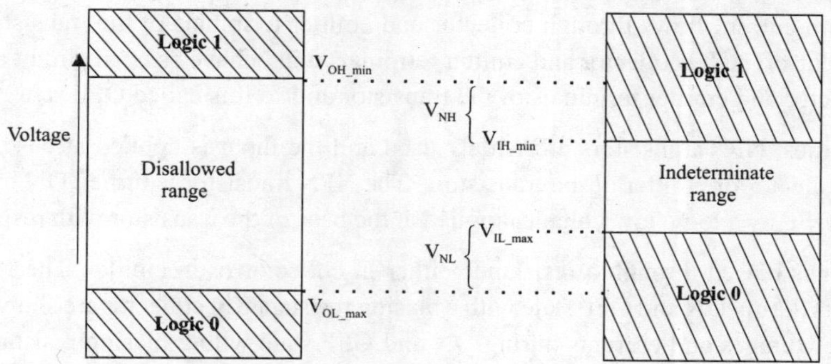

***Fig. a:** Output voltage ranges.* ***Fig. b:** Input voltage ranges.*

***Fig. 1.48:** Noise margin.*

Figure 1.48(b) shows the range of voltages that can be applied at a logic circuit input. The logic circuit will respond to any input greater than V_{IH_min} as a logic 1, and will respond to voltages lower than V_{IL_max} as a logic 0. Voltages in the indeterminate range will produce an unpredictable response and should be avoided.

Based on the minimum voltage range of **high** input and output voltage levels, a high-state noise margin can be defined as shown below:

High-state noise margin, $V_{NH} = V_{OH\_min} - V_{IH\_min}$

Based on the maximum voltage range of **low** input and output voltage levels, a low-state margin can be defined as shown below:

Low-state noise margin, $V_{NL} = V_{IL\_max} - V_{OL\_max}$

Table 1.71: Typical Values of Performance Characteristics of Logic Families

Logic Family	Propagation Delay *(ns)*	Power Dissipation per Gate *(mW)*	Noise Margin *(V)*	Fan-in	Fan-out
RTL	12	30	0.2	5	5
DTL	30	12	0.4	5	8
TTL	9	10	0.4	8	10
ECL	1	50	0.25	5	10
CMOS	20 - 50	0.01	1	10	50

1.9.6 ON-OFF Characteristics of Bipolar Junction Transistors

In logic devices the transistors are used as switches in which they are operated either in saturation region/state or cut-off region/state. In saturation state, collector and emitter terminals will behave as short circuit and maximum current flows through collector and emitter terminals of the transistor and so it is called ON state. In cut-off state, collector and emitter terminals will behave as open circuit and no current flows through collector and emitter terminals of the transistor and so it is called OFF state.

In logic families, NPN transistors are widely used and the input is applied at base and output is taken either from collector or emitter of the transistors. The NPN transistor is turned ON by a logic **high** voltage and turned OFF by a logic **low** voltage, applied at the base of the transistor with respect to emitter.

The collector is biased through a resistance either in collector or in emitter. The NPN transistor voltages and currents during ON and OFF state with a biasing resistance in collector are shown in Fig. 1.49. The NPN transistor voltages and currents during ON and OFF state with a biasing resistance in emitter are shown in Fig. 1.50.

During ON state the transistor is driven to saturation and collector-emitter voltage, $V_{CE} = 0.2\ V$, which is approximately equal to zero and so the collector and emitter terminals will behave as short circuit (S.C.). During OFF state the transistor is driven to cut-off and collector current, I_C is approximately zero and so the collector and emitter terminals will behave as open circuit (O.C.).

When output is obtained from collector the input logic is available as inverted logic in the output. But when the output is obtained from emitter the output logic is same as input logic.

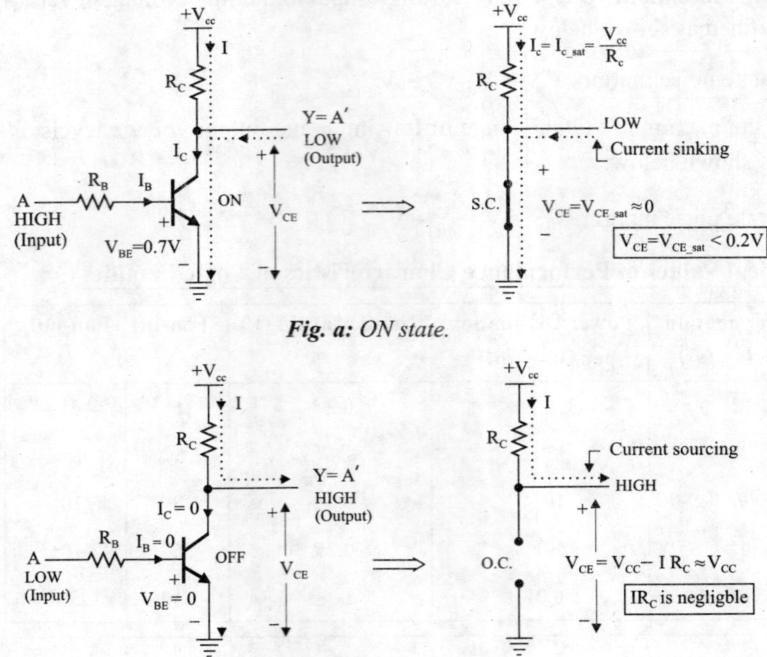

Fig. a: ON state.

Fig. b: OFF state.

***Fig. 1.49:** NPN transistor voltage and current with resistance in collector.*

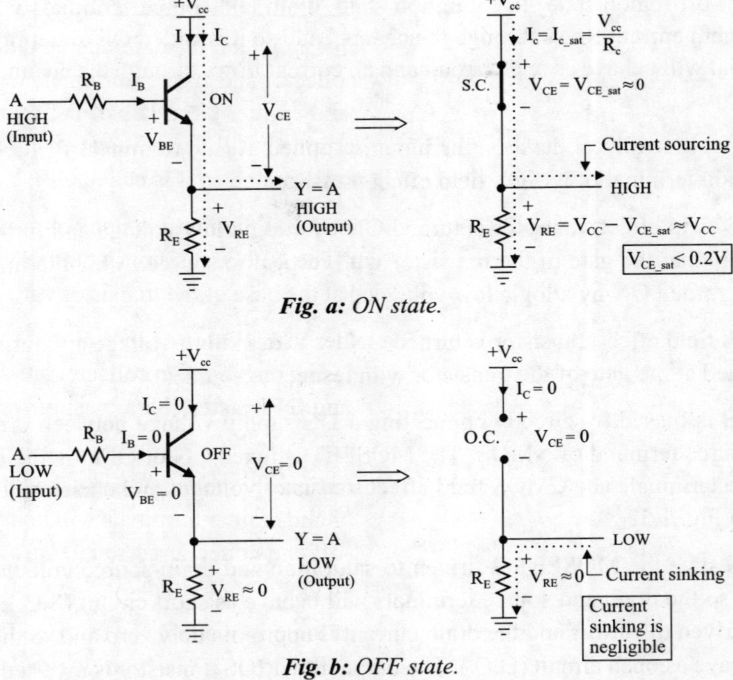

Fig. a: ON state.

Fig. b: OFF state.

***Fig. 1.50:** NPN transistor voltage and current with resistance in emitter.*

1.9.7 ON-OFF Characteristics of CMOS Field Effect Transistors

Enhancement type MOSFETs (Metal Oxide Semiconductor Field Effect Transistor) are the basic switching devices in CMOS (Complementary Metal Oxide Semiconductor) field effect transistors. There are two types of MOSFET devices: PMOS (P-type Metal Oxide Semiconductor) field effect transistor and NMOS (N-type Metal Oxide Semiconductor) field effect transistor. A CMOS field effect transistor is a pair of NMOS and PMOS field effect transistors with their channels connected in series and gate terminals tied together. The symbols of MOSFET devices are shown in Fig. 1.51.

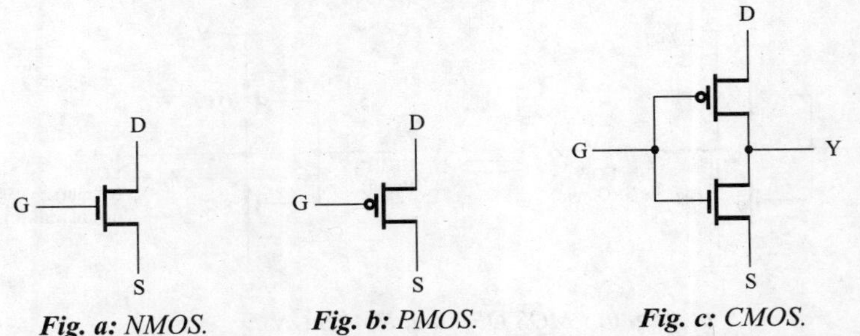

***Fig. a:** NMOS.* ***Fig. b:** PMOS.* ***Fig. c:** CMOS.*

***Fig. 1.51:** Metal Oxide Semiconductor Field Effect Transistor (MOSFET).*

In logic devices the MOSFETs are used as switches in which they are operated either in saturation region/state or cut-off region/state. In saturation state, drain and source terminals will behave as short circuit and maximum current flows through the channel and so it is called ON state. In cut-off state, drain and source terminal will behave as open circuit and no current flows through the channel and so it is called OFF state.

In COMS logic family of devices, the input is applied at gate terminal of MOSFET and output is taken from the drain terminal of NMOS field effect transistor.

The PMOS field effect transistor is turned ON by a logic **low** voltage and turned OFF by a logic **high** voltage, applied at the gate of the transistor with respect to source. A bubble in the gate of PMOS indicates that it is turned ON by a logic **low** voltage.

The NMOS field effect transistor is turned ON by a logic **high** voltage and turned OFF by a logic **low** voltage, applied at the gate of the transistor with respect to source.

The CMOS is biased by directly connecting a DC supply voltage between drain terminal of the PMOS and the source terminal of NMOS. The MOSFET is tuned ON or OFF by the logic voltage input applied at the gate terminal. The CMOS field effect transistor voltages and currents during ON and OFF state are shown in Fig. 1.52.

During ON state the MOSFET is driven to saturation and drain-source voltage is approximately equal to zero and so the drain and source terminals will behave as short circuit (S.C.). During OFF state the MOSFET is driven to cut-off and the drain current is approximately zero and so the drain and source terminals will behave as open circuit (O.C.). The output of CMOS transistor is inverted logic of the input. Hence, CMOS switch basically an inverter.

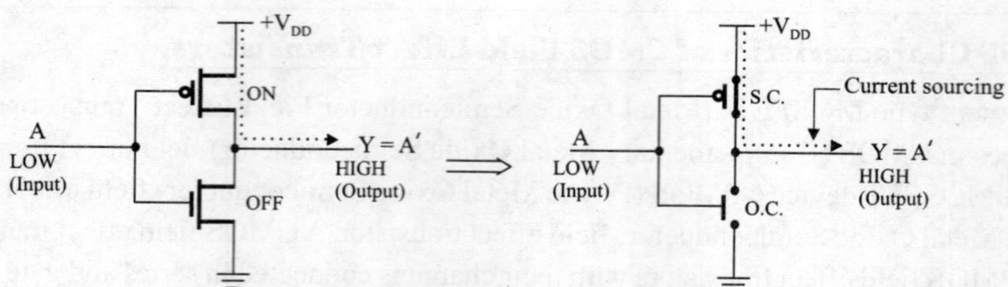

Fig. a: PMOS ON and NMOS OFF.

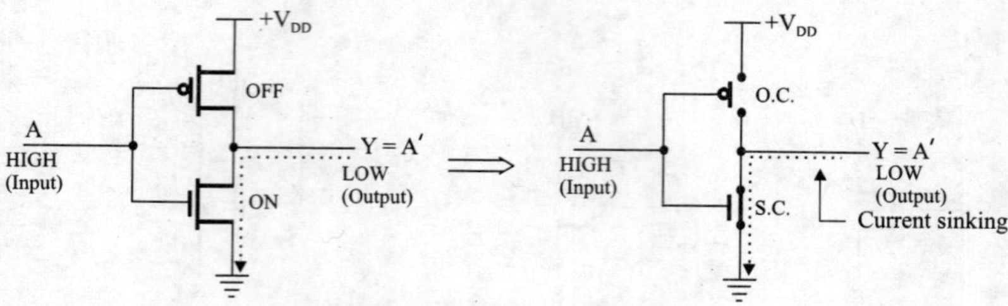

Fig. b: PMOS OFF and NMOS ON.

***Fig. 1.52:** CMOS voltage and current during ON and OFF state.*

1.10 RTL (Resistor-Transistor Logic)

In RTL (Resistor-Transistor Logic) family, input network is made of resistors and switching devices are transistors and so the name RTL. The logic functions are realized by parallel or series connection of switching transistors.

In RTL family, the inputs are applied to base of switching transistors through resistors and the output is obtained from the collector of switching transistor.

The switching transistors operate in cut-off (OFF state) or saturation (ON state). In RTL family the logic **low** is 0 to 0.2 *V* and the logic **high** is 1 *V* to 3.6 *V*.

1.10.1 RTL NOR Gate

A two input NOR gate is shown in Fig. 1.53. Here, two transistors Q_1 and Q_2 are connected in parallel such that collectors are tied together to form the output and emitters are grounded. Inputs are applied to base of the transistors through resistance.

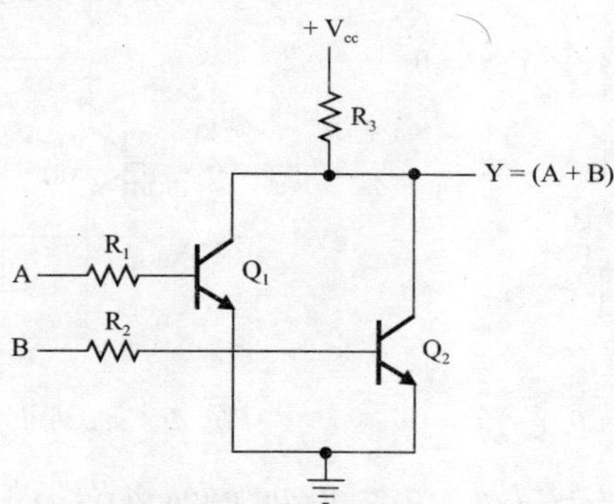

Fig. 1.53: Two input RTL NOR gate.

From the truth table of NOR gate, we get the following four cases of inputs.

Case 1: A = 0, B = 0, Y = 1

Case 2: A = 0, B = 1, Y = 0

Case 3: A = 1, B = 0, Y = 0

Case 4: A = 1, B = 1, Y = 0

Truth Table of NOR Gate

Inputs		Output
A	B	Y
0	0	1
0	1	0
1	0	0
1	1	0

Case 1: A = 0, B = 0, Y = 1

Figure 1.54(a) shows voltage and current condition of NOR gate when both inputs are **low**. Now, the transistors Q_1 and Q_2 are in OFF state and the transistors will behave as open circuit to ground. Hence, a logic **high** voltage appears in the collector. The collector power supply will source current to load at output through the collector resistance.

Case 4: A = 1, B = 1, Y = 0

Figure 1.54(d) shows voltage and current condition of NOR gate when both inputs are **high**. Now, the transistors Q_1 and Q_2 are in ON state (or saturation state) and so collector-emitter terminals of transistors will behave as short circuit. Hence a logic **low** voltage appears in the collector and output load is shorted to ground through the switching transistors and so current will sink to output.

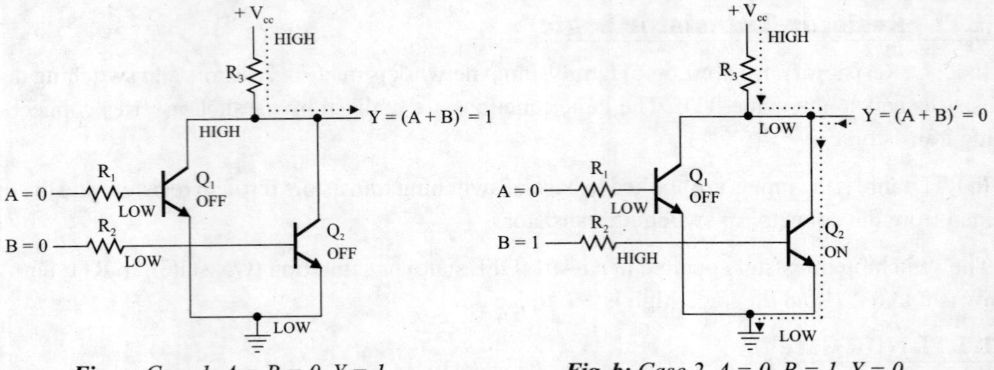

Fig. a: Case 1, A = B = 0, Y = 1. *Fig. b: Case 2, A = 0, B = 1, Y = 0.*

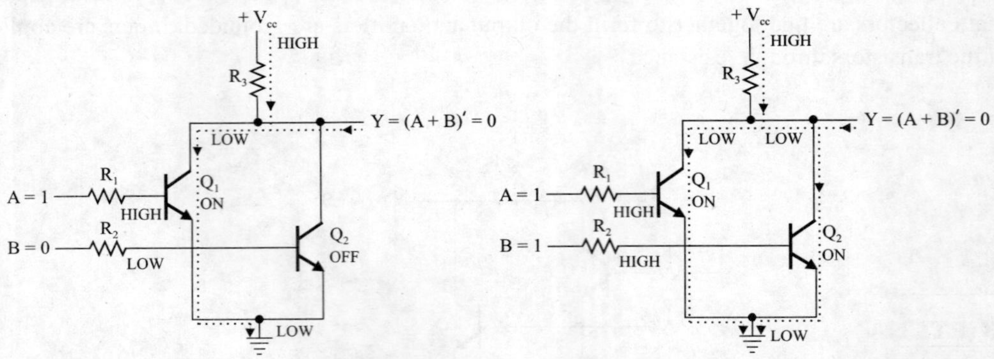

Fig. c: Case 3, A = 1, B = 0, Y = 0. *Fig. d: Case 4, A = B = 1, Y = 0.*

Fig. 1.54: Four cases of operation of RTL NOR gate.

Case 2 and 3: A = 0, B = 1, Y = 0 and A = 1, B = 0, Y = 0

In these two cases one of the base of transistors is **high** and the other is **low**. Hence, one of the transistors will be ON and the other is OFF. The ON transistor will pull the output to **low**. Therefore, the working of NOR gate in these two cases will be similar to case 4. Figures 1.54(b) and 1.54(c) shows voltage and current condition of NOR gate when any one of the inputs is **high**.

1.10.2 RTL OR Gate

The OR logic can be obtained by inverting the NOR logic. The inverted output of NOR gate can be obtained by connecting the biasing resistance on emitter side and taking output from emitter terminal as shown in Fig. 1.55. The voltage and current conditions for various inputs are shown in Figs. 1.56(a) to 1.56(d).

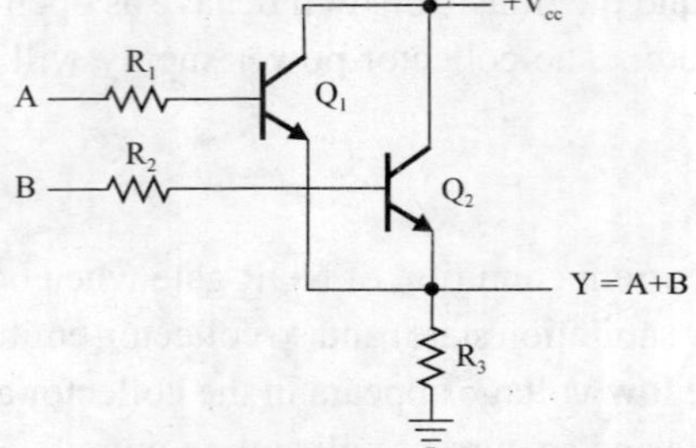

Truth Table of OR Gate

Inputs		Output
A	B	Y
0	0	0
0	1	1
1	0	1
1	1	1

Fig. 1.55: Two input RTL OR gate.

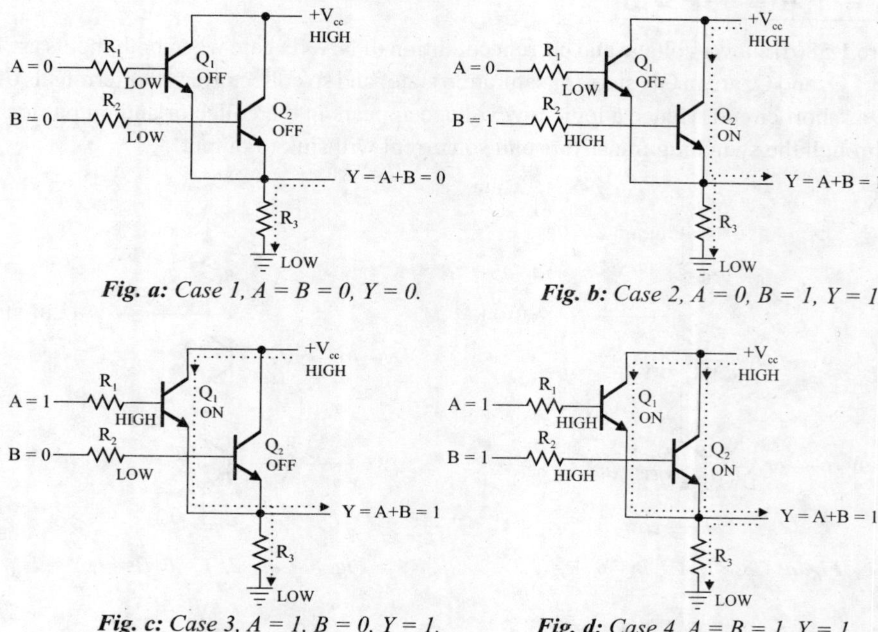

Fig. a: Case 1, A = B = 0, Y = 0.

Fig. b: Case 2, A = 0, B = 1, Y = 1.

Fig. c: Case 3, A = 1, B = 0, Y = 1.

Fig. d: Case 4, A = B = 1, Y = 1.

***Fig. 1.56:** Four cases of operation of RTL OR gate.*

1.10.3 RTL NAND Gate

A two input NAND gate is shown in Fig. 1.57. Here, two transistors Q_1 and Q_2 are connected in series such that emitter of Q_1 is connected to collector of Q_2 and emitter of Q_2 is grounded. Inputs are applied to base of the transistors through resistance. Output is taken from collector of Q_1.

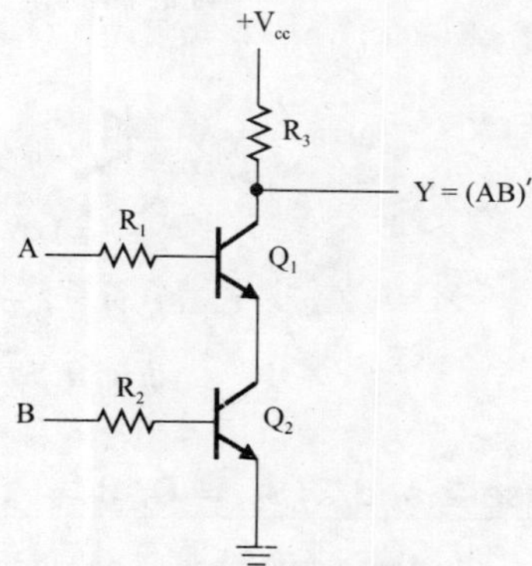

***Fig. 1.57:** Two input RTL NAND gate.*

From the truth table of NAND gate, we get the following four cases of inputs.

Case 1: A = 0, B = 0, Y = 1

Case 2: A = 0, B = 1, Y = 1

Case 3: A = 1, B = 0, Y = 1

Case 4: A = 1, B = 1, Y = 0

Truth Table of NAND Gate

Inputs		Output
A	B	Y
0	0	1
0	1	1
1	0	1
1	1	0

Case 1: A = 0, B = 0, Y = 1

Figure 1.58(a) shows voltage and current condition of NAND gate when both inputs are **low**. Now, the transistors Q_1 and Q_2 are in OFF state and the transistors will behave as open circuit to ground. Hence, logic **high** voltage appears in the collector of transistor Q_1. The collector power supply will source current to load at output through the collector resistance.

Case 4: A = 1, B = 1, Y = 0

Figure 1.58(d) shows voltage and current condition of NAND gate when both inputs are **high**. Now, the transistors Q_1 and Q_2 are in ON state (or saturation state) and so collector-emitter terminals of transistors will behave as short circuit. Hence a logic **low** voltage appears in the collector and output load is shorted to ground through the switching transistors and so current will sink to output.

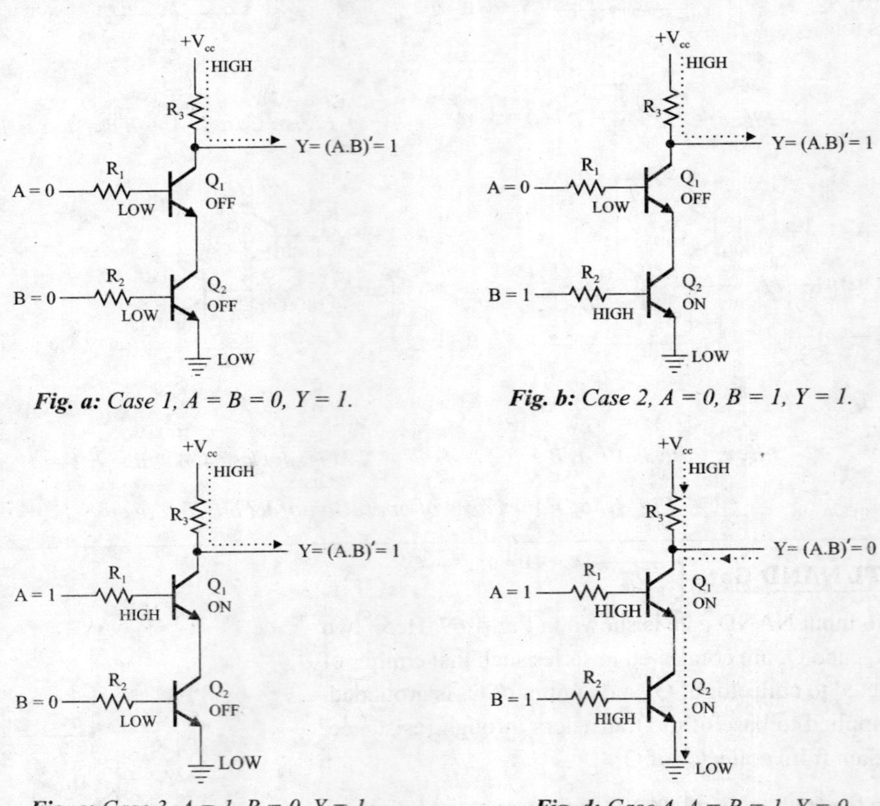

***Fig. a:** Case 1, A = B = 0, Y = 1.*

***Fig. b:** Case 2, A = 0, B = 1, Y = 1.*

***Fig. c:** Case 3, A = 1, B = 0, Y = 1.*

***Fig. d:** Case 4, A = B = 1, Y = 0.*

***Fig. 1.58:** Four cases of operation of RTL NAND gate.*

Case 2 and 3: A = 0, B = 1, Y = 1 and A = 1, B = 0, Y = 1

In these two cases one of the base of transistors is **high** and the other is **low**. Hence, one of th transistors will be ON and the other is OFF. The OFF transistor will behave as open circuit to ground an so no current will flow through transistors and so the output will be **high**. Therefore, the working of NANI gate in these two cases will be similar to case 1. Figures 1.58(b) and 1.58(c) shows voltage and curren condition of NAND gate when any one of the inputs is **high**.

1.10.4 RTL AND Gate

The AND logic can be obtained by inverting the NAND logic. The inverted output of NAND gate can be obtained by connecting the biasing resistance on emitter side and taking output from emitter terminal as shown in Fig. 1.59. The voltage and current conditions for various inputs are shown in Figures 1.60(a) to 1.60(d).

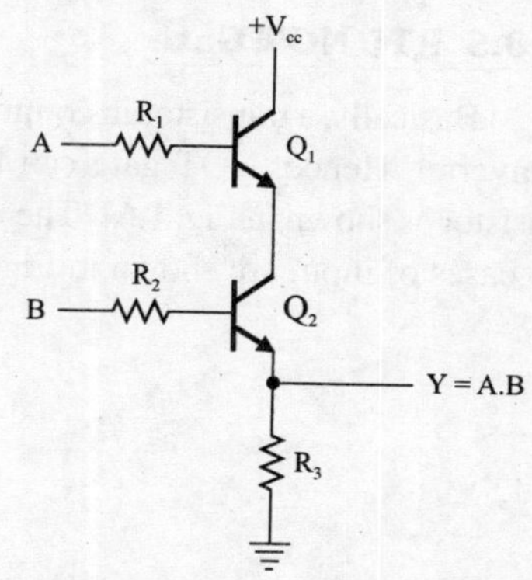

Fig. 1.59: Two input RTL AND gate.

Truth Table of AND Gate

Inputs		Output
A	B	Y
0	0	0
0	1	0
1	0	0
1	1	1

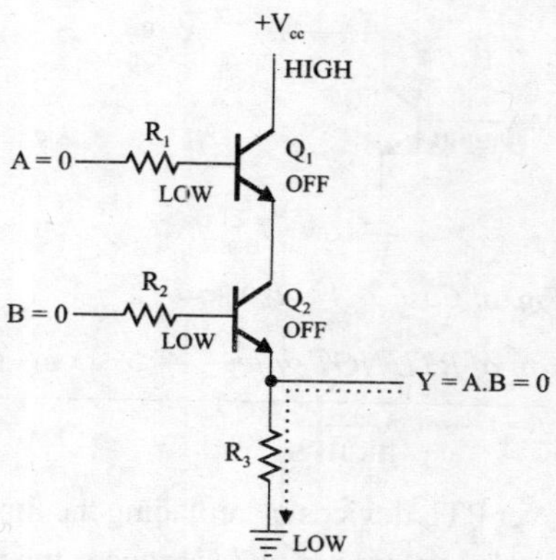

Fig. a: Case 1, $A = B = 0$, $Y = 0$.

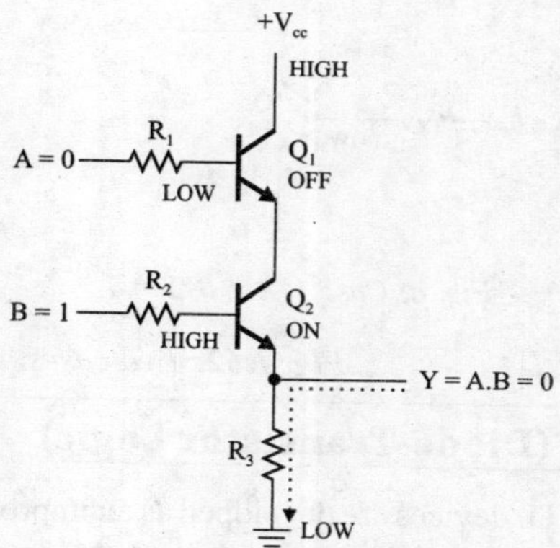

Fig. b: Case 2, $A = 0$, $B = 1$, $Y = 0$.

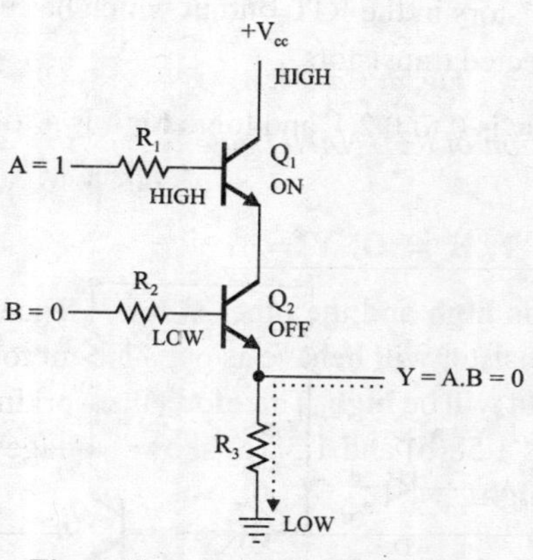

Fig. c: Case 3, $A = 1$, $B = 0$, $Y = 0$.

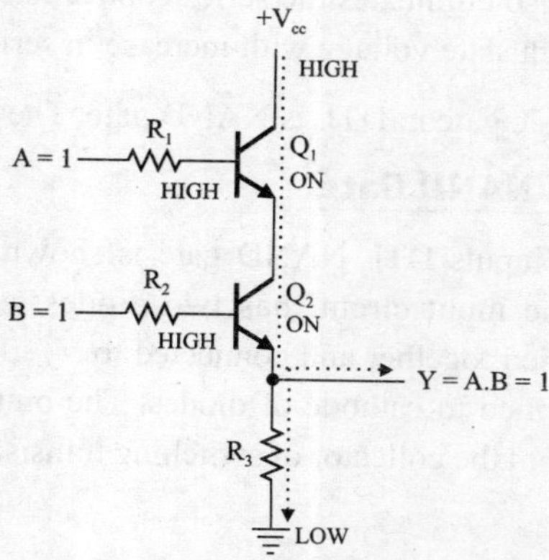

Fig. d: Case 4, $A = B = 1$, $Y = 1$.

Fig. 1.60: Four cases of operation of RTL AND gate.

1.10.5 RTL NOT Gate

(AU, Apr/May'19, 6 Marks)

Basically, a transistor in common emitter configuration will behave as inverter. Hence, NOT gate can be constructed using a single NPN transistor as shown in Fig. 1.61. The voltage and current conditions during two cases of input are shown in Figs. 1.62(a) and 1.62(b).

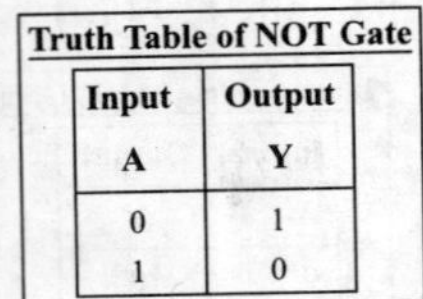

Truth Table of NOT Gate

Input A	Output Y
0	1
1	0

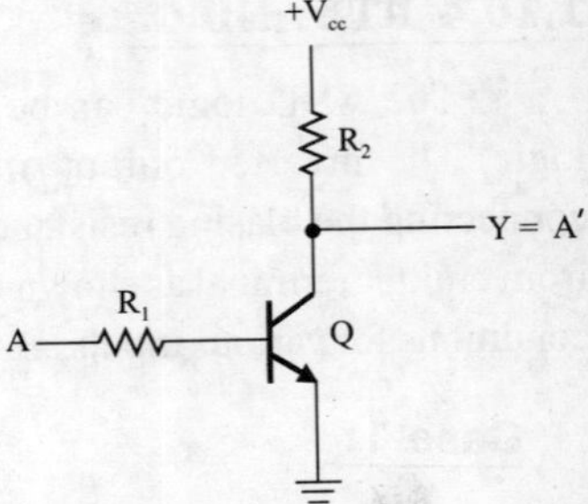

Fig. 1.61: *RTL NOT gate.*

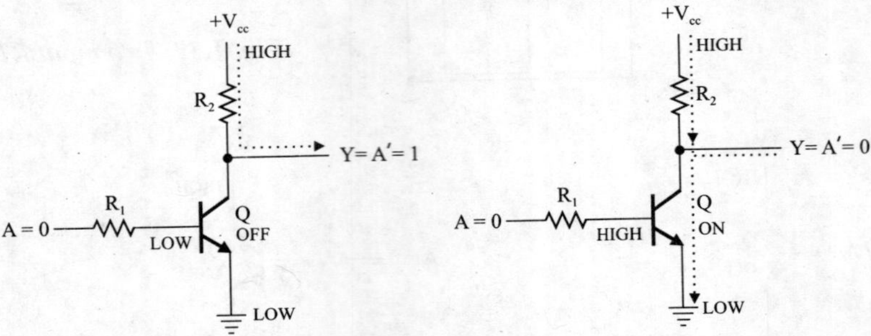

Fig. a: *Case 1, A = 0, Y = 1.* **Fig. b:** *Case 2, A = 1, Y = 0.*

Fig. 1.62: *Two cases of operation of RTL NOT gate.*

1.11 DTL (Diode-Transistor Logic)

The DTL devices are developed as an improvement over RTL devices by replacing the input resistors in RTL devices by diodes in DTL devices. The replacement of resistors by diodes reduces the power loss. Therefore, in DTL devices the diodes are input devices and a single switching transistor is the output device. This also eliminates the series connection of transistors in the RTL output which has a drawback of variable high state voltage with increase in series connected transistors.

The basic gate in DTL is NAND gate. The logic **low** is 0 to 0.2 *V* and logic **high** is 4 to 5 *V*.

1.11.1 DTL NAND Gate

(AU, Nov/Dec'18, 2 Marks)

A two inputs DTL NAND gate is shown in Fig. 1.63. The input circuit has two diodes with their anodes tied together and connected to V_{cc}. The inputs are applied to cathode of diodes. The output is obtained from the collector of switching transistor.

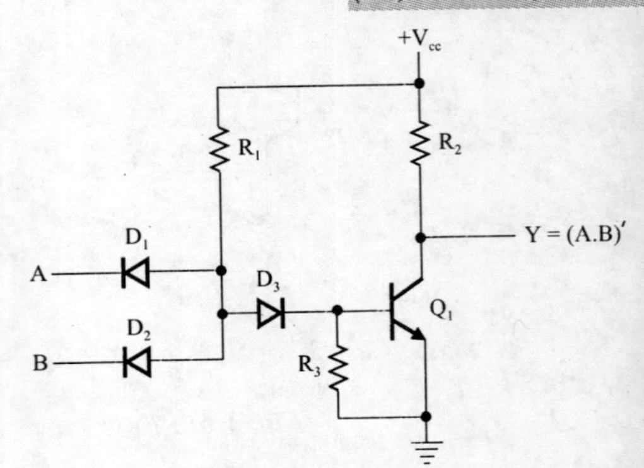

Fig. 1.63: *Two input DTL NAND gate.*

From the truth table of NAND gate we get the following four cases of inputs.

Case 1: A = 0, B = 0, Y = 1

Case 2: A = 0, B = 1, Y = 1

Case 3: A = 1, B = 0, Y = 1

Case 4: A = 1, B = 1, Y = 0

Truth Table of NAND Gate

Inputs		Output
A	B	Y
0	0	1
0	1	1
1	0	1
1	1	0

Case 1: A = 0, B = 0, Y = 1

Figure 1.64(a) shows the voltage and current condition of NAND gate when both the inputs are **low**. The input diodes D_1 and D_2 are forward biased and conduct via input circuit and the voltage at the anode of diodes will be 0.7 *V* which is not sufficient to turn ON the switching transistor. Now, the transistor will be in OFF state and hence the output at the collector of transistor is logic **high**.

Case 2 and 3: A = 0, B = 1, Y = 1 and A = 1, B = 0, Y = 1

In these two cases one of the input diodes is forward biased and conduct via input circuit and the voltage at the anode of diodes will be 0.7 *V* which is not sufficient to turn ON the switching transistor. Now, the transistor will be in OFF state and hence the output at the collector of transistor is logic **high**. Therefore, the working of NAND gate in these two cases will be similar to that of case 1.

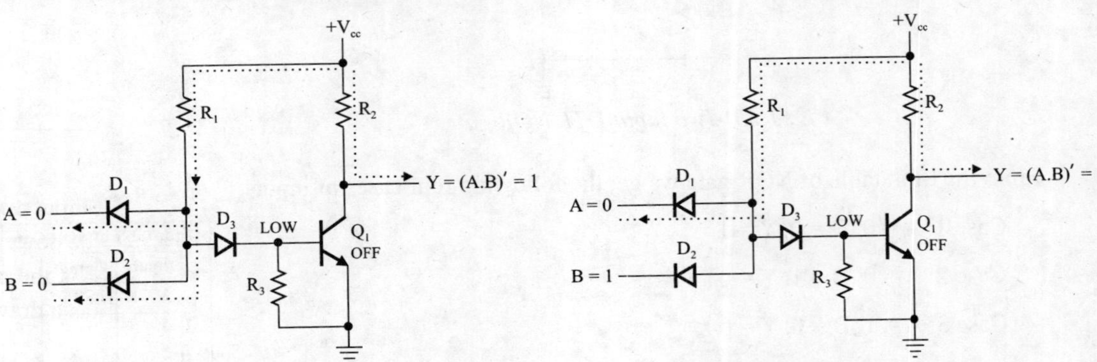

***Fig. a:** Case 1, A = B = 0, Y = 1.*

***Fig. b:** Case 2, A = 0, B = 1, Y = 1.*

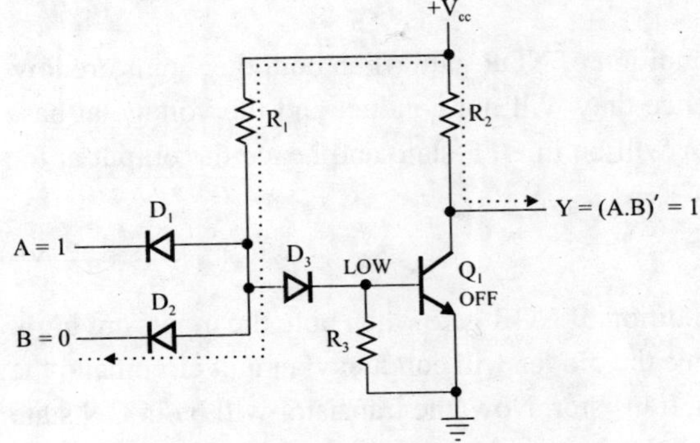

***Fig. c:** Case 3, A = 1, B = 0, Y = 1.*

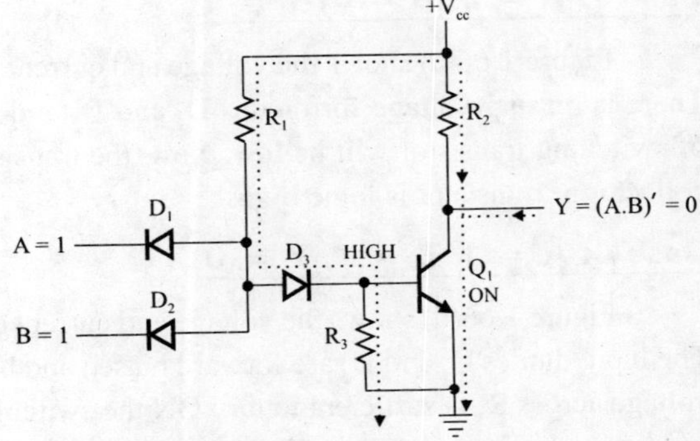

***Fig. d:** Case 2, A = B = 1, Y = 0.*

***Fig. 1.64:** Four cases of operation of DTL NAND gate.*

Case 4: A = 1, B = 1, Y = 0

Figure 1.64(d) shows the voltage and current condition of NAND gate when both the inputs are **high**. The input diodes D_1 and D_2 are reverse biased and hence the diodes will not conduct. The voltage at anode of diode D_3 will be logic **high** and turn ON the switching transistor and hence the output at the collector of transistor is logic **low**.

1.11.2 DTL NOR Gate

A two inputs DTL NOR gate is shown in Fig. 1.65. The input circuit has two diodes with their cathodes tied together and connected to ground via a resistor. The inputs are applied to anode of diodes. The output is obtained from the collector of switching transistor.

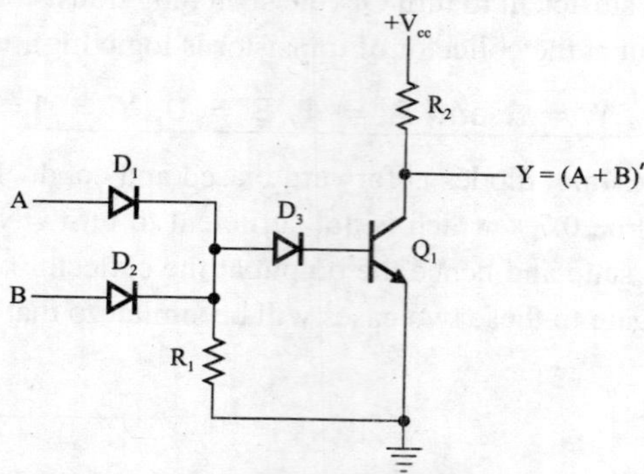

Fig. 1.65: Two input DTL NOR gate.

From the truth table of NOR gate we get the following four cases of inputs.

Case 1: A = 0, B = 0, Y = 1

Case 2: A = 0, B = 1, Y = 0

Case 3: A = 1, B = 0, Y = 0

Case 4: A = 1, B = 1, Y = 0

Truth Table of NOR Gate

Inputs		Output
A	B	Y
0	0	1
0	1	0
1	0	0
1	1	0

Case 1: A = 0, B = 0, Y = 1

Figure 1.66(a) shows the voltage and current condition of NOR gate when both the inputs are **low**. There is biasing voltage for diodes D_1 and D_2 and hence they will not conduct and the voltage at base of switching transistor will be **low**. Now, the transistor will be in OFF state and hence the output at the collector of transistor is logic **high**.

Case 4: A = 1, B = 1, Y = 0

Figure 1.66(d) shows the voltage and current condition of NOR gate when both the inputs are **high**. The input diodes D_1 and D_2 are forward biased and hence the diodes will conduct via input circuit and the voltage across R_3 is sufficient to turn ON the switching transistor. Now, the transistor will be in ON state and hence the output at the collector of transistor is logic **low**.

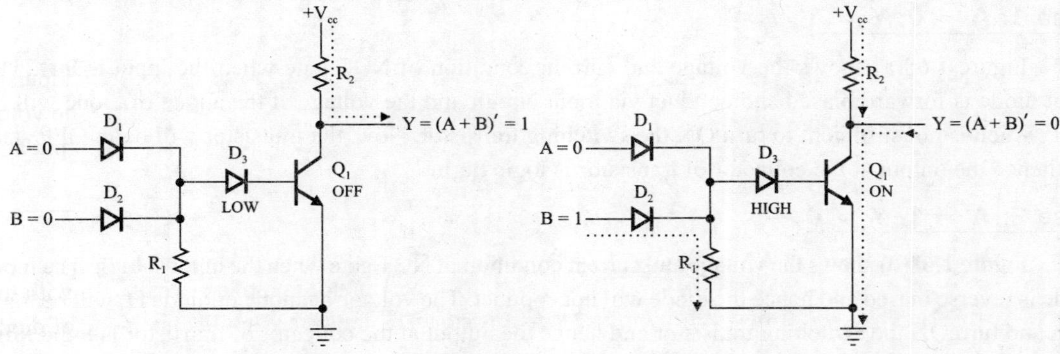

***Fig. a:** Case 1, A = B = 0, Y = 1.*

***Fig. b:** Case 2, A = 0, B = 1, Y = 0.*

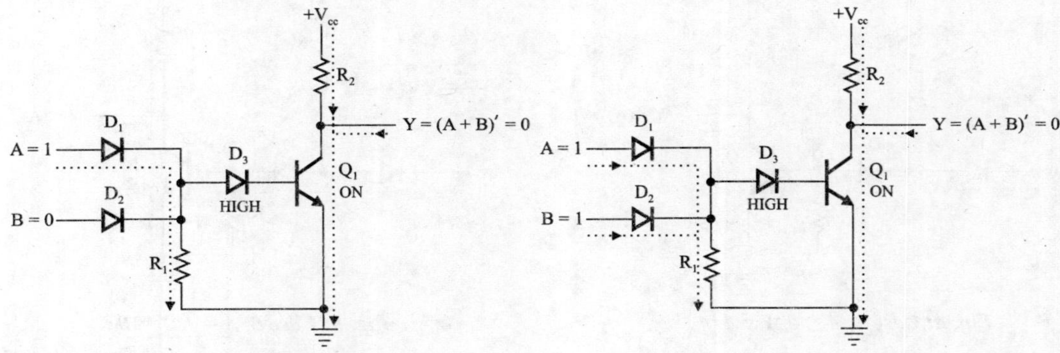

***Fig. c:** Case 1, A = 1, B = 0, Y = 0.*

***Fig. d:** Case 2, A = B = 1, Y = 0.*

***Fig. 1.66:** Four cases of operation of DTL NOR gate.*

Case 2 and 3: A = 0, B = 1, Y = 1 and A = 1, B = 0, Y = 1

In these two cases one of the input diodes is forward biased and conduct via input circuit and the voltage across R_3 is sufficient to turn ON the switching transistor. Now, the transistor will be in ON state and hence the output at the collector of transistor is logic **low**. Therefore, the working of NOR gate in these two cases will be similar to that of case 4.

1.11.3 DTL NOT Gate

A DTL NOT gate is shown in Fig. 1.67. The input circuit has two diodes with anode connected to V_{cc}. The input is applied to cathode of diode. The output is obtained from the collector of switching transistor.

From the truth table of NOT gate we get the following two cases of inputs.

Case 1: A = 0, Y = 1

Case 2: A = 1, Y = 0

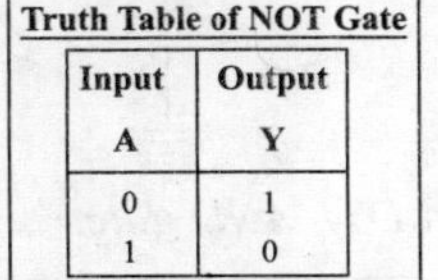

Truth Table of NOT Gate

Input	Output
A	Y
0	1
1	0

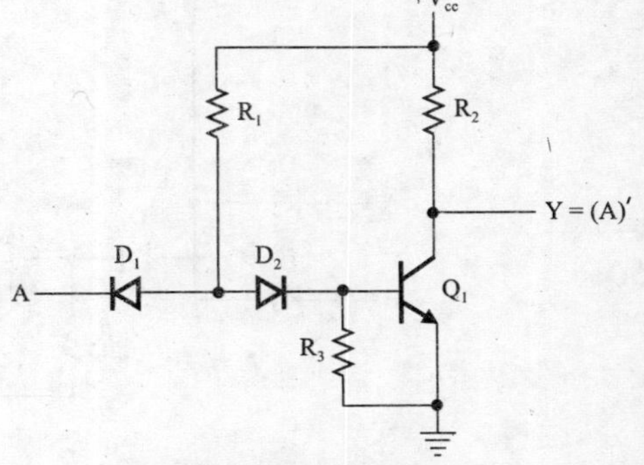

***Fig. 1.67:** DTL NOT gate.*

Case 1: A = 0, Y = 1

Figure 1.68(a) shows the voltage and current condition of NOT gate when the input is **low**. The input diode is forward biased and conduct via input circuit and the voltage at the anode of diode will be 0.7 *V* which is not sufficient to turn ON the switching transistor. Now, the transistor will be in OFF state and hence the output at the collector of transistor is logic **high**.

Case 2: A = 1, Y = 0

Figure 1.68(d) shows the voltage and current condition of NOT gate when the input is **high**. The input diode is reverse biased and hence the diode will not conduct. The voltage at anode of diode D_2 will be logic **high** and turn ON the switching transistor and hence the output at the collector of transistor is logic **low**.

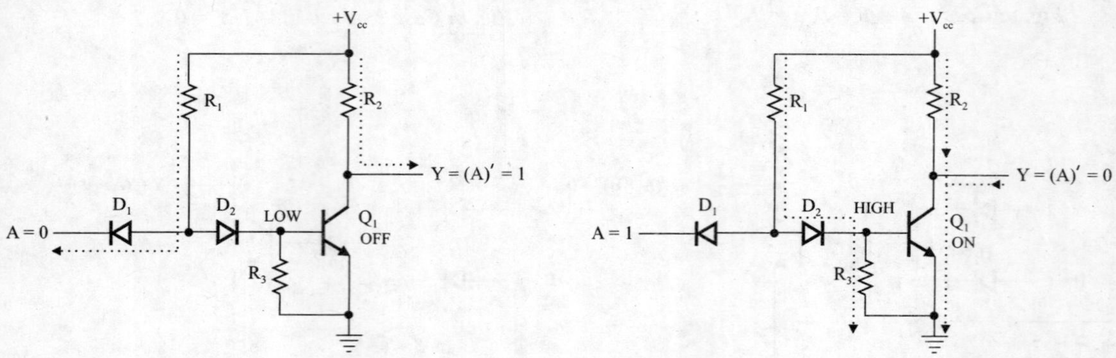

Fig. a: *Case 1, A = 0, Y = 1.* ***Fig. b:*** *Case 2, A = 1, Y = 0.*

Fig. 1.68: *Two cases of operation of DTL NOT gate.*

1.11.4 DTL AND Gate

The AND logic can be obtained by inverting the output of NAND logic. Therefore, the DTL AND gate is obtained by connecting a NOT gate at the output of NAND gate as shown in Fig. 1.69.

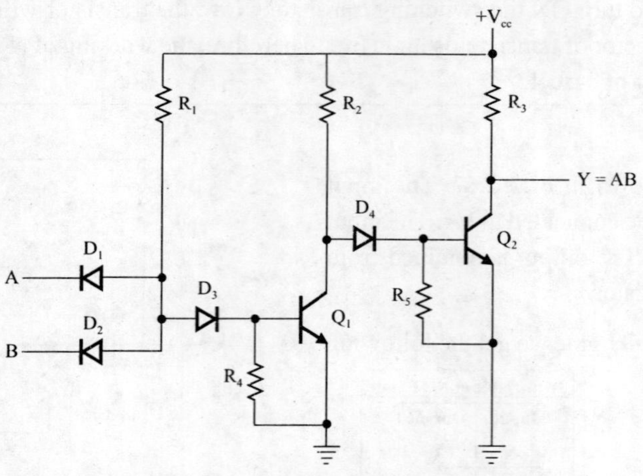

Truth Table of AND Gate

Inputs		Output
A	B	Y
0	0	0
0	1	0
1	0	0
1	1	1

Fig. 1.69: *Two input DTL AND gate.*

1.11.5 DTL OR Gate

The OR logic can be obtained by inverting the output of NOR logic. Therefore, the DTL OR gate is obtained by connecting a NOT gate at the output of NOR gate as shown in Fig. 1.70.

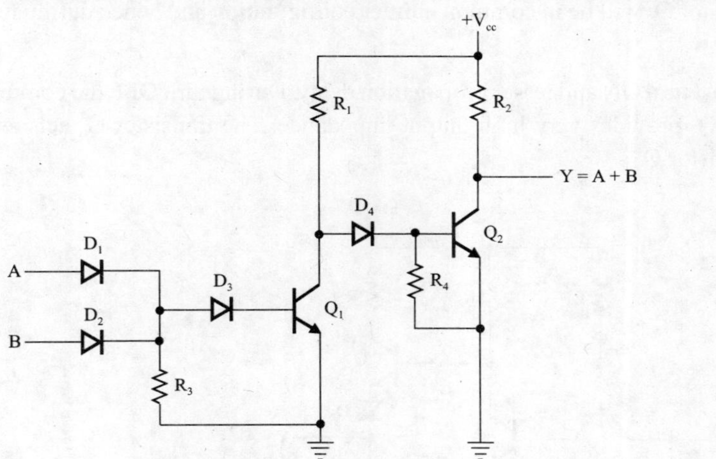

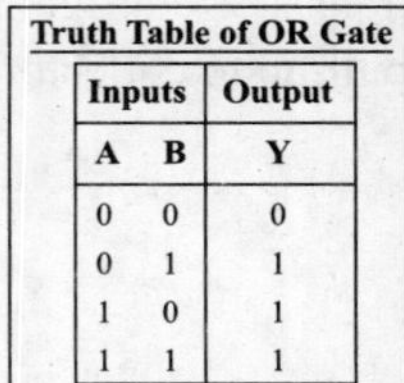

Truth Table of OR Gate

Inputs		Output
A	B	Y
0	0	0
0	1	1
1	0	1
1	1	1

Fig. 1.70: *Two input DTL OR gate.*

1.12 TTL (Transistor-Transistor Logic)

Transistor-transistor logic (TTL) is a class of digital circuits built from bipolar junction transistors (BJT) and resistors. In TTL (Transistor-Transistor Logic) family, input network is made of transistors and switching devices are also transistors and so the name TTL.

In TTL family, the inputs are applied through emitters of a multi-emitter transistor. As every emitter is nothing but a diode, this logic eliminates the use of all diodes. This is the major advantage because transistors ON and OFF much rapidly than a diode and so switching time will be faster.

TTL family of devices are fastest of saturated logic family and so, widely used bipolar logic family. The basic circuit in TTL family is NAND gate. Transistors operate in cut-off (OFF state) or saturation (ON state). In TTL family the logic **low** is 0 *V* to 0.8 *V* and logic **high** is 2 *V* to 5 *V*. In TTL logic devices, there are three different types of outputs namely, totem-pole, open-collector and tri-state output.

1.12.1 Types of Output in TTL Devices

The TTL devices are provided with different types of output circuit configurations. The different types of outputs in TTL devices are,

1. Totem-pole output
2. Open-collector output
3. Three-state (or Tri-state) output

The three output circuit configurations are explained with TTL NOT gate.

Totem-Pole Output

(AU, Nov/Dec'18, 13 Marks)

A TTL NOT gate with totem-pole output is shown in Fig. 1.71. In this configuration the output circuit has two transistors Q_3 and Q_4 such that one sit over the other and hence the name totem-pole. In this arrangement, the transistor Q_3 will be in common emitter configuration and hence during turn ON the output impedance will be **low**.

This will provide a fast turn ON and lesser propagation delay. During turn OFF the common emitter transistor configuration of Q_3 provides very high output impedance. The transistor Q_4 acts as an active pull-up to collector of transistor Q_3.

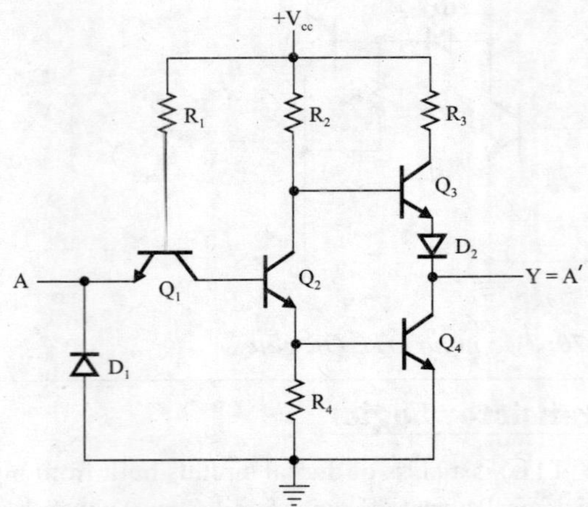

Fig. 1.71: TTL NOT gate with totem-pole output.

Open-Collector Output

A TTL NOT gate with open-collector output is shown in Fig. 1.72. In this configuration the collector of transistor Q_3 is left open and the output is obtained from this open collector. In order to close the output circuit, an external resistor called pull-up resistor should be connected between open collector and power supply V_{cc}.

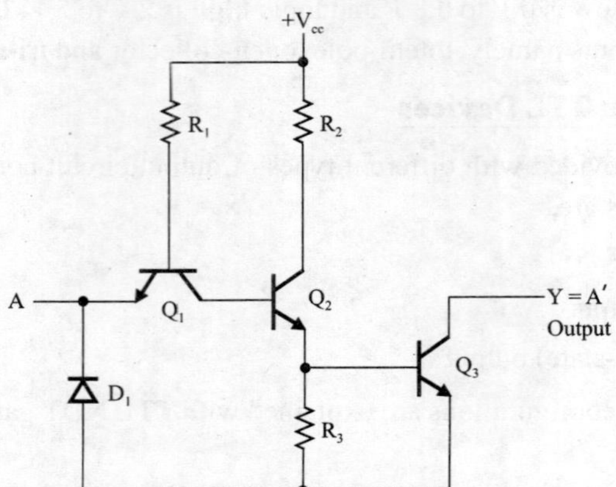

Fig. 1.72: TTL NOT gate with open-collector output.

This will provide flexibility in altering the value of current sourcing and sinking which depends on the value of external resistance. But maximum current is subject to current rating of output transistor. In open-collector configuration, a high value of pull-up resistance will increase the propagation delay. Here, the collector of output transistor is connected to power supply through a resistor and so this connection is called passive pull-up.

Tri-State Output

The tri-state devices will have third state called high impedance state apart from low and high state. The high impedance state is basically electrical open circuit condition so that the device will be isolated when driven to high impedance state. The high impedance is achieved by providing an enable input to TTL device. The totem-pole NOT gate is modified as shown in Fig. 1.73 to convert it to a tri-state device.

Here, when enable input = 1, there is no effect on input and output transistors. Hence, the device will be allowed to accept input and perform logical operation.

When enable input = 0, both the output transistors in totem-pole configuration are turned OFF and hence a floating output is created which is a physical connection with very high impedance. The output will behave as an open circuit. This arrangement helps in connecting the outputs in parallel, without interacting with one another.

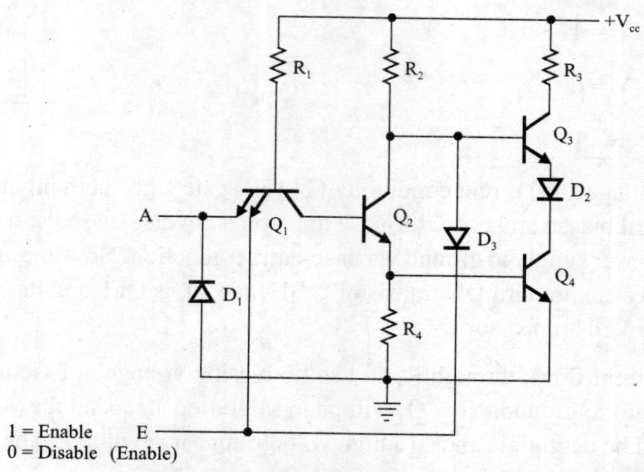

Fig. 1.73: TTL NOT gate with tri-state output.

1.12.2 TTL NAND Gate

(AU, Apr/May'19, 7 Marks)

A two input NAND gate is shown in Fig. 1.74. Transistor Q_1 has two emitters to accept two inputs. The transistor Q_2 is called phase splitter. Transistors Q_3 and Q_4 are totem-pole output transistors. Diodes D_1 and D_2 are provided to bypass input negative spikes to ground. Diode D_3 is provided to prevent simultaneous conduction of Q_3 and Q_4.

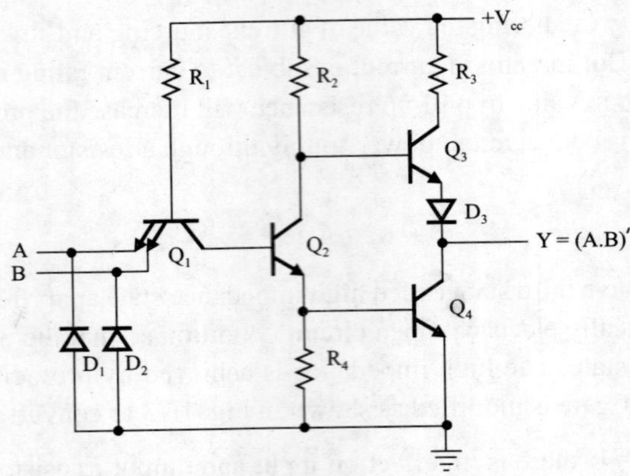

Fig. 1.74: Two input TTL NAND gate.

From the truth table of NAND gate, we get the following four cases of inputs.

Case 1: A = 0, B = 0, Y = 1

Case 2: A = 0, B = 1, Y = 1

Case 3: A = 1, B = 0, Y = 1

Case 4: A = 1, B = 1, Y = 0

Truth Table of NAND Gate

Inputs		Output
A	B	Y
0	0	1
0	1	1
1	0	1
1	1	0

Case 1: A = 0, B = 0, Y = 1

Figure 1.75(a) shows voltage and current condition of NAND gate when both inputs are **low**. Base-emitter junction of Q_1 is forward biased and collector-base junction is reverse biased and so Q_1 turns ON. Current flows from collector power supply to ground via base-emitter junction. Now, the voltage at base of Q_1 will be 0.7 *V* and this voltage cannot turn ON transistor Q_2. Hence, Q_2 is OFF and its collector voltage will be 5 *V* (V_{cc}) which will turn ON transistor Q_3.

Since Q_2 is OFF, no current flows through R_3 and so no biasing voltage at base of Q_4 and hence, Q_4 is OFF and act as open circuit to ground. Now Q_3 will be in saturation state and the voltage across the collector-emitter terminals will be negligible and so a high voltage appears in output. The collector power supply will source current to load at output through Q_3.

Case 4: A = 1, B = 1, Y = 0

Figure 1.75(b) shows voltage and current condition of NAND gate when both inputs are **high**. Base-emitter junction of Q_1 is reverse biased and collector-base junction is forward biased and so Q_1 turns OFF. Current cannot flow through base-emitter junction of Q_1. Now, the voltage at base of Q_1 will be 5 *V* (V_{cc}) and this voltage will turn ON transistor Q_2. Hence, Q_2 is ON and its collector voltage will be V_{sat} which is negligible which will turn OFF transistor Q_3. Since, Q_3 is OFF it will act as open circuit to collector supply.

Since Q_2 is ON, current flows through R_3 and the voltage drop across R_3 will turn ON transistor Q_4 and so Q_4 act as short circuit to ground and so a **low** will appear on the output. Also, the load at output is shorted to ground and so current will sink to output.

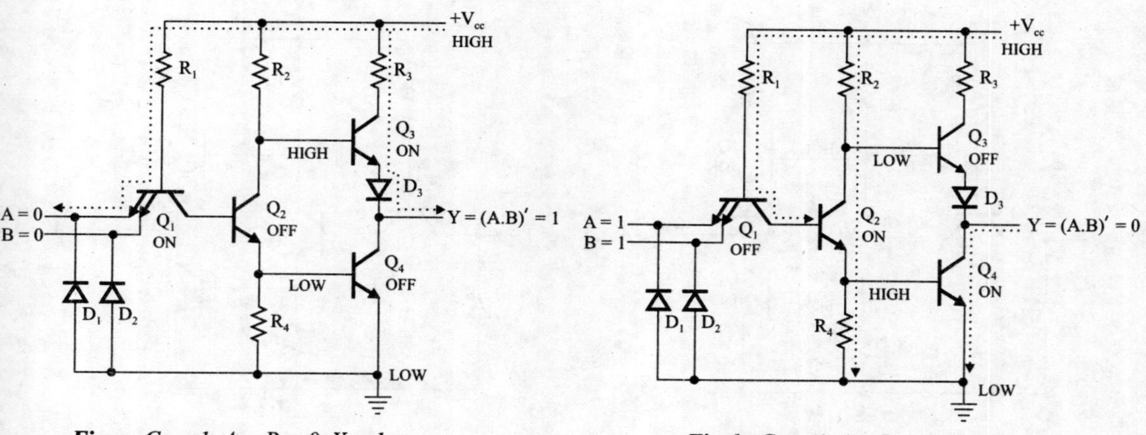

Fig. a: *Case 1, A = B = 0, Y = 1.*

Fig. b: *Case 4, A = B = 1, Y = 0.*

Fig. 1.75: *Two cases of operation of TTL NAND gate.*

Case 2 and 3: A = 0, B = 1, Y = 1 and A = 1, B = 0, Y = 1

In these two cases one of the emitters of transistor Q_1 is **high**, which will turn ON Q_1 and turn OFF Q_2. Therefore, the working of NAND gate in these two cases will be similar to case 1. The voltage and current condition of NAND gate when any one of the input is **high** is similar to Fig. 1.75(a).

1.12.3 TTL NOT Gate

From the truth table of NAND gate, it can be observed that when the inputs are same then the output is complemented. Therefore, when the input emitters of NAND gate are shorted it will behave as NOT gate, which is equivalent to single input NAND gate as shown in Fig. 1.76. The voltage and current conditions of NOT gate during two cases of input are shown in Figs. 1.77(a) and 1.77(b).

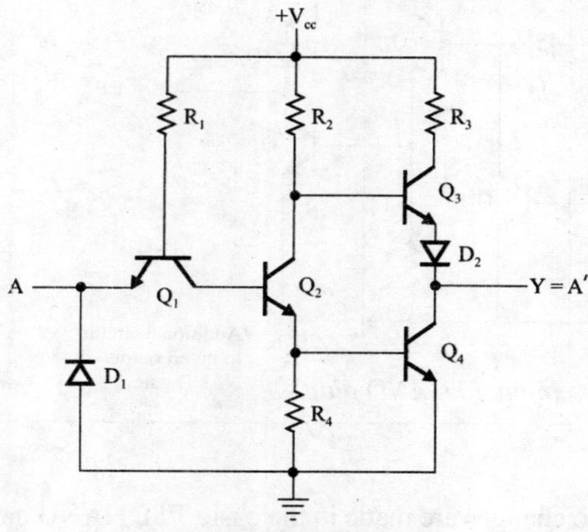

Truth Table of NOT Gate

Input	Output
A	Y
0	1
1	0

Fig. 1.76: *TTL NOT gate.*

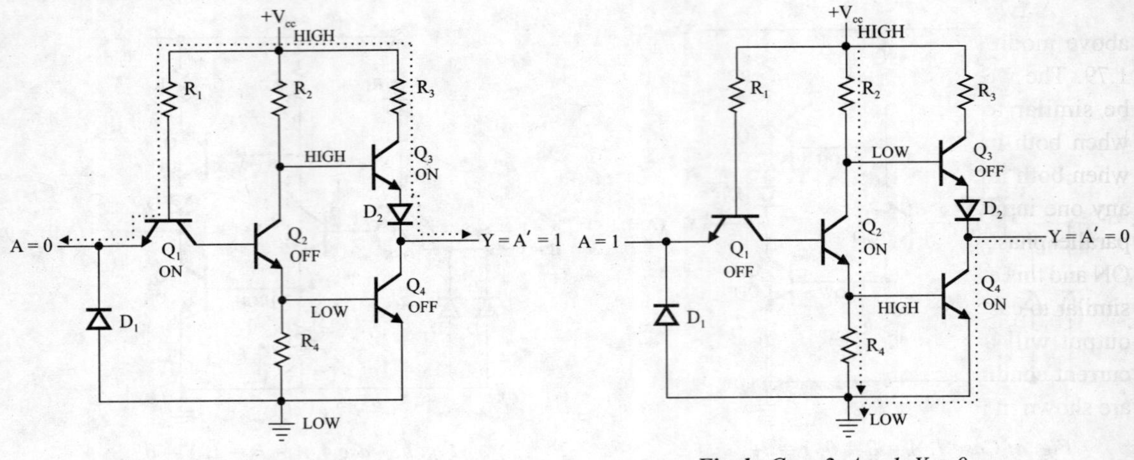

Fig. a: Case 1, A = 0, Y = 1. *Fig. b: Case 2, A = 1, Y = 0.*

Fig. 1.77: Two cases of operation of TTL NOT gate.

1.12.4 TTL AND Gate

The AND logic can be obtained by inverting the NAND logic. On observing the circuit of NAND gate, the NAND logic actual output is available at the collector of Q_2 and so an additional transistor Q_4 can be inserted between Q_2 and totem-pole transistors as shown Fig. 1.78. The transistor Q_4 will invert the output of Q_2 and send to totem pole output transistors. The diode D_3 is provided to prevent reverse current flow and the transistor Q_3 is provided to ensure logic **low** at the base of Q_4 during ON state of Q_2.

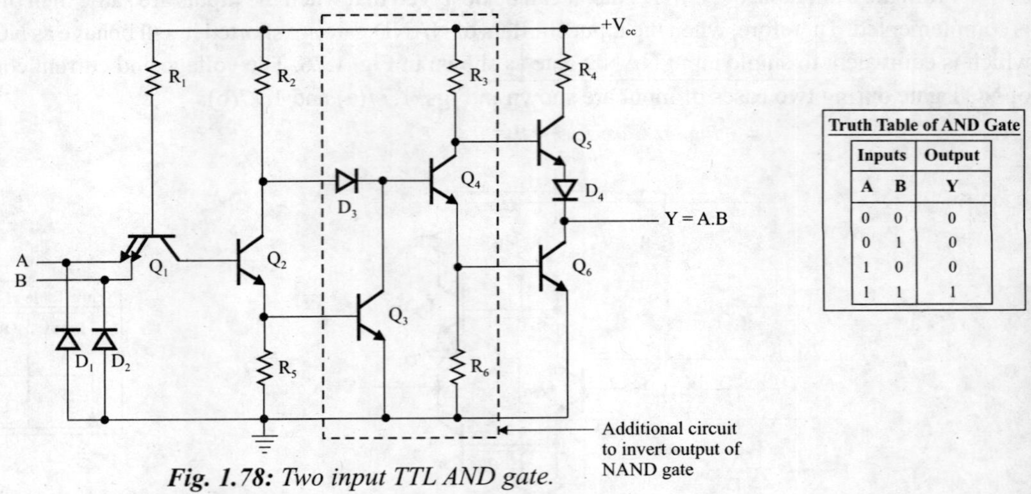

Truth Table of AND Gate

Inputs		Output
A	B	Y
0	0	0
0	1	0
1	0	0
1	1	1

Fig. 1.78: Two input TTL AND gate.

1.12.5 TTL NOR Gate

For NOR logical operation, two changes are made in the basic TTL NAND gate. The multi-emitter transistor at the input is replaced by individual transistors. The phase splitter Q_2 is replaced by parallel transistors.

A two input NOR gate with above modifications is shown in Fig. 1.79. The working of this circuit will be similar to case 1 of NAND gate when both inputs are **low** and case 4 when both inputs are **high**. But, when any one input is **high**, then one of the parallel phase splitter transistors will be ON and this ensures the circuit working similar to case 4 of NAND gate and so output will be **low**. The voltage and current conditions during case 1 and 2 are shown in Figs. 1.80(a) and 1.80(b).

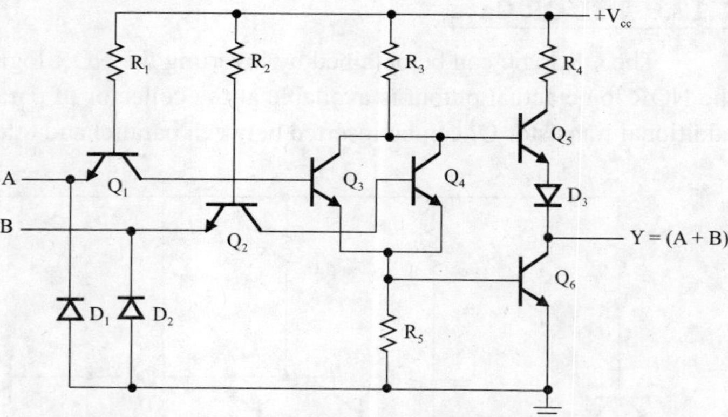

Fig. 1.79: *Two input TTL NOR gate.*

Truth Table of NOR Gate

Inputs		Output
A	**B**	**Y**
0	0	1
0	1	0
1	0	0
1	1	0

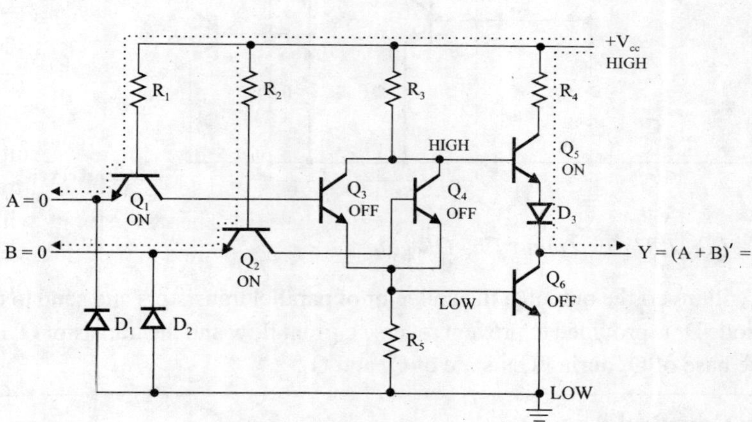

Fig. a: *Case 1, A = B = 0, Y = 1.*

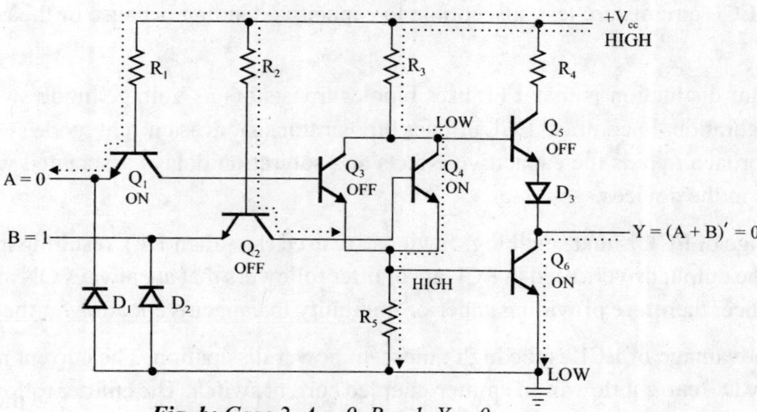

Fig. b: *Case 2, A = 0, B = 1, Y = 0.*

Fig. 1.80: *Two cases of operation of TTL NOR gate.*

1.12.6 TTL OR Gate

The OR logic can be obtained by inverting the NOR logic. On observing the circuit of NOR gate, the NOR logic actual output is available at the collector of parallel combination of Q_3 and Q_4 and so an additional transistor Q_6 can be inserted between parallel and totem-pole transistors as shown Fig. 1.81.

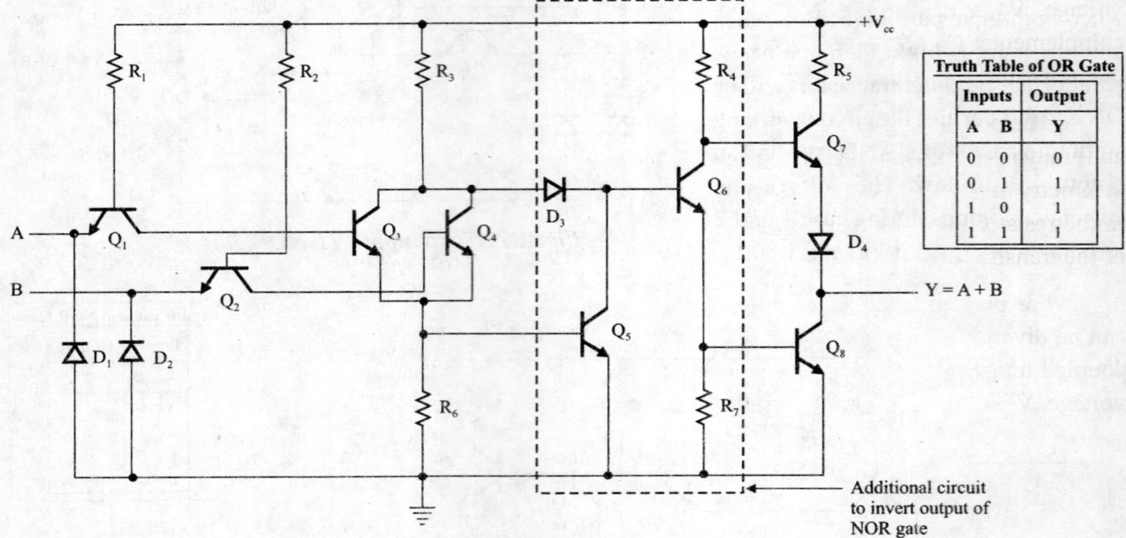

Truth Table of OR Gate

Inputs		Output
A	B	Y
0	0	0
0	1	1
1	0	1
1	1	1

Fig. 1.81: Two input TTL OR gate.

The transistor Q_6 will invert the output at the collector of parallel transistors and send to totem-pole output transistors. The diode D_3 is provided to prevent reverse current flow and the transistor Q_5 is provided to ensure logic low at the base of Q_6 during ON state of Q_3 and Q_4.

1.13 ECL (Emitter Coupled Logic)

Emitter-coupled logic (ECL) circuits are the fastest available bipolar logic gates. When compared to other logic families, ECL circuits are less susceptible to capacitive loading because of their extremely low output impedance.

Here the important distinction is that TTL uses bipolar transistors as voltage mode switches that swing from cut-off to saturation. In contrast, ECL utilizes bipolar transistors as current mode switches that never saturate. This approach avoids the capacitive effects and saturation delays associated with excess minority carrier storage in the devices.

A second advantage of ECL is that small logic swings are used (less than 1 *V*), resulting in short rise and fall times. Finally, the output drivers used in ECL are emitter followers that are always ON and achieve minimal output impedance, therefore providing superior immunity to capacitive loading at the output.

The primary disadvantage of ECL is the high quiescent power dissipation. The current mode logic concept requires a steady DC current flow in an emitter-coupled current switch. The emitter follower output drivers also require large and steady emitter currents.

1.13.1 Emitter Coupled Current Switch

The basic circuit in an ECL gate is the emitter-coupled current switch as shown in Fig. 1.82. Here input is applied to base of transistor Q_1 and a constant reference voltage, V_{ref} is applied to base of transistor, Q_2. Two complementary outputs are available at the collector of the transistors.

The collector is biased by a positive voltage, V_{CC} and emitter is biased by a negative voltage, $-V_{EE}$ and due to symmetry of the switch, a nearly constant current I_E flows in the resistor R_E. Outputs are obtained at the collectors of the transistors.

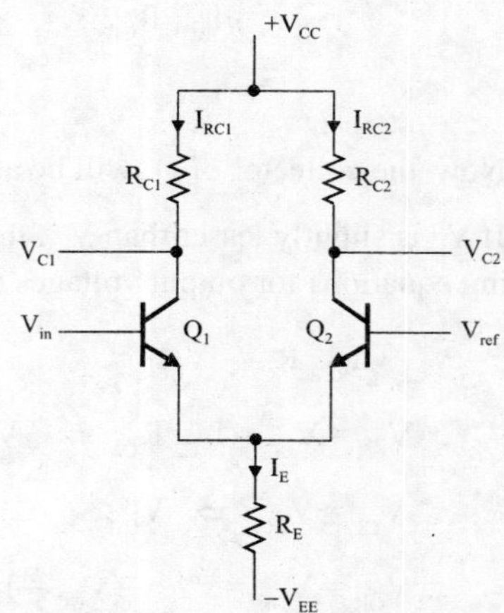

Fig. 1.82: Basic current switch of ECL family.

For performing logical operations, the current I_E can be diverted entirely through one of the two emitter-coupled transistors, depending on the value of the input voltage. V_{in}. Under this situation, the two outputs will be complementary in nature.

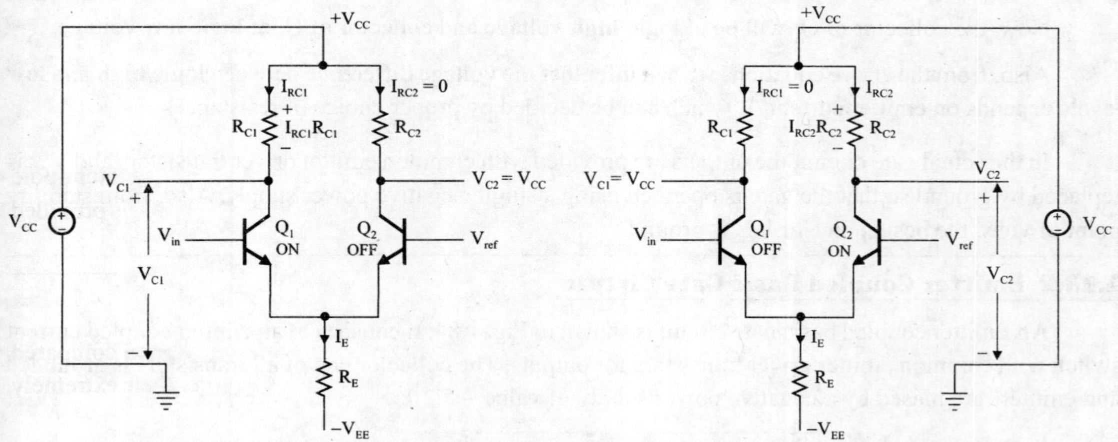

Fig. a: Voltage and current when $V_{in} > V_{ref}$. *Fig. b: Voltage and current when $V_{in} < V_{ref}$.*

Fig. 1.83: Voltage and current in basic current switch of ECL family.

If V_{in} is slightly greater than V_{ref}, then the current I_E flows through Q_1 as shown in Fig. 1.83(a). The following equations for output voltages are obtained using Kirchhoff's Voltage Law (KVL).

$$V_{C1} + I_{RC1}\, R_{C1} = V_{CC}$$

$$\therefore\ V_{C1} = V_{CC} - I_{RC1}\, R_{C1} \;\Rightarrow\; V_{OL}$$

$$V_{C2} = V_{CC} \;\Rightarrow\; V_{OH}$$

$$\therefore V_{OH} - V_{OL} = V_{CC} - (V_{CC} - I_{RC1} R_{C1})$$

$$= I_{RC1} R_{C1}$$

$$= I_E R_{C1} \qquad \boxed{I_E \approx I_{RC1}}$$

Now, the collector of Q_1 will be at logic **low** voltage and collector of Q_2 at logic **high** voltage.

If V_{in} is slightly lesser than V_{ref}, then the current I_E flows through Q_2 as shown in Fig. 1.83(b). The following equations for output voltages are obtained using Kirchhoff's Voltage Law (KVL).

$$V_{C2} + I_{RC2} R_{C2} = V_{CC}$$

$$\therefore V_{C2} = V_{CC} - I_{RC2} R_{C2} \Rightarrow V_{OL}$$

$$V_{C1} = V_{CC} \Rightarrow V_{OH}$$

$$\therefore V_{OH} - V_{OL} = V_{CC} - (V_{CC} - I_{RC2} R_{C2})$$

$$= I_{RC2} R_{C2}$$

$$= I_E R_{C2} \qquad \boxed{I_E \approx I_{RC2}}$$

Now, the collector of Q_1 will be at logic **high** voltage and collector of Q_2 at logic **low** voltage.

Also, from the above equations we can infer that the voltage difference between logic **high** and **low** levels depends on emitter current, I_E which can be decided by proper choice of resistances.

In the actual gate circuit, the outputs are provided with common emitter driver transistors and V_{CC} is replaced by ground so that the gate is operated using a single negative power supply. Also, from stability point of view, the best choice of V_{CC} is ground.

1.13.2 Emitter Coupled Basic Gate Circuit

An Emitter coupled basic gate circuit is shown in Fig. 1.84. It consists of an emitter coupled current switch with common emitter driver transistors for outputs. The collector end of all transistors is grounded and emitters are biased by a negative power supply of value – 5.2 *V*.

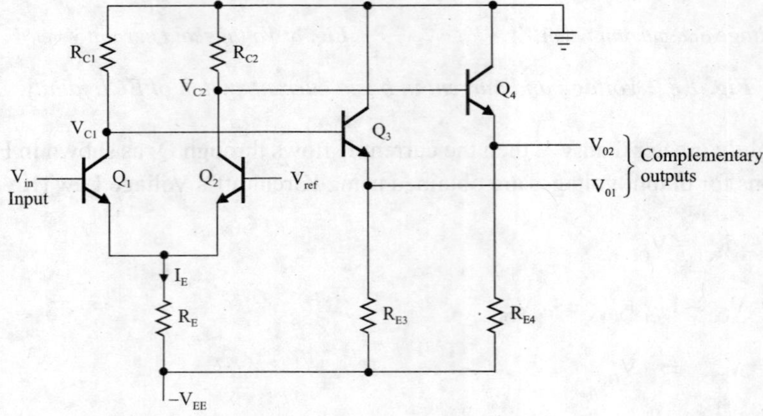

Fig. 1.84: Emitter coupled basic gate circuit.

Emitter coupled transistors operate in active region for both OFF and ON state with a slight difference between logic voltage levels. In ECL family the logic 0 or **low** is –1.5 to –1.7 *V* and logic 1 or **high** is – 0.6 *V* to – 0.9 *V*. The voltages and currents in the ECL basic gate circuit when the inputs are logic **low** and **high** are shown respectively, in Figs. 1.85 and 1.86.

Analysis of ECL basic Gate Circuit when $V_{in} < V_{ref}$

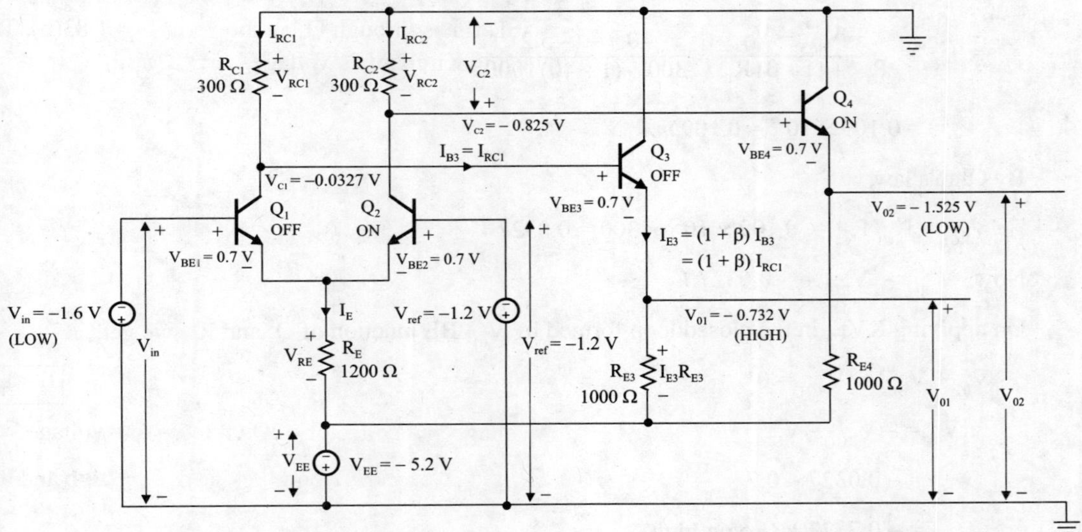

Fig. 1.85: Voltages and currents in emitter coupled basic gate when $V_{in} < V_{ref}$.

In the ECL gate circuit shown in Fig. 1.85, the logic **high** and **low** voltages can be computed using Kirchoff's voltage law and Ohm's law as shown below:

On applying Kirchoff's voltage law (KVL) in the closed loop formed by V_{ref}, BE junction of Q_2, R_E and V_{EE} we can write,

$$V_{ref} = V_{BE2} + V_{RE} + V_{EE}$$

$$\therefore V_{RE} = V_{ref} - V_{BE2} - V_{EE}$$

$$= -1.2 - 0.7 - (-5.2) = -1.2 - 0.7 + 5.2 = 3.3\ V$$

By Ohm's law we can write,

$$V_{RE} = I_E R_E \Rightarrow I_E = \frac{V_{RE}}{R_E} = \frac{3.3}{1200} = 2.75 \times 10^{-3} = 2.75\ mA$$

$$V_{RC2} = I_{RC2} R_{C2} = I_E R_{C2} = 2.75 \times 10^{-3} \times 300 = 0.825\ V \qquad \boxed{I_{RC2} \approx I_E}$$

Now, $V_{C2} = -V_{RC2} = -0.825\ V$

On applying KVL in the closed loop formed by V_{02}, BE junction of Q_4 and V_{C2} we get,

$$V_{02} + V_{BE4} = V_{C2} \Rightarrow V_{02} = V_{C2} - V_{BE4} = -0.825 - 0.7 = -1.525\ V \text{ (Logic } \textbf{low}\text{)}$$

On applying KVL in the closed loop formed by R_{C1}, BE junction of Q_3, R_{E3} and V_{EE} we get,

$$V_{RC1} + V_{BE3} + I_{E3}\, R_{E3} + V_{EE} = 0$$

$$I_{RC1}\, R_{C2} + V_{BE3} + I_{RC1}(1 + \beta)\, R_{E3} + V_{EE} = 0$$

$$I_{RC1}\, R_{C1} + I_{RC1}(1 + \beta)\, R_{E3} = - V_{BE3} - V_{EE}$$

Let, β = 40

$$\therefore\ I_{RC1} = \frac{- V_{BE3} - V_{EE}}{R_{C1} + (1+\beta)\, R_{E3}} = \frac{-0.7 - (-5.2)}{300 + (1+40)\,1000}$$

$$= 0.109 \times 10^{-3} = 0.109\ mA$$

By Ohm's law,

$$V_{RC1} = I_{RC1}\, R_{C1} = 0.109 \times 10^{-3} \times 300 = 0.0327\ V$$

Now, $V_{C1} = - V_{RC1} = - 0.0327\ V$

On applying KVL, in the closed loop formed by V_{01}, BE junction of Q_3 and R_{C1} we get,

$$V_{01} + V_{BE3} + V_{RC1} = 0$$

$$\therefore\ V_{01} = - V_{RC1} - V_{BE3}$$

$$= - 0.0327 - 0.7$$

$$= - 0.7327\ V\ \text{(Logic high)}$$

Analysis of ECL basic Gate Circuit when $V_{in} > V_{ref}$

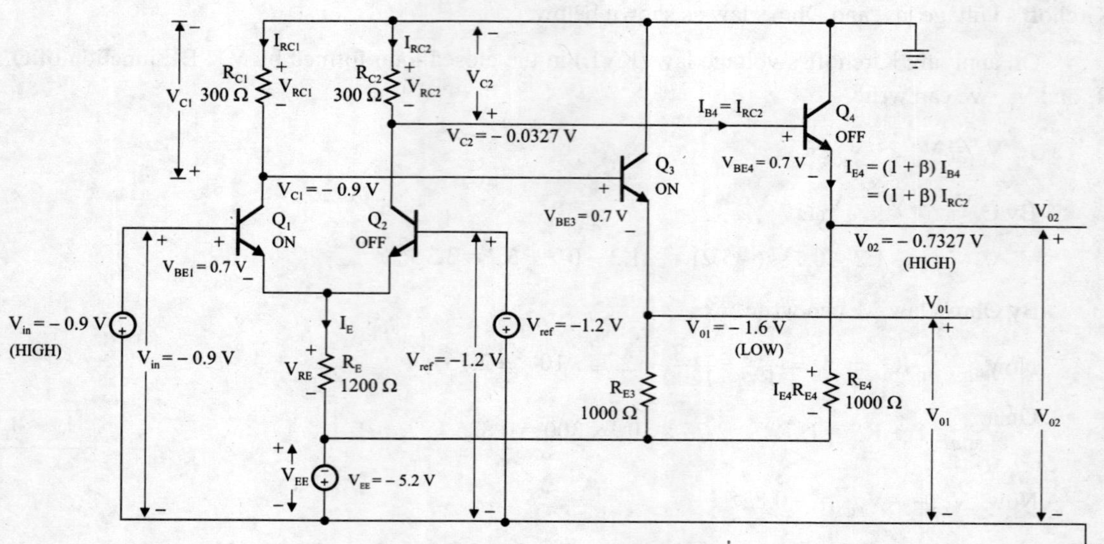

Fig. 1.86: Voltages and currents in emitter coupled basic gate when $V_{in} > V_{ref}$.

In the ECL gate circuit shown in Fig. 1.86, the logic **high** and **low** voltages can be computed using Kirchoff's voltage law and Ohm's law as shown below:

On applying Kirchoff"s voltage law (KVL) in the closed loop formed by V_{in}, BE junction of Q_1, R_E and V_{EE} we can write,

$$V_{in} = V_{BE1} + V_{RE} + V_{EE}$$

$$\therefore V_{RE} = V_{in} - V_{BE1} - V_{EE}$$

$$= -0.9 - 0.7 - (-5.2) = -0.9 - 0.7 + 5.2 = 3.6\ V$$

By Ohm's law we can write,

$$V_{RE} = I_E R_E \Rightarrow I_E = \frac{V_{RE}}{R_E} = \frac{3.6}{1200} = 3 \times 10^{-3} = 3\ mA$$

$$V_{RC1} = I_{RC1} R_{C1} = I_E R_{C1} = 3 \times 10^{-3} \times 300 = 0.9\ V \qquad \boxed{I_{RC1} \approx I_E}$$

Now, $V_{C1} = -V_{RC1} = -0.9\ V$

On applying KVL in the closed loop formed by V_{01}, BE junction of Q_3 and V_{C1} we get,

$V_{01} + V_{BE3} = V_{C1} \Rightarrow V_{01} = V_{C1} - V_{BE3} = -0.9 - 0.7 = -1.6\ V$ (Logic **low**)

On applying KVL in the closed loop formed by R_{C2}, BE junction of Q_4, R_{E4} and V_{EE} we get,

$$V_{RC2} + V_{BE4} + I_{E4} R_{E4} + V_{EE} = 0$$

$$I_{RC2} R_{C2} + V_{BE4} + I_{RC2}(1 + \beta) R_{E4} + V_{EE} = 0$$

$$I_{RC2} R_{C2} + I_{RC2}(1 + \beta) R_{E4} = -V_{BE4} - V_{EE}$$

$$\therefore I_{RC2} = \frac{-V_{BE4} - V_{EE}}{R_{C2} + (1+\beta) R_{E4}} = \frac{-0.7 - (-5.2)}{300 + (1+40)1000} \qquad \boxed{\text{Let, } \beta = 40}$$

$$= 0.109 \times 10^{-3} = 0.109\ mA$$

By Ohm's law,

$$V_{RC2} = I_{RC2} R_{C2} = 0.109 \times 10^{-3} \times 300$$

$$= 0.0327\ V$$

Now, $V_{C2} = -V_{RC2} = -0.0327\ V$

On applying KVL in the closed loop formed by V_{02}, R_{C2} and BE junction of Q_4 we get,

$$V_{02} + V_{BE4} + V_{RC2} = 0$$

$$\therefore V_{02} = -V_{RC2} - V_{BE4}$$

$$= -0.0327 - 0.7$$

$= -0.7327\ V$ (Logic **high**)

1.13.3 ECL NOR Gate

A two input ECL NOR gate can be formed by connecting input transistors in parallel in the basic ECL gate circuit as shown in Fig. 1.87. Transistors Q_2 and Q_3 forms the emitter coupled current switch in which a reference voltage is applied to base of Q_3 and one of the inputs is applied to base of Q_2. An additional input transistor Q_1 is connected in parallel to Q_1. Here, Q_4 is emitter follower output driver transistor and output is obtained from the emitter of Q_4.

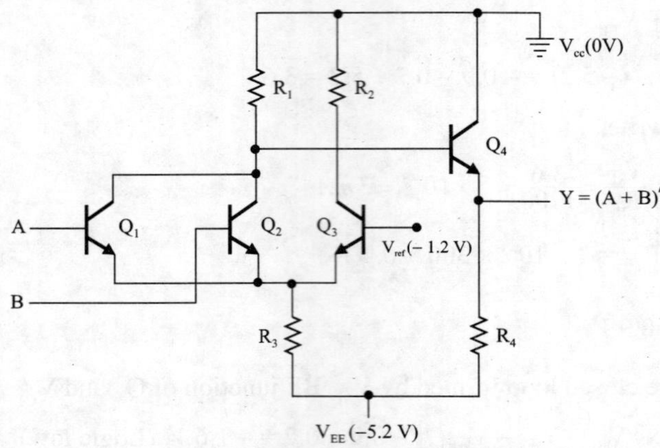

***Fig. 1.87:** Two input ECL NOR gate.*

From the truth table of NOR gate, we get the following four cases of inputs.

Case 1: A = 0, B = 0, Y = 1

Case 2: A = 0, B = 1, Y = 0

Case 3: A = 1, B = 0, Y = 0

Case 4: A = 1, B = 1, Y = 0

Truth Table of NOR Gate

Inputs		Output
A	B	Y
0	0	1
0	1	0
1	0	0
1	1	0

Case 1: A = 0, B = 0, Y = 1

Figure. 1.88(a) shows voltage and current condition of NOR gate when both inputs are **low**. Now, $V_{in} < V_{ref}$ and so forward bias voltage of BE junction of Q_3 is more than Q_2 and so Q_3 turns ON and Q_2 turns OFF. The collector of Q_2 will be at a logic **high** voltage which will turn ON Q_4. Hence a logic **high** voltage appear at the output emitter terminal of Q_4. The collector power supply will source current to load at output through Q_4.

Case 4: A = 1, B = 1, Y = 0

Figure. 1.88(d) shows voltage and current condition of NOR gate when both inputs are **high**. Now, $V_{in} > V_{ref}$ and so forward bias voltage of BE junction of Q_3 is less than Q_2 and so Q_3 turns OFF and Q_2 turns ON. The collector of Q_2 will be at a logic **low** voltage which will turn OFF Q_4. Hence a logic **low** voltage appears at the output emitter terminal of Q_4. The collector power supply will sink current from load at output through emitter resistance R_4.

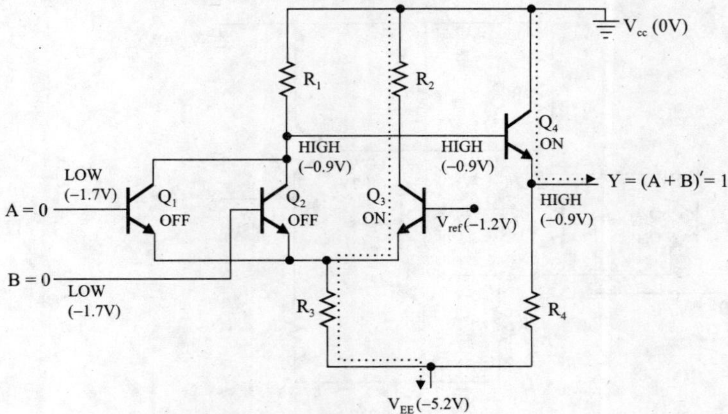

***Fig. a:** Case 1, A = B = 0, Y = 1.*

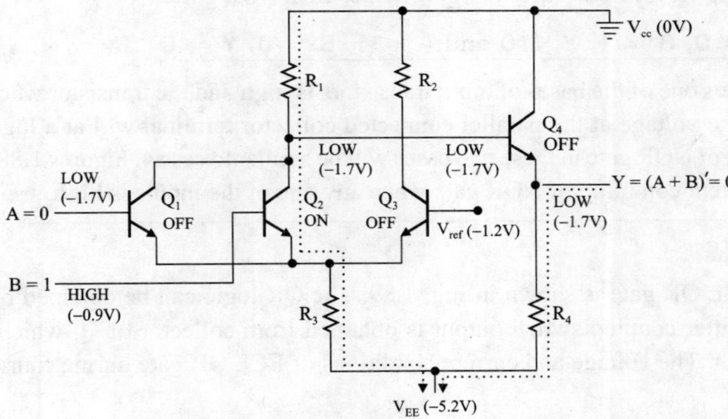

***Fig. b:** Case 2, A = 0, B = 1, Y = 0.*

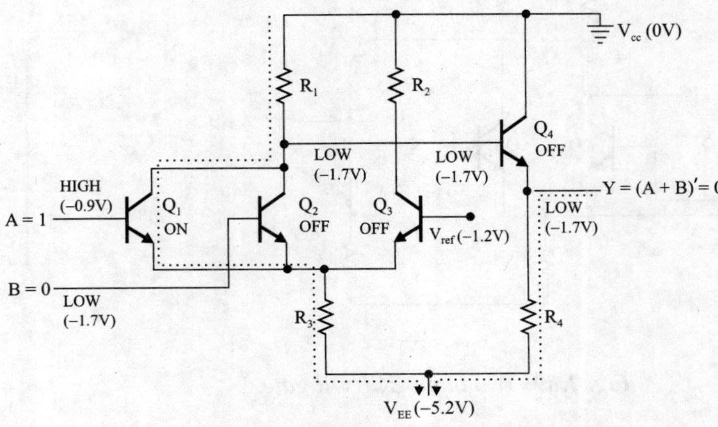

***Fig. c:** Case 3, A = 1, B = 0, Y = 0.*

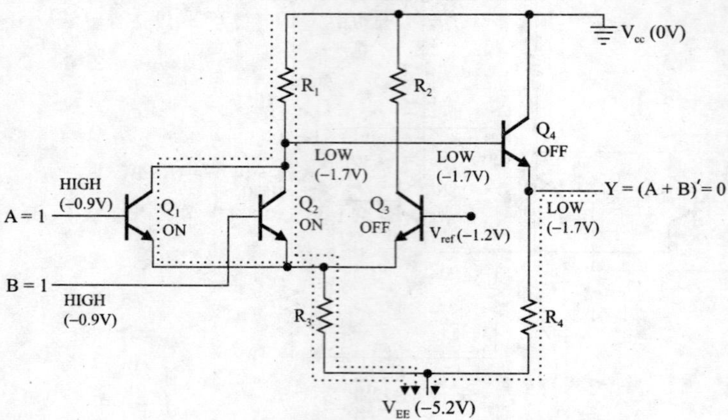

Fig. d: *Case 4, A = B = 1, Y = 0.*

Fig. 1.88: *Four cases of operation of ECL NOR gate.*

Case 2 and 3: A = 0, B = 1, Y = 0 and A = 1, B = 0, Y = 0

In these two cases one of the bases of input transistors is **high** and the transistor whose base is **high** will turn ON so that the voltage at the parallel connected collector terminal will at a logic **low** voltage. Therefore, the working of NOR gate in these two cases will be similar to case 4. Figures 1.88(b) and 1.88(c) shows voltage and current condition of NOR gate when any one of the inputs is **high**.

1.13.4 ECL OR Gate

A two input ECL OR gate is shown in Fig. 1.89. The OR logic can be obtained by inverting the NOR logic and so emitter coupled switch output is obtained from collector of Q_3 which will be invert logic of NOR operation. The voltage and current conditions of ECL OR gate during four cases of inputs is shown in Fig. 1.90.

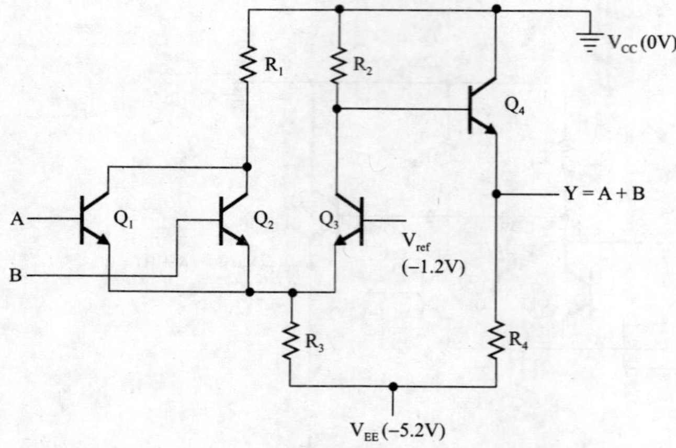

Truth Table of OR Gate

Inputs		Output
A	B	Y
0	0	0
0	1	1
1	0	1
1	1	1

Fig. 1.89: *Two input ECL OR gate.*

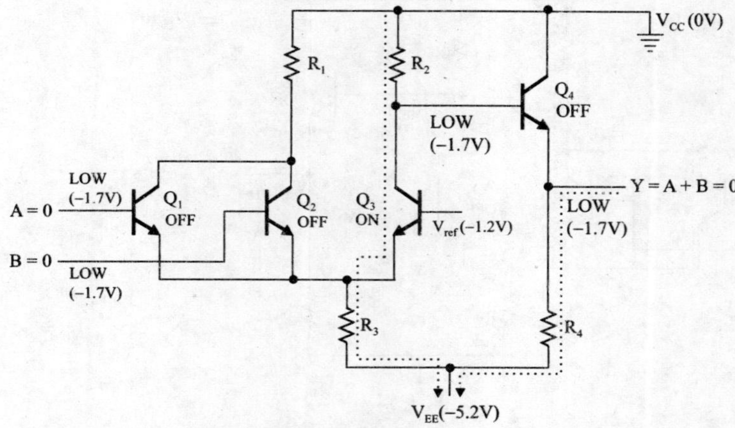

Fig. a: Case 1, A = B = 0, Y = 0.

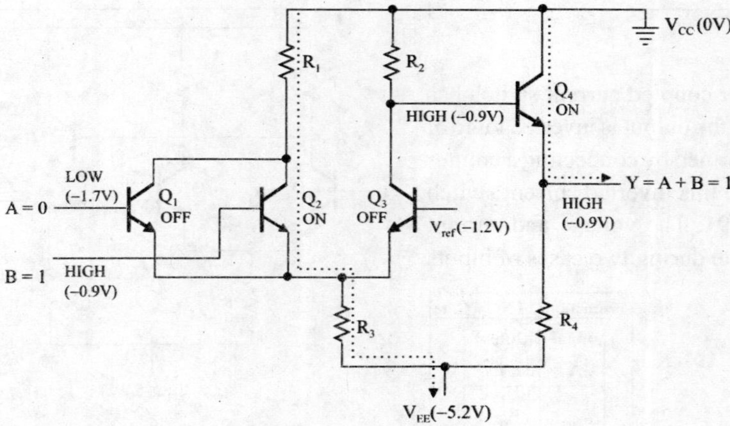

Fig. b: Case 2, A = 0, B = 1, Y = 1.

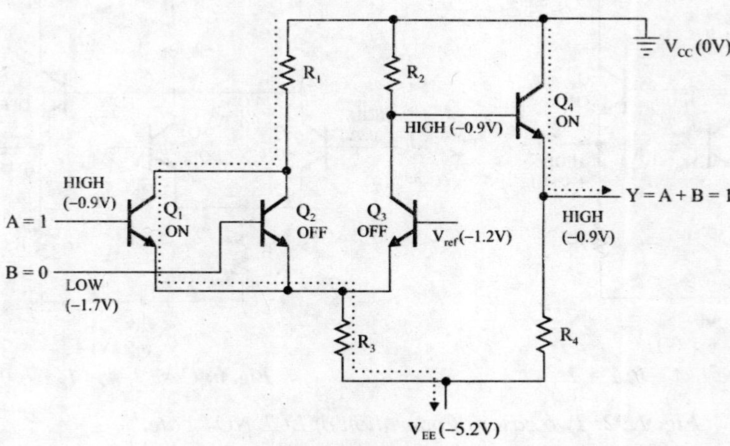

Fig. c: Case 3, A = 1, B = 0, Y = 1.

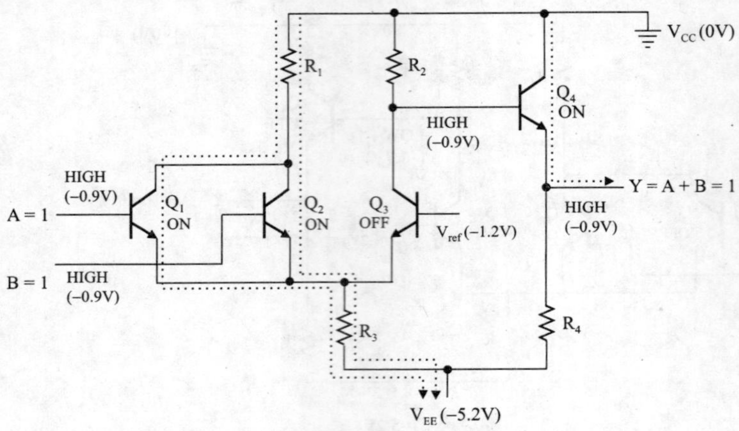

Fig. d: Case 4, A = B = 1, Y = 1.

***Fig. 1.90:** Four cases of operation of ECL OR gate.*

1.13.5 ECL NOT Gate

Basically, the emitter coupled current switch has two outputs in which one of the output is inverted logic of input. Hence NOT gate is formed by connecting a emitter follower driver transistor to this inverted current switch output as shown in Fig. 1.91. The voltage and current conditions of ECL NOT gate during two cases of inputs is shown in Fig. 1.92.

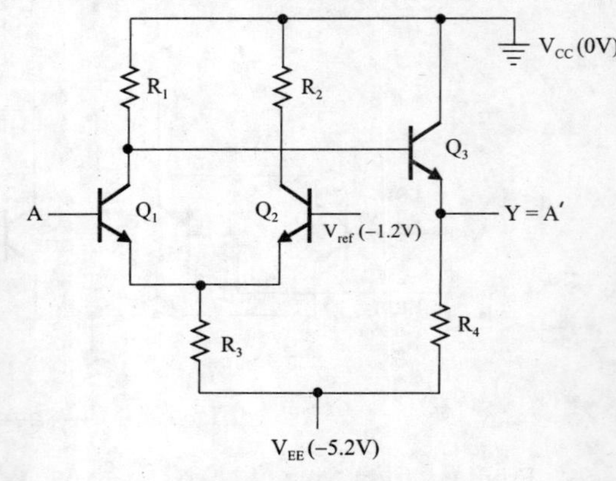

***Fig. 1.91:** ECL NOT gate.*

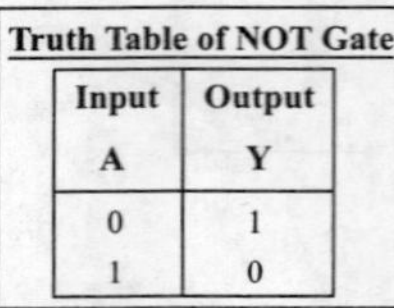

Truth Table of NOT Gate

Input	Output
A	Y
0	1
1	0

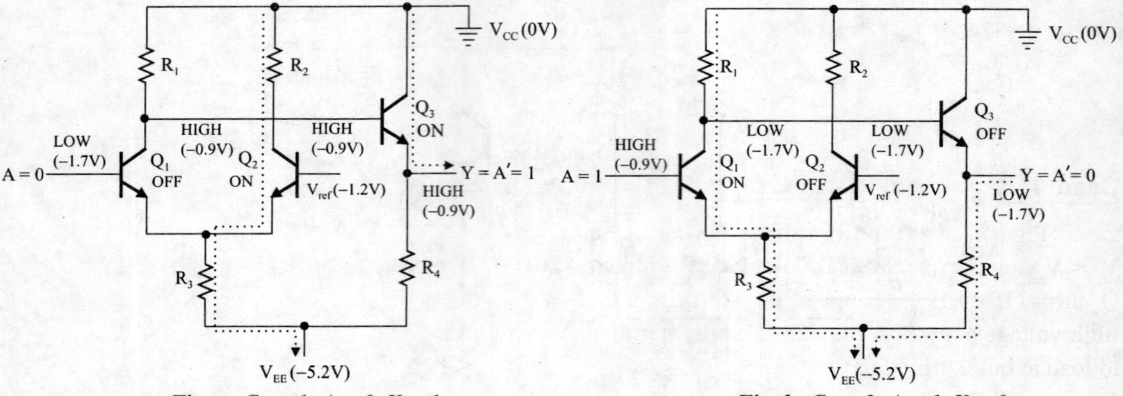

Fig. a: Case 1, A = 0, Y = 1. *Fig. b: Case 2, A = 1, Y = 0.*

***Fig. 1.92:** Two cases of operation of ECL NOT gate.*

1.13.6 ECL NAND Gate

A two input ECL NAND gate can be formed by connecting input transistors in series in the basic ECL gate circuit as shown in Fig. 1.93. Transistors Q_1 and Q_2 forms the emitter coupled current switch in which a reference voltage is applied to base of Q_2 and one of the inputs is applied to base of Q_1. An additional input transistor Q_3 is connected in series with Q_1. The transistor Q_5 is provided to level shift the B input by one base-emitter voltage value.

An additional reference is needed to prevent saturation of Q_2 and Q_3 and so another reference V_{ref2} which is lower by one V_{BE} voltage value than V_{ref1} is added through Q_4. Here, Q_6 is emitter follower output driver transistor and the output is obtained from the emitter of Q_6.

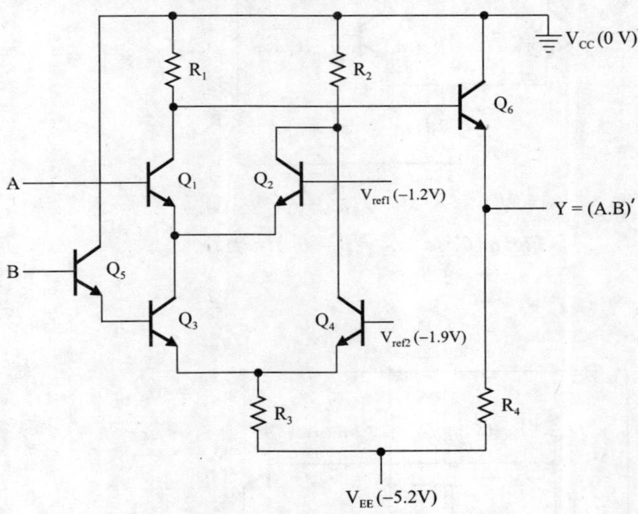

***Fig. 1.93:** Two input ECL NAND gate.*

From the truth table of NAND gate, we get the following four cases of inputs.

Case 1: A = 0, B = 0, Y = 1

Case 2: A = 0, B = 1, Y = 1

Case 3: A = 1, B = 0, Y = 1

Case 4: A = 1, B = 1, Y = 0

Truth Table of NAND Gate

Inputs		Output
A	B	Y
0	0	1
0	1	1
1	0	1
1	1	0

Case 1: A = 0, B = 0, Y = 1

Figure 1.94(a) shows voltage and current condition of NAND gate when both inputs are **low**. Now, $V_{in} < V_{ref}$ and so forward bias voltage of BE junction of Q_2 is more than Q_1 and so Q_2 turns ON and Q_1 and Q_3 turns OFF. The collector of Q_1 will be at a logic **high** voltage which will turn ON Q_6. Hence a logic **high** voltage appears at the output emitter terminal of Q_6. The collector power supply will source current to load at output through Q_6.

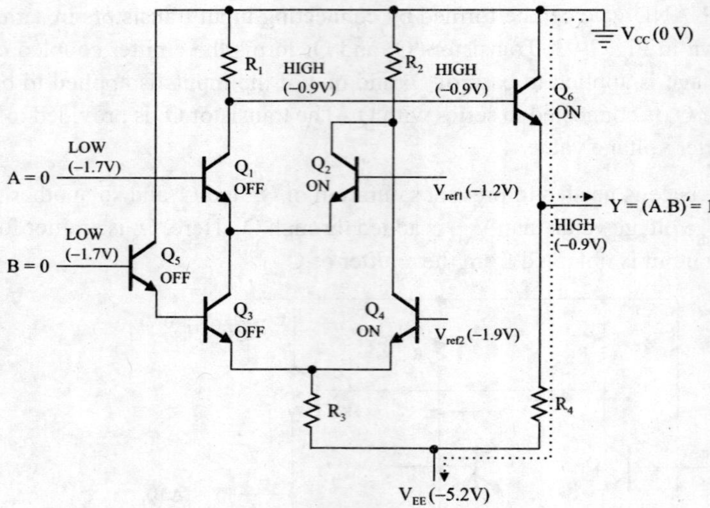

Fig. a: *Case 1, A = B = 0, Y = 1.*

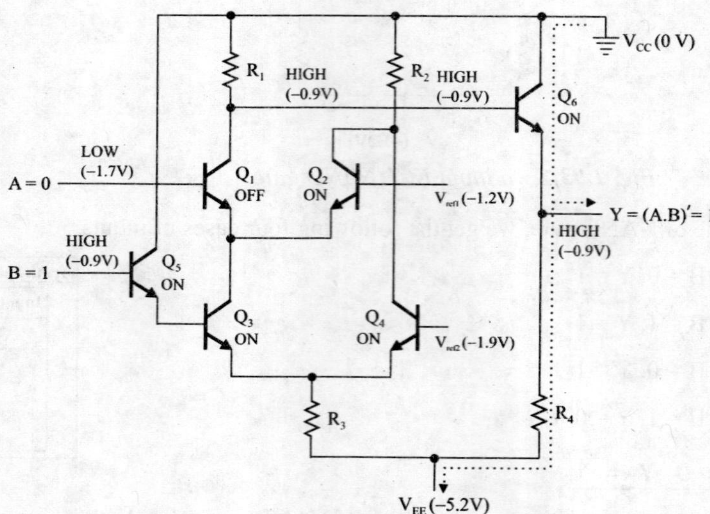

Fig. b: *Case 2, A = 0, B = 1, Y = 1.*

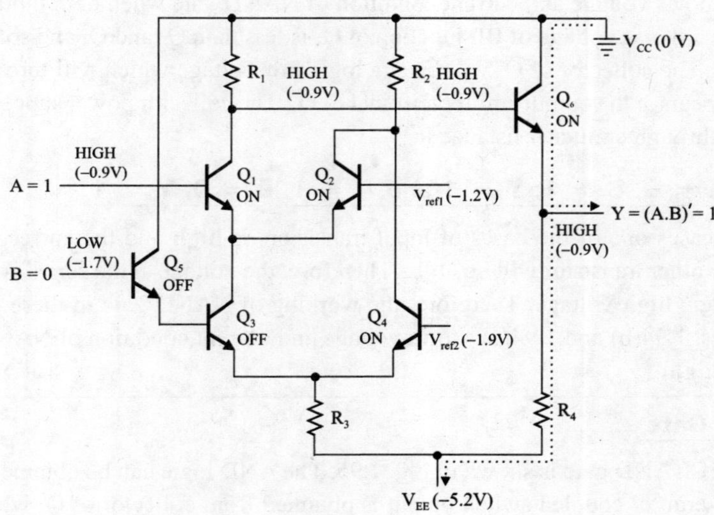

Fig. c: Case 3, A = 1, B = 0, Y = 1.

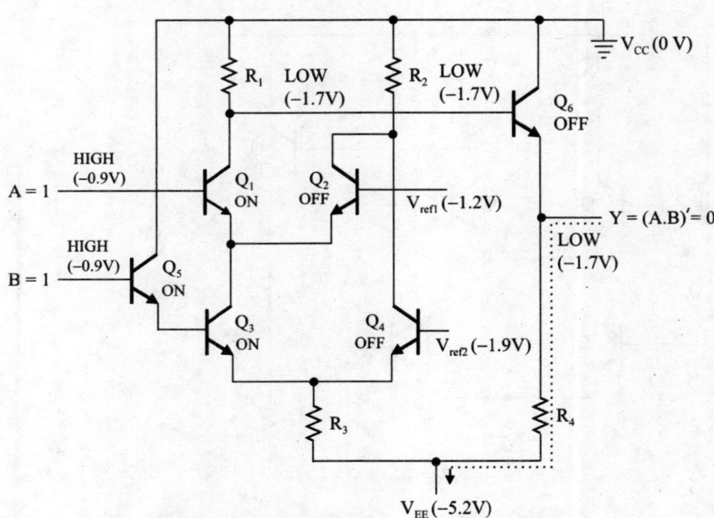

Fig. d: Case 4, A = B = 1, Y = 0.

Fig. 1.94: Four cases of operation of ECL NAND gate.

Case 4: A = 1, B = 1, Y = 0

Fig. 1.94(d) shows voltage and current condition of NAND gate when both inputs are **high**. Now, $V_{in} > V_{ref}$ and so forward bias voltage of BE junction of Q_2 is less than Q_1 and Q_3 and so Q_2 turns OFF and Q_1 and Q_3 turns ON. The collector of Q_1 will be at a logic **low** voltage which will turn OFF Q_6. Hence a logic **low** voltage appears at the output emitter terminal of Q_6. The collector power supply will sink current from load at output through emitter resistance R_4.

Case 2 and 3: A = 0, B = 1, Y = 1 and A = 1, B = 0, Y = 1

In these two cases one of the bases of input transistors is **high** and the corresponding transistor will turn ON and the other transistor will be OFF. Therefore, the voltage at the series connected collector terminal will at a logic **high** voltage. Therefore, the working of NAND gate in these two cases will be similar to case 1 Figs. 1.94(b) and 1.94(c) shows voltage and current condition of NAND gate when any one of the inputs is **high**.

1.13.7 ECL AND Gate

A two input ECL AND gate is shown in Fig. 1.95. The AND logic can be obtained by inverting the NAND logic and so emitter coupled switch output is obtained from collector of Q_2 which will be invert logic of NAND operation. The voltage and current conditions of ECL AND gate during four cases of inputs is shown in Fig. 1.96.

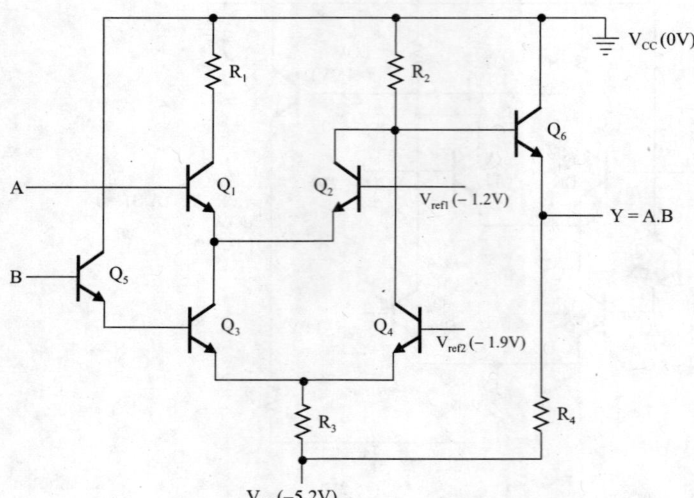

Truth Table of AND Gate

Inputs		Output
A	B	Y
0	0	0
0	1	0
1	0	0
1	1	1

Fig. 1.95: *Two input ECL AND gate.*

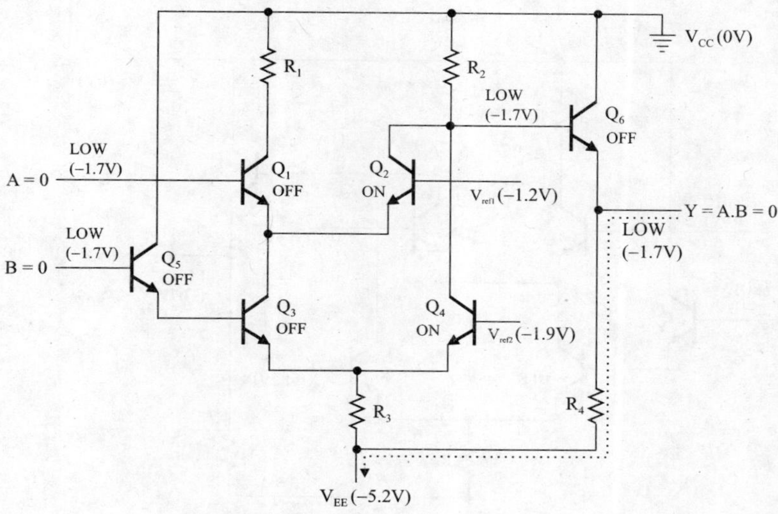

Fig. a: *Case 1, A = B = 0, Y = 0.*

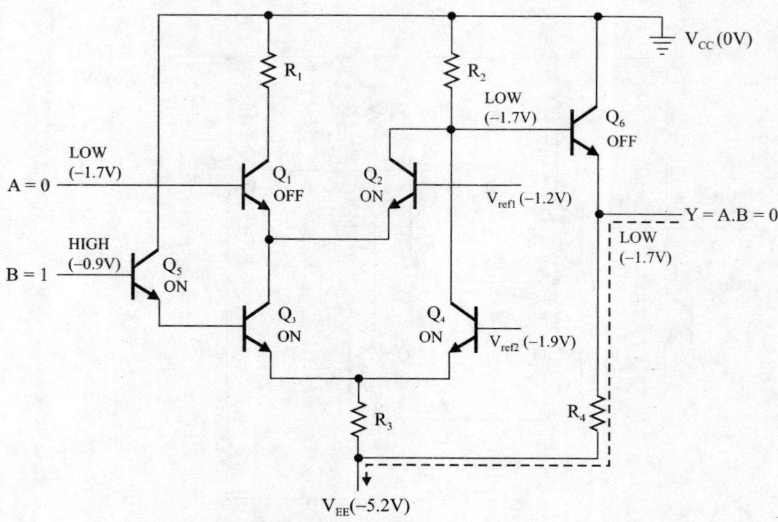

Fig. b: *Case 2, A = 0, B = 1, Y = 0.*

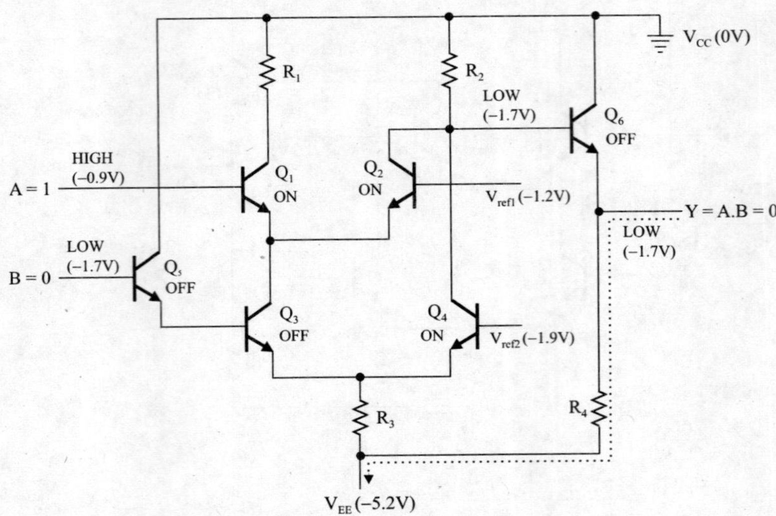

Fig. c: *Case 3, A = 1, B = 0, Y = 0.*

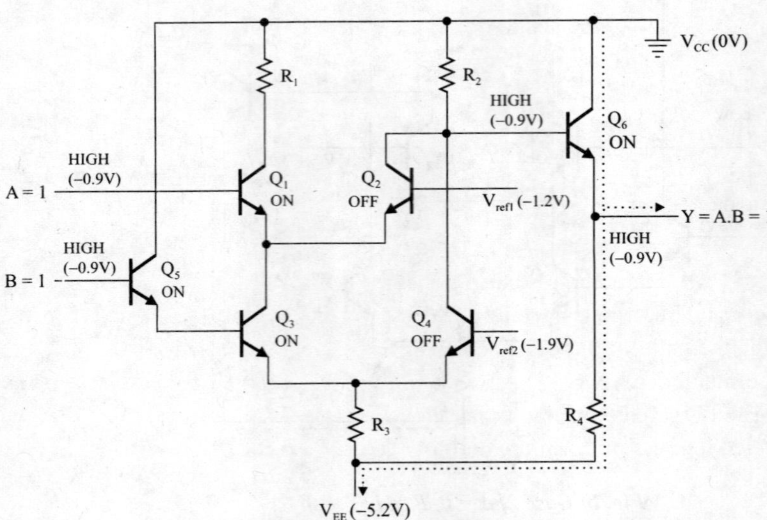

Fig. d: *Case 4, A = B = 1, Y = 1.*

Fig. 1.96: *Four cases of operation of ECL AND gate.*

1.14 CMOS (Complementary MOSFET)

In CMOS logic family of devices, the logic functions are realized by parallel or series connection of PMOS and NMOS field effect transistors.

In CMOS family, the inputs are applied to gate terminal of MOSFET and the output is obtained from the drain terminal of NMOS field effect transistor.

The MOSFETs operate in cut-off (OFF state) or saturation (ON state). In CMOS family, logic **low** is 0 *V* and logic **high** is 5 *V*.

1.14.1 CMOS NOT Gate

Basically, a CMOS field effect transistor will behave as inverter. Hence, NOT gate can be constructed using a CMOS field effect transistor in which a PMOS and NMOS are connected in series with their gates tied together as shown in Fig. 1.97. The input is applied to gate terminal and output is obtained from drain terminal of NMOS. The voltage and current conditions during two cases of inputs are shown in Figs. 1.98(a) and 1.98(b).

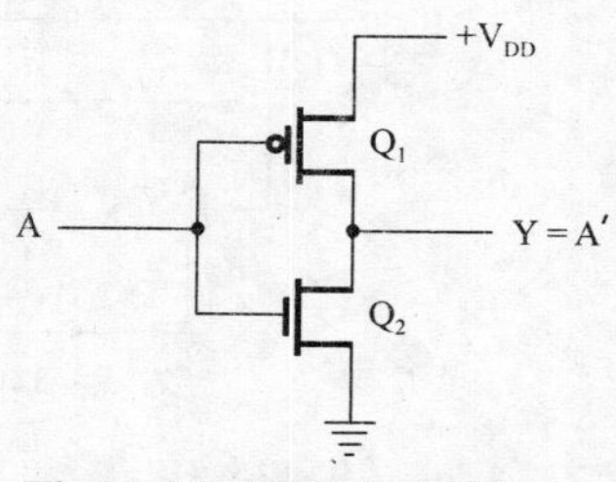

***Fig. 1.97:** CMOS NOT gate.*

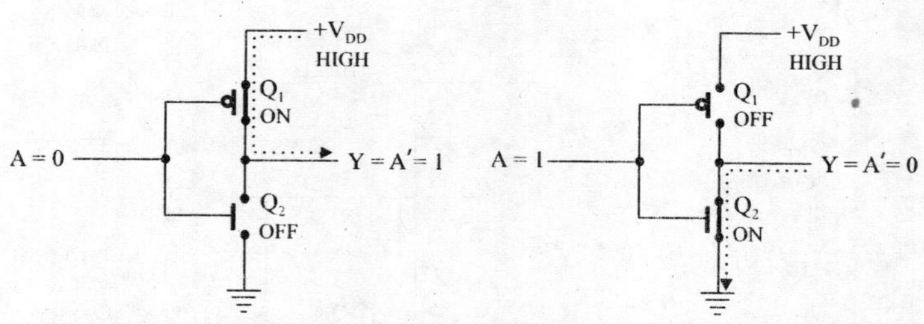

***Fig. a:** Case 1, A = 0, Y = 1.* ***Fig. b:** Case 2, A = 1, Y = 0.*

***Fig. 1.98:** Two cases of operation of CMOS NOT gate.*

Truth Table of NOT Gate

Input	Output
A	Y
0	1
1	0

1.14.2 CMOS NOR Gate *(AU, Apr/May'19, 7 Marks)*

A two input NOR gate is shown in Fig. 1.99. For NOR logical operation two PMOS are connected in series and two NMOS are connected in parallel. The input terminals are formed by combining the gates of one series connected PMOS and one parallel connected NMOS. The two NMOS field effect transistors Q_3 and Q_4 are connected in parallel such that drain terminals are tied together to form the output and source terminals are grounded.

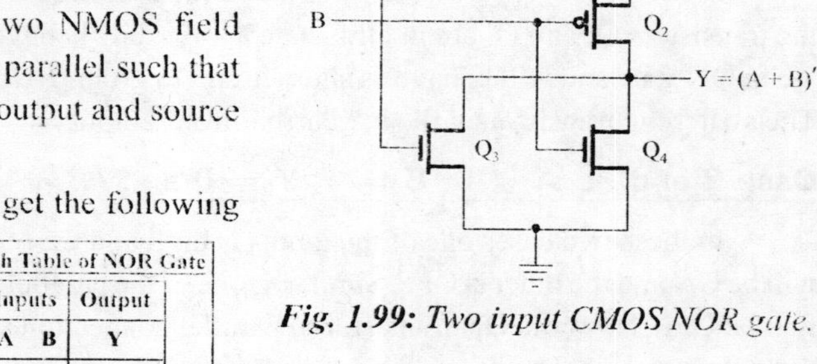

***Fig. 1.99:** Two input CMOS NOR gate.*

From the truth table of NOR gate we get the following four cases of inputs.

Case 1: A = 0, B = 0, Y = 1

Case 2: A = 0, B = 1, Y = 0

Case 3: A = 1, B = 0, Y = 0

Case 4: A = 1, B = 1, Y = 0

Truth Table of NOR Gate

Inputs		Output
A	B	Y
0	0	1
0	1	0
1	0	0
1	1	0

Case 1: A = 0, B = 0, Y = 1

Figure 1.100(a) shows voltage and current condition of NOR gate when both inputs are **low**. Now, the transistors Q_1 and Q_2 are in ON state and the transistors will behave as short circuit and Q_3 and Q_4 are in OFF state and behave as open circuit to ground. Hence, a logic **high** voltage appears in the output. The drain power supply will source current to load at output.

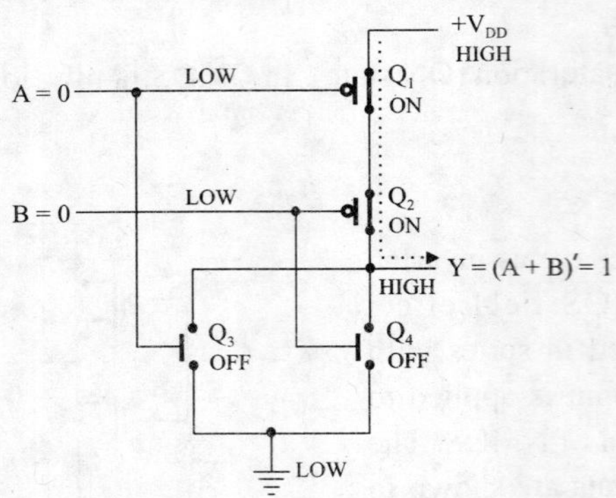

Fig. a: Case 1, A = B = 0, Y = 1.

Fig. b: Case 2, A = 0, B = 1, Y = 0.

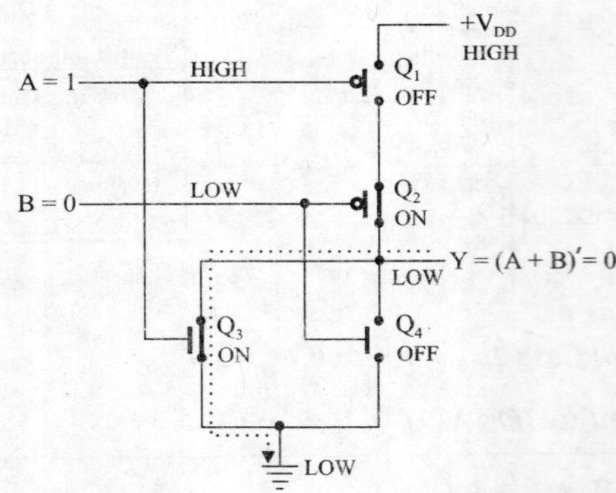

Fig. c: Case 3, A = 1, B = 0, Y = 0.

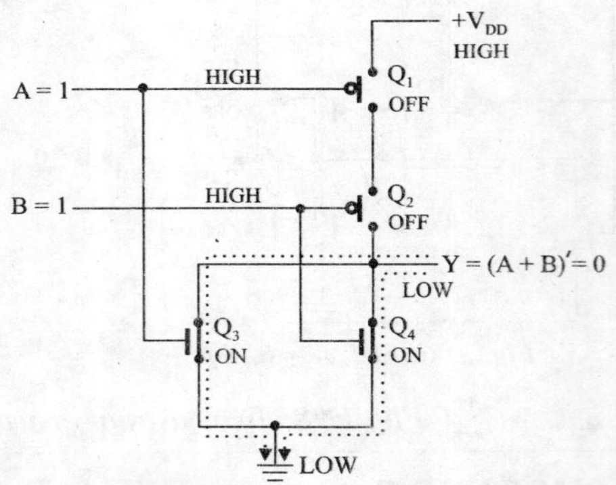

Fig. d: Case 4, A = B = 1, Y = 0.

***Fig. 1.100:** Four cases of operation of CMOS NOR gate.*

Case 4: A = 1, B = 1, Y = 0

Fig. 1.100(d) shows voltage and current condition of NOR gate when both inputs are **high**. Now, the transistors Q_1 and Q_2 are in OFF state and so they behave as open circuit and the transistors Q_3 and Q_4 are in ON state and will behave as short circuit to ground. Hence a logic **low** voltage appears in the output. The switching transistors will sink current from output.

Case 2 and 3: A = 0, B = 1, Y = 0 and A = 1, B = 0, Y = 0

In these two cases one of the inputs is **high** and the other is **low**. Hence, one of the series transistors will be ON and the other is OFF. Similarly, one of the parallel connected transistors will be ON and the other will be OFF. Now, the ON transistor in parallel connection will behave as short circuit to ground and pull the output to **low**. Therefore, the working of NOR gate in these two cases will be similar to case 4. Figs. 1.100(b) and 1.100(c) shows voltage and current condition of NOR gate when any one of the inputs is **high**.

1.14.3 CMOS OR Gate

The OR logic can be obtained by inverting the NOR logic. The inverted output of NOR gate can be obtained by adding another CMOS inverter at the output of NOR gate as shown in Fig. 1.101 which will invert the NOR logic output. The voltage and current conditions for various inputs are shown in Figs. 1.102(a) to 1.102(d).

Truth Table of OR Gate

Inputs		Output
A	**B**	**Y**
0	0	0
0	1	1
1	0	1
1	1	1

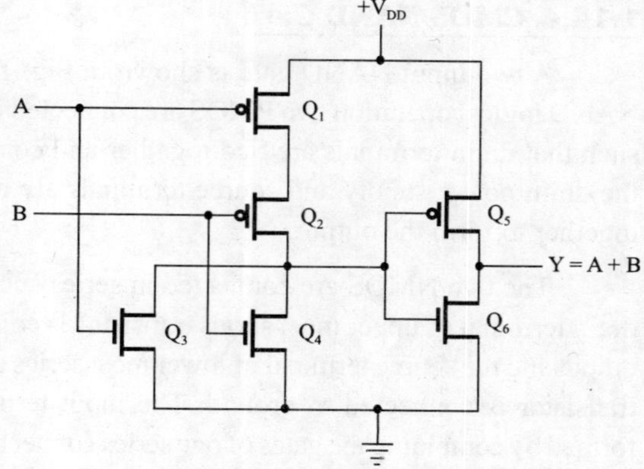

Fig. 1.101: Two input CMOS OR gate.

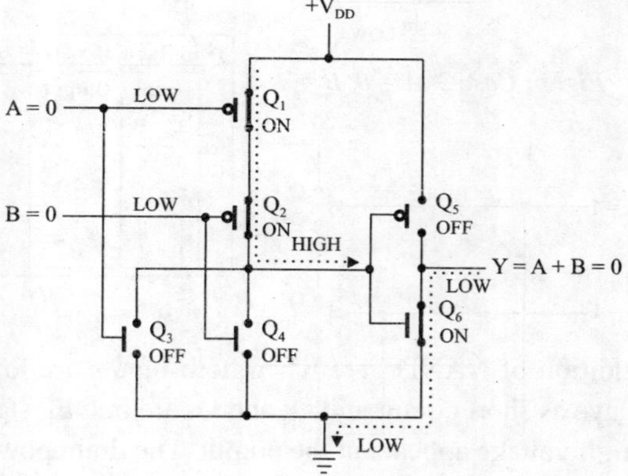

Fig. a: Case 1, A = B = 0, Y = 0.

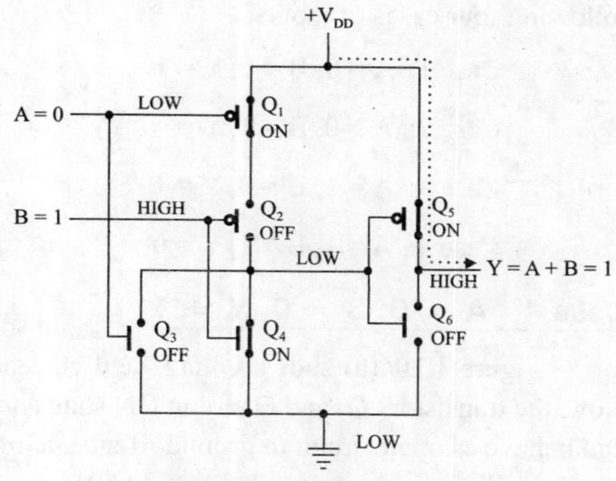

Fig. b: Case 2, A = 0, B = 1, Y = 1.

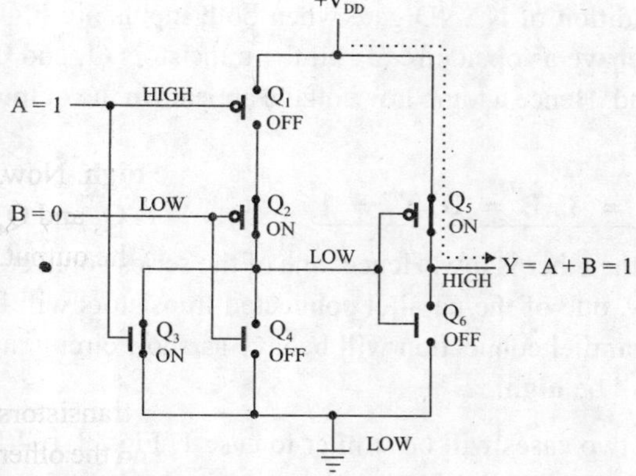

Fig. c: Case 3, A = 1, B = 0, Y = 1.

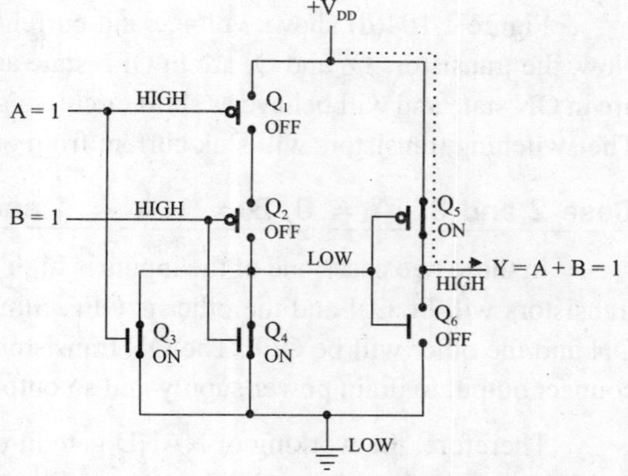

Fig. d: Case 4, A = B = 1, Y = 1.

Fig. 1.102: Four cases of operation of CMOS OR gate.

1.14.4 CMOS NAND Gate

A two input NAND gate is shown in Fig. 1.103. For NAND logical operation two PMOS are connected in parallel such that drain terminals are tied together and connected to the drain power supply and source terminals are connected together to form the output.

The two NMOS are connected in series such that the drain terminal of upper most series transistor is connected to output and the source terminal of lower most series connected transistor is connected to ground. The input terminals are formed by combining the gates of one series connected PMOS and one parallel connected NMOS.

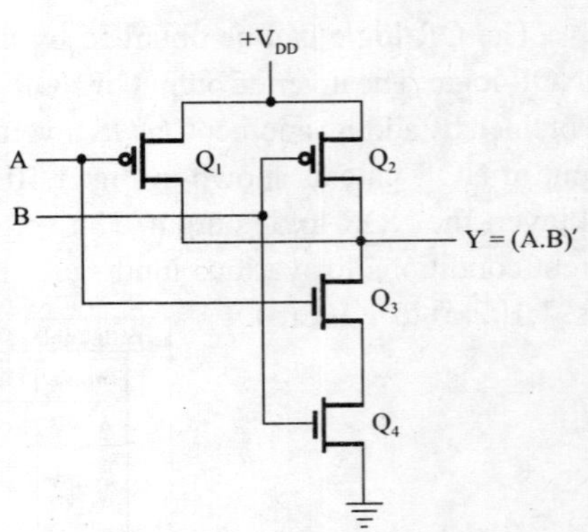

Fig. 1.103: Two input CMOS NAND gate.

From the truth table of NAND gate, we get the following four cases of inputs.

Case 1: A = 0, B = 0, Y = 1

Case 2: A = 0, B = 1, Y = 1

Case 3: A = 1, B = 0, Y = 1

Case 4: A = 1, B = 1, Y = 0

Truth Table of NAND Gate

Inputs		Output
A	B	Y
0	0	1
0	1	1
1	0	1
1	1	0

Case 1: A = 0, B = 0, Y = 1

Figure 1.104(a) shows voltage and current condition of NAND gate when both inputs are **low**. Now, the transistors Q_1 and Q_2 are in ON state and behave as short circuit and Q_3 and Q_4 are in OFF state and behave as open circuit to ground. Hence, a logic **high** voltage appears in the output. The drain power supply will source current to load at output.

Case 4: A = 1, B = 1, Y = 0

Figure 1.104(d) shows voltage and current condition of NAND gate when both inputs are **high**. Now, the transistors Q_1 and Q_2 are in OFF state and behave as open circuit and the transistors Q_3 and Q_4 are in ON state and will behave as short circuit to ground. Hence a logic **low** voltage appears in the output. The switching transistors will sink current from output.

Case 2 and 3: A = 0, B = 1, Y = 1 and A = 1, B = 0, Y = 1

In these two cases one of the inputs is **high** and the other is **low**. Hence, one of the series connected transistors will be ON and the other is OFF. Similarly, one of the parallel connected transistors will be ON and the other will be OFF. The ON transistor in parallel connection will behave as short circuit and connect output to drain power supply and so output will be **high**.

Therefore, the working of NAND gate in these two cases will be similar to case 1. Figs. 1.104(b) and 1.104(c) shows voltage and current condition of NAND gate when any one of the inputs is **high**.

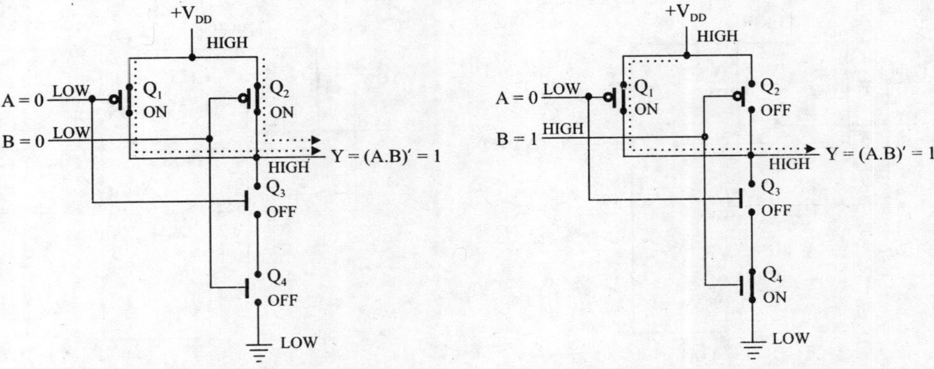

Fig. a: *Case 1, A = B = 0, Y = 1.* ***Fig. b:*** *Case 2, A = 0, B = 1, Y = 1.*

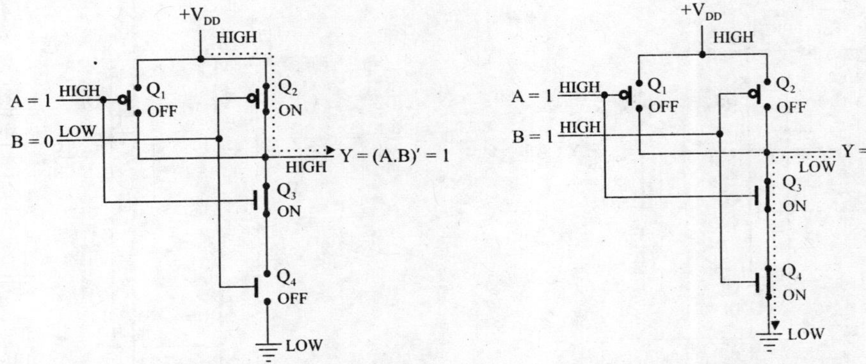

Fig. c: *Case 3, A = 1, B = 0, Y = 1.* ***Fig. d:*** *Case 4, A = B = 1, Y = 0.*

Fig. 1.104: *Four cases of operation of CMOS NAND gate.*

1.14.5 CMOS AND Gate

The AND logic can be obtained by inverting the NAND logic. The inverted output of NAND gate can be obtained by adding another CMOS inverter at the output of NAND gate as shown in Fig. 1.105 which will invert the NAND logic output. The voltage and current conditions for various inputs are shown in Figs. 1.106(a) to 1.106(d).

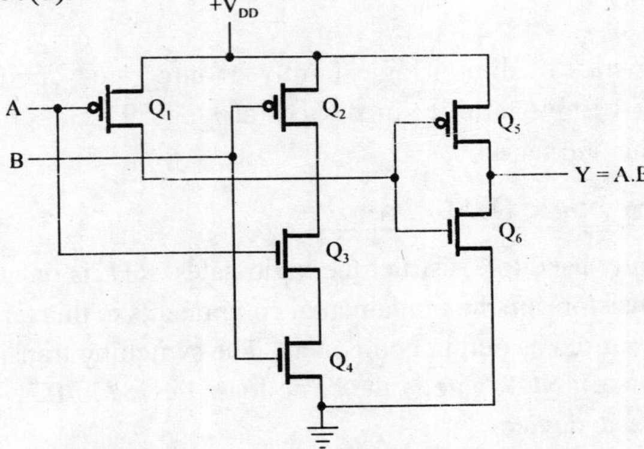

Truth Table of AND Gate

Inputs		Output
A	B	Y
0	0	0
0	1	0
1	0	0
1	1	1

Fig. 1.105: *Two input CMOS AND gate.*

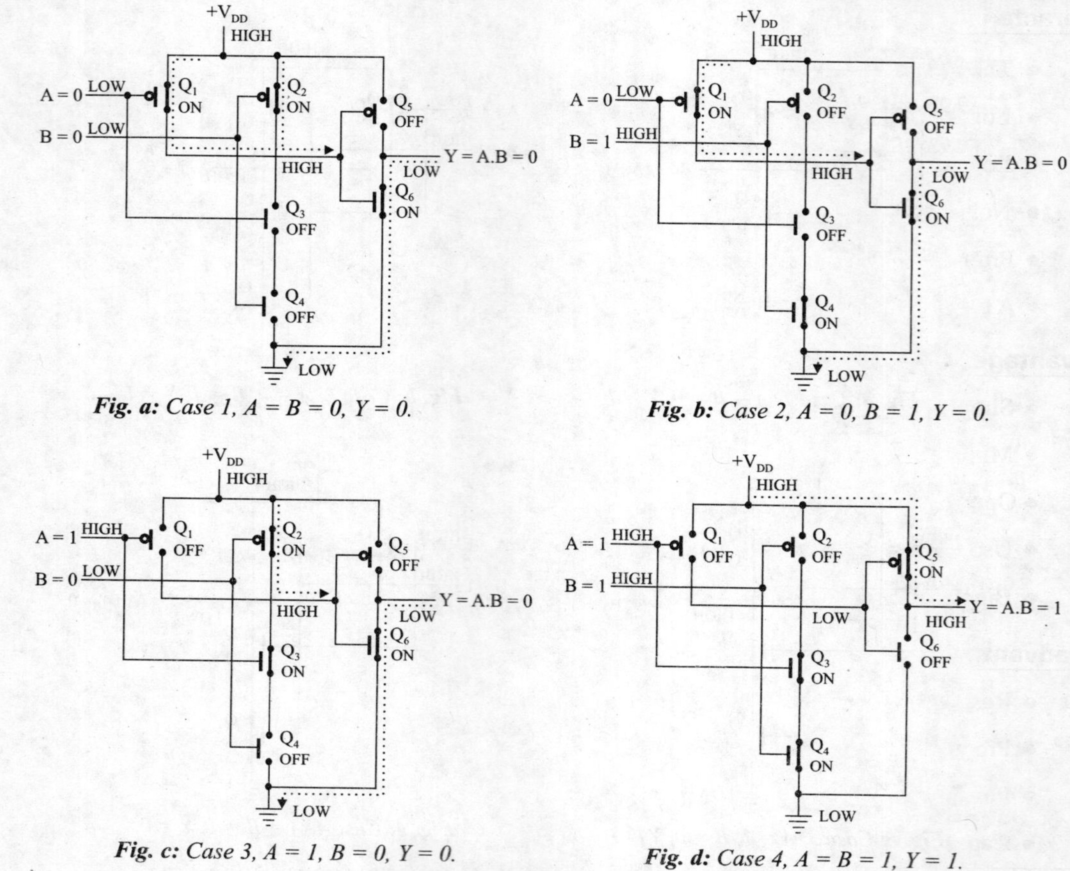

Fig. a: Case 1, A = B = 0, Y = 0.

Fig. b: Case 2, A = 0, B = 1, Y = 0.

Fig. c: Case 3, A = 1, B = 0, Y = 0.

Fig. d: Case 4, A = B = 1, Y = 1.

***Fig. 1.106:** Four cases of operation of CMOS AND gate.*

1.15 Comparison of Logic Families

The popular logic families are RTL, DTL, TTL, ECL and CMOS. The RTL, DTL, TTL and ECL digital logic family of ICs are constructed using bipolar NPN transistors as fundamental switching device. The CMOS digital logic family of devices are constructed using unipolar MOSFET as fundamental switching device.

The important characteristics of digital logic family of integrated circuits (ICs) that are used to compare the performance of various families of devices are logic levels, propagation delay, power dissipation, noise margin, fan-out and fan-in.

1.15.1 Resistor-Transistor Logic (RTL) Family

RTL is the first logic family used to construct the logic gates. RTL is one of the saturated bipolar logic families. Resistors and transistors are the fundamental components in this family. Resistor is used as input component and transistor is used as output component. The switching transistors operate in cut-off (OFF state) or saturation (ON state). NOR gate is the basic logic device in RTL logic family and other logic gates are built using this basic device.

Characteristics of RTL Family

- The output of a RTL device can serve as an input to a maximum of 5 gates.
- Logic **low** voltage for a RTL is defined between 0 - 0.2 *V*.
- Logic **high** voltage for a RTL is at 1 to 3.6 *V*.
- Noise margin is at around 0.2 *V*.
- Propagation delay is about 12 *ns*.
- A typical RTL logic draws a power of about 30 *mW*.

Advantages of RTL Family

- Simple circuit when compared to other logic families.
- Minimum number of transistors is required to implement a logic device.
- Operation is very simple.
- Use silicon bipolar transistor as saturated switches.
- Bipolar transistors exhibit higher trans-conductance than MOSFETs.

Disadvantages of RTL Family

- Response time is poor.
- Propagation delay is high.
- Power dissipation is high due to resistor.
- Capacitive loading reduces fan-out capability.

1.15.2 Diode-Transistor Logic (DTL) Family

DTL is one of the saturated bipolar logic families and it developed to reduce the power dissipation of RTL family by replacing the input resistor by diodes. Diodes are used as input component and transistor used as output component. NAND gate is the basic logic device in RTL logic family and other logic gates are built using this basic device.

Characteristics of DTL Family

- The output of a DTL device can serve as an input to a maximum of 8 gates.
- Logic **low** voltage for a DTL is defined between 0 - 0.2 *V*.
- Logic **high** voltage for a DTL is at 4 to 5 *V*.
- Noise margin is at around 0.4 *V*.
- Propagation delay is about 30 *ns*.
- A typical TTL logic draws a power of about 12 *mW*.

Advantages of DTL Family

- Power loss is lesser than RTL family.
- Minimum number of components is required to implement a logic device.
- Operation is very simple.
- Use silicon bipolar transistor as saturated switches.
- Bipolar transistors exhibit higher trans-conductance than MOSFETs.

Disadvantages of DTL Family

- Propagation delay is more than RTL family due to diode capacitance.
- Usage of DTL logic is limited.
- Capacitive loading increases the OFF time.

1.15.3 Transistor-Transistor Logic (TTL) Family

In TTL family of devices the diodes of DTL family are replaced by transistors to minimize the input capacitance. In TTL, the base-emitter junction is used to receive inputs instead of diodes in DTL. Since transistors are used as input and switching device, this family of devices are called transistor-transistor logic. NAND gate is the basic logic device in TTL logic family and other logic gates are built using this basic device.

Characteristics of TTL Logic Families

- The output of a TTL device can serve as an input to a maximum of 10 gates.
- Logic **low** voltage for a TTL is defined between 0 - 0.2 *V*.
- Logic **high** voltage for a TTL is at 5 *V*.
- Noise margin is at around 0.4 *V*.
- Propagation delay is about 9 *ns*.
- A typical TTL logic draws a power of about 11 *mW*.

Advantages of TTL Logic Families

- TTL has better characteristics than RTL and DTL.
- It is least susceptible to electrical damage.
- Lesser immune to noise when compared to ECL, but more than CMOS.
- Fastest saturation, when compared to other logic families.
- Low output impedance for all states.

Disadvantages of TTL Logic Families

- TTL device dissipates a lot of power, when compared to CMOS devices.
- Not recommended in VLSI chips as it requires more space and isolation.
- Expensive compared to MOSFETs.

1.15.4 Emitter-Coupled Logic (ECL) Family

The basic component in ECL family of devices is emitter coupled differential current switch operating in active region. ECL is the only non-saturated bipolar logic family. OR-NOR gate is the basic logic device in ECL logic family and other logic gates are built using this basic device.

Emitter coupled transistors operate in active region for both OFF and ON state with a slight difference between logic voltage levels.

Characteristics of ECL Logic Families

- Large fan-out among transistor family due to low output impedance.
- The propagation time is about 1ns, making it the fastest logic family.
- A logic **low** voltage for ECL is about – 1.7 to – 1.5 *V*.
- A logic **high** voltage for ECL is about – 0.9 to – 0.6 V.
- It is the fastest among all the logic families.
- It has an average propagation delay time of about 1 *ns*.

Advantages of ECL Logic Families

- Has a fan-out better than the TTL Logic family.
- Offers the highest speed in operation.
- ECL systems produce complementary outputs (OR-NOR, AND-NAND).
- Parameters do not vary much with temperature.

Disadvantages of ECL Logic Families

- Worst noise immunity compared to TTL and CMOS.
- Highest power consumption, when compared with TTL and CMOS.
- VLSI design difficult due to resistors, thus increasing system size.
- Capacitive loading reduces the fan-out capacity.
- Requires negative supply to emitter terminal.

1.15.5 Complementary MOSFET Logic (CMOS) Family

CMOS is a saturated MOSFET family of devices. Enhancement type MOSFETs are the basic components in CMOS. A CMOS field effect transistor is a pair of NMOS and PMOS field effect transistors with their channels connected in series and gate terminals tied together. The CMOS devices are **high** immunity to noise and have **low** static power consumption.

The MOSFETs operate in cut-off (OFF state) or saturation (ON state). NAND gate is the basic logic device in CMOS logic family other logic gates are built using this basic device.

Characteristics of CMOS Logic Families

- CMOS supports a very large fan-out, more than 50.
- It has excellent noise immunity among all families.
- A logic **low** voltage for CMOS is 0 *V*.
- A logic **high** voltage for CMOS is 5 *V*.
- Propagation delay is very large at about 20 to 50 *ns*.

Advantages of CMOS Logic Families

- Has the highest fan-out, when compared with TTL and ECL
- Works well over a wide range of temperature
- Noise immunity is better than TTL and ECL

Disadvantages of CMOS Logic Families

- Large propagation delay

Table 1.72: Comparison of Characteristics of Logic Families

Characteristics	Logic Family				
	RTL	**DTL**	**TTL**	**ECL**	**CMOS**
Basic Component	Resistor, NPN Transistor	Diode, NPN Transistor	NPN Transistor	NPN Transistor	PMOS, NMOS
Basic Gate	NOR	NAND	NAND	OR-NOR	NOR-NAND
Logic Low	0 to 0.2 *V*	0 to 0.2 *V*	0 to 0.8 *V*	– 1.7 to – 1.5 *V*	0 *V*
Logic High	1 to 3.6 *V*	4 to 5 *V*	2 to 5 *V*	–0.9 to – 0.6 *V*	5 *V*
Propagation Delay	12 *ns*	30 *ns*	9 *ns*	1 *ns*	20 to 50 *ns*
Power Dissipation	30 *mW*	12 *mW*	10 *mW*	50 *mW*	0.01 *mW*
Fan-out	5	8	10	10	50
Fan-in	5	5	8	5	10
Noise Margin	0.2 *V*	0.4 *V*	0.4 *V*	0.25 *V*	5 *V*

1.16 Summary of Important Concepts

1. Every number system has a radix, r. The radix, r is also called base.
2. Four types of popular number systems are decimal, binary, octal and hexadecimal.
3. Decimal number system is the standard number system and so the value of a number in any other number system is verified by converting to an equivalent decimal number.
4. The decimal number system is a radix number system in which the radix is 10.
5. The 10 symbols used in decimal number system are 0, 1, 2, 3, 4, 5, 6, 7, 8 and 9.
6. Each digit of a decimal number has a weight which is given by the power of radix, 10.

7. The dot in decimal number system is called decimal point and it separates the integer part and fraction part.
8. The binary number system is a radix number system in which the radix is 2.
9. The two symbols used in binary number system are 0 and 1.
10. The dot in binary number system is called binary point and it separates the integer part and fraction part.
11. Each digit of a binary number has a weight which is given by the power of radix, 2.
12. For binary to decimal conversion multiply each binary digit by its weight and sum up the product.
13. The octal number system is a radix number system in which the radix r is 8.
14. The eight symbols used in octal number system are 0, 1, 2, 3, 4, 5, 6 and 7.
15. Each digit of an octal number has a weight which is given by the power of radix, 8.
16. For octal to decimal conversion multiply each octal digit by its weight and sum up the product.
17. The hexadecimal number system is a radix number system in which the radix r is 16.
18. The 16 symbols used in hexadecimal number system are 0, 1, 2, 3, 4, 5, 6, 7, 8, 9, A, B, C, D, E and F.
19. Each digit of a hexadecimal number has a weight which is given by the power of radix, 16.
20. For hexadecimal to decimal conversion multiply each hexadecimal digit by its weight and sum up the product.
21. A hexadecimal digit can be represented by 4-digit binary and an octal digit by 3-digit binary.
22. An octal number can be directly converted to binary and vice-versa.
23. The three types of signed binary number system are Sign-magnitude form, One's complement form and Two's complement form.
24. The range of decimal number that can be represented by n-bit unsigned binary is 0 to $2^n - 1$.
25. The unsigned binary has only positive numbers. The signed binary has positive and negative numbers.
26. In an n-bit unsigned binary number the digit d_0 has the weight $2^0 = 1$, which is the lowest weight. Hence the digit d_0 is called LSD (Least Significant Digit).
27. In an n-bit unsigned binary number the digit d_{n-1} has the weight 2^{n-1}, which is the highest weight. Hence the digit d_{n-1} is called MSD (Most Significant Digit).
28. In n-bit representation of sign-magnitude form of binary, 1-bit is reserved for sign and n–1 bits are used to represent the magnitude.
29. In sign-magnitude form of binary, for positive number the sign bit is "0" and for negative number the sign bit is "1". The magnitudes are same for a given number with opposite signs.
30. The range of decimal numbers that can be represented in n-bit sign-magnitude form of binary number is, $-(2^{n-1} - 1)$ to $+(2^{n-1} - 1)$.
31. The two types of complements in radix number systems are r's complements or radix complement and (r – 1)'s complement or diminished radix complement.
32. Complements of positive number are used to represent negative numbers in digital system.
33. In one's complement form of binary number representation, the positive numbers will be same as that of unsigned binary. The negative numbers are bit-by-bit complement of the positive numbers.

34. The range of decimal numbers that can be represented in n-bit one's complement form of binary number is, $-(2^{n-1}-1)$ to $+(2^{n-1}-1)$.

35. In two's complement form of binary number representation, the positive numbers will be same as that of unsigned binary. The negative number is obtained by bit-by-bit complementing the positive number and then adding one.

36. The range of decimal numbers that can be represented in n-bit two's complement form of binary number is, $-(2^{n-1})$ to $+(2^{n-1}-1)$.

37. A binary code is a group of n-bit binary used to represent the set of elements of an information.

38. With n-bit binary code it is possible to generate 2^n distinct binary codes.

39. Weighted codes with weight as 8, 4, 2 and 1 for the 4-digits of binary is called BCD.

40. In BCD addition, in order to correct the result, a correction 6_{10} $(0110)_2$ is added to BCD sum of digits whenever sum of BCD digits exceeds 9 or if there is a carry to next addition.

41. Excess-3 code is obtained by adding 4-bit binary value of 3_{10} (0011_2) to 8421 BCD.

42. Different types of BCD codes are, 8421 BCD (Generally called BCD), 2421 code, 5421 code, 8, 4, –2, –1 code, Excess-3 code.

43. 8421, 2421, 5421 and 8, 4, –2, – 1 are weighted codes. In these codes on adding the weights of 1's in the code we get the value of decimal digit represented by the code.

44. The 2421 and Excess-3 are self complementing code.

45. In self complementing code, the complement of the code will give the code for 9's complement representation of decimal digit.

46. In gray code only one bit changes from either "0" to "1" or "1" to "0" between two consecutive codes.

47. ASCII is abbreviated form of **A**merican **S**tandard **C**ode for **I**nformation **I**nterchange. ASCII was developed in the early stage of digital systems in order to standardize the coding system for human and machine interface.

48. ASCII is 7-bit binary code that includes most of the symbols in early type-writing machine.

49. With 7-bit binary code, $2^7 = 128$ ASCII codes are possible. They are assigned as 26 codes for upper case alphabets A to Z, 26 codes for lower case alphabets a to z, 10 codes for numerals 0 to 9, 32 codes for special printable characters and 34 codes for control characters.

50. Parity refer to number of 1's in an n-bit code. If number of 1's is even then the code has even parity and number of 1's is odd then the code has odd parity.

51. Parity of an n-bit code can be modified by appending an extra bit called parity bit. The parity bit is appended at most significant bit position.

52. George Boole developed Boolean algebra in 1854.

53. Boolean algebra is an algebraic structure that includes a set of elements consisting of binary operators " + " and " . ", binary variables x, y, z, ... , binary elements 0 and 1 and Boolean postulates and theorems.

54. The duality principle of Boolean algebra states that every algebraic expression deducible from the postulates of Boolean algebra remains valid if the operator and identity element are interchanged. Here, the operators " + " and " . " and the identity elements are " 0 " and " 1 ".

55. A statement written with Boolean variables, constants and operators is called Boolean expression.

56. Postulates of Boolean algebra are developed by E. V. Huntington in 1904.

57. Binary logic consists of binary variables and a set of logical operations. The binary variables are denoted by letters of alphabet A, B, C, ..., x, y, z, ... Each binary variable can take only two possible values 0 or 1.

58. If, 1 represent **high** voltage and 0 represent **low** voltage then logic system is called positive logic.

59. If, 1 represent **low** voltage and 0 represent **high** voltage then the logic system is called negative logic.

60. Logic gates are electronic devices or circuits that perform logical operations on one or more input logical variables and produce a binary output.

61. The basic logical operations are AND, OR and NOT.

62. In positive logic, the logical AND of two or more variables will be 1 if and only if the value of all the variables are 1.

63. In positive logic, the logical OR of two or more variables will be 1 if the value of one of the variables is 1.

64. NOT operation is same as complement operation.

65. With n variables it is possible to form 2^{2n} Boolean functions. Therefore, with two variables we can form 16 ($2^{2 \times 2} = 16$) Boolean functions.

66. NAND and NOR gates are called universal gates, because any Boolean function can be realized only using NAND gates or only using NOR gates.

67. A Boolean function is described by a Boolean expression which consists of binary variables, binary constants 0 and 1 and logical operators AND, OR and NOT.

68. Each variable within a term of a Boolean expression is called a literal.

69. The complement of a function can be obtained by using duality by replacing "+" by "." and "." by "+" and variables by its complement.

70. Minterms are 2^n possible combinations of AND terms with n variables such that the logical AND of all the variables is 1.

71. Maxterms are 2^n possible combinations of OR terms with n variables such that the logical OR of all the variables is 0.

72. The three types of standard forms of expressing a Boolean function are Sum-of-Products (SOP) form, Product-of-Sums (POS) form and Canonical form.

73. The SOP or POS form is said to be canonical only if all the variables are present in every term of the SOP or POS form.

74. Sum-of-products (SOP) can be realized by AND operation of literals followed by OR operation of output of AND.

75. Product-of-sums (POS) can be realized by OR operation of literals followed by AND operation of output of OR.

76. The realizations in SOP and POS forms are called two level realization of standard forms.

77. It can be observed that complement of sum-of-products will be product-of-sums and vice-versa

78. A Boolean function expressed as a sum of minterms or product of maxterms are said to be in canonical form.

79. K-map is a pictorial form of truth table and used to simplify Boolean functions.
80. A K-map is a diagram made up of squares with each square representing a minterm.
81. Alternatively, maxterm can be used to construct K-map in which each square represent a maxterm.
82. When minterms are considered for simplification using K-map the resultant Boolean function will be in sum-of-products form.
83. When maxterms are considered for simplification using K-map the resultant Boolean function will be in product-of-sums form.
84. For n-variable Boolean function the K-map will have 2^n squares.
85. While arranging the literal/minterms in a K-map, only one change is allowed if we move from one row to next row or from one column to next column.
86. In a K-map, combine adjacent 1's to form prime implicants.
87. A prime implicant with single 1 represent n literal product term. A prime implicant with two 1's represent n – 1 literal product term. A prime implicant with four 1's represent n – 2 literal product term and so on.
88. The literals represented by a prime implicant are the literals that present in all the squares of the prime implicant (or the literals common to all the squares of prime implicants).
89. While forming prime implicants any number of over lapping is allowed in horizontal and vertical directions.
90. While forming prime implicants the K-map can be folded in any direction to get adjacent 1's.
91. When squares with 1's are considered for forming prime implicants, the simplified Boolean function is sum of all the product terms of prime implicant.
92. If all minterms are 1's then all squares of K-map will be filled by 1 and the function value is 1.
93. In K-map if squares with 0's are considered for forming prime implicants then the complement of the resultant sum-of-products form will give product-of-sums form.
94. The undefined function outputs are called don't-care conditions and denoted by ×.
95. In simplification of Boolean functions using K-maps, don't-care conditions can be considered as either 0 or 1.
96. If inclusion of a don't-care condition leads to minimization of a literal then don't-care can be included in the prime implicants.
97. Quine-McCluskey method is a table based method to determine prime implicants and essential prime implicants of a Boolean function and hence to find simplified Boolean expression of a function.
98. Quine-McCluskey method has two phases: Expansion phase and covering phase.
99. The expansion phase is the first phase in which minterms are obtained from truth table and prime implicants are identified after eliminating unwanted/redundant literals.
100. The covering phase is the second phase in which the essential prime implicants are selected from the prime implicants of first phase such that all minterms are covered with minimal number of prime implicants.
101. The collection of different logical devices that perform different logic functions but manufactured using same semiconductor technology is called a logic family.
102. The fundamental devices used in the logic families are either bipolar junction transistors or unipolar CMOS field effect transistors.

103. The RTL, DTL, TTL and ECL digital logic family of ICs are constructed using bipolar NPN transistors as fundamental switching device.

104. The CMOS digital logic family of devices are constructed using unipolar MOSFET as fundamental switching device.

105. The propagation delay is defined as the time required for a change in input to be available in the output.

106. The fan-out is defined as the maximum number of standard load that an output can drive reliably.

107. A standard load is usually defined as the amount of current needed by an input of another gate in the same logic family.

108. Fan-in is the number of inputs provided (or available) in a logic gate and it is decided during manufacturing itself which in turn depends on technology used for manufacturing the IC.

109. A quantitative measure of noise immunity in terms of voltage is called noise margin.

110. The characteristics of logic devices of a logic family depends on the fundamental devices used in that logic family.

111. In logic devices the transistors are operated either in saturation region/state or cut-off region/state.

112. In logic families, NPN transistors are widely used and the input is applied at base and output is taken either from collector or emitter of the transistors.

113. Enhancement type MOSFETs are the basic switching devices in CMOS field effect transistors.

114. A CMOS field effect transistor is a pair of NMOS and PMOS field effect transistors with their channels connected in series and gate terminals tied together.

115. In logic devices the MOSFETs are used as switches in which they are operated either in saturation region/state or cut-off region/state.

116. The RTL (Resistor-Transistor Logic) family of devices uses resistors to receive inputs and transistor as switching device and so the name resistor-transistor logic.

117. DTL (Diode-Transistor Logic) is a class of digital circuits built from diodes, resistors and bipolar junction transistors.

118. In TTL (Transistor-Transistor Logic) family, input network is made of transistors and switching devices are also transistors and so the name TTL.

119. The different types of outputs in TTL devices are totem-pole, open-collector and three-state (or Tri-state) output.

120. The basic circuit in an ECL gate is the emitter-coupled current switch.

121. ECL (Emitter-Coupled Logic) family of logic devices works in active region with a small difference in two logic voltage levels which is decided by the currents.

122. Emitter-coupled logic (ECL) circuits are the fastest available bipolar logic gates.

123. The primary disadvantage of ECL is the high quiescent power dissipation.

124. In CMOS logic family of devices, the logic functions are realized by parallel or series connection of PMOS and NMOS field effect transistors.

1.17 Short-Answer Questions

Q1.1 List the various number systems?

There are four types of popular number systems.

1. Decimal number system
2. Binary number system
3. Octal number system
4. Hexadecimal number system.

Q1.2 What is the largest binary number that can be expressed with 14 bits? Determine the equivalent decima and hexadecimal numbers.

The largest binary number that can be expressed with 14 bits = $11\ 1111\ 1111\ 1111_2$

$$11\ 1111\ 1111\ \ 1111_2 = 0011\ 1111\ 1111\ 1111_2 = 3FFF_{16}$$

$$3FFF_{16} = 3 \times 16^3 + F \times 16^2 + F \times 16^1 + F \times 16^0$$

$$= 3 \times 16^3 + 15 \times 16^2 + 15 \times 16^1 + 15$$

$$= 16383_{10}$$

Q1.3 Convert 101.01_2 to decimal number. **(AU, Apr/May'19, 2 Marks,**

Solution:

$$101.01_2 = 1 \times 2^2 + 0 \times 2^1 + 1 \times 2^0 + 0 \times 2^{-1} + 1 \times 2^{-2}$$

$$= 5.25_{10}$$

Q1.4 Convert the following numbers to decimal 11011.101_2, 5432_8

Solution:

$$11011.101_2 = 1 \times 2^4 + 1 \times 2^3 + 0 \times 2^2 + 1 \times 2^1 + 1 \times 2^0 + 1 \times 2^{-1} + 0 \times 2^{-2} + 1 \times 2^{-3}$$

$$= 27.625_{10}$$

$$5432_8 = 5 \times 8^3 + 4 \times 8^2 + 3 \times 8^1 + 2 \times 8^0$$

$$= 2842_{10}$$

Q1.5 Find the complement of $F = wx + yz$ and then show that $FF' = 0$.

$F = wx + yz$

$F' = (wx + yz)'$

$= (wx)' . (yz)'$

$= (w' + x') . (y' + z')$

$$FF' = (wx + yz)\,.\,(w' + x')\,.\,(y' + z')$$
$$= (wx\,.\,(w' + x') + y\,z\,.\,(w' + x'))\,.\,(y' + z')$$
$$= (wxw' + wxx' + yzw' + yzx')\,.\,(y' + z')$$
$$= (0 + 0 + yzw' + yzx')\,.\,(y' + z')$$
$$= yzw'.\,y' + yzw'z' + yzx'y' + yzx'z'$$
$$= 0 + 0 + 0 + 0 = 0$$

Q1.6 ***Convert the function $Y = A + B'C$ in canonical POS form.***

Solution:

$$Y = A + B'C = (A + B')\,.\,(A + C)$$
$$= ((A + B') + (C.C'))\,.\,((A + C) + (B.B'))$$ Add missing literals

$$= (A + B' + C)\,(A + B' + C')\,(A + B + C)\,(A + B' + C) \quad \text{(Cancel one } M_2\text{)}$$ $x.x' = 1$

$$(M_2) \qquad (M_3) \qquad (M_0) \qquad (M_2)$$

$$= (A + B + C)\,(A + B' + C)\,(A + B' + C')$$
$$= M_0 M_2 M_3$$

Q1.7 ***Perform 1's complement and 2's complement subtraction of fractional numbers 1010_2 and 1000_2 and compare the results with actual subtaction.***

Solution:

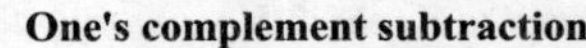

One's complement subtraction

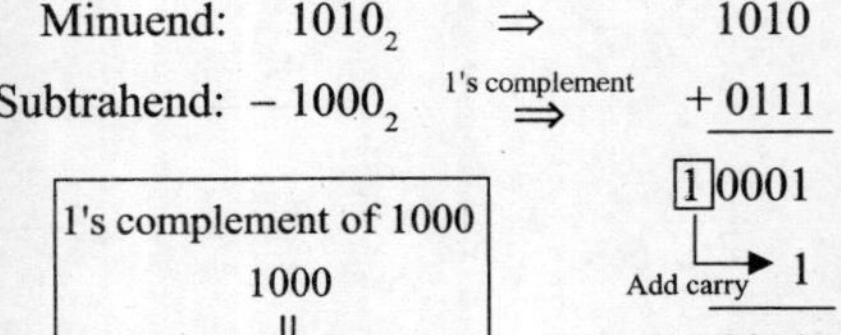

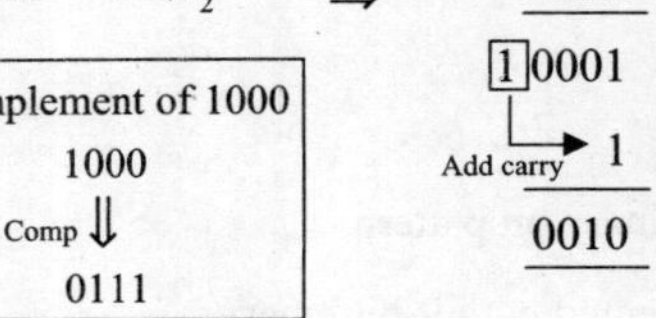

Minuend: 1010_2 ⇒ 1010

Subtrahend: $-\,1000_2$ $\xrightarrow{\text{1's complement}}$ + 0111

[1]0001

Add carry → 1

0010

1's complement of 1000

1000

Comp ⇓

0111

Two's complement subtraction

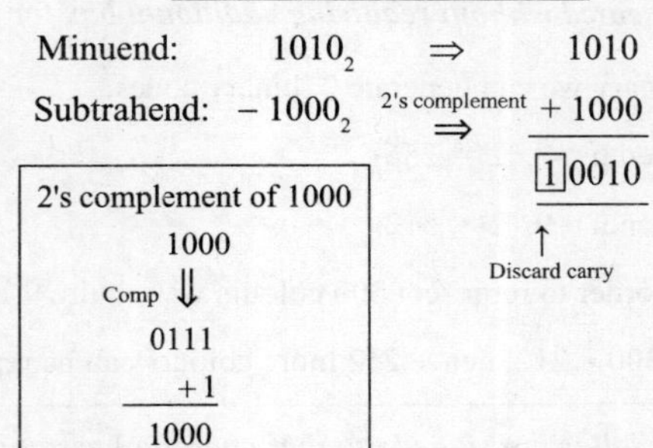

Minuend: 1010_2 ⇒ 1010

Subtrahend: $-\,1000_2$ $\xrightarrow{\text{2's complement}}$ + 1000

[1]0010

↑ Discard carry

2's complement of 1000

1000

Comp ⇓

0111

+1

1000

Actual Subtraction

1010

− 1000

0010

> ***Note:*** *The results are same in direct subtraction, one's complement and two's complement subtraction.*

Q1.8 ***Perform subtraction on the following unsigned binary numbers using the 2's complement of the subtrahend a) $11011_2 - 11001_2$ b) $110100_2 - 10101_2$.***

(AU, Nov/Dec'18, 2 Marks)

Solution:

a) $11011_2 - 11001_2$

Minuend: 11011_2 ⇒ 11011

Subtrahend: -11001_2 $\xrightarrow{\text{2's complement}}$ + 00111

[1] 00010

↑ Discard carry

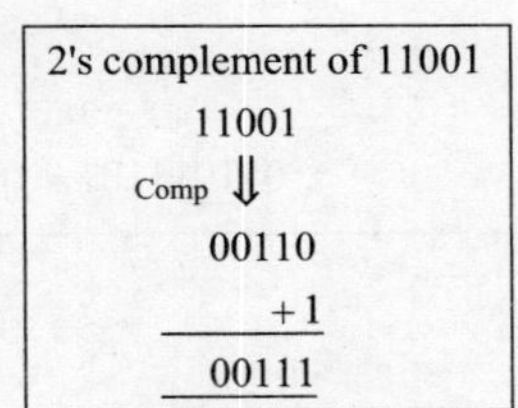

2's complement of 11001

11001

Comp ⇓

00110

+1

00111

$\therefore\ 11011_2 - 11001_2 = 00010_2$

b) $110100_2 - 10101_2$

Minuend: 11011_2 ⇒ 110100

Subtrahend: -10101_2 $\xRightarrow{\text{2's complement}}$ + 101011

[1] 011111

↑ Discard carry

$\therefore 110100_2 - 10101_2 = 011111_2$

2's complement of 10101

10101 $\xrightarrow[\text{6 bits}]{\text{Convert}}$ 010101

Comp ⇓

101010

+1

101011

Q1.9 ***Perform the following arithmetic operation using 2's complement arithmetic***

$11011100_2 - 10011011_2$

Solution:

Two's complement subtraction

Minuend: 11011100_2 ⇒ 11011100

Subtrahend: -10011011_2 $\xRightarrow{\text{2's complement}}$ + 01100101

[1] 01000001

↑ Discard carry

$\therefore 11011100_2 - 10011011_2 = 01000001_2$

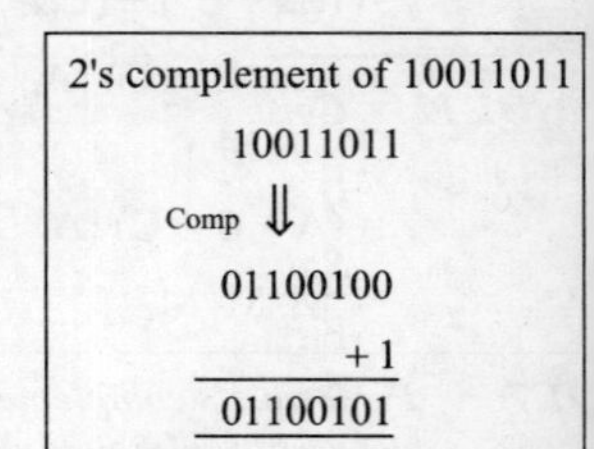

Q1.10 ***Assume that there are 300 colours to be represented in binary digits. If every colour is to be assigned a unique bit pattern, what is the minimum number of bits required to do this? How many more colour's can be represented without requiring additional bits for these colour's unique bit pattern?***

With n-bit binary we can generate 2^n binary codes.

When n = 8, $2^8 = 256_{10}$

When n = 9, $2^9 = 512_{10}$

Therefore in order to represent 300 colours we require 9-bit binary bit pattern.

Here, 512 – 300 = 212, hence 212 more colours can be represented with 9-bit binary.

Q1.11 ***Compute the following signed number operation assuming 2's complement representation.***

i) $0\,1\,1\,0\,1\,1\,1\,0 + 1\,0\,1\,1 = (.....)_2$

ii) $(3B7)_{16} - (5D)_{16} = (.....)_{16}$

Solution:

i) 011011110 ⟶ 01101110

1011 $\xrightarrow[\text{to 8-bits}]{\text{Sign extend}}$ + 11111011

Add [1] 01101001

↑ Discard carry

Note: For mathematical operations signed numbers should be sign extended to same size.

$\therefore 01101110_2 + 1011_2 = 01101001_2$

ii) $(3B7)_{16} - (5D)_{16} = (.....)_{16}$ $\Rightarrow$ $(3B7)_{16} + (-5D)_{16} = (.....)_{16}$

3B7 = 0011 1011 0111

5D = 0101 1101 $\xrightarrow{\text{1's complement}}$ 1010 0010 $\xrightarrow{+1}$ 1010 0011

↓ Sign extend to 12 bits

1111 1010 0011

> ***Note:*** *– 5D is obtained by taking 2's complement of +5D.*

3B7 ⟶ 0011 1011 0111

–5D $\xrightarrow{\text{2's complement of 5D}}$ + 1111 1010 0011

[1] 0011 0101 1010 = $35A_{16}$

↑ Discard carry

> ***Note:*** *In 2's complement arithmetic carry should be discarded.*

$\therefore\ 3B7_{16} - 5D_{16} = 35A_{16}$

Q1.12 Consider a decimal number – 4250. Represent this in each of the specified data type.

i) Signed binary ***ii) 2's complement***

Solution:

First convert 4250 to binary

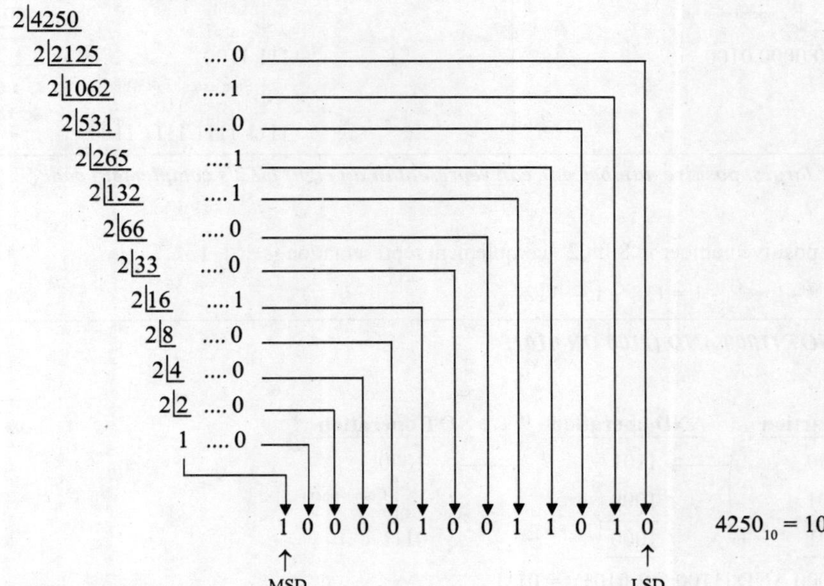

Division	Remainder
2 | 4250	
2 | 2125	0
2 | 1062	1
2 | 531	0
2 | 265	1
2 | 132	1
2 | 66	0
2 | 33	0
2 | 16	1
2 | 8	0
2 | 4	0
2 | 2	0
1	0

1 0 0 0 0 1 0 0 1 1 0 1 0 (MSD ... LSD)

$4250_{10} = 1000010011010_2$

i) Signed binary

Append 1 before MSD to represent negative number. $\therefore\ -4250_{10} =$ [1] 1000010011010_2

↑ Append sign bit '1'

ii) 2's complement

Append 0 before MSD to represent positive number and then convert to 2's complement.

$+ 4250_{10} =$ [0] 1000010011010

↑ Append sign bit '0'   ↓ 1's complement

10111101100101

↓ Add 1

10111101100110 $\therefore\ -4250_{10} = 10111101100110_2$

> ***Note:*** *A negative number is obtained by taking 2's complement of positive number.*

Q1.13 ***a) What is the range of signed integer values in 2's complement representation, that can be represented using 16-bits?***

b) Represent the signed binary number 10000100, as a signed 16-bit binary number as

i) Signed binary

ii) 2's complement binary

Solution:

a) The range of signed integer with 16-bits representation is,

$$-(2^{(16-1)}) \text{ to } (2^{(16-1)} - 1) \quad \Rightarrow \quad -2^{15} \text{ to } +(2^{15} - 1) \quad \Rightarrow \quad -32768_{10} \text{ to } +32767_{10}$$

b) i) Signed binary

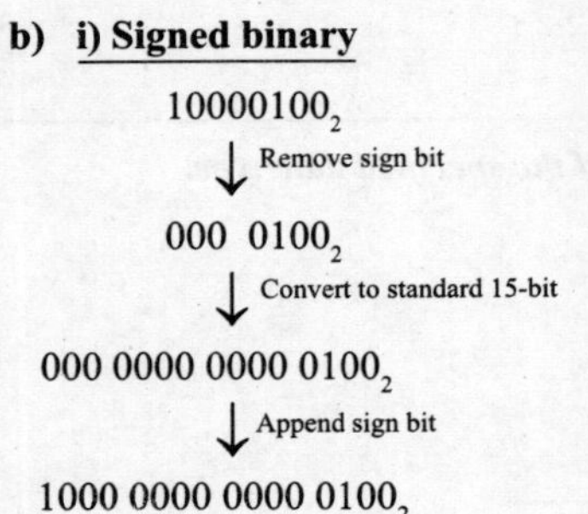

ii) 2's complement binary

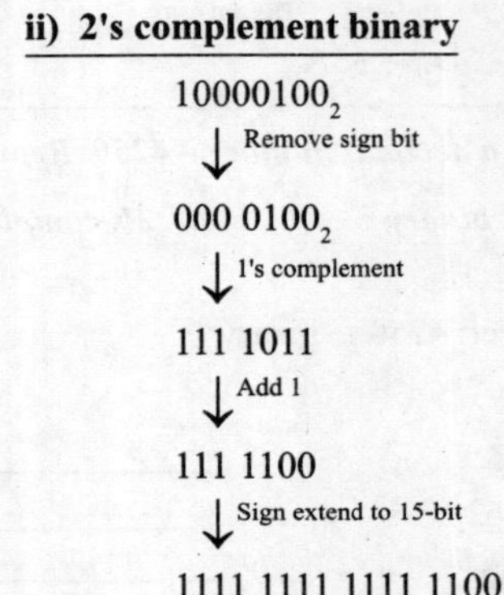

Q1.14 ***What is the largest positive number one can represent in an eight bit 2's complement code?***

Solution:

The largest positive number in 8-bit 2's complement representation is $2^{8-1}-1$

$$2^{8-1}-1 = 2^7 - 1 = 128 - 1 = +127_{10}$$

Q1.15 ***Compute NOT (1000 AND (1100 OR 0101)***

Solution:

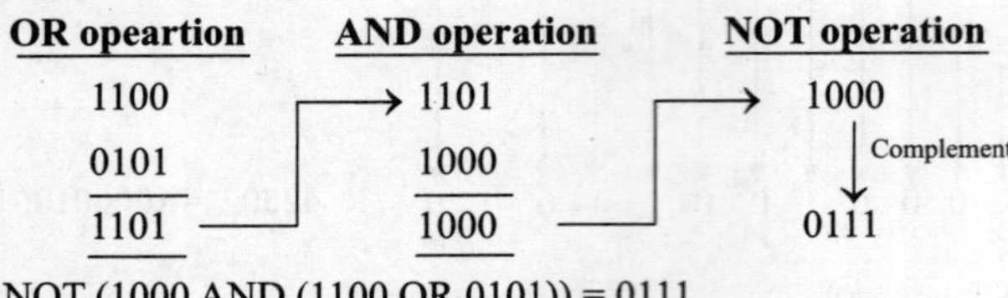

∴ NOT (1000 AND (1100 OR 0101)) = 0111

Q1.16 ***Compute 0101 + 0110. Comment on result if the binary numbers are unsigned binary and 2's complement binary.***

Solution:

```
  0101
+ 0110
------
  1011
```

1011 = 11_{10} in unsigned binary

1011 = – 5_{10} in 2's complement binary

Here, 0101 = 5_{10} and 0110 = 6_{10} and 5_{10} + 6_{10} = 11_{10}

The range of unsigned 4-bit binary is 0 to 15_{10} and so 11_{10} is accommadated in the result and no error in addition.

The range of 4-bit 2's complement numbers are – 8_{10} to +7_{10} and so the result 11_{10} cannot be accommadated in 4-bit 2's complement binary and so the sum has an overflow error.

Q1.17 ***Which of the following binary numbers are even.***

a) 0101100010 ***b) 01000101***

c) 011111110 ***d) 11000001***

Solution:

The binary numbers ending with '0' are even and the binary numbers ending with '1' are odd.

(i.e., if the least significant digit is zero then the number is even otherwise odd).

Therefore, **a)** 0101100010 and **c)** 011111110 are even numbers

and **b)** 01000101 and **d)** 11000001 are odd numbers

Q1.18 ***Write a truth table for the function, Z = (A + B)′ . (A . B + C).***

Solution:

Given that, Z = (A + B)′ . (A . B + C)

Let, (A + B)′ = X

A . B + C = Y

∴ Z = (A + B)′ . (A . B + C) = X . Y

Table 1: Truth Table for Function, Z

Inputs			**Outputs**				
A	**B**	**C**	**A + B**	**(A + B)′ = X**	**A . B**	**(A . B) + C = Y**	**Z = X . Y**
0	0	0	0	1	0	0	0
0	0	1	0	1	0	1	1
0	1	0	1	0	0	0	0
0	1	1	1	0	0	1	0
1	0	0	1	0	0	0	0
1	0	1	1	0	0	1	0
1	1	0	1	0	1	1	0
1	1	1	1	0	1	1	0

Q1.19 ***Demonstrate by means of truth table the validity of the DeMorgan's theorem for three variables: $(xyz)' = x' + y' + z'$***

Solution:

Table 1: Truth Table

Input Variables			Complement of inputs			xyz	(xyz)'	x'+y'+z'
x	y	z	x'	y'	z'			
0	0	0	1	1	1	0	1	1
0	0	1	1	1	0	0	1	1
0	1	0	1	0	1	0	1	1
0	1	1	1	0	0	0	1	1
1	0	0	0	1	1	0	1	1
1	0	1	0	1	0	0	1	1
1	1	0	0	0	1	0	1	1
1	1	1	0	0	0	1	0	0

From the truth table we can say that,

$$(xyz)' = x' + y' + z'$$

Q1.20 ***Write a truth table for, $Q = (X' + (X . Y . Z))'$.***

Solution:

Table 1: Truth Table of Q

Inputs			Outputs				
X	Y	Z	X'	X.Y	X.Y.Z = W	X'+W	Q = (X'+W)'
0	0	0	1	0	0	1	0
0	0	1	1	0	0	1	0
0	1	0	1	0	0	1	0
0	1	1	1	0	0	1	0
1	0	0	0	0	0	0	1
1	0	1	0	0	0	0	1
1	1	0	0	1	0	0	1
1	1	1	0	1	1	1	0

Q1.21 ***In 2's complement representation with 8-bits compute $-30-16$.***

Solution:

$$-30-16 = -30+(-16) = -46_{10}$$

2|30
2|15 - 0
2|7 - 1
2|3 - 1
1 - 1

1 1 1 1 0
MSD LSD

$30 = 11110_2$

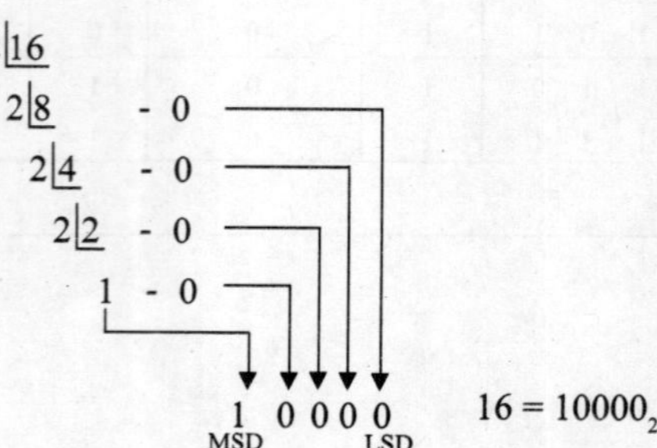

$30_{10} = 11110_2 \xrightarrow{\text{8-bit}} 0001\ 1110_2$ ↓ 1's complement $1110\ 0001 \xrightarrow{+1} 1110\ 0010 \Rightarrow -30_{10}$

$16_{10} = 10000_2 \xrightarrow{\text{8-bit}} 0001\ 0000$ ↓ 1's complement $1110\ 1111 \xrightarrow{+1} 1111\ 0000 \Rightarrow -16_{10}$

$$-30 - 16 \Rightarrow 1110\ 0010$$
$$+\ 1111\ 0000$$
Discard carry ⇒ $\boxed{1}\ 1101\ 0010$

$$1101\ 0010 \xrightarrow{\text{1's complement}} 0010\ 1101 \xrightarrow{+1} 0010\ 1110_2 \xrightarrow{\text{Convert to decimal}} 46_{10}$$

$$\therefore 1101\ 0010 = -46_{10}$$

Q1.22 ***Solve a) $X_{16} + 1334_8 = 2000_8$***

b) $X_8 * X_8 = 10000\ 0000_8$

Solution:

a) $X_{16} + 1334_8 = 2000_8$

$$1334_8 = 1 \times 8^3 + 3 \times 8^2 + 3 \times 8^1 + 4 \times 8^0 = 732_{10}$$
$$2000_8 = 2 \times 8^3 + 0 \times 8^2 + 0 \times 8^1 + 0 \times 8^0$$
$$= 2 \times 512 + 0 + 0 + 0 = 1024_{10}$$
$$X_{16} = 2000_8 - 1334_8 = 1024_{10} - 732_{10} = 292_{10}$$
$$X_{16} = 292_{10} = 124_{16}$$

16 | 292 LSD
16 | 18 - 4
1 - 2
MSD ←

b) $X_8 * X_8 = 10000\ 0000_8$

$$10000\ 0000_8 = (1 \times 8^8)_{10}$$
$$\therefore X_8^2 = (8^8)_{10}$$
$$X_8 = \sqrt{X_8^2} = \sqrt{8^8} = 8^{8 \times \frac{1}{2}} = (8^4)_{10}$$
$$X_8 = 8^4_{10} = 4096_{10} = 10000_8$$

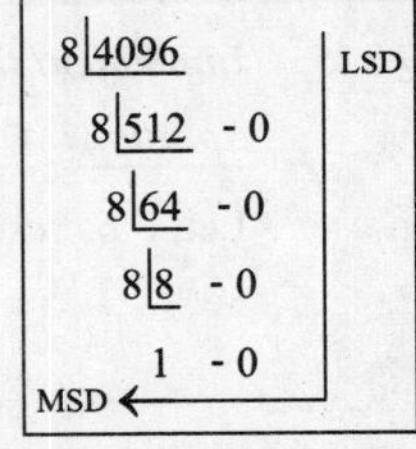
8 | 4096 LSD
8 | 512 - 0
8 | 64 - 0
8 | 8 - 0
1 - 0
MSD ←

Q1.23 ***Draw the NAND gate circuit using NOT, AND and OR gates.*** ***(AU, Nov/Dec'18, 2 Marks)***

Solution:

Table 1:

x	y	Minterm	F = (x . y)′
0	0	m_0	1
0	1	m_1	1
1	0	m_2	1
1	1	m_3	0

$$F = m_0 + m_1 + m_2 = x'y' + x'y + xy'$$

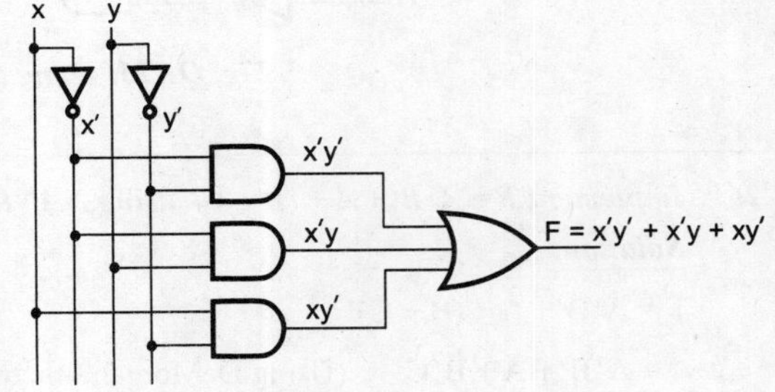

Fig. Q1.23.

Q1.24 Declare a module that describe a circuit that is specified with the following two Boolean expression:

$$E = A + BC + B'D$$

$$F = B'C + BC'D$$

Solution:

The given Boolean equations can be implemented using basic logic gate AND, OR and NOT as shown in Fig. Q1.24.

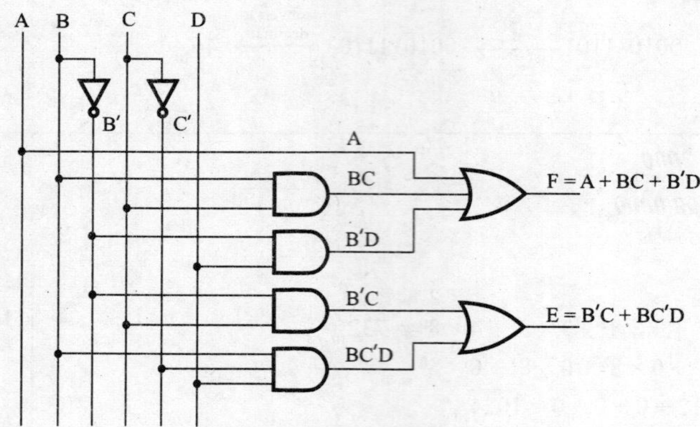

Fig. Q1.24: Logic circuit.

Q1.25 Implement the result of $A + A'D + AC'$. **(AU, Apr/May'19, 2 Marks)**

Solution:

Let, F be result of $A + A'D + AC'$ and it can be obtained from the logic circuit shown in Fig. Q1.25.

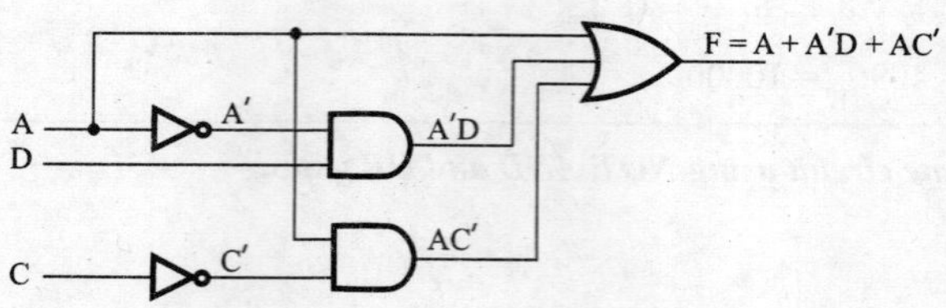

Fig. Q1.25: Logic circuit.

Q1.26 Implement $F = A'B' + A + (B + C)'$ using NAND gates only.

Solution:

$F = A'B' + A + (B + C)'$

$= A'B' + A + B'C'$ (Using DeMorgan's theorem)

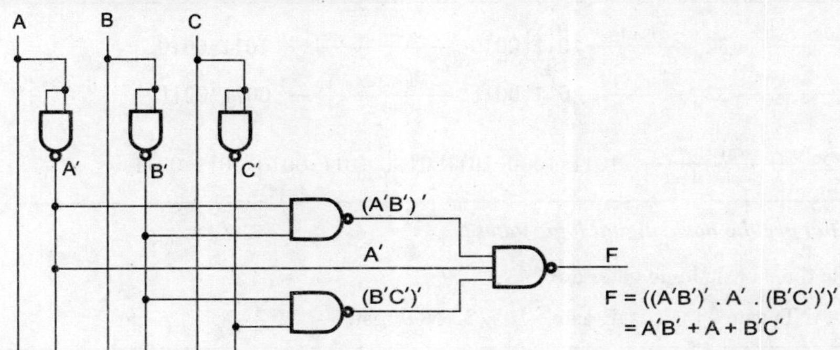

Fig. Q1.26: Logic circuit of F using only NAND gates.

Q1.27 ***i) When is the output of an NAND operation equal to 1?***

ii) When is the output of an NOR operation equal to 1?

From the truth of NAND gate we can say that if any one of the input is 0, then the output of NAND operation is equal to 1.

From the truth table of NOR gate we can say that if all the inputs are 0, then the output of NOR operation is equal to 1.

Table 1: Truth Table of NAND Gates

A	B	(AB)′
0	0	1
0	1	1
1	0	1
1	1	0

Table 2: Truth Table of NOR Gates

A	B	(A + B)′
0	0	1
0	1	0
1	0	0
1	1	0

Q1.28 ***State the classification of binary codes.***

The binary codes are broadly classified into following four categories.

1. Binary coded decimal (BCD)
2. Excess-3 code
3. Gray code
4. Alphanumeric code (ASCII)

Q1.29 ***Represent 3856_{10} in BCD and 2421 code?***

Solution:

$3856_{10} \xrightarrow{\text{BCD}}$ 0011 1000 0101 0110

$3856_{10} \xrightarrow{\text{2421 BCD}}$ 0011 1110 1011 1100

Refer Table 1.20

Q1.30 ***Convert decimal 8723_{10} to both BCD and ASCII with even parity.***

Solution:

8723 $\xrightarrow{\text{BCD}}$ 1000 0111 0010 0011

ASCII is 7-bit binary code with reference to Table 1.29 the ASCII code for 8723_{10} are shown below. Here each decimal digit is represented by 7-bit ASCII code. The addition of even parity bit is also shown below:

8 $\xrightarrow{\text{ASCII}}$ 38_H $\xrightarrow{\text{Binary}}$ 011 1000 $\xrightarrow{\text{Append parity bit}}$ 1011 1000

7 $\xrightarrow{\text{ASCII}}$ 37_H $\xrightarrow{\text{Binary}}$ 011 0111 $\xrightarrow{\text{Append parity bit}}$ 1011 0111

$2 \xrightarrow{\text{ASCII}} 32_H \xrightarrow{\text{Binary}} 011\ 0010 \xrightarrow{\text{Append parity bit}} 1011\ 0010$

$3 \xrightarrow{\text{ASCII}} 33_H \xrightarrow{\text{Binary}} 011\ 0011 \xrightarrow{\text{Append parity bit}} 0011\ 0011$

$8723_{10} \xrightarrow[\text{Append at left}]{\text{ASCII with even parity bit}} 1011\ 1000\ 1011\ 0111\ 1011\ 0010\ 0011\ 0011$

Q1.31 What are the basic digital logic gates?

The three basic logic gates are,

1. AND gate 2. OR gate 3. NOT gate

Q1.32 Simplify the following Boolean function, F = x' y' + xy + x' y.

Solution:

$$F = x'y' + xy + x'y = x'y' + x'y + xy$$

$$= x'(y + y') + xy = x'.1 + xy$$

$$= x' + xy = (x' + x).(x' + y)$$

$$= 1.(x' + y) = x' + y$$

Q1.33 Express F = A + B' C in both canonical SOP and POS form.

Solution:

F = A + B′C

Canonical SOP: It is formed by taking sum of minterms for which F = 1 in truth table.

$\therefore F = \Sigma\,(m_1, m_4, m_5, m_6, m_7)$

$= A'B'C + AB'C' + AB'C + ABC' + ABC$

Canonical POS: It is formed by taking product of maxterms for which F = 0 in truth table.

$\therefore F = \Pi\,(M_0, M_2, M_3)$

$= (A + B + C)\,(A + B' + C)\,(A + B' + C')$

Table 1: Truth Table

A B C	Minterm	Maxterm	B′	B′C	F = A + B′C
0 0 0	m_0	M_0	1	0	0
0 0 1	m_1	M_1	1	1	1
0 1 0	m_2	M_2	0	0	0
0 1 1	m_3	M_3	0	0	0
1 0 0	m_4	M_4	1	0	1
1 0 1	m_5	M_5	1	1	1
1 1 0	m_6	M_6	0	0	1
1 1 1	m_7	M_7	0	0	1

Q1.34 Simplify the function and obtain minimum SOP for F(A, B, C, D) = Σ m(0, 2, 3, 10, 11, 14, 15) + Σ d(8, 12)

Solution:

Minimum SOP form is given by sum of three product terms obtained from K-map.

$\therefore F = B'D' + AC + B'C$

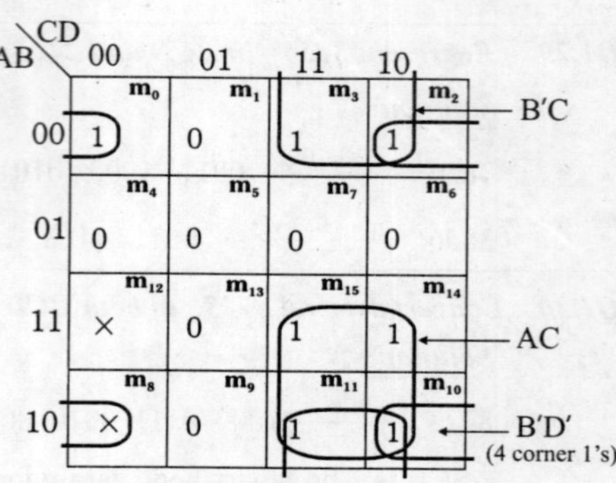

Fig. Q1.34: K-map for F.

Q1.35 ***Find the minimum POS expression for the function F(A, B, C) = Π (2, 6, 7)***

Solution:

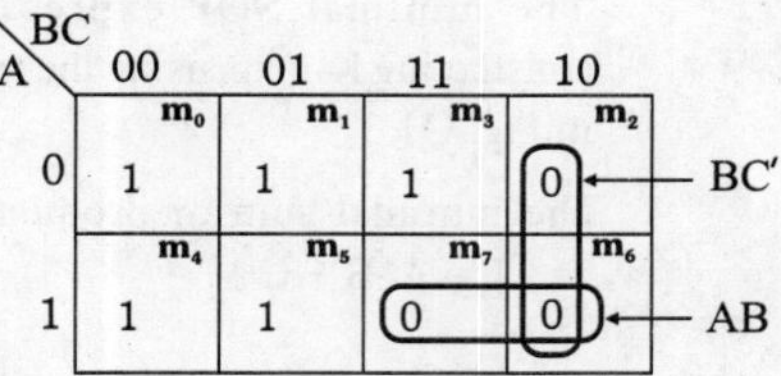

Fig. Q1.35: K-map for F.

Minimum POS form is given by complement of sum of two product terms obtained from K-map.

$\therefore$ F = (BC' + AB)'

= (BC')' (AB)' — Using DeMorgan's theorem

= (B' + C) (A' + B')

Q1.36 ***Obtain the minimal POS for the function, F(A, B, C, D) = Σ m(2, 3, 8, 9, 12, 13)***

Solution:

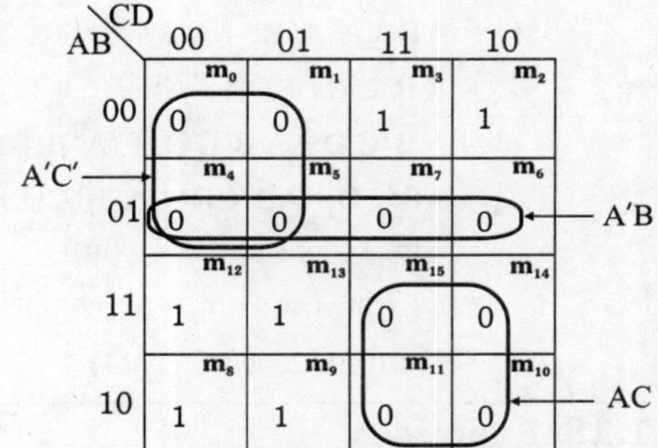

Fig. Q1.36: K-map for F.

Minimum POS form is given by complement of sum of three product terms obtained from K-map.

F(A, B, C, D) = (A'B + A'C' + AC)' — Using DeMorgan's theorem

= (A'B)'(A'C')'(AC)'

= (A + B') (A + C) (A' + C')

Q1.37 ***Given the following boolean function F = A'C + A'B + AB'C + BC.*** **(AU, Nov/Dec'18, 13 Marks)**

a) Express it in sum of minterms.

b) Find the minimal sum of products expression.

Solution:

a) To express as sum of minterms

Table 1: Truth Table

Input Variables			Minterm		Complement of Inputs			Product Terms				Function Output
A	**B**	**C**			**A'**	**B'**	**C'**	**A'C**	**A'B**	**AB'C**	**BC**	**F**
0	0	0	m_0	A'B'C'	1	1	1	0	0	0	0	0
0	0	1	m_1	A'B'C	1	1	0	1	0	0	0	1
0	1	0	m_2	A'BC'	1	0	1	0	1	0	0	1
0	1	1	m_3	A'BC	1	0	0	1	1	0	1	1
1	0	0	m_4	AB'C'	0	1	1	0	0	0	0	0
1	0	1	m_5	AB'C	0	1	0	0	0	1	0	1
1	1	0	m_6	ABC'	0	0	1	0	0	0	0	0
1	1	1	m_7	ABC	0	0	0	0	0	0	1	1

A function can be expressed as a sum of minterms for which the function output is 1. Here, F is 1, when m_1, m_2, m_3, m_5 and m_7 are inputs.

$\therefore$ F = $m_1 + m_2 + m_3 + m_5 + m_7$ = A'B'C + A'BC' + A'BC + AB'C + ABC

b) To determine minimal SOP

The minimal SOP expression can be obtained by constucting K-map, using the truth table (Table 1) as shown in Fig. Q1.37.

The minimal sum of products expression is given by,

F = A'B + C

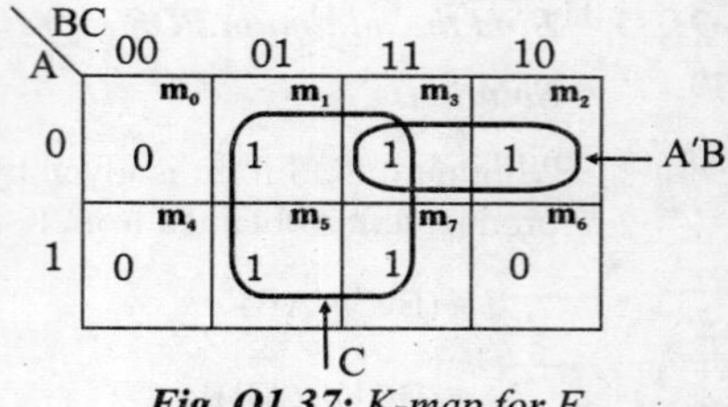

Fig. Q1.37: K-map for F.

Q1.38 Convert the following Boolean expression into standard SOP form:

A B' C + A' B' + A B C' D

Solution:

$$F = AB'C + A'B' + ABC'D$$

$$= AB'C(D + D') + A'B'(C + C') + ABC'D$$

$$= AB'CD + AB'CD' + (A'B'C)(D + D') + (A'B'C')(D + D') + ABC'D$$

Missing literals in first term is D and second term is C and D

$$= \underset{(m_{11})}{AB'CD} + \underset{(m_{10})}{AB'CD'} + \underset{(m_3)}{A'B'CD} + \underset{(m_2)}{A'B'CD'} + \underset{(m_1)}{A'B'C'D} + \underset{(m_0)}{A'B'C'D'} + \underset{(m_{13})}{ABC'D}$$

$$= m_0 + m_1 + m_2 + m_3 + m_{10} + m_{11} + m_{13}$$

$$= \sum m(0, 1, 2, 3, 10, 11, 13)$$

1.18 Exercises

I. Fill in the blanks

1. The unsigned binary has only ______ numbers.
2. The signed binary has ______ and ______ numbers.
3. For positive number the sign bit is ______ and for negative number the sign bit is ______.
4. ASCII was developed in the early stage of digital systems in order to standartize the coding system for ______ and ______ interface.
5. The two voltage levels of logic circuits are called ______ and ______.
6. NAND and NOR gates are called ______.
7. The realizations in SOP and POS forms are called ______.
8. A gate in which any one input must be low to get a high output is called ______.
9. A gate in which any one input must be high to get a low output is called ______.
10. Binary ______ can be performed via addition by using the 1's or 2's complement method.
11. -19_{10} in 2's complement representation is ______.
12. The ______ gate output in high when the inputs are not the same.
13. The output of a 2-input ______ gate is 0 if and only if its inputs are unequal.
14. The most suitable gate for comparing two bits is ______.
15. The ______ of a sum term is equal to the product of the complements of the variables in the sum term.
16. A 3-variable Karnaugh map has ______ cells.
17. The AND operation can be produced with ______ NAND gates.

18. The operational symbol for exclusive-OR operation is ______.
19. AND-OR logic produces the output of an expression in ______ form.
20. To implement the expression AB′C + AC′, it takes ______ OR gate(s), ______ AND gate(s) and ______ inverter(s).
21. _______ is defined as the time required for a change in input to be available in the output.
22. _______ is a class of digital circuits built from diodes, resistors and bipolar junction transistors.
23. The propagation delay of a TTL logic device is _______.
24. The TTL output acts as a current sink in the _______.
25. The basic TTL logic circuit is _______.
26. The noise margin of a TTL gate is about _______.

Answers

1. positive	9. NOR gate	17. 2	25. NAND
2. positive, negative	10. subtraction	18. ⊕	26. 0.4 *V*
3. 0, 1	11. NAND gate	19. SOP	
4. human, machine	12. NOR gate	20. 1, 2, 2	
5. high, low	13. subtraction	21. propagation delay	
6. universal gates	14. 11101101 (Standard 8-bit)	22. DTL	
7. standard form	15. exclusive-OR	23. 9 *ns*	
8. NAND gate	16. XNOR	24. low state	

II. State whether the following statements are True or False

1. The hexadecimal number system is a radix number system in which the radix r is 16.
2. The octal number system is a radix number system in which the radix r is 2.
3. The unsigned binary has only negative numbers.
4. The complements of a number are used to represent negative numbers in digital system.
5. An inverter performs an operation known as complement.
6. The output of NOR gate is high if and only if all its inputs are low.
7. AND gate can be used as an inverter.
8. A + A′B = A + B.
9. The complement of a product term is equal to the sum of the complement of the variables in the product term.
10. An example of a product-of-sums expression is A(B + C) + AC′.
11. On a Karnaugh map grouping the 0's produces AND-OR logic.
12. The expression A′BCD + ABCD′ + AB′C′D cannot be simplified.
13. Fan-in is defined as the maximum number of standard load that an output can drive reliably.
14. A quantitative measure of noise immunity in terms of voltage is called noise margin.
15. The maximum negative voltage that can be applied to a TTL input is −0.2 *V*.
16. The output of a tristate logic state is high.
17. Emitter-coupled logic (ECL) circuits are the fastest available bipolar logic gates.
18. The primary advantage of ECL is the high quiescent power dissipation.

Answers

1. True	5. True	9. True	13. False	17. True
2. False	6. True	10. False	14. True	18. False
3. False	7. False	11. False	15. False	
4. True	8. True	12. True	16. False	

III. Choose the right answer for the following questions

1. ***The Boolean expression A'B' + (AB)' + AB is equivalent to***

a) A' + B'A' **b)** A'B **c)** AB + (AB)' **d)** AB

2. ***The Boolean expression A ⊕ B ⊕ B is equivalent to***

a) AB + A'B' **b)** A'B **c)** A **d)** B

3. ***In the logic circuit shown in Fig. 3, the output F is***

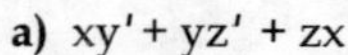

a) xy' + yz' + zx'

b) x + z

c) xy + yz + zx

d) (xy)' + (yz)' + (zx)'

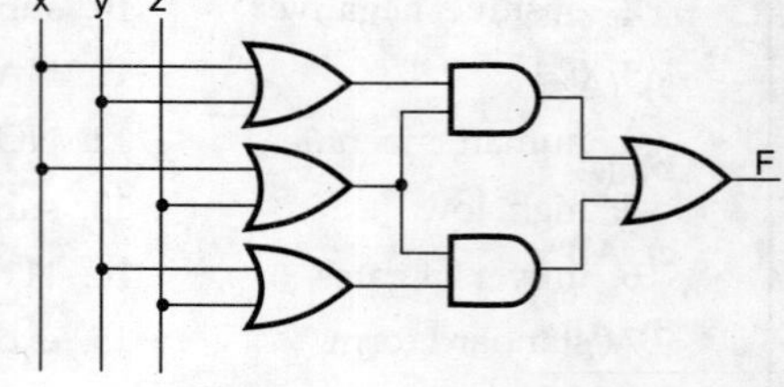

Fig. 3.

4. ***An XOR gate with 6-variables is given by A⊕B⊕C⊕D⊕E⊕F. The number of minterms in the Boolean expression***

a) 6 **b)** 12 **c)** 64 **d)** 32

5. ***Which of the following equation is equivalent to NOT***

a) X NAND X **b)** X NOR X' **c)** X NAND 1 **d)** X NOR 1

6. ***While obtaining minimal SOP expression***

a) All don't cares are ignored

b) All don't cares are treated as logic ones

c) All don't cares are treated as logic zeros

d) Only such don't cares that aid minimization are treated as logic ones

7. ***Y = AB'C' + A'BC' is equivalent to***

a) Y = A **b)** Y = AB **c)** Y = AC **d)** Y = C'(A⊕B)

8. ***The circuit shown in Fig. 8 realizes the function***

a) [(A' + B') + C] . (D' + E')

b) (A+B)C + D + E

c) AB + C + DE

d) AB + C(D +E)

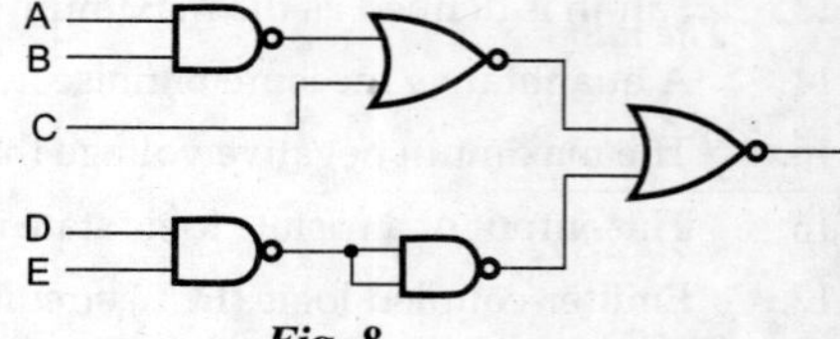

Fig. 8.

9. ***The output Y of the given function in Fig. 9 represents***

a) OR b) AND

c) NAND d) NOR

A B Y

Fig. 9.

10. ***Which of the following is equivalent of AND-OR realization***

a) NAND–NOR realization b) NOR–NOR realization

c) NOR–NAND realization d) NAND–NAND realization

11. ***The number of minterms in the function f(x, y, z) = xy' + z' is***

a) 4 b) 5 c) 3 d) 6

12. ***If X = 1 in the logic equation, (X + Z (Y' + (Z' + XY'))) (X' + Z'(X + Y)) = 1, then***

a) Y = Z b) Y = Z' c) Z = 1 d) Z = 0

13. ***The circuit shown in Fig. 13 realises the function***

a) (A' + B').C + (DE)'

b) (A + B) (C + D + E)

c) AB + C + DE

d) AB + C(D + E)

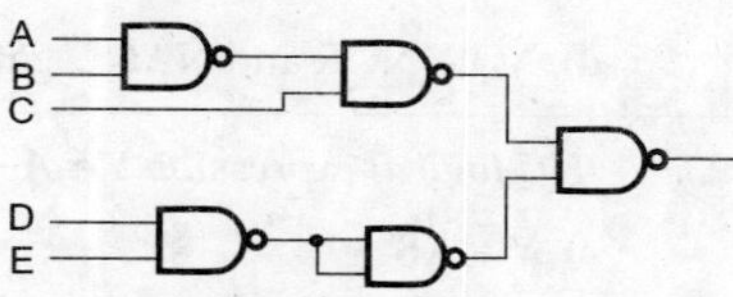

Fig. 13.

14. ***In the Fig. 14 X_0, X_1, X_2 will be 1's complement of ABC if***

a) Y = 1

b) Y = 0

c) Y = A' = B' = C'

d) Y = A = B = C

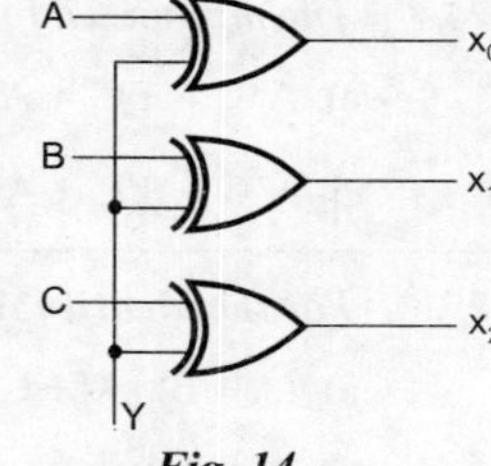

Fig. 14.

15. ***The circuit in Fig. 15 works as***

a) XOR

b) XNOR

c) NOR

d) NAND

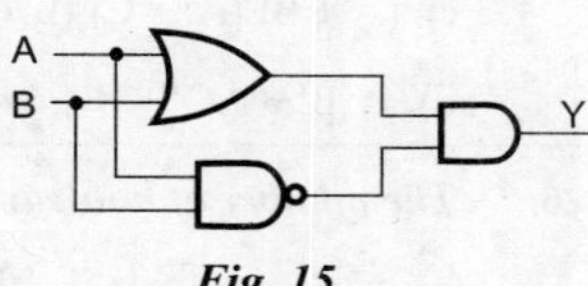

Fig. 15.

16. ***The Boolean function A + BC is a reduced form of***

a) AB + BC b) (A + B) (A + C) c) A'B + AB'C d) (A' + C) B

17. ***The number of distinct Boolean expression of 4 variables is***

a) 16 b) 256 c) 1024 d) 65536

18. ***4 bit 2's complement representation of a decimal number is 1000. The number is***

a) +8 b) 0 c) – 7 d) – 8

19. *The range of signed decimal numbers that can be represented by 6 bit 1's complement number is*

a) – 31 to + 31 b) – 63 to +64 c) – 64 to +63 d) – 32 to 31

20. *The 2's complement representation of – 7 is*

a) 01110 b) 01111 c) 11110 d) 11001

21. *An equivalent 2's complement representation of the 2's complement number 1101 is*

a) 110100 b) 001101 c) 110111 d) 111101

22. *The complete set of only those logic gates designated as universal gates is*

a) NOT, OR and AND gates

b) XNOR, NOR and NAND gates

c) NOR and NAND gates

d) XOR, NOR and NAND gates

23. *The logical expression Y = A + A'B is equivalent to*

a) Y = AB + A b) Y = A'B + A'

c) Y = AB' + B d) Y = AB + A'

24. *The minimized form of logical expression Y = A'B'C' + A'B C' + A'B C + ABC' is*

a) A'C' + BC' + A'B b) AC' + B'C + A'B

c) A'C + B'C + A'B d) AC' + B'C + AB'

25. *For the identity AB + AC + BC = AB + AC the dual form is*

a) (A + B) (A' + C) (B + C) = (A + B) (A' + C)

b) (A' + B') (A' + C') (B' + C') = (A' + B') (A' + C')

c) (A + B) (A' + C) (B + C) = (A' + B') (A + C')

d) A'B' + AC' + B'C' = A'B' + AC'

26. *The number of boolean functions that can be generated by n variable is equal to*

a) 2^{2^n} b) 2^{2n} c) 2^{n-1} d) 2^n

27. *The minimal function that can detect a 8421 BCD code (represented as D_8 D_4 D_2 D_1) divisible by 3 is given by*

a) $D_8D_1 + D_4D_2 + D_3D_2D_1$

b) $D_8D_1 + D_4D_2D_1' + D_4'D_2D_1$

c) $D_4D_1 + D_4D_2 + D_3D_1'D_2'D_1$

d) $D_4 + D_2D_1' + D_4'D_2D_1 + D_8D_4D_2D_1$

28. ***The K-map for a Boolean function is given. The number of essential prime implicants for this function is***

CD \ AB	00	01	11	10
00	m_0 1	m_1 1	m_3 0	m_2 1
01	m_4 0	m_5 0	m_7 0	m_6 1
11	m_{12} 1	m_{13} 0	m_{15} 0	m_{14} 0
10	m_8 1	m_9 0	m_{11} 0	m_{10} 1

Fig. 28.

a) 4

b) 5

c) 6

d) 8

29. ***The Booelan expression AC + BC' is equivalent to***

a) A'C + BC' + AC **b)** B'C + AC + BC' + A'CB'

c) AC + BC' + B'C + ABC **d)** ABC + A'BC' + ABC' + AB'C

30. ***For the logic circuit in Fig. 30 the required input condition (x, y, z) to make the output F = 1 is***

a) 1, 0, 1

b) 0, 0, 1

c) 1, 1, 1

d) 0, 1, 1

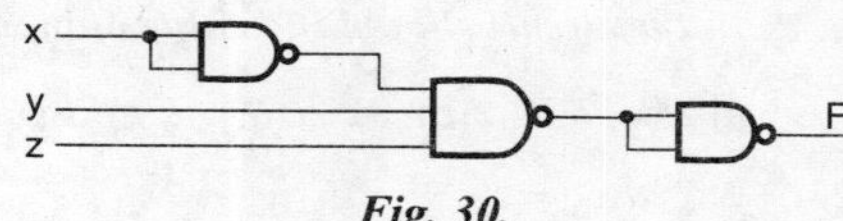

Fig. 30.

31. ***For the logic circuit, the simplified Boolean expression for the output F is***

a) x + y + z

b) x

c) y

d) z

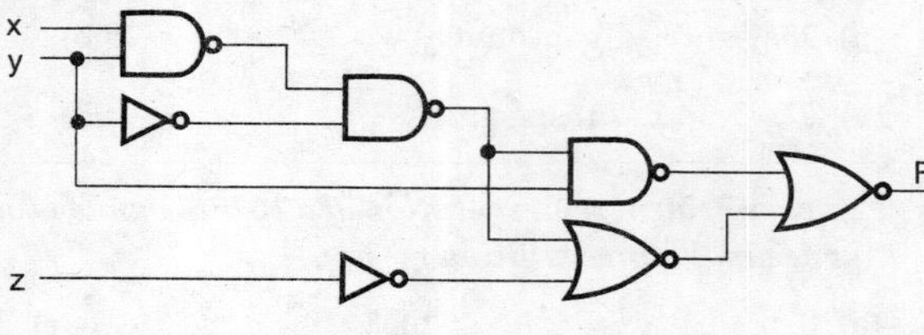

Fig. 31.

32. ***The minimun number of 2 input NAND gates required to implement the Booelan function AB'C assuming that A, B and C are available is***

a) 2 **b)** 3 **c)** 5 **d)** 6

33. ***For the logic circuit shown in Fig. 33 the output F is***

a) (xyz)'

b) x' + y' + z'

c) (xy)' + (yz)' + x' + z'

d) x' + z'

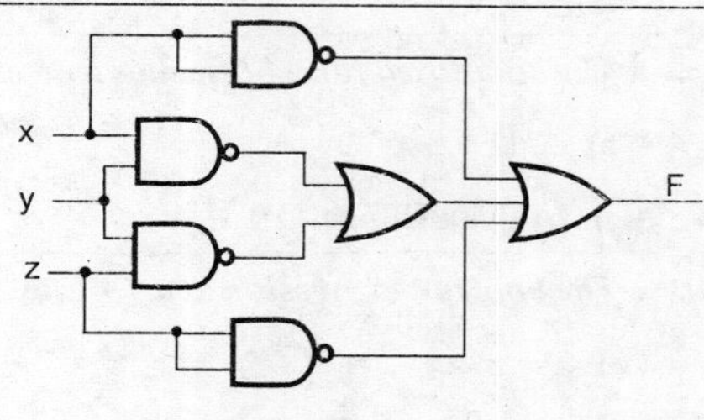

Fig. 33.

34. *Boolean expression for output of XNOR equivalent logic gate in dual form with inputs A and B is*

a) AB' + A'B b) (AB)' + AB

c) (A' + B) (A + B) d) (A' + B') (A + B)

35. *The output of a logic gate is 1 when all its inputs are at logic 0. Then gate is either*

a) a NAND or an XOR gate b) a NOR or an XNOR gate

c) an OR or an XNOR gate d) an AND or an XOR gate

36. *The minimum number of NAND gates required to implement the boolean function A + AB' + AB'C is equal to*

a) 0 b) 1 c) 4 d) 7

37. *The output of the logic gate in Fig. 37 is*

a) 0 b) 1 c) A d) A'

Fig. 37.

38. *What is output Z of an XOR gate whose all inputs are set at A*

a) Z = A b) Z = 1 c) Z = 0 d) Z = A

39. *In the circuit shown the value of input for Y = 0 is*

a) 00 b) 01 c) 10 d) 11

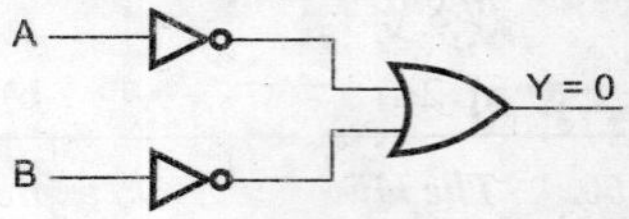

Fig. 39.

40. *11001, 1001 and 111001 corresponding to the 2's complement representation of which one of the following sets of numbers.*

a) 25, 9 and 57 respectively b) – 6, – 6 and – 6 respectively

c) –7, –7 and –7 respectively d) –25, –9 and –57 respectively

41. *2's complement representation of a 16 bits number (one sign and 15 magnitude bits) is FFFF, its magnitude in decimal representation is*

a) 0 b) 1 c) 32,767 d) 65,535

42. *With 2's complement representation, the range of values that can be represented on the data bus of an 8-bit microprocessor is given by*

a) –128 to 127 b) –128 to +128

c) –127 to +128 d) –256 to +256

43. *The simplified form of the boolean expression Y = (A'BC + D) (A'D + B'C') can be written as*

a) A'D + B'C'D b) AD + BC'D

c) (A' + D) (B'C + D') d) AD' + BCD'

44. *The boolean expression x'yz' + (xy)'z + xyz' + xy'z + xyz can be simplified to*

a) x' + z b) y + z

c) y' + z' d) y' + z

45. ***Decimal 43 in Hexadecimal and BCD number system is respectively***

a) B2, 0100 0011 b) 2B, 0100 0011 c) 2B, 0011 0100 d) B2, 0100 0100

46. ***A new Binary Coded Pentary (BCP) number system is proposed in which every digit of a base-5 number is represented by its corresponding 3-bit binary code. For example, the base-5 number 24 will be represented by its BCP code 010100. In this numbering system, the BCP code 100010011001 corresponds to the following number in base-5 system.***

a) 423 b) 1324 c) 2201 d) 4231

47. ***The octal equivalent of the HEX number AB.CD is***

a) 253.314 b) 253.632 c) 526.314 d) 526.632

48. ***X = 0 1 1 1 0 and Y = 1 1 0 0 1 are two 5-bit binary numbers represented in two's complement format. The sum of X and Y represented in two's complement format using 6 bit is***

a) 100111 b) 001000 c) 000111 d) 101001

49. ***The boolean function Y = AB + CD is to be realized using only 2-input NAND gates. The minimum number of gates required is***

a) 2 b) 3 c) 4 d) 5

50. ***The minimum POS expression for K-map given in the Fig. 50 is***

a) (B + C) (C' + D)

b) (B + D) (A + C)

c) (A' + C') (C + D')

d) (B + D) (A + D')

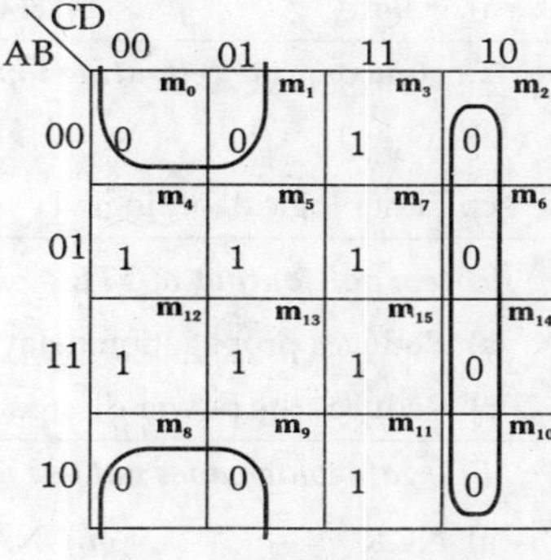

Fig. 50.

51. ***The output F in Fig. 51 is***

a) (A' + B') b) AB

c) (A' . B') d) A + B

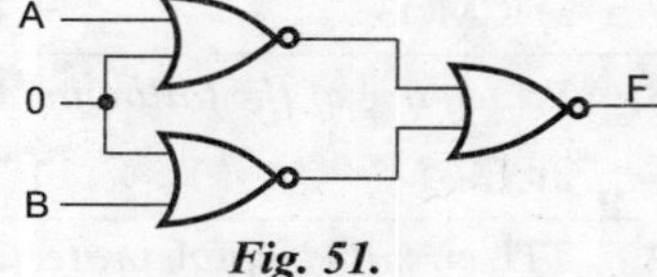

Fig. 51.

52. ***The function F in Fig. 52 is***

a) A + B C + CD

b) A + BC' + C'D'

c) A(B' + C) (C + D)

d) A' (B + C') (C' + D')

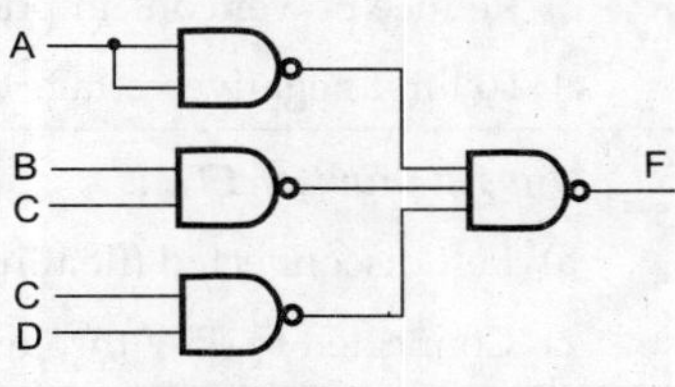

Fig. 52.

53. ***The dual of the function is A(B'C + BC + BC')***

a) A' + (B + C') (B' + C') (B' + C) b) A'(BC' + B'C' + B'C)

c) C (B'A + CA + CA') d) AB'C + ABC + ABC'

54. ***The expression (A + A'B) (B' + BA) can be simplified to***

a) AB' + A'B b) AB' + A'B' c) 1 d) A

55. ***The minimal sum of products form of F = AB'CD + A'BC + A'B'C + BCD is***

a) AC' + BD b) A'C + CD c) AC + B'D d) AB' + CD'

56. ***The circuit in Fig. 56 is works as***

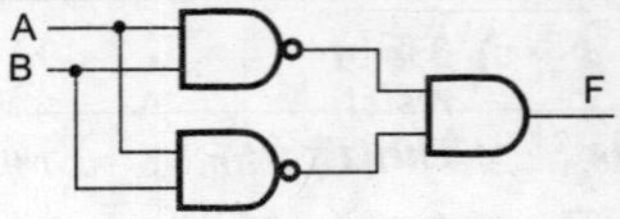

a) XOR b) XNOR

c) NOR d) NAND

A
B
F

Fig. 56.

57. ***Propagation delay t_{PLH} and t_{PHL} of a standard TTL NAND gate is***

a) 10 ns and 6 ns b) 11 ns and 7 ns

c) 11 ns and 6 ns d) 10 ns and 7 ns

58. ***The worst case dc noise margin of a TTL gate is***

a) 600 mV b) 800 mV c) 400 mV d) 100 mV

59. ***The (unconnected) floating input of a TTL circuit acts as a***

a) Logic 0 b) Logic 1

c) Either logic 0 (or) logic 1 d) None of the above

60. ***Totem pole output of TTL***

a) Reduces propagation delay b) Increases the speed

c) Reduces the power dissipation d) Increases the fan out

61. ***TTL gate which does not use multiple emitter is***

a) NOR b) NAND c) AND d) NOR and OR

62. ***Multiple emitter is the characteristics of the***

a) CMOS b) DTL c) TTL d) ECL

63. ***Which one of the following logic families is known as current mode logic (CML)***

a) ECL b) TTL c) DTL d) CMOS

64. ***TTL circuits employ protective shunt diodes on each input to***

a) Reduce power consumption b) Increase the noise margin

c) To limit negative voltage excursions d) None of the above

65. ***Unused input of TTL is***

a) Left disconnected (floating) b) Connected to the used input

c) Connected to +5 V through a 1 kΩ resistor d) All the above

66. ***Basic circuit of CMOS is inverter, which consists of***

a) One p-channel and one n-channel transistor **b)** Both p-channel

c) Both n-channel **d)** None of the above

67. ***TTL open collector output is generally slower than totem pole output. Because,***

a) Open collector output does not use active pull up resistor

b) Open collector output use active pull up resistor

c) Open collector output uses pull down resistor

d) Totem pole output uses pull down resistor

68. ***A 5V battery has an internal resistance of 100 Ω and it is supplying N identical gates each having an input resistance of 3 KΩ. The gates are in parallel. The minimum voltage at each should not fall below 3.5 V. Then N is,***

a) 12 **b)** 10 **c)** 11 **d)** 13

69. ***The unused inputs of an AND gate can be connected to,***

a) used input **b)** to $+V_{CC}$ through 1 kΩ resistance

c) ground **d)** None of the above

70. ***In totem pole arrangement***

a) Only one transistor will be conducting **b)** Both the transistors will not conduct

c) Both the transistors will be conducting **d)** Above all

71. ***Which of the following TTL configuration outputs should not be tied together***

a) Open collector **b)** Totem pole

c) Tristate **d)** None of the above

72. ***In ECL the switching speed is increased by***

a) Reducing the storage delay time **b)** By preventing the transitor to drive in to saturation

c) Using current switching **d)** All the above

73. ***In standard TTL, the totem pole stage refers to***

a) The multi-emitter input stage **b)** The phase splitter

c) The output buffer **d)** Open collector output stage

74. ***Commerically available ECL gate use two ground lines and one negative supply in order to***

a) reduce power dissipation **b)** Increase fan-out

c) reduce loading effect **d)** eliminate the effect of power line glitches or the biasing circuit

75. ***The full form of the abbreviations TTL and CMOS in reference to logic families are***

a) Triple Transistor logic and chip metal oxide semiconductor.

b) Tristate Transistor logic and chip metal oxide semiconductor.

c) Transistor Transistor logic and complementary metal oxide semiconductor.

d) Tristate Transistor logic and complementary metal ocide silicon.

Answers

1. c	11. b	21. d	31. c	41. c	51. b	61. d	71. b
2. c	12. d	22. c	32. c	42. a	52. a	62. c	72. d
3. b	13. a	23. c	33. b	43. a	53. a	63. a	73. c
4. a	14. a	24. a	34. d	44. b	54. d	64. c	74. d
5. a	15. a	25. b	35. b	45. b	55. b	65. b	75. c
6. d	16. b	26. b	36. a	46. d	56. d	66. a	
7. d	17. b	27. b	37. d	47. b	57. b	67. a	
8. a	18. d	28. a	38. c	48. c	58. c	68. a	
9. a	19. a	29. d	39. d	49. b	59. b	69. b	
10. d	20. d	30 d	40. c	50. a	60. a	70. a	

IV. Answer the following questions

1. How many binary bits are needed to represent an octal and hexadecimal digit? Justify.
2. What is the range of decimal numbers that can be represent in n-bit a) Sign-magnitude form of binary number b) Unsigned binary number c) 1's complement binary number d) 2's complement binary number.
3. How is carry handled in complement subtraction?
4. Explain the significance of parity?
5. Explain absorption theorem with truth table.
6. Which are universal gates, why are they called so?
7. What are minterms and maxterms?
8. What is the difference between canonical and SOP/POS form?
9. How will be the resultant function of K-map when minterm and maxterm are used?
10. How do we handle don't-care conditions?
11. What are the methods available to simplify a Boolean expression?

V. Solve the following problems

E1.1 Convert hexadecimal number 2C6B to binary and octal.

E1.2 Convert decimal number 27_{10} to binary in two ways:

a) Directly to binary.

b) Convert first to hexadecimal and then from hexa to binary.

E1.3 Add the following numbers without converting to decimal.

a) Binary numbers 1110 and 001.

b) Hexadecimal numbers 3E and 54.

E1.4 Obtain the 1's and 2's complement of the following binary numbers.

a) 11101111_2 b) 11111111_2 c) 01010101_2 d) 00100101_2

E1.5 Convert the following binary numbers to hexadecimal and to decimal a) 1.11010 b) 100.101

E1.6 Convert decimal +49 and +39 to binary, using the signed 2's complement representation and enough digits to accommodate the numbers. Then perform the binary equivalent of a) (+39) + (–49), b) (–39) + (+49) and c) (– 39) + (– 49). Convert the answers back to decimal and verify that they are correct.

E1.7 Perform the following subtraction by 1's and 2's complement subtraction

a) $100010_2 - 100110_2$ b) $10011_2 - 10010_2$

E1.8 Simplify the following Boolean expression to a minimum number of literals.

a) x'y' + x'y b) (x + y)'(x' + y) c) (x + y' + z') (x' + z')

E1.9 Represent the decimal number 5137 in a) BCD b) Excess-3 code c) 2421 code d) 8, 4, – 2, – 1 code

E1.10 Find the complement of the following expression:

a) x'y + x'y' b) xyz + x'y + xyz'

E1.11 Given two Boolean functions F_1 and F_2, show that (a) the Boolean function E = $F_1 + F_2$ contains the sum of the minterms of F_1 and F_2. (b) The Boolean function G = F_1 F_2 contains only the minterms that are common to F_1 and F_2.

E1.12 List the truth table of the function:

a) F_1 = x'y' + x'y + yz' b) F_2 = b'c' + ac

E1.13 Demonstrate the validity of the following identities by means of truth tables:

Demorgan's theorem for 3 variables: a) (x + y + z)' = x'y'z' b) (xyz)' = x' + y' + z'

The associate law: c) x + (y + z) = (x + y) + z d) x . (y . z) = (x . y) . z

E1.14 Express the complement of the following functions in sum of minterms form,

a) F(A, B, C, D) = $\sum$ m(3, 5, 9, 11, 15)

b) F(x, y, z) = $\prod$ M(2, 4, 5, 7)

E1.15 Convert each of the following expression into sum of products and product of sums.

a) (AB + C) (B + C'D) b) x + x' (x' + y') (y' + z')

E1.16 Show that the dual of the XOR is equal to complement of XOR.

E1.17 Show that the complement of XOR is equal to XNOR.

E1.18 Determine the complement of Boolean expression (b + d) (a' + b' + c) and show that POS becomes SOP.

E1.19 Determine the complement of Boolean expression a'b + a'c' + abc and show that SOP becomes POS.

E1.20 Implement the Boolean function, F = xy + x'y + y'z'

a) With AND, OR and NOT gates. b) With OR and NOT gates.

c) With AND and NOT gates. d) With NAND and NOT gate.

e) With NOR and NOT gates.

E1.21 Obtain the truth table of the following functions and express each function in sum of minterms and product of maxterms form:

a) (xy + z) (y + xz) b) (x + y') (y' + z) c) x'z + wx'y + wyz' + w'y' d) (xy + yz' + x'z) (x + z)

E1.22 For the Boolean function, F = x'yz + xyz' + wx'y' + w'xy + wxy

a) Obtain the truth table of F.

b) Draw the logic diagram, using the original Boolean expressions.

c) Use Boolean algebra to simplify the function to a minimum number of literals.

d) Obtain the truth table of the function from the simplified expression and show that it is the same as the one in part (a).

e) Draw the logic diagram from the simplified expression and compare the total number of gates with the diagram in part (b).

E1.23 Simplify the following Boolean functions using 3-variable K-maps:

a) $F_1(x, y, z) = \Sigma\, m(0, 1, 5, 7)$ b) $F_2(x, y, z) = \Sigma\, m(2, 3, 4, 5)$

E1.24 Simplify the following Boolean functions using K-maps:

a) $F_1(x, y, z) = \Sigma\, m(2, 3, 6, 7)$ b) $F_2(A, B, C, D) = \Sigma\, m(4, 6, 7, 15)$

E1.25 Simplify the following Boolean functions using 4-variable K-map:

a) $F_1(w, x, y, z) = \Sigma\, m(1, 4, 5, 6, 12, 14, 15)$ b) $F_2(A, B, C, D) = \Sigma\, m(2, 3, 6, 7, 12, 13, 14)$

E1.26 Simplify the following Boolean functions using 3-variable K-maps:

a) $F_1 = x'y' + yz' + x'yz'$ b) $F_2 = x'yz + xy'z' + xy'z$

E1.27 Simplify the Boolean function using 4-variable K-maps: F = A'B'C'D' + AC'D' + B'CD' + A'BCD + BC'D

E1.28 Find the minterms of the following Boolean expression by first plotting each function in a K-map:

a) $F_1 = xy + yz + xy'z$ b) $F_2 = C'D + ABC' + ABD' + A'B'D$

E1.29 Using K-map find the minimum sum of products (MSOP) representation for,

$F(A, B, C, D, E) = \Sigma\, m(2, 5, 7, 11, 21, 23, 25, 27) + \Sigma\, d(1, 12, 17, 28)$

Draw the logic circuit of the minimal expression using only NAND gates.

E1.30 Using K-map find the minimum product of sum (MPOS) representation for,

$F(A, B, C, D, E) = \Pi(0, 1, 2, 3, 4, 5, 16, 17, 18, 19, 24, 25) + \Sigma\, d(26, 27)$

Draw the logic circuit of the minimal expression using only NOR gates.

E1.31 Find the prime implicants for the following Boolean function and determine the essential prime implicants.

$F(A, B, C, D) = \Sigma\, m(0, 2, 3, 5, 7, 8, 10, 11, 14, 15)$

Answers

E1.1 $0010\ 1100\ 0110\ 1011_2$, 26153_8

E1.2 a) 11011_2 b) $1B_{16} = 00011011_2$

E1.3 a) 1111_2 b) 92_{16}

E1.4

1's Complement	2's Complement
a) 00010000	a) 00010001
b) 00000000	b) 00000001
c) 10101010	c) 10101011
d) 11011010	d) 11011011

E1.5 **a)** 1.8125_{10}, $1.D_{16}$ **b)** 4.625_{10}, $4.A_{16}$

E1.6 **a)** $-0000\ 1010_2$, -10_{10}

b) $0000\ 1010_2$, $+10_{10}$

c) $-0101\ 1000_2$, -88_{10}

E1.7 **a)** -000100_2 **b)** 00001_2

E1.8 **a)** x′ **b)** x′y′ **c)** z′+x′y′

E1.9 **a)** 0101 0001 0011 0111 **b)** 1000 0100 0110 1010

c) 1011 0001 0011 1101 **d)** 1011 0111 0101 1001

E1.10 **a)** (x + y′) (x + y) **b)** (x′ + y′ + z′) (x + y′) (x′ + y′ + z)

E1.12 a)

Input Variables			Complement of Inputs			Product Terms			Function Output
x	y	z	x′	y′	z′	x′y′	x′y	yz′	F_1
0	0	0	1	1	1	1	0	0	1
0	0	1	1	1	0	1	0	0	1
0	1	0	1	0	1	0	1	1	1
0	1	1	1	0	0	0	1	0	1
1	0	0	0	1	1	0	0	0	0
1	0	1	0	1	0	0	0	0	0
1	1	0	0	0	1	0	0	1	1
1	1	1	0	0	0	0	0	0	0

b)

Input Variables			Complement of Inputs			Product Terms		Function Output
a	b	c	a′	b′	c′	b′c′	ac	F_2
0	0	0	1	1	1	1	0	1
0	0	1	1	1	0	0	0	0
0	1	0	1	0	1	0	0	0
0	1	1	1	0	0	0	0	0
1	0	0	0	1	1	1	0	1
1	0	1	0	1	0	0	1	1
1	1	0	0	0	1	0	0	0
1	1	1	0	0	0	0	1	1

E1.13 a)

x	y	z	x + y + z	(x + y + z)′	x′	y′	z′	x′y′z′
0	0	0	0	1	1	1	1	1
0	0	1	1	0	1	1	0	0
0	1	0	1	0	1	0	1	0
0	1	1	1	0	1	0	0	0
1	0	0	1	0	0	1	1	0
1	0	1	1	0	0	1	0	0
1	1	0	1	0	0	0	1	0
1	1	1	1	0	0	0	0	0

b)

x	y	z	x y z	(x y z)′	x′	y′	z′	x′+ y′+ z′
0	0	0	0	1	1	1	1	1
0	0	1	0	1	1	1	0	1
0	1	0	0	1	1	0	1	1
0	1	1	0	1	1	0	0	1
1	0	0	0	1	0	1	1	1
1	0	1	0	1	0	1	0	1
1	1	0	0	1	0	0	1	1
1	1	1	1	0	0	0	0	0

c)

x	y	z	y + z	x + (y + z)	x + y	(x + y) + z
0	0	0	0	0	0	0
0	0	1	1	1	0	1
0	1	0	1	1	1	1
0	1	1	1	1	1	1
1	0	0	0	1	1	1
1	0	1	1	1	1	1
1	1	0	1	1	1	1
1	1	1	1	1	1	1

d)

x	y	z	y . z	x . (y . z)	x . y	(x . y) . z
0	0	0	0	0	0	0
0	0	1	0	0	0	0
0	1	0	0	0	0	0
0	1	1	1	0	0	0
1	0	0	0	0	0	0
1	0	1	0	0	0	0
1	1	0	0	0	1	0
1	1	1	1	1	1	1

E1.14 a) $\sum m(0, 1, 2, 4, 6, 7, 8, 10, 12, 13, 14)$ **b)** $\sum m(2, 4, 5, 7)$

E1.15 a) AB + BC (SOP), B(A + C) (POS)

b) x + x'y' + x'z' (SOP), x + y' + z' (POS)

E1.18 b'd' + abc' (Sum of products)

E1.19 (a + b') (a + c) (a' + b' + c') (Product of sums)

E1.20 a) F = xy + x'y + y'z'

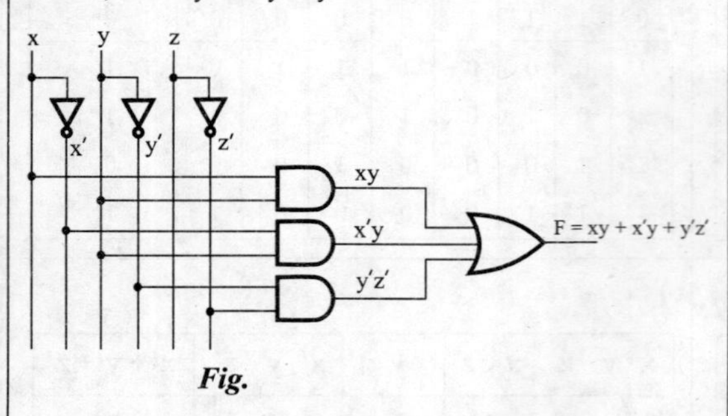

Fig.

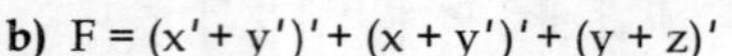

b) F = (x' + y')' + (x + y')' + (y + z)'

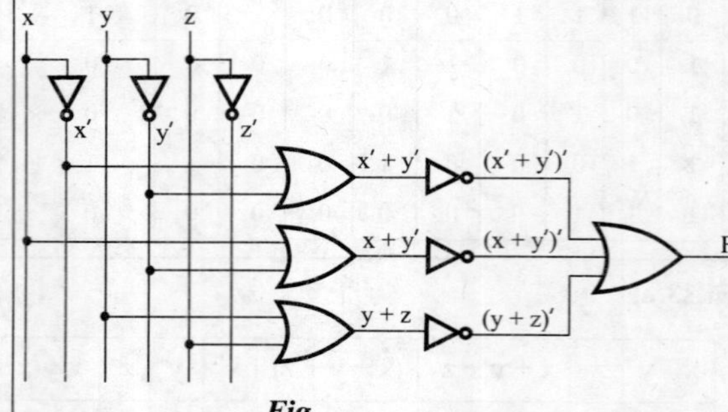

Fig.

c) F = ((xy)'(x'y)'(y'z')')'

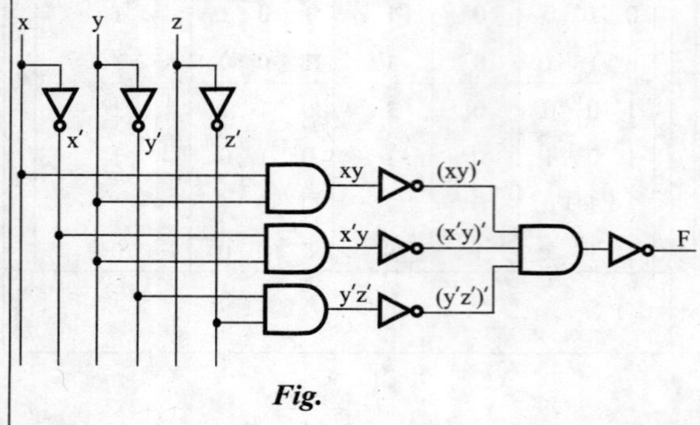

Fig.

d) F = ((xy)'(x'y)'(y'z')')'

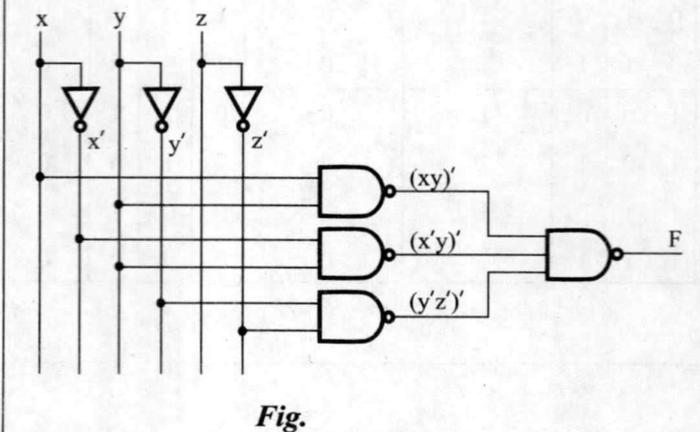

Fig.

e) $F = (x' + y')' + (x + y')' + (y + z)'$

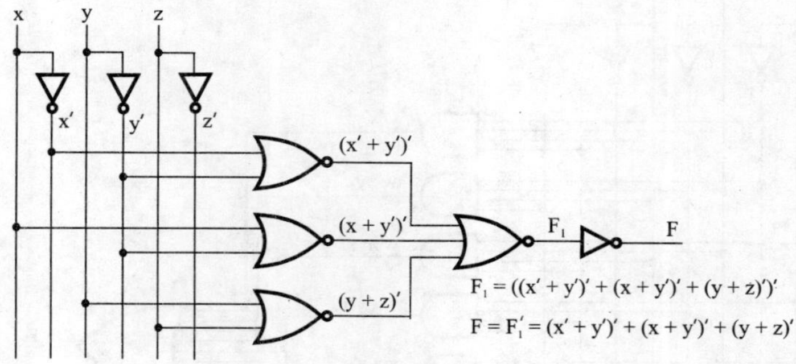

$F_1 = ((x' + y')' + (x + y')' + (y + z)')'$

$F = F_1' = (x' + y')' + (x + y')' + (y + z)'$

Fig.

E1.21 a) $\sum m(3, 5, 6, 7) = \prod M(0, 1, 2, 4)$

b) $\sum m(0, 1, 4, 5, 7) = \prod M(2, 3, 6)$

c) $\sum m(0, 1, 3, 4, 5, 9, 10, 11, 14) = \prod M(2, 6, 7, 8, 12, 13, 15)$

d) $\sum m(1, 3, 6, 7) = \prod M(0, 2, 4, 5)$

E1.22 a)

Input Variables				Complement of Inputs				Product Terms					Function Output
w	x	y	z	w'	x'	y'	z'	x'yz	xyz'	wx'y'	w'xy	wxy	F
0	0	0	0	1	1	1	1	0	0	0	0	0	0
0	0	0	1	1	1	1	0	0	0	0	0	0	0
0	0	1	0	1	1	0	1	0	0	0	0	0	0
0	0	1	1	1	1	0	0	1	0	0	0	0	1
0	1	0	0	1	0	1	1	0	0	0	0	0	0
0	1	0	1	1	0	1	0	0	0	0	0	0	0
0	1	1	0	1	0	0	1	0	1	0	1	0	1
0	1	1	1	1	0	0	0	0	0	0	1	0	1
1	0	0	0	0	1	1	1	0	0	1	0	0	1
1	0	0	1	0	1	1	0	0	0	1	0	0	1
1	0	1	0	0	1	0	1	0	0	0	0	0	0
1	0	1	1	0	1	0	0	1	0	0	0	0	1
1	1	0	0	0	0	1	1	0	0	0	0	0	0
1	1	0	1	0	0	1	0	0	0	0	0	0	0
1	1	1	0	0	0	0	1	0	1	0	0	1	1
1	1	1	1	0	0	0	0	0	0	0	0	1	1

b)

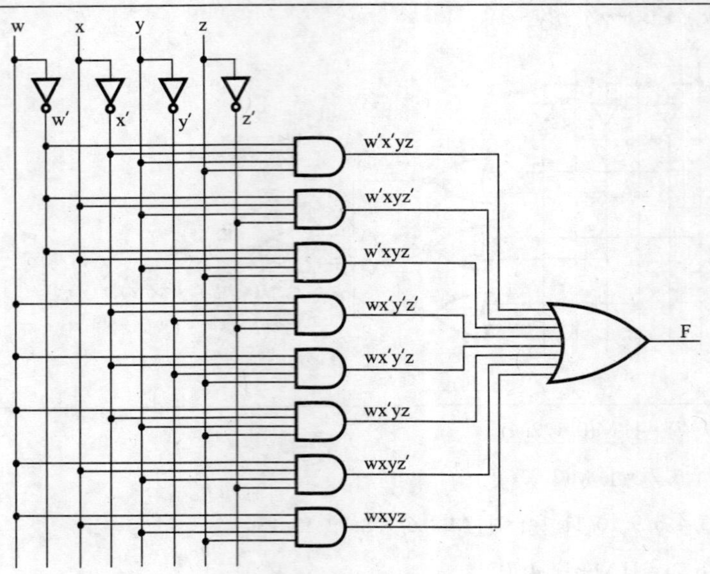

Fig: Logic circuit of F.

c) $F_{simplified} = xy + yz + wx'y'$

d)

Input Variables				Product Terms			Function Output
w	x	y	z	xy	yz	wx′y′	$F_{simplified}$
0	0	0	0	0	0	0	0
0	0	0	1	0	0	0	0
0	0	1	0	0	0	0	0
0	0	1	1	0	1	0	1
0	1	0	0	0	0	0	0
0	1	0	1	0	0	0	0
0	1	1	0	1	0	0	1
0	1	1	1	1	1	0	1
1	0	0	0	0	0	1	1
1	0	0	1	0	0	1	1
1	0	1	0	0	0	0	0
1	0	1	1	0	1	0	1
1	1	0	0	0	0	0	0
1	1	0	1	0	0	0	0
1	1	1	0	1	0	0	1
1	1	1	1	1	1	0	1

e)

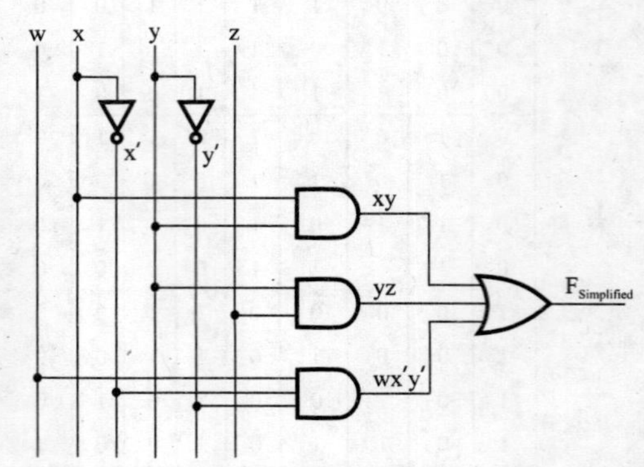

Fig: Logic circuit of $F_{Simplified}$.

E1.23 a) $F_1 = x'y' + xz$ **b)** $F_2 = x'y + xy'$

E1.24 a) $F_1 = y$ **b)** $F_2 = A'BD' + BCD$

E1.25 **a)** $F_1 = w'y'z + wxy + xz'$

b) $F_2 = ABC' + A'C + BCD'$

E1.26 **a)** $F_1 = x'y' + yz'$ **b)** $F_2 = xy' + x'yz$

E1.27 $F = ABC' + A'BD + B'D'$

E1.28 **a)**

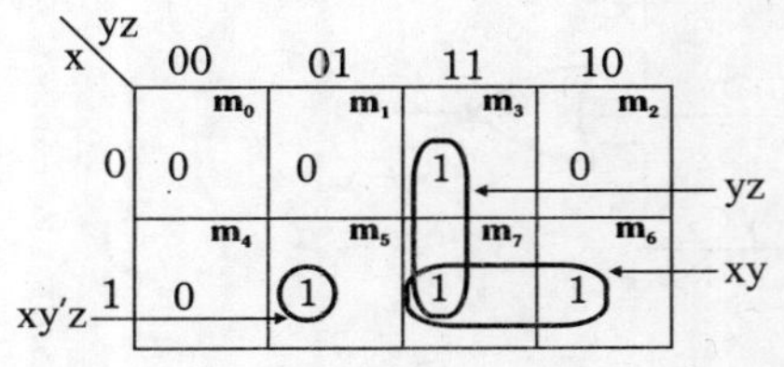

Fig: K-map for F_1.

$F_1 = \Sigma m(3, 5, 6, 7)$

b)

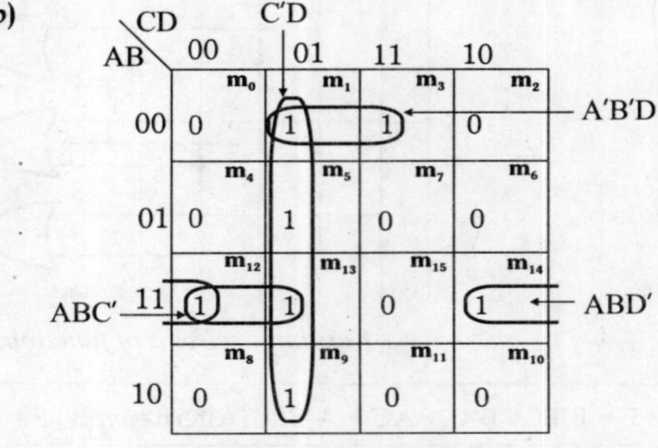

Fig: K-map for F_2.

$F_2 = \Sigma m(1, 3, 5, 9, 12, 13, 14)$

E1.29 $F = A'B'C'DE' + BC'DE + B'CE + ABC'E$

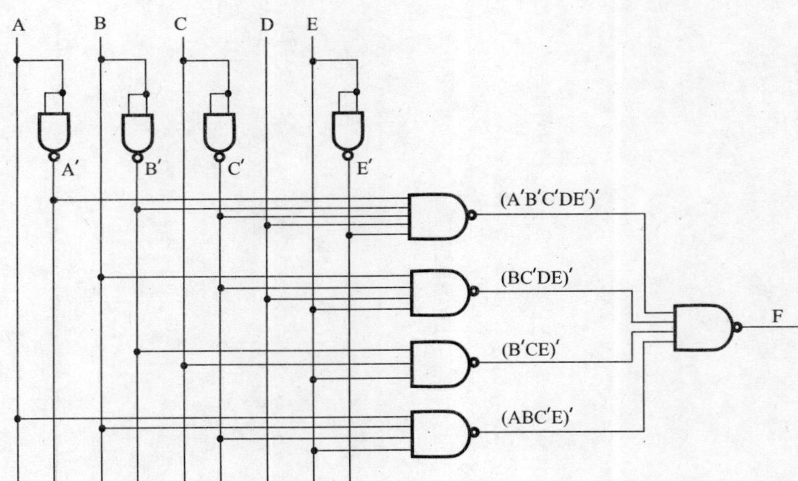

Fig: Logic circuit of function, F using only NAND gates.

E1.30 F = (B + C) (A + B + D) (A′ + C)

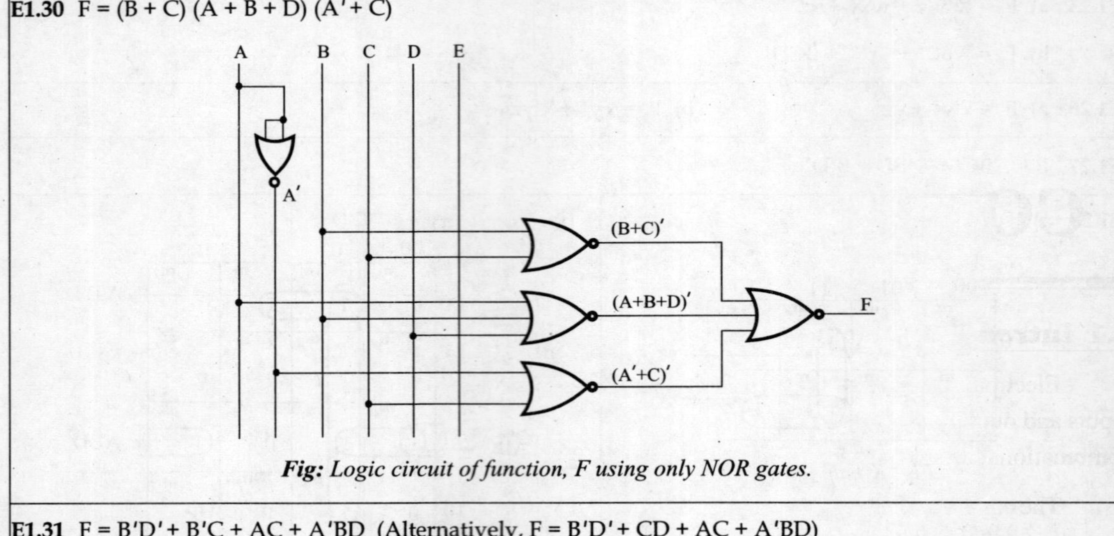

Fig: Logic circuit of function, F using only NOR gates.

E1.31 F = B′D′ + B′C + AC + A′BD (Alternatively, F = B′D′ + CD + AC + A′BD)

CHAPTER 2

COMBINATIONAL CIRCUITS

2.1 Introduction

Electronic circuits constructed using digital logic gates and devices and designed to operate on digital inputs and outputs are called ***digital logic circuits***. The digital logic circuits can be broadly classified into combinational circuits and sequential circuits.

The combinational circuits are digital logic circuits without feedbacks from output to input. Therefore, the outputs of combinational circuit will depend only on present inputs.

Fig. 2.1: Block diagram representation of a combinational circuit.

The sequential circuits are digital logic circuits with feedback from output to input, the sequential circuits are discussed in Chapter-3.

2.2 Combinational Logic

The combinational logic does not depend on feedback signals or previous output. The logical operations are performed using present inputs. The circuits that perform combinational logic operations are called ***combinational logic circuits*** and they are constructed using logic gates. In combinational circuits the output at any time depends on input at that time. In combinational circuits there is no storage element and there is no feedback from output to input.

Therefore, combinational circuits are designed for applications which do not require a record of previous outputs and for applications in which the present outputs do not depend on previous outputs. Some examples of combinational circuits are shown in Fig. 2.2.

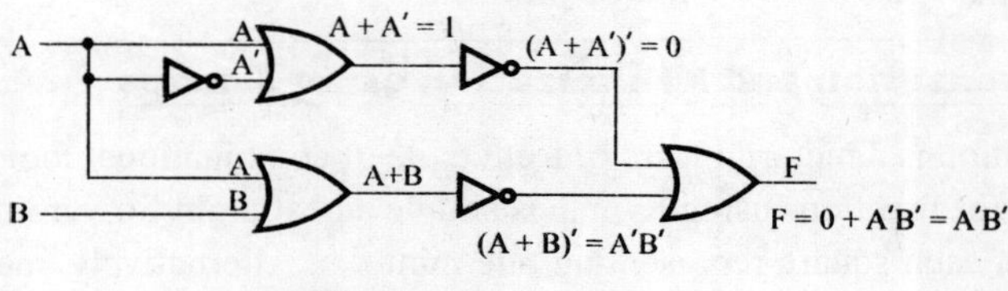

Fig. a.

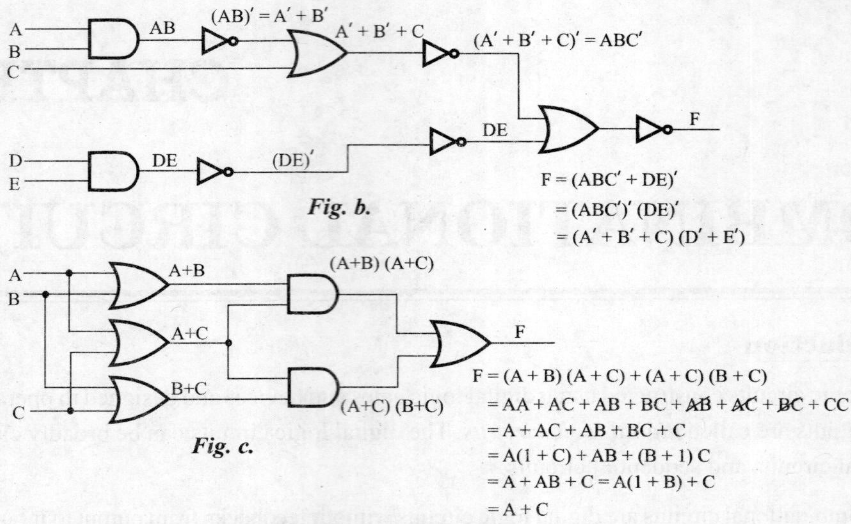

Fig. b.

Fig. c.

Fig. 2.2: Examples of combinational circuit.

2.2.1 Representation of Combinational Logic Functions in SOP and POS Form

The Boolean equations of combinational logic functions can be expressed in SOP and POS standard forms. In SOP form the combinational logic function is expressed as sum of product terms whereas in POS form it is expressed as product of sum terms.

Sum-of-products (SOP) can be realized by AND operation of literals followed by OR operation of output of AND. Here, AND operation of literals are the products and OR operation of output of AND is the sum of product terms. ***Product-of-sums (POS)*** can be realized by OR operation of literals followed by AND operation of output of OR. Here, OR operation of literals are the sums and AND operation of output of OR is the product of sum terms. The realizations in SOP and POS forms are called ***two-level realization of standard forms***.

The two-level realization of SOP form can be obtained by using AND gates to implement product terms followed by OR gate to get sum of product terms. The two-level realization of POS form can be obtained by using OR gates to implement sum terms followed by AND gate to get product of sum terms.

Since NAND and NOR gates are universal gates, the two-level realization of SOP form of a Boolean function of combinational logic can be obtained using only NAND gates. Similarly, two-level realization of POS form of a Boolean function of combinational logic can be obtained using only NOR gates.

It can be observed that complement of sum-of-products will be product-of-sums and vice-versa

i.e, Product-of-sums $\xrightleftharpoons{\text{Complement}}$ Sum-of-products

2.2.2 K-map Representation and Minimization using K-maps

K-map representation is a pictorial form of truth table of combinational logic and used to simplify functions. Simplification of functions using K-map is simple and straight forward. A K-map is a diagram made up of squares with each square representing one minterm. Alternatively, maxterm can be used to construct K-map in which each square represent a maxterm.

When minterms are considered for simplification the resultant combinational logic function will be in sum-of-products form. When maxterms are considered for simplification the resultant combinational logic function will be in product-of-sums form.

For detailed discussion on K-map, refer Chapter-1, Section 1.6.

2.2.3 Simplification and Implementation of Combinational Logic

The combinational circuits are designed to provide hardware based solution for logical problems. The design starts with problem specifications. The procedure to design a combinational circuit is given below:

1. Determine the required inputs and outputs from the problem specifications.
2. Assign a symbol to each input and output.
3. Derive the truth table.
4. Draw K-map for every output and form the prime implicants.
5. Determine the simplified Boolean function for every output from the prime implicants.
6. Implement the Boolean functions of all the outputs as a digital circuit using logic gates.

***Note:** Instead of K-map, alternatively table method can be used.*

Example 2.1

Design a combinational circuit for the following specification.

The output is 1 when the binary value of the inputs for a 3 variables function is less than 3. The output is 0 otherwise.

Solution

The output is 1, when the input is 0, 1 or 2. The output will be 0 if the input is greater than or equal to 3. Let us consider 3-bit binary inputs which can take values from 0 to 7. Let x, y and z be input variables and F be function output. The truth table is shown in Table 1 and the three-variable K-map is shown in Fig. 1.

Table 1: Truth Table

Inputs			Minterm	Output
x	y	z		F
0	0	0	m_0	1
0	0	1	m_1	1
0	1	0	m_2	1
0	1	1	m_3	0
1	0	0	m_4	0
1	0	1	m_5	0
1	1	0	m_6	0
1	1	1	m_7	0

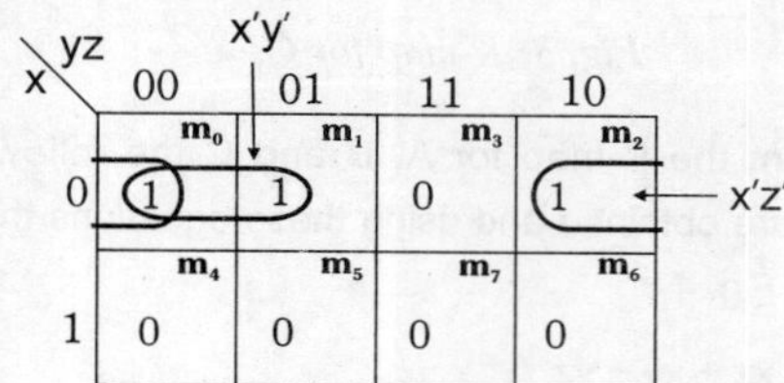

Fig. 1: K-map for F.

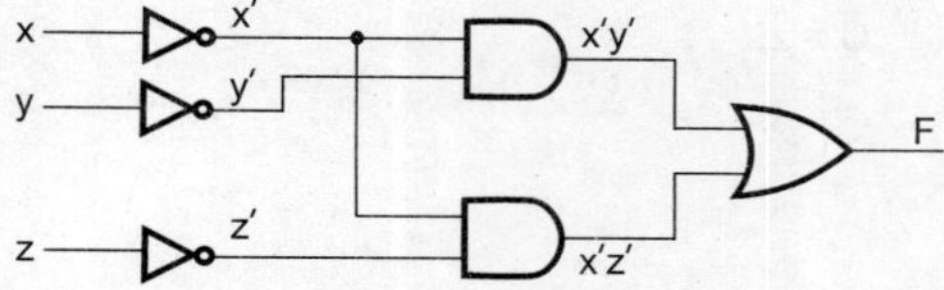

Fig. 2: Logic circuit of F.

In the K-map of Fig. 1, two overlapping prime implicants are formed using two adjacent 1's and the corresponding product terms are x' y' and x' z'. Therefore, the Boolean function, F is sum of these two prime implicants as shown below. The combinational logic circuit is drawn using logic gates as shown in Fig. 2.

$$F = x'y' + x'z'$$

Example 2.2

(AU, Nov/Dec'18, 15 Marks)

Design a combinational circuit with three inputs x, y and z and the 3 outputs, A, B and C. When the binary input is 0, 1, 2 or 3 the binary output is one greater than the input. When the binary input is 4, 5, 6 or 7, the binary output is one less than the input.

Solution

The truth table of the given problem is developed as shown in Table 1. Using the truth table the K-maps are constructed as shown in Figs. 1 to 3.

Table 1: Truth Table

Inputs				Outputs		
x	y	z	Minterm	A	B	C
0	0	0	m_0	0	0	1
0	0	1	m_1	0	1	0
0	1	0	m_2	0	1	1
0	1	1	m_3	1	0	0
1	0	0	m_4	0	1	1
1	0	1	m_5	1	0	0
1	1	0	m_6	1	0	1
1	1	1	m_7	1	1	0

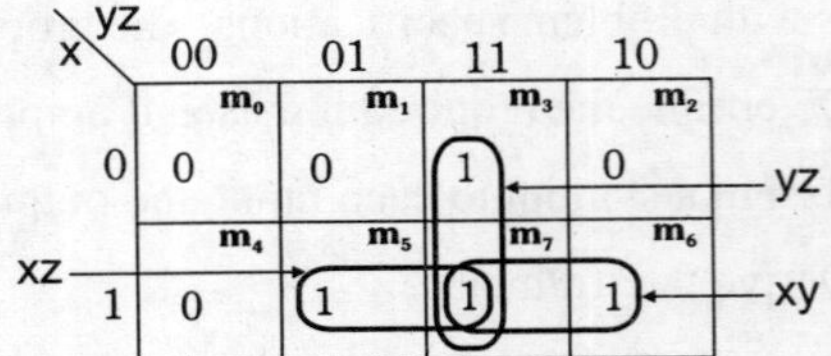

Fig. 1: K-map for A.

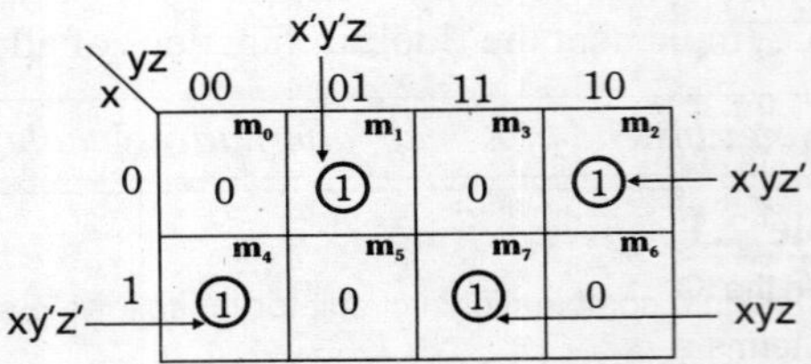

Fig. 2: K-map for B.

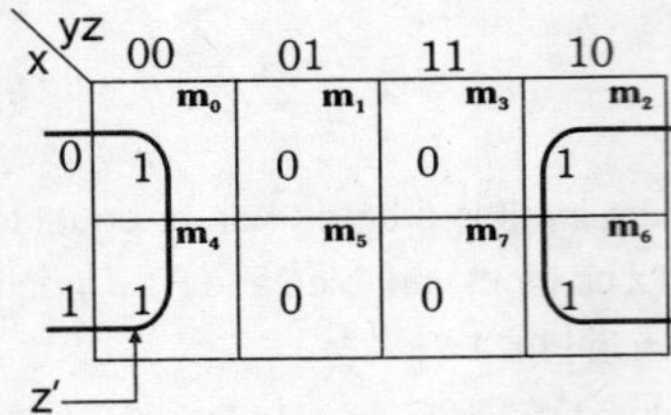

Fig. 3: K-map for C.

From the K-map for A, B and C the following Boolean equations are obtained and using these equations the logic circuit is drawn in Fig. 4.

$$A = xz + xy + yz$$

$$B = xy'z' + xyz + x'y'z + x'yz'$$

$$C = z'$$

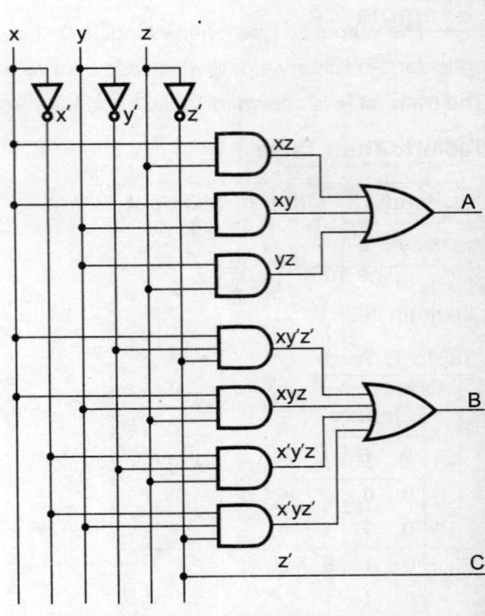

Fig. 4: Logic circuit of A, B and C.

Example 2.3

Design a 3-input majority circuit, in which the output is equal to 1 if the input variables have more 1's than 0's, otherwise, output is 0.

Solution

The truth table of majority circuit described in the problem is shown in Table 1 and the three variable K-map constructed using Table 1 is shown in Fig. 1.

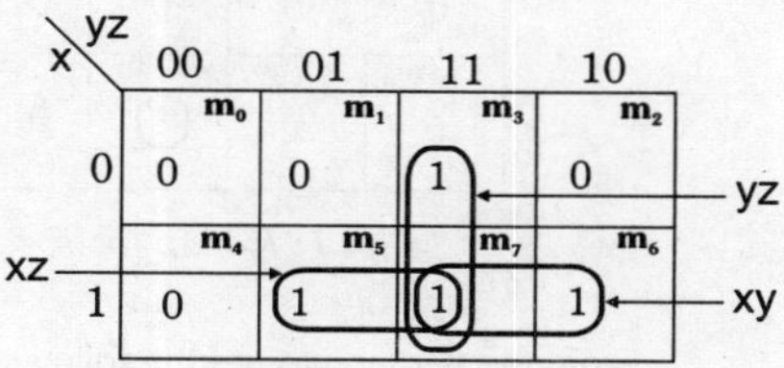

Fig. 1: K-map for F.

Table 1: Truth Table

Inputs			Minterm	Output
x	y	z		F
0	0	0	m_0	0
0	0	1	m_1	0
0	1	0	m_2	0
0	1	1	m_3	1
1	0	0	m_4	0
1	0	1	m_5	1
1	1	0	m_6	1
1	1	1	m_7	1

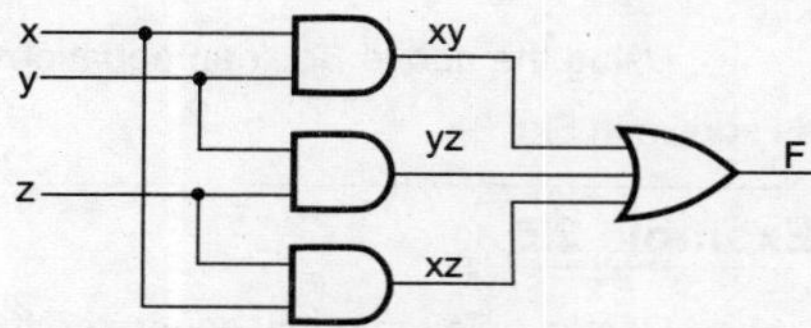

Fig. 2: Logic circuit of F.

In the K-map of Fig. 1, three overlapping prime implicants are formed using two adjacent 1's and the corresponding product terms are xy, yz and xz. Therefore, the Boolean function, F is sum of these three prime implicants as shown below. The combinational logic circuit is drawn using logic gates as shown in Fig. 2.

$$F = xy + yz + xz$$

Example 2.4

(AU, Nov/Dec'18, 13 Marks)

Assume a 3-input AND gate with output F and a 3-input OR gate with output G. Show the signals of the outputs F and G as functions of the three inputs A, B and C. Use all 8 possible combinations of inputs ABC.

Solution

It's given that,

$$F = ABC \quad ; \quad G = A + B + C$$

The truth table for F and G for all possible combinations of A, B and C is developed as shown in Table 1. Using the truth table the K-map for F and G are constructed as shown in Figs. 1 and 2.

Table 1: Truth Table for 3 input AND and OR gate

Inputs				Outputs	
A	B	C	Minterm	F	G
0	0	0	m_0	0	0
0	0	1	m_1	0	1
0	1	0	m_2	0	1
0	1	1	m_3	0	1
1	0	0	m_4	0	1
1	0	1	m_5	0	1
1	1	0	m_6	0	1
1	1	1	m_7	1	1

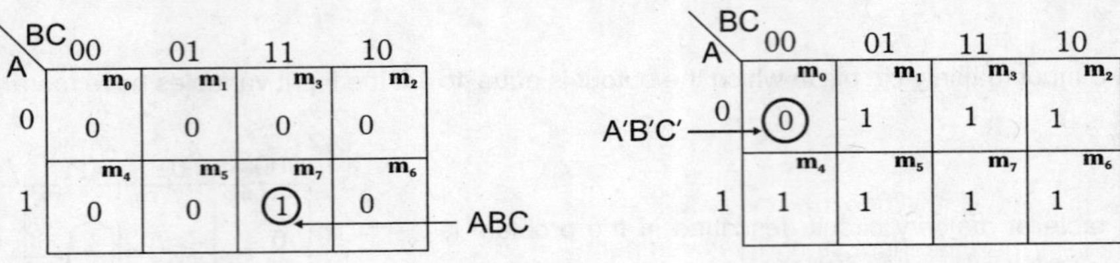

Fig. 1: K-map for F.

Fig. 2: K-map for G.

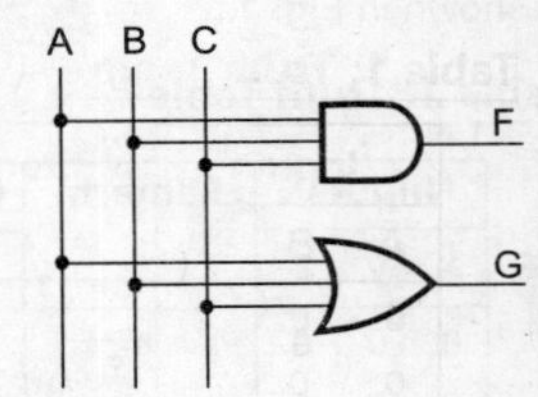

Fig. 3: Logic circuit of F and G.

From the K-map we get the following Boolean equations.

F = ABC

G = (A′B′C′)′ = A + B + C Using DeMorgan's law

Using the above Boolean equations, the logic circuit for F and G is drawn as shown in Fig. 3.

Example 2.5

Design an even parity generator, that generates an even parity bit for every input string of 3-bits.

Solution

Let P be the parity bit. When the input has odd number of 1's parity bit P is assigned 1 in order to convert input string to even parity. When the input string has even number of 1's parity bit is assigned 0. The truth table is shown in Table 1 and using truth table the K-map for P is drawn as shown in Fig. 1.

Table 1: Truth Table

Inputs			Minterm	Output
A	B	C		P
0	0	0	m_0	0
0	0	1	m_1	1
0	1	0	m_2	1
0	1	1	m_3	0
1	0	0	m_4	1
1	0	1	m_5	0
1	1	0	m_6	0
1	1	1	m_7	1

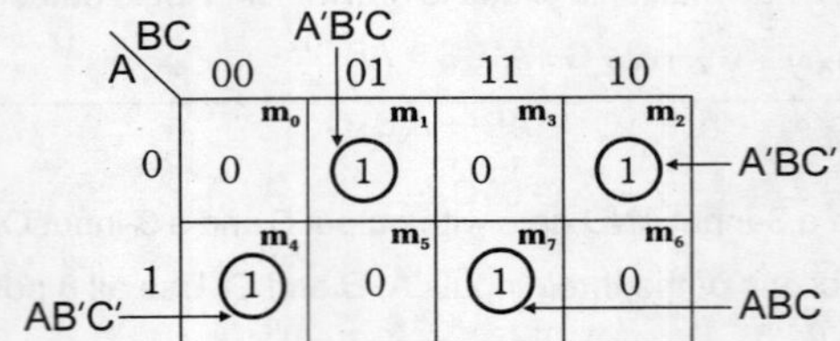

Fig. 1: K-map for parity bit P.

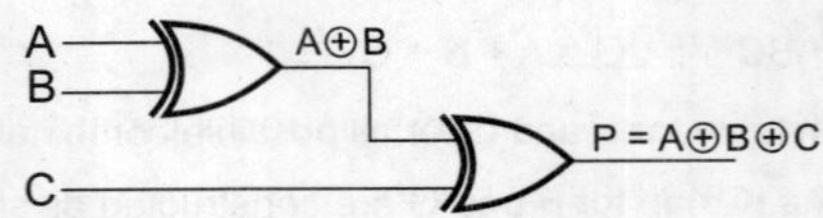

Fig. 2: Logic circuit of even parity bit generator.

From the K-map we get the following Boolean function for P, which is again simplified and using the simplified Boolean function the logic circuit of even parity generator is drawn as shown in Fig. 2.

P = A′B′C + A′BC′ + AB′C′ + ABC

= A′(B′C + BC′) + A(B′C′ + BC)

= A′ (B ⊕ C) + A(B ⊕ C)′

= A′x + Ax′

= A ⊕ x

= A ⊕ (B ⊕ C)

= A ⊕ B ⊕ C

B ⊕ C = B′C + BC′
(B ⊕ C)′ = BC + B′C′

Let, B ⊕ C = x

Example 2.6

Design an odd parity generator that generates an odd parity bit for every input string of 3 bits.

Solution

Let P be the parity bit. When the input has odd number of 1's parity bit P is assigned 0 in order to convert input string to odd parity. When the input string has even number of 1's parity bit is assigned 1. The truth table is shown in Table 1 and using truth table the K-map for P is drawn as shown in Fig. 1.

Table 1: Truth Table

Inputs			Minterm	Output
A	B	C		P
0	0	0	m_0	1
0	0	1	m_1	0
0	1	0	m_2	0
0	1	1	m_3	1
1	0	0	m_4	0
1	0	1	m_5	1
1	1	0	m_6	1
1	1	1	m_7	0

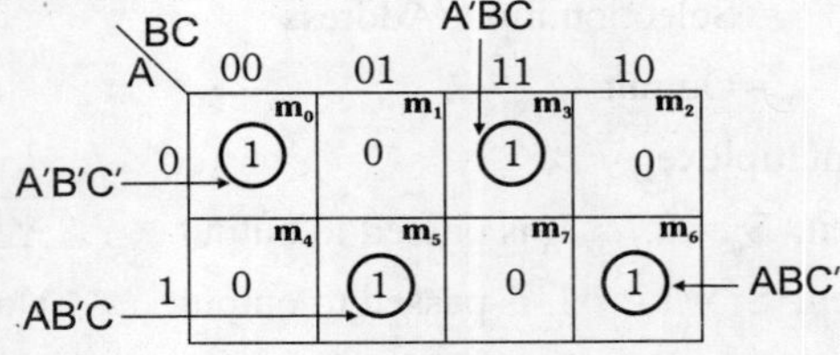

Fig. 1: K-map for parity bit P.

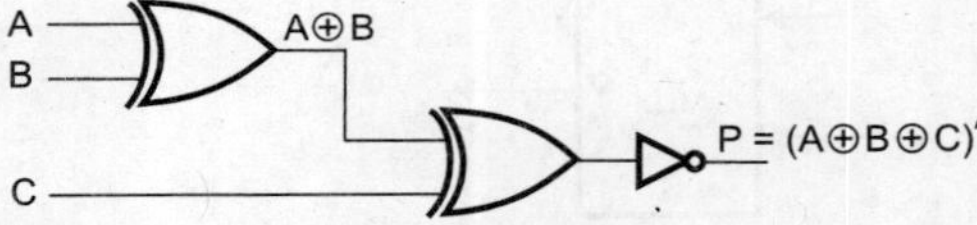

Fig. 2: Logic circuit of odd parity bit generator.

From the K-map we get the following Boolean function for P, which is again simplified and using the simplified Boolean function the logic circuit of odd parity generator is drawn as shown in Fig. 2.

$$A \oplus B = A'B + AB'$$
$$(A \oplus B)' = AB + A'B'$$
Let, $A \oplus B = x$

$$\begin{aligned} P &= A'B'C' + AB'C + A'BC + ABC' \\ &= (AB + A'B')C' + (AB' + A'B)C \\ &= (A \oplus B)'C' + (A \oplus B)C \\ &= x'C' + xC \\ &= (x \oplus C)' \\ &= (A \oplus B \oplus C)' \end{aligned}$$

2.3 Multiplexers

A ***multiplexer*** is a combinational circuit that selects one binary information from many inputs. For this reason a multiplexer is otherwise called a ***data selector***. In general, a multiplexer will have 2^n inputs and one output and so it allows only one input to output at any one time. In order to select one of the 2^n inputs the multiplexer will have n selection lines (or address). In short multiplexer is denoted as MUX.

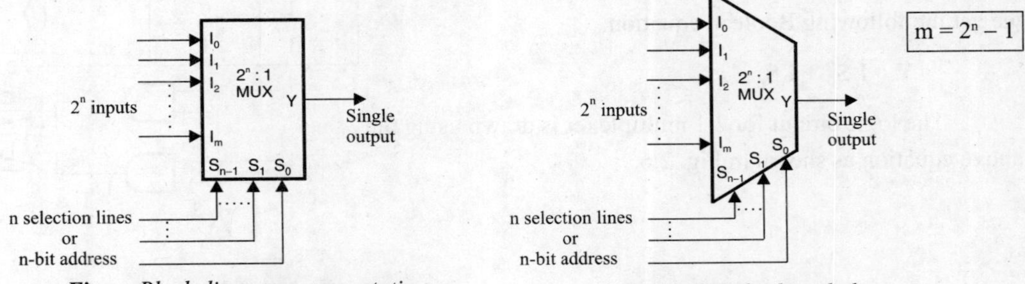

Fig. a: Block diagram representation. *Fig. b: Standard symbol.*

Fig. 2.3: Multiplexer.

The block diagram and symbolic representation of $2^n : 1$ multiplexer with n selection lines to select one of the 2^n inputs are shown in Fig. 2.3.

2:1 Multiplexer

The ***2:1 multiplexer*** is used to select one of the two inputs using 1-bit address or selection input. Since $2 = 2^1$, for 2:1 multiplexer one selection line is required to select one of the two inputs.

Let, I_0, I_1 = Inputs

S_0 = Selection input/Address

Y = Output

In 2:1 multiplexer,

When, $S_0 = 0$, I_0 is passed to output, $\therefore$ $Y = I_0$

When, $S_0 = 1$, I_1 is passed to output, $\therefore$ $Y = I_1$

The above selection logic are listed as truth table in Table 2.1.

Table 2.1: Truth Table of 2:1 Multiplexer

Selection Input S_0	Output Y
0	I_0
1	I_1

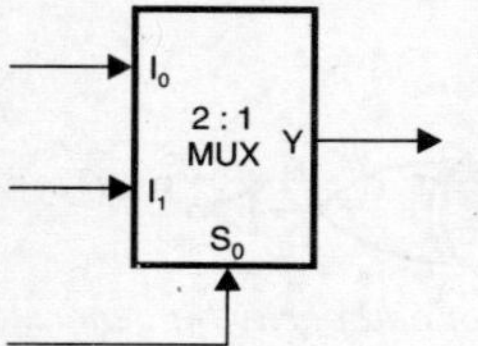

Fig. a: Block diagram representation.

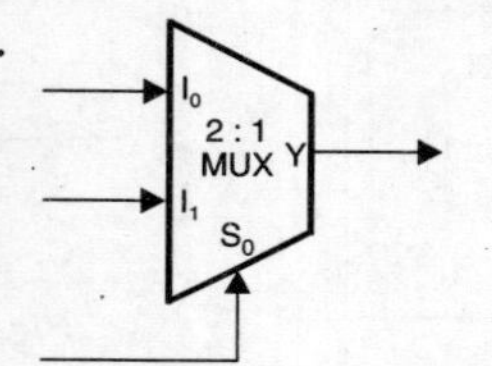

Fig. b: Standard symbol.

***Fig. 2.4:** 2:1 multiplexer.*

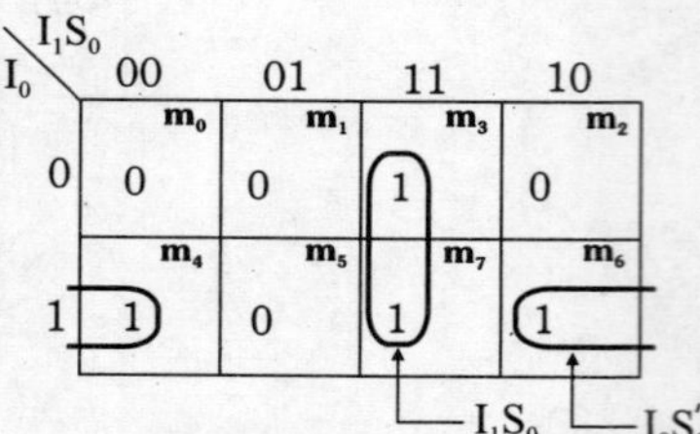

***Fig. 2.5:** K-map for design of 2:1 multiplexer.*

Table 2.2: Truth Table of 2:1 Multiplexer for all Combination of Inputs

Inputs I_0	I_1	Selection Input S_0	Minterm	Output Y
0	0	0	m_0	$Y = I_0 = 0$
0	0	1	m_1	$Y = I_1 = 0$
0	1	0	m_2	$Y = I_0 = 0$
0	1	1	m_3	$Y = I_1 = 1$
1	0	0	m_4	$Y = I_0 = 1$
1	0	1	m_5	$Y = I_1 = 0$
1	1	0	m_6	$Y = I_0 = 1$
1	1	1	m_7	$Y = I_1 = 1$

The truth table of 2:1 multiplexer for all combination of inputs is shown in Table 2.2. The K-map for the design of 2:1 multiplexer is shown in Fig. 2.5 and from the K-map. we get the following Boolean equation.

$$Y = I_0 S_0' + I_1 S_0$$

The logic circuit for 2:1 multiplexer is drawn using the above equation as shown in Fig. 2.6.

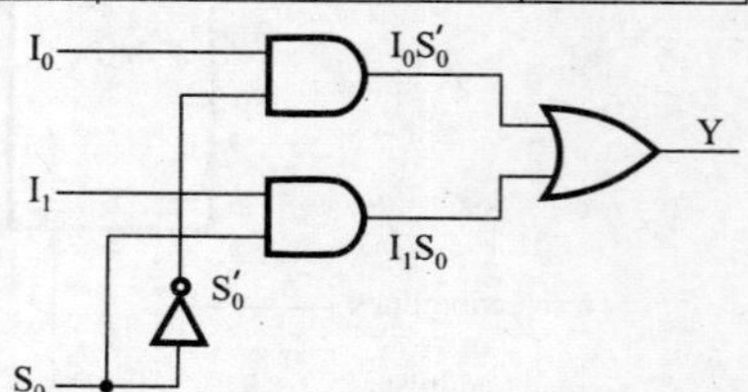

***Fig. 2.6:** Logic circuit of 2:1 multiplexer.*

4:1 Multiplexer

The ***4:1 multiplexer*** is used to select one of the four inputs using 2-bit address or selection input.

Since $4 = 2^2$, for 4:1 multiplexer two selection lines are required to select one of the four inputs.

Let, I_0, I_1, I_2, I_3 = Inputs

S_0, S_1 = Selection inputs/Address

Y = Output

In 4:1 multiplexer,

When, $S_1 = 0$, $S_0 = 0$, I_0 is passed to output, $\therefore Y = I_0$

When, $S_1 = 0$, $S_0 = 1$, I_1 is passed to output, $\therefore Y = I_1$

When, $S_1 = 1$, $S_0 = 0$, I_2 is passed to output, $\therefore Y = I_2$

When, $S_1 = 1$, $S_0 = 1$, I_3 is passed to output, $\therefore Y = I_3$

The above selection logic are listed as truth table in Table 2.3.

The block diagram and symbolic representation of 4:1 multiplexer with two selection lines to select one of the four inputs are shown in Fig. 2.7.

Table 2.3: Truth Table of 4:1 Multiplexer

Selection Input		Output
S_1	S_0	Y
0	0	I_0
0	1	I_1
1	0	I_2
1	1	I_3

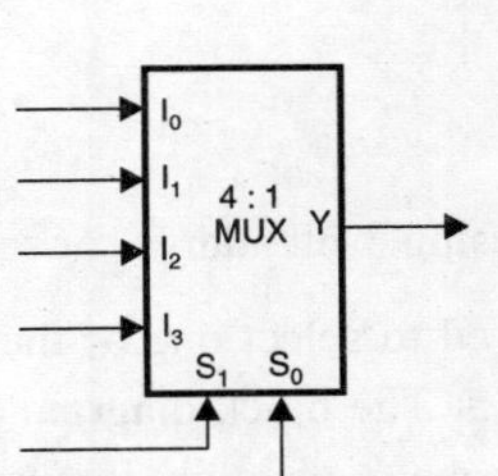

Fig. a: *Block diagram representation.*

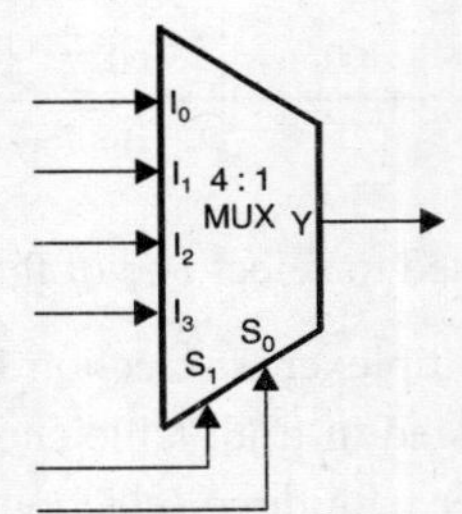

Fig. b: *Standard symbol.*

Fig. 2.7: *4:1 multiplexer.*

The ***dual 4:1 multiplexer*** is available as a standard IC with number 7453. The pin configuration of 7453 is shown in Fig. 2.8.

7453

Signal	Pin	Pin	Signal
$\overline{E}_a$	1	16	V_{cc}(+5V)
S_1	2	15	$\overline{E}_b$
I_{3a}	3	14	S_0
I_{2a}	4	13	I_{3b}
I_{1a}	5	12	I_{2b}
I_{0a}	6	11	I_{1b}
Y_a	7	10	I_{0b}
(0V) GND	8	9	Y_b

Pin	Description
S_0 , S_1	Selection inputs
$\overline{E}_a$, $\overline{E}_b$	Enable (Active low) inputs
I_{0a} - I_{3a}	Multiplexer inputs
I_{0b} - I_{3b}	Multiplexer inputs
Y_a, Y_b	Multiplexer output
V_{cc}	Supply voltage (+5V)
GND	Ground (0V)

Fig. 2.8: *Pin configuration of dual 4:1 multiplexer IC 7453.*

The logic circuit of 4:1 multiplexer can be realized using AND and OR gates as shown in Fig. 2.9. Table 2.4 can be used to check the logic circuit of 4:1 multiplexer.

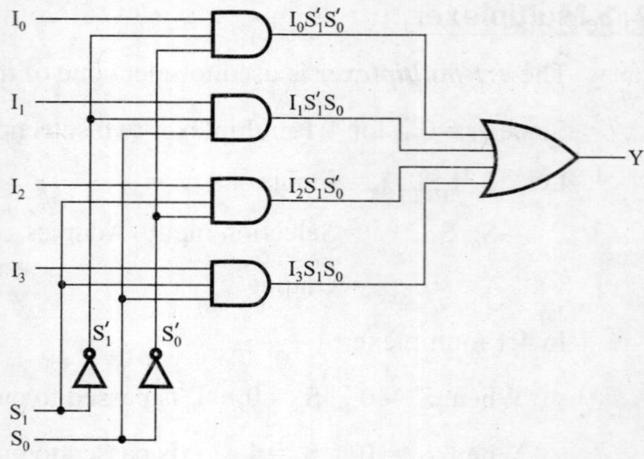

Fig. 2.9: Logic circuit of 4:1 multiplexer.

Table 2.4: Verification of Output of 4:1 MUX shown in Fig. 2.9

S_1	S_0	S_1'	S_0'	$I_0S_1'S_0'$	$I_1S_1'S_0$	$I_2S_1S_0'$	$I_3S_1S_0$	Y
0	0	1	1	I_0	0	0	0	I_0
0	1	1	0	0	I_1	0	0	I_1
1	0	0	1	0	0	I_2	0	I_2
1	1	0	0	0	0	0	I_3	I_3

8:1 Multiplexer

The ***8:1 multiplexer*** is used to select one of the eight inputs using 3-bit address or selection input.

Since $8 = 2^3$, for 8:1 multiplexer 3 selection lines are required to select one of the eight inputs. The input selection logic are listed in truth table shown in Table 2.5. The block diagram and symbolic representation of 8:1 multiplexer with three selection lines to select one of the eight inputs are shown in Fig. 2.10.

Let, $I_0, I_1, I_2, \ldots\ldots I_7$ = Inputs

S_0, S_1, S_2 = Selection inputs/Address

Y = Output

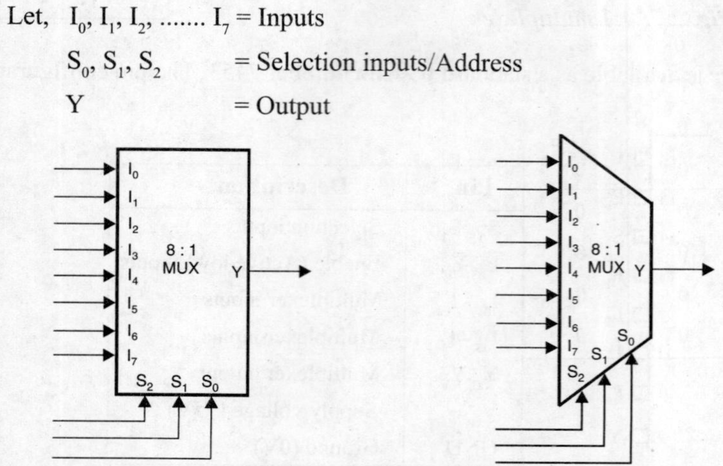

Fig. a: Block diagram representation. *Fig. b: Standard symbol.*

Fig. 2.10: Block diagram representation.

Table 2.5: Truth Table of 8:1 Multiplexer

Selection Inputs			Output
S_2	S_1	S_0	Y
0	0	0	I_0
0	0	1	I_1
0	1	0	I_2
0	1	1	I_3
1	0	0	I_4
1	0	1	I_5
1	1	0	I_6
1	1	1	I_7

The logic circuit of 8:1 multiplexer can be realized using AND and OR gates as shown in Fig. 2.11. Table 2.6 can be used to verify the logic circuit of 8:1 multiplexer.

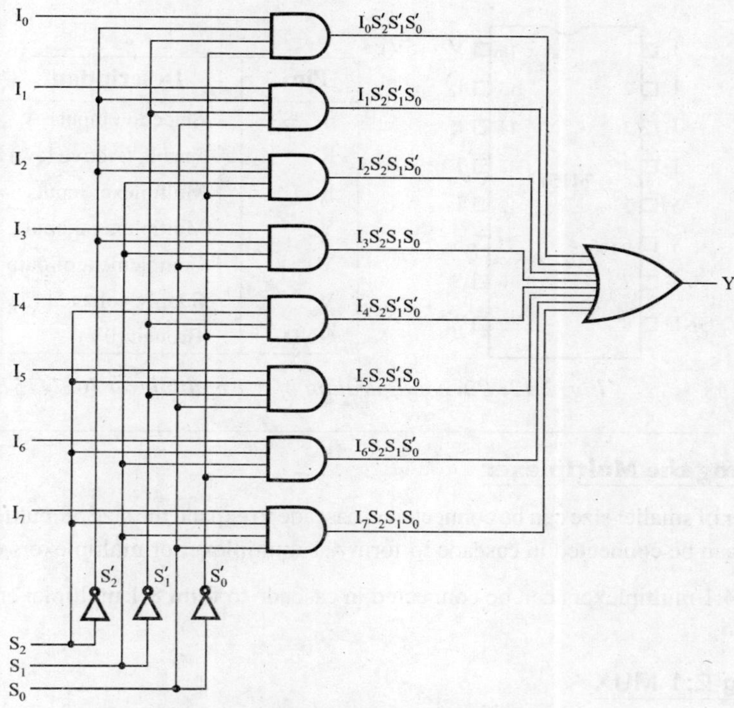

Fig. 2.11: Logic circuit of 8:1 multiplexer.

Table 2.6: Verification of Output of 8:1 Multiplexer of Fig. 2.11

Selection Inputs	Complement of Selection Inputs	Product Terms								Output
S_2 S_1 S_0	S_2' S_1' S_0'	$I_0 S_2' S_1' S_0'$	$I_1 S_2' S_1' S_0$	$I_2 S_2' S_1 S_0'$	$I_3 S_2' S_1 S_0$	$I_4 S_2 S_1' S_0'$	$I_5 S_2 S_1' S_0$	$I_6 S_2 S_1 S_0'$	$I_7 S_2 S_1 S_0$	Y
0 0 0	1 1 1	I_0	0	0	0	0	0	0	0	I_0
0 0 1	1 1 0	0	I_1	0	0	0	0	0	0	I_1
0 1 0	1 0 1	0	0	I_2	0	0	0	0	0	I_2
0 1 1	1 0 0	0	0	0	I_3	0	0	0	0	I_3
1 0 0	0 1 1	0	0	0	0	I_4	0	0	0	I_4
1 0 1	0 1 0	0	0	0	0	0	I_5	0	0	I_5
1 1 0	0 0 1	0	0	0	0	0	0	I_6	0	I_6
1 1 1	0 0 0	0	0	0	0	0	0	0	I_7	I_7

The 8:1 multiplexer is available as a standard IC with number 74151. The pin configuration of 74151 is shown in Fig. 2.12.

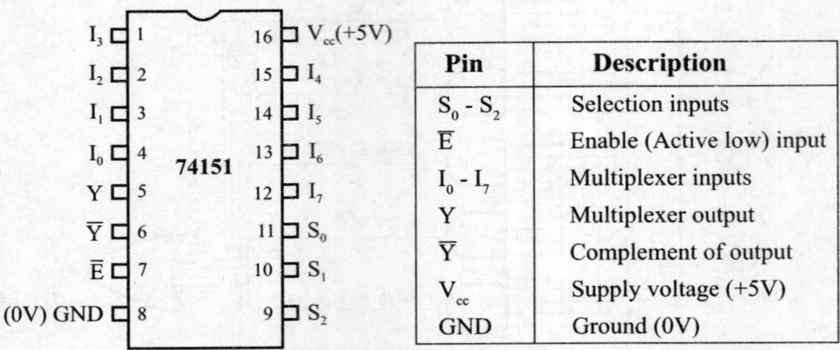

Pin	Description
S_0 - S_2	Selection inputs
$\overline{E}$	Enable (Active low) input
I_0 - I_7	Multiplexer inputs
Y	Multiplexer output
$\overline{Y}$	Complement of output
V_{cc}	Supply voltage (+5V)
GND	Ground (0V)

Fig. 2.12: Pin configuration of 8:1 multiplexer IC 74151.

2.3.1 Expanding the Multiplexer

Multiplexer of smaller size can be connected in cascade to expand the size of multiplexer. For example 2:1 multiplexers can be connected in cascade to form 4:1 multiplexer or multiplexers of higher order.

Similarly, 4:1 multiplexers can be connected in cascade to form 8:1 multiplexer or multiplexers of higher order.

4:1 MUX using 2:1 MUX

An example of constructing 4:1 multiplexer using 2:1 multiplexer is shown in Fig. 2.13.

A ***4:1 multiplexer*** will have 4 inputs I_0, I_1, I_2 and I_3 and two selection inputs S_0 and S_1. Let, Y be the output of 4:1 MUX.

The 4:1 MUX using 2:1 is constructed in two stages with two 2:1 MUX in first stage and one 2:1 MUX in second stage.

The inputs I_0 and I_1 are connected to one of the 2:1 MUX of first stage and another two inputs I_2 and I_3 are connected to another 2:1 MUX of first stage. The outputs of first stage MUXs are connected to inputs of second stage 2:1 MUX.

The selection input S_0 is used to select first stage MUXs and the selection input S_1 is used to select second stage MUX. The output of second stage 2:1 MUX will be same as output of 4:1 MUX which is verified in Table 2.7.

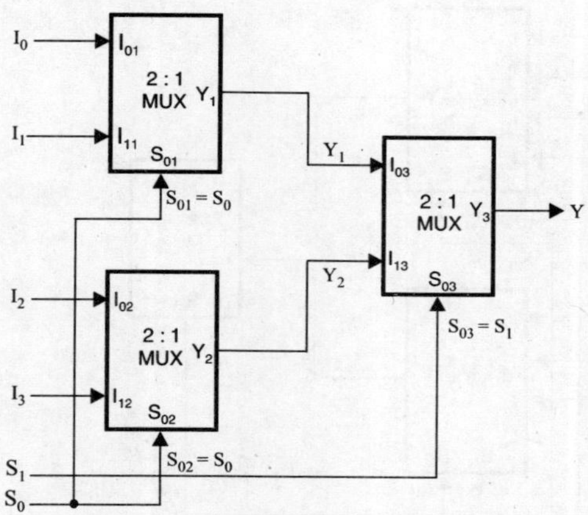

Fig. 2.13: 4:1 MUX using 2:1 MUX.

Table 2.7: Verification of Output of 4:1 Multiplexer of Fig. 2.13

Selection Inputs		2:1 MUX Outputs		4:1 MUX Output
S_1	S_0	Y_1	Y_2	Y
0	0	$Y_1 = I_{01} = I_0$	$Y_2 = I_{02} = I_2$	$Y = I_{03} = Y_1 = I_0$
0	1	$Y_1 = I_{11} = I_1$	$Y_2 = I_{12} = I_3$	$Y = I_{03} = Y_1 = I_1$
1	0	$Y_1 = I_{01} = I_0$	$Y_2 = I_{02} = I_2$	$Y = I_{13} = Y_2 = I_2$
1	1	$Y_1 = I_{11} = I_1$	$Y_2 = I_{12} = I_3$	$Y = I_{13} = Y_2 = I_3$

8:1 MUX using 4:1 and 2:1

An example of constructing 8:1 MUX using 4:1 and 2:1 MUX is shown in Fig. 2.14.

A ***8:1 MUX*** will have eight inputs I_0, I_1, ... I_7 and three selection inputs S_0, S_1 and S_2. Let, Y be the output of 8:1 MUX.

The 8:1 MUX is constructed in two stages with two 4:1 MUX in first stage and one 2:1 MUX in second stage. The inputs I_0, I_1, I_2 and I_3 are connected to inputs of one of the 4:1 MUX of first stage and other 4 inputs I_4, I_5, I_6 and I_7 are connected to input of another 4:1 MUX of first stage. The outputs of first stage MUXs are connected to input of second stage 2:1 MUX.

The selection inputs S_1 and S_0 are used to select first stage MUXs and the selection input S_2 is used to select second stage MUX. The output of second stage 2:1 MUX will be same as output of 8:1 MUX which is verified in Table 2.8.

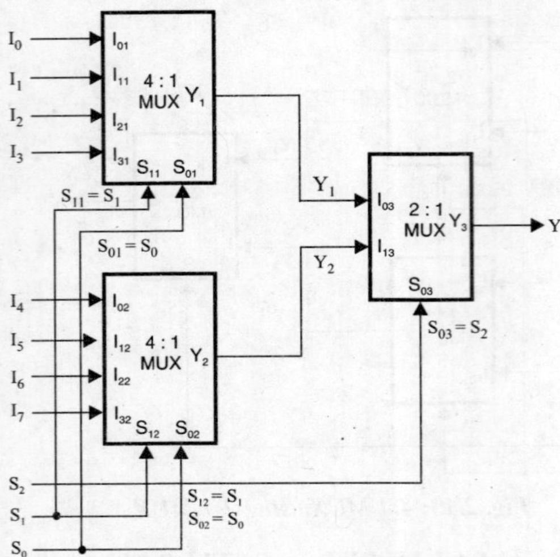

Fig. 2.14: 8:1 MUX using 4:1 and 2:1 MUX.

Table 2.8: Verification of Output of 8:1 MUX using 4:1 and 2:1 MUX of Fig. 2.14

Selection Inputs			4:1 MUX Outputs		8:1 MUX Output
S_2	S_1	S_0	Y_1	Y_2	Y
0	0	0	$Y_1 = I_{01} = I_0$	$Y_2 = I_{02} = I_4$	$Y = I_{03} = Y_1 = I_0$
0	0	1	$Y_1 = I_{11} = I_1$	$Y_2 = I_{12} = I_5$	$Y = I_{03} = Y_1 = I_1$
0	1	0	$Y_1 = I_{21} = I_2$	$Y_2 = I_{22} = I_6$	$Y = I_{03} = Y_1 = I_2$
0	1	1	$Y_1 = I_{31} = I_3$	$Y_2 = I_{32} = I_7$	$Y = I_{03} = Y_1 = I_3$
1	0	0	$Y_1 = I_{01} = I_0$	$Y_2 = I_{02} = I_4$	$Y = I_{13} = Y_2 = I_4$
1	0	1	$Y_1 = I_{11} = I_1$	$Y_2 = I_{12} = I_5$	$Y = I_{13} = Y_2 = I_5$
1	1	0	$Y_1 = I_{21} = I_2$	$Y_2 = I_{22} = I_6$	$Y = I_{13} = Y_2 = I_6$
1	1	1	$Y_1 = I_{31} = I_3$	$Y_2 = I_{32} = I_7$	$Y = I_{13} = Y_2 = I_7$

Example 2.7

Design 32:1 multiplexer using four 8:1 multiplexer and 2-to-4 decoder.

Solution

A 32:1 multiplexer has 32 inputs and 1 output. Let the 32 inputs be I_0, I_1, I_2, ..., I_{31} and the output be Y. The 32:1 multiplexer require 5-bit (2^5 = 32) selection or address input to select any one of the 32 inputs. Let the selection inputs be S_0, S_1, S_2, S_3 and S_4. The truth table of 32:1 multiplexer is shown in Table 1.

A 8:1 multiplexer has 8 inputs, 3 selection inputs and 1 output. To implement 32:1 multiplexer using 8:1 multiplexer we require 4 numbers of 8:1 multiplexer, so that we get 4 × 8 = 32 inputs. But at any one time only one 8:1 multiplexer is allowed to operate and so 4 enable signals are required and they are generated using a 2-to-4 decoder with S_3 and S_4 of 32:1 multiplexer as inputs. The outputs of decoder are used as enables for 8:1 multiplexers.

The selection inputs S_0, S_1 and S_2 of 32:1 multiplexer are used as selection inputs of all the 8:1 multiplexer.

The outputs of 8:1 multiplexers are again connected as inputs of 4:1 MUX so that one of the output of 8:1 MUX is selected as input for 4:1 MUX. The selection input of decoder and 4:1 MUX are same.

Table 1: Truth Table of 32:1 Multiplexer

Selection Inputs					Outputs															
S_4	S_3	S_2	S_1	S_0	Y_0	Y_1	Y_2	Y_3	Y_4	Y_5	Y_6	Y_7	Y_8	Y_9	Y_{10}	Y_{11}	Y_{12}	Y_{13}	Y_{14}	Y_{15}
0	0	0	0	0	I_0	0	0	0	0	0	0	0	0	0	0	0	0	0	0	0
0	0	0	0	1	0	I_1	0	0	0	0	0	0	0	0	0	0	0	0	0	0
0	0	0	1	0	0	0	I_2	0	0	0	0	0	0	0	0	0	0	0	0	0
0	0	0	1	1	0	0	0	I_3	0	0	0	0	0	0	0	0	0	0	0	0
0	0	1	0	0	0	0	0	0	I_4	0	0	0	0	0	0	0	0	0	0	0
0	0	1	0	1	0	0	0	0	0	I_5	0	0	0	0	0	0	0	0	0	0
0	0	1	1	0	0	0	0	0	0	0	I_6	0	0	0	0	0	0	0	0	0
0	0	1	1	1	0	0	0	0	0	0	0	I_7	0	0	0	0	0	0	0	0
0	1	0	0	0	0	0	0	0	0	0	0	0	I_8	0	0	0	0	0	0	0
0	1	0	0	1	0	0	0	0	0	0	0	0	0	I_9	0	0	0	0	0	0
0	1	0	1	0	0	0	0	0	0	0	0	0	0	0	I_{10}	0	0	0	0	0
0	1	0	1	1	0	0	0	0	0	0	0	0	0	0	0	I_{11}	0	0	0	0
0	1	1	0	0	0	0	0	0	0	0	0	0	0	0	0	0	I_{12}	0	0	0
0	1	1	0	1	0	0	0	0	0	0	0	0	0	0	0	0	0	I_{13}	0	0
0	1	1	1	0	0	0	0	0	0	0	0	0	0	0	0	0	0	0	I_{14}	0
0	1	1	1	1	0	0	0	0	0	0	0	0	0	0	0	0	0	0	0	I_{15}

Table 1: Continued....

Selection Inputs					Outputs															
S_4	S_3	S_2	S_1	S_0	Y_{16}	Y_{17}	Y_{18}	Y_{19}	Y_{20}	Y_{21}	Y_{22}	Y_{23}	Y_{24}	Y_{25}	Y_{26}	Y_{27}	Y_{28}	Y_{29}	Y_{30}	Y_{31}
1	0	0	0	0	I_{16}	0	0	0	0	0	0	0	0	0	0	0	0	0	0	0
1	0	0	0	1	0	I_{17}	0	0	0	0	0	0	0	0	0	0	0	0	0	0
1	0	0	1	0	0	0	I_{18}	0	0	0	0	0	0	0	0	0	0	0	0	0
1	0	0	1	1	0	0	0	I_{19}	0	0	0	0	0	0	0	0	0	0	0	0
1	0	1	0	0	0	0	0	0	I_{20}	0	0	0	0	0	0	0	0	0	0	0
1	0	1	0	1	0	0	0	0	0	I_{21}	0	0	0	0	0	0	0	0	0	0
1	0	1	1	0	0	0	0	0	0	0	I_{22}	0	0	0	0	0	0	0	0	0
1	0	1	1	1	0	0	0	0	0	0	0	I_{23}	0	0	0	0	0	0	0	0
1	1	0	0	0	0	0	0	0	0	0	0	0	I_{24}	0	0	0	0	0	0	0
1	1	0	0	1	0	0	0	0	0	0	0	0	0	I_{25}	0	0	0	0	0	0
1	1	0	1	0	0	0	0	0	0	0	0	0	0	0	I_{26}	0	0	0	0	0
1	1	0	1	1	0	0	0	0	0	0	0	0	0	0	0	I_{27}	0	0	0	0
1	1	1	0	0	0	0	0	0	0	0	0	0	0	0	0	0	I_{28}	0	0	0
1	1	1	0	1	0	0	0	0	0	0	0	0	0	0	0	0	0	I_{29}	0	0
1	1	1	1	0	0	0	0	0	0	0	0	0	0	0	0	0	0	0	I_{30}	0
1	1	1	1	1	0	0	0	0	0	0	0	0	0	0	0	0	0	0	0	I_{31}

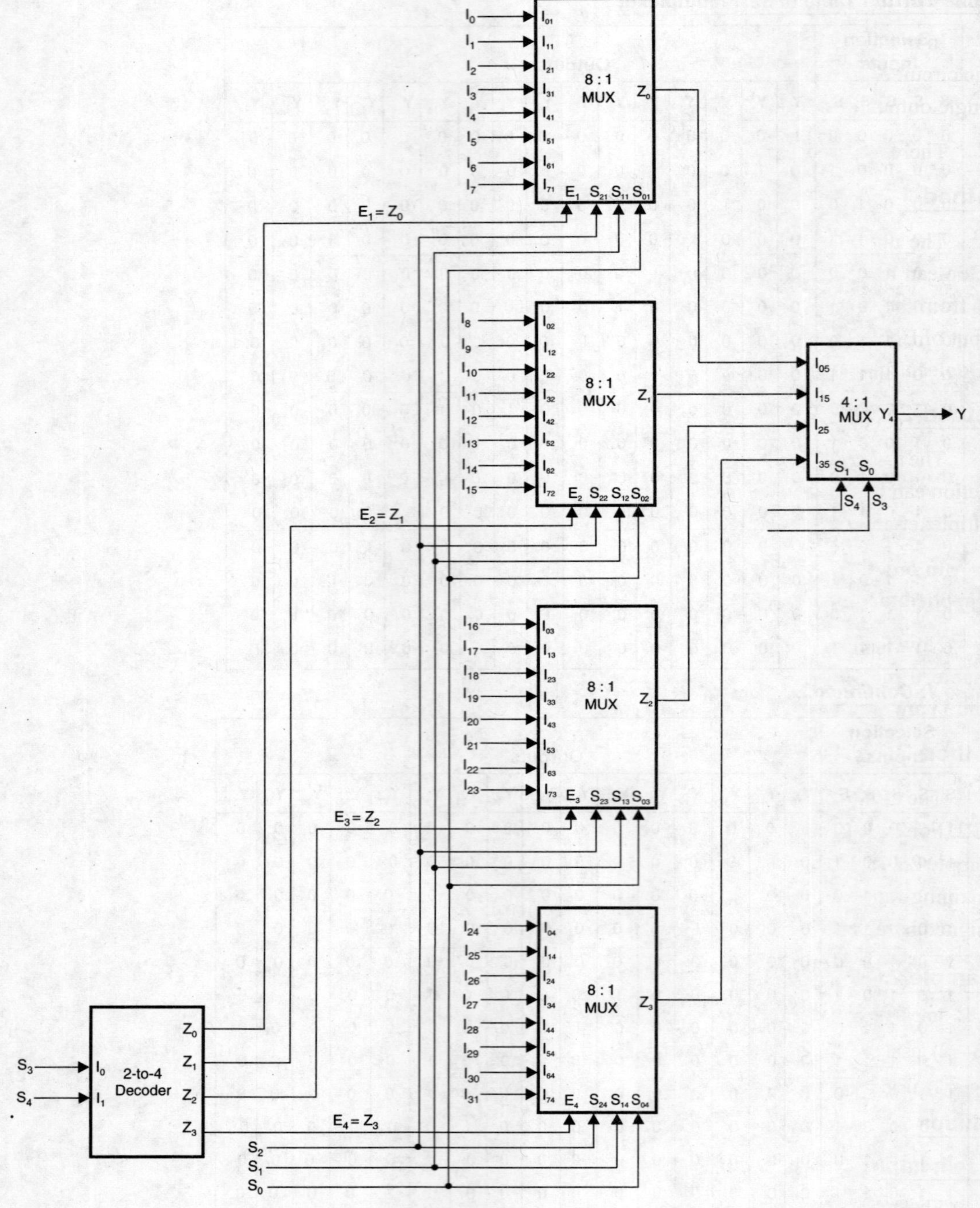

Fig. 1: *32:1 MUX with four 8:1 and the 2-to-4 decoder.*

2.3.2 Combinational Circuit Design using Multiplexer

In multiplexer, the possible binary values of selection inputs are same as minterms. Therefore, the logic circuit of Boolean functions in SOP form can be realized using multiplexer. Since a multiplexer has a single output, a separate multiplexer is required for every output of Boolean function.

There are different methods of implementing combinational circuits using multiplexer.

Method-1

The n-variable Boolean function can be implemented directly by 2^n:1 multiplexer. The n inputs of Boolean function are connected to n selection inputs. The inputs I_0, I_1, I_2, ..., I_{2n-1} are tied to either 0 or 1 from the knowledge of output in the truth table of Boolean function. If a minterm m_p generates 1 in output of Boolean function then input I_p of multiplexer is tied to 1. If a minterm m_q generates 0 in output of Boolean function then input I_q of multiplexer is tied to 0.

Method-2

The n-variable Boolean function can be implemented by 2^{n-1}:1 multiplexer. A three variable Boolean function can be realized using 2:1 multiplexer, a four variable Boolean function can be realized using 8:1 multiplexer and so on.

In order to implement n-variable Boolean function the first n – 1 inputs/variables of the function are applied to selection lines.

Now the function output can be represented by the remaining one variable, complement of this one variable, 0 and 1. The function output in terms of all this four (remaining one variable and its complement, 0 and 1) are applied as input I_0 .... I_{2n-1} of multiplexer.

Method-3

The n-variable Boolean function can be implemented using smaller size multiplexers than method-1 and 2. For example, a four variable Boolean function can be implemented using a 4:1 multiplexer (Refer example 2.24). In this method some of the inputs are connected to selection inputs of multiplexer and the remaining function inputs are applied to inputs of multiplexer through additional logic gates that can be designed by using K-maps.

Example 2.8

Implement the Boolean function $F(A, B, C) = \Sigma(1, 3, 5, 6)$,

i) using 4:1 multiplexer

ii) using 8:1 multiplexer.

Solution

Case i: Implementation using 4:1 Multiplexer

The truth table of the given function is formed as shown in Table 1. The 4:1 multiplexer will have two selection inputs S_0 and S_1. Let us connect the inputs A and B to selection inputs S_1 and S_0 as shown in Fig. 1.

The selection inputs take four different values and so Table 1 is divided into 4 groups based on selection inputs and a separate table is drawn for each group as shown in Tables 2 to 5.

Table 1: Truth Table of F

Inputs			Minterm	Output	Comments
A	B	C		F	
0	0	0	m_0	0	When, A = B = C = 0, F = 0
0	0	1	m_1	1	When, A = B = 0, C = 1, F = 1
0	1	0	m_2	0	When, A = C = 0, B = 1, F = 0
0	1	1	m_3	1	When, A = 0, B = C = 1, F = 1
1	0	0	m_4	0	When, A = 1, B = C = 0, F = 0
1	0	1	m_5	1	When, A = C = 1, B = 0, F = 1
1	1	0	m_6	1	When, A = B = 1, C = 0, F = 1
1	1	1	m_7	0	When, A = B = C = 1, F = 0

Table 2:

Inputs			Selection Inputs		F = C
A	B	C	S_1	S_0	
0	0	0	0	0	0
0	0	1	0	0	1

Table 3:

Inputs			Selection Inputs		F = C
A	B	C	S_1	S_0	
0	1	0	0	1	0
0	1	1	0	1	1

Table 4:

Inputs			Selection Inputs		F = C
A	B	C	S_1	S_0	
1	0	0	1	0	0
1	0	1	1	0	1

Table 5:

Inputs			Selection Inputs		F = C′
A	B	C	S_1	S_0	
1	1	0	1	1	1
1	1	1	1	1	0

The input I_0, I_1, I_2 and I_3 of multiplexer are obtained from Tables 2 to 5.

From Tables 2, 3 and 4 it can be observed that function output is C when selection inputs are 00, 01 and 10 and so I_0, I_1 and I_2 are tied to C as shown in Fig. 1. From Table 5 it can be observed that function output is complement of C when selection input is 11 and so C is inverted to get C′ and connected to I_2 as shown in Fig. 1. The working of 4:1 MUX is verified in Table 6.

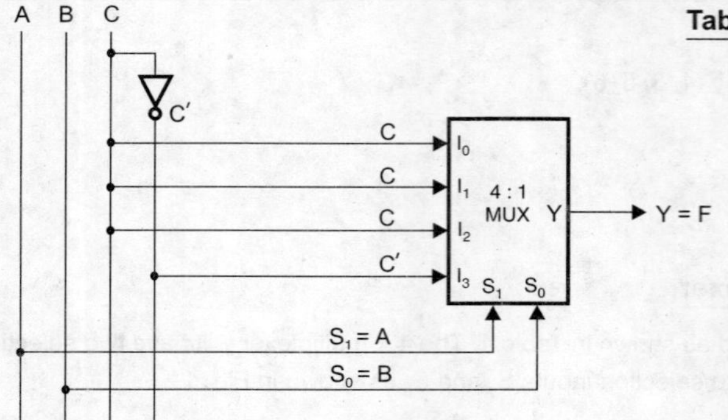

Fig. 1: Implementation of Boolean function using 4:1 MUX.

Table 6: Verification of Logic Circuit of F in Fig. 1

A	B	C	S_1	S_0	F	Y = F
0	0	0	0	0	0	$Y = I_0 = C = 0$
0	0	1	0	0	1	$Y = I_0 = C = 1$
0	1	0	0	1	0	$Y = I_1 = C = 0$
0	1	1	0	1	1	$Y = I_1 = C = 1$
1	0	0	1	0	0	$Y = I_2 = C = 0$
1	0	1	1	0	1	$Y = I_2 = C = 1$
1	1	0	1	1	1	$Y = I_3 = C' = 1$
1	1	1	1	1	0	$Y = I_3 = C' = 0$

Case ii: Implementation using 8:1 Multiplexer

The function F is 3-variable function and so a 2^3:1 multiplexer (8:1 multiplexer) can be used to implement the function as shown in Fig. 2. The function inputs A, B and C are connected to selection inputs S_2, S_1 and S_0. The function output F is 1 for minterms m_1, m_3, m_5 and m_6 and so the MUX inputs I_1, I_3, I_5 and I_6 are permanently tied to 1. The function output F is 0 for minterms m_0, m_2, m_4 and m_7 and so the MUX inputs I_0, I_2, I_4 and I_7 are permanently tied to 0. The working of 8:1 MUX is verified in Table 8.

Table 7: Truth Table of F

Inputs			Minterm	Output	Comments
A	B	C		F	
0	0	0	m_0	0	When, A = B = C = 0, F = 0
0	0	1	m_1	1	When, A = B = 0, C = 1, F = 1
0	1	0	m_2	0	When, A = C = 0, B = 1, F = 0
0	1	1	m_3	1	When, A = 0, B = C = 1, F = 1
1	0	0	m_4	0	When, A = 1, B = C = 0, F = 0
1	0	1	m_5	1	When, A = C = 1, B = 0, F = 0
1	1	0	m_6	1	When, A = B = 1, C = 0, F = 1
1	1	1	m_7	0	When, A = B = C = 1, F = 0

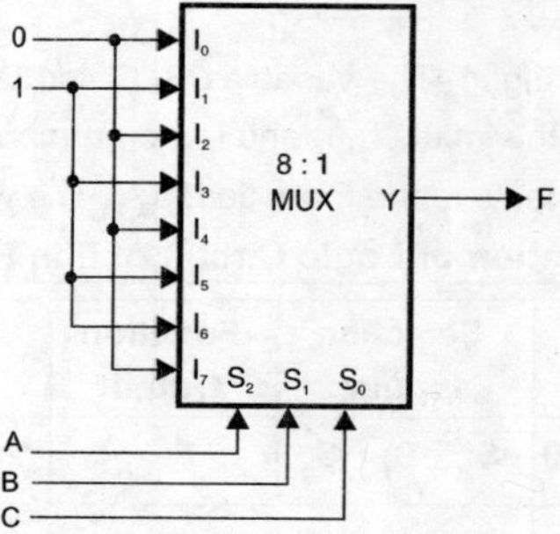

Fig. 2: Realization of F using 8:1 MUX.

Table 8: Verification of Logic Circuit of F in Fig. 2

A	B	C	S_2	S_1	S_0	F	Y = F
0	0	0	0	0	0	0	$Y = I_0 = 0$
0	0	1	0	0	1	1	$Y = I_1 = 1$
0	1	0	0	1	0	0	$Y = I_2 = 0$
0	1	1	0	1	1	1	$Y = I_3 = 1$
1	0	0	1	0	0	0	$Y = I_4 = 0$
1	0	1	1	0	1	1	$Y = I_5 = 1$
1	1	0	1	1	0	1	$Y = I_6 = 1$
1	1	1	1	1	1	0	$Y = I_7 = 0$

Example 2.9

Realize F(A, B, C, D) = Σm(0, 1, 3, 4, 8, 9, 15) using 8:1 MUX by selecting A as an input.

Solution

It is given that function output F is 1 when the inputs are equal to minterms m_0, m_1, m_3, m_4, m_8, m_9 and m_{15}. For all other inputs the function output F is 0.

The given function is analyzed using Table 1 and observed that the function output can be expressed in terms of A, A′, 0 and 1. Therefore, the given 4-variable function can be realized using 2^{4-1}:1 multiplexer (8:1 multiplexer) by taking B, C and D as selection inputs as shown in Fig. 1.

Table 1: Truth Table of F

Inputs				Minterm	Output F	Output in terms of A, A', 0, 1	Comments
A	B	C	D				
0	0	0	0	m_0	1	F = 1	When, B = C = D = 0
1	0	0	0	m_8	1	F = 1	Output, F = 1
0	0	0	1	m_1	1	F = 1	When, B = C = 0, D = 1
1	0	0	1	m_9	1	F = 1	Output, F = 1
0	0	1	0	m_2	0	F = 0	When, B = D = 0, C = 1
1	0	1	0	m_{10}	0	F = 0	Output, F = 0
0	0	1	1	m_3	1	F = A'	When, B = 0, C = D = 1
1	0	1	1	m_{11}	0	F = A'	Output, F = A'
0	1	0	0	m_4	1	F = A'	When B = 1, C = D = 0
1	1	0	0	m_{12}	0	F = A'	Output, F = A'
0	1	0	1	m_5	0	F = 0	When, B = D =1, C = 0
1	1	0	1	m_{13}	0	F = 0	Output, F = 0
0	1	1	0	m_6	0	F = 0	When, B = C = 1, D = 0
1	1	1	0	m_{14}	0	F = 0	Output, F = 0
0	1	1	1	m_7	0	F = A	When, B = C = D = 1
1	1	1	1	m_{15}	1	F = A	Output, F = A

The implementation of function, F using 8:1 MUX is shown in Fig. 1. The variables B, C and D are used as selection inputs S_2, S_1 and S_0 of 8:1 MUX. In 8:1 multiplexer of Fig. 1, the inputs I_2, I_5 and I_6 are permanentely tied to 0 and the inputs I_0 and I_1 are tied to 1. The input I_7 is tied to A and the inputs I_3 and I_4 are tied to A'. The working of 8:1 multiplexer is verified in Table 2.

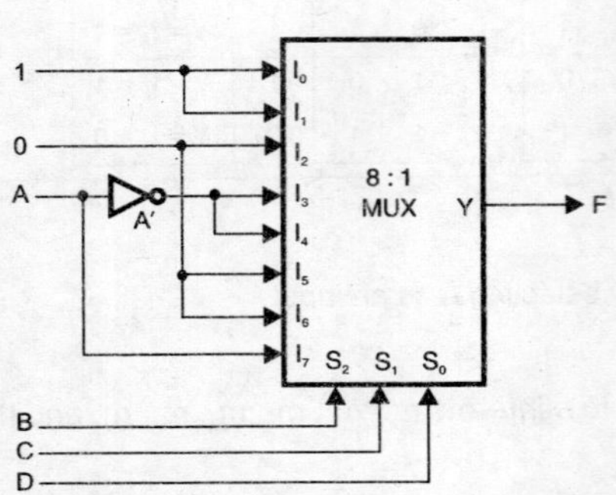

Fig. 1: Realization of F using 8:1 MUX.

Table 2: Verification of Logic Circuit of F in Fig. 1

Inputs				Selection Inputs			Function Output	MUX Output
A	B	C	D	S_2	S_1	S_0	F	Y = F
0	0	0	0	0	0	0	1	$Y = I_0 = 1$
1	0	0	0	0	0	0	1	$Y = I_0 = 1$
0	0	0	1	0	0	1	1	$Y = I_1 = 1$
1	0	0	1	0	0	1	1	$Y = I_1 = 1$
0	0	1	0	0	1	0	0	$Y = I_2 = 0$
1	0	1	0	0	1	0	0	$Y = I_2 = 0$
0	0	1	1	0	1	1	1	$Y = I_3 = A' = 1$
1	0	1	1	0	1	1	0	$Y = I_3 = A' = 0$
0	1	0	0	1	0	0	1	$Y = I_4 = A' = 1$
1	1	0	0	1	0	0	0	$Y = I_4 = A' = 0$
0	1	0	1	1	0	1	0	$Y = I_5 = 0$
1	1	0	1	1	0	1	0	$Y = I_5 = 0$
0	1	1	0	1	1	0	0	$Y = I_6 = 0$
1	1	1	0	1	1	0	0	$Y = I_6 = 0$
0	1	1	1	1	1	1	0	$Y = I_7 = A = 0$
1	1	1	1	1	1	1	1	$Y = I_7 = A = 1$

Example 2.10

Implement the following Boolean function

$F(w, x, y, z) = \sum(2, 3, 5, 6, 11, 14, 15)$

i) using mulitplexer, ii) using decoder, iii) using multiplexer and decoder

Solution

Case i: Implementation using Multiplexer

It is given that function output F is 1 when the inputs are equal to minterms m_2, m_3, m_5, m_6, m_{11}, m_{14} and m_{15}. For all other inputs the function output F is 0.

The given function is analyzed using Table 1 and observed that the function output can be expressed in terms of z, z′, 0 and 1. Therefore, the given 4-variable function can be realized using 2^{4-1}:1 multiplexer (8:1 multiplexer).

Table 1: Truth Table of F

w	x	y	z	Minterm	Output F	Output in terms of z, z′, 0, 1	Comments
0	0	0	0	m_0	0	F = 0	When, w = x = y = 0
0	0	0	1	m_1	0	F = 0	Output, F = 0
0	0	1	0	m_2	1	F = 1	When, w = x = 0, y = 1
0	0	1	1	m_3	1	F = 1	Output, F = 1
0	1	0	0	m_4	0	F = z	When, w = y = 0, x = 1
0	1	0	1	m_5	1	F = z	Output, F = z
0	1	1	0	m_6	1	F = z′	When, w = 0, x = y = 1
0	1	1	1	m_7	0	F = z′	Output, F = z′
1	0	0	0	m_8	0	F = 0	When w = 1, x = y = 0
1	0	0	1	m_9	0	F = 0	Output, F = 0
1	0	1	0	m_{10}	0	F = z	When, w = y =1, x = 0
1	0	1	1	m_{11}	1	F = z	Output, F = z
1	1	0	0	m_{12}	0	F = 0	When, w = x = 1, y = 0
1	1	0	1	m_{13}	0	F = 0	Output, F = 0
1	1	1	0	m_{14}	1	F = 1	When, w = x = y = 1
1	1	1	1	m_{15}	1	F = 1	Output, F = 1

Table 2: Verification of Logic Circuit of F in Fig. 1

w	x	y	z	S_2	S_1	S_0	F	Y = F
0	0	0	0	0	0	0	0	$Y = I_0 = 0$
0	0	0	1	0	0	0	0	$Y = I_0 = 0$
0	0	1	0	0	0	1	1	$Y = I_1 = 1$
0	0	1	1	0	0	1	1	$Y = I_1 = 1$
0	1	0	0	0	1	0	0	$Y = I_2 = z = 0$
0	1	0	1	0	1	0	1	$Y = I_2 = z = 1$
0	1	1	0	0	1	1	1	$Y = I_3 = z' = 1$
0	1	1	1	0	1	1	0	$Y = I_3 = z' = 0$
1	0	0	0	1	0	0	0	$Y = I_4 = 0$
1	0	0	1	1	0	0	0	$Y = I_4 = 0$
1	0	1	0	1	0	1	0	$Y = I_5 = z = 0$
1	0	1	1	1	0	1	1	$Y = I_5 = z = 1$
1	1	0	0	1	1	0	0	$Y = I_6 = 0$
1	1	0	1	1	1	0	0	$Y = I_6 = 0$
1	1	1	0	1	1	1	1	$Y = I_7 = 1$
1	1	1	1	1	1	1	1	$Y = I_7 = 1$

The implementation of function, F using 8:1 MUX is shown in Fig. 1. The variables w, x and y are used as selection inputs S_2, S_1 and S_0 of 8:1 MUX. In 8:1 MUX of Fig. 1, the inputs I_0, I_4 and I_6 are permanently tied to 0 and the inputs I_1 and I_7 are tied to 1. The inputs I_2 and I_5 are tied to z and the input I_3 is tied to z′. The working of 8:1 MUX is verified in Table 2.

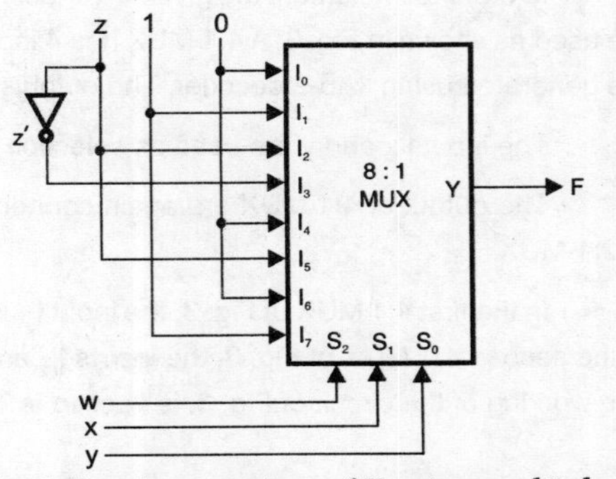

Fig. 1: Implementation of F using multiplexer.

Case ii: Implementation using Decoder

From Table 1 it can be observed that when w = 0, the inputs x, y and z can take 8 combinaitons of binary. Similarly, when w = 1, the inputs x, y and z can take 8 combinations of binary. Hence, two 3-to-8 decoders can be used to implement the given function. One decoder is enabled by w' and other by w.

The inputs of both the decoders are connected to x, y and z. The output of decoder enabled by w' will be same as function output when inputs are minterms m_0 to m_7. The output of decoder enabled by w will be same as function output when inputs are minterms m_8 to m_{15}. The required outputs are logically ORed to get function output as shown in Fig. 2.

On taking into account the minterms for which function outputs are 1, the following Boolean equations are obtained using the logic **high** decoder output.

$F_1 = Y_{21} + Y_{31} + Y_{51} + Y_{61}$; $F_2 = Y_{32} + Y_{62} + Y_{72}$

$F = F_1 + F_2$

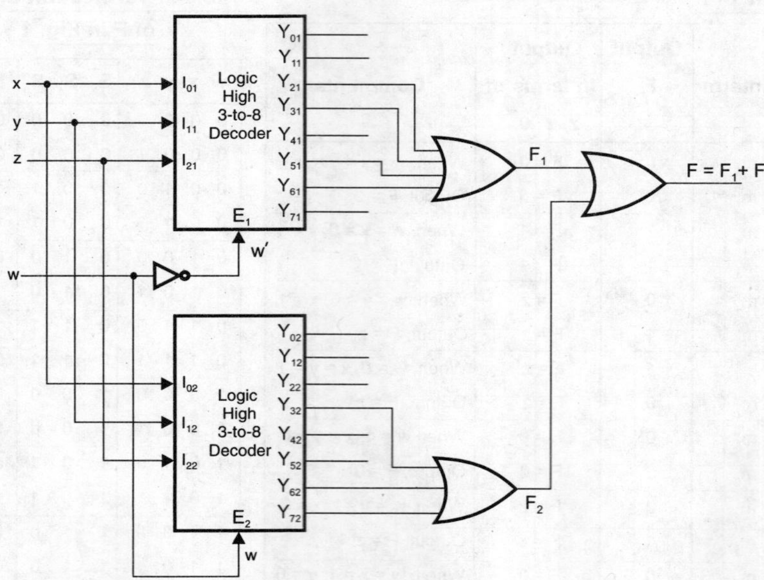

Fig. 2: Implementation of F using decoder.

Case iii: Implementation using Multiplexer and Decoder

In order to implement the given 4-variable Boolean function, one 1-to-2 decoder, two 4:1 MUX and one 2:1 MUX are used as shown in Fig. 3. A 4:1 MUX has 4 inputs, 2 selection inputs,1 enable input and 1 output. The enable inputs are generated using 1-to-2 decoder. The outputs of decoder are used as enable inputs for 4:1 MUX.

The inputs x and y are used as selection inputs S_1 and S_0 of 4:1 MUX.

The output of 4:1 MUX are again connected as inputs of 2:1 MUX. The input w is used as selection input S_{02} of 2:1 MUX.

In the first 4:1 MUX of Fig. 3, the input I_{01} is tied to 0 and I_{11} is tied to 1. The input I_{21} is tied to z and I_{31} is tied to z'. In the second 4:1 MUX of Fig. 3, the inputs I_{02} and I_{22} are tied to 0, the input I_{32} is tied to 1 and the input I_{12} is tied to 1. The working of the circuit of Fig. 3, is verified in Table 3.

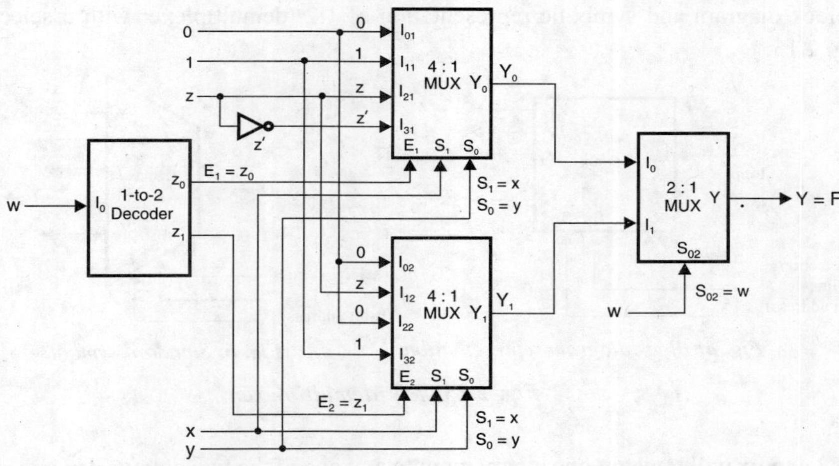

Fig. 3: Implementation of F using decoder and multiplexer.

Table 3: Verification of Logic Circuit of F in Fig. 3

Inputs w x y z	z'	Selection Inputs of 4:1 MUX $S_1 = x$	$S_0 = y$	Selection Input of 2:1 MUX $S_{02} = w$	Function Output F	2:1 MUX Output Y = F	
0 0 0 0	1	0	0	0	0	$Y = I_0 = Y_0$	$Y_0 = I_{01} = 0$
0 0 0 1	0	0	0	0	0		
0 0 1 0	1	0	1	0	1		$Y_0 = I_{11} = 1$
0 0 1 1	0	0	1	0	1		
0 1 0 0	1	1	0	0	0		$Y_0 = I_{21} = z$
0 1 0 1	0	1	0	0	1		
0 1 1 0	1	1	1	0	1		$Y_0 = I_{31} = z'$
0 1 1 1	0	1	1	0	0		
1 0 0 0	1	0	0	1	0	$Y = I_1 = Y_1$	$Y_1 = I_{02} = 0$
1 0 0 1	0	0	0	1	0		
1 0 1 0	1	0	1	1	0		$Y_1 = I_{12} = z$
1 0 1 1	0	0	1	1	1		
1 1 0 0	1	1	0	1	0		$Y_1 = I_{22} = 0$
1 1 0 1	0	1	0	1	0		
1 1 1 0	1	1	1	1	1		$Y_1 = I_{23} = 1$
1 1 1 1	0	1	1	1	1		

2.4 Demultiplexers

A demultiplexer is a combinational circuit that send one binary information to many outputs. For this reason a demultiplexer is otherwise called a ***data distributor***.

A demultiplexer will perform the reverse operation of multiplexer. Therefore, a demultiplexer is a combinational circuit which can transmit a binary input to any one of 2^n output lines using n selection lines or n-bit address. The demultiplexer is shortly called ***demux***.

The block diagram and symbolic representation of $1{:}2^n$ demultiplexer with n selection inputs is shown in Fig. 2.15.

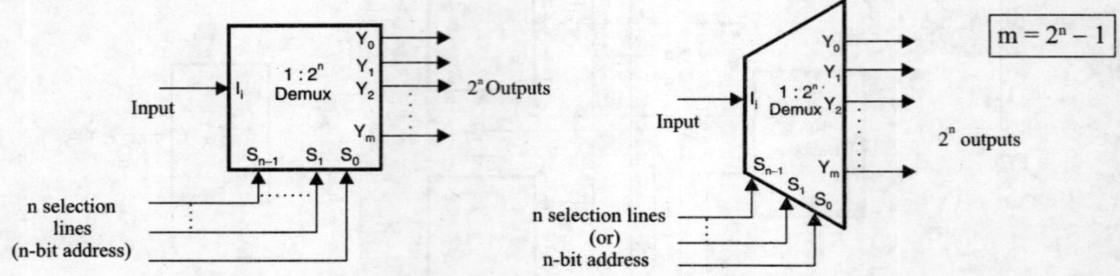

Fig. a: Block diagram representation. *Fig. b: Standard symbol.*

Fig. 2.15: $1{:}2^n$ demultiplexer.

1:2 Demux

The ***1:2 demux*** will transmit one binary input to any one of the two outputs depending on the value of selection input.

Let, I_i = Input

S_0 = Selection input/Address

Y_0, Y_1 = Outputs

In 1:2 demultiplexer,

When, $S_0 = 0$, I_i is passed to Y_0, $\therefore\ Y_0 = I_i$

When, $S_0 = 1$, I_i is passed to Y_1, $\therefore\ Y_1 = I_i$

The above selection logic is listed as truth table in Table 2.9. The block diagram and symbolic representation of 1:2 demux with selection input S_0 are shown in Fig. 2.16.

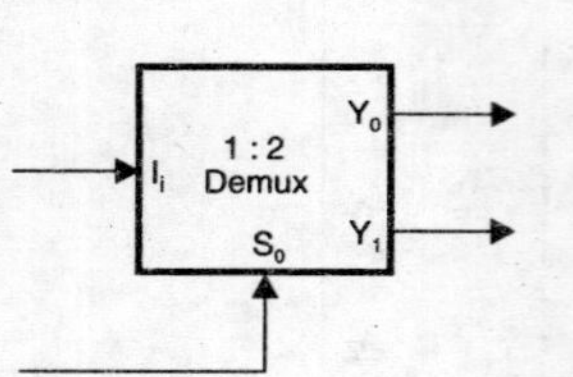

Fig. a: Block diagram representation.

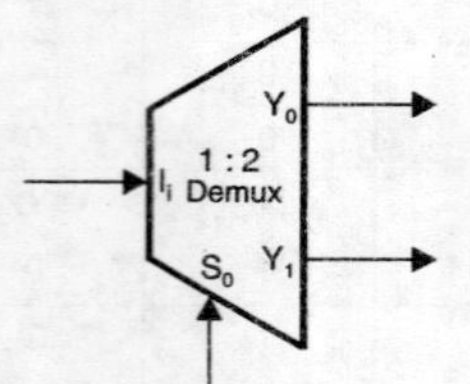

Fig. b: Standard symbol.

Fig. 2.16: 1:2 demultiplexer.

Table 2.9: Truth Table of 1:2 Demultiplexer

S_0	Y_0	Y_1
0	I_i	0
1	0	I_i

The truth table of 1:2 demux for all possible combination of inputs is shown in Table 2.10. The K-maps for design of 1:2 demux are shown in Fig. 2.17 and from the K-map we get the following Boolean equations,

$$Y_0 = I_i\, S_0'$$

$$Y_1 = I_i\, S_0$$

The logic circuit of 1:2 demux is drawn using the above Boolean equations as shown in Fig. 2.18.

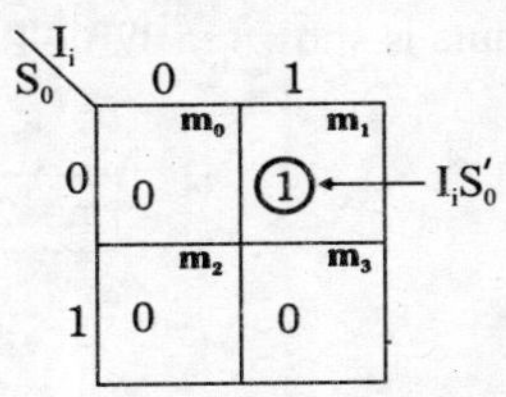

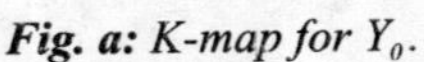

Fig. a: *K-map for Y_0.*

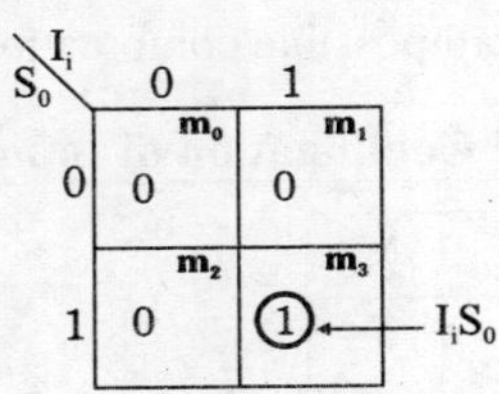

Fig. b: *K-map for Y_1.*

Fig. 2.17: *K-map for design of 1:2 demux.*

Table 2.10: Truth Table of 1:2 Demux for all Combination of Inputs

S_0	I_i	Minterm	Y_0	Y_1
0	0	m_0	$I_i = 0$	0
0	1	m_1	$I_i = 1$	0
1	0	m_2	0	$I_i = 0$
1	1	m_3	0	$I_i = 1$

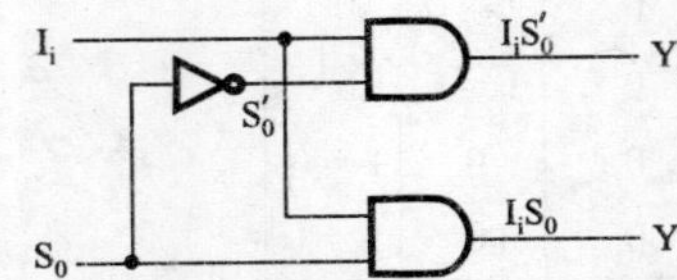

Fig. 2.18: *Logic circuit of 1:2 demux.*

1:4 Demux

The ***1:4 demultiplexer*** will transmit one binary input to any one of the four outputs depending on the value of selection input.

Let, I_i = Input

S_0, S_1 = Selection inputs

Y_0, Y_1, Y_2, Y_3 = Outputs

In 1:4 demultiplexer,

When, $S_1 = 0$ and $S_0 = 0$, I_i is passed to Y_0, $\therefore\ Y_0 = I_i$

When, $S_1 = 0$ and $S_0 = 1$, I_i is passed to Y_1, $\therefore\ Y_1 = I_i$

When, $S_1 = 1$ and $S_0 = 0$, I_i is passed to Y_2, $\therefore\ Y_2 = I_i$

When, $S_1 = 1$ and $S_0 = 1$, I_i is passed to Y_3, $\therefore\ Y_3 = I_i$

The above selection logic is listed as truth table in Table 2.11. The block diagram and symbolic representation of 1:4 demux with selection inputs S_0 and S_i are shown in Fig. 2.19.

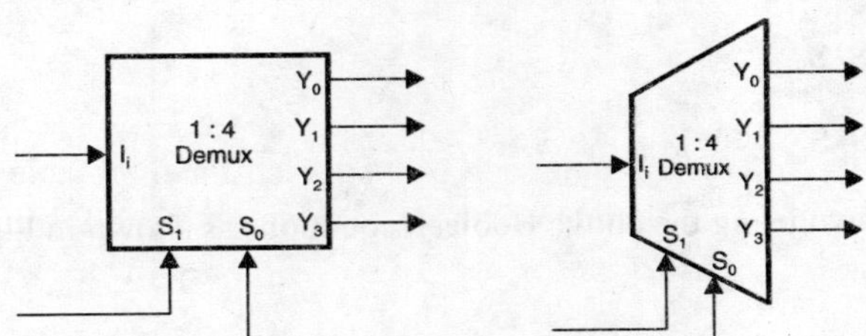

Fig. a: *Block diagram representation.* ***Fig. b:*** *Standard symbol.*

Fig. 2.19: *1:4 demultiplexer.*

Table 2.11: Truth Table of 1:4 Demux

S_1	S_0	Y_0	Y_1	Y_2	Y_3
0	0	I_i	0	0	0
0	1	0	I_i	0	0
1	0	0	0	I_i	0
1	1	0	0	0	I_i

The truth table of 1:4 demultiplexer for all possible combination of inputs is shown in Table 2.12.

Table 2.12: Truth Table of 1:8 Demux for all Combination of Inputs

S_1 S_0 I_i	Minterm	Y_0	Y_1	Y_2	Y_3
0 0 0	m_0	$I_i = 0$	0	0	0
0 0 1	m_1	$I_i = 1$	0	0	0
0 1 0	m_2	0	$I_i = 0$	0	0
0 1 1	m_3	0	$I_i = 1$	0	0
1 0 0	m_4	0	0	$I_i = 0$	0
1 0 1	m_5	0	0	$I_i = 1$	0
1 1 0	m_6	0	0	0	$I_i = 0$
1 1 1	m_7	0	0	0	$I_i = 1$

S_1 \ $S_0 I_i$	00	01	11	10
0	m_0 0	m_1 (1)	m_3 0	m_2 0
1	m_4 0	m_5 0	m_7 0	m_6 0

$S_1' S_0' I_i$

Fig. a: K-map for Y_0.

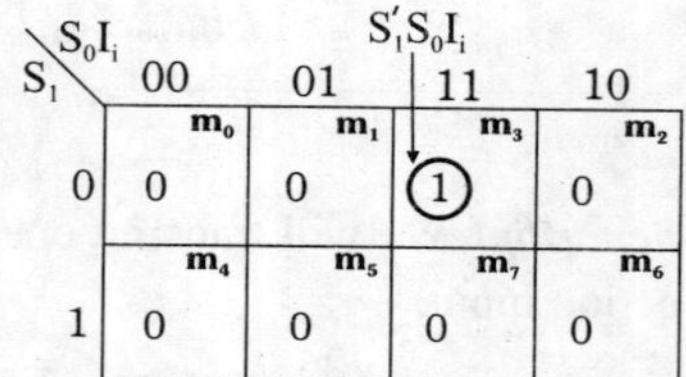

Fig. b: K-map for Y_1.

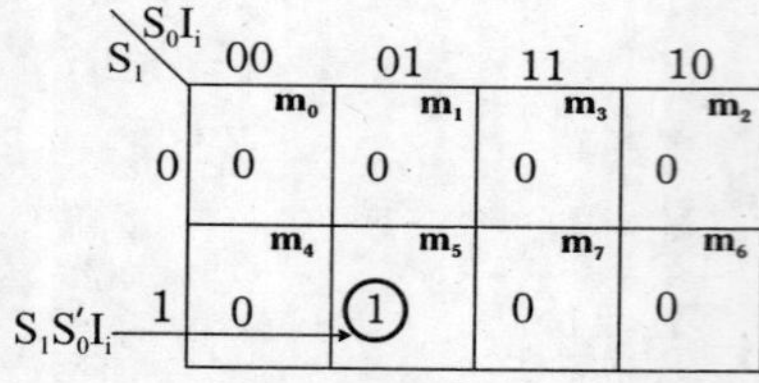

Fig. c: K-map for Y_2.

S_1 \ $S_0 I_i$	00	01	11	10
0	m_0 0	m_1 0	m_3 0	m_2 0
1	m_4 0	m_5 0	m_7 (1)	m_6 0

$S_1 S_0 I_i$

Fig. d: K-map for Y_3.

Fig. 2.20: K-map for design of 1:4 demux.

The K-maps for design of 1:4 demultiplexer are drawn in Fig. 2.20 and from the K-map we get the following Boolean equations,

$$Y_0 = I_i\, S_1'\, S_0' \qquad Y_2 = I_i\, S_1\, S_0'$$

$$Y_1 = I_i\, S_1'\, S_0 \qquad Y_3 = I_i\, S_1\, S_0$$

The logic circuit of 1:4 demux is drawn using the above Boolean equations as shown in Fig. 2.21.

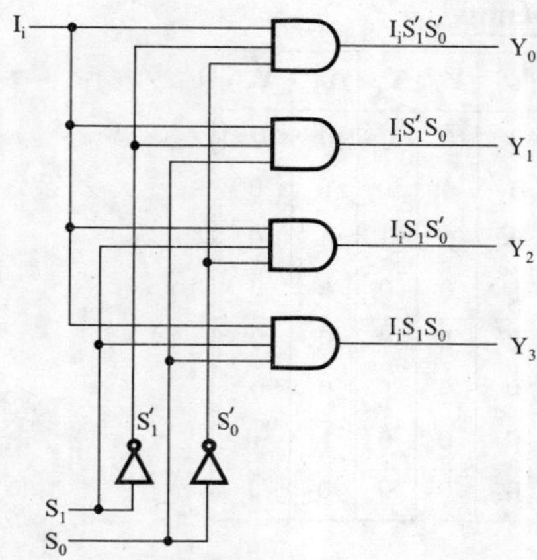

***Fig. 2.21:** Logic circuit of 1:4 demux.*

The dual 1:4 demux is available as a standard IC with number 74155. The pin configuration of 74155 is shown in Fig. 2.22.

74155

Signal	Pin	Pin	Signal
I_{i1}	1	16	V_{cc}(+5V)
E_1	2	15	I_{i2}
S_1	3	14	E_2
$\overline{Y}_{31}$	4	13	S_0
$\overline{Y}_{21}$	5	12	$\overline{Y}_{02}$
$\overline{Y}_{11}$	6	11	$\overline{Y}_{12}$
$\overline{Y}_{01}$	7	10	$\overline{Y}_{22}$
(0V) GND	8	9	$\overline{Y}_{32}$

Pin	Description
I_{i1}, I_{i2}	Inputs
E_1, E_2	Enables
S_0, S_1	Selection inputs
$\overline{Y}_{01}$ - $\overline{Y}_{31}$	Inverted outputs
Y_{02} - Y_{32}	Outputs
V_{cc}	Supply voltage (+5V)
GND	Ground (0V)

***Fig. 2.22:** Pin configuration of dual 1:4 demultiplexer IC 74155.*

1:8 Demux

The ***1:8 demultiplexer*** will transmit one binary input to any one of the eight outputs depending on the value of selection inputs.

Let, I_i = Input

S_0, S_1, S_2 = Selection inputs

Y_0, Y_1, Y_2, ..., Y_7 = Outputs

The logic to pass the input to output in 1:8 demux is listed in Table 2.13.

Table 2.13: Truth Table of 1:8 Demux

S_2	S_1	S_0	Y_0	Y_1	Y_2	Y_3	Y_4	Y_5	Y_6	Y_7
0	0	0	I_i	0	0	0	0	0	0	0
0	0	1	0	I_i	0	0	0	0	0	0
0	1	0	0	0	I_i	0	0	0	0	0
0	1	1	0	0	0	I_i	0	0	0	0
1	0	0	0	0	0	0	I_i	0	0	0
1	0	1	0	0	0	0	0	I_i	0	0
1	1	0	0	0	0	0	0	0	I_i	0
1	1	1	0	0	0	0	0	0	0	I_i

The block diagram and symbolic representation of 1:8 demux is shown in Fig. 2.23.

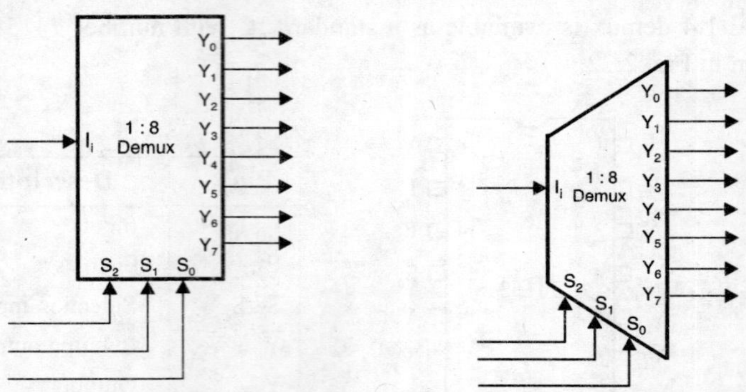

Fig. a: Block diagram representation. *Fig. b: Standard symbol.*

Fig. 2.23: 1:8 demultiplexer.

Table 2.14: Verification of Output of 1:8 Demux of Fig. 2.24

Selection Inputs S_2 S_1 S_0	Complement of Selection Inputs S_2' S_1' S_0'	$Y_0 = I_i S_2' S_1' S_0'$	$Y_1 = I_i S_2' S_1' S_0$	$Y_2 = I_i S_2' S_1 S_0'$	$Y_3 = I_i S_2' S_1 S_0$	$Y_4 = I_i S_2 S_1' S_0'$	$Y_5 = I_i S_2 S_1' S_0$	$Y_6 = I_i S_2 S_1 S_0'$	$Y_7 = I_i S_2 S_1 S_0$
0 0 0	1 1 1	I_i	0	0	0	0	0	0	0
0 0 1	1 1 0	0	I_i	0	0	0	0	0	0
0 1 0	1 0 1	0	0	I_i	0	0	0	0	0
0 1 1	1 0 0	0	0	0	I_i	0	0	0	0
1 0 0	0 1 1	0	0	0	0	I_i	0	0	0
1 0 1	0 1 0	0	0	0	0	0	I_i	0	0
1 1 0	0 0 1	0	0	0	0	0	0	I_i	0
1 1 1	0 0 0	0	0	0	0	0	0	0	I_i

The logic circuit of 1:8 demux can be realized using AND gate as shown in Fig. 2.24 and Table 2.14 can be used to verify the logic circuit of Fig. 2.24.

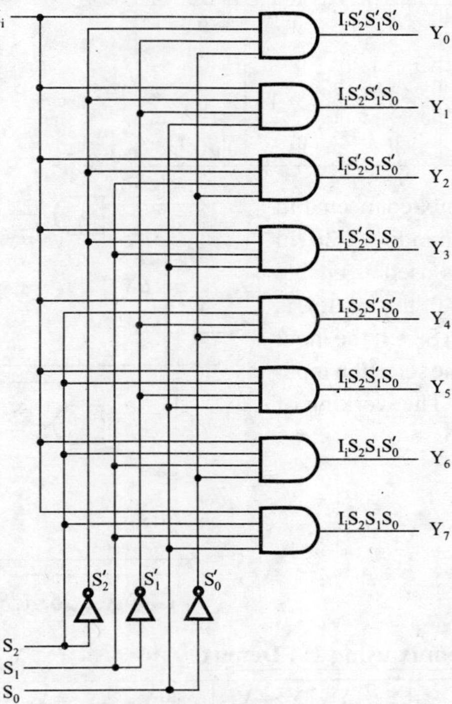

Fig. 2.24: Logic circuit of 1:8 demux.

2.4.1 Expanding the Demultiplexer

Demultiplexers of smaller size can be connected in parallel to expand the size of demultiplexer. For example two 1:2 demultiplexer can be connected in parallel to form 1:4 demultiplexer and two 1:4 demultiplexers can be connected in parallel to form 1:8 demultiplexer and so on.

An example of two 1:2 demux connected in parallel to form 1:4 demux is shown in Fig. 2.11. In this circuit the selection input S_1 is used as enable for the input of demux. When $S_1 = 0$, the input I_i is allowed to first 1:2 demux and when $S_1 = 1$, the input I_i is allowed to second 1:2 demux. The selection input S_0 is common for both 1:2 demux. The working of this demux is verified in Table 2.15.

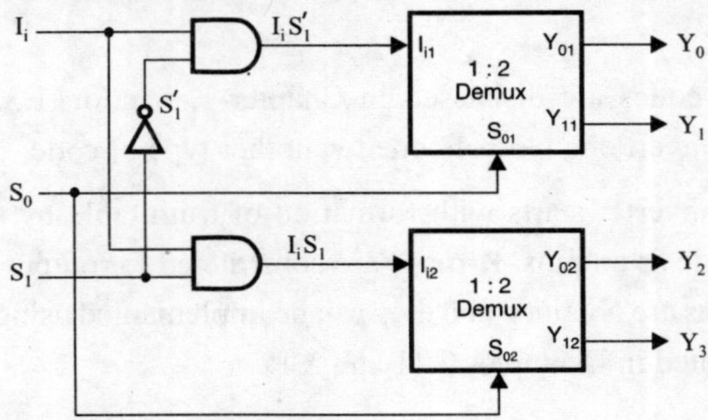

Fig. 2.25: 1:4 demux using 1:2 demux.

Table 2.15: Verification of 1:4 Demux using 1:2 Demux

S_1	S_0	I_iS_1'	I_iS_1	$Y_0 = Y_{01}$	$Y_1 = Y_{11}$	$Y_2 = Y_{02}$	$Y_3 = Y_{12}$
0	0	I_i	0	$I_{i1} = I_i$	0	0	0
0	1	I_i	0	0	$I_{i1} = I_i$	0	0
1	0	0	I_i	0	0	$I_{i2} = I_i$	0
1	1	0	I_i	0	0	0	$I_{i2} = I_i$

An example of two 1:4 demux connected in parallel to form 1:8 demux is shown in Fig. 2.26. In this circuit the selection input S_2 is used as enable for the input of demux. When $S_2 = 0$, the input I_i is allowed to first 1:4 demux and when $S_2 = 1$, the input I_i is allowed to second 1:4 demux. The selection input S_1 is common for both 1:4 demux. The working of this demux is verified in Table 2.16.

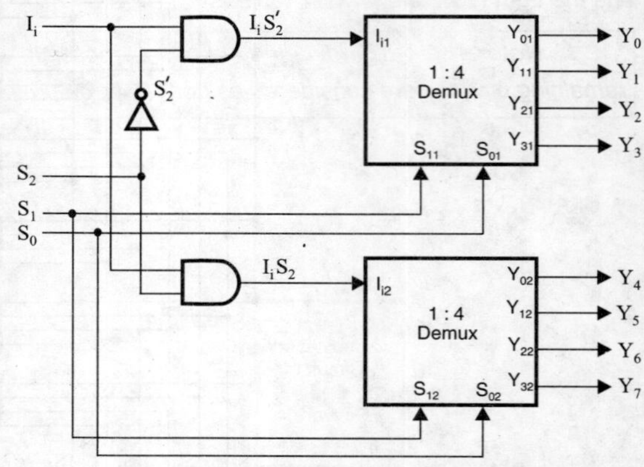

***Fig. 2.26:** 1:8 demux using 1:4 demux.*

Table 2.16: Verification of 1:8 Demux using 1:4 Demux

S_2	S_1	S_0	I_iS_2'	I_iS_2	$Y_0 = Y_{01}$	$Y_1 = Y_{11}$	$Y_2 = Y_{21}$	$Y_3 = Y_{31}$	$Y_4 = Y_{02}$	$Y_5 = Y_{12}$	$Y_6 = Y_{22}$	$Y_7 = Y_{32}$
0	0	0	I_i	0	$I_{i1} = I_i$	0	0	0	0	0	0	0
0	0	1	I_i	0	0	$I_{i1} = I_i$	0	0	0	0	0	0
0	1	0	I_i	0	0	0	$I_{i1} = I_i$	0	0	0	0	0
0	1	1	I_i	0	0	0	0	$I_{i1} = I_i$	0	0	0	0
1	0	0	0	I_i	0	0	0	0	$I_{i2} = I_i$	0	0	0
1	0	1	0	I_i	0	0	0	0	0	$I_{i2} = I_i$	0	0
1	1	0	0	I_i	0	0	0	0	0	0	$I_{i2} = I_i$	0
1	1	1	0	I_i	0	0	0	0	0	0	0	$I_{i2} = I_i$

2.5 Code Converters

The various types of codes are discussed in Chapter-1, Section 1.3. The ***code converters*** are combinational circuits that convert one type of code to another type of code.

The design of code converter starts with formation of truth table by taking one type of code as inputs and another type of code as outputs. K-maps are constructed for output codes and from the K-map simplified Boolean expressions are obtained and they can be implemented using logic gates. Two examples of code conversion are presented in Examples 2.11 and 2.12.

Example 2.11

Design a code converter that converts a decimal digit from the 8, 4, – 2, – 1 code to BCD.

Solution

The 8, 4, – 2, – 1 code and BCD are 4-digit binary codes. Hence the code converter will have 4 inputs and 4 outputs. Let A, B, C, D be inputs and w, x, y, z be outputs. The block diagram of code converter is shown in Fig. 1 and the truth table is shown in Table 2.

> ***Note:*** *The 4-digit binary code has 16 possible codes but here only 10 codes are used and the output for remaining 6 codes are considered as don't-care outputs.*

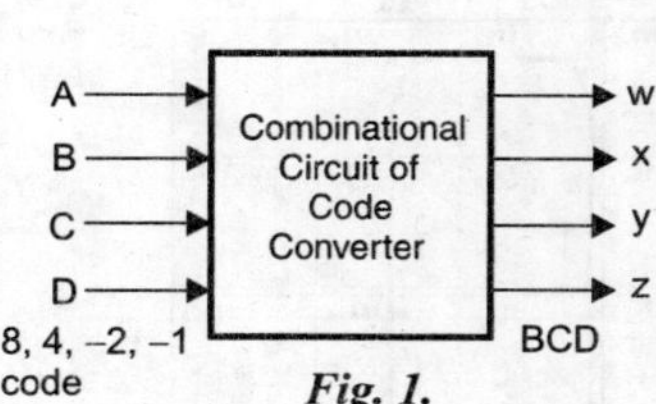

Fig. 1.

Table 1: Input-Output Relation/Truth Table

Decimal Digit	8 A	4 B	–2 C	–1 D	w	x	y	z
0	0	0	0	0	0	0	0	0
1	0	1	1	1	0	0	0	1
2	0	1	1	0	0	0	1	0
3	0	1	0	1	0	0	1	1
4	0	1	0	0	0	1	0	0
5	1	0	1	1	0	1	0	1
6	1	0	1	0	0	1	1	0
7	1	0	0	1	0	1	1	1
8	1	0	0	0	1	0	0	0
9	1	1	1	1	1	0	0	1

Table 2: Truth Table with Don't-care

Decimal Digit	Inputs A	B	C	D	Minterm	Outputs w	x	y	z
0	0	0	0	0	m_0	0	0	0	0
-	0	0	0	1	m_1	×	×	×	×
-	0	0	1	0	m_2	×	×	×	×
-	0	0	1	1	m_3	×	×	×	×
4	0	1	0	0	m_4	0	1	0	0
3	0	1	0	1	m_5	0	0	1	1
2	0	1	1	0	m_6	0	0	1	0
1	0	1	1	1	m_7	0	0	0	1
8	1	0	0	0	m_8	1	0	0	0
7	1	0	0	1	m_9	0	1	1	1
6	1	0	1	0	m_{10}	0	1	1	0
5	1	0	1	1	m_{11}	0	1	0	1
-	1	1	0	0	m_{12}	×	×	×	×
-	1	1	0	1	m_{13}	×	×	×	×
-	1	1	1	0	m_{14}	×	×	×	×
9	1	1	1	1	m_{15}	1	0	0	1

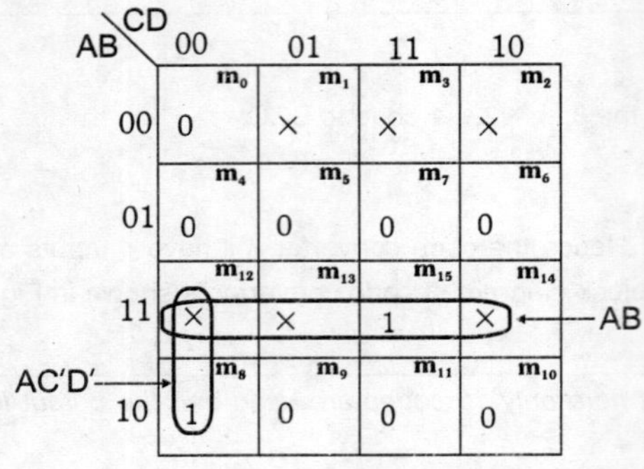

Fig. 2: K-map for w.

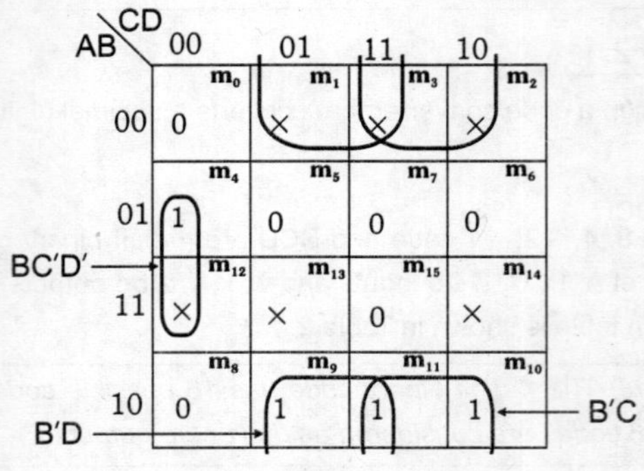

Fig. 3: K-map for x.

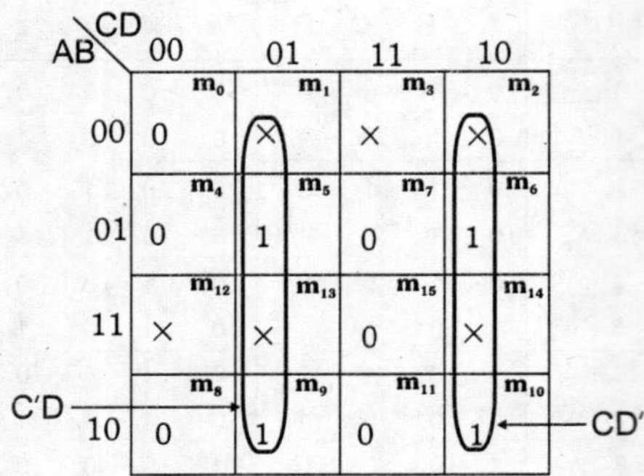

Fig. 4: K-map for y.

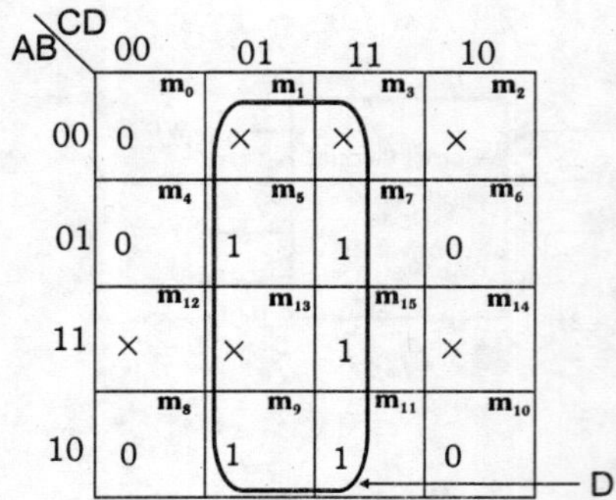

Fig. 5: K-map for z.

In order to find simplified Boolean expressions for output variables, separate 4 variable K-map is drawn for each output variable as shown in Figs. 2 to 5.

The following Boolean functions are obtained from K-map of Figs. 2 to 5.

$$w = AB + AC'D'$$

$$x = B'D + B'C + BC'D'$$

$$y = C'D + CD'$$

$$z = D$$

The logic circuit for 8, 4, – 2, – 1 code to BCD code converter is drawn using the above Boolean equations as shown in Fig. 6.

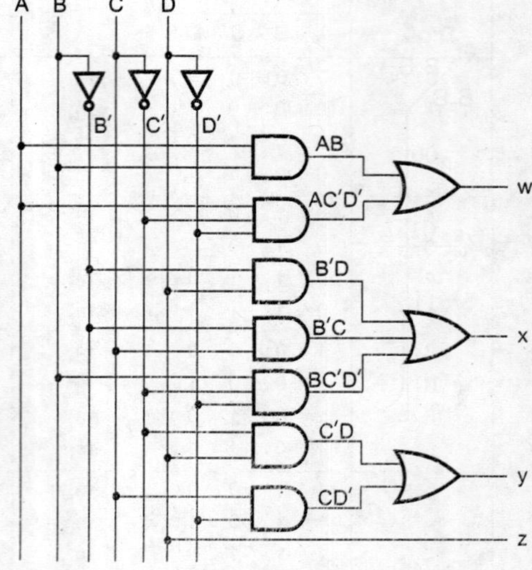

Fig. 6: Logic circuit of 8, 4, –2, –1 code to BCD code converter.

Example 2.12

Design a BCD to excess-3 code converter and explain.

Solution:

The BCD code is 4-bit binary code. In 4-bit binary code, $2^4 = 16$ codes are possible. But the BCD has only 10 valid codes. The excess-3 code is obtained by adding 0011_2 (3_{10}) to BCD code. The excess-3 code will also have only 10 codes and the remaining six codes are don't-care outputs. The truth table of BCD to excess-3 code converter is shown in Table 1.

Using the truth table, K-map for 4 bits of excess-3 code E_0, E_1, E_2 and E_3 are drawn as shown in Figs. 1 to 4.

Table 1: Truth Table

Inputs BCD $B_3\ B_2\ B_1\ B_0$	Minterm	Outputs Excess-3 code $E_3\ E_2\ E_1\ E_0$
0 0 0 0	m_0	0 0 1 1
0 0 0 1	m_1	0 1 0 0
0 0 1 0	m_2	0 1 0 1
0 0 1 1	m_3	0 1 1 0
0 1 0 0	m_4	0 1 1 1
0 1 0 1	m_5	1 0 0 0
0 1 1 0	m_6	1 0 0 1
0 1 1 1	m_7	1 0 1 0
1 0 0 0	m_8	1 0 1 1
1 0 0 1	m_9	1 1 0 0
1 0 1 0	m_{10}	× × × ×
1 0 1 1	m_{11}	× × × ×
1 1 0 0	m_{12}	× × × ×
1 1 0 1	m_{13}	× × × ×
1 1 1 0	m_{14}	× × × ×
1 1 1 1	m_{15}	× × × ×

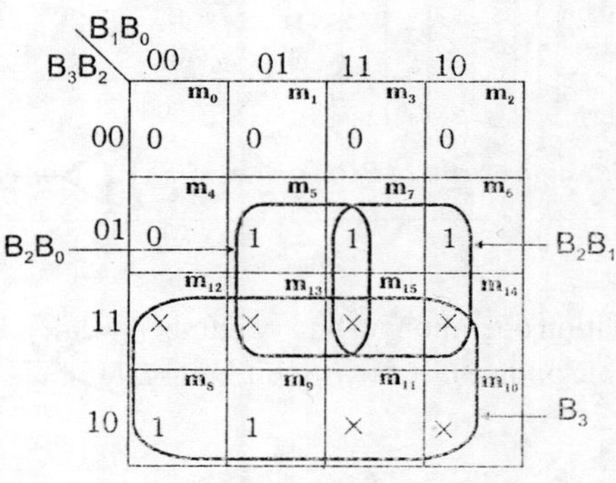

Fig. 1: K-map for E_3.

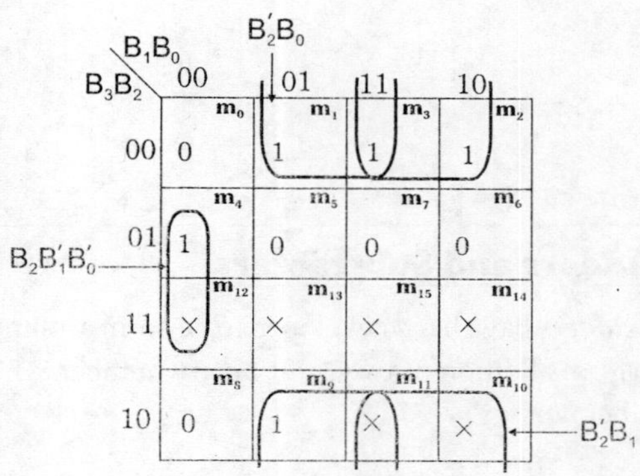

Fig. 2: K-map for E_2.

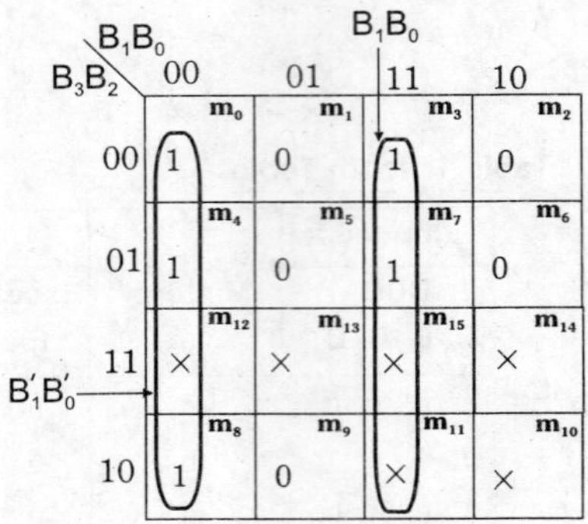

Fig. 3: K-map for E_1.

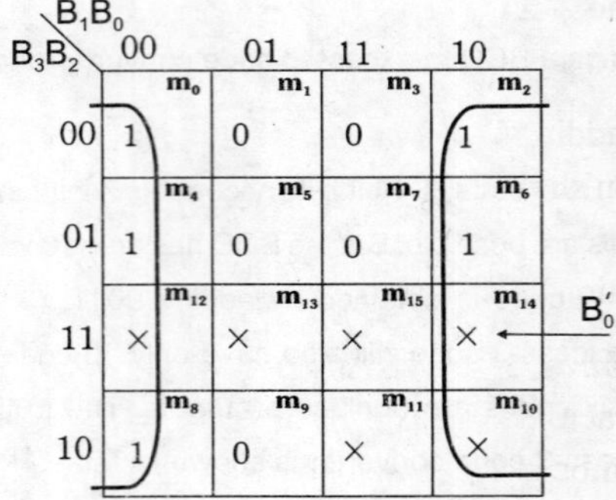

Fig. 4: K-map for E_0.

From the K-map we get the following Boolean equations.

$$E_3 = B_2B_1 + B_2B_0 + B_3$$

$$E_2 = B_2'B_1 + B_2'B_0 + B_2B_1'B_0'$$

$$E_1 = B_1'B_0' + B_1B_0$$

$$E_0 = B_0'$$

Using the above Boolean equations the logic circuit for BCD to excess-3 code converter is drawn as shown in Fig. 5.

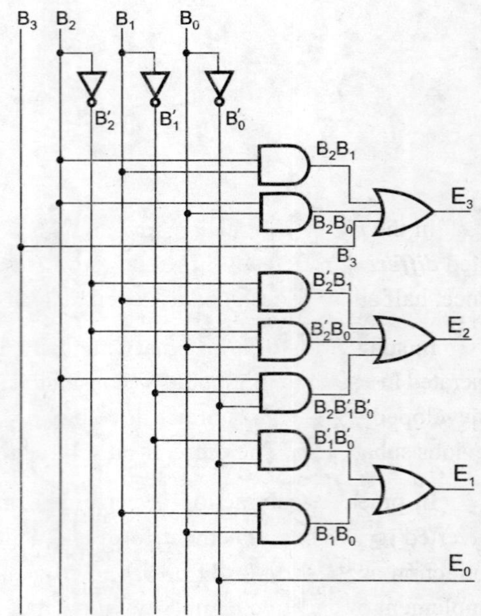

Fig. 5: Logic circuit of BCD to excess-3 code converter.

2.6 Adders and Subtractors

Adders have been developed to perform arithmetic addition operation on binary numbers. Half adder can perform addition of two 1-bit binary number. The possible additions of two 1-bit binary numbers are shown below:

	2-bit standard representation
0 + 0 = 0	0 0
0 + 1 = 1	0 1
1 + 0 = 1	0 1
1 + 1 = 1 0	1 0

(In 1 0: the right bit → Sum, the left bit → Carry)

In the above additions, it can be observed that result is either 1-bit or 2-bit. The 1-bit result is called ***sum***. In the 2-bit result, the first bit is called ***sum*** and the second bit is called ***carry***. Hence, half adder is designed to generate 2-bit standard output.

In addition of n-bit binary numbers, the addition is performed bit-by-bit. In this the carry generated in an addition should be considered in next addition to get correct result and so full adder is developed which can add three binary input bits in which one of the bits is carry generated in previous addition. Therefore, n-bit addition can be performed by using n full adders in parallel.

Subtractors have been developed to perform arithmetic subtraction operation on binary numbers. Half subtractor can perform subtraction of two 1-bit binary number. The possible subtractions of two 1-bit binary numbers are shown below:

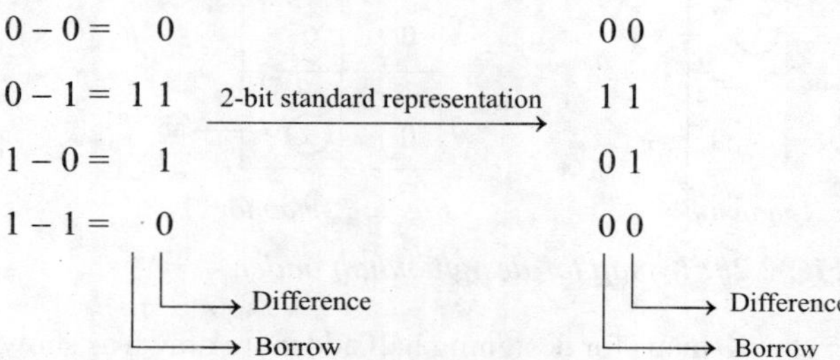

In the above subtractions, it can be observed that result is either 1-bit or 2-bit. The 1-bit result is called ***difference***. In the 2-bit result, the first bit is called ***difference*** and the second bit is called ***borrow***. Hence, half adder is designed to generate 2-bit standard output.

In subtraction of n-bit binary numbers, the subtraction is performed bit-by-bit. In this the borrow generated in a subtraction should be considered in next subtraction to get correct result and so full subtractor is developed which can subtract three binary input bits in which one of the bits is borrow generated in previous subtraction. Therefore, n-bit subtraction can be performed by using n full subtractors in parallel.

In practice, subtraction is performed in two's complement method in which the subtrahend is converted to two's complement form and added with minuend. Therefore, adders are used to perform subtraction as an addition of positive and negative number. The addition of positive number and two's complement of negative number will give the result of subtraction.

2.6.1 Half Adder

Half adder (HA) is a combinational circuit that performs arithmetic sum of two binary bits. The outputs are sum and carry. The truth table of half adder is shown in Table 2.17. The block diagram and symbolic representation of half adder are shown in Figs. 2.27 and 2.30 respectively.

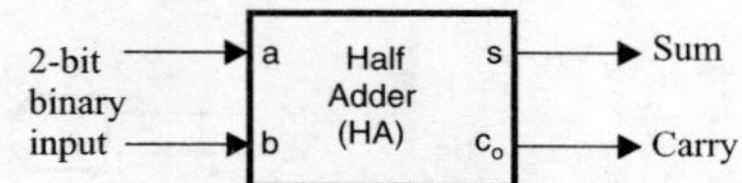

Fig. 2.27: *Block digaram of half adder.*

Table 2.17: Truth Table of Half Adder

Inputs		Minterm	Outputs	
a	b		s	c_0
0	0	m_0	0	0
0	1	m_1	1	0
1	0	m_2	1	0
1	1	m_3	0	1

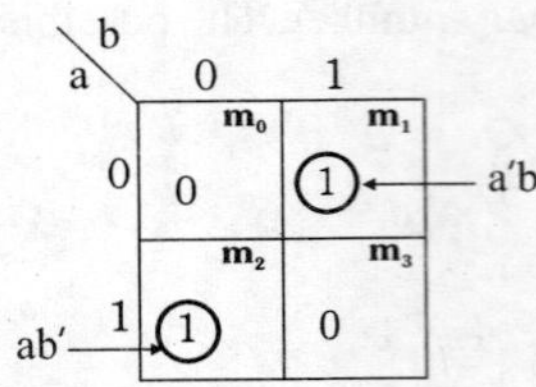

Fig. a: K-map for s.

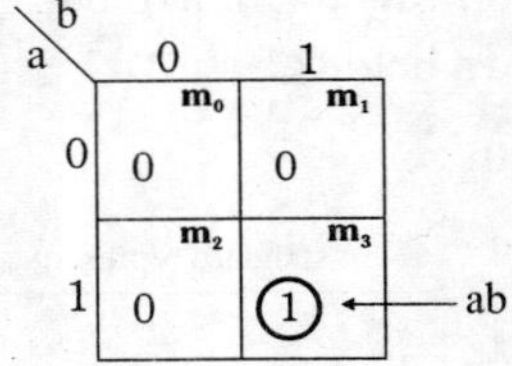

Fig. b: K-map for c_o.

Fig. 2.28: K-map for design of half adder.

Using truth table (Table 2.17) the K-maps for designing half adder are drawn as shown in Fig. 2.28. From the K-map we get the following Boolean equations and using these equations the logic circuit of half adder is drawn as shown in Fig. 2.29.

$$s = a b' + a' b$$

$$c_o = a b$$

$s = ab' + a'b = a \oplus b$

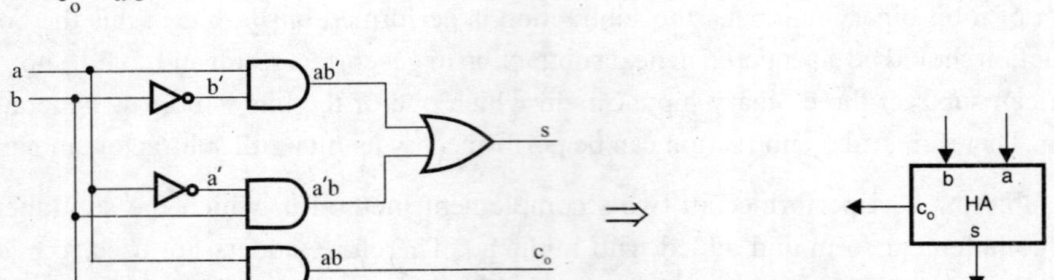

Fig. 2.29: Logic circuit of half adder. *Fig. 2.30: Symbolic representation of half adder.*

Alternatively, the sum can be realized from XOR gate. If we compare the sum and output of XOR gate we observe that both are same. The logic circuit of half adder using XOR gate is shown in Fig. 2.31.

Table 2.18: Truth Table of XOR Gate

a	b	$a \oplus b$
0	0	0
0	1	1
1	0	1
1	1	0

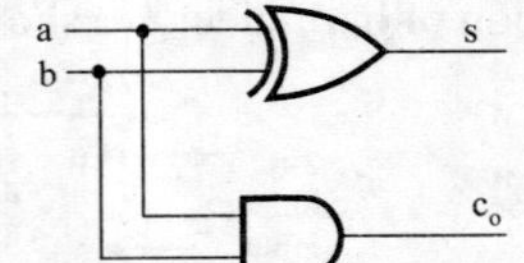

Fig. 2.31: Half adder using XOR gate.

2.6.2 Full Adder

(AU, Apr/May'19, 6 Marks)

Full adder (FA) is a combinational circuit that perform arithmetic sum of three binary bits in which one of the bit is carry generated in previous addition. Hence, full adder will be useful in n-bit addition to add carry of an addition in the next bit addition.

Table 2.19: Truth Table of Full Adder

Inputs			Minterm	Outputs	
a	b	c_i		s	c_o
0	0	0	m_0	0	0
0	0	1	m_1	1	0
0	1	0	m_2	1	0
0	1	1	m_3	0	1
1	0	0	m_4	1	0
1	0	1	m_5	0	1
1	1	0	m_6	0	1
1	1	1	m_7	1	1

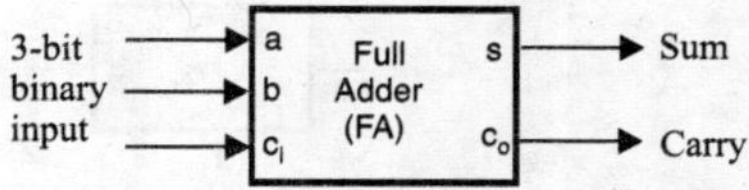

***Fig. 2.32:** Block diagram of full adder.*

The outputs are sum and carry. The truth table of full adder is shown in Table 2.19. The block diagram and symbolic representation of full adder are shown in Figs. 2.32 and 2.37 respectively.

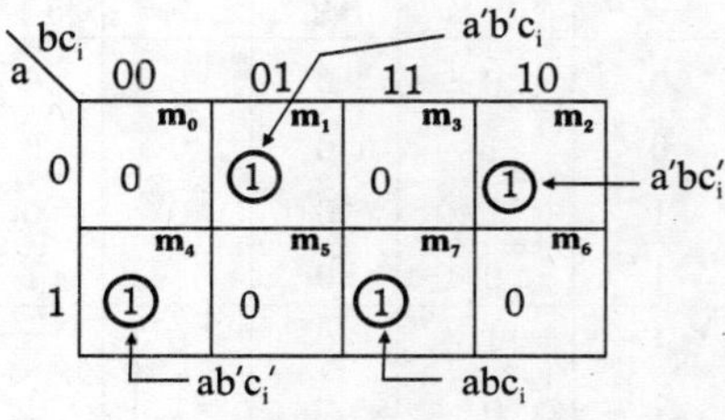

***Fig. 2.33:** K-map for s.*

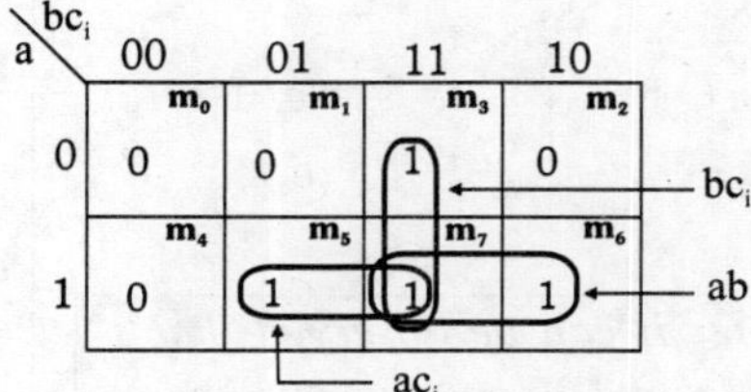

***Fig. 2.34:** K-map for c_o.*

Using truth table (Table 2.19) the K-maps for designing full adder are drawn as shown in Figs. 2.33 and 2.34. From the K-map we get the following Boolean equations and using these equations the logic circuit of full adder is drawn as shown in Fig. 2.35.

$$s = a'b'c_i + a'bc_i' + ab'c_i' + abc_i$$

$$c_o = ab + ac_i + bc_i$$

> ***Note:** With reference to Example 2.5, sum, s can be further simplified to XOR of a, b and c_i.*
>
> $$\therefore s = a \oplus b \oplus c_i$$
>
> a, b, c_i → $s = a \oplus b \oplus c_i$

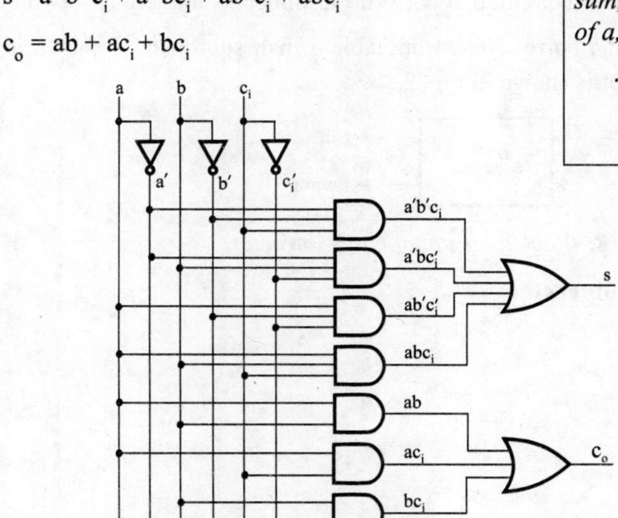

***Fig. 2.35:** Logic circuit of full adder.*

Alternatively, full adder can be realized using two half adders as shown in Fig. 2.36. The first half adder is used to add two binary input bits and the second half adder is used to add sum of first half adder and third binary input.

The carry is obtained by logical OR of the carry of both the half adders.

The combinational circuit of Fig. 2.36 can be verified using the truth table (Table 2.20).

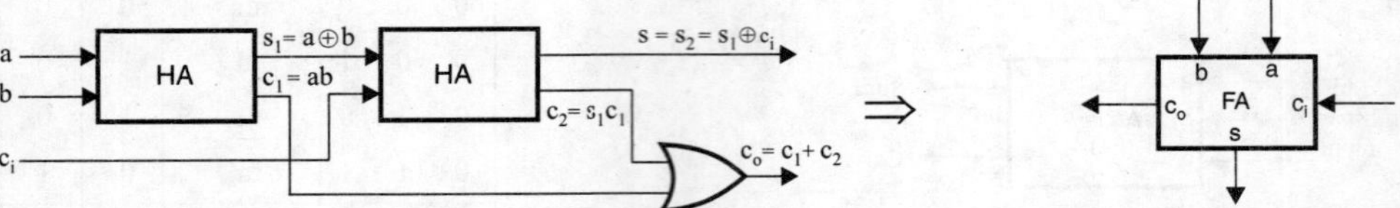

Fig. 2.36: Full adder using two half adders.

Fig. 2.37: Symbolic representation of full adder.

The following equations are obtained from design of half adder in Section 2.6.1, for the sum and carry of the logic circuit of Fig. 2.36.

$s_1 = a \oplus b$	$s = s_2 = s_1 \oplus c_i$
$c_1 = ab$	$c_2 = s_1 \, c_i$

> ***Note:*** *It can be observed that s and c_o of Table 2.20 is same as s and c_o of full adder.*

Table 2.20: Truth Table to Verify Logic Circuit of Fig. 2.36

Inputs			s_1	$s = s_2$	c_1	c_2	c_o
a	b	c_i	$a \oplus b$	$s_1 \oplus c_i$	ab	$s_1 c_i$	$c_1 + c_2$
0	0	0	0	0	0	0	0
0	0	1	0	1	0	0	0
0	1	0	1	1	0	0	0
0	1	1	1	0	0	1	1
1	0	0	1	1	0	0	0
1	0	1	1	0	0	1	1
1	1	0	0	0	1	0	1
1	1	1	0	1	1	0	1

2.6.3 Half Subtractor

Half subtractor is a combination circuit that performs arithmetic subtraction of two binary bits.

The outputs are difference and borrow. The truth table of half subtractor is shown in Table 2.21. The block diagram of half subtractor is shown in Fig. 2.38.

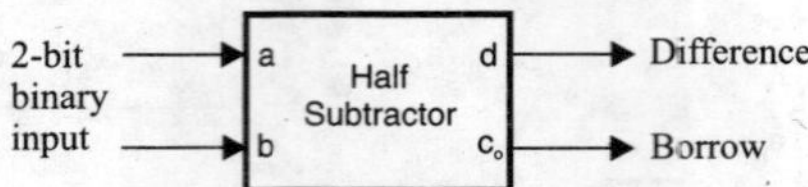

Fig. 2.38: Block diagram of half subtractor.

Table 2.21: Truth Table of Half Subtractor

Inputs		Minterm	Outputs	
a	b		d	c_o
0	0	m_0	0	0
0	1	m_1	1	1
1	0	m_2	1	0
1	1	m_3	0	0

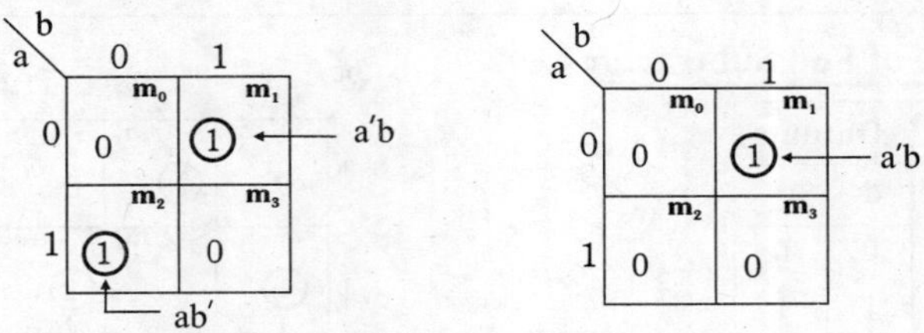

Fig. a: *K-map for d.* ***Fig. b:*** *K-map for c_o.*

Fig. 2.39: *K-map for design of half subtractor.*

Using truth table (Table 2.21) the K-maps for designing half subtractor are drawn as shown in Fig. 2.39.

From the K-map we get the following Boolean equations,

$$d = ab' + a'b$$

$$\boxed{d = ab' + a'b = a \oplus b}$$

$$c_o = a'b$$

Using the above Boolean equations the logic circuit of half subtractor is drawn as shown in Fig. 2.40.

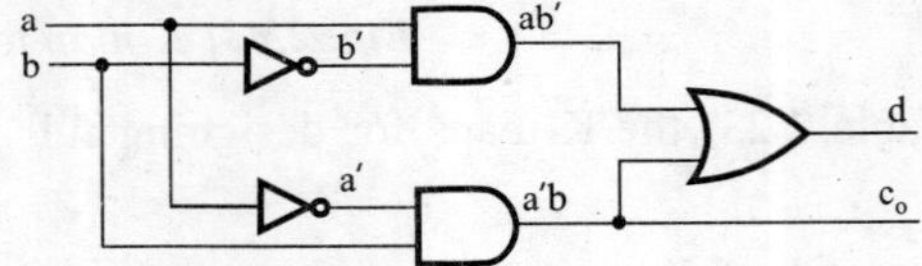

Fig. 2.40: *Logic circuit of half subtractor.*

Alternatively, the difference can be realized from XOR gate. If we compare the difference and output of XOR gate, we observe that both are same. The logic circuit of half subractor using XOR gate is shown in Fig. 2.41.

Table 2.22: Truth Table of XOR Gate

a	b	a ⊕ b
0	0	0
0	1	1
1	0	1
1	1	0

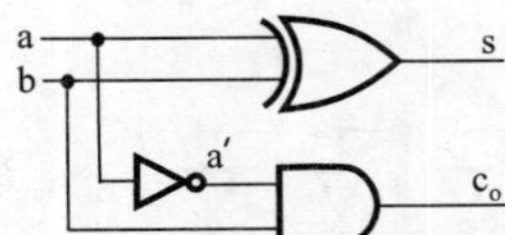

Fig. 2.41: *Half subtractor using XOR gate.*

2.6.4 Full Subtractor

Full subtractor is a combinational circuit that perform arithmetic subtraction of three binary bits in which one of the bit is borrow generated in previous subtraction.

Hence, full subtractor will be useful in n-bit subtraction to subtract borrow of a subtraction in next bit subtraction. The outputs are difference and borrow. The truth table of full subtractor is shown in Table 2.23. The block diagram of full subtractor is shown in Fig. 2.42.

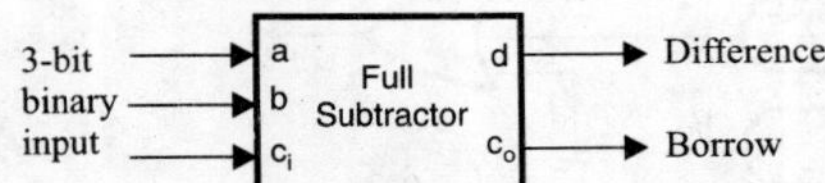

Fig. 2.42: *Block diagram of full subtractor.*

Table 2.23: Truth Table of Full Subtractor

Inputs			Minterm	Outputs	
a	b	c_i		d	c_o
0	0	0	m_0	0	0
0	0	1	m_1	1	1
0	1	0	m_2	1	1
0	1	1	m_3	0	1
1	0	0	m_4	1	0
1	0	1	m_5	0	0
1	1	0	m_6	0	0
1	1	1	m_7	1	1

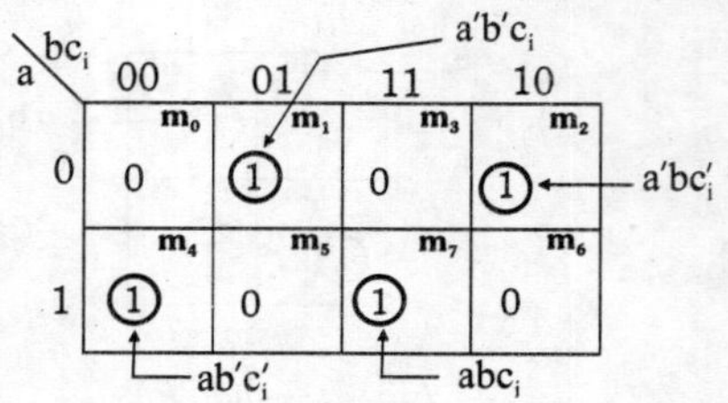

Fig. a: K-map for d.

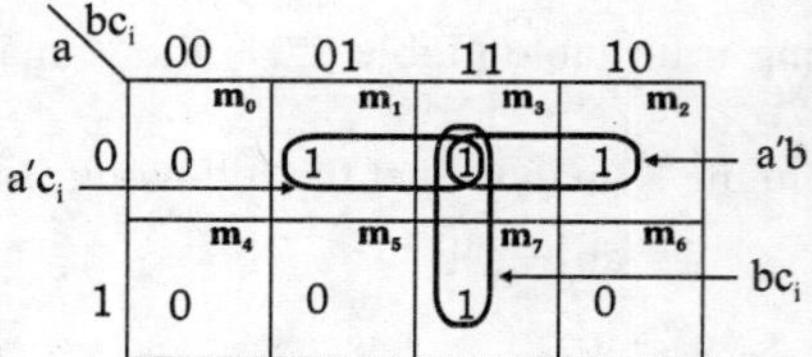

Fig. b: K-map for c_o.

Fig. 2.43: K-map for design of full subtractor.

Using truth table (Table 2.23) the K-maps for designing full subtractor are drawn as shown in Fig. 2.43.

From the K-maps we get the following Boolean equations,

$$d = a'b'c_i + a'bc_i' + ab'c_i' + abc_i$$

$$c_o = a'b + a'c_i + bc_i$$

Using the above Boolean equations the logic circuit of full subtractor is drawn as shown in Fig. 2.44.

> ***Note:*** *With reference to Example 2.5, difference, d can be further simplified to XOR of a, b and c_i.*
>
> $$\therefore d = a \oplus b \oplus c_i$$
>
> a, b, c_i → XOR → $d = a \oplus b \oplus c_i$

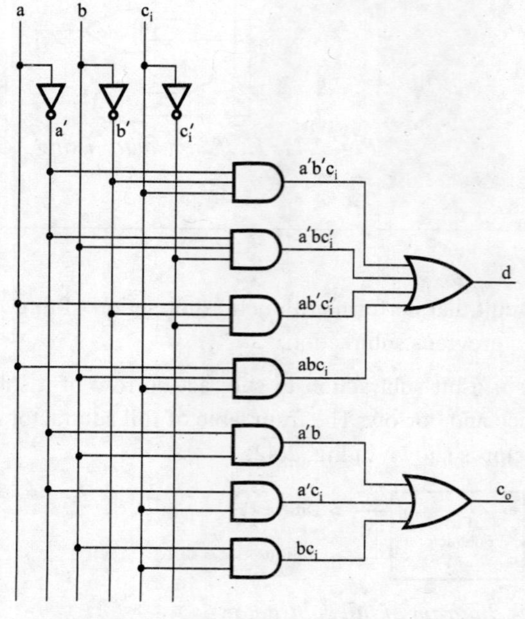

Fig. 2.44: Logic circuit of full subtractor.

2.6.5 Binary Parallel Adder

A ***binary parallel adder*** performs arithmetic sum of two n-bit binary numbers. A full adder can add two 1-bit binary along with previous carry. Hence, in order to add two n-bit binary numbers, n full adders are required. Each full adder will add one bit of binary numbers.

Consider two, 4-bit binary numbers A and B shown below:

Binary number-1: $a_3\ a_2\ a_1\ a_0$ (A)

Binary number-2: $b_3\ b_2\ b_1\ b_0$ (B)

The bit-by-bit addition is performed as follows.

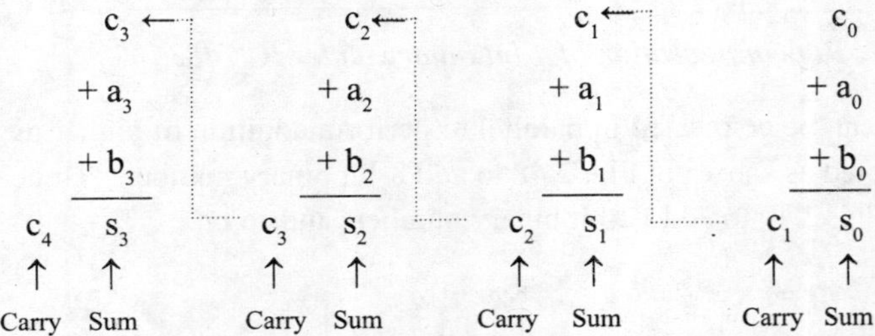

In the above addition the initial carry $c_o = 0$. The carry generated in an addition is considered in the next addition. A 4-bit binary adder to perform above addition is constructed using 4 full adders as shown in Fig. 2.45. The symbolic representations of 4-bit binary adders are shown in Fig. 2.46.

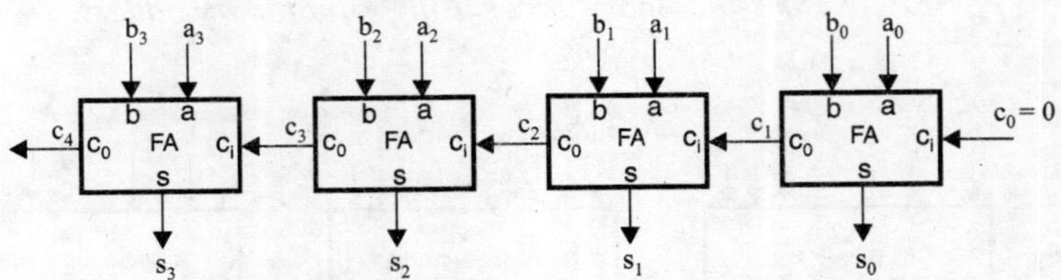

Fig. 2.45: *4-bit binary adder using full adders.*

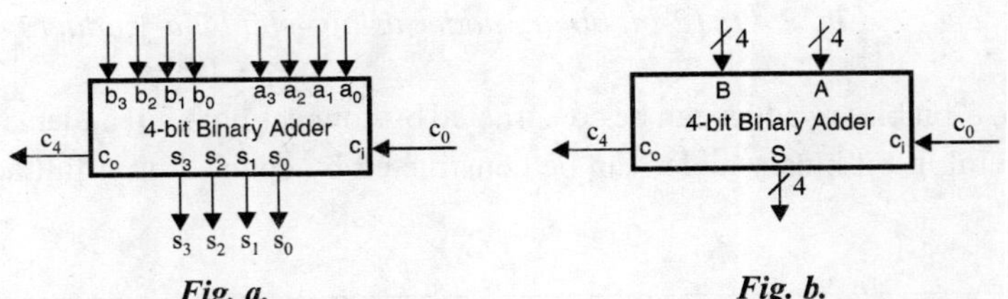

Fig. 2.46: *Symbolic representation of 4-bit binary adder.*

The 4-bit binary adder is available as a standard IC with number 7483. The pin configuration of 7483 is shown in Fig. 2.47.

7483

Pin label	Pin No.	Pin No.	Pin label
s_2	1	16	V_{cc}(+5V)
b_2	2	15	b_3
a_2	3	14	a_3
s_1	4	13	s_3
a_1	5	12	a_4
b_1	6	11	b_4
c_o	7	10	s_4
(0V) GND	8	9	c_4

Pin	Description
a_1 - a_4	Operand a inputs
b_1 - b_4	Operand b inputs
c_o	Carry input
s_1 - s_4	Sum outputs
c_4	Carry output
V_{cc}	Supply voltage (+5V)
GND	Ground (0V)

Fig. 2.47: *Pin configuration of 4-bit binary adder IC 7483.*

The 4-bit binary adders can be connected in parallel to perform addition of higher bit size binary. Two 4-bit adders can be connected as shown in Fig. 2.48 to add 8-bit binary numbers. Three 4-bit adders can be connected as shown in Fig. 2.49 to add 12-bit binary numbers and so on.

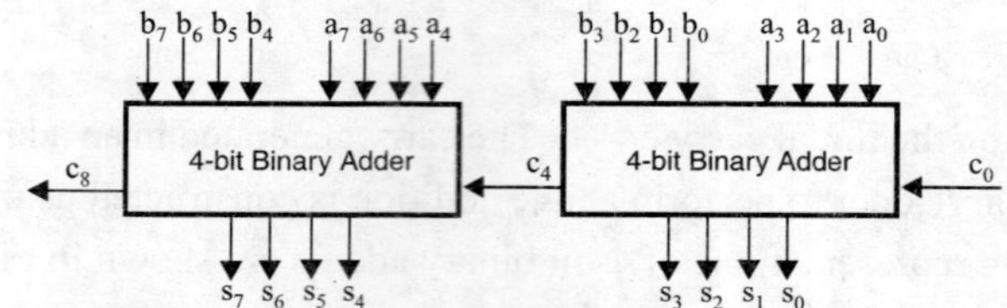

Fig. 2.48: *8-bit binary adder using 4-bit binary adders.*

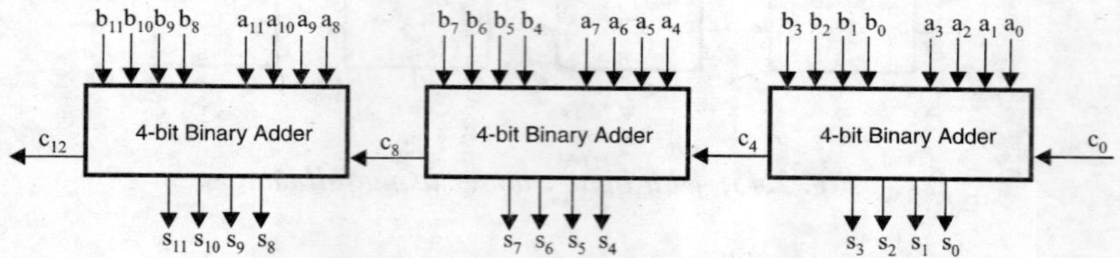

Fig. 2.49: *12-bit binary adder using 4-bit binary adders.*

Alternatively, 8-bit binary adder can be constructed by connecting 8 full adders in parallel as shown in Fig. 2.50. In general, n-bit binary adder can be constructed by connecting n full adders in parallel as shown in Fig. 2.51.

> ***Note:*** *n-bit binary adder is called parallel adder because it add all bits of data simultaneously.*

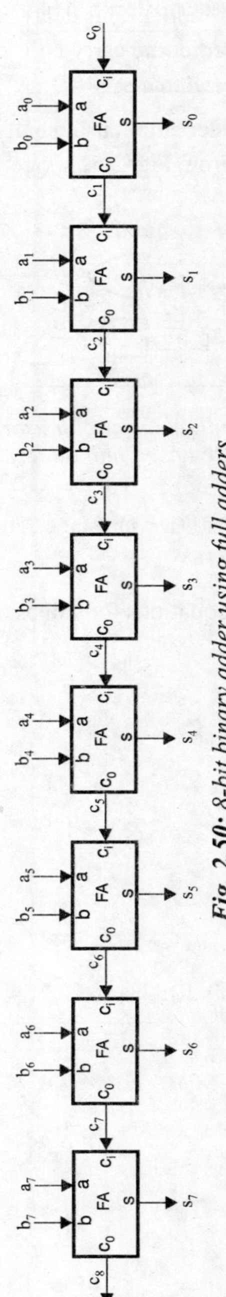

***Fig. 2.50:** 8-bit binary adder using full adders.*

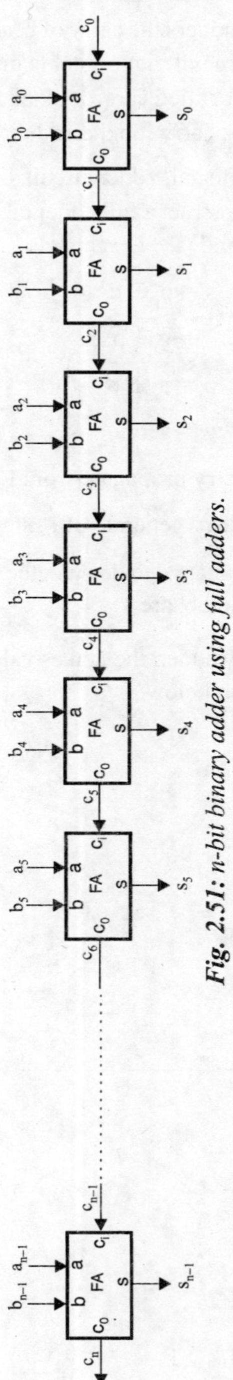

***Fig. 2.51:** n-bit binary adder using full adders.*

2.6.6 Carry Look Ahead Adder

In n-bit binary adders the carry of ith adder propagates to $(i + 1)^{th}$ adder. This creates a delay in getting a stable result. The sum will not be stable until the carry propagates to n^{th} adder. This delay can be avoided if an additional circuit called carry look ahead circuit is designed to predict the carry before addition and the adder that includes carry look ahead circuit is called ***carry look ahead adder***.

With reference to half adder circuit shown in Fig. 2.31 and full adder circuit using half adders shown in Fig. 2.36 we can construct a full adder circuit as shown in Fig. 2.52. From Fig. 2.52 we get the following equations for i^{th} sum and $(i + 1)^{th}$ carry.

$$s_i = p_i \oplus c_i$$

$$c_{i+1} = g_i + p_i\, c_i$$

$$p_i = a_i \oplus b_i$$

$$g_i = a_i\, b_i$$

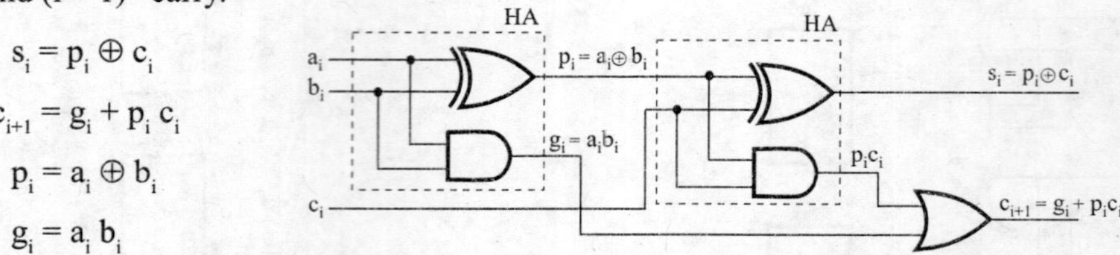

Fig. 2.52: Full adder circuit using two XOR gate based half adders.

Where, p_i = Carry propagate from i_{th} stage

g_i = Carry generate of ith stage

The carry propagate decides whether a carry propagates from i^{th} to $(i + 1)^{th}$ stage. The g_i generates a carry when both a_i and b_i are 1.

In a 4-bit binary adder, the i takes values from 0 to 3. The above equations for values of i = 0, 1, 2, 3 are evaluated as shown below:

When i = 0;

$$p_0 = a_0 \oplus b_0$$

$$g_0 = a_0\, b_0$$

$$s_0 = p_0 \oplus c_0$$

$$c_1 = g_0 + p_0 c_0$$

When i = 1;

$$p_1 = a_1 \oplus b_1$$

$$g_1 = a_1\, b_1$$

$$s_1 = p_1 \oplus c_1$$

$$c_2 = g_1 + p_1\, c_1$$

$$= g_1 + p_1\, (g_0 + p_0\, c_0)$$

$$= g_1 + p_1 g_0 + p_1\, p_0 c_0$$

When i = 2;

$$p_2 = a_2 \oplus b_2$$

$$g_2 = a_2\, b_2$$

$$s_2 = p_2 \oplus c_2$$

$$c_3 = g_2 + p_2\, c_2$$

$$= g_2 + p_2\, (g_1 + p_1\, g_0 + p_1\, p_0 c_0)$$

$$= g_2 + p_2 g_1 + p_2\, p_1 g_0 + p_2\, p_1 p_0 c_0$$

When i = 3;

$$p_3 = a_3 \oplus b_3$$

$$g_3 = a_3\, b_3$$

$$s_3 = p_3 \oplus c_3$$

$$c_4 = g_3 + p_3\, c_3$$

$$= g_3 + p_3\, (g_2 + p_2\, g_1 + p_2\, p_1\, g_0 + p_2\, p_1\, p_0\, c_0)$$

$$= g_3 + p_3 g_2 + p_3 p_2 g_1 + p_3 p_2 p_1 g_0 + p_3 p_2 p_1 p_0 c_0$$

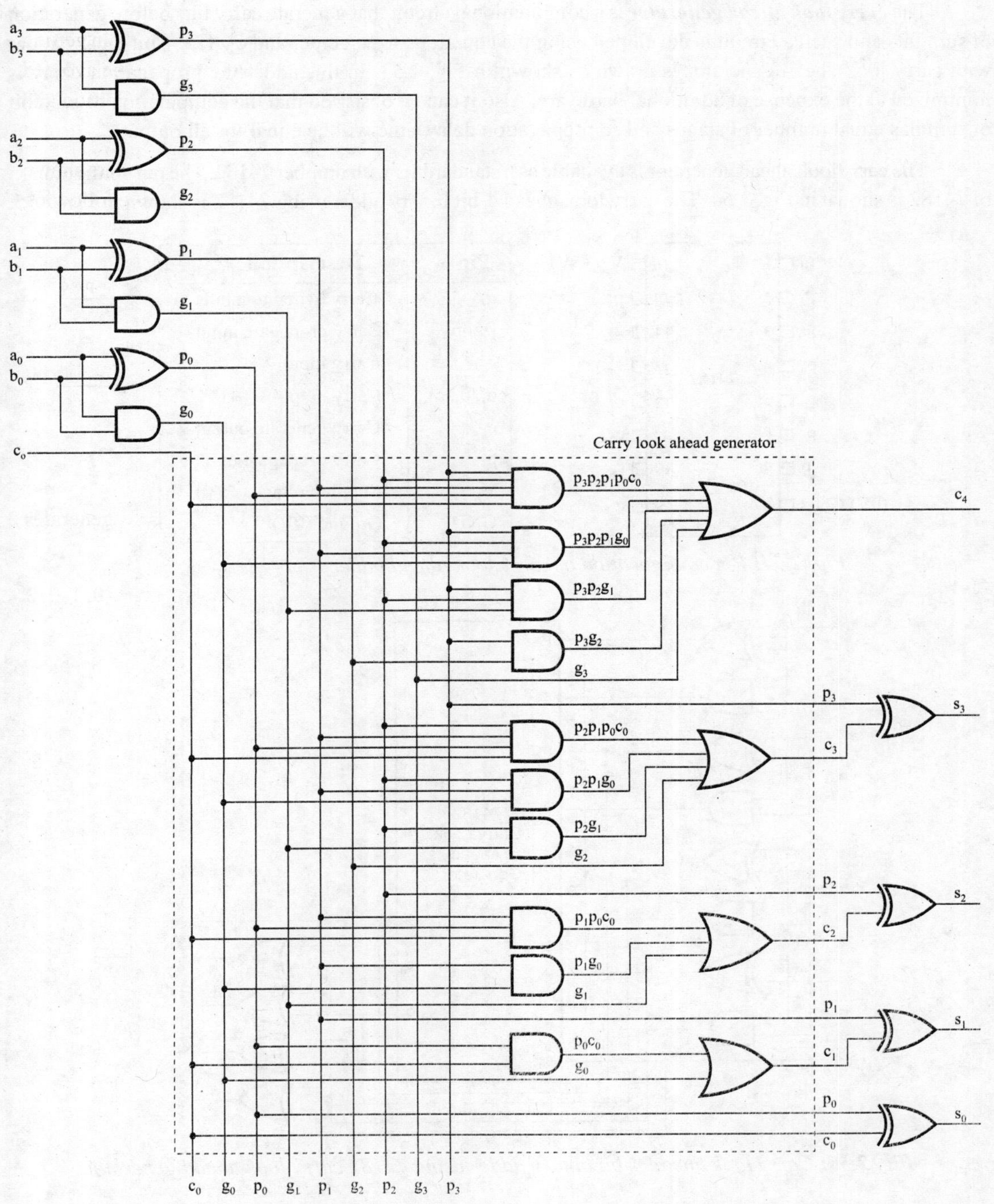

Fig. 2.53: Carry look ahead 4-bit binary adder.

The ***carry look ahead generator*** is a combinational circuit that generate carry bits before generation of sum bits and such a circuit is developed using the equations for c_1, c_2, c_3 and c_4. The 4-bit binary adder with carry look ahead generator is drawn as shown in Fig. 2.53. In this adder the propagation delay is minimized at the expense of additional hardware. Also it can be observed that the computation of each bit of sum has equal number of stages and so propagation delay time will be equal for all outputs.

The carry look ahead generator is available as a standard IC with number 74182. The pin configuration of 74182 is shown in Fig. 2.54. The carry look ahead 4-bit binary adder using 74182 is shown in Fig. 2.55.

74182

Signal	Pin	Pin	Signal
g_1	1	16	V_{cc}(+5V)
p_1	2	15	p_2
g_0	3	14	g_2
p_0	4	13	c_0
g_3	5	12	c_1
p_3	6	11	c_2
p	7	10	c_4
(0V) GND	8	9	c_3

Pin	Description
g_0 - g_3	Carry generate inputs
p_0 - p_3	Carry propagate inputs
c_o	Carry input
c_1, c_2, c_3	Carry output
c_4	Carry generate output
p	Carry propagate output
V_{cc}	Supply voltage (+5V)
GND	Ground (0V)

Fig. 2.54: *Pin configuration of carry look ahead adder IC 74182.*

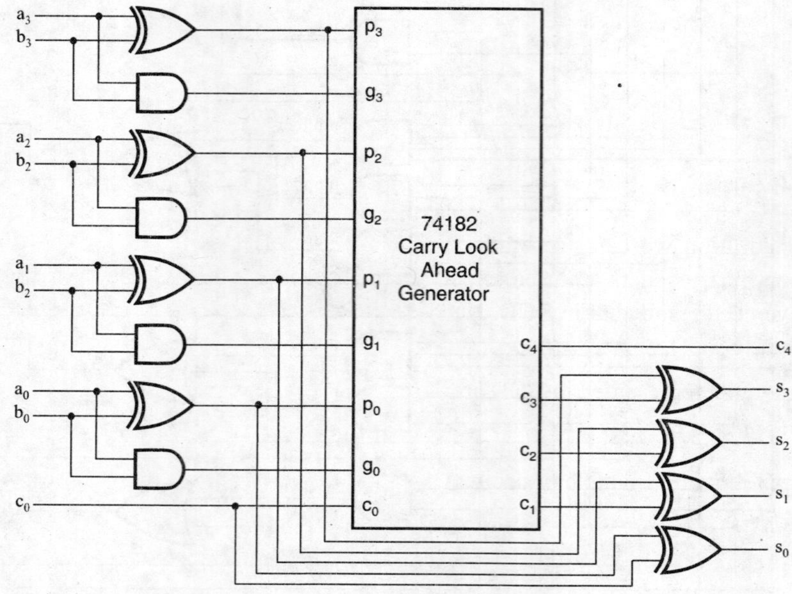

Fig. 2.55: *Carry look ahead 4-bit binary adder using 74182 carry look ahead generator.*

The 4-bit binary full adder with internal carry look ahead generator is available as standard ICs 7483 and 74283. The two ICs are functionally same and differ only in pin configurations. The pin description of IC 7483 and 74283 are shown in Fig. 2.56 and 2.57.

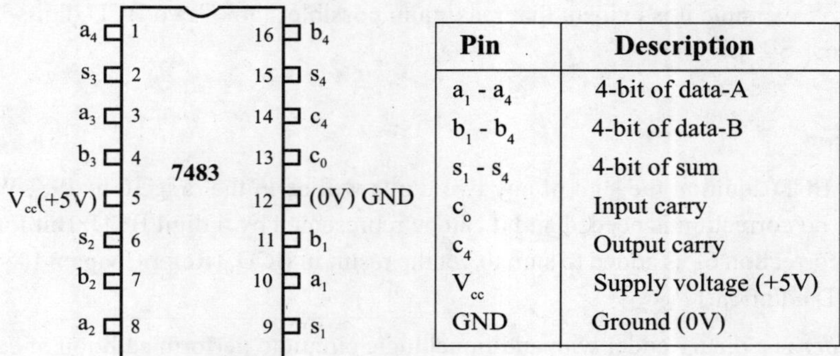

Pin	Description
a_1 - a_4	4-bit of data-A
b_1 - b_4	4-bit of data-B
s_1 - s_4	4-bit of sum
c_o	Input carry
c_4	Output carry
V_{cc}	Supply voltage (+5V)
GND	Ground (0V)

***Fig. 2.56:** Pin description of 7483 4-bit binary adder with internal carry look ahead generator.*

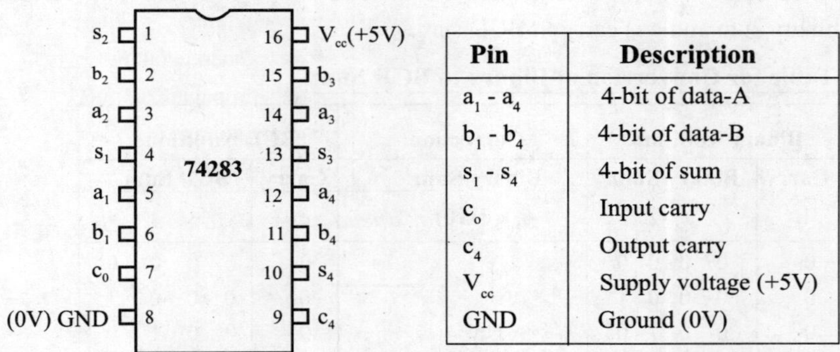

Pin	Description
a_1 - a_4	4-bit of data-A
b_1 - b_4	4-bit of data-B
s_1 - s_4	4-bit of sum
c_o	Input carry
c_4	Output carry
V_{cc}	Supply voltage (+5V)
GND	Ground (0V)

***Fig. 2.57:** Pin description of 74283 4-bit binary adder with internal carry look ahead generator.*

2.6.7 BCD Adder

In decimal addition we can use binary or BCD to represent decimal numbers. In both the representation when binary adders are used to perform addition the result or sum will be in binary. But in BCD addition we need result/sum in BCD. Hence we need additional logic circuit in binary adders to convert binary result to BCD.

The decimal digits represented in BCD are,

0, 1, 2, 3, 4, 5, 6, 7, 8, 9

Consider the following example of decimal addition.

```
      1  1      ← Carry
    4 2  9 8
  + 5 3  9 7
  ----------
    9 6  9 5
  ----------
         ↳ Maximum possible sum = 1 + 9 + 9 = 19
```

From the above example it is evident that maximum possible sum of two BCD digits is,

$$1 + 9 + 9 = 19$$

↑

Previous carry

Therefore in BCD addition the sum of any two digits will be in the range 0 to 19_{10}. When sum is in the range 0 to 9_{10} no correction is needed and it can be represented by 4 digit BCD. But for sum in the range 10_{10} to 19_{10} a correction 6_{10} is added to sum to get the result in BCD. (Refer Chapter-1, Section 1.3.3 for examples of BCD addition.)

The ***BCD adder*** is a binary adder with additional logic circuit to perform addition of correction 6_{10} when the sum of a BCD digit exceeds 9_{10}.

A truth table (Table 2.24) is formed to show the range of binary sum and correction needed to convert binary sum to BCD sum. In Table 2.24, the binary sum is denoted as $z_8\ z_4\ z_2\ z_1$ and BCD sum as $s_8\ s_4\ s_2\ s_1$. Here, k is carry in binary sum and c is carry in BCD sum.

Table 2.24: Truth Table for Conversion of Binary to BCD Sum

Decimal Sum	Binary Addition		Correction	BCD Addition	
	Carry k	Binary Sum $z_8\ z_4\ z_2\ z_1$	Binary Sum $+6_{10}$ (0110)	Carry c	BCD Sum $s_8\ s_4\ s_2\ s_1$
0	0	0 0 0 0	↑	0	0 0 0 0
1	0	0 0 0 1		0	0 0 0 1
2	0	0 0 1 0	No	0	0 0 1 0
3	0	0 0 1 1	Correction	0	0 0 1 1
4	0	0 1 0 0		0	0 1 0 0
5	0	0 1 0 1		0	0 1 0 1
6	0	0 1 1 0		0	0 1 1 0
7	0	0 1 1 1		0	0 1 1 1
8	0	1 0 0 0		0	1 0 0 0
9	0	1 0 0 1	↓	0	1 0 0 1
10	0	1 0 1 0	01010 + 0110	1	0 0 0 0
11	0	1 0 1 1	01011 + 0110	1	0 0 0 1
12	0	1 1 0 0	01100 + 0110	1	0 0 1 0
13	0	1 1 0 1	01101 + 0110	1	0 0 1 1
14	0	1 1 1 0	01110 + 0110	1	0 1 0 0
15	0	1 1 1 1	01111 + 0110	1	0 1 0 1
16	1	0 0 0 0	10000 + 0110	1	0 1 1 0
17	1	0 0 0 1	10001 + 0110	1	0 1 1 1
18	1	0 0 1 0	10010 + 0110	1	1 0 0 0
19	1	0 0 1 1	10011 + 0110	1	1 0 0 1

From the truth table (Table 2.24) we can observe that the correction by adding $+ 6_{10}$ (0110_2) can be achieved using following three conditions.

1. When binary sum is 10_{10} and 11_{10} correction 6 can be added if both z_2 and z_8 are 1.
2. When binary sum is in the range 12_{10} to 15_{10} correction 6 can be added if both z_4 and z_8 are 1.
3. When binary sum is in the range 16_{10} to 19_{10} correction 6 can be added if carry k in binary sum is 1.

Therefore, during BCD addition, first binary addition is performed using 4-bit binary adder then additional logic circuit is used to generate correction 6 for the conditions mentioned above and another 4-bit binary adder is used to add correction 6 to binary sum so that the binary sum is converted to BCD sum.

The logic circuit to perform 1-digit BCD addition using two 4-bit binary adders is shown in Fig. 2.58. The symbolic representation of BCD adder is shown in Fig. 2.59.

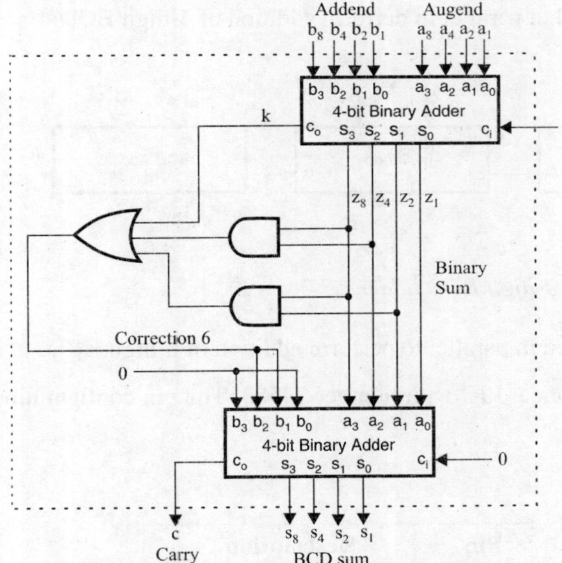

Fig. 2.58: BCD adder using binary adders.

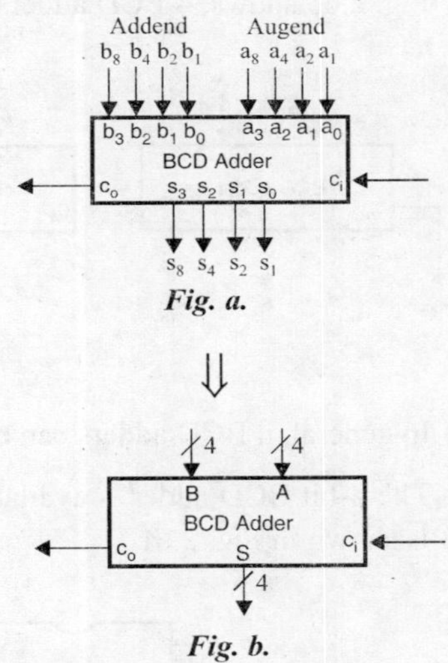

Fig. 2.59: Symbolic representation of BCD adder.

4-digit BCD Adder

The BCD adder shown in Fig. 2.59 is used to perform BCD addition of 1-digit decimal numbers. In order to perform 4-digit BCD addition we need 4 BCD adders shown in Fig. 2.60.

Consider two, 4-digit decimal numbers represented in BCD

Decimal number 1: $A_3 A_2 A_1 A_0$

Decimal number 2: $B_3 B_2 B_1 B_0$

Let, BCD Sum = $c\ S_3\ S_2\ S_1\ S_0$

$\uparrow$ Final carry

Let the format of two number and sum be as follows.

A_3	A_2	A_1	A_0
$a_8\ a_4\ a_2\ a_1$	$a_8\ a_4\ a_2\ a_1$	$a_8\ a_4\ a_2\ a_1$	$a_8\ a_4\ a_2\ a_1$
B_3	B_2	B_1	B_0
$b_8\ b_4\ b_2\ b_1$	$b_8\ b_4\ b_2\ b_1$	$b_8\ b_4\ b_2\ b_1$	$b_8\ b_4\ b_2\ b_1$
S_3	S_2	S_1	S_0
$s_8\ s_4\ s_2\ s_1$	$s_8\ s_4\ s_2\ s_1$	$s_8\ s_4\ s_2\ s_1$	$s_8\ s_4\ s_2\ s_1$

Fig. 2.60 shows, 4 BCD adders connected in parallel to perform addition of 4-digit BCD.

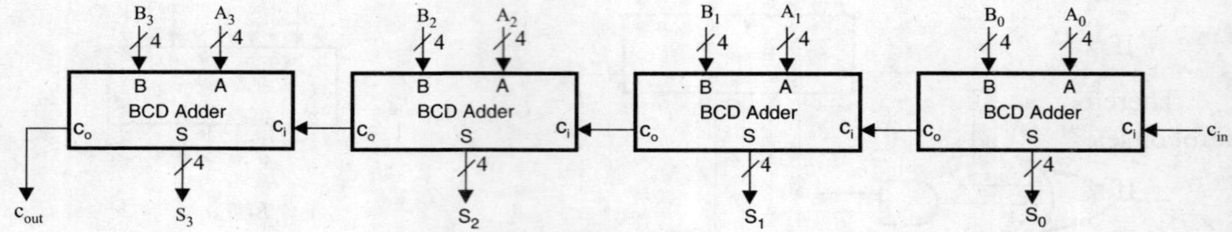

Fig. 2.60: *4-digit BCD adder.*

In general, n BCD adders can be connected in parallel to perform addition of n-digit BCD.

The 4-bit BCD adder is available as a standard IC with number 74583. The pin configuration of 74583 is shown in Fig. 2.61.

74583

Pin no.	Signal	Pin no.	Signal
1	b_1	16	V_{cc}(+5V)
2	b_2	15	a_2
3	b_3	14	a_1
4	a_3	13	a_0
5	c_i	12	b_0
6	c_{out}	11	s_0
7	s_2	10	s_1
8	(0V) GND	9	s_3

Pin	Description
a_0 - a_3	Operand a inputs
b_0 - b_3	Operand b inputs
c_i	Carry input
s_0 - s_3	Sum outputs
c_{out}	Carry output
V_{cc}	Supply voltage (+5V)
GND	Ground (0V)

Fig. 2.61: *Pin configuration of 4-bit BCD adder IC 74583.*

2.6.8 Binary Parallel Adder/Subtractor

Practically subtraction is performed only in 2's complement form in which the subtraction is considered as addition of positive and negative numbers and the negative number is represented in 2's complement form. Therefore, subtractor circuit is not of much use and subtraction can be performed using adder itself.

Therefore, subtraction of two BCD digit A and B is considered as,

$$A - B = A + (-B)$$

↑ 2's complement of B

The complement of binary can be easily obtained using XOR gate. Consider the truth table of XOR gate (Table 2.25).

Table 2.25: Truth Table of XOR Gate

x	y	$x \oplus y$
0	0	0
0	1	1
1	0	1
1	1	0

x	y	$x \oplus y = y$
0	0	0
0	1	1

x	y	$x \oplus y = y'$
1	0	1
1	1	0

With reference to Table 2.25 we can say that,

If $x = 0$, then $x \oplus y = y$

If $x = 1$, then $x \oplus y = y'$

Therefore in XOR gate, input x can be used as control or selection and input y can be data.

If $x = 0$, the data will be output as such.

If $x = 1$, the output will be complement of data.

The n-bit binary adder using full adder can be modified to perform both addition and subtraction as shown in Fig. 2.62.

One input can be given directly to full adders, another input can be given through XOR gate to full adders, so that second data can be sent to full adders with or without complement.

The control input M is used to perform complement operation. When M = 1, the output of XOR gate is complement of the input.

Consider addition and subtraction of two, 4-bit binary numbers.

Binary number-1: $a_3\, a_2\, a_1\, a_0$ (data-A)

Binary number-2: $b_3\, b_2\, b_1\, b_0$ (data-B)

Let us perform, A + B and A − B = A + (−B)

For addition, B is input to full adder without complement as shown in Fig. 2.63.

$\therefore$ For addition, $M = c_o = 0$

For subtraction, B is input to full adder after complement and adding 1 for 2's complement can be performed by making $c_0 = 1$ for subtraction as shown in Fig. 2.64.

$\therefore$ For subtraction, $M = c_o = 1$

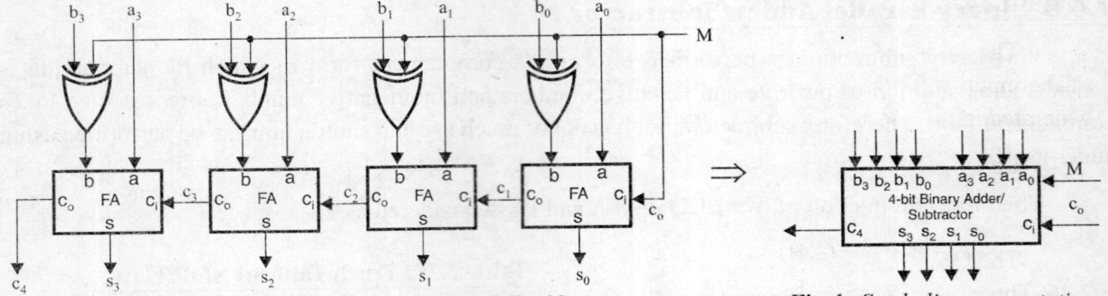

Fig. a: Binary adder/subtractor using full adders.

Fig. b: Symbolic representation.

Fig. 2.62: 4-bit binary/adder subtractor.

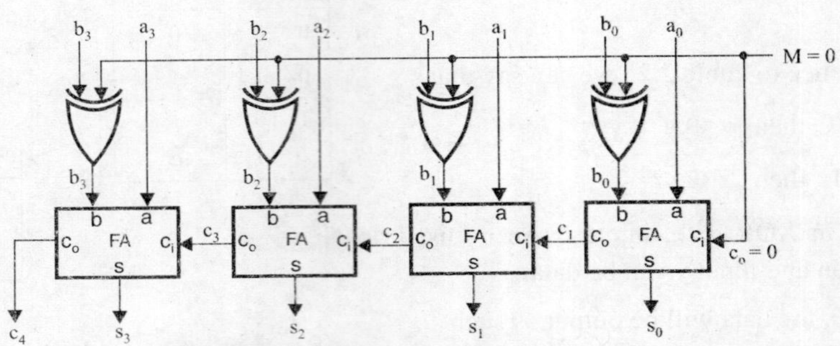

Fig. 2.63: Binary addition using adder/subtractor.

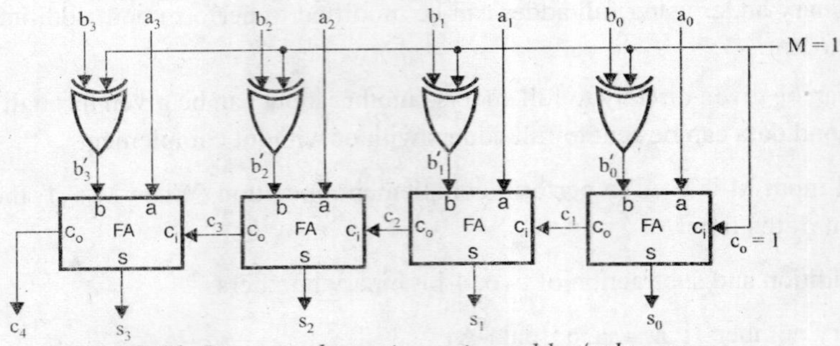

Fig. 2.64: Binary subtraction using adder/subtractor.

> ***Note:*** *Sum is difference and carry is borrow.*

2.7 Encoders

An ***encoder*** is a combinational circuit that performs the inverse operation of a decoder.

In general, an encoder has 2^n input lines and n output lines. When one of 2^n inputs is activated, it generates an unique n-bit output. The block diagram of 2^n-to-n encoder is shown in Fig. 2.65.

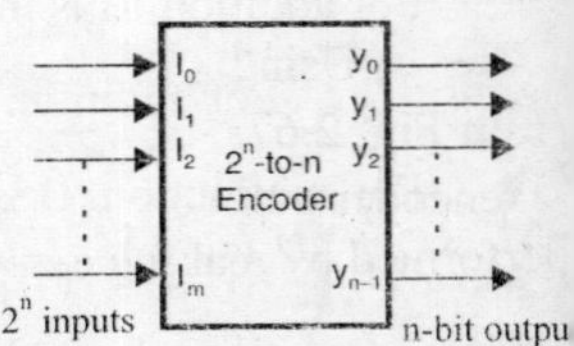

Fig. 2.65: 2^n-to-n encoder.

$m = 2^n - 1$

2.7.1 Logic High 4-to-2 Encoder

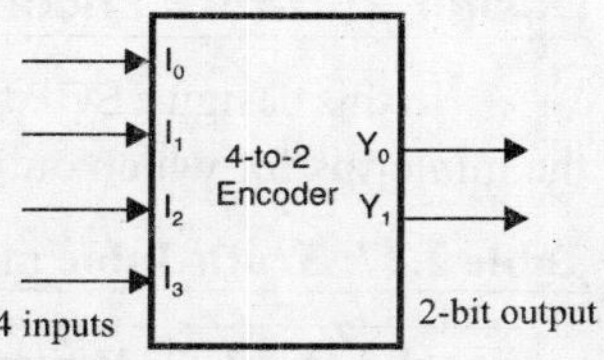

Fig. 2.66: *4-to-2 encoder.*

The ***4-to-2 encoder*** has 4 inputs and 2 outputs. When any one of the inputs is asserted **high** then a corresponding 2-bit binary value is generated in the output.

Let, I_0, I_1, I_2, I_3 = Inputs

Y_0, Y_1 = Outputs

The block diagram of 4-to-2 encoder is shown in Fig. 2.66 and the truth table is shown in Table 2.26. Three different types of encoder design is presented here.

Table 2.26: Truth Table of 4-to-2 Logic High Encoder

Inputs				Minterm	Outputs	
I_0	I_1	I_2	I_3		Y_1	Y_0
1	0	0	0	m_8	0	0
0	1	0	0	m_4	0	1
0	0	1	0	m_2	1	0
0	0	0	1	m_1	1	1

Design 1: 4-to-2 Encoder Design using K-map

Since the 4-to-2 encoder has 4 inputs the possible minterms are $m_0, m_1, m_2, \ldots\ldots, m_{15}$. But only 4 minterms are required for valid outputs as shown in truth table (Table 2.26). The outputs for remaining minterms are don't-care outputs.

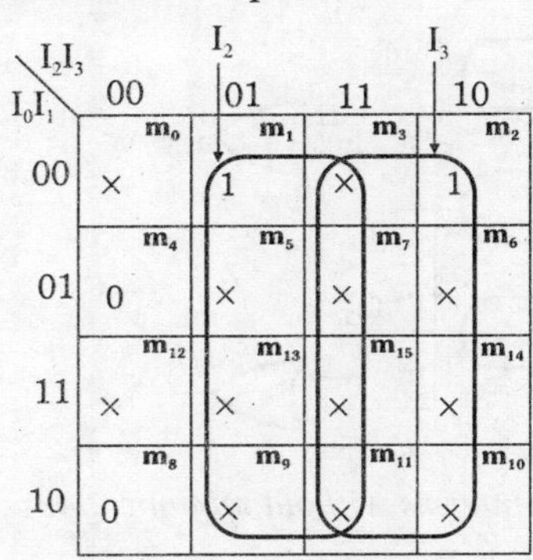

Fig. a: *K-map for Y_1.*

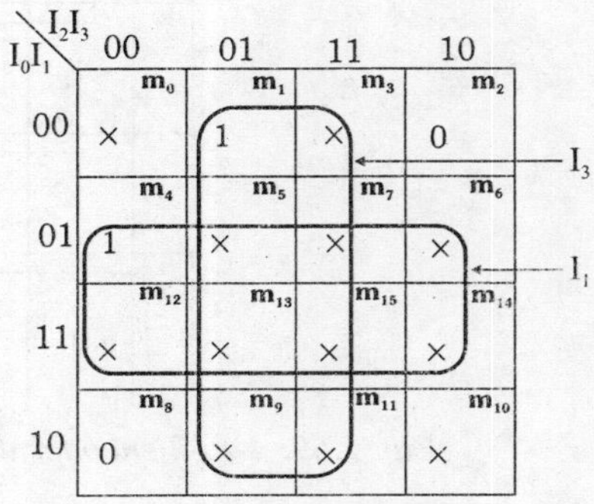

Fig. b: *K-map for Y_0.*

Fig. 2.67: *4-to-2 encoder design using K-map.*

Using Table 2.26, the K-maps for Y_0 and Y_1 are constructed as shown in Fig. 2.67. From the K-map we get the following Boolean equations for encoder outputs.

$$Y_1 = I_2 + I_3 \quad ; \quad Y_0 = I_1 + I_3$$

Using the above Boolean equations the logic circuit of 4-to-2 encoder is drawn as shown in Fig. 2.68.

Fig. 2.68: *4-to-2 encoder.*

> ***Note:*** *Since we use minimum literals I_0 is not used in generating output.*

Design 2: 4-to-2 Encoder Design using Minterms without Minimization

In this design a SOP form of Boolean equation can be directly formed for each output by considering the minterms for which outputs are 1.

Table 2.27: Truth Table of 4-to-2 Encoder

I_0	I_1	I_2	I_3	Minterm	Y_1	Y_0
1	0	0	0	m_8	0	0
0	1	0	0	m_4	0	1
0	0	1	0	m_2	1	0
0	0	0	1	m_1	1	1

The truth table of 4-to-2 encoder is shown in Table 2.27. In this table the minterms for which outputs are 1 are identified and using these minterms the following SOP form of Boolean equations are obtained for outputs.

$$Y_1 = m_2 + m_1 = I_0' I_1' I_2 I_3' + I_0' I_1' I_2' I_3$$

$$Y_0 = m_4 + m_1 = I_0' I_1 I_2' I_3' + I_0' I_1' I_2' I_3$$

Using the above Boolean equations the logic circuit of 4-to-2 encoder is drawn as shown in Fig. 2.69.

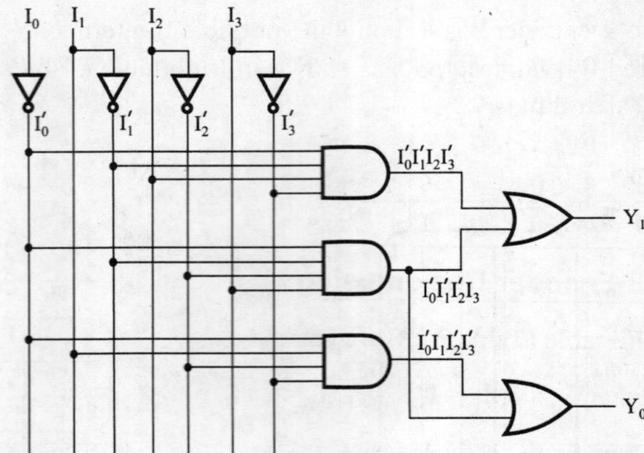

Fig. 2.69: 4-to-2 encoder design using minterms without minimization.

Design 3: 4-to-2 Encoder Design using OR Gate

From the truth table (Table 2.26) we can make following observations.

Y_1 is 1, when $I_2 = 1$ or $I_3 = 1$; $\therefore$ $Y_0 = I_2 + I_3$

Y_0 is 1, when $I_1 = 1$ or $I_3 = 1$; $\therefore$ $Y_1 = I_1 + I_3$

Using the above Boolean equations the logic circuit of 4-to-2 encoder is drawn as shown in Fig. 2.70.

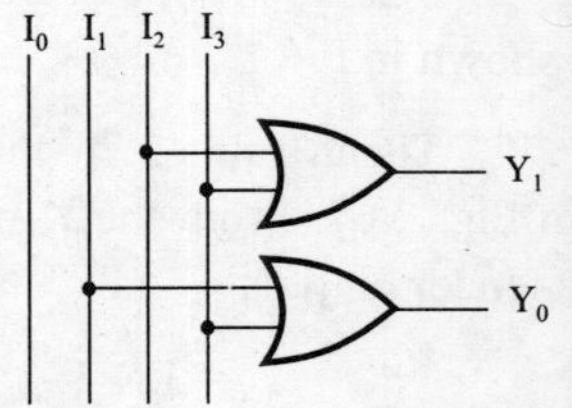

***Fig. 2.70:** 4-to-2 encoder design using OR gate.*

2.7.2 Logic High 8-to-3 Encoder

The ***8-to-3 encoder*** has eight inputs and 3 outputs. When any one of the inputs is asserted **high** then a corresponding 3-bit binary value is generated in the output.

Let, $I_0, I_1, ..., I_7$ = Inputs

$Y_0, Y_1 Y_2$ = Outputs

The block diagram of 8-to-3 encoder is shown in Fig. 2.71 and the truth table is shown in Table 2.28. Two different types of 8-to-3 encoder design is presented here.

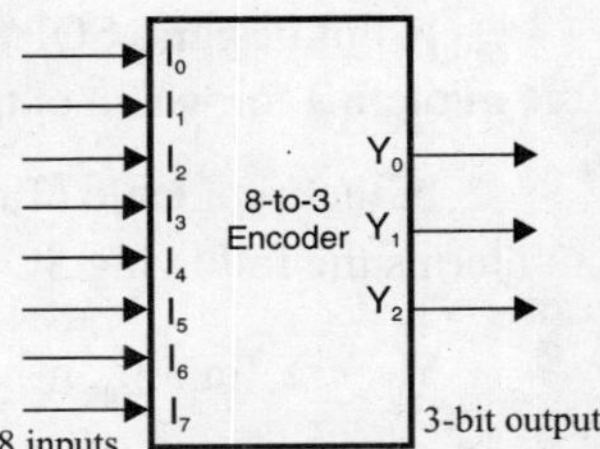

Fig. 2.71: *8-to-3 encoder.*

Since the 8-to-3 encoder has 8 inputs the possible minterms are $m_0, m_1, m_2, ..., m_{255}$. But only 8 minterms are required for valid outputs as shown in Table 2.28. The outputs for remaining minterms are don't-care outputs.

Table 2.28: Truth Table of 8-to-3 Logic High Encoder

Inputs								Minterms	Outputs		
I_0	I_1	I_2	I_3	I_4	I_5	I_6	I_7		Y_2	Y_1	Y_0
1	0	0	0	0	0	0	0	m_{128}	0	0	0
0	1	0	0	0	0	0	0	m_{64}	0	0	1
0	0	1	0	0	0	0	0	m_{32}	0	1	0
0	0	0	1	0	0	0	0	m_{16}	0	1	1
0	0	0	0	1	0	0	0	m_8	1	0	0
0	0	0	0	0	1	0	0	m_4	1	0	1
0	0	0	0	0	0	1	0	m_2	1	1	0
0	0	0	0	0	0	0	1	m_1	1	1	1

Design 1: 8-to-3 Encoder Design using OR gates

From the truth table (Table 2.28) we can make following observations.

Y_0 is 1 when $I_1 = 1, I_3 = 1, I_5 = 1, I_7 = 1$; $\therefore Y_0 = I_1 + I_3 + I_5 + I_7$

Y_1 is 1 when $I_2 = 1, I_3 =1\ I_6 = 1, I_7 = 1$; $\therefore Y_1 = I_2 + I_3 + I_6 + I_7$

Y_2 is 1 when $I_4 = 1, I_5 =1, I_6 =1, I_7 = 1$; $\therefore Y_2 = I_4 + I_5 + I_6 + I_7$

Using the above Boolean equations the logic circuit of 8-to-3 encoder is drawn using OR gates as shown in Fig. 2.72.

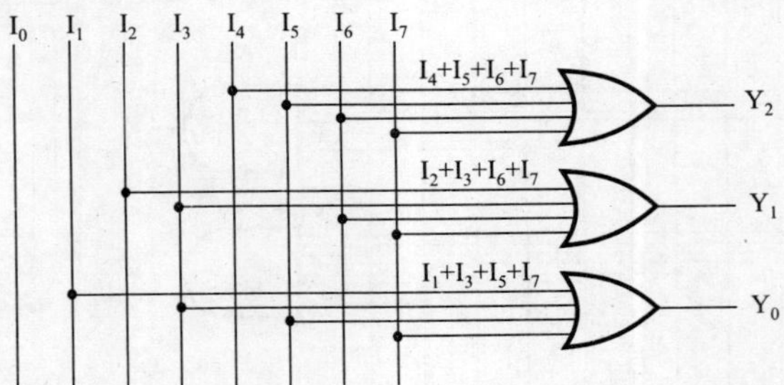

Fig. 2.72: *8-to-3 encoder using OR gates.*

Design 2: 8-to-3 Encoder Design using Minterms and without Minimization

In this design a SOP form of Boolean equation can be directly formed for each output by considering the minterms for which output is 1.

In the truth table (Table 2.28) the minterms for which outputs are 1 are identified and using these minterms the following SOP form of Boolean equations are obtained for encoder outputs.

$$Y_0 = \Sigma (m_1, m_4, m_{16}, m_{64}) = I_0' I_1' I_2' I_3' I_4' I_5' I_6' I_7 + I_0' I_1' I_2' I_3' I_4' I_5 I_6' I_7' + I_0' I_1' I_2' I_3 I_4' I_5' I_6' I_7'$$
$$+ I_0' I_1 I_2' I_3' I_4' I_5' I_6' I_7'$$

$$Y_1 = \Sigma (m_1, m_2, m_{16}, m_{32}) = I_0' I_1' I_2' I_3' I_4' I_5' I_6' I_7 + I_0' I_1' I_2' I_3' I_4' I_5' I_6 I_7' + I_0' I_1' I_2' I_3 I_4' I_5' I_6' I_7'$$
$$+ I_0' I_1' I_2 I_3' I_4' I_5' I_6' I_7'$$

$$Y_2 = \Sigma (m_1, m_2, m_4, m_8) = I_0' I_1' I_2' I_3' I_4' I_5' I_6' I_7 + I_0' I_1' I_2' I_3' I_4' I_5' I_6 I_7' + I_0' I_1' I_2' I_3' I_4' I_5 I_6' I_7'$$
$$+ I_0' I_1' I_2' I_3' I_4 I_5' I_6' I_7'$$

Using the above Boolean equations the logic circuit of 8-to-3 encoder is drawn as shown in Fig. 2.73.

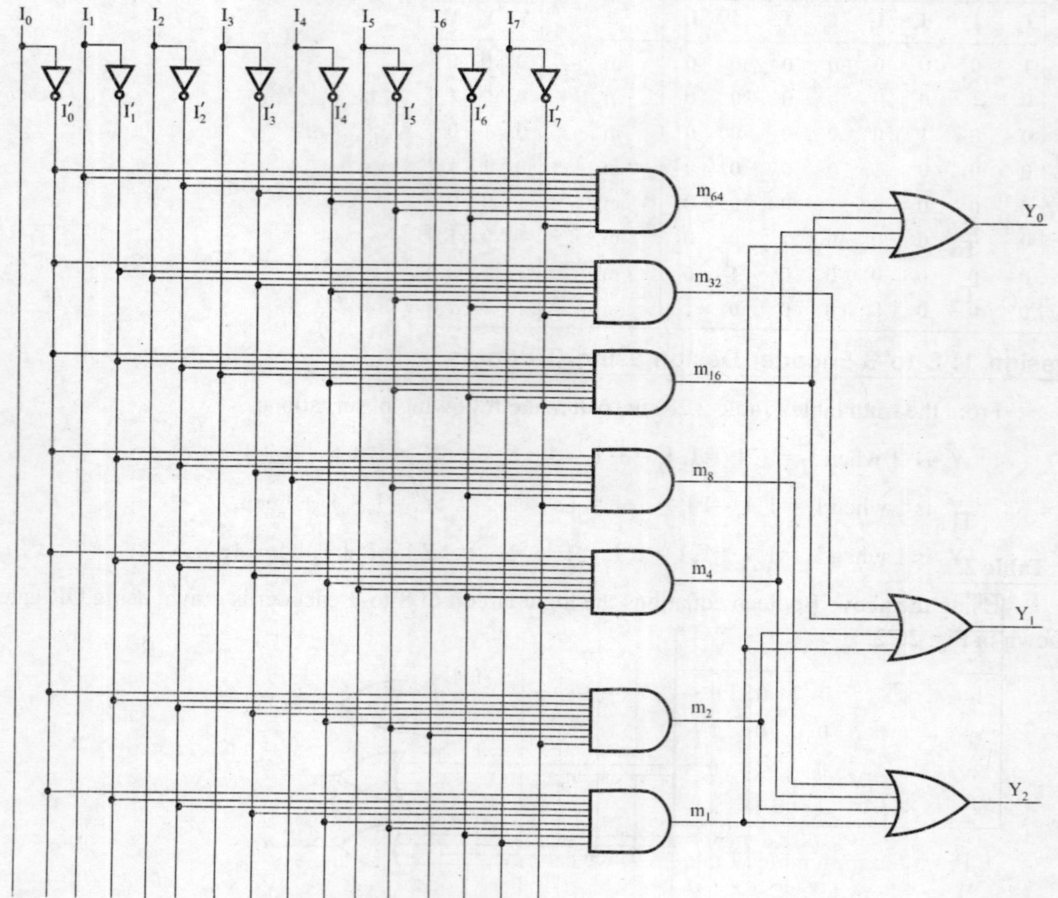

Fig. 2.73: 8-to-3 encoder.

2.8 Priority Encoders

A ***priority encoder*** is an encoder circuit that includes a priority function in order to recognize only one input when multiple inputs are asserted **high**.

The operation of the priority encoder is such that if two or more inputs are asserted **high** (or equal to 1) then the output corresponds to the input having highest priority.

2.8.1 Logic High 4-to-2 Priority Encoder

The 4-to-2 encoder has 4 inputs and 2 outputs.

Let, I_0, I_1, I_2, I_3 = Inputs

Y_0, Y_1 = Outputs

Let the order of priority from highest to lowest be,

I_3 - Highest

I_2

I_1

I_0 - Lowest

Now, the above priority works as follows:

If I_3 and I_2 are set **high** then output will correspond to only I_3.

If I_2 and I_1 are set **high** then output will correspond to only I_2.

If I_1 and I_0 are set **high** then output will correspond to only I_1.

To implement the above priority we have to write a priority function.

The priority functions for the above example are,

$H_3 = I_3$; I_3 is recognized

$H_2 = I_2 I_3'$; I_2 is recognized only if $I_3 = 0$

$H_1 = I_1 I_2' I_3'$; I_1 is recognized only if $I_2 = I_3 = 0$

$H_0 = I_0 I_1' I_2' I_3'$; I_0 is recognized only if $I_1 = I_2 = I_3 = 0$

The truth table of priority encoder with above priority functions is shown in Table 2.29.

Table 2.29: Truth Table of 4-to-2 Priority Encoder

Inputs				Outputs	
I_0	I_1	I_2	I_3	Y_1	Y_0
1	0	0	0	0	0
×	1	0	0	0	1
×	×	1	0	1	0
×	×	×	1	1	1

From the truth table (Table 2.29) we can note that Y_1 is 1 when I_2 and I_3 are 1.

$$\therefore\ Y_1 = H_2 + H_3 = I_2 I_3' + I_3$$

From the truth table (Table 2.29) we can note that Y_0 is 1 when I_1 and I_3 are 1.

$$\therefore\ Y_0 = H_1 + H_3 = I_1\, I_2'\, I_3' + I_3$$

Therefore, the Boolean equations of 4-to-2 priority encoder with priority order from I_3 to I_0 are,

$$Y_1 = I_2\, I_3' + I_3$$

$$Y_0 = I_1\, I_2'\, I_3' + I_3$$

The above equations are verified in Table 2.30 for same possible combinations of inputs.

Table 2.30: Verification of 4-to-2 Priority Encoder

I_0	I_1	I_2	I_3	I_2'	I_3'	$I_2 I_3'$	$I_1 I_2' I_3'$	Y_1	Y_0	Comment
1	0	0	0	1	1	0	0	0	0	Only one input is asserted high and so the output corresponds to the input asserted high.
0	1	0	0	1	1	0	1	0	1	
0	0	1	0	0	1	1	0	1	0	
0	0	0	1	1	0	0	0	1	1	
0	0	1	1	0	0	0	0	1	1	2 and 3 asserted high ; output corresponds to 3
0	1	1	0	0	1	1	0	1	0	1 and 2 asserted high ; output corresponds to 2
1	1	0	0	1	1	0	1	0	1	0 and 1 asserted high ; output corresponds to 1

Usually a priority encoder will have a valid bit v, to indicate that one or more input is set **high**. The valid bit is zero if all inputs are zero. The truth table of 4-to-2 priority encoder with valid bit v is shown in Table 2.31. Since, value of valid bit is 1 if any one input is 1, the equation of valid bit is given by sum of all inputs as shown below:

$$\therefore\ v = I_0 + I_1 + I_2 + I_3$$

Table 2.31: Truth Table of 4-to-2 Priority Encoder with Valid Bit v

Inputs				Outputs		
I_0	I_1	I_2	I_3	Y_1	Y_0	v
0	0	0	0	×	×	0
1	0	0	0	0	0	1
×	1	0	0	0	1	1
×	×	1	0	1	0	1
×	×	×	1	1	1	1

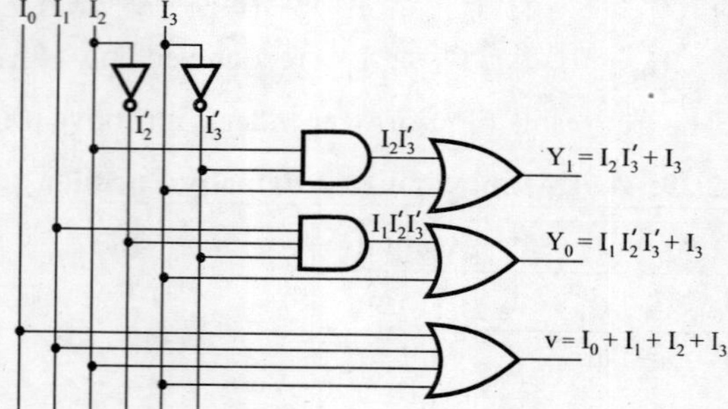

Fig. 2.74: *4-to-2 priority encoder with I_3 having highest priority.*

The logic circuit of 4-to-2 priority encoder with priority order I_3 to I_0 is drawn as shown in Fig. 2.74 using the Boolean equations of outputs and valid bit.

Alternatively, the priority encoder can be designed using K-maps.

2.8.2 Logic High 4-to-2 Priority Encoder Design using K-maps

(AU, Apr/May'19, 6 Marks)

The don't-care inputs are expanded with all possible combination of inputs and listed in Table 2.32.

Table 2.32: Truth Table of Priority Encoder with all Possible Combination of Inputs

Inputs					Outputs		
I_0	I_1	I_2	I_3	Minterm	Y_1	Y_0	v
0	0	0	0	m_0	×	×	0
1	0	0	0	m_8	0	0	1
×	1	0	0	m_4, m_{12}	0	1	1
×	×	1	0	m_2, m_6, m_{10}, m_{14}	1	0	1
×	×	×	1	m_1, m_3, m_5, m_7, m_9, m_{11}, m_{13}, m_{15}	1	1	1

Inputs					Outputs		
I_0	I_1	I_2	I_3	Minterm	Y_1	Y_0	v
0	0	0	0	m_0	×	×	0
1	0	0	0	m_8	0	0	1
0	1	0	0	m_4	0	1	1
1	1	0	0	m_{12}	0	1	1
0	0	1	0	m_2	1	0	1
0	1	1	0	m_6	1	0	1
1	0	1	0	m_{10}	1	0	1
1	1	1	0	m_{14}	1	0	1
0	0	0	1	m_1	1	1	1
0	0	1	1	m_3	1	1	1
0	1	0	1	m_5	1	1	1
0	1	1	1	m_7	1	1	1
1	0	0	1	m_9	1	1	1
1	0	1	1	m_{11}	1	1	1
1	1	0	1	m_{13}	1	1	1
1	1	1	1	m_{15}	1	1	1

The K-map for Y_0, Y_1 and v are constructed using the truth table (Table 2.32) as shown in Fig. 2.75.

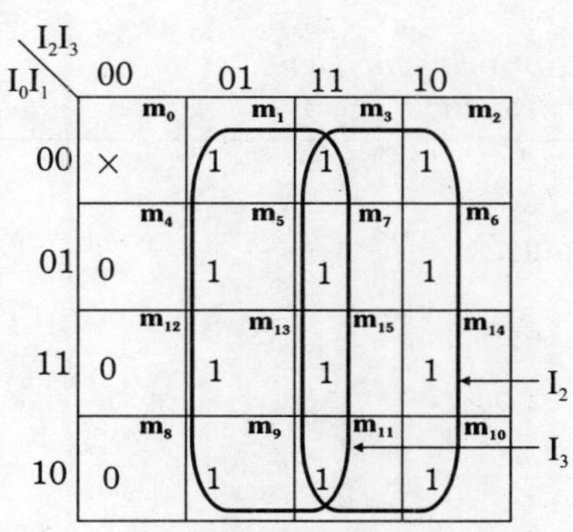

Fig. a: K-map for Y_1.

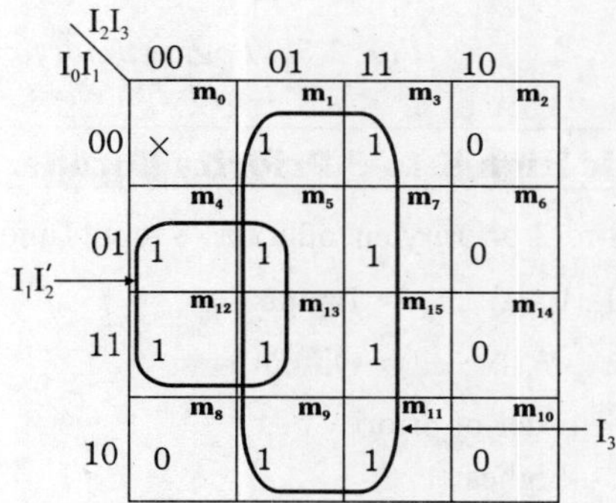

Fig. b: K-map for Y_0.

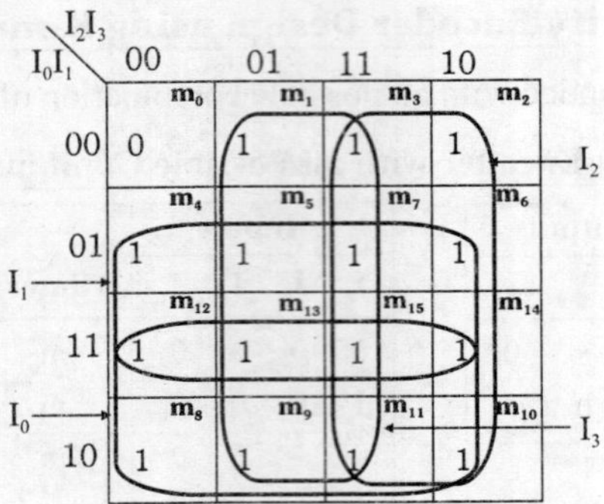

Fig. c: K-map for v.

Fig. 2.75: K-map for design of 4-to-2 encoder.

From the K-map we get the following Boolean equations.

$$Y_1 = I_2 + I_3 \quad ; \quad Y_0 = I_1 I_2' + I_3 \quad ; \quad v = I_0 + I_1 + I_2 + I_3$$

Using the above Boolean equations the logic circuit for 4-to-2 priority encoder is drawn as shown in Fig. 2.76.

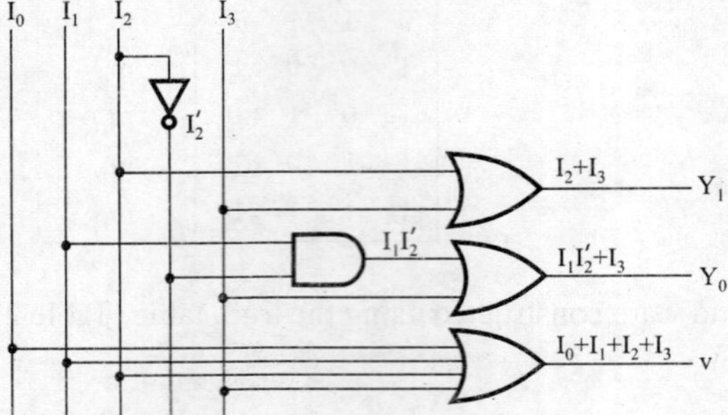

Fig. 2.76: Logic circuit for 4-to-2 priority encoder with I_3 having highest priority.

2.8.3 Logic High 8-to-3 Priority Encoder

The 8-to-3 priority encoder has 8 inputs and 3 outputs.

Let, $I_0, I_1, ..., I_7$ = Inputs

Y_0, Y_1, Y_2 = Outputs

Let the order of priority be,

I_7 - highest

I_6 - .

I_5 - .

I_4 - .

I_3 - .

I_2 - .

I_1 - .

I_0 - lowest

The priority functions for the above priority order are,

$H_7 = I_7$; I_7 is recognized

$H_6 = I_6\ I_7'$; I_6 is recognized only if I_7 is 0

$H_5 = I_5\ I_6'\ I_7'$; I_5 is recognized only if I_6 and I_7 are 0

$H_4 = I_4\ I_5'\ I_6'\ I_7'$; I_4 is recognized only if I_5, I_6 and I_7 are 0

$H_3 = I_3\ I_4'\ I_5'\ I_6'\ I_7'$; I_3 is recognized only if I_4, I_5, I_6 and I_7 are 0

$H_2 = I_2\ I_3'\ I_4'\ I_5'\ I_6'\ I_7'$; I_2 is recognized only if I_3, I_4, I_5, I_6 and I_7 are 0

$H_1 = I_1\ I_2'\ I_3'\ I_4'\ I_5'\ I_6'\ I_7'$; I_1 is recognized only if I_2, I_3, I_4, I_5, I_6 and I_7 are 0

$H_0 = I_0\ I_1'\ I_2'\ I_3'\ I_4'\ I_5'\ I_6'\ I_7'$; I_0 is recognized only if I_1, I_2, I_3, I_4, I_5, I_6 and I_7 are 0

The truth table of priority encoder with above priority functions is shown in Table 2.33.

From the truth table we can note that Y_2 is 1 when, I_4, I_5, I_6 and I_7 are 1.

$$\therefore Y_2 = H_4 + H_5 + H_6 + H_7$$

$$= I_4\ I_5'\ I_6'\ I_7' + I_5\ I_6'\ I_7' + I_6\ I_7' + I_7$$

From the truth table we can note that Y_1 is 1 when I_2, I_3, I_6 and I_7 are 1.

$$\therefore Y_1 = H_2 + H_3 + H_6 + H_7$$

$$= I_2\ I_3'\ I_4'\ I_5'\ I_6'\ I_7' + I_3\ I_4'\ I_5'\ I_6'\ I_7' + I_6\ I_7' + I_7$$

From the truth table we can note that Y_0 is 1 when I_1, I_3, I_5 and I_7 are 1

$$\therefore Y_0 = H_1 + H_3 + H_5 + H_7$$

$$= I_1\ I_2'\ I_3'\ I_4'\ I_5'\ I_6'\ I_7' + I_3\ I_4'\ I_5'\ I_6'\ I_7' + I_5\ I_6'\ I_7' + I_7$$

The valid bit v equation is,

$$v = I_0 + I_1 + I_2 + I_3 + I_4 + I_5 + I_6 + I_7$$

Using the above Boolean equations the logic circuit for 8-to-3 priority encoder is drawn as shown in Fig. 2.77.

Table 2.33: Truth Table of 8-to-3 Priority Encoder

Inputs								Outputs			
I_0	I_1	I_2	I_3	I_4	I_5	I_6	I_7	Y_2	Y_1	Y_0	v
0	0	0	0	0	0	0	0	×	×	×	0
1	0	0	0	0	0	0	0	0	0	0	1
×	1	0	0	0	0	0	0	0	0	1	1
×	×	1	0	0	0	0	0	0	1	0	1
×	×	×	1	0	0	0	0	0	1	1	1
×	×	×	×	1	0	0	0	1	0	0	1
×	×	×	×	×	1	0	0	1	0	1	1
×	×	×	×	×	×	1	0	1	1	0	1
×	×	×	×	×	×	×	1	1	1	1	1

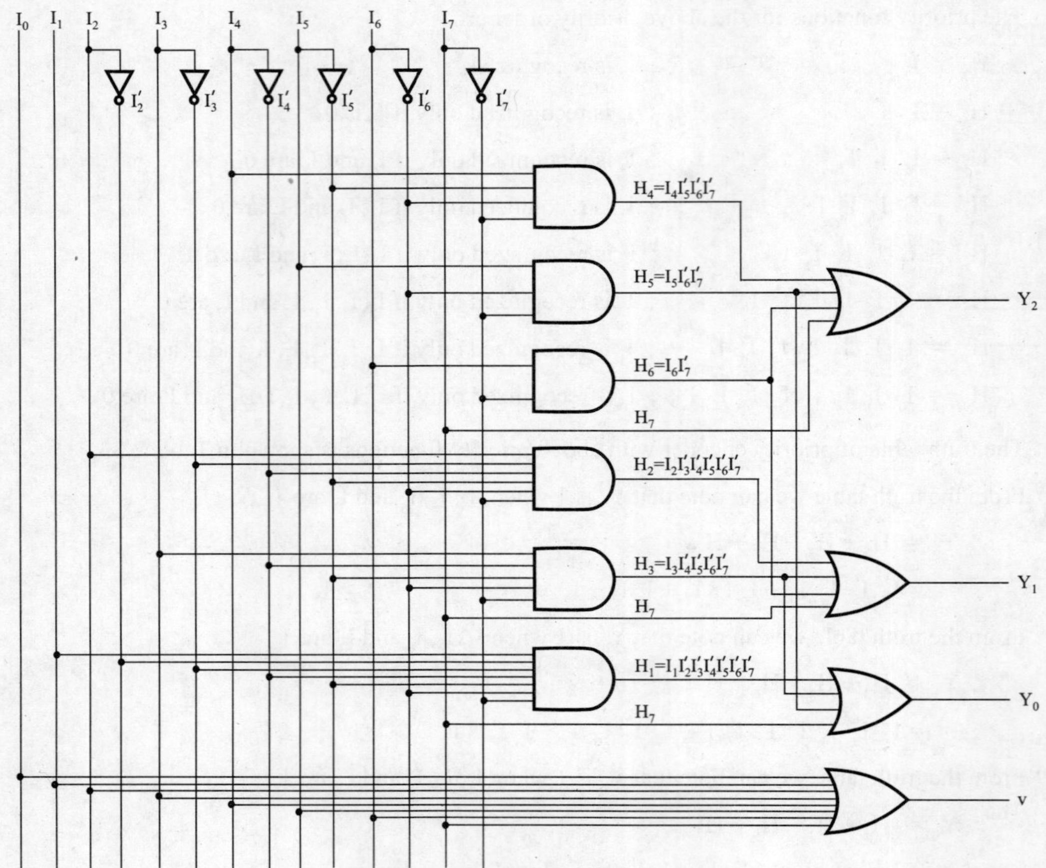

***Fig. 2.77:** 8-to-3 priority encoder with I_7 having highest priority.*

The 8-to-3 priority encoder is available as a standard IC with number 74148. The pin configuration of 74148 is shown in Fig. 2.78.

Left pins (74148)	Right pins
I_4 1	16 V_{cc}(+5V)
I_5 2	15 EO
I_6 3	14 v
I_7 4	13 I_3
EI 5	12 I_2
Y_2 6	11 I_1
Y_1 7	10 I_0
(0V) GND 8	9 Y_0

Pin	Description
I_0 - I_7	Inputs
Y_0 - Y_2	Outputs
v	Output
V_{cc}	Supply voltage (+5V)
GND	Ground (0V)
EI	Input Enable
EO	Output Enable

***Fig. 2.78:** Pin configuration of 8-to-3 priority encoder IC 74148.*

Example 2.13

Design a 4-to-2 priority encoder with input I_0 having the highest priority and input I_3 the lowest priority.

Solution

Let, I_0, I_1, I_2, I_3 = Inputs

Y_0, Y_1, v = Outputs

The truth table of priority encoder is shown in Table 1.

Table 1: Truth Table of 4-to-2 Priority Encoder with I_0 as Highest Priority

Inputs				Outputs		
I_0	I_1	I_2	I_3	Y_1	Y_0	v
0	0	0	0	×	×	0
1	×	×	×	0	0	1
0	1	×	×	0	1	1
0	0	1	×	1	0	1
0	0	0	1	1	1	1

Given that the order of priority highest to lowest are,

I_0 - highest

I_1 - ⋮

I_2 - ⋮

I_3 - lowest

The priority functions are,

$H_0 = I_0$; I_0 is recognized

$H_1 = I_1 I_0'$; I_1 is recognized only if I_0 is 0

$H_2 = I_2 I_1' I_0'$; I_2 is recognized only if I_1 and I_0 are 0

$H_3 = I_3 I_2' I_1' I_0'$; I_3 is recognized only if I_2, I_1 and I_0 are 0

From the truth table we can note that Y_1 is 1 when, I_2 and I_3 are 1.

$$\therefore Y_1 = H_2 + H_3 = I_2 I_1' I_0' + I_3 I_2' I_1' I_0'$$

From the truth table we can note that Y_0 is 1 when I_1 and I_3 are 1.

$$\therefore Y_0 = H_1 + H_3$$

$$= I_1 I_0' + I_3 I_2' I_1' I_0'$$

The equation for valid bit, v is,

$$v = I_0 + I_1 + I_2 + I_3$$

The logic circuit of 4-to-2 priority encoder with priority order I_0 and I_3 is drawn as shown in Fig. 1, using Boolean equations of output and valid bit, v.

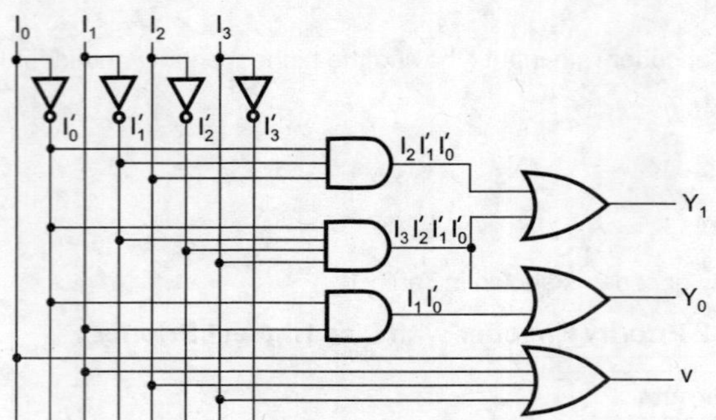

***Fig. 1:** Logic circuit of 4-to-2 priority encoder with I_0 having highest priority.*

The output Boolean equation are verified for some possible combinations of inputs in Table 2.

Table 2: Verification of Output Equations

Inputs				Product Terms						Outputs		Comments
I_0	I_1	I_2	I_3	I_0'	I_1'	I_2'	$I_2 I_1' I_0'$	$I_3 I_2' I_1' I_0'$	$I_1 I_0'$	Y_1	Y_0	
1	0	0	0	0	1	1	0	0	0	0	0	Only one input is asserted high and so the output corresponds to the input asserted high
0	1	0	0	1	0	1	0	0	1	0	1	
0	0	1	0	1	1	0	1	0	0	1	0	
0	0	0	1	1	1	1	0	1	0	1	1	
1	1	0	0	0	0	1	0	0	0	0	0	0 and 1 are asserted high ; output corresponds to 0
0	1	1	0	1	0	0	0	0	1	0	1	1 and 2 are asserted high ; output corresponds to 1
0	0	1	1	1	1	0	1	0	0	1	0	2 and 3 are asserted high ; output corresponds to 2

2.9 Decoders

A ***decoder*** is a combinational logic device that decodes one of the 2^n binary information depending on n-bit binary input. The n-bit binary information is decoded into 2^n binary information.

In other words, for an n-bit binary input one of the 2^n output is activated. The activated output may be logic **high** or logic **low**. In active **high** or logic **high** decoder, one of the output is **high** for an n-bit input and all other outputs are **low**. In active **low** or logic **low** decoder, one of the output is **low** for an n-bit input and all other outputs are **high**.

In general, a decoder is referred to as n-to-2^n decoder and the decoder outputs may be logic **high** or **low**.

For example, consider a 3-to-2^3 (3-to-8) decoder. Let the 8 outputs of decoder be Y_0, Y_1, Y_2, Y_3, Y_4, Y_5, Y_6 and Y_7. In a 3-bit binary input, the possible minterms are m_0, m_1, m_2, m_3, m_4, m_5, m_6 and m_7.

In logic **high** 3-to-8 decoder,

When input is m_0 then output Y_0 is active **high**,

When input is m_1 then output Y_1 is active **high**,

and so on.

Therefore, we can say that output decodes the input binary by asserting the output as **high**.

Similarly, in a 3-bit binay input, the possible maxterms are M_0, M_1, M_2, M_3, M_4, M_5, M_6 and M_7.

In logic **low** 3-to-8 decoder,

When input is M_0 then output Y_0 is active **low**,

When input is M_1 then output Y_1 is active **low**,

and so on.

Therefore, we can say that output decodes the input binary by asserting the output as **low**.

Most of the decoders are provided with an enable signal. The input of decoder is recognized only when enable signal is active. The enable can be active **high** or **low**. In decoder with active **high** enable, the input is recognized only when enable is tied to **high**. In decoder with active **low** enable, the input is recognized only when enable is tied to **low**.

Therefore, six types of n-to-2^n decoders are available as shown in Fig. 2.79. A bubble in the input/output line indicates that the corresponding input/output is active **low**.

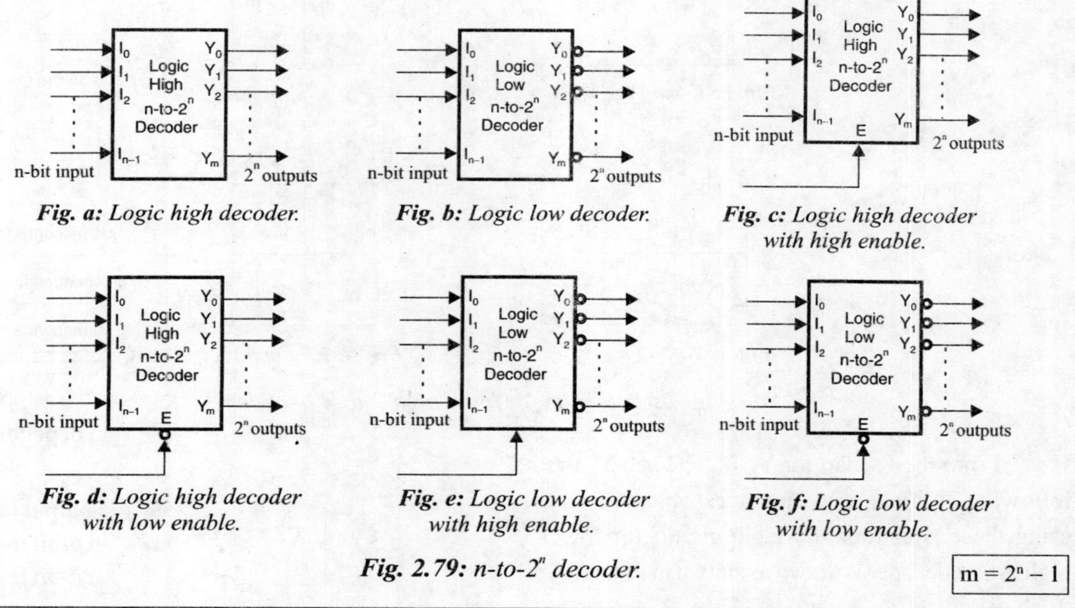

***Fig. a:** Logic high decoder.*

***Fig. b:** Logic low decoder.*

***Fig. c:** Logic high decoder with high enable.*

***Fig. d:** Logic high decoder with low enable.*

***Fig. e:** Logic low decoder with high enable.*

***Fig. f:** Logic low decoder with low enable.*

***Fig. 2.79:** n-to-2^n decoder.*

$m = 2^n - 1$

2.9.1 Logic High 2-to-4 Decoder

A 2-to-4 (2-to-2^2) decoder can generate 4 decoded outputs from the two bit input.

The ***logic high 2-to-4 decoder*** will generate 4 unique outputs in which only one will be **high** at any one time and all other output will be zero. The block diagram representation of logic **high** 2-to-4 decoder is shown in Fig. 2.80.

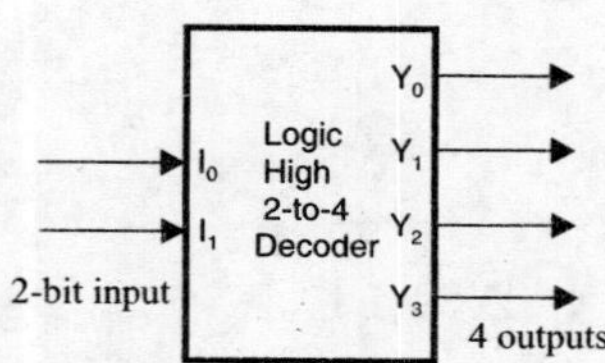

***Fig. 2.80:** Logic high 2-to-4 decoder.*

Table 2.34: Truth Table of Logic High 2-to-4 Decoder

Inputs		Minterm	Outputs			
I_1	I_0		Y_0	Y_1	Y_2	Y_3
0	0	m_0	1	0	0	0
0	1	m_1	0	1	0	0
1	0	m_2	0	0	1	0
1	1	m_3	0	0	0	1

The truth table of logic **high** 2-to-4 decoder is shown in Table 2.34 and using this truth table, the K-maps for the decoder design are drawn as shown in Fig. 2.81.

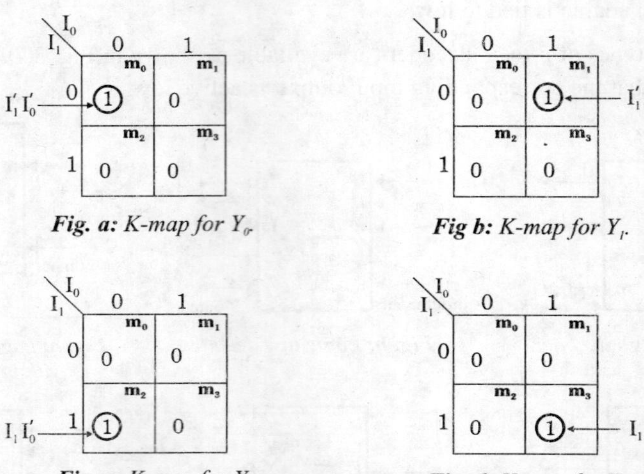

Fig. a: K-map for Y_0. *Fig b: K-map for Y_1.*

Fig. c: K-map for Y_2. *Fig. d: K-map for Y_3.*

***Fig. 2.81:** K-map for logic high 2-to-4 decoder.*

From the K-map for Y_0, Y_1, Y_2 and Y_3 the following Boolean equations are obtained and using these equations the logic circuit for logic **high** 2-to-4 decoder is drawn as shown in Fig. 2.82.

$$Y_0 = I_1' I_0'$$

$$Y_1 = I_1' I_0$$

$$Y_2 = I_1 I_0'$$

$$Y_3 = I_1 I_0$$

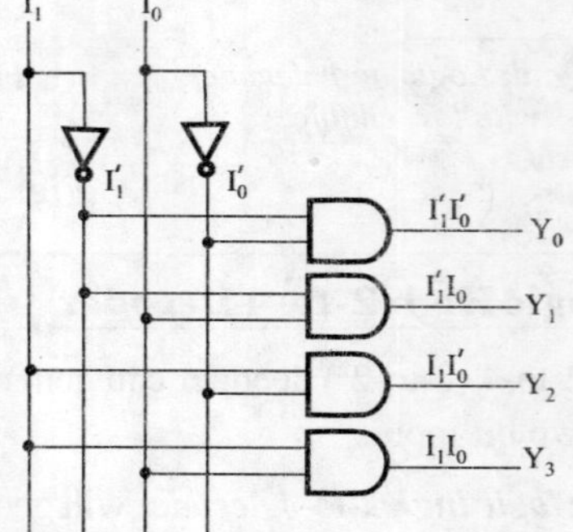

***Fig. 2.82:** Logic circuit for logic high 2-to-4 decoder.*

***Note:** Alternatively, the output equations of logic **high** 2-to-4 decoder can be directly written from the truth table which are minterms for which output is **high**.*

2.9.2 Logic Low 2-to-4 Decoder

(AU, Nov/Dec'18, 13 Marks)

The ***logic low 2-to-4 decoder*** will generate 4 unique outputs in which only one of the output will be **low** at any one time. The output of logic **low** decoder will be complement to that of logic **high** decoder.

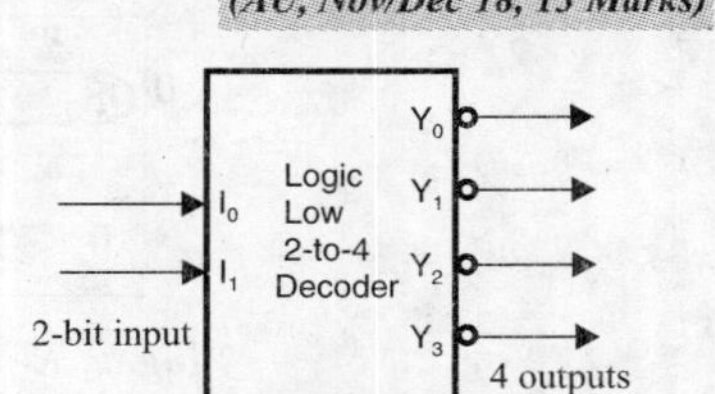

Fig. 2.83: *Logic low 2-to-4 decoder.*

The block diagram representation of logic **low** 2-to-4 decoder is shown in Fig. 2.83. The bubble at the outputs indicate that the outputs are active **low**. The dual 2-to-4 decoder is available as a standard IC with number 74139. The pin configuration of 74139 is shown in Fig. 2.84.

The truth table of logic **low** 2-to-4 decoder is shown in Table 2.35 and using this truth table, the K-maps for the design of decoder are drawn as shown in Fig. 2.85.

Table 2.35: Truth Table of Logic Low 2-to-4 Decoder

Inputs		Minterm	Outputs			
I_1	I_0		Y_0	Y_1	Y_2	Y_3
0	0	m_0	0	1	1	1
0	1	m_1	1	0	1	1
1	0	m_2	1	1	0	1
1	1	m_3	1	1	1	0

Pin	Name	Pin	Name
1	$\overline{E}_a$	16	V_{cc}(+5V)
2	I_{0a}	15	E_b
3	I_{1a}	14	I_{0b}
4	$\overline{Y}_{0a}$	13	I_{1b}
5	$\overline{Y}_{1a}$	12	$\overline{Y}_{0b}$
6	$\overline{Y}_{2a}$	11	$\overline{Y}_{1b}$
7	$\overline{Y}_{3a}$	10	$\overline{Y}_{2b}$
8	(0V) GND	9	$\overline{Y}_{3b}$

74139

Pin	Description
I_{0a}, I_{1a}	Inputs
I_{0b}, I_{1b}	Inputs
$\overline{Y}_{0a}$ - $\overline{Y}_{3a}$	Active low outputs
$\overline{Y}_{0b}$ - $\overline{Y}_{3b}$	Active low outputs
$\overline{E}_a$, $\overline{E}_b$	Enables
V_{cc}	Supply voltage (+5V)
GND	Ground (0V)

Fig. 2.84: *Pin configuration of dual 2-to-4 decoder IC 74139.*

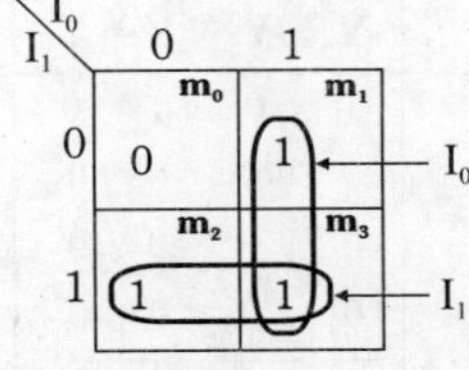

Fig. a: *K-map for Y_0.*

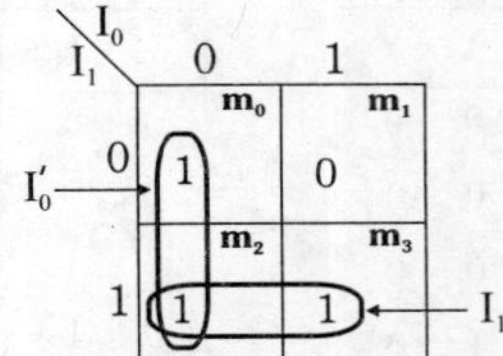

Fig. b: *K-map for Y_1.*

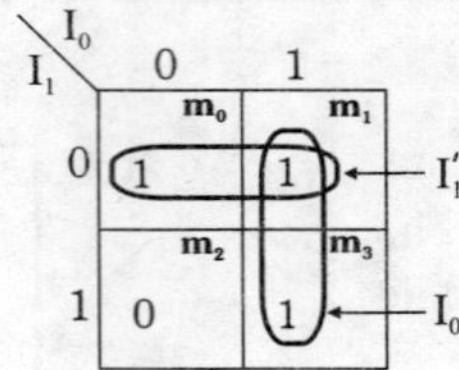

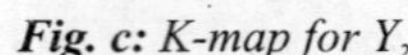
Fig. c: K-map for Y_2.

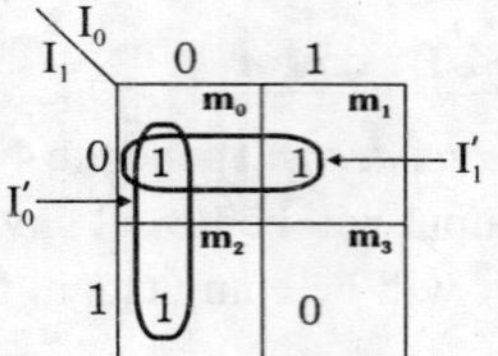

Fig. d: K-map for Y_3.

Fig. 2.85: *K-map for logic low 2-to-4 decoder.*

From the K-map for Y_0, Y_1, Y_2 and Y_3 the following Boolean equations are obtained and using these equations the logic circuit for logic **low** 2-to-4 decoder is drawn in Fig. 2.86.

$$Y_0 = I_1 + I_0$$

$$Y_1 = I_1 + I_0'$$

$$Y_2 = I_1' + I_0$$

$$Y_3 = I_1' + I_0'$$

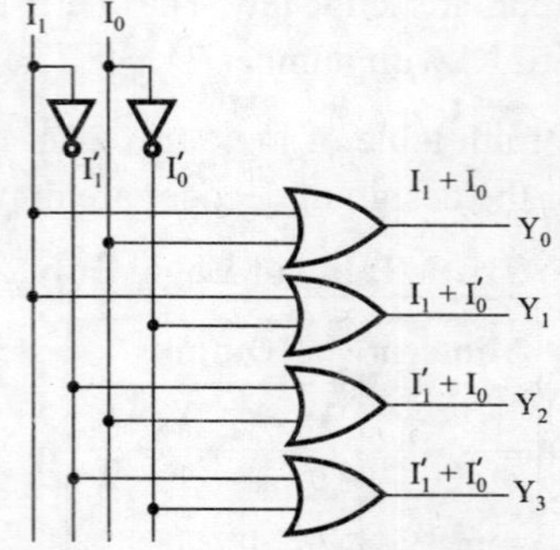

Fig. 2.86: *Logic circuit for logic low 2-to-4 decoder.*

> ***Note:*** *Alternatively, the output equations of logic* ***low*** *2-to-4 decoder can be directly written from the truth table which are maxterms for which output is* ***low****.*

2.9.3 Logic High 2-to-4 Decoder with Logic High Enable

The block diagram representation of logic **high** decoder with logic **high** enable is shown in Fig. 2.87. When enable input is made **low** then inputs are not recognized and so all outputs are zero.

When enable input is made **high** then inputs are recognized and any one of the outputs is set to 1 depending on input.

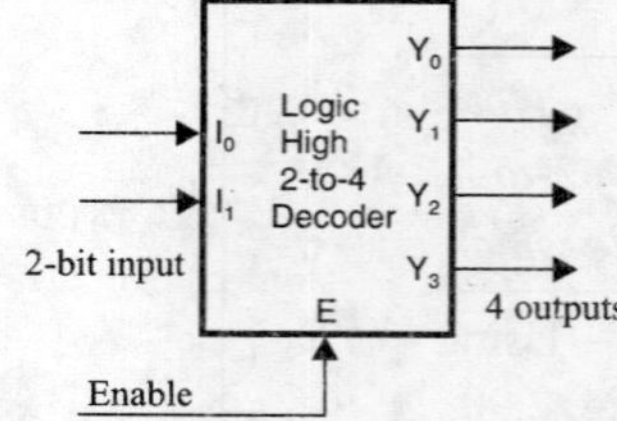

Fig. 2.87: *Logic high 2-to-4 decoder with logic high enable.*

The truth table of decoder for both active and inactive enable is shown in Table 2.36. Using this truth table, 3-variable K-maps are drawn for 4-outputs of decoder as shown in Fig. 2.88.

Table 2.36: Truth Table of Logic High 2-to-4 Decoder with Logic High Enable

Inputs			Outputs			
E	**I_1**	**I_0**	**Y_0**	**Y_1**	**Y_2**	**Y_3**
0	×	×	0	0	0	0
1	0	0	1	0	0	0
1	0	1	0	1	0	0
1	1	0	0	0	1	0
1	1	1	0	0	0	1

Inputs			Minterm	Outputs				Comment
E	**I_1**	**I_0**		**Y_0**	**Y_1**	**Y_2**	**Y_3**	
0	0	0	m_0	0	0	0	0	When enable is not active
0	0	1	m_1	0	0	0	0	
0	1	0	m_2	0	0	0	0	
0	1	1	m_3	0	0	0	0	
1	0	0	m_4	1	0	0	0	When enable is active
1	0	1	m_5	0	1	0	0	
1	1	0	m_6	0	0	1	0	
1	1	1	m_7	0	0	0	1	

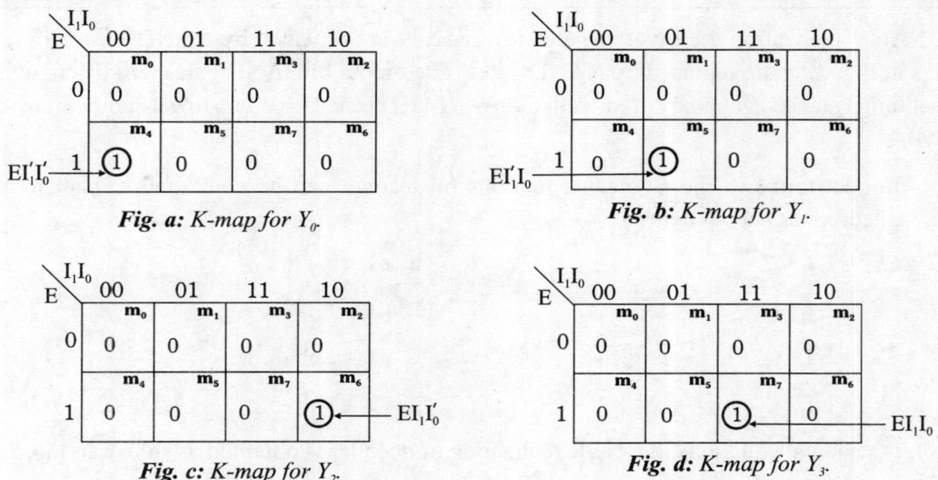

***Fig. 2.88:** K-map for logic high 2-to-4 decoder with logic high enable.*

From the K-map for Y_0, Y_1, Y_2 and Y_3 the following Boolean equations are obtained and using these equations the logic circuit of decoder is drawn in Fig. 2.89.

$$Y_0 = EI_1'I_0'$$

$$Y_1 = EI_1'I_0$$

$$Y_2 = EI_1I_0'$$

$$Y_3 = EI_1I_0$$

> ***Note:** Alternatively, the output equations of logic **high** 2-to-4 decoder can be directly written from the truth table which are minterms for which output is **high**.*

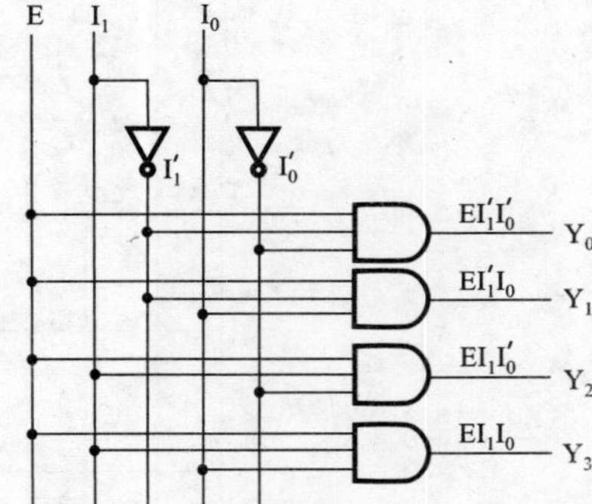

***Fig. 2.89:** Logic circuit for 2-to-4 logic high decoder with logic high enable.*

Using the above equations it is possible to realize the decoder using only NAND gates as shown in Fig. 2.90.

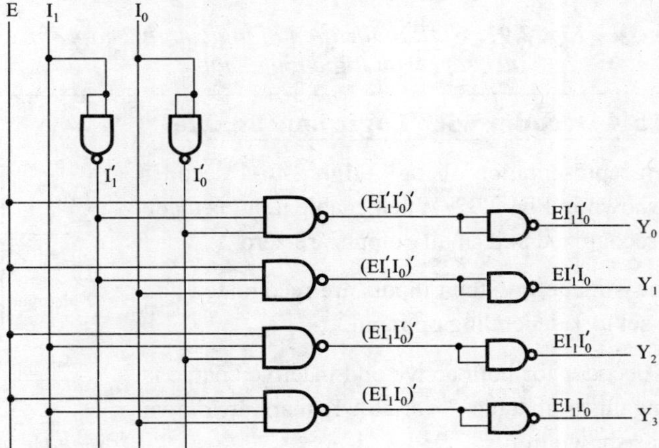

***Fig. 2.90:** NAND realization of logic high 2-to-4 decoder with logic high enable.*

In NAND realization the inverter gates in Fig. 2.89 are replaced by NAND based inverters and AND gates in Fig. 2.89 are replaced by NAND gates. The output of NAND gates will be complement of the product terms and so again they are inverted using NAND based inverters to get right output as shown in Fig. 2.90.

The product terms in the Boolean equations of decoder can be converted to sum terms using DeMorgan's theorem as shown below:

$$Y_0 = E\,I_1'\,I_0' \qquad \xrightarrow{\text{Using DeMorgan's Theorem}} \qquad Y_0 = (E' + I_1 + I_0)'$$

$$Y_1 = E\,I_1'\,I_0 \qquad Y_1 = (E' + I_1 + I_0')'$$

$$Y_2 = E\,I_1\,I_0' \qquad Y_2 = (E' + I_1' + I_0)'$$

$$Y_3 = E\,I_1\,I_0 \qquad Y_3 = (E' + I_1' + I_0')'$$

Using the above equations, the NOR realization of decoder is obtained as shown in Fig. 2.91.

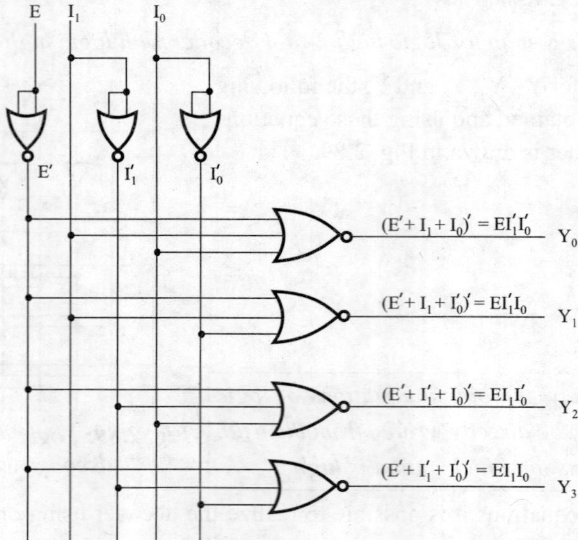

Fig. 2.91: NOR realization of logic high 2-to-4 decoder with logic high enable.

2.9.4 Logic High 2-to-4 Decoder with Logic Low Enable

The block diagram representation of logic **high** 2-to-4 decoder with logic **low** enable is shown in Fig. 2.92. When enable input is made **high** then inputs are not recognized and so all outputs are zero.

When enable input is made **low** then inputs are recognized and any one of the outputs is set to 1 depending on input.

The truth table of decoder for both active and inactive enable is shown in Table 2.37. Using this truth table, 3-variable K-maps are drawn for 4 outputs of decoder as shown in Fig. 2.93.

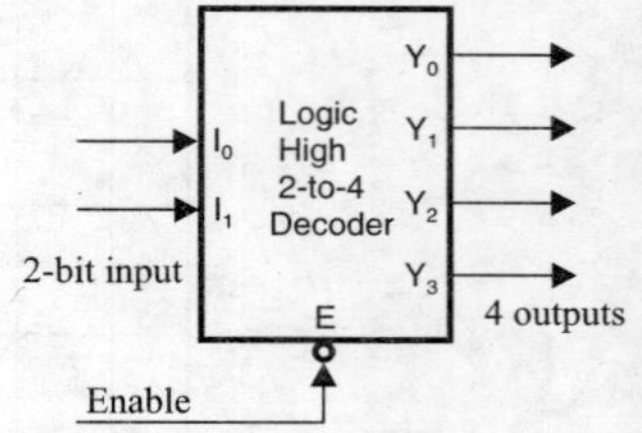

Fig. 2.92: Logic high 2-to-4 decoder with logic low enable.

Table 2.37: Truth Table of Logic High 2-to-4 Decoder with Logic Low Enable

Inputs			Outputs			
E	I_1	I_0	Y_0	Y_1	Y_2	Y_3
1	×	×	0	0	0	0
0	0	0	1	0	0	0
0	0	1	0	1	0	0
0	1	0	0	0	1	0
0	1	1	0	0	0	1

Inputs			Minterm	Outputs				Comment
E	I_1	I_0		Y_0	Y_1	Y_2	Y_3	
0	0	0	m_0	1	0	0	0	Output when enable is active
0	0	1	m_1	0	1	0	0	
0	1	0	m_2	0	0	1	0	
0	1	1	m_3	0	0	0	1	
1	0	0	m_4	0	0	0	0	Output when enable is inactive
1	0	1	m_5	0	0	0	0	
1	1	0	m_6	0	0	0	0	
1	1	1	m_7	0	0	0	0	

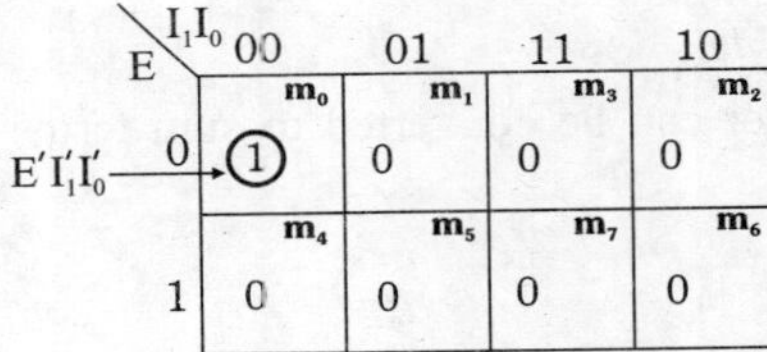

Fig. a: K-map for Y_0.

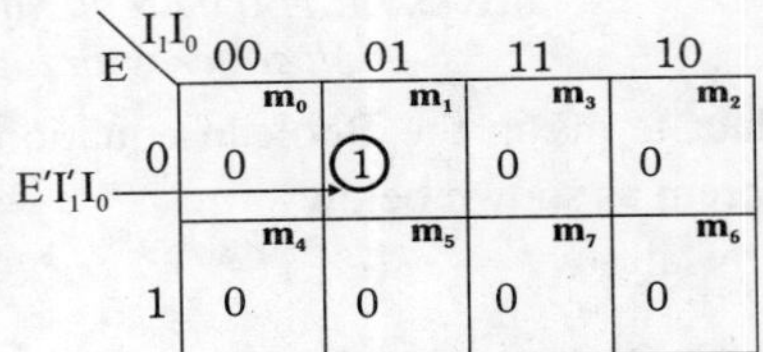

Fig. b: K-map for Y_1.

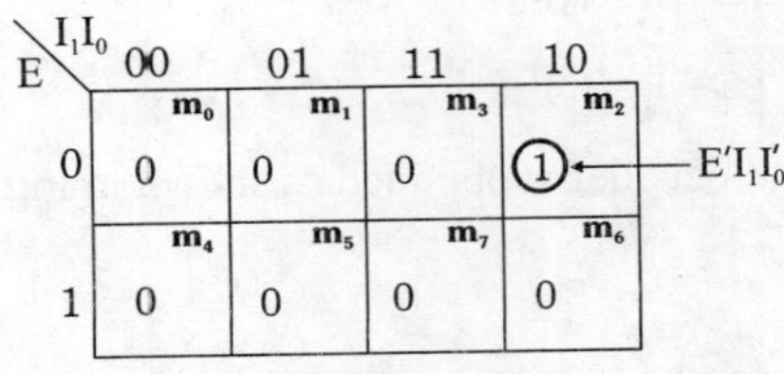

Fig. c: K-map for Y_2.

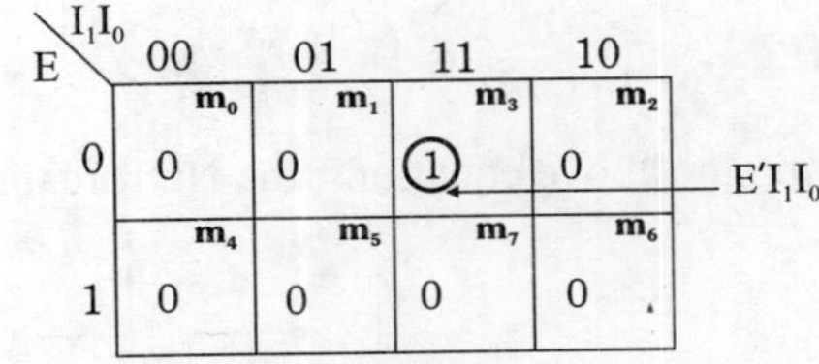

Fig. d: K-map for Y_3.

***Fig. 2.93:** K-map for logic high 2-to-4 decoder with logic low enable.*

From the K-map for Y_0, Y_1, Y_2 and Y_3 the following Boolean equations are obtained and using these equations the logic circuit of decoder is drawn in Fig. 2.94.

$$Y_0 = E' I_1' I_0'$$

$$Y_1 = E' I_1' I_0$$

$$Y_2 = E' I_1 I_0'$$

$$Y_3 = E' I_1 I_0$$

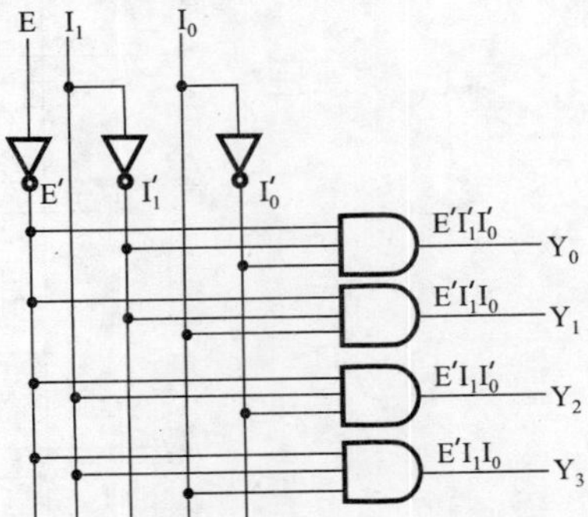

***Fig. 2.94:** Logic circuit for 2-to-4 logic high decoder with logic low enable.*

Using the above equations it is possible to realize the decoder using only NAND gates as shown in Fig. 2.95.

> ***Note:** Alternatively, the output equations of logic **high** 2-to-4 decoder can be directly written from the truth table which are minterms for which output is **high**.*

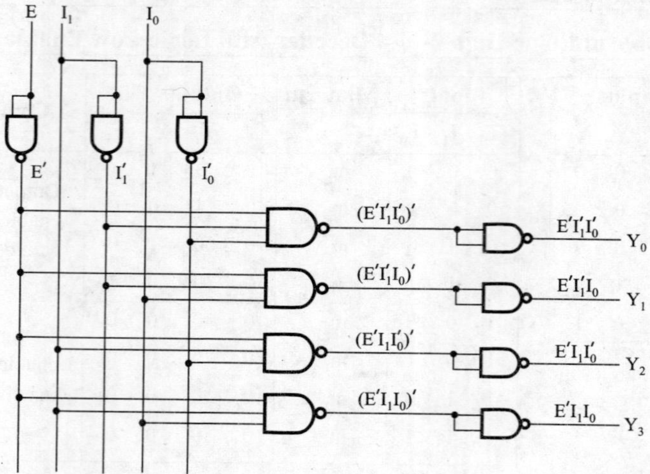

Fig. 2.95: *NAND realization of logic high 2-to-4 decoder with logic low enable.*

The product terms in the Boolean equations of decoder can be converted to sum terms using DeMorgan's theorem as shown below:

$$Y_0 = E' I_1' I_0' \qquad \qquad Y_0 = (E + I_1 + I_0)'$$

$$Y_1 = E' I_1' I_0 \qquad \xrightarrow{\text{Using DeMorgan's Theorem}} \qquad Y_1 = (E + I_1 + I_0')'$$

$$Y_2 = E' I_1 I_0' \qquad \qquad Y_2 = (E + I_1' + I_0)'$$

$$Y_3 = E' I_1 I_0 \qquad \qquad Y_3 = (E + I_1' + I_0')'$$

Using the above equations, the NOR realization of decoder is obtained as shown in Fig. 2.96.

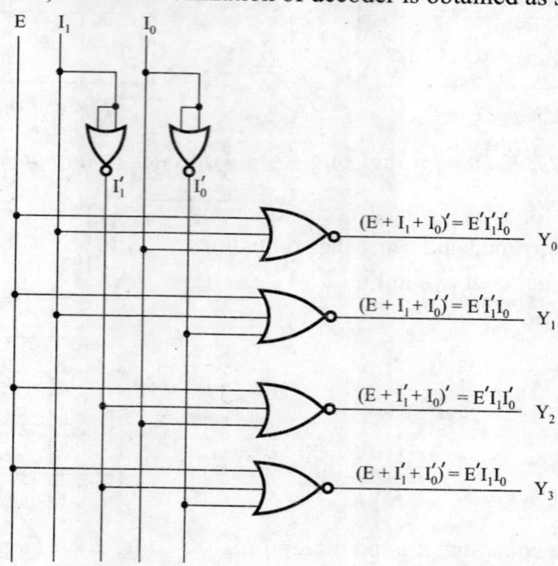

Fig. 2.96: *NOR realization of logic high 2-to-4 decoder with logic low enable.*

2.9.5 Logic Low 2-to-4 Decoder with Logic High Enable

The block diagram representation of logic **low** 2-to-4 decoder with logic **high** enable is shown in Fig. 2.97. When enable input is made **low** then inputs are not recognized and so all outputs are set to 1.

When enable input is made **high** then inputs are recognized and any one of the outputs is made **low** depending on input.

The truth table of decoder for both active and inactive enable is shown in Table 2.38. Using this truth table, 3-variable K-maps are drawn for 4 outputs of decoder as shown in Fig. 2.98.

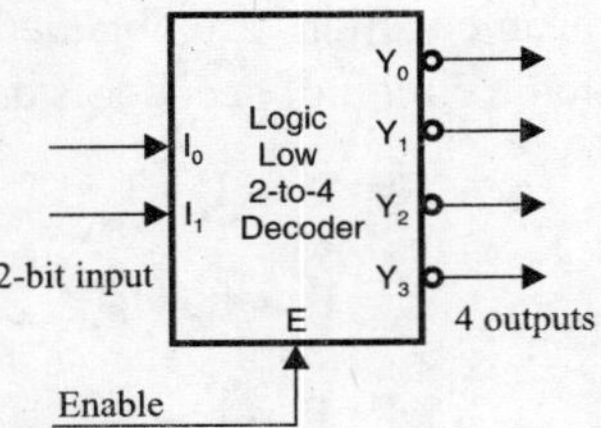

Fig. 2.97: Logic low 2-to-4 decoder with logic high enable.

Table 2.38: Truth Table of Logic low 2-to-4 Decoder with Logic High Enable

Inputs			Minterm	Outputs			
E	I_1	I_0		Y_0	Y_1	Y_2	Y_3
0	×	×	m_0	1	1	1	1
1	0	0	m_4	0	1	1	1
1	0	1	m_5	1	0	1	1
1	1	0	m_6	1	1	0	1
1	1	1	m_7	1	1	1	0

Inputs			Minterm	Outputs				Comment
E	I_1	I_0		Y_0	Y_1	Y_2	Y_3	
0	0	0	m_0	1	1	1	1	Output when enable is inactive
0	0	1	m_1	1	1	1	1	
0	1	0	m_2	1	1	1	1	
0	1	1	m_3	1	1	1	1	
1	0	0	m_4	0	1	1	1	Output when enable is active
1	0	1	m_5	1	0	1	1	
1	1	0	m_6	1	1	0	1	
1	1	1	m_7	1	1	1	0	

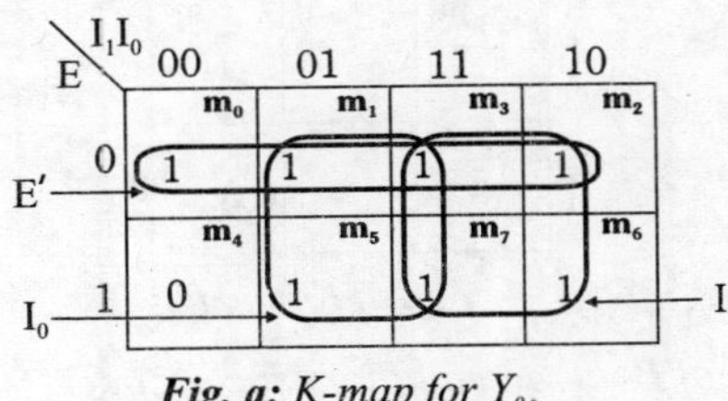

Fig. a: K-map for Y_0.

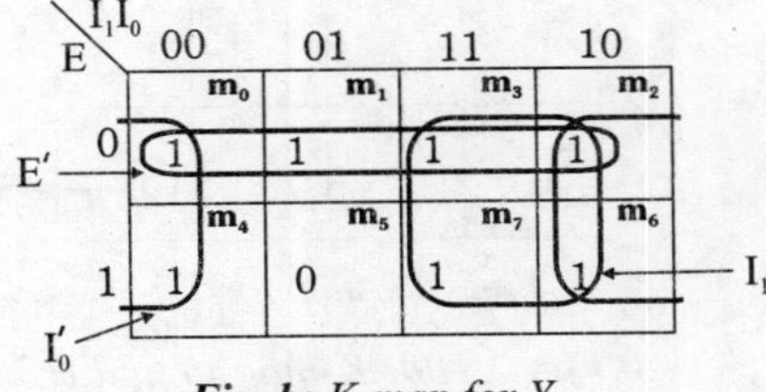

Fig. b: K-map for Y_1.

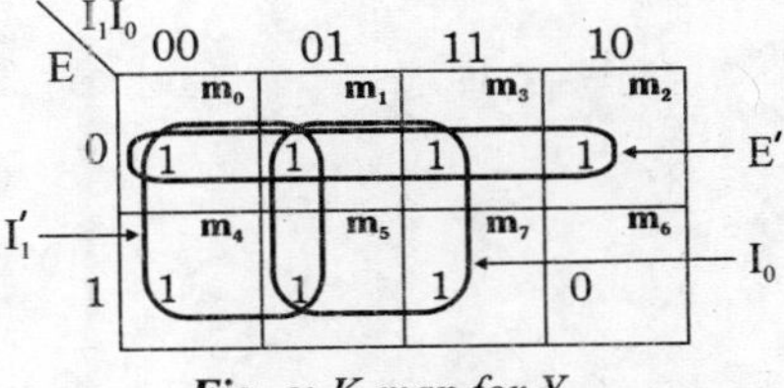

Fig. c: K-map for Y_2.

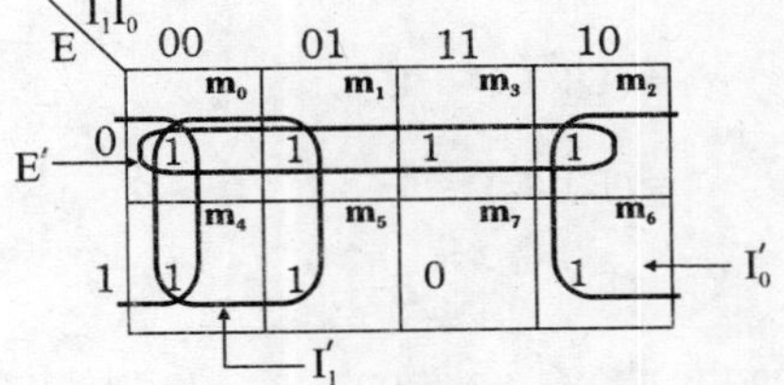

Fig. d: K-map for Y_3.

Fig. 2.98: K-map for logic low 2-to-4 decoder with logic high enable.

From the K-map for Y_0, Y_1, Y_2 and Y_3 the following Boolean equations are obtained and using these equations the logic circuit of decoder is drawn as shown in Fig. 2.99.

$$Y_0 = E' + I_1 + I_0$$

$$Y_1 = E' + I_1 + I_0'$$

$$Y_2 = E' + I_1' + I_0$$

$$Y_3 = E' + I_1' + I_0'$$

> ***Note:** Alternatively, the output equations of logic **low** 2-to-4 decoder can be directly written from the truth table which are maxterms for which output is **low**.*

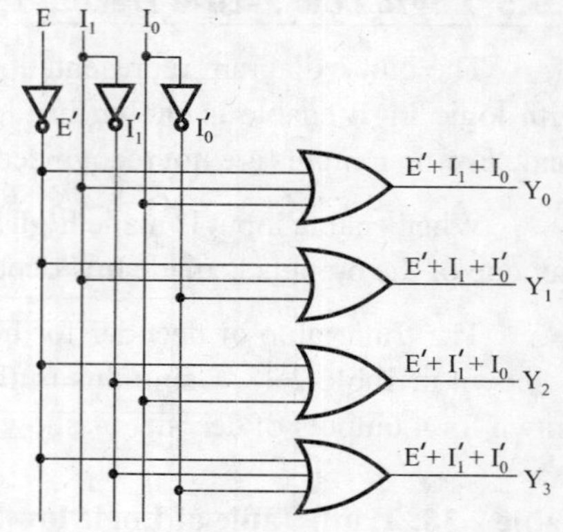

***Fig. 2.99:** Logic circuit for logic low 2-to-4 decoder with logic high enable.*

Alternatively, the following POS form of Boolean equations can be obtained by forming prime implicants using 0's as shown in Fig. 2.100.

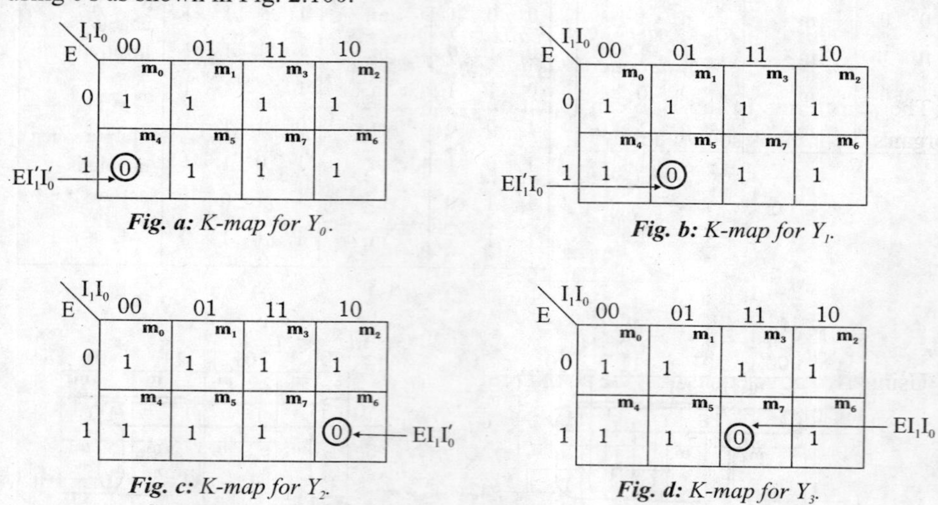

***Fig. a:** K-map for Y_0.*

***Fig. b:** K-map for Y_1.*

***Fig. c:** K-map for Y_2.*

***Fig. d:** K-map for Y_3.*

***Fig. 2.100:** K-map to find POS form of Boolean equations for decoder.*

$$Y_0 = (E\,I_1'\,I_0')' = E' + I_1 + I_0$$

$$Y_1 = (E\,I_1'\,I_0)' = E' + I_1 + I_0'$$

$$Y_2 = (E\,I_1\,I_0')' = E' + I_1' + I_0$$

$$Y_3 = (E\,I_1\,I_0)' = E' + I_1' + I_0'$$

Using the above equations, the NOR realization of decoder is obtained as shown in Fig. 2.101.

In NOR realization the inverter gates in Fig. 2.99 are replaced by NOR based inverters and OR gates in Fig. 2.99 are replaced by NOR gates. The output of NOR gates will be complement of the sum terms and so again they are inverted using NOR based inverters to get right output as shown in Fig. 2.101.

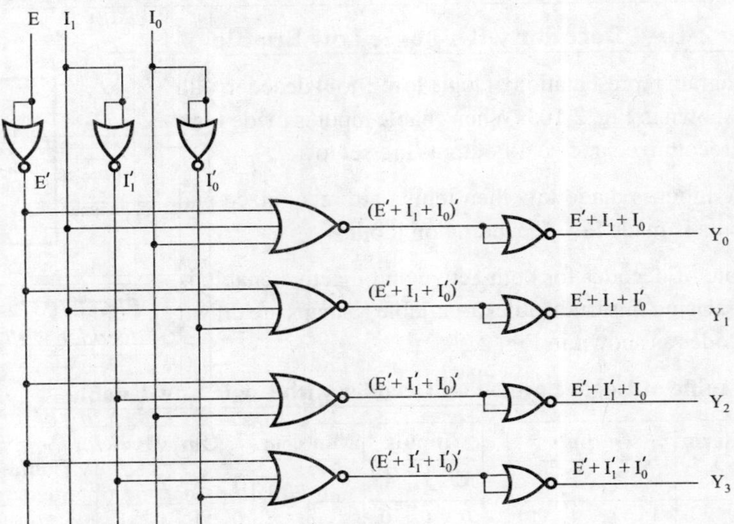

***Fig. 2.101:** NOR realization of logic low 2-to-4 decoder with logic high enable.*

The sum terms in the Boolean equations of decoder can be converted to product terms using DeMorgan's theorem as shown below:

$$Y_0 = E' + I_1 + I_0 \qquad\qquad Y_0 = (E\, I'_1\, I'_0)'$$

$$Y_1 = E' + I_1 + I'_0 \quad \xrightarrow{\text{Using DeMorgan's Theorem}} \quad Y_1 = (E\, I'_1\, I_0)'$$

$$Y_2 = E' + I'_1 + I_0 \qquad\qquad Y_2 = (E\, I_1\, I'_0)'$$

$$Y_3 = E' + I'_1 + I'_0 \qquad\qquad Y_3 = (E\, I_1\, I_0)'$$

Using the above equations, the NAND realization of decoder is obtained as shown in Fig. 2.102.

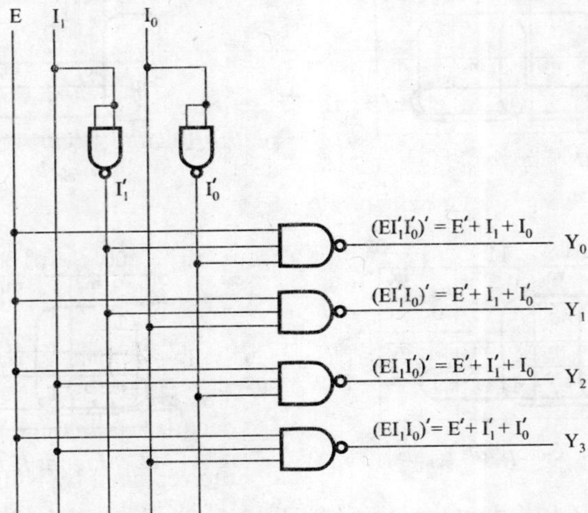

***Fig. 2.102:** NAND realization of logic low 2-to-4 decoder with logic high enable.*

2.9.6 Logic Low 2-to-4 Decoder with Logic Low Enable

The block diagram representation of logic **low** 2-to-4 decoder with logic **low** enable is shown in Fig. 2.103. When enable input is made **high** then inputs are not recognized and so all outputs are set to 1.

When enable input is made **low** then inputs are recognized and any one of the outputs is made **low** depending on input.

The truth table of decoder for both active and inactive enable is shown in Table 2.39. Using this truth table, 3-variable K-maps are drawn for 4 outputs of decoder as shown in Fig. 2.39.

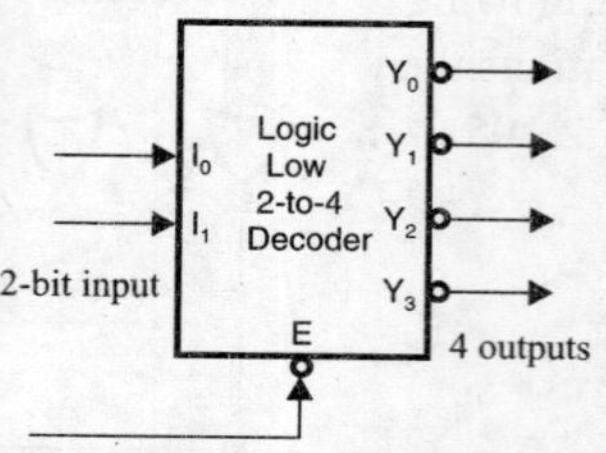

Fig. 2.103: Logic low 2-to-4 decoder with logic low enable.

Table 2.39: Truth Table of Logic Low 2-to-4 Decoder with Logic Low Enable

Inputs			Minterm	Outputs			
E	I_1	I_0		Y_0	Y_1	Y_2	Y_3
1	×	×	m_4	1	1	1	1
0	0	0	m_0	0	1	1	1
0	0	1	m_1	1	0	1	1
0	1	0	m_2	1	1	0	1
0	1	1	m_3	1	1	1	0

Inputs			Minterm	Outputs				Comment
E	I_1	I_0		Y_0	Y_1	Y_2	Y_3	
0	0	0	m_0	0	1	1	1	Output when enable is active
0	0	1	m_1	1	0	1	1	
0	1	0	m_2	1	1	0	1	
0	1	1	m_3	1	1	1	0	
1	0	0	m_4	1	1	1	1	Output when enable is inactive
1	0	1	m_5	1	1	1	1	
1	1	0	m_6	1	1	1	1	
1	1	1	m_7	1	1	1	1	

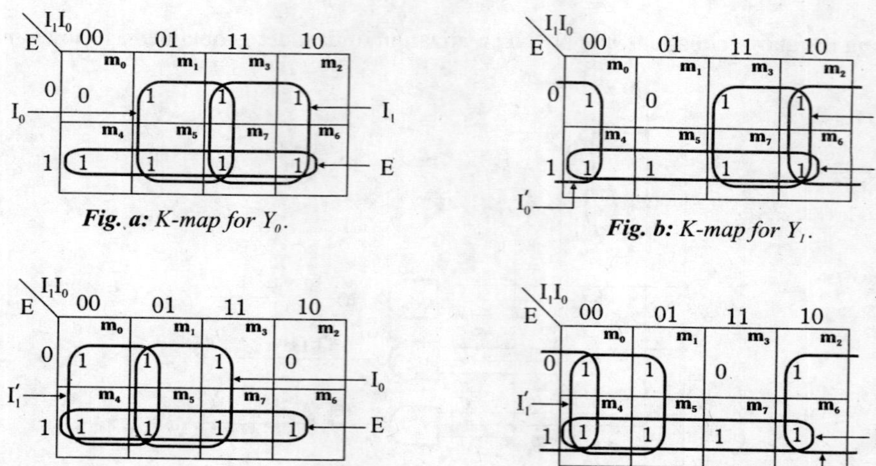

Fig. a: K-map for Y_0.

Fig. b: K-map for Y_1.

Fig. c: K-map for Y_2.

Fig. d: K-map for Y_3.

***Fig. 2.104:** K-map for logic low 2- to-4 decoder with logic low enable.*

From the K-map of Fig. 2.104 for Y_0, Y_1, Y_2 and Y_3 the following Boolean equations are obtained and using these equations the logic circuit of decoder is drawn as shown in Fig. 2.105.

$$Y_0 = E + I_1 + I_0$$

$$Y_1 = E + I_1 + I_0'$$

$$Y_2 = E + I_1' + I_0$$

$$Y_3 = E + I_1' + I_0'$$

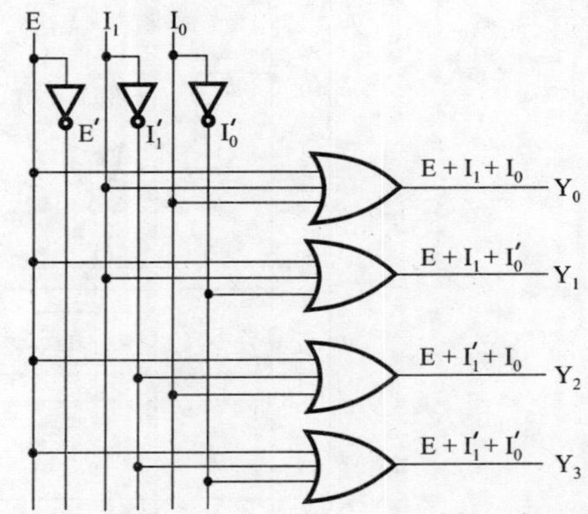

***Fig. 2.105:** Logic circuit for logic low 2-to-4 decoder with logic low enable.*

> ***Note:** Alternatively, the output equations of logic **low** 2-to-4 decoder can be directly written from the truth table which are maxterms for which output is **low**.*

Alternatively, the following POS form of Boolean equations can be obtained by forming prime implicants using 0's as shown in Fig. 2.106.

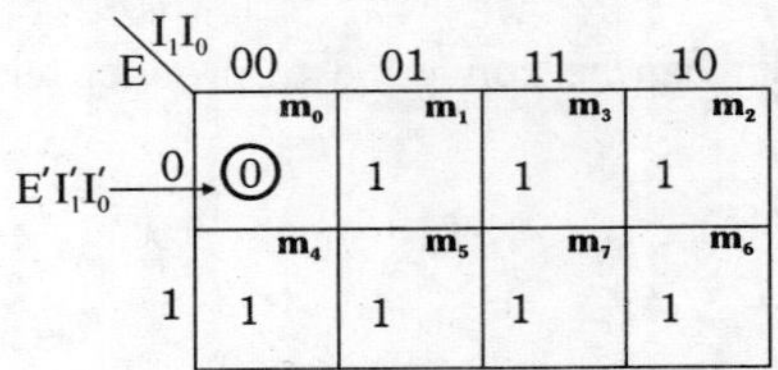

***Fig. a:** K-map for Y_0.*

E \ I_1I_0	00	01	11	10
0	1 (m_0)	(0) (m_1) ← $E'I_1'I_0$	1 (m_3)	1 (m_2)
1	1 (m_4)	1 (m_5)	1 (m_7)	1 (m_6)

***Fig. b:** K-map for Y_1.*

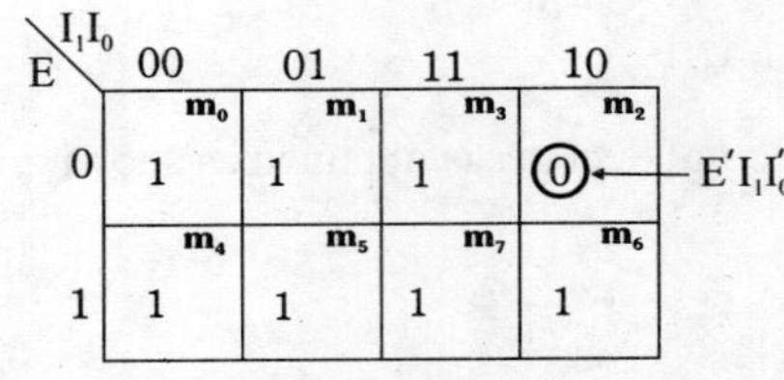

***Fig. c:** K-map for Y_2.*

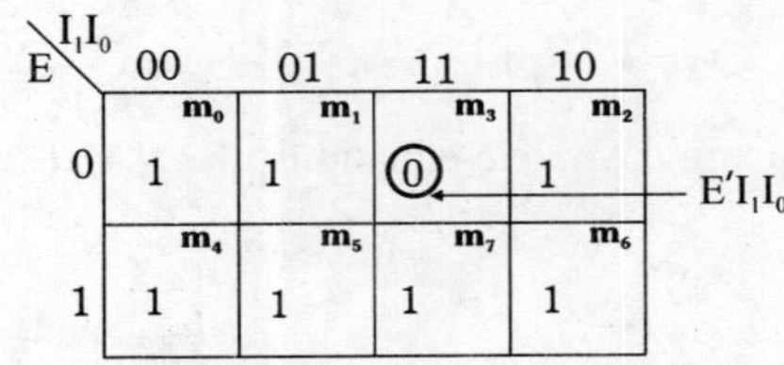

***Fig. d:** K-map for Y_3.*

***Fig. 2.106:** K-map to find POS form of Boolean equations for decoder.*

$$Y_0 = (E'\ I_1'\ I_0')' = E + I_1 + I_0$$

$$Y_1 = (E'\ I_1'\ I_0)' = E + I_1 + I_0'$$

$$Y_2 = (E'\ I_1\ I_0')' = E + I_1' + I_0$$

$$Y_3 = (E'\ I_1\ I_0) \ = \ E + I_1' + I_0'$$

Using the above equations it is possible to realize the decoder using only NOR gates as shown in Fig. 2.107.

In NOR realization the inverter gates in Fig. 2.105 are replaced by NOR based inverters and OR gates in Fig. 2.105 are replaced by NOR gates. The output of NOR gates will be complement of the sum terms and so again they are complemented using NOR based inverters to get right output as shown in Fig. 2.107.

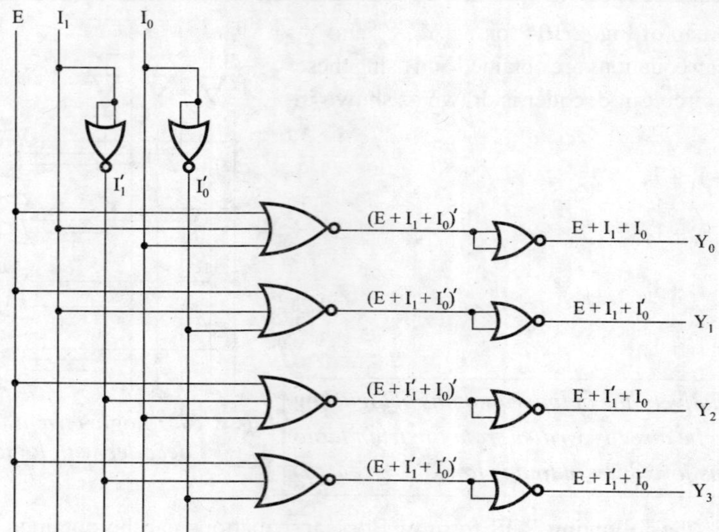

Fig. 2.107: *NOR realization of logic low 2-to-4 decoder with logic low enable.*

The sum terms in the Boolean equations of decoder can be converted to product terms using DeMorgan's theorem as shown below:

$$Y_0 = E + I_1 + I_0 \qquad Y_0 = (E' \, I_1' \, I_0')'$$

$$Y_1 = E + I_1 + I_0' \qquad \xrightarrow{\text{Using DeMorgan's Theorem}} \qquad Y_1 = (E' \, I_1' \, I_0)'$$

$$Y_2 = E + I_1' + I_0 \qquad Y_2 = (E' \, I_1 \, I_0')'$$

$$Y_3 = E + I_1' + I_0' \qquad Y_3 = (E' \, I_1 \, I_0)'$$

Using the above equations, the NAND realization of decoder is obtained as shown in Fig. 2.108.

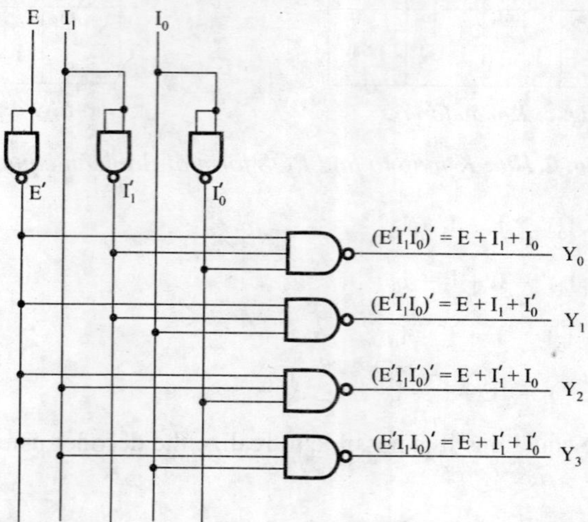

Fig. 2.108: *NAND realization of logic low 2-to-4 decoder with logic low enable.*

2.9.7 Logic High 3-to-8 Decoder

A ***3-to-8 (3-to-2^3) decoder*** can generate 8 decoded outputs from the three bit inputs.

The ***logic high 3-to-8 decoder*** will generate 8 unique outputs in which only one will be **high** at any one time and all other outputs will be **low**.

The block diagram representation of logic **high** 3-to-8 decoder is shown in Fig. 2.109. The truth table of logic **high** 3-to-8 decoder is shown in Table 2.40.

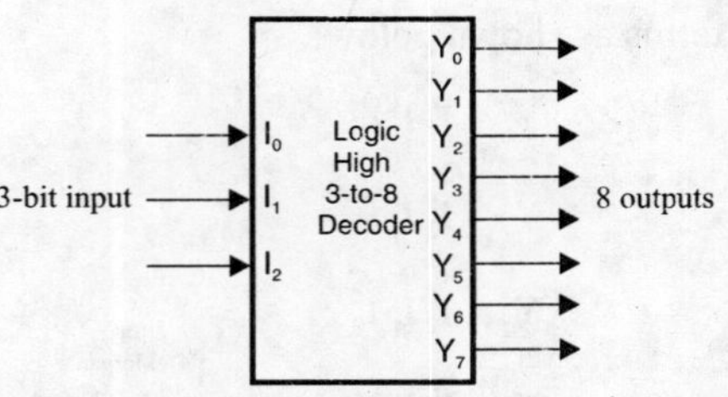

Fig. 2.109: *Logic high 3-to-8 decoder.*

From the truth table we can directly write Boolean equations for outputs of logic **high** 3-to-8 decoder which are the minterms for which output is **high** as shown below:

$$Y_0 = I_2' I_1' I_0'$$

$$Y_1 = I_2' I_1' I_0$$

$$Y_2 = I_2' I_1 I_0'$$

$$Y_3 = I_2' I_1 I_0$$

$$Y_4 = I_2 I_1' I_0'$$

$$Y_5 = I_2 I_1' I_0$$

$$Y_6 = I_2 I_1 I_0'$$

$$Y_7 = I_2 I_1 I_0$$

Table 2.40: Truth Table of Logic High 3-to-8 Decoder

Inputs			**Minterm**	**Outputs**							
I_2	I_1	I_0		Y_0	Y_1	Y_2	Y_3	Y_4	Y_5	Y_6	Y_7
0	0	0	m_0	1	0	0	0	0	0	0	0
0	0	1	m_1	0	1	0	0	0	0	0	0
0	1	0	m_2	0	0	1	0	0	0	0	0
0	1	1	m_3	0	0	0	1	0	0	0	0
1	0	0	m_4	0	0	0	0	1	0	0	0
1	0	1	m_5	0	0	0	0	0	1	0	0
1	1	0	m_6	0	0	0	0	0	0	1	0
1	1	1	m_7	0	0	0	0	0	0	0	1

The logic **high** 3-to-8 decoder is available as a standard IC with number 74237. The pin configuration of 74237 is shown in Fig. 2.110.

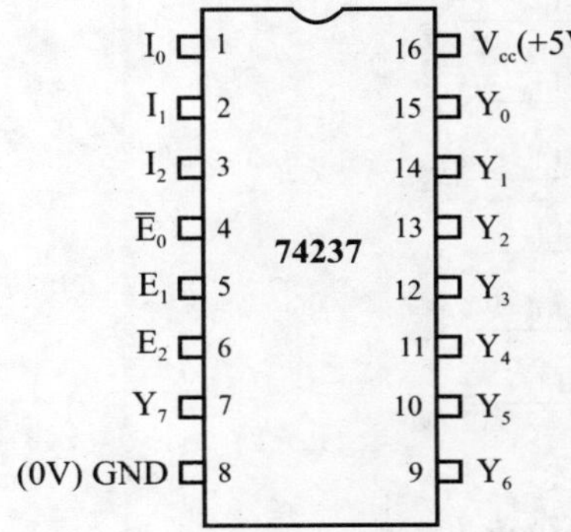

Pin	**Description**
I_0 - I_2	Inputs
$\overline{E}_0$	Logic low enable
E_1, E_2	Logic high enables
Y_0 - Y_7	Active high outputs
V_{cc}	Supply voltage (+5V)
GND	Ground (0V)

Fig. 2.110: *Pin configuration of logic high 3-to-8 decoder IC 74237.*

The product terms in the Boolean equations of logic **high** 3-to-8 decoder can be converted to sum terms as shown below:

(Used for AND Realization)		(Used for NOR Realization)
$Y_0 = I_2' I_1' I_0'$		$Y_0 = (I_2 + I_1 + I_0)'$
$Y_1 = I_2' I_1' I_0$		$Y_1 = (I_2 + I_1 + I_0')'$
$Y_2 = I_2' I_1 I_0'$		$Y_2 = (I_2 + I_1' + I_0)'$
$Y_3 = I_2' I_1 I_0$	Using DeMorgan's Theorem →	$Y_3 = (I_2 + I_1' + I_0')'$
$Y_4 = I_2 I_1' I_0'$		$Y_4 = (I_2' + I_1 + I_0)'$
$Y_5 = I_2 I_1' I_0$		$Y_5 = (I_2' + I_1 + I_0')'$
$Y_6 = I_2 I_1 I_0'$		$Y_6 = (I_2' + I_1' + I_0)'$
$Y_7 = I_2 I_1 I_0$		$Y_7 = (I_2' + I_1' + I_0')'$

Using the above Boolean equations the logic circuit of logic **high** 3-to-8 decoder using AND gates is drawn as shown in Fig. 2.111. It is possible to realize the logic **high** 3-to-8 decoder using only NOR gates as shown in Fig. 2.112.

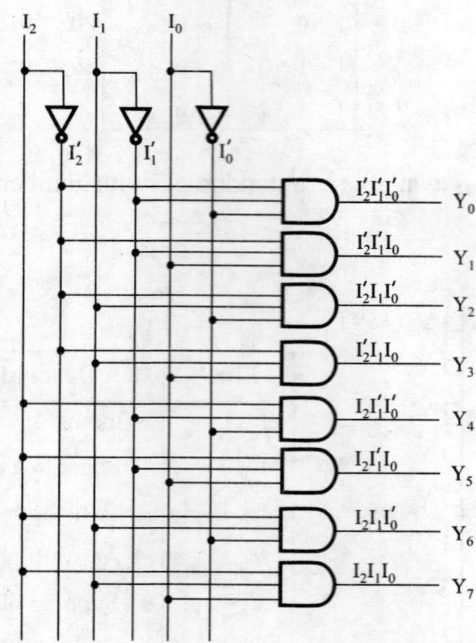

***Fig. 2.111:** AND realization of logic high of 3-to-8 decoder.*

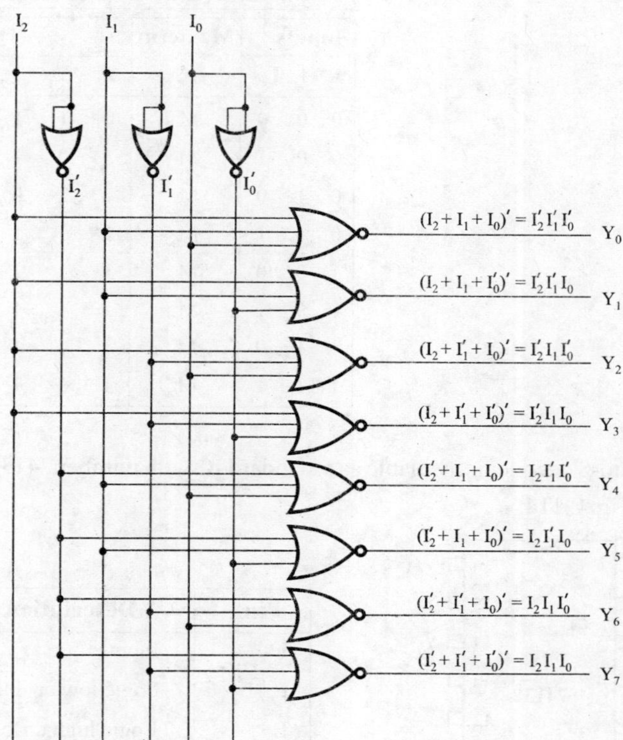

Fig. 2.112: NOR realization of logic high 3-to-8 decoder.

2.9.8 Logic Low 3-to-8 Decoder

The ***logic low 3-to-8 decoder*** will generate 8 unique outputs in which only one will be **low** at any one time and all other outputs will be **high**.

The block diagram representation of logic **low** 3-to-8 decoder is shown in Fig. 2.113. The truth table of logic **low** 3-to-8 decoder is shown in Table 2.41.

From the truth table we can directly write Boolean equations for outputs of logic **low** 3-to-8 decoder which are the maxterms for which output is **low** as shown ahead.

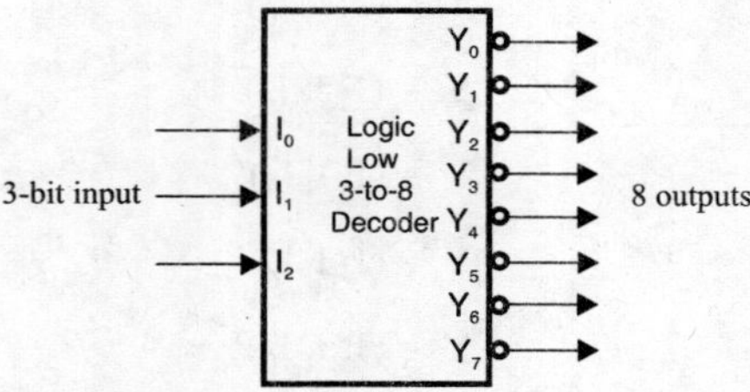

Fig. 2.113: Logic low 3-to-8 decoder.

$Y_0 = I_2 + I_1 + I_0$

$Y_1 = I_2 + I_1 + I_0'$

$Y_2 = I_2 + I_1' + I_0$

$Y_3 = I_2 + I_1' + I_0'$

$Y_4 = I_2' + I_1 + I_0$

$Y_5 = I_2' + I_1 + I_0'$

$Y_6 = I_2' + I_1' + I_0$

$Y_7 = I_2' + I_1' + I_0'$

Table 2.41: Truth Table of Logic Low 3-to-8 Decoder

Inputs			**Maxterm**	**Outputs**							
I_2	I_1	I_0		Y_0	Y_1	Y_2	Y_3	Y_4	Y_5	Y_6	Y_7
0	0	0	M_0	0	1	1	1	1	1	1	1
0	0	1	M_1	1	0	1	1	1	1	1	1
0	1	0	M_2	1	1	0	1	1	1	1	1
0	1	1	M_3	1	1	1	0	1	1	1	1
1	0	0	M_4	1	1	1	1	0	1	1	1
1	0	1	M_5	1	1	1	1	1	0	1	1
1	1	0	M_6	1	1	1	1	1	1	0	1
1	1	1	M_7	1	1	1	1	1	1	1	0

The logic **low** 3-to-8 decoder is available as a standard IC with number 74138. The pin configuration of 74138 is shown in Fig. 2.114.

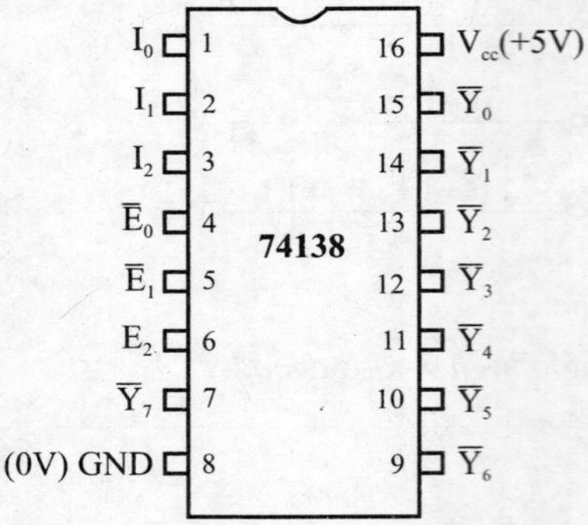

74138

Left pin	No.	No.	Right pin
I_0	1	16	V_{cc}(+5V)
I_1	2	15	$\overline{Y}_0$
I_2	3	14	$\overline{Y}_1$
$\overline{E}_0$	4	13	$\overline{Y}_2$
$\overline{E}_1$	5	12	$\overline{Y}_3$
E_2	6	11	$\overline{Y}_4$
$\overline{Y}_7$	7	10	$\overline{Y}_5$
(0V) GND	8	9	$\overline{Y}_6$

Pin	**Description**
I_0 - I_2	Inputs
$\overline{E}_0$, $\overline{E}_1$	Logic low enables
E_2	Logic high enable
$\overline{Y}_0$ - $\overline{Y}_7$	Active low outputs
V_{cc}	Supply voltage (+5V)
GND	Ground (0V)

Fig. 2.114: *Pin configuration of logic low 3-to-8 decoder IC 74138.*

The sum terms in the Boolean equations of 3-to-8 logic **low** decoder can be converted to product terms as shown below:

(Used for OR Realization)	Using DeMorgan's Theorem	(Used for NAND Realization)
$Y_0 = I_2 + I_1 + I_0$	→	$Y_0 = (I_2' I_1' I_0')'$
$Y_1 = I_2 + I_1 + I_0'$	→	$Y_1 = (I_2' I_1' I_0)'$
$Y_2 = I_2 + I_1' + I_0$	→	$Y_2 = (I_2' I_1 I_0')'$
$Y_3 = I_2 + I_1' + I_0'$	→	$Y_3 = (I_2' I_1 I_0)'$
$Y_4 = I_2' + I_1 + I_0$	→	$Y_4 = (I_2 I_1' I_0')'$
$Y_5 = I_2' + I_1 + I_0'$	→	$Y_5 = (I_2 I_1' I_0)'$
$Y_6 = I_2' + I_1' + I_0$	→	$Y_6 = (I_2 I_1 I_0')'$
$Y_7 = I_2 + I_1 + I_0$	→	$Y_7 = (I_2 I_1 I_0)'$

Using the above Boolean equations the logic circuit of logic **low** 3-to-8 decoder using OR gates is drawn as shown in Fig. 2.115. It is possible to realize the logic **low** 3-to-8 decoder using only NAND gates as shown in Fig. 2.116.

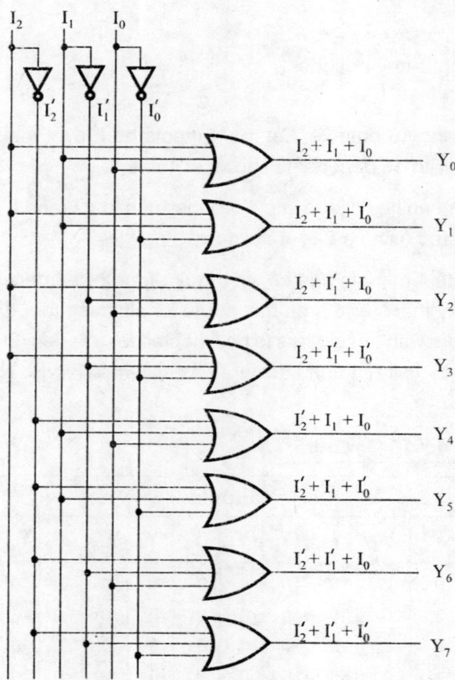

***Fig. 2.115:** OR realization of logic low 3-to-8 decoder.*

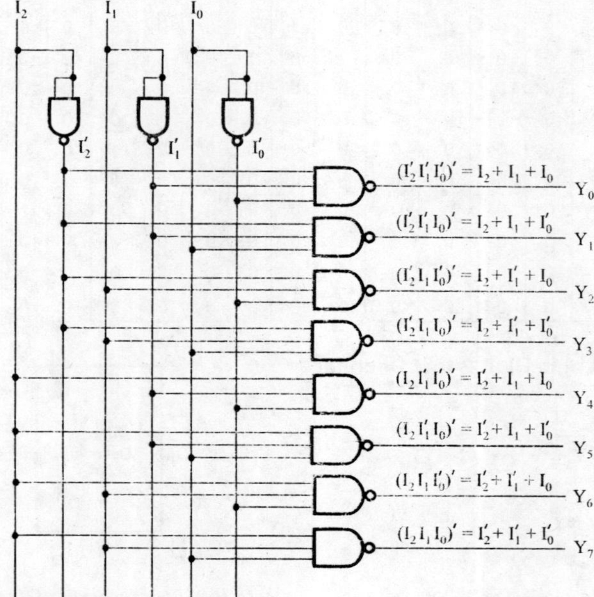

***Fig. 2.116:** NAND realization of logic low 3-to-8 decoder.*

2.9.9 Expanding the Decoder

Decoder of smaller size can be connected in cascade to expand the size of decoder. Three examples of decoder expansion are presented in Examples 2.14, 2.15 and 2.16.

Example 2.14

Construct a 4-to-16 decoder using 2-to-4 decoder.

Solution

A 4-to-16 decoder has 4 inputs and 16 outputs. Let the 4 inputs be I_0, I_1, I_2 and I_3 and the 16 outputs be Y_0, Y_1, ..., Y_{15}. The truth table of logic **high** 4-to-16 decoder is shown in Table 1.

The truth table of 4-to-16 decoder can be divided into 4 groups such that I_2 and I_3 are same in each group. Now, the outputs in each group can be generated using a 2-to-4 decoder.

Each 2-to-4 decoder can generate 4 outputs and so we require four 2-to-4 decoder to generate 16 outputs as shown in Fig. 2. From Table 1 we can say that I_0 and I_1 be the inputs for all these four decoders but only one decoder should be active at a time. Hence decoders with enable has to be selected for this design and another 2-to-4 decoder is required to generate four enable signals by using I_2 and I_3 as input as shown in Fig. 1. The working of 4-to-16 decoder using 2-to-4 decoder is verified in Table 3.

Table 1: Truth Table of Logic High 4-to-16 Decoder

Inputs				Outputs															
I_3	I_2	I_1	I_0	Y_0	Y_1	Y_2	Y_3	Y_4	Y_5	Y_6	Y_7	Y_8	Y_9	Y_{10}	Y_{11}	Y_{12}	Y_{13}	Y_{14}	Y_{15}
0	0	0	0	1	0	0	0	0	0	0	0	0	0	0	0	0	0	0	0
0	0	0	1	0	1	0	0	0	0	0	0	0	0	0	0	0	0	0	0
0	0	1	0	0	0	1	0	0	0	0	0	0	0	0	0	0	0	0	0
0	0	1	1	0	0	0	1	0	0	0	0	0	0	0	0	0	0	0	0
0	1	0	0	0	0	0	0	1	0	0	0	0	0	0	0	0	0	0	0
0	1	0	1	0	0	0	0	0	1	0	0	0	0	0	0	0	0	0	0
0	1	1	0	0	0	0	0	0	0	1	0	0	0	0	0	0	0	0	0
0	1	1	1	0	0	0	0	0	0	0	1	0	0	0	0	0	0	0	0
1	0	0	0	0	0	0	0	0	0	0	0	1	0	0	0	0	0	0	0
1	0	0	1	0	0	0	0	0	0	0	0	0	1	0	0	0	0	0	0
1	0	1	0	0	0	0	0	0	0	0	0	0	0	1	0	0	0	0	0
1	0	1	1	0	0	0	0	0	0	0	0	0	0	0	1	0	0	0	0
1	1	0	0	0	0	0	0	0	0	0	0	0	0	0	0	1	0	0	0
1	1	0	1	0	0	0	0	0	0	0	0	0	0	0	0	0	1	0	0
1	1	1	0	0	0	0	0	0	0	0	0	0	0	0	0	0	0	1	0
1	1	1	1	0	0	0	0	0	0	0	0	0	0	0	0	0	0	0	1

Table 2: Truth Table of Logic High 2-to-4 Decoder

Inputs		Outputs			
I_1	I_0	Y_0	Y_1	Y_2	Y_3
0	0	1	0	0	0
0	1	0	1	0	0
1	0	0	0	1	0
1	1	0	0	0	1

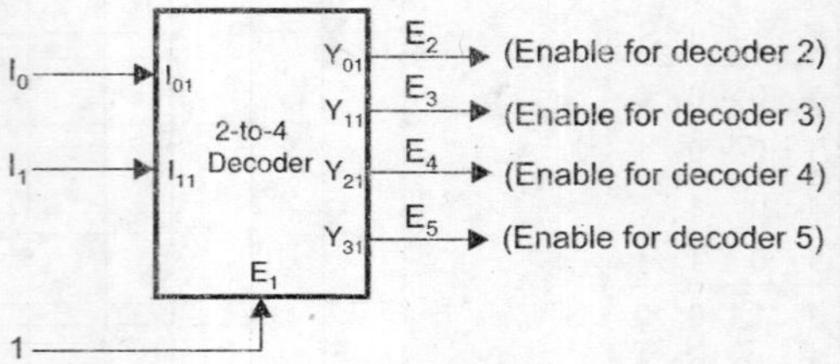

Fig. 1: Generation of enable signal.

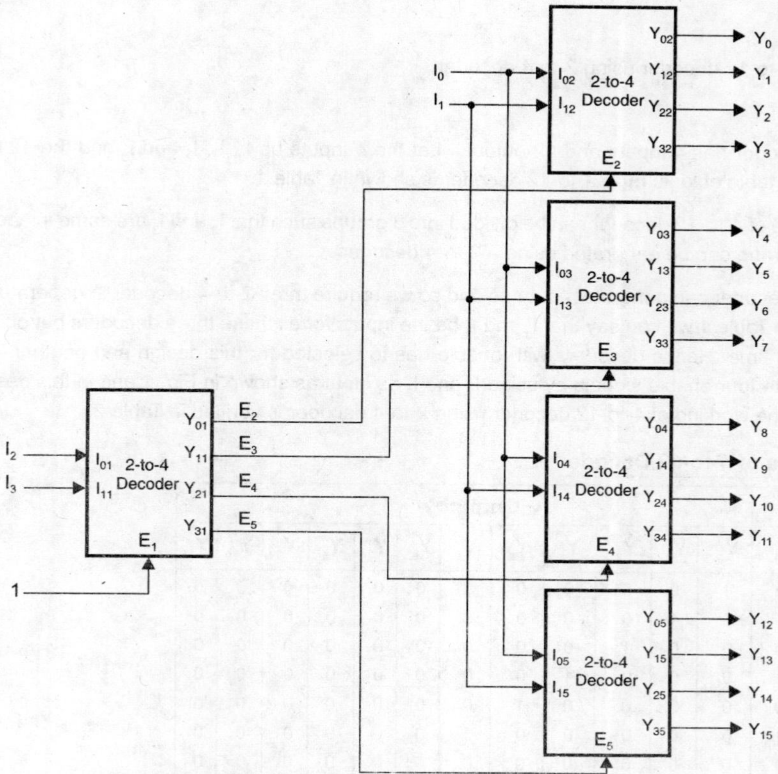

Fig. 2: 4-to-16 decoder using 2-to-4 decoder.

Table 3: Verification of 4-to-16 Decoder Shown in Fig. 2

Inputs				Enables				Outputs															
I_3	I_2	I_1	I_0	$E_2 = Y_{01}$	$E_3 = Y_{11}$	$E_4 = Y_{21}$	$E_5 = Y_{31}$	$Y_0 = Y_{02}$	$Y_1 = Y_{12}$	$Y_2 = Y_{22}$	$Y_3 = Y_{32}$	$Y_4 = Y_{03}$	$Y_5 = Y_{13}$	$Y_6 = Y_{23}$	$Y_7 = Y_{33}$	$Y_8 = Y_{04}$	$Y_9 = Y_{14}$	$Y_{10} = Y_{24}$	$Y_{11} = Y_{34}$	$Y_{12} = Y_{05}$	$Y_{13} = Y_{15}$	$Y_{14} = Y_{25}$	$Y_{15} = Y_{35}$
0	0	0	0	1	0	0	0	1	0	0	0	0	0	0	0	0	0	0	0	0	0	0	0
0	0	0	1	1	0	0	0	0	1	0	0	0	0	0	0	0	0	0	0	0	0	0	0
0	0	1	0	1	0	0	0	0	0	1	0	0	0	0	0	0	0	0	0	0	0	0	0
0	0	1	1	1	0	0	0	0	0	0	1	0	0	0	0	0	0	0	0	0	0	0	0
0	1	0	0	0	1	0	0	0	0	0	0	1	0	0	0	0	0	0	0	0	0	0	0
0	1	0	1	0	1	0	0	0	0	0	0	0	1	0	0	0	0	0	0	0	0	0	0
0	1	1	0	0	1	0	0	0	0	0	0	0	0	1	0	0	0	0	0	0	0	0	0
0	1	1	1	0	1	0	0	0	0	0	0	0	0	0	1	0	0	0	0	0	0	0	0
1	0	0	0	0	0	1	0	0	0	0	0	0	0	0	0	1	0	0	0	0	0	0	0
1	0	0	1	0	0	1	0	0	0	0	0	0	0	0	0	0	1	0	0	0	0	0	0
1	0	1	0	0	0	1	0	0	0	0	0	0	0	0	0	0	0	1	0	0	0	0	0
1	0	1	1	0	0	1	0	0	0	0	0	0	0	0	0	0	0	0	1	0	0	0	0
1	1	0	0	0	0	0	1	0	0	0	0	0	0	0	0	0	0	0	0	1	0	0	0
1	1	0	1	0	0	0	1	0	0	0	0	0	0	0	0	0	0	0	0	0	1	0	0
1	1	1	0	0	0	0	1	0	0	0	0	0	0	0	0	0	0	0	0	0	0	1	0
1	1	1	1	0	0	0	1	0	0	0	0	0	0	0	0	0	0	0	0	0	0	0	1

Example 2.15

Construct a 4-to-12 decoder using 2-to-4 decoder.

Solution

A 4-to-12 decoder has 4 inputs and 12 outputs. Let the 4 inputs be I_0, I_1, I_2 and I_3 and the 12 outputs be Y_0, Y_1, ..., Y_{11}. The truth table of logic **high** 4-to-12 decoder is shown in Table 1.

The truth table of 4-to-12 decoder can be divided into 3 groups such that I_2 and I_3 are same in each group. Now, the outputs in each group can be generated using a 2-to-4 decoder.

Each 2-to-4 decoder can generate 4 outputs and so we require three 2-to-4 decoder to generate 12 outputs as shown in Fig. 2. From Table 1 we can say that I_0 and I_1 be the inputs for all these three decoders but only one decoder should be active at a time. Hence decoders with enable has to selected for this design and another 2-to-4 decoder is required to generate four enable signals by using I_2 and I_3 as input as shown in Fig. 1 and in this design one of the enable is not used. The working of 4-to-12 decoder using 2-to-4 decoder is verified in Table 2.

Table 1: Truth Table of 4-to-12 Decoder

Inputs				Outputs											
I_3	I_2	I_1	I_0	Y_0	Y_1	Y_2	Y_3	Y_4	Y_5	Y_6	Y_7	Y_8	Y_9	Y_{10}	Y_{11}
0	0	0	0	1	0	0	0	0	0	0	0	0	0	0	0
0	0	0	1	0	1	0	0	0	0	0	0	0	0	0	0
0	0	1	0	0	0	1	0	0	0	0	0	0	0	0	0
0	0	1	1	0	0	0	1	0	0	0	0	0	0	0	0
0	1	0	0	0	0	0	0	1	0	0	0	0	0	0	0
0	1	0	1	0	0	0	0	0	1	0	0	0	0	0	0
0	1	1	0	0	0	0	0	0	0	1	0	0	0	0	0
0	1	1	1	0	0	0	0	0	0	0	1	0	0	0	0
1	0	0	0	0	0	0	0	0	0	0	0	1	0	0	0
1	0	0	1	0	0	0	0	0	0	0	0	0	1	0	0
1	0	1	0	0	0	0	0	0	0	0	0	0	0	1	0
1	0	1	1	0	0	0	0	0	0	0	0	0	0	0	1

Table 2: Verification of 4-to-12 Decoder Shown in Fig. 1

Inputs				Enables			Outputs											
I_3	I_2	I_1	I_0	$E_2 = Y_{01}$	$E_3 = Y_{11}$	$E_4 = Y_{21}$	$Y_0 = Y_{02}$	$Y_1 = Y_{12}$	$Y_2 = Y_{22}$	$Y_3 = Y_{32}$	$Y_4 = Y_{03}$	$Y_5 = Y_{13}$	$Y_6 = Y_{23}$	$Y_7 = Y_{33}$	$Y_8 = Y_{04}$	$Y_9 = Y_{14}$	$Y_{10} = Y_{24}$	$Y_{11} = Y_{34}$
0	0	0	0	1	0	0	1	0	0	0	0	0	0	0	0	0	0	0
0	0	0	1	1	0	0	0	1	0	0	0	0	0	0	0	0	0	0
0	0	1	0	1	0	0	0	0	1	0	0	0	0	0	0	0	0	0
0	0	1	1	1	0	0	0	0	0	1	0	0	0	0	0	0	0	0
0	1	0	0	0	1	0	0	0	0	0	1	0	0	0	0	0	0	0
0	1	0	1	0	1	0	0	0	0	0	0	1	0	0	0	0	0	0
0	1	1	0	0	1	0	0	0	0	0	0	0	1	0	0	0	0	0
0	1	1	1	0	1	0	0	0	0	0	0	0	0	1	0	0	0	0
1	0	0	0	0	0	1	0	0	0	0	0	0	0	0	1	0	0	0
1	0	0	1	0	0	1	0	0	0	0	0	0	0	0	0	1	0	0
1	0	1	0	0	0	1	0	0	0	0	0	0	0	0	0	0	1	0
1	0	1	1	0	0	1	0	0	0	0	0	0	0	0	0	0	0	1

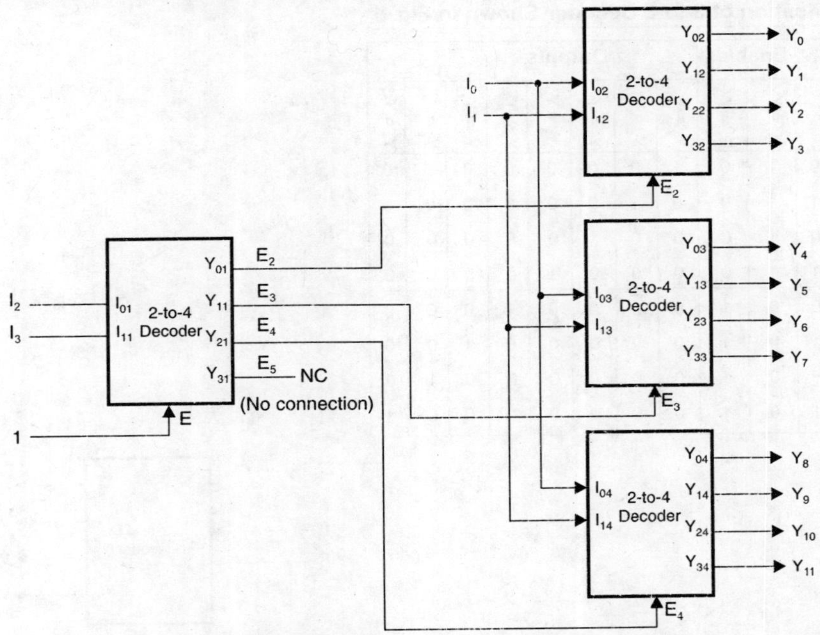

Fig. 1: 4-to-12 decoder using 2-to-4 decoder.

Example 2.16

(AU, Apr/May'19, 7 Marks)

Design 3-to-8 decoder using 2-to-4 decoders. Draw the truth table.

Solution

A 3-to-8 decoder has 3 inputs and 8 outputs. Let the 3 inputs be I_0, I_1 and I_2 and the 8 outputs be Y_0, Y_1, ..., Y_7. The truth table of logic **high** 3-to-8 decoder is shown in Table 1. The truth table of 3-to-8 decoder can be divided into 2 groups such that I_0 and I_1 are same in each group. Now, the outputs in each group can be generated using a 2-to-4 decoder.

Each 2-to-4 decoder can generate 4 outputs and so we require two 2-to-4 decoder to generate 8 outputs as shown in Fig. 1. From Table 1 we can say that I_0 and I_1 be the inputs for all these two decoders but only one decoder should be active at a time. Hence decoders with enable has to be selected for this design and another 2-to-4 decoder is required to generate four enable signals by using I_2 and 0 as input as shown in Fig. 1 and in this two enables are not used. The working of 3-to-8 decoder using 2-to-4 decoder is verified in Table 2.

Table 1: Truth Table of 3-to-8 Decoder

Inputs			Outputs							
I_2	I_1	I_0	Y_0	Y_1	Y_2	Y_3	Y_4	Y_5	Y_6	Y_7
0	0	0	1	0	0	0	0	0	0	0
0	0	1	0	1	0	0	0	0	0	0
0	1	0	0	0	1	0	0	0	0	0
0	1	1	0	0	0	1	0	0	0	0
1	0	0	0	0	0	0	1	0	0	0
1	0	1	0	0	0	0	0	1	0	0
1	1	0	0	0	0	0	0	0	1	0
1	1	1	0	0	0	0	0	0	0	1

Table 2: Verification of 3-to-8 Decoder Shown in Fig. 1

Inputs			Enables		Outputs							
I_2	I_1	I_0	$E_2 = Y_{01}$	$E_3 = Y_{11}$	$Y_0 = Y_{02}$	$Y_1 = Y_{12}$	$Y_2 = Y_{22}$	$Y_3 = Y_{32}$	$Y_4 = Y_{03}$	$Y_5 = Y_{13}$	$Y_6 = Y_{23}$	$Y_7 = Y_{33}$
0	0	0	1	0	1	0	0	0	0	0	0	0
0	0	1	1	0	0	1	0	0	0	0	0	0
0	1	0	1	0	0	0	1	0	0	0	0	0
0	1	1	1	0	0	0	0	1	0	0	0	0
1	0	0	0	1	0	0	0	0	1	0	0	0
1	0	1	0	1	0	0	0	0	0	1	0	0
1	1	0	0	1	0	0	0	0	0	0	1	0
1	1	1	0	1	0	0	0	0	0	0	0	1

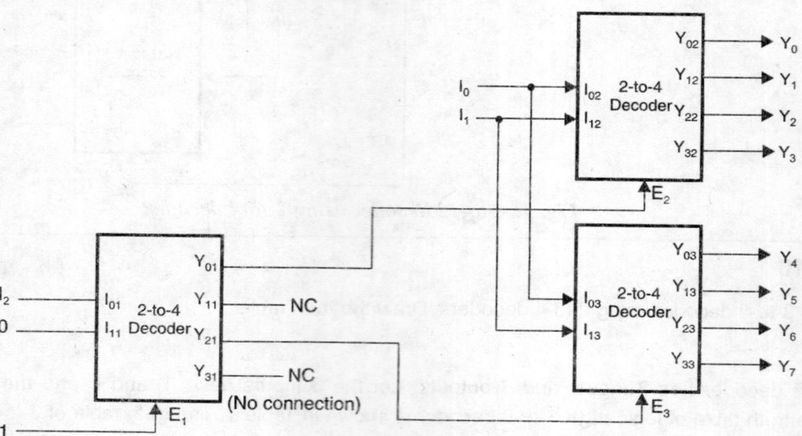

Fig. 1: 3-to-8 decoder using 2-to-4 decoder.

2.9.10 Combinational Circuit Design using Decoder

The logic **high** decoder generate output 1 for only one particular minterm as input. Therefore, a Boolean function expressed as a sum of minterms can be obtained by logical OR of the outputs of decoder for which the minterms of function generates logic **high** or 1.

The complement of logic **low** decoder output will be same as output of logic **high** decoder. Hence, the complement of output of logic **low** decoder can be used to realize SOP form of Boolean function.

Similarly, the logic **low** decoder generates output 0 for only one particular maxterm as input. Therefore, a Boolean function expressed as a product of maxterms can be obtained by logical AND of the outputs of decoder for which the maxterms of function generates logic **low** or 0.

The complement of logic **high** decoder output will be same as output of logic **low** decoder. Hence the complement of output of logic **high** decoder can be used to realize POS form of Boolean function.

Example 2.17

Realize the following functions,

$F_1 = \Sigma\, m(0, 1, 3, 6)$

$F_2 = \Pi\, M(0, 2, 4, 7)$

(a) Using logic **high** 3-to-8 decoder and external OR gates only

(b) Using logic **high** 3-to-8 decoder and external NOR gates only.

Solution

Given that, $F_1 = \Sigma\, m(0, 1, 3, 6)$

$F_2 = \Pi\, M(0, 2, 4, 7) = \Sigma\, m(1, 3, 5, 6)$

Since the given functions has minterms m_0 to m_7, the given functions are 3-variable functions and so can be realized using a 3-to-8 decoder. The 3-to-8 decoder has three inputs and the minterms are formed from all possible combination of inputs as shown in Table 1. Here, x, y and z are input variables of the functions.

The truth table of F_1 and F_2 is shown in Table 2.

Table 1: Truth Table of Logic High 3-to-8 Decoder

Inputs				Outputs							
x	y	z	Minterm	Y_0	Y_1	Y_2	Y_3	Y_4	Y_5	Y_6	Y_7
0	0	0	m_0	1	0	0	0	0	0	0	0
0	0	1	m_1	0	1	0	0	0	0	0	0
0	1	0	m_2	0	0	1	0	0	0	0	0
0	1	1	m_3	0	0	0	1	0	0	0	0
1	0	0	m_4	0	0	0	0	1	0	0	0
1	0	1	m_5	0	0	0	0	0	1	0	0
1	1	0	m_6	0	0	0	0	0	0	1	0
1	1	1	m_7	0	0	0	0	0	0	0	1

Table 2: Truth Table of F_1 and F_2

Minterm	F_1	F_2
m_0 = 0 0 0	1	0
m_1 = 0 0 1	1	1
m_2 = 0 1 0	0	0
m_3 = 0 1 1	1	1
m_4 = 1 0 0	0	0
m_5 = 1 0 1	0	1
m_6 = 1 1 0	1	1
m_7 = 1 1 1	0	0

(a) Using Logic High 3-to-8 Decoder and External OR Gates Only

In function F_1, the output is "1" for minterms 0, 1, 3 and 6. In logic **high** decoder the output Y_n is "1" when the input corresponds to minterm m_n and so the outputs Y_0, Y_1, Y_3 and Y_6 are logically ORed for OR implementation.

Similarly, for function, F_2 the decoder outputs Y_1, Y_3, Y_5 and Y_6 are logically ORed for OR implementation.

The Boolean equations of functions F_1 and F_2 in terms of decoder output are given below:

$$F_1 = Y_0 + Y_1 + Y_3 + Y_6 \quad .....(1)$$

$$F_2 = Y_1 + Y_3 + Y_5 + Y_6 \quad .....(2)$$

The equations (1) and (2) are used to implement functions F_1 and F_2 using logic **high** 3-to-8 decoder and external OR gates as shown in Fig. 1.

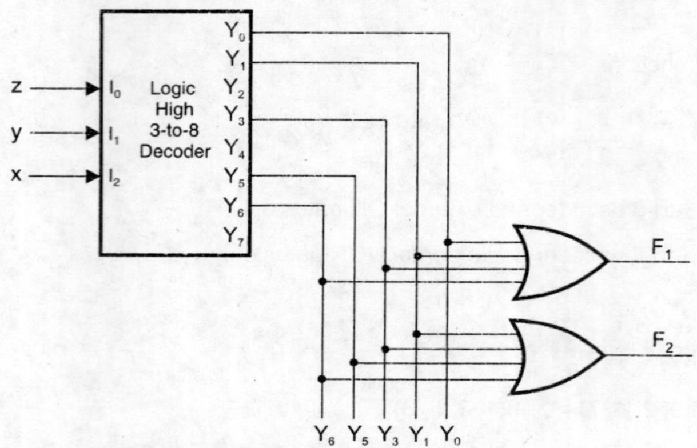

Fig. 1: OR implementation of F_1 and F_2.

(b) Using Logic High 3-to-8 Decoder and External NOR Gates Only

The equations (1) and (2) can be modified for NOR implementation as shown below:

$$\therefore F_1 = Y_0 + Y_1 + Y_3 + Y_6$$

$$= \left((Y_0 + Y_1 + Y_3 + Y_6)'\right)' = (F_1')' \quad .....(3)$$

$$F_2 = Y_1 + Y_3 + Y_5 + Y_6$$

$$= \left((Y_1 + Y_3 + Y_5 + Y_6)'\right)' = (F_1')' \quad .....(4)$$

The equations (3) and (4) are used to implement functions F_1 and F_2 using logic **high** 3-to-8 decoder and external NOR gates as shown in Fig. 2.

In NOR implementation the inverters are constructed using NOR gates and the OR gate is constructed using NOR gate followed by an inverter as shown in Fig. 2.

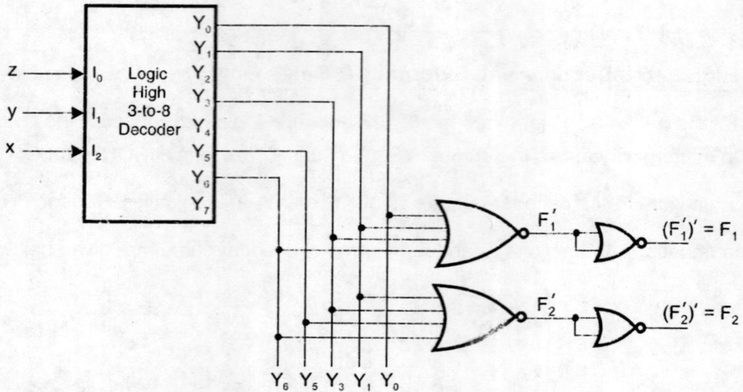

Fig. 2: NOR implementation of F_1 and F_2.

Example 2.18

Using decoder and external gates, design the combination circuit defined by the following Boolean functions:

i) $F_1 = (y + x')z$

ii) $F_2 = yz + x'y + y'z$

iii) $F_3 = (x + y')z$

Solution

The truth table of Boolean functions F_1, F_2 and F_3 are shown in Table 1. Here, x, y and z are input variables of the functions. From the truth table the following Boolean equations in SOP form are obtained. The function is given by sum of minterms for which output is 1.

$$F_1 = m_1 + m_3 + m_7$$

$$F_2 = m_1 + m_2 + m_3 + m_5 + m_7$$

$$F_3 = m_1 + m_5 + m_7$$

Table 1: Truth Table for F_1, F_2 and F_3

x y z	Minterm	Maxterm	x' y' z'	y + x'	F_1	yz	x'y	y'z	F_2	x + y'	F_3
0 0 0	m_0	M_0	1 1 1	1	0	0	0	0	0	1	0
0 0 1	m_1	M_1	1 1 0	1	1	0	0	1	1	1	1
0 1 0	m_2	M_2	1 0 1	1	0	0	1	0	1	0	0
0 1 1	m_3	M_3	1 0 0	1	1	1	1	0	1	0	0
1 0 0	m_4	M_4	0 1 1	0	0	0	0	0	0	1	0
1 0 1	m_5	M_5	0 1 0	0	0	0	0	1	1	1	1
1 1 0	m_6	M_6	0 0 1	1	0	0	0	0	0	1	0
1 1 1	m_7	M_7	0 0 0	1	1	1	0	0	1	1	1

On taking into account the minterms for which function outputs are 1 the following Boolean equations are obtained using the logic **high** decoder output.

$F_1 = Y_1 + Y_3 + Y_7$; $F_2 = Y_1 + Y_2 + Y_3 + Y_5 + Y_7$; $F_3 = Y_1 + Y_5 + Y_7$

Using the above Boolean equations the logic circuit for the given functions using logic **high** 3-to-8 decoder and OR gate is drawn as shown in Fig. 1.

From the truth table the following Boolean equations in POS form are obtained. The function is given by product of maxterms for which function output is 0.

$F_1 = M_0 M_2 M_4 M_5 M_6$; $F_2 = M_0 M_4 M_6$; $F_3 = M_0 M_2 M_3 M_4 M_6$

On taking into account the maxterms for which the functions outputs are 0, the following Boolean equations are obtained using logic **low** decoder output.

$F_1 = Y_0 Y_2 Y_4 Y_5 Y_6$; $F_2 = Y_0 Y_4 Y_6$; $F_3 = Y_0 Y_2 Y_3 Y_4 Y_6$

Using the above Boolean equations the logic circuit for the given functions using logic **low** 3-to-8 decoder and AND gate is drawn as shown in Fig. 2.

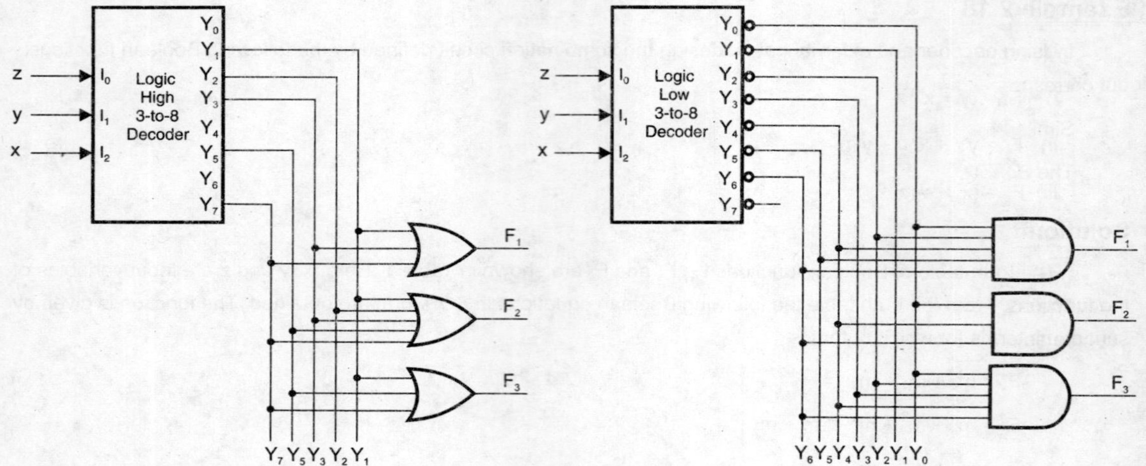

Fig. 1: Logic circuit of F_1, F_2, F_3 using logic high decoder and OR gate.

Fig. 2: Logic circuit of F_1, F_2, F_3 using logic low decoder and AND gate.

Example 2.19

Realize the following functions,

$F_1 = \Sigma\, m(0, 1, 3, 6)$

$F_2 = \Pi\, M(0, 2, 4, 5)$

(a) Using logic **low** 3-to-8 decoder and external AND gates only

(b) Using logic **low** 3-to-8 decoder and external NAND gates only.

Solution

Given that, $F_1 = \Sigma\, m(0, 1, 3, 6) = \Pi\, M(2, 4, 5, 7)$

$F_2 = \Pi\, M(0, 2, 4, 5)$

Since the given function has maxterms M_0 to M_7 the given functions are 3-variable functions and so can be realized using a 3-to-8 decoder. The 3-to-8 decoder has three inputs and the maxterms are formed from all possible combination of inputs as shown in Table 1. Here, x, y and z are input variables of the functions.

The truth table of F_1 and F_2 is shown in Table 2.

Table 1: Truth Table of Logic Low 3-to-8 Decoder

Inputs				Outputs							
x	y	z	Maxterm	Y_0	Y_1	Y_2	Y_3	Y_4	Y_5	Y_6	Y_7
0	0	0	M_0	0	1	1	1	1	1	1	1
0	0	1	M_1	1	0	1	1	1	1	1	1
0	1	0	M_2	1	1	0	1	1	1	1	1
0	1	1	M_3	1	1	1	0	1	1	1	1
1	0	0	M_4	1	1	1	1	0	1	1	1
1	0	1	M_5	1	1	1	1	1	0	1	1
1	1	0	M_6	1	1	1	1	1	1	0	1
1	1	1	M_7	1	1	1	1	1	1	1	0

Table 2: Truth Table of F_1 and F_2

Maxterm	F_1	F_2
$M_0 = 0 + 0 + 0$	1	0
$M_1 = 0 + 0 + 1'$	1	1
$M_2 = 0 + 1' + 0$	0	0
$M_3 = 0 + 1' + 1'$	1	1
$M_4 = 1' + 0 + 0$	0	0
$M_5 = 1' + 0 + 1'$	0	0
$M_6 = 1' + 1' + 0$	1	1
$M_7 = 1' + 1' + 1'$	0	1

(a) Using Logic Low 3-to-8 Decoder and External AND Gates Only

In function, F_1 the output is "0" for maxterms 2, 4, 5 and 7. In logic **low** decoder the output Y_n is "0" when the input corresponds to maxterm M_n and so the outputs Y_2, Y_4, Y_5 and Y_7 are logically ANDed in order to realize F_1.

Similarly, for function, F_2 the decoder outputs Y_0, Y_2, Y_4 and Y_5 are logically ANDed in order to realize F_2.

The Boolean equations of functions F_1 and F_2 in terms of logic **low** decoder output are given below:

$$F_1 = \prod M(2, 4, 5, 7) = Y_2\, Y_4\, Y_5\, Y_7 \qquad .....(1)$$

$$F_2 = \prod M(0, 2, 4, 5) = Y_0\, Y_2\, Y_4\, Y_5 \qquad .....(2)$$

The equations (1) and (2) are used to implement functions F_1 and F_2 using logic **low** 3-to-8 decoder and external AND gate as shown in Fig. 1.

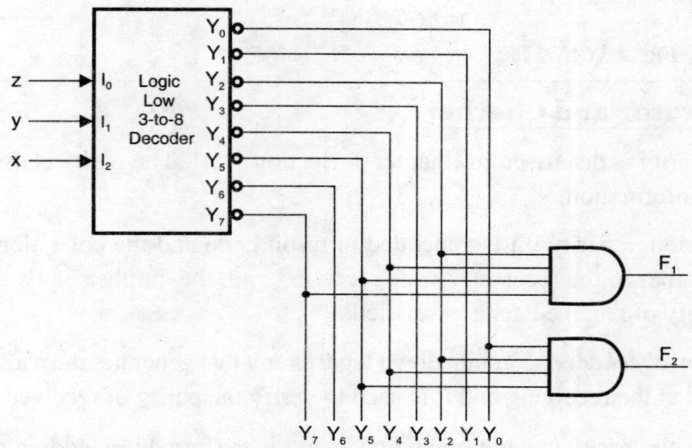

Fig. 1: AND implementation of F_1 and F_2.

(b) Using Logic Low 3-to-8 Decoder and External NAND Gates Only

The equations (1) and (2) can be modified for NAND implementation as shown below:

$$F_1 = \prod M(2, 4, 5, 7) = Y_2\, Y_4\, Y_5\, Y_7 = ((Y_2\, Y_4\, Y_5\, Y_7)')' = (F_1')' \qquad .....(3)$$

$$F_2 = \prod M(0, 2, 4, 5) = Y_0\, Y_2\, Y_4\, Y_5 = ((Y_0\, Y_2\, Y_4\, Y_5)')' = (F_2')' \qquad .....(4)$$

The equations (3) and (4) are used to implement functions F_1 and F_2 using logic **low** 3-to-8 decoder and external NAND gate as shown in Fig. 2.

In NAND implementation the inverters are constructed using NAND gates and AND gate is constructed using NAND gate followed by an inverter as shown in Fig. 2.

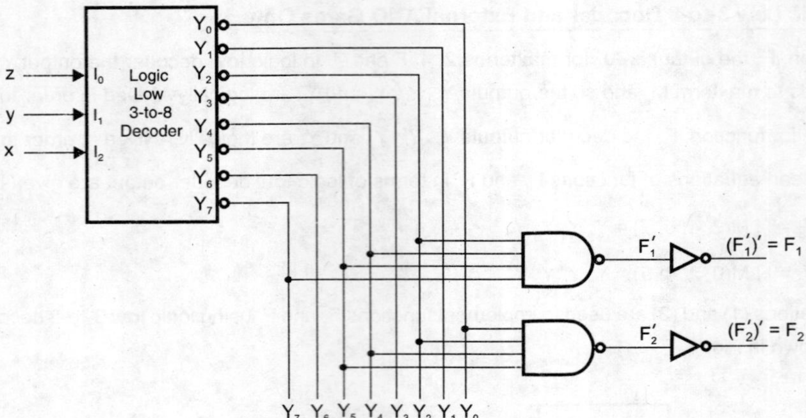

Fig. 2: NAND implementation of F_1 and F_2.

2.10 Parity Generator and Checker

The concept of parity is discussed in Chapter-1, Section 1.2.1. The parity is used to check error in transmission of digital information.

During transmission, a parity bit is appended to m-bit code and the code along with parity bit is transmitted from transmitter to the receiver. At the receiving end, the number of 1's in the code received are counted and the parity of received code is verified.

A ***parity generator/checker*** is a combinational logic circuit that generates the parity bit to be appended during transmission and at the receiving end it is used to verify the parity of received information.

For a m-bit code, the parity generation and checking is performed by adding all the m-bits of the code. In addition, the sum bit (or simply sum) is the least significant bit of the result and the other bits are called ***carry***.

When an m-bit code has even parity then sum of all the bits will be 0. When the m-bit code has odd parity then sum of all the bits will be 1.

When ***even parity*** is adopted and if the m-bit code already has even parity then parity bit is 0 and if the m-bit code already has odd parity then parity bit is 1. Therefore, in even parity system the sum of all the bits can be used as parity bit. The sum of 4-bit code is generated using three XOR gates as shown in Fig. 2.117.

When ***odd parity*** is adopted the parity bit of circuit in Fig. 2.117 is inverted and used as parity bit for odd parity system as shown in Fig. 2.118.

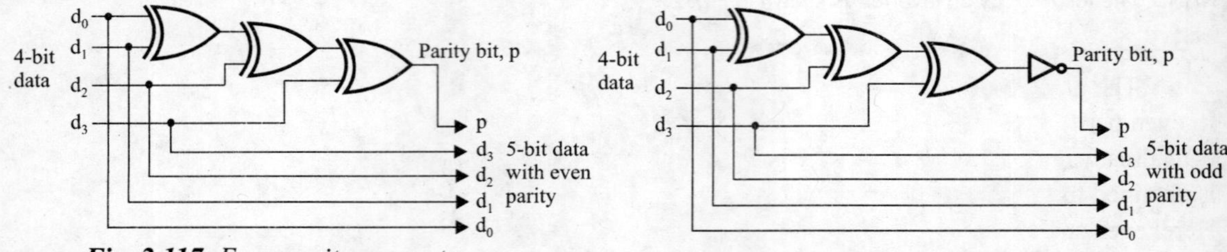

Fig. 2.117: *Even parity generator.*

Fig. 2.118: *Odd parity generator.*

On the receiver side the error can be checked by same technique as that of parity generation by adding all the bits of codes and verify the sum. When even parity is adopted the sum of received bits will be 0 and if odd parity is adopted then sum of received bits will be 1.

A parity checker for 5-bit code with parity is shown in Fig. 2.119. In this circuit 4 XOR gates are used to get sum of 5 bits. In even parity system if sum is 0, then no error and in odd parity system if sum is 1, then no error.

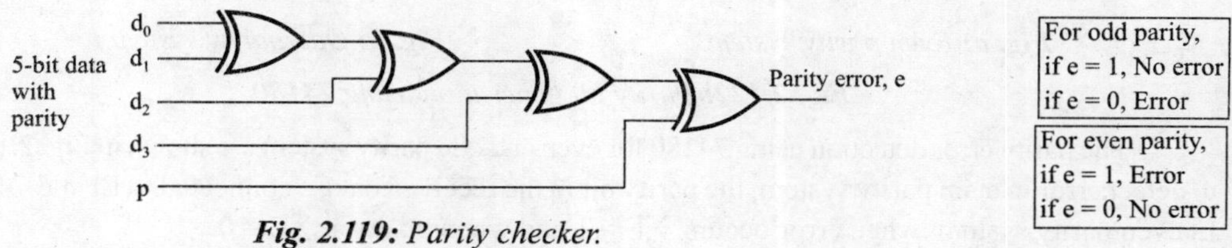

Fig. 2.119: Parity checker.

A standard IC 74180 is available to generate parity bit or verify parity of the system when the information is transmitted as 9-bit binary code consisting of 8 data bits and 1 parity bit. The function table of 74180 is shown in Table 2.42.

74180 Parity Bit Generator/ Checker

Pin	Left side	Pin	Right side
1	d_6	14	V_{cc}(+5V)
2	d_7	13	d_5
3	EI	12	d_4
4	OI	11	d_3
5	ΣE	10	d_2
6	ΣO	9	d_1
7	(0V) GND	8	d_0

Pin	Description
d_0 - d_7	8-bit data input
ΣE	Even sum output
ΣO	Odd sum output
EI	Even parity select input
OI	Odd parity select input
V_{cc}	Supply voltage (+5V)
GND	Ground (0V)

Fig. 2.120: Pin configuration of IC 74180.

Table 2.42: Function Table of 74180

Number of 1's in Data Bits d_0 - d_7	Inputs EI	Inputs OI	Outputs ΣE	Outputs ΣO	Comment
EVEN	1	0	1	0	Even parity system
ODD	1	0	0	1	
EVEN	0	1	0	1	Odd parity system
ODD	0	1	1	0	
×	1	1	0	0	Not allowed
×	0	0	1	1	

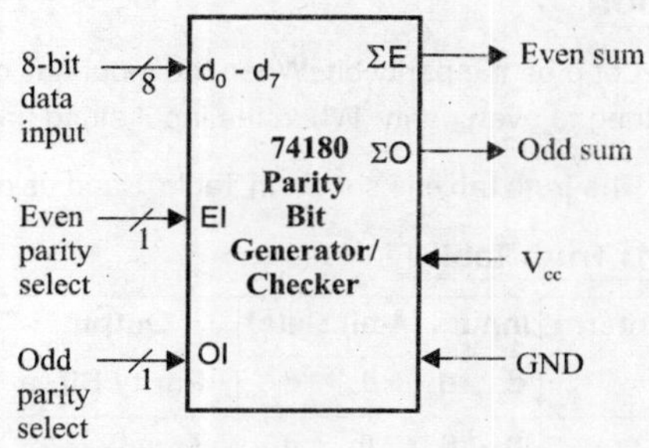

Fig. 2.121: Functional grouping of IC 74180.

The parity bit generation using IC 74180 for even and odd parity system are shown in Fig. 2.122. In even parity bit generation system, EI is tied to 1 and OI is tied to 0. Here, odd sum output is 0 if the data already has even parity and odd sum ouput is 1 if the data already has odd parity and so the parity bit is obtained from odd sum output, ΣO. In odd parity bit generation system, EI is tied to 0 and OI is tied 1. Here, even sum output is 1 if the data already has odd parity and even sum output is 0 if the data already has even parity and so the parity bit is obtained from even sum output, ΣE.

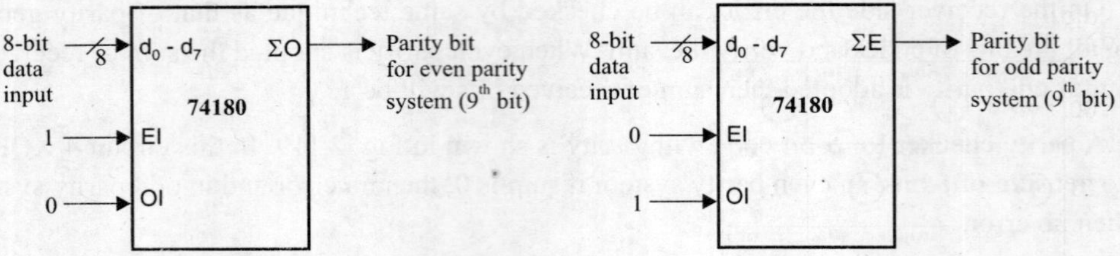

Fig. a: Even parity system. *Fig. b: Odd parity system.*

Fig. 2.122: Parity bit generation using 74180.

The parity error detection using 74180 for even and odd parity system are shown in Fig. 2.123. In order to detect error in even parity system, the parity bit in the received data is connected to EI and OI is tied to 0. In even parity system, when error occurs, $\Sigma E = 1$ and when no error, $\Sigma E = 0$.

In order to detect error in odd parity system, the parity bit in the received data is connected to OI and EI is tied to 0. In odd parity system, when error occurs, $\Sigma O = 0$ and when no error, $\Sigma O = 1$.

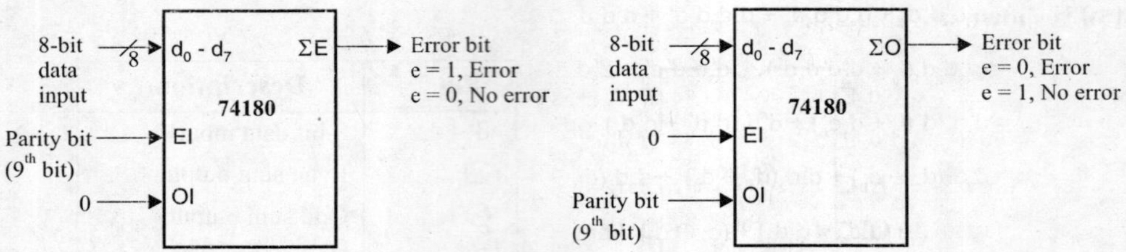

Fig. a: Even parity system. *Fig. b: Odd parity system.*

Fig. 2.123: Parity error detection using 74180.

Example 2.20

Design an even parity generator, that generates an even parity bit for every input string of 4-bits.

Solution

Let p be the parity bit. When the input has odd number of 1's then parity bit p is assigned 1 in order to convert input string to even parity. When the input string has even number of 1's then parity bit is assigned 0.

The truth table is shown in Table 1 and using truth table the K-map for p is drawn as shown in Fig. 1.

Table 1: Truth Table

Minterm	Inputs (4-bit Data)				Output	Minterm	Inputs (4-bit Data)				Output
	d_3	d_2	d_1	d_0	Parity Bit, p		d_3	d_2	d_1	d_0	Parity Bit, p
m_0	0	0	0	0	0	m_8	1	0	0	0	1
m_1	0	0	0	1	1	m_9	1	0	0	1	0
m_2	0	0	1	0	1	m_{10}	1	0	1	0	0
m_3	0	0	1	1	0	m_{11}	1	0	1	1	1
m_4	0	1	0	0	1	m_{12}	1	1	0	0	0
m_5	0	1	0	1	0	m_{13}	1	1	0	1	1
m_6	0	1	1	0	0	m_{14}	1	1	1	0	1
m_7	0	1	1	1	1	m_{15}	1	1	1	1	0

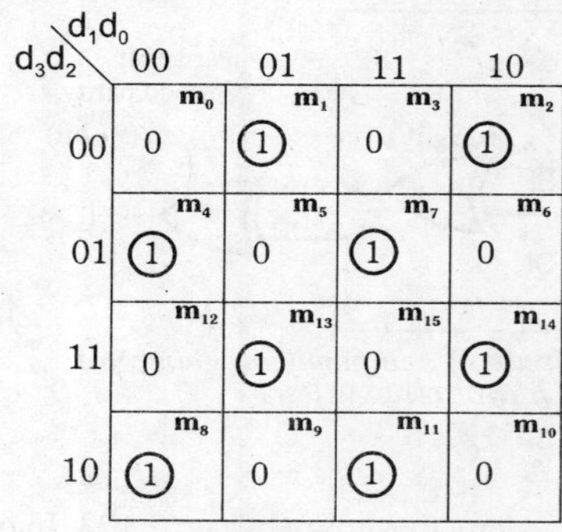

Fig. 1: K-map for p.

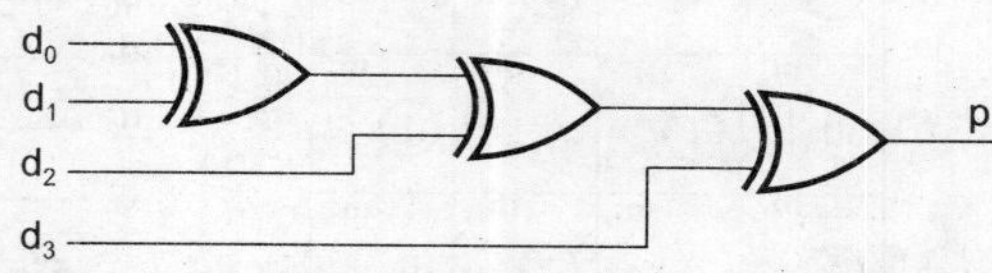

Fig. 2: Logic circuit of even parity bit generator.

All the prime implicants in the K-map are single 1's and so simplification using K-map is not possible. Hence the Boolean equation for p is given by sum of all the minterms for which output is 1.

$$\therefore p = m_1 + m_2 + m_4 + m_7 + m_8 + m_{11} + m_{13} + m_{14}$$

$$= d_3'd_2'd_1'd_0 + d_3'd_2'd_1d_0' + d_3'd_2d_1'd_0' + d_3'd_2d_1d_0 + d_3d_2'd_1'd_0' + d_3d_2'd_1d_0 + d_3d_2d_1'd_0 + d_3d_2d_1d_0'$$

$$= (d_3'd_2'd_1'd_0 + d_3'd_2'd_1d_0') + (d_3'd_2d_1'd_0' + d_3'd_2d_1d_0) + (d_3d_2d_1'd_0 + d_3d_2d_1d_0') + (d_3d_2'd_1'd_0' + d_3d_2'd_1d_0)$$

$$= d_3'd_2'(d_1'd_0 + d_1d_0') + d_3'd_2(d_1'd_0' + d_1d_0) + d_3d_2(d_1'd_0 + d_1d_0') + d_3d_2'(d_1'd_0' + d_1d_0)$$

$$= d_3'd_2'(d_1 \oplus d_0) + d_3'd_2(d_1 \oplus d_0)' + d_3d_2(d_1 \oplus d_0) + d_3d_2'(d_1 \oplus d_0)'$$

$a'b + ab' = a \oplus b$
$ab + a'b' = (a \oplus b)'$

$$= (d_1 \oplus d_0)(d_3'd_2' + d_3d_2) + (d_1 \oplus d_0)'(d_3'd_2 + d_3d_2')$$

$$= (d_1 \oplus d_0)(d_3 \oplus d_2)' + (d_1 \oplus d_0)'(d_3 \oplus d_2)$$

$$= d_3 \oplus d_2 \oplus d_1 \oplus d_0$$

Using the above simplified Boolean function the logic circuit of even parity generator is drawn as shown in Fig. 2.

Example 2.21

Design an odd parity generator that generates an odd parity bit for every input string of 4 bits.

Solution

Let p be the parity bit. When the input has odd number of 1's then parity bit p is assigned 0 in order to convert input string to odd parity. When the input string has even number of 1's then parity bit is assigned 1. The truth table is shown in Table 1 and using truth table the K-map for p is drawn as shown in Fig. 1.

Table 1: Truth Table

Minterm	Inputs (4-bit Data)				Output	Minterm	Inputs (4-bit Data)				Output
	d_3	d_2	d_1	d_0	Parity Bit, p		d_3	d_2	d_1	d_0	Parity Bit, p
m_0	0	0	0	0	1	m_8	1	0	0	0	0
m_1	0	0	0	1	0	m_9	1	0	0	1	1
m_2	0	0	1	0	0	m_{10}	1	0	1	0	1
m_3	0	0	1	1	1	m_{11}	1	0	1	1	0
m_4	0	1	0	0	0	m_{12}	1	1	0	0	1
m_5	0	1	0	1	1	m_{13}	1	1	0	1	0
m_6	0	1	1	0	1	m_{14}	1	1	1	0	0
m_7	0	1	1	1	0	m_{15}	1	1	1	1	1

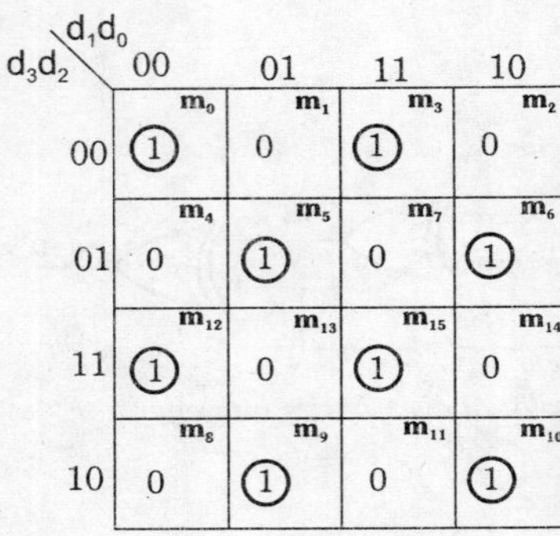

Fig. 1: K-map for F.

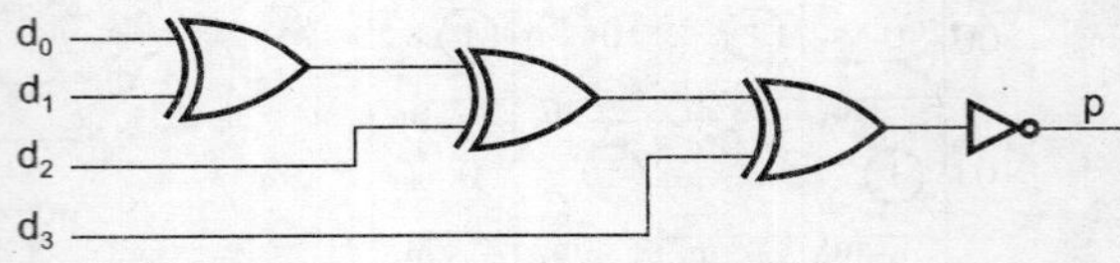

Fig. 2: Logic circuit of odd parity bit generator.

All the prime implicants in the K-map are single 1's and so simplication using K-map is not possible. Hence the Boolean equation for p is given by sum of all the minterms for which output is 1.

$$\therefore\ p = m_0 + m_3 + m_5 + m_6 + m_9 + m_{10} + m_{12} + m_{15}$$

$$= d_3'd_2'd_1'd_0' + d_3'd_2'd_1d_0 + d_3'd_2d_1'd_0 + d_3'd_2d_1d_0' + d_3d_2'd_1'd_0 + d_3d_2'd_1d_0' + d_3d_2d_1'd_0' + d_3d_2d_1d_0$$

$$= (d_3'd_2'd_1'd_0' + d_3'd_2'd_1d_0) + (d_3d_2d_1'd_0' + d_3d_2d_1d_0) + (d_3'd_2d_1'd_0 + d_3'd_2d_1d_0') + (d_3d_2'd_1'd_0 + d_3d_2'd_1d_0')$$

$$= d_3'd_2'(d_1'd_0' + d_1d_0) + d_3d_2(d_1'd_0' + d_1d_0) + d_3'd_2(d_1'd_0 + d_1d_0') + d_3d_2'(d_1'd_0 + d_1d_0')$$

$$= d_3'd_2'(d_1 \oplus d_0)' + d_3d_2(d_1 \oplus d_0)' + d_3'd_2(d_1 \oplus d_0) + d_3d_2'(d_1 \oplus d_0)$$

$$= (d_1 \oplus d_0)'(d_3'd_2' + d_3d_2) + (d_1 \oplus d_0)(d_3'd_2 + d_3d_2')$$

$$= (d_1 \oplus d_0)'(d_3 \oplus d_2)' + (d_1 \oplus d_0)(d_3 \oplus d_2)$$

$$= (d_3 \oplus d_2 \oplus d_1 \oplus d_0)'$$

$a'b + ab' = a \oplus b$
$ab + a'b' = (a \oplus b)'$

Using the above simplified Boolean function the logic circuit of odd parity generator is drawn as shown in Fig. 2

2.11 Implementation of Logic Design using Standard Logic Gate IC's

A combinational logic design is represented by a set of Boolean equations or functions. A Boolean function basically involves product of variables, sum of variables and complement of variables. Therefore the straight forward implementation of Boolean function can be made by using basic gates AND, OR and NOT. The complement of a variable is obtained using NOT gate. The product term is realized using AND gate. The sum term is realized using OR gate. The AND, OR and NOT gates that are available as standard ICs for implementing the Boolean functions are listed in Appendix-1.

Example 2.22

Implement the following functions using basic logic gates available as standard ICs.

a) $F_1 = A'B + C'$ b) $F_2 = x'y'z + x'yz + xy'z$ c) $F_3 = xy' + x'yz$

Solution

a) $F_1 = A'B + C'$

The given function can be implemented using NOT gate IC 7404, 2-input AND gate IC 7408 and 2-input OR gate IC 7432.

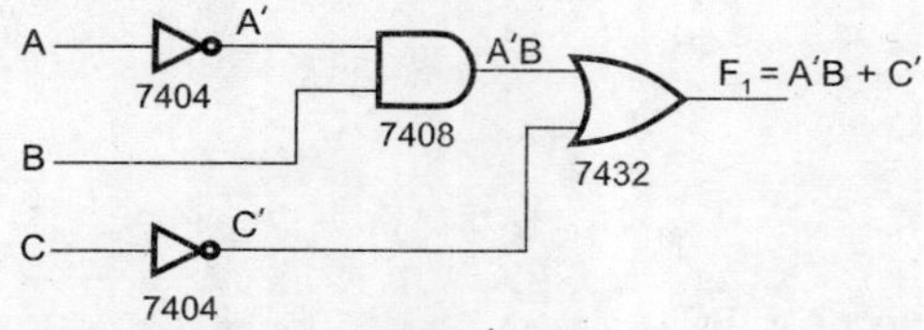

Fig. 1: Logic circuit for function, F_1.

b) $F_2 = x'y'z + x'yz + xy'z$

The given function can be implemented using NOT gate IC 7404, 3-input AND gate IC 7411 and 3-input OR gate IC 744075.

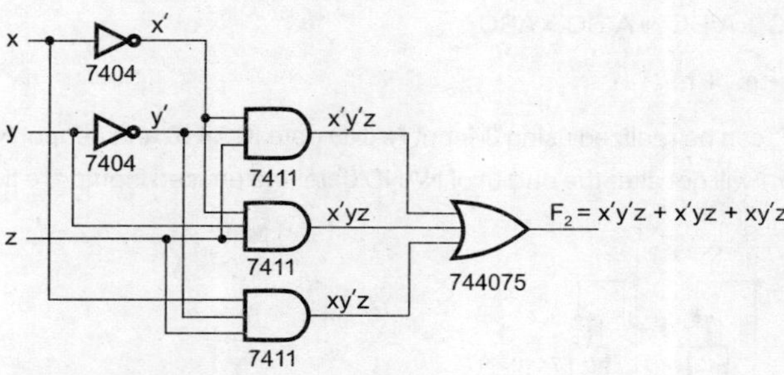

Fig. 2: Logic circuit for function, F_2.

c) $F_3 = xy' + x'yz$

The given function can be implemented using NOT gate IC 7404, 2-input AND gate IC 7408 and 2-input OR gate IC 7432.

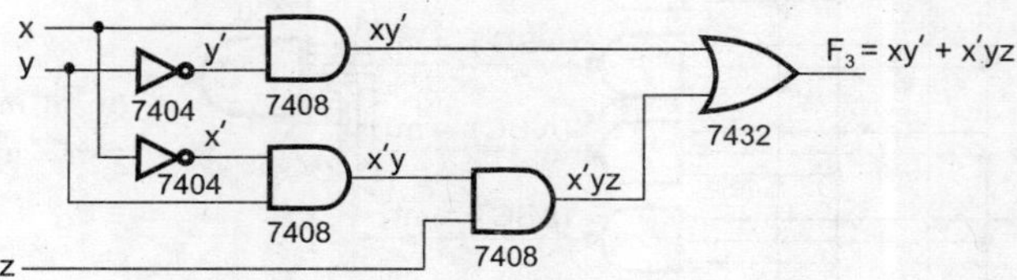

Fig. 3: Logic circuit for function, F_3.

2.12 Implementation of Logic Design using Standard Universal Gate IC's

The NAND and NOR gates are called ***universal gates***. Combinational logic design represented by SOP or POS form of Boolean functions can be implemented using universal gates. When a Boolean function is expressed in SOP form, it can be realized using only NAND gates. When a Boolean function is expressed in POS form, it can be realized using only NOR gates. The NAND and NOR gates that are available as standard ICs for implementation of SOP and POS form of Boolean functions are listed in Appendix-1.

Example 2.23

Express the following Boolean function in canonical form and implement the Boolean functions using either only NAND or NOR gates available as standard ICs.

a) $F_1 = A' + BC$

b) $F_2 = A'C + AB' + BC$

Solution

a) $F_1 = A' + BC$

Case i: SOP

$F_1 = A' + BC$

$= A' . (B + B') . (C + C') + (A + A') . BC$

$= (A'B + A'B') . (C + C') + ABC + A'BC$

$= A'BC + A'BC' + A'B'C + A'B'C' + ABC + A'BC$

$(m_3) \quad (m_2) \quad (m_1) \quad (m_0) \quad (m_7) \quad (m_3)$

$= A'B'C' + A'B'C + A'BC' + A'BC + ABC$

$= m_0 + m_1 + m_2 + m_3 + m_7$

Missing literals in first term are A and B and in second term is A.

$a + a' = 1$

$a . 1 = a$

Repeated minterms are considered only one time.

The SOP form of F_1 can be realized using 3-input NAND gate IC 7410 and 8-input NAND gate IC 7430 as shown in Fig. 1. Since NAND with 1 will not alter the output of NAND gate, the unused inputs are tied to 1 in 8-input NAND gate.

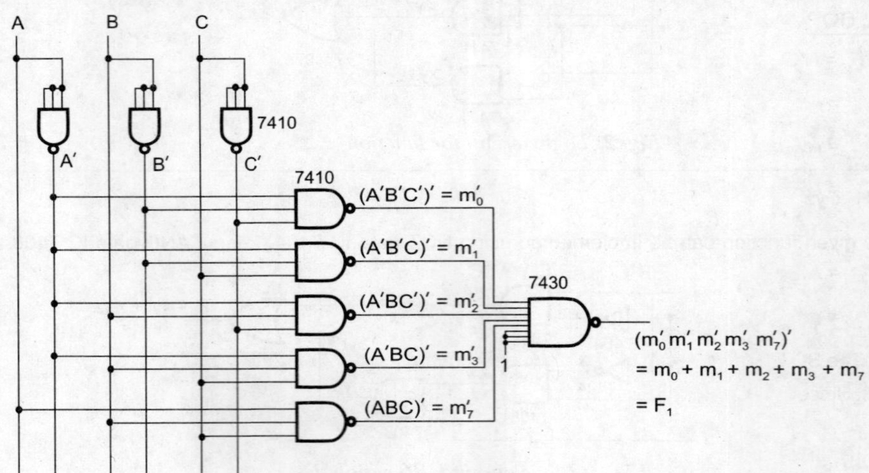

Fig. 1: Logic circuit of F_1 using only NAND gates.

Case ii: POS

$F_1 = A' + BC$

$= (A' + B) . (A' + C) = (A' + B + (C . C')) . (A' + C + (B . B'))$

$= (A' + B + C) . (A' + B + C') . (A' + C + B) . (A' + C + B')$

$= (A' + B + C) . (A' + B + C') . (A' + B + C) . (A' + B' + C)$

$(M_4) \quad (M_5) \quad (M_4) \quad (M_6)$

$= (A' + B + C) (A' + B + C') (A' + B' + C)$

$= M_4 M_5 M_6$

$a . a' = 0$

$a + 0 = a$

Missing literal in first term is C and in second term is B.

Repeated maxterms are considered only one time.

The POS form of F_1 can be realized using 3-input NOR gate IC 7427 as shown in Fig. 2.

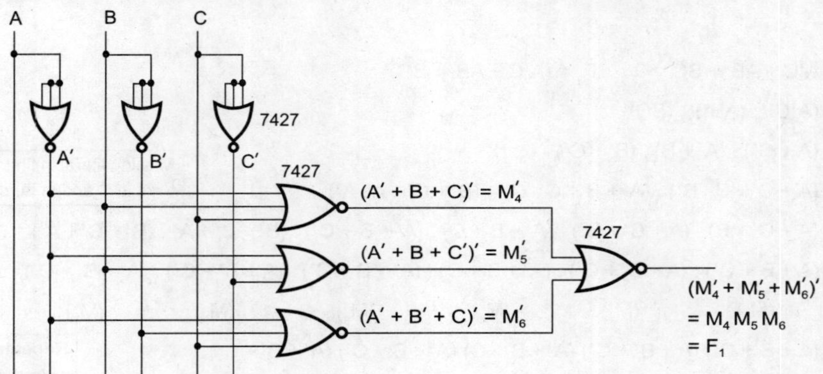

Fig. 2: Logic circuit of F_1 using only NOR gates.

b) $F_2 = A'C + AB' + BC$

Case i: SOP

$F_2 = A'C + AB' + BC$

$= A'C . (B + B') + AB' . (C + C') + BC . (A + A')$

Missing literal in first term is B, second term is C and third term is A.

$= A'CB + A'CB' + AB'C + AB'C' + BCA + BCA'$

a + a′ = 1
a . 1 = a

$= A'BC + A'B'C + AB'C + AB'C' + ABC + A'BC$

$(m_3) \quad (m_1) \quad (m_5) \quad (m_4) \quad (m_7) \quad (m_3)$

$= A'B'C + A'BC + AB'C' + AB'C + ABC$

Repeated minterms are considered only one time.

$= m_1 + m_3 + m_4 + m_5 + m_7$

The SOP form of F_2 can be realized using 3-input NAND gate IC 7410 and 8-input NAND gate IC 7430 as shown in Fig. 3. Since NAND with 1 will not alter the output of NAND gate, the unused inputs are tied to 1 in 8-input NAND gate.

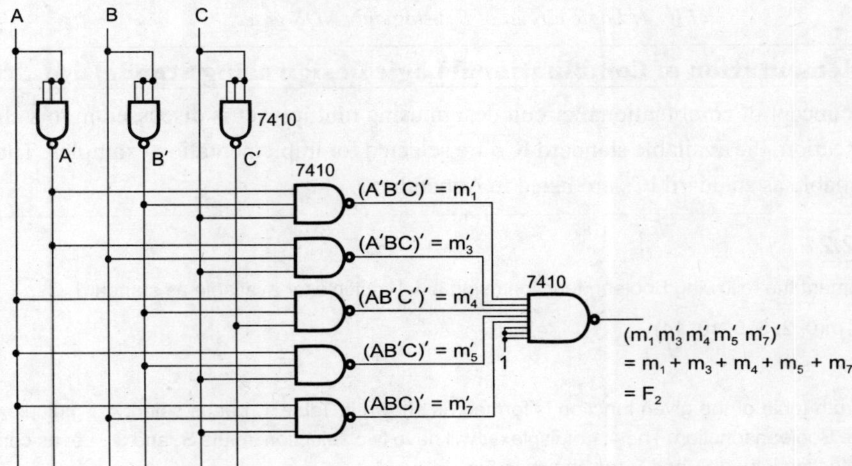

Fig. 3: Logic circuit of F_2 using only NAND gates.

Case ii: POS

$$F_2 = A'C + AB' + BC \quad \Rightarrow \quad F_2 = (A'C + AB' + BC)'$$

$$\therefore F_2 = (A'C)' . (AB')' . (BC)'$$

$$= (A + C') . (A' + B) . (B' + C')$$

$$= (A + C' + B . B') . (A' + B + C . C') . (B' + C' + A . A')$$

a . a' = 0

a + 0 = a

Missing literal in first term is B, second term is C and in third term is A.

$$= (A + C' + B) . (A + C' + B') . (A' + B + C) . (A' + B + C') . (B' + C' + A) . (B' + C' + A')$$

$$= (A + B + C') . (A + B' + C') . (A' + B + C) . (A' + B + C') . (A + B' + C') . (A' + B' + C')$$

$$(M_1) \quad (M_3) \quad (M_4) \quad (M_5) \quad (M_3) \quad (M_7)$$

$$= (A + B + C') (A + B' + C') (A' + B + C) (A' + B + C') (A' + B' + C')$$

Repeated maxterms are considered only one time.

$$= M_1 M_3 M_4 M_5 M_7$$

$$\therefore F_2 = (M_1 M_3 M_4 M_5 M_7)' = M_0 M_2 M_6$$

$$= (A + B + C) (A + B' + C) (A' + B' + C)$$

The POS form of F_2 can be realized using 3-input NOR gate IC 7427 as shown in Fig. 4.

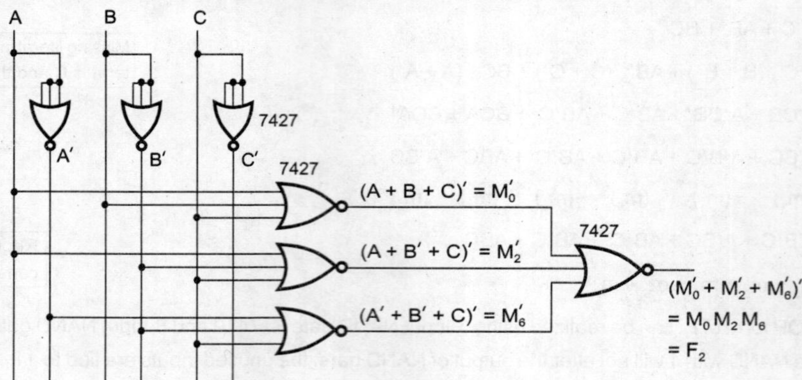

Fig. 4: Logic circuit of F_2 using only NOR gates.

2.13 Implementation of Combinational Logic Design using Standard Multiplexer IC's

The concept of combinational circuit design using multiplexer is discussed in this chapter, section 2.3. In this section, the available standard ICs are selected for implementation examples. The multiplexers that are available as standard ICs are listed in Appendix-1.

Example 2.24

Implement the following Boolean function using a 4:1 multiplexer available as standard ICs.

$$F = \sum m(0, 2, 5, 8, 10, 14)$$

Solution

The truth table of the given function is formed as shown in Table 1. Let us select 4:1 multiplexer IC 74153 to implement the Boolean function. The 4:1 multiplexer will have two selection inputs S_0 and S_1. Let us connect the inputs A and B to selection inputs S_1 and S_0 as shown in Fig. 1.

The selection inputs take four different values and so Table 1 is divided into 4 groups based on selection inputs and a separate table is drawn for each group as shown in Tables 2 to 5.

Table 1: Truth Table

A	B	C	D	Minterm	Output F	A	B	C	D	Minterm	Output F
0	0	0	0	m_0	1	1	0	0	0	m_8	1
0	0	0	1	m_1	0	1	0	0	1	m_9	0
0	0	1	0	m_2	1	1	0	1	0	m_{10}	1
0	0	1	1	m_3	0	1	0	1	1	m_{11}	0
0	1	0	0	m_4	0	1	1	0	0	m_{12}	0
0	1	0	1	m_5	1	1	1	0	1	m_{13}	0
0	1	1	0	m_6	0	1	1	1	0	m_{14}	1
0	1	1	1	m_7	0	1	1	1	1	m_{15}	0

Table 2: Truth Table When Selection Input 00

Inputs A B	C	D	Selection Inputs S_1	S_0	D′	F = D′
0 0	0	0	0	0	1	1
0 0	0	1	0	0	0	0
0 0	1	0	0	0	1	1
0 0	1	1	0	0	0	0

Table 3: Truth Table When Selection Input 01

Inputs A B	C	D	Selection Inputs S_1	S_0	C′	C′D	F = C′D
0 1	0	0	0	1	1	0	0
0 1	0	1	0	1	1	1	1
0 1	1	0	0	1	0	0	0
0 1	1	1	0	1	0	0	0

Table 4: Truth Table When Selection Input 10

Inputs A B	C	D	Selection Inputs S_1	S_0	D′	F = D′
1 0	0	0	1	0	1	1
1 0	0	1	1	0	0	0
1 0	1	0	1	0	1	1
1 0	1	1	1	0	0	0

Table 5: Truth Table When Selection Input 11

Inputs A B	C	D	Selection Inputs S_1	S_0	D′	CD′	F = CD′
1 1	0	0	1	1	1	0	0
1 1	0	1	1	1	0	0	0
1 1	1	0	1	1	1	1	1
1 1	1	1	1	1	0	0	0

Note: *The Boolean equations for F in Tables 2 to 5 can be obtained by drawing 2-variable K-maps using C and D as inputs.*

The input I_0, I_1, I_2 and I_3 of multiplexer are obtained from Tables 2 to 5. The NOT and AND gates are needed to generate inputs for I_0 and I_1 as shown in Fig. 1. The NOT gate IC 7404 and 2-input AND gate IC 7408 are used in this design.

From Tables 2 and 4 it can be observed that function output is D′ when selection inputs are 00 and 10 and so I_0 and I_2 are tied to D′ in Fig. 1. From Table 3 it can be observed that function output is logical AND of C′ and D when selection inputs are 01 and so I_1 is tied to C′D in Fig. 1. From Table 5 it can be observed that function output is logical AND of C and D′ when selection inputs are 11 and so I_3 is tied to CD′ in Fig. 1. The working of 4:1 MUX is verified in Table 6.

Fig. 1: Implementation of Boolean function using 4:1 MUX.

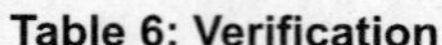

Table 6: Verification

A	B	C	D	S_1	S_0	F	Y = F
0	0	0	0	0	0	1	$Y = I_0 = D' = 1$
0	0	0	1	0	0	0	$Y = I_0 = D' = 0$
0	0	1	0	0	0	1	$Y = I_0 = D' = 1$
0	0	1	1	0	0	0	$Y = I_0 = D' = 0$
0	1	0	0	0	1	0	$Y = I_1 = C'D = 0$
0	1	0	1	0	1	1	$Y = I_1 = C'D = 1$
0	1	1	0	0	1	0	$Y = I_1 = C'D = 0$
0	1	1	1	0	1	0	$Y = I_1 = C'D = 0$
1	0	0	0	1	0	1	$Y = I_2 = D' = 1$
1	0	0	1	1	0	0	$Y = I_2 = D' = 0$
1	0	1	0	1	0	1	$Y = I_2 = D' = 1$
1	0	1	1	1	0	0	$Y = I_2 = D' = 0$
1	1	0	0	1	1	0	$Y = I_3 = CD' = 0$
1	1	0	1	1	1	0	$Y = I_3 = CD' = 0$
1	1	1	0	1	1	1	$Y = I_3 = CD' = 1$
1	1	1	1	1	1	0	$Y = I_3 = CD' = 0$

Example 2.25

Implement the following Boolean function using multiplexers available as standard ICs.

i) $F_1(x, y) = \Sigma\, m(0, 2)$

ii) $F_2(A, B, C) = \Sigma\, m(1, 3, 5, 7)$

iii) $F_3(A, B, C, D) = \Sigma\, m(1, 5, 8, 9, 11, 12)$

Solution

i) $F_1(x, y) = \Sigma\, m(0, 2)$

The function F_1 is 2-variable function and so a 2^2:1 multiplexer (4:1 multiplexer IC 74153) can be used to implement the function. The function inputs x and y are connected to selection inputs S_1 and S_0 as shown in Fig. 1. The function output F_1 is 1 for minterms m_0 and m_2 and so the inputs I_0 and I_2 of MUX are permanently tied to 1 as shown in Fig. 1. The function output F_1 is 0 for minterms m_1 and m_3 and so the inputs I_1 and I_3 are permanently tied to 0 as shown in Fig. 1. The working of 4:1 MUX is verified in Table 1.

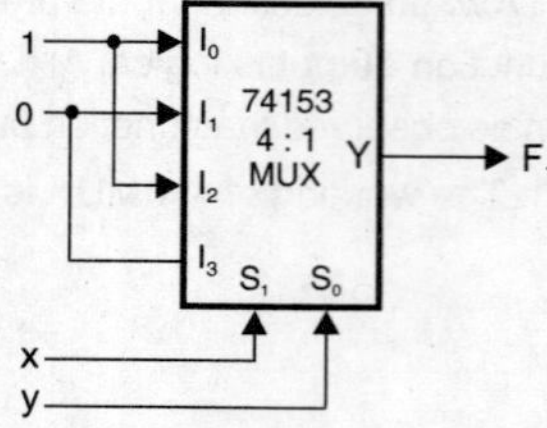

Fig. 1: Realization of F_1 using 4:1 MUX.

Table 1: Verification

x	y	Minterm	S_1	S_0	F_1	$Y = F_1$
0	0	m_0	0	0	1	$Y = I_0 = 1$
0	1	m_1	0	1	0	$Y = I_1 = 0$
1	0	m_2	1	0	1	$Y = I_2 = 1$
1	1	m_3	1	1	0	$Y = I_3 = 0$

ii) $F_2(A, B, C) = \sum m(1, 3, 5, 7)$

The function F_2 is 3-variable function and so a 2^3:1 multiplexer (8:1 multiplexer IC 74151) can be used to implement the function. The function inputs A, B and C are connected to selection inputs S_2, S_1 and S_0 as shown in Fig. 2. The function output F_2 is 1 for minterms m_1, m_3, m_5 and m_7 and so the MUX inputs I_1, I_3, I_5 and I_7 are permanently tied to 1 as shown in Fig. 2. The function output F_2 is 0 for minterms m_0, m_2, m_4 and m_6 and so the MUX inputs I_0, I_2, I_4 and I_6 are permanently tied to 0 as shown in Fig. 2. The working of 8:1 MUX is verified in Table 2.

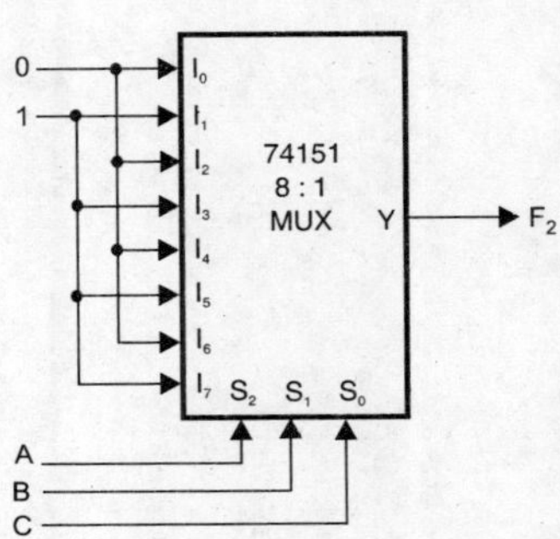

Fig. 2: Realization of F_2 using 8:1 MUX.

Table 2: Verification

A	B	C	Minterm	S_2	S_1	S_0	F_2	$Y = F_2$
0	0	0	m_0	0	0	0	0	$Y = I_0 = 0$
0	0	1	m_1	0	0	1	1	$Y = I_1 = 1$
0	1	0	m_2	0	1	0	0	$Y = I_2 = 0$
0	1	1	m_3	0	1	1	1	$Y = I_3 = 1$
1	0	0	m_4	1	0	0	0	$Y = I_4 = 0$
1	0	1	m_5	1	0	1	1	$Y = I_5 = 1$
1	1	0	m_6	1	1	0	0	$Y = I_6 = 0$
1	1	1	m_7	1	1	1	1	$Y = I_7 = 1$

iii) $F_3(A, B, C, D) = \sum m(1, 5, 8, 9, 11, 12)$

The function F_3 is 4-variable function and so a 2^4:1 multiplexer (16:1 multiplexer IC 74150) can be used to implement the function. The function inputs A, B, C and D are connected to selection inputs S_3, S_2, S_1 and S_0 of MUX as shown in Fig. 3.

The function output F_3 is 1 for minterms m_1, m_5, m_8, m_9, m_{11} and m_{12} and so the MUX inputs I_1, I_5, I_8, I_9, I_{11} and I_{12} are permanently tied to 1 as shown in Fig. 3. The function output F_3 is 0 for minterms m_0, m_2, m_3, m_4, m_6, m_7, m_{10}, m_{13}, m_{14} and m_{15} and so the MUX inputs I_0, I_2, I_3, I_4, I_6, I_7, I_{10}, I_{13}, I_{14} and I_{15} are permanently tied to 0 as shown in Fig. 3. The working of 16:1 MUX is verified in Table 3.

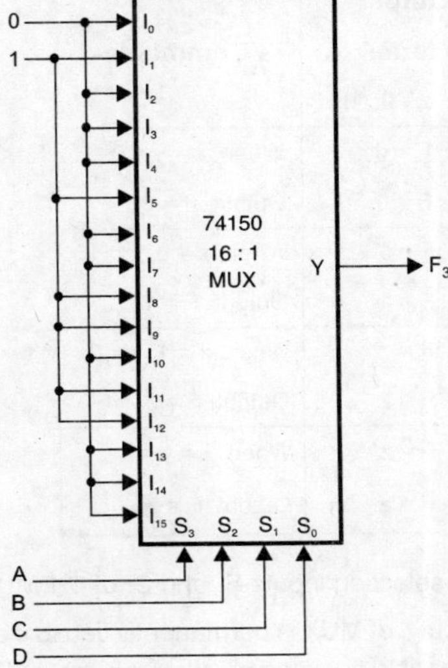

Fig. 3: Realization of F_3 using 16:1 MUX.

Table 3: Verification

A	B	C	D	Minterm	S_3	S_2	S_1	S_0	F_3	$Y = F_3$
0	0	0	0	m_0	0	0	0	0	0	$Y = I_0 = 0$
0	0	0	1	m_1	0	0	0	1	1	$Y = I_1 = 1$
0	0	1	0	m_2	0	0	1	0	0	$Y = I_2 = 0$
0	0	1	1	m_3	0	0	1	1	0	$Y = I_3 = 0$
0	1	0	0	m_4	0	1	0	0	0	$Y = I_4 = 0$
0	1	0	1	m_5	0	1	0	1	1	$Y = I_5 = 1$
0	1	1	0	m_6	0	1	1	0	0	$Y = I_6 = 0$
0	1	1	1	m_7	0	1	1	1	0	$Y = I_7 = 0$
1	0	0	0	m_8	1	0	0	0	1	$Y = I_8 = 1$
1	0	0	1	m_9	1	0	0	1	1	$Y = I_9 = 1$
1	0	1	0	m_{10}	1	0	1	0	0	$Y = I_{10} = 0$
1	0	1	1	m_{11}	1	0	1	1	1	$Y = I_{11} = 1$
1	1	0	0	m_{12}	1	1	0	0	1	$Y = I_{12} = 1$
1	1	0	1	m_{13}	1	1	0	1	0	$Y = I_{13} = 0$
1	1	1	0	m_{14}	1	1	1	0	0	$Y = I_{14} = 0$
1	1	1	1	m_{15}	1	1	1	1	0	$Y = I_{15} = 0$

Example 2.26

Implement the Boolean function, $F(x, y, z) = \Sigma(m_2, m_3, m_5, m_6)$ using multiplexer available as standard ICs.

Solution

It is given that function output F is 1, when the inputs are equal to minterms m_2, m_3, m_5 and m_6. Function output F is 0 for all other inputs. The given function is analyzed using Table 1 and observed that the output F can be expressed in terms of z, z′, 0 and 1. Therefore, the given 3-variable function can be realized using 2^{3-1}:1 multiplexer (4:1 multiplexer IC 7453).

Table 1: Truth Table

Inputs			Minterm	Output	Output	Comments
x	y	z		F	in terms of z, z′, 0, 1	
0	0	0	m_0	0	F = 0	When, x = y = 0
0	0	1	m_1	0	F = 0	Output, F = 0
0	1	0	m_2	1	F = 1	When, x = 0, y = 1
0	1	1	m_3	1	F = 1	Output, F = 1
1	0	0	m_4	0	F = z	When, x = 1, y = 0
1	0	1	m_5	1	F = z	Output, F = z
1	1	0	m_6	1	F = z′	When, x = y = 1
1	1	1	m_7	0	F = z′	Output, F = z′

Let the variables x and y be used as selection inputs S_1 and S_0 of 4:1 MUX as shown in Fig. 1.

In 4:1 multiplexer of Fig. 1, the first input I_0 of MUX is permanently tied to 0 and the second input I_1 is permanently tied to 1. The third variable z and its complement z′ are used as other two inputs I_2 and I_3 respectively. The complement of z is obtained by using a standard NOT gate IC 7404. The working of 4:1 multiplexer is verified in Table 2.

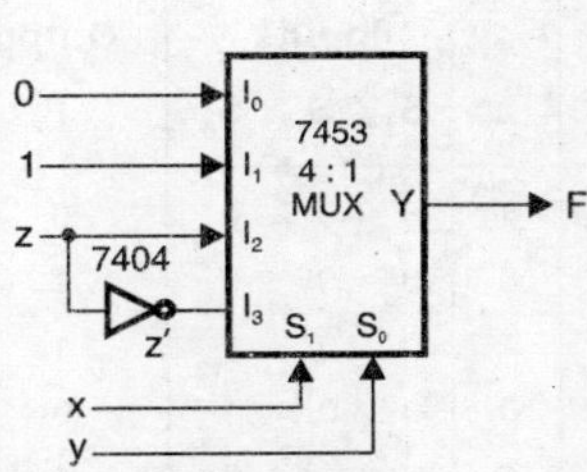

Fig. 1: Implementation of Boolean function using 4:1 MUX.

Table 2: Verification

Inputs				Selection Inputs		Function Output	Output
x	y	z	z'	S_1	S_0	F	Y = F
0	0	0	1	0	0	0	$Y = I_0 = 0$
0	0	1	0	0	0	0	$Y = I_0 = 0$
0	1	0	1	0	1	1	$Y = I_1 = 1$
0	1	1	0	0	1	1	$Y = I_1 = 1$
1	0	0	1	1	0	0	$Y = I_2 = z = 0$
1	0	1	0	1	0	1	$Y = I_2 = z = 1$
1	1	0	1	1	1	1	$Y = I_3 = z' = 1$
1	1	1	0	1	1	0	$Y = I_3 = z' = 0$

Example 2.27

Implement the Boolean function, $F(w, x, y, z) = \Sigma(m_1, m_2, m_4, m_6, m_7, m_{10}, m_{13})$.

Solution

It is given that function output F is 1 when the inputs are equal to minterms m_1, m_2, m_4, m_6, m_7, m_{10} and m_{13}. For all other inputs the function output F is 0.

The given function is analyzed using Table 1 and observed that the function output can be expressed in terms of z, z', 0 and 1. Therefore, the given 4-variable function can be realized using 2^{4-1}:1 multiplexer (8:1 multiplexer IC 74151).

Table 1: Truth Table

Inputs				Minterm	Output F	Output in terms of z, z', 0, 1	Comments
w	x	y	z				
0	0	0	0	m_0	0	F = z	When, w = x = y = 0
0	0	0	1	m_1	1	F = z	Output, F = z
0	0	1	0	m_2	1	F = z'	When, w = x = 0, y = 1
0	0	1	1	m_3	0	F = z'	Output, F = z'
0	1	0	0	m_4	1	F = z'	When, w = y = 0, x = 1
0	1	0	1	m_5	0	F = z'	Output, F = z'
0	1	1	0	m_6	1	F = 1	When, w = 0, x = y = 1
0	1	1	1	m_7	1	F = 1	Output, F = 1
1	0	0	0	m_8	0	F = 0	When w = 1, x = y = 0
1	0	0	1	m_9	0	F = 0	Output, F = 0
1	0	1	0	m_{10}	1	F = z'	When, w = y =1, x = 0
1	0	1	1	m_{11}	0	F = z'	Output, F = z'
1	1	0	0	m_{12}	0	F = z	When, w = x = 1, y = 0
1	1	0	1	m_{13}	1	F = z	Output, F = z
1	1	1	0	m_{14}	0	F = 0	When, w = x = y = 1
1	1	1	1	m_{15}	0	F = 0	Output, F = 0

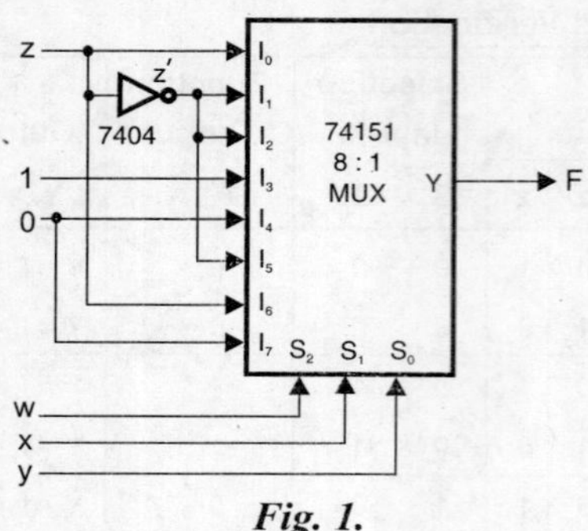

Fig. 1.

Let the variables w, x and y be used as selection inputs S_2, S_1 and S_0 of 8:1 MUX as shown in Fig. 1. In 8:1 multiplexer of Fig. 1, the inputs I_4 and I_7 are permanentely tied to 0 and the input I_3 is tied to 1. The input I_0 and I_6 are tied to z and the inputs I_1, I_2 and I_5 are tied to z′. The complement of z is obtained by using a standard NOT gate IC 7404. The working of 8:1 multiplexer is verified in Table 2.

Table 2: Verification

Inputs					Selection Inputs			Function Output	MUX Output
w	x	y	z	z′	S_2	S_1	S_0	F	Y = F
0	0	0	0	1	0	0	0	0	$Y = I_0 = z = 0$
0	0	0	1	0	0	0	0	1	$Y = I_0 = z = 1$
0	0	1	0	1	0	0	1	1	$Y = I_1 = z' = 1$
0	0	1	1	0	0	0	1	0	$Y = I_1 = z' = 0$
0	1	0	0	1	0	1	0	1	$Y = I_2 = z' = 1$
0	1	0	1	0	0	1	0	0	$Y = I_2 = z' = 0$
0	1	1	0	1	0	1	1	1	$Y = I_3 = 1$
0	1	1	1	0	0	1	1	1	$Y = I_3 = 1$
1	0	0	0	1	1	0	0	0	$Y = I_4 = 0$
1	0	0	1	0	1	0	0	0	$Y = I_4 = 0$
1	0	1	0	1	1	0	1	1	$Y = I_5 = z' = 1$
1	0	1	1	0	1	0	1	0	$Y = I_5 = z' = 0$
1	1	0	0	1	1	1	0	0	$Y = I_6 = z = 0$
1	1	0	1	0	1	1	1	1	$Y = I_6 = z = 1$
1	1	1	0	1	1	1	1	0	$Y = I_7 = 0$
1	1	1	1	0	1	1	1	0	$Y = I_7 = 0$

Example 2.28

Implement the following Boolean function using 2 : 1 multiplexer available as standard ICs.

$F = \Sigma\, m(0, 2, 3, 5)$

Solution

The truth table of given function can be formed as shown in Table 1. Let us select a standard 2:1 multiplexer IC 74157 to implement the Boolean function. The 2:1 multiplexer will have one selection input S_0. Let us connect input x to selection input S_0 as shown in Fig. 1. The selection input take two different values and so Table 1 is divided into two groups based on selection input and a separate table is drawn for each group as shown in Tables 2 and 3.

Table 1: Truth Table of Function, F

Inputs				Output
x	y	z	Minterm	F
0	0	0	m_0	1
0	0	1	m_1	0
0	1	0	m_2	1
0	1	1	m_3	1
1	0	0	m_4	0
1	0	1	m_5	1
1	1	0	m_6	0
1	1	1	m_7	0

Table 2: Truth Table When Selection Input 0

Inputs			Selection Input	y + z′	F = y + z′
x	y	z	S_0		
0	0	0	0	1	1
0	0	1	0	0	0
0	1	0	0	1	1
0	1	1	0	1	1

Table 3: Truth Table When Selection Input 1

Inputs			Selection Input	y'z	F = y'z
x	y	z	S_0		
1	0	0	1	0	0
1	0	1	1	1	1
1	1	0	1	0	0
1	1	1	1	0	0

Table 4: Verification

x	y	z	S_0	F	Y = F
0	0	0	0	1	$Y = I_0 = y + z' = 1$
0	0	1	0	0	$Y = I_0 = y + z' = 0$
0	1	0	0	1	$Y = I_0 = y + z' = 1$
0	1	1	0	1	$Y = I_0 = y + z' = 1$
1	0	0	1	0	$Y = I_1 = y'z = 0$
1	0	1	1	1	$Y = I_1 = y'z = 1$
1	1	0	1	0	$Y = I_1 = y'z = 0$
1	1	1	1	0	$Y = I_1 = y'z = 0$

Note: *The Boolean equations for F in Tables 2 and 3 can be obtained by drawing 2-variable K-maps using y and z as inputs.*

The input I_0 and I_1 for 2:1 MUX are obatined from Tables 2 and 3. The AND, OR and NOT gates are required to generate inputs for I_0 as are shown in Fig. 1. The AND gate IC 7408, OR gate IC 7432 and NOT gate IC 7404 are used in this design.

From Table 2 we get, when $S_0 = 0$, $F = y + z'$ and so I_0 is tied to $y + z'$ in Fig. 1.

From Table 3 we get, when $S_0 = 1$, $F = y'z$ and so I_1 is tied to $y'z$ in Fig. 1.

The working of 2:1 MUX is verified in Table 4.

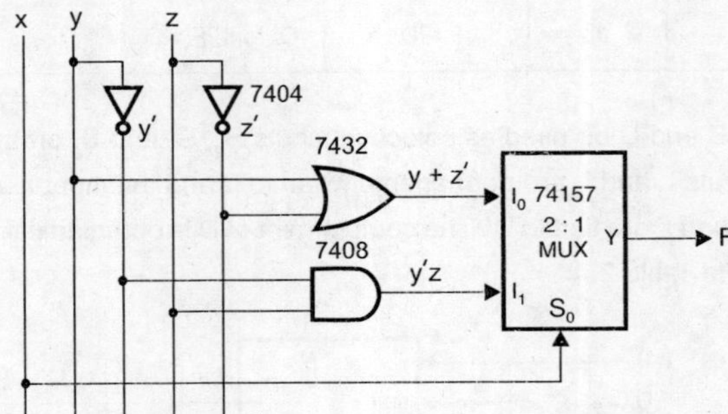

Fig. 1: Implementation of Boolean function using 2:1 MUX.

Example 2.29

Implement the following boolean function with an 8-to-1 line multiplexer and an inverter.

$F(A, B, C, D) = \sum m(2, 4, 6, 9, 10, 11, 15)$

Solution

It is given that function output F is 1 when the inputs are equal to minterms m_2, m_4, m_6, m_9, m_{10}, m_{11} and m_{15}. For all other inputs the function output F is 0.

The given function is analyzed using Table 1 and observed that the function output can be expressed in terms of D, D', 0 and 1. Therefore, the given 4-variable function can be realized using 2^{4-1}:1 multiplexer (8:1 multiplexer IC 74151).

Table 1: Truth Table

Inputs				Minterm	Output	Output	Comments
A	B	C	D		F	in terms of D, D′, 0, 1	
0	0	0	0	m_0	0	F = 0	When, A = B = C = 0
0	0	0	1	m_1	0	F = 0	Output, F = 0
0	0	1	0	m_2	1	F = D′	When, A = B = 0, C = 1
0	0	1	1	m_3	0	F = D′	Output, F = D′
0	1	0	0	m_4	1	F = D′	When, A = C = 0, B = 1
0	1	0	1	m_5	0	F = D′	Output, F = D′
0	1	1	0	m_6	1	F = D′	When, A = 0, B = C = 1
0	1	1	1	m_7	0	F = D′	Output, F = D′
1	0	0	0	m_8	0	F = D	When A = 1, B = C = 0
1	0	0	1	m_9	1	F = D	Output, F = D
1	0	1	0	m_{10}	1	F = 1	When, A = C =1, B = 0
1	0	1	1	m_{11}	1	F = 1	Output, F = 1
1	1	0	0	m_{12}	0	F = 0	When, A = B = 1, C = 0
1	1	0	1	m_{13}	0	F = 0	Output, F = 0
1	1	1	0	m_{14}	0	F = D	When, A = B = C = 1
1	1	1	1	m_{15}	1	F = D	Output, F = D

Let the variables A, B and C be used as selection inputs S_2, S_1 and S_0 of 8:1 MUX as shown in Fig. 1. In 8:1 multiplexer of Fig. 1, the inputs I_0 and I_6 are permanentely tied to 0 and the input I_5 is tied to 1. The input I_4 and I_7 are tied to D and the inputs I_1, I_2 and I_3 are tied to D′. The complement of D is obtanied using standard IC 7404. The working of 8:1 multiplexer is verified in Table 2.

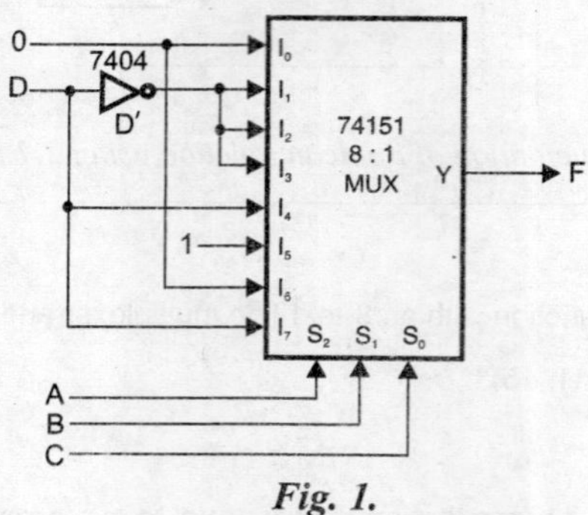

Fig. 1.

Table 2: Verification

Inputs					Selection Inputs			Function Output	MUX Output
A	B	C	D	D′	S_2	S_1	S_0	F	Y = F
0	0	0	0	1	0	0	0	0	$Y = I_0 = 0$
0	0	0	1	0	0	0	0	0	$Y = I_0 = 0$
0	0	1	0	1	0	0	1	1	$Y = I_1 = D' = 1$
0	0	1	1	0	0	0	1	0	$Y = I_1 = D' = 0$
0	1	0	0	1	0	1	0	1	$Y = I_2 = D' = 1$
0	1	0	1	0	0	1	0	0	$Y = I_2 = D' = 0$
0	1	1	0	1	0	1	1	1	$Y = I_3 = D' = 1$
0	1	1	1	0	0	1	1	0	$Y = I_3 = D' = 0$
1	0	0	0	1	1	0	0	0	$Y = I_4 = D = 0$
1	0	0	1	0	1	0	0	1	$Y = I_4 = D = 1$
1	0	1	0	1	1	0	1	1	$Y = I_5 = 1$
1	0	1	1	0	1	0	1	1	$Y = I_5 = 1$
1	1	0	0	1	1	1	0	0	$Y = I_6 = 0$
1	1	0	1	0	1	1	1	0	$Y = I_6 = 0$
1	1	1	0	1	1	1	1	0	$Y = I_7 = D = 0$
1	1	1	1	0	1	1	1	1	$Y = I_7 = D = 1$

2.14 Implementation of Combinational Logic Design using Standard Decoder IC's

The concept of combinational circuit design using decoders is discussed in this chapter, section 2.9. In this section, the available standard ICs are selected for implementation examples. The decoders that are available as standard ICs are listed in Appendix-1.

Example 2.30

Design a full adder using decoder:

a) using logic **high** decoder b) using logic **low** decoder.

Solution

Table 1: Truth Table of Full Adder

Inputs			Minterm	Maxterm	Outputs	
a	b	c_i			Carry c_o	Sum s
0	0	0	m_0	M_0	0	0
0	0	1	m_1	M_1	0	1
0	1	0	m_2	M_2	0	1
0	1	1	m_3	M_3	1	0
1	0	0	m_4	M_4	0	1
1	0	1	m_5	M_5	1	0
1	1	0	m_6	M_6	1	0
1	1	1	m_7	M_7	1	1

The truth table of full adder along with minterms and maxterms is shown in Table 1.

a) Full Adder using Logic High Decoder

From the truth table of full adder we can write the following SOP form of Boolean equations for sum and carry of full adder.

$$\text{Sum, } s = m_1 + m_2 + m_4 + m_7$$

$$\text{Carry, } c_o = m_3 + m_5 + m_6 + m_7$$

Since the full adder has three inputs, a 3-to-8 decoder can be selected to implement the full adder. The truth table of logic **high** 3-to-8 decoder is shown in Table 2.

On taking into account the minterms for which sum and carry are 1, the following Boolean equations can be obtained for sum and carry using the logic high decoder output.

$$s = Y_1 + Y_2 + Y_4 + Y_7 \quad ; \quad c_o = Y_3 + Y_5 + Y_6 + Y_7$$

Using the above Boolean equations the logic circuit of full adder using logic **high** 3-to-8 decoder is drawn as shown in Fig. 1.

The above Boolean equations are implemented using logic **high** 3-to-8 decoder IC 74237 and 4-input OR gate IC 744072 as shown in Fig. 1.

Table 2: Truth Table of Logic High 3-to-8 Decoder

Inputs			Minterm	Outputs							
a	b	c_i		Y_0	Y_1	Y_2	Y_3	Y_4	Y_5	Y_6	Y_7
0	0	0	m_0	1	0	0	0	0	0	0	0
0	0	1	m_1	0	1	0	0	0	0	0	0
0	1	0	m_2	0	0	1	0	0	0	0	0
0	1	1	m_3	0	0	0	1	0	0	0	0
1	0	0	m_4	0	0	0	0	1	0	0	0
1	0	1	m_5	0	0	0	0	0	1	0	0
1	1	0	m_6	0	0	0	0	0	0	1	0
1	1	1	m_7	0	0	0	0	0	0	0	1

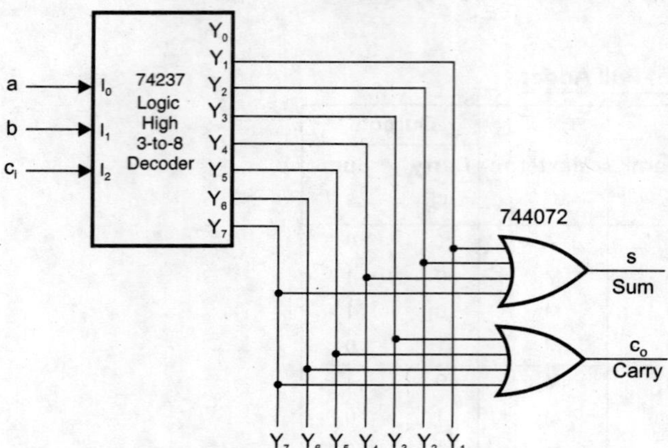

Fig. 1: Full adder using logic high 3-to-8 decoder.

b) Full Adder using Logic Low Decoder

From the truth table of full adder we can write the following POS form of Boolean equations for sum and carry of full adder.

$$s = M_0\, M_3\, M_5\, M_6 \quad ; \quad c = M_0\, M_1\, M_2\, M_4$$

The truth table of logic **low** 3-to-8 decoder is shown in Table 3.

On taking into account the maxterms for which sum and carry are 0, the following Boolean equations can be obtained for sum and carry using the logic **high** decoder output.

$$s = Y_0\, Y_3\, Y_5\, Y_6 \quad ; \quad c = Y_0\, Y_1\, Y_2\, Y_4$$

Using the above Boolean equations the logic circuit of full adder using logic low 3-to-8 decoder is drawn as shown in Fig. 1.

The above Boolean equations are implemented using logic **low** 3-to-8 decoder IC 74138 and 4-input AND gate IC 7421 as shown in Fig. 2.

Table 3: Truth Table of Logic Low 3-to-8 Decoder

Inputs			Maxterm	Outputs							
a	**b**	**c_i**		**Y_0**	**Y_1**	**Y_2**	**Y_3**	**Y_4**	**Y_5**	**Y_6**	**Y_7**
0	0	0	M_0	0	1	1	1	1	1	1	1
0	0	1	M_1	1	0	1	1	1	1	1	1
0	1	0	M_2	1	1	0	1	1	1	1	1
0	1	1	M_3	1	1	1	0	1	1	1	1
1	0	0	M_4	1	1	1	1	0	1	1	1
1	0	1	M_5	1	1	1	1	1	0	1	1
1	1	0	M_6	1	1	1	1	1	1	0	1
1	1	1	M_7	1	1	1	1	1	1	1	0

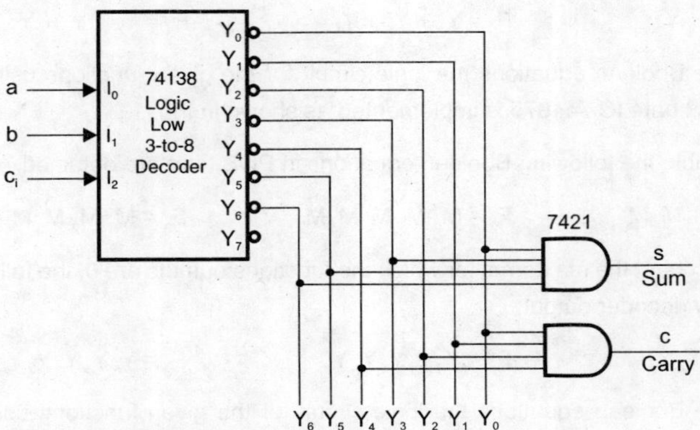

Fig. 2: Full adder using logic low 3-to-8 decoder.

Example 2.31

Using a decoder and external gates, design the combinational circuit defined by the following three Boolean functions:

i) $F_1 = a'bc' + ac$ **ii)** $F_2 = ab'c + a'b$ **iii)** $F_3 = a'b'c' + ab$

Solution

The truth table of Boolean functions F_1, F_2 and F_3 are shown in Table 1: Here, a, b and c are input variables of the functions. From the truth table the following Boolean equations in SOP form are obtained.

$$F_1 = m_2 + m_5 + m_7$$

$$F_2 = m_2 + m_3 + m_5$$

$$F_3 = m_0 + m_6 + m_7$$

Table 1: Truth Table for F_1, F_2 and F_3

a	b	c	Minterm	Maxterm	a'	b'	c'	a'bc'	ac	F_1	ab'c	a'b	F_2	a'b'c'	ab	F_3
0	0	0	m_0	M_0	1	1	1	0	0	0	0	0	0	1	0	1
0	0	1	m_1	M_1	1	1	0	0	0	0	0	0	0	0	0	0
0	1	0	m_2	M_2	1	0	1	1	0	1	0	1	1	0	0	0
0	1	1	m_3	M_3	1	0	0	0	0	0	0	1	1	0	0	0
1	0	0	m_4	M_4	0	1	1	0	0	0	0	0	0	0	0	0
1	0	1	m_5	M_5	0	1	0	0	1	1	1	0	1	0	0	0
1	1	0	m_6	M_6	0	0	1	0	0	0	0	0	0	0	1	1
1	1	1	m_7	M_7	0	0	0	0	1	1	0	0	0	0	1	1

On taking into account the minterms for which function outputs are 1 the following Boolean equations are obtained using the logic **high** decoder output.

$$F_1 = Y_2 + Y_5 + Y_7 \quad ; \quad F_2 = Y_2 + Y_3 + Y_5 \quad ; \quad F_3 = Y_0 + Y_6 + Y_7$$

Using the above Boolean equations the logic circuit for the given functions using logic **high** 3-to-8 decoder IC 74237 and 3-input OR gate IC 744075 is implemented as shown in Fig. 1.

From the truth table the following Boolean equations in POS form are obtained.

$$F_1 = M_0\,M_1\,M_3\,M_4\,M_6 \quad ; \quad F_2 = M_0\,M_1\,M_4\,M_6\,M_7 \quad ; \quad F_3 = M_1\,M_2\,M_3\,M_4\,M_5$$

On taking into account the maxterms for which the functions outputs are 0, the following Boolean equations are obtained using logic **low** decoder output.

$$F_1 = Y_0Y_1Y_3Y_4Y_6 \quad ; \quad F_2 = Y_0Y_1Y_4Y_6Y_7 \quad ; \quad F_3 = Y_1Y_2Y_3Y_4Y_5$$

Using the above Boolean equations the logic circuit for the given functions using logic **low** 3-to-8 decoder IC 74138 and 3-input AND gate IC 7411 is implemented as shown in Fig. 2.

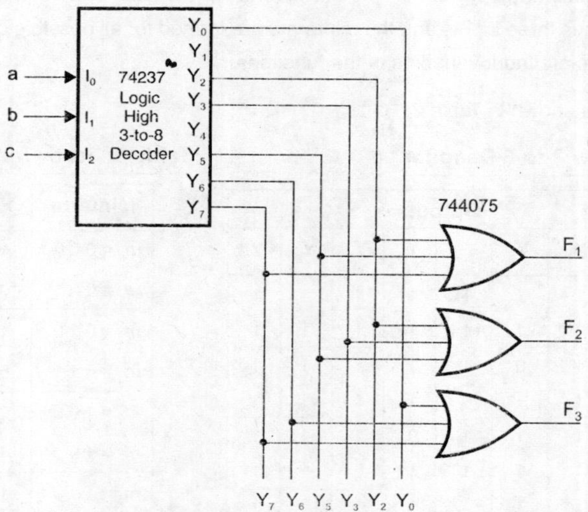

***Fig. 1:** Logic circuit of F_1, F_2, F_3 using logic high decoder and OR gate.*

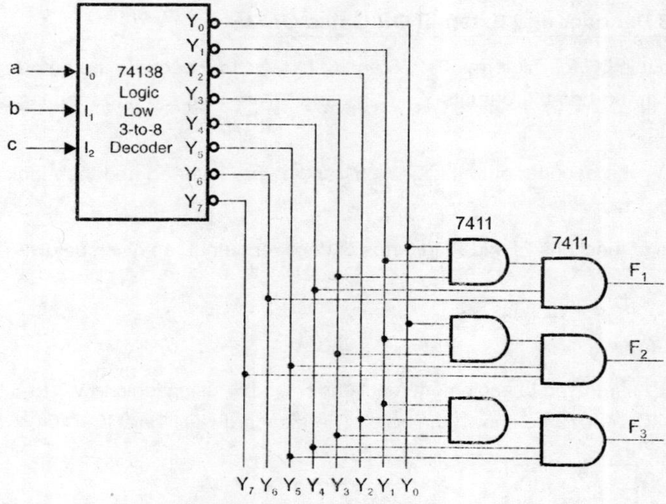

***Fig. 2:** Logic circuit of F_1, F_2, F_3 using logic low decoder and AND gate.*

Example 2.32

Realize the following functions,

$F_1 = \Sigma m(0, 2, 5, 6)$ $F_2 = \Pi M(0, 1, 3, 7)$

(a) Using logic **low** 3-to-8 decoder and external OR gates only

(b) Using logic **low** 3-to-8 decoder and external NOR gates only.

Solution

Given that, $F_1 = \Sigma m(0, 2, 5, 6)$

$F_2 = \Pi M(0, 1, 3, 7) = \Sigma m(2, 4, 5, 6)$

Since the given function has minterms m_0 to m_7 they are 3-variable functions and so can be realized using a 3-to-8 decoder. The 3-to-8 decoder has three inputs and the minterms are formed for all possible combination of inputs as shown in Table 1. Here, a, b and c are input variables of the functions.

The truth table of F_1 and F_2 is shown in Table 2.

Table 1: Truth Table of Logic Low 3-to-8 Decoder

Inputs				Outputs							
a	b	c	Minterm	Y_0	Y_1	Y_2	Y_3	Y_4	Y_5	Y_6	Y_7
0	0	0	m_0	0	1	1	1	1	1	1	1
0	0	1	m_1	1	0	1	1	1	1	1	1
0	1	0	m_2	1	1	0	1	1	1	1	1
0	1	1	m_3	1	1	1	0	1	1	1	1
1	0	0	m_4	1	1	1	1	0	1	1	1
1	0	1	m_5	1	1	1	1	1	0	1	1
1	1	0	m_6	1	1	1	1	1	1	0	1
1	1	1	m_7	1	1	1	1	1	1	1	0

Table 2: Truth Table of F_1 and F_2

Minterm	F_1	F_2
m_0 = 0 0 0	1	0
m_1 = 0 0 1	0	0
m_2 = 0 1 0	1	1
m_3 = 0 1 1	0	0
m_4 = 1 0 0	0	1
m_5 = 1 0 1	1	1
m_6 = 1 1 0	1	1
m_7 = 1 1 1	0	0

(a) Using Logic Low 3-to-8 Decoder and External OR Gates Only

In function, F_1 the output is "1" for minterms 0, 2, 5 and 6. In logic **low** decoder the output Y_n is "0" when the input corresponds to minterm m_n and so the outputs Y_0, Y_2, Y_5 and Y_6 are inverted and then logically ORed for OR implementation as shown in Fig. 1.

Similarly, for function, F_2 the decoder outputs Y_2, Y_4, Y_5 and Y_6 are inverted and then logically ORed for OR implementation as shown in Fig. 1.

The Boolean equations of functions F_1 and F_2 in terms of decoder output are given below:

$$F_1 = Y_0' + Y_2' + Y_5' + Y_6' \qquad .....(1)$$

$$F_2 = Y_2' + Y_4' + Y_5' + Y_6' \qquad .....(2)$$

The equations (1) and (2) are used to implement functions F_1 and F_2 using logic **low** 3-to-8 decoder IC 74138 and external 4-input OR gate IC 744072 as shown in Fig. 1. The standard NOT gate IC 7404 is used to invert the decoder outputs.

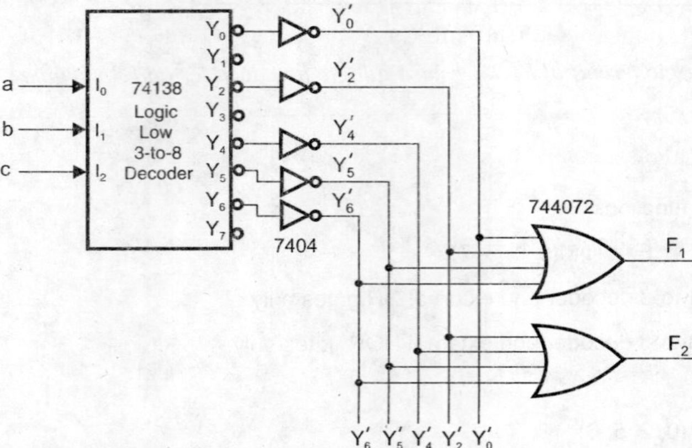

Fig. 1: OR implementation of F_1 and F_2.

(b) Using Logic Low 3-to-8 Decoder and External NOR Gates Only

In NOR implementation the inverters are constructed using NOR gates and OR gate is constructed using NOR gate followed by an inverter as shown in Fig. 2.

The equations (1) and (2) can be modified for NOR implementation as shown below:

$$\therefore F_1 = Y_0' + Y_2' + Y_5' + Y_6'$$

$$= \left(\left(Y_0' + Y_2' + Y_5' + Y_6'\right)'\right)' \quad .....(3)$$

$$F_2 = Y_2' + Y_4' + Y_5' + Y_6'$$

$$= \left(\left(Y_2' + Y_4' + Y_5' + Y_6'\right)'\right)' \quad .....(4)$$

The equations (3) and (4) are used to implement functions F_1 and F_2 using logic **low** 3-to-8 decoder IC 74138 and external 4-input NOR gate IC 7429 as shown in Fig. 2. The inverters are implemented using 2-input NOR gate IC 7402.

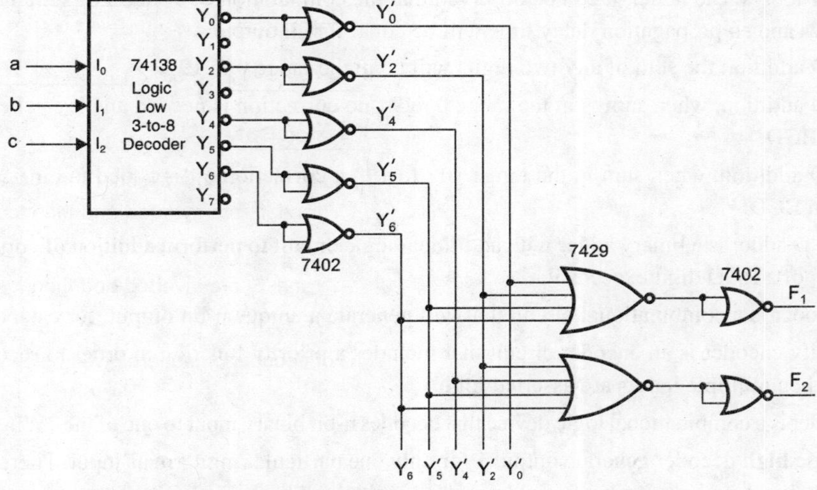

Fig. 2: NOR implementation of F_1 and F_2.

2.15 Summary of Important Concepts

1. Electronic circuits constructed using digital logic gates and devices and designed to operate on digital inputs and outputs are called digital logic circuits.
2. The digital logic circuits can be broadly classified into combinational circuits and sequential circuits.
3. The combinational circuits are digital logic circuits without feedbacks from output to input.
4. The sequential circuits are digital logic circuits with feedback from output to input.
5. Combinational logic circuits and they are constructed using logic gates.
6. A multiplexer is a combinational circuit which can select one of the 2^n inputs as output using n selection lines.
7. In multiplexer the possible binary values of selection inputs are same as minterms. Therefore, the logic circuit of Boolean functions in SOP form can be realized using multiplexer.
8. Multiplexer of smaller size can be connected in cascade to expand the size of multiplexer.
9. A demultiplexer will perform the reverse operation of multiplexer.

10. A demultiplexer is a combinational circuit which can transmit a binary input to any one of 2^n output lines using n selection lines or n-bit address.
11. Demultiplexers of smaller size can be connected in parallel to expand the size of demultiplexer.
12. Adders have been developed to perform arithmetic addition operation on binary numbers.
13. Half adder can perform addition of two 1-bit binary number.
14. Subtractors have been developed to perform arithmetic subtraction operation on binary numbers.
15. Half subtractor can perform subtraction of two 1-bit binary number.
16. The outputs of half and full adders are sum and carry.
17. Full adder is a combinational circuit that perform the arithmetic sum of three binary bits.
18. Full subtractor is a combinational circuit that perform arithmetic subtraction of three binary bits.
19. Full adder can also be realized using two half adders.
20. A binary parallel adder performs arithmetic sum of two n-bit binary numbers.
21. In carry look ahead adder, it can be observed that the computation of each bit of sum has equal number of stages and so propagation delay time will be equal for all outputs.
22. In BCD addition the sum of any two digits will be in the range 0 to 19_{10}.
23. In BCD addition, when sum is in the range 0 to 9_{10} no correction is needed and it can be represented by 4 digit BCD.
24. In BCD addition, when sum in the range 10_{10} to 19_{10} a correction 6_{10} is added to binary sum to get the result in BCD.
25. The BCD adder is a binary adder with additional logic circuit to perform addition of correction 6_{10} when the sum of a BCD digit exceeds 9_{10}.
26. An encoder is a combinational circuit that will generate a unique n-bit output, for each of the 2^n inputs.
27. A priority encoder is an encoder circuit that includes a priority function in order to recognize only one input when multiple inputs are asserted **high**.
28. A decoder is a combinational logic device that decodes n-bit binary input to one of the 2^n binary information.
29. The logic **high** decoder generate output 1 for only one particular minterm as input. Therefore, a Boolean function expressed as a sum of minterms can be obtained by logical OR of the outputs of decoder for which the minterms of function generates logic **high** or 1.
30. The logic **low** decoder generates output 0 for only one particular maxterm as input. Therefore, a Boolean function expressed as a product of maxterms can be obtained by logical AND of the outputs of decoder for which the maxterms of function generates logic **low** or 0.
31. A parity generator/checker is a combinational logic circuit that generates the parity bit to be appended during transmission and at the receiving end it is used to verify the parity of received information.

2.16 Short-Answer Questions

Q2.1 ***Mention the dependency of output in combinational circuits.*** *(AU, Nov/Dec'18, 2 Marks)*

The output depends only on the preset or current value of input. Since there is no feedback of output to input in combinational circuits, the present output will not depend on previous output.

Q2.2 ***Define digtal logic circuits.***

Electronic circuits constructed using digital logic gates and devices and designed to operate on digital inputs and outputs are called digital logic circuits.

Q2.3 ***What are the types of digtal logic circuits.***

The digital logic circuits can be broadly classified into,

1. Combinational circuits
2. Sequential circuits.

Q2.4 ***Explain the design procedure for combinational circuits.***

1. Determine the required inputs and outputs from the problem specifications.
2. Assign a symbol to each input and output.
3. Derive the truth table.
4. Draw K-map for every output and form the prime implicants.
5. Determine the simplified Boolean function for every output from the prime implicants.
6. Implement the Boolean functions of all the outputs as a digital circuit using logic gates.

Q2.5 ***Define Half adder and Full adder.***

Half adder is a combinational circuit that performs arithmetic sum of 2-bit binary.

Full adder is a combinational circuit that performs arithmetic sum of 3-bit binary.

Q2.6 ***Determine the exact number of half adders and full adders required for performing the addition of two binary numbers of 5 bits length each.*** *(AU, Apr/May'19, 2 Marks)*

When addition is performed using half adders, we require two half adders to add one bit binary number. Therefore, for addition of 5-bit binary numbers 10 half adders are required.

When addition is performed using full adders we require one full adder to add one bit binary number. Therefore, for addition of 5-bit binary numbers we require 5 full adders.

Q2.7 ***Define Decoder.***

A decoder is a combinational circuit that converts binary information from n input lines to a maximum of 2^n unique output lines. In a decoder, for an n-bit binary input one of the 2^n output is activated.

Q2.8 ***How multiplexer differ from decoder?***

A decoder accept n-bit binary input to activate one of the 2^n outputs whereas the multiplexer accept n-bit address to allow one of the 2^n input on the output line.

Q2.9 ***Write the truth table of 2-to-4 line decoder and draw its logic diagram.***

Solution:

Table 1: Truth Table

Inputs		Outputs			
I_1	I_0	Y_0	Y_1	Y_2	Y_3
0	0	1	0	0	0
0	1	0	1	0	0
1	0	0	0	1	0
0	1	0	0	0	1

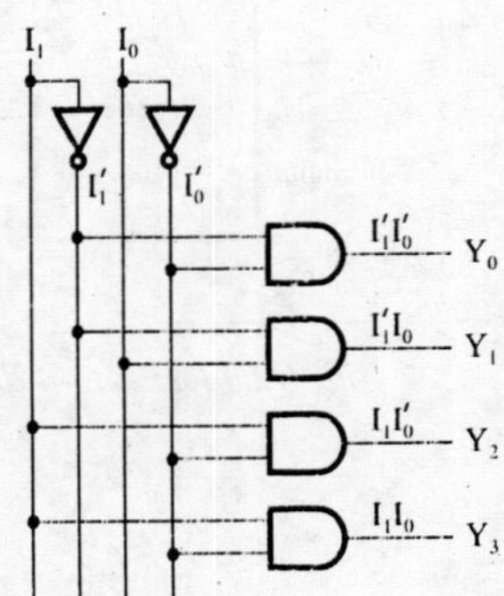

***Fig. Q2.9:** Logic diagram.*

Boolean equations:

$Y_0 = I_1' I_0'$; $Y_1 = I_1' I_0$; $Y_2 = I_1 I_0'$; $Y_3 = I_1 I_0$

Q2.10 Define Encoder.

An encoder is a combinational circuit that perform the inverse operation of a decoder. An encoder has 2^n input lines and n output lines. When one of 2^n inputs is activated, it generates an unique n-bit output.

Q2.11 What is priority Encoder.

A priority encoder is an encoder circuit that includes the priority function. In priority encoder, if two or more inputs are asserted high (or equal to 1) then the output corresponds to the input having highest priority.

Q2.12 What is Demultiplexer.

Demultiplexer is a combinational circuit which can transmit a binary input to any one of 2^n output lines using n selection lines or n-bit address.

Q2.13 An odd parity operation results in 1 if an odd number of bits in a given sequence of bits are 1. Write a truth table for 4-input odd parity operation.

Solution:

Inputs				Output	Inputs				Output
a	b	c	d		a	b	c	d	
0	0	0	0	0	1	0	0	0	1
0	0	0	1	1	1	0	0	1	0
0	0	1	0	1	1	0	1	0	0
0	0	1	1	0	1	0	1	1	1
0	1	0	0	1	1	1	0	0	0
0	1	0	1	0	1	1	0	1	1
0	1	1	0	0	1	1	1	0	1
0	1	1	1	1	1	1	1	1	0

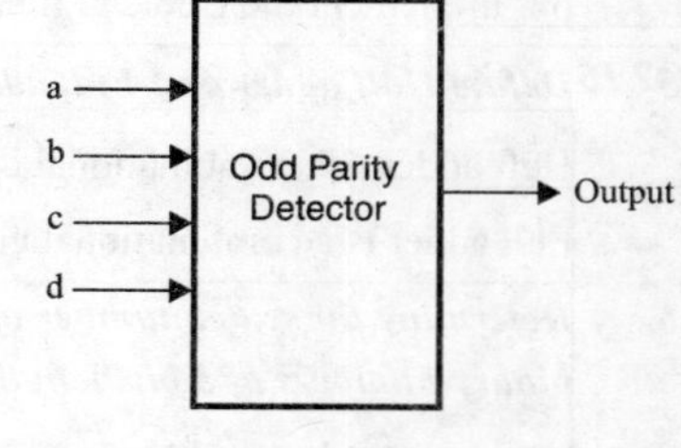

Fig. Q2.13.

Output = 1, When number of 1's odd
Output = 0, When number of 1's even

Q2.14 Convert a 2-to-4 decoder with enable input to 1:4 demux.

Solution:

The truth table of 2-to-4 decoder working as 1:4 MUX is shown in Table 1.

When the 2 inputs I_0, I_1 of decoder are used as selection inputs S_0 and S_1 of demux and enable of decoder used as demux input, a 2-to-4 decoder will work as a 1:4 demux.

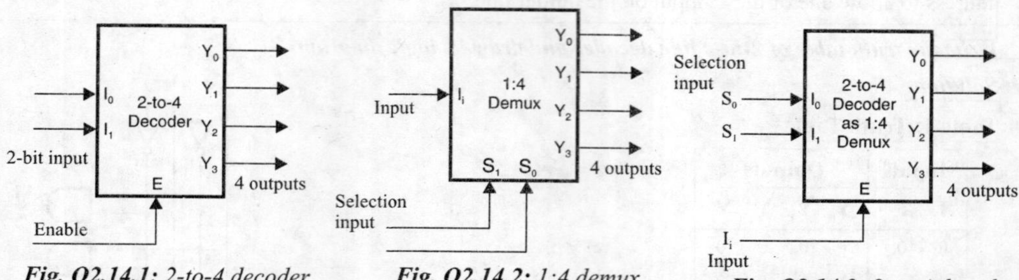

Fig. Q2.14.1: 2-to-4 decoder. *Fig. Q2.14.2: 1:4 demux.* *Fig. Q2.14.3: 2-to-4 decoder as 1:4 demux.*

Table 1: Truth Table of 2-to-4 Decoder with 1:4 Demux

Inputs	Selection Inputs		Outputs			
$I_i = E$	S_1	S_0	Y_0	Y_1	Y_2	Y_3
$I_i = 0$	0	0	0	0	0	0
$I_i = 0$	0	1	0	0	0	0
$I_i = 0$	1	0	0	0	0	0
$I_i = 0$	1	1	0	0	0	0
$I_i = 1$	0	0	I_i	0	0	0
$I_i = 1$	0	1	0	I_i	0	0
$I_i = 1$	1	0	0	0	I_i	0
$I_i = 1$	1	1	0	0	0	I_i

Q2.15 ***Compare demux and decoder.***

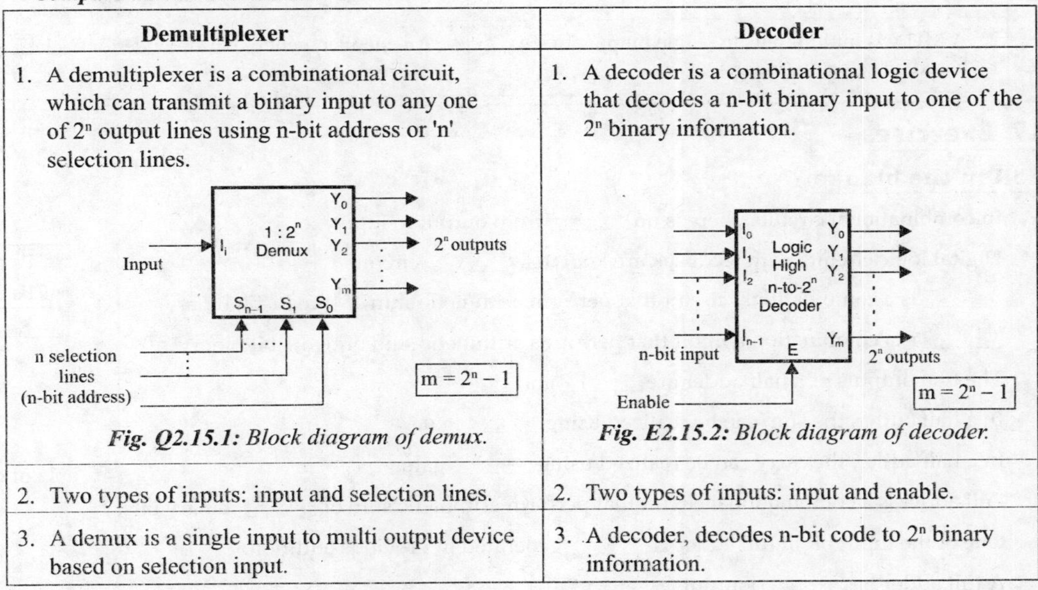

Demultiplexer	Decoder
1. A demultiplexer is a combinational circuit, which can transmit a binary input to any one of 2^n output lines using n-bit address or 'n' selection lines.	1. A decoder is a combinational logic device that decodes a n-bit binary input to one of the 2^n binary information.
Fig. Q2.15.1: Block diagram of demux.	*Fig. E2.15.2: Block diagram of decoder.*
2. Two types of inputs: input and selection lines.	2. Two types of inputs: input and enable.
3. A demux is a single input to multi output device based on selection input.	3. A decoder, decodes n-bit code to 2^n binary information.

Q2.16 ***What do you mean by parity checker?***

A parity bit is an extra bit added to the binary code to make the number of 1's either odd or even. It is mainly used for detecting the errors in transmission of binary information. Parity checker is the logic circuit that checks the parity of binary code received by the receiver. In even parity system, the parity checker will check for even parity in the received codes.

In odd parity system, the parity checker will check for odd parity in the received codes. If the parity of a codes is not correct then an error signal is generated to indicate error in transmitted code.

Q2.17 Compare multiplexer and encoder.

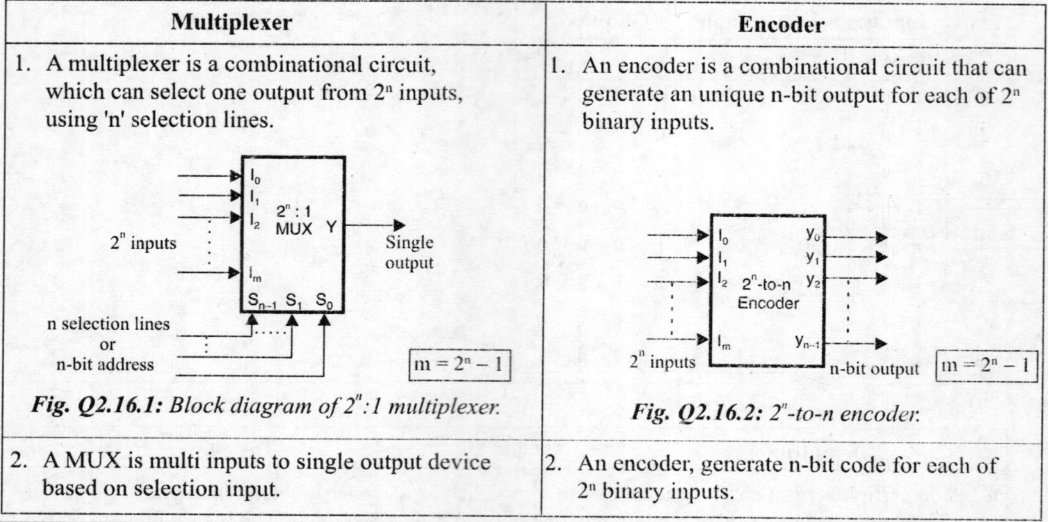

Multiplexer	Encoder
1. A multiplexer is a combinational circuit, which can select one output from 2^n inputs, using 'n' selection lines.	1. An encoder is a combinational circuit that can generate an unique n-bit output for each of 2^n binary inputs.
Fig. Q2.16.1: Block diagram of 2^n:1 multiplexer.	*Fig. Q2.16.2: 2^n-to-n encoder.*
2. A MUX is multi inputs to single output device based on selection input.	2. An encoder, generate n-bit code for each of 2^n binary inputs.

2.17 Exercises

I. Fill in the blanks

1. In combinational circuits, there is no _______ from output to input.
2. Digital logic circuits with feedback are known as _______ circuits.
3. _______ is a combinational circuit that perform arithmetic sum of three 1-bit binary.
4. _______ is a combinational ircuit that performs arithmetic sum of two 1-bit binary.
5. The two outputs of a half adder are _______ and _______.
6. In a half adder, the sum can be realized using _______ gate.
7. In a half adder, the carry can be realized using _______ gate.
8. Full adder is a combinational circuit that perform arithmetic sum of _______ binary bits.
9. One of the input bit in full adder is _______ generated in previous addition.
10. A full adder has _______ outputs.
11. Practically _______ is considered as addition of positive and negative numbers.
12. _______ is a combinational circuit used to compare two binary numbers.
13. A decoder decodes n-bit binary information into _______ binary information.
14. An encoder performs the _______ operation of a decoder.
15. The digital multiplexer is basically a combination logic to perform the _______ operation.
16. The multiplexer is otherwise called _______.
17. A multiplexer will have _______ inputs and _______ output.
18. A demultiplexer will perform the _______ operation of multiplexer.
19. A demultiplexer is shortly called as _______.
20. Demultiplexers of smaller size can be connected in _______ to expand the size of demux.

Answers

1. feedback	6. XOR	11. subtraction	16. data selector
2. sequential	7. AND	12. magnitude comparator	17. 2^n, one
3. full adder	8. three	13. 2^n	18. reverse
4. half adder	9. carry	14. inverse	19. Demux
5. sum, carry	10. two	15. AND-OR	20. parallel

II. State whether the following statements are True or False

1. A logic gate is a data storage device.
2. A full adder is characterized by three inputs and two outputs.
3. A 4-bit parallel adder can add two 4-bit binary numbers.
4. In general, a multiplexer has several data inputs, several data outputs and selection inputs.
5. A comparator compares for the equality of two input numbers.
6. A demultiplexer can transmit a binary input to any one of 2^n output lines using n-bit address.
7. Demultiplexer of smaller size can be connected in series to expand the size of demultiplexer.
8. In a decoder, n-bit binary information is decoded into 2^n binary information.
9. A n-to-2^n decoder outputs may be logic **high** or **low**.
10. An encoder is a combinational circuit that performs the reverse of decoder function.
11. A priority encoder is a decoder.
12. A demultiplexer is a data distributor.
13. The code 10011000 exhibits even parity.
14. Half adder has two inputs and two outputs.

Answers

1. False	6. True	11. False
2. True	7. False	12. True
3. True	8. True	13. False
4. False	9. True	14. True
5. True	10. True	

III. Choose the right answer for the following questions

1. ***A combinational circuit is one in which the output depends on the***

   a) Input combination at that time

   b) Input combination and the previous output

   c) Input combination at that time and previous input combination

   d) Present output and previous output

2. *The following switching functions are to be implemented using a decoder*

$f_1 = \Sigma m(1, 2, 4, 8, 10, 14)$; $f_2 = \Sigma m(2, 5, 9, 11)$; $f_3 = \Sigma m(2, 4, 5, 6, 7)$

The minimum configuration of the decoder should be

a) 2 to 4 line **b)** 3 to 8 line **c)** 4 to 16 line **d)** 5 to 32 line

3. *The circuit of Fig. 3 is equivalent to,*

a) AND gate **b)** OR gate

c) XOR gate **d)** XNOR gate

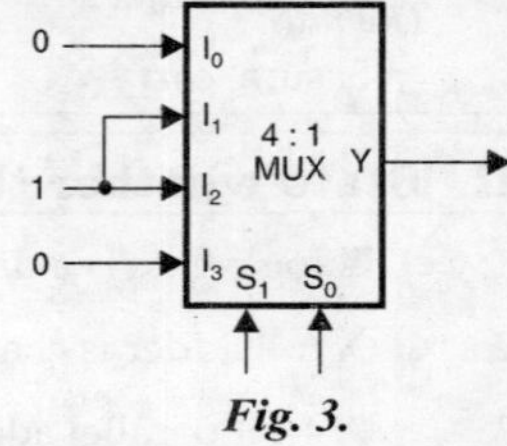

Fig. 3.

4. *The minimum number of 2×1 MUX needed (with no added gates) to form a 3×1 MUX is*

a) 2 **b)** 3 **c)** 4 **d)** 1

5. *The circuit in Fig. 5 works as*

a) binary to gray converter

b) gray to binary converter

c) odd parity generator

d) even parity generator

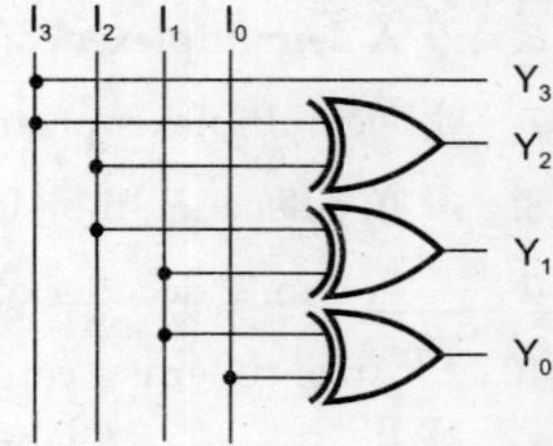

Fig. 5.

6. *The function "F" implemented by the multiplexer shown in Fig. 6 is*

a) A

b) B

c) A′B

d) AB′ + A′B

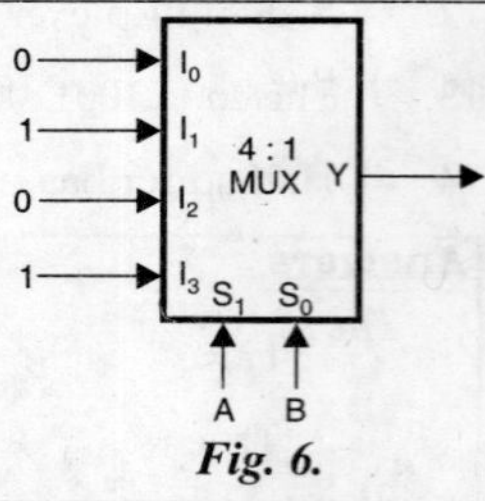

Fig. 6.

7. *The given logic diagram in Fig. 7 represents a*

a) Multiplexer

b) Full adder

c) Half adder

d) None of the above

A
B
Sum
Carry

Fig. 7.

8. *The MUX in Fig. 8 implements the function*

a) F = BC′

b) F = BC

c) F = B′C′

d) F = B′C

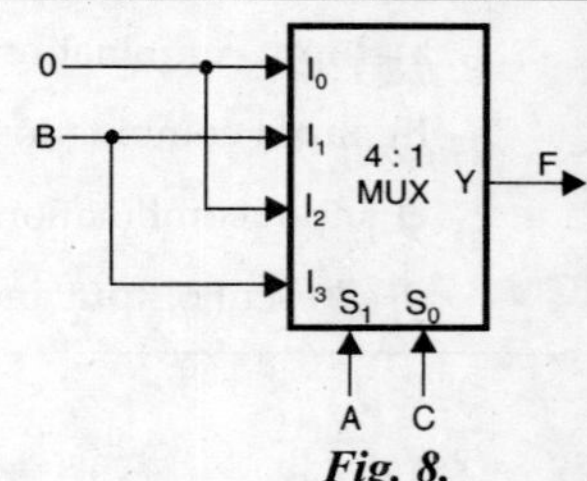

Fig. 8.

9. ***The circuit shown in Fig. 9 is a***

a) Odd parity generator b) Even parity generator

c) Full adder d) Comparator

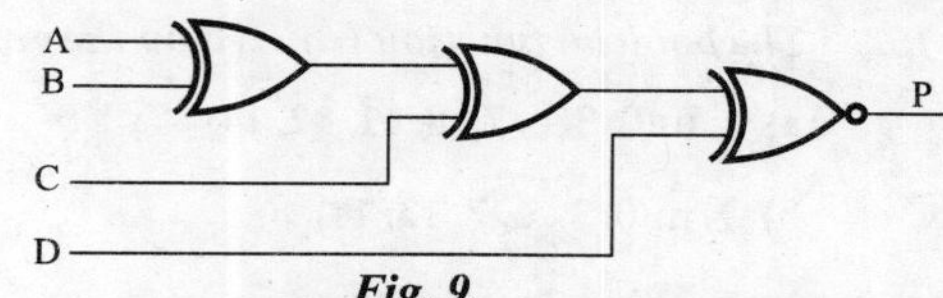

Fig. 9.

10. ***The multiplexer in Fig. 10 implements the function***

a) Y = AB

b) Y = A′B + B′C

c) Y = A + BC

d) Y = BC

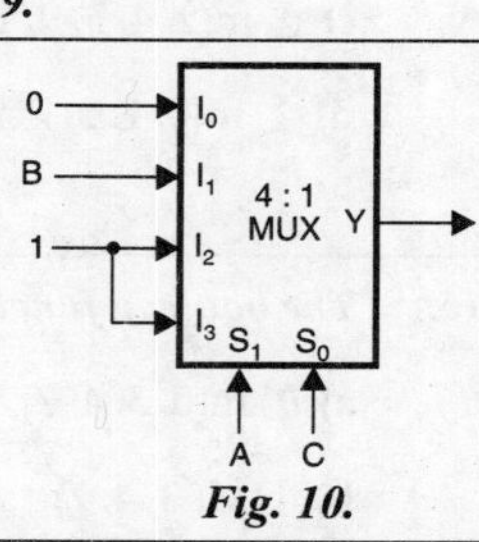

Fig. 10.

11. ***The MUX in Fig. 11 implements the function***

a) $Y = \sum m(1, 3, 5, 7)$

b) $Y = \sum m(1, 2, 3, 6)$

c) $Y = \sum m(2, 5, 6, 7)$

d) $Y = \sum m(1, 4, 5, 6)$

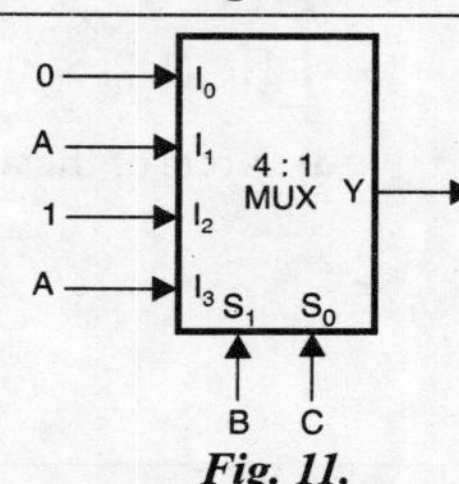

Fig. 11.

12. ***The MUX in Fig. 12 implements the function***

a) B′

b) B

c) $B \oplus C$

d) $B \odot C$

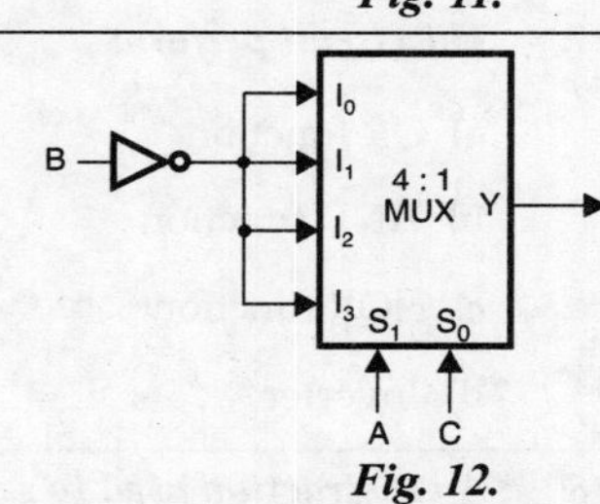

Fig. 12.

13. ***The output of the 4:1 MUX shown in Fig. 13 is***

a) x′y′ + x

b) x + y

c) x′ + y′

d) xy + x′

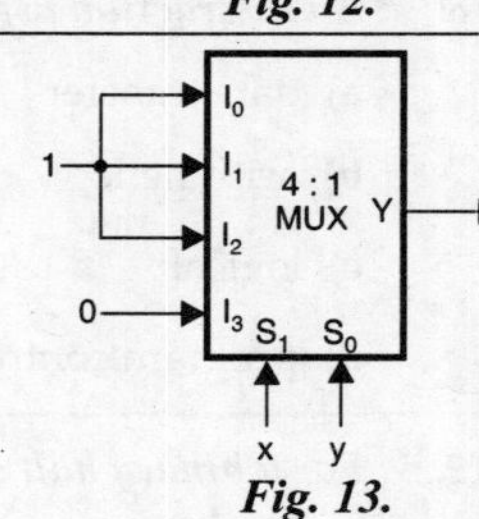

Fig. 13.

14. ***The inputs to the following circuit is***

A: 1 0 1 0 0 1

B: 0 1 1 1 1 0

The outputs X and Y are

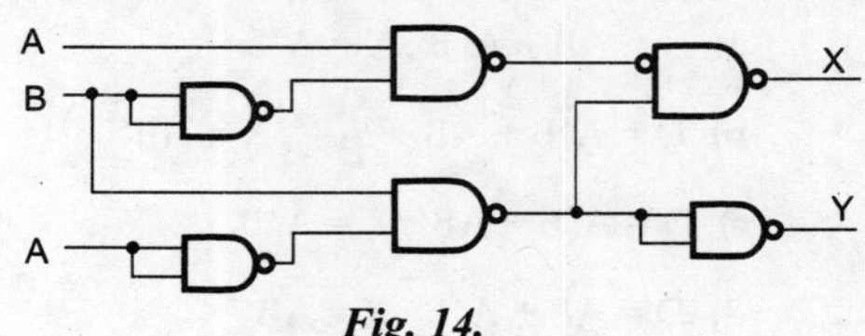

Fig. 14.

a) X: 0 1 1 1 1 0
 Y: 0 1 0 1 1 0

b) X: 0 1 0 1 1 0
 Y: 1 1 0 1 1 1

c) X: 0 1 0 1 1 0
 Y: 0 1 0 1 1 0

d) X: 1 1 0 1 1 1
 Y: 0 0 1 0 0 0

15. *The boolean function realized by the given multiplexer is given by,*

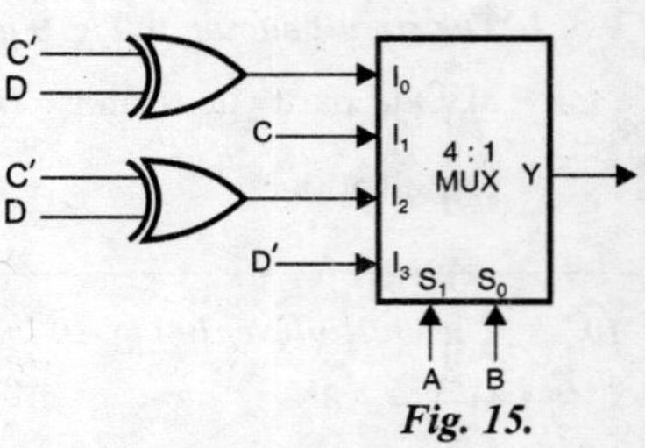

Fig. 15.

a) Σ m(0, 3, 6, 7, 8, 11, 12, 14)

b) Σ m(0, 3, 6, 7, 12, 14)

c) Σ m(2, 4, 5, 9, 10, 13, 15)

d) Σ m(0, 1, 5, 7, 9, 12)

16. *The boolean function implemented by the MUX is given as,*

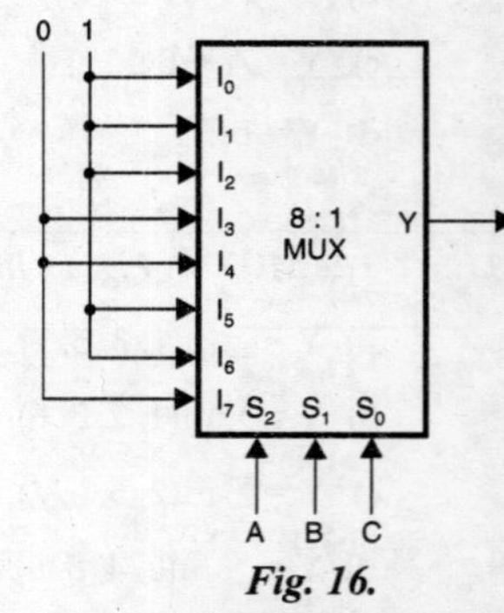

Fig. 16.

a) Σ m(0, 3, 4, 7)

b) ΠM(3, 4, 7)

c) ΠM(1, 2, 5, 6)

d) None of the above

17. *The circuit perform*

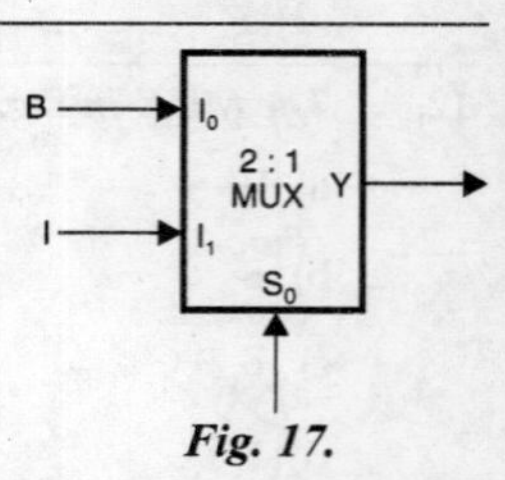

Fig. 17.

a) OR function

b) AND function

c) NOR function

d) Inverter

18. *An instruction used to set the carry flag in a computer can be classified as*

a) data transfer

b) arithmetic

c) logical

d) program control

19. *For a binary half subtractor having 2 input A and B, the correct set of logical expression for the outputs D = (A – B) and X = (Borrow) are,*

a) D = AB + A'B ; X = A'B

b) D = A'B + AB'+ AB'; X = AB'

c) D = A'B + AB'; X = A'B

d) D = AB + A'B ; X = AB'

20. ***Without any additional circuitry, an 8 : 1 MUX can be used to obtain***

a) Some but not all boolean function of 3 variables

b) All functions of 3 variables but none of 4 variables

c) All functions of 3 variables and some but not all of 4 variables

d) All functions of 4 variables

---

21. ***The minimum number of 2:1 MUX required to realize a 4:1 MUX is***

a) 1 **b)** 2 **c)** 3 **d)** 4

---

22. ***A digital system is required to amplify a binary encoded audio signal, the user should be able to control the gain of the amplifier from a minimum to a maximum in 248 increments. The minimum number of bits required to encode in straight binary is***

a) 8 **b)** 6 **c)** 5 **d)** 7

---

23. ***The logic realized by the circuit of Fig. 23 is***

a) F = AC

b) $F = A \oplus C$

c) F = BC

d) F = B + C

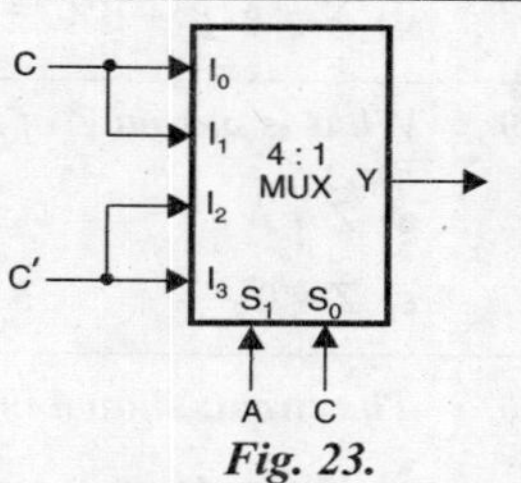

Fig. 23.

---

24. ***Two 2's complement number having sign bit x and y are added and the sign bit of the result is 2. If z is input carry, then the occurance of overflow is indicated by the Boolean function,***

a) xyz

b) x'y'z'

c) x'y'z + xyz'

d) xy + yz + zx

---

25. ***A 4 : 1 MUX is used to implement a 3-input Boolean function as shown in Fig. 25. The Boolean function F(A, B, C) implemented is***

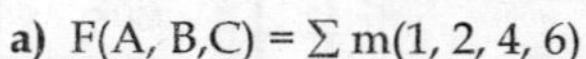

a) $F(A, B,C) = \Sigma\, m(1, 2, 4, 6)$

b) $F(A, B,C) = \Sigma\, m(1, 2, 6)$

c) $F(A, B,C) = \Sigma\, m(2, 4, 5, 6)$

d) $F(A, B,C) = \Sigma\, m(1, 5, 6)$

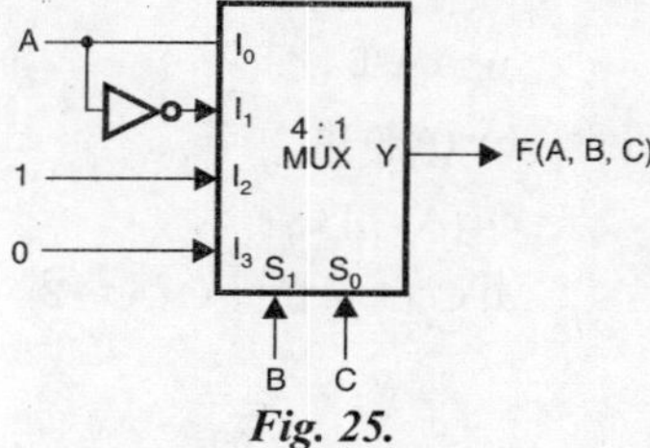

Fig. 25.

---

26. ***A, B, C and D are input bits and Y is the output bit in the XOR gate circuit of Fig. 26. Which of the following statements about the sum S of A, B, C, D is correct?***

a) S is always either zero or odd

b) S is always either zero or even

c) S = 1 only if the sum of A, B, C and D is even

d) S = 1 only if the sum of A, B, C and D is odd

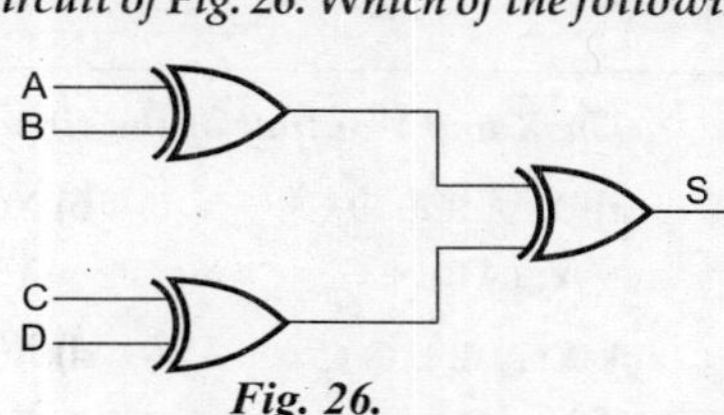

Fig. 26.

27. *The Boolean function realized by the logic circuit shown in Fig. 27 is*

a) $F = \Sigma m(0, 1, 3, 5, 9, 10, 14)$

b) $F = \Sigma m(2, 3, 5, 7, 8, 12, 13)$

c) $F = \Sigma m(1, 2, 4, 5, 11, 14, 15)$

d) $F = \Sigma m(2, 3, 5, 7, 8, 9, 12)$

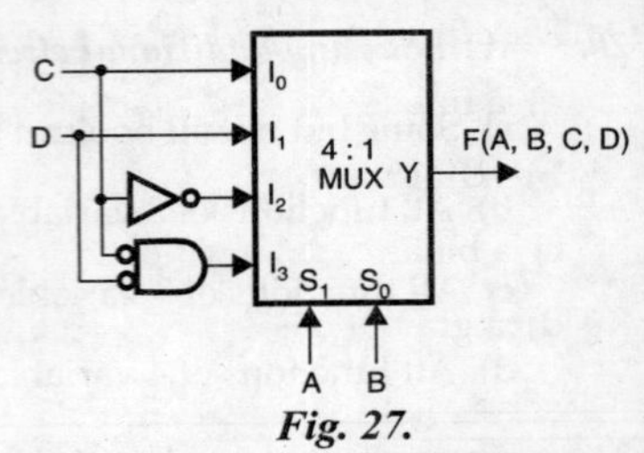

Fig. 27.

28. *In the circuit of Fig. 28, X is given by*

a) $X = AB'C + A'BC' + A'B'C + ABC$

b) $X = A'BC + AB'C + ABC' + A'B'C'$

c) $X = AB + BC + AC$

d) $X = A'B' + B'C' + A'C'$

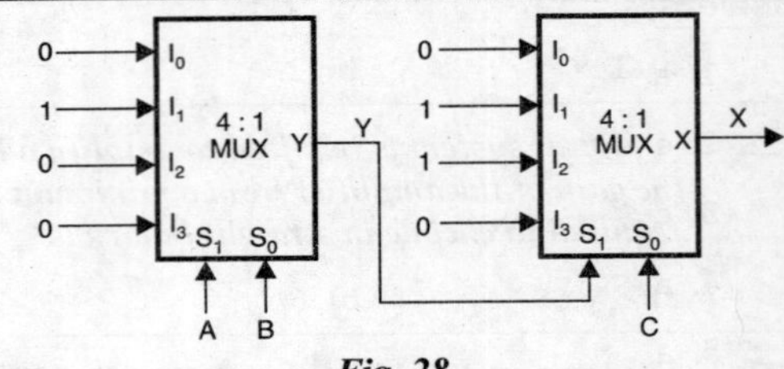

Fig. 28.

29. *What is output Z of an XOR gate whose all inputs are set at A*

a) $Z = A$

b) $Z = 1$

c) $Z = 0$

d) $Z = A$

30. *The circuit shown in Fig. 30 works as*

a) binary to gray converter

b) gray to binay converter

c) binary to its 2's complement converter

d) a parity generator

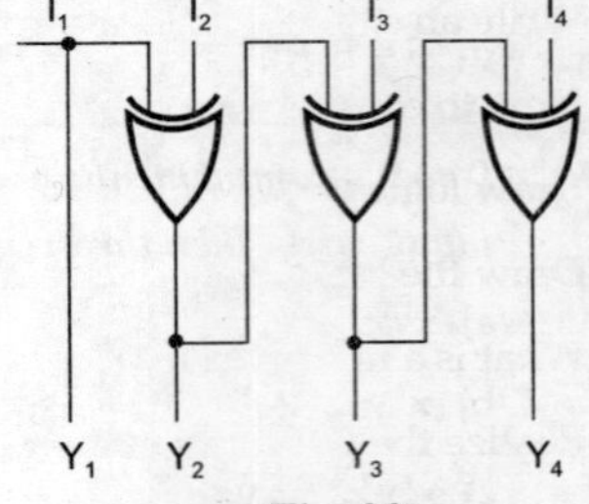

Fig. 30.

31. *The output of MUX shown in Fig. 31 is*

a) $A \oplus B$

b) $(A \oplus B)'$

c) $(A . B) C$

d) $C'(A \oplus B)' + C (A \oplus B)$

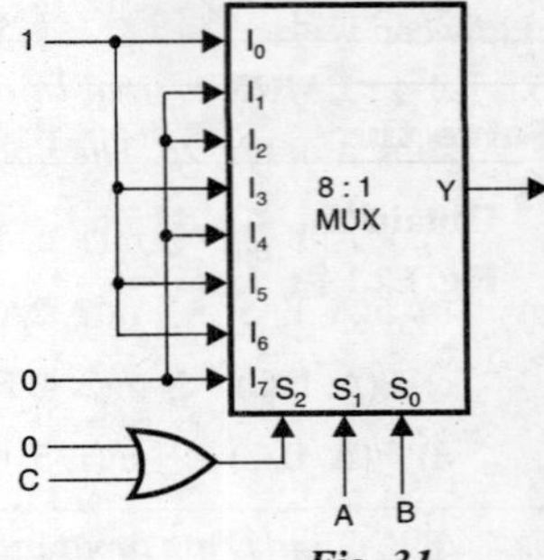

Fig. 31.

32. *The X and Y output of the circuit in Fig. 32 is*

a) X: $A \oplus B \oplus C$
Y: $(AB)' + C$

b) X: $A \oplus B \oplus C$
Y: ABC

c) X: $A \oplus B \oplus C$
Y: $AB + (A \oplus B)C$

d) X: $A \oplus B \oplus C$
Y: $(A \oplus B)C'$

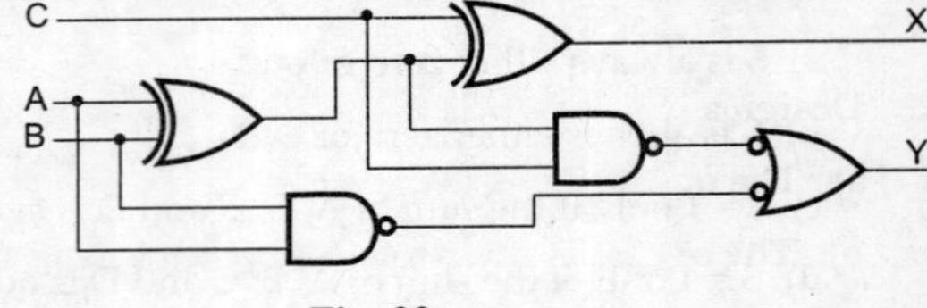

Fig. 32.

33. ***The circuit in Fig. 33 works as***

    a) a full subractor

    b) a full adder

    c) a binary to gray converter

    d) a gray to binary converter

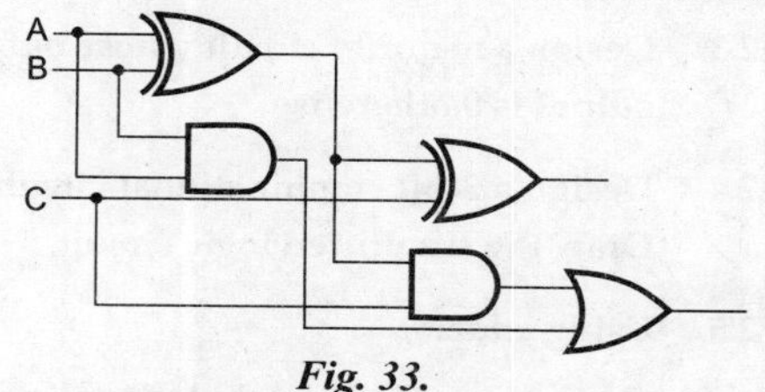

Fig. 33.

Answers

1. a	6. b	11. c	16. b	21. c	26. d	31. d
2. c	7. c	12. a	17. a	22. a	27. d	32. c
3. c	8. b	13. c	18. b	23. b	28. a	33. b
4. a	9. a	14. a	19. c	24. d	29. c	
5. a	10. c	15. a	20. d	25. a	30. b	

IV. Answer the following questions

1. Give examples for combinational circuit.
2. Draw a full adder using two half adders.
3. Write an expression for borrow and difference in a full subtractor.
4. How to convert binary adder to BCD adder?
5. Draw logic circuit of half subtractor using basic gates.
6. Draw the logic block diagram of n-bit binary adder using full adders.
7. What is a multiplexer? How it is called otherwise? Why?
8. Realize the Boolean function F(A, B, C, D) = $\Sigma(0, 1, 5, 6)$ using appropriate multiplexer.
9. What is the purpose of enable input in a decoder?
10. How can we derive a priority encoder from an encoder?

V. Solve the following problems

E2.1 **Obtain the simplified Boolean expression for output F interms of the input variables in the circuit of Fig. E2.1.**

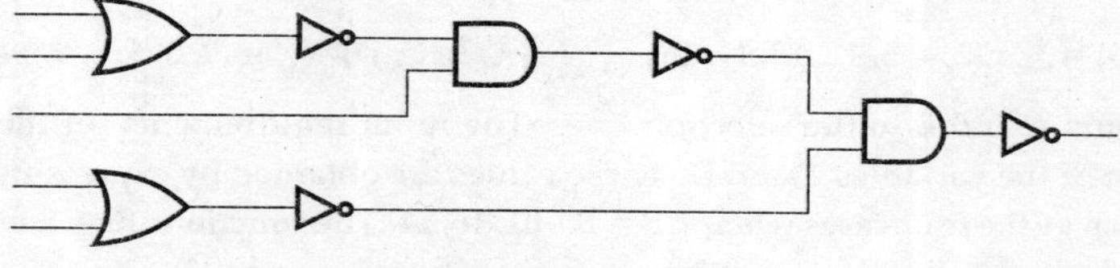

Fig. E2.1.

E2.2 **Design a Boolean function with 3 inputs and 1 output.**

a) **The output is 0 when the binary value of the input is less than 3. The output is 1 otherwise.**

b) **The output is 1 when the binary value of the input is an odd number.**

E2.3 Design a majority circuit whose output is equal to 1 if the input variables have more 0's than 1's. The output is 0 otherwise.

E2.4 Design a 3-bit combinational circuit that generates the 2's complement of the input binary number. Draw the simplified logic circuit.

E2.5 Using adders,

a) Design a 3-bit combinational circuit incrementer (a circuit that adds 1 to a 3-bit binary number).

b) Design a 3-bit combinational circuit decrementer (a circuit that subtracts 1 from a 3-bit binary number).

E2.6 A combinational circuit is specified by the following Boolean functions.

$F_1(A, B, C) = \sum m(3, 5, 6)$

$F_2(A, B, C) = \sum m(1, 4)$

$F_3(A, B, C) = \sum m(2, 3, 5, 6, 7)$

Implement the circuit with a logic low decoder and NAND or AND gates connected to the decoder outputs. Use a block diagram of the decoder. Minimize the number of inputs in the external gates.

E2.7 Implement a half adder with a decoder and NAND gates. The adder inputs are A and B. The adder produces output s and c_o.

E2.8 Implement a half subtractor with a decoder and NAND gates. The subtractor inputs A, B. The subtractor produces output d and c_o.

E2.9 Using decoder and external gates, design the combination circuit defined by the following Boolean functions.

$F_1 = x'y'z' + xz + yz, \quad F_2 = y'z' + xy' + yz', \quad F_3 = x'y'z + xy$

E2.10 Construct a 16:1 multiplexer with two 8:1 and a 2:1 MUX.

E2.11 Design a combinational circuit that compares two 3-bit numbers to check if they are equal. The circuit output is 0 if the two numbers are equal and 1 otherwise.

E2.12 Design a code converter that converts a BCD to gray code.

E2.13 Implement the Boolean function with a MUX

a) $F(A, B, C, D) = \sum m(0, 2, 5, 7, 11, 14)$ b) $F(A, B, C, D) = \prod M(3, 8, 12, 15)$

E2.14 Implement a following Boolean function with 4:1 MUX and gates.

$F_1 (A, B, C, D) = \sum m(2, 4, 5, 12, 13, 14, 15)$ $F_2 (A, B, C, D) = \sum m(2, 3, 6, 8, 9, 14, 15)$

Connect inputs A and B to the selection lines. The input requirements for the four data lines will be a function of the variables C and D. These values are obtained by expressing F as a function of C and D for each of the four cases when AB = 00, 01, 10, 11. The functions may have to be implemented with external gates and with connections to power (1) and ground (0).

E2.15 Implement a full adder with two 4:1 MUX.

E2.16 Implement a full subtractor with two 4:1 MUX.

Answers

E2.1 $F = A + B + C' + D + E$

E2.2 **a)** $x + yz$ **b)** z

E2.3 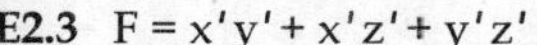

$F = x'y' + x'z' + y'z'$

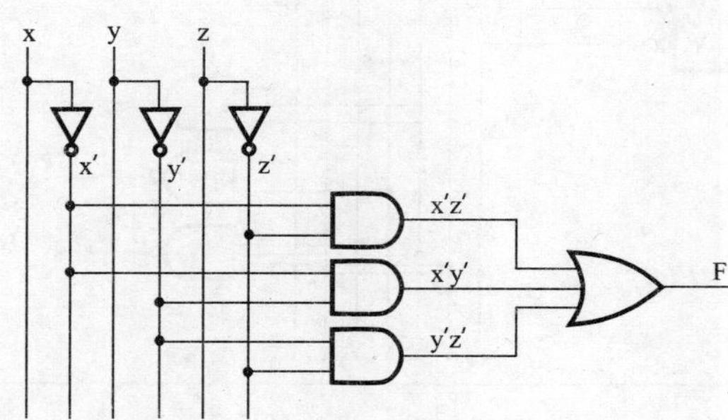

Fig: Logic circuit of F.

E2.4

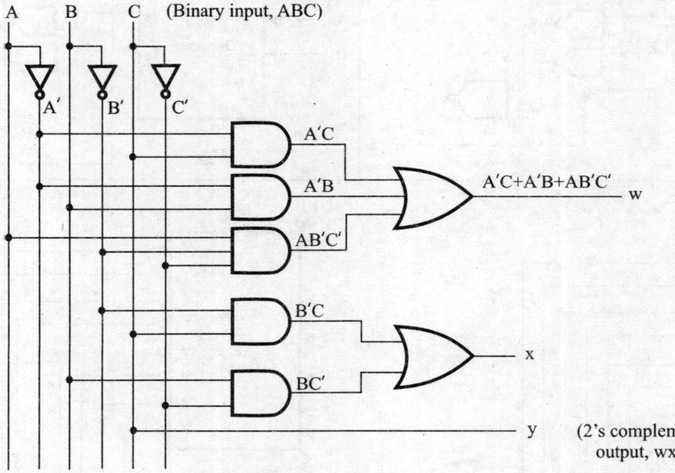

Fig: Logic circuit.

E2.5 **a)**

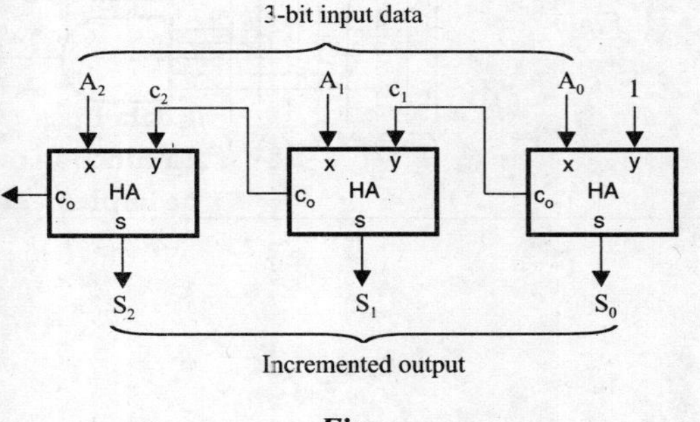

Fig.

b)

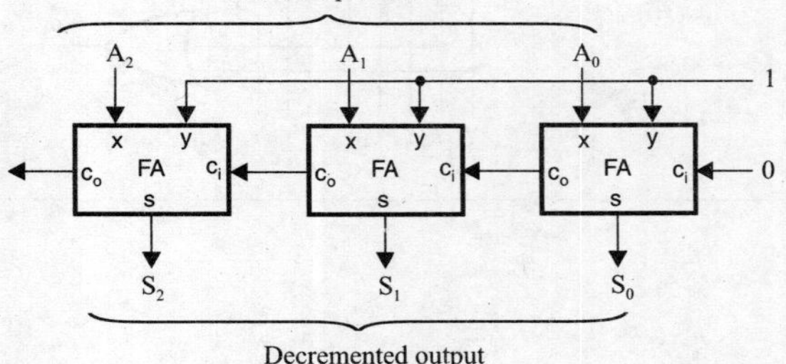

Fig.

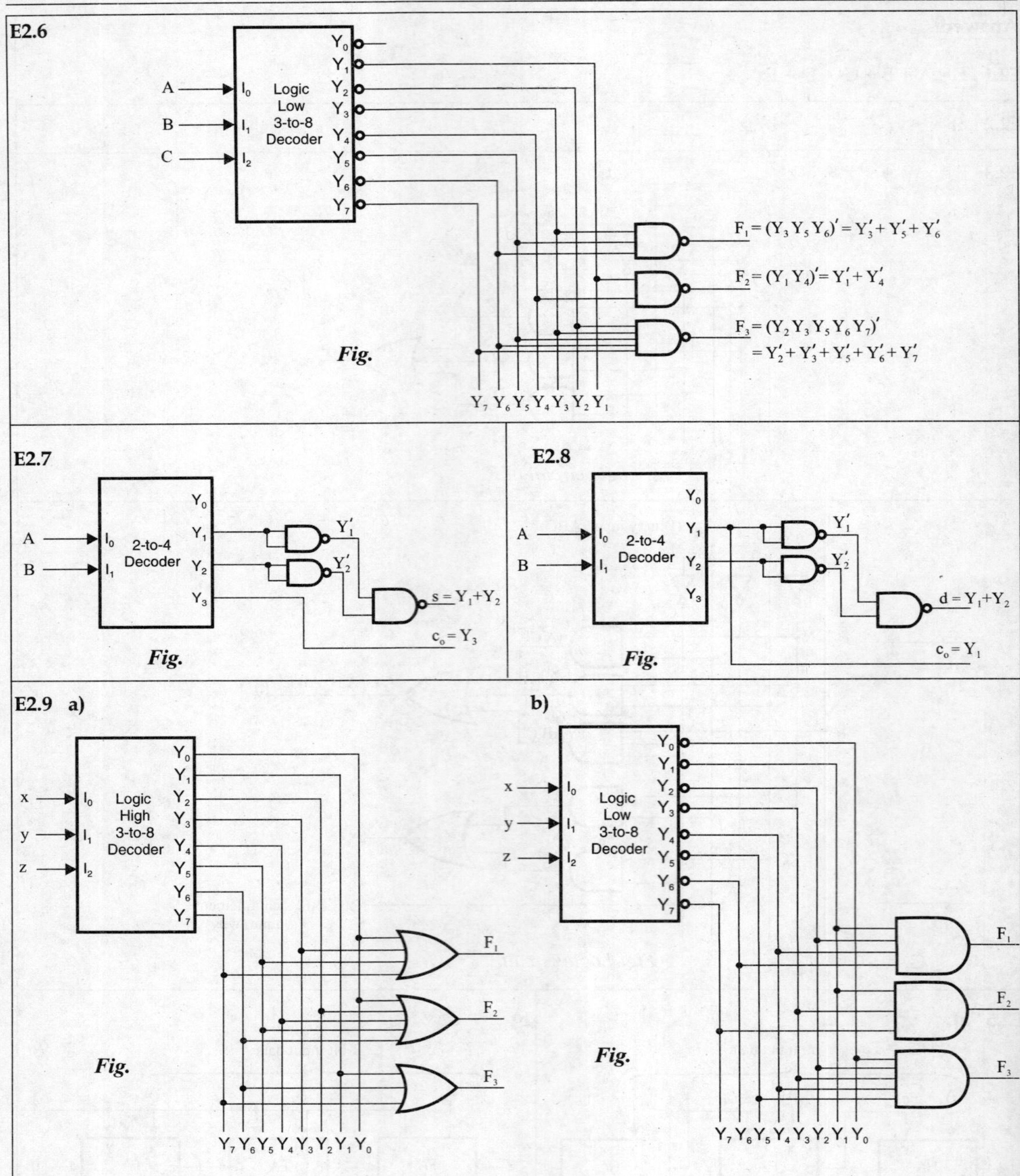
E2.6
A
B
C
I_0
I_1
I_2
Logic Low 3-to-8 Decoder
Y_0 Y_1 Y_2 Y_3 Y_4 Y_5 Y_6 Y_7
$F_1 = (Y_3 Y_5 Y_6)' = Y_3' + Y_5' + Y_6'$
$F_2 = (Y_1 Y_4)' = Y_1' + Y_4'$
$F_3 = (Y_2 Y_3 Y_5 Y_6 Y_7)'$
$= Y_2' + Y_3' + Y_5' + Y_6' + Y_7'$
Fig.
Y_7 Y_6 Y_5 Y_4 Y_3 Y_2 Y_1
E2.7
A
B
I_0
I_1
2-to-4 Decoder
Y_0 Y_1 Y_2 Y_3
Y_1'
Y_2'
$s = Y_1 + Y_2$
$c_o = Y_3$
Fig.
E2.8
A
B
I_0
I_1
2-to-4 Decoder
Y_0 Y_1 Y_2 Y_3
Y_1'
Y_2'
$d = Y_1 + Y_2$
$c_o = Y_1$
Fig.
E2.9 a)
x
y
z
I_0
I_1
I_2
Logic High 3-to-8 Decoder
Y_0 Y_1 Y_2 Y_3 Y_4 Y_5 Y_6 Y_7
F_1
F_2
F_3
Fig.
Y_7 Y_6 Y_5 Y_4 Y_3 Y_2 Y_1 Y_0
b)
x
y
z
I_0
I_1
I_2
Logic Low 3-to-8 Decoder
Y_0 Y_1 Y_2 Y_3 Y_4 Y_5 Y_6 Y_7
F_1
F_2
F_3
Fig.
Y_7 Y_6 Y_5 Y_4 Y_3 Y_2 Y_1 Y_0

E2.10

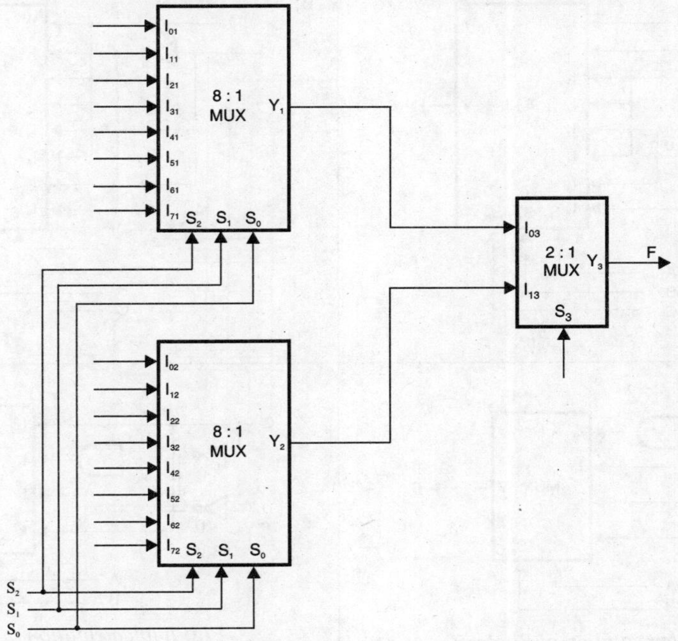

Fig: *16:1 MUX with two 8:1 and one 2:1 MUX.*

E2.11

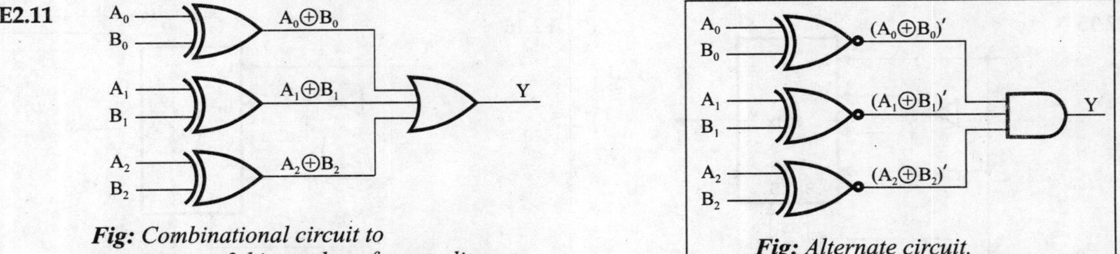

Fig: *Combinational circuit to compare two 3-bit numbers for equality.*

Fig: *Alternate circuit.*

E2.12

$G_3 = B_3$

$G_2 = B_2 + B_3$

$G_1 = B_2 B_1' + B_2' B_1$

$G_0 = B_1' B_0 + B_1 B_0'$

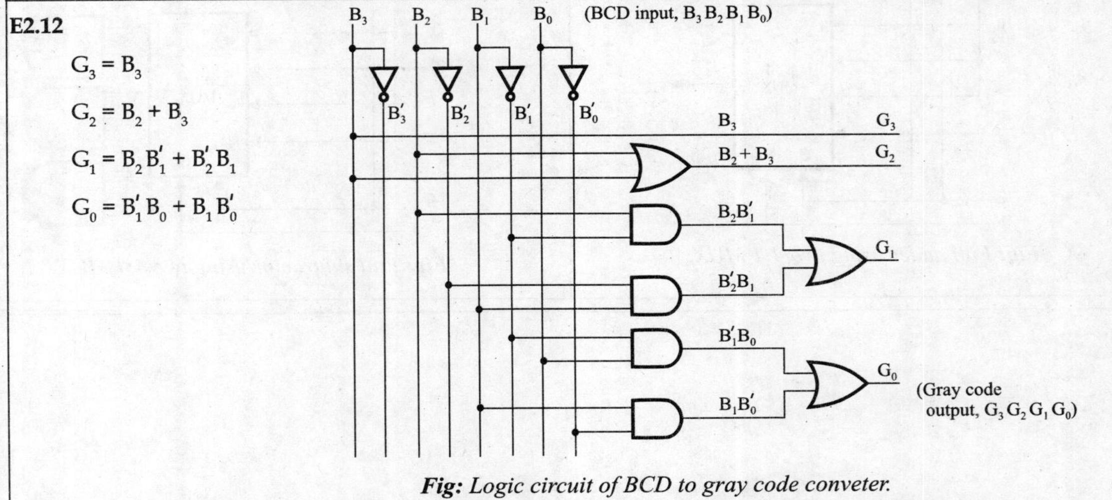

Fig: *Logic circuit of BCD to gray code conveter.*

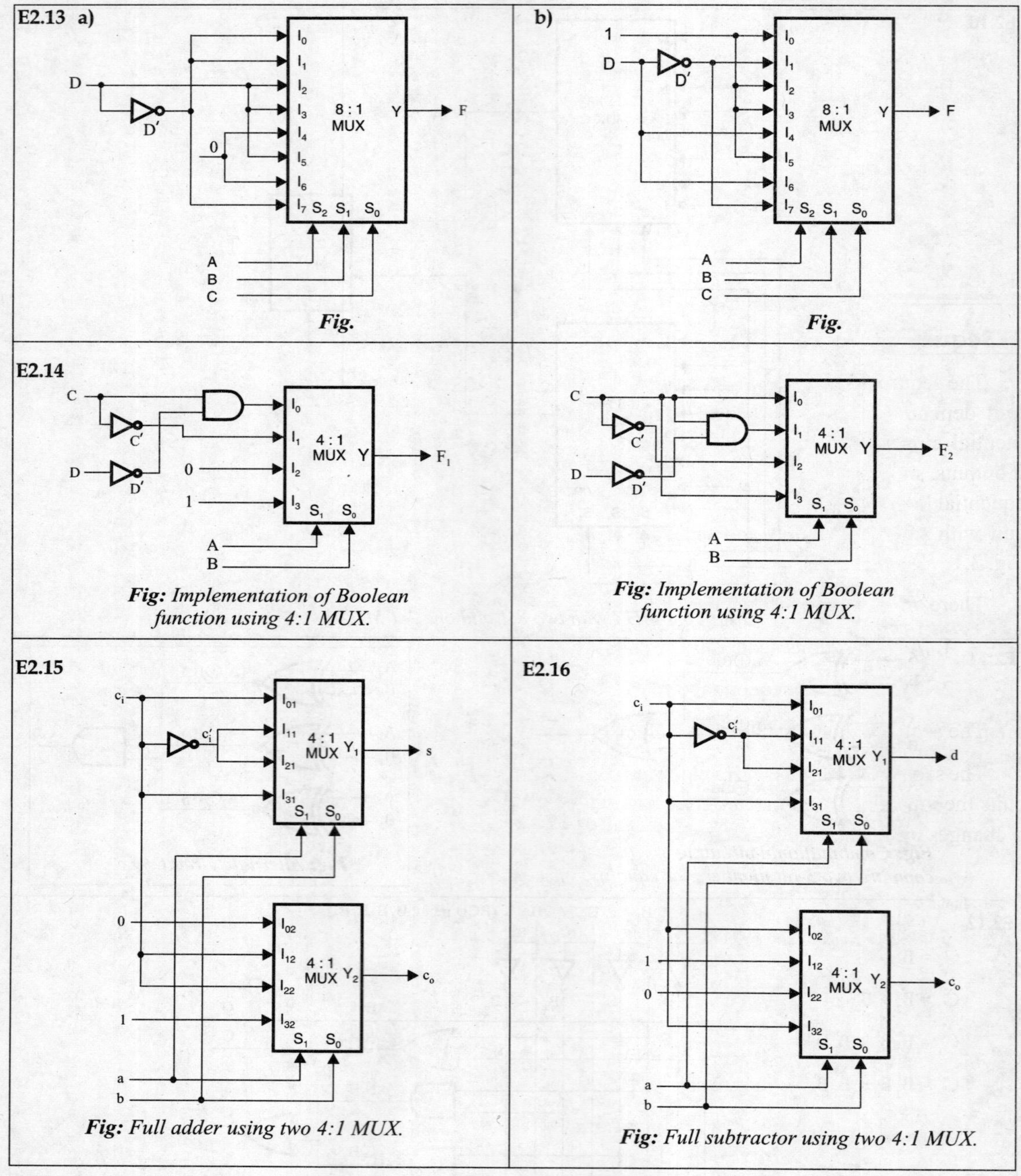

Fig.

Fig.

Fig: *Implementation of Boolean function using 4:1 MUX.*

Fig: *Implementation of Boolean function using 4:1 MUX.*

Fig: *Full adder using two 4:1 MUX.*

Fig: *Full subtractor using two 4:1 MUX.*

CHAPTER 3

SYNCHRONOUS SEQUENTIAL CIRCUITS

3.1 Sequential Logic

The sequential logic circuits are digital circuits whose output depends on present inputs and past outputs. The sequential circuits will have storage elements to store the past outputs, so that they can be fed back to input. Therefore, a sequential logic circuit can be expressed as a combinational circuit with storage elements and feedback paths as shown in Fig. 3.1.

Inputs
Outputs
Combinational circuit
Storage elements
Feedback signals
(Outputs fed back as inputs)

Fig. 3.1: Sequential logic circuit.

There are two types of sequential circuits.

1. Asynchronous sequential circuits
2. Synchronous sequential circuits

The sequential circuits without a clock are called ***asynchronous sequential circuits***.

The sequential circuits with a clock are called ***synchronous sequential circuits***. In clocked sequential circuits the inputs are recognized only when clock signal is applied otherwise ignored. Therefore, output also changes only when clock is applied.

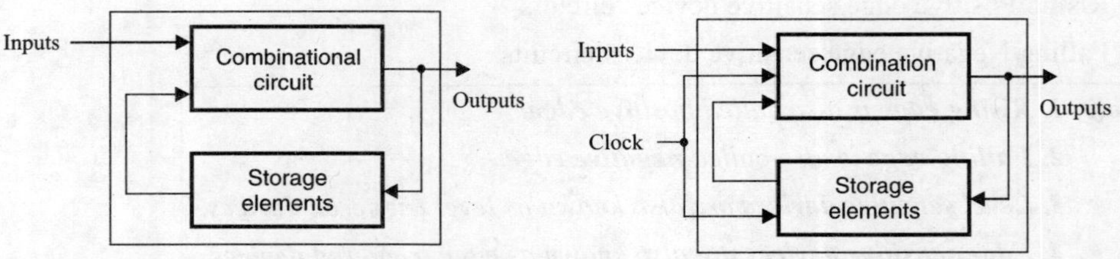

Fig. a: Asynchronous sequential circuit. *Fig. b: Synchronous sequential circuit.*

Fig. 3.2: Asynchronous and synchronous sequential circuits.

The basic storage elements in sequential circuits are latches and flip-flops. Latches are sequential circuits without clock. Flip-flops are sequential circuits with clock.

3.1.1 Clock

The clock (or clock signal) is basically a periodic square pulse signal. Usually the clock pulses will have equal ON and OFF time (or **high** and **low** time). The sum of ON and OFF time is called ***one period of clock***. The inverse of clock period is called ***clock frequency***. In today's technology, typical clock frequencies are in the order of GHz and so clock periods are in the order of nano-seconds. A clock signal will have a definite rise and fall time as shown in Fig. 3.3(b). But for simplicity, the clock signal can be expressed as shown in Fig. 3.3(a).

A clock has three edges.

1. Rising edge (Positive edge)
2. Falling edge (Negative edge)
3. Level edge

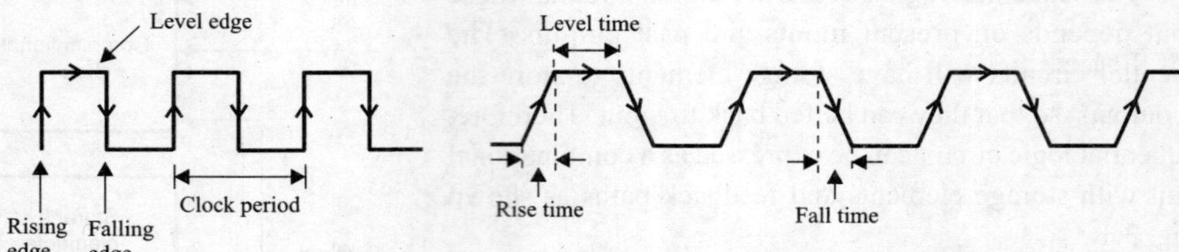

Fig. a: *Various edges of a clock signal.*

Fig. b: *Practical clock signal.*

Fig. 3.3: *Clock signal.*

Clock is used in sequential logic circuits in order to have precise timing control over changes in logic circuits. Usually, the clocked sequential circuits or synchronous sequential circuits are designed as edge sensitive device. In edge sensitive device or circuits the timing of signal (or input) recognization is more precise. Inputs are recognized only at the specified edge of the clock.

Therefore, the synchronous sequential circuits can be classified as,

- Level sensitive devices/circuits
- Rising/Positive edge sensitive devices/circuits
- Falling/Negative edge sensitive devices/circuits

> ***Note: 1. Rising edge*** *is also called* ***positive edge.***
>
> ***2. Falling edge*** *is also called* ***negative edge.***
>
> ***3. Level sensitive devices*** *are also known as* ***level triggered devices.***
>
> ***4. Edge sensitive devices*** *are also known as* ***edge triggered devices.***

Specific clock symbol is used to denote positive/negative/level edge sensitivity in sequential devices as shown in Fig. 3.4.

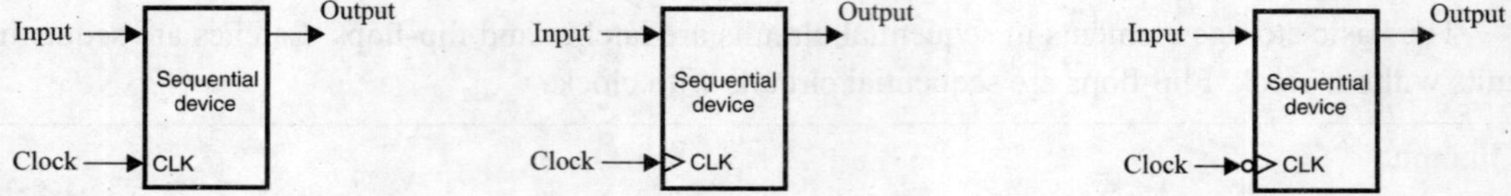

Fig. a: *Level sensitive device.* ***Fig. b:*** *Positive edge sensitive device.* ***Fig. c:*** *Negative edge sensitive device.*

Fig. 3.4: *Clock symbol for clock sensitivity in sequential device.*

3.2 Latches

A ***latch*** is the basic storage device to store a binary bit. A latch is designed with two stable states in order to store two possible values of a binary bit 1 and 0. Latches are designed with complementary inputs and outputs so that positive or negative logic can be implemented easily. The latches are sequential circuits without a clock. The two types of basic latches are SR latch and D latch.

3.2.1 SR Latch

The ***SR latch*** is a digital circuit designed to store 1 or 0 applied as two inputs called set and reset and it is provided with an output and complement of the output.

Set is an input used to set the output to 1 and it is denoted by S.

Reset is an input used to clear the output to 0 and it is denoted by R.

The two outputs are denoted as Q and Q′. Here, Q and Q′ are complements. The standard symbol of SR latch is shown in Fig. 3.5.

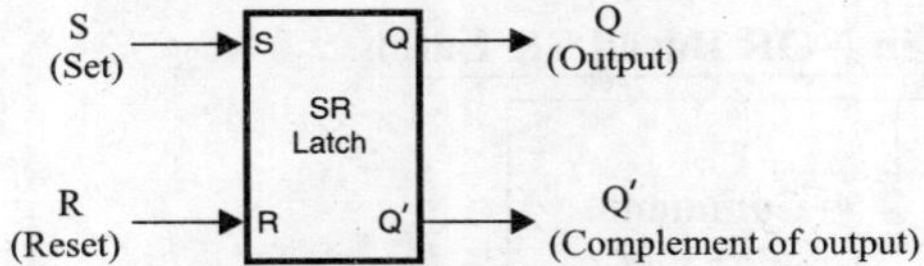

Fig. 3.5: *Symbol of SR latch.*

Table 3.1: Stable States of SR Latch

Set State	Reset State
Q = 1	Q = 0
Q′= 0	Q′= 1

The SR latch is designed to have two stable states as shown in Table 3.1. A transition from one state to another can be initiated by applying appropriate input. The SR latch can be constructed using two cross-coupled NOR gates or using two cross-coupled NAND gates as shown in Fig. 3.6.

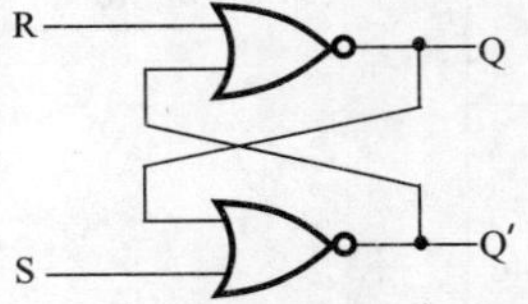

Fig. a: *SR latch using NOR gates.*

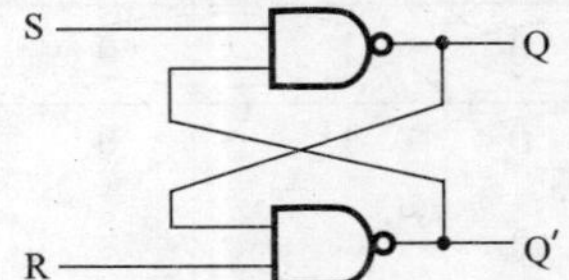

Fig. b: *SR latch using NAND gates.*

Fig. 3.6: *SR latch using NOR and NAND gates.*

3.2.2 Analysis of SR Latch using NOR Gates

(AU, Nov/Dec'18, 13 Marks)

Consider SR latch using NOR gates as shown in Fig. 3.7. The analysis of the SR latch can be carried for driving the output from set to reset state and vice-versa.

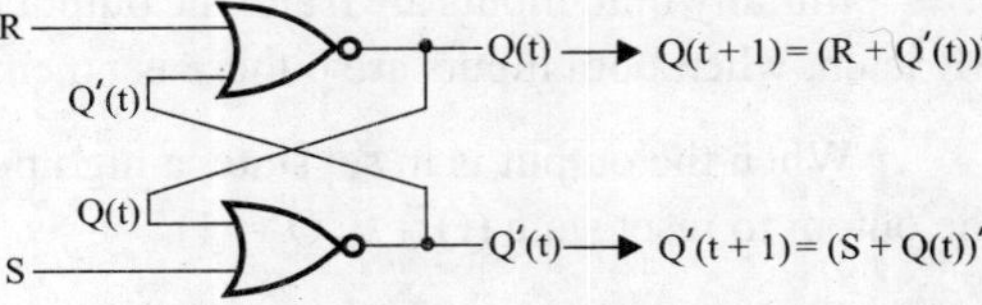

Fig. 3.7: *SR latch using NOR gates.*

Let, Q(t) and Q′(t) be present state output.

Let, Q(t + 1) and Q′(t + 1) be next state output.

A ***stable state*** is the state when present state output and next state output are same after a change in input.

Case i: Drive the output from reset to set state

Initially both inputs are 0 and the output is in reset state which is one of the stable state. In NOR based SR latch, when both inputs are 0 there is no change in output.

When the output is in reset state, a high pulse (or momentary 1) applied at set input (S = 1) drives the output to set state (Q = 1, Q′= 0).

From Fig. 3.7 we get the following equations for next state output.

$$Q(t+1) = (R + Q'(t))'$$

$$Q'(t+1) = (S + Q(t))'$$

The various transitions that occur in the NOR based SR latch when it is driven from reset to set state are listed in Table 3.2.

When the output settle at stable set state after a momentary 1 applied at S, the inputs can remain at both 0 and this will maintain the previous state (set state) and will not make any change in output.

Table 3.2: Transition from Reset to Set State in NOR Based SR Latch

Inputs		Outputs				Comment
S	R	Present State		Next State		
				Q(t + 1)	Q′(t + 1)	
		Q(t)	Q′(t)	(R + Q′(t))′	(S + Q(t))′	
0	0	0	1	-	-	Reset state
1	0	0	1	0	0	S →1
1	0	0	0	1	0	Transition
1	0	1	0	1	0	Stable set state
0	0	1	0	1	0	S → 0 No change Remain in set state

S should remain at 1 until a stable set state is achieved.

Note: Q(t) = Previous Q(t + 1) ; Q′(t) = Previous Q′(t + 1)

Case ii: Drive the output from set to reset state

Initially both inputs are 0 and the output is in set state which is one of the stable state. In NOR based SR latch, when both inputs are 0 there is no change in output.

When the output is in set state, a high pulse (or momentary 1) applied at reset input (R = 1) drives the output to reset state (Q = 0, Q′= 1).

The equations for next state output are same as that of case i.

The various transitions that occur in the NOR based SR latch when it is driven from set to reset state are listed in Table 3.3.

When the output settle at stable reset state after a momentary 1 applied at R, the inputs can remain at both 0 and this will maintain the previous state (reset state) and will not make any change in output.

Table 3.3: Transition from Set to Reset State in NOR Based SR Latch

Inputs		Outputs				Comment
S	R	Present State		Next State		
		Q(t)	Q′(t)	Q(t + 1) (R + Q′(t))′	Q′(t + 1) (S + Q(t))′	
0	0	1	0	-	-	Set state
0	1	1	0	0	0	R →1
0	1	0	0	0	1	Transition
0	1	0	1	0	1	Stable reset state
0	0	0	1	0	1	R → 0 No change Remain in reset state

R should remain at 1 until a stable reset state is achieved.

Note: Q(t) = Previous Q(t + 1) ; Q′(t) = Previous Q′(t + 1)

Based on the two cases of analysis of NOR based SR latch, the truth table of SR latch using NOR gates is obtained as shown in Table 3.4.

***Note:** Applying both R = 1 and S = 1 is illegal and lead to indeterminate or unpredictable output.*

Table 3.4: Truth Table of SR latch using NOR Gates

Inputs		Outputs	
S	R	Q	Q′
0	0	No change	
0	1	0	1
1	0	1	0
1	1	Indeterminate	

3.2.3 Analysis of SR Latch using NAND Gates

(AU, Nov/Dec'18, 13 Marks)

Consider SR latch using NAND gates as shown in Fig. 3.8. The analysis of the SR latch can be carried for driving the output from set to reset state and vice-versa.

Let, Q(t) and Q′(t) be present state output.

Let, Q(t + 1) and Q′(t + 1) be next state output.

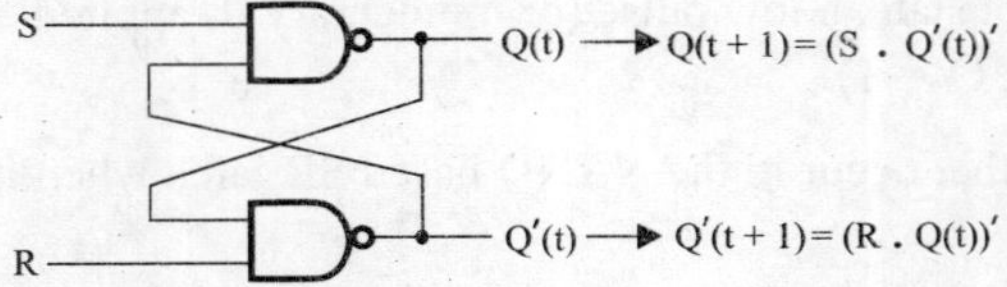

Fig. 3.8: SR latch using NAND gates.

Case i: Drive the output from reset to set state

Initially both inputs are 1 and the output is in reset state which is one of the stable state. In NAND based SR latch, when both inputs are 1 there is no change in output.

When the output is in reset state, a low pulse (or momentary 0) applied at set input (S = 0) drives the output to set state (Q = 1, Q' = 0).

From Fig. 3.8 we get the following equations for next state output.

$$Q(t+1) = (S \,.\, Q'(t))'$$

$$Q'(t+1) = (R \,.\, Q(t))'$$

The various transitions that occur in the NAND based SR latch when it is driven from reset to set state are listed in Table 3.5.

When the output settle at stable set state after a momentary 0 applied at S, the inputs can remain at both 1, and this will maintain the previous state (set state) and will not make any change in output.

Table 3.5: Transition from Reset to Set State in NAND Based SR Latch

Inputs		Outputs				Comment
S	R	Present State		Next State		
		Q(t)	Q'(t)	Q(t + 1) (S . Q'(t))'	Q'(t + 1) (R . Q(t))'	
1	1	0	1	-	-	Reset state
0	1	0	1	1	1	S → 0
0	1	1	1	1	0	Transition
0	1	1	0	1	0	Stable set state
1	1	1	0	1	0	S → 1 No change Remain in set state

S should remain at 0 until a stable set state is achieved.

Note: Q(t) = Previous Q(t + 1) ; Q'(t) = Previous Q'(t + 1)

Case ii: Drive the output from set to reset state

Initially both inputs are 1 and the output is in set state which is one of the stable state. In NAND based SR latch, when both inputs are 1 there is no change in output.

When the output is in set state, a low pulse (or momentary 0) applied at reset input (R = 0) drives the output to reset state (Q = 0, Q' = 1).

The various transitions that occur in the NAND based SR latch when it is driven from set to reset state are listed in Table 3.6.

When the output settle at stable reset state after a momentary 0 applied at R, the inputs can remain at both 1, and this will maintain the previous state (reset state) and will not make any change in output.

Table 3.6: Transition from Set to Reset State in NAND Based SR Latch

Inputs		Outputs				Comment
S	R	Present State		Next State		
		Q(t)	Q'(t)	Q(t + 1) (S . Q'(t))'	Q'(t + 1) (R . Q(t))'	
1	1	1	0	-	-	Set state
1	0	1	0	1	1	R → 0
1	0	1	1	0	1	Transition
1	0	0	1	0	1	Stable reset state
1	1	0	1	0	1	R → 1 No change Remain in reset state

R should remain at 0 until a stable reset state is achieved.

Note: Q(t) = Previous Q(t + 1) ; Q'(t) = Previous Q'(t + 1)

Based on the two cases of analysis of NAND based SR latch, the truth table of SR latch using NAND gates is obtained as shown in Table 3.7. The NAND based SR latch is available as a standard IC 74279. The pin configuration of 74279 is shown in Fig. 3.9.

Table 3.7: Truth Table of SR Latch using NAND Gates

Inputs		Outputs	
S	R	Q	Q'
0	0	Indeterminate	
0	1	1	0
1	0	0	1
1	1	No change	

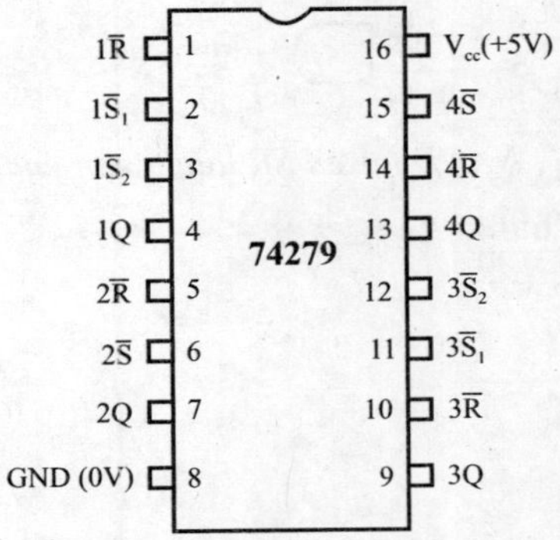

Note: *Applying both R = 0 and S = 0 is illegal and lead to indeterminate or unpredictable output.*

Pin			Pin
$1\overline{R}$	1	16	V_{cc}(+5V)
$1\overline{S}_1$	2	15	$4\overline{S}$
$1\overline{S}_2$	3	14	$4\overline{R}$
1Q	4	13	4Q
$2\overline{R}$	5	12	$3\overline{S}_2$
$2\overline{S}$	6	11	$3\overline{S}_1$
2Q	7	10	$3\overline{R}$
GND (0V)	8	9	3Q

74279

Pin	Description
$1\overline{S}_1$, $1\overline{S}_2$, $3\overline{S}_1$, $3\overline{S}_2$	Dual set inputs
$2\overline{S}$, $4\overline{S}$	Set inputs
$1\overline{R}$, $2\overline{R}$, $3\overline{R}$, $4\overline{R}$	Reset inputs
1Q, 2Q, 3Q, 4Q	Outputs
V_{cc}	Supply voltage (+5V)
GND	Ground (0V)

Note: *Latches 1 and 3 has two set inputs $\overline{S}_1$, $\overline{S}_2$.*

Fig. 3.9: *Pin configuration of quad SR latch IC 74279.*

Enable Circuit for SR Latch using NAND Gate

An enable circuit can be added to NAND based SR latch such that the inputs are allowed only when enable is active. The NAND operation of each input with a logic **high** enable signal, E as shown in Fig. 3.10 can be used as enable circuit. The truth table of enable circuit is shown in Table 3.8.

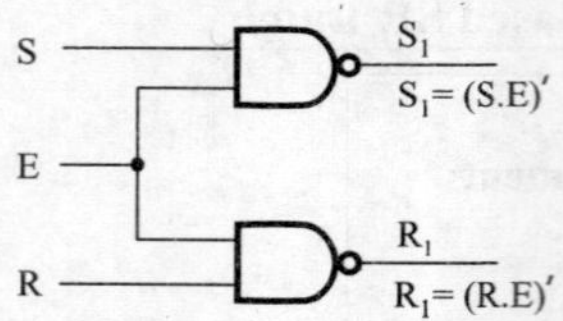

Fig. 3.10: Enable circuit for SR latch.

Table 3.8: Truth Table of Enable Circuit

E	S	R	S_1	R_1
0	0	0	1	1
0	0	1	1	1
0	1	0	1	1
0	1	1	1	1
1	0	0	1	1
1	0	1	1	0
1	1	0	0	1
1	1	1	0	0

From Table 3.8 it can be observed that,

As long as E = 0, both S_1 and R_1 are both 1.

As long as E = 1, both S_1 and R_1 are complement of S and R.

On comparing Table 3.7 and 3.8 we can conclude that on connecting enable circuit to NAND based SR latch as shown in Fig. 3.11, there is no change in output as long as E = 0 and output is allowed to change only when E = 1.

SR Latch using NAND Gates and with Enable Circuit

The NAND based SR latch with logic **high** enable input is shown in Fig. 3.11. From the knowledge of Tables 3.7 and 3.8, the truth table of SR latch with enable is formed as shown in Table 3.9.

When E = 0, $S_1 = R_1 = 1$; no change in output

When E = 1, $S_1 = S', R_1 = R'$; inputs are complemented

When E = 1, the next state output equations are,

$$Q(t+1) = (S_1 \cdot Q'(t))'$$

$$Q'(t+1) = (R_1 \cdot Q(t))'$$

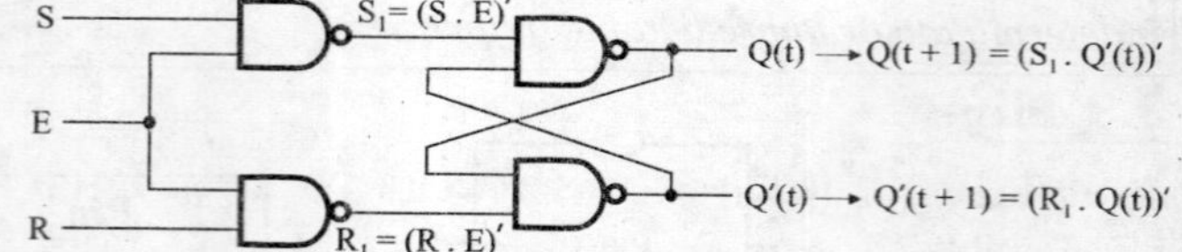

Fig. 3.11: NAND based SR latch with enable.

Table 3.9: Truth Table of NAND based SR Latch with Enable

Inputs			Intermediate Signal		Outputs
E	S	R	S_1	R_1	Q Q'
0	×	×	1	1	No change
1	0	0	1	1	No change
1	0	1	1	0	0 1 (Reset state)
1	1	0	0	1	1 0 (Set state)
1	1	1	0	0	Indeterminate

On comparing Tables 3.4 and 3.9, it can be observed that the truth table of NAND based SR latch with enable is same as NOR based SR latch without enable. Therefore, working of SR latch of Fig. 3.11 is same as working of NOR based SR latch of Fig. 3.7.

3.2.4 D Latch

A single input latch can be developed from SR latch, if the inputs to SR latch are always complementary inputs. The complementary input conditions can be achieved by adding an inverter to one of the inputs of SR latch. The input D is applied to S and complement of D is applied to R as shown in Fig. 3.12. Now SR latch is single input latch called D latch.

Table 3.10: Stable States of D Latch

Inputs	Outputs	State
D = S = 1, R = 0	Q = 1, Q′ = 0	Set state
D = S = 0, R = 1	Q = 0, Q′ = 1	Reset state

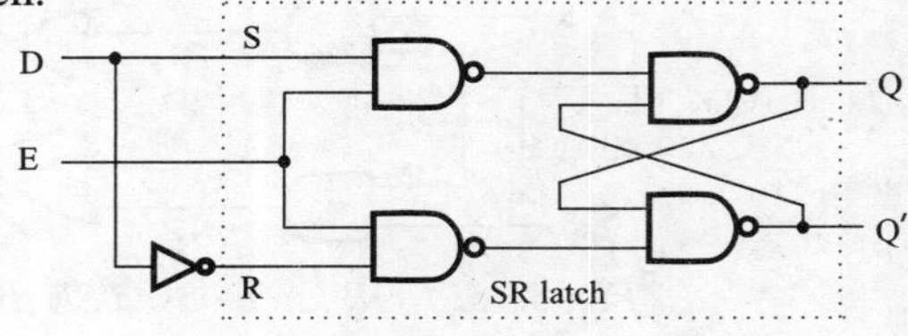

Fig. 3.12: D latch.

Therefore, D latch is a single input latch with two stable states which are used to store two different values 0 and 1 of a binary data. Hence, the D latch is used as a basic storage device in digital systems.

The stable states of D latch are shown in Table 3.10. The truth table of D latch is shown in Table 3.11 and the standard symbol of D latch is shown in Fig. 3.13.

Table 3.11: Truth Table of D Latch

E	D	Q	Q′	Comment
0	×	No change		-
1	0	0	1	Reset state
1	1	1	0	Set state

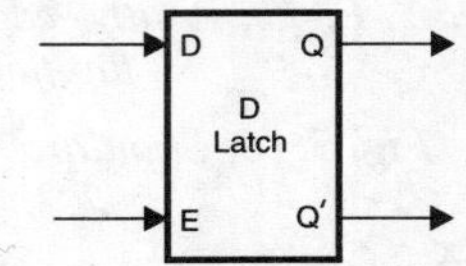

Fig. 3.13: Symbol of D latch.

3.3 Flip-Flops

In latches with enable, it is observed that inputs are recognized only when enable is active. Therefore, it is possible to replace enable by a momentary pulse called ***clock***, so that inputs can be recognized only at a specified time. Such a device is called a flip-flop.

Flip-flop is a storage device constructed using latch and controlled by enable signal called clock. Flip-flops are made sensitive to level/positive edge/negative edge clock.

The different types of flip-flops are,

- SR flip-flop
- JK flip-flop
- T flip-flop (or Toggle flip-flop)
- D flip-flop (or Data flip-flop)

3.3.1 SR Flip-Flop

The ***SR flip-flop*** is a digital device with two stable states: set and reset states. Hence it can be used as a switch or as a storage device. The SR latch with enable shown in Fig. 3.11 becomes a SR flip-flop, if the enable is replaced by a clock as shown in Fig. 3.14.

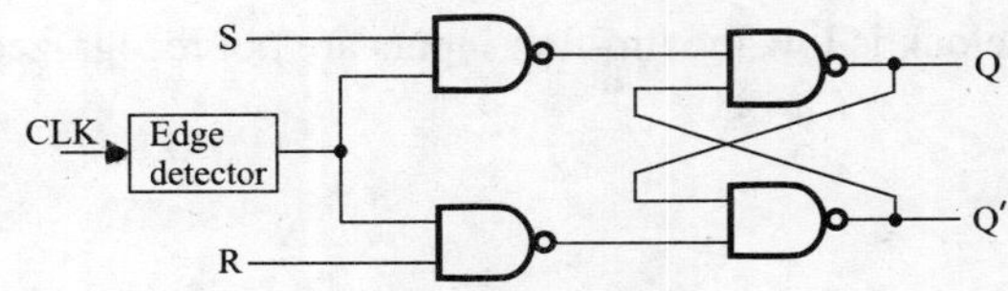

Fig. 3.14: NAND latch based SR flip-flop.

Similarly, if an AND gate based clock circuit is added to NOR based SR latch of Fig. 3.7 it will work as SR flip-flop. The NOR latch based SR flip-flop is shown in Fig. 3.15. The truth table of SR flip-flop is shown in Table 3.12 and standard symbols are shown in Fig. 3.16. The drawback in SR flip-flop is that it has an indeterminate output state.

Fig. 3.15: NOR latch based SR flip-flop.

Table 3.12: Truth Table of SR Flip-Flop

Inputs		Outputs	
S	**R**	**Q**	**Q′**
0	0	No change	
0	1	0	1
1	0	1	0
1	1	Indeterminate	

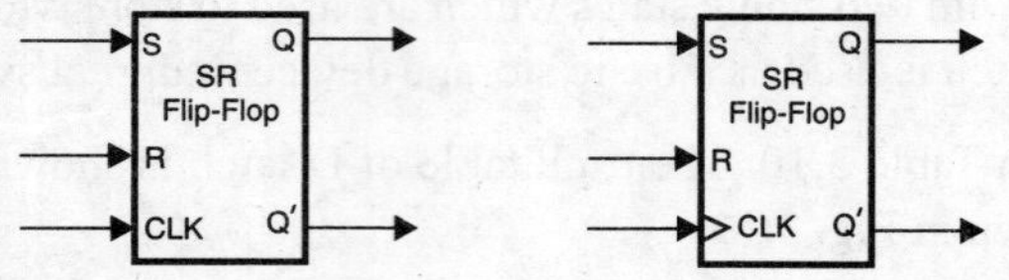

Fig. a: Level triggered SR flip-flop. *Fig. b: Positive edge triggered SR flip-flop.* *Fig. c: Negative edge triggered SR flip-flop.*

Fig. 3.16: Standard symbols of SR flip-flop.

Verification of SR Flip-Flop Truth Table

The input-output relations of SR flip-flop shown in truth table are verified by taking following six cases of inputs and initial outputs.

Case i: S = R = 0, Q = 0, Q′ = 1

Case ii: S = R = 0, Q = 1, Q′ = 0

Case iii: S = 0, R = 1, Q = 1, Q′ = 0

Case iv: S = 1, R = 0, Q = 0, Q′ = 1

Case v: S = R = 1, Q = 0, Q′ = 1

Case vi: S = R = 1, Q = 1, Q′ = 0

The various state transitions in the above six cases of inputs and outputs are diagrammatically shown in Figs. 3.17 to 3.22. In each case, the condition of flip-flop when a clock pulse is applied is shown in these figures. The condition of flip-flop before and after clock are also shown in these figures.

The clock is a narrow pulse with duration of **high** level, equal to a transition time of a NAND gate, so that the clock is valid only for a short time to allow the flip-flop inputs to reach the input of latch. When clock is **low** the flip-flop inputs are not recognized.

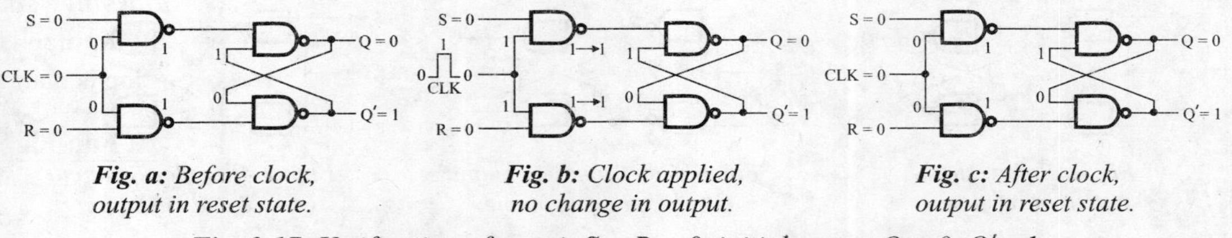

Fig. a: *Before clock, output in reset state.*

Fig. b: *Clock applied, no change in output.*

Fig. c: *After clock, output in reset state.*

Fig. 3.17: *Verification of case i: S = R = 0, initial output Q = 0, Q' = 1.*

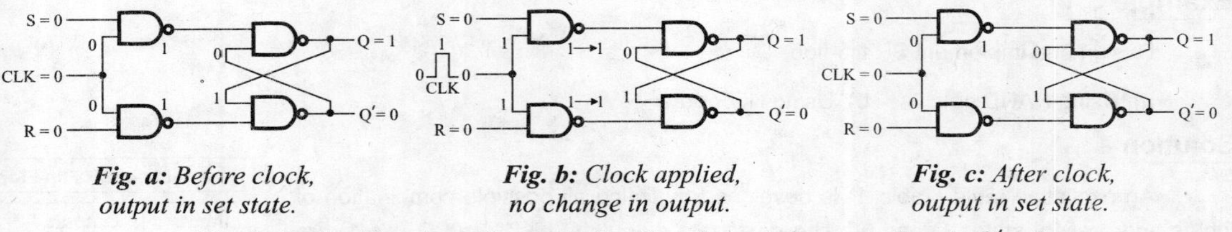

Fig. a: *Before clock, output in set state.*

Fig. b: *Clock applied, no change in output.*

Fig. c: *After clock, output in set state.*

Fig. 3.18: *Verification of case ii: S = R = 0, initial output Q = 1, Q' = 0.*

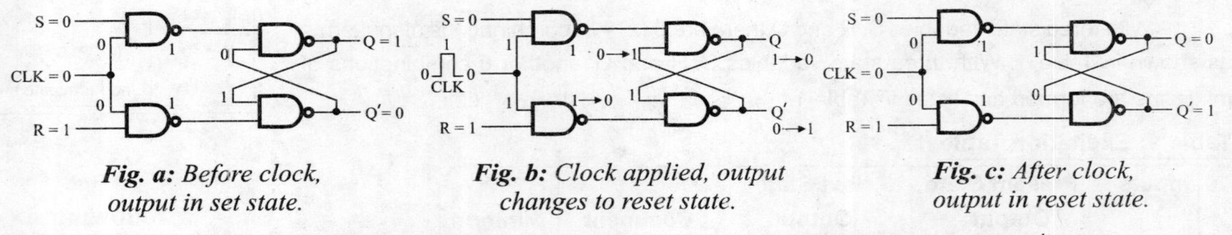

Fig. a: *Before clock, output in set state.*

Fig. b: *Clock applied, output changes to reset state.*

Fig. c: *After clock, output in reset state.*

Fig. 3.19: *Verification of case iii: S = 0, R = 1, initial output Q = 1, Q' = 0.*

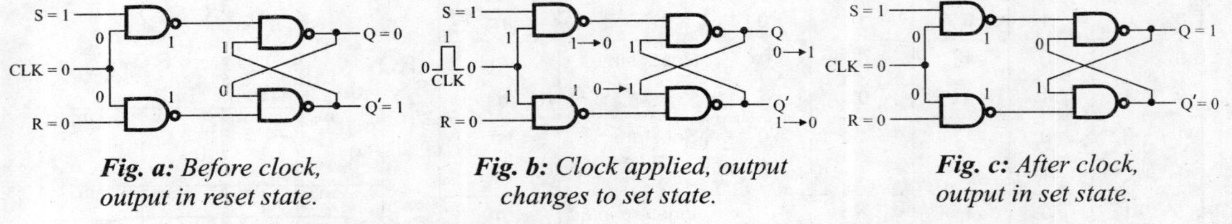

Fig. a: *Before clock, output in reset state.*

Fig. b: *Clock applied, output changes to set state.*

Fig. c: *After clock, output in set state.*

Fig. 3.20: *Verification of case iv: S = 1, R = 0, initial output Q = 0, Q' = 1.*

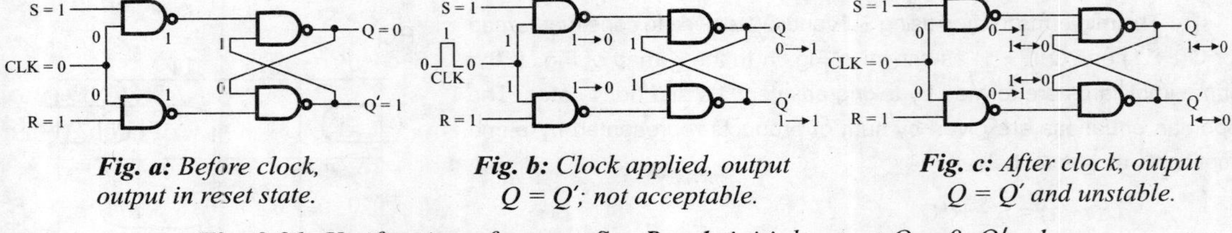

Fig. a: *Before clock, output in reset state.*

Fig. b: *Clock applied, output Q = Q'; not acceptable.*

Fig. c: *After clock, output Q = Q' and unstable.*

Fig. 3.21: *Verification of case v: S = R = 1, initial output Q = 0, Q' = 1.*

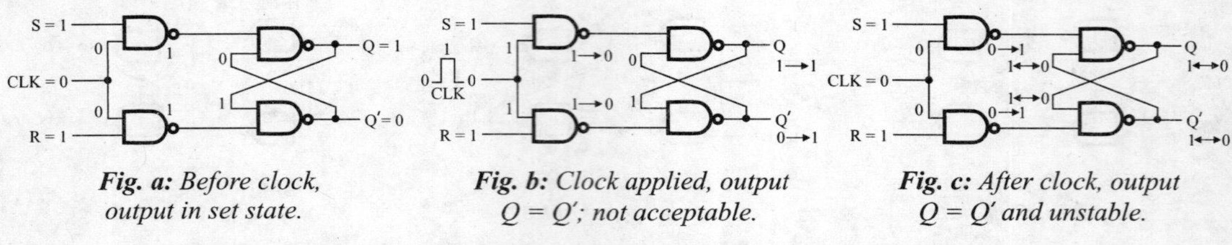

Fig. a: Before clock, output in set state. *Fig. b: Clock applied, output Q = Q′; not acceptable.* *Fig. c: After clock, output Q = Q′ and unstable.*

Fig. 3.22: Verification of case vi: S = R = 1, initial output Q = 1, Q′= 0.

Example 3.1

Design and implement SR flip-flop.

a) Using NAND gate b) Using NOR gate

Solution

An excitation table (Table 1) is developed by listing all possible combination of inputs and present states Q and Q′. The next state outputs Q(t + 1) and Q′(t + 1) are determined for all the combinations of S, R, Q and Q′ from the knowledge of truth table of SR flip-flop and listed in Table 1.

Truth Table of SR Flip-Flop

Inputs		Outputs	
S	R	Q	Q′
0	0	No change	
0	1	0	1
1	0	1	0
1	1	Indeterminate	

With three state variables S, R and Q there are 8 ($2^3 = 8$) combinations of minterms as shown in Table 1. With three state variables S, R and Q′ another 8 combinations of minterms are formed and listed in Table 1.

Table 1: Excitation Table

Inputs		Present State Output		Next State Output		Comment	Minterm
S	R	Q	Q′	Q(t + 1)	Q′(t+1)		SRQ
0	0	0	1	0	1	No change	m_0
0	0	1	0	1	0		m_1
0	1	0	1	0	1	Reset state	m_2
0	1	1	0	0	1		m_3
1	0	0	1	1	0	Set state	m_4
1	0	1	0	1	0		m_5
1	1	0	1	×	×	Indeterminate	m_6
1	1	1	0	×	×		m_7

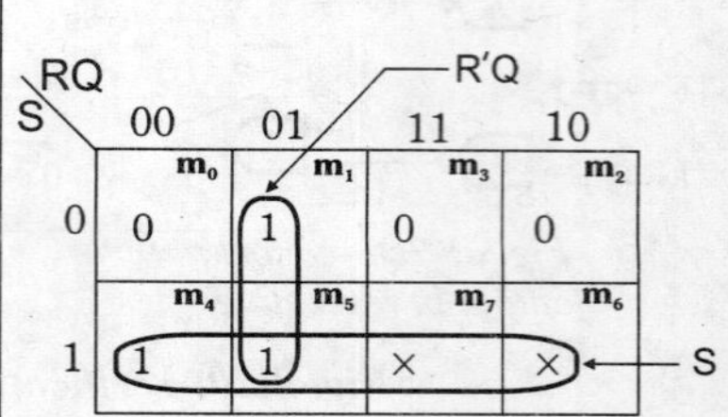

Fig. a: K-map for Q(t + 1).

a) Design using NAND Gate

The minterms formed using S, R and Q are used to construct K-map for Q(t + 1) and Q′(t + 1) as shown in Fig. 1. In the K-map of Fig. 1, the prime implicants are formed by taking adjacent 1's and don't-cares. The Boolean equations are given by sum of products represented by prime implicants as shown below:

$$Q(t+1) = S + R'Q$$

$$Q'(t+1) = R + S'Q'$$

The SR flip-flop that satisfy the above equations is shown in Fig. 2. The SR flip-flop output is verified in Table 2.

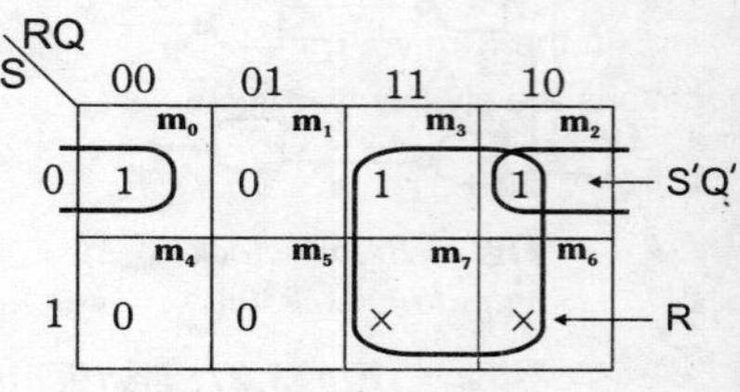

Fig. b: K-map for Q′(t + 1).

Fig. 1: K-map for SOP form of Q(t + 1) and Q′(t + 1).

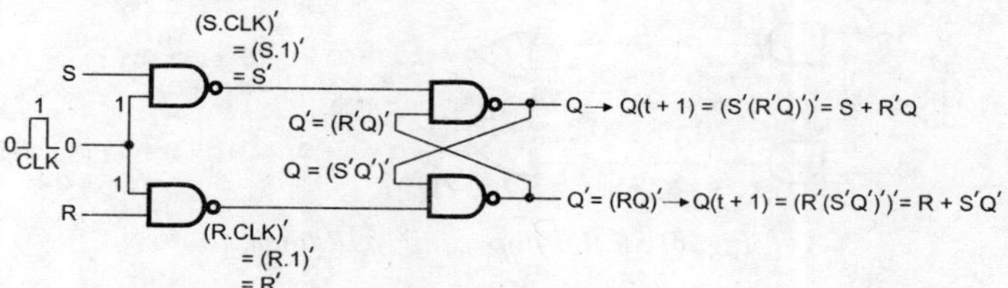

Fig. 2: SR flip-flop using NAND latch.

Table 2: Verification of SR Flip-Flop Outputs

Inputs		Present State Output						Next State Output		Comment
S	R	Q	Q′	S′	R′	R′Q	S′Q′	Q(t + 1) = S + R′Q	Q′(t + 1) = R + S′Q′	
0	0	0	1	1	1	0	1	0	1	No change
0	0	1	0	1	1	1	0	1	0	
0	1	0	1	1	0	0	1	0	1	Reset state
0	1	1	0	1	0	0	0	0	1	
1	0	0	1	0	1	0	0	1	0	Set state
1	0	1	0	0	1	1	0	1	0	
1	1	0	1	0	0	0	0	1	1	Not acceptable
1	1	1	0	0	0	0	0	1	1	

b) Design using NOR Gate

The excitation table (Table 1) formed for NAND based SR latch design is also used for NOR based SR latch. The minterms formed using S, R and Q are used to construct K-map for Q(t + 1) and Q′(t + 1) as shown in Fig. 3.

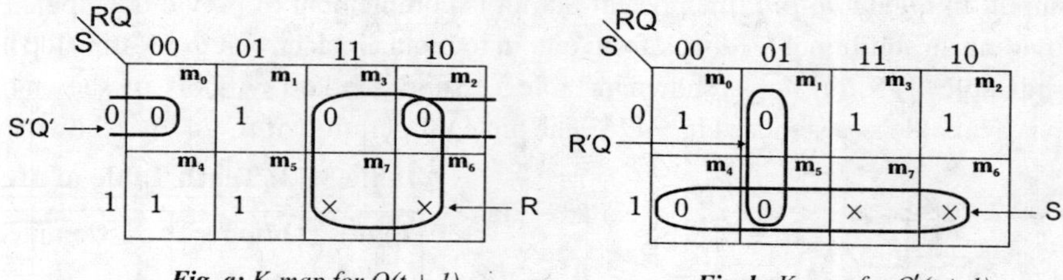

Fig. a: K-map for Q(t + 1). *Fig. b: K-map for Q′(t + 1).*

Fig. 3: K-map for POS form of Q(t + 1) and Q′(t + 1).

In the K-map of Fig. 3, the prime implicants are formed by taking adjacent 0's and don't-cares. The Boolean equations are given by complement of sum of products represented by prime implicants as shown below:

Q(t + 1) = (R + S′Q′)′	Q′(t + 1) = (S + R′Q)′
= R′(S′Q′)′	= S′(R′Q)′
= R′(S + Q)	= S′(R + Q′)

The SR flip-flop that satisfy the above equations is shown in Fig. 4. The SR flip-flop output is verified in Table 3.

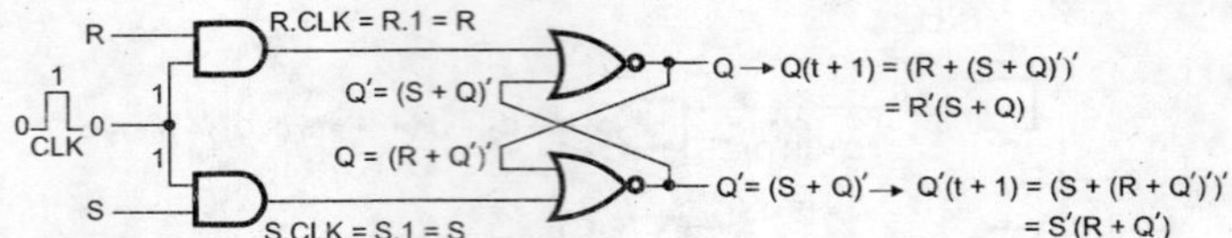

Fig. 4: SR flip-flop using NOR latch.

Table 3: Verification of SR Flip-Flop Outputs

Inputs		Present State Output						Next State Output		Comment
S	R	Q	Q′	R′	S + Q	S′	R + Q′	Q(t + 1) = R′(S + Q)	Q′(t + 1) = S′(R + Q′)	
0	0	0	1	1	0	1	1	0	1	No change
0	0	1	0	1	1	1	0	1	0	
0	1	0	1	0	0	1	1	0	1	Reset state
0	1	1	0	0	1	1	1	0	1	
1	0	0	1	1	1	0	1	1	0	Set state
1	0	1	0	1	1	0	0	1	0	
1	1	0	1	0	1	0	1	0	0	Not acceptable
1	1	1	0	0	1	0	1	0	0	

3.3.2 JK Flip-Flop

The JK flip-flop is a digital device with two stable states: set and reset states, similar to SR flip-flop but without indeterminate output.

In JK flip-flop, the indeterminate output of SR flip-flop is converted to another useful output called toggle output. In ***toggle output***, the present output is complement of previous output. This can be achieved by providing an additional feedback from output to input clock circuit in SR flip-flop as shown in Fig. 3.23. The truth table of JK flip-flop is shown in Table 3.13 and standard symbols are shown in Fig. 3.24. The JK flip-flop is available as a standard IC 7476. The pin configuration of IC 7476 is shown in Fig. 3.25.

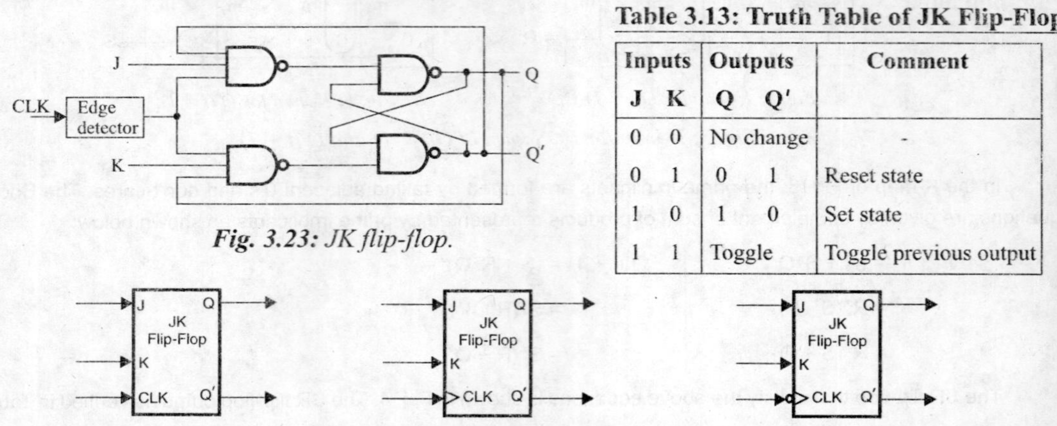

Table 3.13: Truth Table of JK Flip-Flop

Inputs		Outputs		Comment
J	K	Q	Q′	
0	0	No change		-
0	1	0	1	Reset state
1	0	1	0	Set state
1	1	Toggle		Toggle previous output

Fig. 3.23: JK flip-flop.

Fig. a: Level triggered. *Fig. b: Positive edge triggered.* *Fig. c: Negative edge triggered.*

Fig. 3.24: Standard symbols of JK flip-flop.

Pin	Label		Pin	Label
1	CLK1	7476	16	K_1
2	$\overline{PR1}$		15	Q_1
3	$\overline{CLR1}$		14	$\overline{Q}_1$
4	J_1		13	GND (0V)
5	V_{cc} (+5V)		12	K_2
6	CLK2		11	Q_2
7	$\overline{PR2}$		10	$\overline{Q}_2$
8	$\overline{CLR2}$		9	J_2

Pin	Description
J, K	Inputs
$\overline{CLR}_1$, $\overline{CLR}_2$	Clear
CLK_1, CLK_2	Clock
$\overline{PR}_1$, $\overline{PR}_2$	Set output
Q_1, Q_2	Outputs
$\overline{Q}_1$, $\overline{Q}_2$	Complement of outputs
V_{cc}	Supply voltage (+5V)
GND	Ground (0V)

***Fig. 3.25:** Pin configuration of dual JK flip-flop IC 7476.*

Verification of JK Flip-Flop Truth Table

The input-output relations of JK flip-flop shown in truth table are verified by taking following six cases of inputs and initial outputs.

Case i: J = K = 0, Q = 0, Q′ = 1

Case ii: J = K = 0, Q = 1, Q′ = 0

Case iii: J = 0, K = 1, Q = 1, Q′ = 0

Case iv: J = 1, K = 0, Q = 0, Q′ = 1

Case v: J = K = 1, Q = 0, Q′ = 1

Case vi: J = K = 1, Q = 1, Q′ = 0

The various state transitions in the above six cases of inputs and outputs are diagrammatically shown in Figs. 3.26 to 3.31. In each case, the condition of flip-flop when a clock pulse is applied is shown in these figures. The condition of flip-flop before and after clock are also shown in these figures. The clock is a narrow pulse with duration of **high** level, equal to a transition time of a NAND gate, so that the clock is valid only for a short time to allow the flip-flop inputs to reach the input of latch. When clock is **low** the flip-flop inputs are not recognized.

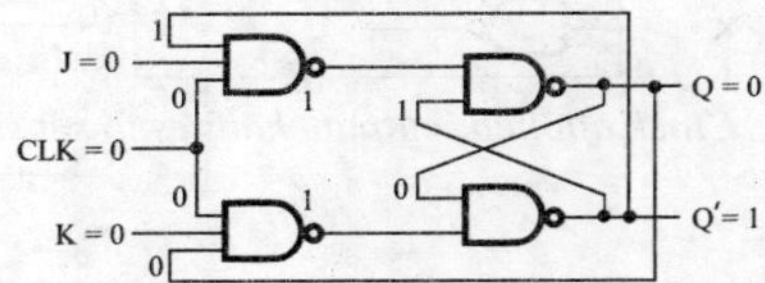

***Fig. a:** Before clock, output in reset state.*

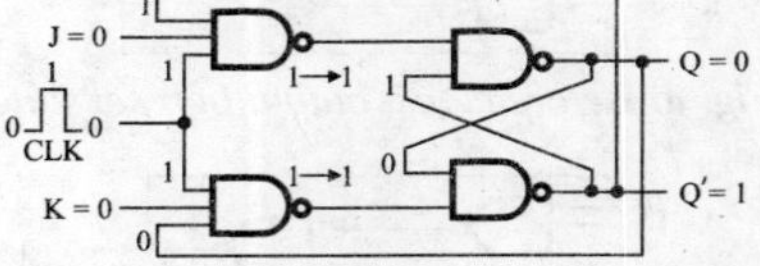

***Fig. b:** Clock applied, no change in output.*

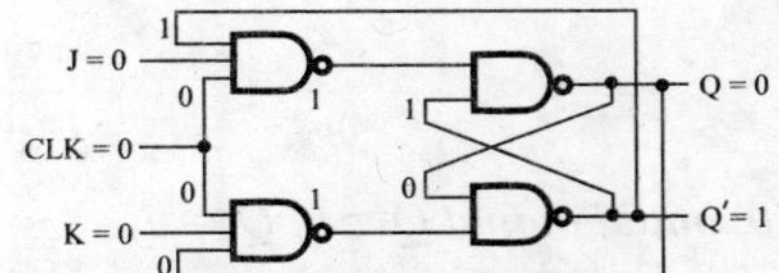

***Fig. c:** After clock, output in reset state.*

***Fig. 3.26:** Verification of case i: J = K = 0, initial output Q = 0, Q′ = 1.*

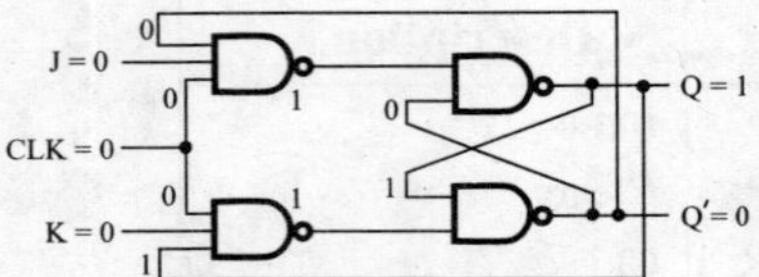

Fig. a: *Before clock, output in set state.*

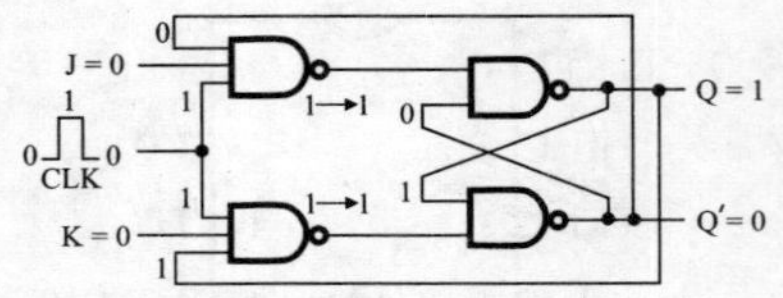

Fig. b: *Clock applied, no change in output.*

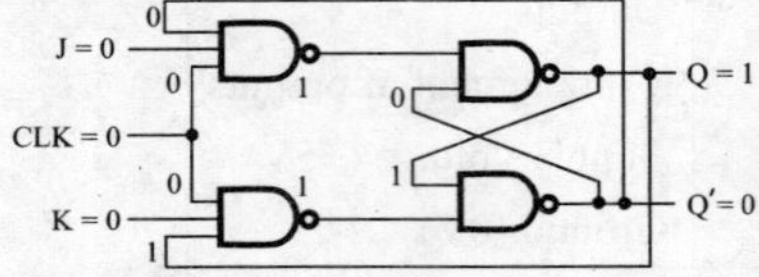

Fig. c: *After clock, output in set state.*

Fig. 3.27: *Verification of case ii: J = K = 0, initial output Q = 1, Q′ = 0.*

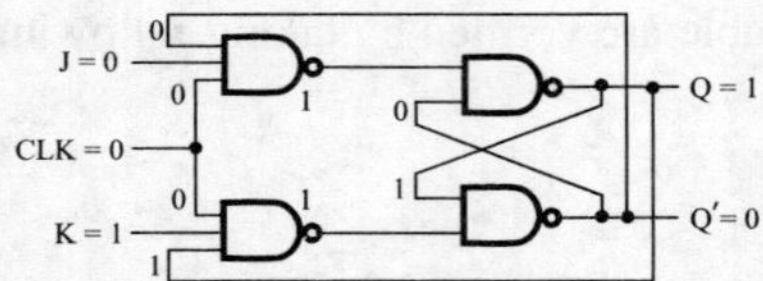

Fig. a: *Before clock, output in set state.*

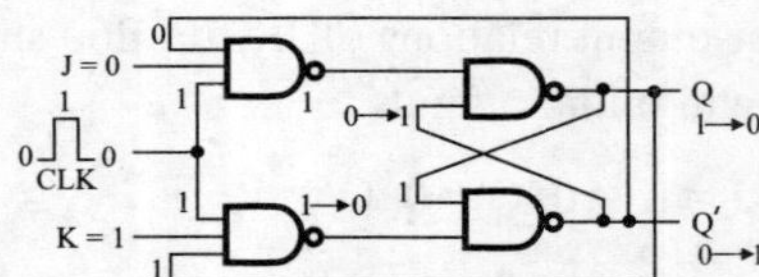

Fig. b: *Clock applied, output changes to reset state.*

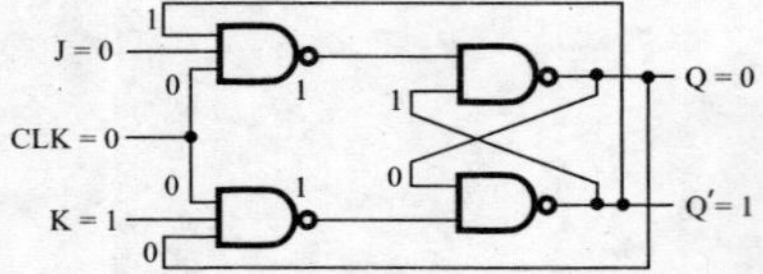

Fig. c: *After clock, output in reset state.*

Fig. 3.28: *Verification of case iii: J = 0, K = 1, initial output Q = 1, Q′ = 0.*

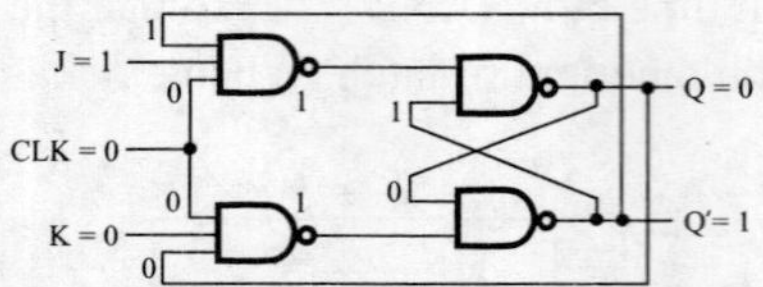

Fig. a: *Before clock, output in reset state.*

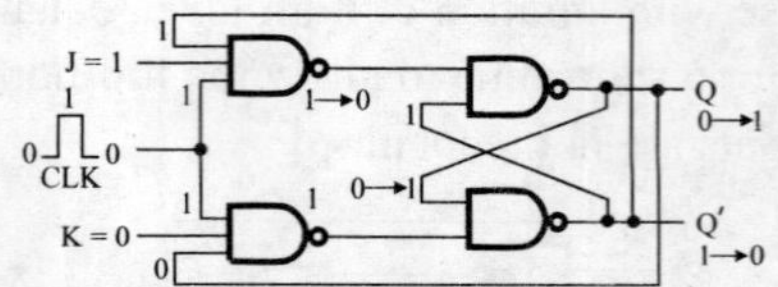

Fig. b: *Clock applied, output changes to set state.*

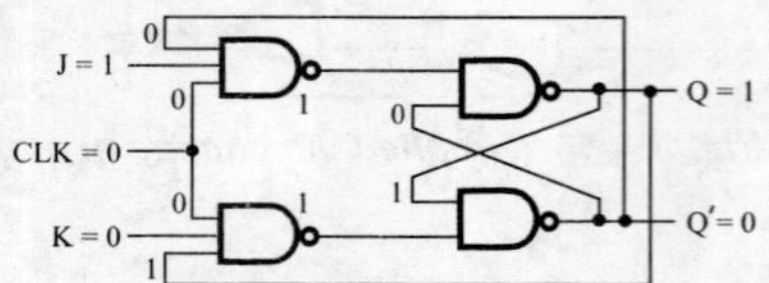

Fig. c: *After clock, output in set state.*

Fig. 3.29: *Verification of case iv: J = 1, K = 0, initial output Q = 0, Q′ = 1.*

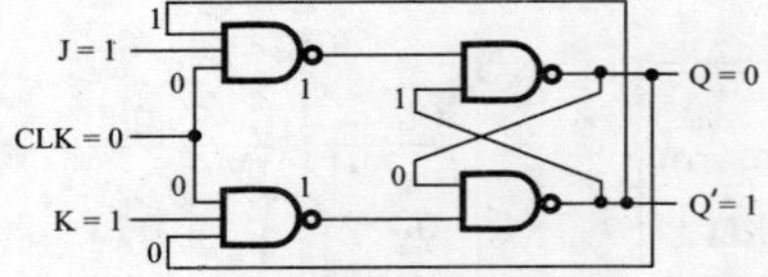

Fig. a: *Before clock, output in reset state.*

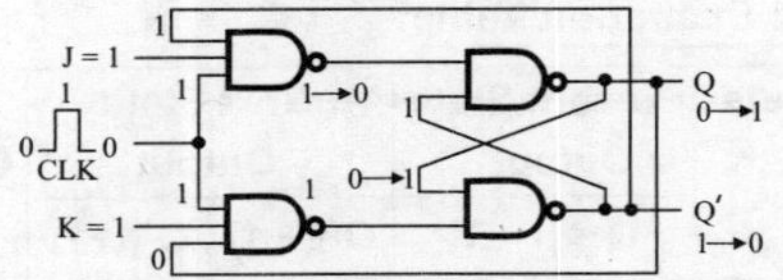

Fig. b: *Clock applied, output toggles.*

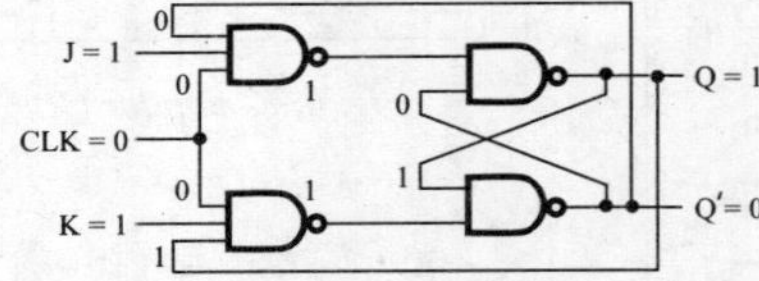

Fig. c: *After clock, output in set state.*

Fig. 3.30: *Verification of case v: J = K = 1, initial output Q = 0, Q′ = 1.*

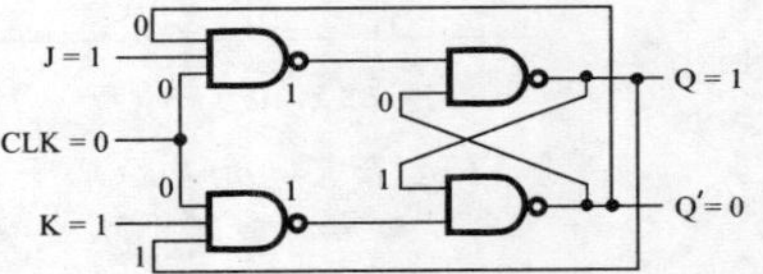

Fig. a: *Before clock, output in set state.*

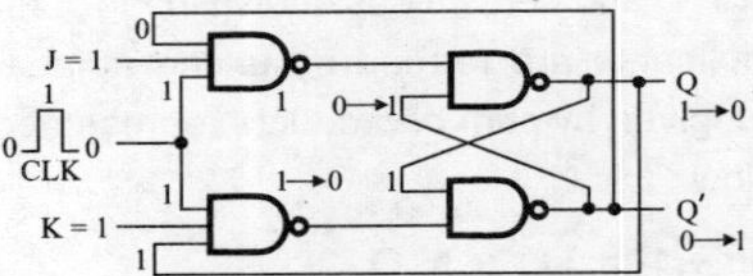

Fig. b: *Clock applied, output toggles.*

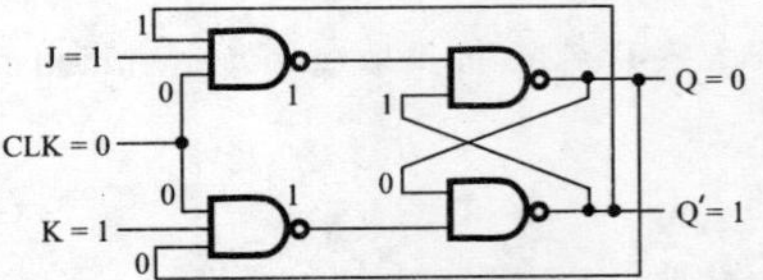

Fig. c: *After clock, output in reset state.*

Fig. 3.31: *Verification of case vi: J = K = 1, initial output Q = 1, Q′ = 0.*

Example 3.2

Design and implement JK flip-flop.

Solution

An excitation table (Table 1) is developed by listing all possible combination of inputs and present states Q and Q′. The next state outputs Q(t + 1) and Q′(t + 1) are determined for all the combinations of J, K, Q and Q′ from the knowledge of truth table of JK flip-flop and listed in Table 1.

With three state variables J, K and Q there are 8 ($2^3 = 8$) combinations of minterms as shown in Table 1. With three state variables J, K and Q′ another 8 combinations of minterms are formed and listed in Table 1.

Truth Table of JK Flip-Flop

Inputs		Outputs	
J	**K**	**Q**	**Q′**
0	0	No change	
0	1	0	1
1	0	1	0
1	1	Toggle	

Table 1: Excitation Table

Inputs		Present State Output		Next State Output		Comment	Minterm
J	K	Q	Q′	Q(t + 1)	Q′(t+1)		JKQ
0	0	0	1	0	1	No change	m_0
0	0	1	0	1	0		m_1
0	1	0	1	0	1	Reset state	m_2
0	1	1	0	0	1		m_3
1	0	0	1	1	0	Set state	m_4
1	0	1	0	1	0		m_5
1	1	0	1	1	0	Toggle	m_6
1	1	1	0	0	1		m_7

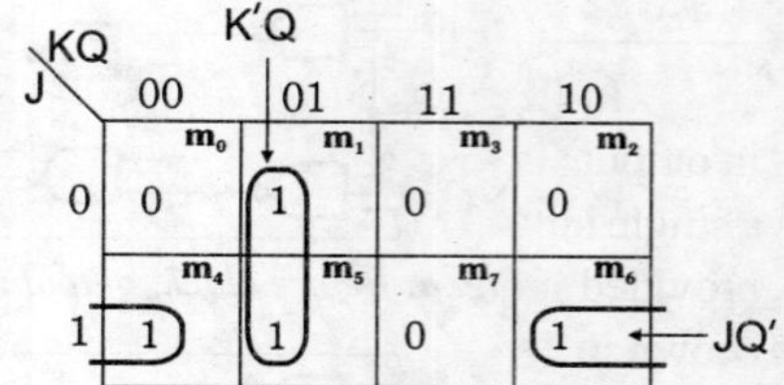

Fig. a: K-map for Q(t + 1).

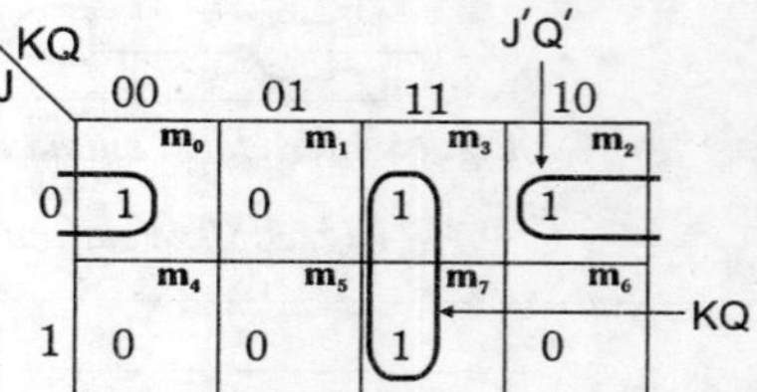

Fig. b: K-map for Q′(t + 1).

***Fig. 1:** K-map for Q(t + 1) and Q′(t + 1).*

The minterms formed using J, K and Q are used to construct K-map for Q(t + 1) and Q′(t + 1) as shown in Fig. 1. In the K-map of Fig. 1, the prime implicants are formed by taking adjacent 1's. The Boolean equations are given by sum of products represented by prime implicants as shown below:

$$Q(t + 1) = J\,Q' + K'Q$$

$$Q'(t + 1) = J'Q' + KQ$$

The JK flip-flop that satisfy the above equations is shown in Fig. 2. The JK flip-flop output is verified in Table 2.

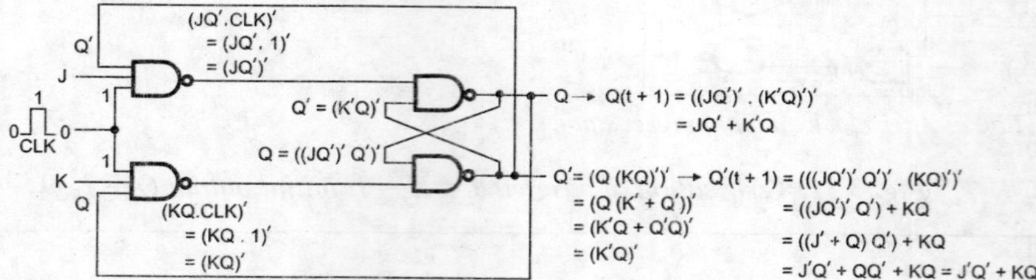

Fig. 2: JK flip-flop using NAND latch.

Table 2: Verification of JK Flip-Flop Outputs

Inputs		Present State Output								Next State Output		
J	K	Q	Q′	K′	JQ′	K′Q	J′	J′Q′	KQ	Q(t + 1) = JQ′ + K′Q	Q′(t + 1) = J′Q′ + KQ	Comment
0	0	0	1	1	0	0	1	1	0	0	1	No change
0	0	1	0	1	0	1	1	0	0	1	0	
0	1	0	1	0	0	0	1	1	0	0	1	Reset state
0	1	1	0	0	0	0	1	0	1	0	1	
1	0	0	1	1	1	0	0	0	0	1	0	Set state
1	0	1	0	1	0	1	0	0	0	1	0	
1	1	0	1	0	1	0	0	0	0	1	0	Toggle
1	1	1	0	0	0	0	0	0	1	0	1	

3.3.3 T Flip-Flop (Toggle Flip-Flop)

(AU, Nov/Dec'18, 13 Marks)

In JK flip-flop when J = K = 1 the output toggles for every clock input and when J = K = 0 no change in output. This property can be used to construct T flip-flop as shown in Fig. 3.32. Therefore, T flip-flop is a single input flip-flop in which the present output is complement of previous output when clock is applied, provided input is 1. The T flip-flops are used to design binary counters. The truth table of T flip-flop is shown in Table 3.14 and standard symbols of T flip-flop are shown in Fig. 3.33.

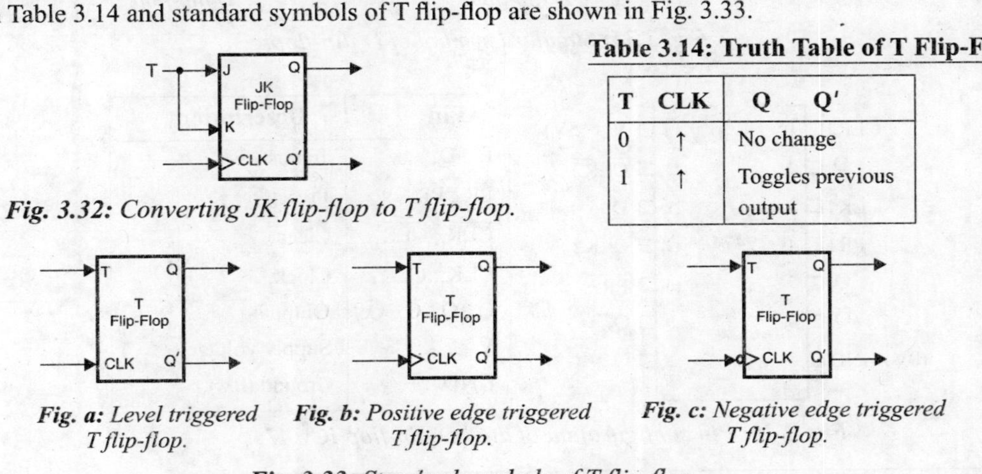

Fig. 3.32: Converting JK flip-flop to T flip-flop.

Table 3.14: Truth Table of T Flip-Flop

T	CLK	Q Q'
0	↑	No change
1	↑	Toggles previous output

Fig. a: Level triggered T flip-flop. *Fig. b: Positive edge triggered T flip-flop.* *Fig. c: Negative edge triggered T flip-flop.*

Fig. 3.33: Standard symbols of T flip-flop.

3.3.4 D Flip-Flop (Data Flip-Flop)

A single input flip-flop can be developed from SR flip-flop, if the input to SR flip-flops are always complementary inputs, so that if input is 1 then output is also 1 and if input is 0 then output is also 0. Therefore, D flip-flop is a single input flip-flop in which the input is stored in the output when clock is applied.

The D latch with enable shown in Fig. 3.12 becomes a D flip-flop if the enable is replaced by a clock as shown in Fig. 3.34. The truth table of D flip-flop is shown in Table 3.15 and standard symbols are shown in Fig. 3.35. The D flip-flop is available as a standard IC 7474. The pin configuration of IC 7474 is shown in Fig. 3.36.

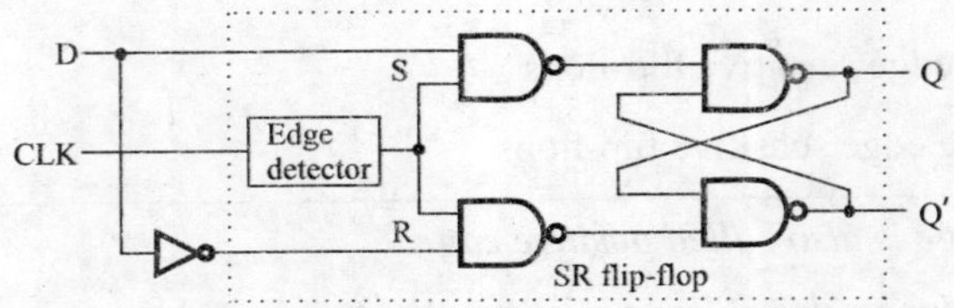

Fig. 3.34: D flip-flop.

Table 3.15: Truth Table of D Flip-Flop

Input	Outputs	
D	Q	Q'
0	0	1
1	1	0

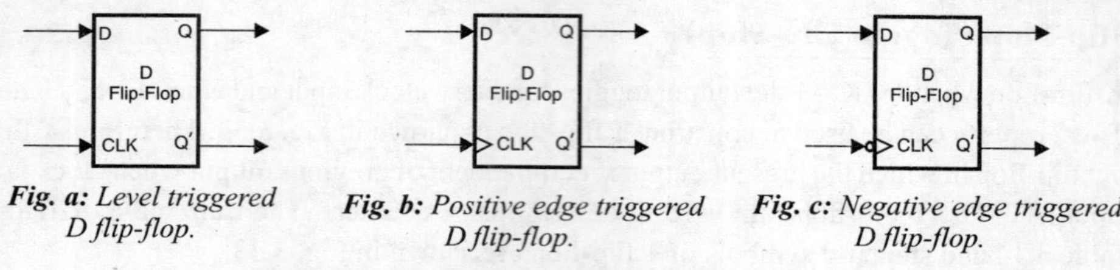

Fig. a: Level triggered D flip-flop. *Fig. b: Positive edge triggered D flip-flop.* *Fig. c: Negative edge triggered D flip-flop.*

***Fig. 3.35:** Standard symbol of D flip-flop.*

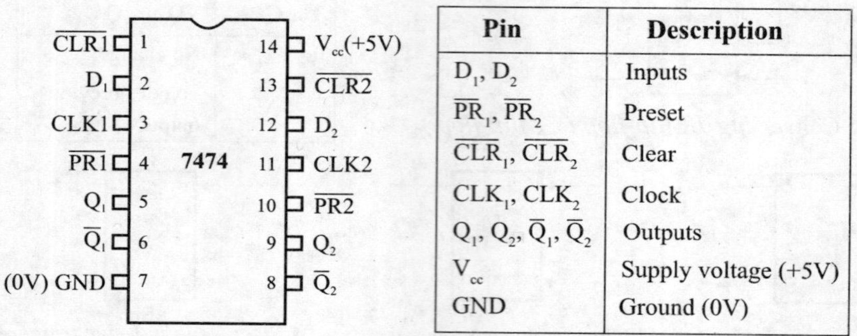

Pin	Description
D_1, D_2	Inputs
$\overline{PR}_1$, $\overline{PR}_2$	Preset
$\overline{CLR}_1$, $\overline{CLR}_2$	Clear
CLK_1, CLK_2	Clock
Q_1, Q_2, $\overline{Q}_1$, $\overline{Q}_2$	Outputs
V_{cc}	Supply voltage (+5V)
GND	Ground (0V)

***Fig. 3.36:** Pin configuration of dual D flip-flop IC 7474.*

3.3.5 Level and Edge Triggering of Flip-Flops

In flip-flops the process of applying a control signal to change from one stable state to another stable state is called ***triggering***. Usually, the control signal is a clock pulse discussed in Section 3.1.1.

Clock is used in flip-flops in order to have precise timing control over changes in flip-flops. Usually, the clocked flip-flops are designed as edge sensitive flip-flop. In edge sensitive flip-flop or circuits the timing of signal (or input) recognization is more precise. Inputs are recognized only at the specified edge of the clock.

Therefore, the flip-flops can be classified as,

- Level sensitive flip-flops
- Rising/Positive edge sensitive flip-flops
- Falling/Negative edge sensitive flip-flops

> ***Note:* 1. *Rising edge* is also called *positive edge.***
>
> **2. *Falling edge* is also called *negative edge.***
>
> **3. *Level sensitive flip-flops* are also known as *level triggered flip-flops.***
>
> **4. *Edge sensitive flip-flops* are also known as *edge triggered flip-flops.***

Specific clock symbol is used to denote positive/negative/level edge sensitivity in flip-flops as shown in Fig. 3.37.

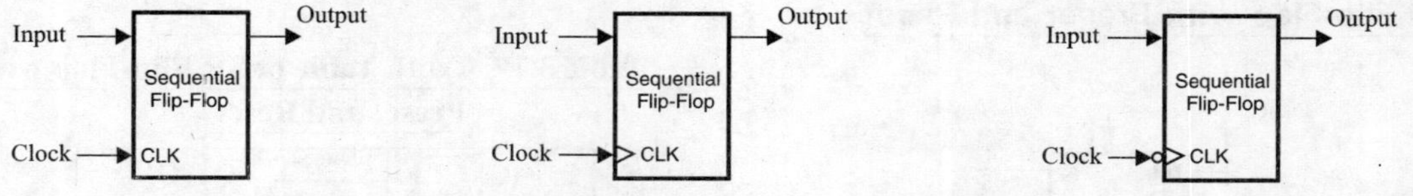

Fig. a: Level sensitive flip-flop. *Fig. b: Positive edge sensitive flip-flop.* *Fig. c: Negative edge sensitive flip-flop.*

***Fig. 3.37:** Clock symbol for clock sensitivity in flip-flop.*

Mostly the triggering is performed by edge sensitive clock. A typical clock edge detector circuit is shown in Fig. 3.38 which generates a narrow pulse during a positive edge or negative edge. In this circuit a small propagation delay for clock pulse to travel through inverter will create a time difference between CLK and $\overline{\text{CLK}}$ which is used to generate a very narrow clock pulse.

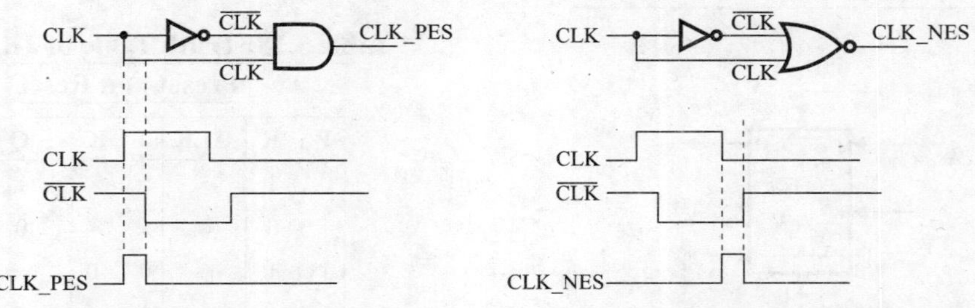

Fig. a: Positive edge detection. *Fig. b: Negative edge detection.*

***Fig. 3.38:** Clock edge detector circuits.*

3.3.6 Flip-Flop with Preset and Reset Facility

When power is switched ON flip-flop outputs will be set or reset at random.

In order to ensure a definite set state, preset facility is provided in flip-flops.

In order to ensure a definite reset state, reset facility is provided in flip-flops.

The preset and reset are logic **low** input signals denoted by P and R respectively. As long as preset is 0, the output is set to 1 (set state). Only when P is made 1, the input and clock are allowed to change the output. As long as reset is 0 the output is reset to 0 (reset state). Only when R is made 1, the input and clock are allowed to change the output. Therefore, for output to change depending on input and clock, preset and reset should be tied to 1.

***Note:** Both P = 0 and R = 0 is illegal.*

D Flip-Flop with Preset and Reset

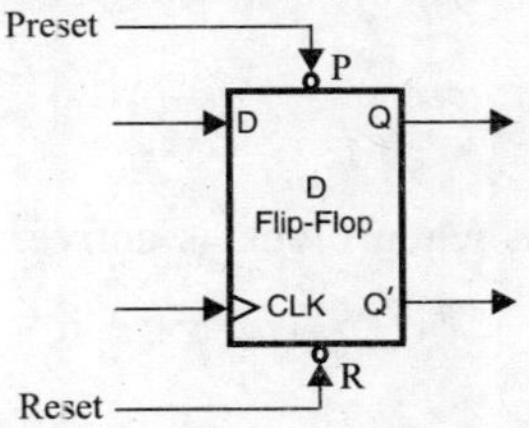

***Fig. 3.39:** D flip-flop with preset and reset.*

Table 3.16: Truth Table D Flip-Flop with Preset and Reset

P	R	CLK	D	Q	Q'
0	1	×	×	1	0
1	0	×	×	0	1
1	1	↑	0	0	1
1	1	↑	1	1	0

SR Flip-Flop with Preset and Reset

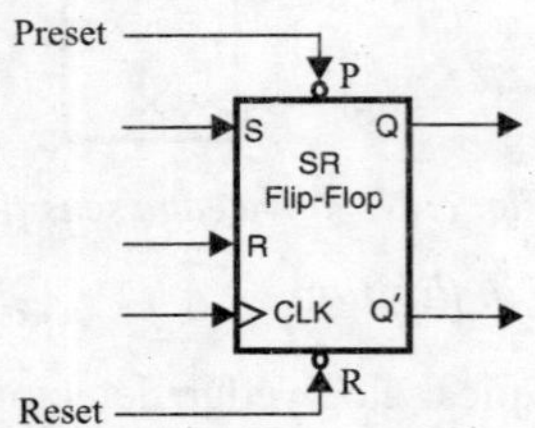

Fig. 3.40: SR flip-flop with preset and reset.

Table 3.17: Truth Table of SR Flip-Flop with Preset and Reset

P	R	CLK	S	R	Q	Q'
0	1	×	×	×	1	0
1	0	×	×	×	0	1
1	1	↑	0	0	No change	
1	1	↑	0	1	0	1
1	1	↑	1	0	1	0
1	1	↑	1	1	Indeterminate	

JK Flip-Flop with Preset and Reset

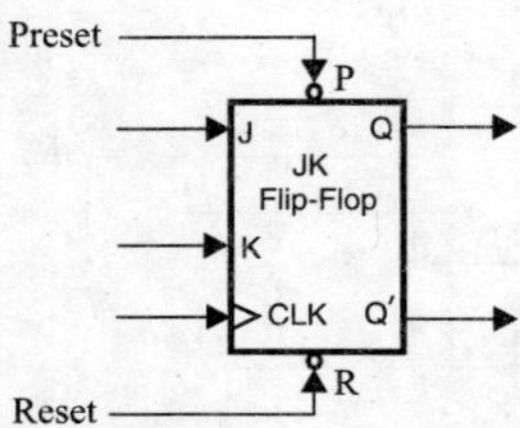

Fig. 3.41: JK flip-flop with preset and reset.

Table 3.18: Truth Table of JK Flip-Flop with Preset and Reset

P	R	CLK	J	K	Q	Q'
0	1	×	×	×	1	0
1	0	×	×	×	0	1
1	1	↑	0	0	No change	
1	1	↑	0	1	0	1
1	1	↑	1	0	1	0
1	1	↑	1	1	Toggle previous output	

T Flip-Flop with Preset and Reset

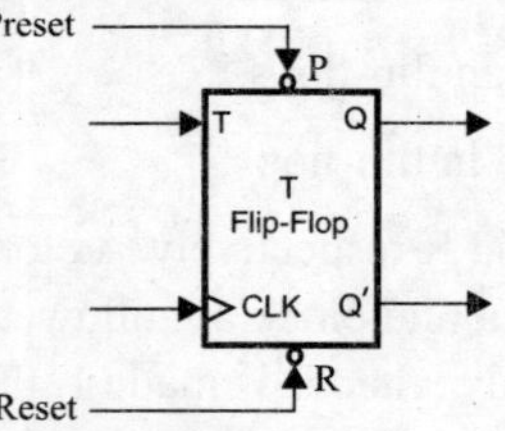

Fig. 3.42: T flip-flop with preset and reset.

Table 3.19: Truth Table of T Flip-Flop with Preset and Reset

P	R	CLK	T	Q	Q'
0	1	×	×	1	0
1	0	×	×	0	1
1	1	↑	0	1	0
1	1	↑	1	0	1

3.4 Excitation Table and Characteristic Equation of Flip-Flops

The design using flip-flops require the knowledge of inputs to flip-flops for the next state output. The flip-flop inputs can be obtained from excitation table. The excitation table lists the required flip-flop inputs for a specified change of state in output.

3.4.1 Excitation Table of D Flip-Flop

In D flip-flop the next state output is given by the input when clock is active. Therefore, the characteristic equation of D flip-flop is,

$$Q(t+1) = D$$

From the characteristic equation we can infer that the input directly determine the next state output. Therefore, in the design using D flip-flop the flip-flop inputs can be directly obtained from state table from the knowledge of next state output and so excitation table is not required for design using D flip-flop.

However, the excitation table for D flip-flop can be formed as shown in Table 3.20. There are four possible transitions from present state to next state. When present state is 0, the next state can be 0 or 1. When present state is 1, the next state can be 0 or 1.

Table 3.20: Excitation Table of D Flip-Flop

Present State Q(t)	Next State Q(t+1)	Flip-Flop Input D
0	0	0
0	1	1
1	0	0
1	1	1

The explanation for formation of each row of excitation table is given below:

Truth Table of D Flip-Flop

Input	Outputs	
D	Q	Q′
0	0	1
1	1	0

Row 1 of Excitation Table: Present and Next States are 0

Now, $Q(t) = 0$; $Q(t+1) = 0$

Output in reset state. No change from present state to next state.

From truth table of D flip-flop we can say that,

if $D = 0$, then there is no change in output

Therefore, input to flip-flop is,

$D = 0$

Row 2 of Excitation Table: Present State is 0 and Next State is 1

Now, $Q(t) = 0$; $Q(t+1) = 1$

Output changes from reset to set state.

From the truth table of D flip-flop we can say that,

if $D = 1$, then the output will be driven to set state.

Therefore, input to flip-flop is,

$D = 1$

Row 3 of Excitation Table: Present State is 1 and Next State is 0

Now, $Q(t) = 1$; $Q(t+1) = 0$

Output changes from set to reset state.

From the truth table of D flip-flop we can say that,

if $D = 0$, then the output will be driven to reset state.

Therefore, input to flip-flop is,

$D = 0$

Row 4 of Excitation Table: Present State and Next States are 1

Now, $Q(t) = 1$; $Q(t+1) = 1$

Output in set state. No change from present state to next state.

From the truth table of D flip-flop we can say that,

if $D = 1$, there is no change in output.

Therefore, input to flip-flop is,

$D = 1$

The excitation table (Table 3.20) can be used to develop characteristic equation of D flip-flop by forming minterms using present state, Q(t) and input, D as shown below and constructing K-map for next state Q(t + 1) as shown in Fig. 3.43.

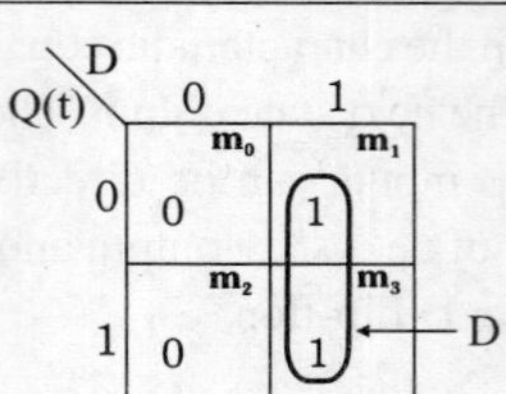

Fig. 3.43: K-map for Q(t + 1) of D flip-flop.

Q(t) D	Minterm	Q(t + 1)
0 0	m_0	0
0 1	m_1	1
1 0	m_2	0
1 1	m_3	1

From the K-map we get the following characteristic equation for D flip-flop.

$Q(t + 1) = D$

Using the characteristic equation the characteristic table of D flip-flop is formed as shown in Table 3.21.

Table 3.21: Characteristic Table of D Flip-Flop

Input D	Present State Q	Next State Q(t+1)
0	0	0
0	1	0
1	0	1
1	1	1

3.4.2 Excitation Table of JK Flip-Flop

In the design using JK flip-flops an excitation table is formed to determine the inputs of flip-flops for next state output.

The excitation table of JK flip-flop is shown in Table 3.22. There are four possible transitions from present state to next state. When present state is 0 the next state can be 0 or 1. When present state is 1 the next state can be 0 or 1.

The explanation for formation of each row of excitation table is given below:

Table 3.22: Excitation Table of JK Flip-Flop

Present State Q(t)	Next State Q(t+1)	Flip-Flop Inputs J	K
0	0	0	×
0	1	1	×
1	0	×	1
1	1	×	0

Truth Table of JK Flip-Flop

Inputs J	K	Outputs Q	Q′
0	0	No change	
0	1	0	1 (Reset)
1	0	1	0 (Set)
1	1	Toggle	

Row 1 of Excitation Table: Present and Next States are 0

Now, $Q(t) = 0$; $Q(t + 1) = 0$

Output in reset state. No change from present state to next state.

From truth table of JK flip-flop we can say that,

if J = 0 and K = 0, then there is no change in output

Also, if J = 0 and K = 1 then output remain in reset state.

Therefore, inputs to flip-flop are,

$J = 0 \Rightarrow J = 0$

$K = 0$ or 1 (don't-care) $\quad K = \times$

Row 2 of Excitation Table: Present State is 0 and Next State is 1

Now, $Q(t) = 0$; $Q(t + 1) = 1$

Output changes from reset to set state.

From the truth table of JK flip-flop we can say that,

if J = 1 and K = 0, then the output will be driven to set state.

Also, if J = 1 and K = 1 then the output toggles. Here present state 0 will toggle to next state 1.

Therefore, inputs to flip-flop are,

J = 1 ⇒ J = 1

K = 0 or 1 (don't-care) K = ×

Row 3 of Excitation Table: Present State is 1 and Next State is 0

Now, Q(t) = 1 ; Q(t + 1) = 0

Output changes from set to reset state.

From the truth table of JK flip-flop we can say that,

if J = 0 and K = 1, then the output will be driven to reset state.

Also, if J = 1 and K = 1 then the output toggles. Here present state 1 will toggle to next state 0.

Therefore, inputs to flip-flop are,

J = 0 or 1 (don't-care) ⇒ J = ×

K = 1 K = 1

Row 4 of Excitation Table: Present State and Next States are 1

Now, Q(t) = 1 ; Q(t + 1) = 1

Output in set state. No change from present state to next state.

From the truth table of JK flip-flop we can say that,

if J = 0 and K = 0, there is no change in output.

Also, if J = 1 and K = 0 then output remain in set state.

Therefore, inputs to flip-flop are,

J = 0 or 1 (don't-care) ⇒ J = ×

K = 0 K = 0

The excitation table (Table 3.22) can be used to develop characteristic equation of JK flip-flop by forming minterms using Q(t), J and K as shown below and constructing K-map for next state Q(t + 1) as shown in Fig. 3.44.

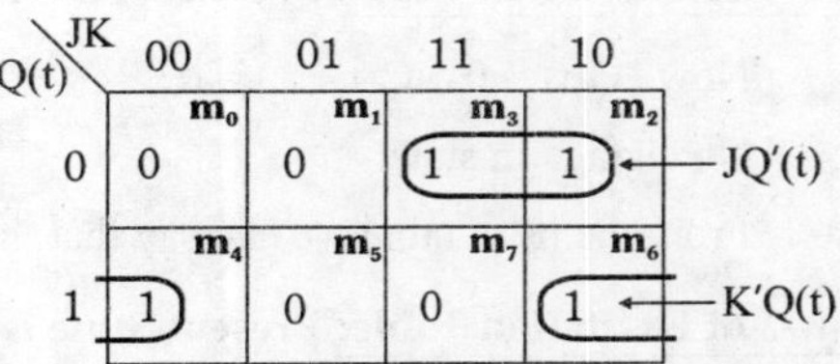

Fig. 3.44: K-map for Q(t + 1) of JK flip-flop.

Q(t)	J	K	Minterm	Q(t + 1)	Q(t)	J	K	Minterm	Q(t + 1)
0	0	0	m_0	0	1	0	1	m_5	0
0	0	1	m_1	0	1	1	1	m_7	0
0	1	0	m_2	1	1	0	0	m_4	1
0	1	1	m_3	1	1	1	0	m_6	1

> ***Note:*** *Two possible values 0 and 1 are taken for don't-care, × of J and K in excitation table.*

From the K-map we get the following characteristic equation for JK flip-flop.

$$Q(t+1) = JQ'(t) + K'Q(t)$$

Using the characteristic equation the characteristic table of JK flip-flop is formed as shown in Table 3.23.

Table 3.23: Characteristic Table of JK Flip-Flop *(AU, Apr/May'19, 2 Marks)*

Inputs J K	Present State Q	Next State Q(t+1)
0 0	0	0
0 0	1	1
0 1	0	0
0 1	1	0
1 0	0	1
1 0	1	1
1 1	0	1
1 1	1	0

3.4.3 Excitation Table of T Flip-Flop

In design using T flip-flops an excitation table is formed to determine the inputs to flip-flops for the next state output.

The excitation table of T flip-flop is shown in Table 3.24. There are four possible transitions from present to next state. When present state is 0 the next state can be 0 or 1. When present state is 1 the next state can be 0 or 1. The explanation for formation of each row of excitation table is given below:

Table 3.24: Excitation Table of T Flip-Flop

Present State Q(t)	Next State Q(t + 1)	Flip-Flop Input T
0	0	0
0	1	1
1	0	1
1	1	0

Row 1 of Excitation Table: Present and Next States are 0

Now, $Q(t) = 0$; $Q(t+1) = 0$

No change in state.

From the truth table we can say that, input T = 0 for this output condition.

Truth Table of T Flip-Flop

Input T	Outputs Q Q'
0	No change
1	Toggle

Row 2 of Excitation Table: Present State is 0 and Next State are 1

Now, $Q(t) = 0$; $Q(t+1) = 1$

Output toggles.

From the truth table we can say that, input T = 1 for this output condition.

Row 3 of Excitation Table: Present State is 1 and Next State are 0

Now, $Q(t) = 1$; $Q(t+1) = 0$

Output toggles.

From the truth table we can say that, input T = 1 for this output condition.

Row 4 of Excitation Table: Present and Next States are 1

Now, Q(t) = 1 ; Q(t + 1) = 1

No change in state.

From the truth table we can say that, input T = 0 for this output condition.

The excitation table (Table 3.24) can be used to develop characteristic equation of T flip-flop by forming minterms using present state, Q(t) and input, T as shown below and constructing K-map for next state Q(t + 1) as shown in Fig. 3.45.

Q(t)	T	Minterm	Q(t + 1)
0	0	m_0	0
0	1	m_1	1
1	0	m_2	1
1	1	m_3	0

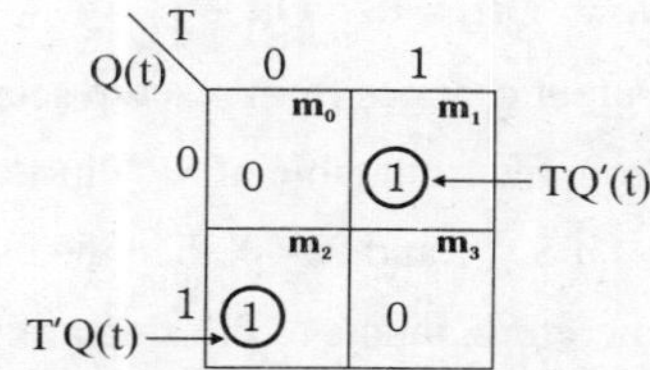

Fig. 3.45: K-map for Q(t + 1) of T flip-flop.

From the K-map we get the following characteristic equation for T flip-flop.

$$Q(t+1) = TQ'(t) + T'Q(t)$$

Using the characteristic equation the characteristic table of T flip-flop is formed as shown in Table 3.25.

Table 3.25: Characteristic Table of T Flip-Flop

Input T	Present State Q	Next State Q(t+1)
0	0	0
0	1	1
1	0	1
1	1	0

3.4.4 Excitation Table of SR Flip-Flop

In the design using SR flip-flops an excitation table is formed to determine the inputs of flip-flops for next state output.

The excitation table of SR flip-flop is shown in Table 3.26. There are four possible transitions from present state to next state. When present state is 0 the next state can be 0 or 1. When present state is 1 the next state can be 0 or 1.

The explanation for formation of each row of excitation table is given below:

Table 3.26: Excitation Table of SR Flip-Flop

Present State Q(t)	Next State Q(t+1)	Flip-Flop Inputs S	R
0	0	0	×
0	1	1	0
1	0	0	1
1	1	×	0
0	×	1	1
1	×	1	1

Row 1 of Excitation Table: Present and Next States are 0

Now, Q(t) = 0 ; Q(t + 1) = 0

Output in reset state. No change from present state to next state.

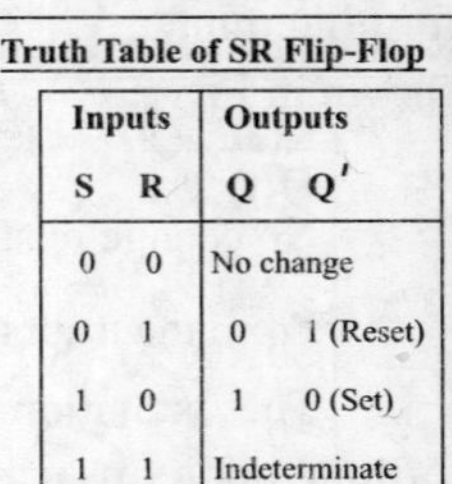

Truth Table of SR Flip-Flop

Inputs		Outputs	
S	R	Q	Q'
0	0	No change	
0	1	0	1 (Reset)
1	0	1	0 (Set)
1	1	Indeterminate	

From truth table of SR flip-flop we can say that,

if S = 0 and R = 0, then there is no change in output

Also, if S = 0 and R = 1 then output remain in reset state.

Therefore, inputs to flip-flop are,

S = 0 ⇒ S = 0

R = 0 or 1 (don't-care) R = ×

Row 2 of Excitation Table: Present State is 0 and Next State is 1

Now, $Q(t) = 0$; $Q(t+1) = 1$

Output changes from reset to set state.

From the truth table of SR flip-flop we can say that,

if S = 1 and R = 0, then the output will be driven to set state.

Therefore, inputs to flip-flop are,

S = 1, R = 0

Row 3 of Excitation Table: Present State is 1 and Next State is 0

Now, $Q(t) = 1$; $Q(t+1) = 0$

Output changes from set to reset state.

From the truth table of SR flip-flop we can say that,

if S = 0 and R = 1, then the output will be driven to reset state.

Therefore, inputs to flip-flop are,

S = 0, R = 1

Row 4 of Excitation Table: Present State and Next States are 1

Now, $Q(t) = 1$; $Q(t+1) = 1$

Output in set state. No change from present state to next state.

From the truth table of SR flip-flop we can say that,

if S = 0 and R = 0, there is no change in output.

Also, if S = 1 and R = 0 then output remain in set state.

Therefore, inputs to flip-flop are,

S = 0 or 1 (don't-care) ⇒ S = ×

R = 0 R = 0

Rows 5 and 6 of Excitation Table: Indeterminate Next State

Now, $Q(t) = 0$ or 1 ; $Q(t+1) = \times$

Output in indeterminate state.

From the truth table of SR flip-flop we can say that,

if S = 1 and R = 1, then the output will be indeterminate.

Therefore, inputs to flip-flop are,

S = 1, R = 1

The excitation table (Table 3.26) can be used to develop characteristic equation of SR flip-flop by forming minterms using Q(t), S and R as shown below and constructing K-map for next state Q(t + 1) as shown in Fig. 3.46.

Q(t)	S	R	Minterm	Q(t + 1)	Q(t)	S	R	Minterm	Q(t + 1)
0	0	0	m_0	0	1	0	0	m_4	1
0	0	1	m_1	0	1	1	0	m_6	1
0	1	0	m_2	1	0	1	1	m_3	×
1	0	1	m_5	0	1	1	1	m_7	×

Note: *Two possible values 0 and 1 are taken for don't-care, × in S and R in excitation table.*

From the K-map we get the following characteristic equation for SR flip-flop.

$$Q(t+1) = S + R'Q(t)$$

Using the characteristic equation the characteristic table of SR flip-flop is formed as shown in Table 3.27.

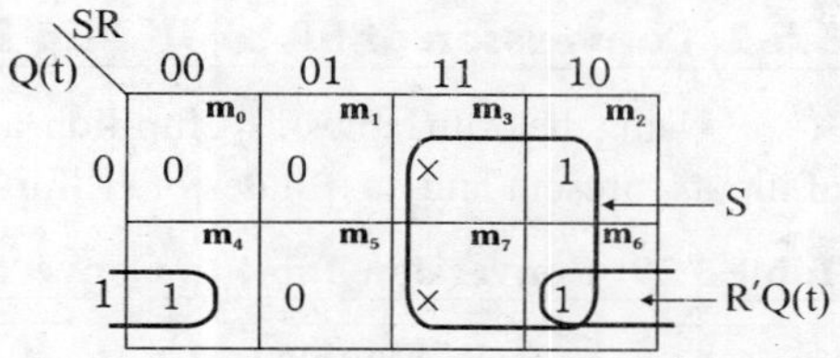

Fig. 3.46: K-map for Q(t + 1) of SR flip-flop.

Table 3.27: Characteristic Table of SR Flip-Flop

Inputs S R	Present State Q	Next State Q(t+1)
0 0	0	0
0 0	1	1
0 1	0	0
0 1	1	0
1 0	0	1
1 0	1	1
1 1	0	×
1 1	1	×

Table 3.28: Summary of Characteristic Equation of Flip-Flops

Flip-Flop	Characteristic Equation
D	$Q(t+1) = D$
JK	$Q(t+1) = JQ'(t) + K'Q(t)$
T	$Q(t+1) = TQ'(t) + T'Q(t)$
SR	$Q(t+1) = S + R'Q(t)$

3.5 Conversion of Flip-Flops

The different types of flip-flops are,

- SR flip-flop
- JK flip-flop
- D flip-flop (or Data flip-flop)
- T flip-flop (or Toggle flip-flop)

It is possible to convert one type of flip-flop to any other type of flip-flop. For example, SR flip-flop can be converted to JK flip-flop or D flip-flop or T flip-flop.

The flip-flop outputs are commonly denoted as Q and Q′. The different types of inputs will drive the flip-flop to get different values of Q and Q′. Therefore, the conversion of flip-flop requires design of a combinational circuit to supply the required inputs to a flip-flop in order to get the output of another desired flip-flop.

The design of combinational circuit requires a table that lists possible combinations of inputs, present state and next state of the desired flip-flop. These informations are obtained from truth table of desired flip-flop. Using the excitation table of flip-flop to be converted, the inputs of desired flip-flop to drive the output to next state are determined and listed in same table.

The minterms are formed for all possible combinations of input and present state of desired flip-flop. Using these minterms the K-maps are constructed for input of flip-flop to be converted. From the K-map the simplified equations are obtained for flip-flop inputs.

3.5.1 Conversion of SR to JK Flip-Flop

Using the truth table of JK flip-flop, a conversion table is formed by listing the possible combinations of inputs, present and next states of JK flip-flop as shown in Table 3.29.

Table 3.29: Conversion Table to Convert SR to JK Flip-Flop

JK Flip-Flop					SR Flip-Flop Inputs		Minterm
Inputs		Present State	Next State	Comment			
J	K	Q	Q(t+1)		S	R	JKQ
0	0	0	0	No change	0	×	m_0
0	0	1	1		×	0	m_1
0	1	0	0	Reset state	0	×	m_2
0	1	1	0		0	1	m_3
1	0	0	1	Set state	1	0	m_4
1	0	1	1		×	0	m_5
1	1	0	1	Toggle	1	0	m_6
1	1	1	0		0	1	m_7

Truth Table of JK Flip-Flop

Inputs		Outputs	
J	K	Q	Q′
0	0	No change	
0	1	0	1 (Reset)
1	0	1	0 (Set)
1	1	Toggles	

Excitation Table of SR Flip-Flop

Present State	Next State	Flip-Flop Inputs	
Q(t)	Q(t + 1)	S	R
0	0	0	×
0	1	1	0
1	0	0	1
1	1	×	0
0	×	1	1
1	×	1	1

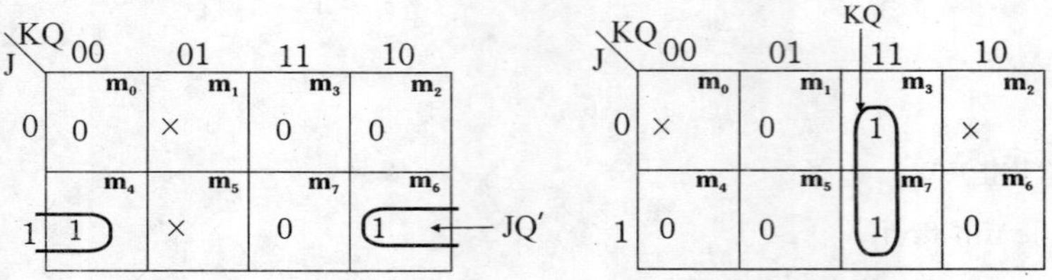

In order to drive the present state to next state listed in Table 3.29, the SR flip-flop inputs are obtained using the excitation table of SR flip-flop and listed in Table 3.29. Minterms are formed using J, K and Q and the K-maps are constructed for S and R as shown in Fig. 3.47.

J \ KQ	00	01	11	10
0	m_0: 0	m_1: ×	m_3: 0	m_2: 0
1	m_4: 1	m_5: ×	m_7: 0	m_6: 1

JQ′

Fig. a: K-map for S.

J \ KQ	00	01	11	10
0	m_0: ×	m_1: 0	m_3: 1	m_2: ×
1	m_4: 0	m_5: 0	m_7: 1	m_6: 0

KQ

Fig. b: K-map for R.

Fig. 3.47: K-map for SR flip-flop inputs.

From the K-map we get the following Boolean equations for SR flip-flop inputs and using these equations, the JK flip-flop logic circuit is drawn as shown in Fig. 3.48.

$$S = JQ' \quad ; \quad R = KQ$$

The working of JK flip-flop shown in Fig. 3.48 is verified for all possible combinations of inputs and present state as shown in Table 3.30. The next state output of JK flip-flop is obtained from SR flip-flop characteristic table.

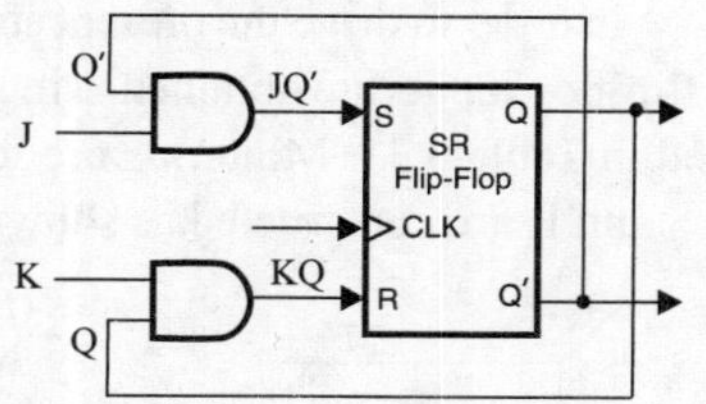

Fig. 3.48: Logic circuit of JK flip-flop using SR flip-flop.

Table 3.30: Verification of JK Flip-Flop of Fig. 3.48

JK Flip-Flop				SR Flip-Flop Inputs		JK Flip-Flop		
Inputs		Present State				Next State		Comment
J	K	Q	Q′	S = JQ′	R = KQ	Q(t+1)	Q′(t+1)	
0	0	0	1	0	0	0	1	No change
0	0	1	0	0	0	1	0	
0	1	0	1	0	0	0	1	Reset state
0	1	1	0	0	1	0	1	
1	0	0	1	1	0	1	0	Set state
1	0	1	0	0	0	1	0	
1	1	0	1	1	0	1	0	Toggle
1	1	1	0	0	1	0	1	

Characteristic Table of SR Flip-Flop

Inputs		Present State	Next State
S	R	Q	Q(t+1)
0	0	0	0
0	0	1	1
0	1	0	0
0	1	1	0
1	0	0	1
1	0	1	1
1	1	0	×
1	1	1	×

3.5.2 Conversion of SR to D Flip-Flop

From the truth table of SR flip-flop we can observe that when the inputs to SR flip-flop are always complementary then the output will be same as D flip-flop. Therefore, when the D input is directly applied to S and complement of D is applied to R as shown in Fig. 3.49, then the SR flip-flop will behave as D flip-flop.

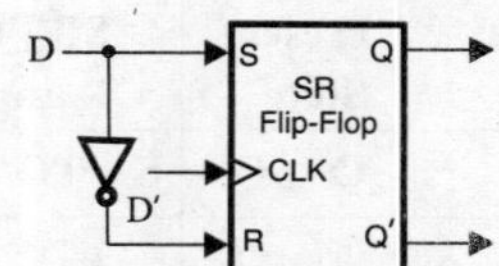

Fig. 3.49: Logic circuit of D flip-flop using SR flip-flop.

3.5.3 Conversion of SR to T Flip-Flop

Using the truth table of T flip-flop, a conversion table is formed by listing the possible combinations of input, present and next states of T flip-flop as shown in Table 3.31.

Table 3.31: Conversion Table to Convert SR to T Flip-Flop

T Flip-Flop				SR Flip-Flop Inputs		Minterm
Input	Present State	Next State	Comment			
T	Q	Q(t+1)		S	R	TQ
0	0	0	No change	0	×	m_0
0	1	1		×	0	m_1
1	0	1	Toggle	1	0	m_2
1	1	0		0	1	m_3

Truth Table of T Flip-Flop

Input	Outputs
T	Q Q′
0	No change
1	Toggle

In order to drive the present state to next state listed in Table 3.31, the SR flip-flop inputs are obtained using the excitation table of SR flip-flop and listed in Table 3.31. Minterms are formed using T and Q and the K-maps for S and R are constructed as shown in Fig. 3.50.

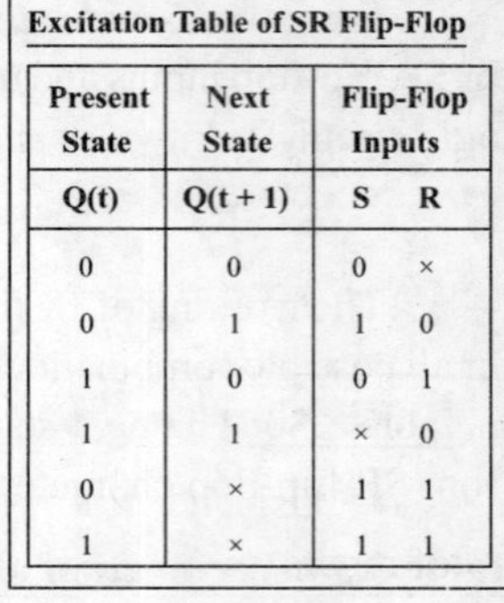

Excitation Table of SR Flip-Flop

Present State	Next State	Flip-Flop Inputs	
Q(t)	Q(t + 1)	S	R
0	0	0	×
0	1	1	0
1	0	0	1
1	1	×	0
0	×	1	1
1	×	1	1

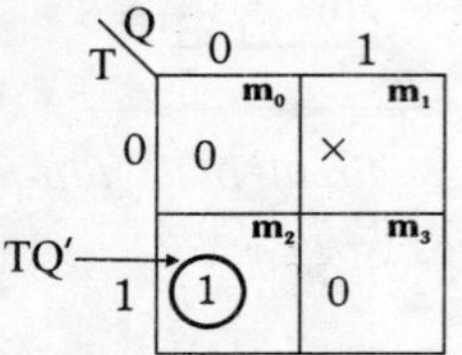

Fig. a: K-map for S.

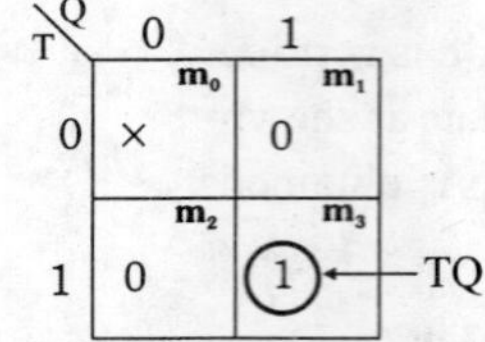

Fig. b: K-map for R.

***Fig. 3.50:** K-map for SR flip-flop inputs.*

From the K-map we get the following Boolean equations for SR flip-flop inputs and using these equations, the T flip-flop logic circuit is drawn as shown in Fig. 3.51.

$$S = TQ' \quad ; \quad R = TQ$$

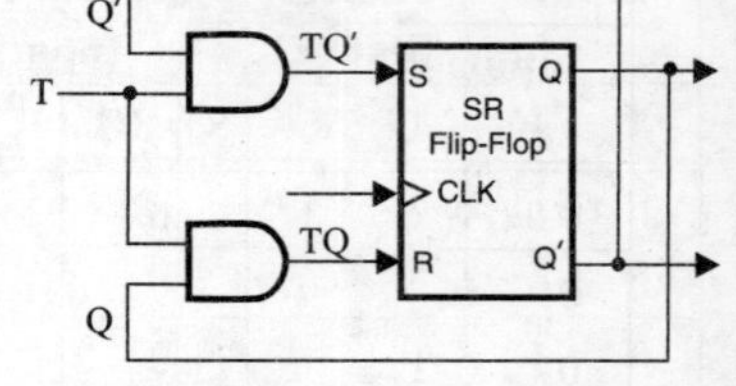

***Fig. 3.51:** Logic circuit of T flip-flop using SR flip-flop.*

The working of T flip-flop shown in Fig. 3.51 is verified for all possible combinations of input and present state as shown in Table 3.32. The next state output of T flip-flop is obtained from SR flip-flop characteristic table.

Characteristic Table of SR Flip-Flop

Inputs	Present State	Next State
S R	Q	Q(t+1)
0 0	0	0
0 0	1	1
0 1	0	0
0 1	1	0
1 0	0	1
1 0	1	1
1 1	0	×
1 1	1	×

Table 3.32: Verification of T Flip-Flop of Fig. 3.51

T Flip-Flop Input	T Flip-Flop Present State		SR Flip-Flop Inputs		T Flip-Flop Next State		Comment
T	Q	Q′	S = TQ′	R = TQ	Q(t+1)	Q′(t+1)	
0	0	1	0	0	0	1	No change
0	1	0	0	0	1	0	
1	0	1	1	0	1	0	Toggle
1	1	0	0	1	0	1	

3.5.4 Conversion of JK Flip-Flop to SR, D and T Flip-Flops

The difference between JK and SR flip-flop is the type of output when the inputs are same. In JK flip-flop the output toggles when the inputs are same, whereas in SR flip-flop the output is indeterminate when the inputs are same. Practically, the indeterminate state is not used in SR flip-flop and so if we ignore the toggle state of JK flip-flop, then the behaviour of JK flip-flop will be same as SR flip-flop.

From the truth table of JK flip-flop we can observe that when the inputs to JK flip-flop are always complementary then the output will be same as D flip-flop. Therefore, when the D input is directly applied to J and complement of D is applied to K then the JK flip-flop will behave as D flip-flop.

From the truth table of JK flip-flop we can observe that when the inputs of JK flip-flop are always same then the output will be same as T flip-flop. Therfore, when the T input is directly applied to J and K then the JK flip-flop will behave as T flip-flop.

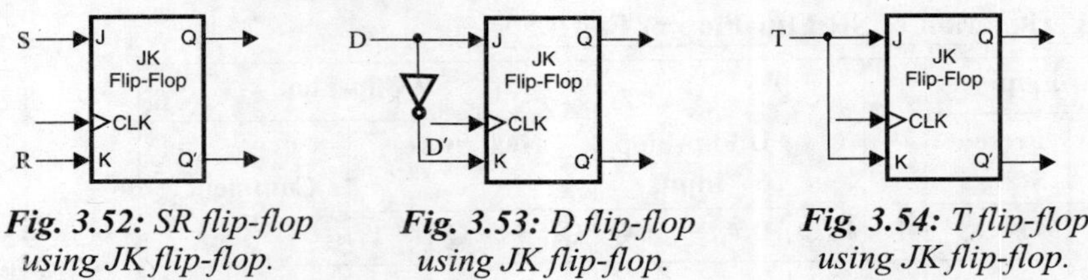

Fig. 3.52: SR flip-flop using JK flip-flop.

Fig. 3.53: D flip-flop using JK flip-flop.

Fig. 3.54: T flip-flop using JK flip-flop.

3.5.5 Conversion of D to SR Flip-Flop

Using the truth table of SR flip-flop, a conversion table is formed by listing the possible combinations of inputs, present and next states of SR flip-flop as shown in Table 3.33.

Table 3.33: Conversion Table to Convert D to SR Flip-Flop

SR Flip-Flop					D Flip-Flop Input	Minterm
Inputs		Present State	Next State	Comment		
S	R	Q	Q(t+1)			SRQ
0	0	0	0	No change	0	m_0
0	0	1	1		1	m_1
0	1	0	0	Reset state	0	m_2
0	1	1	0		0	m_3
1	0	0	1	Set state	1	m_4
1	0	1	1		1	m_5
1	1	0	×	Indeterminate	×	m_6
1	1	1	×		×	m_7

Truth Table of SR Flip-Flop

Inputs		Outputs	
S	R	Q	Q′
0	0	No change	
0	1	0	1 (Reset)
1	0	1	0 (Set)
1	1	Indeterminate	

Excitation Table of D Flip-Flop

Present State	Next State	Flip-Flop Input
Q(t)	Q(t + 1)	D
0	0	0
0	1	1
1	0	0
1	1	1

In order to drive the present state to next state listed in Table 3.33, the D flip-flop input is obtained using the excitation table of D flip-flop and listed in Table 3.33. Minterms are formed using S, R and Q and the K-map is constructed for D as shown in Fig. 3.55.

S \ RQ	00	01	11	10
0	m_0: 0	m_1: 1	m_3: 0	m_2: 0
1	m_4: 1	m_5: 1	m_7: ×	m_6: ×

(Groupings: R′Q — column 01; S — row 1)

Fig. 3.55: K-map for D flip-flop input.

From the K-map we get the following Boolean equation for D flip-flop input and using this equation, the SR flip-flop logic circuit is drawn as shown in Fig. 3.56.

$$D = S + R'Q$$

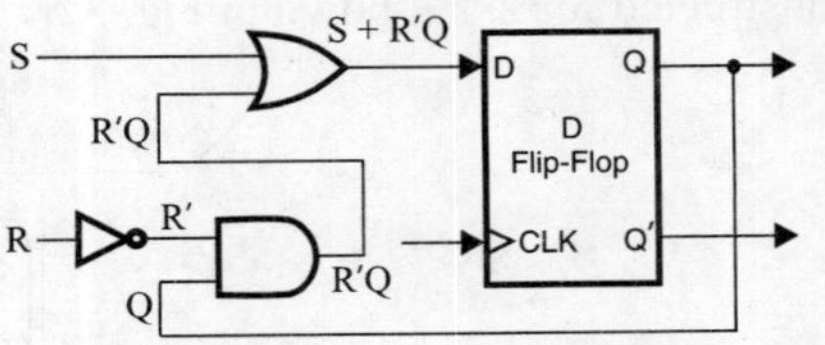

Fig. 3.56: SR flip-flop using D flip-flop.

The working of SR flip-flop shown in Fig. 3.56 is verified for all possible combinations of inputs and present state as shown in Table 3.34. The next state output of SR flip-flop is obtained from D flip-flop characteristic table.

Table 3.34: Verification of SR Flip-Flop of Fig. 3.56

SR Flip-Flop							SR Flip-Flop		
Inputs		Present State				D Flip-Flop Input	Next State		Comment
S	R	Q	Q′	R′	R′Q	D = S + R′Q	Q(t+1)	Q′(t+1)	
0	0	0	1	1	0	0	0	1	No change
0	0	1	0	1	1	1	1	0	
0	1	0	1	0	0	0	0	1	Reset state
0	1	1	0	0	0	0	0	1	
1	0	0	1	1	0	1	1	0	Set state
1	0	1	0	1	1	1	1	0	
1	1	0	1	0	0	1	1	0	Not acceptable
1	1	1	0	0	0	1	1	0	

Characteristic Table of D Flip-Flop

Input	Present State	Next State
D	Q	Q(t+1)
0	0	0
0	1	0
1	0	1
1	1	1

3.5.6 Conversion of D to JK Flip-Flop

Using the truth table of JK flip-flop, a conversion table is formed by listing the possible combinations of inputs, present and next states of JK flip-flop as shown in Table 3.35.

Table 3.35: Conversion Table to Convert D to JK Flip-Flop

JK Flip-Flop						
Inputs		Present State	Next State	Comment	D Flip-Flop Input	Minterm
J	K	Q	Q(t+1)			JKQ
0	0	0	0	No change	0	m_0
0	0	1	1		1	m_1
0	1	0	0	Reset state	0	m_2
0	1	1	0		0	m_3
1	0	0	1	Set state	1	m_4
1	0	1	1		1	m_5
1	1	0	1	Toggle	1	m_6
1	1	1	0		0	m_7

In order to drive the present state to next state listed in Table 3.35, the D flip-flop input is obtained using the excitation table of D flip-flop and listed in Table 3.35. Minterms are formed using J, K and Q and the K-map is constructed for D as shown in Fig. 3.57.

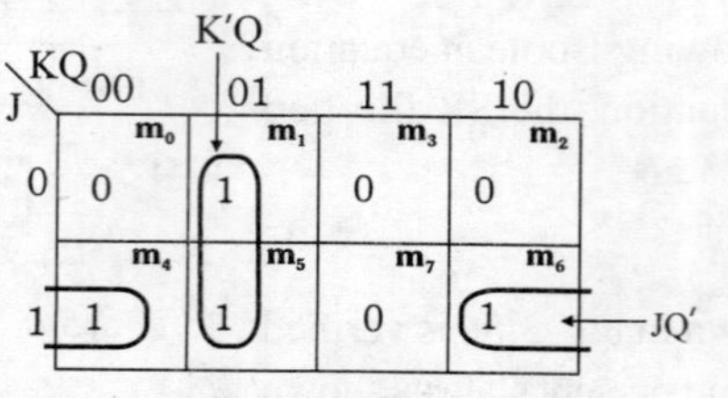

Truth Table of JK Flip-Flop

Inputs		Outputs	
J	K	Q	Q′
0	0	No change	
0	1	0	1 (Reset)
1	0	1	0 (Set)
1	1	Toggle	

Excitation Table of D Flip-Flop

Present State	Next State	Flip-Flop Input
Q(t)	Q(t + 1)	D
0	0	0
0	1	1
1	0	0
1	1	1

J \ KQ	00	01	11	10
0	m_0: 0	m_1: 1	m_3: 0	m_2: 0
1	m_4: 1	m_5: 1	m_7: 0	m_6: 1

K′Q

JQ′

Fig. 3.57: K-map for D flip-flop input.

From the K-map we get the following Boolean equation for D flip-flop input and using this equation, the JK flip-flop logic circuit is drawn as shown in Fig. 3.58.

$$D = JQ' + K'Q$$

The working of JK flip-flop shown in Fig. 3.58 is verified for all possible combinations of inputs and present state as shown in Table 3.36. The next state output of JK flip-flop is obtained from D flip-flop characteristic table.

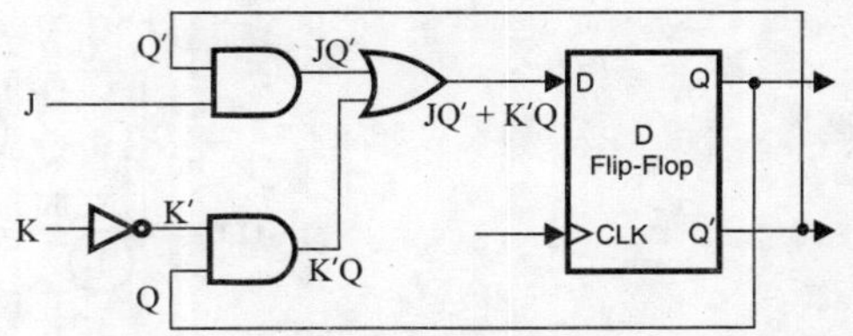

Fig. 3.58: JK flip-flop using D flip-flop.

Table 3.36: Verification of JK Flip-Flop of Fig. 3.58

JK Flip-Flop								JK Flip-Flop		
Inputs		Present State					D Flip-Flop Input	Next State		Comment
J	K	Q	Q'	K'	JQ'	K'Q	D = JQ' + K'Q	Q(t+1)	Q'(t+1)	
0	0	0	1	1	0	0	0	0	1	No change
0	0	1	0	1	0	1	1	1	0	
0	1	0	1	0	0	0	0	0	1	Reset state
0	1	1	0	0	0	0	0	0	1	
1	0	0	1	1	1	0	1	1	0	Set state
1	0	1	0	1	0	1	1	1	0	
1	1	0	1	0	1	0	1	1	0	Toggle
1	1	1	0	0	0	0	0	0	1	

Characteristic Table of D Flip-Flop

Input	Present State	Next State
D	Q	Q(t+1)
0	0	0
0	1	0
1	0	1
1	1	1

3.5.7 Conversion of D to T Flip-Flop

Using the truth table of T flip-flop, a conversion table is formed by listing the possible combinations of input, present and next states of T flip-flop as shown in Table 3.37.

Table 3.37: Conversion Table to Convert D to T Flip-Flop

T Flip-Flop					
Input	Present State	Next State	Comment	D Flip-Flop Input	Minterm
T	Q	Q(t+1)			TQ
0	0	0	No change	0	m_0
0	1	1		1	m_1
1	0	1	Toggle	1	m_2
1	1	0		0	m_3

Truth Table of T Flip-Flop

Input	Outputs
T	Q Q'
0	No change
1	Toggle

In order to drive the present state to next state listed in Table 3.37, the D flip-flop input is obtained using the excitation table of D flip-flop and listed in Table 3.37. Minterms are formed using T and Q and the K-map is constructed for D as shown in Fig. 3.59.

Excitation Table of D Flip-Flop

Present State	Next State	Flip-Flop Input
Q(t)	Q(t + 1)	D
0	0	0
0	1	1
1	0	0
1	1	1

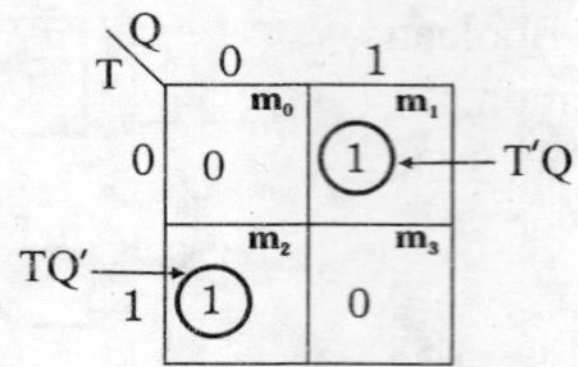

Fig. 3.59: K-map for D flip-flop input.

From the K-map we get the following Boolean equation for D flip-flop input and using this equation, the T flip-flop logic circuit is drawn as shown in Fig. 3.60.

$$D = TQ' + T'Q$$

The working of T flip-flop shown in Fig. 3.60 is verified for all possible combinations of input and present state as shown in Table 3.38. The next state output of T flip-flop is obtained from D flip-flop characteristic table.

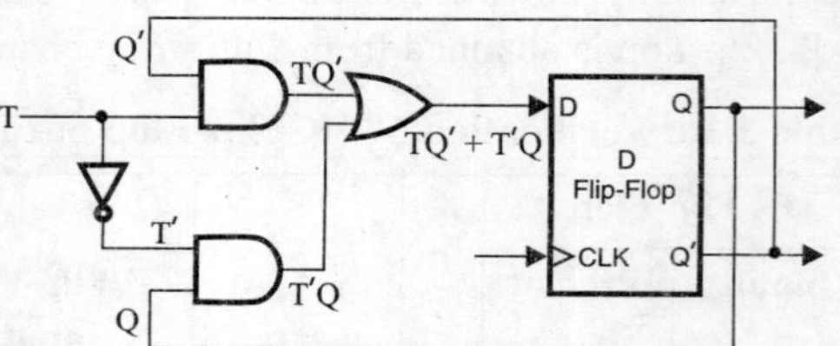

Fig. 3.60: T flip-flop using D flip-flop.

Table 3.38: Verification of T Flip-Flop of Fig. 3.60

T Flip-Flop Input	T Flip-Flop Present State					D Flip-Flop Input	T Flip-Flop Next State		Comment
T	Q	Q'	T'	TQ'	T'Q	D = TQ' + T'Q	Q(t+1)	Q'(t+1)	
0	0	1	1	0	0	0	0	1	No change
0	1	0	1	0	1	1	1	0	
1	0	1	0	1	0	1	1	0	Toggle
1	1	0	0	0	0	0	0	1	

Characteristic Table of D Flip-Flop

Input	Present State	Next State
D	Q	Q(t+1)
0	0	0
0	1	0
1	0	1
1	1	1

3.5.8 Conversion of T to SR Flip-Flop

Using the truth table of SR flip-flop, a conversion table is formed by listing the possible combinations of inputs, present and next states of SR flip-flop as shown in Table 3.39.

Table 3.39: Conversion Table to Convert T to SR Flip-Flop

SR Flip-Flop Inputs		Present State	Next State	Comment	T Flip-Flop Input	Minterm
S	R	Q	Q(t+1)			SRQ
0	0	0	0	No change	0	m_0
0	0	1	1		0	m_1
0	1	0	0	Reset state	0	m_2
0	1	1	0		1	m_3
1	0	0	1	Set state	1	m_4
1	0	1	1		0	m_5
1	1	0	×	Indeterminate	×	m_6
1	1	1	×		×	m_7

Truth Table of SR Flip-Flop

Inputs		Outputs	
S	R	Q	Q'
0	0	No change	
0	1	0	1 (Reset)
1	0	1	0 (Set)
1	1	Indeterminate	

In order to drive the present state to next state listed in Table 3.39, the T flip-flop input is obtained using the excitation table of T flip-flop and listed in Table 3.39. Minterms are formed using S, R and Q and the K-map is constructed for T as shown in Fig. 3.61.

Excitation Table of T Flip-Flop

Present State Q(t)	Next State Q(t + 1)	Flip-Flop Input T
0	0	0
0	1	1
1	0	1
1	1	0

S \ RQ	00	01	11	10
0	m_0: 0	m_1: 0	m_3: 1	m_2: 0
1	m_4: 1	m_5: 0	m_7: ×	m_6: ×

RQ ; SQ'

Fig. 3.61: K-map for T flip-flop input.

From the K-map we get the following Boolean equation for T flip-flop input and using this equation, the SR flip-flop logic circuit is drawn as shown in Fig. 3.62.

$T = SQ' + RQ$

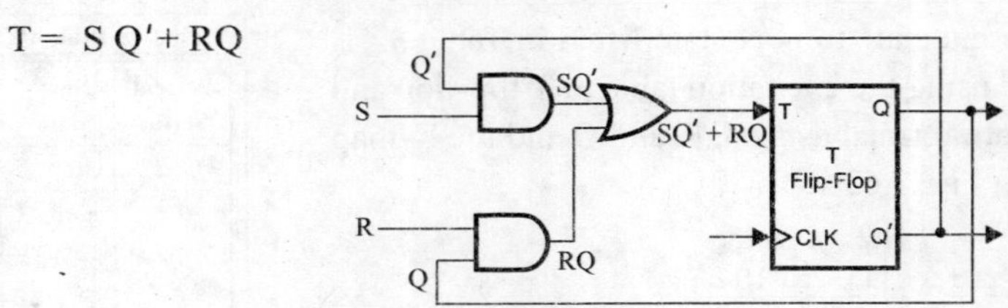

Fig. 3.62: SR flip-flop using T flip-flop.

The working of SR flip-flop shown in Fig. 3.62 is verified for all possible combinations of inputs and present state as shown in Table 3.40. The next state output of SR flip-flop is obtained from T flip-flop characteristic table.

Table 3.40: Verification of SR Flip-Flop of Fig. 3.62

SR Flip-Flop							SR Flip-Flop		
Inputs		Present State				T Flip-Flop Input	Next State		Comment
S	R	Q	Q'	SQ'	RQ	T = SQ' + RQ	Q(t+1)	Q'(t+1)	
0	0	0	1	0	0	0	0	1	No change
0	0	1	0	0	0	0	1	0	
0	1	0	1	0	0	0	0	1	Reset state
0	1	1	0	0	1	1	0	1	
1	0	0	1	1	0	1	1	0	Set state
1	0	1	0	0	0	0	1	0	
1	1	0	1	1	0	1	1	0	Toggle
1	1	1	0	0	1	1	0	1	

Characteristic Table of T Flip-Flop

Input T	Present State Q	Next State Q(t+1)
0	0	0
0	1	1
1	0	1
1	1	0

3.5.9 Conversion of T to JK Flip-Flop

Using the truth table of JK flip-flop, a conversion table is formed by listing the possible combinations of inputs, present and next states of JK flip-flop as shown in Table 3.41.

Table 3.41: Conversion Table to Convert T to JK Flip-Flop

JK Flip-Flop					T Flip-Flop Input	Minterm
Inputs		Present State	Next State	Comment		
J	K	Q	Q(t+1)			JKQ
0	0	0	0	No change	0	m_0
0	0	1	1		0	m_1
0	1	0	0	Reset state	0	m_2
0	1	1	0		1	m_3
1	0	0	1	Set state	1	m_4
1	0	1	1		0	m_5
1	1	0	1	Toggle	1	m_6
1	1	1	0		1	m_7

Truth Table of JK Flip-Flop

Inputs		Outputs	
J	K	Q	Q′
0	0	No change	
0	1	0	1 (Reset)
1	0	1	0 (Set)
1	1	Toggle	

Excitation Table of T Flip-Flop

Present State	Next State	Flip-Flop Input
Q(t)	Q(t + 1)	T
0	0	0
0	1	1
1	0	1
1	1	0

In order to drive the present state to next state listed in Table 3.41, the T flip-flop input is obtained using the excitation table of T flip-flop and listed in Table 3.41. Minterms are formed using J, K and Q and the K-map is constructed for T as shown in Fig. 3.63.

J \ KQ	00	01	11	10
0	m_0 0	m_1 0	m_3 1	m_2 0
1	m_4 1	m_5 0	m_7 1	m_6 1

(KQ ← group of m_3, m_7; JQ′ ← group of m_4, m_6)

Fig. 3.63: K-map for T flip-flop input.

From the K-map we get the following Boolean equation for T flip-flop input and using this equation, the JK flip-flop logic circuit is drawn as shown in Fig. 3.64.

$$T = JQ' + KQ$$

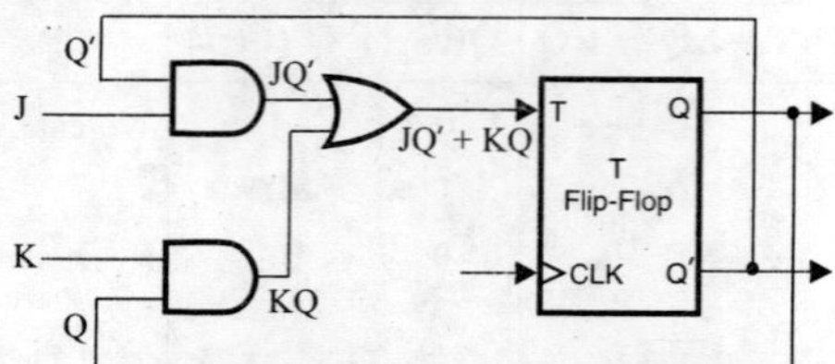

Fig. 3.64: JK flip-flop using T flip-flop.

The working of JK flip-flop shown in Fig. 3.64 is verified for all possible combinations of inputs and present state as shown in Table 3.42. The next state output of JK flip-flop is obtained from T flip-flop characteristic table.

Table 3.42: Verification of JK Flip-Flop of Fig. 3.64

JK Flip-Flop							SR Flip-Flop		
Inputs		Present State				T Flip-Flop Input	Next State		Comment
J	K	Q	Q′	JQ′	KQ	T = JQ′ + KQ	Q(t+1)	Q′(t+1)	
0	0	0	1	0	0	0	0	1	No change
0	0	1	0	0	0	0	1	0	
0	1	0	1	0	0	0	0	1	Reset state
0	1	1	0	0	1	1	0	1	
1	0	0	1	1	0	1	1	0	Set state
1	0	1	0	0	0	0	1	0	
1	1	0	1	1	0	1	1	0	Toggle
1	1	1	0	0	1	1	0	1	

Characteristic Table of T Flip-Flop

Input	Present State	Next State
T	Q	Q(t+1)
0	0	0
0	1	1
1	0	1
1	1	0

3.5.10 Conversion of T to D Flip-Flop

Using the truth table of D flip-flop, a conversion table is formed by listing the possible combinations of input, present and next states of D flip-flop as shown in Table 3.43.

Table 3.43: Conversion Table to Convert T to D Flip-Flop

D Flip-Flop					
Input	Present State	Next State	Comment	T Flip-Flop Input	Minterm
D	Q	Q(t+1)			TQ
0	0	0	Next state output is same as input	0	m_0
0	1	0		1	m_1
1	0	1		1	m_2
1	1	1		0	m_3

Truth Table of D Flip-Flop

Input	Outputs	
D	Q	Q′
0	0	1
1	1	0

In order to drive the present state to next state listed in Table 3.43, the T flip-flop input is obtained using the excitation table of T flip-flop and listed in Table 3.43. Minterms are formed using D and Q and the K-map is constructed for T as shown in Fig. 3.65.

Excitation Table of T Flip-Flop

Present State	Next State	Flip-Flop Input
Q(t)	Q(t + 1)	T
0	0	0
0	1	1
1	0	1
1	1	0

D \ Q	0	1
0	m_0: 0	m_1: (1) ← D′Q
1	DQ′ → m_2: (1)	m_3: 0

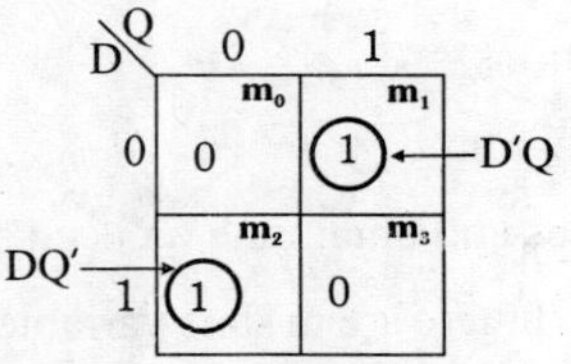

Fig. 3.65: K-map for T flip-flop input.

From the K-map we get the following Boolean equation for T flip-flop input and using this equation, the D flip-flop logic circuit is drawn as shown in Fig. 3.66.

$$T = DQ' + D'Q$$

The working of D flip-flop shown in Fig. 3.66 is verified for all possible combinations of input and present state as shown in Table 3.44. The next state output of D flip-flop is obtained from T flip-flop characteristic table.

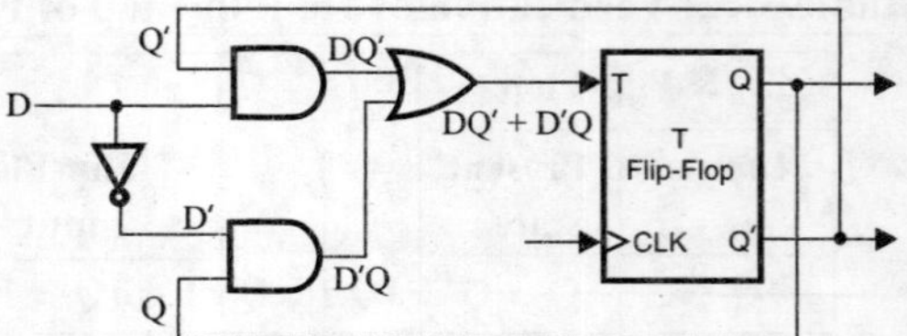

Fig. 3.66: D flip-flop using T flip-flop.

Table 3.44: Verification of D Flip-Flop of Fig. 3.66

D Flip-Flop							D Flip-Flop		
Input		Present State				T Flip-Flop Input	Next State		Comment
D	D'	Q	Q'	DQ'	D'Q	T = DQ' + D'Q	Q(t+1)	Q'(t+1)	
0	1	0	1	0	0	0	0	1	Next state output is same as input
0	1	1	0	0	1	1	0	1	
1	0	0	1	1	0	1	1	0	
1	0	1	0	0	0	0	1	0	

Characteristic Table of T Flip-Flop

Input	Present State	Next State
T	Q	Q(t+1)
0	0	0
0	1	1
1	0	1
1	1	0

3.6 Design of Synchronous Sequential Circuits

The sequential circuits can be designed using flip-flops with combinational circuits at the input and output of flip-flops. The sequential circuit involves feedback from output to input. The circuit can be designed using either D or JK or T flip-flop. Usually the circuit is designed with only any one of the type of flip-flops.

The T flip-flops do not have set state and reset state and so they cannot be used to design general storage devices or sequential circuits like D flip-flop and JK flip-flop. But T flip-flops are popularly used to design counters which is also a type of sequential circuits. The combinational circuit at the input is designed by constructing K-map for flip-flop inputs. The combinational circuit at the output is designed by constructing K-map for outputs.

The design starts with specifications which includes a word description, inputs and outputs of the circuit.

The steps in design of sequential circuits are,

1. Draw the state diagram from specifications.

2. Determine the state table from state diagram.

3. If necessary, reduce the number of states and obtain the reduced state table.

4. Assign binary values or codes to states. If there are m state variables and n inputs then 2^{m+n} binary codes are needed to code the states.

5. Form the binary coded state table.

6. Choose the type of flip-flop to be used in design (either D or JK or T). The number of flip-flops is equal to number of state variables.

7. The flip-flops inputs are same as next state outputs, when D flip-flops are used for design. In order to design using JK or T flip-flops determine the flip-flop inputs using excitation table.
8. Form the expanded state table with flip-flop inputs.
9. Derive the simplified flip-flop input and output equations using K-map.
10. Draw the logic diagram using the input and output equations.

3.6.1 Mealy and Moore Models

(AU, Apr/May'19, 7 Marks)

The sequential circuits consists of inputs, states and outputs. The outputs of sequential circuits may be functions of only states or functions of both inputs and states.

When the outputs of a sequential circuit are functions of inputs and states (or present states) then the sequential circuit is called ***Mealy type*** or ***Mealy model*** of sequential circuits. The sequential circuits in Examples 3.5, 3.6, 3.8, 3.9 and 3.10 are Mealy models.

When the outputs of a sequential circuit are functions of states (or present states) alone, then the sequential circuit is called ***Moore type*** or ***Moore model*** of sequential circuits. The sequential circuits in Examples 3.3, 3.4, 3.11, 3.14 and 3.15 are Moore models.

Since the output depends only on states in Moore model of sequential circuits, the state diagram of Moore type models will have circles with state and associated output. The lables on the directed lines will have only inputs that initiate transitions from one state to another. (Refer examples 3.3, 3.4 and 3.11 for state diagram of Moore model sequential circuits)

The functional behaviour of sequential circuits can be represented using a finite number of states. Hence the sequential circuits are also called ***finite state machines (FSM)***. Therefore, the Mealy model is called ***Mealy FSM*** or ***Mealy Machine***. The moore model is called ***Moore FSM*** or ***Moore Machine***.

The general block diagrams of Mealy and Moore Machine are shown in Figs. 3.67 and 3.68 respectively. In the block diagram the state registers are flip-flops that are used to store the value of states.

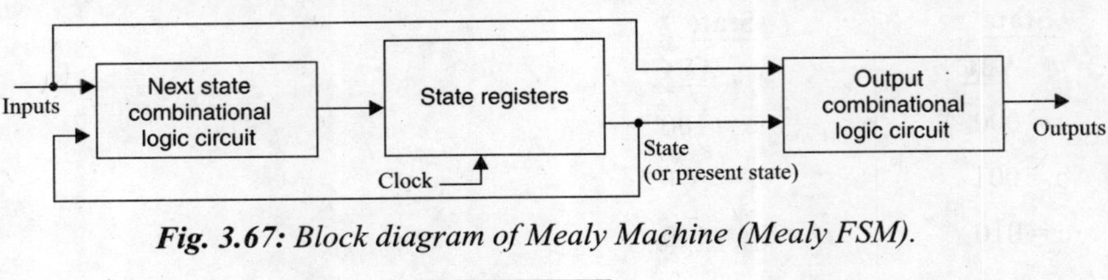

Fig. 3.67: *Block diagram of Mealy Machine (Mealy FSM).*

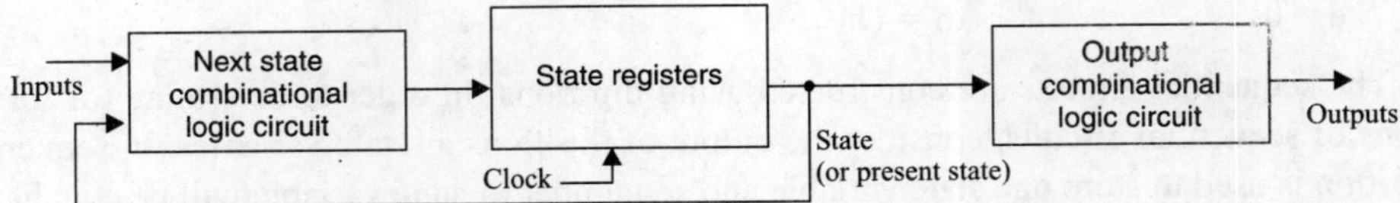

Fig. 3.68: *Block diagram of Moore Machine (Moore FSM).*

The logic circuits on the input side of flip-flops are called next state combination logic circuit, because they are needed to develop the inputs for flip-flops before arrival of clock to change the state from present state to next state. The logic circuits on the output side of flip-flops are used to construct outputs, and so they are called output combinational logic circuit.

In a Moore model, the outputs are synchronized with clock. But in a Mealy model the inputs may change any time and so outputs are not synchronized with clock. The synchronization of inputs with clock in Mealy models is necessary to ensure right outputs and to overcome the difference in delays in input changes and state changes.

3.6.2 State and State Equations

A ***state*** is the condition of a sequential circuit described by a set of variables called ***state variables***. Flip-flops are used to store state variables and so the output of flip-flops constitute the states of a sequential circuit. With m state variables we can form 2^m different states.

The ***state equations*** (or transition equations) are a set of equations that describe the next state as a function of the present states and inputs.

Let, x, y, z, ... are used to denote inputs.

X, Y, Z, ... are used to denote outputs of a sequential circuit.

a, b, c, ... are used to denote states

A, B, C, ... are used to denote the state variables or flip-flop outputs.

Since the inputs, outputs and state variables are all functions of time we can differentiate the present and next state variables by two different notations as function of time.

Present State Variables: A(t), B(t), C(t) ......

Next State Variables: A(t + 1), B(t + 1), C(t + 1) ....

Note: *For simplicity the " t " and " t+1" can be dropped and variables in present and next state can be represented in different columns in state table, refer examples or solved problems.*

Example: Let A, B and C be three state variables. Now, with 3 state variables, number of different states are 2^3 = 8 states. Let the 8 states be denoted as a, b, c, d, e, f, g and h. The states and state variables are related by assigning 3-bit code for state variables as shown below:

State ABC	State ABC
a = 000	e = 100
b = 001	f = 101
c = 010	g = 110
d = 011	h = 111

The sequential circuits are constructed using flip-flops. In order to determine (or form) the state equations of sequential circuits, consider the output of flip-flops as state variables. In sequential circuits one flip-flop is used to store one state variable and so number of state variable will be equal to number of flip-flops.

The present state is the output of flip-flops before clock pulse and next state is the output of flip-flop after clock pulse. The next state can be obtained from the truth table of flip-flop or characteristic equations of flip-flops. Thus state equations are obtained from characteristic equations.

Refer examples 3.5 and 3.6 for the formation of state equations of sequential circuits.

3.6.3 State Table and State Diagram

The ***state table*** is listing of outputs and next states for all possible combinations of inputs and present states. In general, if there are m state variables and n inputs then possible combinations of inputs and states are 2^{m+n}. The state table list the outputs and next states for 2^{m+n} combinations of inputs and states.

The ***state diagram*** is graphical representation of all transitions from present state to next state for all possible combination of inputs and outputs. Therefore, the state diagram is the graphical representation of information available in a state stable.

The state diagram consists of circles and directed lines. The states are represented by circles and directed lines indicate the transition from one state to another. The associated inputs and outputs of every state transition is marked as label in the directed lines. The format of label is input/output.

Refer examples 3.5 and 3.6 for the formation of state table and state diagram of sequential circuits.

3.6.4 State Minimization

The synthesis or design of sequential circuit start from a state diagram. There is a possibility of realizing the same input-output relationship of the given state diagram with an equivalent or alternate state diagram with reduced number of states. An example of reduced state diagram is shown in Fig. 3.69.

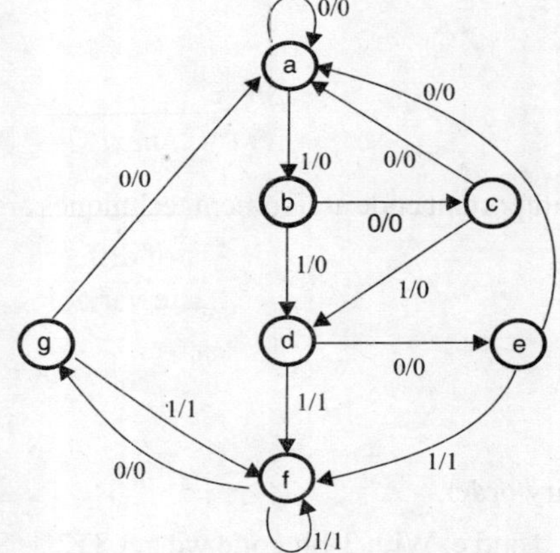

Fig. a: *State diagram with seven states.*

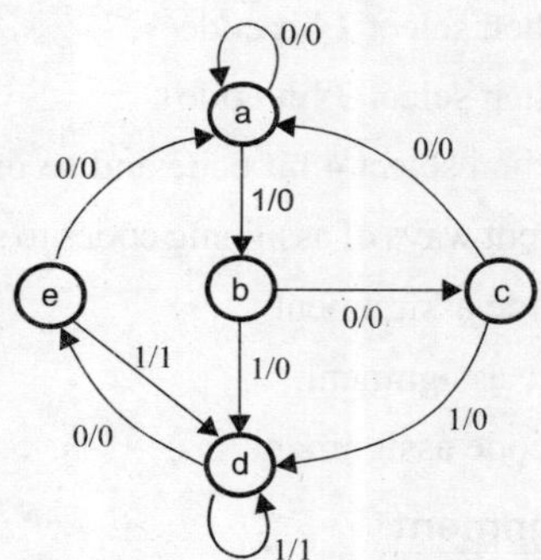

Fig. b: *Reduced equivalent state diagram with 5 states.*

The reduction is explained in the Example 3.7.

Fig. 3.69: *Example of state diagram and reduced state diagram.*

In order to check the possibility of reducing the number of states in a given state diagram, first form the state table. Then check for equivalent states in state table. Two states are said to be equivalent if the input-output relationship are same with identical present state and changes in input, which drive the present state to identical next state. When two states are equivalent, one of them can be removed without altering the input-output relationship.

In the state reduction problem first check for a pair of equivalent states in the state table and remove or eliminate one of the identical state and form a reduced state table. Then again check for a pair of equivalent states in the reduced state table and remove or eliminate one of the identical state and again form another reduced state table.

The above procedure is repeated until all the equivalent pair of states are identified and one of the state is removed in each pair. The original (or specified or given) state diagram and the reduced state diagram are verified using an input-output sequence. If both the original and reduced state diagram satisfy the input-output sequence then the reduced state diagram is equivalent to specified state diagram.

The advantage in reducing the states is that the sequential circuit can be realized with less number of flip-flops, so that the hardware cost will be reduced.

3.6.5 State Assignment

In the design specifications when the states are specified by alphabets it is necessary to assign binary codes for practical implementations. Therefore, an n-bit binary code is assigned to each state. Assigning binary codes to states represented by alphabets is called ***state assignment*** (or ***code assignment***).

With n-bits we can form 2^n binary codes.

When $n = 2$, $2^n = 2^2 = 4$

When $n = 3$, $2^n = 2^3 = 8$

When $n = 4$, $2^n = 2^4 = 16$ and so on.

Let, m = Number of state variables.

2^m = Number of states

If, $2^m \leq 4$, then select 2 bit code

If, $2^m \leq 8$, then select 3 bit code

If, $2^m \leq 16$, then select 4 bit code and so on.

There are different ways of assigning codes to states. Three important code assignment techniques are,

1. Binary code assignment
2. Gray code assignment
3. One hot code assignment

Binary Code Assignment

In binary code assignment, the codes are assigned in binary order.

Consider assigning three bit binary code to 5 states a, b, c, d and e. With 3-bit code we get 8 ($2^3 = 8$) binary codes. The first 5 codes are assigned to states a, b, c, d and e as shown below and the remaining 3 codes are unused or don't-care codes.

State		Three bit binary code		
a	⟶	0 0 0	Assigned codes	
b	⟶	0 0 1	Assigned codes	
c	⟶	0 1 0	Assigned codes	
d	⟶	0 1 1	Assigned codes	
e	⟶	1 0 0	Assigned codes	
		1 0 1	Don't-care codes	
		1 1 0	Don't-care codes	
		1 1 1	Don't-care codes	

Gray Code Assignment

Consider assigning three bit gray code to 5 states a, b, c, d and e. With 3-bit gray code we get 8 ($2^3 = 8$) gray codes. The first 5 codes are assigned to states a, b, c, d and e as shown below and the remaining 3 codes are unused or don't-care codes.

State		Three bit gray code	
a	→	0 0 0	Assigned codes
b	→	0 0 1	
c	→	0 1 1	
d	→	0 1 0	
e	→	1 1 0	
		1 1 1	Don't-care codes
		1 0 1	
		1 0 0	

One Hot Code Assignment

The one-hot assignment will not satisfy 2^n logic. Here for n state assignments, an n-bit code is chosen. In n-bit code only one digit is 1 and remaining are zeros. The 1 is assigned starting from lower order bit.

The assignment of 5 bit one-hot code to 5 states a, b, c, d and e is shown below:

State		One-hot code
a	→	0 0 0 0 1
b	→	0 0 0 1 0
c	→	0 0 1 0 0
d	→	0 1 0 0 0
e	→	1 0 0 0 0

3.6.6 Lock-out Condition Circuit Implementation

In sequential circuits with some don't-care or unused states, the sequential circuit may find itself in some unused state during starting or switch ON. If next state of the unused state is again some unused one and if the sequential circuit cycle through only unused states and never arrives at in a used state then this condition is called ***lock-out***. An example of lock-out state diagram is shown in Fig. 3.70. In the state diagram of Fig. 3.70, the sequential circuit has to detect the sequence 1, 2, 4 and 7. Here, unused states are 0, 3, 5 and 6.

During switch ON if the sequential circuit enter an unused state and it cycles through only unused states as shown in Fig. 3.70(a) then the circuit will be in lock-out condition.

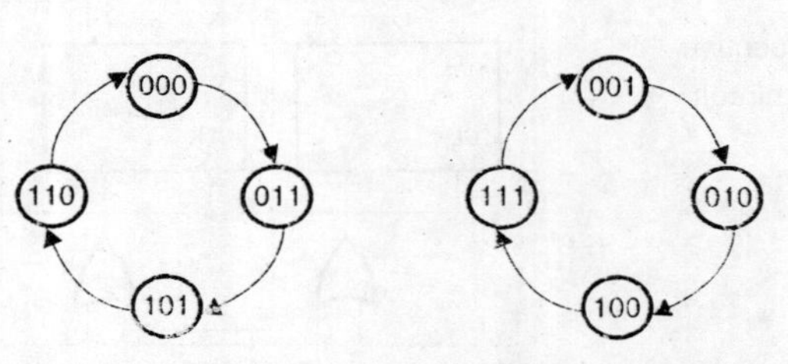

Fig. a: Unused states. *Fig. b: Used states.*

Fig. 3.70: *State diagram with lock-out condition.*

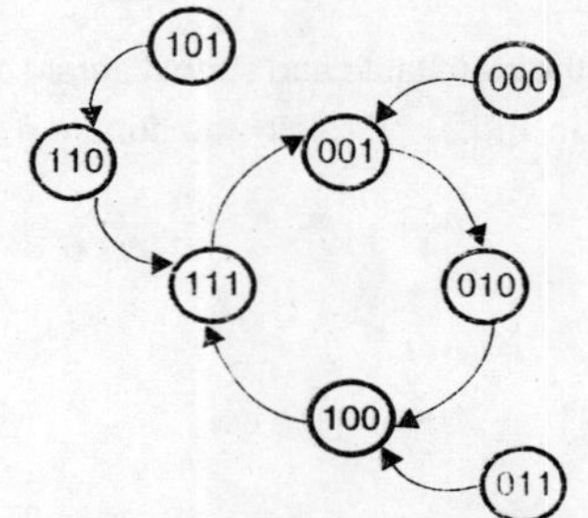

Fig. 3.71: *State diagram to prevent lock-out condition.*

In order to ensure that lock-out does not occur, the counter should be designed such that if the initial state is an unused state, then the state transition occurs from unused state to a used state. Therefore, every unused state is introduced in front of a used state. The example of Fig. 3.70, is modified as shown in Fig. 3.71, to prevent lock-out condition. The solution of this state diagram is discussed in Example 3.18. Alternatively, special circuit can be designed to reset the counter whenever it got struck into an unused state.

Example 3.3

(AU, Apr/May'19, 6 Marks)

Draw the state stable for the following state diagram.

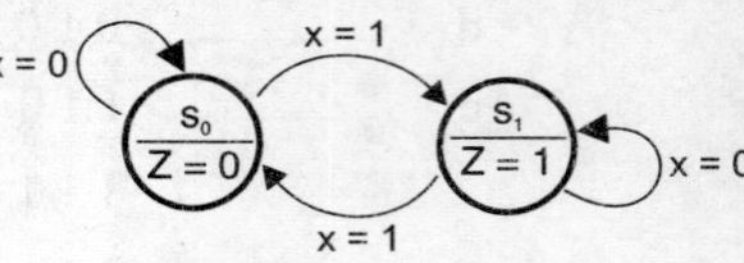

Fig. 1: State diagram.

Solution

From the knowledge of state transitions shown in Fig. 1, the state table is constructed as shown in Table 1.

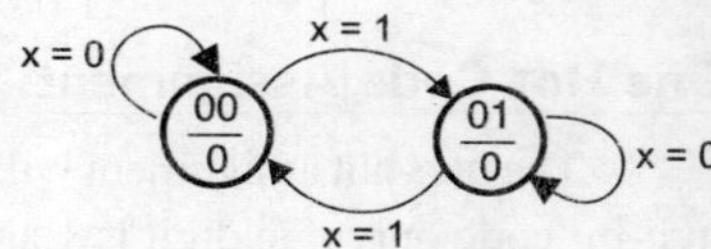

Fig. 2: State diagram with binary code assignment.

Let us choose two state variables A and B and assign two bit binary code for state variables.

∴ Possible states, AB = 0 0

0 1

1 0

1 1

Let, S_0 = 00 and S_1 = 01

Here, 10 and 11 are unused states.

Table 1: State Table

Present States	Next States		Outputs	
A B	x = 0	x = 1	x = 0	x = 1
	A(t+1)	B(t+1)	A(t+1)	B(t+1)
S_0	S_0	S_1	0	0
S_1	S_1	S_0	1	1

The state diagram with binary code assignment is shown in Fig. 2 and using this state diagram the state table is constructed as shown in Table 2.

Table 2: State Table with Code Assignment

Present States		Input	Next States		Output
A	B	x	A(t+1)	B(t + 1)	Z
0	0	0	0	0	0
0	0	1	0	1	0
0	1	0	0	1	1
0	1	1	0	0	1
1	0	0	×	×	×
1	0	1	×	×	×
1	1	0	×	×	×
1	1	1	×	×	×

Here, the output depends only on states and so it is a Moore model sequential circuit.

Example 3.4

Derive the state table and state diagram of the sequential circuit shown in Fig. 1. Explain the function that the circuit performs.

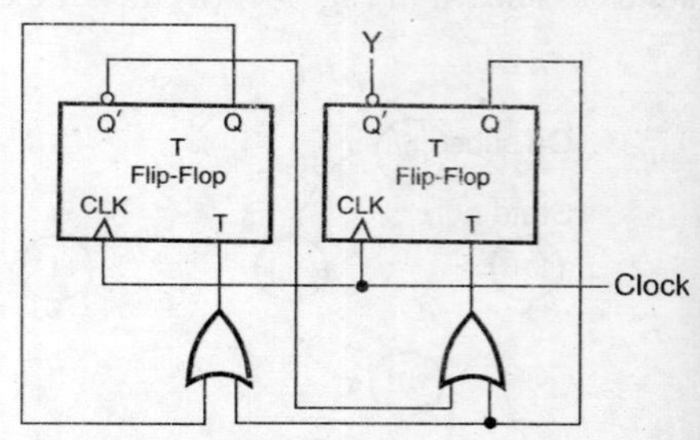

Fig. 1: Sequential circuit.

Solution

The sequential circuit is redrawn as shown in Fig. 2.

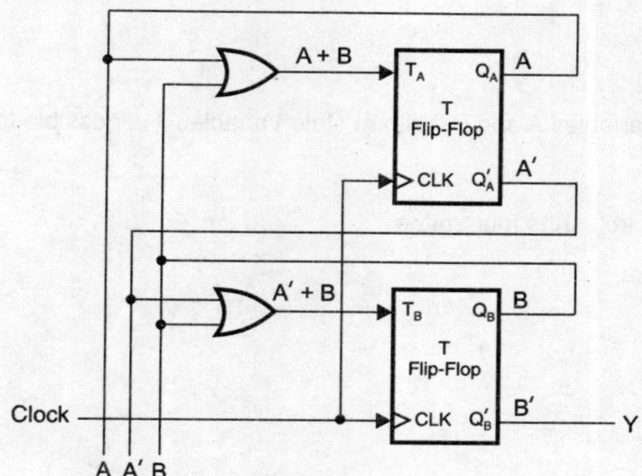

Fig. 2: Sequential circuit.

In the given sequential circuit, the output depends on only present states and so it is a Moore model.

The sequential circuit is constructed using two T flip-flops.

Since there are two flip-flops let us take two state variables A and B and equate the state variables to flip-flop output.

$\therefore \; A = Q_A \;\Rightarrow\; A(t+1) = Q_A(t+1)$

$B = Q_B \;\Rightarrow\; B(t+1) = Q_B(t+1)$

The characteristic equations of the two T flip-flops are,

$Q_A(t+1) = T_A A' + T'_A A$

$Q_B(t+1) = T_B B' + T'_B B$

Characteristic equation of T flip-flop,
$Q(t+1) = TQ' + T'Q$

Now, the characteristic equations, in terms of state variables are,

$A(t+1) = T_A A' + T'_A A$

$B(t+1) = T_B B' + T'_B B$

From Fig. 2 we get the following Boolean equations for flip-flop inputs.

$T_A = A + B$

$T_B = A' + B$

On substituting the above Boolean equations in the characteristic equations we get the following state equations.

State equations: $A(t+1) = T_A A' + T'_A A$

$= (A+B)A' + (A+B)'A = AA' + A'B + A'B'A$

$= 0 + A'B + 0 = A'B$

$B(t+1) = T_B B' + T'_B B$

$= (A'+B)B' + (A'+B)'B = A'B' + BB' + AB'B$

$= A'B' + 0 + 0 = A'B'$

Here, Y = Output

From Fig. 2 we get the output equation as,

Output equation : Y = B′

State Table

The circuit has two state variables A and B. With m state variables it is possible to have 2^m different states.

Here, m = 2, $\therefore 2^2 = 4$

Therefore, the sequential circuit has four states.

The four possible states are,

AB = 0 0

= 0 1

= 1 0

= 1 1

The state table is listing of output and next states for 4 possible present states.

Table 1: State Table

Present States		Complement of Present States		Next States		Output
A	B	A′	B′	A(t + 1)	B(t + 1)	Y
0	0	1	1	0	1	1
0	1	1	0	1	0	0
1	0	0	1	0	0	1
1	1	0	0	0	0	0

A(t + 1) = A′B

B(t + 1) = A′B′

Y = B′

State Diagram

In the state diagram the states are represented by circles. The transition from one state to another is represented by directed lines (i.e, lines with arrow). The arrow indicates the transition from one state to another.

In this example, there are 4 possible states and so the state diagram has four circles with values of state variables inside the circle as shown below:

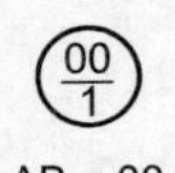

AB = 00

$\frac{01}{0}$ AB = 01

$\frac{10}{1}$ AB = 10

$\frac{11}{0}$ AB = 11

The state diagram is drawn using state table as shown in Fig. 3.

Fig. 3: State diagram.

Working of Circuit

From the state diagram we can say that the output is 1 when both states are reset to 0 or A is set when B is reset and in all other cases the output is 0. Therefore, the circuit detects a condition whether both states are reset or A alone is set after a reset.

Example 3.5

Construct the state table and state diagram for the sequential circuit using D flip-flops shown in Fig. 1.

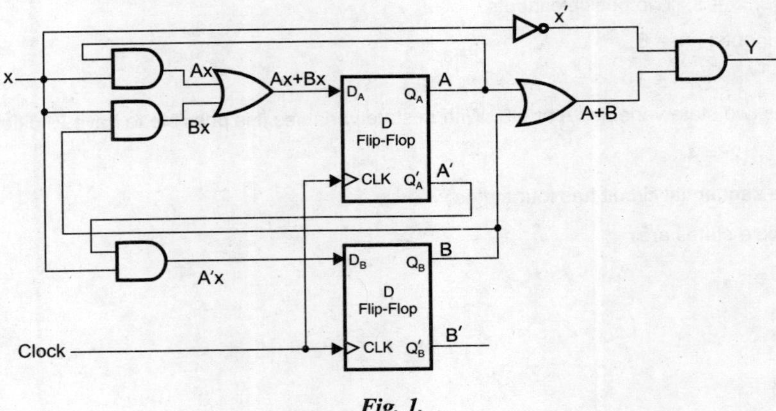

Fig. 1.

Solution

In the given sequential circuit, the output depends on input and states and so it is a Mealy model.

The sequential circuit is constructed using two D flip-flops.

Since there are two flip-flops let us take two state variables A and B and equate the state variables to flip-flop output.

$\therefore\ A = Q_A \Rightarrow A(t+1) = Q_A(t+1)$

$B = Q_B \Rightarrow B(t+1) = Q_B(t+1)$

The characteristic equations of the two D flip-flops are,

Characteristic equation of D flip-flop, $Q(t+1) = D$

$Q_A(t+1) = D_A$

$Q_B(t+1) = D_B$

Now, the characteristic equations, in terms of state variables are,

$A(t+1) = D_A$

$B(t+1) = D_B$

From Fig. 1 we get the following Boolean equations for flip-flop inputs.

$D_A = Ax + Bx$

$D_B = A'x$

On substituting the above Boolean equations in the characteristic equations we get the following state equations.

State equations: $A(t+1) = Ax + Bx$

$B(t+1) = A'x$

Here, Y = Output

From Fig. 1 we get the output equation as,

Output equation : $Y = (A + B)\,x'$

State Table

With n-bit input it is possible to have 2^n different inputs.

Here, n = 1, $\therefore 2^1 = 2$, Two possible inputs.

Two possible inputs: x = 0

x = 1

The circuit has two state variables A and B. With m state variables it is possible to have 2^m different states.

Here, m = 2, $\therefore 2^2 = 4$

Therefore, the sequential circuit has four states.

The four possible states are,

AB = 0 0

= 0 1

= 1 0

= 1 1

Here, when x = 0, AB can take four possible states.

Similarly, when x = 1, AB can take four possible states.

i.e, when, x = 0, AB = 0 0	when x = 1, AB = 0 0
= 0 1	= 0 1
= 1 0	= 1 0
= 1 1	= 1 1

When there are m state variables and n inputs then possible combinations of inputs and states are 2^{m+n}.

Here, m = 2, n = 1

$\therefore 2^{m+n} = 2^{2+1} = 2^3 = 8$

The evaluation of next state and output are shown in Table 1.

The state table is listing of outputs and next states for 8 combinations of inputs and present states as shown in Table 2 which can be represented in the simplified form as shown in Table 3.

Table 1: Evaluation of Next State and Output

Present States			Input		Product and Sum Terms				Next States		Output
A	B	A′	x	x′	Ax	Bx	A′x	A + B	A(t + 1)	B(t + 1)	Y
0	0	1	0	1	0	0	0	0	0	0	0
0	0	1	1	0	0	0	1	0	0	1	0
0	1	1	0	1	0	0	0	1	0	0	1
0	1	1	1	0	0	1	1	1	1	1	0
1	0	0	0	1	0	0	0	1	0	0	1
1	0	0	1	0	1	0	0	1	1	0	0
1	1	0	0	1	0	0	0	1	0	0	1
1	1	0	1	0	1	1	0	1	1	0	0

$A(t + 1) = Ax + Bx$

$B(t + 1) = A'x$

$Y = (A + B)x'$

Table 2: State Table

Present States		Input	Next States		Output
A	B	x	A(t+1)	B (t+1)	Y
0	0	0	0	0	0
0	0	1	0	1	0
0	1	0	0	0	1
0	1	1	1	1	0
1	0	0	0	0	1
1	0	1	1	0	0
1	1	0	0	0	1
1	1	1	1	0	0

Table 3: Simplified form of State Table

Present States		Next States				Output	
		x = 0		x = 1		x = 0	x = 1
A	B	A(t+1)	B(t+1)	A(t+1)	B(t+1)	Y	Y
0	0	0	0	0	1	0	0
0	1	0	0	1	1	1	0
1	0	0	0	1	0	1	0
1	1	0	0	1	0	1	0

State Diagram

In the state diagram the states are represented by circles. The transition from one state to another is represented by directed lines (i.e, lines with arrow). The arrow indicates the transition from one state to another.

Every transition has associated inputs and outputs, those inputs and outputs are marked as labels on the directed lines. The format of label is "input/output".

In this example, there are 4 possible states and so the state diagram has four circles with values of state variables inside the circle as shown below:

(00)	(01)	(10)	(11)
AB = 00	AB = 01	AB = 10	AB = 11

In this example, input = x

output = Y

Therefore, format of label on the directed lines is x/Y.

In this example there is a possibility of two transitions from one state to another for x = 0 and x = 1.

The possible transitions are shown in Fig. 2.

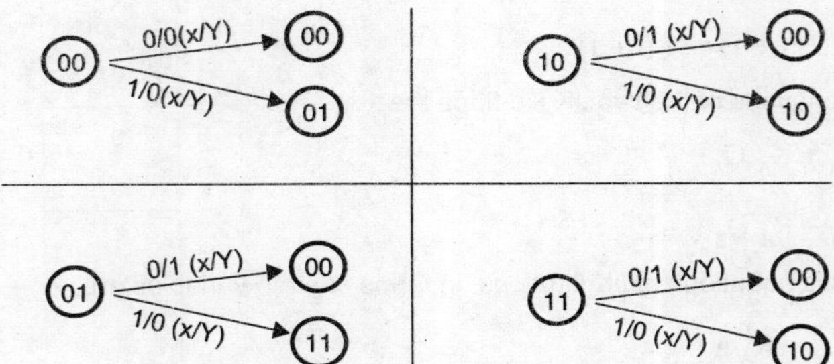

Fig. 2: Transitions from one state to another state.

The possible transitions shown in Fig. 2 has four states and 8 directed lines. On combining all the state transitions into a single diagram, we get the state diagram as shown in Fig. 3.

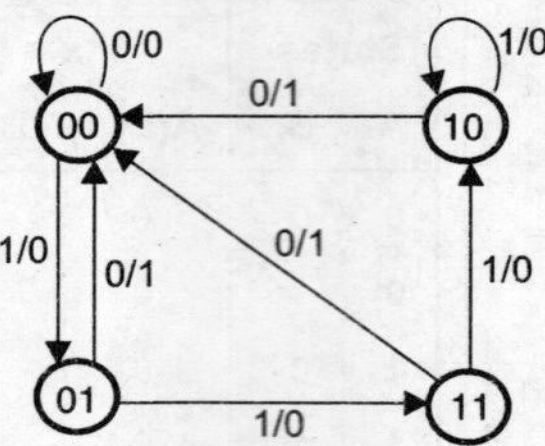

Fig. 3: State diagram.

Example 3.6

Develop the state table and diagram for the sequential circuit using JK flip-flop shown in Fig. 1.

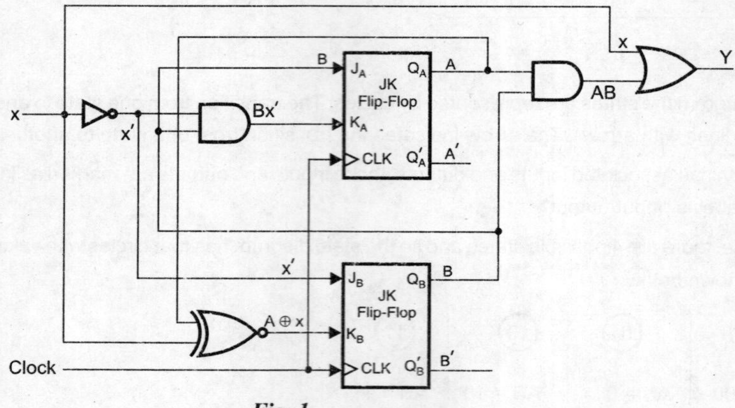

Fig. 1.

Solution

In the given sequential circuit, the output depends on input and states and so it is a Mealy model.

The sequential circuit is constructed using two JK flip-flops.

Since there are two flip-flops let us take two state variables A and B and equate the state variables to flip-flop output.

$$\therefore \; A = Q_A \;\Rightarrow\; A(t+1) = Q_A(t+1)$$

$$B = Q_B \;\Rightarrow\; B(t+1) = Q_B(t+1)$$

The characteristic equations of the two JK flip-flops are,

$$Q_A(t+1) = J_A\,Q_A' + K_A'\,Q_A$$

$$Q_B(t+1) = J_B\,Q_B' + K_B'\,Q_B$$

Characteristic equation of JK flip-flop, $Q(t+1) = JQ' + K'Q$

Now, the characteristic equations of JK flip-flops, in terms of state variables are,

$$A(t+1) = J_A\,A' + K_A'\,A$$

$$B(t+1) = J_B\,B' + K_B'\,B$$

From Fig. 1, we get the following Boolean equations for flip-flop inputs.

$$J_A = B$$

$K_A = B\,x'$

$J_B = x'$

$K_B = A \oplus x$

On substituting the above Boolean equations in characteristic equations we get the following state equations.

$$A(t+1) = J_A\,A' + K_A'\,A$$

$$= B\,A' + (B\,x')'A$$ Using DeMorgan's law

$$= B\,A' + (B' + x)\,A$$

$$= A'B + A\,B' + Ax$$

$$B(t+1) = J_B\,B' + K_B'\,B$$

$$= x'B' + (A \oplus x)'B$$ $(a \oplus b)' = a \odot b = ab + a'b'$

$$= x'B' + (Ax + A'x')\,B$$

$$= B'x' + A\,B\,x + A'B\,x'$$

$\therefore$ State equations: $A(t+1) = A'B + A\,B' + Ax$

$B(t+1) = B'x' + A\,B\,x + A'B\,x'$

From Fig. 1, we get the output equation as,

Output equation: $Y = AB + x$

State Table

Here we have 1 input and 2 state variables.

$\therefore\ n = 1,\ m = 2$

Number of inputs = $2^n = 2^1 = 2$ ($x = 0$ and $x = 1$)

Number of states = $2^m = 2^2 = 4$

Combinations of inputs and states = $2^{n+m} = 2^{1+2} = 2^3 = 8$

The 8 combinations of inputs and states are,

$x = 0,\ A\,B = 0\,0$

$= 0\,1$

$= 1\,0$

$= 1\,1$

$x = 1,\ A\,B = 0\,0$

$= 0\,1$

$= 1\,0$

$= 1\,1$

The next states are decided by the state equations and the output is determined by the output equation. The next states and output are evaluated for all possible combinations of inputs and states and listed in Table 1 and the state table is shown in Table 2.

Table 1: Evaluation of Next State and Output

Present States				Input		Product Terms							Next States		Output
A	B	A′	B′	x	x′	A′B	AB′	Ax	B′x′	AB x	A′B x′	A B	A(t+1)	B(t+1)	Y
0	0	1	1	0	1	0	0	0	1	0	0	0	0	1	0
0	0	1	1	1	0	0	0	0	0	0	0	0	0	0	1
0	1	1	0	0	1	1	0	0	0	0	1	0	1	1	0
0	1	1	0	1	0	1	0	0	0	0	0	0	1	0	1
1	0	0	1	0	1	0	1	0	1	0	0	0	1	1	0
1	0	0	1	1	0	0	1	1	0	0	0	0	1	0	1
1	1	0	0	0	1	0	0	0	0	0	0	1	0	0	1
1	1	0	0	1	0	0	0	1	0	1	0	1	1	1	1

$A(t+1) = A'B + AB' + Ax$; $B(t+1) = B'x' + ABx + A'Bx'$; $Y = AB + x$

Table 2: State Table

Present States		Input	Next States		Output
A	B	x	A(t+1)	B(t+1)	Y
0	0	0	0	1	0
0	0	1	0	0	1
0	1	0	1	1	0
0	1	1	1	0	1
1	0	0	1	1	0
1	0	1	1	0	1
1	1	0	0	0	1
1	1	1	1	1	1

State Diagram

This example has 4 states. Therefore, the state diagram has four circles with values of state variables inside the circle as shown below:

(00)	(01)	(10)	(11)
AB = 00	AB = 01	AB = 10	AB = 11

In this example, input = x

output = Y

Therefore, format of label on the directed lines is x/Y.

There is a possibility of two transitions from one state to another for x = 0 and x = 1. The possible transitions are shown in Fig. 2.

The possible transitions shown in Fig. 2 has four states and 8 directed lines. On combining all the transitions into one single diagram we get the state diagram as shown in Fig. 3.

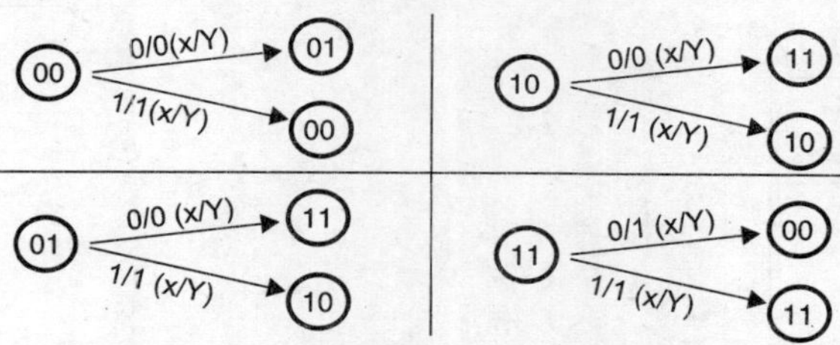

Fig. 2: Transitions from one state to another state.

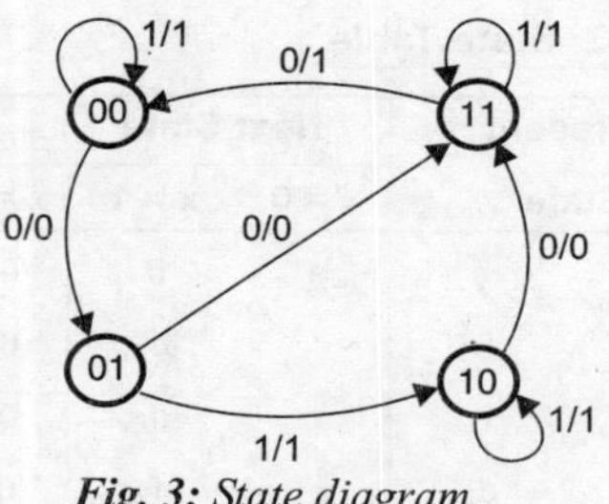

Fig. 3: State diagram.

Alternate Method for State Equations

K-maps are constructed for A(t+1), B(t+1) and Y using the values listed in Table 1. From the K-map we get the following state equations.

$$A(t+1) = A'B + AB' + Ax$$

$$B(t+1) = B'x' + ABx + A'Bx'$$

$$Y = AB + x$$

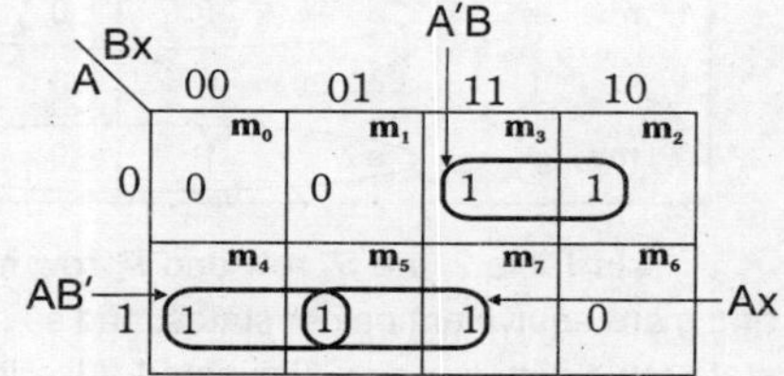

Fig. 4: K-map for A(t+1).

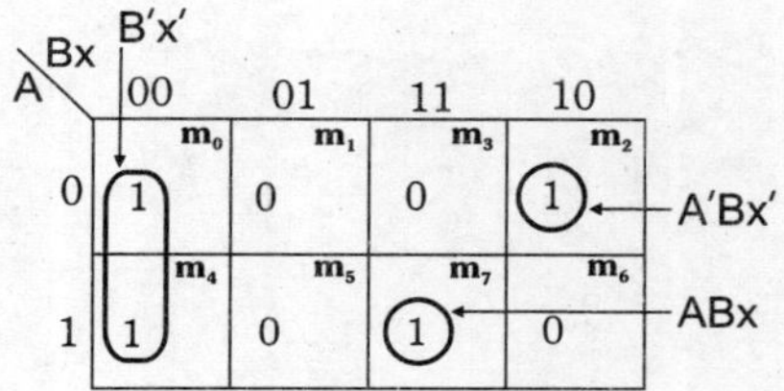

Fig. 5: K-map for B(t+1).

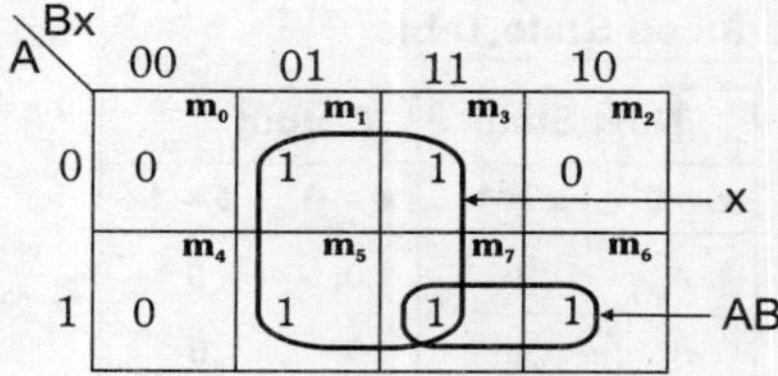

Fig. 6: K-map for Y.

Example 3.7

Reduce the states in the state diagram shown in Fig. 1 and draw the reduced state diagram. Verify the state diagram with the following input sequence. Also assign binary code to states in reduced state diagram.

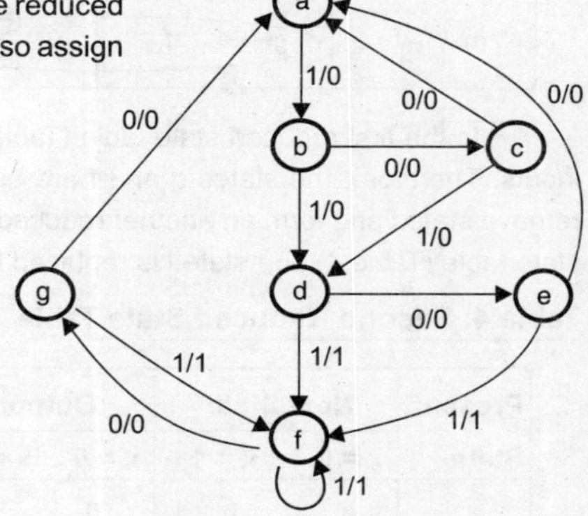

Fig 1: State diagram.

Table 1: Input Sequence for Verification

	Column Number											
	1	**2**	**3**	**4**	**5**	**6**	**7**	**8**	**9**	**10**	**11**	
State	a	a	b	c	d	e	f	f	g	f	g	a
Input	0	1	0	1	0	1	1	0	1	0	0	
Output	0	0	0	0	0	1	1	0	1	0	0	

Solution

In the given state diagram the following state transitions are identified.

a —0/0→ a | b —0/0→ c | c —0/0→ a | d —0/0→ e

a —1/0→ b | b —1/0→ d | c —1/0→ d | d —1/1→ f

e —0/0→ a | f —0/0→ g | g —0/0→ a

e —1/1→ f | f —1/1→ f | g —1/1→ f

Using the above transitions, the state table is constructed as shown in Table 2.

Table 2: State Table

Present State	Next State x = 0	Next State x = 1	Output x = 0	Output x = 1
a	a	b	0	0
b	c	d	0	0
c	a	d	0	0
d	e	f	0	1
5th row, e	a	f	0	1
f	g	f	0	1
7th row, g	a	f	0	1

Equivalent pair

In Table 2, the 5th row and 7th row has identical next states and outputs for same input, Therefore, the states e and g are equivalent pair of states, and so one of the state can be removed. Let us remove state g and form a reduced state table with last row eliminated. Also in the reduced state table, the state g is replaced by e in all entries in the retained rows.

Table 3: First Reduced State Table

Present State	Next State x = 0	Next State x = 1	Output x = 0	Output x = 1
a	a	b	0	0
b	c	d	0	0
c	a	d	0	0
4th row, d	e	f	0	1
e	a	f	0	1
6th row, f	g → e	f	0	1

Equivalent pair

In the first reduced state table (Table 3) the 4th row and 6th row has identical next states and outputs for same inputs. Therefore, the states d and f are equivalent pair of states, and so one of the state can be removed. Let us remove state f and form an another reduced state table (Table 4), with last row eliminated. Also in the second reduced state table (Table 4), the state f is replaced by d in all entries in the retained rows.

Table 4: Second Reduced State Table

Present State	Next State x = 0	Next State x = 1	Output x = 0	Output x = 1
a	a	b	0	0
b	c	d	0	0
c	a	d	0	0
d	e	f → d	0	1
e	a	f → d	0	1

The reduced state diagram is drawn using the second reduced state table (Table 4) as shown below:

In the second reduced state diagram the following state transitions are identified.

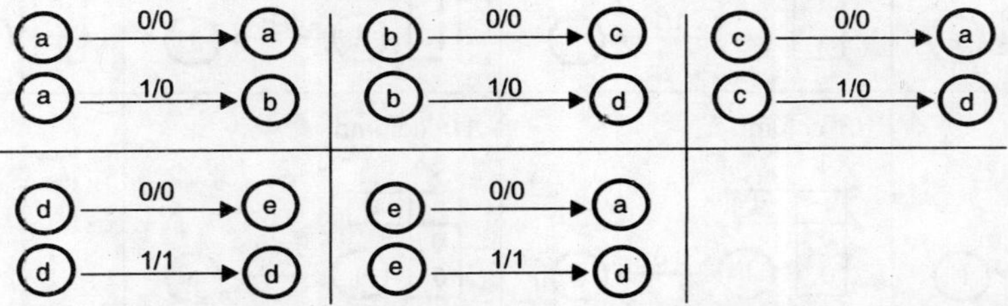

On combining the above state transitions the reduced state diagram is drawn as shown in Fig. 2.

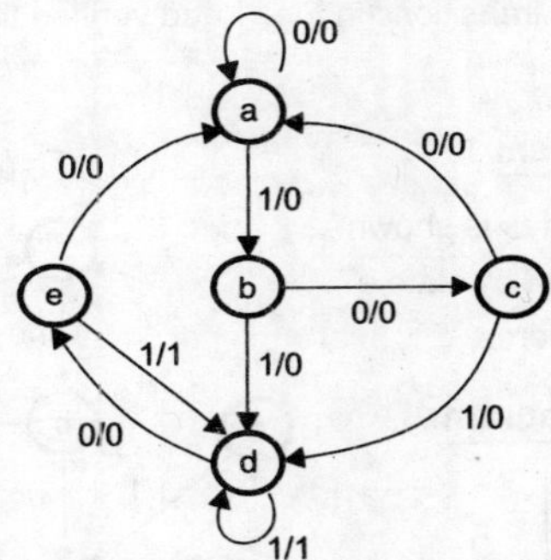

Fig. 2: Reduced state diagram.

Verification of Given Input Sequence in Original State Diagram

For convenience, the Table 1 and Fig. 1 are reproduced here. In Table 1, each column represent a state transition due to change in input.

Table 1: Input Sequence for Verification

	Column Number											
	1	2	3	4	5	6	7	8	9	10	11	
State	a	a	b	c	d	e	f	f	g	f	g	a
Input	0	1	0	1	0	1	1	0	1	0	0	
Output	0	0	0	0	0	1	1	0	1	0	0	

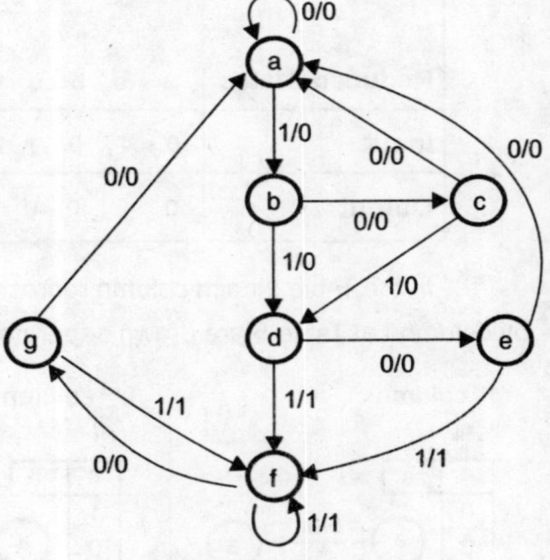

Fig 1: State diagram.

The state transitions represented by columns of Table 1 are drawn separately as shown below:

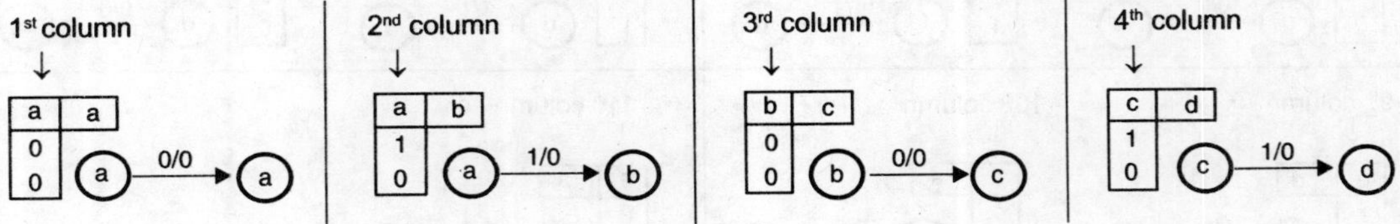

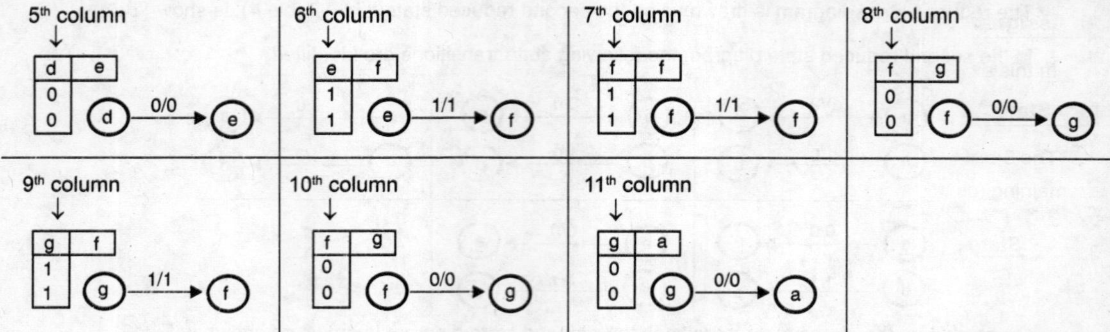

The above state transitions are compared with possible state transitions in Fig. 1 and verified that all the state transitions of Table 1 are valid transitions.

Verification of Given Input Sequence in Reduced State Diagram

The sequence to be verified after reducing the number of states is shown in Table 5. Here the state f is replaced by state d and state g is replaced by state e. For convenience the reduced state diagram of Fig. 2 is reproduced here.

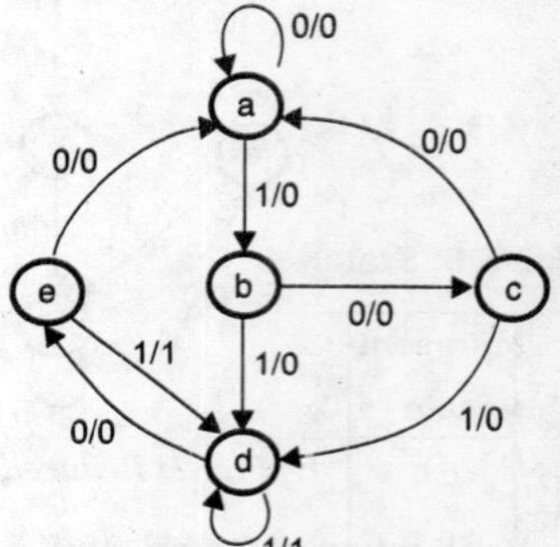

Fig. 2: Reduced state diagram.

Table 5: Input Sequence for Verification in Reduced State Diagram

States	a	a	b	c	d	e	f	f	g	f	g	a
							↓	↓	↓	↓	↓	
Reduced States	a	a	b	c	d	e	d	d	e	d	e	a
Input	0	1	0	1	0	1	1	0	1	0	0	
Output	0	0	0	0	0	1	1	0	1	0	0	

In the Table 5 each column represent a state transition due to change in input. The state transition represented by columns of Table 5 are drawn separately as shown below:

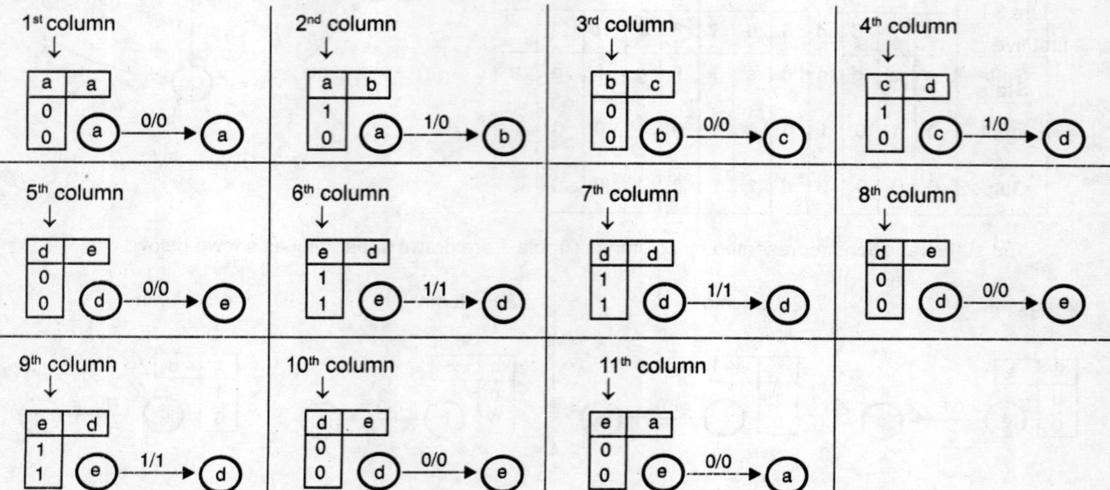

The above state transitions are compared with possible state transitions in Fig. 2 and verified that all the state transitions of Table 5 are valid transitions.

State Assignments

In this example the reduced state diagram has 5 states and so let us assign 3 bit code for states.

Binary State Assignments

The 3-bit binary codes are 000, 001, 010, 011, 100, 101, 110, 111. Let us assign first five codes to a, b, c, d and e. The remaining (or last three) are don't-cares.

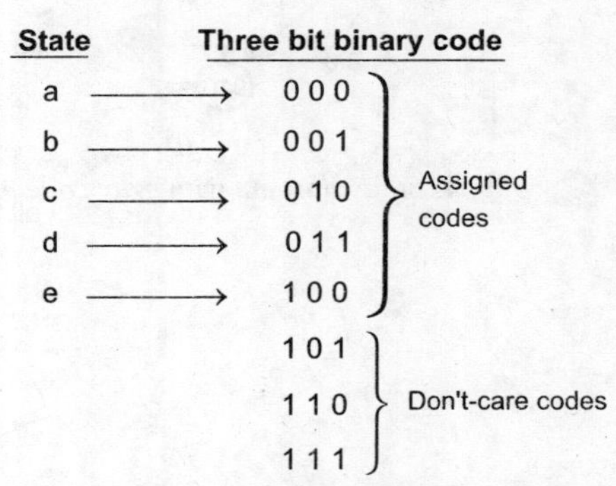

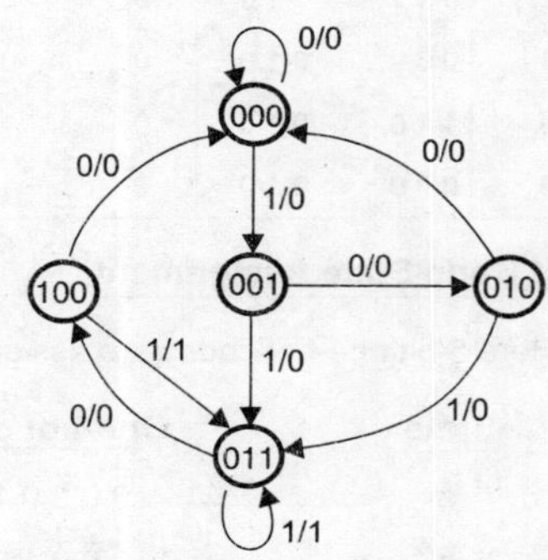

Fig. 3: State diagram with binary code assignment.

Table 6: State Table with Binary Code Assignment

Present State	Next State x = 0	Next State x = 1	Output x = 0	Output x = 1
000	000	001	0	0
001	010	011	0	0
010	000	011	0	0
011	100	011	0	1
100	000	011	0	1

Gray Code State Assignment

Here 3-bit gray codes are assigned to states. There are eight possible combinations of three bit gray code and in this first five codes are assigned to states a, b, c, d and e. The remaining are don't-cares.

State	Three bit gray code	
a →	000	Assigned codes
b →	001	
c →	011	
d →	010	
e →	110	
	111	Don't-care codes
	101	
	100	

Table 7: State Table with Gray Code Assignment

Present State	Next State x = 0	Next State x = 1	Output x = 0	Output x = 1
000	000	001	0	0
001	011	010	0	0
011	000	010	0	0
010	110	010	0	1
110	000	010	0	1

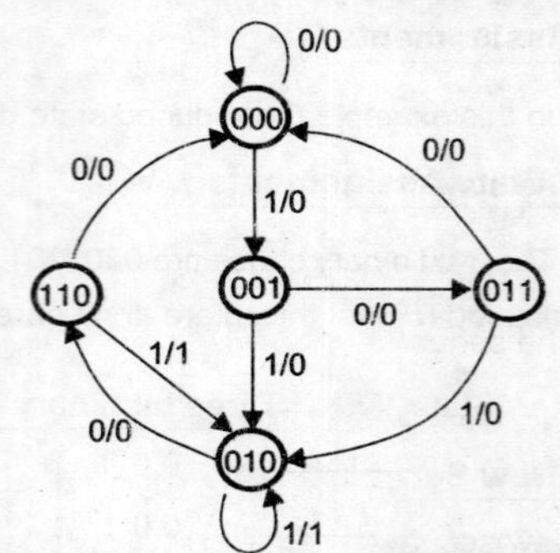

Fig. 4: State diagram with gray code assignement

One-hot Code State Assignment

Here 5-bit one-hot codes are assigned to states.

State		One-hot code
a	⟶	00001
b	⟶	00010
c	⟶	00100
d	⟶	01000
e	⟶	10000

Table 8: StateTable with One-hot Code Assignement

Present State	Next State x = 0	Next State x = 1	Output x = 0	Output x = 1
00001	00001	00010	0	0
00010	00100	01000	0	0
00100	00001	01000	0	0
01000	10000	01000	0	1
10000	00001	01000	0	1

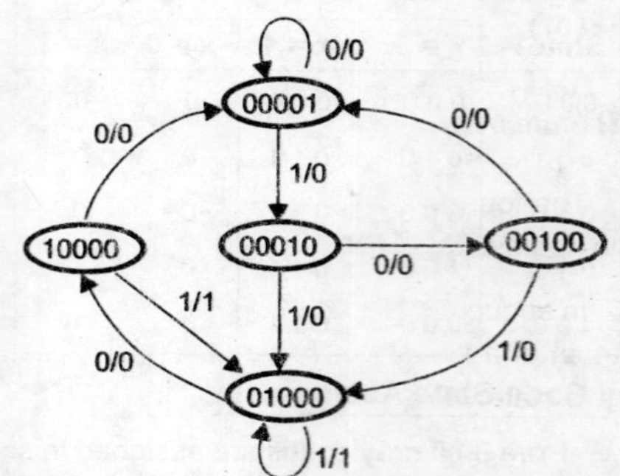

Fig. 5: State diagram with one-hot code assignment.

Example 3.8

The state diagram of a digital lock with password and two keys 0 and 1 is shown in Fig. 1.

a) Determine the password of the digital lock.

b) Modify the design by adding an additional security feature in digital lock.

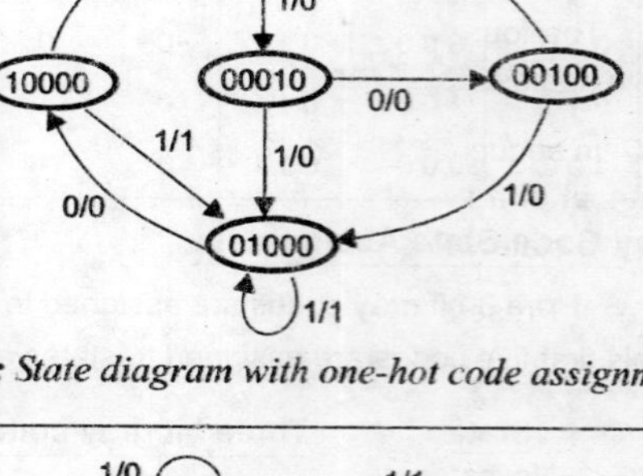

Fig. 1.

Solution

a) To determine the password

The given state diagram has 4 states, single input and output. The two keys in the lock is provided to input two possible values 0 and 1 of single input.

In the state diagram there are 8 transitions and the output is 1 only in one transition, when state changes from 11 to 00. Therefore, we have to start from 00 and come back to 00 with output 1.

This is possible only if we go through following state transitions,

$$00 \xrightarrow[0/0]{} 01 \xrightarrow[1/0]{} 10 \xrightarrow[1/0]{} 11 \xrightarrow[1/1]{} 00$$

The state transitions with input alone are shown below:

$$00 \xrightarrow[\text{Input } 0]{} 01 \xrightarrow[\text{Input } 1]{} 10 \xrightarrow[\text{Input } 1]{} 11 \xrightarrow[\text{Input } 1]{} 00$$

The sequence of inputs in the state transitions is the password.

$\therefore$ Password = 0111

b) Add New Security Feature

New security feature can be added by introducing an additional output. Additional or second output is defined such that output 1 in all wrong path and output 0 in all right path.

The state transitions with two outputs in the right path and wrong path are shown separately in Figs. 2 and 3 respectively. The state diagram with two output is shown in Fig. 4.

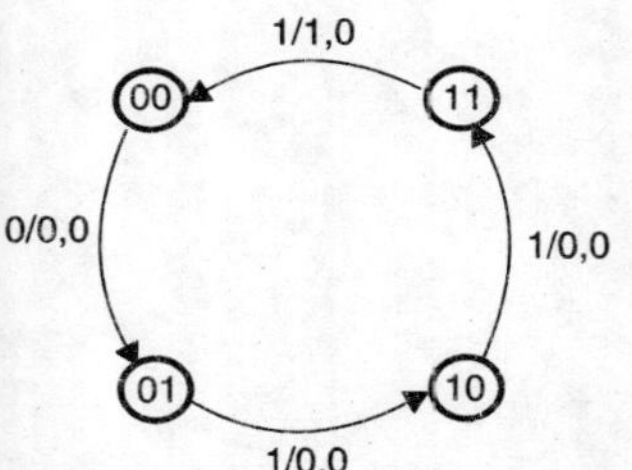

Fig. 2: State transitions in right path.

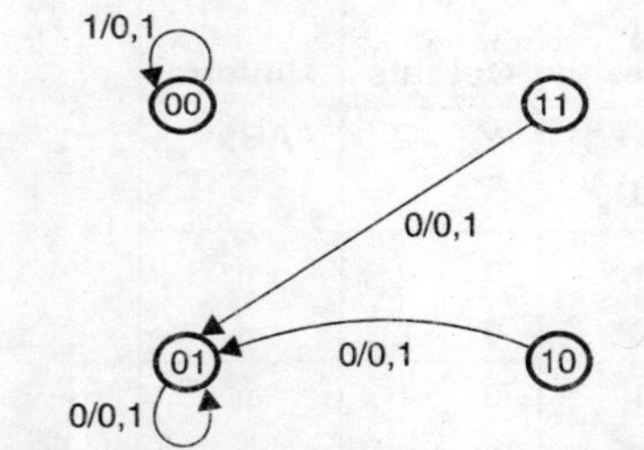

Fig. 3: State transitions in wrong path.

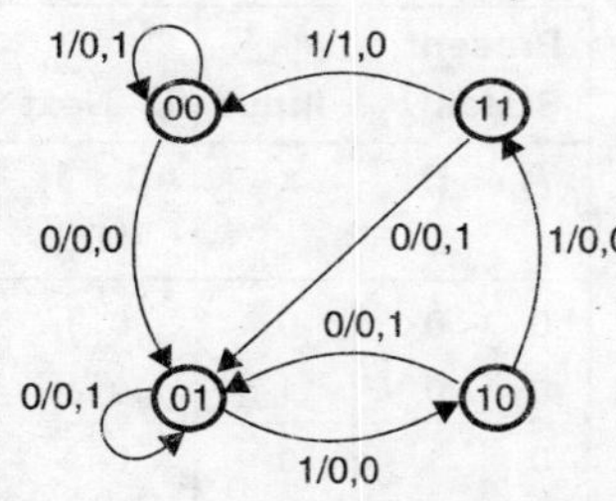

Fig. 4: State diagram with additional output.

The four ($2^2 = 4$) states can be obtained from two state variables. Let A and B be the two state variables.

In sequential circuit, each state variable is stored in the output of a flip-flop. Hence, choose two D flip-flops to store two state variables A and B as shown in Fig. 6.

$\therefore$ Present state variables : A(t), B(t)

⇓

A, B (For simplicity)

Next state variables : A(t + 1), B(t + 1)

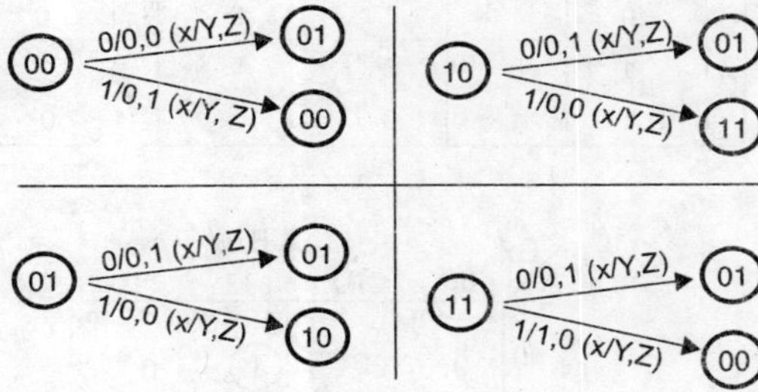

Fig. 5: Transitions from one state to another state.

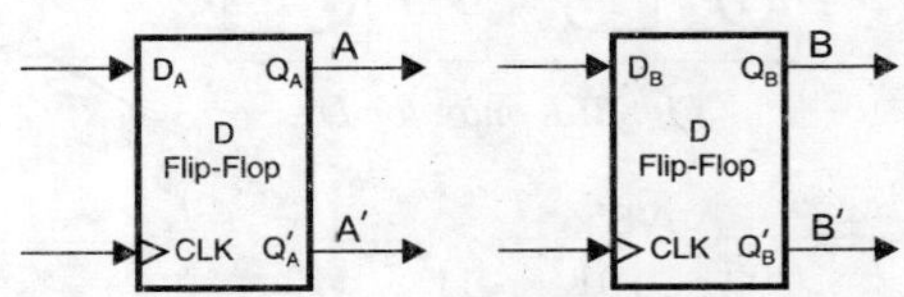

Fig. a: Flip-flop A. *Fig. b: Flip-flop B.*

Fig. 6: D flip-flop to store state variables.

The characteristic equations of two D flip-flops are,

$$Q_A(t + 1) = D_A$$

$$Q_B(t + 1) = D_B$$

Characteristic equation of D flip-flop.
$Q(t + 1) = D$

The two state variables can be equated to flip-flop outputs,

$\therefore\ A = Q_A \Rightarrow A(t + 1) = Q_A(t + 1)$

$B = Q_B \Rightarrow B(t + 1) = Q_B(t + 1)$

Now the characteristic equations of D flip-flop in terms of state variables are,

$A(t + 1) = D_A$

$B(t + 1) = D_B$

From the above equations we can say that the D flip-flop input is same as next state output.

Let, x = Input

Y, Z = Outputs

Using the state transitions shown in Fig. 5 and the characteristic equations, the state table is constructed as shown in Table 1.

Table 1: State Table

Present States		Input	Next States		Outputs		Minterm
A	B	x	A(t + 1) = D_A	B(t + 1) = D_B	Y	Z	ABx
0	0	0	0	1	0	0	m_0
0	0	1	0	0	0	1	m_1
0	1	0	0	1	0	1	m_2
0	1	1	1	0	0	0	m_3
1	0	0	0	1	0	1	m_4
1	0	1	1	1	0	0	m_5
1	1	0	0	1	0	1	m_6
1	1	1	0	0	1	0	m_7

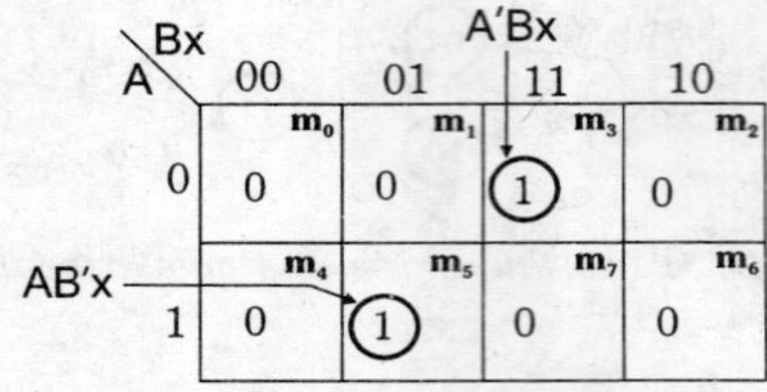

Fig. 7: K-map for D_A.

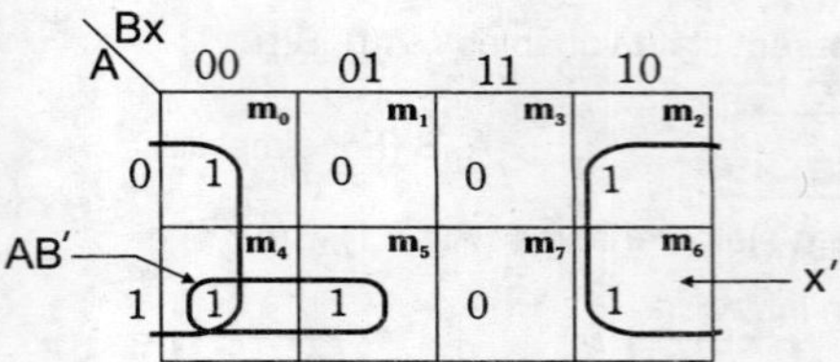

Fig. 8: K-map for D_B.

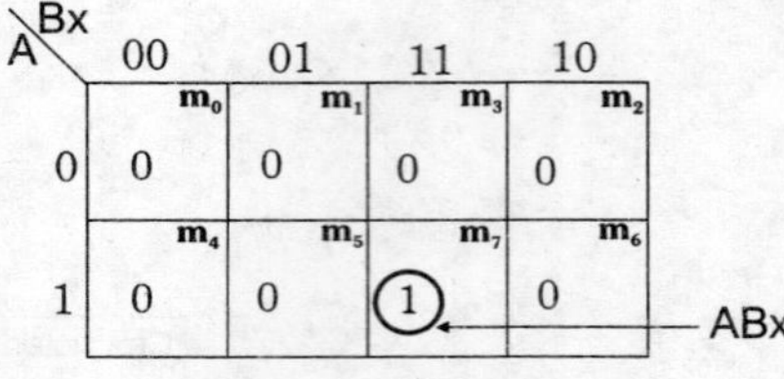

Fig. 9: K-map for Y.

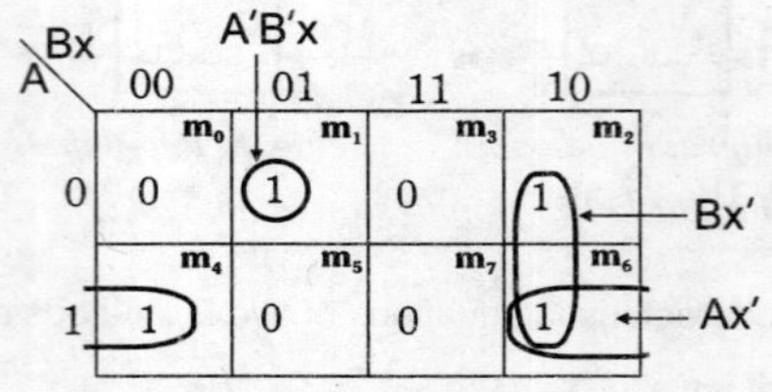

Fig. 10: K-map for Z.

Using the state table, the K-maps for D_A, D_B, Y and Z are constructed as shown in Figs. 7 to 10. From the K-map we get the following Boolean equations for D_A, D_B, Y and Z.

$$D_A = A'Bx + AB'x$$

$$D_B = AB' + x'$$

$$Y = ABx$$

$$Z = A'B'x + Ax' + Bx'$$

Using the Boolean equations for D_A, D_B, Y and Z the sequential logic circuit of digital lock with two outputs is drawn as shown in Fig. 11. The output depends on input and states and so this model is Mealy model.

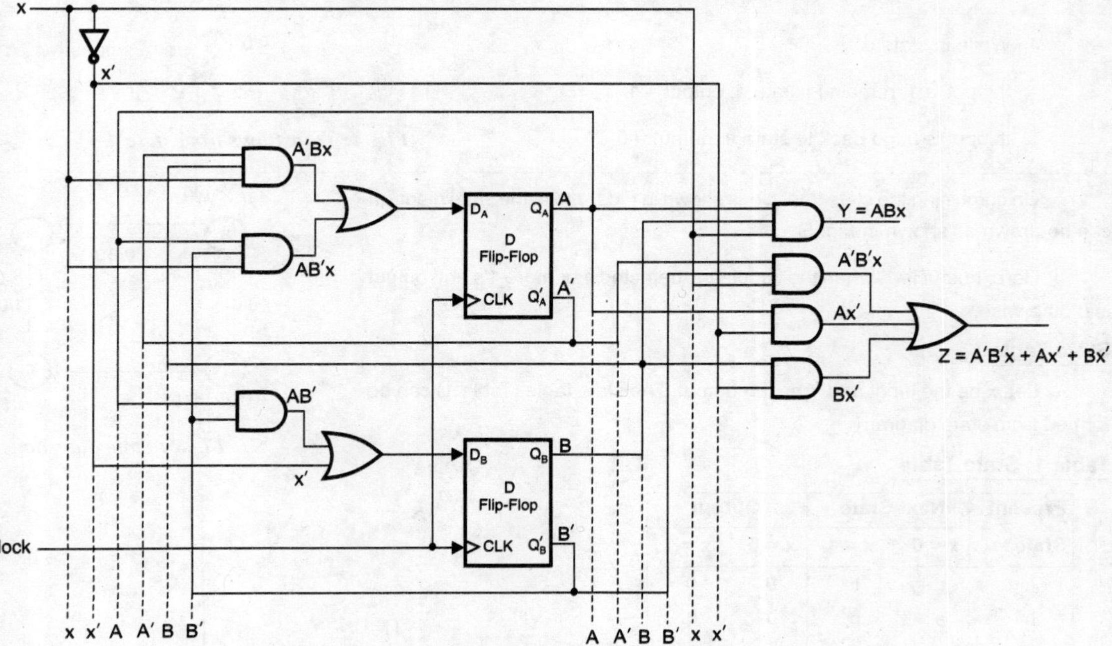

Fig. 11: Sequential circuit to implement digital lock with two outputs.

Example 3.9

Design a sequential circuit to detect a sequence of four or more consecutive 1's in a string of bits coming through an input line.

Case 1: Design using D flip-flop

Case 2: Design using JK flip-flop

Solution

The sequential circuit is designed to detect four or more consecutive 1's. Hence we can choose four states a, b, c and d. The possible state transitions from one state to other states are shown in Figs. 1 to 4.

1. While in state a

   if input = 1, go to next state b, output = 0

   if input = 0, remain in state a, output = 0

0/0
a
b
1/0

Fig. 1: Transitions from state a.

2. While in state b

   if input = 1, go to next state c output = 0

   if input = 0, go back to state a, output = 0

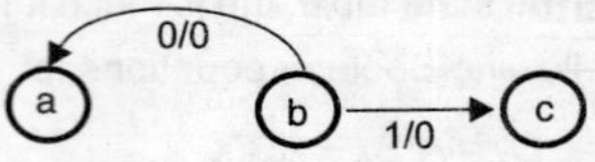

Fig. 2: Transitions from state b.

3. While in state c

   if input = 1, go to next state d, output = 0

   if input = 0, go back to state a, output = 0

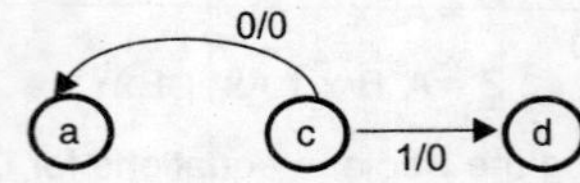

Fig. 3: Transitions from state c.

4. While in state d

   if input is 1, remain in state d, output = 1

   if input is 0, go back to state a, output = 0

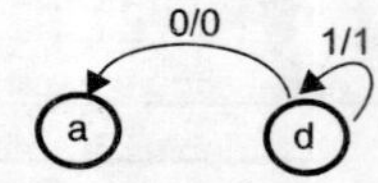

Fig. 4: Transitions from state d.

On combining the state transitions shown in Figs. 1 to 4, the state diagram can be drawn as shown in Fig. 5.

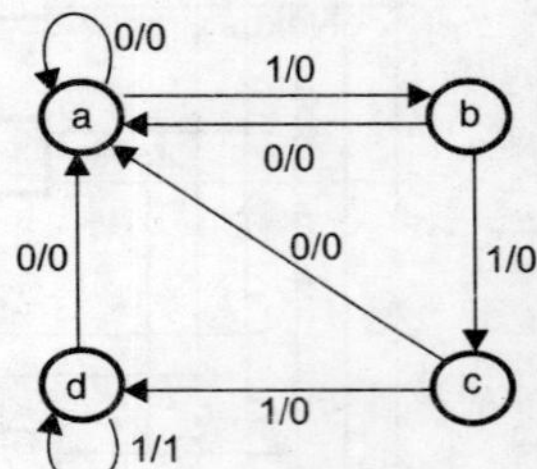

Fig. 5: State diagram.

Here, output is 1, when the circuit has detected 4 or more 1's and output is 0 otherwise.

State Table

Let, x be the input and Y be the output. The state table (Table 1) can be formed from state diagram.

Table 1: State Table

Present State	Next State x = 0	Next State x = 1	Output x = 0	Output x = 1
a	a	b	0	0
b	a	c	0	0
c	a	d	0	0
d	a	d	0	1

There is no identical or equivalent pair of states in the state table with same output and same next state, and so there is no need for state reduction. The four ($2^2 = 4$) states can be formed with two state variables. Let A and B be the two state variables.

Since there are four states, let us choose 2-bit code for states. The state diagram is redrawn as shown in Fig. 6 and state table is modified as shown in Table 2 with following state assignment.

a ⟶ 0 0
b ⟶ 0 1
c ⟶ 1 0
d ⟶ 1 1

} Assigned codes

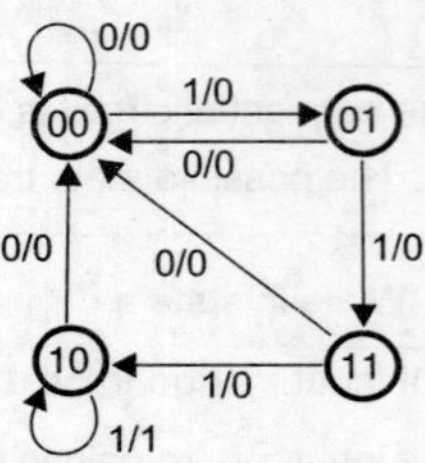

Fig. 6: State diagram.

Table 2: State Table with Binary Code Assignment for States

Present States		Next States				Output	
		x = 0		x = 1		x = 0	x = 1
A	B	A(t+1)	B(t+1)	A(t+1)	B(t+1)	Y	Y
0	0	0	0	0	1	0	0
0	1	0	0	1	0	0	0
1	0	0	0	1	1	0	0
1	1	0	0	1	1	0	1

Case i: Design using D Flip-Flop

In the sequential circuit, each state variable is stored in the output of a flip-flop. Hence, choose two D flip-flops to store the two state variables A and B as shown in Fig. 7.

$\therefore$ Present state variables : A(t), B(t) $\Rightarrow$ A, B (For simplicity)

Next state variables : A(t + 1), B(t + 1)

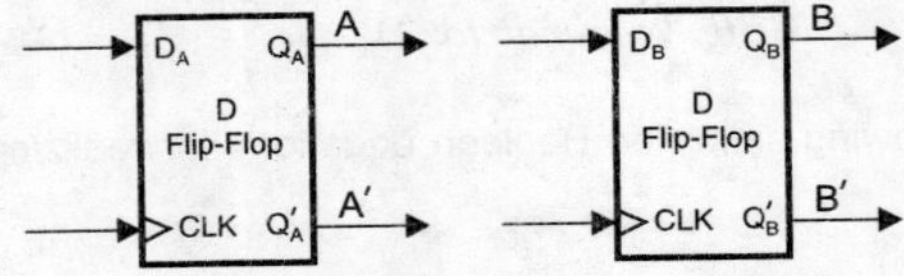

Fig. a: Flip-flop A. *Fig. b: Flip-flop B.*

Fig. 7: D flip-flop to store state variables.

The characteristic equation of two D flip-flops are,

> Characteristic equation of D flip-flop.
> Q(t + 1) = D

$Q_A(t + 1) = D_A$

$Q_B(t + 1) = D_B$

The two state variables can be equated to flip-flop outputs.

$\therefore A = Q_A \Rightarrow A(t + 1) = Q_A(t + 1)$

$B = Q_A \Rightarrow B(t + 1) = Q_B(t + 1)$

Now the characteristic equations in terms of state variables are,

$A(t + 1) = D_A$

$B(t + 1) = D_B$

Using the above equations the flip-flop inputs are determined from the next state and listed in Table 3. The D flip-flop the input is same as next state output.

Table 3: Expanded State Table

Present States		Input	Next States		Flip-Flop Inputs		Output	Minterm
A	B	x	A(t+1)	B(t+1)	D_A	D_B	Y	ABx
0	0	0	0	0	0	0	0	m_0
0	0	1	0	1	0	1	0	m_1
0	1	0	0	0	0	0	0	m_2
0	1	1	1	0	1	0	0	m_3
1	0	0	0	0	0	0	0	m_4
1	0	1	1	1	1	1	0	m_5
1	1	0	0	0	0	0	0	m_6
1	1	1	1	1	1	1	1	m_7

Here, D_A, D_B, Y are functions of A, B, x. Using Table 3, the Boolean equations for D_A, D_B and Y are determined as sum of minterms as shown below:

$$\therefore D_A = \Sigma m(3, 5, 7)$$

$$D_B = \Sigma m(1, 5, 7)$$

$$Y = \Sigma m(7)$$

Let us draw K-map for D_A, D_B and Y as shown in Figs. 8 to 10 to develop simplified Boolean equations for D_A, D_B and Y.

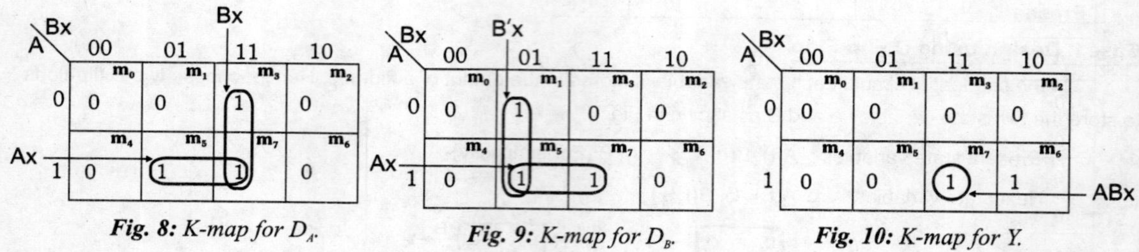

Fig. 8: K-map for D_A. *Fig. 9: K-map for D_B.* *Fig. 10: K-map for Y.*

From the K-map we get the following simplified Boolean equations for realizing sequence detector using D flip-flops.

$$D_A = Ax + B\,x$$

$$D_B = Ax + B'x$$

$$Y = ABx$$

The logic diagram of sequence detector using the above equations is drawn as shown in Fig. 11. Here, the output depends on input and states and so this model is Mealy model.

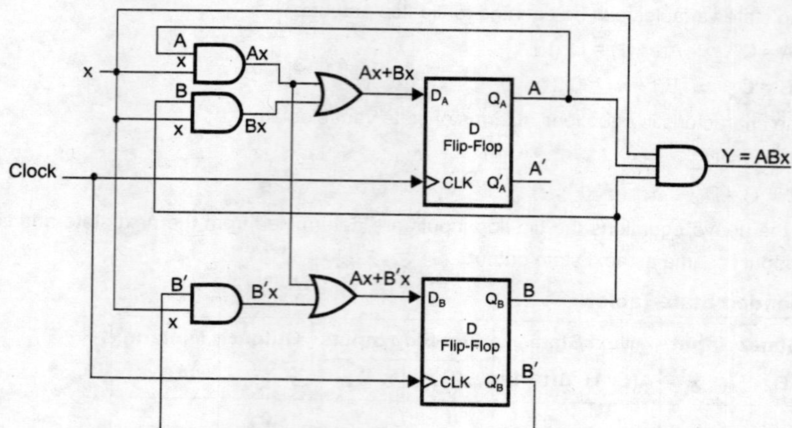

Fig. 11: Logic diagram of sequence detector using D flip-flop.

Case ii: Design using JK Flip-Flop

The state table and state assignment in the design using D flip-flop can be used for design using JK flip-flops. But the expanded state table require two columns for flip-flop inputs. Let us choose two JK flip-flops to store two state variables A and B as shown in Fig. 12. The state table is shown in Table 4.

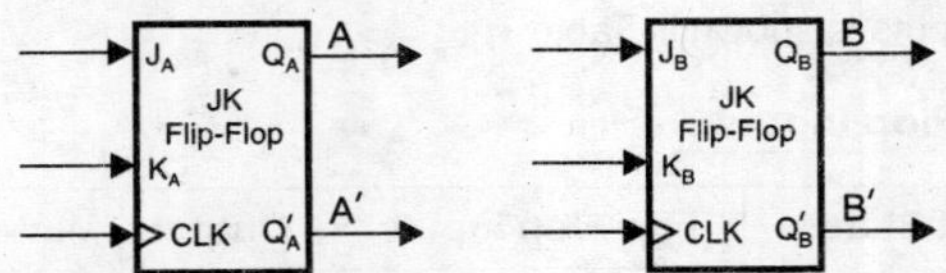

***Fig. a:** Flip-flop A.* ***Fig. a:** Flip-flop B.*

***Fig. 12:** JK flip-flops to store state variables.*

Table 4: State Table

Present States		Input	Next States	
A	**B**	**x**	**A(t+1)**	**B(t + 1)**
0	0	0	0	0
0	0	1	0	1
0	1	0	0	0
0	1	1	1	0
1	0	0	0	0
1	0	1	1	1
1	1	0	0	0
1	1	1	1	1

Excitation Table of JK Flip-Flop

Q(t)	Q(t + 1)	J	K
0	0	0	×
0	1	1	×
1	0	×	1
1	1	×	0

The inputs J_A, J_B, K_A, K_B of flip-flops can be determined using excitation table as shown below:

State A

For state A the Q(t), Q(t + 1), J and K of excitation table are considered as A, A(t + 1), J_A and K_A.

$$\therefore Q(t) \rightarrow A \quad ; \quad Q(t+1) \rightarrow A(t+1)$$

$$J \rightarrow J_A \quad ; \quad K \rightarrow K_A$$

The flip-flop inputs J_A and K_A that drives the present state A to next states A(t + 1) are obtained from excitation table and listed in Table 5.

Table 5:

A	A(t+1)	J_A	K_A
0	0	0	×
0	0	0	×
0	0	0	×
0	1	1	×
1	0	×	1
1	1	×	0
1	0	×	1
1	1	×	0

State B

For state B the Q(t), Q(t + 1), J and K of excitation table are considered as B, B(t + 1), J_B and K_B.

$$\therefore Q(t) \rightarrow B \quad ; \quad Q(t+1) \rightarrow B(t+1)$$

$$J \rightarrow J_B \quad ; \quad K \rightarrow K_B$$

The flip-flop inputs J_B and K_B that drives the present state B to next states B(t + 1) are obtained from excitation table and listed in Table 6.

Table 6:

B	B(t+1)	J_B	K_B
0	0	0	×
0	1	1	×
1	0	×	1
1	0	×	1
0	0	0	×
0	1	1	×
1	0	×	1
1	1	×	0

The state table with flip-flop inputs is shown in Table 7.

Table 7: State Table with JK Flip-Flop Inputs

Present States		Input	Next States		Flip-Flop Inputs				Output	Minterm
A	B	x	A(t+1)	B(t+1)	J_A	K_A	J_B	K_B	Y	ABx
0	0	0	0	0	0	×	0	×	0	m_0
0	0	1	0	1	0	×	1	×	0	m_1
0	1	0	0	0	0	×	×	1	0	m_2
0	1	1	1	0	1	×	×	1	0	m_3
1	0	0	0	0	×	1	0	×	0	m_4
1	0	1	1	1	×	0	1	×	0	m_5
1	1	0	0	0	×	1	×	1	0	m_6
1	1	1	1	1	×	0	×	0	1	m_7

***Note:** Output is same as in D flip-flop design.*

The inputs J_A, K_A, J_B, K_B are functions of present state and input x. Therefore, minterms are same as in design using D flip-flops.

J_A, K_A, J_B, K_B are functions of A, B, x and using Table 7, J_A, K_A, J_B, K_B are expressed as sum of minterms and don't-cares as shown below:

$$J_A = \Sigma m(3) + \Sigma d(4, 5, 6, 7)$$

$$K_A = \Sigma m(4, 6) + \Sigma d(0, 1, 2, 3)$$

$$J_B = \Sigma m(1, 5) + \Sigma d(2, 3, 6, 7)$$

$$K_B = \Sigma m(2, 3, 6) + \Sigma d(0, 1, 4, 5)$$

Let us draw K-map for J_A, K_A, J_B, K_B as shown in Figs. 13 to 16 to develop simplified equations for J_A, K_A, J_B, K_B. The equation for Y and K-map for Y are same as in design using D flip-flop.

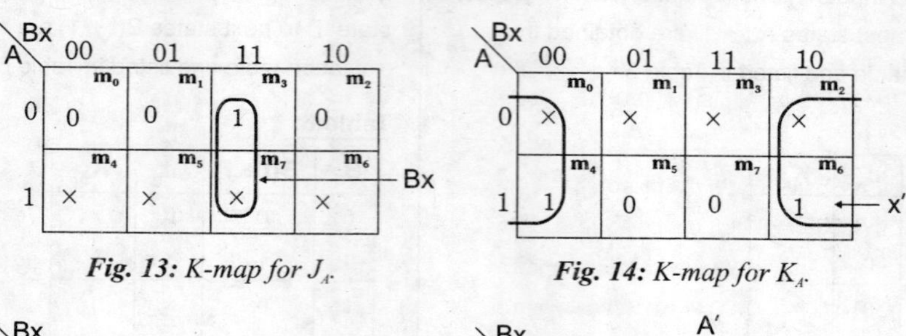

Fig. 13: K-map for J_A.

Fig. 14: K-map for K_A.

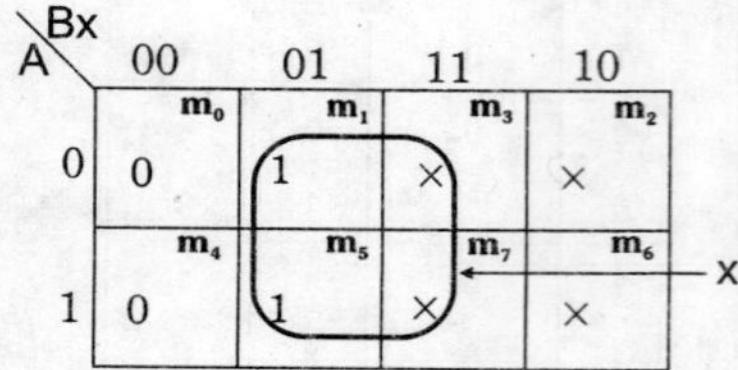

Fig. 15: K-map for J_B.

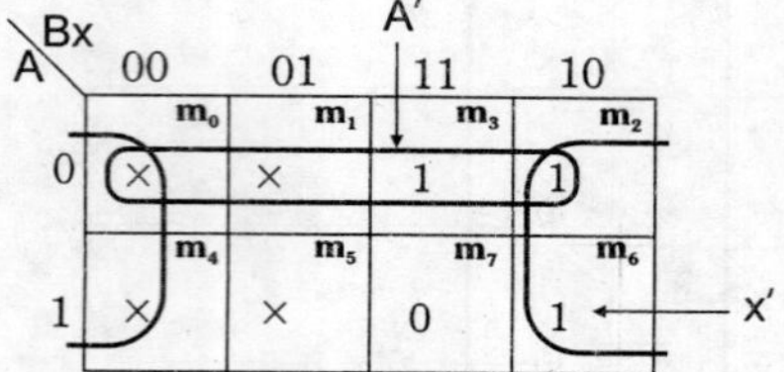

Fig. 16: K-map for K_B.

From the K-map we get the following simplified Boolean equations for realizing sequence detector using JK flip-flops.

$J_A = Bx$	$J_B = x$	$Y = ABx$
$K_A = x'$	$K_B = A' + x'$	

The logic diagram of sequence detector using above equations is drawn as shown in Fig. 17. Here, the output depends on input and states and so this model is Mealy model.

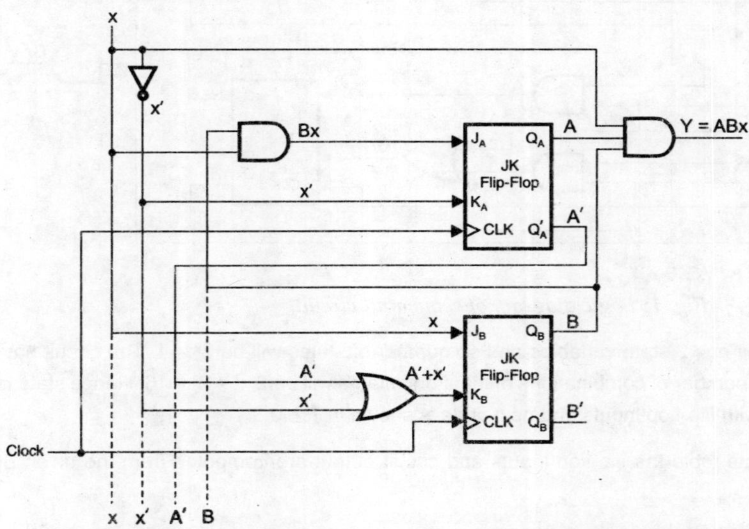

Fig. 17: Logic diagram of sequence detector using JK flip-flop.

Example 3.10

A sequential circuit has two JK flip-flops A and B, two inputs x and y and one output Z. The flip-flop input equations and circuit output equation are,

$J_A = Bx + B'y'$

$K_A = B'xy'$

$J_B = A'x$

$K_B = A + xy'$

$Z = Axy + Bx'y'$

Draw the logic diagram and state table of the circuit. Also derive the state equations for A and B.

Solution

The input and output equations are directly implemented using AND, OR and NOT gates as shown in Fig. 1. The sequential circuit has two flip-flops and so the circuit has two state variables A and B which are output of flip-flops. From the logic diagram of sequential circuit in Fig. 1, the output depends on inputs and states and so this model is Mealy model.

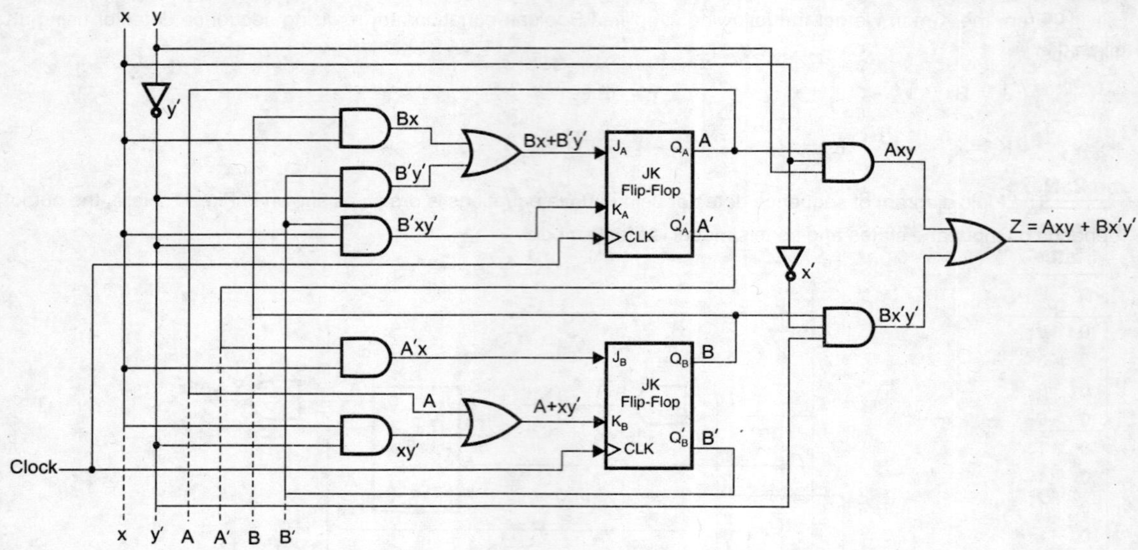

Fig. 1: Logic diagram of sequential circuit.

The circuit has 2 state variables and so number of states will be $2^2 = 4$. The circuit have 2 inputs and 2 state variables and so number of combinations of input and states will be $2^{2+2} = 2^4 = 16$. Hence state table will have 16 rows. The state table with flip-flop inputs and next state is shown in Table 3.

In the state table the flip-flop inputs and circuit output are computed from the given Boolean equations and listed in Table 1.

$J_A = Bx + B'y'$ | $K_A = B'xy'$

$J_B = A'x$ | $K_B = A + xy'$

$Z = Axy + Bx'y'$

Table 1: Flip-Flop Inputs and Circuit Output

Present States		Complement of Present States		Inputs		Complement of Inputs		Product Terms							Flip-Flop Inputs				Output
A	B	A'	B'	x	y	x'	y'	Bx	B'y'	B'xy'	A'x	xy'	Axy	Bx'y'	J_A	K_A	J_B	K_B	Z
0	0	1	1	0	0	1	1	0	1	0	0	0	0	0	1	0	0	0	0
0	0	1	1	0	1	1	0	0	0	0	0	0	0	0	0	0	0	0	0
0	0	1	1	1	0	0	1	0	1	1	1	1	0	0	1	1	1	1	0
0	0	1	1	1	1	0	0	0	0	0	1	0	0	0	0	0	1	0	0
0	1	1	0	0	0	1	1	0	0	0	0	0	0	1	0	0	0	0	1
0	1	1	0	0	1	1	0	0	0	0	0	0	0	0	0	0	0	0	0
0	1	1	0	1	0	0	1	1	0	0	1	1	0	0	1	0	1	1	0
0	1	1	0	1	1	0	0	1	0	0	1	0	0	0	1	0	1	0	0
1	0	0	1	0	0	1	1	0	1	0	0	0	0	0	1	0	0	1	0
1	0	0	1	0	1	1	0	0	0	0	0	0	0	0	0	0	0	1	0
1	0	0	1	1	0	0	1	0	1	1	0	1	0	0	1	1	0	1	0
1	0	0	1	1	1	0	0	0	0	0	0	0	1	0	0	0	0	1	1
1	1	0	0	0	0	1	1	0	0	0	0	0	0	1	0	0	0	1	1
1	1	0	0	0	1	1	0	0	0	0	0	0	0	0	0	0	0	1	0
1	1	0	0	1	0	0	1	1	0	0	0	1	0	0	1	0	0	1	0
1	1	0	0	1	1	0	0	1	0	0	0	0	1	0	1	0	0	1	1

The next states A(t + 1) and B(t + 1) are determined using the following JK flip-flop characteristics equations and listed in Table 2.

> Characteristic equation of JK flip-flop.
> $Q(t+1) = JQ' + K'Q$

$$A(t+1) = J_A A' + K_A' A$$

$$B(t+1) = J_B B' + K_B' B$$

Table 2: Next State Flip-Flop Outputs

Present States		Complement of Present States		Flip-Flop Inputs				Complement of K_A, K_B		Product Terms				Next States	
A	B	A′	B′	J_A	K_A	J_B	K_B	K_A'	K_B'	$J_A A'$	$K_A' A$	$J_B B'$	$K_B' B$	A(t + 1)	B(t + 1)
0	0	1	1	1	0	0	0	1	1	1	0	0	0	1	0
0	0	1	1	0	0	0	0	1	1	0	0	0	0	0	0
0	0	1	1	1	1	1	1	0	0	1	0	1	0	1	1
0	0	1	1	0	0	1	0	1	1	0	0	1	0	0	1
0	1	1	0	0	0	0	0	1	1	0	0	0	1	0	1
0	1	1	0	0	0	0	0	1	1	0	0	0	1	0	1
0	1	1	0	1	0	1	1	1	0	1	0	0	0	1	0
0	1	1	0	1	0	1	0	1	1	1	0	0	1	1	1
1	0	0	1	1	0	0	1	1	0	0	1	0	0	1	0
1	0	0	1	0	0	0	1	1	0	0	1	0	0	1	0
1	0	0	1	1	1	0	1	0	0	0	0	0	0	0	0
1	0	0	1	0	0	0	1	1	0	0	1	0	0	1	0
1	1	0	0	0	0	0	1	1	0	0	1	0	0	1	0
1	1	0	0	0	0	0	1	1	0	0	1	0	0	1	0
1	1	0	0	1	0	0	1	1	0	0	1	0	0	1	0
1	1	0	0	1	0	0	1	1	0	0	1	0	0	1	0

Table 3: State Table with JK Flip-Flop Inputs

Present States		Inputs		Next States		Flip-Flop Inputs				Output
A	B	x	y	A(t + 1)	B(t + 1)	J_A	K_A	J_B	K_B	Z
0	0	0	0	1	0	1	0	0	0	0
0	0	0	1	0	0	0	0	0	0	0
0	0	1	0	1	1	1	1	1	1	0
0	0	1	1	0	1	0	0	1	0	0
0	1	0	0	0	1	0	0	0	0	1
0	1	0	1	0	1	0	0	0	0	0
0	1	1	0	1	0	1	0	1	1	0
0	1	1	1	1	1	1	0	1	0	0
1	0	0	0	1	0	1	0	0	1	0
1	0	0	1	1	0	0	0	0	1	0
1	0	1	0	0	0	1	1	0	1	0
1	0	1	1	1	0	0	0	0	1	1
1	1	0	0	1	0	0	0	0	1	1
1	1	0	1	1	0	0	0	0	1	0
1	1	1	0	1	0	1	0	0	1	0
1	1	1	1	1	0	1	0	0	1	1

From the knowledge of inputs, outputs and present and next states obtained from state table, the state diagram is drawn as shown in Fig. 2

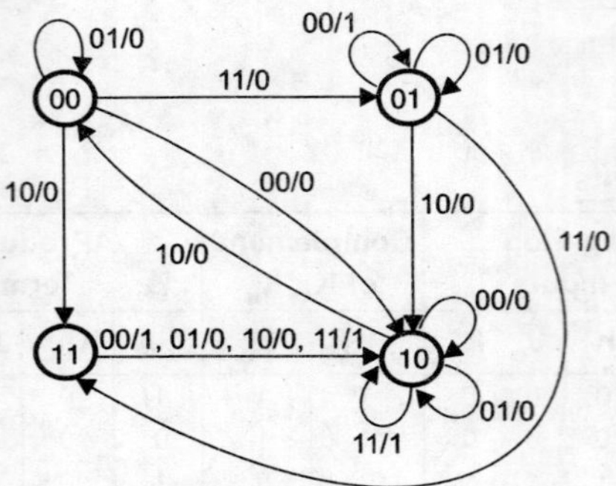

Fig. 2: State diagram.

The state equations can be obtained by substituting the Boolean equations of flip-flop inputs in the characteristic equations as shown below:

$$A(t+1) = J_A A' + K_A' A$$

$$= (Bx + B'y')A' + ((B'xy')')A = (Bx + B'y')A' + (B + x' + y)A$$

$$= A'(Bx + B'y') + A(B + x' + y)$$

Using DeMorgan's theorem

$$B(t+1) = J_B B' + K_B' B$$

$$= (A'x)B' + ((A + xy')')B$$

$$= A'xB' + (A'(xy')')B$$

$$= A'xB' + B(A'(x' + y))$$

$$= B'(A'x) + B(A'x' + A'y)$$

Note: *Alternatively the state equations can be obtained by constructing K-maps for A(t+1) and B(t+1). Here, minterms are formed using ABxy.*

State equations: $A(t+1) = A'(Bx + B'y) + A(B + x' + y)$

$$B(t+1) = B'(A'x) + B(A'x' + A'y)$$

Example 3.11

Draw the state diagram for the sequential circuit whose state table is shown in Table 1. Also, design a logic circuit using JK flip-flop.

Table 1: State Table of a Sequential Circuit

Present States		Input	Next States		Output
A	**B**	**x**	**A(t + 1)**	**B(t + 1)**	**Y**
0	0	0	0	0	0
0	0	1	0	1	0
0	1	0	1	0	0
0	1	1	0	1	0
1	0	0	1	0	0
1	0	1	1	1	0
1	1	0	1	1	1
1	1	1	0	0	1

Solution

In the given state stable it can be observed that the output does not depends on input and so output can be associated with states. Hence the given model is Moore model.

The various state transitions listed in Table 1 are diagrammatically shown in Fig. 1.

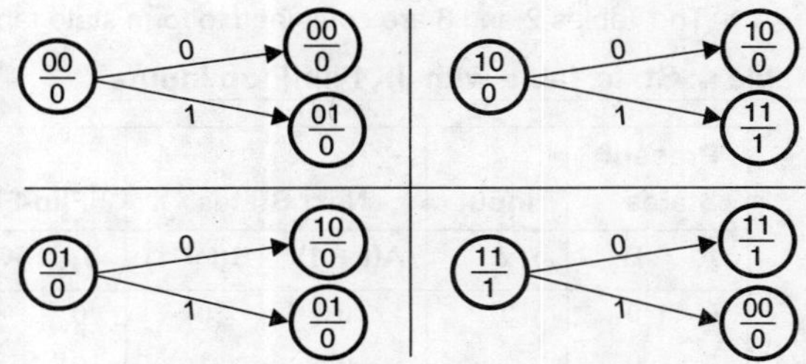

Fig. 1: Transitions from one state to another state.

On combining the state transitions of Fig. 1 into single diagram we get the Moore model state diagram shown in Fig. 2.

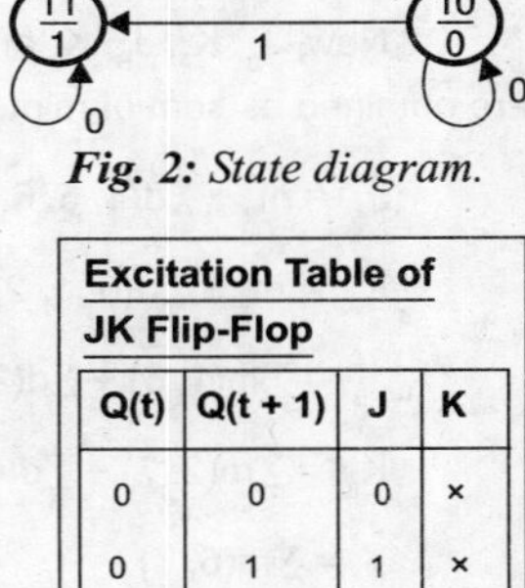

Fig. 2: State diagram.

The given state stable has two state variables A and B. In sequential circuits, each state variable is stored in a flip-flop. Hence, choose two JK flip-flops to store two state variables A and B as shown in Fig. 3.

$\therefore$ Present state variables : A(t), B(t) $\Rightarrow$ A, B (For simplicity)

Next state variables : A(t + 1), B(t + 1)

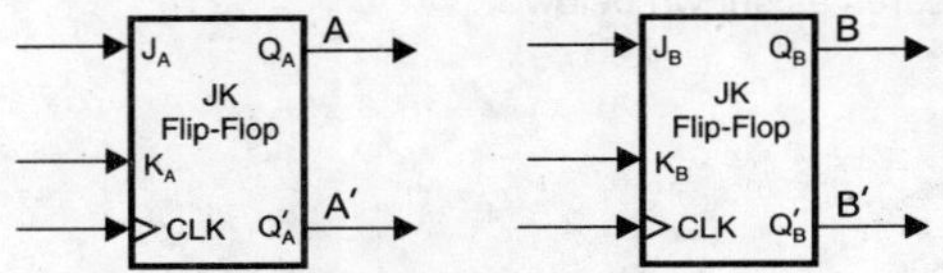

Fig. a: Flip-flop A. *Fig. b: Flip-flop B.*

Fig. 3: JK flip-flops to store state variables.

Excitation Table of JK Flip-Flop

Q(t)	Q(t + 1)	J	K
0	0	0	×
0	1	1	×
1	0	×	1
1	1	×	0

The flip-flops inputs J_A, K_A, J_B, K_B can be obtained from excitation table as shown below:

State A

For state A the Q(t), Q(t + 1), J and K of excitation table are considered as A, A(t + 1), J_A and K_A.

$\therefore$ $Q(t) \rightarrow A$; $Q(t+1) \rightarrow A(t+1)$

$J \rightarrow J_A$; $K \rightarrow K_A$

The flip-flop inputs J_A and K_A that drives the present state A to next state A(t + 1) are obtained from excitation table and listed in Table 2.

Table 2:

A	A(t + 1)	J_A	K_A
0	0	0	×
0	0	0	×
0	1	1	×
0	0	0	×
1	1	×	0
1	1	×	0
1	1	×	0
1	0	×	1

State B

For state B the Q(t), Q(t + 1), J and K of excitation table are considered as B, B(t + 1), J_B and K_B.

$\therefore$ $Q(t) \rightarrow B$; $Q(t+1) \rightarrow B(t+1)$

$J \rightarrow J_B$; $K \rightarrow K_B$

The flip-flop inputs J_B and K_B that drives the present state B to next state B(t + 1) are obtained from excitation table and listed in Table 3.

Table 3:

B	B(t + 1)	J_B	K_B
0	0	0	×
0	1	1	×
1	0	×	1
1	1	×	0
0	0	0	×
0	1	1	×
1	1	×	0
1	0	×	1

The Tables 2 and 3 are combined to form state table with flip-flop inputs as shown in Table 4.

Table 4: State Table with JK Flip-Flop Inputs

Present States		Input	Next States		Flip-Flop Inputs				Output	Minterm
A	B	x	A(t + 1)	B(t + 1)	J_A	K_A	J_B	K_B	Y	ABx
0	0	0	0	0	0	×	0	×	0	m_0
0	0	1	0	1	0	×	1	×	0	m_1
0	1	0	1	0	1	×	×	1	0	m_2
0	1	1	0	1	0	×	×	0	0	m_3
1	0	0	1	0	×	0	0	×	0	m_4
1	0	1	1	1	×	0	1	×	0	m_5
1	1	0	1	1	×	0	×	0	1	m_6
1	1	1	0	0	×	1	×	1	1	m_7

Now, J_A, K_A, J_B, K_B and Y are functions of A, B, x. Using Table 4, the Boolean equations of J_A, K_A, J_B, K_B and Y are obtained as sum of minterms and don't-cares as shown below:

$$J_A = m_2 + \Sigma d(4, 5, 6, 7)$$

$$K_A = m_7 + \Sigma d(0, 1, 2, 3)$$

$$J_B = \Sigma m(1, 5) + \Sigma d(2, 3, 6, 7)$$

$$K_B = \Sigma m(2, 7) + \Sigma d(0, 1, 4, 5)$$

$$Y = \Sigma m(6, 7)$$

Let us draw K-map for J_A, K_A, J_B, K_B and Y as shown in Figs. 4 to 8 and develop simplified Boolean equations to implement logic diagram.

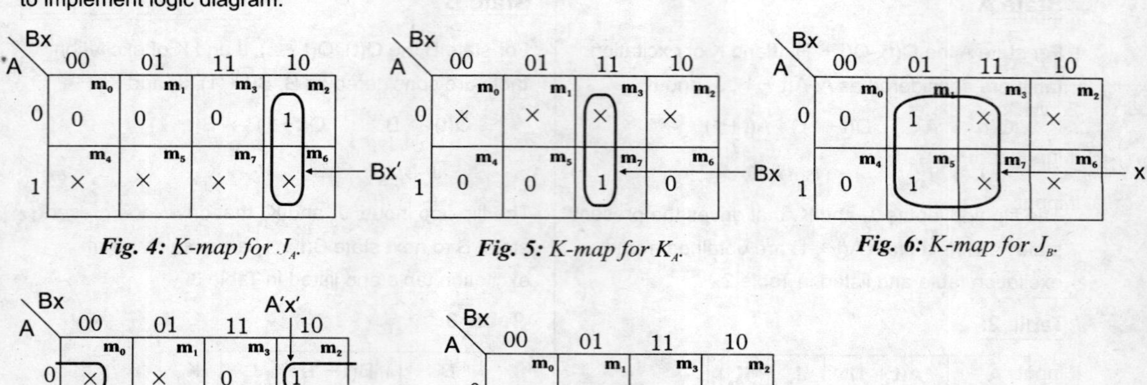

Fig. 4: K-map for J_A. *Fig. 5: K-map for K_A.* *Fig. 6: K-map for J_B.*

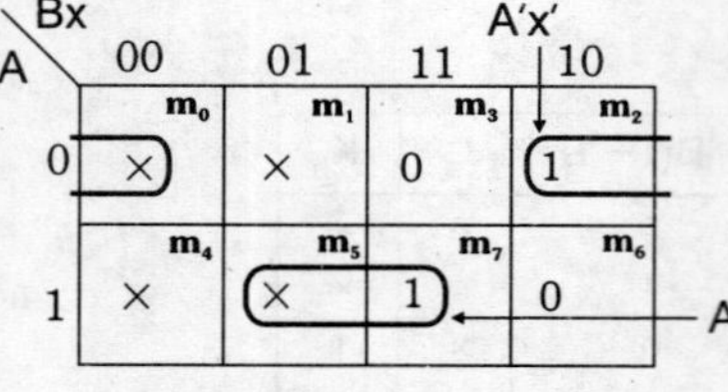

Fig. 7: K-map for K_B.

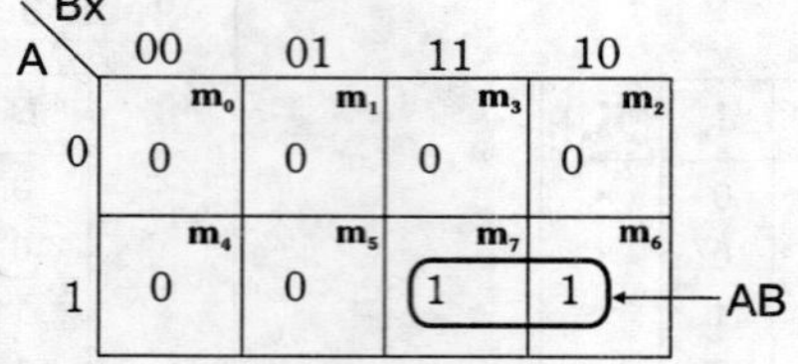

Fig. 8: K-map for Y.

From the K-maps we get the following simplified Boolean equations for realizing the given state diagrams using JK flip-flops,

$J_A = Bx'$

$K_A = Bx$

$J_B = x$

$K_B = A'x' + Ax$

$Y = AB$

The logic diagram using above equations is drawn as shown in Fig. 9.

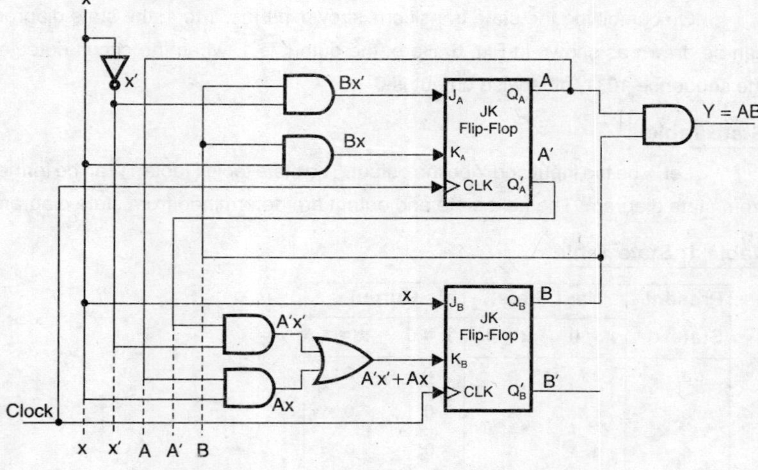

***Fig. 9:** Logic diagram of sequential circuit.*

Example 3.12

Design a synchronous sequence detector that produces an output 1, whenever the non overlapping sequence 1011 is detected.

Solution

In order to detect 1011, four states a, b, c and d are required as shown in Fig. 1. The possible state transitions from one state to other states are shown in Figs. 2 to 5.

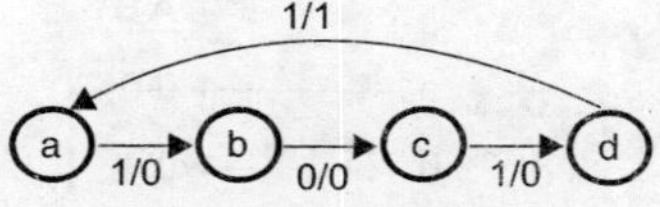

***Fig. 1:** States to detect 1011.*

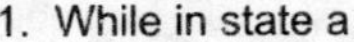

1. While in state a

   if input = 1, go to next state b, output = 0

   if input = 0, remain in state a, output = 0

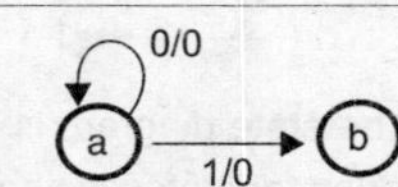

***Fig. 2:** Transitions from state a.*

2. While in state b

   if input = 1, go to state a, output = 0

   if input = 0, go to next state c, output = 0

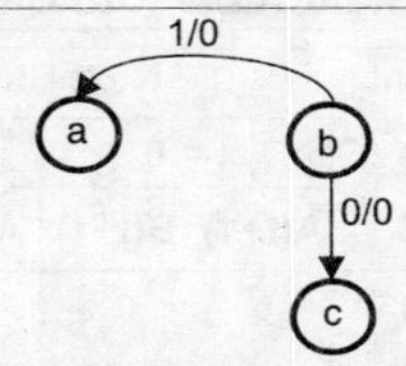

***Fig. 3:** Transitions from state b.*

3. While in state c

   if input = 1, go to next state d, output = 0

   if input = 0, go to state a, output = 0

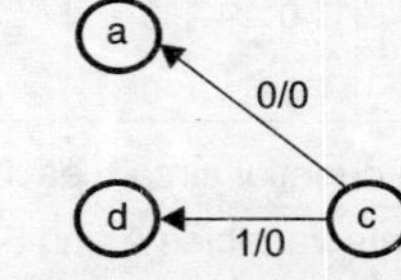

***Fig. 4:** Transitions from state c.*

4. While in state d

   if input = 1, go to state a, output = 1

   if input = 0, go to state a, output = 0

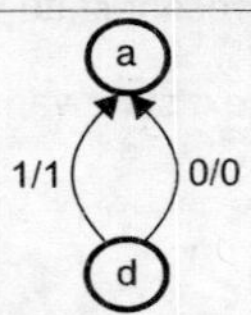

***Fig. 5:** Transitions from state d.*

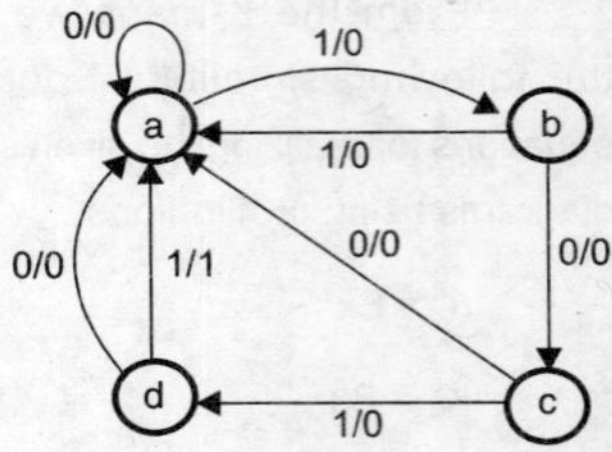

Fig. 6: State diagram.

On combining the state transitions shown in Figs. 1 to 4, the state diagram can be drawn as shown in Fig. 6. Here, the output is 1, when the circuit detected the sequence 1011, otherwise output is 0.

State Table

Let, x be the input and Y be the output. The state table (Table 1) can be formed from state diagram. The next state and output are determined from state diagram.

Table 1: State Table

Present	Next State		Output	
State	x = 0	x = 1	x = 0	x = 1
a	a	b	0	0
b	c	a	0	0
c	a	d	0	0
d	a	a	0	1

There is no identical or equivalent pair of states in the state table with same output and same next state and so, there is no need for state reduction. The four ($2^2 = 4$) states can be formed with two state variables. Let A and B be the two state variables. Since there are four states, let us choose 2-bit code for states as shown below:

	AB	
a ⟶	0 0	Assigned codes
b ⟶	0 1	
c ⟶	1 0	
d ⟶	1 1	

The state diagram is redrawn as shown in Fig. 7 with the above code assignments. Also the state table with binary codes for states is shown in Table 2.

Table 2: State Table with Binary Code Assignment for States

Present		Next States				Output	
States		x = 0		x = 1		x = 0	x = 1
A	B	A(t+1)	B(t+1)	A(t+1)	B(t+1)	Y	Y
0	0	0	0	0	1	0	0
0	1	1	0	0	0	0	0
1	0	0	0	1	1	0	0
1	1	0	0	0	0	0	1

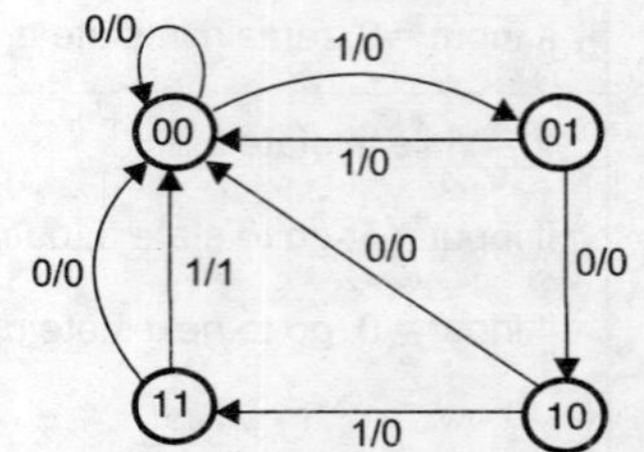

Fig. 7: State diagram with state assignment.

In sequential circuit, each state variable is stored in the output of a flip-flop. Hence, choose two D flip-flops to store two state variables A and B as shown in Fig. 8.

∴ Present state variables : A(t), B(t) ⇒ A, B (for simplicity)

Next state variables : A(t + 1), B(t + 1)

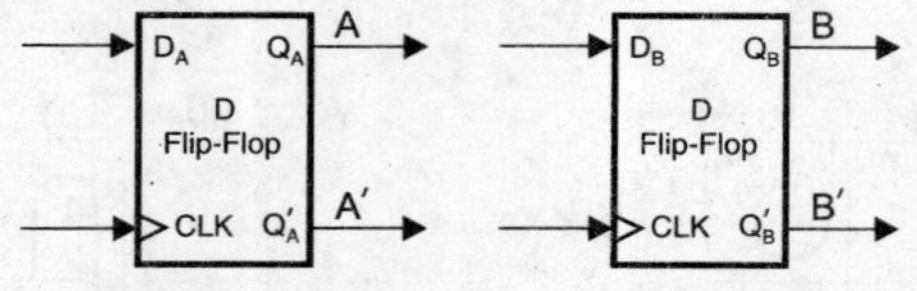

Fig. a: Flip-flop A. *Fig. b: Flip-flop B.*

***Fig. 8:** D flip-flop to store state variables.*

The characteristic equation of two D flip-flops are,

> Characteristic equation of D flip-flop.
> $Q(t + 1) = D$

$Q_A(t + 1) = D_A$

$Q_B(t + 1) = D_B$

The two state variables can be equated to flip-flop outputs.

$\therefore A = Q_A \Rightarrow A(t + 1) = Q_A(t + 1)$

$B = Q_B \Rightarrow B(t + 1) = Q_B(t + 1)$

Now the characteristic equations in terms of state variables are,

$A(t + 1) = D_A$

$B(t + 1) = D_B$

Using the above equations the flip-flop inputs are determined for the next state and listed in Table 3. In D flip-flop, the input is same as next state output.

Table 3: Expanded State Table

Present States		Input	Next States		Flip-Flop Inputs		Output	Minterm
A	**B**	**x**	**A(t+1)**	**B(t+1)**	D_A	D_B	**Y**	**ABx**
0	0	0	0	0	0	0	0	m_0
0	0	1	0	1	0	1	0	m_1
0	1	0	1	0	1	0	0	m_2
0	1	1	0	0	0	0	0	m_3
1	0	0	0	0	0	0	0	m_4
1	0	1	1	1	1	1	0	m_5
1	1	0	0	0	0	0	0	m_6
1	1	1	0	0	0	0	1	m_7

Using Table 3, the Boolean equations for D_A, D_B and Y are determined as sum of minterms as shown below:

$D_A = \Sigma m(2, 5)$

$D_B = \Sigma m(1, 5)$

$Y = m_7$

Let us draw K-map for D_A, D_B and Y to develop simplified Boolean equations for D_A, D_B and Y.

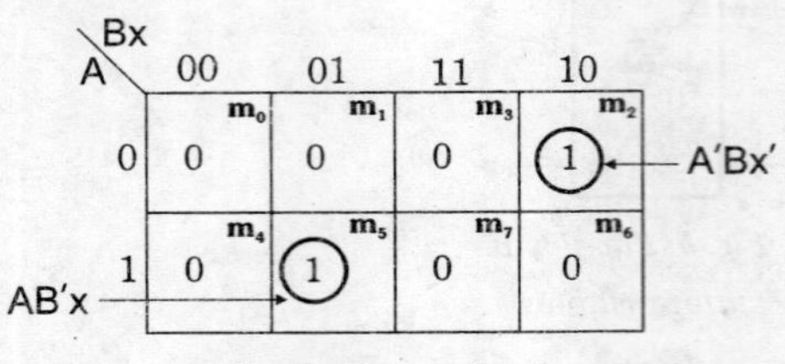

Fig. 9: K-map for D_A.

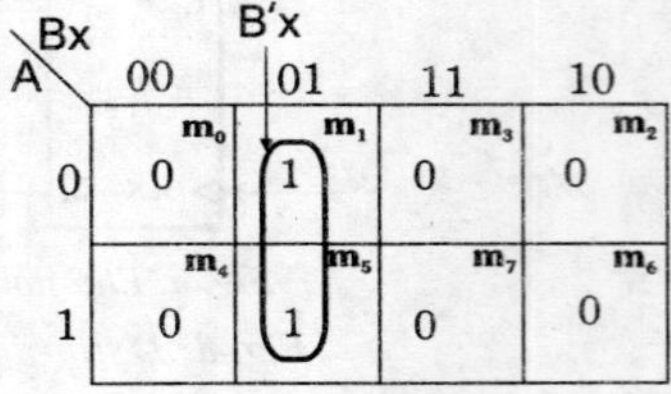

Fig. 10: K-map for D_B.

From the K-map we get the following simplified Boolean equations for realizing sequence detector using D flip-flops,

$D_A = A'Bx' + AB'x$

$D_B = B'x$

$Y = ABx$

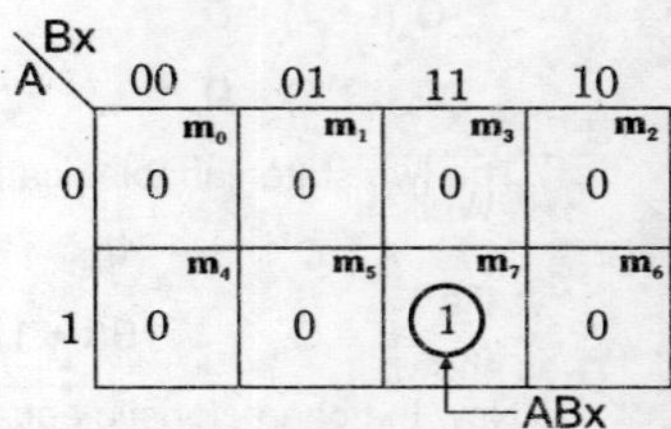

Fig. 11: K-map for Y.

The logic diagram of sequence detector using the above equations is drawn as shown in Fig. 12. Here, the output depends on input and states and so this model is Mealy model.

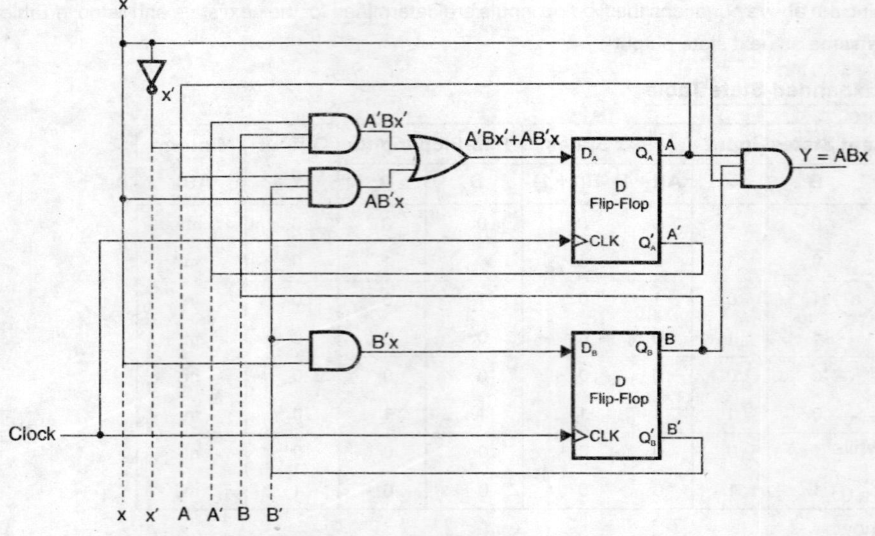

Fig. 12: Logic diagram of sequence detector.

Example 3.13

A clocked sequential circuit is provided with a single input x and a single output Z. Whenever the input produces a string of pulses 111 or 000 output a 1, otherwise output a 0.

Obtain the state diagram, the state table and design the logic circuit of sequence detector.

i) When overlapping not allowed

ii) When overlapping allowed

Solution

i) When Overlapping Not Allowed

In order to detect 000 three states are required as shown in Fig. 1.

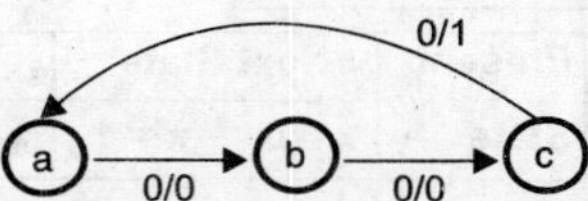

Fig. 1: States to deduct 000.

In order to detect 111 three states are required as shown in Fig. 2. The other possible transitions are shown in Figs. 3 to 6.

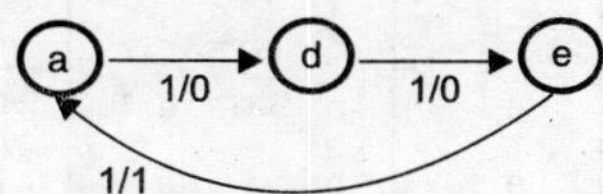

Fig. 2: States to deduct 111.

While in state b,

if a 1 is detected then the circuit can go through states b, d, e and a to detect 111 as shown in Fig. 3.

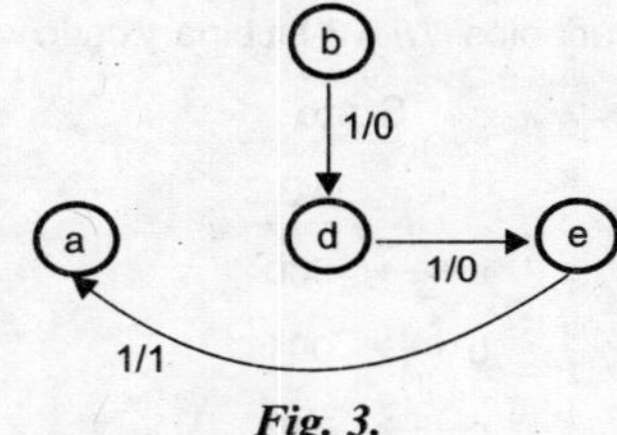

Fig. 3.

While in state c,

if a 1 is detected then the circuit can go through states c, d, e and a to detect 111 as shown in Fig. 4.

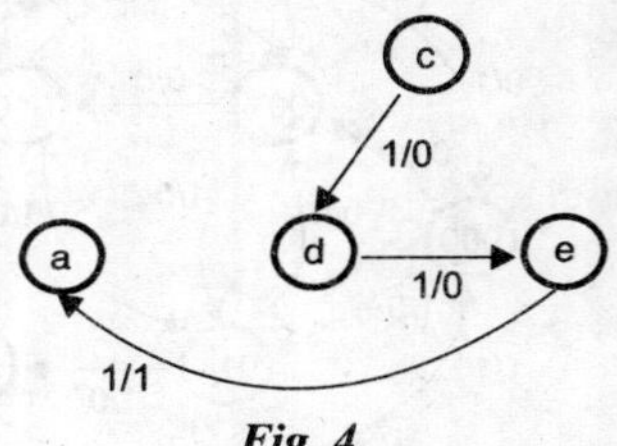

Fig. 4.

While in state d,

if a 0 is detected then the circuit can go through states d, b, c and a to detect 000 as shown in Fig. 5.

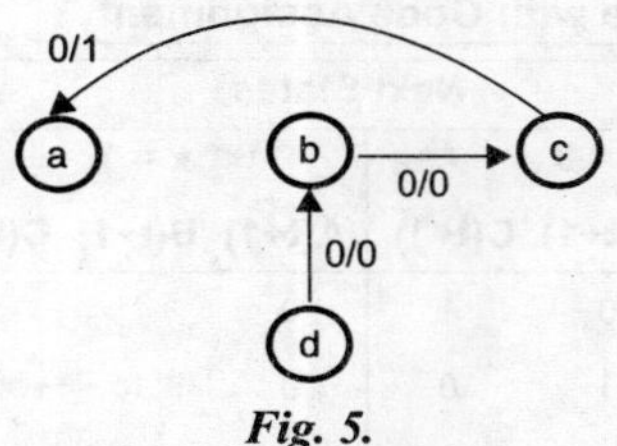

Fig. 5.

While in state e,

if a 0 is detected then the circuit can go through states e, b, c and a to detect 000 as shown in Fig. 6.

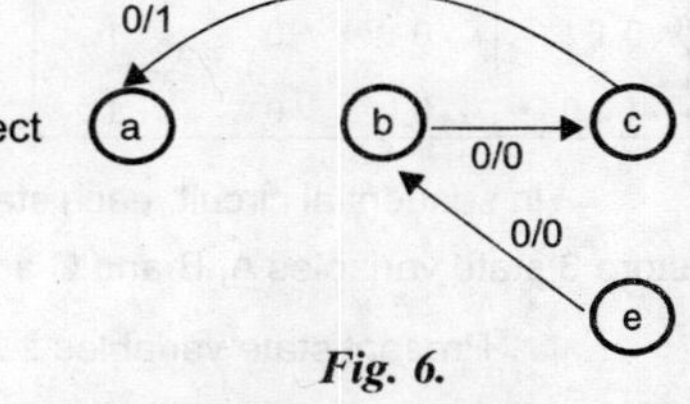

Fig. 6.

The state transitions shown in Fig. 7 are combined to draw the state diagram as shown in Fig. 8. The state table is constructed from state diagram as shown in Table 1.

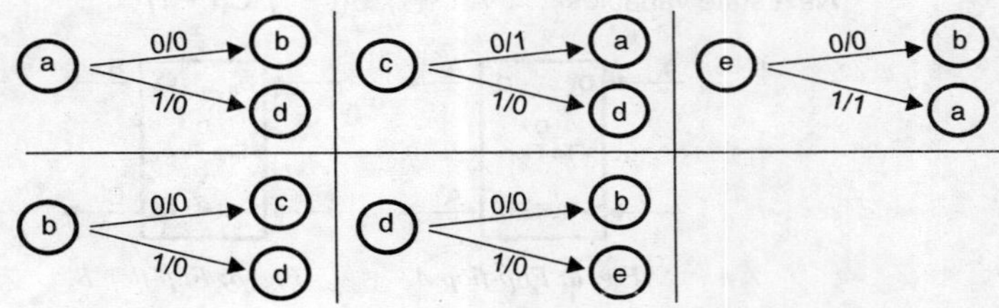

Fig. 7: Summary of state transitions from one state to another state.

Table 1: State Table

Present State	Next State		Output	
	x = 0	x = 1	x = 0	x = 1
a	b	d	0	0
b	c	d	0	0
c	a	d	1	0
d	b	e	0	0
e	b	a	0	1

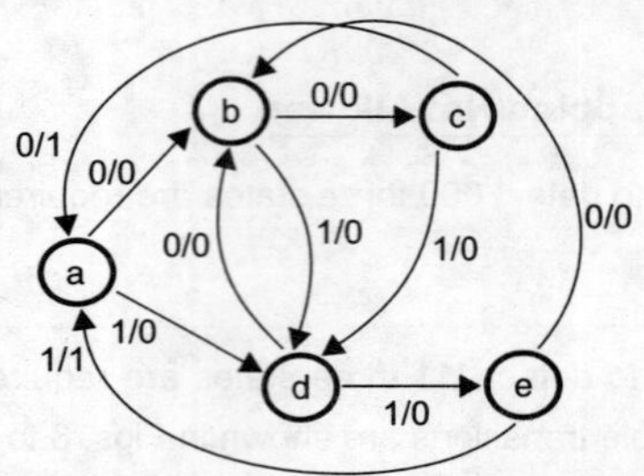

Fig. 8: State diagram when overlapping not allowed.

The sequential circuit has 5 states and they can be formed using three state variables. Let A, B and C be the state variables. With 3-bit binary code we get $2^3 = 8$ different codes and in this 5 codes are assigned to states as shown below:

State

ABC

a → 000

b → 001

c → 010

d → 011

e → 100

The state diagram with assigned codes is shown in Fig. 9. The state table with code assignment is shown in Table 2.

Table 2: State Table with Code Assignment

Present State	Next States						Outputs	
	x = 0			x = 1			x = 0	x = 1
A B C	A(t+1)	B(t+1)	C(t+1)	A(t+1)	B(t+1)	C(t+1)	Y	Y
0 0 0	0	0	1	0	1	1	0	0
0 0 1	0	1	0	0	1	1	0	0
0 1 0	0	0	0	0	1	1	1	0
0 1 1	0	0	1	1	0	0	0	0
1 0 0	0	0	1	0	0	0	0	1

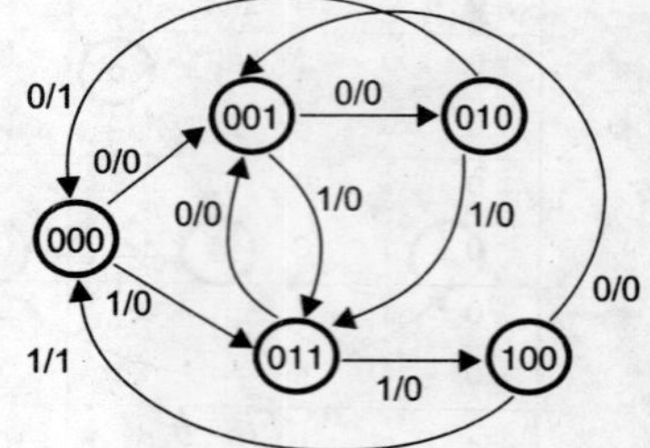

Fig. 9: State diagram with code assignment.

In sequential circuit, each state variable is stored in the output of a flip-flop. Hence, choose three D flip-flops to store 3 state variables A, B and C as shown in Fig. 10.

∴ Present state variables : A(t), B(t), C(t) ⇒ A, B, C (For simplicity)

Next state variables : A(t + 1), B(t + 1), C(t + 1)

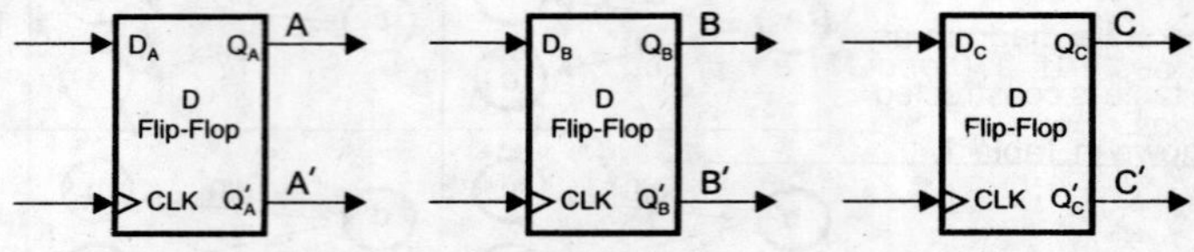

Fig. a: Flip-flop A. *Fig. b: Flip-flop B.* *Fig. c: Flip-flop C.*

Fig. 10: D flip-flop to store state variables.

The characteristic equation of two D flip-flops are,

$Q_A(t + 1) = D_A$

$Q_B(t + 1) = D_B$

$Q_C(t + 1) = D_C$

> Characteristic equation of D flip-flop.
> $Q(t + 1) = D$

The three state variables can be equated to flip-flop outputs.

$\therefore A = Q_A \quad \Rightarrow \quad A(t + 1) = Q_A(t + 1)$

$B = Q_B \quad \Rightarrow \quad B(t + 1) = Q_B(t + 1)$

$C = Q_C \quad \Rightarrow \quad C(t + 1) = Q_C(t + 1)$

Now the characteristic equations in terms of state variables are,

$A(t + 1) = D_A$

$B(t + 1) = D_B$

$C(t + 1) = D_C$

Using the above equations the flip-flop inputs are determined for the next state and listed in Table 3. In D flip-flop the input is same as next state output.

Table 3: Expanded State Table

Present States			Input	Next States			Flip-Flop Inputs			Output	Minterm
A	B	C	x	A(t+1)	B(t+1)	C(t+1)	D_A	D_B	D_C	Y	ABCx
0	0	0	0	0	0	1	0	0	1	0	m_0
0	0	0	1	0	1	1	0	1	1	0	m_1
0	0	1	0	0	1	0	0	1	0	0	m_2
0	0	1	1	0	1	1	0	1	1	0	m_3
0	1	0	0	0	0	0	0	0	0	1	m_4
0	1	0	1	0	1	1	0	1	1	0	m_5
0	1	1	0	0	0	1	0	0	1	0	m_6
0	1	1	1	1	0	0	1	0	0	0	m_7
1	0	0	0	0	0	1	0	0	1	0	m_8
1	0	0	1	0	0	0	0	0	0	1	m_9
1	0	1	0	×	×	×	×	×	×	×	m_{10}
1	0	1	1	×	×	×	×	×	×	×	m_{11}
1	1	0	0	×	×	×	×	×	×	×	m_{12}
1	1	0	1	×	×	×	×	×	×	×	m_{13}
1	1	1	0	×	×	×	×	×	×	×	m_{14}
1	1	1	1	×	×	×	×	×	×	×	m_{15}

> **Note:** *Binary codes 101, 110 and 111 are unused codes and so next states and output are considered as don't-care for these codes.*

Here, D_A, D_B, D_C and Y are functions of A, B, C and x. Using Table 3, the Boolean equations for D_A, D_B, D_C and Y are determined as sum of minterms and don't-cares as shown below:

$$D_A = m_7 + \Sigma d(10, 11, 12, 13, 14, 15)$$

$$D_B = \Sigma m(1, 2, 3, 5) + \Sigma d(10, 11, 12, 13, 14, 15)$$

$$D_C = \Sigma m(0, 1, 3, 5, 6, 8) + \Sigma d(10, 11, 12, 13, 14, 15)$$

$$Y = \Sigma m(4, 9) + \Sigma d(10, 11, 12, 13, 14, 15)$$

Using the above equations K-maps are drawn for D_A, D_B, D_C and Y in order to obtain simplified Boolean equations.

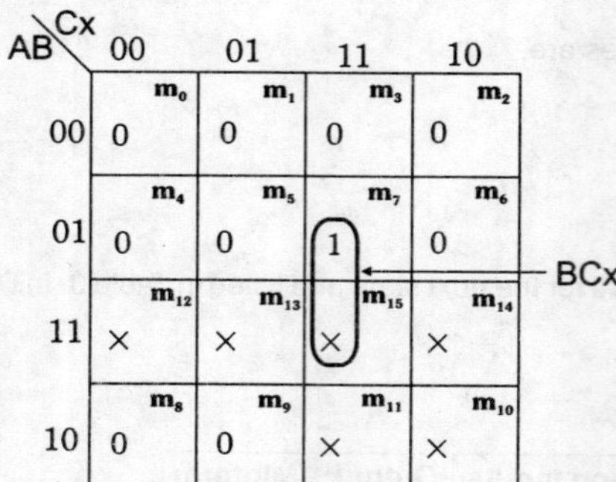

Fig. 11: K-map for D_A.

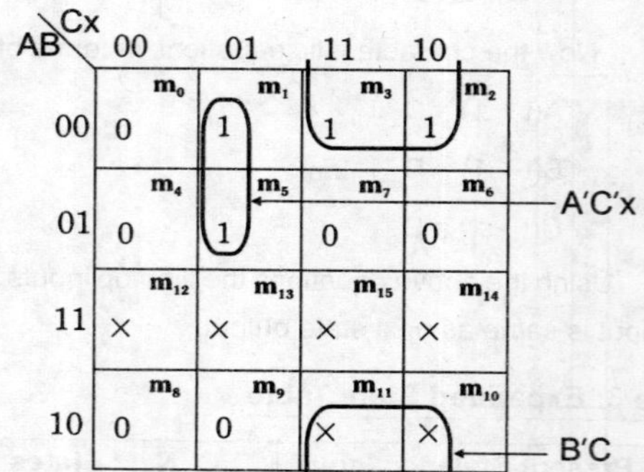

Fig. 12: K-map for D_B.

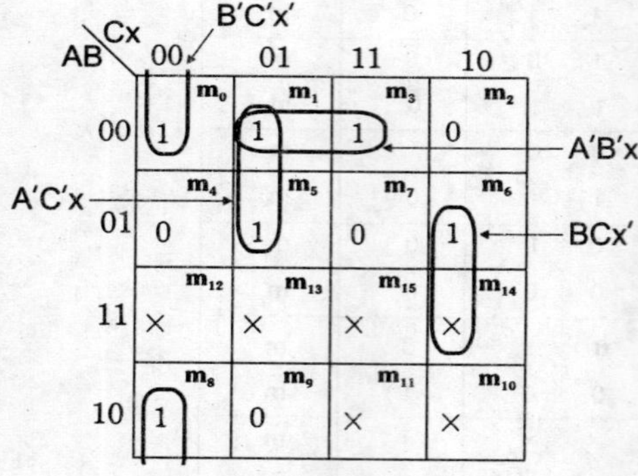

Fig. 13: K-map for D_C.

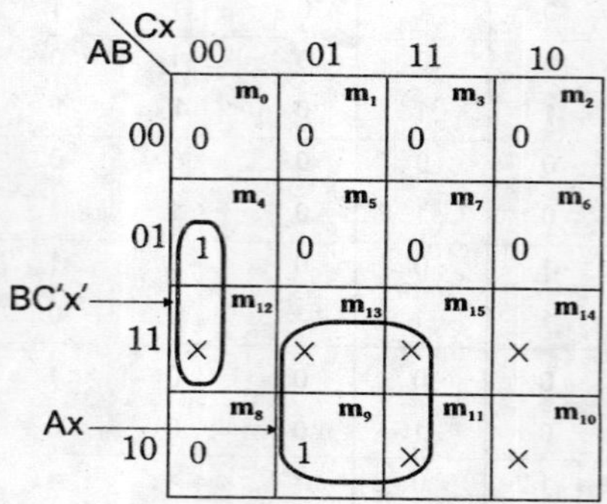

Fig. 14: K-map for Y.

From the K-map we get the following simplified Boolean equations for detecting the sequence 111 or 000 when overlapping not allowed using D flip-flops,

$$D_A = BCx$$

$$D_B = B'C + A'C'x$$

$$D_C = B'C'x' + A'C'x + A'B'x + BCx'$$

$$Y = BC'x' + Ax$$

Using the above equations the logic diagram of sequence detector is drawn as shown in Fig. 15. Here, the output depends on input and states and so this model is Mealy model.

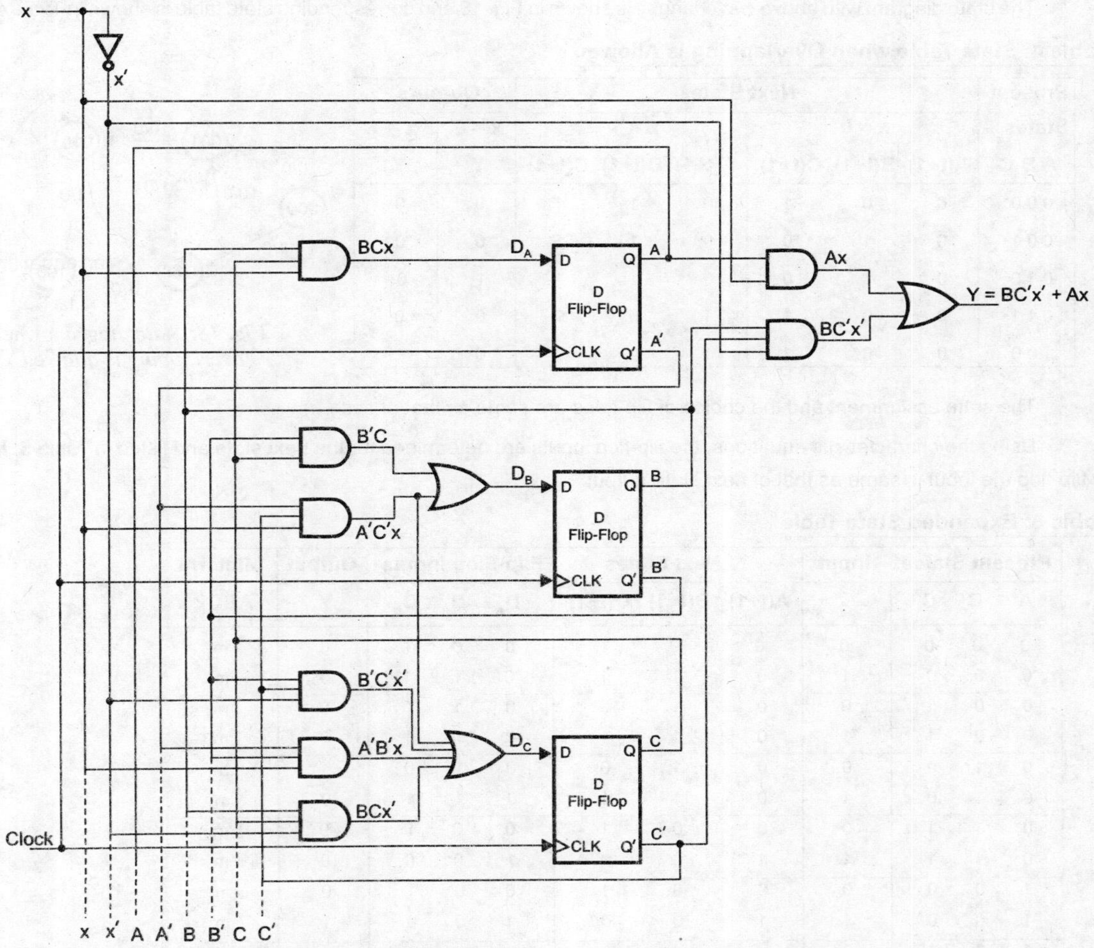

Fig. 15: Logic diagram to detect 111 or 000 when overlapping not allowed.

ii) When Overlapping Allowed

When overlapping is allowed the following two changes are to be made in state diagram of Fig. 9 and redrawn as in Fig. 18.

1. With reference to Fig. 1, while in state c (010), if input is 0 then output 1 and remain in state c itself instead of going to state a (000) as shown in Fig. 16.

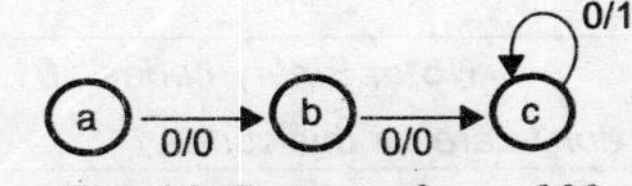

Fig. 16: States to detect 000 in overlapping sequence.

2. With reference to Fig. 2, while in state e (100), if input is 1 then output 1 and remain in state e itself instead of going to state a (000) as shown in Fig. 17.

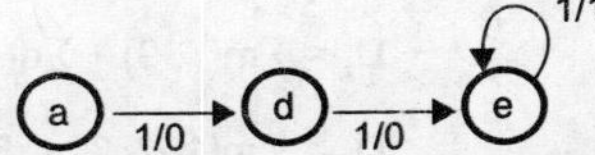

Fig. 17: States to detect 111 in overlapping sequence.

The state diagram with above two changes is shown in Fig. 18 and corresponding state table is shown in Table 4.

Table 4: State Table when Overlapping is Allowed

Present States	Next States						Outputs	
	x = 0			x = 1			x = 0	x = 1
ABC	A(t+1)	B(t+1)	C(t+1)	A(t+1)	B(t+1)	C(t+1)	Y	Y
000	0	0	1	0	1	1	0	0
001	0	1	0	0	1	1	0	0
010	0	1	0	0	1	1	1	0
011	0	0	1	1	0	0	0	0
100	0	0	1	1	0	0	0	1

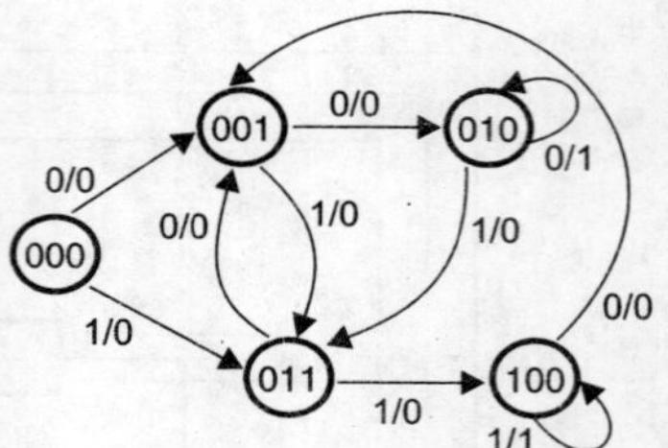

Fig. 18: State diagram when overlapping is allowed.

The state assignment and the choice of flip-flops are same as that of case (i).

Using the characteristic equations, the flip-flop inputs are determined for the next state and listed in Table 5. In D flip-flop the input is same as that of next state output.

Table 5: Expanded State Table

Present States			Input	Next States			Flip-Flop Inputs			Output	Minterm
A	B	C	x	A(t+1)	B(t+1)	C(t+1)	D_A	D_B	D_C	Y	ABCx
0	0	0	0	0	0	1	0	0	1	0	m_0
0	0	0	1	0	1	1	0	1	1	0	m_1
0	0	1	0	0	1	0	0	1	0	0	m_2
0	0	1	1	0	1	1	0	1	1	0	m_3
0	1	0	0	0	1	0	0	1	0	1	m_4
0	1	0	1	0	1	1	0	1	1	0	m_5
0	1	1	0	0	0	1	0	0	1	0	m_6
0	1	1	1	1	0	0	1	0	0	0	m_7
1	0	0	0	0	0	1	0	0	1	0	m_8
1	0	0	1	1	0	0	1	0	0	1	m_9
1	0	1	0	×	×	×	×	×	×	×	m_{10}
1	0	1	1	×	×	×	×	×	×	×	m_{11}
1	1	0	0	×	×	×	×	×	×	×	m_{12}
1	1	0	1	×	×	×	×	×	×	×	m_{13}
1	1	1	0	×	×	×	×	×	×	×	m_{14}
1	1	1	1	×	×	×	×	×	×	×	m_{15}

> ***Note:*** *Binary codes 101, 110 and 111 are unused codes and so next states and output are considered as don't-care for this codes.*

Here, D_A, D_B, D_C and Y are functions of A, B, C and x. Using Table 5, the Boolean equations for D_A, D_B, D_C and Y are determined as sum of minterms and don't-cares as shown below:

$$D_A = \Sigma m(7, 9) + \Sigma d(10, 11, 12, 13, 14, 15)$$

$$D_B = \Sigma m(1, 2, 3, 4, 5) + \Sigma d(10, 11, 12, 13, 14, 15)$$

$$D_C = \Sigma m(0, 1, 3, 5, 6, 8) + \Sigma d(10, 11, 12, 13, 14, 15)$$

$$Y = \Sigma m(4, 9) + \Sigma d(10, 11, 12, 13, 14, 15)$$

The equations for D_C and Y are same as in case (i), hence K-maps are constructed for D_A and D_B as shown in Figs. 19 and 20 to get simplified equations for D_A and D_B.

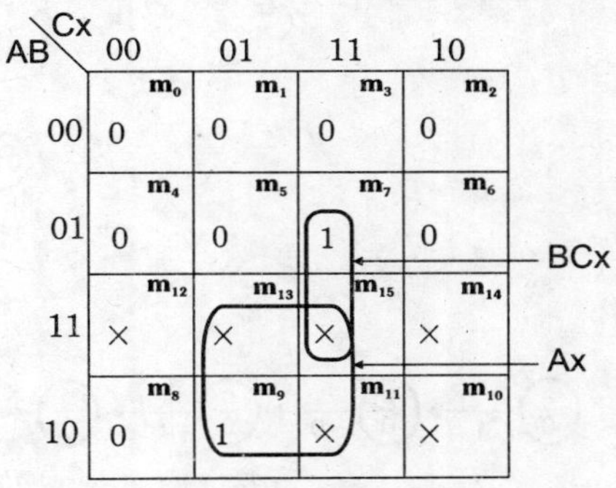

Fig. 19: *K-map for D_A.*

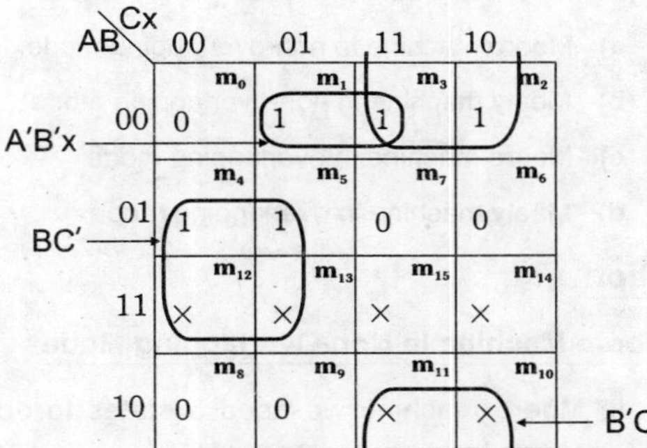

Fig. 20: *K-map for D_B.*

From the K-map we get the following simplified Boolean equations for D_A, D_B, D_C and Y.

$D_A = BCx + Ax$

$D_B = A'B'x + B'C + BC'$

$D_C = B'C'x' + A'C'x + A'B'x + BCx'$

$Y = BC'x' + Ax$

Using the above equations the logic circuit using D flip-flops to detect 111 or 000 when overlapping is allowed is drawn as shown in Fig. 21. Here, the output depends on input and states and so this model is Mealy model.

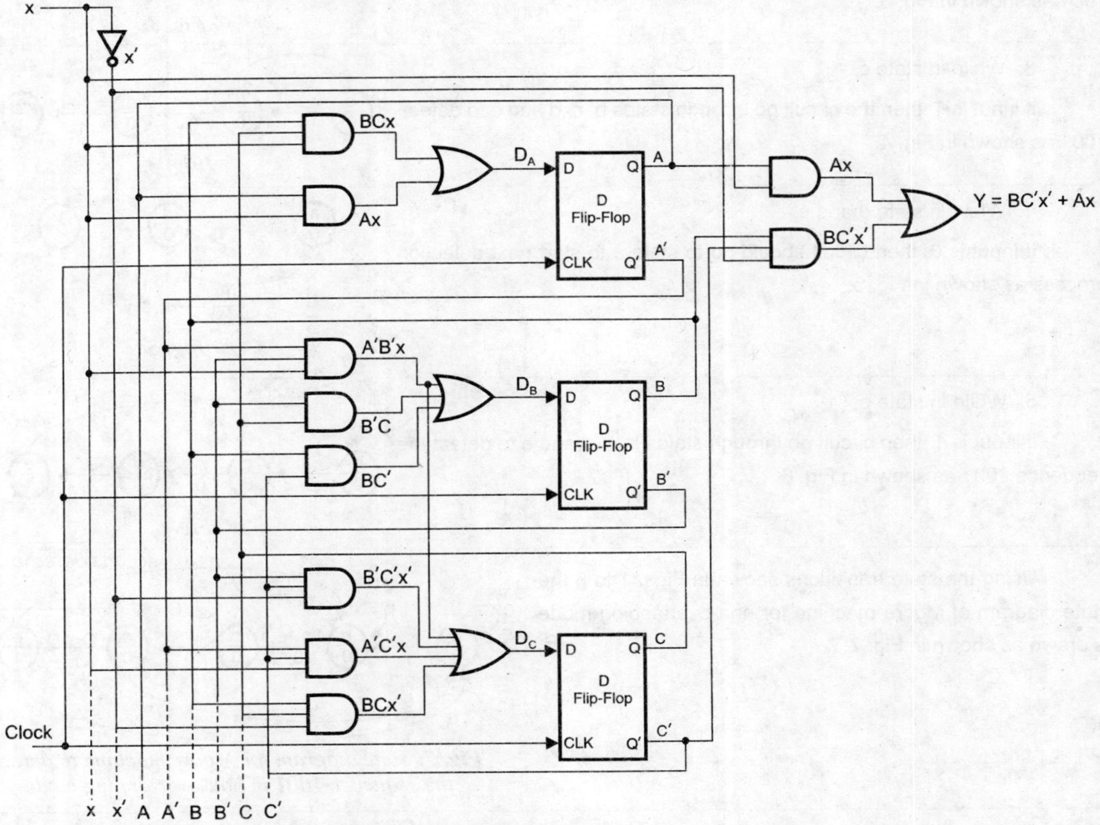

Fig. 21: *Logic diagram to detect 111 or 000 when overlapping allowed.*

Example 3.14

Draw the state diagram for sequence detector 1001 for the following cases.

a) Moore machine in non-overlapping mode

b) Mealy machine in non-overlapping mode

c) Moore machine in overlapping mode

d) Mealy machine in overlapping mode

Solution

a) Moore Machine in Non-Overlapping Mode

In Moore machine we need 5 states to detect the sequence 1001 in non-overlapping mode as shown in Fig. 1.

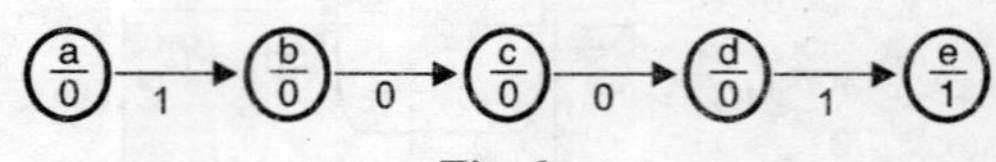

Fig. 1.

1. While in state a,

if input is 0, then circuit should remain in state a itself as shown in Fig. 2.

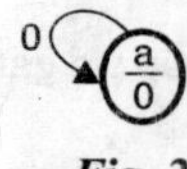

Fig. 2.

2. While in state b,

if input is 1, then the circuit go through states b, c, d and e to detect 1001 as shown in Fig. 3.

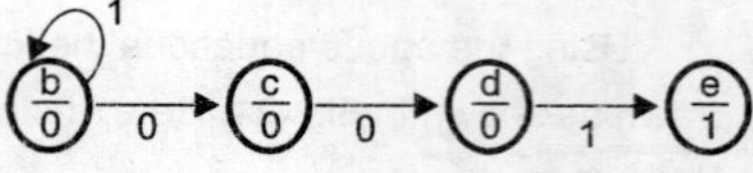

Fig. 3.

3. While in state c,

if input is 1, then the circuit go through states b, c, d and e to detect 1001 as shown in Fig. 4.

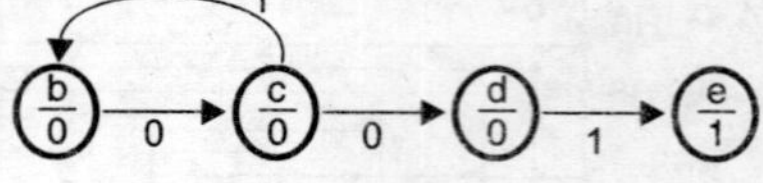

Fig. 4.

4. While in state d and e,

if input is 0, then circuit should go to state a to start next detection process as shown in Fig. 5.

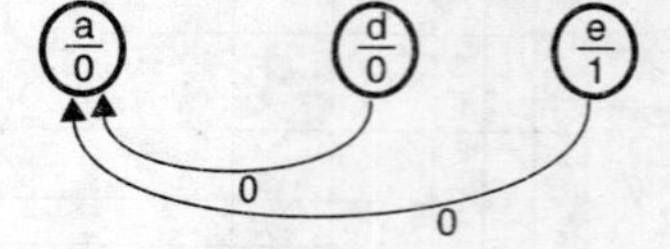

Fig. 5.

5. While in state e,

if input is 1, then circuit go through states b, c, d and e to detect the sequence 1001 as shown in Fig. 6.

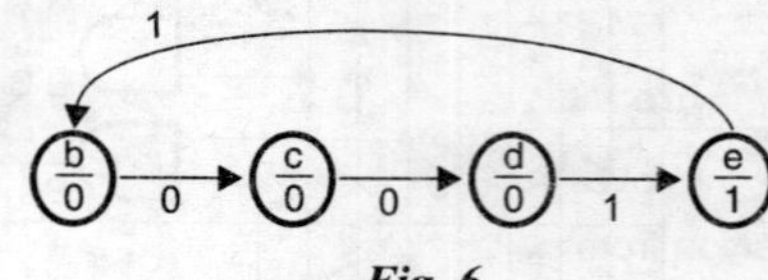

Fig. 6.

Using the state transitions shown in Figs. 1 to 6 the state diagram of Moore machine for non-overlapping mode is drawn as shown in Fig. 7.

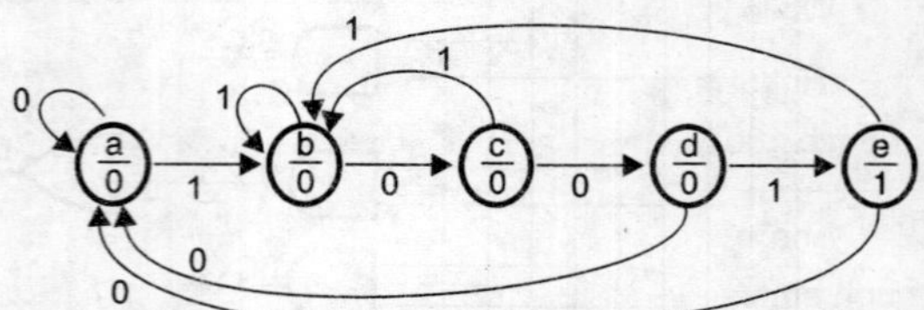

Fig. 7: State diagram for Moore machine to detect the sequence 1001 in non-overlapping mode.

b) Mealy Machine in Non-Overlapping Mode

In Mealy machine we need 4 states to detect the sequence 1001 in non-overlapping mode as shown in Fig. 8.

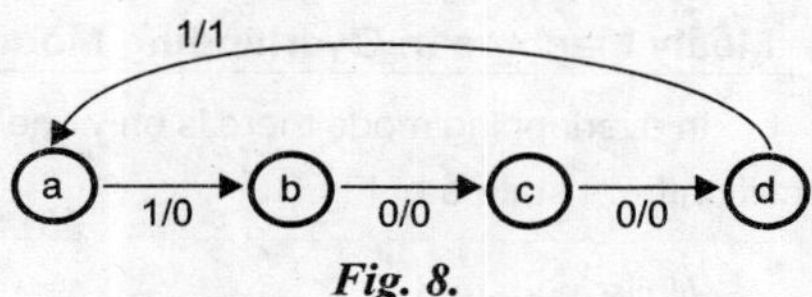

Fig. 8.

1. While in state a,

if input is 0, then circuit should remain in state a itself as shown in Fig. 9.

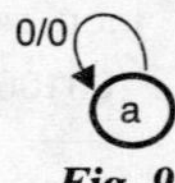

Fig. 9.

2. While in state b,

if input is 1, then the circuit go through states b, c, d and a to detect 1001 as shown in Fig. 10.

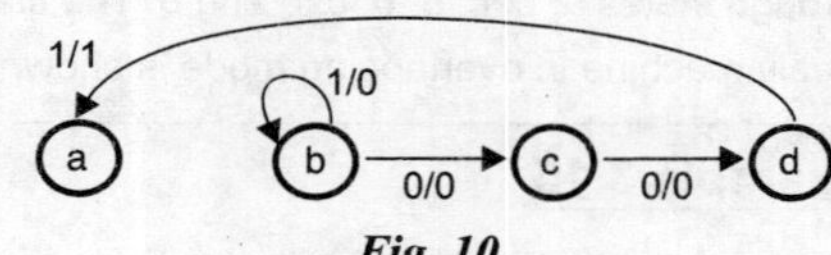

Fig. 10.

3. While in state c,

if input is 1, then the circuit go through states b, c, d and a to detect 1001 as shown in Fig. 11.

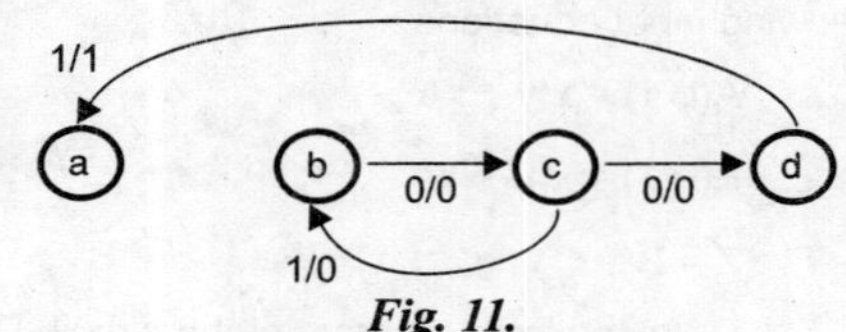

Fig. 11.

4. While in state d,

if input is 0, then circuit should go to state a to start next detection process as shown in Fig. 12.

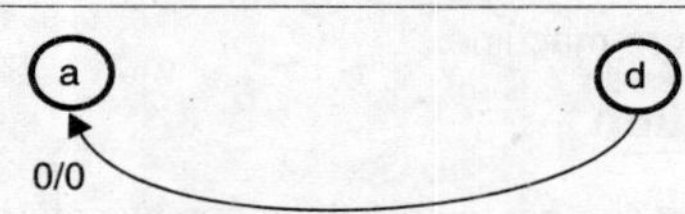

Fig. 12.

Using the state transitions shown in Figs. 8 to 12 the state diagram of Mealy machine for non-overlapping mode is drawn as shown in Fig. 13.

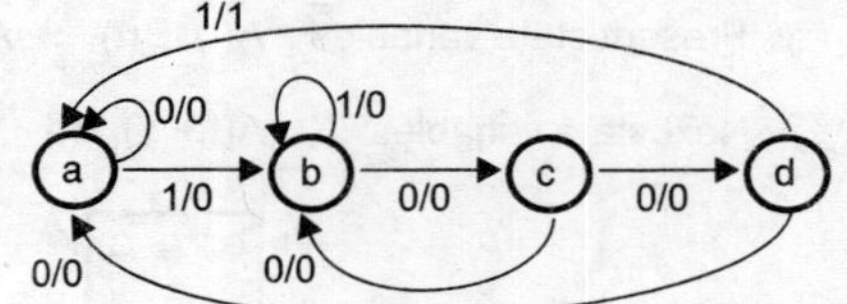

***Fig. 13:** State diagram for Mealy machine to detect the sequence 1001 in non-overlapping mode.*

c) Moore Machine in Overlapping Mode

In overlapping mode there is only one change in state transition from state e of Fig. 7.

While in state e,

if input is 0, then circuit go to state c to detect overlapped sequence 1001001.

The overlapped sequence is detected when passing through states a, b, c, d, e, c, d and e. The state diagram for Moore machine in overlapping mode is shown in Fig. 14.

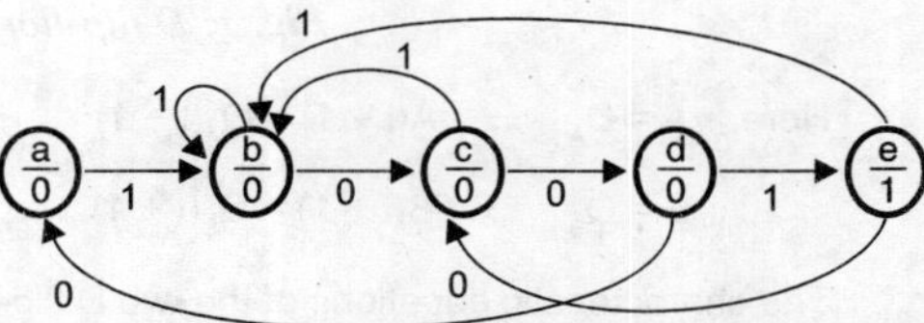

***Fig. 14:** State diagram for Moore machine to detect the sequence 1001 in overlapping mode.*

d) Mealy Machine in Overlapping Mode

In overlapping mode there is only one change in state transition from state d of Fig. 13.

While in state d,

if input is 1, then circuit go to state b to detect overlapped sequence 1001001.

The overlapped sequence is detected when passing through states a, b, c, d, b, c, d and b. The state diagram for Mealy machine in overlapping mode is shown in Fig. 15.

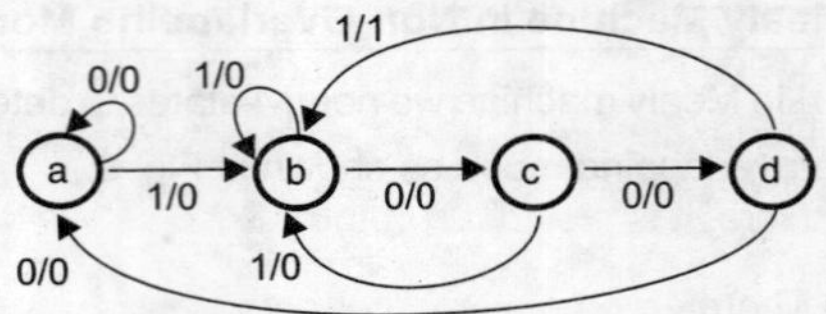

Fig. 15: State diagram for Mealy machine to detect the sequence 1001 in overlapping mode.

Example 3.15

A sequential circuit with two D filp-flops A and B and two inputs x and y, and one output Z is specified by the following input equations :

$A(t+1) = x'y + xA$

$B(t+1) = x'B + xA$

$Z = B$

Draw the logic diagram of the circuit. Derive the state table and state diagram and state whether it is a Mealy or Moore machine.

Solution

The given equations has two state variables A and B. Hence the sequential circuit can be designed using two D flip-flops shown in Fig. 1 to store two state variables.

∴ Present state variables : A(t), B(t) ⇒ A, B (For simplicity)

Next state variables : A(t + 1), B(t + 1)

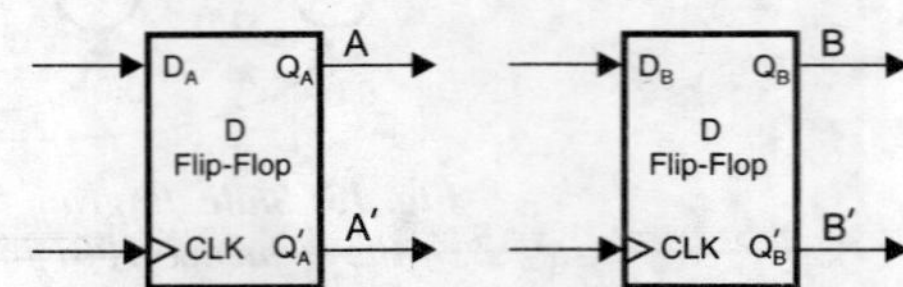

Fig. a: Flip-flop A. *Fig. b: Flip-flop B.*

Fig. 1: D flip-flop to store state variables.

Here, $A = Q_A \Rightarrow A(t+1) = Q_A(t+1)$

$B = Q_B \Rightarrow B(t+1) = Q_B(t+1)$

The characteristic equations of the two D flip-flops are,

$Q_A(t+1) = D_A$

$Q_B(t+1) = D_B$

It is given that,

$A(t+1) = x'y + x'A$

$B(t+1) = x'B + xA$

Characteristic equation of D flip-flop,

$Q(t+1) = D$

From the knowledge of above equations we get the following equations for flip-flop input.

$A(t+1) = D_A \Rightarrow D_A = x'y + xA$

$B(t+1) = D_B \Rightarrow D_B = x'B + xA$

Using the flip-flop input equations and given output equation the logic diagram of sequential circuit is drawn as shown in Fig. 2.

Here, it is given that output, Z = B

Therefore, the output depends only on states and so it is a Moore machine.

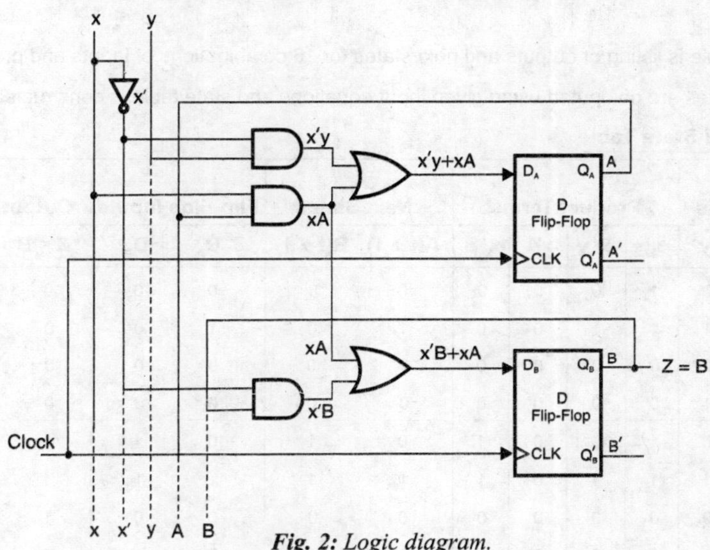

Fig. 2: Logic diagram.

State Table

With n-bit input it is possible to have 2^n different inputs.

Here, n = 2, $\therefore$ $2^2 = 4$, Four possible inputs.

Four possible inputs: xy = 0 0

= 0 1

= 1 0

= 1 1

The circuit has two state variables A and B. With m state variables it is possible to have 2^m different states.

Here, m = 2, $\therefore$ $2^2 = 4$

Therefore, the sequential circuit has four states.

The four possible states are, A B = 0 0

= 0 1

= 1 0

= 1 1

When there are m state variables and n inputs then possible combinations of inputs and states are 2^{m+n}.

Here, m = 2, n = 2

$\therefore$ $2^{m+n} = 2^{2+2} = 2^4 = 16$

The 16 combinations of inputs and states are given below:

when, xy = 0 0, AB = 0 0	when xy = 0 1, AB = 0 0	when xy = 1 0, AB = 0 0	when xy = 1 1, AB = 0 0
= 0 1	= 0 1	= 0 1	= 0 1
= 1 0	= 1 0	= 1 0	= 1 0
= 1 1	= 1 1	= 1 1	= 1 1

The state table is listing of outputs and next states for 16 combinations of inputs and present states.

The next states are computed using given input equations and state table is constructed as shown in Table 1.

Table 1: Expanded State Table

Present States		Inputs		Product Terms				Next States		Flip-Flop Inputs		Output
A	B	x	y	x′	x′y	xA	x′B	A(t + 1)	B(t + 1)	D_A	D_B	Z = B
0	0	0	0	1	0	0	0	0	0	0	0	0
0	0	0	1	1	1	0	0	1	0	1	0	0
0	0	1	0	0	0	0	0	0	0	0	0	0
0	0	1	1	0	0	0	0	0	0	0	0	0
0	1	0	0	1	0	0	1	0	1	0	1	1
0	1	0	1	1	1	0	1	1	1	1	1	1
0	1	1	0	0	0	0	0	0	0	0	0	1
0	1	1	1	0	0	0	0	0	0	0	0	1
1	0	0	0	1	0	0	0	0	0	0	0	0
1	0	0	1	1	1	0	0	1	0	1	0	0
1	0	1	0	0	0	1	0	1	1	1	1	0
1	0	1	1	0	0	1	0	1	1	1	1	0
1	1	0	0	1	0	0	1	0	1	0	1	1
1	1	0	1	1	1	0	1	1	1	1	1	1
1	1	1	0	0	0	1	0	1	1	1	1	1
1	1	1	1	0	0	1	0	1	1	1	1	1

$A(t + 1) = x'y + xA$

$B(t + 1) = x'B + xA$

$D_A = A(t + 1)$

$D_B = B(t + 1)$

Note: *In D flip-flop the inputs are same as next state flip-flop outputs.*

State Diagram

The possible state transitions from one state to other states are obtained from Table 1 and state transitions are diagrammatically shown in Figs. 3 to 6.

1. While in state 00,

   if input = 00, 10,11, remain in state 00.

   if input = 01, go to state 10.

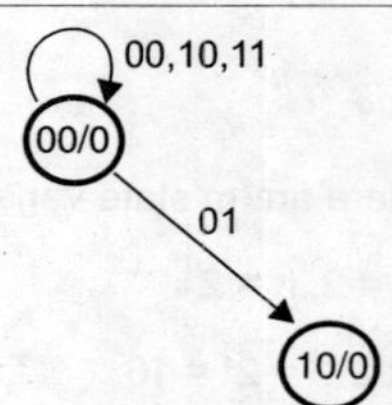

Fig. 3: *Transitions from state 00.*

2. While in state 01,

   if input = 00, remain in state state 01.

   if input = 10,11, go to state 00.

   if input = 01, go to state 11.

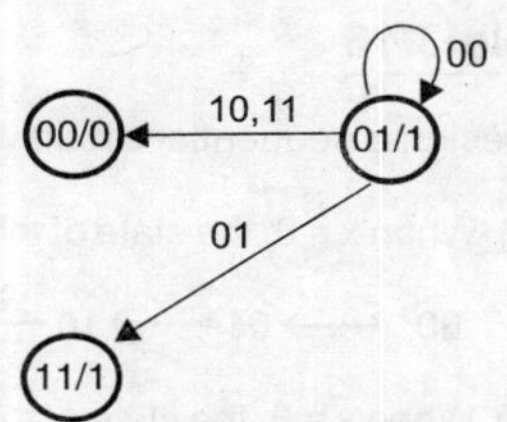

Fig. 4: Transitions from state 01.

3. While in state 10,

   if input = 00, go to state 00

   if input = 01, go to state 10

   if input = 10, 11, go to state 11

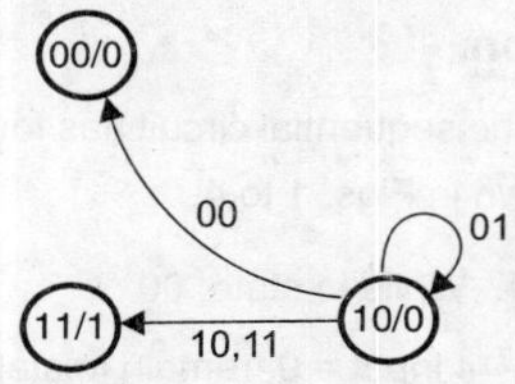

Fig. 5: Transitions from state 10.

4. While in state 11,

   if input = 00, go to state 01

   if input = 01, 10,11, remain in state 11

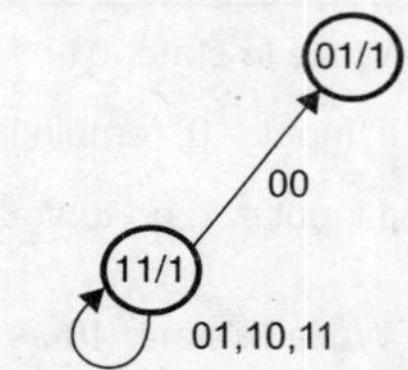

Fig. 6: Transitions from state 11.

On combining the state transitions shown in Figs. 3 to 6, the state diagram can be drawn as shown in Fig. 7. The standard form of state stable is shown in Table 2.

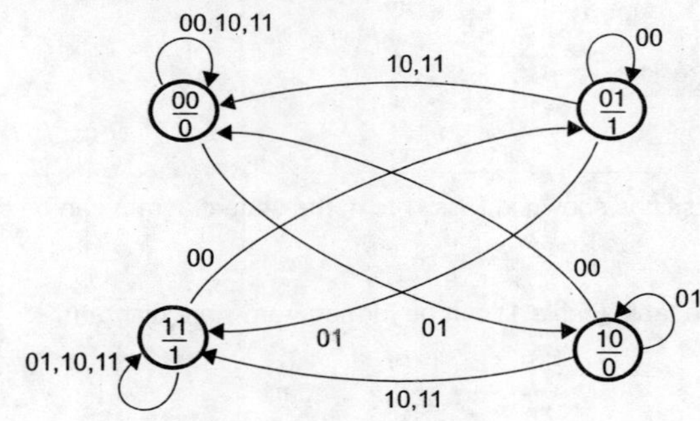

Fig. 7: State diagram.

Table 2: State Table

Present States		Next States								Output			
		xy = 00		xy = 01		xy = 10		xy = 11		xy = 00	xy = 01	xy = 10	xy = 11
A	B	A(t + 1)	B(t + 1)	A(t + 1)	B(t + 1)	A(t + 1)	B(t + 1)	A(t + 1)	B(t + 1)	Z	Z	Z	Z
0	0	0	0	1	0	0	0	0	0	0	0	0	0
0	1	0	1	1	1	0	0	0	0	1	1	1	1
1	0	0	0	1	0	1	1	1	1	0	0	0	0
1	1	0	1	1	1	1	1	1	1	1	1	1	1

Example 3.16

(AU, Nov/Dec'19, 15 Marks)

Design a sequential circuit with two D flip-flops A and B and one input x.

a) When x = 0, the state of the circuit remains the same. When x = 1, the circuit goes through the state transistions

00 ⟶ 01 ⟶ 10 ⟶ 11 ⟶ 00

b) When x = 0, the state of the circuit remains the same. When x = 1, the circuit goes through the state transitions

00 ⟶ 11 ⟶ 01 ⟶ 10 ⟶ 00

Solution

a) The sequential circuit has four states 00, 01, 10, 11. The possible state transitions from one state to other states are shown in Figs. 1 to 4.

1. While in state, 00

   if input = 0, remain in state 00

   if input = 1, go to state 01

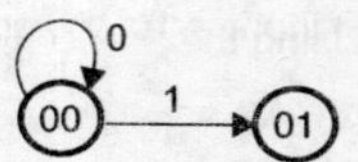

Fig. 1: Transitions from state 00.

2. While in state, 01

   if input = 0, remain in state 01

   if input = 1, go to state 10

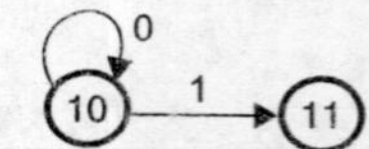

Fig. 2: Transitions from state 01.

3. While in state, 10

   if input = 0, remain in state 10

   if input = 1, go to state 11

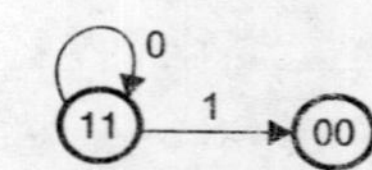

Fig. 3: Transitions from state 10.

4. While in state, 11

   if input = 0, remain in state 11

   if input = 1, go to state 00

Fig. 4: Transitions from state 11.

On combining the state transitions shown in Figs. 1 to 4, the state diagram can be drawn as shown in Fig. 5.

State Table

Let, x be the input. The state table (Table 1) can be formed from state diagram.

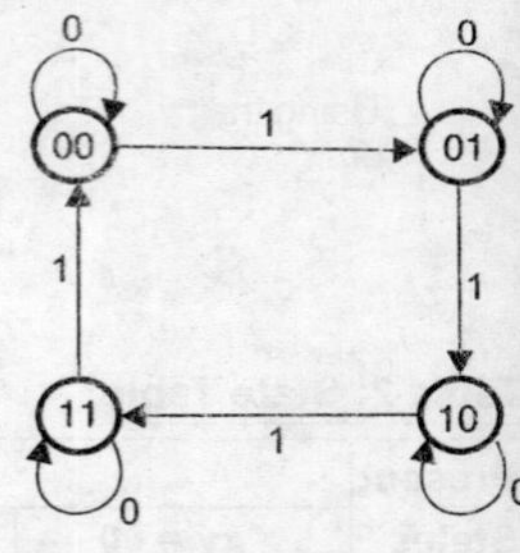

Fig. 5: State diagram.

Table 1: State Table

Present States	Next State x = 0	Next State x = 1
0 0	0 0	0 1
0 1	0 1	1 0
1 0	1 0	1 1
1 1	1 1	0 0

Design using D Flip-Flop

Here, there are two state variables. Let A and B be the two state variables.

∴ Present state variables : A(t), B(t) ⇒ A, B (For simplicity)

∴ Next state variables : A(t + 1), B(t + 1)

Let us choose two flip-flops to store two state variables A and B as shown in Fig. 6. Here, A and B are output of flip-flops.

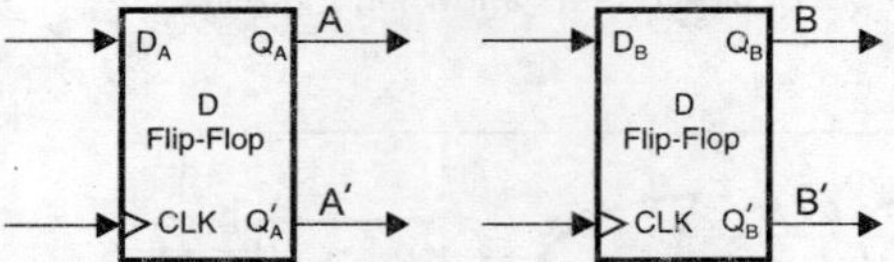

Fig. a: Flip-flop A. *Fig. b: Flip-flop B.*

Fig. 6: D flip-flop to store state variables.

Here, $A = Q_A \Rightarrow A(t+1) = Q_A(t+1)$; $B = Q_B \Rightarrow B(t+1) = Q_B(t+1)$

Characteristic equation of D flip-flop, $Q(t+1) = D$

The characteristic equation of two D flip-flops are,

$Q_A(t+1) = D_A$; $Q_B(t+1) = D_B$

Using the above equation we can write,

$A(t+1) = D_A$; $B(t+1) = D_B$

Using the above equations the flip-flop inputs are determined for the next state and listed in Table 2. Here, the flip-flop inputs are same as next state outputs.

Table 2: Expanded State Table

Present States		Input	Next States		Flip-Flop inputs		Minterm
A	**B**	**x**	**A(t+1)**	**B(t+1)**	**D_A**	**D_B**	**ABx**
0	0	0	0	0	0	0	m_0
0	0	1	0	1	0	1	m_1
0	1	0	0	1	0	1	m_2
0	1	1	1	0	1	0	m_3
1	0	0	1	0	1	0	m_4
1	0	1	1	1	1	1	m_5
1	1	0	1	1	1	1	m_6
1	1	1	0	0	0	0	m_7

Here, D_A and D_B are functions of A, B, x and so minterms are formed using A, B and x. Using Table 2, the Boolean equations for D_A and D_B are determined as sum of minterms as shown below:

$\therefore D_A = \Sigma m(3, 4, 5, 6)$; $D_B = \Sigma m(1, 2, 5, 6)$

Using the above equations, let us draw K-map for D_A and D_B to develop simplified Boolean equations for D_A and D_B.

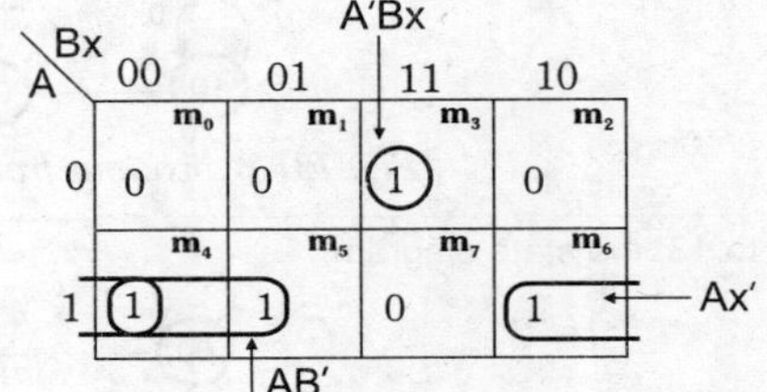

Fig. 7: K-map for D_A.

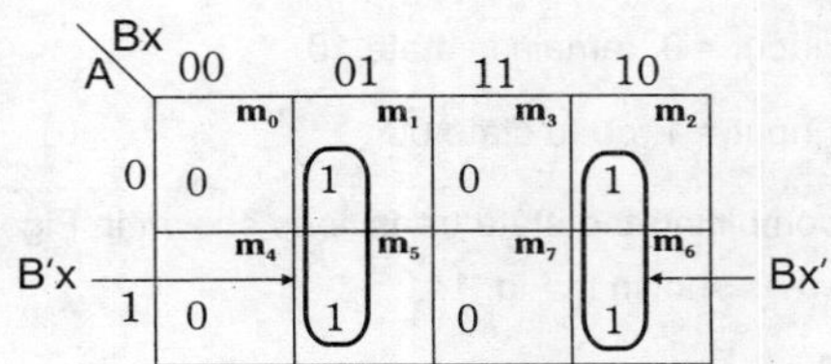

Fig. 8: K-map for D_B.

From the K-map we get the following simplified Boolean equations for realizing sequence detector using D flip-flops,

$D_A = Ax' + AB' + A'Bx$; $D_B = B'x + Bx'$

The logic diagram of sequence detector using the above equations is drawn as shown in Fig. 9. Here the sequence detector output is the present state output of flip-flops as shown in Fig. 9.

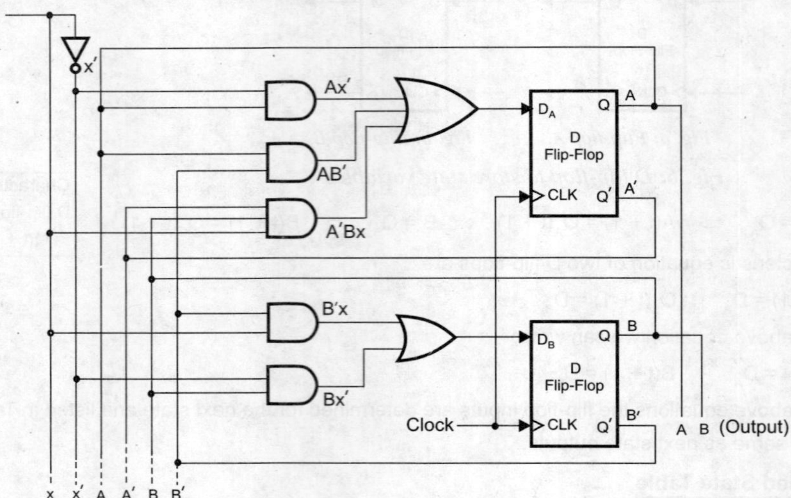

Fig. 9: Logic diagram using D flip-flops.

b) The sequential circuit has four states 00, 11, 01, 10. The possible state transitions from one state to other states are shown in Figs. 10 to 13.

1. While in state, 00

   if input = 0, remain in state 00

   if input = 1, go to state 11

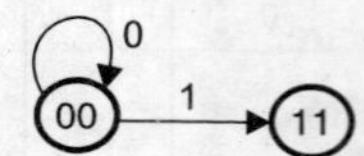

Fig. 10: Transitions from state 00.

2. While in state, 11

   if input = 0, remain in state 11

   if input = 1, go to state 01

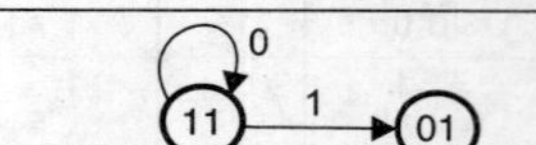

Fig. 11: Transitions from state 11.

3. While in state, 01

   if input = 0, remain in state 01

   if input = 1, go to state 10

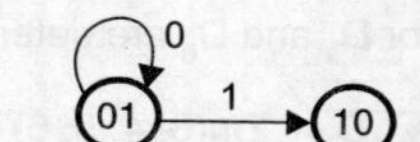

Fig. 12: Transitions from state 01.

4. While in state, 10

   if input = 0, remain in state 10

   if input = 1, go to state 00

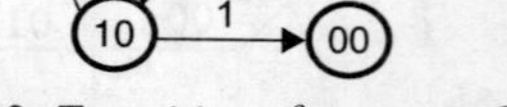

Fig. 13: Transitions from state 10.

On combining the state transitions shown in Figs. 10 to 13, the state diagram can be drawn as shown in Fig. 14.

State Table

Let, x be the input. The state table (Table 1) can be formed from state diagram.

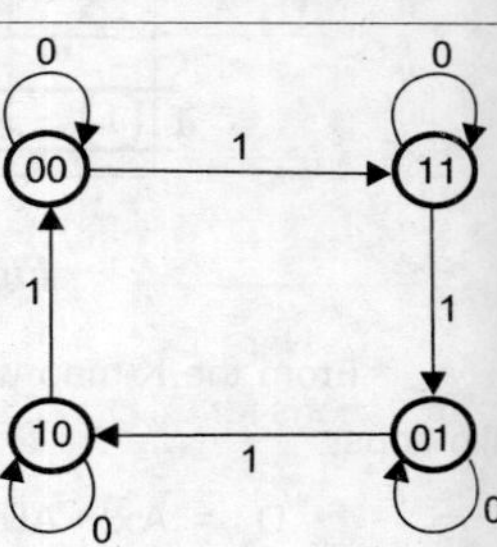

Fig. 14: State diagram.

Table 1: State Table

Present States	Next State x = 0	Next State x = 1
0 0	0 0	1 1
0 1	0 1	1 0
1 0	1 0	0 0
1 1	1 1	0 1

Design using D Flip-Flop

Here, there are two state variables. Let A and B be the two state variables.

∴ Present state variables : A(t), B(t) ⇒ A, B (For simplicity)

Next state variables : A(t + 1), B(t + 1)

Let us choose two flip-flops to store two state variables A and B as shown in Fig. 15. Here, A and B are output of flip-flops.

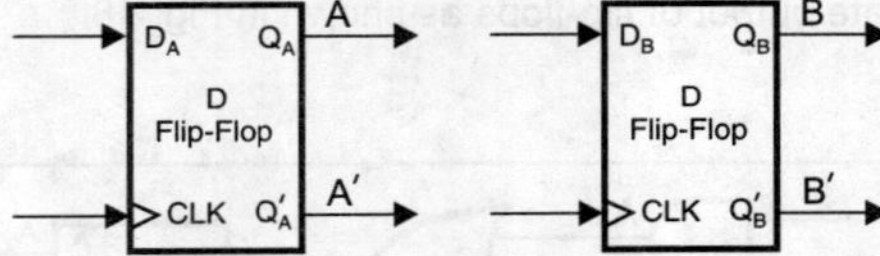

***Fig. a:** Flip-flop A.* ***Fig. b:** Flip-flop B.*

***Fig. 15:** D flip-flop to store state variables.*

> Characteristic equation of D flip-flop,
> Q(t + 1) = D

Here, $A = Q_A \Rightarrow A(t+1) = Q_A(t+1)$; $B = Q_B \Rightarrow B(t+1) = Q_B(t+1)$

The characteristic equation of two D flip-flops are,

$Q_A(t+1) = D_A$; $Q_B(t+1) = D_B$

Using the above equation we can write,

$A(t+1) = D_A$; $B(t+1) = D_B$

Using the above equations the flip-flop inputs are determined for the next state and listed in Table 2. Here, the flip-flop inputs are same as next state outputs.

Table 2: Expanded State Table

Present States A	Present States B	Input x	Next States A(t+1)	Next States B(t+1)	Flip-Flop inputs D_A	Flip-Flop inputs D_B	Minterm ABx
0	0	0	0	0	0	0	m_0
0	0	1	1	1	1	1	m_1
0	1	0	0	1	0	1	m_2
0	1	1	1	0	1	0	m_3
1	0	0	1	0	1	0	m_4
1	0	1	0	0	0	0	m_5
1	1	0	1	1	1	1	m_6
1	1	1	0	1	0	1	m_7

Here, D_A and D_B are functions of A, B, x and so minterms are formed using A,B and x. Using Table 2, the Boolean equations for D_A and D_B are determined as sum of minterms as shown below:

$\therefore D_A = \Sigma m(1, 3, 4, 6)$; $D_B = \Sigma m(1, 2, 6, 7)$

Using the above equations, let us draw K-map for D_A and D_B to develop simplified Boolean equations for D_A and D_B.

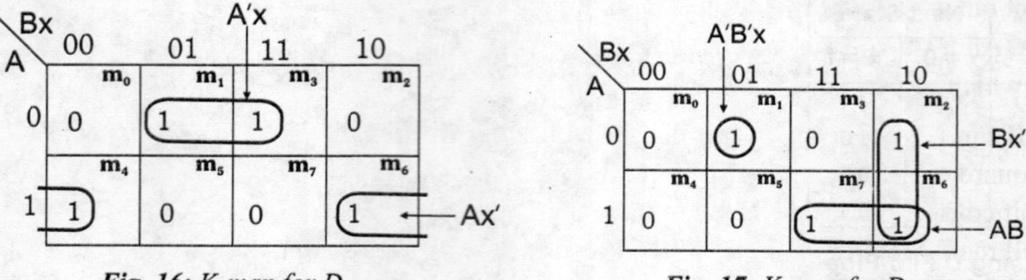

Fig. 16: K-map for D_A.

Fig. 17: K-map for D_B.

From the K-map we get the following simplified Boolean equations for realizing sequence detector using D flip-flops,

$$D_A = Ax' + A'x \quad ; \quad D_B = AB + Bx' + A'B'x$$

The logic diagram of sequence detector using the above equations is drawn as shown in Fig. 18. Here the sequence detector output is the present state output of flip-flops as shown in Fig. 18.

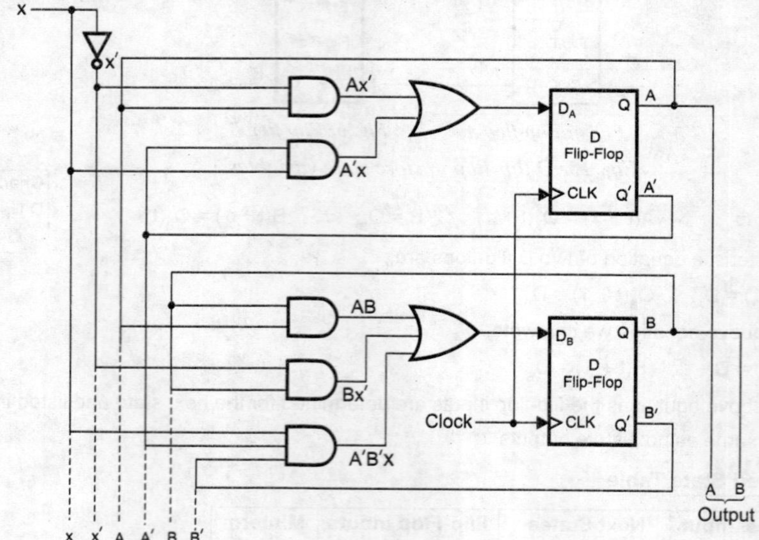

Fig. 18: Logic diagram using D flip-flops.

3.7 Counters

A register that goes through prescribed (or predefined) sequence of states upon the application of clock pulses is called a ***counter***. The counters are used to generate or go through a sequence of binary code known as ***count sequence***. A counter that follows a binary count sequence is called ***binary counter***. An n-bit binary counter consists of n flip-flops and can count binary from 0 to $2^n - 1$. In binary up counter, the output will increment for every clock pulse and in binary down counter the output will decrement for every clock pulse. Mostly, counters will have a reset facility to clear the output at any desired time and also will have a preload facility to load a known value during starting.

The modulus of the counter refers to number states a counter will pass through before coming back to initial state. An n-bit counter will have 2^n possible states and if the n-bit counter passes through all possible states before coming to initial state then it is called ***MOD-2^n counter***. A 3-bit binary counter has

8 possible states and so it is called MOD-8 counter. Also, the counter divides the input clock frequency. An n-bit binary counter divides the clock frequency by 2^n and so n-bit counter is also called divide-by-2^n counter. An n-bit counter is constructed using n flip-flops and the output of each flip-flop is the counter output which is also a clock.

If f is the frequency of the clock then frequency of outputs of flip-flop is f/2, f/4, f/8, ..., $f/2^n$. A 3-bit counter will have three flip-flops and the frequency of output clock of flip-flops is f/2, f/4 and f/8. An n-bit counter can be designed to have modulus less than 2^n. Such counters use less number of states than full modulus counter. Let, MOD-N counter be a reduced modulus n-bit counter, where $N<2^n$.

Now n-bit MOD-N counter uses only N states in the 2^n possible states of n-bit counter and the remaining states are unused states. Also, the MOD-N counter divides the clock frequency by N and so it is also called ***divide-by-N counter***.

The counters can be broadly classified into two categories: Asynchronous and synchronous counters.

The asynchronous counters are also called ***ripple counters***.

In a ripple counter, a flip-flop output will act as clock for another flip-flop.

In synchronous counters all the flip-flops are triggered simultaneously by a common clock.

3.8 Asynchronous or Ripple Counter

A binary ripple counter is constructed using a series of complementing flip-flops. The binary ripple counter can be constructed using any flip-flop, that are connected to generate complement of present state as next state output. The pulses to be counted or the clock signal is applied to clock of first flip-flop. The output of first flip-flop is connected to clock of second flip-flop. The output of second flip-flop is connected to clock of third and so on. In general, the output of low order flip-flop is connected to clock of next higher order flip-flop.

3.8.1 Up Counter

In up counter, the output will increment for every clock pulse and in n-bit up counter, the count sequence will be from 0 to $2^n - 1$. In generally, counter refers to up counter.

Binary Ripple Counter using T Flip-Flop

In T-flip-flop, if the input is tied to 1 then the output will complement on every clock pulse and so in counter using T-flip-flop the input of all flip-flops are tied to 1. The count sequence or clock signal is applied as clock to first flip-flop. The output of low order flip-flop is connected to clock of next higher order flip-flop. Also, the flip-flop outputs are counter outputs. A 4-bit binary ripple counter using T flip-flop is shown in Fig. 3.72.

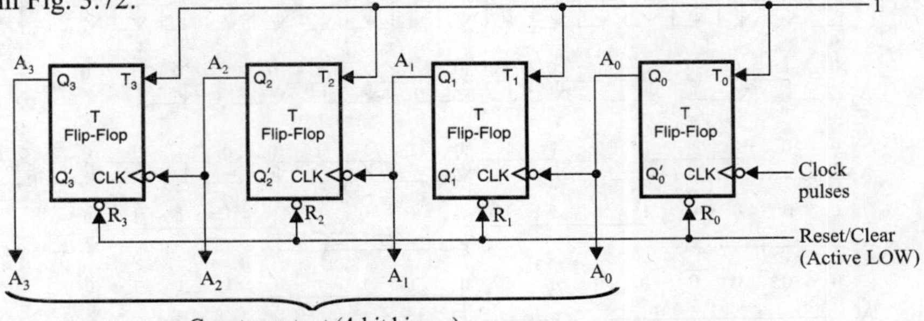

***Fig. 3.72:** 4-bit binary ripple counter using T flip-flop.*

Binary Ripple Counter using JK Flip-Flop

When the inputs J and K are tied to 1 then the JK flip-flop will behave as T-flip-flop. Therefore, in ripple counter using JK flip-flops, the J and K inputs of all flip-flops are tied to 1 and all other are same as T flip-flop based ripple counter. A 4-bit binary ripple counter using JK flip-flop is shown in Fig. 3.73.

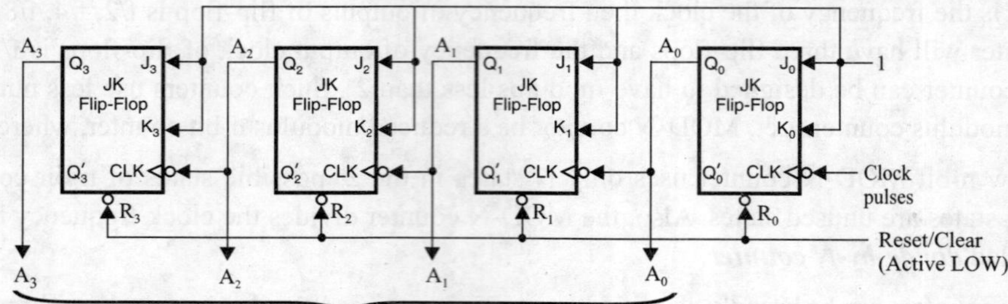

Fig. 3.73: 4-bit binary ripple counter using JK flip-flop.

Binary Ripple Counter using D Flip-Flop

In D flip-flop, after a reset, Q = 0, and Q′ = 1, and so if Q′ is connected to D, then for every clock pulse output toggles. This concept is used in construction of binary ripple counter using D flip-flop. A 4-bit binary ripple counter using D flip-flop is shown in Fig. 3.74.

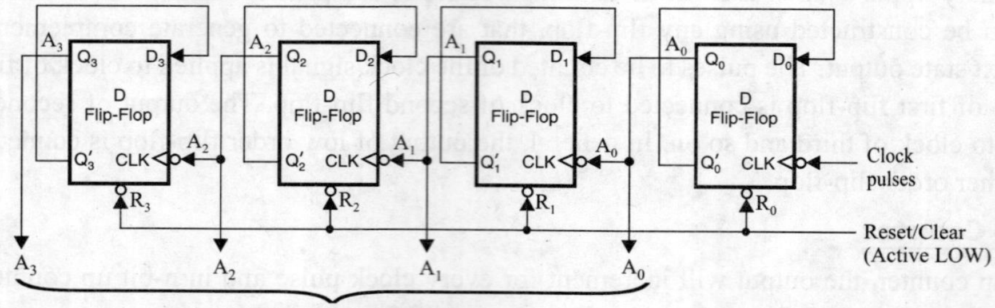

Fig. 3.74: 4-bit binary counter using D flip-flop.

Working of 4-bit Binary Ripple Counter

In binary ripple counter, the input to flip-flops are tied to 1. The clock is negative edge triggered and so on every negative edge of clock, the output of flip-flop toggles.

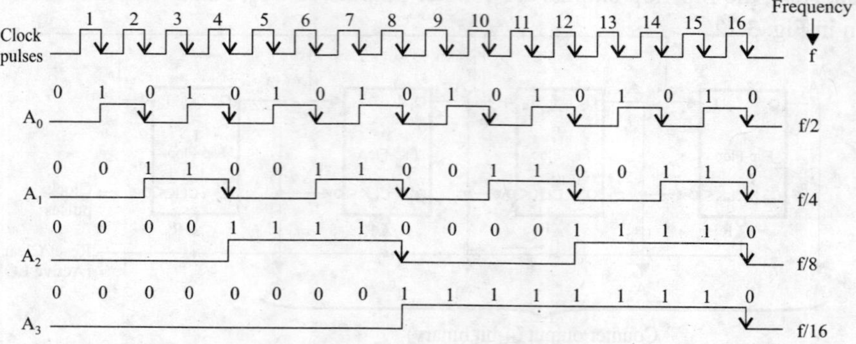

Fig. 3.75: Clock and output signals of 4-bit binary ripple counter.

Initially all the outputs are made zero using reset/clear signal. Then pulses to be counted or clock pulse is applied to clock of first flip-flop. So first flip-flop output A_0 complements or toggle on negative edge of every clock pulse.

Since A_0 is connected as clock of second flip-flop, the output A_1 of second flip-flop complements on every negative edge of A_0.

Since A_1 is connected as clock of third flip-flop, the output A_2 of third flip-flop complements on every negative edge of A_1.

Since A_2 is connected as clock of fourth flip-flop, the output A_3 of fourth flip-flop, complements on every negative edge of A_2. The clock and output signals of 4-bit binary ripple counter are shown in Fig. 3.75.

In Fig. 3.75, it can be observed that after 15th count pulses, all the outputs become zero. Therefore the count sequence again start after 16th clock pulse.

Hence, the 4-bit binary ripple counter can count from 0 to 15. The output of 4-bit binay counter is tabulated in Table 3.45.

The counter outputs are also known as divide-by-2^n clock signal for n = 1, 2, 3, 4. The frequency of counter output is shown in Fig. 3.75. Here, A_0 is divide-by-2 clock, A_1 is divide-by-4 clock, A_2 is divide-by-8 clock and A_3 is divide-by-16 clock.

Table 3.45: Output of 4-bit Binary Ripple Counter

Count or Clock Pulse	Counter Outputs			
	A_3	A_2	A_1	A_0
-	0	0	0	0
1	0	0	0	1
2	0	0	1	0
3	0	0	1	1
4	0	1	0	0
5	0	1	0	1
6	0	1	1	0
7	0	1	1	1
8	1	0	0	0
9	1	0	0	1
10 (A)	1	0	1	0
11 (B)	1	0	1	1
12 (C)	1	1	0	0
13 (D)	1	1	0	1
14 (E)	1	1	1	0
15 (F)	1	1	1	1
16	0	0	0	0

The 4-bit binary ripple counter is available as a standard IC 74293. The pin configuration of the IC 74293 is shown in Fig. 3.76. Here, Q_A, Q_B, Q_C and Q_D are output of internal flip-flops A, B, C and D respectively. Separate clock signals can be applied for flip-flops A and B. The flip-flop B, C and D are internally connected to function as 3-bit ripple counter.

The flip-flop A can function as divide-by-2 or MOD-2 counter. The flip-flop output Q_A should be connected to CLK-B externally for 4-bit ripple counter. The two reset inputs are internally NANDed and applied as logic **low** reset for flip-flops. Hence, any one reset should be tied **low** for counter operation.

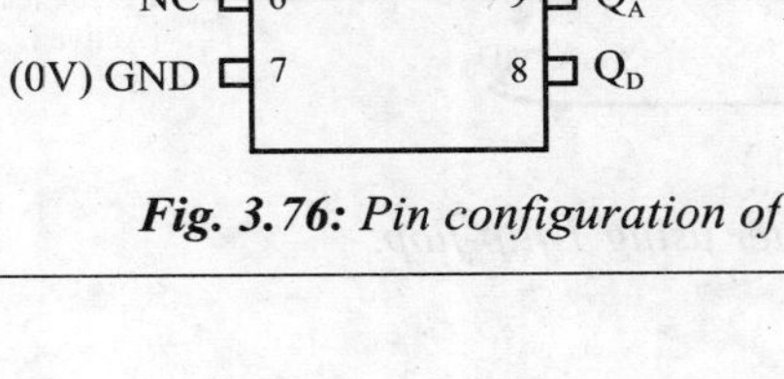

Pin (left)	No.	74293	No.	Pin (right)
NC	1		14	V_{cc}(+5V)
NC	2		13	R_2
NC	3		12	R_1
Q_C	4		11	CLK-B
Q_B	5		10	CLK-A
NC	6		9	Q_A
(0V) GND	7		8	Q_D

Pin	Description
NC	No connection
R_1, R_2	Reset
Q_A, Q_B, Q_C, Q_D	Counter outputs
CLK-A, CLK-B	A and B clock pulse
V_{cc}	Supply voltage (+5V)
GND	Ground (0V)

Fig. 3.76: *Pin configuration of 4-bit binary counter IC 74293.*

3.8.2 Down Counter

In down counter, the counter output decrement for every clock pulse and in an n-bit down counter the count sequence will be from 2^{n-1} to 0. The complement of up counter output will be the down count. A 4-bit down counter using T flip-flop is shown in Fig. 3.77. The up counter and complement of up counter output signals are shown in Fig. 3.78. The up and down counters outputs are tabultated in Table 3.46.

The clock of down counter is similar to up counter, but output is taken from complement of output Q′ of each flip-flop.

The down counter using JK and D flip-flops are same as up counter except that output is taken from Q′ of every flip-flop.

Table 3.46: Up and Down Counter Outputs

Count or Clock Pulse	Up Counter Outputs				Down Counter Outputs			
	A_3	A_2	A_1	A_0	A'_3 D_3	A'_2 D_2	A'_1 D_1	A'_0 D_0
-	0	0	0	0	1	1	1	1
1	0	0	0	1	1	1	1	0
2	0	0	1	0	1	1	0	1
3	0	0	1	1	1	1	0	0
4	0	1	0	0	1	0	1	1
5	0	1	0	1	1	0	1	0
6	0	1	1	0	1	0	0	1
7	0	1	1	1	1	0	0	0
8	1	0	0	0	0	1	1	1
9	1	0	0	1	0	1	1	0
10 (A)	1	0	1	0	0	1	0	1
11 (B)	1	0	1	1	0	1	0	0
12 (C)	1	1	0	0	0	0	1	1
13 (D)	1	1	0	1	0	0	1	0
14 (E)	1	1	1	0	0	0	0	1
15 (F)	1	1	1	1	0	0	0	0
16	0	0	0	0	1	1	1	1

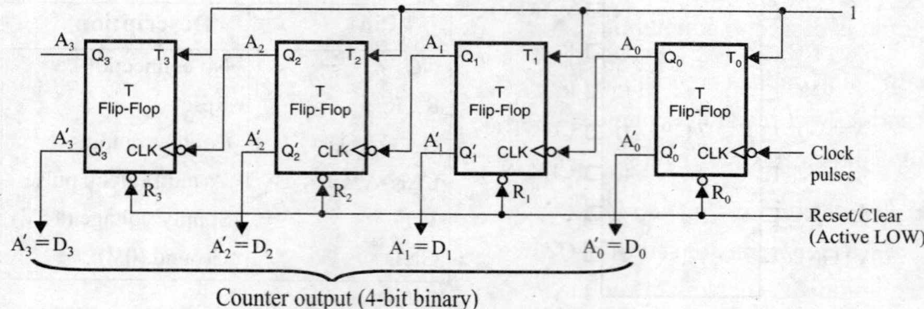

Fig. 3.77: 4-bit down counter using T flip-flop.

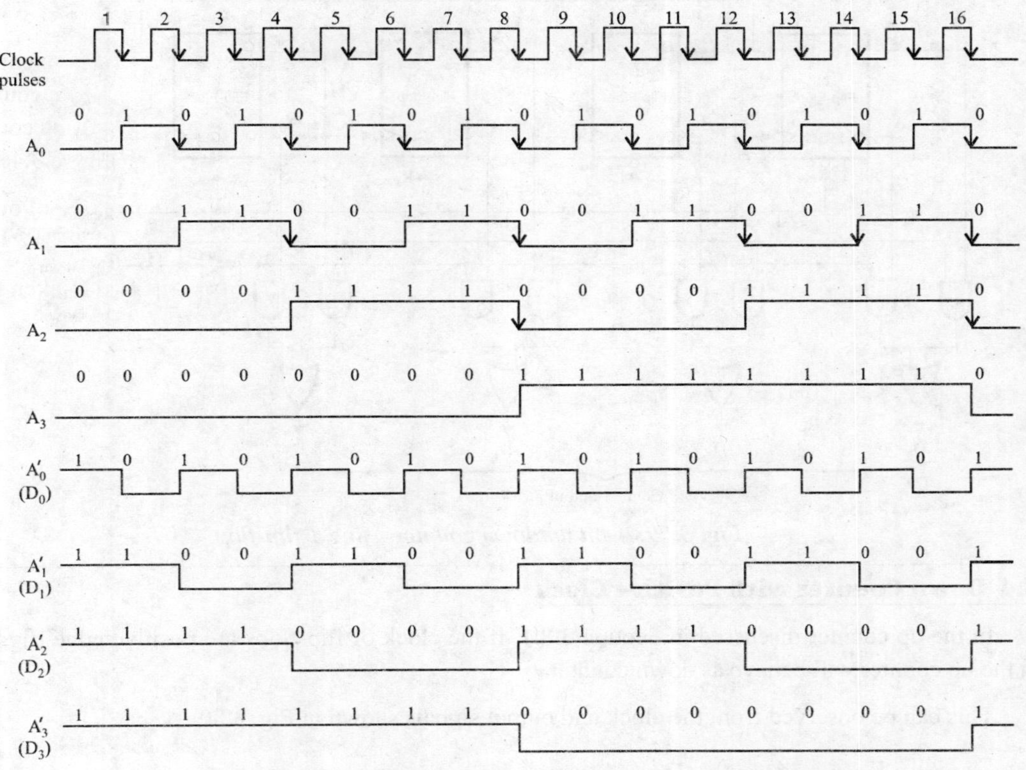

Fig. 3.78: Output signals of up and down counter.

3.8.3 Up/Down Counter using T Flip-Flop

A binary counter can be used to perform both up/down counting by adding a selection logic in the output as shown in Fig. 3.79.

When up/down = 1, only A_0, A_1, A_2 and A_3 are allowed to reach output. When up/down = 0 only A_0', A_1', A_2' and A_3' will reach output. The output and the complement of output of flip-flops are separately ANDed with up/down control signal and then output of both AND gates are logically ORed. In this arrangement, at any one time only one of the input to OR gate is valid and the other input will be 0, so that the active input is passed to counter output.

The flip-flop outputs are logically ANDed with up/down control signal so that the flip-flop outputs A_0, A_1, A_2 and A_3 will reach the counter outputs D_0, D_1, D_2 and D_3 when up/down = 1. Now the counter behave as up counter.

The complement of flip-flop outputs are logically ANDed with complement of up/down control signal so that the complement of flip-flop outputs A_0', A_1', A_2' and A_3' will reach counter outputs D_0, D_1, D_2 and D_3 when up/down = 0. Now the counter behave as down counter.

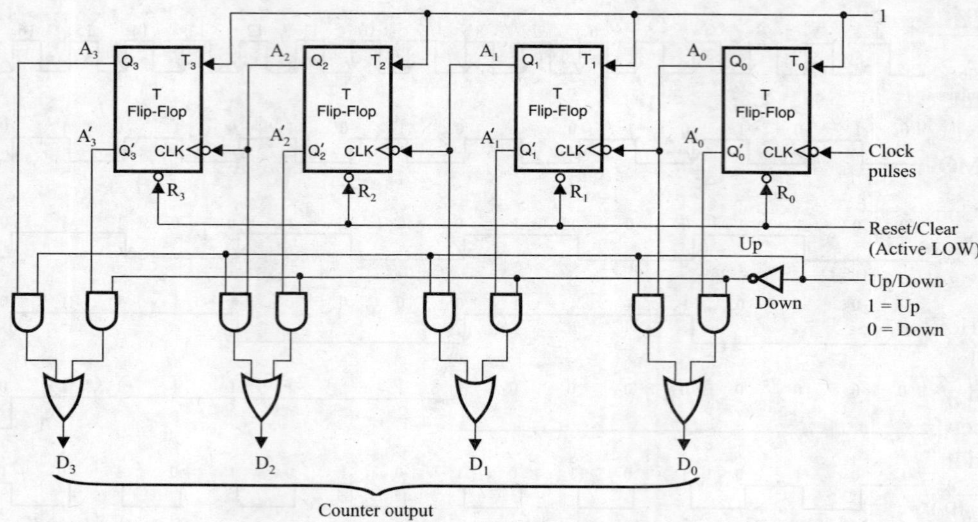

Fig. 3.79: *4-bit up/down counter using T flip-flop.*

3.8.4 Down Counter with Positive Clock

In the up counter discussed in Section 3.8.1, if the clock of flip-flops are positive edge triggered then the up counter will behave as down counter.

This can be observed from the clock and output signals shown in Fig. 3.80.

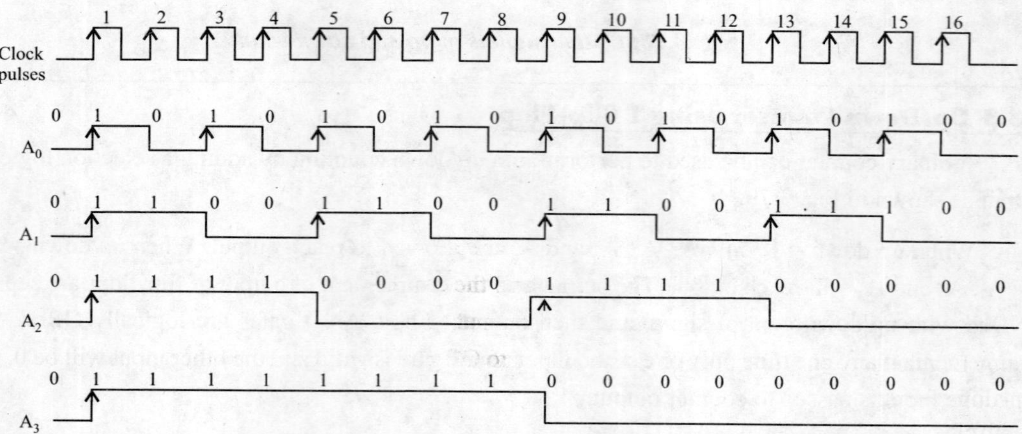

Fig. 3.80: *Output signals of down counter with positive edge triggered clock.*

Also it can be observed that the complements of A_0, A_1, A_2 and A_3 will be up count.

Therefore the outputs A_0', A_1', A_2' and A_3' will be output of up counter, when the clock in positive edge triggered.

This concept is applicable for binary ripple counter using, T or JK or D flip-flops.

3.8.5 BCD Counter

A BCD counter has to count only from 0 to 9, and when the count reaches 10_{10}, the output has to be reset to all zero, so that next count sequence will start. The BCD counter is also called MOD-10 counter.

Consider the 4-bit output of ripple counter for 10 clock pulses shown in Table 3.47. From Table 3.47, it can be observed that A_1 and A_3 are 1 on tenth clock pulse.

Here, the logical NAND gate of A_1 and A_3, will be "0" on tenth clock pulse. The output of NAND gate can be logically ANDed with Reset/clear signal of flip-flops, so that the counter will reset on every tenth clock pulse. The output of BCD counter is listed in Table 3.47.

> ***Note:** For count operation reset is tied to 1.*

Table 3.47: Output of BCD Counter

Clock Pulse	Counter Outputs			
	A_3	A_2	A_1	A_0
	d_8	d_4	d_2	d_1
-	0	0	0	0
1	0	0	0	1
2	0	0	1	0
3	0	0	1	1
4	0	1	0	0
5	0	1	0	1
6	0	1	1	0
7	0	1	1	1
8	1	0	0	0
9	1	0	0	1
10	1	0	1	0
	↓ Reset			
	0	0	0	0

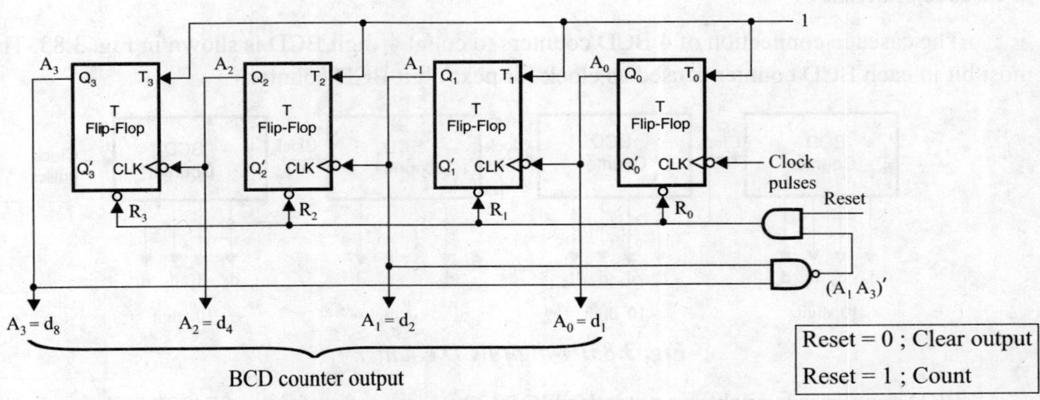

Fig. 3.81: BCD counter using T flip-flops.

A BCD counter using T flip-flops is shown in Fig. 3.81. Since a BCD counter needs 4-bit output, 4 flip-flops are used to construct a BCD counter. Here the arrangement is same as binary up counter, but on tenth clock pulse the logical NAND of A_1 and A_3 will reset all the output bit to zero. So that next count sequence starts from next clock pulse.

The output signals of BCD counter are shown in Fig. 3.82. The BCD counter using JK and D flip-flop are same as that of binary counter with addition of additional NAND gate to reset the output on tenth clock pulse.

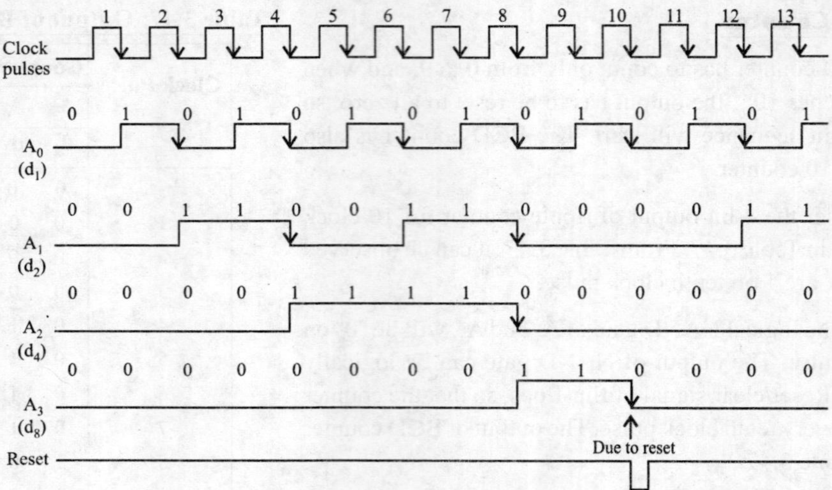

Fig. 3.82: Output signals of BCD counter.

The BCD counter shown in Fig. 3.81 can count one digit of BCD in the range 0 to 9. In order to count m digit BCD we need m BCD counters shown in Fig. 3.81. The BCD counters should be connected in cascade or series.

The cascade connection of 4 BCD counters to count 4-digit BCD is shown in Fig. 3.83. The upper most bit in each BCD counter is used as clock for next digit BCD counter.

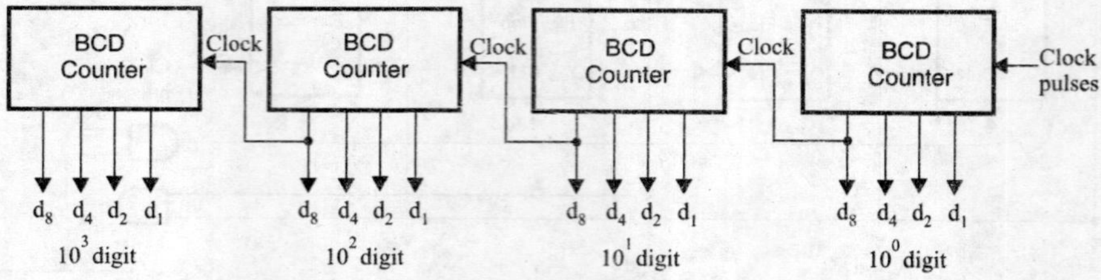

Fig. 3.83: 4-digit BCD counter.

BCD counter is available as a standard IC 74290. Pin configuration of 74290 is shown in Fig. 3.84.

74290

- $R9_1$ 1 — 14 V_{cc}(+5V)
- NC 2 — 13 R_1
- $R9_2$ 3 — 12 R_2
- Q_C 4 — 11 CLK-A
- Q_B 5 — 10 CLK-B
- NC 6 — 9 Q_A
- (0V) GND 7 — 8 Q_B

Pin	Description
NC	No connection
R_1, R_2	Reset
Q_A, Q_B, Q_C, Q_D	Counter outputs
CLK-A, CLK-B	A and B clock pulse
$R9_1$, $R9_2$	Enable to set output to 9
V_{cc}	Supply voltage (+5V)
GND	Ground (0V)

Fig. 3.84: Pin configuration of BCD counters IC 74290.

Here, Q_A, Q_B, Q_C and Q_D are output of internal flip-flops A, B, C and D respectively. Separate clock signals can be applied for flip-flops A and B. The flip-flops B, C and D are internally connected to function as 3-bit BCD ripple counter.

The flip-flop A can work independently as divide-by-2 or MOD-2 counter. The flip-flop output Q_D should be connected to CLK-A externally and counter clock should be connected to CLK-B for fout digit BCD count. When $R9_1$ and $R9_2$ are both tied **high**, the output will be set to 9 which will can be used in 9's complement arithmetic. Similar to that of 74293, any one reset input should be tied **low** for count operation.

3.8.6 Modulo Counter

An n-bit binary counter will have 2^n states. MOD-N counters can be designed using n-bit binary counter by using only N states of n-bit counter provided $N < 2^n$.

Consider 3-bit binary counter which has $2^3 = 8$ states. Here, counter output and states are same. The output of 3-bit binary counter is shown in Table 3.48. The 3-bit binary counter is shown in Fig. 3.85. The 3-bit counter can be modified to work as MOD-2, MOD-3, MOD-4, MOD-5 and MOD-6 counters as shown in Figs. 3.86 to 3.95.

Table 3.48: Output of 3-bit Binary Counter

Clock Pulse	Counter Outputs or States		
	A_2	A_1	A_0
-	0	0	0
1	0	0	1
2	0	1	0
3	0	1	1
4	1	0	0
5	1	0	1
6	1	1	0
7	1	1	1

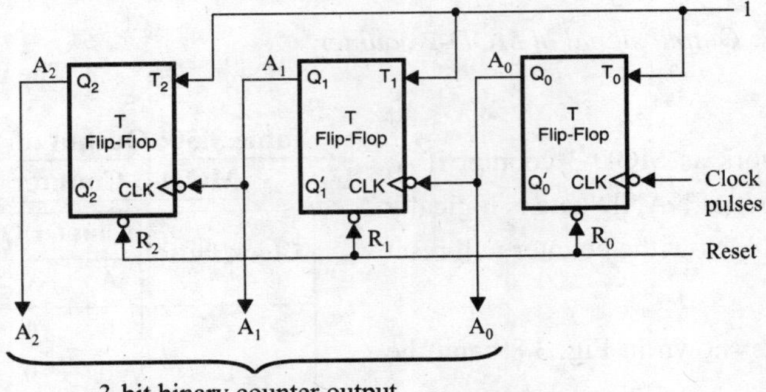

***Fig. 3.85:** 3-bit binary counter.*

MOD-2 Counter

The 3-bit binary counter will work as MOD-2 counter if A_1 is inverted and logically ANDed with reset as shown in Fig. 3.86. Now the counter will reset in every second clock pulse.

The clock and output signal are shown in Fig. 3.87 and the counter outputs are listed in Table 3.49.

Table 3.49: Output of MOD-2 Counter

Clock Pulse	Counter Outputs		
	A_2	A_1	A_0
-	0	0	0
1	0	0	1
2	0	0	0
3	0	0	1
4	0	0	0
5	0	0	1
6	0	0	0

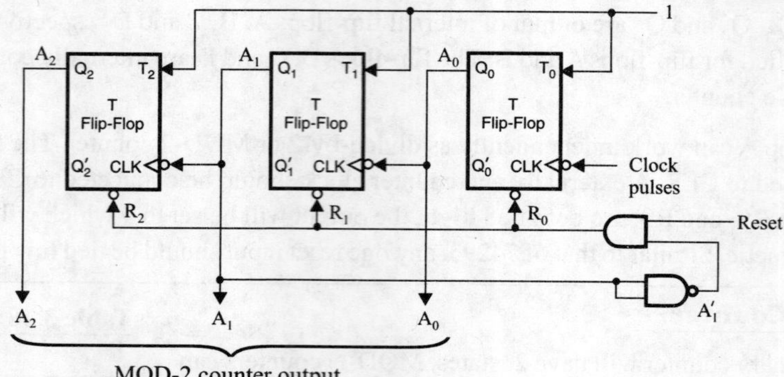

***Fig. 3.86:** MOD-2 counter.*

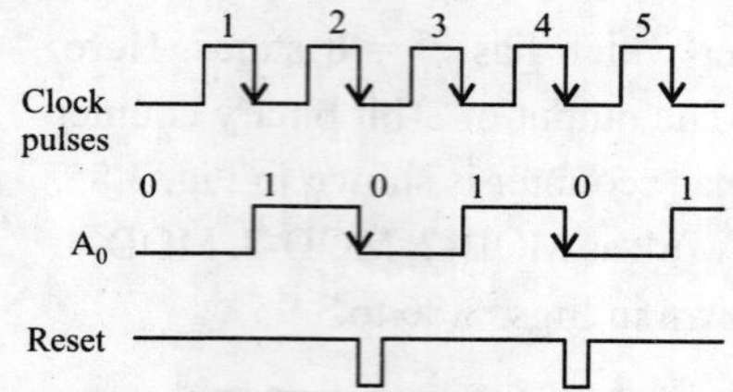

***Fig. 3.87:** Output signal of MOD-2 counter.*

MOD-3 Counter

The 3-bit binary counter will work as MOD-3 counter if A_1 and A_0 are logically NANDed and output of NAND gate is logically ANDed with reset as shown in Fig. 3.88. Now the counter will reset in every third clock pulse.

The clock and output signals are shown in Fig. 3.89 and the counter outputs are listed in Table 3.50.

Table 3.50: Output of MOD-3 Counter

Clock Pulse	Counter Outputs		
	A_2	A_1	A_0
-	0	0	0
1	0	0	1
2	0	1	0
3	0	0	0
4	0	0	1
5	0	1	0
6	0	0	0
7	0	0	1
8	0	1	0
9	0	0	0

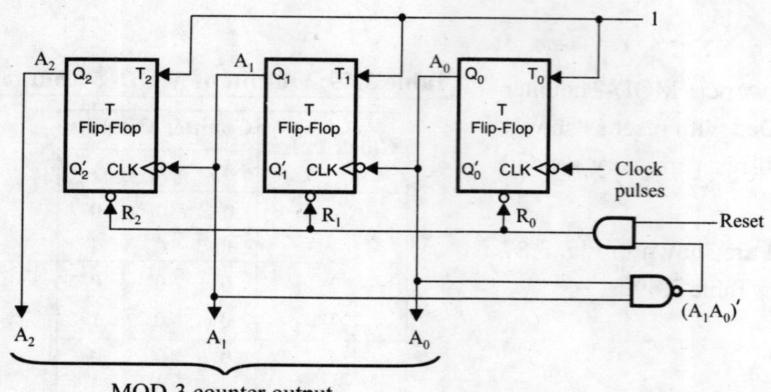

***Fig. 3.88:** MOD-3 counter.*

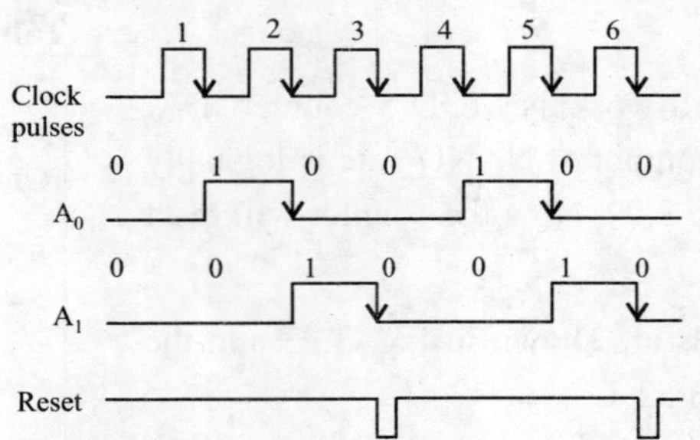

Fig. 3.89: *Output signals of MOD-3 counter.*

MOD-4 Counter

The 3-bit binary counter will work as MOD-4 counter if A_2 is inverted and logically ANDed with reset as shown in Fig. 3.90. Now the counter will reset in every fourth clock pulse.

The clock and output signals are shown in Fig. 3.91 and the counter outputs are listed in Table 3.51.

Table 3.51: Output of MOD-4 Counter

Clock Pulse	Counter Outputs		
	A_2	A_1	A_0
-	0	0	0
1	0	0	1
2	0	1	0
3	0	1	1
4	0	0	0
5	0	0	1
6	0	1	0
7	0	1	1
8	0	0	0

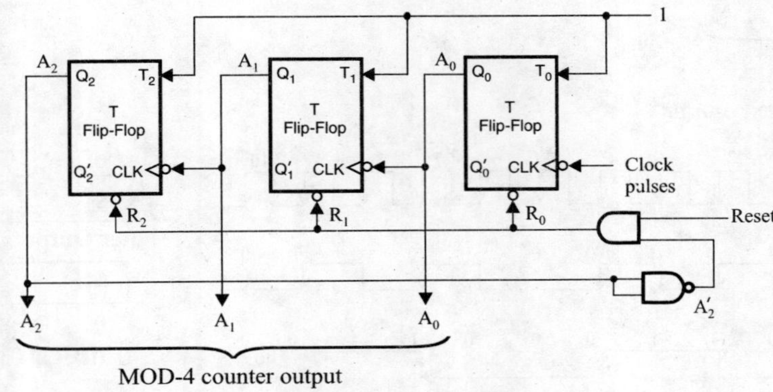

Fig. 3.90: *MOD-4 counter.*

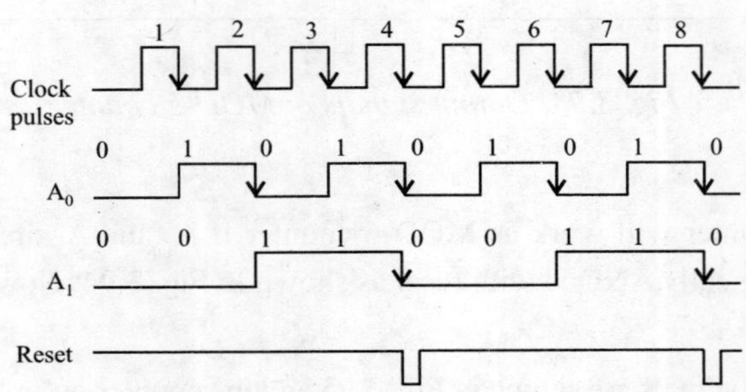

Fig. 3.91: *Outputs signal of MOD-4 counter.*

MOD-5 Counter

The 3-bit binary counter will work as MOD-5 counter if A_0 and A_2 are logically NANDed and output of NAND gate is logically ANDed with reset as shown in Fig. 3.92. Now the counter will reset in every fifth clock pulse.

The clock and output signals are shown in Fig. 3.93 and the counter outputs are listed in Table 3.52.

Table 3.52: Output of MOD-5 Counter

Clock Pulse	Counter Outputs		
	A_2	A_1	A_0
-	0	0	0
1	0	0	1
2	0	1	0
3	0	1	1
4	1	0	0
5	0	0	0
6	0	0	1
7	0	1	0
8	0	1	1
9	1	0	0
10	0	0	0

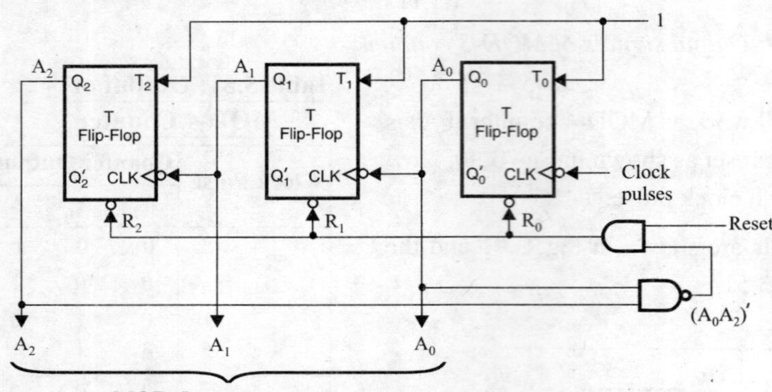

Fig. 3.92: MOD-5 counter.

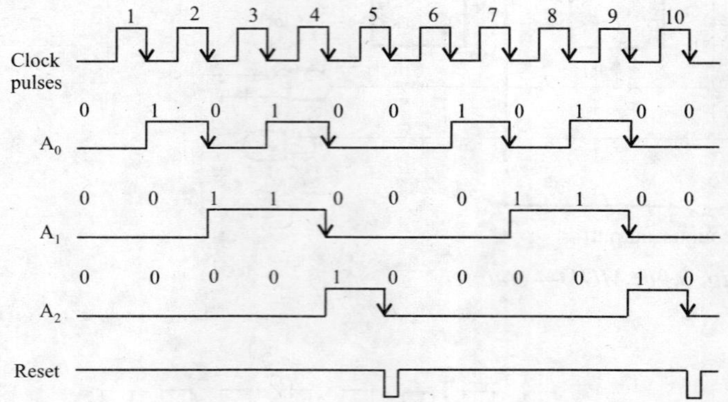

Fig. 3.93: Output signals of MOD-5 counter.

MOD-6 Counter

(AU, Apr/May'19, 7 Marks)

The 3-bit binary counter will work as MOD-6 counter if A_1 and A_2 are logically NANDed and output of NAND gate is logically ANDed with reset as shown in Fig. 3.94. Now the counter will reset in every sixth clock pulse.

The clock and output signals are shown in Fig. 3.95 and the counter outputs are listed in Table 3.53.

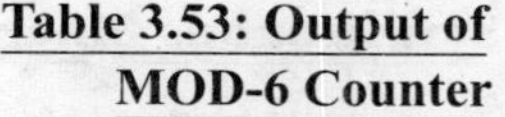

Table 3.53: Output of MOD-6 Counter

Clock Pulse	Counter Outputs		
	A_2	A_1	A_0
-	0	0	0
1	0	0	1
2	0	1	0
3	0	1	1
4	1	0	0
5	1	0	1
6	0	0	0
7	0	0	1
8	0	1	0
9	0	1	1
10	1	0	0
11	1	0	1
12	0	0	0

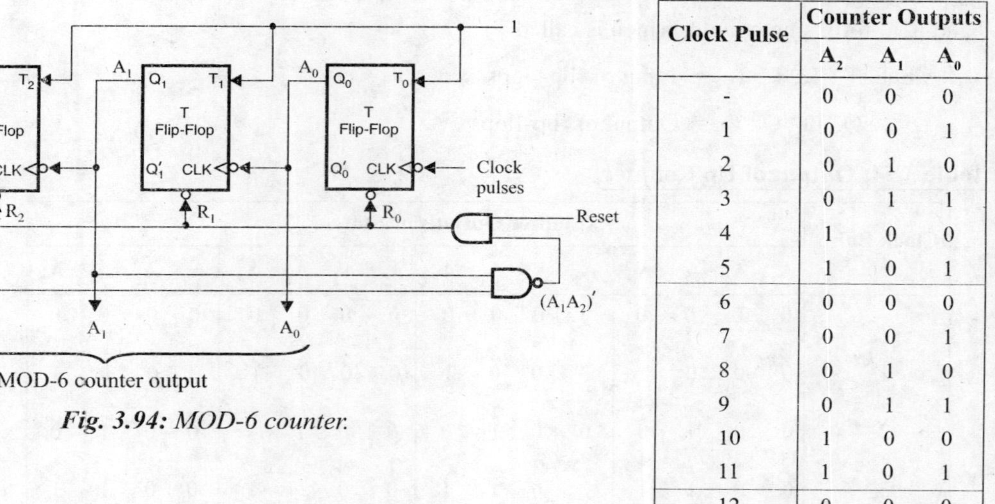

Fig. 3.94: MOD-6 counter.

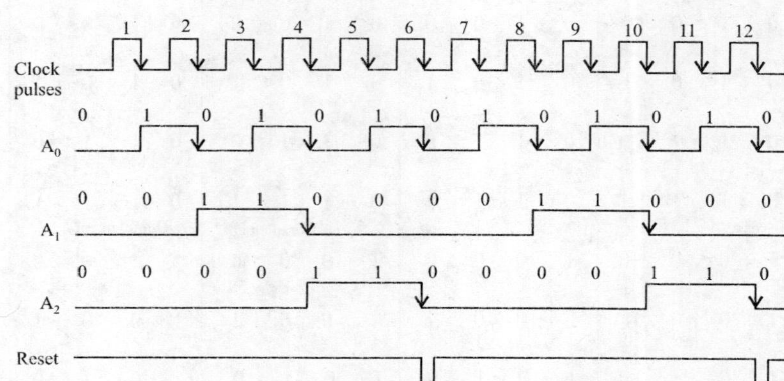

Fig. 3.95: Output signals of MOD-6 counter.

3.9 Synchronous Counter

The synchronous counters are designed using complementing flip-flops similar to ripple counters, but in synchronous counters, the clock is applied simultaneously to all the flip-flops and the inputs to the flip-flops are applied through some combinational logic circuits.

The combinational logic circuits will consider the output of one or more flip-flops to decide the input to flip-flops for next state outputs. Since the counters are also sequential circuits, the design of counters are same as that of design of sequential circuits. In design of counters, the counter outputs are considered as states.

3.9.1 Synchronous Binary Up Counter

Let us consider 4-bit binary counter using four T flip-flops. (JK flip-flops can also be used by joining J and K to form single input, which is called T).

Let, T_0, T_1, T_2, T_3 = Input of flip-flops

Q_0, Q_1, Q_2, Q_3 = Output of flip-flops

Table 3.54: Output of Up Counter

Clock Pulse	Counter Outputs or States															
	A_3	A_2	A_1	A_0	A_3	A_2	A_1	A_0	A_3	A_2	A_1	A_0	A_3	A_2	A_1	A_0
-	0	0	0	0	0	0	0	0	0	0	0	0	0	0	0	0
				↓												
1	0	0	0	1	0	0	0	1	0	0	0	1	0	0	0	1
				↓			↓									
2	0	0	1	0	0	0	1	0	0	0	1	0	0	0	1	0
				↓												
3	0	0	1	1	0	0	1	1	0	0	1	1	0	0	1	1
				↓			↓			↓ Complement						
4	0	1	0	0	0	1	0	0	0	1	0	0	0	1	0	0
				↓												
5	0	1	0	1	0	1	0	1	0	1	0	1	0	1	0	1
				↓			↓									
6	0	1	1	0	0	1	1	0	0	1	1	0	0	1	1	0
				↓												
7	0	1	1	1	0	1	1	1	0	1	1	1	0	1	1	1
				↓			↓			↓ Complement			↓ Complement			
8	1	0	0	0	1	0	0	0	1	0	0	0	1	0	0	0
				↓												
9	1	0	0	1	1	0	0	1	1	0	0	1	1	0	0	1
				↓			↓									
10	1	0	1	0	1	0	1	0	1	0	1	0	1	0	1	0
				↓												
11	1	0	1	1	1	0	1	1	1	0	1	1	1	0	1	1
				↓			↓			↓ Complement						
12	1	1	0	0	1	1	0	0	1	1	0	0	1	1	0	0
				↓												
13	1	1	0	1	1	1	0	1	1	1	0	1	1	1	0	1
				↓			↓									
14	1	1	1	0	1	1	1	0	1	1	1	0	1	1	1	0
				↓												
15	1	1	1	1	1	1	1	1	1	1	1	1	1	1	1	1
				↓			↓			↓ Complement			↓ Complement			
16	0	0	0	0	0	0	0	0	0	0	0	0	0	0	0	0

Note: *1. The down arrow (↓) indicates complement.*

2. Here the counter output is listed 4 times to show the complement operation of A_0, A_1, A_2 and A_3 separately.

Here, A_3, A_2, A_1 and A_0 be the output of binary counter which are same as flip-flop outputs. The output of binary counter for 16 clock pulses are listed in Table 3.54. Here, counter outputs and states are same. The count sequence of 4-bit up counter is from 0 to $2^4 - 1$ (i.e, 0 to 15_{10}).

In Table 3.54 the following observations are made,

The bit A_0 is complemented for every clock pulse.

The bit A_1 is complemented when A_0 is 1 in previous state.

The bit A_2 is complemented when both A_0 and A_1 are 1 in previous state.

The bit A_3 is complemented when A_0, A_1 and A_2 are 1 in previous state.

A 4-bit synchronous binary up counter using T flip-flop is shown in Fig. 3.96.

Since the counter output is 4-bit, 4 flip-flops are selected. A control signal called count enable is provided to start and stop the count operation. When count enable is 0, the flip-flop inputs are 0 and so no change in flip-flop outputs. The counter will start counting only when count enable is tied to 1.

The clock and output signals of binary counter are shown in Fig. 3.97. The synchronous binary up counter is same as binary ripple up counter except the following two differences.

1. The clock is directly applied to all flip-flops.
2. The inputs to flip-flops are applied through a combinational circuit.

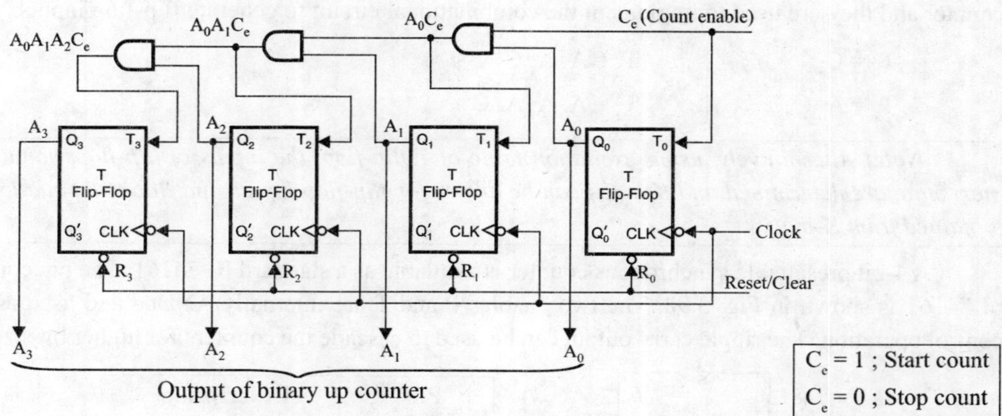

Fig. 3.96: *4-bit synchronous binary up counter.*

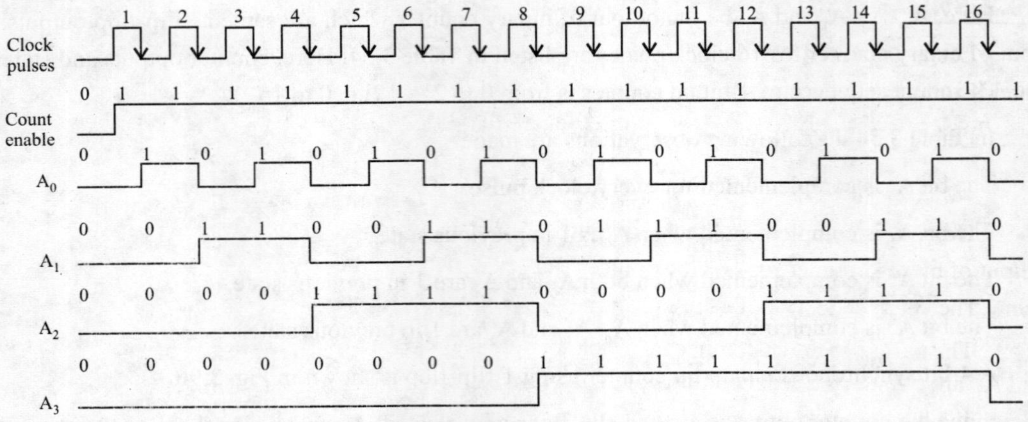

Fig. 3.97: Signals of synchronous binary counter.

The logic circuit of counter is designed as follows,

Count enable signal is applied directly to first flip-flop.

A_0 and count enable are logically ANDed and applied as input to second flip-flop.

A_0, A_1 and count enable are logically ANDed and applied as input to third flip-flop.

A_0, A_1, A_2 and count enable are logically ANDed and applied as input to fourth flip-flop.

Clock signal is applied simultaneously to all the flip-flops.

Based on the above discussion, we get the following Boolean equations for flip-flop inputs of up counter and they are used to implement the combinational circuit to generate flip-flop inputs.

$$T_0 = C_e \qquad T_2 = A_0 A_1 C_e$$

$$T_1 = A_0 C_e \qquad T_3 = A_0 A_1 A_2 C_e$$

> ***Note:** Alternatively, using excitation table of T flip-flop, the inputs of flip-flops for the desired next state are determined and K-maps can be drawn for flip-flop inputs and Boolean equations can be obtained from K-map.*

A 4-bit presettable synchronous counter is available as a standard IC 74161. The pin configuration of 74161 is shown in Fig. 3.98. The two enables P and T are internally ANDed and used as start/stop control operation. The ripple carry output can be used to cascade the counters for higher bit size counters.

74161

Signal	Pin	Pin	Signal
$\overline{R}$	1	16	V_{cc}(+5V)
$\overline{CLK}$	2	15	RCO
d_0	3	14	Q_0
d_1	4	13	Q_1
d_2	5	12	Q_2
d_3	6	11	Q_3
P	7	10	T
(0V) GND	8	9	$\overline{Load}$

Pin	Description
Q_0, Q_1, Q_2, Q_3	Counter outputs
d_0, d_1, d_2, d_3	Preload count inputs
$\overline{R}$	Reset
$\overline{CLK}$	Clock
P, T	Enable
RCO	Ripple carry output
V_{cc}	Supply voltage (+5V)
GND	Ground (0V)

Fig. 3.98: Pin configuration of synchronous 4-bit counters IC 74161.

3.9.2 Synchronous Binary Down Counter

(AU, Apr/May'19, 7 Marks)

Let us consider 4-bit binary counter using four T flip-flops. (JK flip-flops can also be used by joining J and K to form single input, which is called T).

Let, T_0, T_1, T_2, T_3 = Input of flip-flops

Q_0, Q_1, Q_2, Q_3 = Output of flip-flops

Here, A_3, A_2, A_1 and A_0 be the output of binary counter which are same as flip-flop outputs. The output of binary counter for 16 clock pulses are listed in Table 3.55. Here, counter outputs and states are same. The count sequence of 4-bit up counter is from $2^4 - 1$ to 0 (i.e, 15_{10} to 0).

The output of down counter for 16 clock pulses are listed in Table 3.55, and in this table A_0, A_1, A_2 and A_3 are output of down counter and A_0', A_1', A_2' and A_3' are complement of output which are used to design the combinational circuit to generate flip-flop inputs.

Table 3.55: Output of Down Counter

Clock Pulse	Down Counter Outputs				Complement of Outputs			
	A_3	A_2	A_1	A_0	A_3'	A_2'	A_1'	A_0'
-	0	0	0	0	1	1	1	1
1	1	1	1	1	0	0	0	0
2	1	1	1	0	0	0	0	1
3	1	1	0	1	0	0	1	0
4	1	1	0	0	0	0	1	1
5	1	0	1	1	0	1	0	0
6	1	0	1	0	0	1	0	1
7	1	0	0	1	0	1	1	0
8	1	0	0	0	0	1	1	1
9	0	1	1	1	1	0	0	0
10	0	1	1	0	1	0	0	1
11	0	1	0	1	1	0	1	0
12	0	1	0	0	1	0	1	1
13	0	0	1	1	1	1	0	0
14	0	0	1	0	1	1	0	1
15	0	0	0	1	1	1	1	0
16	0	0	0	0	1	1	1	1

In Table 3.55, the following observations are made.

A_0' complemented for every clock pulse.

A_1' is complemented when A_0' is 1 in previous state.

A_2' is complemented when A_0' and A_1' are 1 in previous state.

A_3' is complemented when A_0', A_1' and A_2' are 1 in previous state.

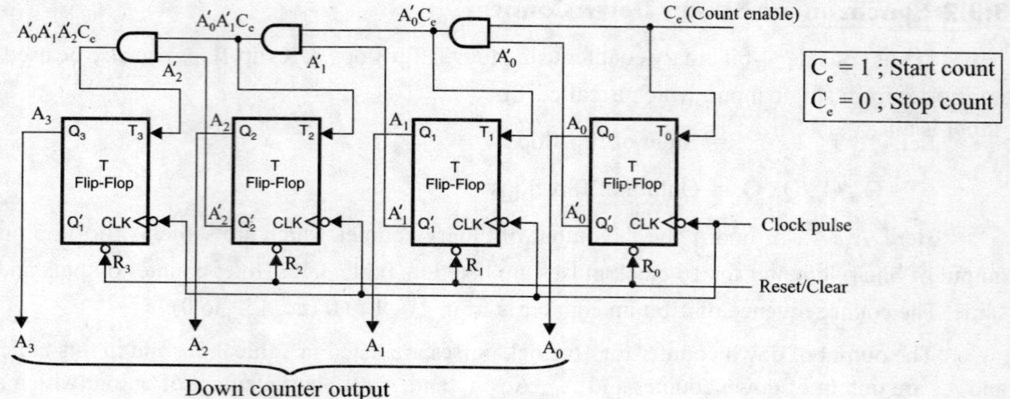

Fig. 3.99: 4-bit synchronous binary down counter.

A 4-bit synchronous binary down counter is shown in Fig. 3.99. Since the counter output is 4-bit, 4 flip-flops are selected. A count enable is provided to similar to up counter. The output signals of synchronous down counter will be same as D_0, D_1, D_2 and D_3 shown in Fig. 3.78.

The synchronous binary down counter is same as binary ripple down counter except the following three differences.

1. Clock is directly applied to all flip-flops.
2. The inputs to flip-flops are applied through a combinational circuit.
3. The counter output is obtained from flip-flop outputs similar to up counter.

The logic circuit of down counter is designed as follows,

Count enable is applied directly to first flip-flop.

A_0' and count enable are logically ANDed and applied as input to second flip-flop.

A_0', A_1' and count enable are logically ANDed and applied as input to third flip-flop.

A_0', A_1', A_2' and count enable are logically ANDed and applied as input to fourth flip-flop.

Based on the above discussion, we get the following Boolean equations for flip-flop inputs of down counter and they are used to implement the combinational circuit to generate flip-flop inputs in Fig. 3.99.

$$T_0 = C_e$$

$$T_1 = A_0' \, C_e$$

$$T_2 = A_0' \, A_1' \, C_e$$

$$T_3 = A_0' \, A_1' \, A_2' \, C_e$$

Note: *Alternatively, using excitation table of T flip-flop, the inputs of flip-flops for the desired next state are determined and K-maps can be drawn for flip-flop inputs and Boolean equations can be obtained from K-map.*

3.9.3 Synchronous Up/Down Counter

The up and down count operation can achieved in a single counter by logical OR operation of input to flip-flops in up and down counter and with two control inputs as shown in Fig. 3.100, so that only one input is allowed to flip-flops at any one time.

Let, C_u = Up count enable

C_d = Down count enable

Now, the control logic for up/down count is,

$C_u = 1$ and $C_d = 0$ for up count

$C_u = 0$ and $C_d = 1$ for down count

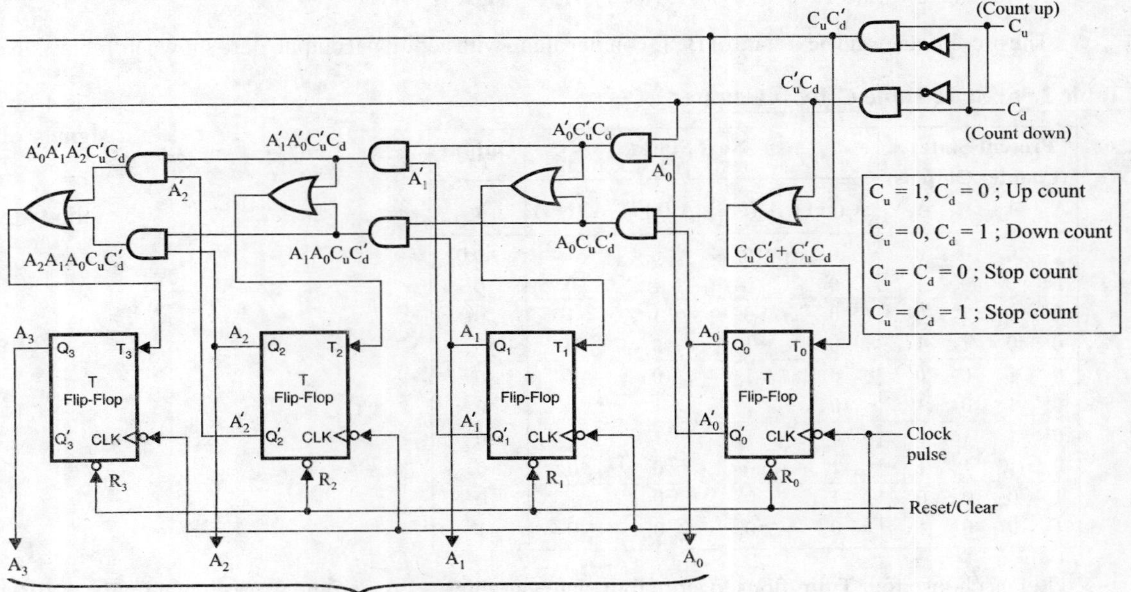

Fig. 3.100: Synchronous 4-bit up/down counter.

The Boolean equations of up/down counter are given by sum of Boolean equations of up and down counters after replacing C_e of up counter by $C_u C_d'$ and C_e of down counter by $C_u' C_d$.

The following are Boolean equations for input of flip-flops of up/down counter and they are used to implement the combinational circuit to generate flip-flop inputs.

$$T_0 = C_u C_d' + C_u' C_d$$

$$T_1 = A_0 C_u C_d' + A_0' C_u' C_d$$

$$T_2 = A_0 A_1 C_u C_d' + A_0' A_1' C_u' C_d$$

$$T_3 = A_0 A_1 A_2 C_u C_d' + A_0' A_1' A_2' C_u' C_d$$

3.9.4 Synchronous BCD Counter

BCD counter will count only from 0 to 9 for 9 clock pulses and upon reaching tenth pulse output is reset to 0, so that next count sequence will start. Also, an additional output 1 is needed to use as carry to next stage.

The design of synchronous BCD counter using T flip-flop is presented in this section.

The flip-flops inputs to achieve the desired next state are determined from the excitation table of flip-flop. The combinational circuit to generate flip-flops are designed using K-maps.

Let, A_8, A_4, A_2 and A_1 be 4 state variables of BCD counter.

$\therefore$ Present state variables : $A_8(t), A_4(t), A_2(t), A_1(t) \Rightarrow A_8, A_4, A_2, A_1$ (For simplicity)

Next state variables : $A_8(t+1), A_4(t+1), A_2(t+1), A_1(t+1)$

The present state and next state of BCD counter along with additional output Y are shown in Table 3.56.

Table 3.56: State Table of BCD Counter

Present States (Counter Outputs)				Next States				Output
A_8	A_4	A_2	A_1	$A_8(t+1)$	$A_4(t+1)$	$A_2(t+1)$	$A_1(t+1)$	Y
0	0	0	0	0	0	0	1	0
0	0	0	1	0	0	1	0	0
0	0	1	0	0	0	1	1	0
0	0	1	1	0	1	0	0	0
0	1	0	0	0	1	0	1	0
0	1	0	1	0	1	1	0	0
0	1	1	0	0	1	1	1	0
0	1	1	1	1	0	0	0	0
1	0	0	0	1	0	0	1	0
1	0	0	1	0	0	0	0	1

Let us choose four T flip-flops to store four state variables A_1, A_2, A_4 and A_8 as shown in Fig. 3.101. Here, A_1, A_2, A_4 and A_8 are output of flip-flops.

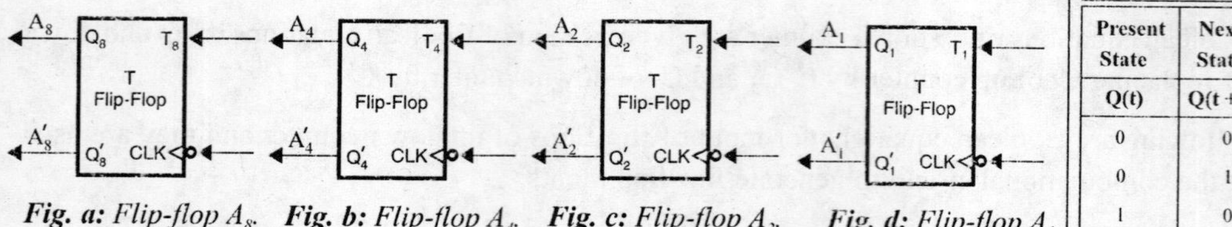

Fig. a: Flip-flop A_8. *Fig. b: Flip-flop A_4.* *Fig. c: Flip-flop A_2.* *Fig. d: Flip-flop A_1.*

Excitation Table of T Flip-Flop

Present State	Next State	Flip-Flop Input
Q(t)	Q(t + 1)	T
0	0	0
0	1	1
1	0	1
1	1	0

Fig. 3.101: T flip-flops to store state variables.

The flip-flop inputs can be obtained from excitation table as shown below:

State A_8

For state A_8, the Q(t), Q(t + 1) and T of excitation table are considered as A_8, $A_8(t + 1)$ and T_8.

$\therefore Q(t) \rightarrow A_8$; $Q(t+1) \rightarrow A_8(t+1)$

$T \rightarrow T_8$

The flip-flop input T_8 that drives the present state A_8 to next state is obtained from excitation table and listed in Table 3.57.

State A_4

For state A_4, the Q(t), Q(t + 1) and T of excitation table are considered as A_4, $A_4(t + 1)$ and T_4.

$\therefore Q(t) \rightarrow A_4$; $Q(t+1) \rightarrow A_4(t+1)$

$T \rightarrow T_4$

The flip-flop input T_4 that drives the present state A_4 to next state is obtained from excitation table and listed in Table 3.58.

State A_2

For state A_2, the Q(t), Q(t + 1) and T of excitation table are considered as A_2, $A_2(t + 1)$ and T_2.

$\therefore Q(t) \rightarrow A_2$; $Q(t+1) \rightarrow A_2(t+1)$

$T \rightarrow T_2$

The flip-flop input T_2 that drives the present state A_2 to next state is obtained from excitation table and listed in Table 3.59.

State A_1

For state A_1, the Q(t), Q(t + 1) and T of excitation table are considered as A_1, $A_1(t + 1)$ and T_1.

$\therefore Q(t) \rightarrow A_1$; $Q(t+1) \rightarrow A_1(t+1)$

$T \rightarrow T_1$

The flip-flop input T_1 that drives the present state A_1 to next state is obtained from excitation table and listed in Table 3.60.

Table 3.57:

A_8	$A_8(t+1)$	T_8
0	0	0
0	0	0
0	0	0
0	0	0
0	0	0
0	0	0
0	0	0
0	1	1
1	1	0
1	0	1

Table 3.58:

A_4	$A_4(t+1)$	T_4
0	0	0
0	0	0
0	0	0
0	1	1
1	1	0
1	1	0
1	1	0
1	0	1
0	0	0
0	0	0

Table 3.59:

A_2	$A_2(t+1)$	T_2
0	0	0
0	1	1
1	1	0
1	0	1
0	0	0
0	1	1
1	1	0
1	0	1
0	0	0
0	0	0

Table 3.60:

A_1	$A_1(t+1)$	T_1
0	1	1
1	0	1
0	1	1
1	0	1
0	1	1
1	0	1
0	1	1
1	0	1
0	1	1
1	0	1

The tables 3.56 to 3.60 are combined to form state table with flip-flop inputs as shown in Table 3.61.

Table 3.61: State Table with Flip-Flop Inputs

Present States (Counter Outputs)				Next States				Output	Flip-Flop Inputs				Minterm
A_8	A_4	A_2	A_1	$A_8(t+1)$	$A_4(t+1)$	$A_2(t+1)$	$A_1(t+1)$	Y	T_8	T_4	T_2	T_1	$A_8A_4A_2A_1$
0	0	0	0	0	0	0	1	0	0	0	0	1	m_0
0	0	0	1	0	0	1	0	0	0	0	1	1	m_1
0	0	1	0	0	0	1	1	0	0	0	0	1	m_2
0	0	1	1	0	1	0	0	0	0	1	1	1	m_3
0	1	0	0	0	1	0	1	0	0	0	0	1	m_4
0	1	0	1	0	1	1	0	0	0	0	1	1	m_5
0	1	1	0	0	1	1	1	0	0	0	0	1	m_6
0	1	1	1	1	0	0	0	0	1	1	1	1	m_7
1	0	0	0	1	0	0	1	0	0	0	0	1	m_8
1	0	0	1	0	0	0	0	1	1	0	0	1	m_9
1	0	1	0	×	×	×	×	×	×	×	×	×	m_{10}
1	0	1	1	×	×	×	×	×	×	×	×	×	m_{11}
1	1	0	0	×	×	×	×	×	×	×	×	×	m_{12}
1	1	0	1	×	×	×	×	×	×	×	×	×	m_{13}
1	1	1	0	×	×	×	×	×	×	×	×	×	m_{14}
1	1	1	1	×	×	×	×	×	×	×	×	×	m_{15}

***Note:** 1. The states 1010 to 1111 are unused states and so next states, output and flip-flop inputs are considered as don't-care for these states.*

2. Minterms are formed using state variables A_8, A_4, A_2 and A_1.

Here, T_8, T_4, T_2, T_1 and Y are functions of A_8, A_4, A_2 and A_1. Therefore, the Boolean equations for T_8, T_4, T_2, T_1 and Y are obtained from Table 3.61 as sum of minterms and don't-cares as shown below:

$$T_8 = \Sigma m(7, 9) + \Sigma d(10, 11, 12, 13, 14, 15)$$

$$T_4 = \Sigma m(3, 7) + \Sigma d(10, 11, 12, 13, 14, 15)$$

$$T_2 = \Sigma m(1, 3, 5, 7) + \Sigma d(10, 11, 12, 13, 14, 15)$$

$$T_1 = \Sigma m(0, 1, 2, 3, 4, 5, 6, 7, 8, 9) + \Sigma d(10, 11, 12, 13, 14, 15)$$

$$Y = m_9 + \Sigma d(10, 11, 12, 13, 14, 15)$$

Let us draw K-maps for T_8, T_4, T_2, T_1 and Y in order to determine the simplified Boolean equations for flip-flop inputs and output Y as shown in Figs. 3.102 to 3.106.

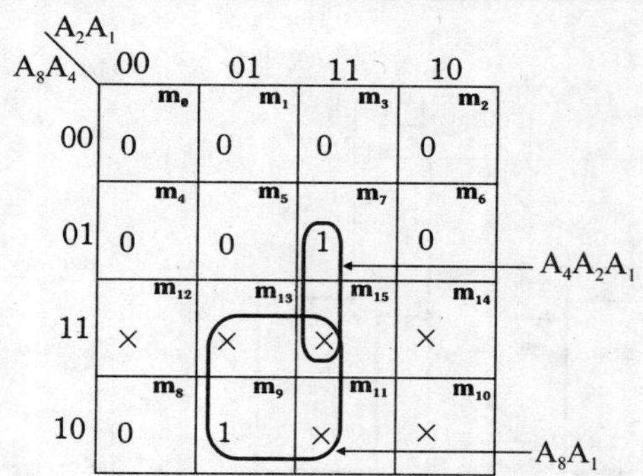

Fig. 3.102: K-map for T_8.

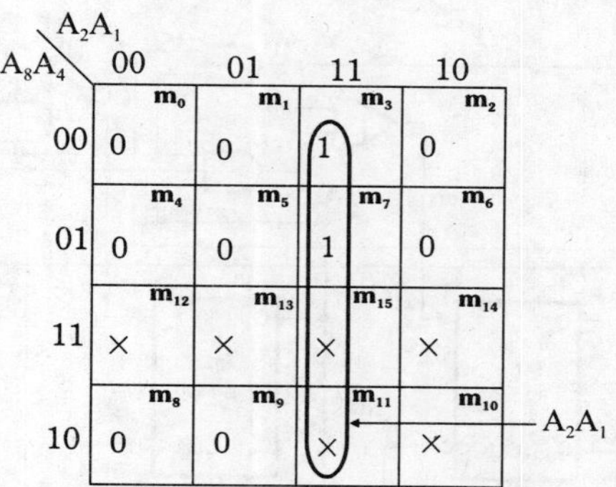

Fig. 3.103: K-map for T_4.

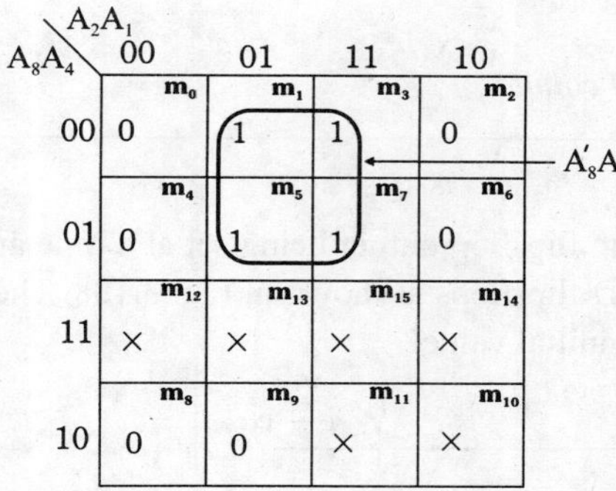

Fig. 3.104: K-map for T_2.

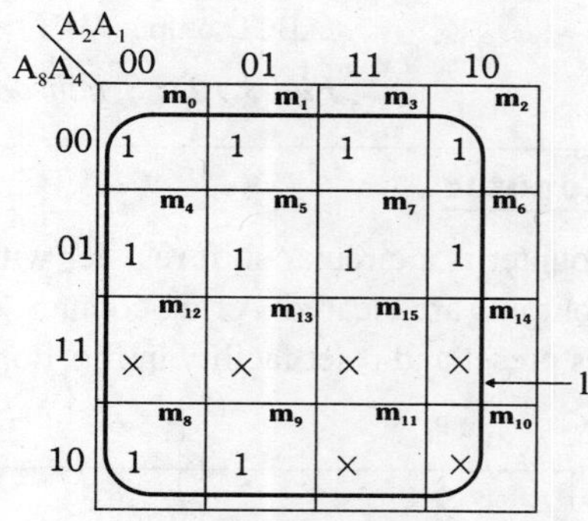

Fig. 3.105: K-map for T_1.

From the K-map we get the following simplified Boolean equations for T_8, T_4, T_2, T_1 and Y. Using these equations the logic circuit of synchronous BCD counter is drawn as shown in Fig. 3.107.

$$T_8 = A_8A_1 + A_4A_2A_1$$

$$T_4 = A_2A_1$$

$$T_2 = A'_8A_1$$

$$T_1 = 1$$

$$Y = A_8A_1$$

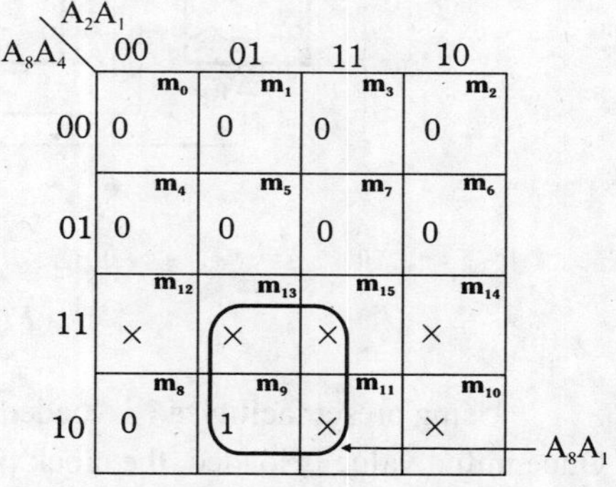

Fig. 3.106: K-map for Y.

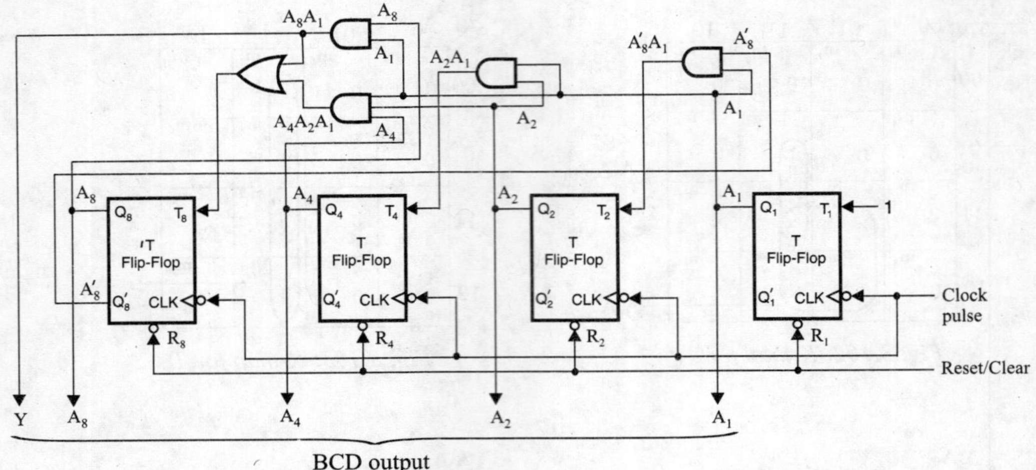

Fig. 3.107: Synchronous BCD counter.

3.9.5 Ring Counter

A ring counter is a circular shift register with only one flip-flop output being set at a time and all other flip-flop outputs are cleared. A ring counter with four D flip-flops is shown in Fig. 3.108. The ring counter requires preset and reset facility in flip-flops to load initial value.

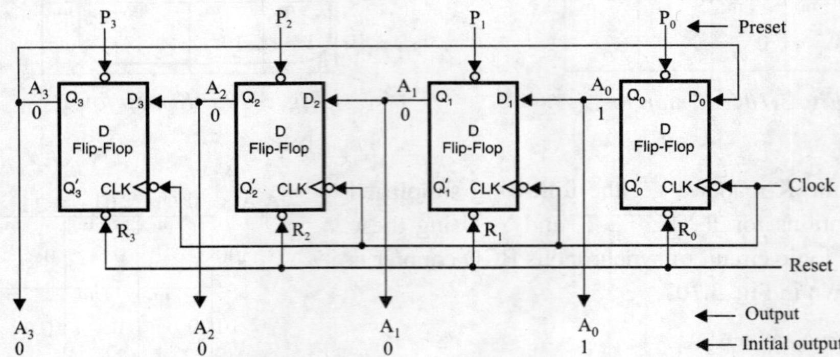

Fig. 3.108: Ring counter.

Using preset facility, a 1 is loaded in one flip-flop output and all other flip-flop outputs are loaded 0. Once initial value is loaded, the clock pulses can be applied to start the counting operation. Here, all the flip-flops are clocked simultaneosly. For every clock pulse, the flip-flop input is passed to output. Here, only one flip-flop input is 1 and all other flip-flop inputs are 0 and so 1 will shift by one position.

This produces a sequence of timing signals that are shifted by one clock time period. The time delay between start of any two consecutive output will be one clock time period. The ring counter output for 12 clock pulses are listed in Table 3.62. The output signals of ring counter are shown in Fig. 3.109.

Table 3.62: Ring Counter Output

Clock Pulse	Counter Outputs A_3	A_2	A_1	A_0
-	0	0	0	1
1	0	0	1	0
2	0	1	0	0
3	1	0	0	0
4	0	0	0	1
5	0	0	1	0
6	0	1	0	0
7	1	0	0	0
8	0	0	0	1
9	0	0	1	0
10	0	1	0	0
11	1	0	0	0
12	0	0	0	1

Fig. 3.109: Clock and output signals of ring counter.

3.9.6 Johnson Counter

Johnson counter is basically a ring counter with complement of last flip-flop connected to input of first flip-flop. This arrangement is called ***switch tail ring counter.***

Johnson counter using D flip-flops is shown in Fig. 3.110. Johnson counter operation starts with clear operation. When the flip-flops are cleared or reset all the outputs will be zero, and complement of outputs will be one. So that input to first flip-flop is 1 due to the connection of last flip-flop complemented output to input of first flip-flop.

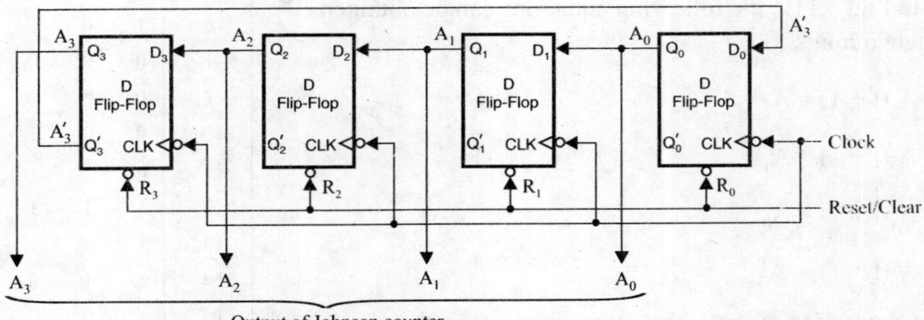

Fig. 3.110: Johnson counter using D flip-flop.

When clock pulses are applied, the counter generates 8 distinguishable outputs and this output repeats for every 8 clock pulses. The output signals of Johnson counter are shown in Fig. 3.111. The Johnson counter output for 16 clock pulses are listed in Table 3.63.

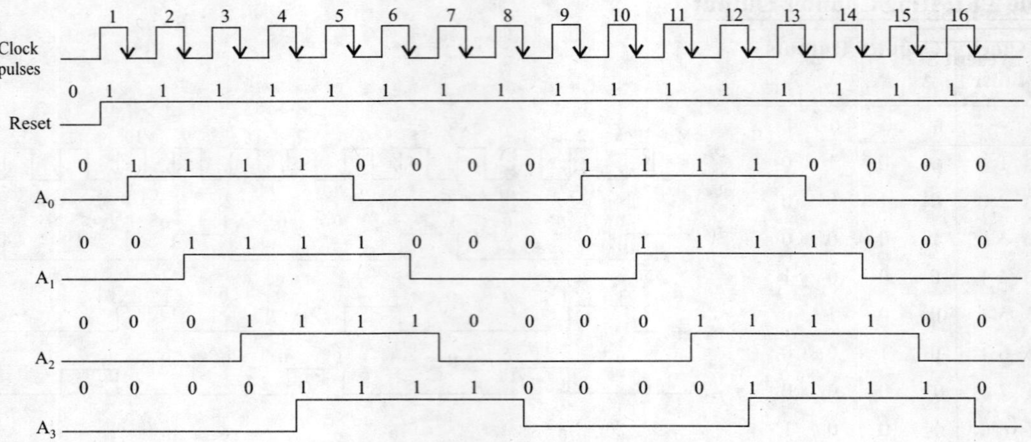

Fig. 3.111: Clock and output signals of Johnson counter.

The counter output are state variables. Hence the counter has 4 state variables A_3, A_2, A_1 and A_0.

$\therefore$ Present state variables : $A_3(t), A_2(t), A_1(t), A_0(t)$

$$\Downarrow$$

$$A_3, A_2, A_1, A_0 \quad \text{(For simplicity)}$$

Next state variables : $A_3(t + 1), A_2(t + 1), A_1(t + 1), A_0(t + 1)$

In D flip-flops the next state output is given by input of the flip-flop at the time of applying clock pulse. Hence with reference to Fig. 3.110, the following equations can be obtained for next state outputs.

$$A_3(t + 1) = A_2$$

$$A_2(t + 1) = A_1$$

$$A_1(t + 1) = A_0$$

$$A_0(t + 1) = A_3'$$

The next state outputs are determined using the above equations and the state table of Johnson counter is constructed by taking counter output as present states as shown in Table 3.64.

Table 3.63: Johnson Counter Output

Clock Pulse	Counter Outputs A_3	A_2	A_1	A_0
-	0	0	0	0
1	0	0	0	1
2	0	0	1	1
3	0	1	1	1
4	1	1	1	1
5	1	1	1	0
6	1	1	0	0
7	1	0	0	0
8	0	0	0	0
9	0	0	0	1
10	0	0	1	1
11	0	1	1	1
12	1	1	1	1
13	1	1	1	0
14	1	1	0	0
15	1	0	0	0
16	0	0	0	0

Table 3.64: State Table of Johnson Counter

Present States					Next States			
A_3	A_2	A_1	A_0		$A_3(t+1) = A_2$	$A_2(t+1) = A_1$	$A_1(t+1) = A_0$	$A_0(t+1) = A_3'$
0	0	0	0	8 distinguishable states	0	0	0	1
0	0	0	1		0	0	1	1
0	0	1	1		0	1	1	1
0	1	1	1		1	1	1	1
1	1	1	1		1	1	1	0
1	1	1	0		1	1	0	0
1	1	0	0		1	0	0	0
1	0	0	0		0	0	0	0

The 4-bit Johnson counter generates 8 distinguishable states. In general, k bit Johnson counter will generate 2k distinguishable states.

The 8 distinguishable states of 4-bit Johnson counter can be decoded to generate 8 unique output signals as shown in Table 3.65. The required decoded signals with respect to clock signal is shown in Fig. 3.112.

Table 3.65: Decoded Output Signals of Johnson Counter

Present States				Decoded Outputs								Minterm
A_3	A_2	A_1	A_0	Y_7	Y_6	Y_5	Y_4	Y_3	Y_2	Y_1	Y_0	$A_3A_2A_1A_0$
0	0	0	0	0	0	0	0	0	0	0	1	m_0
0	0	0	1	0	0	0	0	0	0	1	0	m_1
0	0	1	1	0	0	0	0	0	1	0	0	m_3
0	1	1	1	0	0	0	0	1	0	0	0	m_7
1	1	1	1	0	0	0	1	0	0	0	0	m_{15}
1	1	1	0	0	0	1	0	0	0	0	0	m_{14}
1	1	0	0	0	1	0	0	0	0	0	0	m_{12}
1	0	0	0	1	0	0	0	0	0	0	0	m_8

Note: *The states corresponding to minterms m_2, m_4, m_5, m_6, m_9, m_{10}, m_{11} and m_{13} are unused states and so corresponding decoded outputs are don't-cares.*

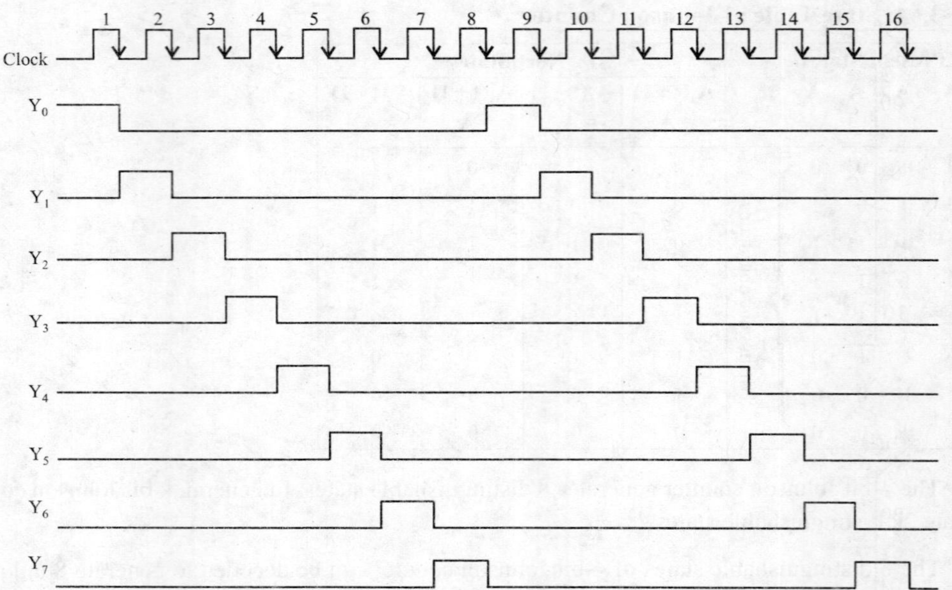

***Fig. 3.112:** Decoded signals of 8 states of Johnson counter.*

In Table 3.65, the minterms are formed using present state variables A_3, A_2, A_1 and A_0. From Table 3.65 the following Boolean equations are obtained for the decoded output of Johnson counter a sum of minterms and don't-cares. Using these equations, K-maps are constructed as shown in Figs. 3.113 to 3.120 to get simplified Boolean equations.

$$Y_0 = m_0 + \Sigma d(2, 4, 5, 6, 9, 10, 11, 13)$$
$$Y_1 = m_1 + \Sigma d(2, 4, 5, 6, 9, 10, 11, 13)$$
$$Y_2 = m_3 + \Sigma d(2, 4, 5, 6, 9, 10, 11, 13)$$
$$Y_3 = m_7 + \Sigma d(2, 4, 5, 6, 9, 10, 11, 13)$$
$$Y_4 = m_{15} + \Sigma d(2, 4, 5, 6, 9, 10, 11, 13)$$
$$Y_5 = m_{14} + \Sigma d(2, 4, 5, 6, 9, 10, 11, 13)$$
$$Y_6 = m_{12} + \Sigma d(2, 4, 5, 6, 9, 10, 11, 13)$$
$$Y_7 = m_8 + \Sigma d(2, 4, 5, 6, 9, 10, 11, 13)$$

A_3A_2 \ A_1A_0	00	01	11	10
00	m_0: 1	m_1: 0	m_3: 0	m_2: ×
01	m_4: ×	m_5: ×	m_7: 0	m_6: ×
11	m_{12}: 0	m_{13}: ×	m_{15}: 0	m_{14}: 0
10	m_8: 0	m_9: ×	m_{11}: ×	m_{10}: ×

$\leftarrow A_3'A_0'$

***Fig. 3.113:** K-map for Y_0.*

A_3A_2 \ A_1A_0	00	01	11	10
00	m_0: 0	m_1: 1	m_3: 0	m_2: ×
01	m_4: ×	m_5: ×	m_7: 0	m_6: ×
11	m_{12}: 0	m_{13}: ×	m_{15}: 0	m_{14}: 0
10	m_8: 0	m_9: ×	m_{11}: ×	m_{10}: ×

$A_1'A_0 \rightarrow$

***Fig. 3.114:** K-map for Y_1.*

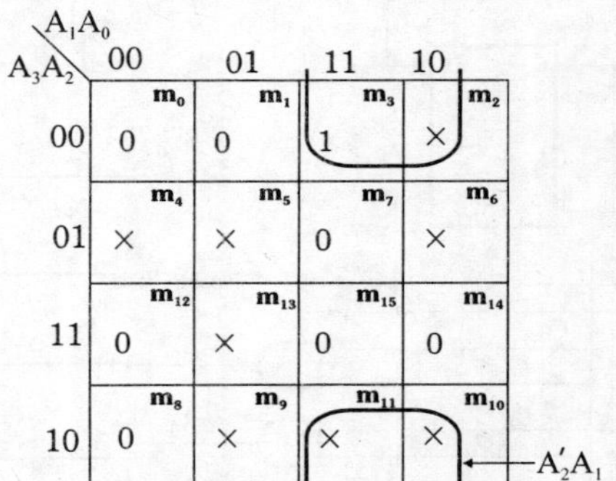

Fig. 3.115: K-map for Y_2.

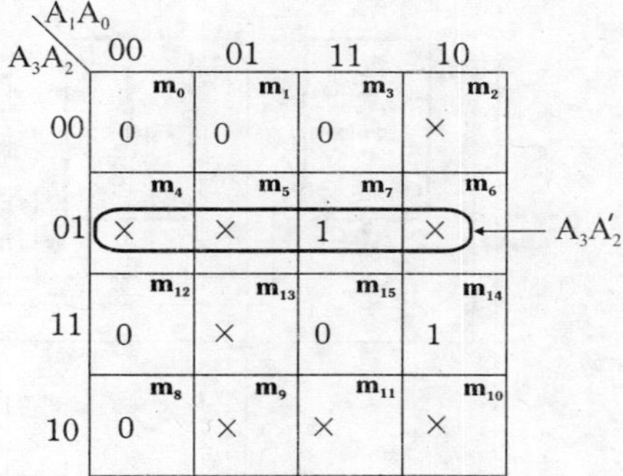

Fig. 3.116: K-map for Y_3.

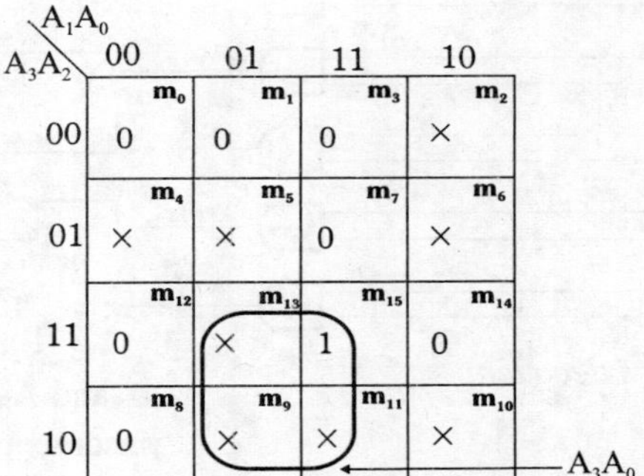

Fig. 3.117: K-map for Y_4.

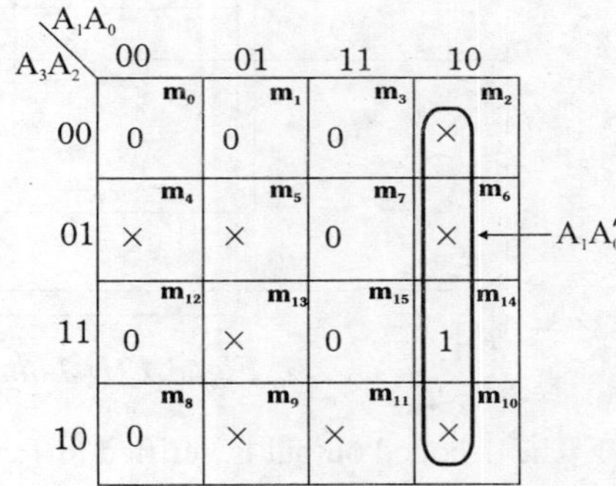

Fig. 3.118: K-map for Y_5.

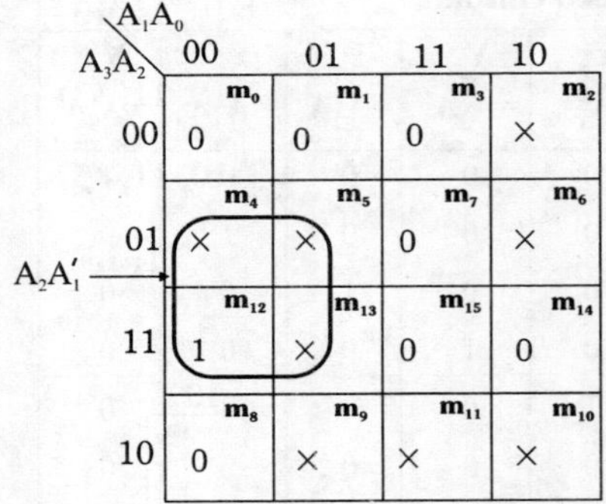

Fig. 3.119: K-map for Y_6.

A_3A_2 \ A_1A_0	00	01	11	10
00	m_0 0	m_1 0	m_3 0	m_2 ×
01	m_4 ×	m_5 ×	m_7 0	m_6 ×
11	m_{12} 0	m_{13} ×	m_{15} 0	m_{14} 0
10	m_8 1	m_9 ×	m_{11} ×	m_{10} ×

A_3A_2'

Fig. 3.120: K-map for Y_7.

From the K-map the following simplified Boolean equations are obtained for decoded output Y_0 to Y_7. Using these equations the logic circuits of Johnson counter as decoder is drawn as shown in Fig. 3.121.

$$Y_0 = A_3'A_0' \qquad Y_4 = A_3A_0$$

$$Y_1 = A_1'A_0 \qquad Y_5 = A_1A_0'$$

$$Y_2 = A_2'A_1 \qquad Y_6 = A_2A_1'$$

$$Y_3 = A_3A_2' \qquad Y_7 = A_3A_2'$$

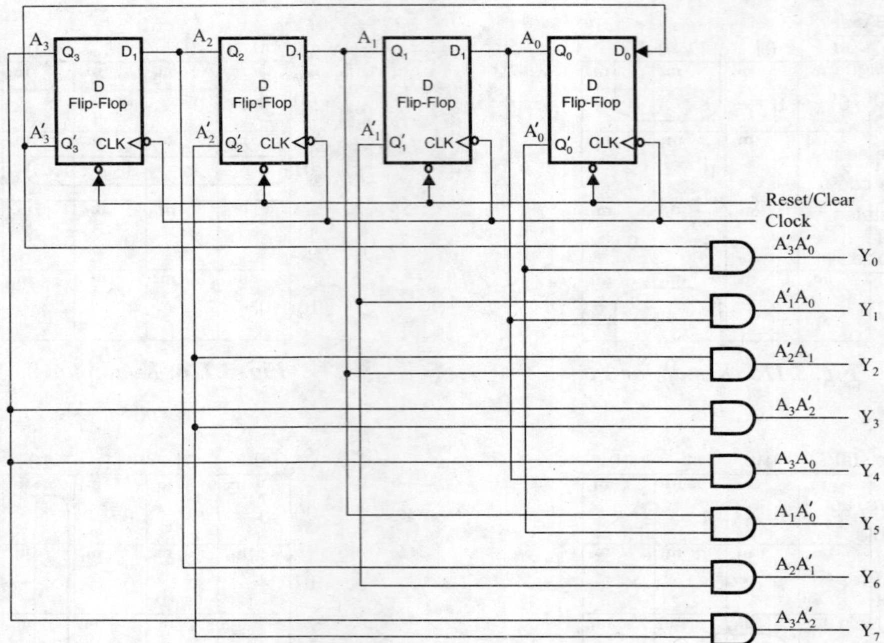

Fig. 3.121: Johnson counter as decoder.

The decoded output is verified in Table 3.66.

Table 3.66: Verification of Decoding of Johnson Counter Output

Counter Outputs				Complement of Counter Outputs				Decoded Outputs							
								Y_7	Y_6	Y_5	Y_4	Y_3	Y_2	Y_1	Y_0
A_3	A_2	A_1	A_0	A'_3	A'_2	A'_1	A'_0	$=A_3A'_2$	$=A_2A'_1$	$=A_1A'_0$	$=A_3A_0$	$=A_3A'_2$	$=A'_2A_1$	$=A'_1A_0$	$=A'_3A'_0$
0	0	0	0	1	1	1	1	0	0	0	0	0	0	0	1
0	0	0	1	1	1	1	0	0	0	0	0	0	0	1	0
0	0	1	1	1	1	0	0	0	0	0	0	0	1	0	0
0	1	1	1	1	0	0	0	0	0	0	0	1	0	0	0
1	1	1	1	0	0	0	0	0	0	0	1	0	0	0	0
1	1	1	0	0	0	0	1	0	0	1	0	0	0	0	0
1	1	0	0	0	0	1	1	0	1	0	0	0	0	0	0
1	0	0	0	0	1	1	1	1	0	0	0	0	0	0	0

Example 3.17

Design a counter using T flip-flop that goes through states 1, 2, 4, 7, 1. Is the counter is self starting?

Solution

The sequential circuit has to detect decimal values in the range 0 to 8. Let us assign binary value of decimal number as code for states. Hence we need 3-bit binary code to represent the decimal sequence. With 3-bit code 8 (2^3 = 8) states are possible and in this 4 states are used states and 4 states are unused states.

Binary Code Assignment		**Unused Binary Codes**	
1	001	0	000
2	010	3	011
4	100	5	101
7	111	6	110

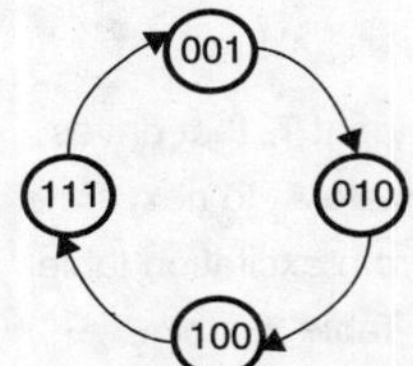

Fig. 1: State diagram with only used states.

The state diagram using only used states is shown in Fig. 1.

Here, each bit of binary code is a state variable. Hence the sequence detector has 3 state variables. Let the state variables be A_0, A_1 and A_2.

$\therefore$ Present state variablels : $A_0(t), A_1(t), A_2(t) \Rightarrow A_0, A_1, A_2$ (For simplicity)

Next state variablels : $A_0(t + 1), A_1(t + 1), A_2(t + 1)$

Using the state diagram of Fig. 1, the state table is constructed as shown in Table 1. The next state for unused states are considered as don't-cares.

Table 1: State Table

Present States			Next States		
A_2	A_1	A_0	$A_2(t+1)$	$A_1(t+1)$	$A_0(t+1)$
0	0	0	×	×	×
0	0	1	0	1	0
0	1	0	1	0	0
0	1	1	×	×	×
1	0	0	1	1	1
1	0	1	×	×	×
1	1	0	×	×	×
1	1	1	0	0	1

Let us choose three T flip-flops to store three state variables A_0, A_1 and A_2 as shown in Fig. 2. Here, A_0, A_1 and A_2 are output of flip-flops.

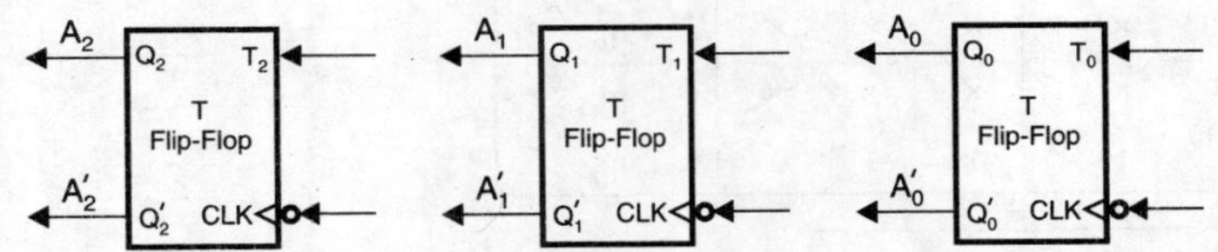

Fig. a: Flip-flop A_2. *Fig. b: Flip-flop A_1.* *Fig. c: Flip-flop A_0.*

Fig. 2: T flip-flops to store state variables.

Excitation Table of T Flip-Flop

Present State Q(t)	Next State Q(t + 1)	Flip-Flop Input T
0	0	0
0	1	1
1	0	1
1	1	0

The flip-flop inputs can be obtained from excitation table as shown below:

State A_2

For state A_2, the Q(t), Q(t + 1) and T of excitation table are considered as A_2, $A_2(t+1)$ and T_2.

$\therefore Q(t) \rightarrow A_2$; $Q(t+1) \rightarrow A_2(t+1)$

$T \rightarrow T_2$

The flip-flop input T_2 that drives the present state A_2 to next state is obtained from excitation table and listed in Table 2.

State A_1

For state A_1, the Q(t), Q(t + 1) and T of excitation table are considered as A_1, $A_1(t+1)$ and T_1.

$\therefore Q(t) \rightarrow A_1$; $Q(t+1) \rightarrow A_1(t+1)$

$T \rightarrow T_1$

The flip-flop input T_1 that drives the present state A_1 to next state is obtained from excitation table and listed in Table 3.

State A_0

For state A_0, the Q(t), Q(t + 1) and T of excitation table are considered as A_0, $A_0(t+1)$ and T_0.

$\therefore Q(t) \rightarrow A_0$; $Q(t+1) \rightarrow A_0(t+1)$

$T \rightarrow T_0$

The flip-flop input T_0 that drives the present state A_0 to next state is obtained from excitation table and listed in Table 4.

Table 2:

A_2	$A_2(t+1)$	T_2
0	×	×
0	0	0
0	1	1
0	×	×
1	1	0
1	×	×
1	×	×
1	0	1

Table 3:

A_1	$A_1(t+1)$	T_1
0	×	×
0	1	1
1	0	1
1	×	×
0	1	1
0	×	×
1	×	×
1	0	1

Table 4:

A_0	$A_0(t+1)$	T_0
0	×	×
1	0	1
0	0	0
1	×	×
0	1	1
1	×	×
0	×	×
1	1	0

The Tables 1 to 4 are combined to form state table with flip-flop inputs as shown in Table 5.

Table 5: Expanded State Table

Present States			Next States			Flip-Flop Inputs			Minterm
A_2	A_1	A_0	$A_2(t+1)$	$A_1(t+1)$	$A_0(t+1)$	T_2	T_1	T_0	$A_2\,A_1\,A_0$
0	0	0	×	×	×	×	×	×	m_0
0	0	1	0	1	0	0	1	1	m_1
0	1	0	1	0	0	1	1	0	m_2
0	1	1	×	×	×	×	×	×	m_3
1	0	0	1	1	1	0	1	1	m_4
1	0	1	×	×	×	×	×	×	m_5
1	1	0	×	×	×	×	×	×	m_6
1	1	1	0	0	1	1	1	0	m_7

Note: *Here, present state $A_2\,A_1\,A_0$ is the counter output.*

The Boolean equations for T_2, T_1 and T_0 are obtained from Table 5 by taking present state as minterms.

$$T_2 = \Sigma m(2, 7) + \Sigma d(0, 3, 5, 6) \quad ; \quad T_1 = \Sigma m(1, 2, 4, 7) + \Sigma d(0, 3, 5, 6) \quad ; \quad T_0 = \Sigma m(1, 4) + \Sigma d(0, 3, 5, 6)$$

The simplified Boolean equations are obtained by constructing K-map using the above equations for T_2, T_1 and T_0 as shown in Figs. 3 to 6.

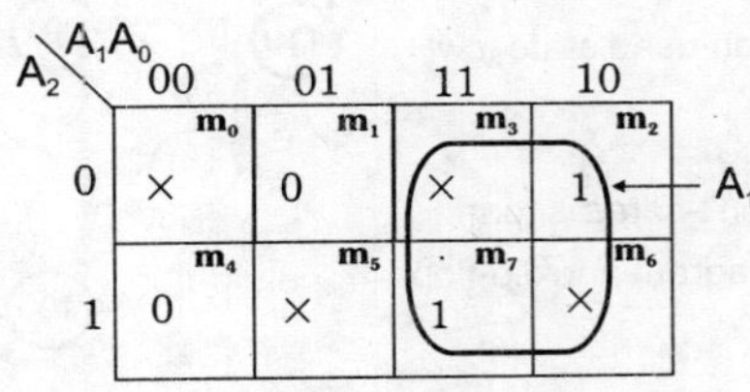

Fig. 3: K-map for T_2.

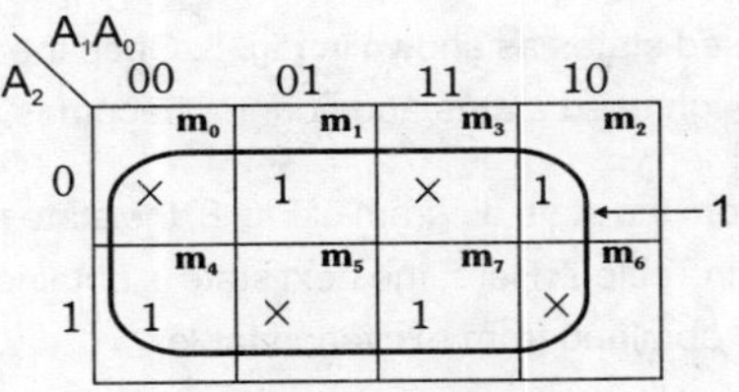

Fig. 4: K-map for T_1.

From the K-map we get the following simplified Boolean equations to realize 3-bit binary counter using T flip-flops and they are used to construct the logic diagram of 3-bit counter as shown in Fig. 6.

$$T_2 = A_1 \quad ; \quad T_1 = 1 \quad ; \quad T_0 = A_1'$$

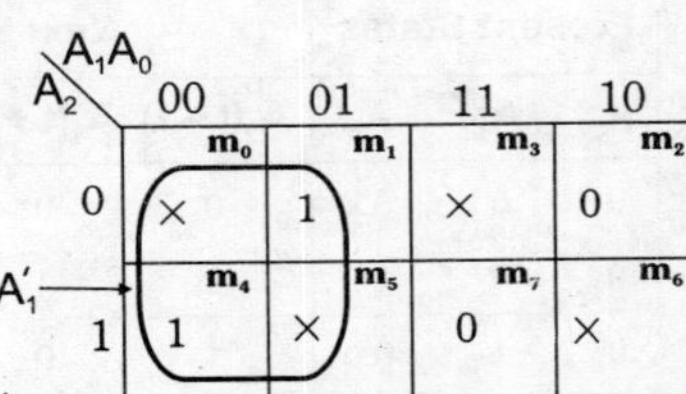

Fig. 5: K-map for T_0.

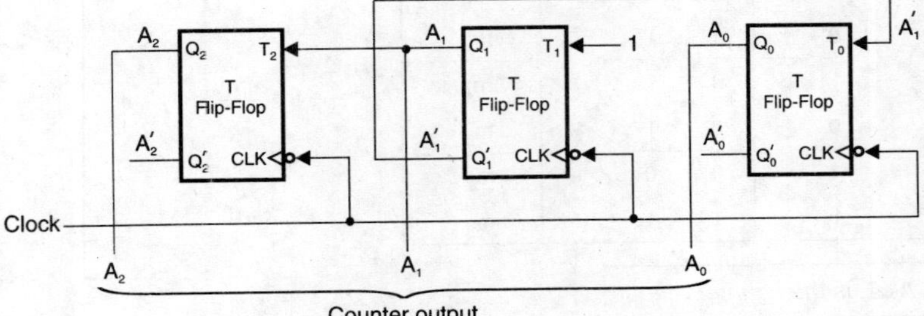

Fig. 6: Logic diagram.

Check for Self Starting

In order to check self starting, the flip-flop inputs are determined for each unused state using simplified equations of flip-flop inputs and listed in Table 6. The next state is obtained using characteristic table of T flip-flop.

Characteristic Table of T Flip-Flop

Input T	Present State Q	Next State Q(t+1)
0	0	0
0	1	1
1	0	1
1	1	0

Table 6: Check for Self Starting

Unused Present States			Flip-Flop Inputs			Next States		
			T_2	T_1	T_0	$A_2(t+1)$	$A_1(t+1)$	$A_0(t+1)$
A_2	A_1	A_0	$= A_1$	$= 1$	$= A_1'$			
0	0	0	0	1	1	0	1	1
0	1	1	1	1	0	1	0	1
1	0	1	0	1	1	1	1	0
1	1	0	1	1	0	0	0	0

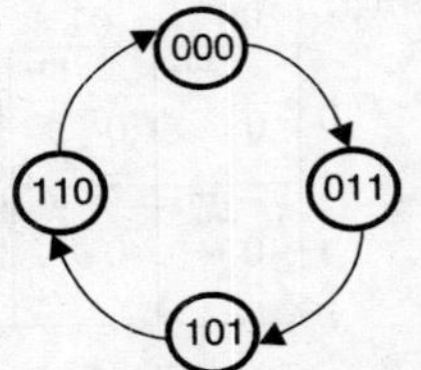

Fig. 7: State diagram of unused states.

Using Table 6 we can construct the state diagram of unused states as shown in Fig. 7 and from this state diagram we can observe that when the counter enter an unused state then it enter a lock-out condition so that the counter will cycle through only unused states and will never enter a used state. Hence the counter is not self starting.

Counter Design for Self Starting

In order to make the counter self starting the unused states can be connected to enter used states as shown in Fig. 8. Once the counter enter an used state it will circle through used states and counter becomes self starting.

From the state diagram of Fig. 8, the state table without don't-cares is formed as shown in Table 7. Here, the next state is obtained from state diagram and flip-flop inputs are obtained from excitation table.

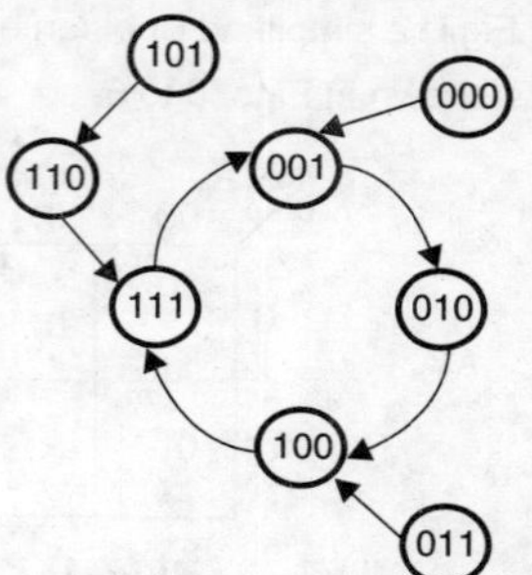

Fig. 8: State diagram to prevent lock-out condition.

Table 7: Expanded State Table without don't-cares

Present States			Next States			Flip-Flop Inputs			Minterm
A_2	A_1	A_0	$A_2(t+1)$	$A_1(t+1)$	$A_0(t+1)$	T_2	T_1	T_0	$A_2 A_1 A_0$
0	0	0	0	0	1	0	0	1	m_0
0	0	1	0	1	0	0	1	1	m_1
0	1	0	1	0	0	1	1	0	m_2
0	1	1	1	0	0	1	1	1	m_3
1	0	0	1	1	1	0	1	1	m_4
1	0	1	1	1	0	0	1	1	m_5
1	1	0	1	1	1	0	0	1	m_6
1	1	1	0	0	1	1	1	0	m_7

***Note:** Here, present state $A_2 A_1 A_0$ is the counter output.*

In Table 7, the minterms are formed using present states. The Boolean equations for T_2, T_1 and T_0 are obtained as sum of minterms as shown below:

$T_2 = \Sigma m(2, 3, 7)$

$T_1 = \Sigma m(1, 2, 3, 4, 5, 7)$

$T_0 = \Sigma m(0, 1, 3, 4, 5, 6)$

The simplified Boolean equations are obtained by constructing K-map using the above equations for T_2, T_1 and T_0 as shown in Figs. 9 to 11.

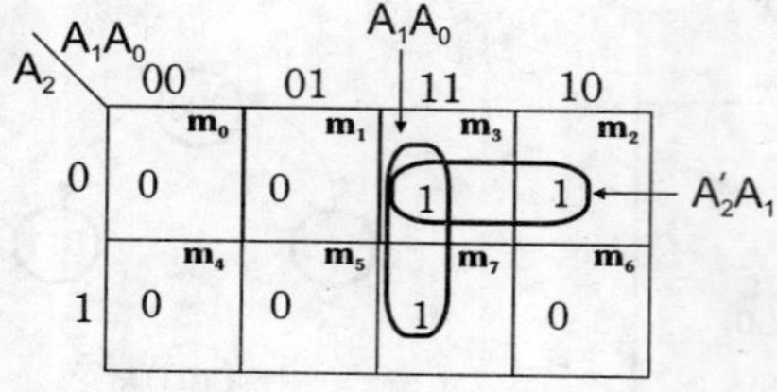

Fig. 9: K-map for T_2.

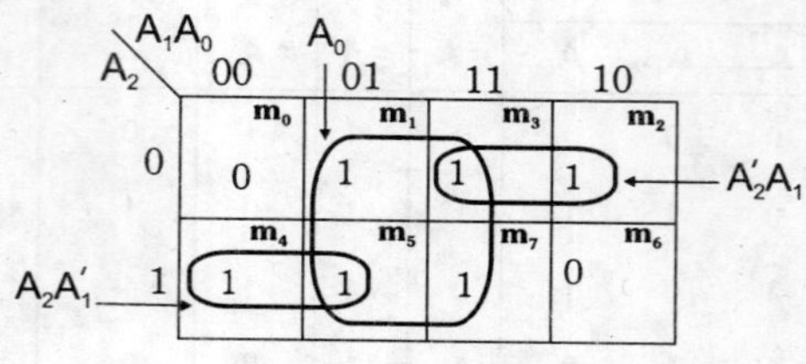

Fig. 10: K-map for T_1.

From the K-map we get the following simplified Boolean equations to realize 3-bit binary counter using T flip-flops and they are used to realize the logic diagram of 3-bit counter as shown in Fig. 12.

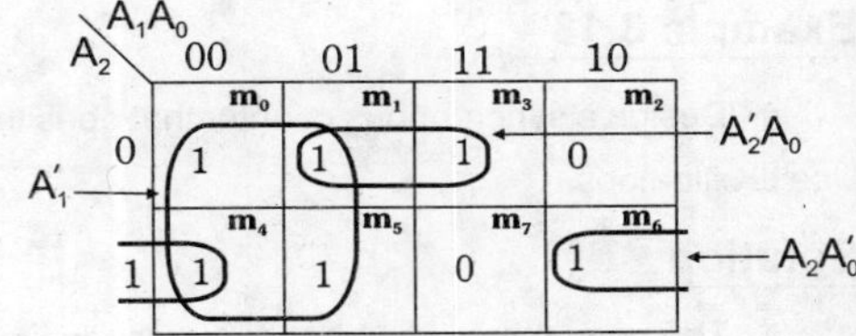

Fig. 11: K-map for T_0.

$$T_2 = A_2' A_1 + A_1 A_0 \qquad T_0 = A_1' + A_2' A_0 + A_2 A_0'$$

$$T_1 = A_0 + A_2 A_1' + A_2' A_1$$

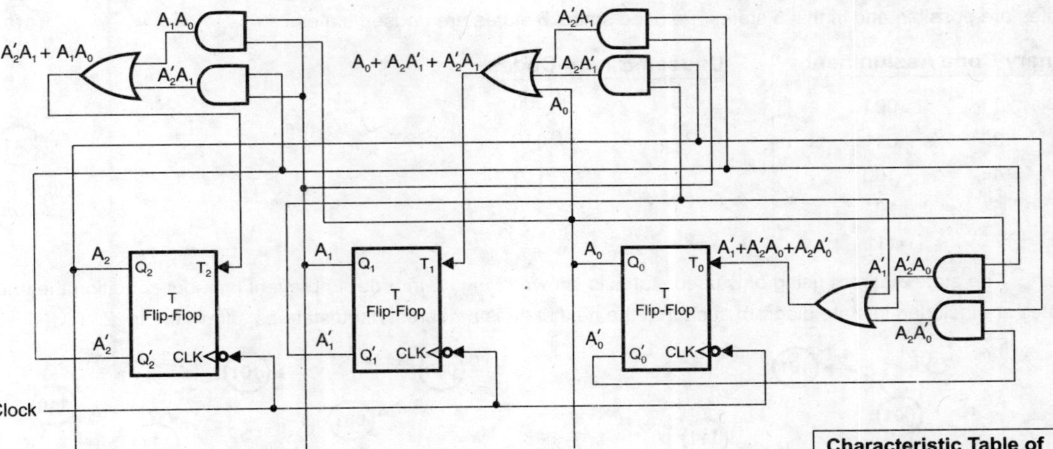

Fig. 12: Logic diagram.

Characteristic Table of T Flip-Flop

Input T	Present State Q	Next State Q(t+1)
0	0	0
0	1	1
1	0	1
1	1	0

Check for Self Starting

In order to check for self starting, the flip-flop inputs are determined for each unused state using simplified equations of flip-flop inputs and listed in Table 8. The next states are obtained from characteristic table of T flip-flop.

Table 8: Check for Self Starting

Unused Present States			Flip-Flop Inputs			Next States		
			T_2	T_1	T_0	$A_2(t+1)$	$A_1(t+1)$	$A_0(t+1)$
A_2	A_1	A_0	$= A_2' A_1 + A_1 A_0$	$= A_0 + A_2 A_1' + A_2' A_1$	$= A_1' + A_2' A_0 + A_2 A_0'$			
0	0	0	0	0	1	0	0	1
0	1	1	1	1	1	1	0	0
1	0	1	0	1	1	1	1	0
1	1	0	0	0	1	1	1	1

From the Table 8 we can observe the following state transitions from unused state to next state.

000 → 001 ; When the counter enters unused state 000 then in next transition it go to a used state 001.

011 → 100 ; When the counter enters unused state 011 then in next transition it go to unused state 100

101 → 110 → 111 ; When the counter enters unused state 101 then in next two transitions it go to a used state 111.

110 → 111 ; When the counter enters unused state 110 then in next transition it go to a used state 111.

The unused states enters used state in the next transition as shown above and so the counter is self starting.

Example 3.18

Design a synchronous counter that goes through states 5 → 7 → 4 → 3 → 1 → 5 by avoiding lock-out condition. Use JK flip-flops.

Solution

The sequential circuit has to detect decimal values in the range 0 to 8. Let us assign binary value of decimal number as code for states. Hence we need 3-bit binary code to represent the decimal sequence. With 3-bit code 8 ($2^3 = 8$) states are possible and in this 5 states are used states 3 states are unused states.

Binary Code Assignment		**Unused Binary Codes**	
1	001	0	000
3	011	2	010
4	100	6	110
5	101		
7	111		

The state diagram using only used states is shown in Fig. 1. In order to prevent lock-out condition the unused states are included in state diagram such that the next state is a nearest used state as shown in Fig. 2.

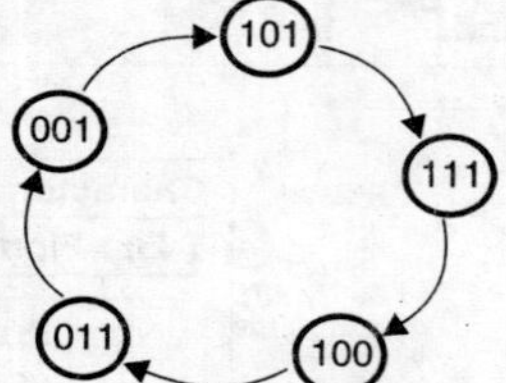

Fig. 1: State diagram with only used states.

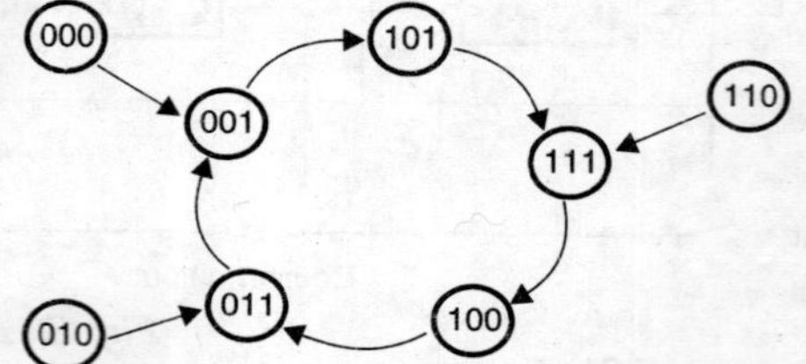

Fig. 2: State diagram to prevent lock-out condition.

Here, each bit of binary code is a state variable. Hence the sequence detector has 3 state variables. Let the state variables be A_0, A_1, A_2.

∴ Present state variables : $A_0(t)$, $A_1(t)$, $A_2(t)$ ⇒ A_0, A_1, A_2 (For simplicity)

Next state variables : $A_0(t + 1)$, $A_1(t + 1)$, $A_2(t + 1)$

Using the state diagram of Fig. 2, the state table is constructed as shown in Table 1.

Table 1: State Table

Present States			**Next States**		
A_2	A_1	A_0	$A_2(t+1)$	$A_1(t+1)$	$A_0(t+1)$
0	0	0	0	0	1
0	0	1	1	0	1
0	1	0	0	1	1
0	1	1	0	0	1
1	0	0	0	1	1
1	0	1	1	1	1
1	1	0	1	1	1
1	1	1	1	0	0

Let us choose three JK flip-flops to store three state variables A_0, A_1 and A_2 as shown in Fig. 3. Here, A_0, A_1 and A_2 are output of flip-flops.

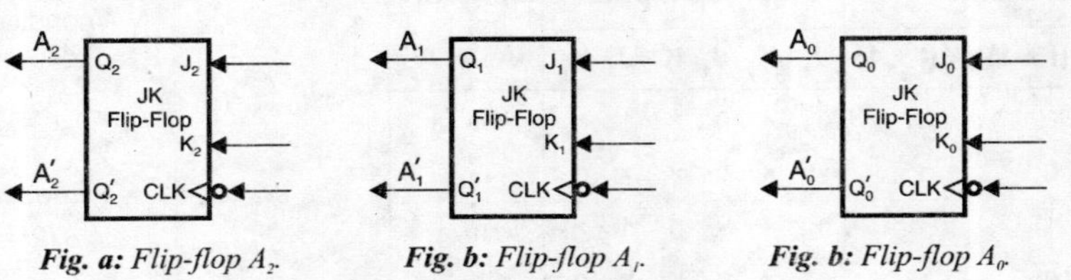

Fig. a: Flip-flop A_2. *Fig. b: Flip-flop A_1.* *Fig. b: Flip-flop A_0.*

Fig. 3: JK flip-flops to store state variables.

Excitation Table of JK Flip-Flop

Present State Q(t)	Next State Q(t + 1)	Flip-Flop Inputs J	K
0	0	0	×
0	1	1	×
1	0	×	1
1	1	×	0

The flip-flop inputs can be obtained from excitation table as shown below:

State A_2

For state A_2, the Q(t), Q(t + 1), J and K of excitation table are considered as A_2, $A_2(t + 1)$, J_2 and K_2.

$\therefore Q(t) \rightarrow A_2$; $Q(t+1) \rightarrow A_2(t+1)$

$J \rightarrow J_2$; $K \rightarrow K_2$

The flip-flop inputs J_2 and K_2 that drives the present state A_2 to next state is obtained from excitation table and listed in Table 2.

Table 2:

A_2	$A_2(t+1)$	J_2	K_2
0	0	0	×
0	1	1	×
0	0	0	×
0	0	0	×
1	0	×	1
1	1	×	0
1	1	×	0
1	1	×	0

State A_1

For state A_1, the Q(t), Q(t + 1), J and K of excitation table are considered as A_1, $A_1(t + 1)$, J_1 and K_1.

$\therefore Q(t) \rightarrow A_1$; $Q(t+1) \rightarrow A_1(t+1)$

$J \rightarrow J_1$; $K \rightarrow K_1$

The flip-flop inputs J_1 and K_1 that drives the present state A_1 to next state is obtained from excitation table and listed in Table 3.

Table 3:

A_1	$A_1(t+1)$	J_1	K_1
0	0	0	×
0	0	0	×
1	1	×	0
1	0	×	1
0	1	1	×
0	1	1	×
1	1	×	0
1	0	×	1

State A_0

For state A_0, the Q(t), Q(t + 1), J and K of excitation table are considered as A_0, $A_0(t + 1)$, J_0 and K_0.

$\therefore Q(t) \rightarrow A_0$; $Q(t+1) \rightarrow A_0(t+1)$

$J \rightarrow J_0$; $K \rightarrow K_0$

The flip-flop inputs J_0 and K_0 that drives the present state A_0 to next state is obtained from excitation table and listed in Table 4.

Table 4:

A_0	$A_0(t+1)$	J_0	K_0
0	1	1	×
1	1	×	0
0	1	1	×
1	1	×	0
0	1	1	×
1	1	×	0
0	1	1	×
1	0	×	1

The Tables 1 to 4 are combined to form state table with flip-flop inputs as shown in Table 5.

Table 5: Expanded State Table

Present States			Next States			Flip-Flop Inputs						Minterm
A_2	A_1	A_0	$A_2(t+1)$	$A_1(t+1)$	$A_0(t+1)$	J_2	K_2	J_1	K_1	J_0	K_0	$A_2 A_1 A_0$
0	0	0	0	0	1	0	×	0	×	1	×	m_0
0	0	1	1	0	1	1	×	0	×	×	0	m_1
0	1	0	0	1	1	0	×	×	0	1	×	m_2
0	1	1	0	0	1	0	×	×	1	×	0	m_3
1	0	0	0	1	1	×	1	1	×	1	×	m_4
1	0	1	1	1	1	×	0	1	×	×	0	m_5
1	1	0	1	1	1	×	0	×	0	1	×	m_6
1	1	1	1	0	0	×	0	×	1	×	1	m_7

***Note:** Here, present state $A_2 A_1 A_0$ is the counter output.*

The flip-flop inputs are functions of present state and so minterms are formed using present state and the Boolean equations for J_2, K_2, J_1, K_1, J_0 and K_0 are obtained from Table 5 as sum of minterms and don't-cares as shown below:

$J_2 = m_1 + \Sigma d(4, 5, 6, 7)$

$K_2 = m_4 + \Sigma d(0, 1, 2, 3)$

$J_1 = \Sigma m(4, 5) + \Sigma d(2, 3, 6, 7)$

$K_1 = \Sigma m(3, 7) + \Sigma d(0, 1, 4, 5)$

$J_0 = \Sigma m(0, 2, 4, 6) + \Sigma d(1, 3, 5, 7)$

$K_0 = m_7 + \Sigma d(0, 2, 4, 6)$

The simplified Boolean equations are obtained by constructing K-map using the above equations for J_2, K_2, J_1, K_1, J_0 and K_0 as shown in Figs. 4 to 9.

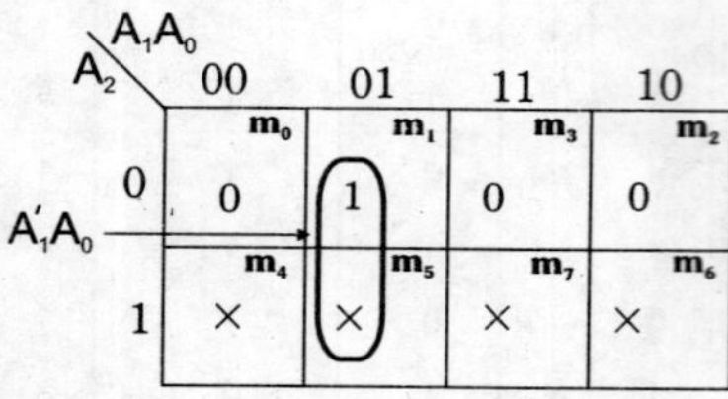

Fig. 4: K-map for J_2.

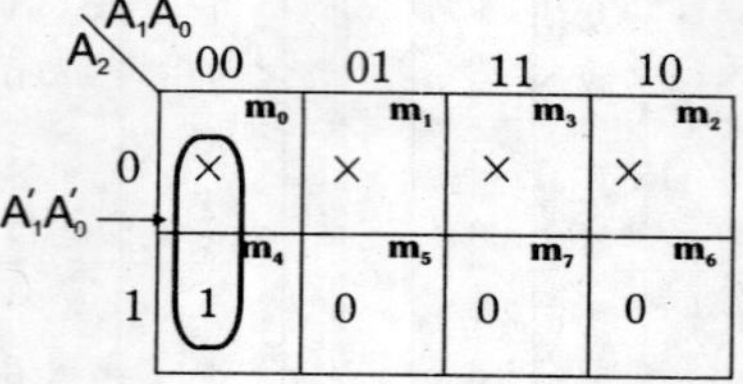

Fig. 5: K-map for K_2.

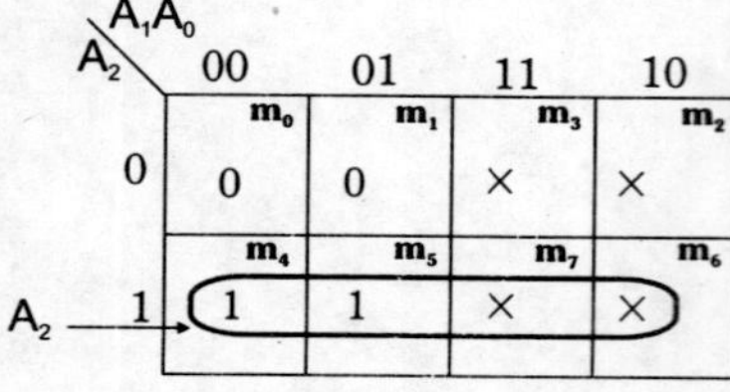

Fig. 6: K-map for J_1.

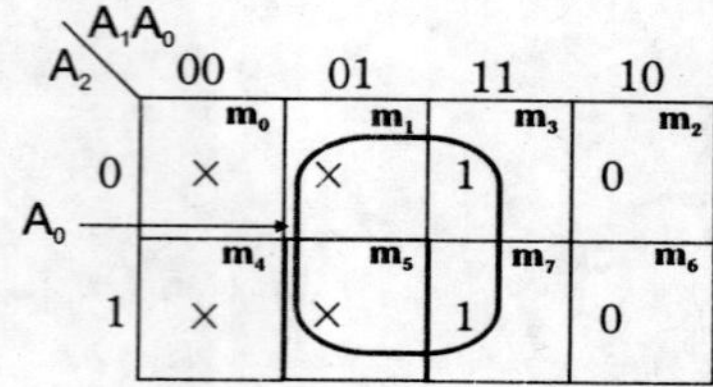

Fig. 7: K-map for K_1.

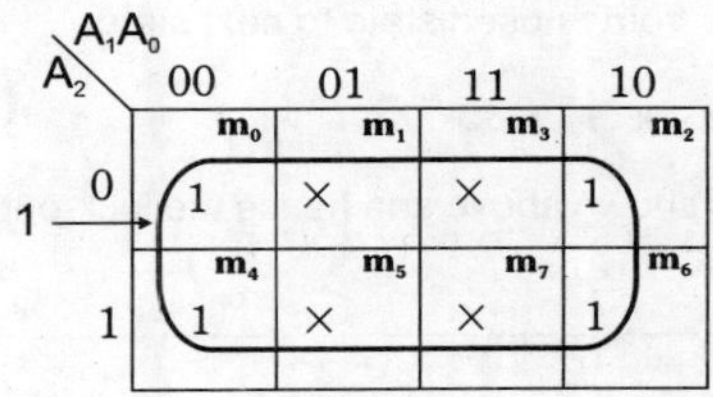

Fig. 8: K-map for J_0.

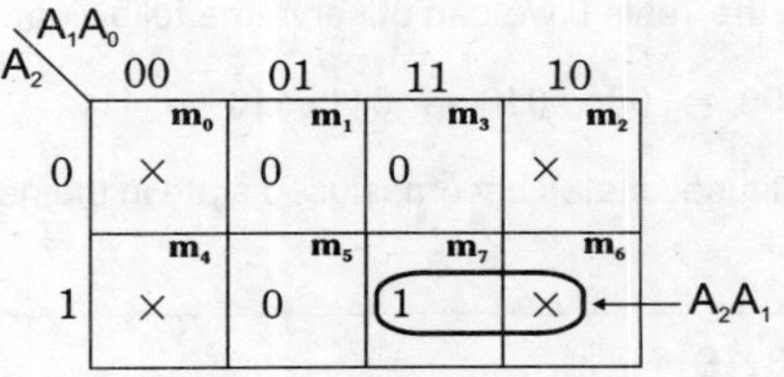

Fig. 9: K-map for K_0.

From the K-map we get the following simplified Boolean equations to realize 3-bit binary counter using JK flip-flops.

$J_2 = A_1' A_0$ $\quad$ $K_2 = A_1' A_0'$

$J_1 = A_2$ $\quad$ $K_1 = A_0$

$J_0 = 1$ $\quad$ $K_0 = A_2 A_1$

Using the simplified Boolean equations obtained from K-map the logic diagram of 3-bit counter is drawn as shown in Fig. 10.

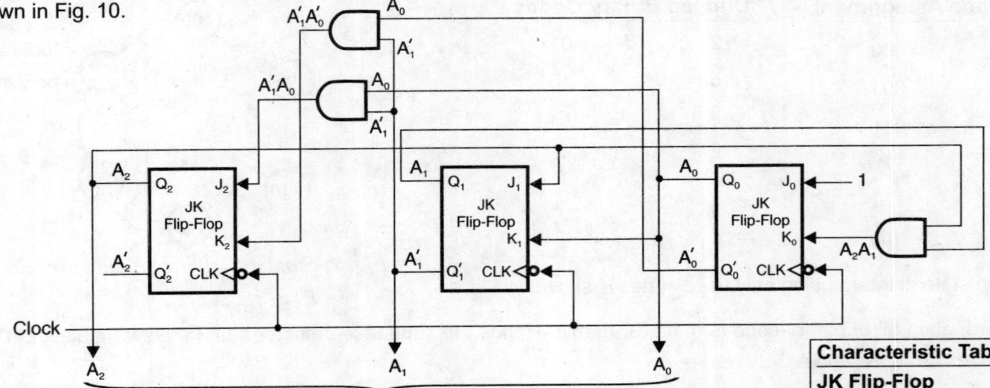

Fig. 10: Logic diagram.

Characteristic Table of JK Flip-Flop

Input J	Input K	Present State Q	Next State Q(t+1)
0	0	0	0
0	0	1	1
0	1	0	0
0	1	1	0
1	0	0	1
1	0	1	1
1	1	0	1
1	1	1	0

Check for Lock-out Condition

In order to check for lock-out condition, the flip-flop inputs are determined for each unused state using simplified equations of flip-flop inputs and listed in Table 6. The next states are obtained from characteristic table of JK flip-flop.

Table 6: Check for Lock-out Condition

Unused Present States			Flip-Flop Inputs						Next States		
A_2	A_1	A_0	$J_2 = A_1'A_0$	$K_2 = A_1'A_0'$	$J_1 = A_2$	$K_1 = A_0$	$J_0 = 1$	$K_0 = A_2A_1$	$A_2(t+1)$	$A_1(t+1)$	$A_0(t+1)$
0	0	0	0	1	0	0	1	0	0	0	1
0	1	0	0	0	0	0	1	0	0	1	1
1	1	0	0	0	1	0	1	1	1	1	1

From the Table 6 we can observe the following state transitions from unused state to next state.

000 → 001, 010 → 011, 110 → 111

The unused states becomes used state in the next transition as shown above and hence the lock-out condition is avoided.

Example 3.19

Design a counter with T flip-flops that goes through the binary sequence : 0, 1, 3, 7, 6, 4. Show that when binary states 010 and 101 are considered as don't-care conditions, the counter may not operate properly. Find a way to correct the design.

Solution

The sequential circuit has to detect decimal values in the range 0 to 7. Let us assign binary value of decimal number as code for states. Hence we need 3-bit binary code to represent the decimal sequence. With 3-bit code 8 ($2^3 = 8$) states all possible and in this 6 states are used states 2 states are unused states.

Binary Code Assignment

0	000
1	001
3	011
7	111
6	110
4	100

Unused Binary Codes

2	011
5	101

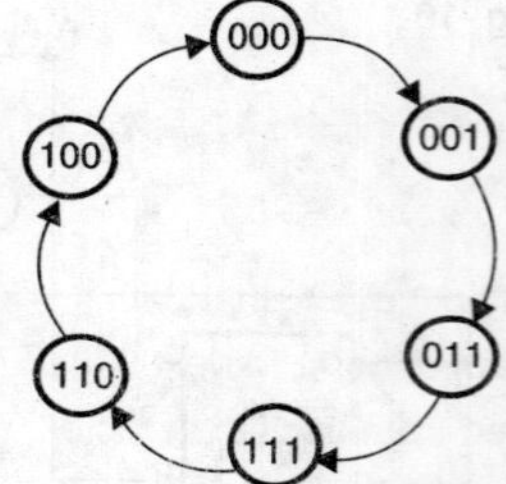

Fig. 1: State diagram with only used states.

The state diagram using only used states is shown in Fig. 1.

Here, each bit of binary code is a state variable. Hence the sequence detector has 3 state variables. Let the state variables be A_0, A_1 and A_2.

∴ Present state variables : $A_0(t), A_1(t), A_2(t) \Rightarrow A_0, A_1, A_2$ (For simplicity)

Next state variables : $A_0(t + 1), A_1(t + 1), A_2(t + 1)$

Using the state diagram of Fig. 1, the state table is constructed as shown in Table 1. The next state for unused states are considered as don't-cares.

Table 1: State Table

Present States			Next States		
A_2	A_1	A_0	$A_2(t+1)$	$A_1(t+1)$	$A_0(t+1)$
0	0	0	0	0	1
0	0	1	0	1	1
0	1	0	×	×	×
0	1	1	1	1	1
1	0	0	0	0	0
1	0	1	×	×	×
1	1	0	1	0	0
1	1	1	1	1	0

Let us choose three T flip-flops to store three state variables A_0, A_1 and A_2 as shown in Fig. 2. Here, A_0, A_1 and A_2 are output of flip-flops.

Excitation Table of T Flip-Flop

Present State Q(t)	Next State Q(t + 1)	Flip-Flop Input T
0	0	0
0	1	1
1	0	1
1	1	0

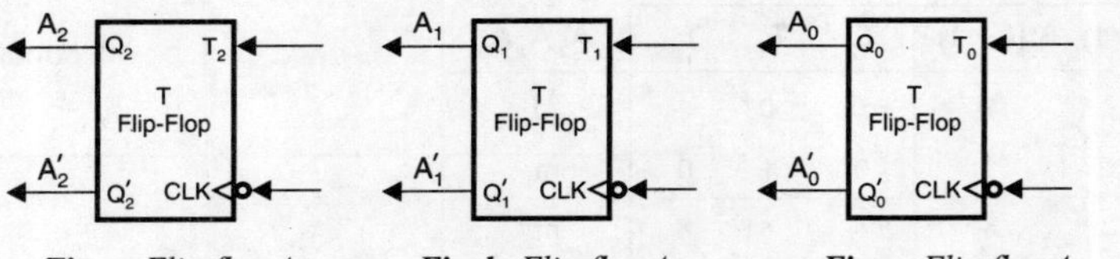

***Fig. a:** Flip-flop A_2.* ***Fig. b:** Flip-flop A_1.* ***Fig. c:** Flip-flop A_0.*

***Fig. 2:** T flip-flops to store state variables.*

The flip-flop inputs can be obtained from excitation table as shown below:

State A_2

For state A_2, the Q(t), Q(t + 1) and T of excitation table are considered as A_2, $A_2(t + 1)$ and T_2.

$\therefore Q(t) \rightarrow A_2$; $Q(t+1) \rightarrow A_2(t+1)$

$T \rightarrow T_2$

The flip-flop input T_2 that drives the present state A_2 to next state is obtained from excitation table and listed in Table 2.

Table 2:

A_2	$A_2(t+1)$	T_2
0	0	0
0	0	0
0	×	×
0	1	1
1	0	1
1	×	×
1	1	0
1	1	0

State A_1

For state A_1, the Q(t), Q(t + 1) and T excitation table are considered as A_1, $A_1(t + 1)$ and T_1.

$\therefore Q(t) \rightarrow A_1$; $Q(t+1) \rightarrow A_1(t+1)$

$T \rightarrow T_1$

The flip-flop input T_1 that drives the present state A_1 to next state is obtained from excitation table and listed in Table 3.

Table 3:

A_1	$A_1(t+1)$	T_1
0	0	0
0	1	1
1	×	×
1	1	0
0	0	0
0	×	×
1	0	1
1	1	0

State A_0

For state A_0, the Q(t), Q(t + 1) and T of excitation table are considered as A_0, $A_0(t + 1)$ and T_0.

$\therefore Q(t) \rightarrow A_0$; $Q(t+1) \rightarrow A_0(t+1)$

$T \rightarrow T_0$

The flip-flop input T_0 that drives the present state A_0 to next state is obtained from excitation table and listed in Table 4.

Table 4:

A_0	$A_0(t+1)$	T_0
0	1	1
1	1	0
0	×	×
1	1	0
0	0	0
1	×	×
0	1	0
1	1	1

The Tables 1 to 4 are combined to form state table with flip-flop inputs as shown in Table 5.

Table 5: Expanded State Table

Present States			Next States			Flip-Flop Inputs			Minterm
A_2	A_1	A_0	$A_2(t+1)$	$A_1(t+1)$	$A_0(t+1)$	T_2	T_1	T_0	$A_2A_1A_0$
0	0	0	0	0	1	0	0	1	m_0
0	0	1	0	1	1	0	1	0	m_1
0	1	0	×	×	×	×	×	×	m_2
0	1	1	1	1	1	1	0	0	m_3
1	0	0	0	0	0	1	0	0	m_4
1	0	1	×	×	×	×	×	×	m_5
1	1	0	1	0	0	0	1	0	m_6
1	1	1	1	1	0	0	0	1	m_7

***Note:** Here, present state $A_2\,A_1\,A_0$ is the counter output.*

The Boolean equations for T_2, T_1 and T_0 are obtained from Table 5 by taking present state as minterms.

$$T_2 = \Sigma m(3, 4) + \Sigma d(2, 5)$$

$$T_1 = \Sigma m(1, 6) + \Sigma d(2, 5)$$

$$T_0 = \Sigma m(0, 7) + \Sigma d(2, 5)$$

The simplified Boolean equations are obtained by constructing K-map using the above equation for T_2, T_1 and T_0 as shown in Figs. 3 to 5.

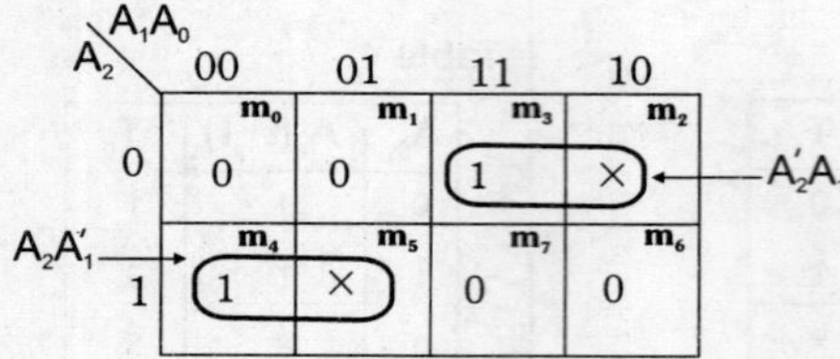

Fig. 3: K-map for T_2.

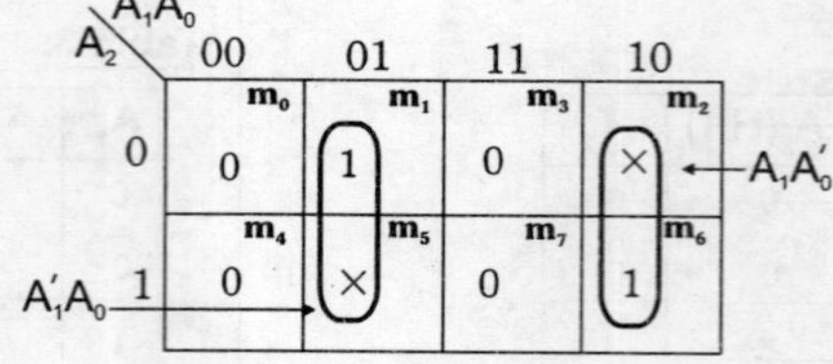

Fig. 4: K-map for T_1.

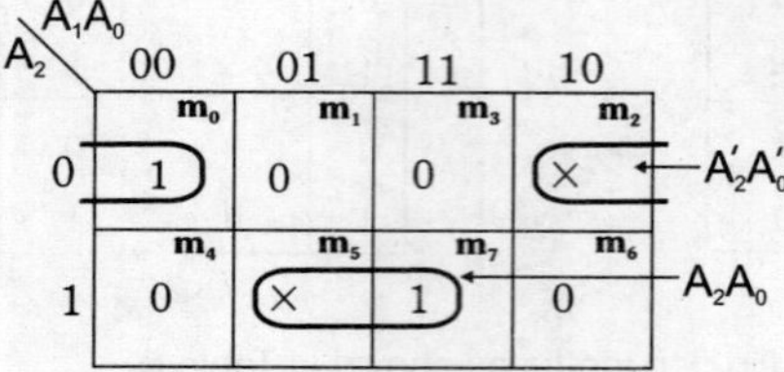

Fig. 5: K-map for T_0.

From the K-map we get the following simplified Boolean equations to realize 3-bit binary counter using T flip-flops and they are used to construct the logic diagram of 3-bit counter as shown in Fig. 6.

$$T_2 = A_2'\,A_1 + A_2A_1'$$

$$T_1 = A_1'A_0 + A_1A_0'$$

$$T_0 = A_2'A_0' + A_2A_0$$

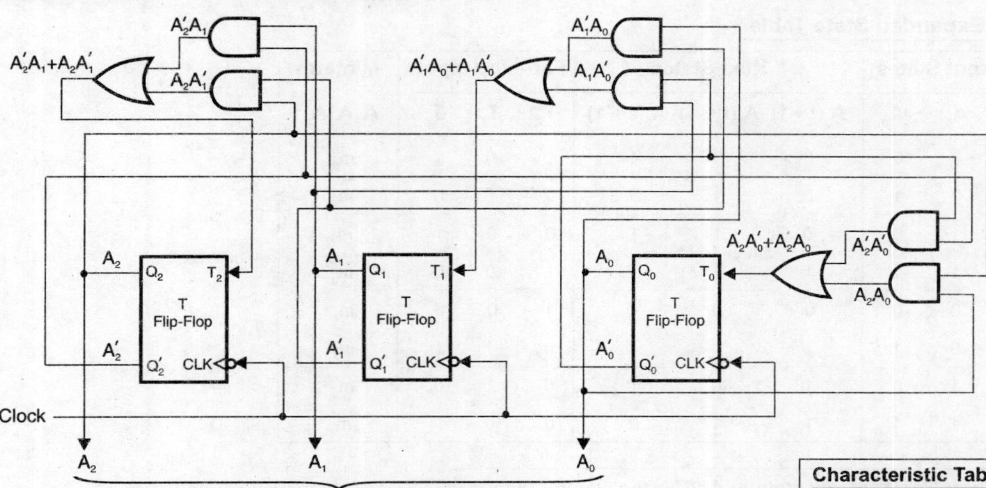

Fig. 6: Logic diagram.

Characteristic Table of T Flip-Flop

Input T	Present State Q	Next State Q(t+1)
0	0	0
0	1	1
1	0	1
1	1	0

Check for Self Starting

In order to check for self starting, the flip-flop inputs are determined for each unused state using simplified equations of flip-flop inputs and listed in Table 6. The next states are obtained from characteristic table of T flip-flop.

Table 6: Check for Self Starting

Unused Present States			Flip-Flop Inputs			Next States		
			T_2	T_1	T_0	$A_2(t+1)$	$A_1(t+1)$	$A_0(t+1)$
A_2	A_1	A_0	$= A_2'A_1 + A_2A_1'$	$= A_1'A_0 + A_1A_0'$	$= A_2'A_0' + A_2A_0$			
0	1	0	1	1	1	1	0	1
1	0	1	1	1	1	0	1	0

From Table 6 we can construct the state diagram of unused states as shown in Fig. 7. Therefore, when the counter enter an unused state then it enter a lock-out condition so that the counter will cycle through only unused states and will never enter a used state. Hence the counter is not self starting.

Fig. 7: State diagram of unused states.

In order to make the counter self starting the unused states can be connected to enter used states as shown in Fig. 8. Once the counter enter an used state it will circle through used states and counter becomes self starting.

From the state diagram of Fig. 8, the state table without don't-cares is formed as shown in Table 7. Here, the next state is obtained from state diagram and flip-flop inputs are obtained from excitation table.

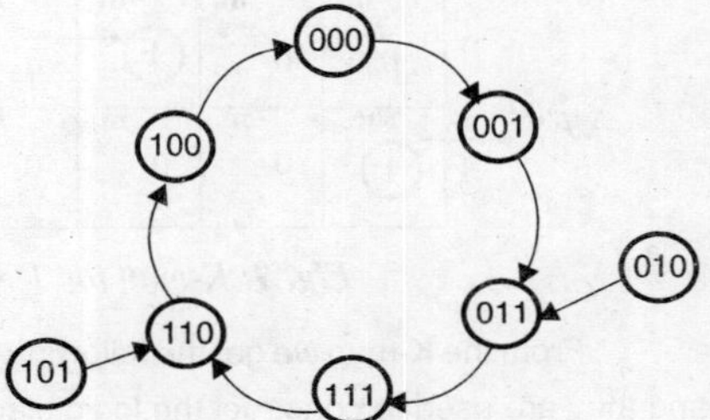

Fig. 8: State diagram to prevent lock-out condition.

Table 7: Expanded State Table

Present States			Next States			Flip-Flop Inputs			Minterm
A_2	A_1	A_0	$A_2(t+1)$	$A_1(t+1)$	$A_0(t+1)$	T_2	T_1	T_0	$A_2A_1A_0$
0	0	0	0	0	1	0	0	1	m_0
0	0	1	0	1	1	0	1	0	m_1
0	1	0	0	1	1	0	0	1	m_2
0	1	1	1	1	1	1	0	0	m_3
1	0	0	0	0	0	1	0	0	m_4
1	0	1	1	1	0	0	1	1	m_5
1	1	0	1	0	0	0	1	0	m_6
1	1	1	1	1	0	0	0	1	m_7

Note: *Here, present state $A_2 A_1 A_0$ is the counter output.*

The Boolean equations for T_2, T_1 and T_0 are obtained from Table 7 by taking present state as minterms.

$$T_2 = \Sigma m(3, 4)$$

$$T_1 = \Sigma m(1, 5, 6)$$

$$T_0 = \Sigma m(0, 2, 5, 7)$$

The simplified Boolean equations are obtained by constructing K-map using the above equation for T_2, T_1 and T_0 as shown in Figs. 9 to 11.

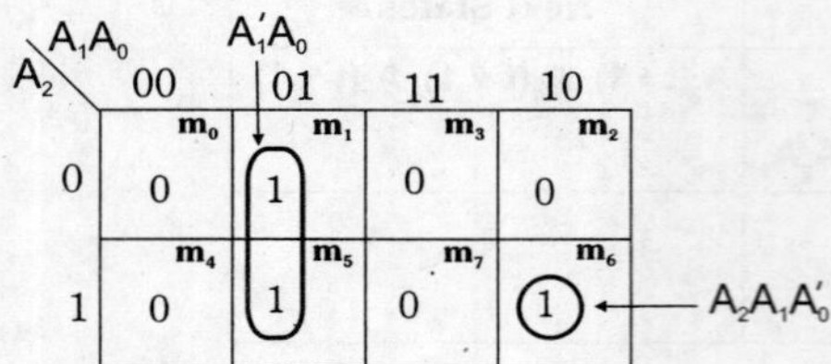

Fig. 10: *K-map for T_1.*

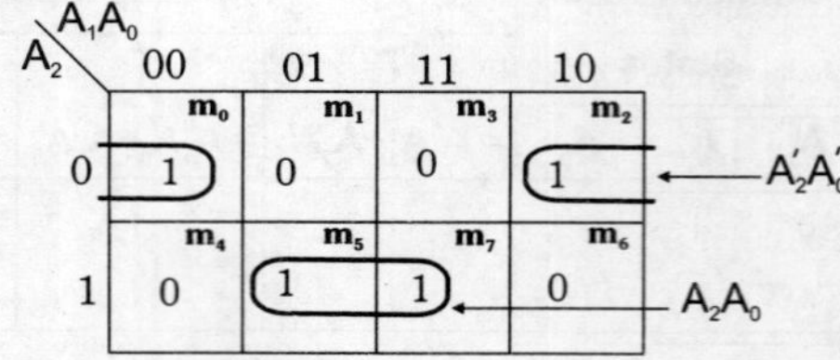

Fig. 11: *K-map for T_0.*

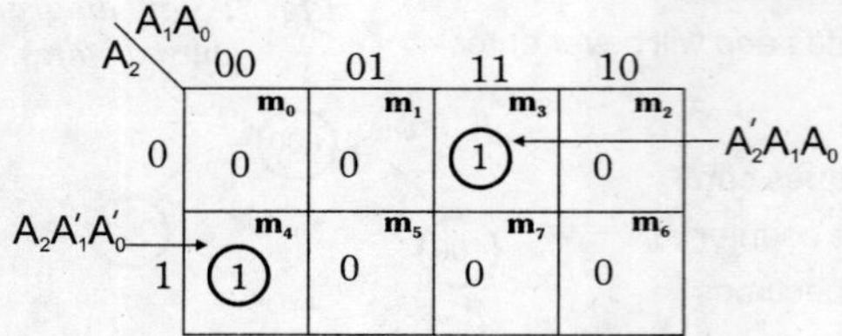

Fig. 9: *K-map for T_2.*

From the K-map we get the following simplified Boolean equations to realize 3-bit binary counter using T flip-flops and they are used to construct the logic diagram of 3-bit counter as shown in Fig. 12.

$$T_2 = A_2' A_1 A_0 + A_2 A_1' A_0'$$

$$T_1 = A_1' A_0 + A_2 A_1 A_0'$$

$$T_0 = A_2' A_0' + A_2 A_0$$

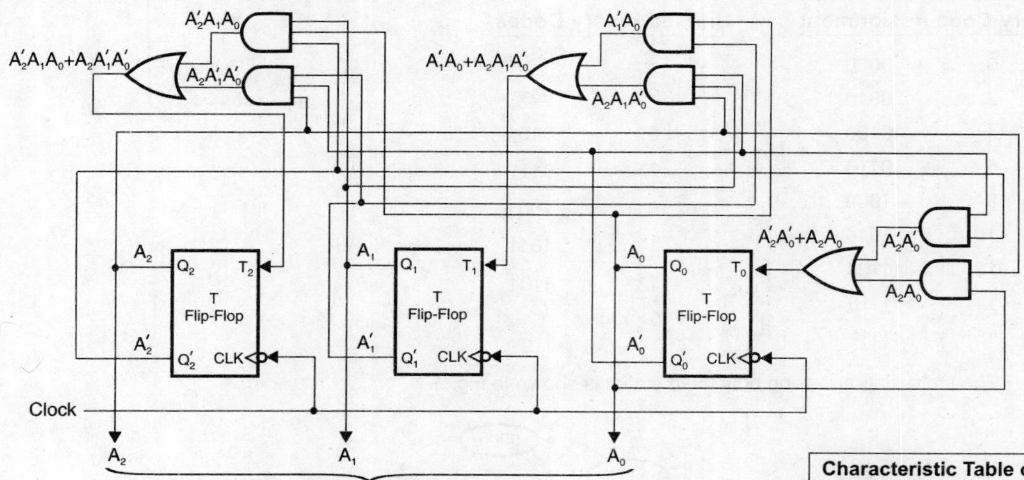

Fig. 12: Logic diagram.

Characteristic Table of T Flip-Flop

Input T	Present State Q	Next State Q(t+1)
0	0	0
0	1	1
1	0	1
1	1	0

Check for Self Starting

In order to check for self starting, the flip-flop inputs are determined for each unused state using simplified equations of flip-flop inputs and listed in Table 8. The next states are obtained from characteristic table of T flip-flop.

Table 8: Check for Self Starting

Unused Present States			Flip-Flop Inputs			Next States		
			T_2	T_1	T_0	$A_2(t+1)$	$A_1(t+1)$	$A_0(t+1)$
A_2	A_1	A_0	$= A_2'A_1A_0 + A_2A_1'A_0'$	$= A_1'A_0 + A_2A_1A_0'$	$= A_2'A_0' + A_2A_0$			
0	1	0	0	0	1	0	1	1
1	0	1	0	1	1	1	1	0

From the Table 8 we can observe the following state transitions from unused state to next state.

010 → 011 ; When the counter enters unused state 010 then in next transition it go to a used state 011.

101 → 110 ; When the counter enters unused state 101 then in next transition it go to unused state 110.

The unused states enters used state in the next transition as shown above and so the counter is self starting.

Example 3.20

(AU, Apr/May'19, 15 Marks)

Design a synchronous sequential circuit that goes through the sequence 0, 2, 4, 6, 8, 10, 12, 14, 15 repeatedly. Use D flip-flops for your design.

Solution

The sequential circuit has to detect decimal values in the range 0 to 15. Let us assign the binary value of decimal number as code for states. Hence we need 4-bit binary code to represent the given decimal sequence.

With 4-bit code 16 ($2^4 = 16$) states are possible and in this 9 states are used states and 7 states are unused states.

Binary Code Assignment		Unused Binary Codes	
0	0000	1	0001
2	0010	3	0011
4	0100	5	0101
6	0110	7	0111
8	1000	9	1001
10	1010	11	1011
12	1100	13	1101
14	1110		
15	1111		

The state diagram using only used states is shown in Fig. 1.

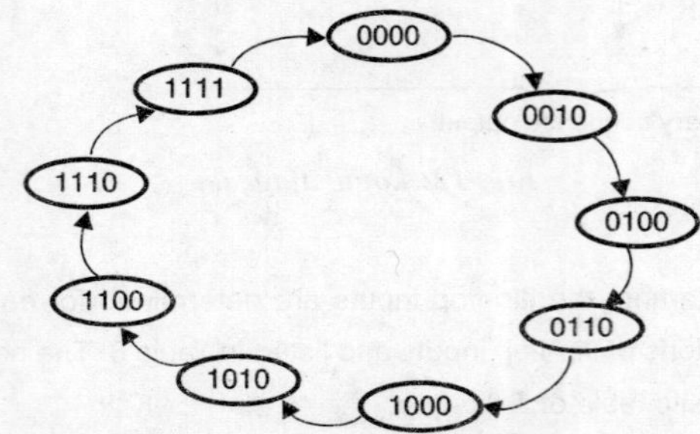

Fig. 1: State diagram with used state.

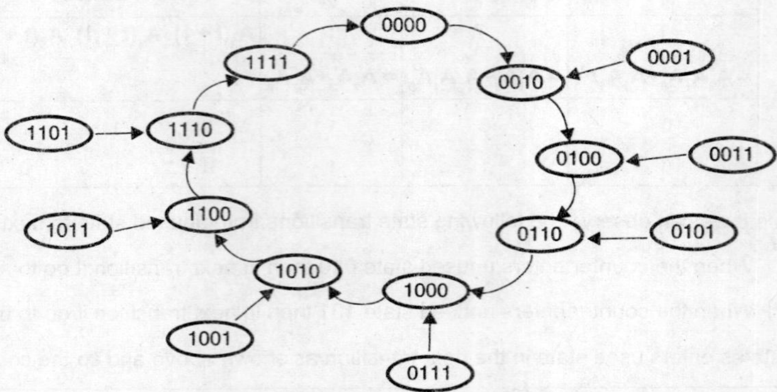

Fig. 2: State diagram to prevent lock-out condition.

Here, each bit of binary code is a state variable. Hence the sequence counter has 4 state variables. Let the state variables be A_0, A_1, A_2 and A_3.

$\therefore$ Present state variables : $A_0(t), A_1(t), A_2(t), A_3(t) \Rightarrow A_0, A_1, A_2, A_3$ (For simplicity)

Next state variables : $A_0(t+1), A_1(t+1), A_2(t+1), A_3(t+1)$

Using the state diagram of Fig. 2, the state table is constructed as shown in Table 1.

Table 1: State Table

Present States				Next States			
A_3	A_2	A_1	A_0	$A_3(t+1)$	$A_2(t+1)$	$A_1(t+1)$	$A_0(t+1)$
0	0	0	0	0	0	1	0
0	0	0	1	0	0	1	0
0	0	1	0	0	1	0	0
0	0	1	1	0	1	0	0
0	1	0	0	0	1	1	0
0	1	0	1	0	1	1	0
0	1	1	0	1	0	0	0
0	1	1	1	1	0	0	0
1	0	0	0	1	0	1	0
1	0	0	1	1	0	1	0
1	0	1	0	1	1	0	0
1	0	1	1	1	1	0	0
1	1	0	0	1	1	1	0
1	1	0	1	1	1	1	0
1	1	1	0	1	1	1	1
1	1	1	1	0	0	0	0

Let us choose four D flip-flops to store four state variables A_0, A_1, A_2 and A_3 as shown in Fig. 3. Here, A_0, A_1, A_2 and A_3 are output of flip-flops.

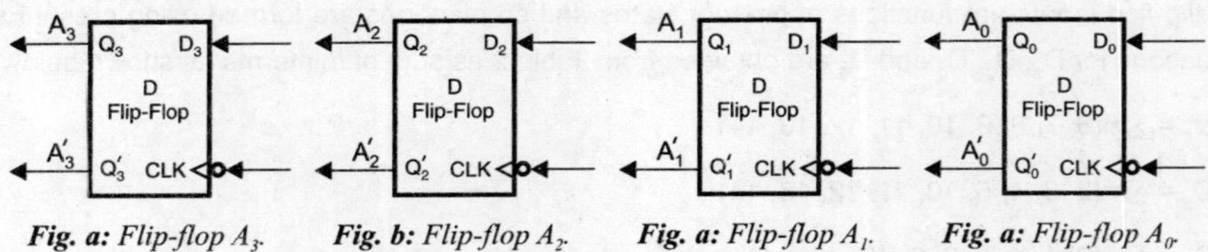

Fig. a: Flip-flop A_3. *Fig. b: Flip-flop A_2.* *Fig. a: Flip-flop A_1.* *Fig. a: Flip-flop A_0.*

Fig. 3: D flip-flop to store state variables.

Here, $A_0 = Q_0$ $\quad$ $A_0(t+1) = Q_0(t+1)$

$A_1 = Q_1$ $\Rightarrow$ $A_1(t+1) = Q_1(t+1)$

$A_2 = Q_2$ $\quad$ $A_2(t+1) = Q_2(t+1)$

$A_3 = Q_3$ $\quad$ $A_3(t+1) = Q_3(t+1)$

The characteristic equation of four D flip-flop are,

$Q_0(t+1) = D_0$

$Q_1(t+1) = D_1$

$Q_2(t+1) = D_2$

$Q_3(t+1) = D_3$

Characteristic equation of D flip-flop,
$Q(t+1) = D$

Using the above equations we can write,

$A_3(t+1) = D_3$ $\quad$ $A_1(t+1) = D_1$

$A_2(t+1) = D_2$ $\quad$ $A_0(t+1) = D_0$

Using the above equations the flip-flop inputs are determined for the next state and listed in Table 2. Here, the flip-flop inputs are same as next state outputs.

Table 2: Expanded State Table

Present States				Next States				Flip-Flop Inputs				Minterm
A_3	A_2	A_1	A_0	$A_3(t+1)$	$A_2(t+1)$	$A_1(t+1)$	$A_0(t+1)$	D_3	D_2	D_1	D_0	$A_3 A_2 A_1 A_0$
0	0	0	0	0	0	1	0	0	0	1	0	m_0
0	0	0	1	0	0	1	0	0	0	1	0	m_1
0	0	1	0	0	1	0	0	0	1	0	0	m_2
0	0	1	1	0	1	0	0	0	1	0	0	m_3
0	1	0	0	0	1	1	0	0	1	1	0	m_4
0	1	0	1	0	1	1	0	0	1	1	0	m_5
0	1	1	0	1	0	0	0	1	0	0	0	m_6
0	1	1	1	1	0	0	0	1	0	0	0	m_7
1	0	0	0	1	0	1	0	1	0	1	0	m_8
1	0	0	1	1	0	1	0	1	0	1	0	m_9
1	0	1	0	1	1	0	0	1	1	0	0	m_{10}
1	0	1	1	1	1	0	0	1	1	0	0	m_{11}
1	1	0	0	1	1	1	0	1	1	1	0	m_{12}
1	1	0	1	1	1	1	0	1	1	1	0	m_{13}
1	1	1	0	1	1	1	1	1	1	1	1	m_{14}
1	1	1	1	0	0	0	0	0	0	0	0	m_{15}

Note: *Here, present state $A_3\, A_2\, A_1\, A_0$ is the sequence detector output.*

The flip-flop inputs are functions of present states and so minterms are formed using present states and the Boolean equations for D_3, D_2, D_1 and D_0 are obtained from Table 2 as sum of minterms as shown below:

$$D_3 = \Sigma m(6, 7, 8, 9, 10, 11, 12, 13, 14)$$

$$D_2 = \Sigma m(2, 3, 4, 5, 10, 11, 12, 13, 14)$$

$$D_1 = \Sigma m(0, 1, 4, 5, 8, 9, 12, 13, 14)$$

$$D_0 = m_{14}$$

The simplified Boolean equations are obtained by constructing K-map using the above equations for D_3, D_2, D_1 and D_0 as shown in Figs. 4 to 7.

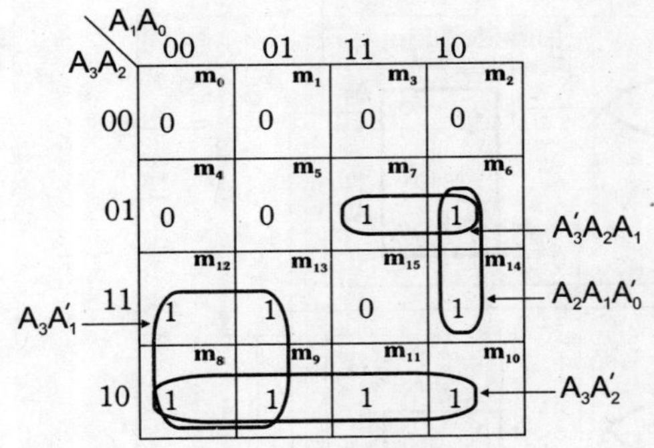

Fig. 4: K-map for D_3.

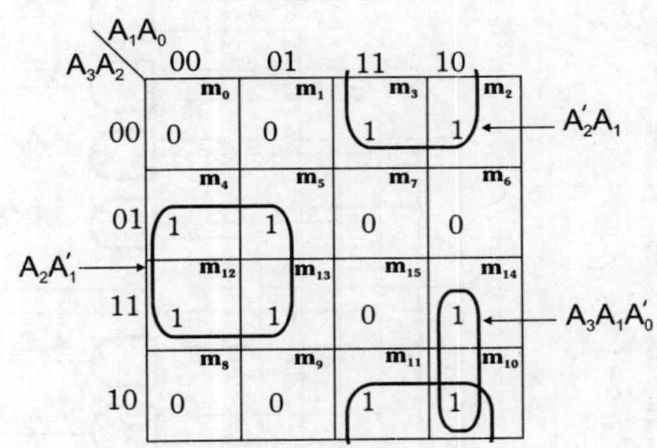

Fig. 5: K-map for D_2.

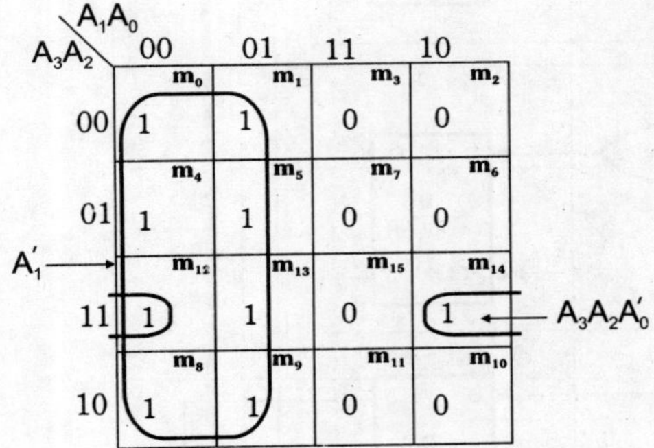

Fig. 6: K-map for D_1.

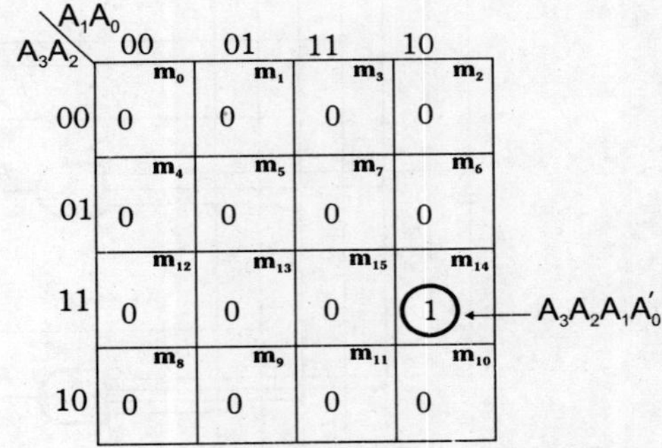

Fig. 7: K-map for D_0.

From the K-map we get the following simplified Boolean equations to realize 4-bit binary sequence counter using D flip-flops and they are used to realize the sequence counter as shown in Fig. 8.

$$D_3 = A_3A_1' + A_3A_2' + A_3'A_2A_1 + A_2A_1A_0'$$

$$D_2 = A_2A_1' + A_2'A_1 + A_3A_1A_0'$$

$$D_1 = A_1' + A_3A_2A_0'$$

$$D_0 = A_3A_2A_1A_0'$$

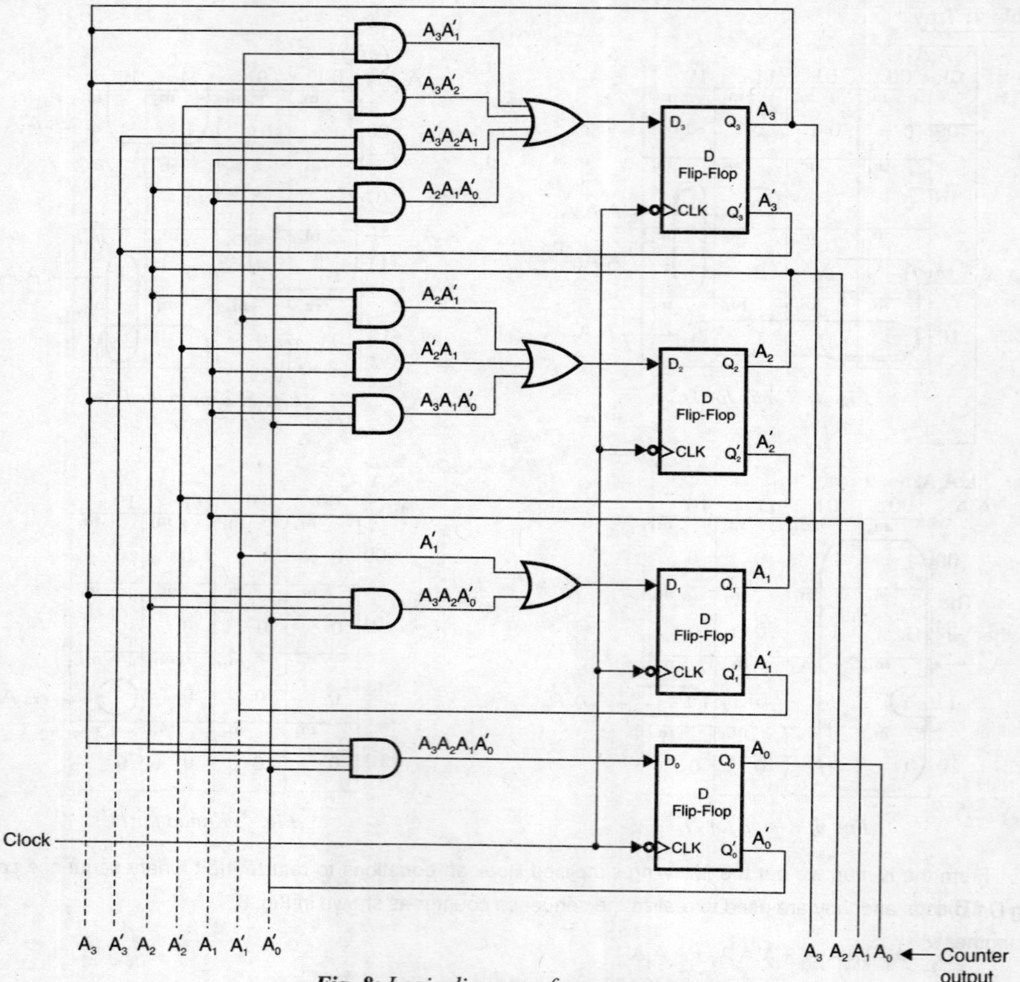

Fig. 8: Logic diagram of sequence counter.

Example 3.21

Design of 3-bit synchronous up/down counter using JK flip-flops.

Solution

Binary output size = 3 bits.

The 3 bits are represented by three state variables A_2, A_1 and A_0.

$\therefore$ Present state variables : $A_2(t), A_1(t), A_0(t) \Rightarrow A_2, A_1, A_0$ (For simplicity)

Next state variables : $A_2(t+1), A_1(t+1), A_0(t+1)$

With 3 state variables it is possible to form 8 ($2^3 = 8$) states.

The 8 states are 000, 001, 010, 011, 100, 101, 110 and 111.

The truth table of up and down count are shown in Tables 1 and 2 respectively. Here, the present states are the counter outputs.

Table 1: Truth Table of Up Count

Clock	Counter Outputs		
	A_2	A_1	A_0
-	0	0	0
↓	0	0	1
↓	0	1	0
↓	0	1	1
↓	1	0	0
↓	1	0	1
↓	1	1	0
↓	1	1	1
↓	0	0	0

Table 2: Truth Table of Down Count

Clock	Counter Outputs		
	A_2	A_1	A_0
-	1	1	1
↓	1	1	0
↓	1	0	1
↓	1	0	0
↓	0	1	1
↓	0	1	0
↓	0	0	1
↓	0	0	0
↓	1	1	1

Let, M be a mode control bit to select up and down count.

∴ When, M = 1, select up count

When, M = 0, select down count

The state diagram of up counter is drawn as shown in Fig. 1. Here, M = 1 is the input that drives one state to another when clock is applied.

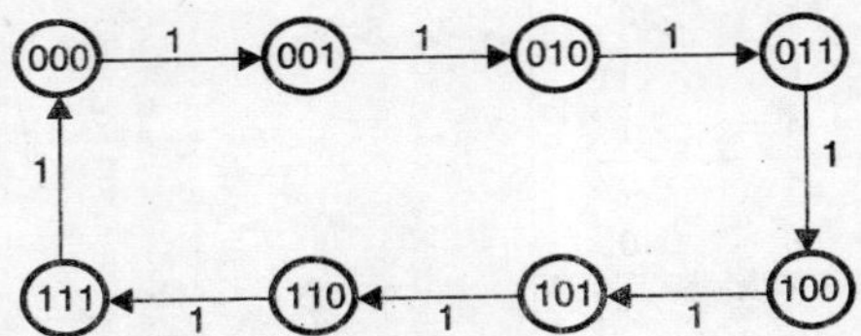

Fig. 1: State diagram for up count.

The state diagram of down counter is drawn as shown in Fig. 2. Here, M = 0 is the input that drives one state to another when clock is applied.

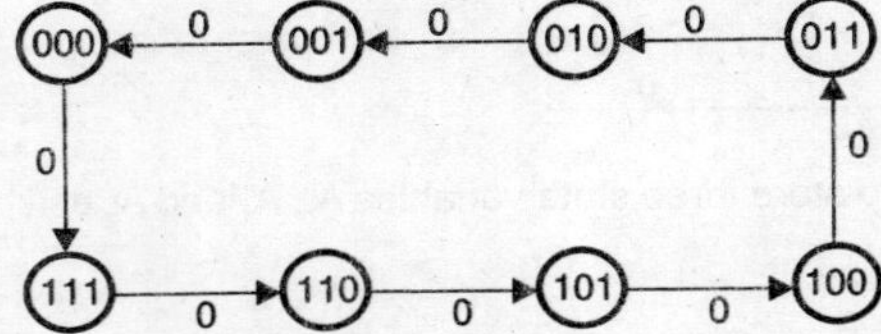

Fig. 2: State diagram for down count.

The state diagram of Figs. 1 and 2 are combined to form the state diagram of up/down counter as shown in Fig. 3.

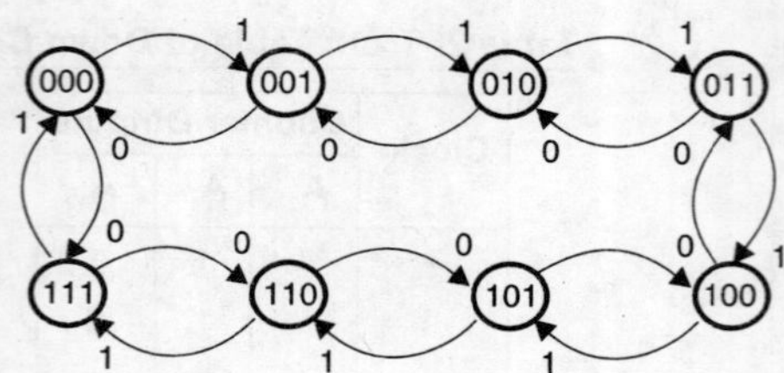

Fig. 3: State diagram for up/down count.

> ***Note:*** *Mode control can be changed at any state (Need not be changed only at initial state).*
>
> ***Example:*** *If mode is changed from 1 to 0, when circuit is in state 101, then it will start down count starting from 101.*

From the knowledge of state transitions in the state diagram of Fig. 3, the state table is formed as shown in Table 3.

Table 3: State Table

Present States			Mode Control	Next States		
A_2	A_1	A_0	M	$A_2(t+1)$	$A_1(t+1)$	$A_0(t+1)$
0	0	0	0	1	1	1
0	0	0	1	0	0	1
0	0	1	0	0	0	0
0	0	1	1	0	1	0
0	1	0	0	0	0	1
0	1	0	1	0	1	1
0	1	1	0	0	1	0
0	1	1	1	1	0	0
1	0	0	0	0	1	1
1	0	0	1	1	0	1
1	0	1	0	1	0	0
1	0	1	1	1	1	0
1	1	0	0	1	0	1
1	1	0	1	1	1	1
1	1	1	0	1	1	0
1	1	1	1	0	0	0

Let us choose three JK flip-flops to store three state variables A_0, A_1 and A_2 as shown in Fig. 4. Here, A_0, A_1 and A_2 are output of flip-flops.

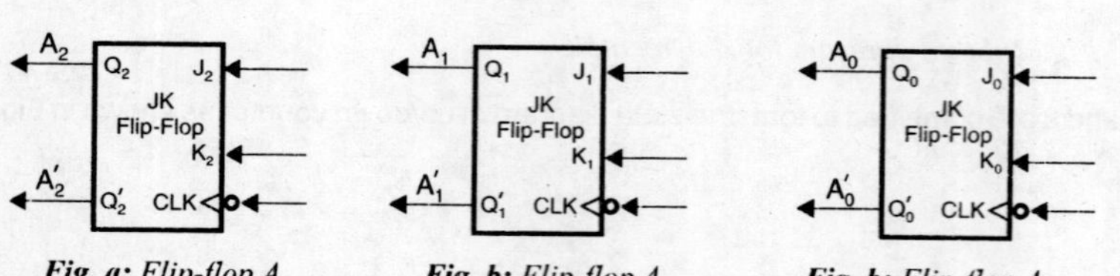

Fig. a: *Flip-flop A_2.* ***Fig. b:*** *Flip-flop A_1.* ***Fig. b:*** *Flip-flop A_0.*

Fig. 4: *JK flip-flops to store state variables.*

Excitation Table of JK Flip-Flop

Present State Q(t)	Next State Q(t + 1)	Flip-Flop Inputs J	K
0	0	0	×
0	1	1	×
1	0	×	1
1	1	×	0

The flip-flop inputs can be obtained from excitation table as shown below:

State A_2

For state A_2, the Q(t), Q(t + 1), J and K of excitation table are considered as A_2, $A_2(t + 1)$, J_2 and K_2.

$$\therefore Q(t) \rightarrow A_2 \; ; \; Q(t+1) \rightarrow A_2(t+1)$$

$$J \rightarrow J_2 \; ; \; K \rightarrow K_2$$

The flip-flop inputs J_2 and K_2 that drives the present state A_2 to next state is obtained from excitation table and listed in Table 4.

Table 4:

A_2	$A_2(t+1)$	J_2	K_2
0	1	1	×
0	0	0	×
0	0	0	×
0	0	0	×
0	0	0	×
0	0	0	×
0	0	0	×
0	1	1	×
1	0	×	1
1	1	×	0
1	1	×	0
1	1	×	0
1	1	×	0
1	1	×	0
1	1	×	0
1	0	×	1

State A_1

For state A_1, the Q(t), Q(t + 1), J and K of excitation table are considered as A_1, $A_1(t + 1)$, J_1 and K_1.

$$\therefore Q(t) \rightarrow A_1 \; ; \; Q(t+1) \rightarrow A_1(t+1)$$

$$J \rightarrow J_1 \; ; \; K \rightarrow K_1$$

The flip-flop inputs J_1 and K_1 that drives the present state A_1 to next state is obtained from excitation table and listed in Table 5.

Table 5:

A_1	$A_1(t+1)$	J_1	K_1
0	1	1	×
0	0	0	×
0	0	0	×
0	1	1	×
1	0	×	1
1	1	×	0
1	1	×	0
1	0	×	1
0	1	1	×
0	0	0	×
0	0	0	×
0	1	1	×
1	0	×	1
1	1	×	0
1	1	×	0
1	0	×	1

State A_0

For state A_0, the Q(t), Q(t + 1), J and K of excitation table are considered as A_0, $A_0(t + 1)$, J_0 and K_0.

$$\therefore Q(t) \rightarrow A_0 \; ; \; Q(t+1) \rightarrow A_0(t+1)$$

$$J \rightarrow J_0 \; ; \; K \rightarrow K_0$$

The flip-flop inputs J_0 and K_0 that drives the present state A_0 to next state is obtained from excitation table and listed in Table 6.

Table 6:

A_0	$A_0(t+1)$	J_0	K_0
0	1	1	×
0	1	1	×
1	0	×	1
1	0	×	1
0	1	1	×
0	1	1	×
1	0	×	1
1	0	×	1
0	1	1	×
0	1	1	×
1	0	×	1
1	0	×	1
0	1	1	×
0	1	1	×
1	0	×	1
1	0	×	1

The Tables 3 to 6 are combined to form state table with flip-flop inputs as shown in Table 7.

Table 7: Expanded State Table

Present States			Mode Control	Next States			Flip-Flop Inputs						Minterm
A_2	A_1	A_0	M	$A_2(t+1)$	$A_1(t+1)$	$A_0(t+1)$	J_2	K_2	J_1	K_1	J_0	K_0	$A_2A_1A_0M$
0	0	0	0	1	1	1	1	×	1	×	1	×	m_0
0	0	0	1	0	0	1	0	×	0	×	1	×	m_1
0	0	1	0	0	0	0	0	×	0	×	×	1	m_2
0	0	1	1	0	1	0	0	×	1	×	×	1	m_3
0	1	0	0	0	0	1	0	×	×	1	1	×	m_4
0	1	0	1	0	1	1	0	×	×	0	1	×	m_5
0	1	1	0	0	1	0	0	×	×	0	×	1	m_6
0	1	1	1	1	0	0	1	×	×	1	×	1	m_7
1	0	0	0	0	1	1	×	1	1	×	1	×	m_8
1	0	0	1	1	0	1	×	0	0	×	1	×	m_9
1	0	1	0	1	0	0	×	0	0	×	×	1	m_{10}
1	0	1	1	1	1	0	×	0	1	×	×	1	m_{11}
1	1	0	0	1	0	1	×	0	×	1	1	×	m_{12}
1	1	0	1	1	1	1	×	0	×	0	1	×	m_{13}
1	1	1	0	1	1	0	×	0	×	0	×	1	m_{14}
1	1	1	1	0	0	0	×	1	×	1	×	1	m_{15}

Note: *Here, present state $A_2\,A_1\,A_0$ is the counter output.*

The flip-flop inputs are functions of present states and mode control bit, M. Therefore, minterms are formed using present states and M and the Boolean equations for flip-flop inputs are obtained as sum of minterms and don't-cares as shown below:

$J_2 = \Sigma m(0, 7) + \Sigma d(8, 9, 10, 11, 12, 13, 14, 15)$

$J_1 = \Sigma m(0, 3, 8, 11) + \Sigma d(4, 5, 6, 7, 12, 13, 14, 15)$

$J_0 = \Sigma m(0, 1, 4, 5, 8, 9, 12, 13) + \Sigma d(2, 3, 6, 7, 10, 11, 14, 15)$

$K_2 = \Sigma m(8, 15) + \Sigma d(0, 1, 2, 3, 4, 5, 6, 7)$

$K_1 = \Sigma m(4, 7, 12, 15) + \Sigma d(0, 1, 2, 3, 8, 9, 10, 11)$

$K_0 = \Sigma m(2, 3, 6, 7, 10, 11, 14, 15) + \Sigma d(0, 1, 4, 5, 8, 9, 12, 13)$

The simplified Boolean equations for flip-flop inputs are obtained by constructing K-map using the above equations as shown in Figs. 5 to 8.

Here, J_0 and K_0 are always 1 or × and so no need of K-map and they can be permanently tied to 1.

$\therefore J_0 = K_0 = 1$

From the K-map we get the following Boolean equations for flip-flop inputs.

$J_1 = K_1 = A_0M + A_0'M'$

$J_2 = K_2 = A_1A_0M + A_1'A_0'M'$

Using the above Boolean equations the logic diagram of 3-bit counter is drawn as shown in Fig. 9.

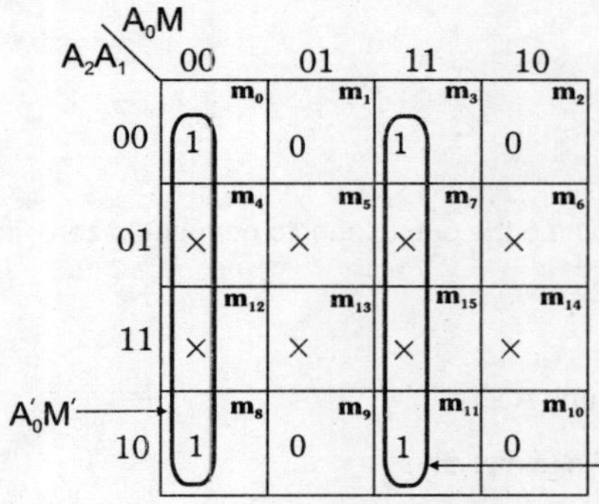

$A_2A_1 \backslash A_0M$	00	01	11	10
00	m_0: 1	m_1: 0	m_3: 1	m_2: 0
01	m_4: ×	m_5: ×	m_7: ×	m_6: ×
11	m_{12}: ×	m_{13}: ×	m_{15}: ×	m_{14}: ×
10	m_8: 1	m_9: 0	m_{11}: 1	m_{10}: 0

$A_0'M'$; A_0M

***Fig. 5:** K-map for J_1.*

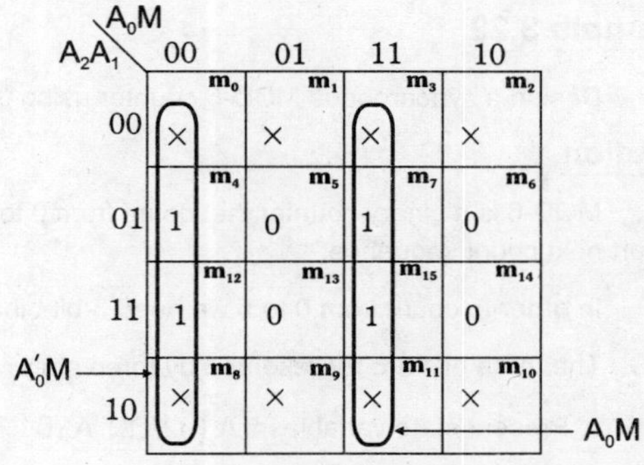

$A_2A_1 \backslash A_0M$	00	01	11	10
00	m_0: ×	m_1: ×	m_3: ×	m_2: ×
01	m_4: 1	m_5: 0	m_7: 1	m_6: 0
11	m_{12}: 1	m_{13}: 0	m_{15}: 1	m_{14}: 0
10	m_8: ×	m_9: ×	m_{11}: ×	m_{10}: ×

$A_0'M'$; A_0M

***Fig. 6:** K-map for K_1.*

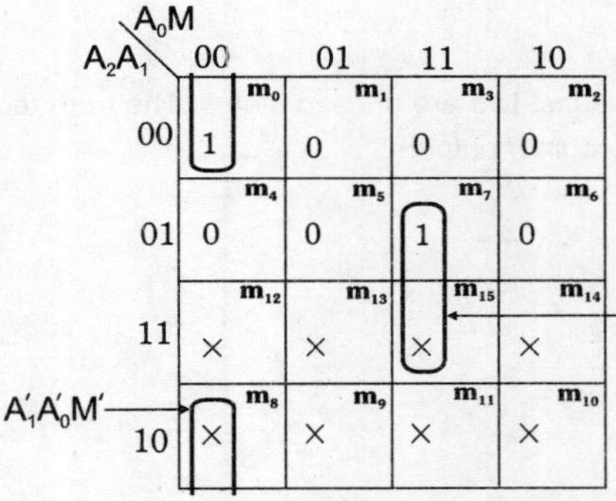

$A_2A_1 \backslash A_0M$	00	01	11	10
00	m_0: 1	m_1: 0	m_3: 0	m_2: 0
01	m_4: 0	m_5: 0	m_7: 1	m_6: 0
11	m_{12}: ×	m_{13}: ×	m_{15}: ×	m_{14}: ×
10	m_8: ×	m_9: ×	m_{11}: ×	m_{10}: ×

A_1A_0M ; $A_1'A_0'M'$

***Fig. 7:** K-map for J_2.*

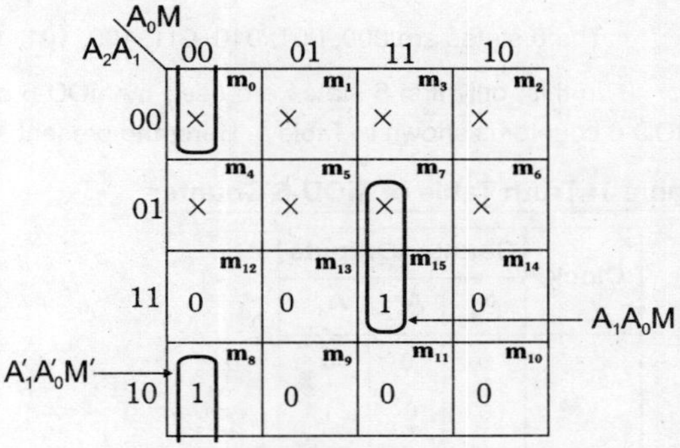

$A_2A_1 \backslash A_0M$	00	01	11	10
00	m_0: ×	m_1: ×	m_3: ×	m_2: ×
01	m_4: ×	m_5: ×	m_7: ×	m_6: ×
11	m_{12}: 0	m_{13}: 0	m_{15}: 1	m_{14}: 0
10	m_8: 1	m_9: 0	m_{11}: 0	m_{10}: 0

A_1A_0M ; $A_1'A_0'M'$

***Fig. 8:** K-map for K_2.*

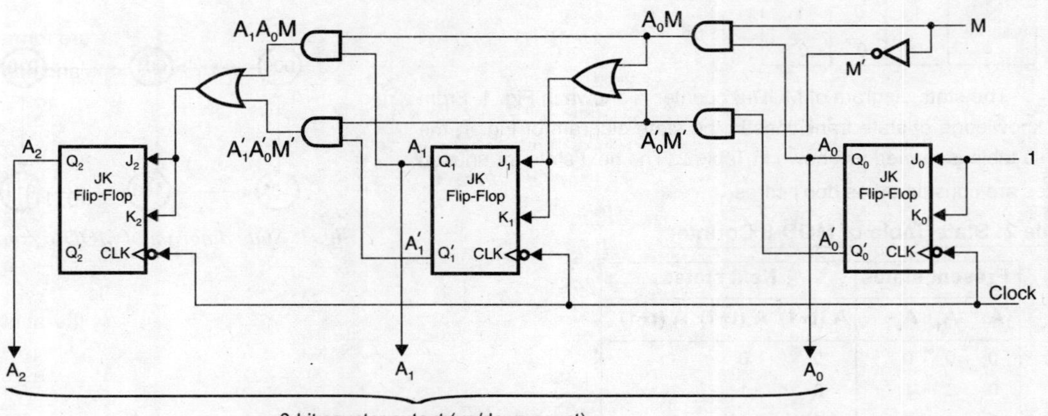

***Fig. 9:** Logic diagram of 3-bit synchronous up/down counter.*

Example 3.22

Design a synchronous MOD-6 counter using JK flip-flops.

Solution

MOD-6 is a binary counter that count from 0 to 5 and on reaching 6 the output should be reset to zero, in order to start next count sequence.

In order to count from 0 to 5 we need 3-bit binary output.

The three bits are represented by three state variables A_2, A_1 and A_0.

$\therefore$ Present state variables : $A_2(t), A_1(t), A_0(t) \Rightarrow A_2, A_1, A_0$ (For simplicity)

Next state variables : $A_2(t+1), A_1(t+1), A_0(t+1)$

With 3 state variables it is possible to form 8 ($2^3 = 8$) states.

The 8 states are 000, 001, 010, 011, 100, 101, 110 and 111.

In this, only first 6 states are used by MOD-6 counter and the last two are unused states. The truth table of MOD-6 counter is shown in Table 1. Here, the present states are the counter outputs.

Table 1: Truth Table of MOD-6 Counter

Clock	Counter Outputs		
	A_2	A_1	A_0
-	0	0	0
↓	0	0	1
↓	0	1	0
↓	0	1	1
↓	1	0	0
↓	1	0	1
↓	0	0	0

The state diagram of MOD-6 counter is shown in Fig. 1. From the knowledge of state transitions in the state diagram of Fig. 1, the stable table is formed as shown in Table 2. The next state for unused states are considered as don't cares.

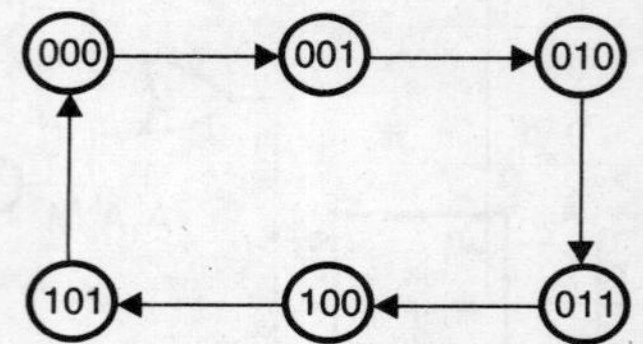

Fig. 1: State diagram of MOD-6 counter.

Table 2: State Table of MOD-6 Counter

Present states			Next states		
A_2	A_1	A_0	$A_2(t+1)$	$A_1(t+1)$	$A_0(t+1)$
0	0	0	0	0	1
0	0	1	0	1	0
0	1	0	0	1	1
0	1	1	1	0	0
1	0	0	1	0	1
1	0	1	0	0	0
1	1	0	×	×	×
1	1	1	×	×	×

Let us choose three JK flip-flops to store three state variables A_0, A_1 and A_2 as shown in Fig. 2. Here, A_0, A_1 and A_2 be output of flip-flops.

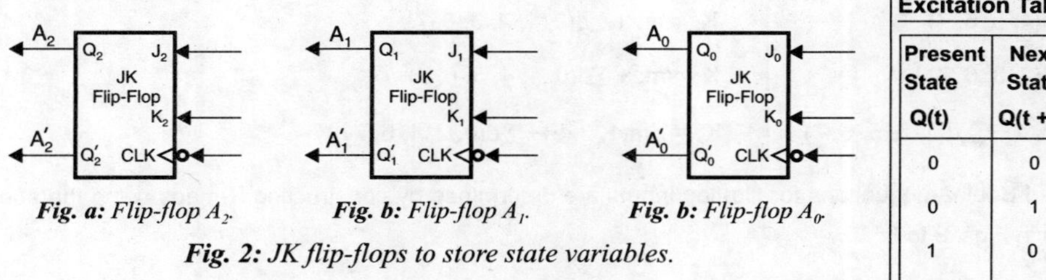

***Fig. a:** Flip-flop A_2.* ***Fig. b:** Flip-flop A_1.* ***Fig. b:** Flip-flop A_0.*

***Fig. 2:** JK flip-flops to store state variables.*

Excitation Table of JK Flip-Flop

Present State	Next State	Flip-Flop Inputs	
Q(t)	Q(t + 1)	J	K
0	0	0	×
0	1	1	×
1	0	×	1
1	1	×	0

The flip-flop inputs can be obtained from excitation table as shown below:

State A_2

For state A_2, the Q(t), Q(t + 1), J and K of excitation table are K of considered as A_2, $A_2(t + 1)$, J_2 and K_2.

$$\therefore Q(t) \rightarrow A_2 \;\; ; \;\; Q(t+1) \rightarrow A_2(t+1)$$

$$J \rightarrow J_2 \;\; ; \;\; K \rightarrow K_2$$

The flip-flop inputs J_2 and K_2 that drives the present state A_2 to next state is obtained from excitation table and listed in Table 3.

Table 3:

A_2	$A_2(t + 1)$	J_2	K_2
0	0	0	×
0	0	0	×
0	0	0	×
0	1	1	×
1	1	×	0
1	0	×	1

State A_1

For state A_1, the Q(t), Q(t + 1), J and excitation table are considered as A_1, $A_1(t + 1)$, J_1 and K_1.

$$\therefore Q(t) \rightarrow A_1 \;\; ; \;\; Q(t+1) \rightarrow A_1(t+1)$$

$$J \rightarrow J_1 \;\; ; \;\; K \rightarrow K_1$$

The flip-flop inputs J_1 and K_1 that drives the present state A_1 to next state is obtained from excitation table and listed in Table 4.

Table 4:

A_1	$A_1(t + 1)$	J_1	K_1
0	0	0	×
0	1	1	×
1	1	×	0
1	0	×	1
0	0	0	×
0	0	0	×

State A_0

For state A_0, the Q(t), Q(t + 1), J and K of excitation table are considered as A_0, $A_0(t + 1)$, J_0 and K_0.

$$\therefore Q(t) \rightarrow A_0 \;\; ; \;\; Q(t+1) \rightarrow A_0(t+1)$$

$$J \rightarrow J_0 \;\; ; \;\; K \rightarrow K_0$$

The flip-flop inputs J_0 and K_0 that drives the present state A_0 to next state is obtained from excitation table and listed in Table 5.

Table 5:

A_0	$A_0(t + 1)$	J_0	K_0
0	1	1	×
1	0	×	1
0	1	1	×
1	0	×	1
0	1	1	×
1	0	×	1

The tables 2 to 5 are combined to form state table with flip-flop inputs as shown in Table 6.

Table 6: Expanded State Table

Present States			Next States			Flip-Flop Inputs						Minterm
A_2	A_1	A_0	$A_2(t+1)$	$A_1(t+1)$	$A_0(t+1)$	J_2	K_2	J_1	K_1	J_0	K_0	$A_2 A_1 A_0$
0	0	0	0	0	1	0	×	0	×	1	×	m_0
0	0	1	0	1	0	0	×	1	×	×	1	m_1
0	1	0	0	1	1	0	×	×	0	1	×	m_2
0	1	1	1	0	0	1	×	×	1	×	1	m_3
1	0	0	1	0	1	×	0	0	×	1	×	m_4
1	0	1	0	0	0	×	1	0	×	×	1	m_5
1	1	0	×	×	×	×	×	×	×	×	×	m_6
1	1	1	×	×	×	×	×	×	×	×	×	m_7

The flip-flop inputs are functions of present states and so minterms are formed using present states and the Boolean equations for flip-flop inputs are obtained as sum of minterms and don't-cares as shown below:

$J_2 = m_3 + \Sigma d(4, 5, 6, 7)$ | $K_2 = m_5 + \Sigma d(0, 1, 2, 3, 6, 7)$

$J_1 = m_1 + \Sigma d(2, 3, 6, 7)$ | $K_1 = m_3 + \Sigma d(0, 1, 4, 5, 6, 7)$

$J_0 = \Sigma m(0, 2, 4) + \Sigma d(1, 3, 5, 6, 7)$ | $K_0 = \Sigma m(1, 3, 5) + \Sigma d(0, 2, 4, 6, 7)$

The simplified Boolean equations for flip-flop inputs are determined by constructing K-maps using the above equations as shown in Figs. 3 to 6.

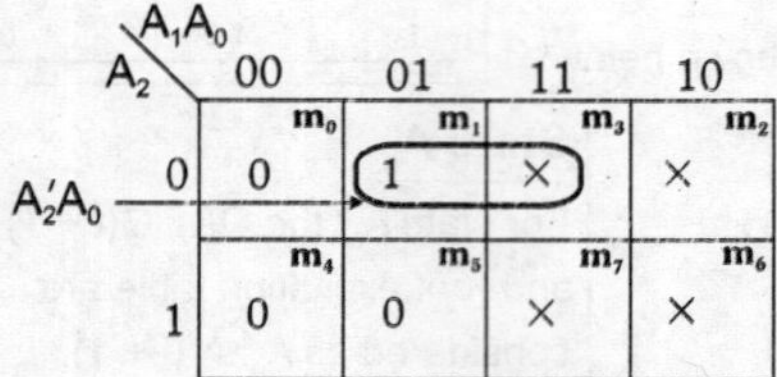

Fig. 3: K-map for J_1.

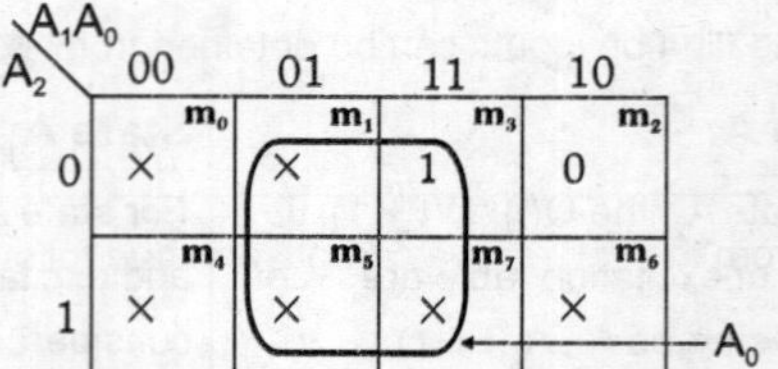

Fig. 4: K-map for K_1.

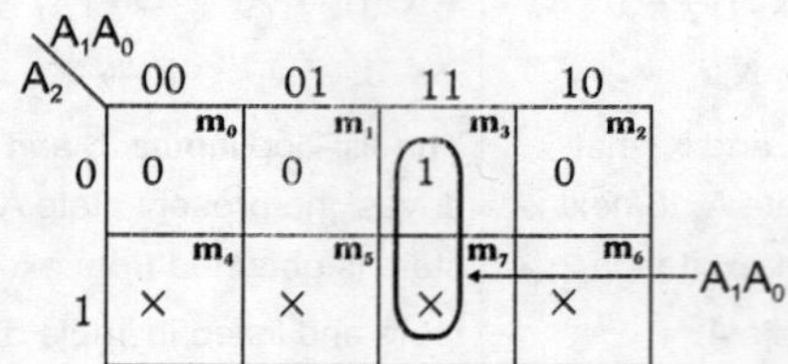

Fig. 5: K-map for J_2.

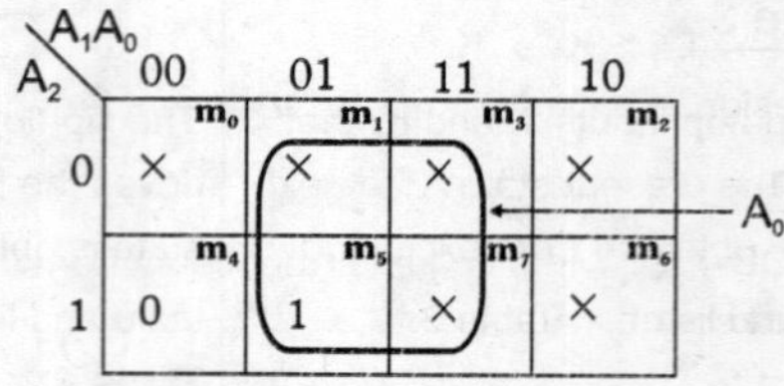

Fig. 6: K-map for K_2.

Here, J_0 and K_0 are always 1 or × and so no need of K-map and they can be permanently tied to 1.

$\therefore J_0 = K_0 = 1$

From the K-map we get the following Boolean equations for flip-flop inputs and using these equation the logic circuit of MOD-6 counter is drawn as shown in Fig. 7.

$J_1 = A_2' A_0$ | $K_1 = A_0$

$J_2 = A_1 A_0$ | $K_2 = A_0$

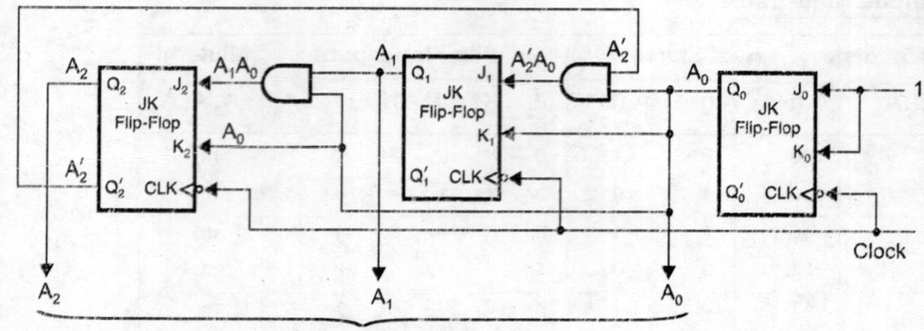

Fig. 7: Logic diagram of MOD-6 synchronous counter.

Check for Self Starting

In order to check for self starting, the flip-flop inputs are determined for each unused state using simplified equations of flip-flop inputs and listed in Table 7. The next state is obtained using characteristic table of JK flip-flop.

Characteristic Table of JK Flip-Flop

Input J K	Present State Q	Next State Q(t+1)
0 0	0	0
0 0	1	1
0 1	0	0
0 1	1	0
1 0	0	1
1 0	1	1
1 1	0	1
1 1	1	0

Table 7: Check for Self Starting

Unused Present States			Flip-Flop Inputs						Next States		
			J_2	K_2	J_1	K_1	J_0	K_0	$A_2(t+1)$	$A_1(t+1)$	$A_0(t+1)$
A_2	A_1	A_0	$= A_2' A_0$	$= A_0$	$= A_1 A_0$	$= A_0$	$= 1$	$= 1$			
1	1	0	0	0	0	0	1	1	1	1	1
1	1	1	0	1	1	1	1	1	0	0	0

From the Table 7 we can observe the following state transitions from unused state to next state.

110 → 111 → 000 ; 111 → 000

The unused states enters used state in the next transition as shown above. Hence the counter is self-starting.

Example 3.23

Design and implement a MOD-7 synchronous binary counter using T flip-flops.

Solution

MOD-7 is a binary counter that count from 0 to 6 and on reaching 7 the output should be reset to zero, in order to start next count sequence.

In order to count from 0 to 6 we need 3-bit binary output.

The three bits are represented by three state variables A_2, A_1 and A_0.

∴ Present state variables : $A_2(t), A_1(t), A_0(t) \Rightarrow A_2, A_1, A_0$ (For simplicity)

Next state variables : $A_2(t+1), A_1(t+1), A_0(t+1)$

With 3 state variables it is possible to form 8 ($2^3 = 8$) states.

The 8 states are 000, 001, 010, 011, 100, 101, 110 and 111.

In this except the state 111 all other states are used by MOD-7 counter. The truth table of MOD-7 counter is shown in Table 1. Here, the present states are the counter outputs.

Table 1: Truth Table of MOD-7 Counter

Clock	Counter Outputs		
	A_2	A_1	A_0
-	0	0	0
↓	0	0	1
↓	0	1	0
↓	0	1	1
↓	1	0	0
↓	1	0	1
↓	1	1	0
↓	0	0	0

The state diagram of MOD-7 counter is shown in Figs. 1 and 2. In Fig. 2 the unused state is connected to an used state to prevent lock-out condition.

From the knowledge of state transitions in the state diagram of Fig. 2, the stable table is formed as shown in Table 2.

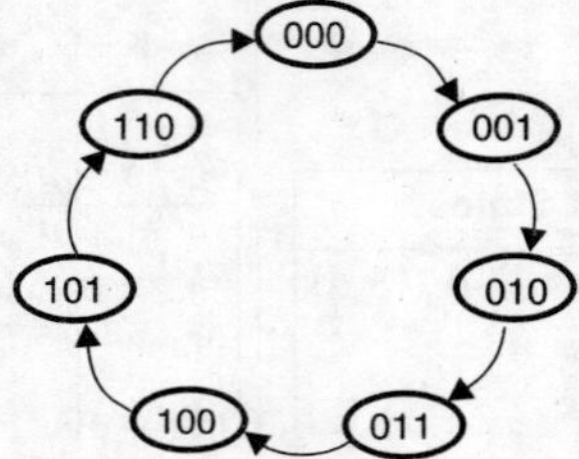

Fig. 1: State diagram of MOD-7 counter.

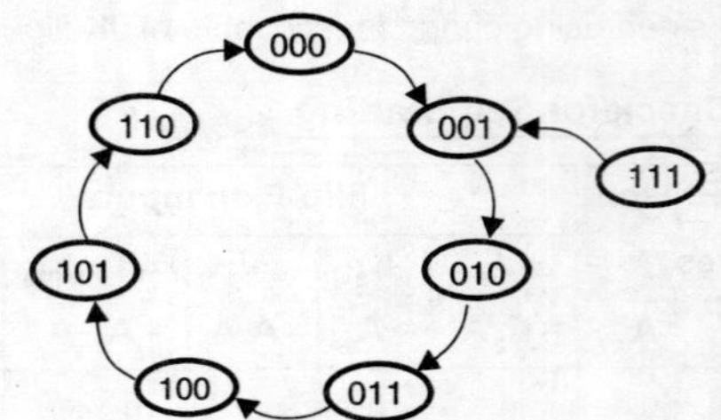

Fig. 2: State diagram of MOD-7 counter to prevent lock-out condition.

Table 2: State Table

Present States			Next States		
A_2	A_1	A_0	$A_2(t+1)$	$A_1(t+1)$	$A_0(t+1)$
0	0	0	0	0	1
0	0	1	0	1	0
0	1	0	0	1	1
0	1	1	1	0	0
1	0	0	1	0	1
1	0	1	1	1	0
1	1	0	0	0	0
1	1	1	0	0	1

Let us choose three T flip-flops to store three state variables A_0, A_1 and A_2 as shown in Fig. 3. Here, A_0, A_1 and A_2 are output of T flip-flops.

Excitation Table of T Flip-Flop

Present State Q(t)	Next State Q(t + 1)	Flip-Flop Input T
0	0	0
0	1	1
1	0	1
1	1	0

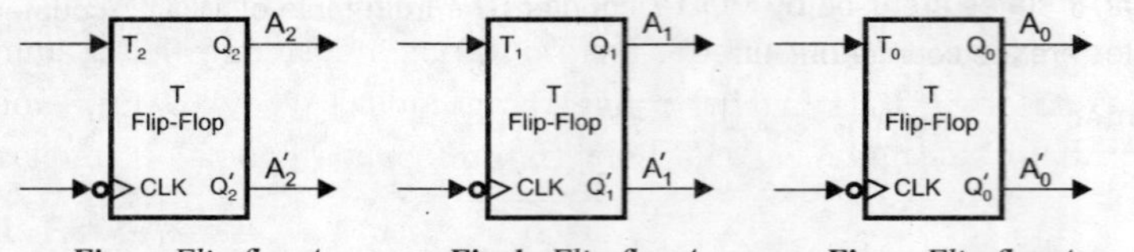

Fig. a: Flip-flop A_2. *Fig. b: Flip-flop A_1.* *Fig. c: Flip-flop A_0.*

Fig. 3: T flip-flop to store state variables.

The flip-flop inputs can be determined from excitation table of T flip-flop as shown below:

State A_2

For state A_2, the Q(t), Q(t + 1) and T of excitation table are considered as A_2, A_2(t + 1) and T_2.

$\therefore Q(t) \rightarrow A_2$; $Q(t+1) \rightarrow A_2(t+1)$

$T \rightarrow T_2$

The flip-flop input T_2 that drives the present state A_2 to next states is obtained from excitation table and listed in Table 3.

Table 3:

A_2	A_2(t+1)	T_2
0	0	0
0	0	0
0	0	0
0	1	1
1	1	0
1	1	0
1	0	1
1	0	1

State A_1

For state A_1, the Q, Q(t + 1) and T of excitation table are considered as A_1, A_1(t + 1) and T_1.

$\therefore Q(t) \rightarrow A_1$; $Q(t+1) \rightarrow A_1(t+1)$

$T \rightarrow T_1$

The flip-flop input T_1 that drives the present state T_1 to next states is obtained from excitation table and listed in Table 4.

Table 4:

A_1	A_1(t+1)	T_1
0	0	0
0	1	1
1	1	0
1	0	1
0	0	0
0	1	1
1	0	1
1	0	1

State A_0

For state A_0, the Q(t), Q(t + 1) and T of excitation table are considered as A_0, A_0(t + 1) and T_0.

$\therefore Q(t) \rightarrow A_0$; $Q(t+1) \rightarrow A_0(t+1)$

$T \rightarrow T_0$

The flip-flop input T_0 that drives the present state A_0 to next states is obtained from excitation table and listed in Table 5.

Table 5:

A_0	A_0(t+1)	T_0
0	1	1
1	0	1
0	1	1
1	0	1
0	1	1
1	0	1
0	0	0
1	1	0

The tables 2 to 5 are combined to form expanded state table with flip-flop inputs as shown in Table 6.

Table 6: Expanded State Table

Present States			Next States			Flip-Flop Inputs			Minterm
A_2	A_1	A_0	A_2(t+1)	A_1(t+1)	A_0(t+1)	T_2	T_1	T_0	$A_2 A_1 A_0$
0	0	0	0	0	1	0	0	1	m_0
0	0	1	0	1	0	0	1	1	m_1
0	1	0	0	1	1	0	0	1	m_2
0	1	1	1	0	0	1	1	1	m_3
1	0	0	1	0	1	0	0	1	m_4
1	0	1	1	1	0	0	1	1	m_5
1	1	0	0	0	0	1	1	0	m_6
1	1	1	0	0	1	1	1	0	m_7

Note: *Here, present state $A_2 A_1 A_0$ is the counter output.*

The flip-flop inputs are functions of present states. Hence, minterms are formed using present states and the Boolean equations for T_2, T_1 and T_0 are obtained from Table 6 as sum as minterms as shown below:

$$T_2 = \Sigma m(3, 6, 7)$$

$$T_1 = \Sigma m(1, 3, 5, 6, 7)$$

$$T_0 = \Sigma m(0, 1, 2, 3, 4, 5)$$

The simplified Boolean equations are obtained by constructing K-map for T_2, T_1 and T_0 as shown in Figs. 4 to 6.

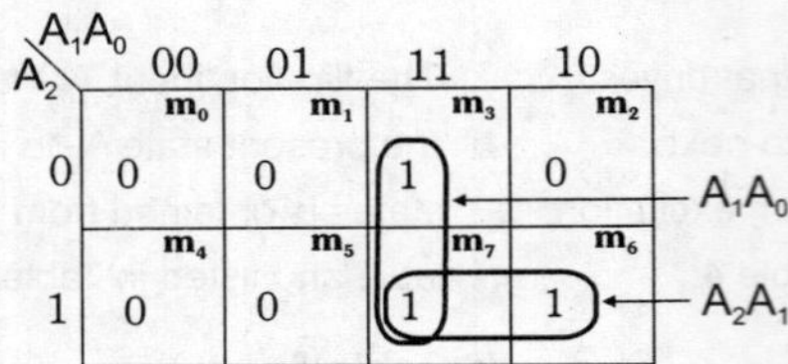

Fig. 4: K-map for T_2.

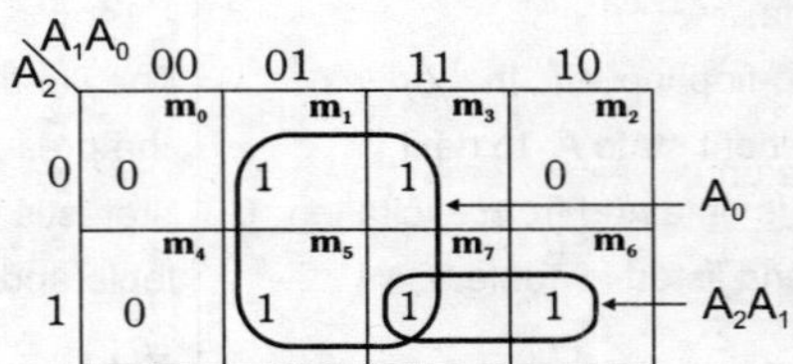

Fig. 5: K-map for T_1.

From the K-map we get the following simplified Boolean equations to realize 4-bit binary counter using T flip-flops and they are used to construct the logic circuit of synchronous MOD-7 counter as shown in Fig. 7.

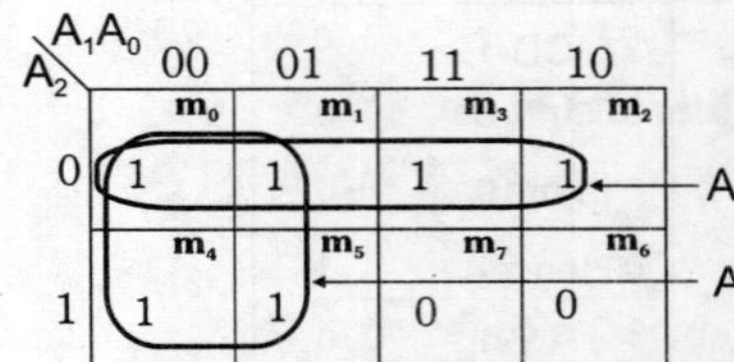

Fig. 6: K-map for T_0.

$$T_2 = A_2A_1 + A_1A_0$$

$$T_1 = A_2A_1 + A_0$$

$$T_0 = A_2' + A_1'$$

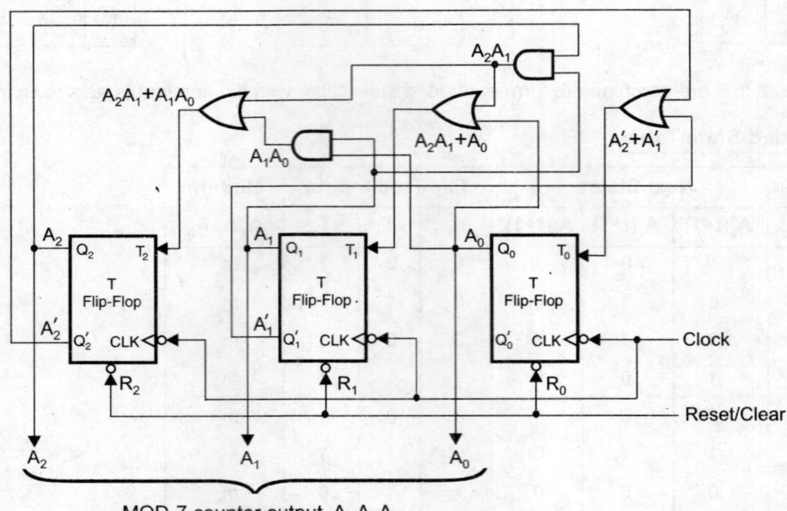

Fig. 7: Logic diagram of MOD-7 synchronous counter using T flip-flop.

Check for Lock-out Condition

In order to check lock-out condition the flip-flop inputs are determined for each unused state using simplified equation of flip-flop inputs and listed in Table 7. The next state are obtained from characteristic table of T flip-flop.

Table 7: Check for Lock-out Condition

Unused Present States			Flip-Flop Inputs			Next States		
			T_2	T_1	T_0	$A_2(t+1)$	$A_1(t+1)$	$A_0(t+1)$
A_2	A_1	A_0	$= A_2A_1 + A_1A_0$	$= A_2A_1 + A_0$	$= A_2' + A_1'$			
1	1	1	1	1	0	0	0	1

Characteristic Table of T Flip-Flop

Input T	Present State Q	Next State Q(t+1)
0	0	0
0	1	1
1	0	1
1	1	0

From the Table 7 we can observe the following state transition from unused state to next state.

$111 \rightarrow 001$

The unused state enter an used state in the next transition and so the counter is self starting.

Example 3.24

Design a MOD-12 synchronous binary up counter using T flip-flops.

Solution

MOD-12 is a binary counter that count from 0 to 11_{10} and on reaching 12_{10} the output should be reset to zero in order to start next count sequence.

In order to count from 0 to 11_{10} we need 4-bit binary output.

Hence we need 4 bit binary code to represent the decimal sequence. Let us assign binary value of decimal number as code for states. With 4-bit code 16 ($2^4 = 16$) states are possible and in this 12 states are used states and remaining 4 states are unused states.

The code assignment for states are as follows,

Binary code assignment

0	0000
1	0001
2	0010
3	0011
4	0100
5	0101
6	0110
7	0111
8	1000
9	1001
10	1010
11	1011

Unused binary code

12	1100
13	1101
14	1110
15	1111

The four bits of binary codes are represented by four state variables A_3, A_2, A_1 and A_0.

$\therefore$ Present state variables : $A_3(t), A_2(t), A_1(t), A_0(t) \Rightarrow A_3, A_2, A_1, A_0$ (For simplicity)

Next state variables : $A_3(t+1), A_2(t+1), A_1(t+1), A_0(t+1)$

The truth table of MOD-12 counter is shown in Table 1. Here, the present states are counter outputs. The state diagram using only used states is shown in Fig. 1.

Table 1: Truth Table of MOD-12 Counter

Clock	Counter Outputs			
	A_3	A_2	A_1	A_0
-	0	0	0	0
↓	0	0	0	1
↓	0	0	1	0
↓	0	0	1	1
↓	0	1	0	0
↓	0	1	0	1
↓	0	1	1	0
↓	0	1	1	1
↓	1	0	0	0
↓	1	0	0	1
↓	1	0	1	1
↓	0	0	0	0

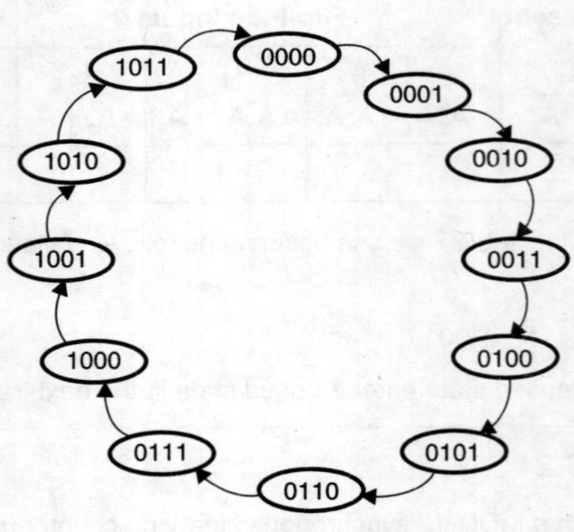

Fig. 1: Used states.

From the knowledge of state transitions in the state diagram of Fig. 1, the state table is constructed as shown in Table 2. Here, the next state for unused states are considered as don't-cares.

Table 2: State Table

Present States				Next States			
A_3	A_2	A_1	A_0	$A_3(t+1)$	$A_2(t+1)$	$A_1(t+1)$	$A_0(t+1)$
0	0	0	0	0	0	0	1
0	0	0	1	0	0	1	0
0	0	1	0	0	0	1	1
0	0	1	1	0	1	0	0
0	1	0	0	0	1	0	1
0	1	0	1	0	1	1	0
0	1	1	0	0	1	1	1
0	1	1	1	1	0	0	0
1	0	0	0	1	0	0	1
1	0	0	1	1	0	1	0
1	0	1	0	1	0	1	1
1	0	1	1	0	0	0	0
1	1	0	0	×	×	×	×
1	1	0	1	×	×	×	×
1	1	1	0	×	×	×	×
1	1	1	1	×	×	×	×

Excitation Table of T Flip-Flop

Present State Q(t)	Next State Q(t + 1)	Flip-Flop Input T
0	0	0
0	1	1
1	0	1
1	1	0

Let us choose four T flip-flops to store four state variables A_0, A_1, A_2 and A_3 as shown in Fig. 2. Here, A_0, A_1, A_2 and A_3 are output of T flip-flops.

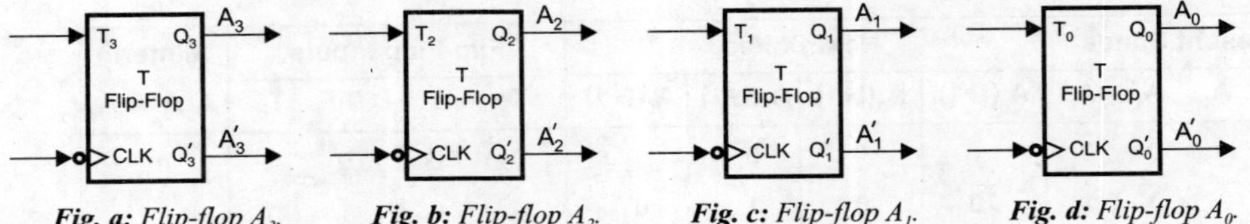

***Fig. a:** Flip-flop A_3.* ***Fig. b:** Flip-flop A_2.* ***Fig. c:** Flip-flop A_1.* ***Fig. d:** Flip-flop A_0.*

***Fig. 2:** T flip-flop to store state variables.*

The flip-flop inputs are determined using excitation table of T flip-flop as shown below:

State A_3

For state A_3, the Q(t), Q(t + 1) and T of excitation table are considered as A_3, $A_3(t + 1)$ and T_3.

$$\therefore Q(t) \rightarrow A_3 \quad ; \quad Q(t+1) \rightarrow A_3(t+1) \quad ; \quad T \rightarrow T_3$$

The flip-flop input T_3 that drives the present state A_3 to next states is obtained from excitation table and listed in Table 3.

State A_2

For state A_2, the Q(t), Q(t + 1) and T of excitation table are considered as A_2, $A_2(t + 1)$ and T_2.

$$\therefore Q(t) \rightarrow A_2 \quad ; \quad Q(t+1) \rightarrow A_2(t+1) \quad ; \quad T \rightarrow T_2$$

The flip-flop input T_2 that drives the present state A_2 to next states is obtained from excitation table and listed in Table 4.

State A_1

For state A_1, the Q(t), Q(t + 1) and T of excitation table are considered as A_1, $A_1(t + 1)$ and T_1.

$$\therefore Q(t) \rightarrow A_1 \quad ; \quad Q(t+1) \rightarrow A_1(t+1) \quad ; \quad T \rightarrow T_1$$

The flip-flop input T_1 that drives the present state A_1 to next states is obtained from excitation table and listed in Table 5.

State A_0

For state A_0, the Q(t), Q(t + 1) and T of excitation table are considered as A_0, $A_0(t + 1)$ and T_0.

$$\therefore Q(t) \rightarrow A_0 \quad ; \quad Q(t+1) \rightarrow A_0(t+1) \quad ; \quad T \rightarrow T_0$$

The flip-flop input T_0 that drives the present state A_0 to next states is obtained from excitation table and listed in Table 6.

Table 3:

A_3	$A_3(t+1)$	T_3
0	0	0
0	0	0
0	0	0
0	0	0
0	0	0
0	0	0
0	0	0
0	1	1
1	1	0
1	1	0
1	1	0
1	0	1
1	×	×
1	×	×
1	×	×
1	×	×

Table 4:

A_2	$A_2(t+1)$	T_2
0	0	0
0	0	0
0	0	0
0	1	1
1	1	0
1	1	0
1	1	0
1	0	1
0	0	0
0	0	0
0	0	0
0	0	0
1	×	×
1	×	×
1	×	×
1	×	×

Table 5:

A_1	$A_1(t+1)$	T_1
0	0	0
0	1	1
1	1	0
1	0	1
0	0	0
0	1	1
1	1	0
1	0	1
0	0	0
0	1	1
1	1	0
1	0	1
0	×	×
0	×	×
1	×	×
1	×	×

Table 6:

A_0	$A_0(t+1)$	T_0
0	1	1
1	0	1
0	1	1
1	0	1
0	1	1
1	0	1
0	1	1
1	0	1
0	1	1
1	0	1
0	1	1
1	0	1
0	×	×
1	×	×
0	×	×
1	×	×

The Tables 2 to 6 are combined to form expanded state table with flip-flop inputs as shown in Table 7.

Table 7: Expanded State Table

Present States				Next States				Flip-Flop Inputs				Minterm
A_3	A_2	A_1	A_0	$A_3(t+1)$	$A_2(t+1)$	$A_1(t+1)$	$A_0(t+1)$	T_3	T_2	T_1	T_0	$A_3A_2A_1A_0$
0	0	0	0	0	0	0	1	0	0	0	1	m_0
0	0	0	1	0	0	1	0	0	0	1	1	m_1
0	0	1	0	0	0	1	1	0	0	0	1	m_2
0	0	1	1	0	1	0	0	0	1	1	1	m_3
0	1	0	0	0	1	0	1	0	0	0	1	m_4
0	1	0	1	0	1	1	0	0	0	1	1	m_5
0	1	1	0	0	1	1	1	0	0	0	1	m_6
0	1	1	1	1	0	0	0	1	1	1	1	m_7
1	0	0	0	1	0	0	1	0	0	0	1	m_8
1	0	0	1	1	0	1	0	0	0	1	1	m_9
1	0	1	0	1	0	1	1	0	0	0	1	m_{10}
1	0	1	1	0	0	0	0	1	0	1	1	m_{11}
1	1	0	0	×	×	×	×	×	×	×	×	m_{12}
1	1	0	1	×	×	×	×	×	×	×	×	m_{13}
1	1	1	0	×	×	×	×	×	×	×	×	m_{14}
1	1	1	1	×	×	×	×	×	×	×	×	m_{15}

***Note:** Here, present state $A_3\,A_2\,A_1\,A_0$ is the counter output.*

The flip-flop inputs are functions of present states and so minterms are formed using present state and the Boolean equations for T_3, T_2, T_1 and T_0 are obtained from Table 7 as sum of minterms and don't-cares as shown below:

$$T_3 = \Sigma m(7, 11) + \Sigma d(12, 13, 14, 15)$$

$$T_2 = \Sigma m(3, 7) + \Sigma d(12, 13, 14, 15)$$

$$T_1 = \Sigma m(1, 3, 5, 7, 9, 11) + \Sigma d(12, 13, 14, 15)$$

$$T_0 = \Sigma m(0, 1, 2, 3, 4, 5, 6, 7, 8, 9, 10, 11) + \Sigma d(12, 13, 14, 15)$$

The simplified Boolean equations are obtained by constructing K-map for T_3, T_2, T_1 and T_0 as shown in Figs. 3 to 6.

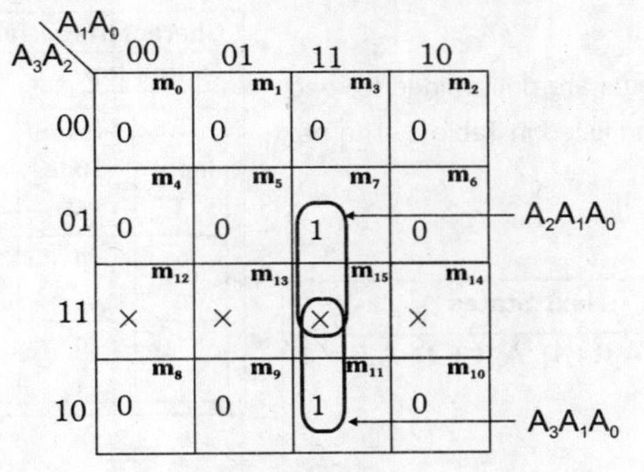

Fig. 3: K-map for T_3.

A_3A_2 \ A_1A_0	00	01	11	10
00	m_0 0	m_1 0	m_3 1	m_2 0
01	m_4 0	m_5 0	m_7 1	m_6 0
11	m_{12} ×	m_{13} ×	m_{15} ×	m_{14} ×
10	m_8 0	m_9 0	m_{11} 0	m_{10} 0

$\leftarrow A'_3A_1A_0$

Fig. 4: K-map for T_2.

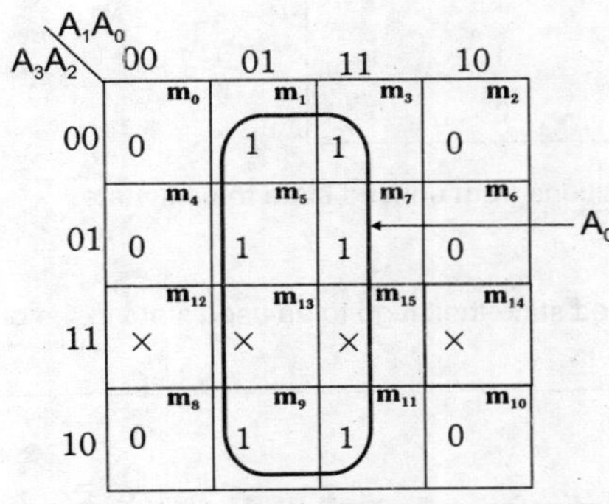

Fig. 5: K-map for T_1.

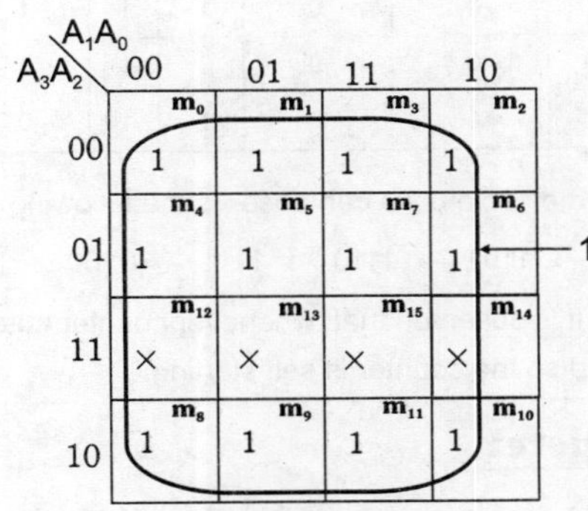

Fig. 6: K-map for T_0.

From the K-map we get the following simplified Boolean equations to realize MOD-12 counter using T flip-flops. Using the following equations, the MOD-12 synchronous counter is implemented as shown in Fig. 7.

$$T_3 = A_2A_1A_0 + A_3A_1A_0 \qquad T_1 = A_0$$

$$T_2 = A'_3A_1A_0 \qquad T_0 = 1$$

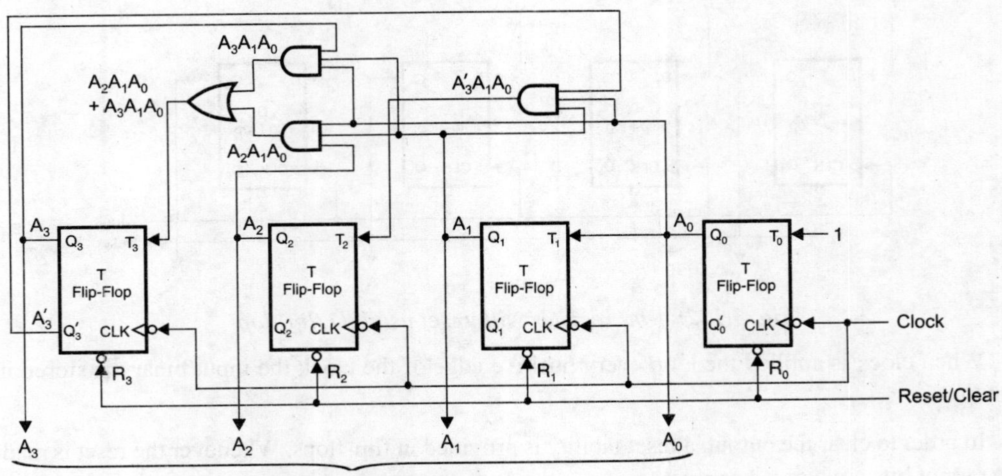

Fig. 7: Logic diagram of MOD-12 synchronous counter using T flip-flop.

Check for Lock-out Condition

In order to check lock-out condition the flip-flop inputs are determined for each unused state using simplified equation of flip-flop inputs and listed in Table 8. The next state are obtained from characteristic table of T flip-flop.

Characteristic Table of T Flip-Flop

Input T	Present State Q	Next State Q(t+1)
0	0	0
0	1	1
1	0	1
1	1	0

Table 8: Check for Lock-out Condition

Unused Present States $A_3\ A_2\ A_1\ A_0$	Flip-Flop Inputs $T_3 = A_2A_1A_0 + A_3A_1A_0$	$T_2 = A_3'A_1A_0$	$T_1 = A_0$	$T_0 = 1$	Next States $A_3(t+1)$	$A_2(t+1)$	$A_1(t+1)$	$A_0(t+1)$
1 1 0 0	0	0	0	1	1	1	0	1
1 1 0 1	0	0	1	1	1	1	1	0
1 1 1 0	0	0	0	1	1	1	1	1
1 1 1 1	1	0	1	1	0	1	0	0

From the Table 8 we can observe the following state transitions from unused state to next state.

1100 → 1101 → 1110 → 1111 → 0100

Here, it is observed that, whenever counter enter an unused state then it go to an used state in the one to four transitions and so the counter is self starting.

3.10 Register

A ***register*** is a group of flip-flops that can store binary information. Each flip-flop in a register can store one bit binary and so, in order to store n-bit binary the register should have n flip-flops.

All the flip-flops in a register are controlled or synchronized to a common clock. A simple 4-bit register using D flip-flops is shown in Fig. 3.122.

Let, I_3, I_2, I_1 and I_0 are 4-bit input and A_3, A_2, A_1 and A_0 are 4-bit output.

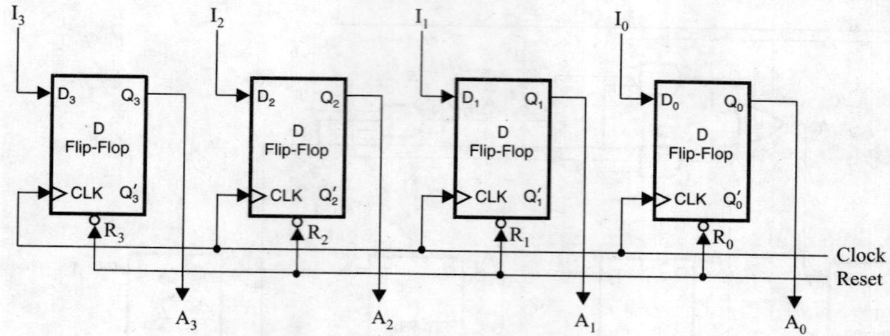

Fig. 3.122: 4-bit register with reset using D flip-flop.

When clock is applied then, on every positive edge of the clock the input binary is stored in flip-flop output.

In order to clear the output, a reset facility is provided in flip-flops. Whenever the reset is made low, all the output bits are cleared or reset to zero.

Register with Parallel Load Control

(AU, Nov/Dec'18, 13 Marks)

A 4-bit register with logic **high** load control is shown in Fig. 3.123. Generally in a digital system the clock will be used by or all the devices in the system. Therefore, an individual control is needed for every register in order to load data or store data. A practical register will have a load control (or write control) so that the input data will be stored in the register only when load control is activated.

Also, the D flip-flop require a facility to rewrite the output data to input so that the output data is preserved on every clock pulse when the load control is inactive.

Let, L = Load control input

L′= Complement of load control input

The load control L is logically ANDed with input data and complement of load control L′is logically ANDed with flip-flop output separately. The output of both AND gates are logically ORed and applied as input to D flip-flop.

When L = 1 and L′= 0 then input data is passed to flip-flop input and this logic is illustrated in Fig. 3.124.

When L = 0 and L′= 1 then flip-flop output is rewritten as input and this logic is illustrated in Fig. 3.125.

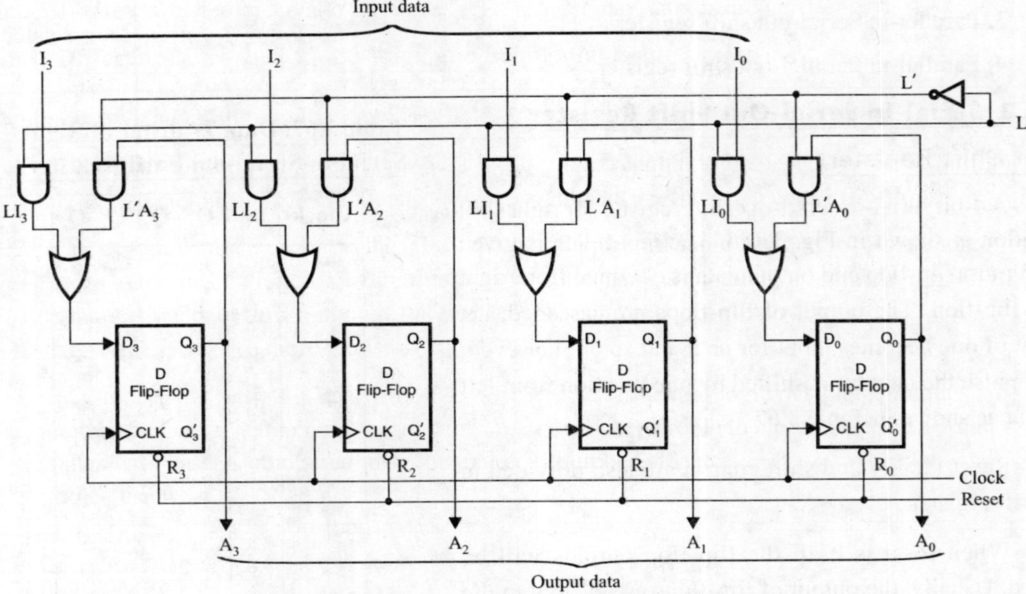

Fig. 3.123: 4-bit register with high parallel load control.

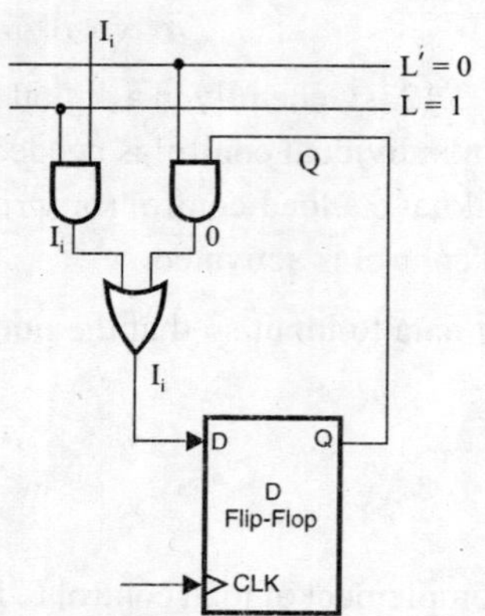

Fig. 3.124: Load input data.

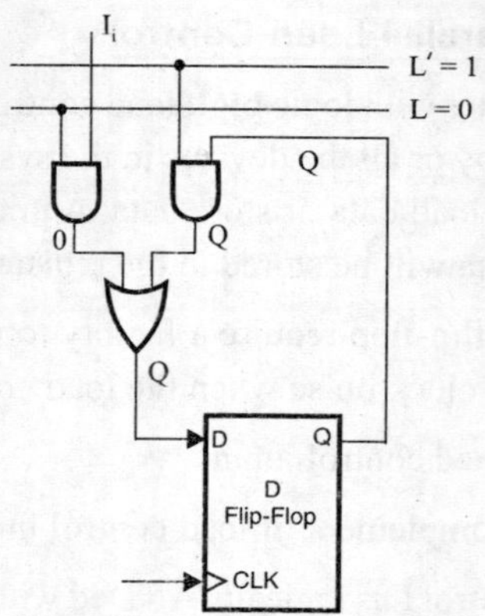

Fig. 3.125: Rewriting output to input.

3.11 Shift Register

A register capable of shifting the binary information from one flip-flop to another (or one cell to another) is called ***shift register***.

There are four types of shift registers.

1. Serial-in Serial-out shift register
2. Serial-in Parallel-out shift register
3. Parallel-in Serial-out shift register
4. Parallel-in Parallel-out shift register.

3.11.1 Serial-In Serial-Out Shift Register

Right Shift Register

A 4-bit serial-in serial-out shift register for right shift operation is shown in Fig. 3.126. The input data is given to left most flip-flop and output data is obtained from right most flip-flop. The output of flip-flops are cascaded, i.e, output of one becomes input for next and so on. For every clock pulse the data bit is shifted by one position from left to right as shown in Table 3.67

Reset facility and shift control are also provided in the shift register.

When reset is 0 all the flip-flop outputs will be cleared. Usually, the output of flip-flops are cleared in the beginning by making reset **low** and then reset is tied to **high**.

Table 3.67: Data Transfer in Right Shift Serial-in Serial-out Shift Register

Clock	D_{in}	Q_3	Q_2	Q_1	Q_0	D_{out}
-	d_0	×	×	×	×	×
↑	d_0	d_0	×	×	×	×
↑	d_1	d_1	d_0	×	×	×
↑	d_2	d_2	d_1	d_0	×	×
↑	d_3	d_3	d_2	d_1	d_0	d_0
↑	×	×	d_3	d_2	d_1	d_1
↑	×	×	×	d_3	d_2	d_2
↑	×	×	×	×	d_3	d_3

The shift control is provided to allow or disallow the clock. When shift control is 1, clock is allowed. When shift control is 0, clock is not allowed. The data will shift only when clock is allowed and so shift operation is performed as long as shift control is 1.

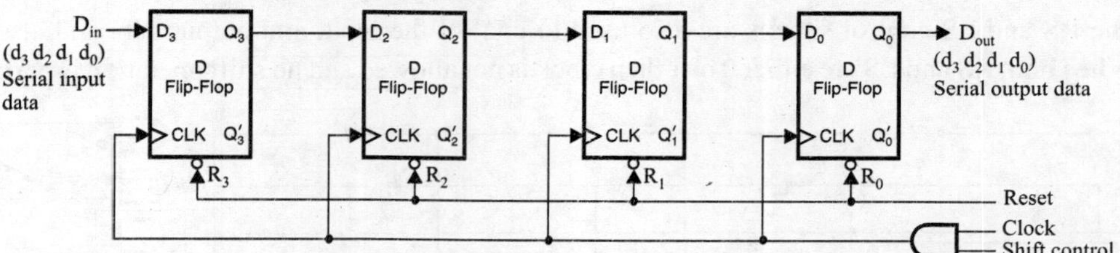

Fig. 3.126: Serial-in serial-out shift register for right shift (shift left to right).

Left Shift Register

For left shift operation the shifting is performed from right to left and the connections in Fig. 3.126 should be modified as shown in Fig. 3.127.

The input data is given to right most flip-flop and the output data is obtained from left most flip-flop. The shift operation from right to left is shown in Table 3.68. The reset and shift control are similar to left to right shift register.

Table 3.68: Data Transfer in Left Shift Serial-in Serial-out Shift Register

Clock	D_{out}	Q_3	Q_2	Q_1	Q_0	D_{in}
-	×	×	×	×	×	d_0
↑	×	×	×	×	d_0	d_0
↑	×	×	×	d_0	d_1	d_1
↑	×	×	d_0	d_1	d_2	d_2
↑	d_0	d_0	d_1	d_2	d_3	d_3
↑	d_1	d_1	d_2	d_3	×	×
↑	d_2	d_2	d_3	×	×	×
↑	d_3	d_3	×	×	×	×

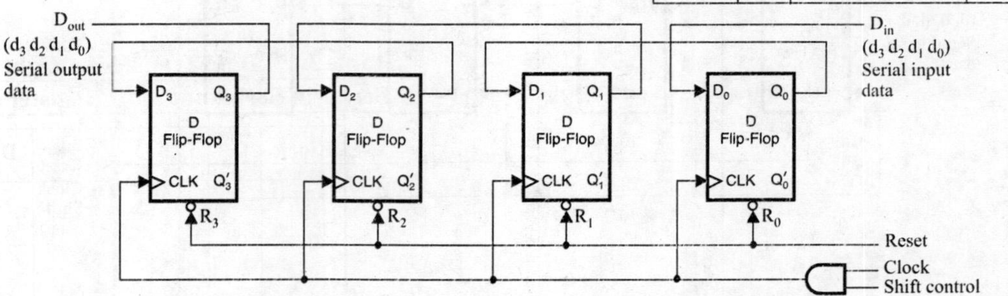

Fig. 3.127: Serial-in serial-out shift register for left shift (shift right to left).

Shift Register for Right and Left Shift Operation

The left to right and right to left shift operation can be performed in a single serial-in serial-out shift register with additional logic circuits and two shift control signals as shown in Fig. 3.128.

The input of flip-flops are applied through AND-OR combinational circuit. The flip-flop input for left to right shift is logically ANDed with RS control signal. The flip-flop input for right to left shift is logically ANDed with LS control signal. The output of both AND gates are logically ORed and applied as flip-flop inputs. The arrangement will allow only one shift data to flip-flop inputs.

When RS = 1 and LS = 0, the left to right shift data is allowed to flip-flop inputs to perform right shift operation. When RS = 0 and LS = 1, the right to left shift data is allowed to flip-flop inputs to perform left shift operation.

The RS and LS control signals are also used to control the input and output data and also clock signal. When both RS and LS are either 0 or 1 then clock is not allowed and no shift operation is performed.

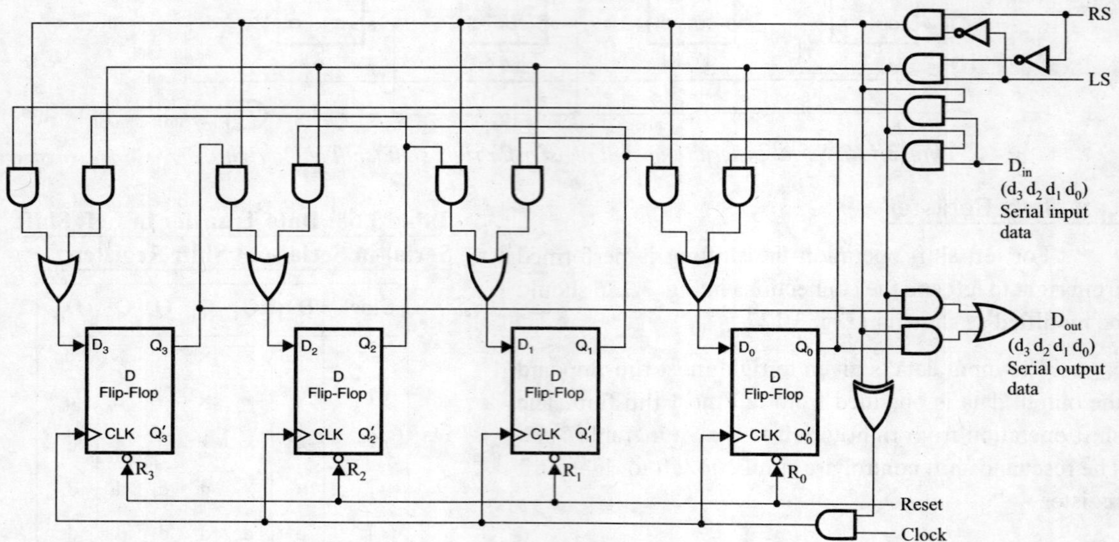

Fig. 3.128: Serial-in serial-out shift register for both right and left shift.

3.11.2 Serial-In Parallel-Out Shift Register

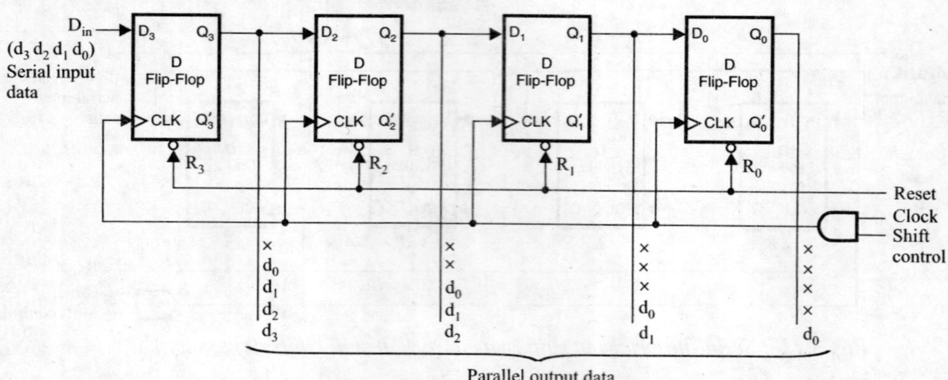

Fig. 3.129: Serial-in parallel-out shift register.

A 4-bit serial-in parallel-out shift register is shown in Fig. 3.129. The input data is applied to left most flip-flop. The reset and shift control are similar to serial-in serial-out shift register. Here, 4 clock pulses are needed to load the data in the shift register by right shift operation.

Here input data bit is shifted one position from left to right for every clock pulse provided shift control is **high**. But the all the 4-bit output are available for read operation at any one time. The left shift of input data is possible if input connections are modified as shown in left shift register of Fig. 3.127.

3.11.3 Parallel-In Serial-Out Shift Register

A 4-bit parallel-in serial-out shift register is shown in Fig. 3.130. Here, the flip-flops are connected to output data by right shift operation. The shift register require two control inputs for proper operation. One is parallel load control and the other is shift control. Here, the complement of shift control can be used for load control.

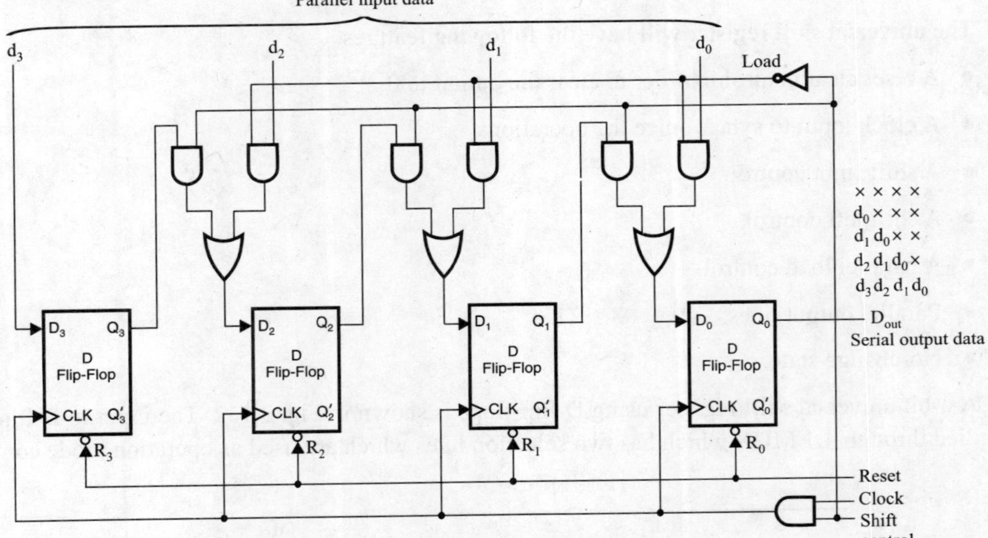

Fig. 3.130: Parallel-in serial-out shift register.

When shift control is 0, the data can be loaded in parallel and clock also not allowed. When shift control is 1, the data is shifted left to right for every positive edge of clock. The data output by left shift is possible if connections are modified for left shift of output as shown in Fig. 3.127.

3.11.4 Parallel-In Parallel-Out Shift Register

A 4-bit parallel-in parallel-out shift register is shown in Fig. 3.131. Here, a load control signal is logically ANDed with clock to allow or disallow the clock. When load control is 1, the clock is allowed and input data is loaded as the output. When load control is 0, the clock is not allowed and input data is not loaded in output.

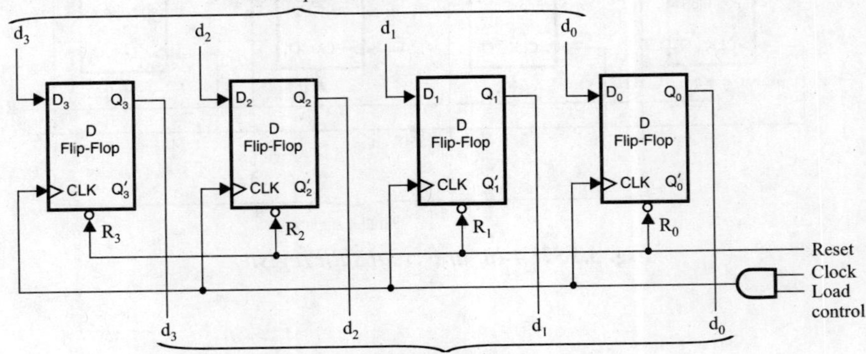

Fig. 3.131: Parallel-in parallel-out shift register.

3.11.5 Universal Shift Register

(AU, Apr/May'19, 6 Marks)

A shift register that can accept serial or parallel input data and output data in parallel along with right or left shift operation is called ***universal shift register***.

Therefore, an universal shift register can be used to implement any type of shift register with right/left shift selection.

The universal shift register will have the following features.

- A reset/clear control in order to clear the output to 0.
- A clock input to synchronize the operations.
- A shift right control.
- A shift left control.
- A parallel load control.
- Parallel output lines.
- No-change state.

A 4-bit universal shift register using D flip-flops is shown in Fig. 3.132. The input to D flip-flops are applied through 4:1 MUX which has two selection lines which are used as operation mode control.

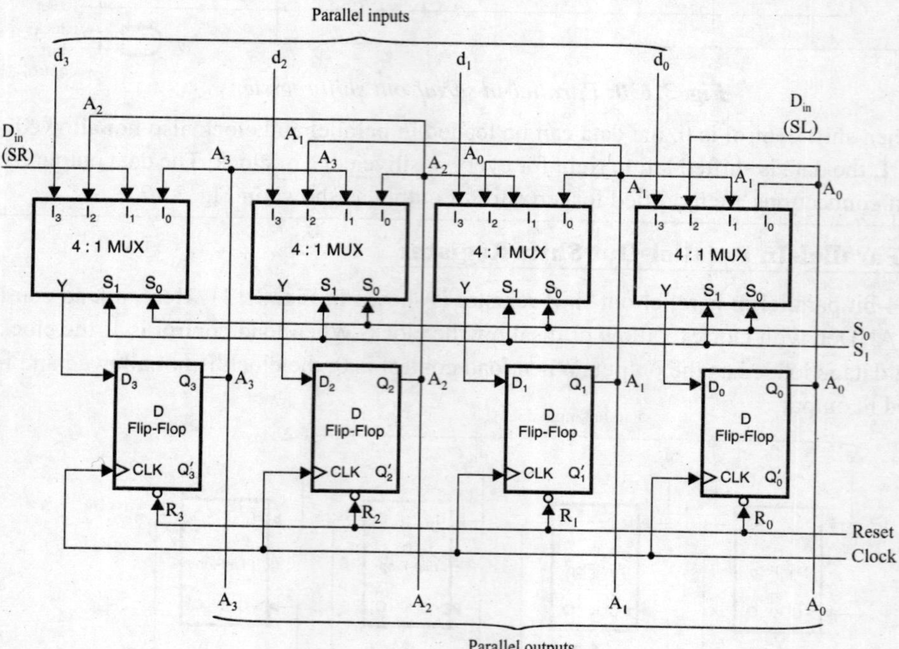

Fig. 3.132: 4-bit universal shift register.

Table 3.69: Operation Modes of Universal Shift Register

Operation Mode	Mode Selection S_1	S_0	Operation
Mode-0	0	0	No change
Mode-1	0	1	Shift right
Mode-2	1	0	Shift left
Mode-3	1	1	Parallel load

Mode-0 : $S_1 = S_0 = 0$, No Change in Output

In the universal shift register the, output of each flip-flop is connected to input 4:1 MUX that is used to input data to D flip-flop as shown in Fig. 3.133.

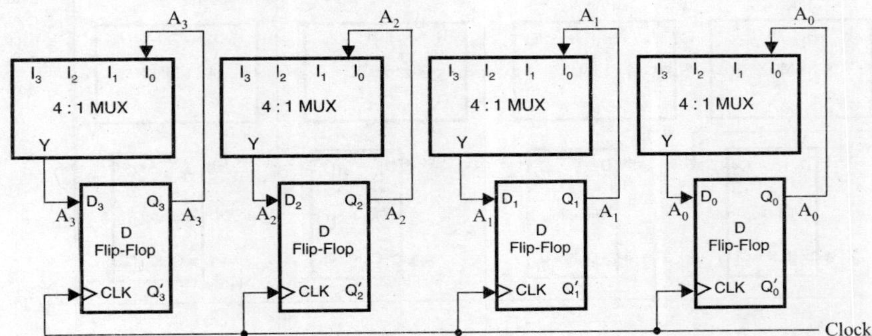

Fig. 3.133: Mode-0 (no change) operation of universal shift register.

Therefore, when $S_1 = S_0 = 0$, the MUX input I_0 is passed to output Y of MUX and so the input of D flip-flops are same as outputs. Hence, for every clock the output is rewritten as input and so there is no change in output.

Mode-1 : $S_1 = 0$, $S_0 = 1$, Shift Right

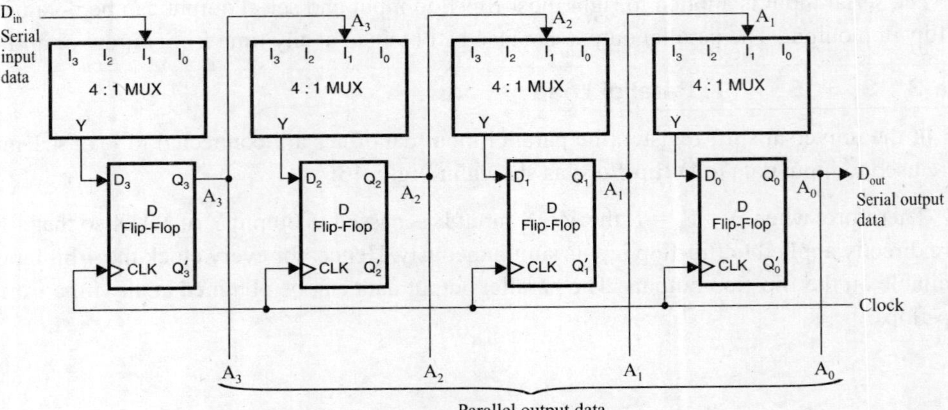

Fig. 3.134: Mode-1 (shift right) operation of universal shift register.

In the universal shift register, the output of a higher order flip-flop is connected to input I_1 of 4:1 MUX that is used to input data to D flip-flop as shown in Fig. 3.134.

Therefore, when $S_1 = 0$ and $S_0 = 1$, the MUX input I_1 is passed to output Y of MUX so that the output of higher order flip-flop is applied as input to lower order flip-flop. Hence, for every clock the output is shifted right by one position.

The serial input is applied to left most flip-flop input and serial output can be obtained from right most flip-flop output. The parallel output can also be obtained at any time from output of all flip-flops.

Mode-2 : $S_1 = 1, S_0 = 0$, Shift Left

In the universal shift register, the output of a lower order flip-flop is connected to input I_2 of 4:1 MUX that is used to input data to D flip-flop as shown in Fig. 3.135.

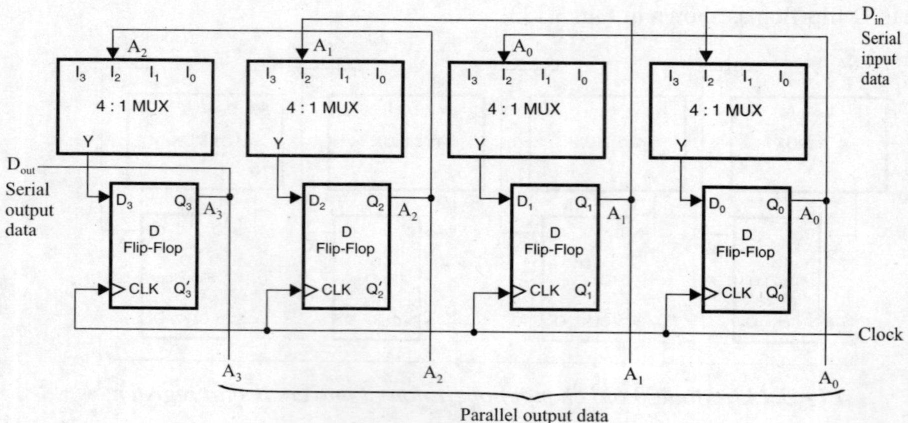

Fig. 3.135: Mode-2 (shift left) operation of universal shift register.

Therefore, when $S_1 = 1$ and $S_0 = 0$, the MUX input I_2 is passed to output Y of MUX so that the output of a lower order flip-flop is applied as input to higher order flip-flop. Hence, for every clock the output is shifted left by one position.

The serial input is applied to right most flip-flop input and serial output can be obtained from left most flip-flop output. The parallel output can also be obtained at any time from output of all flip-flops.

Mode-3 : $S_1 = S_0 = 1$, Parallel Load

In the universal shift register, the parallel input data lines are connected to I_3 of 4:1 multiplexers that are used to input data to D flip-flops as shown in Fig. 3.136.

Therefore, when $S_1 = S_0 = 1$, the MUX input I_3 is passed to output Y of MUX so that all input data bits are directly applied to flip-flop inputs simultaneously. Hence, for every clock the 4-bit input data will be available on the flip-flop outputs. The parallel output data can be obtained at any time from output of all flip-flops.

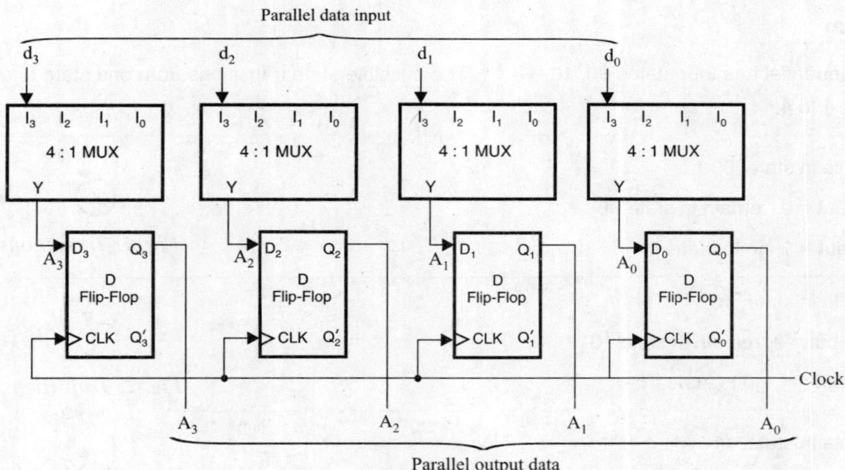

Fig. 3.136: *Mode-3 (parallel load) operation of universal shift register.*

The universal shift register discussed in this section is available as a standard IC 74194. The pin configuration of IC 74194 is shown in Fig. 3.137. The operation modes are same as that of Table 3.69.

74194

Pin	Signal	Pin	Signal
1	$\overline{CLR}$	16	V_{cc}(+5V)
2	SR_{in}	15	Q_A
3	A	14	Q_B
4	B	13	Q_C
5	C	12	Q_D
6	D	11	CP
7	SL_{in}	10	S_1
8	(0V) GND	9	S_0

Pin	Description
A, B, C, D	Parallel inputs
Q_A, Q_B, Q_C, Q_D	Parallel outputs
SR_{in}	Serial data input for right shift
SL_{in}	Serial data input for left shift
S_0, S_1	Operation mode selection inputs
$\overline{CLR}$	Clear/Reset output
V_{cc}	Supply voltage (+5V)
GND	Ground (0V)

Fig. 3.137: *Pin configuration of 4-bit high speed shift register IC 74194.*

3.12 Implementation of Sequential Logic Design using Standard IC's

The concept of sequential circuit design using flip-flops and logic gates is discussed in this chapter. In this section, the available standard ICs are selected for implementation examples. The flip-flops and logic gates that are available as standard ICs are listed in Appendix-1.

Example 3.25

Design a sequential circuit with two D filp-flops A and B and one input x. When x = 0, the state of the circuit remains the same. When x = 1, the circuit goes through the state transistions from 00 to 10 to 11 to 01, back to 00 and repeats. Implement the logic circuit using available standard ICs.

Solution

State Diagram

The sequential has four states 00, 10, 11, 01. The possible state transitions from one state to other states are shown in Figs. 1 to 4.

1. While in state, 00

   if input = 0, remain in state 00

   if input = 1, go to state 10

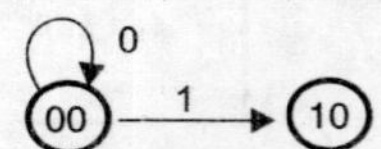

Fig. 1: Transitions from state 00.

2. While in state, 10

   if input = 0, remain in state 10

   if input = 1, go to state 11

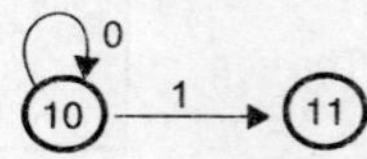

Fig. 2: Transitions from state 10.

3. While in state, 11

   if input = 0, remain in state 11

   if input = 1, go to state 01

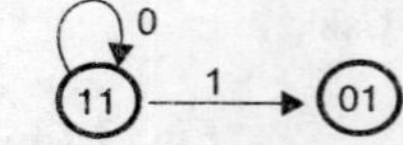

Fig. 3: Transitions from state 11.

4. While in state, 01

   if input = 0, remain in state 01

   if input = 1, go to state 00

Fig. 4: Transitions from state 01.

On combining the state transitions shown in Figs. 1 to 4, the state diagram can be drawn as shown in Fig. 5.

State Table

Let, x be the input. The state table (Table 1) can be formed from state diagram.

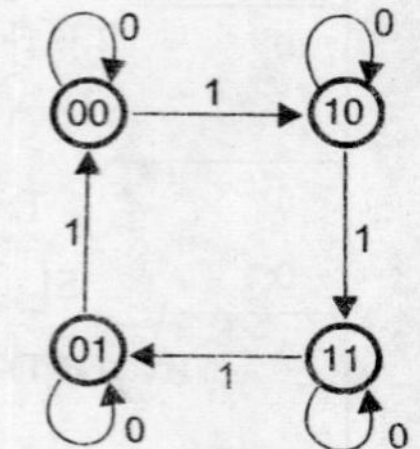

Fig. 5: State diagram.

Table 1: State Table

Present States	Next State	
	x = 0	x = 1
0 0	0 0	1 0
0 1	0 1	0 0
1 0	1 0	1 1
1 1	1 1	0 1

Design using D Flip-Flop

Here, there are two state variables. Let A and B be the two state variables.

∴ Present state variables : A(t), B(t) ⇒ A, B (For simplicity)

∴ Next state variables : A(t + 1), B(t + 1)

Let us choose a standard dual D flip-flop IC 7474 to store two state variables A and B as shown in Fig. 6. Here, A and B are output of flip-flops.

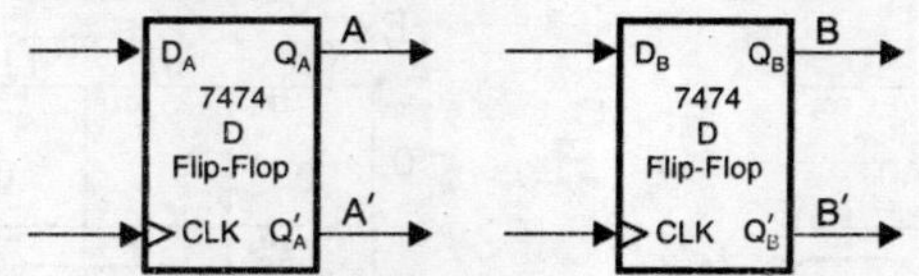

Fig. a: *Flip-flop A.* ***Fig. b:*** *Flip-flop B.*

Fig. 6: *D flip-flop to store state variables.*

Here, $A = Q_A \Rightarrow A(t+1) = Q_A(t+1)$

$B = Q_B \Rightarrow B(t+1) = Q_B(t+1)$

The characteristic equation of two D flip-flops are,

Characteristic equation of D flip-flop,
$Q(t+1) = D$

$Q_A(t+1) = D_A$

$Q_B(t+1) = D_B$

Using the above equation we can write,

$A(t+1) = D_A$

$B(t+1) = D_B$

Using the above equations the flip-flop inputs are determined for the next state and listed in Table 3. Here, the flip-flop inputs are same as next state outputs.

Table 3: Expanded State Table

Present States		Input	Next States		Flip-Flop Inputs		Minterm
A	**B**	**x**	**A(t+1)**	**B(t+1)**	D_A	D_B	**ABx**
0	0	0	0	0	0	0	m_0
0	0	1	1	0	1	0	m_1
0	1	0	0	1	0	1	m_2
0	1	1	0	0	0	0	m_3
1	0	0	1	0	1	0	m_4
1	0	1	1	1	1	1	m_5
1	1	0	1	1	1	1	m_6
1	1	1	0	1	0	1	m_7

Here, D_A and D_B are functions of A, B, x. Using Table 3, the Boolean equations for D_A and D_B are determined as sum of minterms as shown below:

$\therefore D_A = \Sigma m(1, 4, 5, 6)$

$D_B = \Sigma m(2, 5, 6, 7)$

Using the above equations, let us draw K-map for D_A and D_B to develop simplified Boolean equations for D_A and D_B.

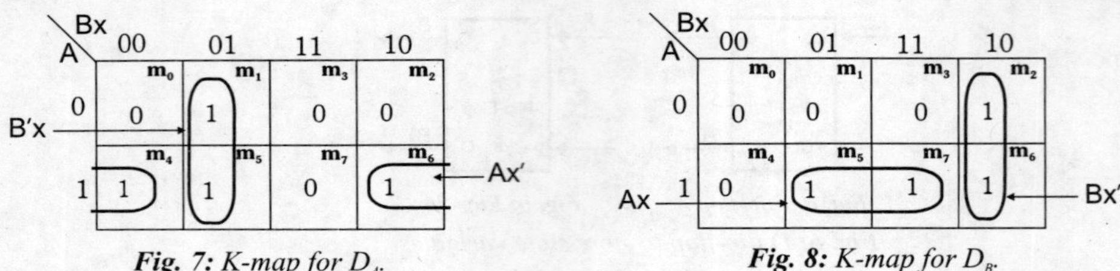

Fig. 7: K-map for D_A. *Fig. 8: K-map for D_B.*

From the K-map we get the following simplified Boolean equations for realizing sequence detector using D flip-flops,

$D_A = Ax' + B'x$; $D_B = Ax + Bx'$

The logic diagram of sequence detector using the above equations is realized using NOT gate IC 7404, 2-input AND gate IC 7408 and OR gate IC 7432 and dual D flip-flop IC 7474 as shown in Fig. 9. Here the sequence detector output is the present state output of flip-flops as shown in Fig. 9.

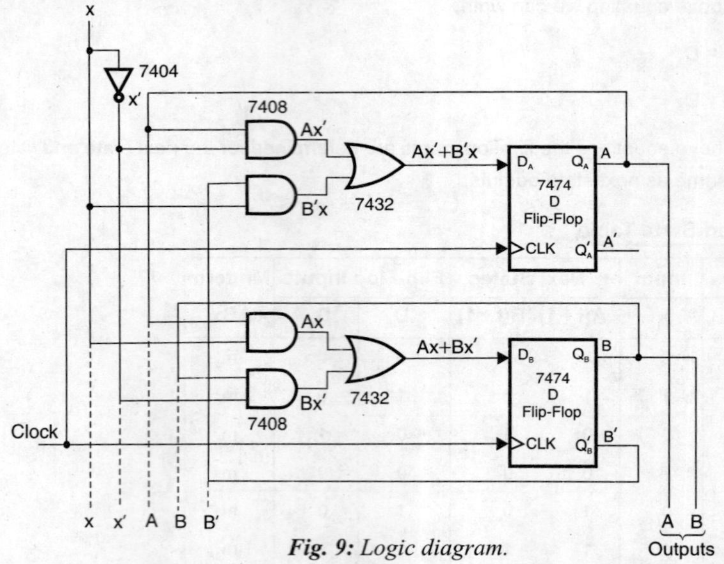

Fig. 9: Logic diagram.

Example 3.26

Design a synchronous counter that goes through states 4 → 6 → 7 → 3 → 1 → 4 by avoiding lock-out condition. Use JK flip-flop. Implement the logic circuit using standard ICs.

Solution

The sequential circuit has to detect decimal values in the range 0 to 8. Let us assign binary value of decimal number as code for states. Hence we need 3-bit binary code to represent the decimal sequence. With 3-bit code 8 ($2^3 = 8$) states are possible and in this 5 states are used states 3 states are unused states.

Binary Code Assignment

1	001
3	011
4	100
6	110
7	111

Unused Binary Codes

0	000
2	010
5	101

The state diagram using only used states is shown in Fig. 1. In order to prevent lock-out condition the unused states are included in state diagram such that the next state is a nearest used state as shown in Fig. 2.

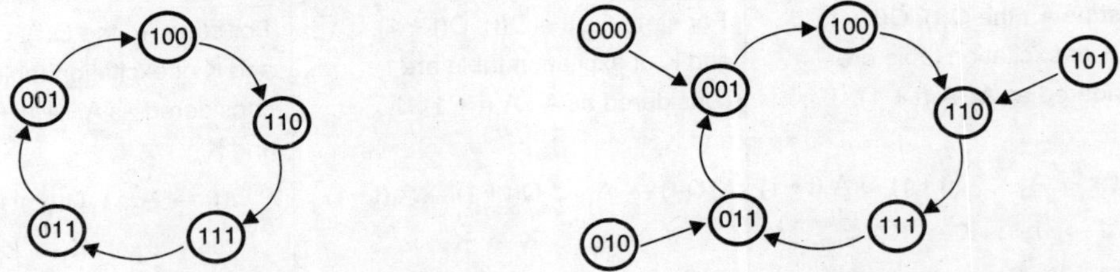

Fig. 1: State diagram with only used states.

Fig. 2: State diagram to prevent lock-out condition.

Here, each bit of binary code is a state variable. Hence the sequence counter has 3 state variables. Let the state variables be A_0, A_1 and A_2.

$\therefore$ Present state variables : $A_0(t), A_1(t), A_2(t) \Rightarrow A_0, A_1, A_2$ (For simplicity)

Next state variables : $A_0(t+1), A_1(t+1), A_2(t+1)$

Using the state diagram of Fig. 2, the state table is constructed as shown in Table 1.

Table 1: State Table

Present States			Next States		
A_2	A_1	A_0	$A_2(t+1)$	$A_1(t+1)$	$A_0(t+1)$
0	0	0	0	0	1
0	0	1	1	0	0
0	1	0	0	1	1
0	1	1	0	0	1
1	0	0	1	1	0
1	0	1	1	1	0
1	1	0	1	1	1
1	1	1	0	1	1

Let us choose two numbers of standard dual JK flip-flop IC 74112 to store three state variables A_0, A_1 and A_2 as shown in Fig. 3. Here, A_0, A_1 and A_2 are output of flip-flops and one of the flip-flop in one 74112 is not used.

Excitation Table of JK Flip-Flop

Present State Q(t)	Next State Q(t + 1)	Flip-Flop Inputs J	Flip-Flop Inputs K
0	0	0	×
0	1	1	×
1	0	×	1
1	1	×	0

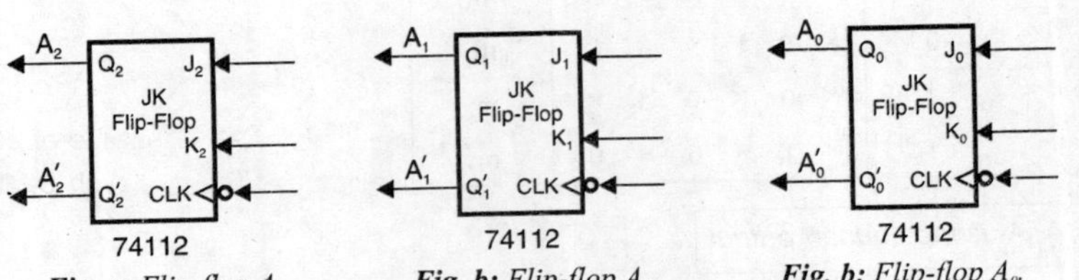

Fig. a: Flip-flop A_2. *Fig. b: Flip-flop A_1.* *Fig. b: Flip-flop A_0.*

Fig. 3: JK flip-flops to store state variables.

The flip-flop inputs can be obtained from excitation table as shown below:

State A_2

For state A_2, the Q(t), Q(t + 1), J and K of excitation table are considered as A_2, $A_2(t + 1)$, J_2 and K_2.

$$\therefore Q(t) \rightarrow A_2 \ ; \ Q(t+1) \rightarrow A_2(t+1)$$
$$J \rightarrow J_2 \ ; \ K \rightarrow K_2$$

The flip-flop inputs J_2 and K_2 that drives the present state A_2 to next state is obtained from excitation table and listed in Table 2.

Table 2:

A_2	$A_2(t+1)$	J_2	K_2
0	0	0	×
0	1	1	×
0	0	0	×
0	0	0	×
1	1	×	0
1	1	×	0
1	1	×	0
1	0	×	1

State A_1

For state A_1, the Q(t), Q(t + 1), J and K of excitation table are considered as A_1, $A_1(t + 1)$, J_1 and K_1.

$$\therefore Q(t) \rightarrow A_1 \ ; \ Q(t+1) \rightarrow A_1(t+1)$$
$$J \rightarrow J_1 \ ; \ K \rightarrow K_1$$

The flip-flop inputs J_1 and K_1 that drives the present state A_1 to next state is obtained from excitation table and listed in Table 3.

Table 3:

A_1	$A_1(t+1)$	J_1	K_1
0	0	0	×
0	0	0	×
1	1	×	0
1	0	×	1
0	1	1	×
0	1	1	×
1	1	×	0
1	1	×	0

State A_0

For state A_0, the Q(t), Q(t + 1), J and K of excitation table are considered as A_0, $A_0(t + 1)$, J_0 and K_0.

$$\therefore Q(t) \rightarrow A_0 \ ; \ Q(t+1) \rightarrow A_0(t+1)$$
$$J \rightarrow J_0 \ ; \ K \rightarrow K_0$$

The flip-flop inputs J_0 and K_0 that drives the present state A_0 to next state is obtained from excitation table and listed in Table 4.

Table 4:

A_0	$A_0(t+1)$	J_0	K_0
0	1	1	×
1	0	×	1
0	1	1	×
1	1	×	0
0	0	0	×
1	0	×	1
0	1	1	×
1	1	×	0

The tables 1 to 4 are combined to form state table with flip-flop inputs as shown in Table 5.

Table 5: Expanded State Table

Present States			Next States			Flip-Flop Inputs						Minterm
A_2	A_1	A_0	$A_2(t+1)$	$A_1(t+1)$	$A_0(t+1)$	J_2	K_2	J_1	K_1	J_0	K_0	$A_2 A_1 A_0$
0	0	0	0	0	1	0	×	0	×	1	×	m_0
0	0	1	1	0	0	1	×	0	×	×	1	m_1
0	1	0	0	1	1	0	×	×	0	1	×	m_2
0	1	1	0	0	1	0	×	×	1	×	0	m_3
1	0	0	1	1	0	×	0	1	×	0	×	m_4
1	0	1	1	1	0	×	0	1	×	×	1	m_5
1	1	0	1	1	1	×	0	×	0	1	×	m_6
1	1	1	0	1	1	×	1	×	0	×	0	m_7

Note: *Here, present state $A_2 A_1 A_0$ is the counter output.*

The flip-flop inputs are functions of present state and so minterms are formed using present states and the Boolean equations for J_2, K_2, J_1, K_1, J_0 and K_0 are obtained from Table 5 as sum of minterms and don't-cares as shown below:

$$J_2 = m_1 + \Sigma d(4, 5, 6, 7)$$

$$K_2 = m_7 + \Sigma d(0, 1, 2, 3)$$

$J_1 = m(4, 5) + \Sigma d(2, 3, 6, 7)$

$K_1 = m_3 + \Sigma d(0, 1, 4, 5)$

$J_0 = \Sigma m(0, 2, 6) + \Sigma d(1, 3, 5, 7)$

$K_0 = \Sigma m(1, 5) + \Sigma d(0, 2, 4, 6)$

The simplified Boolean equations are obtained by constructing K-map using the above equations for J_2, K_2, J_1, K_1, J_0 and K_0 as shown in Figs. 4 to 9.

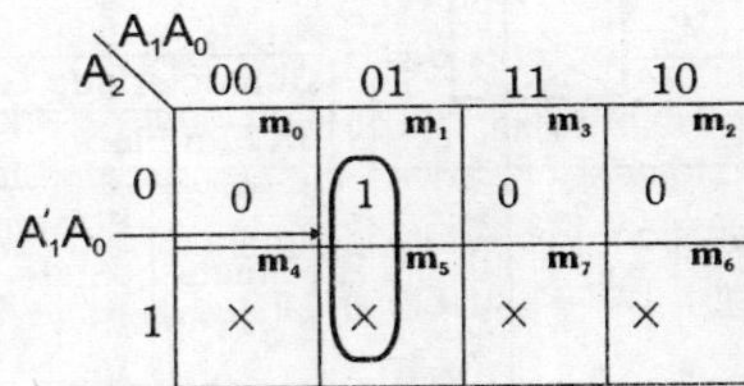

Fig. 4: K-map for J_2.

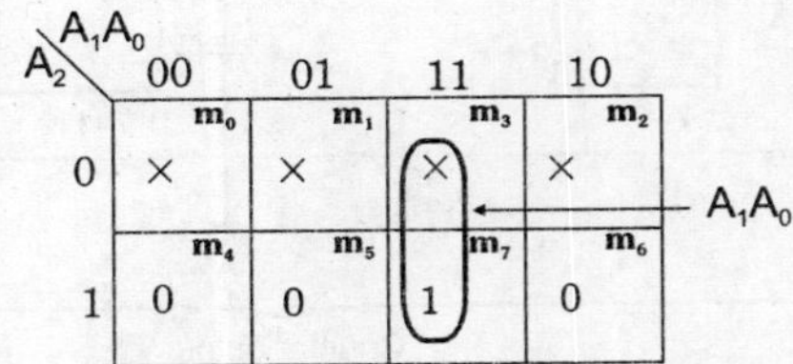

Fig. 5: K-map for K_2.

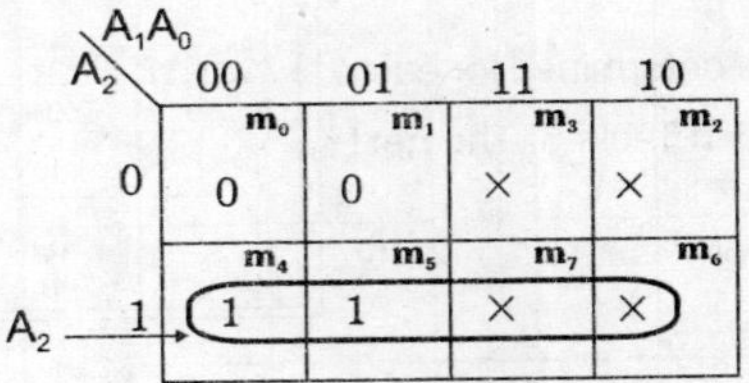

Fig. 6: K-map for J_1.

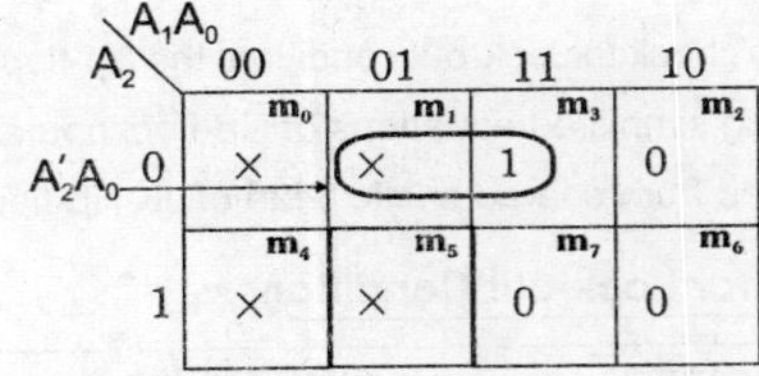

Fig. 7: K-map for K_1.

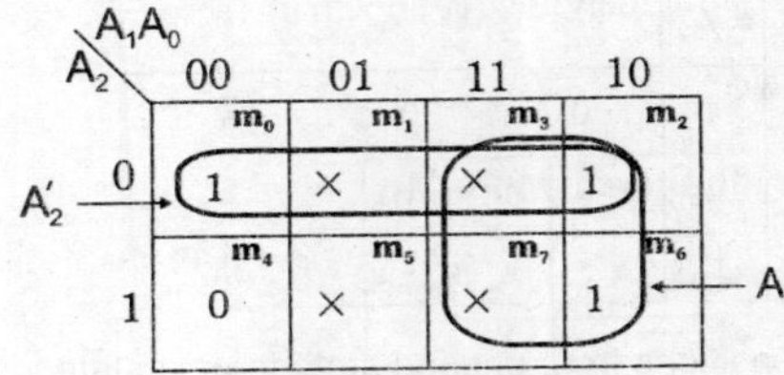

Fig. 8: K-map for J_0.

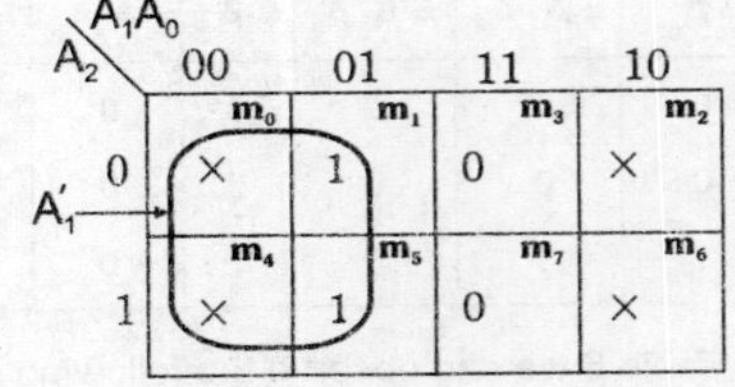

Fig. 9: K-map for K_0.

From the K-map we get the following simplified Boolean equations to realize 3-bit binary counter using JK flip-flops.

$J_2 = A_1' A_0$	$K_2 = A_1 A_0$
$J_1 = A_2$	$K_1 = A_2' A_0$
$J_0 = A_2' + A_1$	$K_0 = A_1'$

Using the simplified Boolean equations obtained from K-map the logic diagram of 3-bit counter is drawn as shown in Fig. 10. The logic circuit is implemented using 2-input AND gate IC 7408 and OR gate IC 7432 and dual JK flip-flop IC 74112.

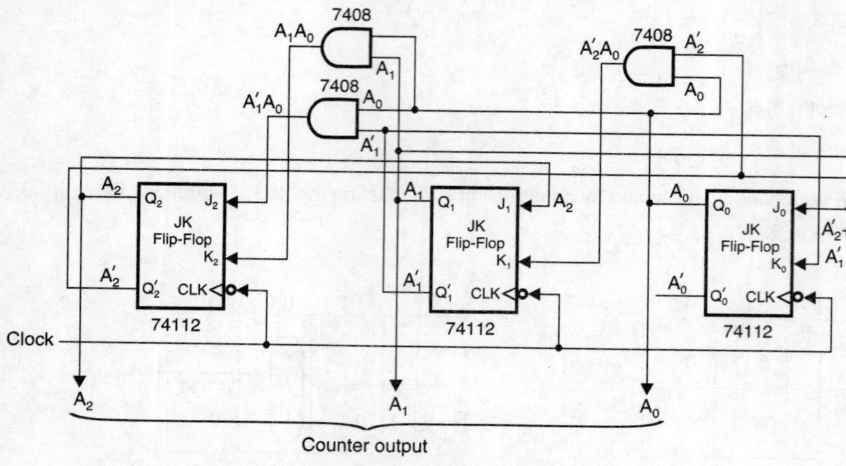

Fig. 10: Logic diagram.

Characteristic Table of JK Flip-Flop

Inputs J K	Present State Q	Next State Q(t+1)
0 0	0	0
0 0	1	1
0 1	0	0
0 1	1	0
1 0	0	1
1 0	1	1
1 1	0	1
1 1	1	0

Check for Lock-out Condition

In order to check for lock-out condition, the flip-flop inputs are determined for each unused state using simplified equations of flip-flop inputs and listed in Table 6. The next states are obtained from characteristic table of JK flip-flop

Table 6: Check for Lock-out Condition

Unused Present States			Flip-Flop Inputs						Next States		
A_2	A_1	A_0	$J_2 = A'_1A_0$	$K_2 = A_1A_0$	$J_1 = A_2$	$K_1 = A'_2A_0$	$J_0 = A'_2+A_1$	$K_0 = A'_1$	$A_2(t+1)$	$A_1(t+1)$	$A_0(t+1)$
0	0	0	0	0	0	0	1	1	0	0	1
0	1	0	0	0	0	0	1	0	0	1	1
1	0	1	1	0	1	0	0	1	1	1	0

From the Table 6 we can observe the following state transitions from unused state to next state.

000 → 001, 010 → 011, 101 → 110

The unused states enter used state in the next transition as shown above and hence the lock-out condition is avoided.

Example 3.27

Design a sequential circuit for a 3-bit synchronous binary counter that counts clock pulses and generate output in the range 000_2 to 111_2. Implement the logic circuit using standard ICs.

Solution

In counters the states are directly taken as outputs.

Since 3-bit output is needed, there are eight possible combination of states formed using 3 state variables. Let the three state variables be A_0, A_1 and A_2.

∴ Present state variables : $A_0(t), A_1(t), A_2(t) \Rightarrow A_0, A_1, A_2$ (For simplicity)

Next state variables : $A_0(t+1), A_1(t+1), A_2(t+1)$

Since T flip-flops are not available as standard IC, the JK flip-flop is converted to T flip-flop by connecting the J and K inputs to function as T input. The standard dual JK flip-flop IC 74112 is selected for storing state variables A_0, A_1 and A_2 as shown in Fig. 1. Here A_0, A_1 and A_2 are output of flip-flops and J and K inputs are connected to function as T input.

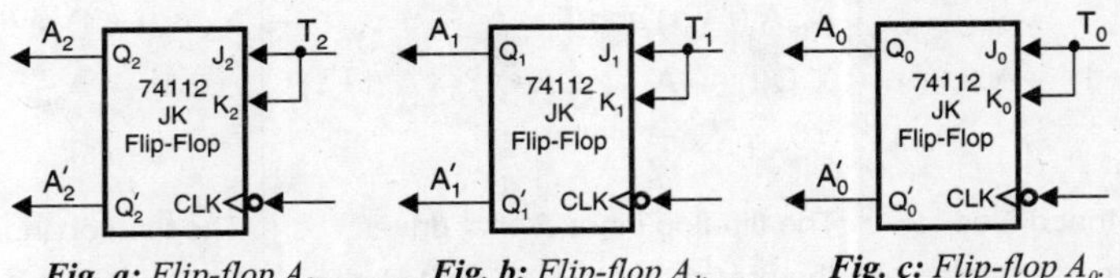

Fig. a: *Flip-flop A_2.* ***Fig. b:*** *Flip-flop A_1.* ***Fig. c:*** *Flip-flop A_0.*

Fig. 1: *T flip-flops to store state variables.*

The binary counter has to increment by 1 for every clock pulse. The initial output is 000, then for every clock pulse it has to increment by 1. So that the output sequence is,

000, 001, 010, 011, 100, 101, 110, 111

Upon reaching the upper most count 111 the count should start from 000 for next clock pulse. Using this logic, the state diagram is drawn as shown in Fig. 2.

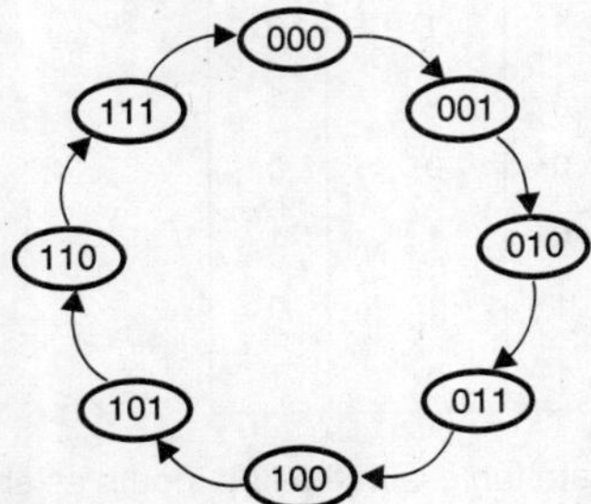

Fig. 2: *State diagram of 3-bit binary counter.*

The state table is formed using state diagram as shown in Table 1. The flip-flop inputs can be obtained from excitation table.

Table 1: State Table

Present States			Next States		
A_2	A_1	A_0	$A_2(t+1)$	$A_1(t+1)$	$A_0(t+1)$
0	0	0	0	0	1
0	0	1	0	1	0
0	1	0	0	1	1
0	1	1	1	0	0
1	0	0	1	0	1
1	0	1	1	1	0
1	1	0	1	1	1
1	1	1	0	0	0

Excitation Table of T Flip-Flop

Present State Q(t)	Next State Q(t + 1)	Flip-Flop Input T
0	0	0
0	1	1
1	0	1
1	1	0

State A_2

For state A_2, the Q(t), Q(t + 1) and T of excitation table are considered as A_2, $A_2(t + 1)$ and T_2.

$\therefore Q(t) \rightarrow A_2$; $Q(t+1) \rightarrow A_2(t+1)$

$T \rightarrow T_2$

The flip-flop input T_2 that drives the present state A_2 to next states is obtained from excitation table and listed in Table 2.

Table 2:

A_2	$A_2(t+1)$	T_2
0	0	0
0	0	0
0	0	0
0	1	1
1	1	0
1	1	0
1	1	0
1	0	1

State A_1

For state A_1, the Q(t), Q(t + 1) and T of excitation table are considered as A_1, $A_1(t + 1)$ and T_1.

$\therefore Q(t) \rightarrow A_1$; $Q(t+1) \rightarrow A_1(t+1)$

$T \rightarrow T_1$

The flip-flop input T_1 that drives the present state A_1 to next states is obtained from excitation table and listed in Table 3.

Table 3:

A_1	$A_1(t+1)$	T_1
0	0	0
0	1	1
1	1	0
1	0	1
0	0	0
0	1	1
1	1	0
1	0	1

State A_0

For state A_0, the Q(t), Q(t + 1) and T of excitation table are considered as A_0, $A_0(t + 1)$ and T_0.

$\therefore Q(t) \rightarrow A_0$; $Q(t+1) \rightarrow A_0(t+1)$

$T \rightarrow T_0$

The flip-flop input T_0 that drives the present state A_0 to next states is obtained from excitation table and listed in Table 4.

Table 4:

A_0	$A_0(t+1)$	T_0
0	1	1
1	0	1
0	1	1
1	0	1
0	1	1
1	0	1
0	1	1
1	0	1

The tables 1 to 4 are combined to form state table with flip-flop inputs as shown in Table 5.

Table 5: Expanded State Table

Present States			Next States			Flip-Flop Inputs			Minterm
A_2	A_1	A_0	$A_2(t+1)$	$A_1(t+1)$	$A_0(t+1)$	T_2	T_1	T_0	$A_2A_1A_0$
0	0	0	0	0	1	0	0	1	m_0
0	0	1	0	1	0	0	1	1	m_1
0	1	0	0	1	1	0	0	1	m_2
0	1	1	1	0	0	1	1	1	m_3
1	0	0	1	0	1	0	0	1	m_4
1	0	1	1	1	0	0	1	1	m_5
1	1	0	1	1	1	0	0	1	m_6
1	1	1	0	0	0	1	1	1	m_7

> ***Note:*** *Here, present state $A_2A_1A_0$ is the counter output.*

The flip-flop inputs are functions of present states and so minterms are formed using present states and Boolean equations for flip-flop inputs are obtained as sum of minterms as shown below:

$$T_2 = \Sigma m(3, 7) \quad ; \quad T_1 = \Sigma m(1, 3, 5, 7) \quad ; \quad T_0 = \Sigma m(0, 1, 2, 3, 4, 5, 6, 7)$$

The simplified Boolean equations are obtained by constructing K-map for T_2, T_1 and T_0 as shown in Figs. 3 to 5.

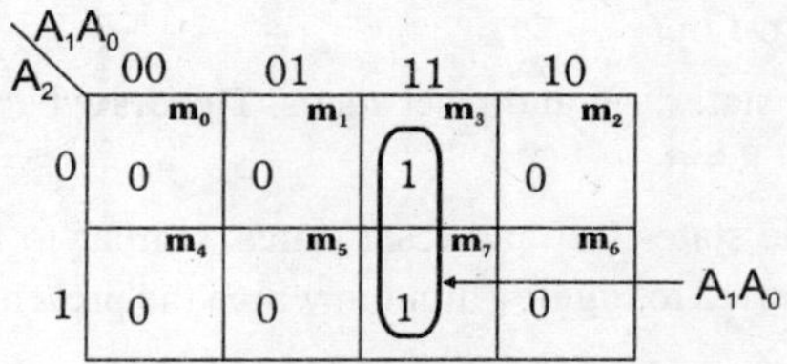

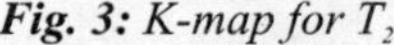

Fig. 3: K-map for T_2.

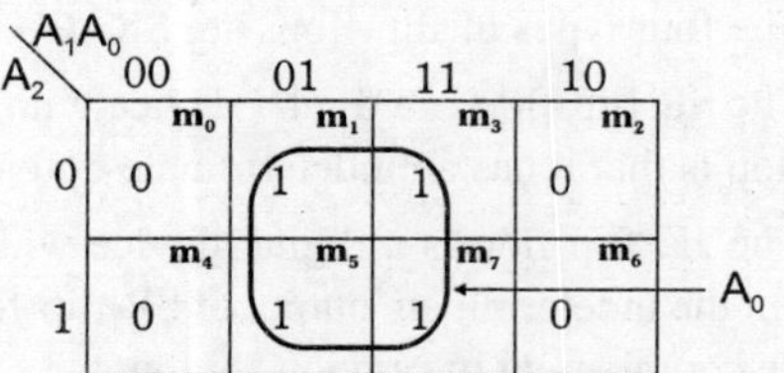

Fig. 4: K-map for T_1.

From the K-map we get the following simplified Boolean equations to realize 3-bit binary counter using T flip-flops.

$T_2 = A_1 A_0$

$T_1 = A_0$

$T_0 = 1$

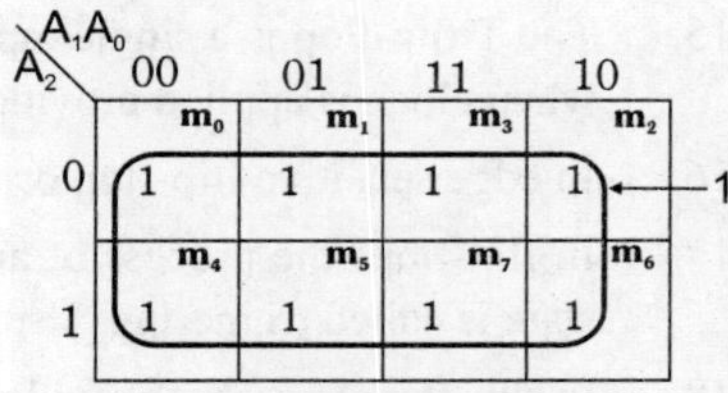

Fig. 5: K-map for T_0.

Using the above equations the logic diagram of 3-bit counter is drawn as shown in Fig. 6. Here, 2-input AND gate IC 7408 and dual JK flip-flop ICs 74112 with J and K inputs connected to work as T input are used to implement the logic circuit.

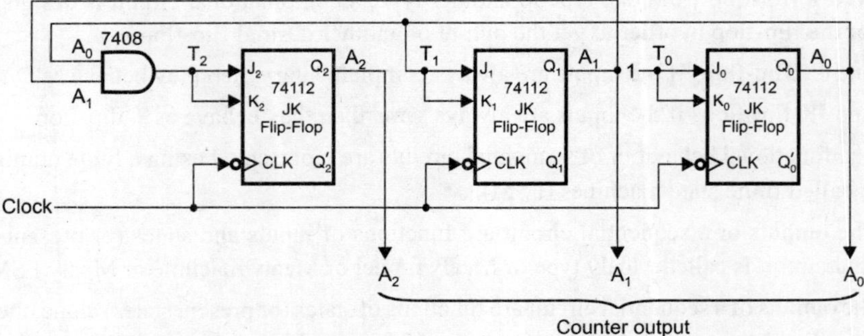

Fig. 6: Logic diagram of 3-bit synchronous binary counter.

3.13 Summary of Important Concepts

1. Sequential logic circuits are digital circuits whose output depends on present inputs and past outputs.
2. A sequential logic circuit is a combinational circuit with storage elements and feedback paths.
3. The sequential circuits without a clock are called asynchronous sequential circuits.
4. The sequential circuits with a clock are called synchronous sequential circuits.
5. A latch is the basic storage device to store a binary bit.
6. A latch is designed with two stable states in order to store two possible values of a binary bit 1 and 0.
7. The SR latch is a digital circuit designed to store 1 or 0 applied as two inputs called set and reset and it is provided with an output and complement of the output.
8. A stable state is the state when present state output and next state output are same after a change in input.
9. D latch is a single input latch with two stable states which are used to store two different values 0 and 1 of a binary data.

10. Flip-flop is a storage device constructed using latch and controlled by enable signal called clock.
11. The four types of flip-flops are SR, JK, D and T flip-flops.
12. The SR flip-flop is a digital device with two stable states: set and reset states. The draw back in SR flip flop is that it has an indeterminate output when S = R = 1.
13. The JK flip-flop is a digital device with two stable states: set and reset states, similar to SR flip-flop but the indeterminate output of SR flip-flop is converted to toggle output in which the present output will be complement of previous output.
14. The D flip-flop is a single input flip-flop in which the input is stored in the output when clock is applied.
15. The T flip-flop is a single input flip-flop in which the present output is complement of previous output when clock is applied provided input is 1.
16. In edge sensitive flip-flop or circuits the timing of signal (or input) recognization is more precise.
17. In flip-flops the process of applying a control signal to change from one stable state to another stable state is called triggering. Usually, the control signal is a clock pulse.
18. In flip-flops preset is provided to set the output and reset is provided to reset the output.
19. In flip-flops, when preset is 0, the output is set to 1 and when reset is 0, the output is reset to 0.
20. In flip-flops, the preset and reset should be tied 1, for input and clock to make changes in output.
21. In flip-flops, the excitation table lists the required inputs for a specified change of state in output.
22. To convert a flip-flop from one type to another type, a combinational circuit is designed to supply the inputs of the flip-flop in order to get the output of another desired flip-flop.
23. In SR and JK flip-flops if the inputs are always complementary then they behave as D flip-flop.
24. In SR and JK flip-flops if the inputs are always same then they behave as T flip-flop.
25. Since, the functional behaviour of sequential circuits are represented using a finite number of states they are also called finite state machines (FSM).
26. When the outputs of a sequential circuit are functions of inputs and states (or present states) then the sequential circuit is called Mealy type or Mealy model or Mealy machine or Mealy FSM.
27. When the outputs of a sequential circuit are functions of states (or present states) alone, then the sequential circuit is called Moore type or Moore model or Moore machine or Moore FSM.
28. A state is the condition of a sequential circuit described by a set of variables called state variables.
29. The state equations (or transition equations) are a set of equations that describe the next state as a function of the present states and inputs.
30. State table is listing of outputs and next states for all possible combinations of inputs and present states.
31. If there are m state variables and n inputs then the state table list the outputs and next states for 2^{m+n} combinations of inputs and states.
32. State diagram is graphical representation of all transitions from present state to next state for all possible combination of inputs and outputs. Therefore, the state diagram is the graphical representation of information available in a state stable.
33. Two states are said to be equivalent if the input-output relationship are same with identical present state and changes in input, which drive the present state to identical next state.
34. Assigning binary codes to states represented by alphabets is called state assignment (or code assignment).
35. In sequential circuits with unused states, when clock pulses are applied, if the sequential circuit cycle through only unused states and never arrives at in a used state then this condition is called lock-out.

36. A register that goes through prescribed (or predefined) sequence of states upon the application of clock pulses is called a counter.
37. A counter that follows a binary count sequence is called binary counter.
38. An n-bit binary counter consists of n flip-flops and can count binary from 0 to $2^n - 1$.
39. The modulus of the counter refers to number states a counter will pass through before coming back to initial state.
40. An n-bit counter will have 2^n possible states and if the n-bit counter passes through all possible states before coming to initial state then it is called MOD-2^n counter.
41. An n-bit binary counter divides the clock frequency by 2^n and so n-bit counter is also called divide-by-2^n counter
42. The n-bit MOD-N counter uses only N states in the 2^n possible states of n-bit counter and the remaining states are unused states.
43. MOD-N counters can be designed using n-bit binary counter by using only N states of n-bit counter provided $N < 2^n$
44. The MOD-N counter divides the clock frequency by N and so it is also called divide-by-N counter.
45. The counters can be broadly classified into two categories: Ripple counters and synchronous counters.
46. In a ripple counter, a flip-flop output will act as clock for another flip-flop.
47. In synchronous counters all the flip-flops are triggered simultaneously by a common clock.
48. A binary ripple counter is constructed using a series of complementing flip-flops.
49. In up counter, the output will increment for every clock pulse and the count sequence will be from 0 to 2^n-1.
50. In down counter, the output will decrement for every clock pulse and counter the count sequence will be from $2^n - 1$ to 0.
51. A BCD counter has to count only from 0 to 9, and when the count reaches 1010, the output has to be reset to all zero, so that next count sequence will start.
52. The BCD counter is also called MOD-10 counter.
53. The synchronous counters are designed using complementing flip-flops similar to ripple counters, but the clock is applied simultaneously to all the flip-flops.
54. In synchronous counters, the flip-flop inputs are applied through some combinational logic circuits.
55. A ring counter is a circular shift register with only one flip-flop output being set at a time and all other flip-flop outputs are cleared.
56. A ring counter produces a sequence of timing signals that are shifted by one clock time period.
57. Johnson counter is basically a ring counter with complement of last flip-flop connected to input of first flip-flop. This arrangement is called switch tail ring counter.
58. In general, k bit Johnson counter will generate 2k distinguishable states and they decoded to generate 2k unique output signals that resembles the output of a decoder.
59. A register is a group of flip-flops that can store binary information.
60. Each flip-flop in a register can store one bit binary and so, in order to store n-bit binary the register should have n flip-flops.
61. A register capable of shifting the binary information from one flip-flop to another (or one cell to another) is called shift register.

62. The four types of shift registers are serial-in serial-out, serial-in parallel-out, parallel-in serial out and parallel-in parallel out shift register.

63. A shift register that can accept serial or parallel input data and output data in parallel along with right or left shift is called universal shift register.

3.14 Short-Answer Questions

Q3.1 Define clock and clock frequency.

The clock is basically a periodic square pulse signal. The sum of ON and OFF time (or **high** and **low** time) is called one period of clock. The inverse of clock period is called clock frequency. The flip-flop is possible to replace enable by a momentary pulse called clock.

Q3.2 Define latch and flip-flop.

Latch: A latch is the basic storage device to store a binary bit. A latch is designed with two stable states in order to store two possible values of a binary bit 1 and 0. Latches are designed with complementary inputs and outputs so that positive or negative logic can be implemented easily. The latches are sequential circuits without a clock.

Flip-flop: Flip-flop is a storage device constructed using latch and controlled by enable signal called clock. Flip-flops are made sensitive to level/positive edge/negative edge clock.

Q3.3 State the difference between latches and flip-flops.

The main difference between latches and flip-flops is that the flip-flops are clocked storage devices but the latches are unclocked storage devices.

Q3.4 Write the function of set and reset inputs of a flip-flop.

Set is an input used to set the output to 1 and it is denoted by S.

Reset is an input used to clear the output to 0 and it is denoted by R.

Q3.5 What is meant by stable state.

A ***stable state*** is the state when present state output and next state output are same after a change in input.

Q3.6 Define data flip-flop and toggle flip-flop.

D flip-flop is a single input flip-flop in which the input is stored in the output when clock is applied. It is also called as data flip-flop.

T flip-flop is a single input flip-flop in which the present output is complement of previous output when clock is applied, provided input is 1. It is also called as toggle flip-flop.

Q3.7 What is mean by triggering of flip-flops?

In flip-flops the process of applying a control signal to change from one stable state to another stable state is called triggering. Mostly the triggering is performed by edge sensitive clock.

Q3.8 Define Moore and Melay model of sequential circuits.

When the outputs of a sequential circuit are functions of inputs and states (or present states) then the sequential circuit is called Mealy type or Mealy model of sequential circuits.

When the outputs of a sequential circuit are functions of states (or present states) alone, then the sequential circuit is called Moore type or Moore model of sequential circuits.

Q3.9 What is the difference between Melay and Moore state machines?

In Mealy machine, the output depends on inputs and present states but in Moore machine, the outputs depends only present states and does not depend on input.

Q3.10 Define FSM. ***(AU, Apr/May'19, 2 Marks)***

The functional behaviour of sequential circuits can be represented using a finite number of states. Hence the sequential circuits are also called finite state machines (FSM). Therefore, the Mealy model is called Mealy FSM or Mealy Machine. The moore model is called Moore FSM or Moore Machine.

Q3.11 Define state and state equations.

A state is the condition of a sequential circuit described by a set of variables called state variables. Flip-flops are used to store state variables and so the output of flip-flops constitute the states of a sequential circuit. With m state variables we can form 2^m different states.

The state equations (or transition equations) are a set of equations that describe the next state as a function of the present states and inputs.

Q3.12 What is state diagram?

The state diagram is graphical representation of all transitions from present state to next state for all possible combination of inputs and outputs. Therefore, the state diagram is the graphical representation of information available in a state stable.

Q3.13 What is state table?

The state table is listing of outputs and next states for all possible combinations of inputs and present states. In general, if there are m state variables and n inputs then possible combinations of inputs and states are 2^{m+n}. The state table list the outputs and next states for 2^{m+n} combinations of inputs and states.

Q3.14 Define state assignment.

In the design specifications when the states are specified by alphabets it is necessary to assign binary codes for practical implementations. Therefore, an n-bit binary code is assigned to each state. Assigning binary codes to states represented by alphabets is called state assignment (or code assignment).

Q3.15 Define lock-out condition?

In sequential circuits with some don't-care or unused states, the sequential circuit may find itself in some unused state during starting or switch ON. If next state of the unused state is again some unused one and if the sequential circuit cycle through only unused states and never arrives at in a used state then this condition is called lock-out.

Q3.16 Define counter and its classifications.

A register that goes through prescribed (or predefined) sequence of states upon the application of clock pulses is called a counter. The counters are used to generate or go through a sequence binary code known as count sequence.

The counters can be broadly classified into two categories: Ripple counters and synchronous counters.

In a ripple counter, a flip-flop output will act as clock for another flip-flop.

In synchronous counters all the flip-flops are triggered simultaneously by a common clock.

Q3.17 Define binary counter.

A counter that follows a binary count sequence is called binary counter. An n-bit binary counter consists of n flip-flops and can count binary from 0 to $2^n - 1$.

Q3.18 What is switch tail ring counter?

Johnson counter is basically a ring counter with complement of last flip-flop connected to input of first flip-flop. This arrangement is called switch tail ring counter.

Q3.19 *Explain ring counter.*

A ring counter is a circular shift register with only one flip-flop output being set at a time and all other flip-flop outputs are cleared. Using preset facility, a 1 is loaded in one flip-flop output and all other outputs are loaded 0. When clock pulses are applied the 1 will shift by one position. This produces a sequence of timing signals that are shifted by one clock time period.

Q3.20 *How to make the counter self starting?*

In order to make the counter self starting the unused states can be connected to enter used states. Once the counter enter an used state it will circle through used states and counter becomes self starting.

Q3.21 *How many fip-flops will be complemented in a 10-bit ripple counter to reach the next count after the count value of 1001100111?*

Add 1 to the binary value 1001100111 in order to get next count value.

```
1001100111
       + 1
----------
1001101000
```

In the result of above addition we observe that 4 bits are complemented. In a counter, each bit is stored in a flip-flop. Therefore, 4 flip-flops will be complemented in a 10-bit ripple counter to reach the next count after the count of 1001100111.

Q3.22 *Define BCD or MOD 10 counter.*

A BCD counter has to count only from 0 to 9, and when the count reaches 10_{10}, the output has to be reset to all zero, so that next count sequence will start. The BCD counter is also called MOD-10 counter.

Q3.23 *Comment about a presettable counter and ripple counter.* ***(AU, Nov/Dec'18, 2 Marks)***

A presettable counter will have a preload facility to load a count value before the start of the count and count sequence starts from this preloaded value. But in a ripple counter without a preload facility the count sequence will be all possible combinations of binary.

Q3.24 *Define register.*

A register is a group of flip-flops that can store binary information. Each flip-flop in a register can store one bit binary and so, in order to store n-bit binary the register should have n flip-flops.

Q3.25 *Define shift register.* ***(AU, Nov/Dec'18, 2 Marks)***

A register capable of shifting the binary information from one flip-flop to another (or one cell to another) is called shift register.

Q3.26 *Write a types of shift register.*

There are four types of shift registers.

1. Serial-in Serial-out shift register
2. Serial-in Parallel-out shift register
3. Parallel-in Serial-out shift register
4. Parallel-in Parallel-out shift register

3.15 Exercises

I. Fill in the blanks

1. The SR latch is a digital circuit designed to store 1 or 0 applied as two inputs called _____ and _______.
2. A JK flip-flop is in toggle condition when ______.
3. A state is the condition of a sequential circuit described by a set of variables called ______ .
4. The graphical representation of information available in state stable is called _____.
5. A register capable of shifting the binary information from one flip-flop to another is called ______.
6. A counter that follows a binary counter sequence is called _____.
7. A gated D latch does not have ______.
8. A register is a group of _______ .
9. The number of flip-flops needed to construct a MOD-17 synchronous counter is _______ .
10. The BCD counter is also called _____ .
11. A _____ is a group of flip-flops that can store binary information.
12. A _____ counter produces a sequence of timing signals that are shifted by one clock time period.
13. MOD-N counters can be designed using n-bit binary counter by using only N states of n-bit counter provided _____ .
14. A _____ is constructed using a series of complementing flip-flops.
15. A n-bit binary counter consists of n flip-flops and can count binary from _____ .

Answers

1. set, reset
2. J = 1, K = 1
3. state variables
4. state diagram
5. shift register
6. binary counter
7. a clock input
8. flip-flops
9. 5
10. MOD-10 counter
11. register
12. ring
13. $N < 2^n$
14. binary ripple counter
15. 0 to $2^n - 1$

II. State whether the following statements are True or False

1. The serial-in parallel-out shift registers transfers data from one parallel data bus to another parallel data bus by transferring one bit at a time across a single line.
2. A ripple counter is an asynchronous counter.
3. Counters are common components in digital clocks.
4. An SR latch is a one bit memory element.
5. A pulse train can be delayed by a finite number of clock periods using a serial in serial out shift register.
6. A ring counter consisting of seven flip-flops will have 5 states.
7. A BCD counter has to count only from 0 to 8, when the count reaches 1010, the output has to be reset to all zero, so that next count sequence will start.
8. In up counter, the output will increment for every clock pulse and the count sequence will be from 0 to $2^n - 1$.
9. State table is listing of outputs and next states for all possible combinations of inputs and present states.
10. A stable state is the state when present state output and next state output are same after a change in input.
11. In flip-flops, the preset and reset should be tied 0, for input and clock to make changes in output.
12. In SR and JK flip-flops if the inputs are always complementary then they behave as T flip-flop.
13. In flip-flops, the excitation table lists the required inputs for a specified change of state in output.
14. A latch is the basic storage device to store a binary bit.
15. A latch is designed with three stable states in order to store three possible values of a binary bit 1 and 0.

Answers

1. False	5. False	9. True	13. True
2. True	6. False	10. True	14. True
3. True	7. False	11. False	15. False
4. True	8. True	12. False	

III. Choose the right answer for the following questions

1. *A 0 to 6 counter consists of 3 flip-flops and a combination circuit of 2 input gate. The combination circuits consists of*

   a) one AND gate b) one OR gate
   c) one AND gate and one OR gate d) two AND gate

2. *A 4 bit ripple counter and a 4 bit synchronous counter are made using flip-flops having a propagation delay of 10ns each. If the worst case delay in the ripple counter and synchronous counter be R and S respectively then.*

   a) R = 10 ns, S = 40 ns b) R = 40 ns, S = 10 ns
   c) R = 10 ns, S = 30 ns d) R = 30 ns, S = 10 ns

3. *A master slave flip-flop has the characteristics that*

   a) Change in the input is immediately reflected in the output
   b) Change in the output occurs when the state of the master is affected
   c) Change in the output occurs when the state of the slave is affected
   d) both the master and the slave states are affected at the same time.

4. *A ring counter consisting of seven flip-flops will have*

   a) 5 states b) 7 states c) 32 states d) Infinie states

5. *The initial state of a MOD-16 down counter is 0110. After 37 clock pulses, the state of the counter will be*

   a) 1011 b) 0110 c) 0101 d) 0001

6. *A 4-bit binary ripple counter uses flip-flops with a propagation delay time of 20 ns each. The maximum possible time required for change of state will be,*

   a) 25 ns b) 80 ns c) 75 ns d) 100 ns

7. *The minimum number of flip-flops needed to make MOD-2 counter is*

   a) 1 b) 2 c) 3 d) 4

8. *The characteristics equation of a JK flip-flop is*

   a) JQ' + K'Q b) JQ + K'Q c) J'Q + K'Q d) JQ + K'Q

9. *Symmetrical square wave of time period 100 μs can be obtained from square wave of time period 10 μs by using a*

   a) divide by 5 counter b) divide by 2 counter
   c) divide by 7 counter d) divide by 10 counter

10. *A master slave flip-flop is made up of*

a) two flip-flop connected in series **b)** two flip-flop connected in parallel

c) debouncer circuit **d)** D-latch

11. *A decade counter has a maximum of*

a) 8 states **b)** 10 states **c)** 9 states **d)** 7 states

12. *The input clock frequency of the MOD-5 counter is 25 KHz. Then the output has the frequency of*

a) 5 KHz **b)** 2 KHz **c)** 0.2 KHz **d)** 1 KHz

13. *The output of a toggle flip-flop is _____ the frequency of its input.*

a) twice **b)** half **c)** thrice **d)** one-third

14. *The number of flip-flops needed to construct a MOD-15 synchronous counter is*

a) 9 **b)** 4 **c)** 3 **d)** 2

15. *Assume that initially Q = 1 in Fig. 15, with clock pulses being given, the subsequence states of Q will be*

a) 10101

b) 0000

c) 1111

d) 01010

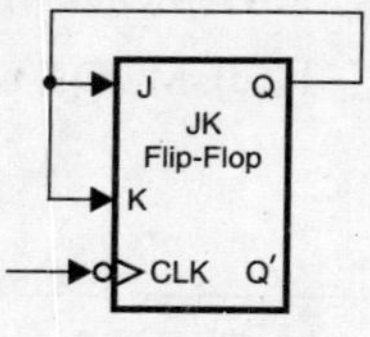

Fig. 15.

16. *The Q output of a JK flip-flop is '1'. The output doesn't change when a clock pulse is applied. The inputs J and K will be respectively (where X is don't care input)*

a) 0 and X **b)** X and 0 **c)** 1 and 0 **d)** 0 and 1

17. *The circuit shown in Fig. 17 behave as*

a) T flip-flop

b) D flip-flop

c) JK flip-flop with K connected to ground and X connected to J

d) none of the above

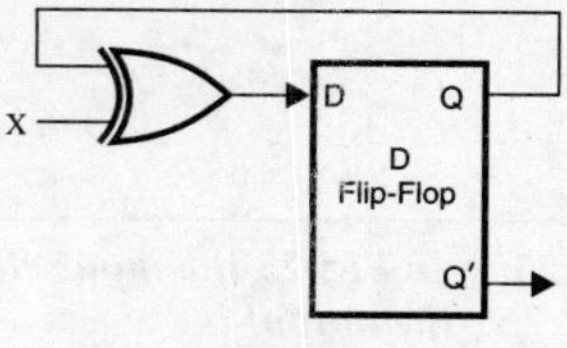

Fig. 17.

18. *The modulus of the counter shown in Fig. 18*

a) 4

b) 3

c) 2

d) 1

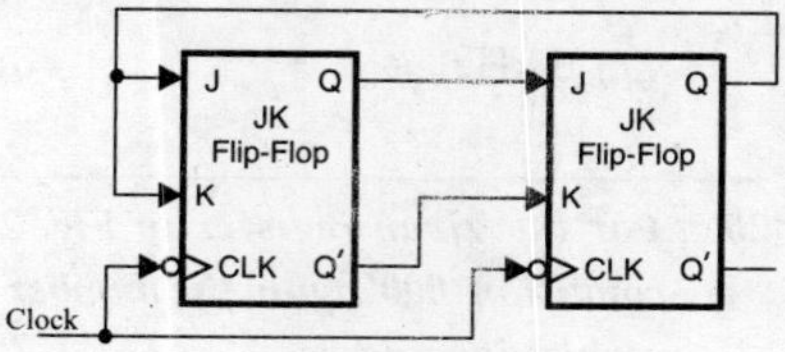

Fig. 18.

19. *The number of flip-flops needed to construct a MOD-17 synchronous counter is*

a) 9 **b)** 5 **c)** 3 **d)** 4

20. *A switch-tail ring counter is made by using a single D flip-flop. The resulting circuit is a*

a) SR flip-flop **b)** JK flip-flop **c)** D flip-flop **d)** T flip-flop

21. ***In a JK flip-flop we have J = Q' and K = 1. (Refer Fig. 21) Assuming the flip-flop was initially cleared and then clocked for 6 pulses. The sequence at the Q output will be.***

a) 010000

b) 011001

c) 010010

d) 010101

Fig. 21.

22. ***The present output Q(t) of an edge triggered JK flip-flop is logic 0. If J = 1, then Q(t+1)***

a) Cannot be determined b) will be logic 0

c) will be logic 1 d) will race around

23. ***The counter shown in Fig. 23 is***

a) MOD-3 ripple counter

b) MOD-7 synchronous counter

c) MOD-5 ripple counter

d) MOD-7 ripple counter

Fig. 23.

24. ***The modulus of the counter in Fig. 24 is***

a) 2

b) 3

c) 4

d) 5

Fig. 24.

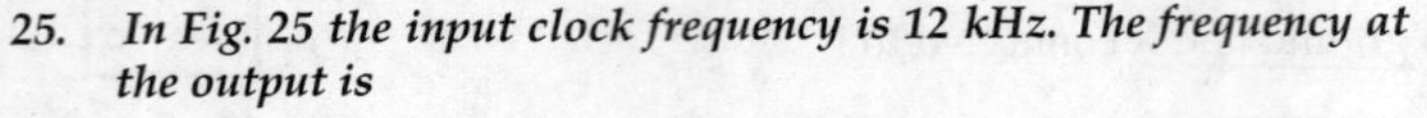

25. ***In Fig. 25 the input clock frequency is 12 kHz. The frequency at the output is***

a) 24 kHz b) 12 kHz

c) 3 kHz d) 6 kHz

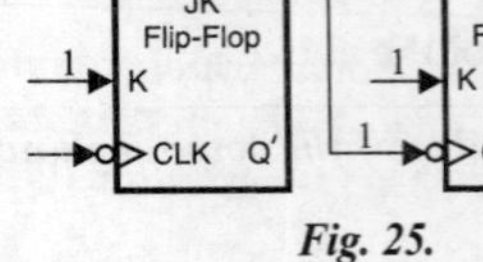

Fig. 25.

26. ***For the given register in Fig. 26 to have the condition '000' again the number of clock pulses required would be***

a) 3 b) 8

c) 6 d) 16

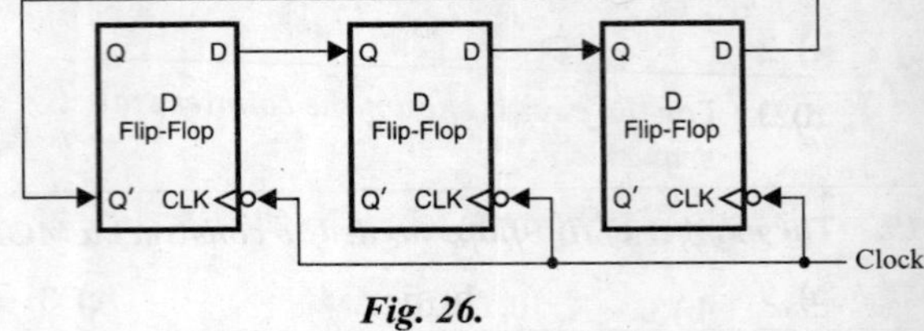

Fig. 26.

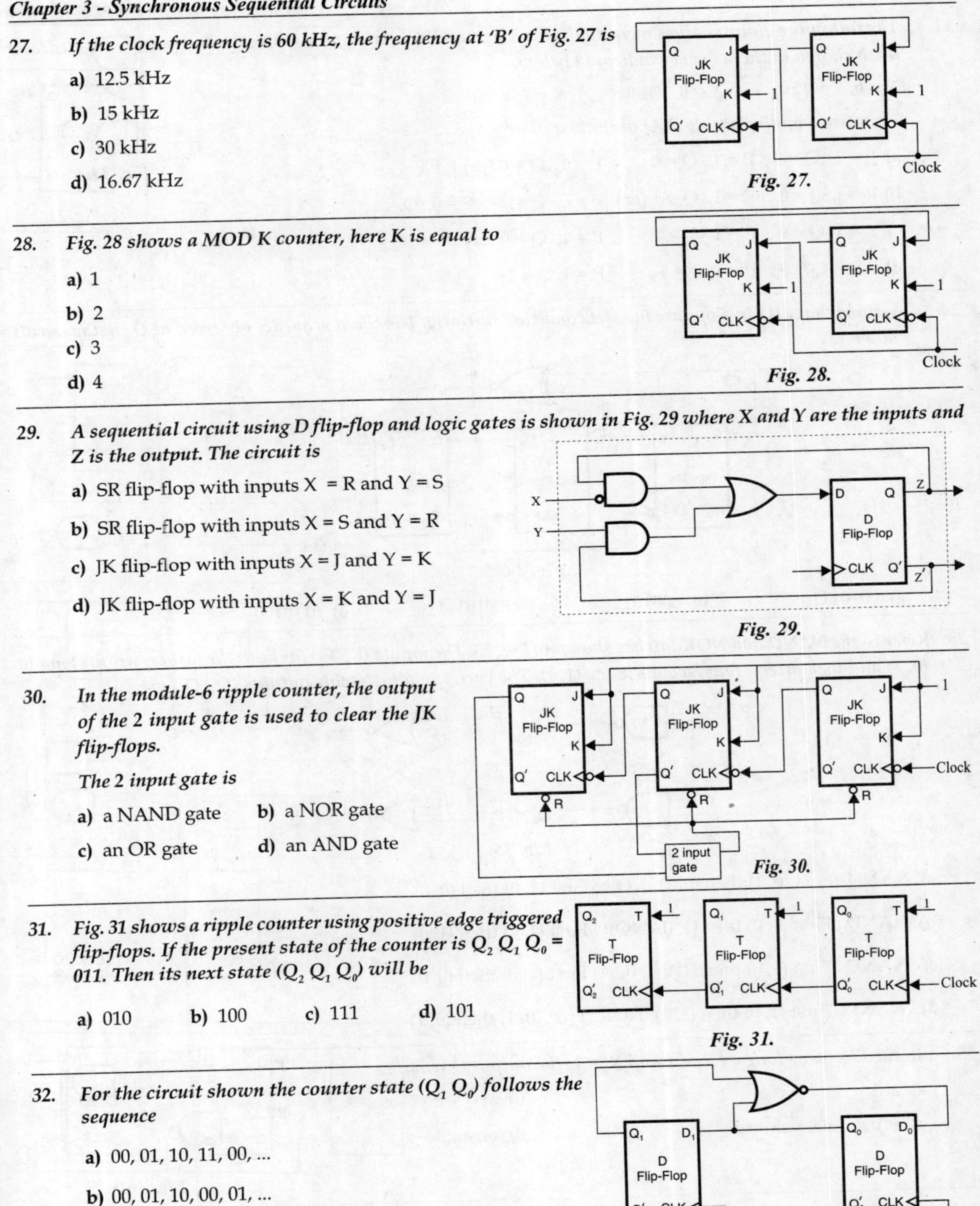

27. *If the clock frequency is 60 kHz, the frequency at 'B' of Fig. 27 is*

a) 12.5 kHz

b) 15 kHz

c) 30 kHz

d) 16.67 kHz

Fig. 27.

28. *Fig. 28 shows a MOD K counter, here K is equal to*

a) 1

b) 2

c) 3

d) 4

Fig. 28.

29. *A sequential circuit using D flip-flop and logic gates is shown in Fig. 29 where X and Y are the inputs and Z is the output. The circuit is*

a) SR flip-flop with inputs X = R and Y = S

b) SR flip-flop with inputs X = S and Y = R

c) JK flip-flop with inputs X = J and Y = K

d) JK flip-flop with inputs X = K and Y = J

Fig. 29.

30. *In the module-6 ripple counter, the output of the 2 input gate is used to clear the JK flip-flops.*

The 2 input gate is

a) a NAND gate b) a NOR gate

c) an OR gate d) an AND gate

Fig. 30.

31. *Fig. 31 shows a ripple counter using positive edge triggered flip-flops. If the present state of the counter is $Q_2\,Q_1\,Q_0$ = 011. Then its next state ($Q_2\,Q_1\,Q_0$) will be*

a) 010 b) 100 c) 111 d) 101

Fig. 31.

32. *For the circuit shown the counter state ($Q_1\,Q_0$) follows the sequence*

a) 00, 01, 10, 11, 00, ...

b) 00, 01, 10, 00, 01, ...

c) 00, 01, 11, 00, 11, ...

d) 00, 10, 11, 00, 10, ...

Fig. 32.

33. ***The following binary values were applied to the X and Y inputs of the NAND latch shown in the sequence indicated below.***

X = 0, Y = 1 ; X = 0, Y = 0 ; X = 1, Y = 1

The corresponding stable P, Q outputs will be

a) P = 1, Q = 0 ; P = 1, Q = 0 ; P = 1, Q = 0 (or) P = 0, Q = 1

b) P = 1, Q = 0 ; P = 1, Q = 1 (or) P = 0, Q = 1 ; P = 0, Q = 1

c) P = 1, Q = 0 ; P = 1, Q = 1 ; P = 1, Q = 0 (or) P = 0, Q = 1

d) P = 1, Q = 0 ; P = 1, Q = 1 ; P = 1, Q = 1

Fig. 33.

34. ***Assume that all flip-flops are in reset condition initially. The count sequence observed at Q_A in the circuit shown is***

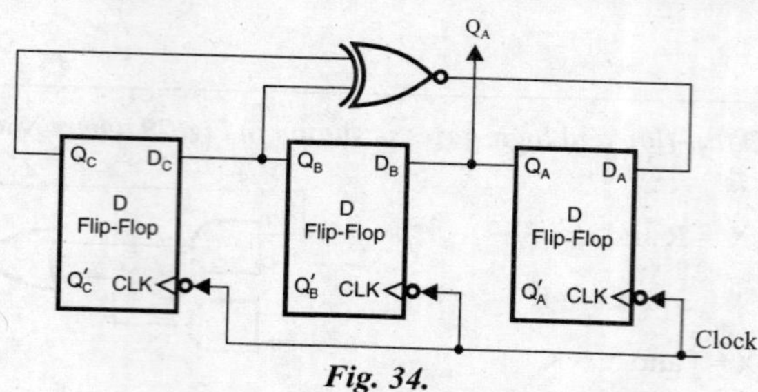

Fig. 34.

a) 0010111 **b)** 0001011 **c)** 0101111 **d)** 0110100

35. ***Refer to the NAND and NOR latches shown in Fig. 35. The inputs (P_1, P_2) for both the lathces are first made (0, 1) and then after a few seconds made (1, 1). The corresponding stable outputs (Q_1, Q_2) are***

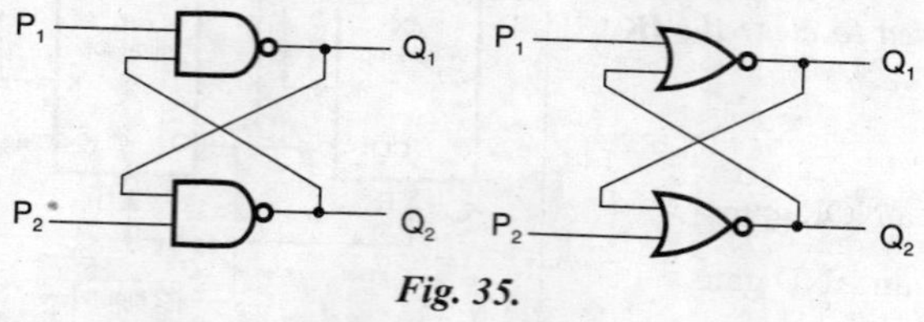

Fig. 35.

a) NAND : First (0, 1) then (0, 1) NOR : First (1, 0) then (0, 0)

b) NAND : First (1, 0) then (1, 0) NOR : First (1, 0) then (1, 0)

c) NAND : First (1, 0) then (1, 0) NOR : First (1, 0) then (0, 0)

d) NAND : First (1, 0) then (1, 1) NOR : First (0, 1) then (0, 1)

36. ***In the Fig. 36 as long as X_1 = 1 and X_2 = 1, the output Q remains***

a) at 1 **b)** at 0

c) at its initial value **d)** unstable

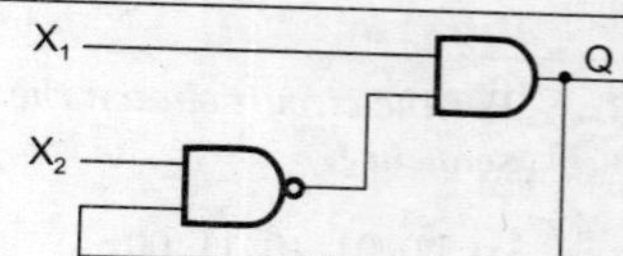

Fig. 36.

Answers

1. d	8. a	15. b	22. c	29. d	36. c
2. b	9. d	16. d	23. d	30. a	
3. c	10. a	17. a	24. c	31. a	
4. b	11. b	18. a	25. c	32. b	
5. d	12. a	19. b	26. c	33. d	
6. b	13. b	20. c	27. c	34. d	
7. a	14. b	21. d	28. b	35. a	

V. Answer the following questions

1. Draw the symbol of SR latch with using NAND and OR gates?
2. How does the D latch developed from SR latch and draw the symbol of D latch?
3. Write a few applications of flip-flops.
4. List the difference between level triggeried and edge triggered flip-flop.
5. Write down the characteristic equation and excitation table for SR, JK , T and D flip-flop.
6. Define universal shift register and its operation mode.
7. Explain the preset and reset facility in a flip-flop.
8. Explain the up and down counter.
9. Define MOD-2^n counter.
10. Write down the output table of 3-bit binary counter.
11. Write down the output table of 4-bit binary counter.
12. Explain synchronous counter.
13. Draw the logic diagram and timing sequence of a 4-bit ring counter?

VI. Solve the following problems

E3.1 **A sequential circuit with two D flip-flops A and B, two inputs x and y and one output Z is specified by the following next state and output equations.**

$$A(t+1) = xy' + xB \quad ; \quad B(t+1) = xA + xB' \quad ; \quad Z = A$$

a) Draw the logic diagram of the circuit.

b) List the state table for the sequential circuit.

c) Draw the corresponding state diagram.

E3.2 **A sequential circuit has one D flip-flop, two inputs x and y, and one output d. The full-subtractor output carry c_o is connected to the D flip-flop, as shown in Fig. E3.2. Derive the state table and state diagram of the sequential circuit.**

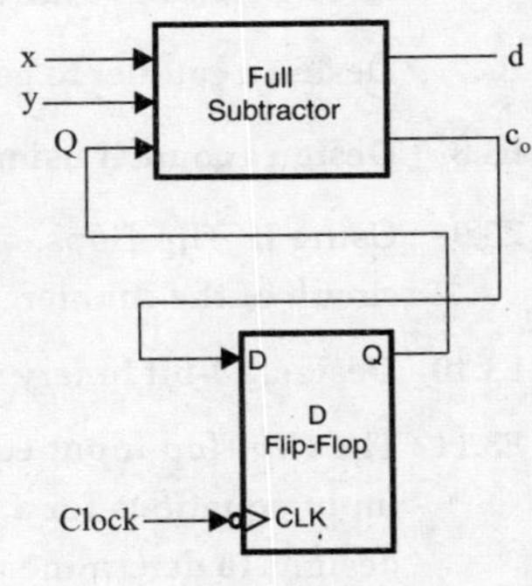

Fig. E3.2.

E3.3 **For the state table shown in Table E3.3,**

a) Draw the corresponding state diagram.

b) Tabulate the reduced state table.

c) Draw the state diagram corresponding to the reduced state table.

Table E3.3: State Table

Present State	Next State		Output	
Q(t)	x = 0	x = 1	x = 0	x = 1
a	f	b	0	0
b	d	c	0	0
c	f	e	0	0
d	g	a	1	0
e	d	c	0	0
f	f	b	1	1
g	g	h	0	1
h	g	a	1	0

E3.4 **A sequential circuit has two JK flip-flops A and B and one input x. The circuit is described by the following flip-flop input equations.**

$J_A = x \quad ; \quad K_A = B' \quad ; \quad J_B = x \quad ; \quad K_B = A$

a) Derive the state equations A(t + 1) and B(t + 1) by substituting the input equations for the J and K variables.

b) Draw the state diagram of the circuit.

E3.5 **A sequential circuit has three flip-flops A, B, and C one input x and one output Y. The state diagram is shown in Fig. E3.5. The circuit is to be designed by treating the unused states as don't-care conditions. Analyze the circuit obtained from the design to determine the effect of the unused states.**

a) Use JK flip-flops in the design. **b) Use T flip-flops in the design.**

c) Use D flip-flops in the design.

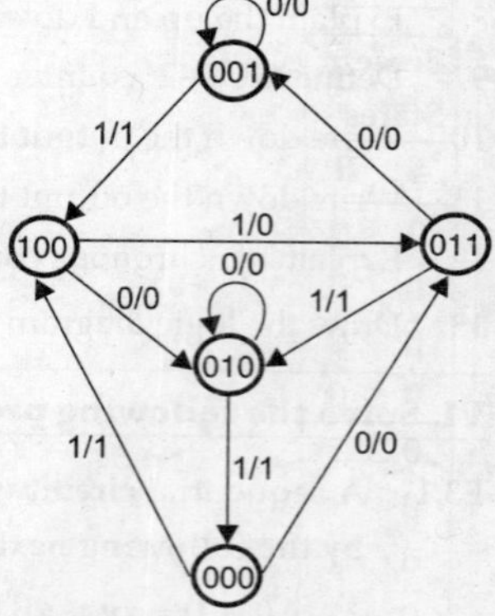

Fig. E3.5: State diagram.

E3.6 **Develop the state diagram for a Mealy state machine that detects a sequence of three or more consecutive 1's in a string of bits coming through an input line.**

E3.7 **Design a counter to generate the binary sequence : 0, 1, 2, 4, 6 using D flip-flops.**

E3.8 **Design a counter using T flip-flops that goes through states 0, 2, 4, 6, 7, 0. Is the counter is self starting?**

E3.9 **Using JK flip-flops, design a counter to generate the binary sequence 0, 1, 2, 4, 5, 6. Draw the logic circuit of the counter.**

E3.10 **Design a 4-bit binary synchronous counter using D flip-flops.**

E3.11 **The flip-flop input equations for a BCD counter using T flip-flops are given in below. Obtain the input equations for a BCD counter that uses a) JK flip-flops and b) D flip-flops. Compare the three designs to determine which is most efficient.**

$T_{A1} = 1 \quad ; \quad T_{A2} = A_8'A_1$

$T_{A4} = A_2A_1 \quad ; \quad T_{A8} = A_8A_1 + A_4A_2A_1$

Answers

E3.1 a)

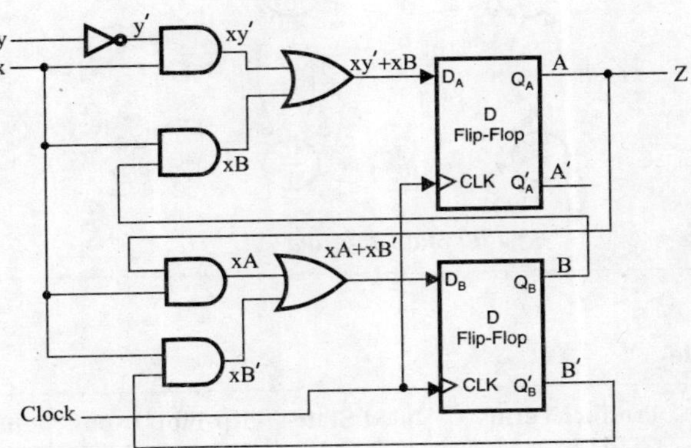

***Fig:** Logic diagram.*

b)

Table 1: Expanded State Table

Present States		Complement of Present State B	Inputs			Product terms				Next States		Flip-Flop Inputs		Output
A	B	B′	x	y	y′	xy′	xB	xA	xB′	A(t+1)	B(t+1)	D_A	D_B	Z = A
0	0	1	0	0	1	0	0	0	0	0	0	0	0	0
0	0	1	0	1	0	0	0	0	0	0	0	0	0	0
0	0	1	1	0	1	1	0	0	1	1	1	1	1	0
0	0	1	1	1	0	0	0	0	1	0	1	0	1	0
0	1	0	0	0	1	0	0	0	0	0	0	0	0	0
0	1	0	0	1	0	0	0	0	0	0	0	0	0	0
0	1	0	1	0	1	1	1	0	0	1	0	1	0	0
0	1	0	1	1	0	0	1	0	0	1	0	1	0	0
1	0	1	0	0	1	0	0	0	0	0	0	0	0	1
1	0	1	0	1	0	0	0	0	0	0	0	0	0	1
1	0	1	1	0	1	1	0	1	1	1	1	1	1	1
1	0	1	1	1	0	0	0	1	1	0	1	0	1	1
1	1	0	0	0	1	0	0	0	0	0	0	0	0	1
1	1	0	0	1	0	0	0	0	0	0	0	0	0	1
1	1	0	1	0	1	1	1	1	0	1	1	1	1	1
1	1	0	1	1	0	0	1	1	0	1	1	1	1	1

c)

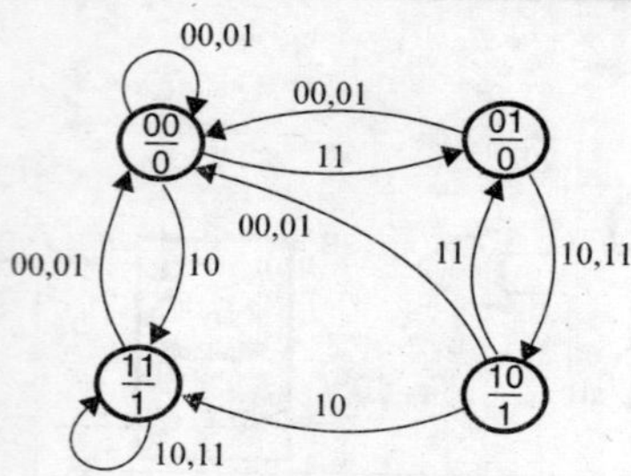

Fig: State diagram.

E3.2 a)

Table 1: Expanded State Table

Present State	Input		Product Terms				Next State	Flip-Flop Input	Output
A	x	y	x'	x'y	x'A	yA	A(t+1)	D	d
0	0	0	1	0	0	0	0	0	0
0	0	1	1	1	0	0	1	1	1
0	1	0	0	0	0	0	0	0	1
0	1	1	0	0	0	0	0	0	0
1	0	0	1	0	1	0	1	1	1
1	0	1	1	1	1	1	1	1	0
1	1	0	0	0	0	0	0	0	0
1	1	1	0	0	0	1	1	1	1

b)

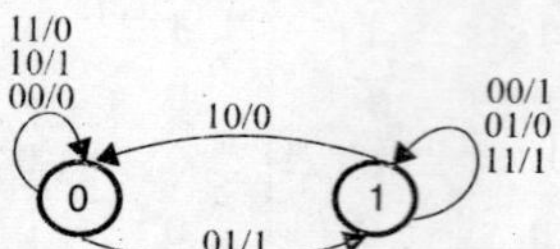

Fig: State diagram.

E3.3 a)

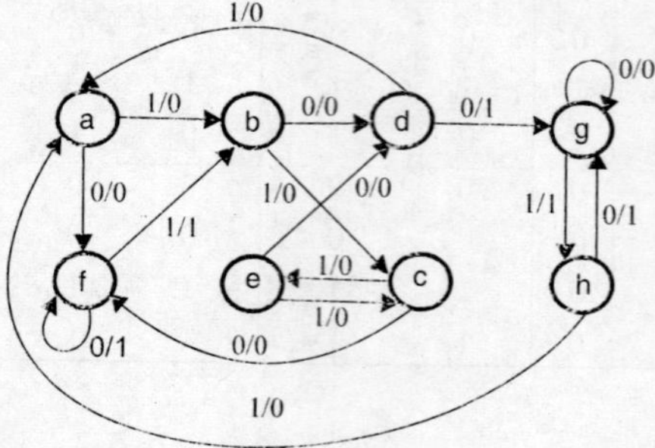

Fig: State diagram.

b)

Table: Third Reduced State Table

Present State	Next State		Output	
Q(t)	x = 0	x = 1	x = 0	x = 1
a	f	b	0	0
b	d	c → a	0	0
d	g	a	1	0
f	f	b	1	1
g	g	d	0	1

c)

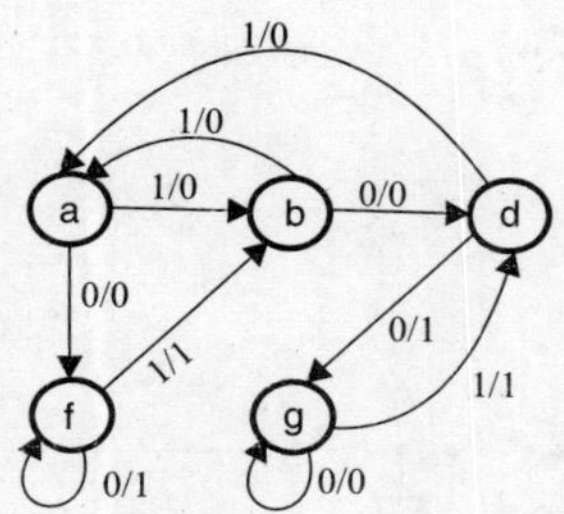

Fig: *State diagram with reduced states.*

E3.4 a) State equations: $A(t+1) = A'x + AB$

$$B(t+1) = B'x + A'B$$

b)

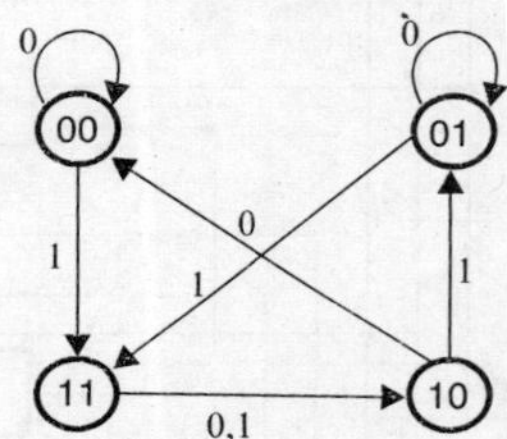

Fig: *State diagram.*

E3.5 a)

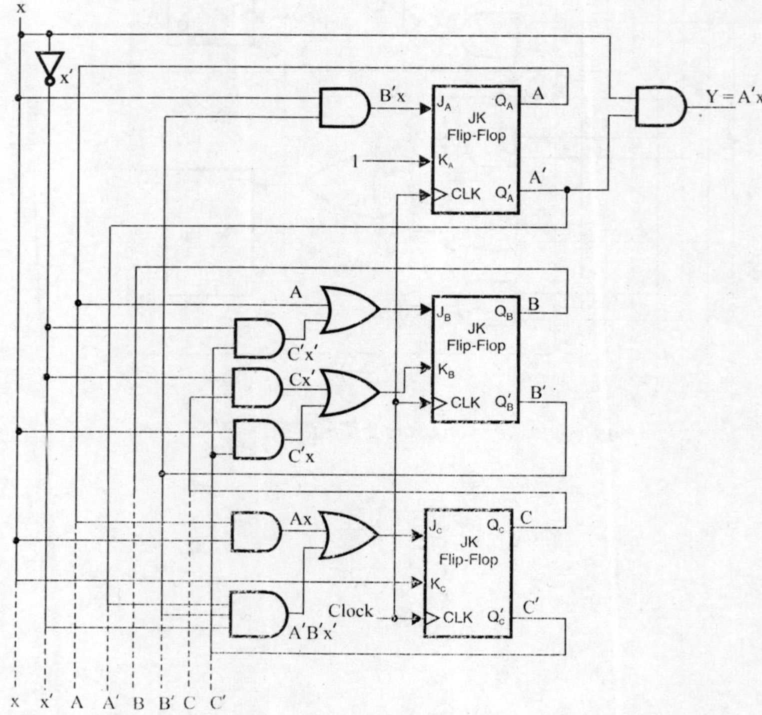

Fig: Logic diagram using JK flip-flops.

b)

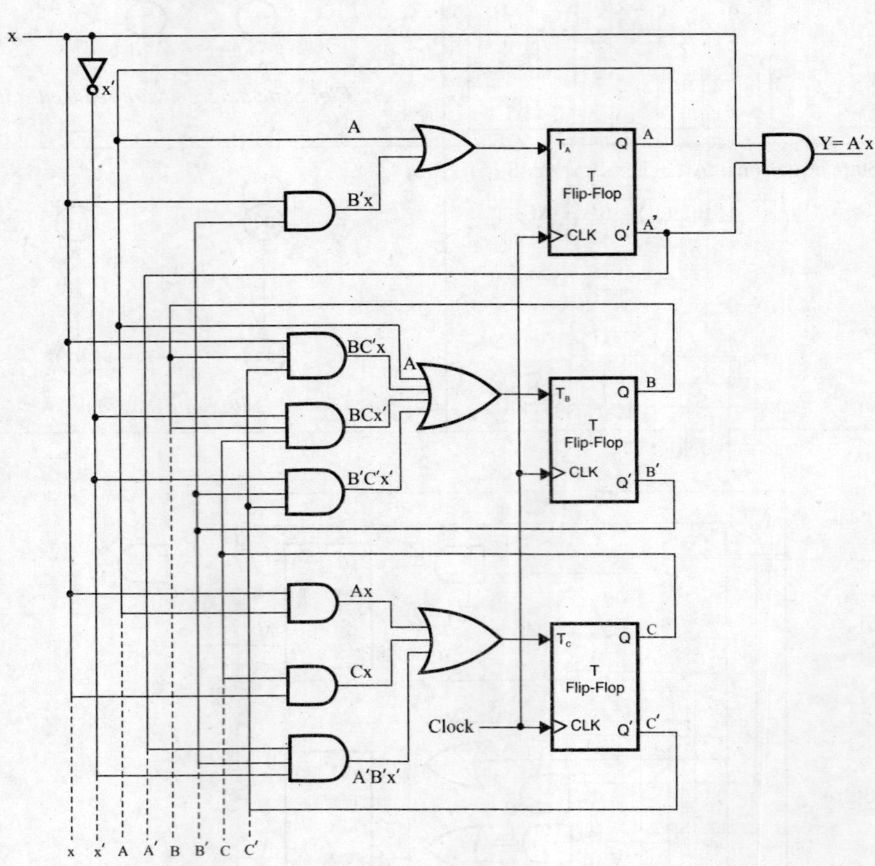

Fig: *Logic diagram using T flip-flops.*

c)

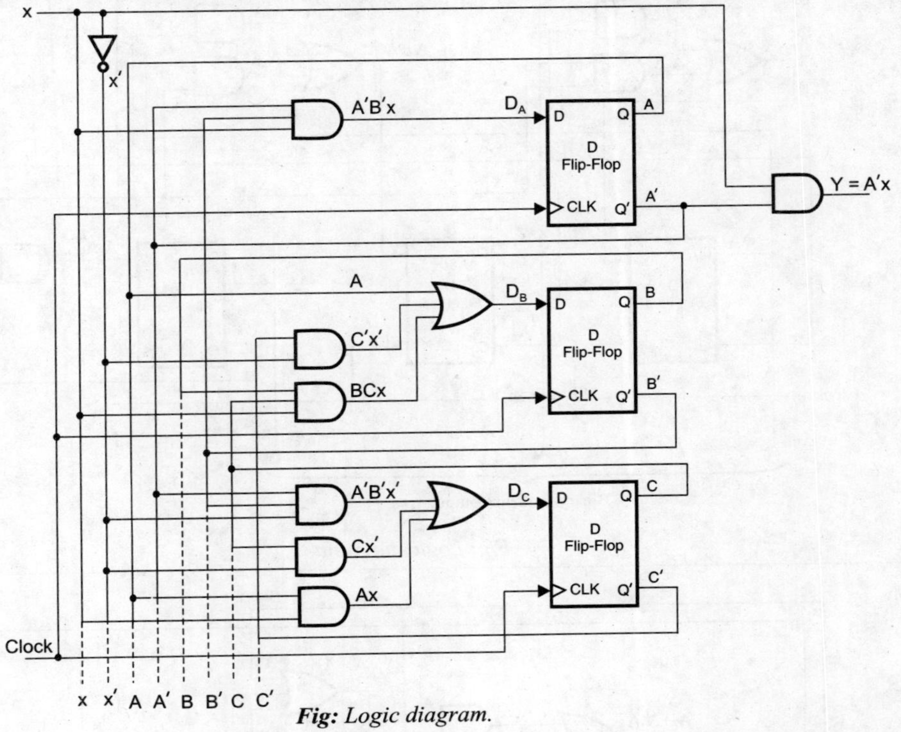

x
x′
A′B′x
D_A
D Q A
D
Flip-Flop
CLK Q′ A′
Y = A′x
A
D_B
D Q B
D
Flip-Flop
C′x′
BCx
CLK Q′ B′
A′B′x′
D_C
D Q C
D
Flip-Flop
Cx′
Ax
CLK Q′ C′
Clock
x x′ A A′ B B′ C C′

***Fig:** Logic diagram.*

E3.6

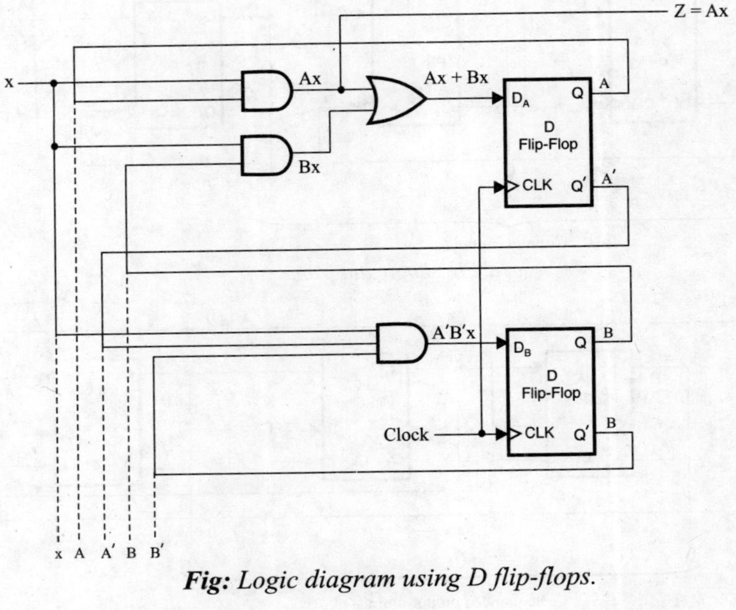

Z = Ax
x
Ax
Ax + Bx
D_A Q A
D
Flip-Flop
Bx
CLK Q′ A′
A′B′x
D_B Q B
D
Flip-Flop
Clock
CLK Q′ B
x A A′ B B′

***Fig:** Logic diagram using D flip-flops.*

E3.7

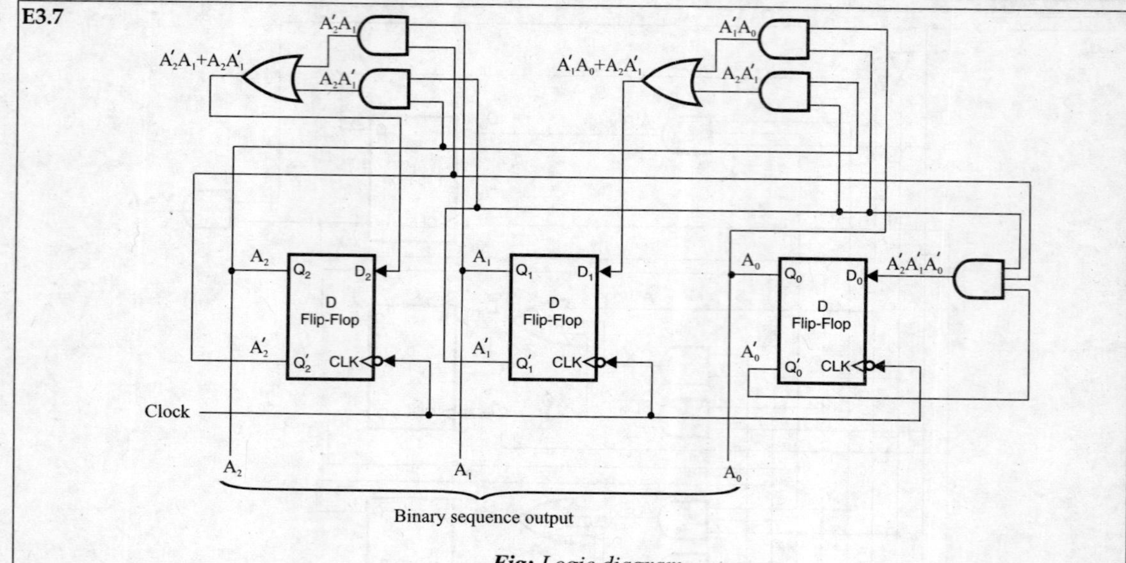

Fig: Logic diagram.

E3.8

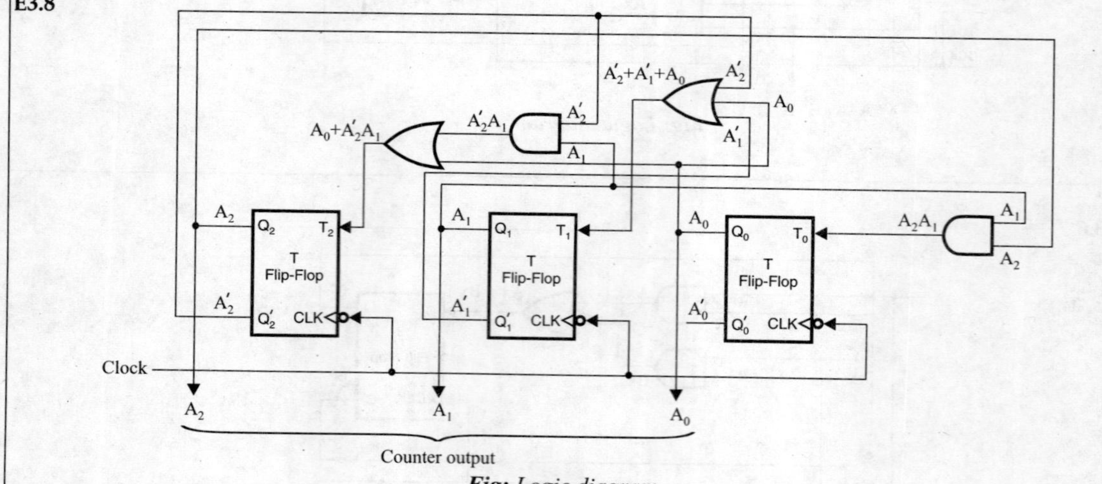

Fig: Logic diagram.

E3.9

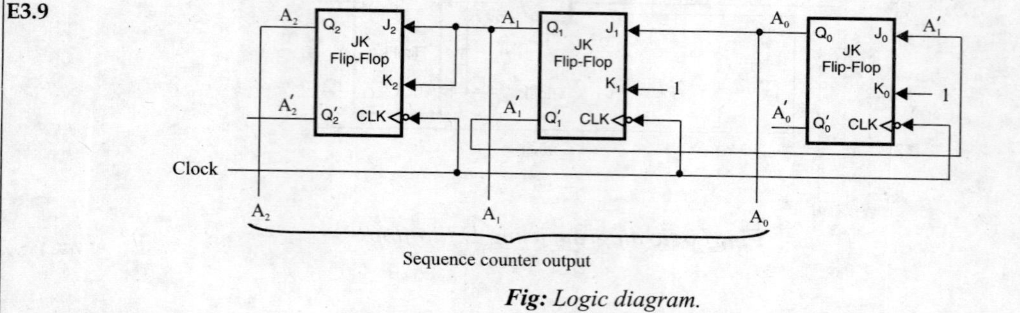

Fig: Logic diagram.

E3.10

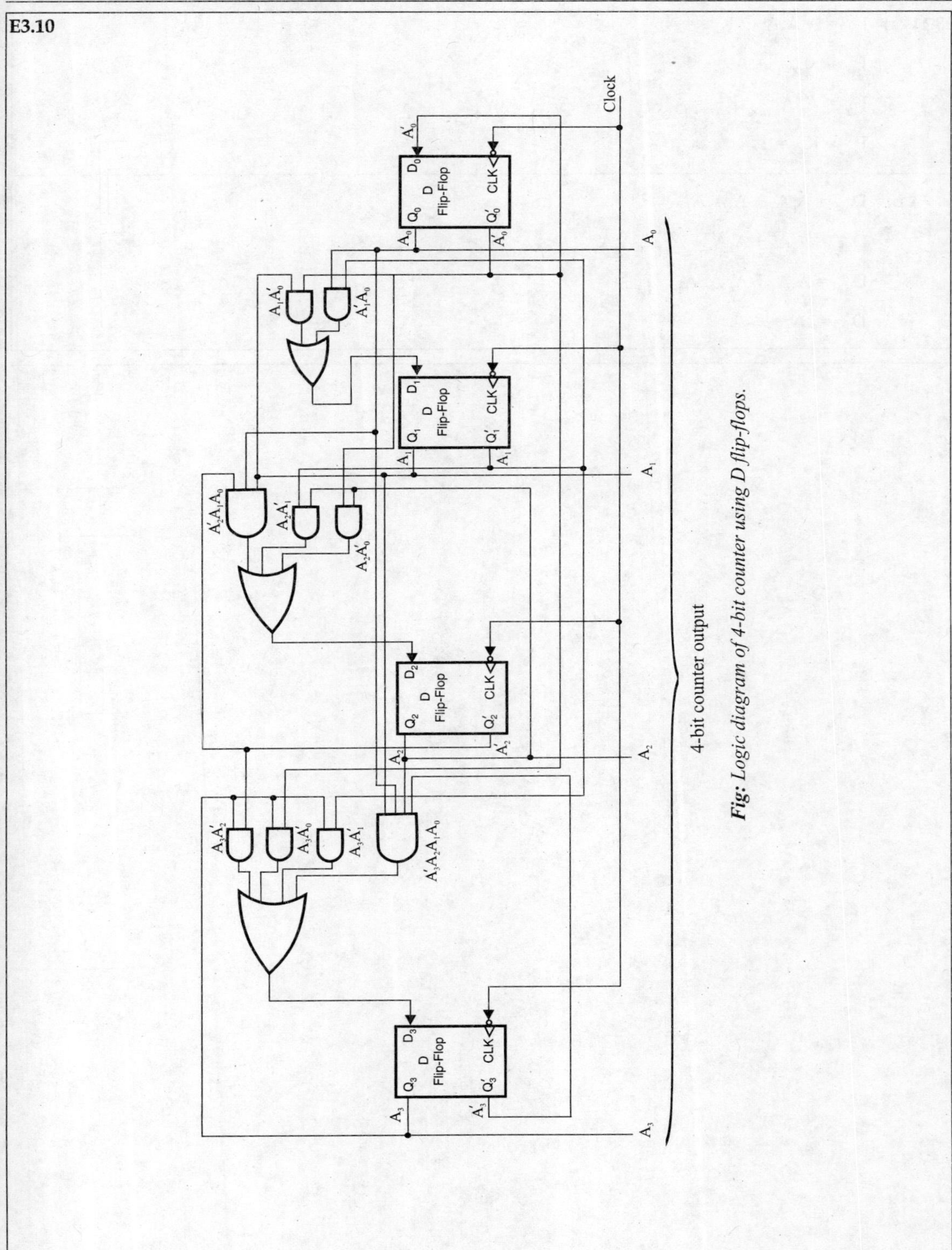

***Fig:** Logic diagram of 4-bit counter using D flip-flops.*

E3.11 a) $J_{A8} = A_4A_2A_1$; $K_{A8} = A_1$

$J_{A4} = K_{A4} = A_2A_1$

$J_{A2} = A'_8 A_1$; $K_{A2} = A_1$

$J_{A1} = K_{A1} = 1$

b) $D_{A8} = A_8A'_1 + A_4A_2A_1$

$D_{A4} = A_4A'_2 + A_4A'_1 + A'_8 A'_4 A_2A_1$

$D_{A2} = A_2A'_1 + A'_8 A'_2 A_1$

$D_{A1} = A'_1$

CHAPTER 4

ASYNCHRONOUS SEQUENTIAL CIRCUITS AND PROGRAMMABILITY LOGIC DEVICES

4.1 Asynchronous Sequential Logic Circuits

The combinational logic circuits with feedback and without clock are called ***asynchronous sequential circuits***. Since there is no clock, the changes in internal states occur immediately when there is a change in input and so the response in an asynchronous sequential circuit will be fast. In the design of asynchronous sequential circuit sufficient care must be taken to ensure that every change in state leads to a stable condition.

Table 4.1: Differences between Synchronous and Asynchronous Sequential Circuits

Synchronous Sequential Circuit	Asynchronous Sequential Circuit
1. Provided with clock	1. Do not have a clock
2. Changes in internal states occur in response to clock.	2. Changes in internal states occur when there is a change in input variables.
3. Memory elements are clocked flip-flops.	3. Memory elements are either latches (unclocked flip-flops) or time-delay elements.

Basically, an asynchronous sequential circuit can be expressed as a combinational circuit with feedback paths having delay elements. The asynchronous sequential circuits will have input, output and state variables. The present and next state variables of asynchronous sequential circuit are called secondary and excitation variables respectively. The block diagram of asynchronous sequential circuit is shown in Fig. 4.1.

Here, $x_1, x_2, \ldots, x_n$ = n input variables

$Z_1, Z_2, \ldots, Z_m$ = m output variables

$Y_1, Y_2, \ldots, Y_k$ = k excitation variables (next state variables)

$y_1, y_2, \ldots, y_k$ = k secondary variables (present state variables)

In asynchronous circuits the state transitions occur when there is a change in input. The circuit will attain a stable condition when excitation (or next state) variables are same as secondary (or present) state variables. Therefore, a change in input is allowed only after the circuit attains a stable state due to previous change in input. Also, when there are multiple inputs, only one input is allowed to change at any one time in order to ensure proper operation of circuit and this is called fundamental mode of operation. Therefore, the ***fundamental mode*** of operation is defined as the operating mode in which only one input is allowed to change when the circuit is in stable state.

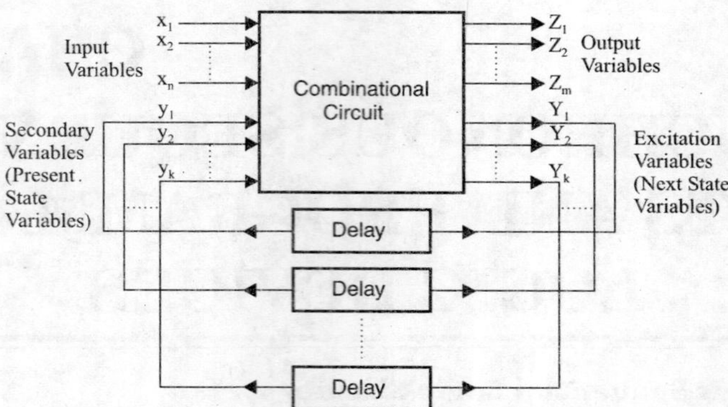

Fig. 4.1: Block diagram representation of an asynchronous sequential circuit.

4.1.1 Fundamental and Pulse Mode Sequential Circuits

In fundamental mode asynchronous sequential circuits the inputs are level triggered inputs, whereas in pulse mode asynchronous sequential circuits the inputs are pulses of sufficient duration for the output to change for a given input.

Design of Fundamental Mode Sequential Circuits

The design of asynchronous sequential logic circuit is designing a race-free and hazard free logic circuit from the given system specifications. The following procedural steps can be followed for the design.

- Develop a primitive flow table from the word description of asynchronous circuit
- Minimize the number of states in the flow table.
- Assign race-free binary codes to states.
- Construct the transition table.
- Construct excitation maps and find Boolean equations for a hazard free realization.
- Check for essential hazards.
- Draw the logic circuit by implementing methods to eliminate essential hazards.

The examples presented in sections 4.1 to 4.6 in this chapter are related to design of fundamental mode asynchronous circuits.

4.2 Transition and Flow Table and Stability

4.2.1 Transition Table

In asynchronous sequential circuit, the state table is called ***transition table***. Practically, the state table itself is listing of all possible transitions between the states of a circuit. In asynchronous circuit state table, there should be at least one next state that is equal to present state in each row.

The procedure to form transition table is as follows,

1. Identify all feedback paths in the circuit. Let the circuit has k number of feedback paths.
2. Assign secondary (or present state) and excitation (or next state) variables to each feedback path. Let, y_1, y_2, …, y_k be secondary variables and Y_1, Y_2, ……, Y_k be excitation variables.

3. Now, the excitation variables are functions of secondary variables and circuit external inputs. Determine the Boolean equations for excitation variables in terms of secondary variables and inputs. Note that, the secondary variables are inputs to a combinational circuit and excitation variables are output from the combinational circuit. The excitation variables are given as inputs to feedback paths.
4. Draw maps similar to K-maps for every excitation variable, Y_1, Y_2, ……., Y_k. In these maps, the rows are formed by secondary variables and columns by inputs.
5. Now, there will be k number of maps. Combine all the k number of maps into a single map, by combining the corresponding entries in each square of individual map. Now each square will have k-bit binary representing next state.
6. In each row mark the next states ($Y_1 Y_2 \ldots Y_k$) that are same as present state ($y_1 y_2 \ldots y_k$) by circles. The circled entries represent stable states and uncircled entries represent unstable states.

The ***transition table*** lists the possible state transitions for all possible changes in inputs and so it can be used to analyze the circuit behavior.

Consider the asynchronous sequential circuit shown in Fig. 4.2. This circuit has one input and output and two feedback paths. Let us assume that the delay in feedback is due to propagation delay in the conducting wires that connect the output and input.

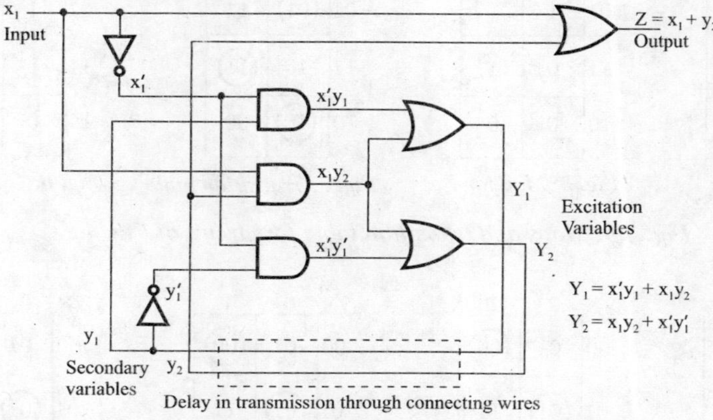

Fig. 4.2: Example of an asynchronous sequential circuit.

Let, x_1 = Input

y_1, y_2 = Secondary variables

Y_1, Y_2 = Excitation variables

Z = Output

Now, with reference to Fig. 4.2, we can write the following Boolean equations for excitation variables, Y_1 and Y_2 and output Z.

$$Y_1 = x_1' y_1 + x_1 y_2$$

$$Y_2 = x_1 y_2 + x_1' y_1'$$

$$Z = x_1 + y_2$$

Using the above Boolean equations the excitation variables, Y_1 and Y_2 are evaluated for all possible combinations input x_1 and secondary variables y_1 and y_2 and listed in Table 4.2.

Table 4.2: Evaluation of Excitation Variables and Output

y_1	y_2	x_1	y_1'	y_2'	x_1'	$x_1'y_1$	x_1y_2	$x_1'y_1'$	Y_1	Y_2	$Z = x_1 + y_2$
0	0	0	1	1	1	0	0	1	0	1	0
0	0	1	1	1	0	0	0	0	0	0	1
0	1	0	1	0	1	0	0	1	0	1	1
0	1	1	1	0	0	0	1	0	1	1	1
1	0	0	0	1	1	1	0	0	1	0	0
1	0	1	0	1	0	0	0	0	0	0	1
1	1	0	0	0	1	1	0	0	1	0	1
1	1	1	0	0	0	0	1	0	1	1	1

y_1y_2 \ x_1	0	1
00	0	0
01	0	1
11	1	1
10	1	0

Fig. a: Map for Y_1.

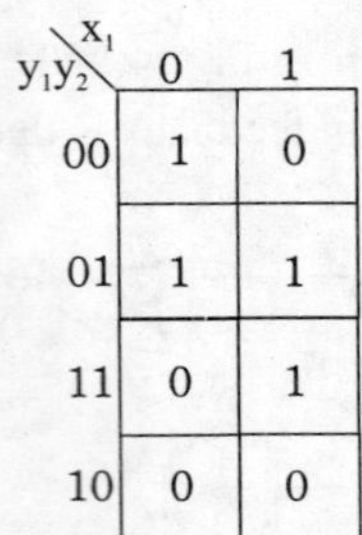

Fig. b: Map for Y_2.

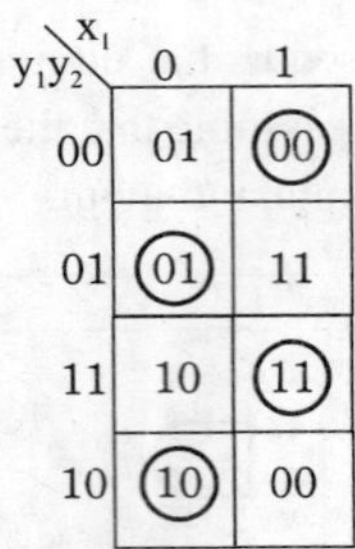

Fig. c: Transition table.

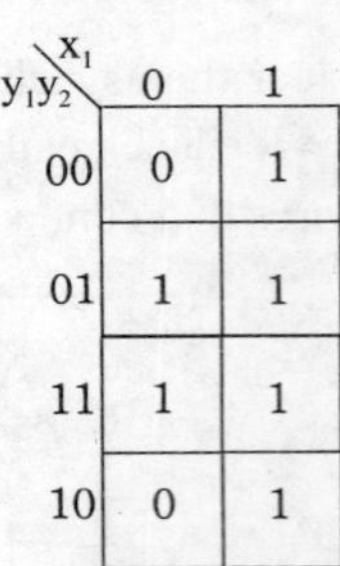

Fig. d: Output map.

Fig. 4.3: Maps and transition table for circuit of Fig. 4.2.

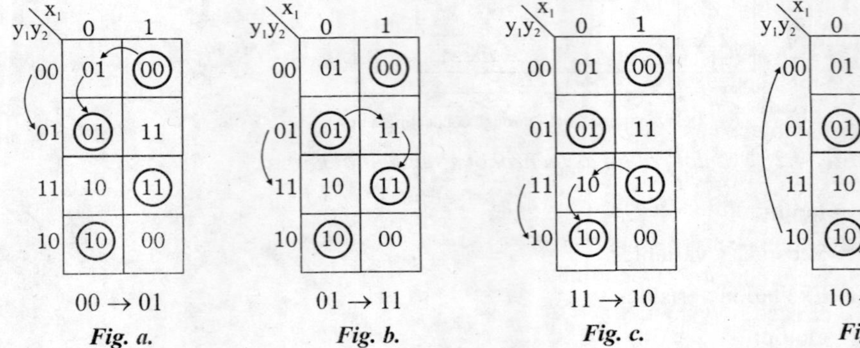

Fig. 4.4: State transitions due to change in input.

Maps are constructed for Y_1 and Y_2 by taking possible values of secondary variables y_1 and y_2 on rows and input x_1 on columns as shown in Figs. 4.3(a) and (b) and the squares in the maps are filled with values of Y_1 and Y_2 listed in Table 4.2. The transition table is formed by combining the maps of Figs. 4.3(a) and (b) into a single map as shown in Fig. 4.3(c). The output map is shown in Fig. 4.3(d).

4.2.2 Stable and Unstable States

A state is a circuit condition represented by a set of variables called ***state variables***.

The states are represented either by letter symbols a, b, c, d, ... or by binary codes. The circuit of Fig. 4.2 has two state variables and so states are represented by 2-bit binary codes 00, 01, 11 and 10.

The circuit will change from one state to another state when there is a change in input. During a state transition when the next state is same as present state then the circuit attains a stable condition and the corresponding state is called ***stable state***. On the other hand in a state transition, if the next state is not equal to present state then the corresponding state is called ***unstable state***.

In the transition table shown in Fig. 4.3(c), the entries in each square represents next state. Therefore, in each row, the squares that have identical values with present state are marked with circles. The circled entries represent stable states and uncircled entries represent unstable states.

Normally, an asynchronous sequential circuit will remain in a stable state for a given input. The transition from one stable state to another stable state occurs due to changes in input.

Consider the circuit shown in Fig. 4.2 and its transition table shown in Fig. 4.3. In this circuit, the following four state transitions due to changes in input are possible as shown in Fig. 4.4.

When in state 00, if the input changes from 1 to 0, then state transitions occur as shown in Fig. 4.4(a).

When in state 01, if the input changes from 0 to 1, then state transitions occur as shown in Fig. 4.4(b).

When in state 11, if the input changes from 1 to 0, then state transitions occur as shown in Fig. 4.4(c).

When in state10, if the input changes from 0 to 1, then state transitions occur as shown in Fig. 4.4(d).

In an asynchronous circuit, the state changes occurs immediately when there is a change in input, and so the input value can be combined with state to declare a condition called total state of the circuit. All possible combinations of state variables and inputs are called ***total states***. The total state in the circuit of Fig. 4.2 is given by the possible binary values of $y_1y_2x_1$. With three variables 8 different total states can be formed, in which the four states that are circled in transition table are stable states and the other four states are unstable.

Stable total states: 001, 010, 111, 100

Unstable total states: 000, 011, 110, 101

Using the transition table, the state table for asynchronous circuit similar to that of a state table of synchronous circuit can be constructed as shown in Table 4.3. In state table of asynchronous circuit, at least one entry of each row should have same present state and next state.

Table 4.3: State Table of the Circuit of Fig. 4.2

Present State		Next State				Output	
		$x_1 = 0$		$x_1 = 1$		$x_1 = 0$	$x_1 = 1$
y_1	y_2	Y_1	Y_2	Y_1	Y_2	Z	Z
0	0	0	1	0	0	0	1
0	1	0	1	1	1	1	1
1	0	1	0	0	0	0	1
1	1	1	0	1	1	1	1

Unstable Condition in Asynchronous Circuit

Consider the transition table of an asynchronous circuit shown in Fig. 4.5(a). In this transition table, there, is no stable state when input is 11. Hence, when input changes from 01 to 11 or 10 to 11 the circuit will enter an unstable condition with next state keeps on changing from 1 to 0 and 0 to 1 continuously as shown in Fig. 4.5(b) and (c).

While designing asynchronous sequential circuits such unstable conditions should be avoided and this can be achieved by ensuring one stable state for every input.

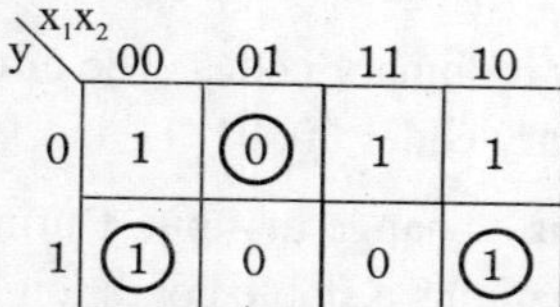

y \ x_1x_2	00	01	11	10
0	1	(0)	1	1
1	(1)	0	0	(1)

Fig. a: Transition table.

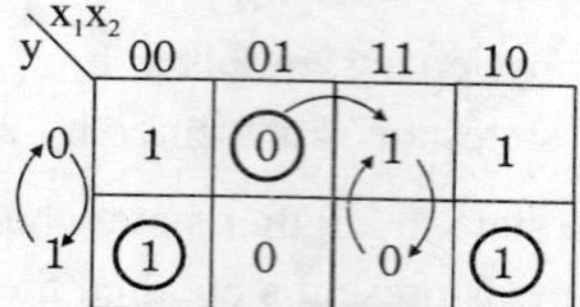

y \ x_1x_2	00	01	11	10
0	1	(0)	1	1
1	(1)	0	0	(1)

Fig. b: Unstable condition when input change from 01 to 11.

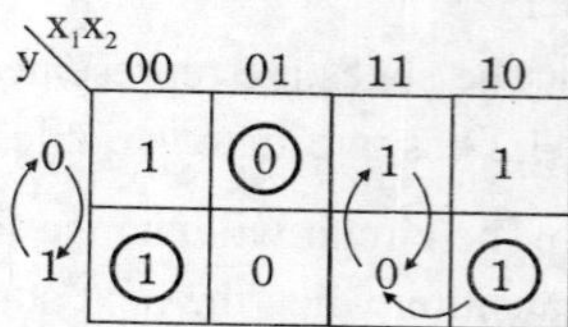

y \ x_1x_2	00	01	11	10
0	1	(0)	1	1
1	(1)	0	0	(1)

Fig. c: Unstable condition when input change from 10 to 11.

Fig. 4.5: Example of transition table of unstable circuit.

The continuous change between two unstable states is also called ***metastability*** which refers to an equilibrium condition in an unstable circuit.

4.2.3 Flow Table

The ***flow table*** lists all state transitions from present state to next state along with output for all possible changes in input. It is similar to transition table, but has two differences. One is the states are represented by letter symbols and not by binary values. Another difference is that the output is also listed in each square along with next state.

A flow table may have one or more stable states in a row. When there is only one stable state in each row of a flow table then it is called a ***primitive flow table***. The flow table for the circuit of Fig. 4.2 is shown in Fig. 4.6. Here, the following letter symbols are assigned to binary values of states.

$y_1y_2 \rightarrow y$
$00 \rightarrow a$
$01 \rightarrow b$
$11 \rightarrow c$
$10 \rightarrow d$

y \ x_1	0	1
a	b, 0	(a),1
b	(b),1	c, 1
c	d, 1	(c),1
d	(d),0	a, 1

Fig. 4.6: Flow table of circuit shown in Fig. 4.2.

4.2.4 Cycles and Races

In an asynchronous sequential circuit, when there are changes in two or more state variables due to change in input, a ***race condition*** may exist, in which changes in variables may occur at different timings due to difference in delay in the feedback paths. This leads to unpredictable transition paths during a change in state.

For example, consider the state change from 011 to 100. Some of the possible transition paths are shown below:

$011 \rightarrow 111 \rightarrow 101 \rightarrow 100$

$011 \rightarrow 010 \rightarrow 000 \rightarrow 100$

$011 \rightarrow 001 \rightarrow 101 \rightarrow 100$

The race condition may be classified into critical and noncritical race. In a state transition, if the final stable state depends on the order of change in state variables, then the race is called ***critical race***. On the other hand, in a state transition, if the final stable state does not depends on the order of change in state variables, then the race is called ***noncritical race***.

Examples of noncritical race are shown in Fig. 4.7 and examples of critical race are shown in Fig. 4.8.

Races are identified by tracing possible state transitions through unstable states with one bit change while going from one stable state to another stable state for a change in input. The actual race is through internal states or secondary variables.

In the transition table shown in Fig. 4.7(a), while in stable state 00, if the input changes from 0 to 1, there are two possible transition paths and in both the transition paths the circuit go to same stable next state 11 as shown in Fig. 4.7(b). Hence they are noncritical race. Similarly, two more state transitions from stable state 01 to 11 and 10 to 11 for change in input from 0 to 1 are shown in Figs. 4.7(c) and (d). In these transitions also races are noncritical.

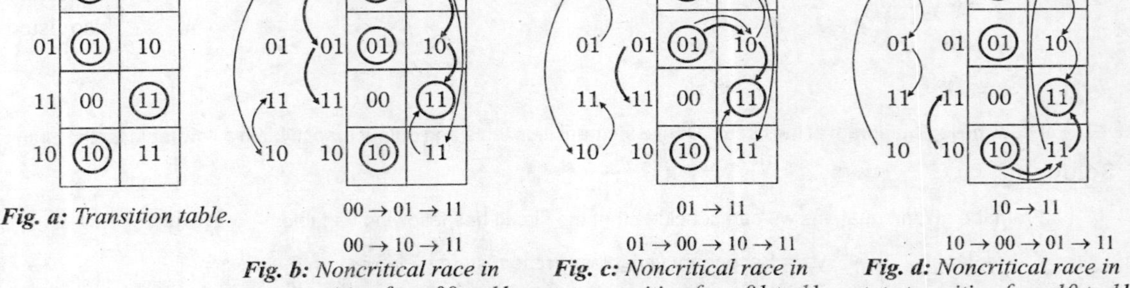

Fig. a: *Transition table.* ***Fig. b:*** *Noncritical race in state transition from 00 to 11.* ***Fig. c:*** *Noncritical race in state transition from 01 to 11.* ***Fig. d:*** *Noncritical race in state transition from 10 to 11.*

Fig. 4.7: *Examples of noncritical race when input change from 0 to 1.*

In the transition table shown in Fig. 4.8(a), while in stable state 00, if the input changes from 0 to 1, there are three possible state transition paths and in each transition path the circuit go to different stable next state as shown in Fig. 4.8(b). Hence they are critical race.

Also, it can be observed that when a column has only one stable state the critical race do not occur.

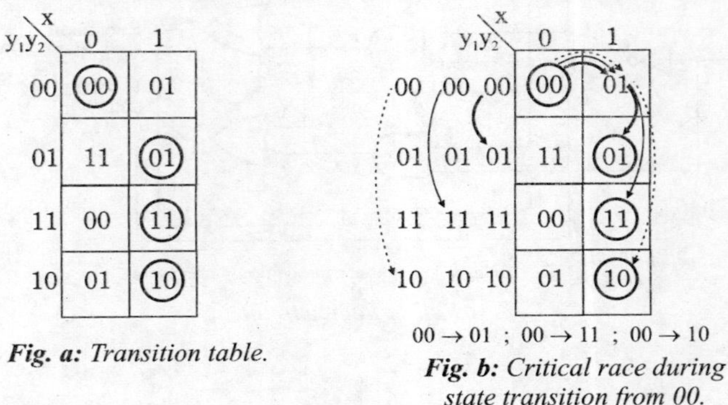

Fig. a: *Transition table.* ***Fig. b:*** *Critical race during state transition from 00.*

Fig. 4.8: *Examples of critical race when input change from 0 to 1.*

Races can be avoided by proper choice of binary values to states. While assiging binary values to states, if the values are assigned such that only one variable change at a time during a state transition in a flow table then the races can be avoided. The race free state assignment is discussed in this chapter in Section 4.3.

When the state transitions in a circuit go through a unique sequence of unstable states then that condition is called a ***cycle***. Races can also be avoided by using cycles, which terminate on a stable state. When a cycle does not terminate on a stable state then the entire circuit will be unstable.

The following state transitions shown in Fig. 4.7 is also examples of cycles because while going from one stable state to another stable state, the transitions go through unstable states.

$00 \rightarrow 01 \rightarrow 11$	$01 \rightarrow 00 \rightarrow 10 \rightarrow 11$	$10 \rightarrow 00 \rightarrow 01 \rightarrow 11$
$00 \rightarrow 10 \rightarrow 11$	$01 \rightarrow 10 \rightarrow 11$	$10 \rightarrow 01 \rightarrow 11$
$00 \rightarrow 01 \rightarrow 10 \rightarrow 11$		

Example 4.1

The Boolean equations for excitation variables and output of an asynchronous circuit are,

$$Y_1 = y_1 y_2 + y_2 x'$$

$$Y_2 = y_1 + y_2 x$$

$$Z = y_1 y_2$$

Draw the logic diagram of the circuit. Derive the transition table and output map. Obtain a flow table for the circuit.

Solution

From the given equations we can conclude that the circuit has following variables.

x = Input	y_1, y_2 = Secondary variables (Present state variables)
Z = Output	Y_1, Y_2 = Excitation variables (Next state variables)

Using the given Boolean equations the logic circuit is drawn as shown in Fig. 1 and the excitation variables Y_1 and Y_2 is evaluated for possible values of input, x and secondary variables, y_1 and y_2 and listed in Table 1.

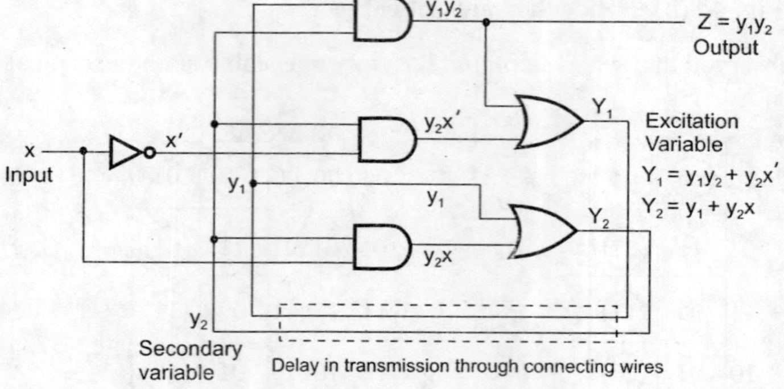

Fig. 1: Logic circuit.

Table 1: Evaluation of Excitation Variables and Output

y_1	y_2	x	y_1'	y_2'	x'	y_1y_2	y_2x'	y_2x	Y_1	Y_2	$Z = y_1y_2$
0	0	0	1	1	1	0	0	0	0	0	0
0	0	1	1	1	0	0	0	0	0	0	0
0	1	0	1	0	1	0	1	0	1	0	0
0	1	1	1	0	0	0	0	1	0	1	0
1	0	0	0	1	1	0	0	0	0	1	0
1	0	1	0	1	0	0	0	0	0	1	0
1	1	0	0	0	1	1	1	0	1	1	1
1	1	1	0	0	0	1	0	1	1	1	1

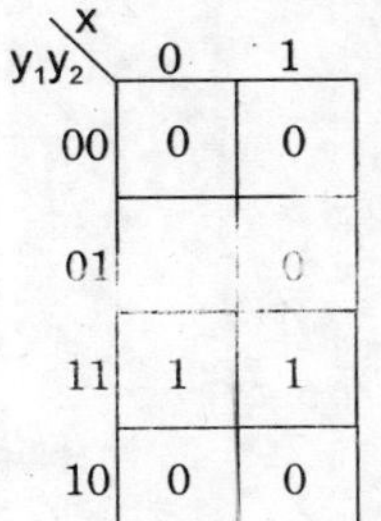

Fig. a: Map for Y_1.

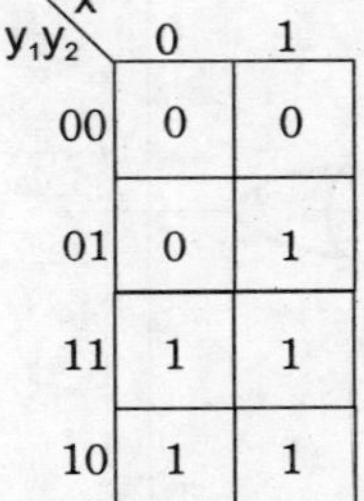

Fig. b: Map for Y_2.

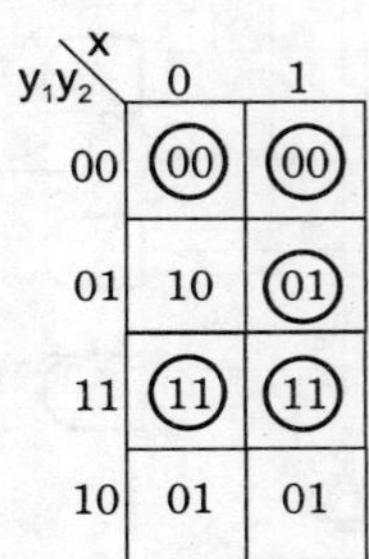

Fig. c: Transition table.

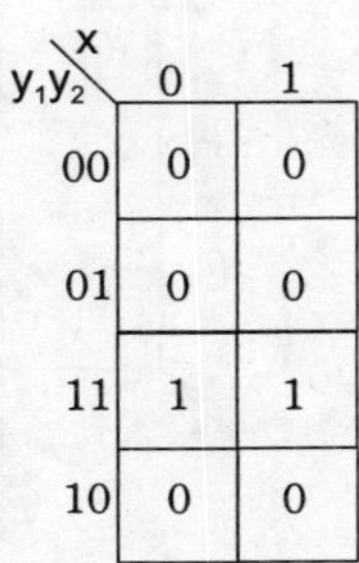

Fig. d: Output map.

Fig. 2: Maps and transition table.

Maps are constructed for Y_1 and Y_2 by taking possible values of secondary variable y_1 and y_2 on rows and input x on columns as shown in Fig. 2 (a) and (b) and the squares in the maps are filled with values of Y_1 and Y_2 listed in Table 1. The transition table is formed by combining the maps of Figs. 2(a) and (b) into a single map as shown in Fig. 2(c). The output map is constructed as shown in Fig. 2(d) and the squares are filled with values of Z.

In each row of the transition table, the squares that have identical values with secondary values are marked with circles. The circled entries represent stable states and uncircled entries represent unstable states.

In order to draw flow table let us assign letter symbols to binary value of states as shown below:

$y_1 y_2 \rightarrow y$

$00 \rightarrow a$

$01 \rightarrow b$

$11 \rightarrow c$

$10 \rightarrow d$

Using the above symbol assignment for states the flow table is obtained by combining the transition table and output map as shown in Fig. 3.

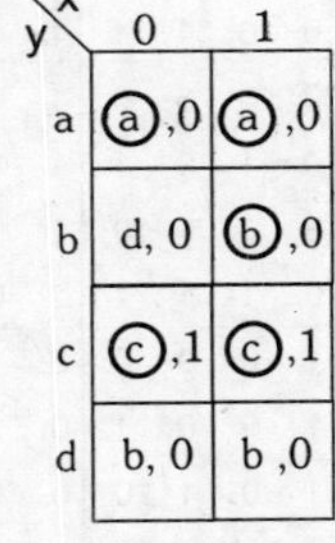

Fig. 3: Flow table.

Example 4.2

The Boolean equations for excitation variables and output of an asynchronous circuit are,

$$Y_1 = x_1x_2 + x_1y_2' + x_2'y_1 \quad ; \quad Y_2 = x_2 + x_1y_1'y_2 + x_1'y_1 \quad ; \quad Z = x_2 + y_1$$

Draw the logic diagram of the circuit. Derive the transition table and output map. Obtain a flow table for the circuit.

Solution

From the given equations we can conclude that the circuit has following variables.

x = Input | y_1, y_2 = Secondary variables (Present state variables)

Z = Output | Y_1, Y_2 = Excitation variables (Next state variables)

Using the given Boolean equations the logic circuit is drawn as shown in Fig. 1 and the excitation variables Y_1 and Y_2 are evaluated for all possible combinations for inputs (x_1 and x_2) and secondary variables (y_1 and y_2) and listed in Table 1.

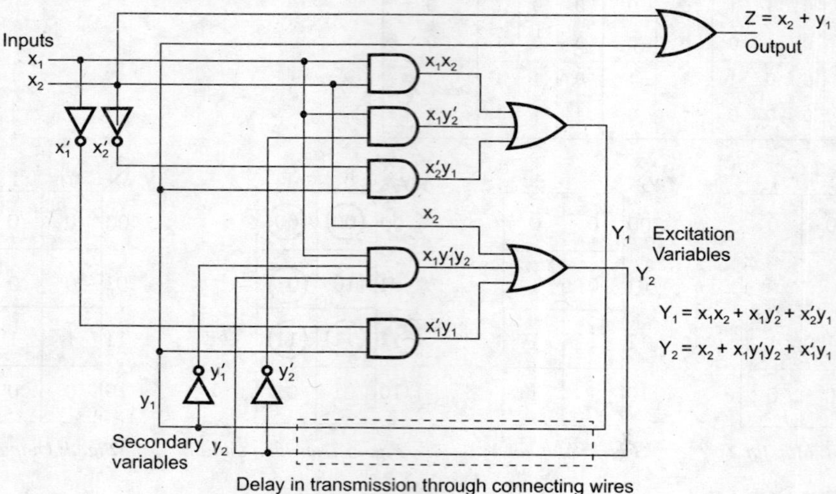

Fig. 1: Logic circuit.

Table 1: Evaluation of Excitation Variables and Output

y_1	y_2	x_1	x_2	y_1'	y_2'	x_1'	x_2'	x_1x_2	x_1y_2'	$x_2'y_1$	$x_1y_1'y_2$	$x_1'y_1$	Y_1	Y_2	$Z = x_2 + y_1$
0	0	0	0	1	1	1	1	0	0	0	0	0	0	0	0
0	0	0	1	1	1	1	0	0	0	0	0	0	0	1	1
0	0	1	0	1	1	0	1	0	1	0	0	0	1	0	0
0	0	1	1	1	1	0	0	1	1	0	0	0	1	1	1
0	1	0	0	1	0	1	1	0	0	0	0	0	0	0	0
0	1	0	1	1	0	1	0	0	0	0	0	0	0	1	1
0	1	1	0	1	0	0	1	0	0	0	1	0	0	1	0
0	1	1	1	1	0	0	0	1	0	0	1	0	1	1	1
1	0	0	0	0	1	1	1	0	0	1	0	1	1	1	1
1	0	0	1	0	1	1	0	0	0	0	0	1	0	1	1
1	0	1	0	0	1	0	1	0	1	1	0	0	1	0	1
1	0	1	1	0	1	0	0	1	1	0	0	0	1	1	1
1	1	0	0	0	0	1	1	0	0	1	0	1	1	1	1
1	1	0	1	0	0	1	0	0	0	0	0	1	0	1	1
1	1	1	0	0	0	0	1	0	0	1	0	0	1	0	1
1	1	1	1	0	0	0	0	1	0	0	0	0	1	1	1

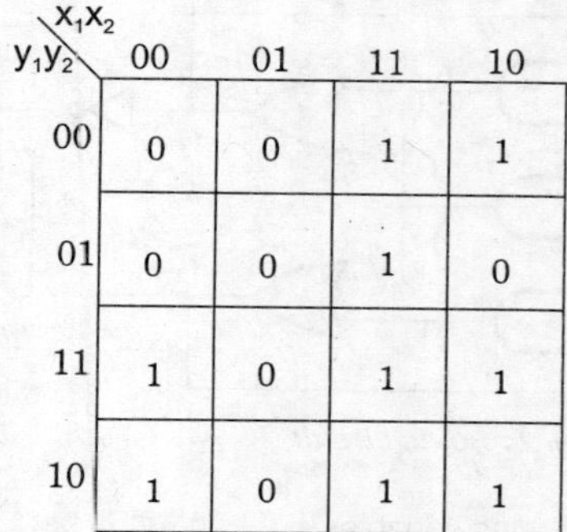

y_1y_2 \ x_1x_2	00	01	11	10
00	0	0	1	1
01	0	0	1	0
11	1	0	1	1
10	1	0	1	1

Fig. 2: *Map for Y_1.*

y_1y_2 \ x_1x_2	00	01	11	10
00	0	1	1	0
01	0	1	1	1
11	1	1	1	0
10	1	1	1	0

Fig. 3: *Map for Y_2.*

y_1y_2 \ x_1x_2	00	01	11	10
00	(00)	01	11	10
01	00	(01)	11	(01)
11	(11)	01	(11)	10
10	11	01	11	(10)

Fig. 4: *Transition table.*

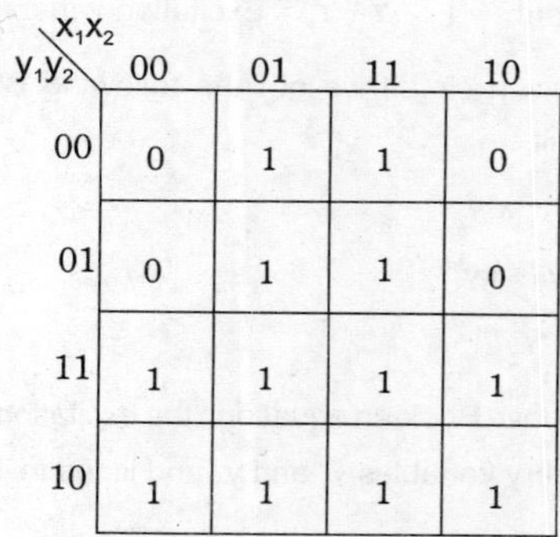

y_1y_2 \ x_1x_2	00	01	11	10
00	0	1	1	0
01	0	1	1	0
11	1	1	1	1
10	1	1	1	1

Fig. 5: *Output map.*

Maps are constructed for Y_1 and Y_2 by taking possible values of secondary variables y_1 and y_2 on rows and input x_1 and x_2 on columns as shown in Figs. 2 and 3 and the squares in the maps are filled with values of Y_1 and Y_2 listed in Table 1. The transition table is formed by combining the maps of Figs. 2 and 3 into a single map as shown in Fig. 4. The output map is constructed as shown in Fig. 5 and the squares are filled with values of Z.

In each row of the transition table, the squares that have identical values with secondary values are marked with circles. The circled entries represent stable states and uncircled entries represent unstable states.

In order to draw flow table let us assign letter symbols to binary value of states as shown below:

$y_1y_2 \rightarrow y$

$00 \rightarrow a$

$01 \rightarrow b$

$11 \rightarrow c$

$10 \rightarrow d$

Using the above symbol assignment for states the flow table is obtained by combining the transition table and output map as shown in Fig. 6.

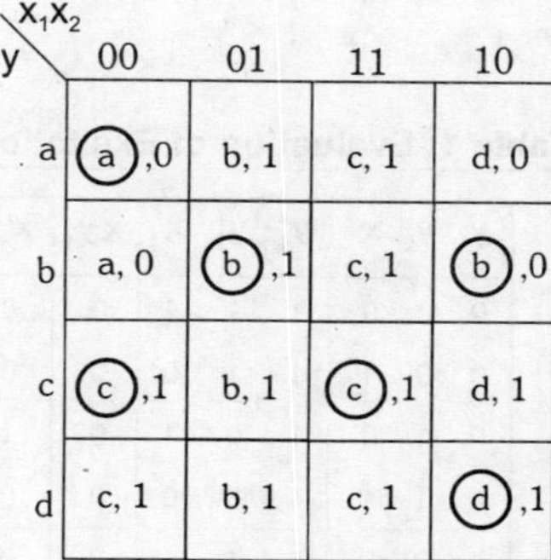

y \ x_1x_2	00	01	11	10
a	(a),0	b, 1	c, 1	d, 0
b	a, 0	(b),1	c, 1	(b),0
c	(c),1	b, 1	(c),1	d, 1
d	c, 1	b, 1	c, 1	(d),1

Fig. 6: *Flow table.*

Example 4.3

Write logical equation and construct transition table for the circuit output in terms of the circuit input and secondary variables. Show the state transitions for changes in input and hence find the races and cycles. Also, construct the state and flow table.

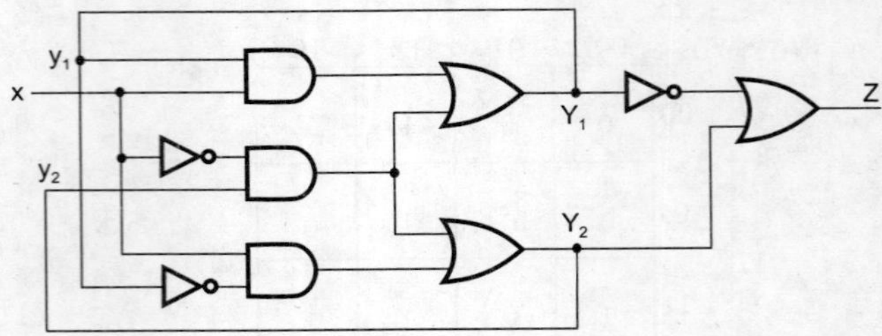

Fig. 1: Logic circuit.

Solution

From the given circuit we can conclude that the circuit has following variables.

x = Input	y_1, y_2 = Secondary variables (Present state variables)
Z = Output	Y_1, Y_2 = Excitation variables (Next state variables)

From the given circuit we get the following Boolean equations for excitation variables and output of the asynchronous circuit.

$$Y_1 = xy_1 + x'y_2$$

$$Y_2 = x'y_2 + xy_1'$$

$$Z = Y_1' + Y_2$$

Using the above Boolean equations the excitation variables, Y_1 and Y_2 are evaluated for all possible combinations input x and secondary variables y_1 and y_2 and listed in Table 1.

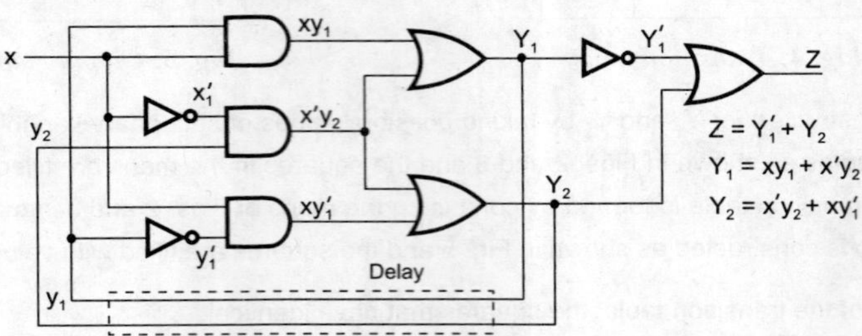

Fig. 2.

Table 1: Evaluation of Excitation Variables and Output

y_1	y_2	x	y_1'	y_2'	x'	xy_1	$x'y_2$	xy_1'	Y_1	Y_2	Y_1'	$Z = Y_1' + Y_2$
0	0	0	1	1	1	0	0	0	0	0	1	1
0	0	1	1	1	0	0	0	1	0	1	1	1
0	1	0	1	0	1	0	1	0	1	1	0	1
0	1	1	1	0	0	0	0	1	0	1	1	1
1	0	0	0	1	1	0	0	0	0	0	1	1
1	0	1	0	1	0	1	0	0	1	0	0	0
1	1	0	0	0	1	0	1	0	1	1	0	1
1	1	1	0	0	0	1	0	0	1	0	0	0

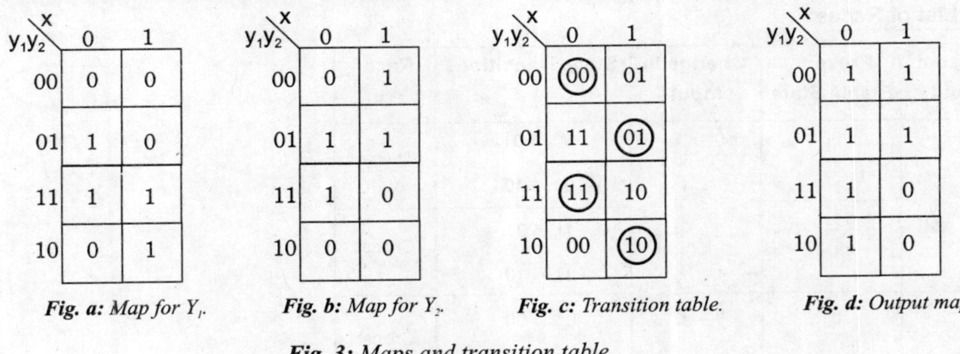

Fig. a: Map for Y_1. *Fig. b:* Map for Y_2. *Fig. c:* Transition table. *Fig. d:* Output map.

Fig. 3: Maps and transition table.

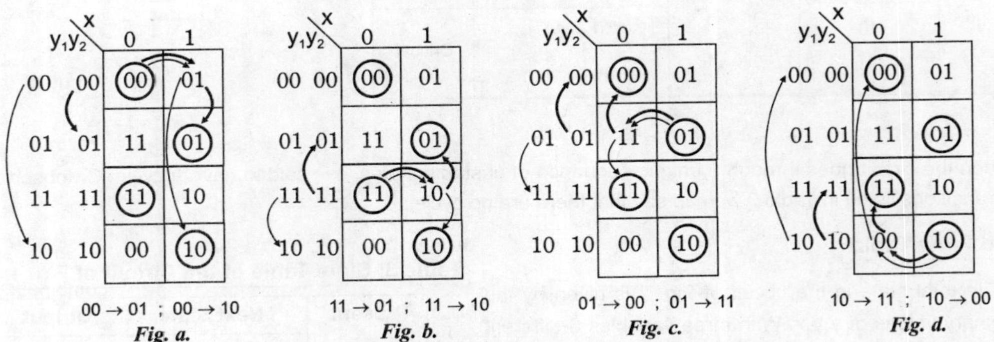

Fig. a. *Fig. b.* *Fig. c.* *Fig. d.*

Fig. 4: State transitions due to change in input.

Maps are constructed for Y_1 and Y_2 by taking possible values of secondary variables y_1 and y_2 on rows and input x on columns as shown in Figs. 3(a) and (b) and the squares in the maps are filled with values of Y_1 and Y_2 listed in Table 1. The transition table is formed by combining the maps of Figs. 3(a) and (b) into a single map as shown in Fig. 3(c). The output map is constructed as shown in Fig. 3(d) and the squares are filled with values of Z.

In each row of the transition table, the squares that have identical values with secondary values are marked with circles. The circled entries represent stable states and uncircled entries represent unstable states.

In the transition table it can be observed that each column has two stable states for every input. But, for a given input the circuit will remain in one of the stable state.

The transition from one stable state to another stable state occurs due to changes in input.

Since there are two stable states for each input the circuit can go to any one of the stable state for a change in input as shown in Fig. 4 and hence every change in input leads to a critical race.

When in state 00, if the input changes from 0 to 1, then state transitions occur as shown in Fig. 4(a).

When in state 11, if the input changes from 0 to 1, then state transitions occur as shown in Fig. 4(b).

When in state 01, if the input changes from 1 to 0, then state transitions occur as shown in Fig. 4(c).

When in state10, if the input changes from 1 to 0, then state transitions occur as shown in Fig. 4(d).

The state transitions in Fig. 4 are also listed in Table 2.

Table 2: List of Races

Present Input	Present Stable State	Change in Input	State Transitions	Race
0	00	0 → 1	00 → 01 00 → 10	Critical
	11		11 → 01 11 → 10	Critical
1	01	1 → 0	01 → 00 01 → 11	Critical
	10		10 → 00 10 → 11	Critical

Cycles

When the circuit goes through a unique sequence of unstable states, it is said to have a cycle. On observing the state transitions listed in Table 2 we can say that there are no cycles.

State and Flow Table

The total state in the circuit of Fig. 2 is given by the possible binary values of y_1y_2x. With three variables 8 different total states can be formed, in which the four states that are circled in transition table are stable states and the other four states are unstable.

Stable states: 000, 011, 110, 101

Unstable states: 001, 010, 111, 100

Using the transition table, the state table for asynchronous circuit similar to that of a state table of synchronous circuit can be constructed as shown in Table 3.

Table 3: State Table of the Circuit of Fig. 1

Present State		Next State				Output	
		x = 0		x = 1		x = 0	x = 1
y_1	y_2	Y_1	Y_2	Y_1	Y_2	Z	Z
0	0	0	0	0	1	1	1
0	1	1	1	0	1	1	1
1	0	0	0	1	0	1	0
1	1	1	1	1	0	1	0

In order to draw flow table let us assign letter symbols to binary value of states as shown below:

$y_1y_2 \rightarrow y$

00 → a
01 → b
11 → c
10 → d

Using the above symbol assignment for states the flow table is obtained by combining the transition table and output map as shown in Fig. 5.

y \ x	0	1
a	(a),1	b, 1
b	c, 1	(b),1
c	(c),1	d, 0
d	a, 1	(d),0

Fig. 5: Flow table.

Example 4.4

An asynchronous sequential circuit is described by the excitation functions, $Y_1 = y_1 y_2' + (y_1 + y_2') x$ and $Y_2 = y_2' x + y_1 x$ and the output function, $Z = Y_2$. Draw the logic diagram of the circuit. Derive the transition table and output map. Also discuss about the behaviour of the circuit.

Solution

From the given equations we can conclude that the circuit has following variables.

x = Inputs	y_1, y_2 = Secondary variables (Present state variables)
Z = Output	Y_1, Y_2 = Excitation variables (Next state variables)

Given that, $Y_1 = y_1 y_2' + (y_1 + y_2') x$

$$Y_2 = y_2' x + y_1 x$$

$$Z = Y_2$$

Using the given Boolean equations the logic circuit is drawn as shown in Fig. 1 and the excitation variables, Y_1 and Y_2 are evaluated for all possible combinations of input x and secondary variables y_1 and y_2 and listed in Table 1.

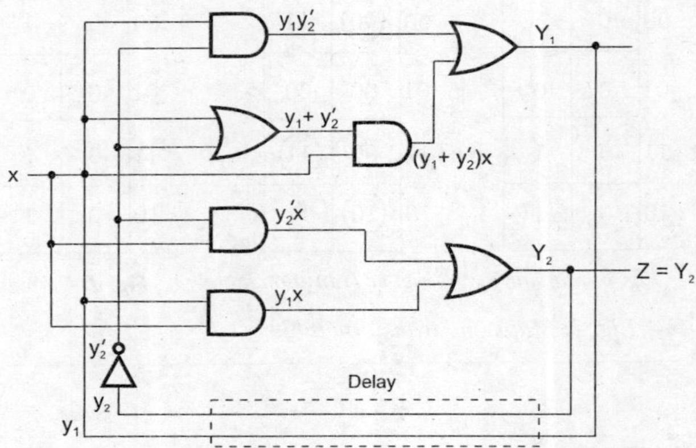

Fig. 1: Logic circuit.

Table 1: Evaluation of Excitation Variables and Output

y_1	y_2	x	y_1'	y_2'	x'	$y_1 y_2'$	y_1+y_2'	$(y_1+y_2')x$	$y_2'x$	$y_1 x$	Y_1	Y_2	$Z = Y_2$
0	0	0	1	1	1	0	1	0	0	0	0	0	0
0	0	1	1	1	0	0	1	1	1	0	1	1	1
0	1	0	1	0	1	0	0	0	0	0	0	0	0
0	1	1	1	0	0	0	0	0	0	0	0	0	0
1	0	0	0	1	1	1	1	0	0	0	1	0	0
1	0	1	0	1	0	1	1	1	1	1	1	1	1
1	1	0	0	0	1	0	1	0	0	0	0	0	0
1	1	1	0	0	0	0	1	1	0	1	1	1	1

The transition table is formed by taking possible values of secondary variable on rows and input x on columns as shown in Fig. 2(c) and the squares in the transition table are filled with values of Y_1 and Y_2 listed in Table 1. The output map is constructed as shown in Fig. 2(d) and the squares are filled with values of Z.

In each row of the transition table, the squares that have identical values with secondary values are maked with circles. The circled entires represent stable states and uncircled entries represent unstable states.

In order to draw flow table let us assign letter symbols to binary value of states as shown below:

$y_1 y_2 \rightarrow y$

$00 \rightarrow a$

$01 \rightarrow b$

$11 \rightarrow c$

$10 \rightarrow d$

Using the above symbol assignment for states the flow table is obtained by combining the transition table and output map as shown in Fig. 3.

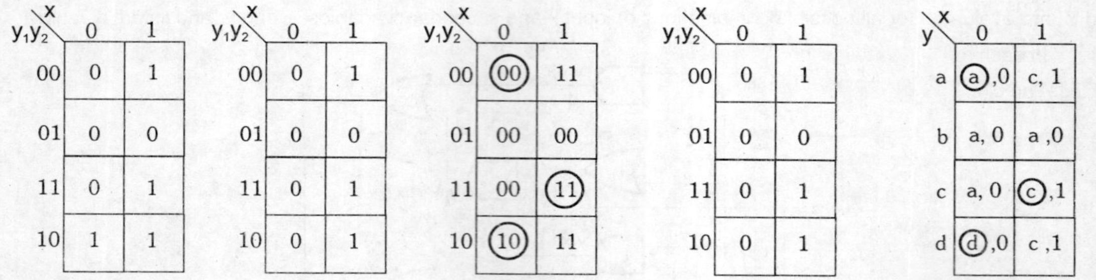

y_1y_2 \ x	0	1
00	0	1
01	0	0
11	0	1
10	1	1

Fig. a: Map for Y_1.

y_1y_2 \ x	0	1
00	0	1
01	0	0
11	0	1
10	0	1

Fig. b: Map for Y_2.

y_1y_2 \ x	0	1
00	(00)	11
01	00	00
11	00	(11)
10	(10)	11

Fig. c: Transition table.

y_1y_2 \ x	0	1
00	0	1
01	0	0
11	0	1
10	0	1

Fig. d: Output map.

y \ x	0	1
a	(a),0	c, 1
b	a, 0	a ,0
c	a, 0	(c),1
d	(d),0	c ,1

Fig. 3: Flow table.

Fig. 2: Maps and transition table.

Example 4.5

An asynchronous sequential circuit is described by the excitation function, $Y = x_1 x_2' + (x_1 + x_2')y$ and the output function, Z = Y. Draw the logic diagram of the circuit. Derive the transition table and output map. Also discuss about the behaviour of the circuit.

Solution

From the given equations we can conclude that the circuit has following variables.

x_1, x_2 = Inputs | y = Secondary variable (Present state variable)

Z = Output | Y = Excitation variable (Next state variable)

Given that, $Y = x_1 x_2' + (x_1 + x_2')y$

$Z = Y$

Using the given Boolean equations the logic circuit is drawn as shown in Fig. 1 and the excitation variable, Y is evaluated for all possible combinations of input x and secondary variables y_1 and y_2 and listed in Table 1.

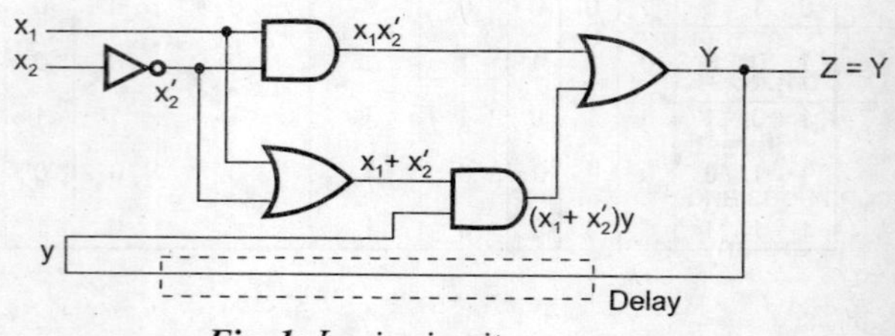

Fig. 1: Logic circuit.

Table 1: Evaluation of Excitation Variables and Output

y	x_1	x_2	y′	x_1'	x_2'	x_1x_2'	x_1+x_2'	$(x_1+x_2')y$	Y	Z = Y
0	0	0	1	1	1	0	1	0	0	0
0	0	1	1	1	0	0	0	0	0	0
0	1	0	1	0	1	1	1	0	1	1
0	1	1	1	0	0	0	1	0	0	0
1	0	0	0	1	1	0	1	1	1	1
1	0	1	0	1	0	0	0	0	0	0
1	1	0	0	0	1	1	1	1	1	1
1	1	1	0	0	0	0	1	1	1	1

The transition table is formed by taking possible values of secondary variable on rows and inputs x_1 and x_2 on columns as shown in Fig. 2 and the squares in the transition table are filled with values of Y listed in Table 1. The output map is constructed as shown in Fig. 3 and the squares are filled with values of Z.

In each row of the transition table, the squares that have identical values with secondary values are maked with circles. The circled entires represent stable states and uncircled entries represent unstable states.

The flow table is formed by combining transition table and output map as shown in Fig. 4.

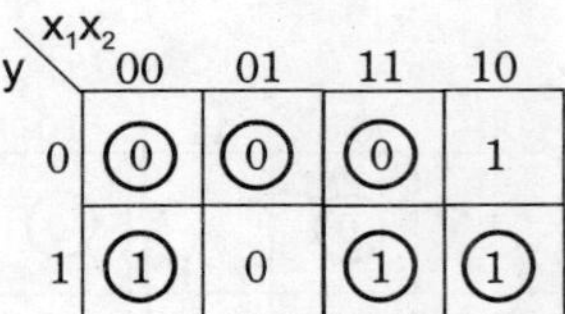

y \ x_1x_2	00	01	11	10
0	(0)	(0)	(0)	1
1	(1)	0	(1)	(1)

Fig. 2: Transition table.

y \ x_1x_2	00	01	11	10
0	0	0	0	1
1	1	0	1	1

Fig. 3: Output map.

y \ x_1x_2	00	01	11	10
0	(0),0	(0),0	(0),0	1, 1
1	(1),1	0, 0	(1),1	(1),1

Fig. 4: Flow table.

y \ x_1x_2	00	01	11	10
0	(0),0	(0),0	(0),0	1, 1
1	(1),1	0, 0	(1),1	(1),1

Fig. 5: State transitions when the input changes to either 00 or 11.

Comment : From the flow table we can conclude that when the input is 01, the output is 0. When the input is 10 and the output is 1. Whenever the input assumes one of the other two combinations (either 00 or 11), the output retains its previous value as shown in Fig. 5.

Example 4.6

Investigate the transition table of Fig. 1, and determine all race conditions and whether they are critical or noncritical. Determine also whether there are any cycles.

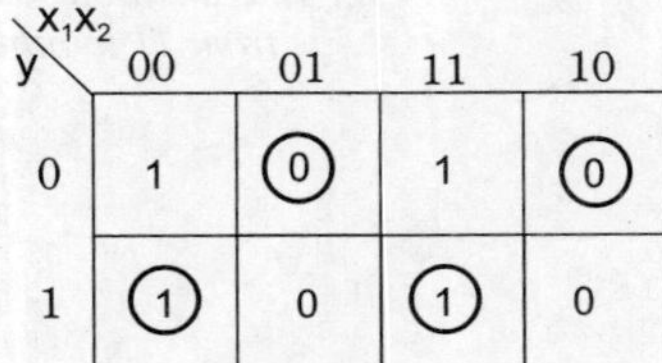

y \ x_1x_2	00	01	11	10
0	1	(0)	1	(0)
1	(1)	0	(1)	0

Fig. 1: Transition table.

Solution

State Transitions Due to Change in Input

Here, each input has one stable state. State transitions occur only when there is a change in input. Normally, when the input is stable or when no change in input the circuit remain in the stable state corresponding to the present input.

Here, there are four possible combinations of inputs. When the circuit remain in a stable state at any one input then state transitions will occur when the input changes to other possible inputs with one bit change.

When input is 00, the circuit remain in state 1. The input can change from 00 to 01 or 00 to 10 and the corresponding state transitions will occur as shown in Fig. 2.

When input is 01, the circuit remain in state 0. The input can change from 01 to 11 or 01 to 00 and the corresponding state transitions will occur as shown in Fig. 3.

When input is 11, the circuit remain in state 1. The input can change from 11 to 10 or 11 to 01 and the corresponding state transitions will occur as shown in Fig. 4.

When input is 10, the circuit remain in state 0. The input can change from 10 to 00 or 10 to 11 and the corresponding state transitions will occur as shown in Fig. 5.

The state transitions shown in Figs. 2 to 5 are listed in Table 1. It can be observed that there are no races and cycles in the state transitions.

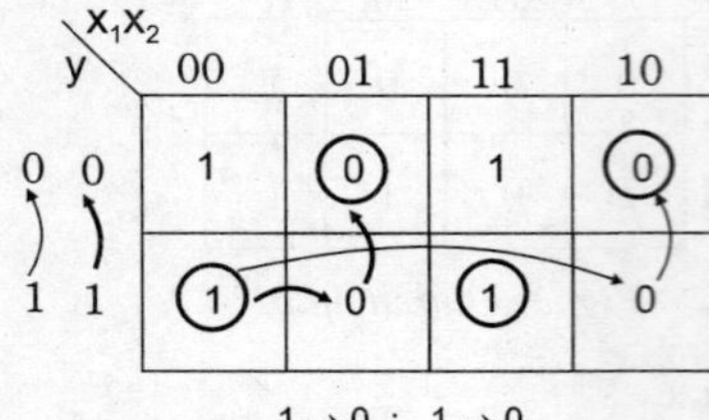

Fig. 2: Transitions due to change in input from 00 to other possible inputs.

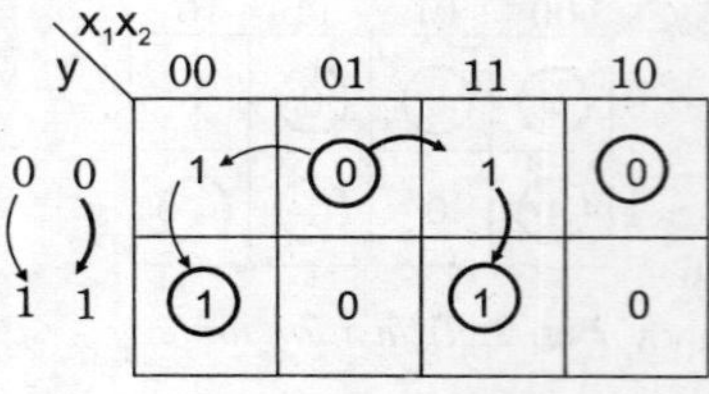

Fig. 3: Transitions due to change in input from 01 to other possible inputs.

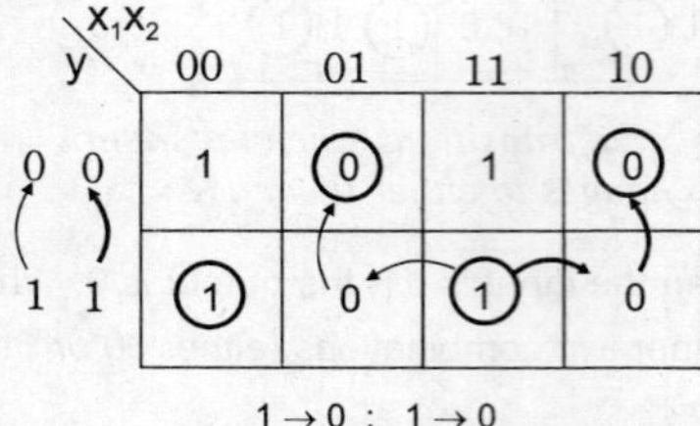

Fig. 4: Transitions due to change in input from 11 to other possible inputs.

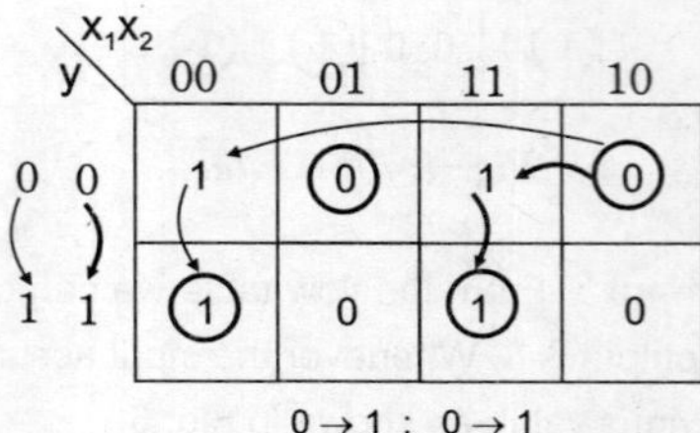

Fig. 5: Transitions due to change in input from 10 to other possible inputs.

Table 1: List of Races

Present Input	Present Stable State	Change in Input	State Transitions
00	1	00 → 01	1 → 0
		00 → 10	1 → 0
01	0	01 → 00	0 → 1
		01 → 11	0 → 1
11	1	11 → 01	1 → 0
		11 → 10	1 → 0
10	0	10 → 11	0 → 1
		10 → 00	0 → 1

Example 4.7

Investigate the transition table of Fig. 1, and determine all race conditions and whether they are critical or noncritical. Determine also whether there are any cycles.

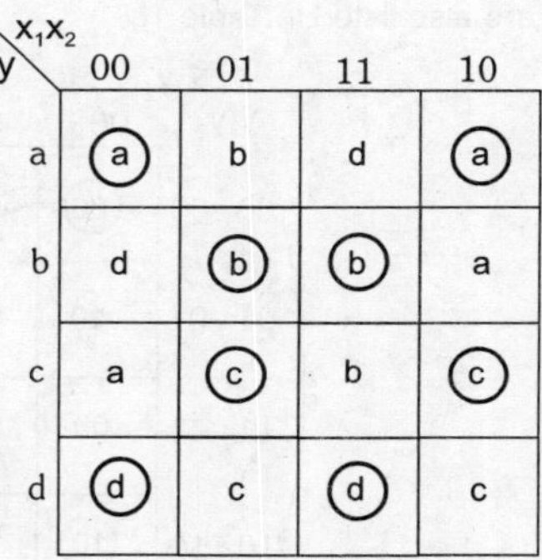

y \ x_1x_2	00	01	11	10
a	(a)	b	d	(a)
b	d	(b)	(b)	a
c	a	(c)	b	(c)
d	(d)	c	(d)	c

Fig. 1: Transition table.

Solution

Binary State Assignment

Let us assign binary codes to states as shown below. Here there are 4 states and hence 2-bit binary codes are selected, so that we get, $2^2 = 4$ codes.

Let the two bits of the binary codes be represented by the two state variables y_1 and y_2.

$$a \rightarrow 00 \quad c \rightarrow 11 \quad y = y_1 y_2$$
$$b \rightarrow 01 \quad d \rightarrow 10$$

The transition table with binary code assignment is shown in Fig. 2.

y_1y_2 \ x_1x_2	00	01	11	10
00	(00)	01	10	(00)
01	10	(01)	(01)	00
11	00	(11)	01	(11)
10	(10)	11	(10)	11

Fig. 2: Transition table with binary code assignment.

State Transitions Due to Change in Input

State transitions occur only when there is a change in input. Normally, when the input is stable or when no change in input the circuit remain in one of the stable state corresponding to the present input.

Here, there are four possible combinations of inputs. When the circuit remain in a stable state at any one input then state transitions will occur when the input changes to other possible inputs with one bit change.

Each input has two stable states. Hence at any one input the circuit can remain in one of the stable state.

When input is 00, the circuit remain in either state 00 or 10. The input can change from 00 to 01 or 00 to 10 and the corresponding state transitions will occur as shown in Fig. 3.

When input is 01, the circuit remain in either state 01 or 11. The input can change from 01 to 11 or 01 to 00 and the corresponding state transitions will occur as shown in Fig. 4.

When input is 11, the circuit remain in either state 00 or 10. The input can change from 11 to 10 or 11 to 01 and the corresponding state transitions will occur as shown in Fig. 5.

When input is 10, the circuit remain in either state 00 or 11. The input can change from 10 to 00 or 10 to 11 and the corresponding state transitions will occur as shown in Fig. 6.

The state transitions shown in Figs. 3 to 6 are listed in Table 1. The races and cycles involved in state transitions are also listed in Table 1.

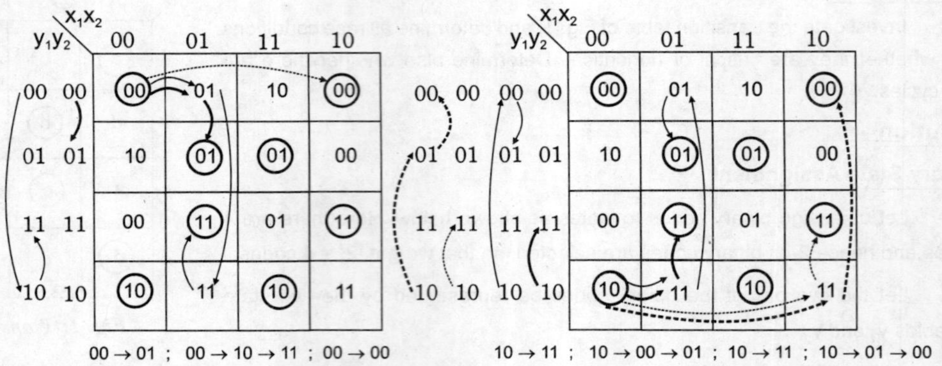

Fig. a: Transitions from stable state 00. *Fig. b: Transitions from stable state 10.*

Fig. 3: Transitions due to change in input from 00 to other possible inputs.

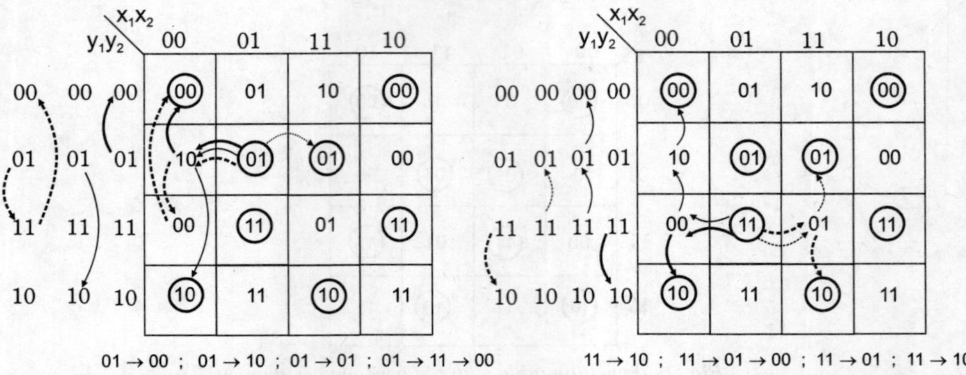

Fig. a: Transitions from stable state 01. *Fig. b: Transitions from stable state 11.*

Fig. 4: Transitions due to change in input from 01 to other possible inputs.

Table 1: List of Races and Cycles

Present Input	Present Stable State	Change in Input	State Transitions	Race	Cycles
00	00	00 → 01	00 → 01 00 → 10 → 11	Critical	- 00 → 10 → 11
		00 → 10	00 → 00	-	-
	10	00 → 01	10 → 11 10 → 00 → 01	Critical	- 10 → 00 → 01
		00 → 10	10 → 11 10 → 01 → 00	Critical	- 10 → 01 → 00
01	01	01 → 00	01 → 00 01 → 11 → 10	Critical	- 01 → 11 → 10
		01 → 11	01 → 01	-	-
	11	01 → 00	11 → 10 11 → 01 → 00	Critical	- 11 → 01 → 00
		01 → 11	11 → 01 11 → 10	Critical	-
11	01	11 → 01	01 → 01	-	-
		11 → 10	01 → 00 01 → 11	Critical	- -
	10	11 → 01	10 → 11 10 → 00 → 01	Critical	- 10 → 00 → 01
		11 → 10	10 → 11 10 → 00	Critical	-
10	00	10 → 00	00 → 00	-	-
		10 → 11	00 → 01 00 → 10	Critical	- -
	11	10 → 00	11 → 01 → 00 11 → 10	Critical	11 → 01 → 00 -
		10 → 11	11 → 10 11 → 01	Critical	- -

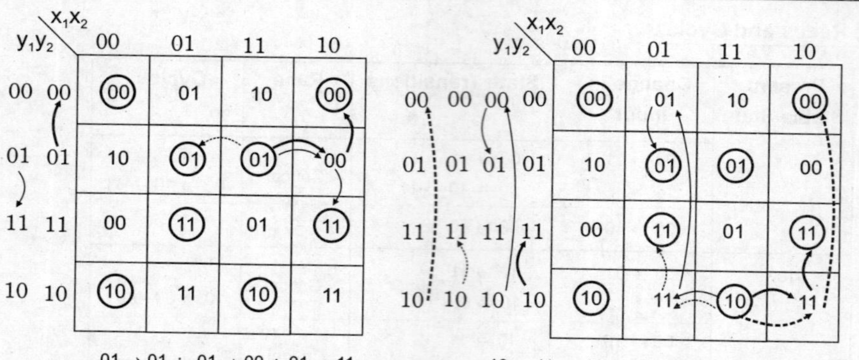

01 → 01 ; 01 → 00 ; 01 → 11

***Fig. a:** Transitions from stable state 01.*

10 → 11 ; 10 → 00 → 01 ; 10 → 11 ; 10 → 00

***Fig. b:** Transitions from stable state 10.*

***Fig. 5:** Transitions due to change in input from 11 to other possible inputs.*

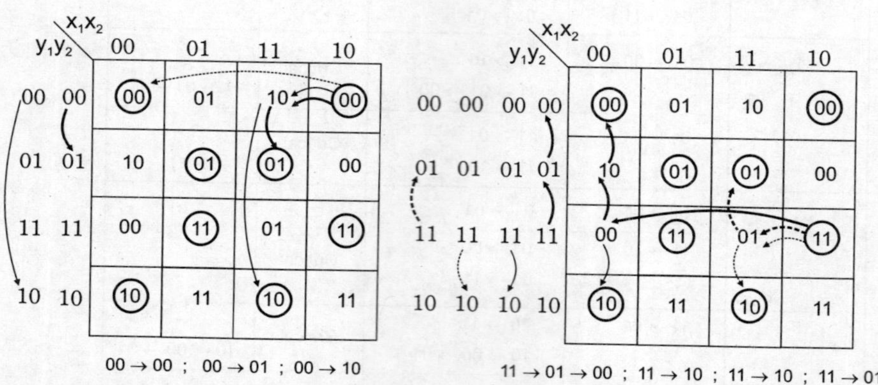

00 → 00 ; 00 → 01 ; 00 → 10

***Fig. a:** Transitions from stable state 00.*

11 → 01 → 00 ; 11 → 10 ; 11 → 10 ; 11 → 01

***Fig. b:** Transitions from stable state 11.*

***Fig. 6:** Transitions due to change in input from 10 to other possible inputs.*

Cycles

When the circuit goes through a unique sequence of unstable states, it is said to have a cycle and they are listed in Table 1.

The following cycles are observed during state transition for changes in inputs,

Cycles during change in input from 00.	Cycles during change in input from 01.	Cycle during change in input from 11.	Cycle during change in input from 10.
00 → 10 → 11	01 → 11 → 10	10 → 00 → 01	11 → 01 → 00
10 → 00 → 01	11 → 01 → 00		

Example 4.8

Investigate the transition table of Fig. 1, and determine all race conditions and whether they are critical or noncritical. Determine also whether there are any cycles.

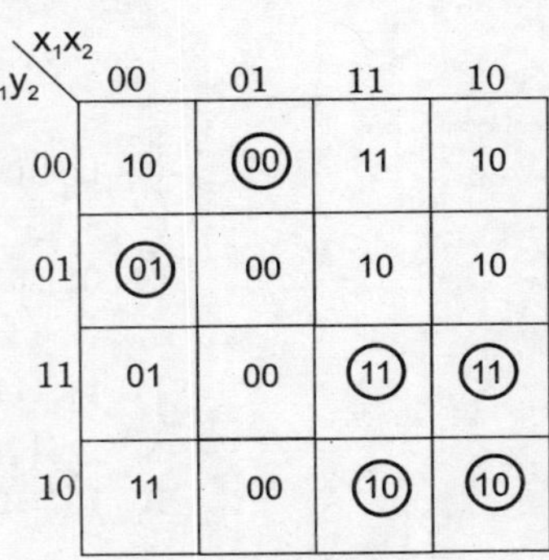

y_1y_2 \ x_1x_2	00	01	11	10
00	10	(00)	11	10
01	(01)	00	10	10
11	01	00	(11)	(11)
10	11	00	(10)	(10)

***Fig. 1:** Transition table.*

Solution

State Transitions Due to Change in Input

State transitions occur only when there is a change in input. Normally, when the input is stable or when no change in input the circuit remain in one of the stable state corresponding to the input.

Here, there are four possible combinations of inputs. When the circuit remain in a stable state at any one input then state transitions will occur when the input changes to other possible inputs with one bit change.

Here, two inputs has one stable state each and the other two inputs has two stable states each. Hence at any one input the circuit can remain in one of the stable state.

When input is 00, the circuit remain in state 01. The input can change from 00 to 01 or 00 to 10 and the corresponding state transitions will occur as shown in Fig. 2.

When input is 01, the circuit remain in state 00. The input can change from 01 to 11 or 01 to 00 and the corresponding state transitions will occur as shown in Fig. 3.

When input is 11, the circuit remain in either state 11 or 10. The input can change from 11 to 10 or 11 to 01 and the corresponding state transitions will occur as shown in Fig. 4.

When input is 10, the circuit remain in either state 11 or 10. The input can change from 10 to 00 or 10 to 11 and the corresponding state transitions will occur as shown in Fig. 5.

The state transitions shown in Figs. 2 to 5 are listed in Table 1. The races and cycles that are involved in state transitions are also listed in Table 1.

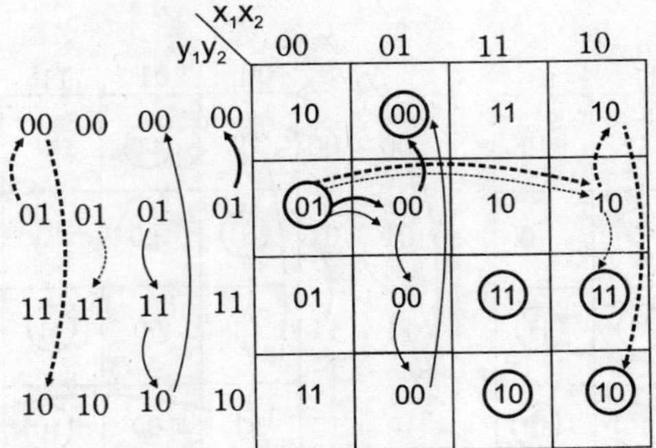

01 → 00 ; 01 → 11 → 10 → 00 ; 01 → 11 ; 01 → 00 → 10

***Fig. 2:** Transitions due to change in input from 00 to other possible inputs.*

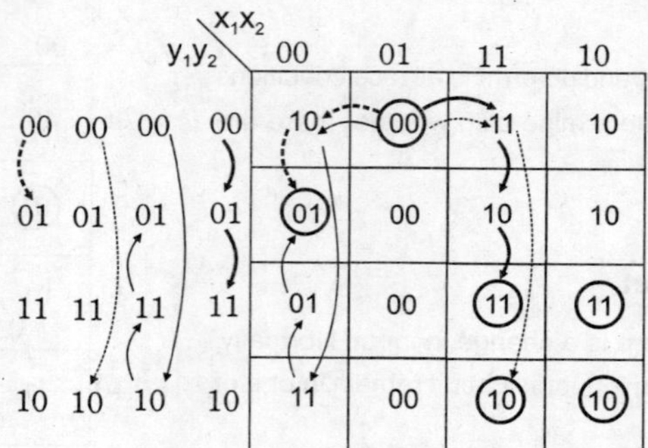

00 → 01 ; 00 → 10 → 11 → 01 ; 00 → 10 ; 00 → 01 → 11

***Fig. 3:** Transitions due to change in input from 01 to other possible inputs.*

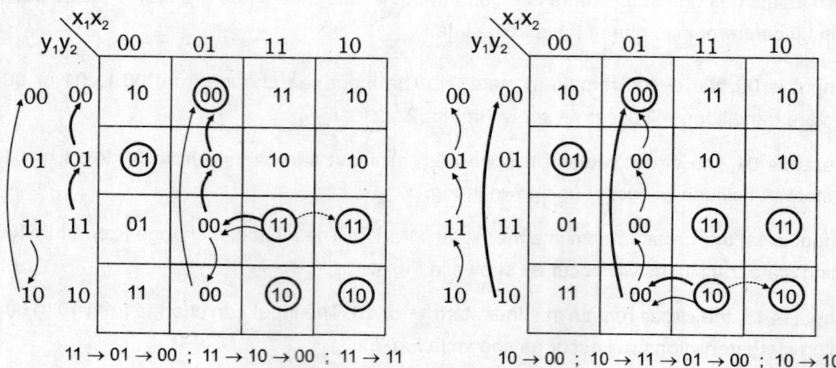

***Fig. a:** Transitions from stable state 11.* ***Fig. b:** Transitions from stable state 10.*

***Fig. 4:** Transitions due to change in input from 11 to other possible inputs.*

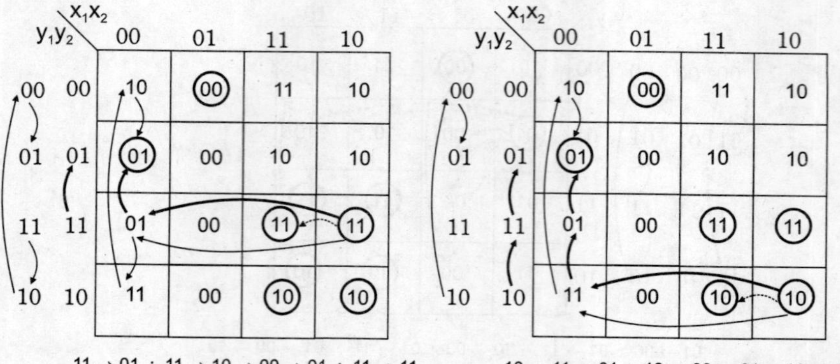

11 → 01 ; 11 → 10 → 00 → 01 ; 11 → 11 10 → 11 → 01 ; 10 → 00 → 01 ; 10 → 10

***Fig. a:** Transitions from stable state 11.* ***Fig. b:** Transitions from stable state 10.*

***Fig. 5:** Transitions due to change in input from 10 to other possible inputs.*

Table 1: List of Races and Cycles

Present Input	Present Stable State	Change in Input	State Transitions	Race	Cycles
00	01	00 → 01	01 → 00	Noncritical	-
			01 → 11 → 10 → 00		01 → 11 → 10 → 00
		00 → 10	01 → 11	Critical	-
			01 → 00 → 10		01 → 00 → 10
01	00	01 → 00	00 → 01	Noncritical	-
			00 → 10 → 11 → 01		00 → 10 → 11 → 01
		01 → 11	00 → 10	Critical	-
			00 → 01 → 11		00 → 01 → 11
11	11	11 → 01	11 → 01 → 00	Noncritical	11 → 01 → 00
			11 → 10 → 00		11 → 10 → 00
		11 → 10	11 → 11	-	-
	10	11 → 01	10 → 00	Noncritical	-
			10 → 11 → 01 → 00		10 → 11 → 01 → 00
		11 → 10	10 → 10	-	-
10	11	10 → 00	11 → 01	Noncritical	-
			11 → 10 → 00 → 01		11 → 10 → 00 → 01
		10 → 11	11 → 11	-	-
	10	10 → 00	10 → 11 → 00	Noncritical	10 → 11 → 01
			10 → 00 → 01		10 → 00 → 01
		10 → 11	10 → 10	-	-

Cycles

When the circuit goes through a unique sequence of unstable states, it is said to have a cycle and they are listed in Table 1.

The following cycles are observed during state transitions for changes in inputs,

Cycles during change in input from 00.	Cycles during change in input from 01.	Cycles during change in input from 11.	Cycles during change in input from 10.
01 → 11 → 10 → 00	00 → 10 → 11 → 01	11 → 01 → 00	11 → 10 → 00 → 01
01 → 00 → 10	00 → 01 → 11	11 → 10 → 00	10 → 11 → 01
		10 → 11 → 01 → 00	10 → 00 → 01

4.3 Race-Free Assignment

In order to avoid critical races, the state assignment or binary code assignment should be such that only one variable change at a time when a state transition occurs in a flow table. Two binary codes are said to be adjacent if they differ by only one variable. For example the 3-bit code 011 and 001 are adjacent, because they differ by only one variable. In order to avoid critical races, the two states between which a transition occurs should be assigned adjacent codes. Assigning adjacent codes to states to avoid critical races is called ***race-free assignment***. Sometimes, race free assignment needs addition of extra rows in flow table which can be achieved by the following two methods.

Method 1: Additional Rows with Extra States

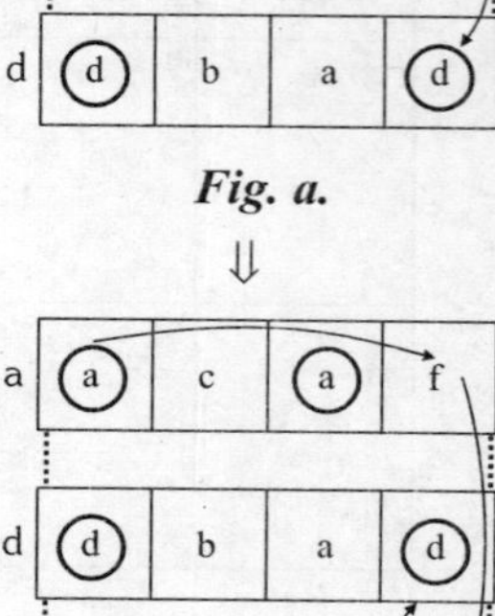

Fig. a.

Fig. b.

Fig. 4.9.

Using the flow table, a transition diagram can be constructed that represents all possible state transitions. Binary code assignment for states can be made in the transition diagram such that states that are involved in every transition are assigned adjacent codes.

When adjacent code assignment is not possible for a pair of states involved in a state transition insert an additional state such that the state transition between the pair of states go via the additional states. The concept of inserting additional state can be extended to all state pair involved in state transitions that do not have adjacent state assignment.

The insertion of additional states leads to extra rows in flow table. For every extra state one row is added in flow table.

Consider a state transition between two rows a and d as shown in Fig. 4.9(a).

Let a and d are not adjacent. Now an additional state f is added such that the transition from a to d go via state f as shown below:

$$a \rightarrow d \quad \Rightarrow \quad a \rightarrow f \rightarrow d$$

In order to assure this transition an additional state named state f is selected which leads to an additional row for state f in flow table table as shown in Fig. 4.9(b). Now, the state d in row a is replaced by f and the corresponding column in row f is filled by d and all other columns in row f are don't-cares.

Method 2: Multiple Row with Complement States

In multiple row method each row in a flow table is represented by two or more equivalent rows. In two rows representation, each row is represented by two rows and they are known as original row and equivalent row.

The rows corresponding to original flow table are assigned binary codes such that they are adjacent. The complement of binary codes are assigned to equivalent rows. The stable states are filled with subscript 1 in original rows and subscript 2 in the equivalent rows. The unstable states are filled such that they are adjacent to stable states. (Refer Example 4.9(b) for multiple row method).

4.3.1 Transition Diagram

The transition table lists all possible state transitions from present to next state. The state transitions between unstable states in each row of transition table can be converted to transition diagram which will be more useful for race-free state assignment. The diagrammetic way of representing all possible state transitions between unstable states in every row of transition table is called ***transition diagram***.

An example of drawing the transition diagram from a transition table is shown in Fig. 4.10.

From the transition table shown in Fig. 4.10(a) we can observe the following state transition.

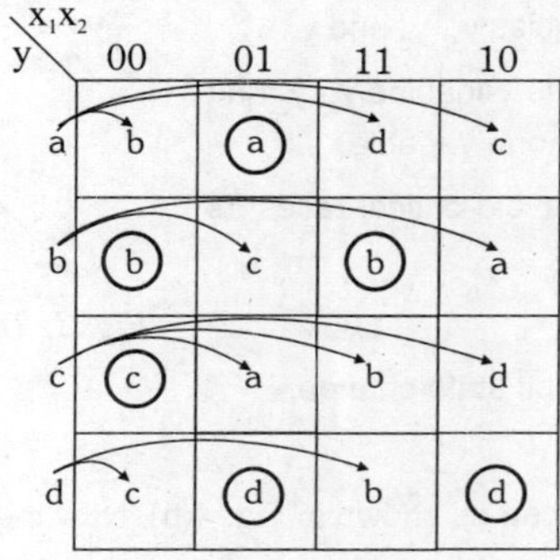

Fig. a: Transition table.

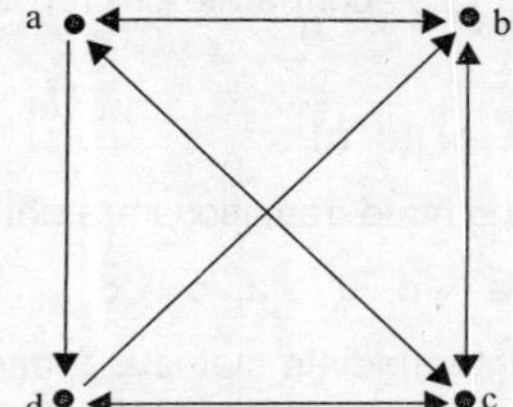

Fig. b: Transition diagram.

***Fig. 4.10:** Example of drawing transition diagram from a transition table.*

From state a	From state b	From state c	From state d
a → b	b → c	c → a	d → c
a → d	b → a	c → b	d → b
a → c		c → d	

The above state transitions can be combined as a single state transition diagram shown in Fig. 4.10(b).

Example 4.9

Consider the reduced flow table shown in Fig. 1.

a) Obtain the transition diagram, and show that three state variables are needed for a race-free binary state assignment.

b) Obtain the expanded flow table, using the multiple row method of assignment.

y \ x_1x_2	00	01	11	10
a	(a)	c	(a)	d
b	a	(b)	c	(b)
c	(c)	(c)	(c)	d
d	(d)	b	a	(d)

***Fig. 1:** Reduced flow table.*

Solution

a) Transition Diagram and Race-Free State Assignment

The following transitions from present state to next state are observed in transition table in Fig. 2.

From a	From b	From c	From d
a → c	b → a	c → d	d → b
a → d	b → c		d → a

The above transitions are combined as a single diagram in the transition diagram shown in Fig. 3.

For a critical race-free state assignment the state transitions in Fig. 2 should have adjacent state assignments. Here, there are 4 states and so a minimum of 4 binary codes are required for state assignment. But for race free state assignment additional rows are required and so let us choose 8 binary codes which is possible with 3-bit code (2^3 = 8 possible codes). The possible 3-bit codes are,

000, 001, 010, 011, 100, 101, 110, 111

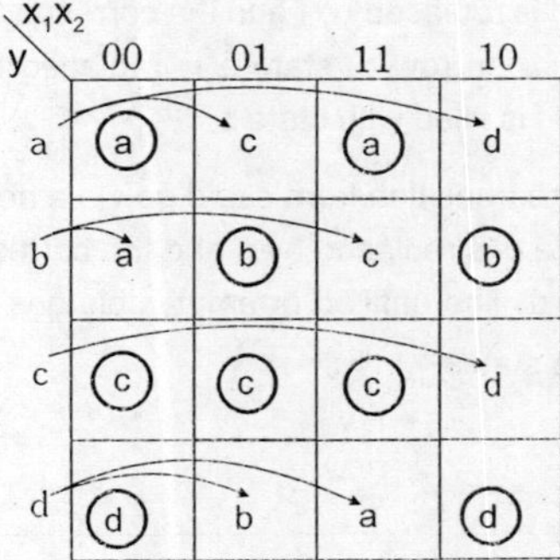

***Fig. 2:** Transitions from present state to next state.*

Let us represent the three binary bits by three state variables y_1, y_2 and y_3.

Let us construct a binary assignment table with three state variables y_3, y_2 and y_1. Two binary values are said to be adajcent if they differ in only one variable.

Let us assign adjacent states to following state pair to avoid critical races as shown in Fig. 4(a).

(b, a), (b, c), (d, b)

The following state transitions are not possible with one bit state change,

$a \to c$, $a \to d$, $d \to a$, $c \to d$

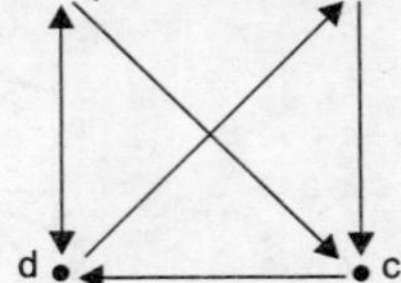

***Fig. 3:** Transition diagram.*

Therefore, intermediate states e, f, and g can be introduced as shown in Fig. 4(b). Now the above transitions will be as follows,

$a \to c \Rightarrow a \to e \to c$ | $a \to d \Rightarrow a \to f \to d$

$c \to d \Rightarrow c \to g \to d$ | $d \to a \Rightarrow d \to f \to a$

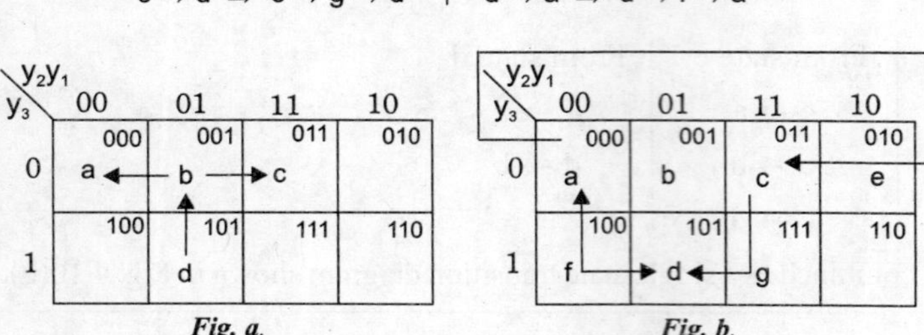

***Fig. 4:** Binary state assignment.*

a = 000
b = 001
e =010
f = 100
d =101
c = 011
g = 111

***Fig. 5:** Transition diagram with binary assignment.*

The transition diagram with critical race-free binary assignment is shown in Fig. 5. The introduction of three additional states e, f and g leads to three extra rows in transition table so that 4 rows table is modified to 7 rows table as shown in Fig. 6. The rows are arranged such that their state assignments are adjacent. Here, one the of code is not used and so there is a row with all don't-care entries in columns. The state transition from state b does not require additional state and so the correspond row in modified table is same as original table.

The transition from a to c go via a new state e. Therefore, in row a, c is replaced by e and in row e the corresponding column is filled with c.

The transition from a to d and d to a go via a new state f. Therefore, in row a, d is replaced by f and the corresponding column in row f is filled with d. Also, in row d, state a is replaced by f and the corresponding column in f is filled with state a.

The transition from c to d go via a new state g. Therefore, in row c, the state d is replaced by g and the corresponding column in row g is filled with d. The unfilled or empty columns in Fig. 6 are considered as don't-care states.

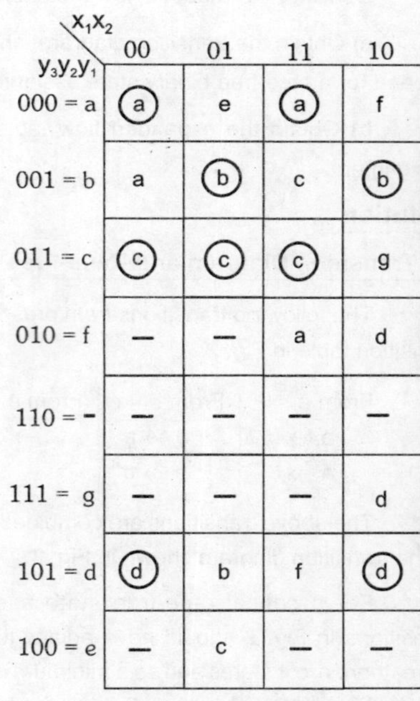

$y_3y_2y_1$ \ x_1x_2	00	01	11	10
000 = a	(a)	e	(a)	f
001 = b	a	(b)	c	(b)
011 = c	(c)	(c)	(c)	g
010 = f	—	—	a	d
110 = –	—	—	—	—
111 = g	—	—	—	d
101 = d	(d)	b	f	(d)
100 = e	—	c	—	—

***Fig. 6:** Modified flow table with race free state assignment.*

b) Multiple Row Method

In multpile row method each row in original flow table is represented by two or more equivalent rows. Let us represent each row by two equivalent rows so that,

- a is represented by a_1 and a_2
- b is represented by b_1 and b_2
- c is represented by c_1 and c_2
- d is represented by d_1 and d_2

Now, there are 8 states and so a 3-bit code is required for 8 binary codes ($2^3 = 8$). The 8 possible codes are,

000, 001, 010, 011, 100,101, 110, 111

Let us represent three bits of binary code by three state variables y_1, y_2 and y_3.

Let us assign 4 adjacent codes to a_1, b_1, c_1 and d_1 as shown in Fig. 7(a). In multiple row method the equivalent states are assigned with complement codes. Hence complement of codes for a_1, b_1, c_1 and d_1 are assigned to states a_2, b_2, c_2 and d_2 as shown in Fig. 7(b).

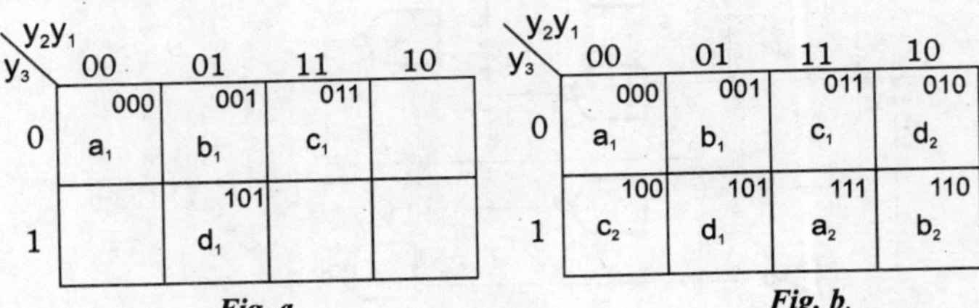

Fig. a. *Fig. b.*

Fig. 7: Binary state assignment in multiple row method.

$y_3y_2y_1$ \ x_1x_2	00	01	11	10
000 = a_1	(a_1)		(a_1)	
111 = a_2	(a_2)		(a_2)	
001 = b_1		(b_1)		(b_1)
110 = b_2		(b_2)		(b_2)
011 = c_1	(c_1)	(c_1)	(c_1)	
100 = c_2	(c_2)	(c_2)	(c_2)	
101 = d_1	(d_1)			(d_1)
010 = d_2	(d_2)			(d_2)

Fig. a.

$y_3y_2y_1$ \ x_1x_2	00	01	11	10
000 = a_1	(a_1)	c_2	(a_1)	d_2
111 = a_2	(a_2)	c_1	(a_2)	d_1
001 = b_1	a_1	(b_1)	c_1	(b_1)
110 = b_2	a_2	(b_2)	c_2	(b_2)
011 = c_1	(c_1)	(c_1)	(c_1)	d_2
100 = c_2	(c_2)	(c_2)	(c_2)	d_1
101 = d_1	(d_1)	b_1	a_2	(d_1)
010 = d_2	(d_2)	b_2	a_1	(d_2)

Fig. b.

Fig. 8: Expanded flow table.

The expanded flow table is formed by representing each row of original flow table by their two equivalent rows as shown in Fig. 8.

First fill the table shown in Fig. 8 with equivalent stable states as shown in Fig. 8(a). The equivalent stable states are same as original stable states but with subscript 1 and 2. Next, the unstable states are marked with reference to state assignment table such that they are adjacent.

In rows a_1 and a_2 the unstable states c_1, c_2, d_1 and d_2 are filled such that they are adjacent to a_1 and a_2.

In rows b_1 and b_2 the unstable states a_1, a_2, c_1 and c_2 are filled such that they are adjacent to b_1 and b_2.

In rows c_1 and c_2 the unstable states d_1 and d_2 are filled such that they are adjacent to c_1 and c_2.

In rows d_1 and d_2 the unstable states a_1, a_2, b_1 and b_2 are filled such that they are adjacent to d_1 and d_2.

Example 4.10

Derive the transition table for the asynchronous sequential circuit shown in Fig. 1. Determine the sequence of the internal states $Y_1 Y_2$ for the following sequence of inputs x_1x_2 : 00, 10, 11, 01, 11, 00.

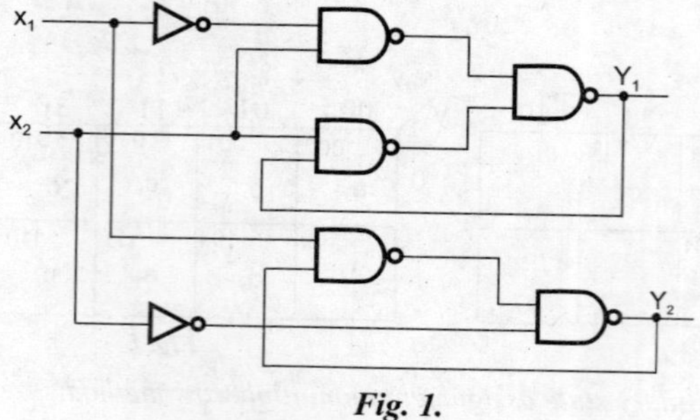

Fig. 1.

Solution

From the given circuit we can conclude that the circuit has following variables.

x_1, x_2 = Input

y_1, y_2 = Secondary variables (Present state variables)

Y_1, Y_2 = Excitation variables (Next state variables)

From the asynchronous sequential circuit shown in Fig. 2, we get the following equations,

$$Y_1 = ((x_1\,x_2)'\,.\,(x_2y_1)')' = x_1x_2 + x_2y_1$$

$$Y_2 = ((x_1y_2)'\,.\,x_2)' = x_1y_2 + x_2$$

Using the above equations the excitation variables Y_1 and Y_2 are evaluated for all possible combinations for inputs (x_1 and x_2) and secondary variables (y_1 and y_2) and listed in Table 1.

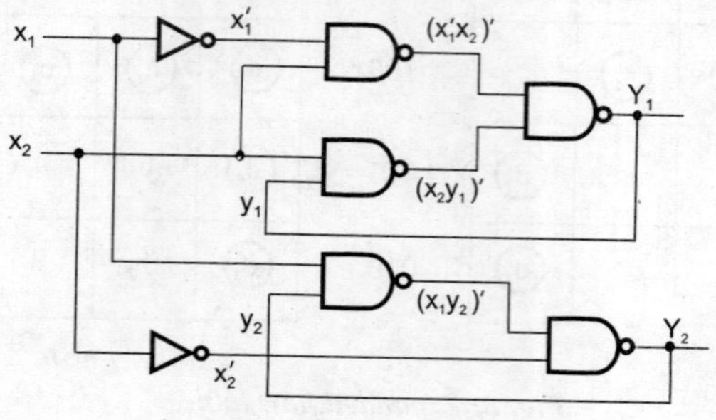

Fig. 2.

Table 1: Evaluation of Excitation Variables

Present states				Product Terms			Next States	
y_1	y_2	x_1	x_2	$x_1'x_2$	x_2y_1	x_1y_2	Y_1	Y_2
0	0	0	0	0	0	0	0	0
0	0	0	1	1	0	0	1	1
0	0	1	0	0	0	0	0	0
0	0	1	1	0	0	0	0	1
0	1	0	0	0	0	0	0	0
0	1	0	1	1	0	0	1	1
0	1	1	0	0	0	1	0	1
0	1	1	1	0	0	1	0	1
1	0	0	0	0	0	0	0	0
1	0	0	1	1	1	0	1	1
1	0	1	0	0	0	0	0	0
1	0	1	1	0	1	0	1	1
1	1	0	0	0	0	0	0	0
1	1	0	1	1	1	0	1	1
1	1	1	0	0	0	1	0	1
1	1	1	1	0	1	1	1	1

Maps are constructed for Y_1 and Y_2 by taking possible values of secondary variables y_1 and y_2 on rows and input x_1 and x_2 on columns as shown in Figs. 3(a) and (b) and the squares in the maps are filled with values of Y_1 and Y_2 listed in Table 1. The transition table is formed by combining the maps of Figs. 3(a) and (b) into a single map as shown in Fig. 4.

In the transition table of each row, the squares that have identical values with secondary values are marked with circles. The circled entries represent stable states and uncircled entries represent unstable states.

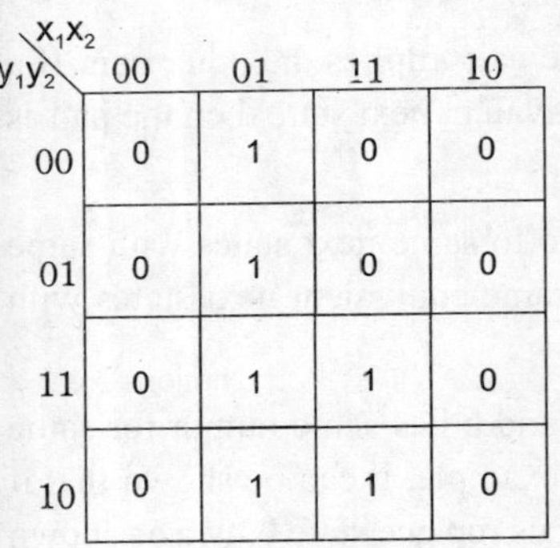

y_1y_2 \ x_1x_2	00	01	11	10
00	0	1	0	0
01	0	1	0	0
11	0	1	1	0
10	0	1	1	0

Fig. a: K-map for Y_1.

y_1y_2 \ x_1x_2	00	01	11	10
00	0	1	1	0
01	0	1	1	1
11	0	1	1	1
10	0	1	1	0

Fig. b: K-map for Y_2.

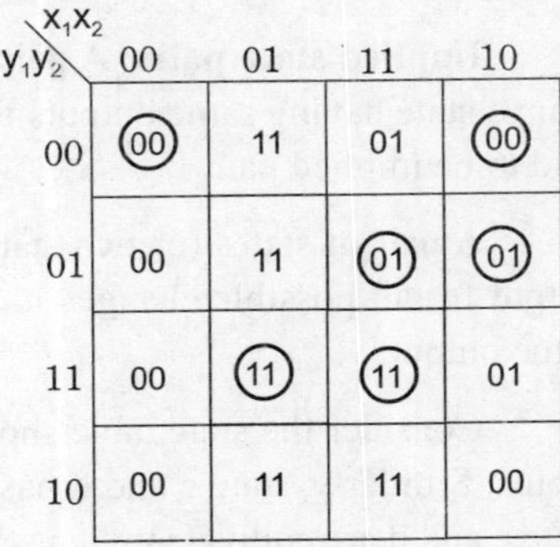

y_1y_2 \ x_1x_2	00	01	11	10
00	(00)	11	01	(00)
01	00	11	(01)	(01)
11	00	(11)	(11)	01
10	00	11	11	00

Fig. 4: Transition table.

Fig. 3: Map for two state variables Y_1, Y_2.

It is given that the sequence of input changes is,

$$00 \rightarrow 10 \rightarrow 11 \rightarrow 01 \rightarrow 11 \rightarrow 00$$

The state transitions for the above input changes are shown in Fig. 5. The present and next state for the sequence of changes in input are also listed in Table 1.

Table 1:

Change in Input x_1 x_2	Present State y_1 y_2	State Transitions	Next State Y_1 Y_2
0 0	0 0	–	0 0
00 → 10	0 0	00 → 00	0 0
10 → 11	0 0	00 → 01 → 01	0 1
11 → 01	0 1	01 → 11 → 11	1 1
01 → 11	1 1	11 → 11	1 1
11 → 00	1 1	11 → 00 → 00	0 0

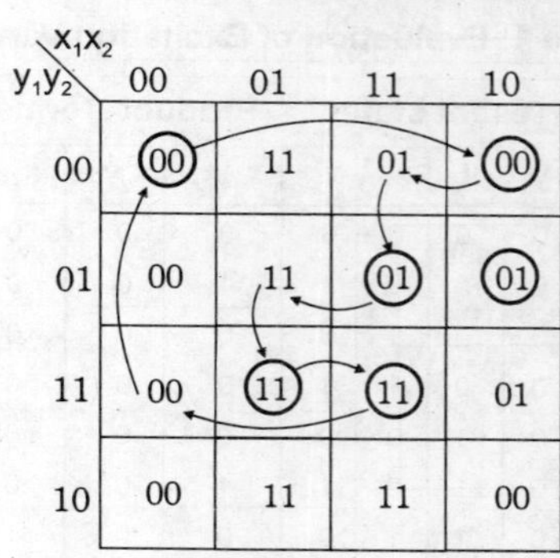

Fig. 5: Transitions through the sequence of inputs x_1 x_2.

Conclusion: For the input sequence of 00, 10, 11, 01, 11 and 00, the sequence of Y_1Y_2 is,

Y_1Y_2 : 00, 00, 01, 11, 11, 00.

4.4 State Reduction

In asynchronous sequential circuits, the state reduction is performed via an implication table. In the original state table, the equivalent states are identified and each pair of equivalent states is merged as a single state.

The rows of state tables can be classified in to following three groups.

Equivalent States: State pairs that have same next state and output for the same input are equivalent states.

Not equivalent states: State pairs having different next state and output for same input are not equivalent states.

Implied state pairs: A pair of states can be equivalent only if one pair implies the other pair. If a pair of state having same outputs for the same inputs and go to same equivalent next state then the pair is said to be implied pair.

A pair of states (or two states) are said to be equivalent if they go to same next states with same output for all possible changes in inputs. Alternatively, they can go to same equivalent next states with same output.

Consider the state table shown in Table 4.4. In this table, state a and b has same output for same inputs. Similarly, state c and d has same output for same inputs. In this example, it can be shown that if state c and d are equivalent then state a and b will also be equivalent. Let us replace state b by a as shown in Table 2. Now, in Table 4.5, if state d is equal to c then state a and b will be equivalent as well as state c and d are also equivalent. Hence we can say that (a, b) implies (c, d).

The state table with state d replaced by state c is shown in Table 4.6 and from this table we can say that state a and b are equivalent and also, state c and d are equivalent. Hence, one of the equivalent states can be removed and reduced state table can be formed as shown in Table 4.7.

Table 4.4: State Table

Present State	Next State		Output	
	x = 0	x = 1	x = 0	x = 1
a	d	a	1	0
b	c	b	1	0
c	b	d	0	1
d	a	d	0	1

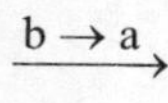
b → a

Table 4.5: State Table with State b Replaced by State a

Present State	Next State		Output	
	x = 0	x = 1	x = 0	x = 1
a	d	a	1	0
a (b → a)	c	a (b → a)	1	0
c	a (b → a)	d	0	1
d	a	d	0	1

d → c

Table 4.6: State Table with State d Replaced by State c

Present State	Next State		Output	
	x = 0	x = 1	x = 0	x = 1
a	c (d → c)	a	1	0
a	c	a	1	0
c	a	c (d → c)	0	1
c (d → c)	a	c (d → c)	0	1

Reduced state table ←

Table 4.7: Reduced State Table

Present State	Next State		Output	
	x = 0	x = 1	x = 0	x = 1
a	c	a	1	0
c	a	c	0	1

4.4.1 Implication Table

When the state table has large number of rows it is difficult to find equivalent and implied states and in this situation implication table is used to find equivalent and implied states and then form the reduced state table.

The implication table is constructed by listing the states on rows and columns in order to form squares at the intersection of every row and column. Each square represent a pair of state. While forming rows, the first state is not considered and while forming column the last state is not considered so that only state pairs are represented by squares. The implication table is also known as ***merger table***.

Consider the state table of asynchronous sequential circuit with 7 states a, b, c, d, e, f and g shown in Table 4.8. Now, the implication table is formed by listing states b, c, d, e, f and g on rows and states a, b, c, d, e, and f on columns in order to form squares at the intersection of each row and column as shown in Fig. 4.11(a). The state pair represented by each square is shown in Fig. 4.11(b).

Table 4.8: State Table

Present State	Next State x = 0	Next State x = 1	Output x = 0	Output x = 1
a	e	a	0	0
b	d	b	0	0
c	g	f	1	0
d	a	d	0	1
e	a	d	0	1
f	b	c	0	0
g	a	e	0	1

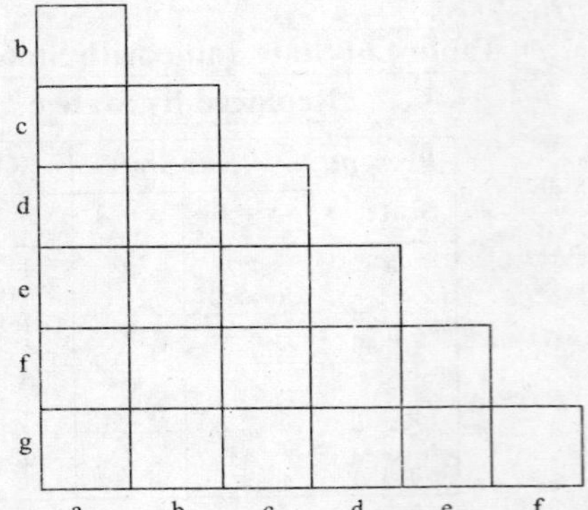

Fig. a: Listing of states in rows and columns.

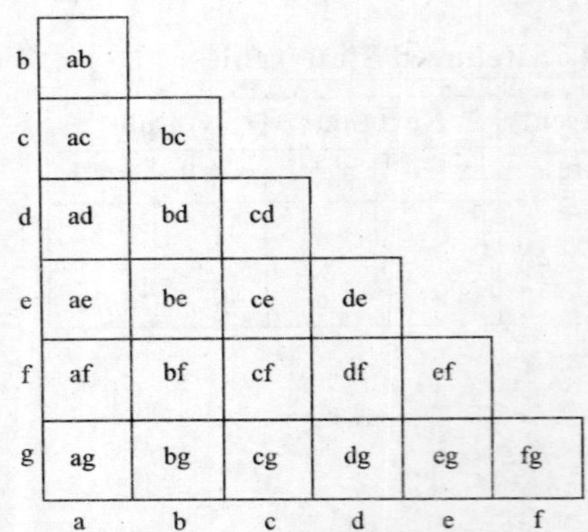

Fig. b: Pair of states represented by squares.

***Fig. 4.11:** Structure of implication table.*

First mark × in the squares in which the pair of states do not have same output for same input as shown in Fig. 4.12(a). The state pairs marked by × are not equivalent pairs. Next mark a tick (√) in the squares in which the state pairs are equivalent as shown in Fig. 4.12(b). Here, the state pair (d, e) i equivalent because they go to same next state and output for same input.

The next step is to identify the implied pairs among the state pairs that are not marked by × and tic (√) in Figs. 4.12(a) and (b). Mark the next state pairs corresponding to the state pair represented by th squares that are blank in Figs. 4.12(a) and (b), as shown in Fig. 4.12(c).

When next state pair and present state pair are same then those state pairs can be ignored in Fig 4.12(c). Similarly, when a present state pair go to same next state then those next states can also be ignore in Fig. 4.12(c).

The state pair entries in Fig. 4.12(c), are marked × if the pair is not equivalent and marked wit tick (√) if the state pair is equivalent. The three implication tables shown in Figs. 4.12(a), (b) and (c) ar combined to form the implication table shown in Fig. 4.12(d). Now, the equivalent state pairs are determine from implication table of Fig. 4.12(d).

When a square representing a state pair has an equivalent state pair entry then that state pair is also equivalent. In Fig. 4.12(d), the squares representing (a, b), (d, g) and (e, g) has equivalent state pair entires and so (a, b), (d, g) and (e, g) are equivalent state pairs.

When a square representing a state pair has not equivalent state pair entry then that state pair is also not equivalent. In Fig. 4.12(d), the squares representing (a, f) and (b, f) has not equivalent state pairs and so (a, f) and (b, f) are not equivalent state pairs.

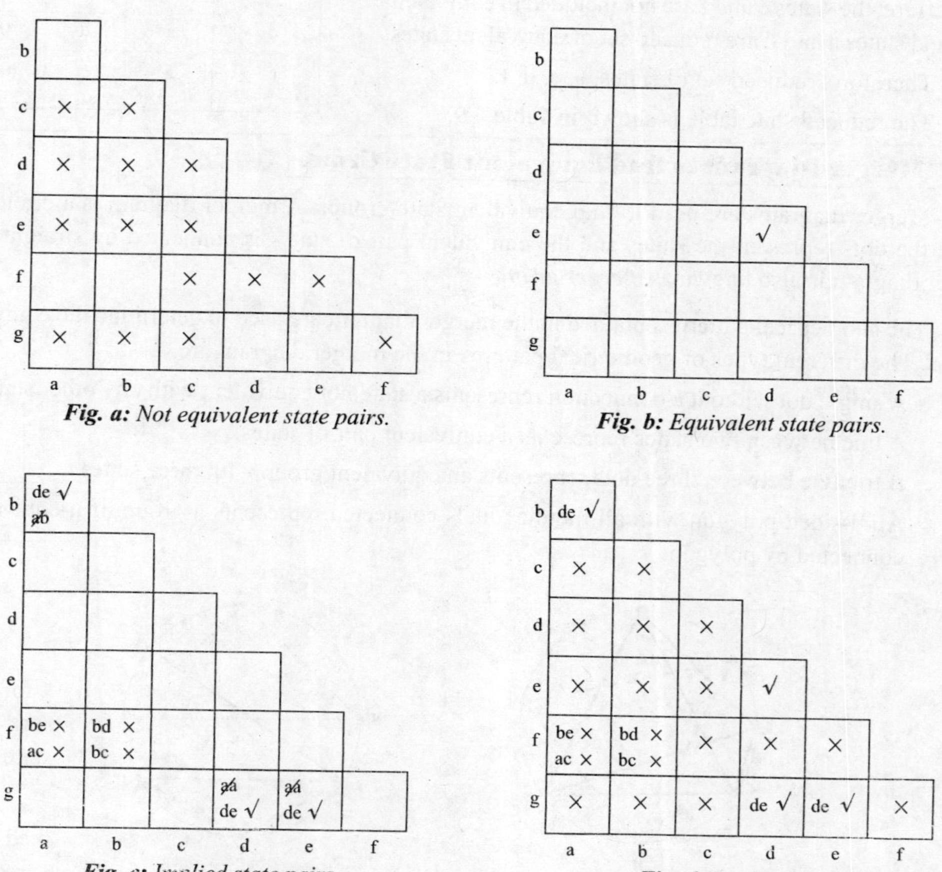

Fig. a: Not equivalent state pairs.

Fig. b: Equivalent state pairs.

Fig. c: Implied state pairs.

Fig. d: Implication table.

Fig. 4.12: Formation of implication table.

Now, the equivalent state pairs are,

(a, b), (d, e), (d, g), (e, g)

The state pairs (d, e), (d, g) and (e, g) can be merged to larger group (d, e, g). Now, the equivalent state groups are,

(a, b), (d, e, g)

In the equivalent state groups shown above only one state is selected in each group. Let us select state a in group (a, b) and state d in group (d, e, g).

Now the reduced states are given by one of the state in each equivalent group and the states not listed in the equivalent pairs.

Here, the states c and f are not included in equivalent pairs and states a and d are reduced set of equivalent states.

Therefore, reduced set of states: a, c, d, f.

The reduced state table is shown in Table 4.9.

Table 4.9: Reduced State Table

Present State	Next State		Output	
	x = 0	x = 1	x = 0	x = 1
a	d (e → d)	a	0	0
c	d (g → d)	f	1	0
d	a	d	0	1
f	a (b → a)	c	0	0

4.4.2 Merger Diagram to find Equivalent State Groups

Merger diagrams are used to find equivalent state groups. A merger diagram is a circle with dots where the dots represent the states and the equivalent pair of states is connected by straight lines. The merger diagram is also known as ***merger graph***.

The geometrical patterns obtained in the merger diagram are used to determine the equivalent state groups. The different types of geometrical patterns in the merger diagram are,

- A single dot without a connection represents a state not equivalent with any other state.
- A line between two states represents a equivalent pair of states.
- A triangle between three dots represents an equivalent group with three states.
- An n-sided polygon with all the diagonals connected represents a group of n-equivalent states connected by polygon.

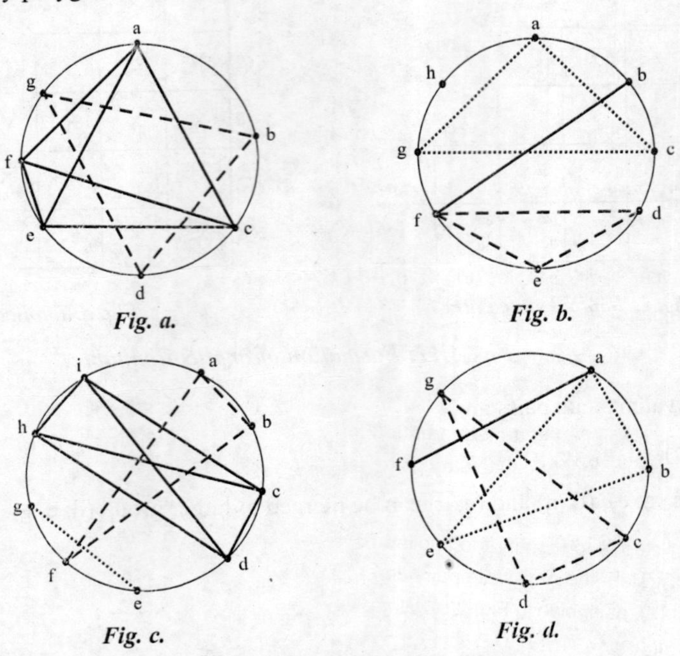

Fig. 4.13: Examples of merger diagrams.

Some examples of merger diagrams with different groups of equivalent states are shown in Fig. 4.13.

Equivalent state groups in Fig. 4.13(a) : (b, d, g), (a, c, e, f)

Equivalent state groups in Fig. 4.13(b) : (b, f), (a, c, g), (d, e, f)

Equivalent state groups in Fig. 4.13(c) : (e, g), (a, b, f), (c, d, h, i)

Equivalent state groups in Fig. 4.13(d) : (a, f), (a, b, e), (c, d, g)

Example 4.11

Reduce the state table shown in Table 1 to a minimum number of states using the implication table method.

Table 1: State Table

Present State	Next State		Output	
	x = 0	x = 1	x = 0	x = 1
a	a	e	0	0
b	c	b	0	1
c	a	f	0	0
d	c	b	0	1
e	f	e	0	0
f	a	f	0	0

Solution

The given state table (Table 1) of asynchronous sequential circuit has 6 states a, b, c, d, e and f. Now, the implication table is formed by listing states b, c, d, e and f on rows and states a, b, c, d, and e on columns and forming squares at the intersection of each row and column as shown in Fig. 1(a). The state pair represented by each square is shown in Fig. 1(b).

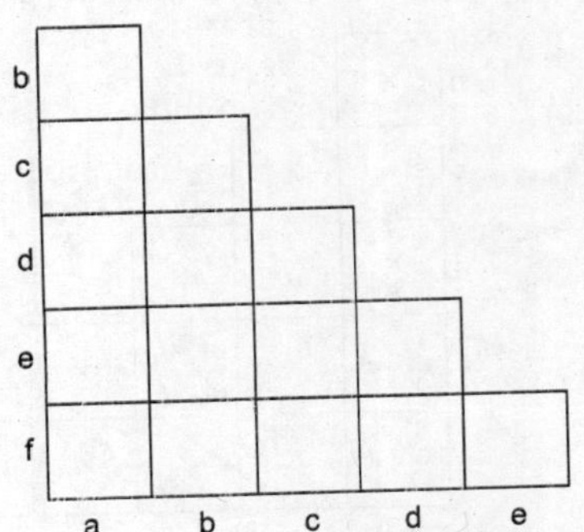

Fig. a: Listing of states in rows and columns.

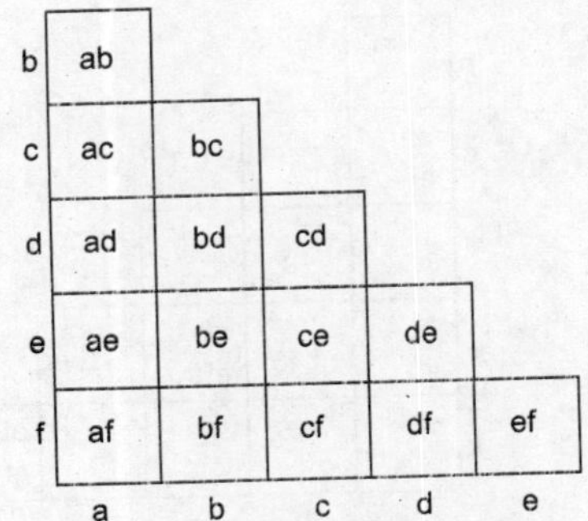

Fig. b: Pair of states represented by squares.

Fig. 1: Structure of implication table.

Here, three implication tables (Fig. 2(a), (b) and (c)) are drawn to separately show the not equivalent, equivalent and implied state pairs.

First mark × in the squares in Fig. 2(a) in which the pair of states do not have same output for same inputs. The state pairs marked by × are not equivalent pairs. The not equivalent state pairs are listed in Table 2.

Next mark a tick (√) in the squares in Fig. 2(b) in which the state pairs are equivalent. Here, the state pair (b, d) and (c, f) is equivalent because they go to same next state and output for same input. The equivalent state pairs are listed in Table 3.

The next step is to identify the implied pairs among the state pairs that are not marked by × and tick (√) in Figs. 2(a) and (b). Mark the next state pairs corresponding to the state pair represented by the squares that are blank in Figs. 2(a) and (b), as shown in Fig. 2(c). When a present state pair go to same next state then those next states can be ignored in Fig. 2(c).

Table 2: Not Equivalent State Pairs

Present State	Next State x = 0	Next State x = 1	Output x = 0	Output x = 1	Pair
a	a	e	0	0	(a, b)
b	c	b	0	1	
a	a	e	0	0	(a, d)
d	c	b	0	1	
b	c	b	0	1	(b, c)
c	a	f	0	0	
b	c	b	0	1	(b, e)
e	f	e	0	0	
b	c	b	0	1	(b, f)
f	a	f	0	0	
c	a	f	0	0	(c, d)
d	c	b	0	1	
d	c	b	0	1	(d, e)
e	f	e	0	0	
d	c	b	0	1	(d, f)
f	a	f	0	0	

Table 3: Equivalent State Pairs

Present State	Next State x = 0	Next State x = 1	Output x = 0	Output x = 1	Pair
b	c	b	0	1	(b, d)
d	c	b	0	1	
c	a	f	0	0	(c, f)
f	a	f	0	0	

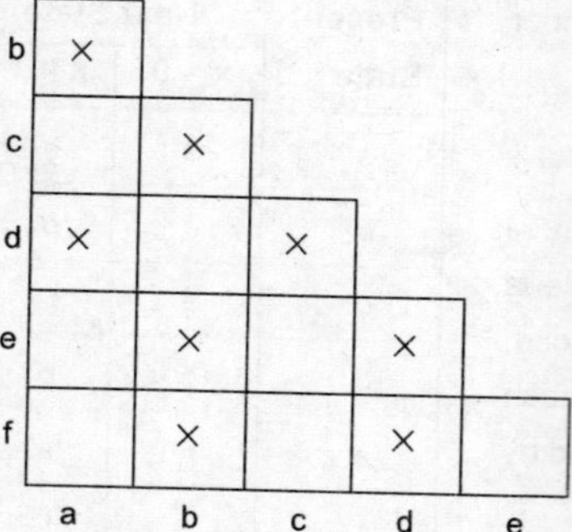

Fig. a: Not equivalent state pairs.

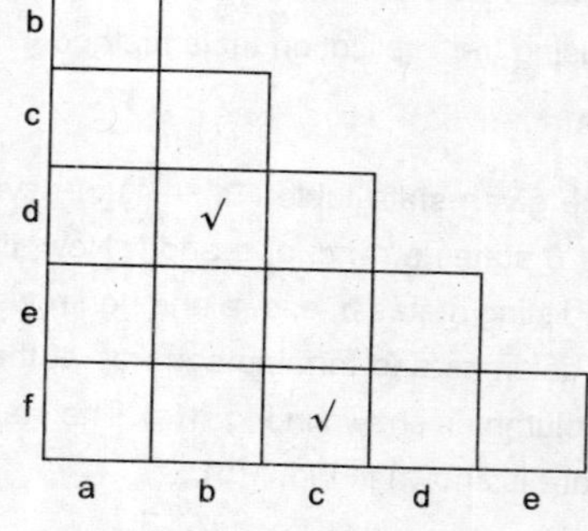

Fig. b: Equivalent state pairs.

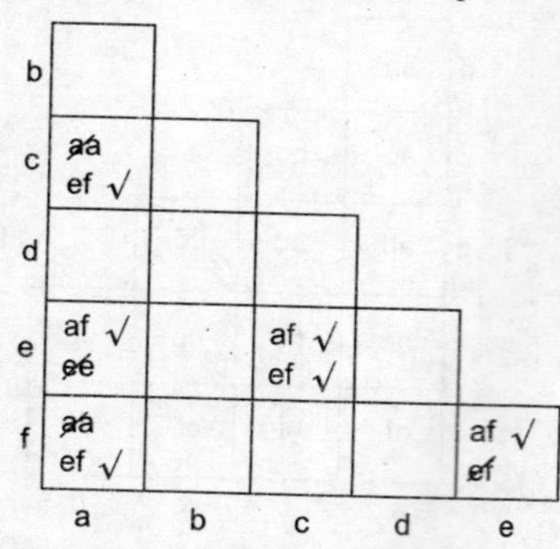

Fig. c: Implied state pairs.

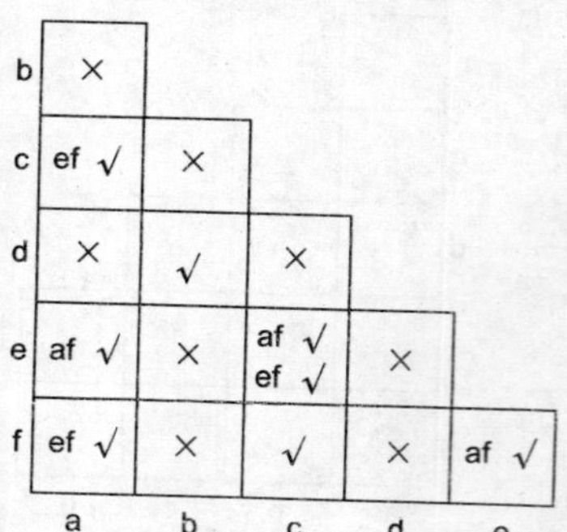

Fig. d: Implication table.

***Fig. 2:** Formation of implication table.*

Here, all entries in Fig. 2(c) are either state pair (a, f) or (e, f). The state pair (a, f) go to same next state for same input. Similarly, state pair (e, f) go to same next state for same input. Hence, state pairs (a, f) and (e, f) are implied pairs.

With reference to Tables 4 and 5 when f = a then state pair (a, f) implies (e, f). Hence state pairs (a, f) and (e, f) are equivalent pairs and they are marked with tick (√).

Table 4: State Table for State Pairs (a, f) and (e, f)

Present State	Next State x = 0	Next State x = 1	Output x = 0	Output x = 1
a	a	e	0	0
f	a	f	0	0
e	f	e	0	0
f	a	f	0	0

f → a

Table 5: Table 2 with State f Replaced by State a

Present State	Next State x = 0	Next State x = 1	Output x = 0	Output x = 1
a	a	e	0	0
a (f → a)	a	a (f → a)	0	0
e	a (f → a)	e	0	0
a (f → a)	a	a (f → a)	0	0

The three implication tables in Fig. 2(a), (b) and (c) are combined as a single implication table in Fig. 2(d) and the equivalent state pairs are determined from Fig. 2(d).

When a square representing a state pair has an equivalent state pair entry then that state pair is also equivalent. In Fig. 2(d), the squares representing (a, c), (a, e), (a, f), (c, e) and (e, f) has equivalent state pair entires and so (a, c), (a, e), (a, f), (c, e) and (e, f) are equivalent state pairs.

With reference to Fig. 2(d), the equivalent state pairs are,

(a, c), (a, e), (a, f), (c, e) (e, f), (c, f), (b, d)

The state pairs (a, c), (a, e), (a, f), (c, e) (e, f) and (c, f) can be merged to larger group (a, c, e, f). Now, the equivalent state groups are,

(a, c, e, f), (b, d)

Alternatively, the above equivalent groups can be obtained from the merger diagram shown in Fig. 3. In this diagram a 4 sided polygon with diagonals connected form an equivalent group (a, c, e, f) and a line betweem two states form an equivalent pair (b, d)

In the equivalent state groups shown above only one state is selected in each group. Let us select state a in group (a, c, e, f) and state b in group (b, d).

Now the reduced states are given by one of the state in each equivalent group and the states not listed in the equivalent pairs. Here, all states are included in equivalent state groups and so the states a and b are reduced set of equivalent states.

Therefore, reduced set of states: a, b.

The reduced state table is shown in Table 6.

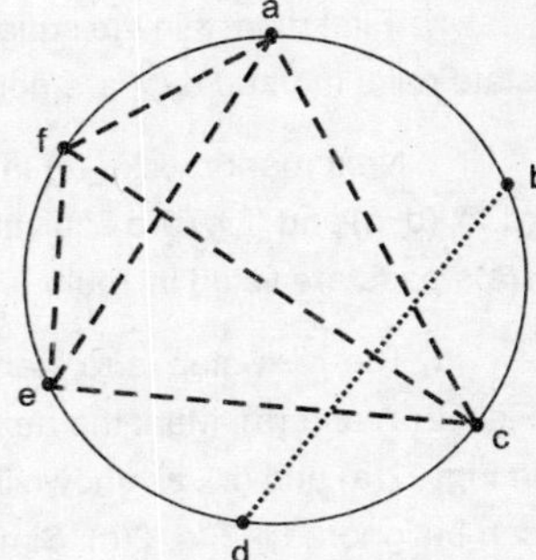

Fig. 3: Merger diagram.

Table 6: Reduced State Table

Present State	Next State x = 0	Next State x = 1	Output x = 0	Output x = 1
a	a	a (e → a)	0	0
b	a (c → a)	b	0	1

Example 4.12

Reduce the state table shown in Table 1 to a minimum number of states using the implication table method.

Solution

The given state table (Table 1) of asynchronous sequential circuit has 7 states a, b, c, d, e, f and g. Now, the implication table is formed by listing states b, c, d, e, f and g on rows and states a, b, c, d, e and f on columns and forming squares at the intersection of each row and column as shown in Fig. 1(a). The state pair represented by each square is shown in Fig. 1(b).

Table 1: State Table

Present State	Next State x = 0	Next State x = 1	Output x = 0	Output x = 1
a	d	b	0	0
b	f	a	0	0
c	g	f	0	1
d	a	d	1	0
e	c	b	0	0
f	a	d	1	0
g	a	d	1	0

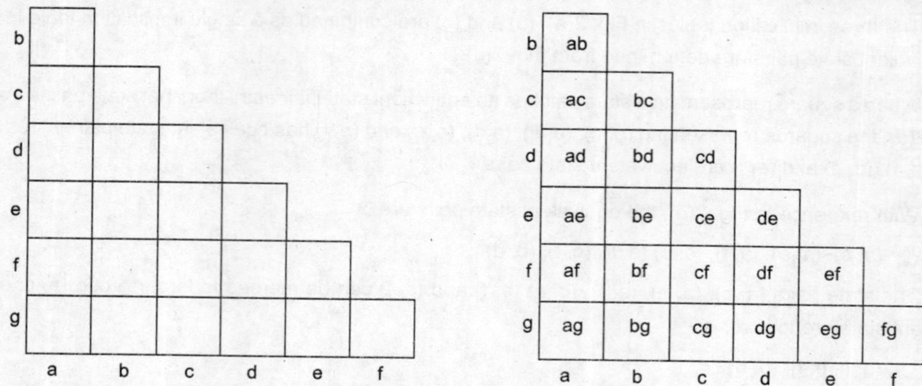

Fig. a: Listing of states in rows and columns.

Fig. b: Pair of states represented by squares.

Fig. 1: Structure of implication table.

Here, three implication tables (Figs. 2(a), (b) and (c)) are drawn to separately show the not equivalent, equivalent and implied state pairs.

First mark × in the squares in Fig. 2(a) in which the pair of states do not have same output for same inputs. The state pairs marked by × are not equivalent pairs. The not equivalent state pairs are listed in Table 2.

Next mark a tick (√) in the squares in Fig. 2(b) in which the state pairs are equivalent. Here, the state pairs (d, f), (d, g) and (f, g) are equivalent pairs because they go to same next state and output for same input. The equivalent state pairs are listed in Table 3.

The next step is to identify the implied pairs among the state pairs that are not marked by × and tick (√) in Figs. 2(a) and (b). Mark the next state pairs corresponding to the state pair represented by the squares that are blank in Figs. 2(a) and (b), as shown in Fig. 2(c). When next state pair and present state pair are same then those state pairs can be ignored in Fig. 2(c). Similarly, when a present state pair go to same next state then those next states can also be ignored in Fig. 2(c).

The state pair entries in Fig. 2(c), are marked × if the pair is not equivalent and marked with tick (√) if the state pair is equivalent.

The three implication tables in Fig. 2(a), (b) and (c) are combined as a single implication table in Fig. 2(d) and the equivalent state pairs are determined from Fig. 2(d).

When a square representing a state pair has an equivalent state pair entry then that state pair is also equivalent. In Fig. 2(d), the square representing (a, b) has equivalent state pair entry and so (a, b) is equivalent state pair. When a square representing a state pair has not equivalent state pair entry then that state pair is also not equivalent. In Fig. 2(d), the squares representing (a, e) and (b, e) have not equivalent state pairs and so (a, e) and (b, e) are not equivalent state pairs.

Table 2: Not Equivalent State Pairs

State	Next states	Outputs	Pair
a	d b	0 0	(a, c)
c	g f	0 1	
a	d b	0 0	(a, d)
d	a d	1 0	
a	d b	0 0	(a, f)
f	a d	1 0	
a	d b	0 0	(a, g)
g	a d	1 0	
b	f a	0 0	(b, c)
c	g f	0 1	
b	f a	0 0	(b, d)
d	a d	1 0	
b	f a	0 0	(b, f)
f	a d	1 0	
b	f a	0 0	(b, g)
g	a d	1 0	
c	g f	0 1	(c, d)
d	a d	1 0	
c	g f	0 1	(c, e)
e	c b	0 0	
c	g f	0 1	(c, f)
f	a d	1 0	
c	g f	0 1	(c, g)
g	a d	1 0	
d	a d	1 0	(d, e)
e	c b	0 0	
e	c b	0 0	(e, f)
f	a d	1 0	
e	c b	0 0	(e, g)
g	a d	1 0	

Table 3: Equivalent State Pairs

State	Next states	Outputs	Pair
d	a d	1 0	(d, f)
f	a d	1 0	
d	a d	1 0	(d, g)
g	a d	1 0	
f	a d	1 0	(f, g)
g	a d	1 0	

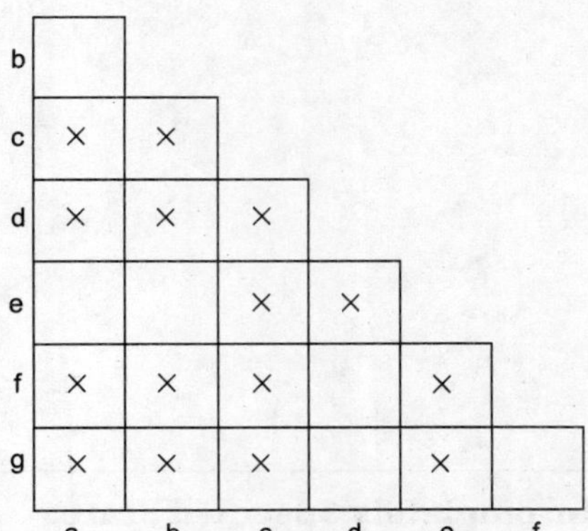

Fig. a: Not equivalent state pairs.

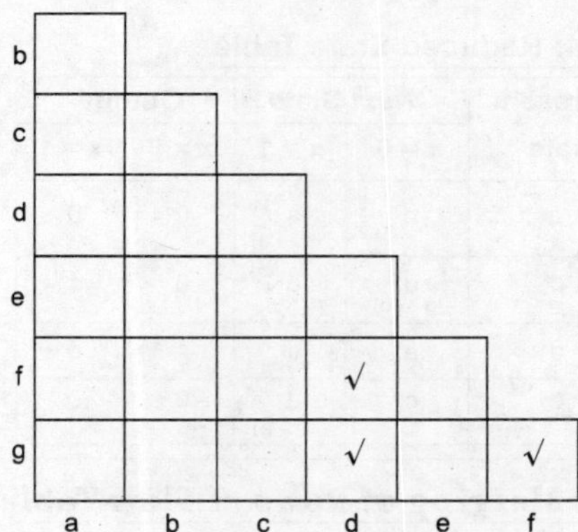

Fig. b: Equivalent state pairs.

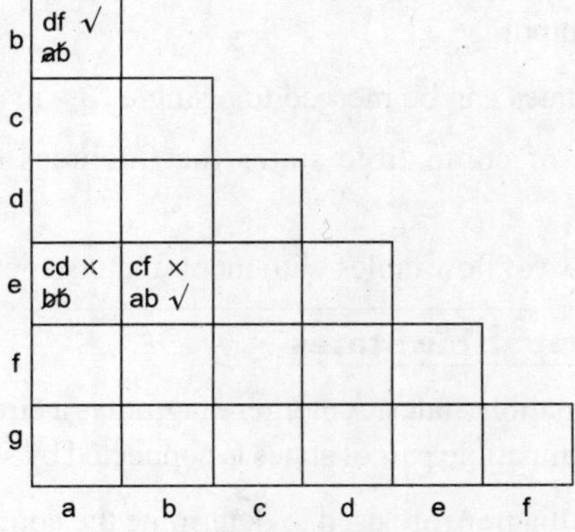

Fig. c: Implied state pairs.

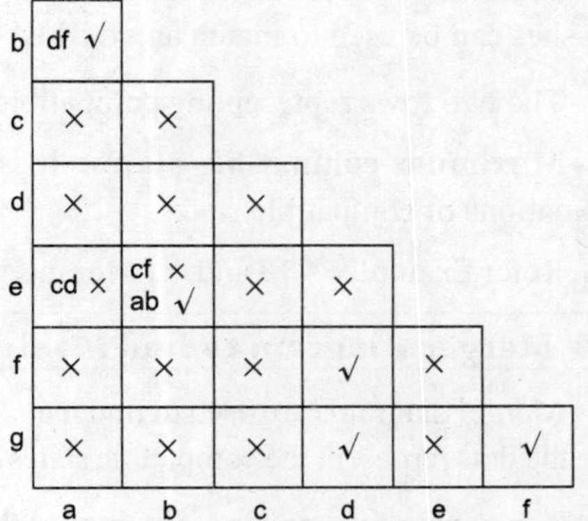

Fig. d: Implication table.

Fig. 2: Formation of implication table.

With reference to Fig. 2(d), the equivalent state pairs are,

(a, b), (d, f), (d, g), (f, g)

The state pairs (d, f), (d, g) and (f, g) can be merged to larger group (d, f, g). Now, the equivalent state groups are,

(a, b), (d, f, g)

Alternatively, the above equivalent states can be obtained from merger diagram shown in Fig. 3. In this diagram, a triangle form an equivalent group (d, f, g) and a line between two states form an equivalent pair (a, b). In the equivalent state groups shown above only one state is selected in each group. Let us select state a in group (a, b) and state d in group (d, f, g).

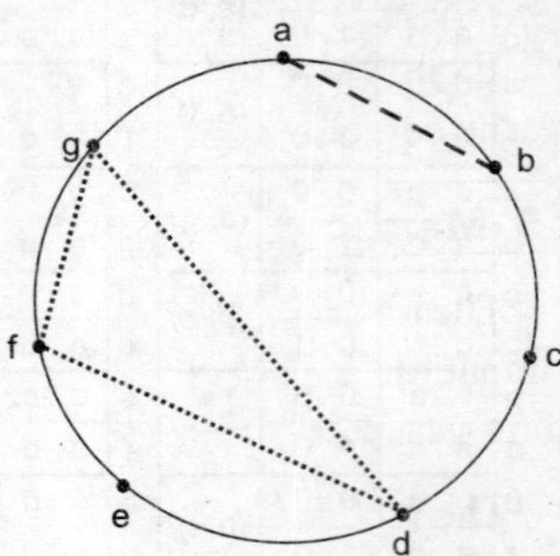

Fig. 3: Merger diagram.

Now the reduced states are given by one of the state in each equivalent group and the states not listed in the equivalent pairs. Here, the states selected from equivalent groups are a and d. The states not included in state groups are c and e.

Therefore, reduced set of states: a, c, d, e.

The reduced state table with 4 states is shown in Table 4.

Table 4: Reduced State Table

Present State	Next State x = 0	Next State x = 1	Output x = 0	Output x = 1
a	d	a (b → a)	0	0
c	d (g → d)	d (f → d)	0	1
d	a	d	1	0
e	c	a (b → a)	0	0

4.4.3 Merging of Rows of Flow Tables with Incompletely Specified States

In asynchronous circuits with incompletely specified states, the states can be reduced by identifying compatible pairs. Two states are said to be a compatible pair if the specified outputs and next states are same. Here, dashes are used to represent unspecified states or outputs. While identifying compatible pairs the dashes can be used to match any desired state or output.

The two rows representing compatible pair of states can be merged to a single row in a flow table.

Maximum compatible states: It is a group of compatible states that includes all possible combinations of compatible states.

Refer Examples 4.13 and 4.14 for merging of rows of flow tables with incompletely specified states.

4.4.4 Merger Diagram to find Maximum Compatible States

Merger diagrams are used to find maximum compatible states. A merger diagram is a circle with dots where the dots represent the compatible states and the compatible pair of states is connected by straight lines.

The geometrical patterns obtained in the merger diagram are used to determine the compatible state groups. The different types of geometrical patterns in the merger diagram are,

- A single dot without a connection represents a state not compatible with any other state.

- A line between two states represents a compatible pair of states.
- A triangle between three dots represents a state compatibility with the three connected states.
- An n-sided polygon with all the diagonals connected represents a group of n-compatible states connected by polygon.

Examples of merger diagram shown in Fig. 4.13 are applicable for determining compatible groups. Here, the equivalent groups are called ***compatible groups***.

4.4.5 Maximum Compatible

The maximum compatible is a group of compatible states that includes all the possible combinations of compatible states. The maximum compatible states can be determined from merger diagram. The states forming various geometrical patterns in the merger diagram are the maximum compatibles.

The reduced flow table can be formed by taking one member of each maximum compatible set as a row in reduced flow table.

In many cases, the maximum compatibles will not be a minimal set of compatibles. A minimum set of compatibles can be determined from closed-covering condition.

4.4.6 Closed-Covering Condition

In implication table, the compatible pairs may or may not have implied pairs. Hence the closed condition can be defined for two different types of implication table.

1. In implication table, if the compatible pairs does not have implied pairs then the closed-condition refers to a set of compatibles that does not include implied states.

2. In implication table, when the compatible pairs has implied states then the implied pairs should be included in selected compatible pairs.

The covering-condition refers to inclusion of all states of original state table in the compatible sets. The closed-covering condition refers to set of compatibles that covers all the states and satisfy covering condition. The set of compatibles that satisfy the closed-covering condition form the minimal set of compatibles. The reduced flow table can be formed by taking one element of each of the minimal compatible set to form the rows of reduced flow table.

4.4.7 Assigning Outputs in Flow Table with Incompletely Specified States

In flow tables with incompletely specified states, every stable state will be associated with an output. Normally, output will not specified for unstable states. But in order to find compatible pairs and reduce the states, output is necessary for unstable states too. Hence, output has to be assigned to all unstable states such that no transient outputs are generated during a transition from one stable state to another stable state. Therefore, the following rules are followed to assign output to unstable states in a flow table.

Rule 1 : Assign 0 to an output of an unstable state which is transient state during a transition between two stable states that have a 0 as output.

Rule 2 : Assign 1 to an output of an unstable state which is transient state during a transition between two stable states that have a 1 as output.

Rule 3 : Assign a don't-care condition to an output of an unstable state which is transient state during a transition between two stable states that have a different values (0 and 1 or 1 and 0) as output.

Examples of output assignment using the above rules are shown in Fig. 4.14.

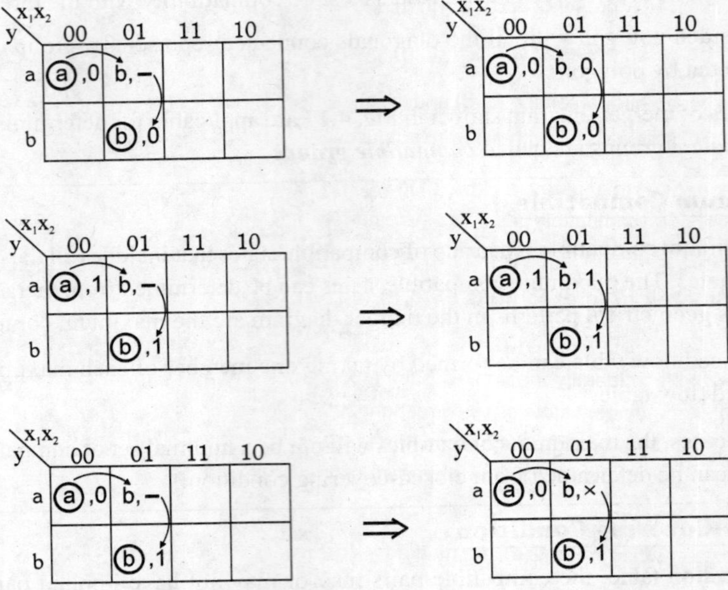

Fig. 4.14: Examples of output assignment.

Refer Examples 4.16, 4.18 and 4.23 for output assignment.

Example 4.13

Merge the primitive flow table shown in Fig. 1. Find all compatible pairs by means of an implication table. Find the maximal compatibles by means of a merger diagram. Find a minimal set of compatibles that covers all the states and is closed.

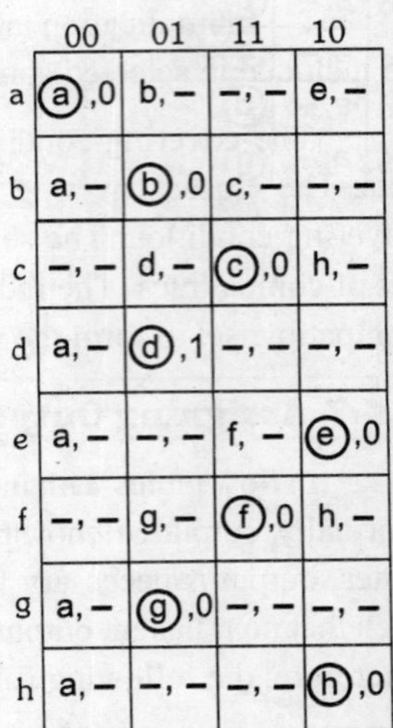

	00	01	11	10
a	(a),0	b, –	–, –	e, –
b	a, –	(b),0	c, –	–, –
c	–, –	d, –	(c),0	h, –
d	a, –	(d),1	–, –	–, –
e	a, –	–, –	f, –	(e),0
f	–, –	g, –	(f),0	h, –
g	a, –	(g),0	–, –	–, –
h	a, –	–, –	–, –	(h),0

Fig. 1: Primitive flow table.

Solution

The given flow table (Table 1) has 8 states a, b, c, d, e, f, g and h. The states b, c, d, e, f, g and h are listed to form rows and the states a, b, c, d, e, f and g are listed to form columns in implication table.

Here, three implication tables (Figs. 2(a), (b) and (c)) are drawn to separately show the not compatible, compatible and implied pairs.

The given flow table has 4 combinations of inputs. Hence compatibility of states for all the four inputs should be marked in each square of implication table.

The dashes in the given flow table represents unspecified states or outputs. While identifying compatible pairs the dashes can be used to match any desired state or output.

First mark × in squares in Fig. 2(a) in which the state pairs do not have same outputs for same inputs.

- When input is 01, present state pair (b, d), go to next state (b, d) with different output.
- When input is 01, present state pair (d, g), go to next state (d, g) with different output.

Hence the squares representing (b, d) and (d, g) pairs are marked × in Fig. 2(a). The not compatible state pairs that has transitions to stable states with different output are listed in Table 1.

Next mark tick (√) in squares in Fig. 2(b) in which state pairs are compatible. Here, dashes are matched with states and outputs to form compatible pairs. The state pair (a, b) goes to same next state with same output when dashes are considered as same state and output and so (a, b) is compatible pair.

Similarly, (a, e), (b, h), (c, d), (c, h), (d, e), (d, h), (e, g), (f, g), (f, h) and (g, h) are compatible pairs. The compatible state pairs are listed in Table 2.

The next step is to identify the implied pairs among the state pairs that are not marked by × and tick (√) in Figs. 2(a) and (b). Mark the next state pairs corresponding to the state pair represented by the squares that are blank in Figs. 2(a) and (b), as shown in Fig. 2(c). When next state pair and present state pair are same then those state pairs can be ignored in Fig. 2(c). Similarly, when a present state pair goes to same next state then those next states can also be ignored in Fig. 2(c).

The state pair entries in Fig. 2(c), are marked × if the pair is not compatible and marked with tick (√) if the state pair is compatible.

Table 1: Not Compatible State Pairs

	00	01	11	10	
b	a, –	ⓑ,0	c, –	–, –	(b, d)
d	a, –	ⓓ,1	–, –	–, –	
d	a, –	ⓓ,1	–, –	–, –	(d, g)
g	a, –	ⓖ,0	–, –	–, –	

Table 2: Compatible State Pairs

	00	01	11	10	
a	ⓐ,0	b, 0	c, –	e, –	(a, b)
b	a, 0	ⓑ,0	c, –	e, –	
a	ⓐ,0	b, –	f, –	e, 0	(a, e)
e	a, 0	b, –	f, –	ⓔ,0	
b	a, –	ⓑ,0	c, –	h, 0	(b, h)
h	a, –	b, 0	c, –	ⓗ,0	
c	a, –	d, 1	ⓒ,0	h, –	(c, d)
d	a, –	ⓓ,1	c, 0	h, –	
c	a, –	d, –	ⓒ,0	h, 0	(c, h)
h	a, –	d, –	c, 0	ⓗ,0	
d	a, –	ⓓ,1	f, –	e, 0	(d, e)
e	a, –	d, 1	f, –	ⓔ,0	

	00	01	11	10	
d	a, –	ⓓ,1	–, –	h, 0	(d, h)
h	a, –	d, 1	–, –	ⓗ,0	
e	a, –	g, 0	f, –	ⓔ,0	(e, g)
g	a, –	ⓖ,0	f, –	e, 0	
f	a, –	g, 0	ⓕ,0	h, –	(f, g)
g	a, –	ⓖ,0	f, 0	h, –	
f	a, –	g, –	ⓕ,0	h, 0	(f, h)
h	a, –	g, –	f, 0	ⓗ,0	
g	a, –	ⓖ,0	–, –	h, 0	(g, h)
h	a, –	g, 0	–, –	ⓗ,0	

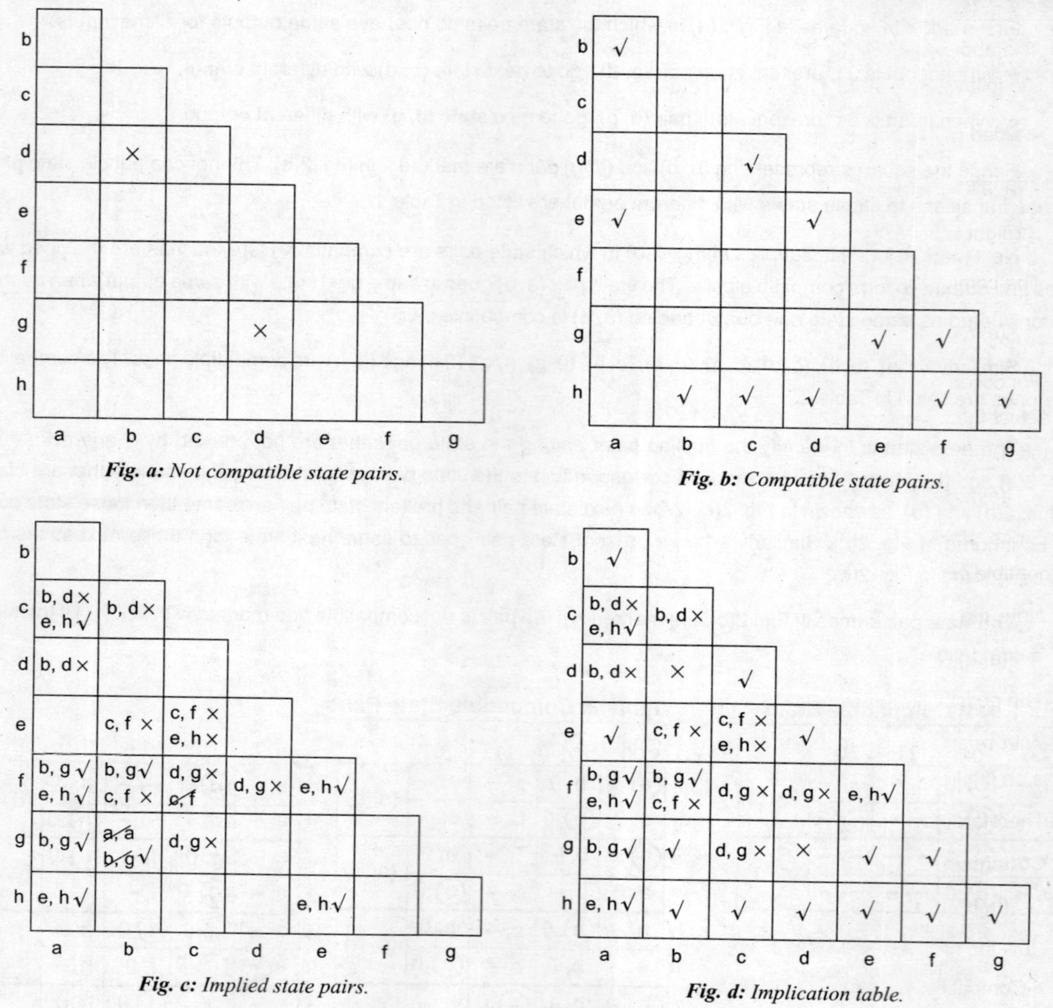

Fig. a: Not compatible state pairs.

Fig. b: Compatible state pairs.

Fig. c: Implied state pairs.

Fig. d: Implication table.

Fig. 2: *Formation of implication table.*

The three implication tables in Figs. 2(a), (b) and (c) are combined as a single implication table in Fig. 2(d) and the compatible state pairs are determined from Fig. 2(d).

When a square representing a state pair has a compatible state pair entry then that state pair is also compatible. In Fig. 2(d), the square representing (a, f), (a, g), (a, h), (e, f) and (e, h) has compatible state pair entry and so they are compatible state pairs. Here, (b, g) implies (b, g) itself and so (b, g) is also a compatible state pair.

When a square representing a state pair has not compatible state pair entry then that state pair is also not compatible. In Fig. 2(d), the squares representing (a, c), (a, d), (b, c), (b, e), (b, f), (c, e), (c, f), (c, g) and (d, f) have not compatible state pairs and so they are not compatible state pairs.

The compatible pairs obtained from implication table of Fig. 2(d) are,

(a, b), (a, e), (a, f), (a, g), (a, h), (b, g), (b, h), (c, d), (c, h), (d, e), (d, h), (e, f), (e, g), (e, h), (f, g), (f, h), (g, h)

The above compatible pairs can be merged to form maximum compatibles using a merger diagram shown in Fig. 3. The following compatible groups are obtained from merger diagram.

4-sided polygon : (e, f, g, h)

Triangles : (a, b, h), (c, d, h)

Straight lines : (a, g), (a, f), (a, e), (b, g), (e, d)

The compatible groups obtained from merger diagram are,

(e, f, g, h), (a, b, h), (c, d, h), (a, g), (a, f), (a, e), (b, g), (e, d)

For covering condition all states should be included in compatible groups.

Let us select the following compatible groups to cover all the 8 states a, b, c, d, e, f, g and h.

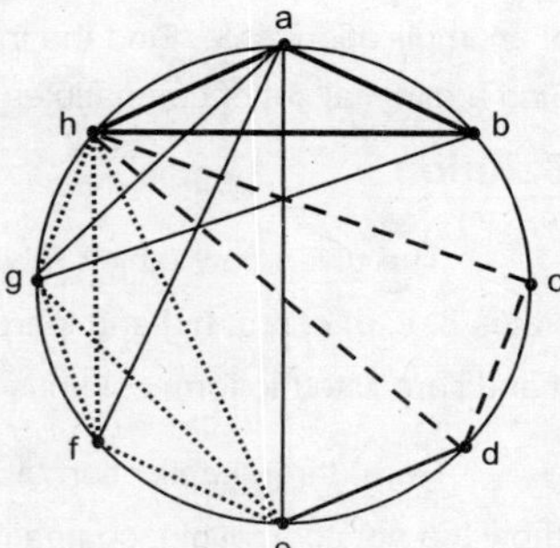

Fig. 3: Merger diagram.

(e, f, g, h), (a, b, h), (c, d, h)

Since, state h is covered in (e, f, g, h), the state h can be eliminated in the groups (a, b, h) and (c, d, h) to form the compatible pairs (a, b) and (c, d).

Now, the modified compatible groups that satisfy covering condition are,

(a, b) (c, d) (e, f, g, h)

The possible compatible pairs in the above group are (a, b), (c, d), (e, f), (e, g), (e, h), (f, g), (f, h) and (g, h).

With reference to Fig. 2(d) we can say that the compatible pairs has no implied pair except (e, f). The pair (e, f) has an implied pair (e, h) which is included in (e, f, g, h).

Therefore, the closure condition is also satisfied by the selected compatible pairs.

Conclusion: The compatible pairs that satisfy closed-covering condition are,

(a, b), (c, d), (e, f, g, h)

Now the reduced flow table can be obtained by merging the rows in the compatible group.

- Compatible group (a, b) : Rows b is merged to row a.
- Compatible group (c, d) : Row d is merged to row c.
- Compatible group (e, f, g, h) : Row f, g and h are merged to row e.

The reduced flow table with only three rows is shown in Fig. 4.

	00	01	11	10
a, b	(a),0	(b),0	c, –	e, –
c, d	a, –	(d),1	(c),0	h, –
e, f, g, h	a, –	(g),0	(f),0	(h),0

Fig. a.

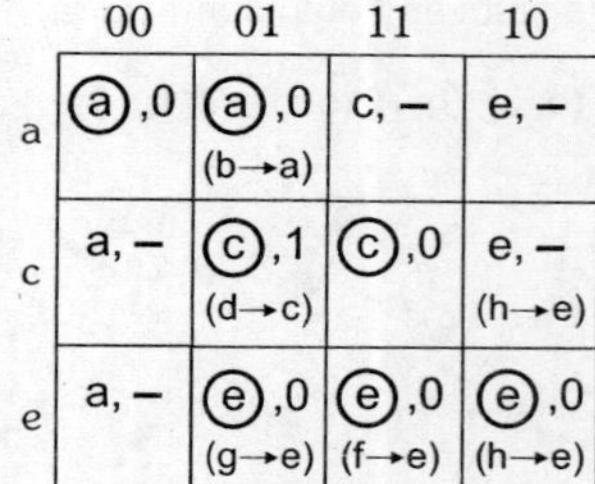

	00	01	11	10
a	(a),0	(a),0 (b→a)	c, –	e, –
c	a, –	(c),1 (d→c)	(c),0	e, – (h→e)
e	a, –	(e),0 (g→e)	(e),0 (f→e)	(e),0 (h→e)

Fig. b.

Fig. 4: Reduced flow table.

Example 4.14

Merge the primitive flow table shown in Fig. 1. Find all compatible pairs by means of an implication table. Find the maximal compatibles by means of a merger diagram. Find a minimal set of compatibles that covers all the states and is closed.

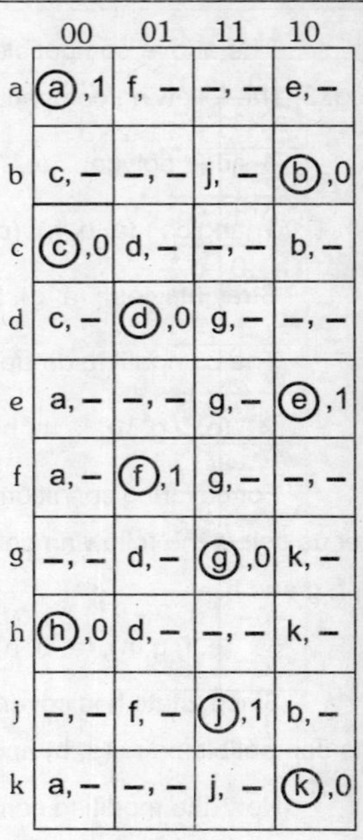

Fig. 1: Primitive flow table.

Solution

The given flow table (Table 1) has 10 states a, b, c, d, e, f, g, h, j and k. The states b, c, d, e, f, g, h, j and k are listed to form rows and the states a, b, c, d, e, f, g, h and j are listed to form columns in implication table.

Here, three implication tables (Figs. 2(a), (b) and (c)) are drawn to separately show the not compatible, compatible and implied pairs.

The given flow table has 4 combinations of inputs. Hence compatibility of states for all the four inputs should be marked in each square of implication table.

The dashes in the given flow table represents unspecified states or outputs. While identifying compatible pairs the dashes can be used to match any desired state or output.

First mark × in squares in Fig. 2(a) in which the state pairs do not have same outputs for same inputs.

- When input is 00, present state pair (a, c), go to next state (a, c) with different output.
- When input is 00, present state pair (a, h), go to next state (a, h) with different output.
- When input is 01, present state pair (d, f), go to next state (d, f) with different output.
- When input is 11, present state pair (g, j), go to next state (g, j) with different output.
- When input is 10, present state pair (b, e), go to next state (b, e) with different output.
- When input is 10, present state pair (e, k), go to next state (e, k) with different output.

Hence squares representing (a, c), (a, h), (b, e), (d, f), (e, k) and (g, j) are marked × in Fig. 2(a). The not compatible state pairs that has transitions to stable states with different output are listed in Table 1.

Next mark tick (√) in squares in Fig. 2(b) in which state pairs are compatible. Here, dashes are matched with states and outputs to form compatible pairs. The state pair (a, e) goes to same next state with same output when dashes are considered as same state and output and so (a, e) is compatible pair.

Similarly, (a, f), (b, c), (b, j), (c, d), (d, g), (e, f) and (g, h) are compatible state pairs. The compatible state pairs are listed in Table 2.

Table 1: Not Compatible State Pairs

	00	01	11	10	
a	ⓐ,1	f, –	–, –	e, –	(a, c)
c	ⓒ,0	d, –	–, –	b, –	
a	ⓐ,1	f, –	–, –	e, –	(a, h)
h	ⓗ,0	d, –	–, –	k, –	
b	c, –	–, –	j, –	ⓑ,0	(b, e)
e	a, –	–, –	g, –	ⓔ,1	
d	c, –	ⓓ,0	g, –	–, –	(d, f)
f	a, –	ⓕ,1	g, –	–, –	
e	a, –	–, –	g, –	ⓔ,1	(e, k)
k	a, –	–, –	j, –	ⓚ,0	
g	–, –	d, –	ⓖ,0	k, –	(g, j)
j	–, –	f, –	ⓙ,1	b, –	

Table 2: Compatible State Pairs

	00	01	11	10	
a	ⓐ,1	f, –	g, –	e, 1	(a, e)
e	a, 1	f, –	g, –	ⓔ,1	
a	ⓐ,1	f, 1	g, –	e, –	(a, f)
f	a, 1	ⓕ,1	g, –	e, –	
b	c, 0	d, –	j, –	ⓑ,0	(b, c)
c	ⓒ,0	d, –	j, –	b, 0	
b	c, –	f, –	j, 1	ⓑ,0	(b, j)
j	c, –	f, –	ⓙ,1	b, 0	
c	ⓒ,0	d, 0	g, –	b, –	(c, d)
d	c, 0	ⓓ,0	g, –	b, –	
d	c, –	ⓓ,0	g, 0	k, –	(d, g)
g	c, –	d, 0	ⓖ,0	k, –	
e	a, –	f, 1	g, –	ⓔ,1	(e, f)
f	a, –	ⓕ,1	g, –	e, 1	
g	h, 0	d, –	ⓖ,0	k, –	(g, h)
h	ⓗ,0	d, –	g, 0	k, –	

The next step is to identify the implied pairs among the state pairs that are not marked by × and tick (√) in Figs. 2(a) and (b). Mark the next state pairs corresponding to the state pair represented by the squares that are blank in Figs. 2(a) and (b), as shown in Fig. 2(c). When next state pair and present state pair are same then those state pairs can be ignored in Fig. 2(c). Similarly, when a present state pair goes to same next state then those next states can also be ignored in Fig. 2(c). The state pair entries in Fig. 2(c), are marked × if the pair is not compatible and marked with tick (√) if the state pair is compatible.

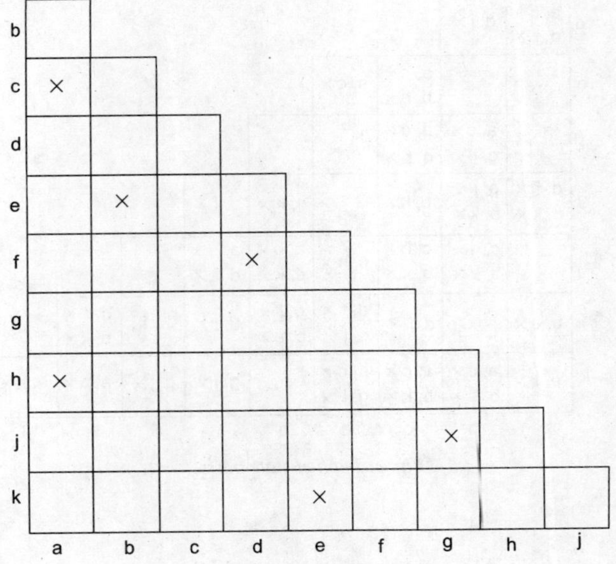

***Fig. a:** Not compatible state pairs.*

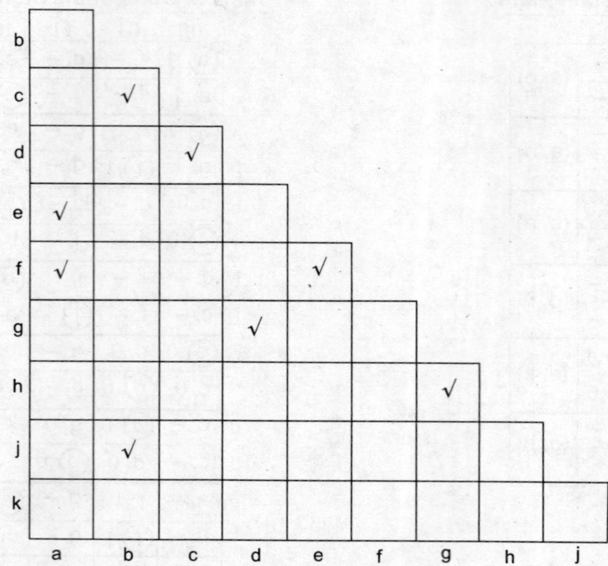

Fig. b: Compatible state pairs.

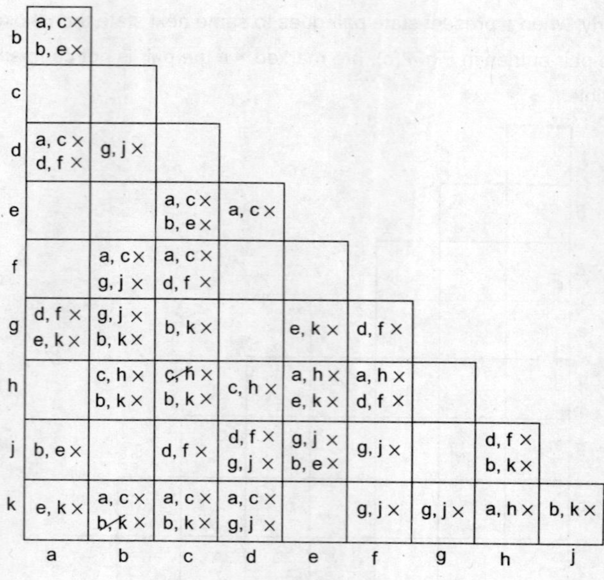

Fig. c: Implied state pairs.

b	a, c × b, e ×								
c	×	√							
d	a, c × d, f ×	g, j ×	√						
e	√	×	a, c × b, e ×	a, c ×					
f	√	a, c × g, j ×	a, c × d, f ×	×	√				
g	d, f × e, k ×	g, j × b, k ×	b, k ×	√	e, k ×	d, f ×			
h	×	c, h × b, k ×	b, k ×	c, h ×	a, h × e, k ×	a, h × d, f ×	√		
j	b, e ×	√	d, f ×	d, f × g, j ×	g, j × b, e ×	g, j ×	×	d, f × b, k ×	
k	e, k ×	a, c ×	a, c × b, k ×	a, c × g, j ×	×	g, j ×	g, j ×	a, h ×	b, k ×
	a	b	c	d	e	f	g	h	j

Fig. d: Implication table.

***Fig. 2:** Formation of implication table.*

The three implication tables in Figs. 2(a), (b) and (c) are combined as a single implication table in Fig. 2(d) and the compatible state pairs are determined from Fig. 2(d).

When a square representing a state pair has not compatible state pair entry then that state pair is also not compatible. In Fig. 2(d), the squares representing (a, b), (a, d), (a, g), (a, j), (a, k), (b, d), (b, f), (b, g), (b, h), (b, k), (c, e), (c, f), (c, g), (c, h), (c, j), (c, k), (d, e), (d, h), (d, j), (d, k), (e, g), (e, h), (e, j), (f, g), (f, h), (f, j), (f, k), (g, k), (h, j), (h, k) and (j, k) have not compatible state pairs and so they are not compatible state pairs.

The compatible pairs obtained from implication table of Fig. 2(d) are,

(a, e), (a, f), (b, c), (b, j), (c, d), (d, g), (e, f), (g, h)

The above compatible pairs can be merged to form maximum compatibles using a merger diagram shown in Fig. 3. The following compatible groups are obtained from merger diagram.

Triangle : (a, e, f)

Straight lines : (b, j), (b, c), (c, d), (d, g), (g, h)

The compatible groups obtained from merger diagram are,

(a, e, f), (b, j), (b, c), (c, d) (d, g), (g, h), (k)

For covering condition all states should be included in compatible groups. Let us select the following compatible groups to cover all the 10 states a, b, c, d, e, f, g, h, j and k.

(a, e, f), (b, j), (c, d), (g, h), (k)

The possible compatible pairs in the above group are (a, e), (a, f), (e, f), (b, j), (c, d) and (g, h). With reference to Fig. 2(d) we can say that the compatible pairs has no implied pair. Therefore, the closure condition is also satisfied by the selected compatible pairs.

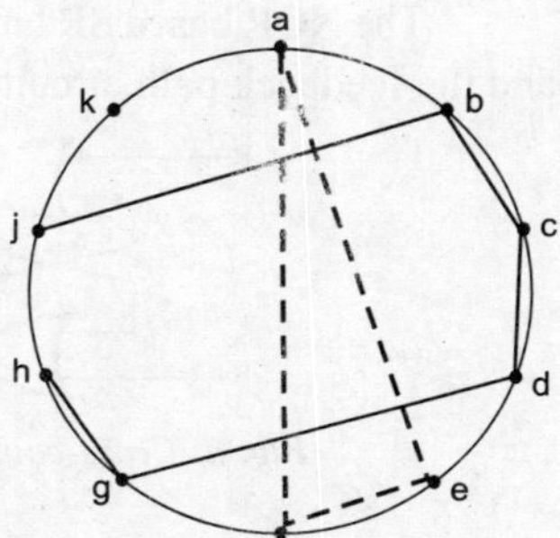

***Fig. 3:** Merger diagram.*

Conclusion: The compatible pairs that satisfy closed-covering condition are,

(a, e, f), (b, j), (c, d), (g, h), (k)

Now the reduced flow table can be obtained by merging the rows of each compatible group. Here, the following row mergers are made to obtain reduced flow table.

- Compatible group (a, e, f) : Rows e and f are merged to row a.
- Compatible group (b, j) : Row j is merged to row b.
- Compatible group (c, d) : Row d is merged to row c.
- Compatible group (g, h) : Row h is merged to row g.

The reduced flow table with five rows is shown in Fig. 4.

	00	01	11	10
a, e, f	(a),1	(f),1	g, –	(e),1
b, j	c, –	f, –	(j),1	(b),0
c, d	(c),0	(d),0	g, –	b, –
g, h	(h),0	d, –	(g),0	k, –
k	a, –	–,–	j, –	(k),0

Fig. a.

	00	01	11	10
a	(a),1	(a),1 (f→a)	g, –	(a),1 (e→a)
b	c, –	a, – (f→a)	(b),1 (j→b)	(b),0
c	(c),0	(c),0 (d→c)	g, –	b, –
g	(g),0 (h→g)	c, – (d→c)	(g),0	k, –
k	a, –	–,–	b, – (j→b)	(k),0

Fig. b.

Fig. 4: Reduced flow table.

4.5 Asynchronous Circuits with SR Latches

The latches are used as memory elements in asynchronous circuits. Therefore, in an n-variable asynchronous circuit, n number of SR latches are employed to store the n variables. There are two types of SR latches: one type is constructed using two cross coupled NOR gate and another type is constructed using two cross coupled NAND gates. Both these two latches has two inputs S and R. The working of SR latch is discussed in detail in Chapter-3, Section 3.2.1.

4.5.1 NOR Based SR Latch

The NOR based SR latch is constructed using two cross coupled NOR gates as shown in Fig. 4.15(a) and the feedback path circuit is shown in Fig. 4.15(b).

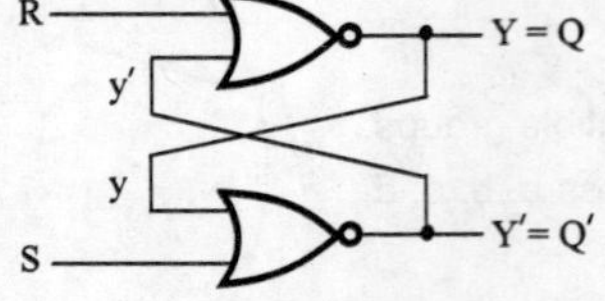

Fig. a: Cross-coupled NOR gates.

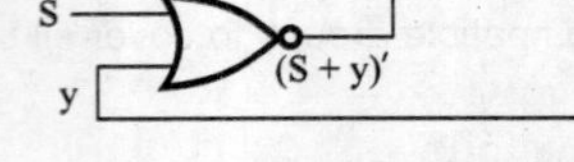

Fig. b: Feedback circuit.

Fig. 4.15: SR latch using NOR gates.

Let, y = Present state

Y = Next state

From the feedback circuit of Fig. 4.15(b), we get the following equation for next state or output of NOR based SR latch.

$$Y = (R + (S + y)')' = R'(S + y) = R'S + R'y$$

$$\therefore Y = SR' + R'y$$

Truth Table of NOR Based SR Latch

Inputs		Outputs	
S	R	Q	Q'
0	0	No change	
0	1	0	1
1	0	1	0
1	1	Indeterminate	

From the truth table of NOR based SR latch we can say that both S and R should not be equal to 1 simultaneously and this can be transferred to a condition that logical AND of S and R should not be 1 or should be always 0. Therefore, in NOR based SR latch, S . R = 0.

Consider the Boolean expression, SR' + SR.

Now, SR' + SR = S(R' + R)

Since R' + R = 1, we can write, SR' + SR = S

From the above equation we can say that if SR = 0, then SR' = S.

On substituting, SR' = S in the next state equation of SR latch we get the following simplified equation for NOR based SR latch.

$$Y = SR' + R'y \quad \Rightarrow \quad Y = S + R'y$$

(Simplified equation of SR latch)

The simplified equation of SR latch can be used to analyze and design the asynchronous circuit using NOR based SR latch when SR = 0.

4.5.2 NAND Based SR Latch

The NAND based SR latch is constructed using two cross coupled NAND gates as shown in Fig. 4.16(a) and the feedback path circuit is shown in Fig. 4.16(b).

Let, y = Present state

Y = Next state

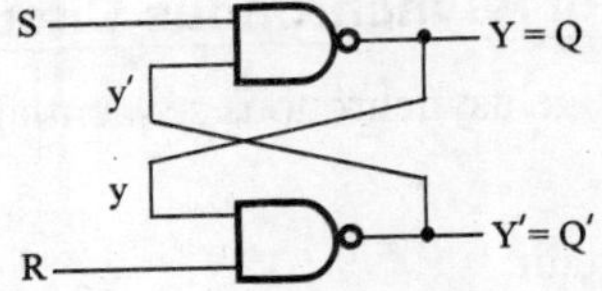

Fig. a: Cross-coupled NAND gates.

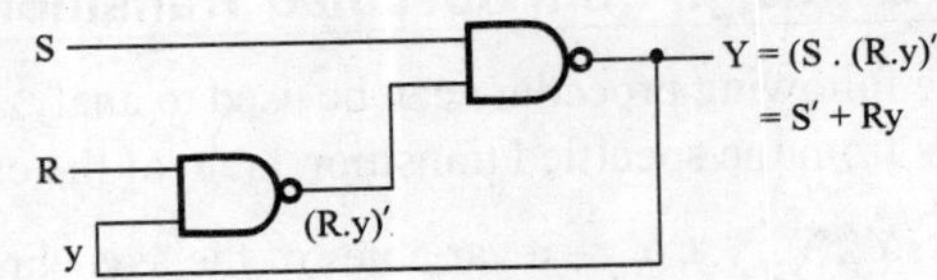

Fig. b: Feedback circuit.

Fig. 4.16: SR latch using NAND gates.

Truth Table of NAND Based SR Latch

Inputs		Outputs	
S	R	Q	Q'
0	0	Indeterminate	
0	1	1	0
1	0	0	1
1	1	No change	

From the truth table of NAND based SR latch we can say that both S and R should not be equal to 0 simultaneously. This can be transferred to a condition that logical AND of S' and R' should not be 1 or should be always 0. Therefore, in NAND based SR latch, S' . R' = 0. From the feedback circuit of Fig. 4.16(b), we get the following equation for next state or output of NAND based SR latch.

$$Y = S' + Ry$$

On comparing the above equation with NOR based SR latch we can observe that S is replaced by S' and R' is replaced by R and so the above equation will be valid only when S'R'=0. Therefore, the above equation can be used to analyze and design the asynchronous circuit using NAND based SR latch when S'R' = 0.

4.5.3 Design using SR Latches

The design of asynchronous circuits can start with either Boolean equations of asynchronous circuit or with transition table.

Method-1: Design from Specified Boolean Equations of Asynchronous Circuit

The following procedure can be used to analyze and design asynchronous sequential circuits using SR latches from the specified Boolean equations of the circuit.

Let, $Y_1, Y_2, \ldots Y_n$ = n variables of the asynchronous circuit.

Now, $y_1, y_2, \ldots y_n$ are present state variables and $Y_1, Y_2, \ldots Y_n$ are next state variables.

In order to store n variables, choose n number of NOR or NAND based SR latches.

Determine the Boolean functions for the R and S input of each latch by comparing the given equations with following next state equation of SR latch.

For NOR based SR latch: $Y_j = S_j + R_j' y_j$; for j = 1, 2, … n.

For NAND based SR latch: $Y_j = S_j' + R_j y_j$; for j = 1, 2, … n.

When NOR based SR latch is selected, check for $S_j R_j = 0$ for j = 1, 2, … n.

When NAND based SR latch is selected, check for $S_j' R_j' = 0$ for for j = 1, 2, … n.

When the above condition is not satisfied, the circuit designed using given equation may not work properly and so design is not possible. When the above condition is satisfied, draw the logic circuit using equations of Y_j for j = 1, 2, … n.

The asynchronous circuit can be analyzed for cycles and races by constructing transition table using the equations of Y_j for j = 1, 2, … n.

Method-2: Design from Specified Transition Table of Asynchronous Circuit

The following procedure can be used to analyze and design asynchronous sequential circuits using SR latches from the specified transition table of the circuit.

Let, $Y_1, Y_2, \ldots Y_n$ = n variables of the asynchronous circuit.

Now, $y_1, y_2, \ldots y_n$ are present state variables and $Y_1, Y_2, \ldots Y_n$ are next state variables.

In order to store n variables, choose n number of NOR or NAND based SR latches.

When the states are denoted by letter symbols, assign race-free binary codes for states.

The design starts from the transition table with binary codes for states. The entries in the squares of transition tables are binary values for next states $Y_1, Y_2, \ldots, Y_n$. From the knowledge of present state and next state values, the excitation table of SR latch is used to determine the binary values of S and R inputs of all SR latches and list in a table.

Then construct K-map for every S and R inputs and determine the simplified Boolean equations for latch inputs.

It can be observed that the S and R inputs in the excitation table of NOR and NAND latches are complements. The equations obtained from K-map for NOR implemented can be complemented and used for NAND implementation.

Table 4.10: Excitation Table of NOR based SR Latch

Present State	Next State	Latch Inputs	
y	Y	S	R
0	0	0	×
0	1	1	0
1	0	0	1
1	1	×	0

Table 4.11: Excitation Table of NAND based SR Latch

Present State	Next State	Latch Inputs	
y	Y	S	R
0	0	1	×
0	1	0	1
1	0	1	0
1	1	×	1

When NOR based SR latch is selected, check for $S_j R_j = 0$ for $j = 1, 2, \ldots n$.

When NAND based SR latch is selected, check for $S'_j R'_j = 0$ for for $j = 1, 2, \ldots n$.

When the above condition is not satisfied, the circuit designed using given equation may not work properly and so design is not possible. When the above condition is satisfied, draw the combinational logic circuit to generate latch inputs, using equations of S_j and R_j for $j = 1, 2, \ldots n$. The complete logic circuit is given by this combinational logic circuit connected to SR latches.

Short-cut Method for Deriving SR Latch Input Equations

The following theorem is stated without proof.

In realization of a flow table with race-free state assignment using SR latches there exists equations for inputs S_i and R_i of latch-i independent of y_i.

The above theorem can be used to find equations for latch inputs from the maps of next state variables. For example, consider a flow table with two state varaibles Y_1 and Y_2. The maps of Y_1 and Y_2 are shown in Figs. 4.17 and 4.18.

From the map of Y_1 we can get equations for S_1 and R_1 independent of y_1. Hence divide the map of Y_1 into two maps: One map when $y_1 = 0$ and another map when $y_1 = 1$.

In the map of Fig. 4.17(b), the prime implicants are formed by considering adjacent 1's. The sum of prime implicants gives the equation for S_1. In the map of Fig. 4.17(c), the prime implicants are formed by considering adjacent 0's. The complement of sum of prime implicants gives the equation for R'_1.

Similarly, from the map of Y_2 we get the equations for S_2 and R_2 independent of y_2. Here, the map of Y_2 is divided into two: One map when $y_2 = 0$ and another map when $y_2 = 1$. The formation of equations for S_2 and R'_2 is similar to that of formation of equations for S_1 amd R'_1.

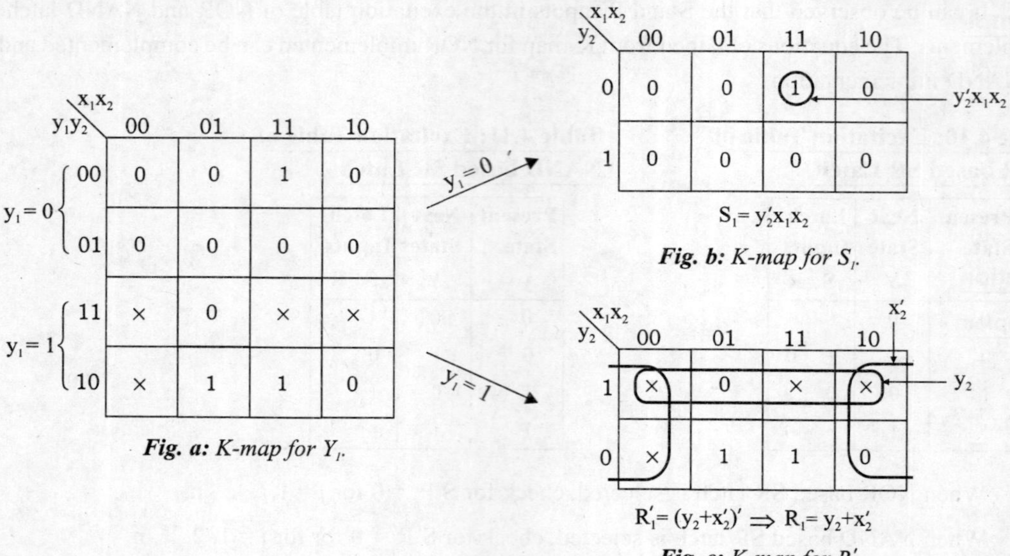

Fig. a: K-map for Y_1.

Fig. b: K-map for S_1.

Fig. c: K-map for R_1'.

Fig. 4.17: K-maps to find equations for S_1 and R_1.

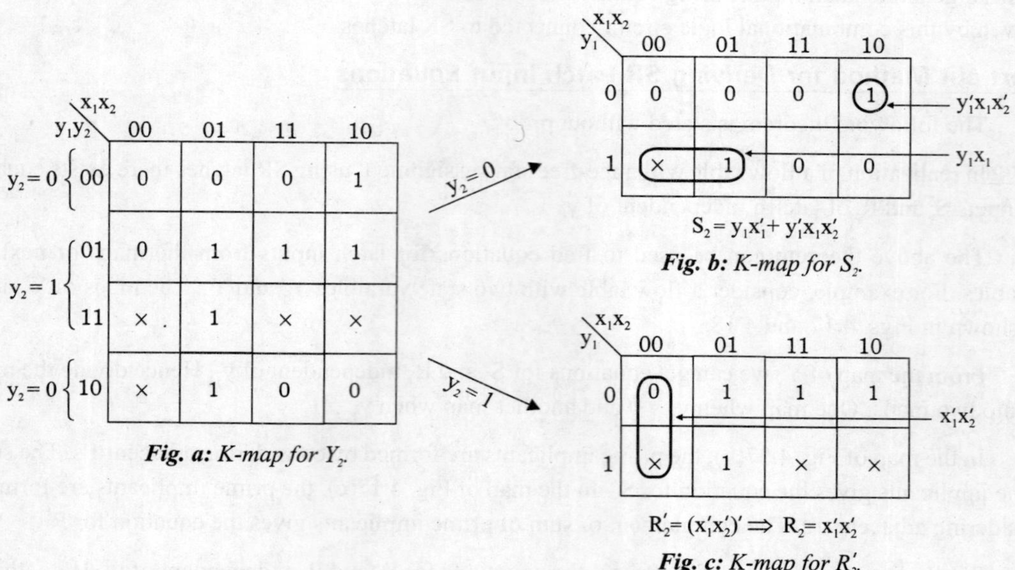

Fig. a: K-map for Y_2.

Fig. b: K-map for S_2.

Fig. c: K-map for R_2'.

Fig. 4.18: K-maps to find equations for S_2 and R_2.

Example 4.15

An asynchronous sequential circuit is described by the following excitation and output functions

$Y = x_1x_2' + (x_1 + x_2')\,y \quad ; \; Z = Y$

Implement the above equations in a logic circuit

a) using NOR latch

b) using NAND latch

Solution

a) Implementation using NOR Latch

The excitation function of the asynchronous sequential circuit is,

$$Y = x_1x_2' + (x_1 + x_2')\,y \qquad .....(1)$$

The excitation function of SR latch using NOR gate is,

$$Y = S + R'y \qquad .....(2)$$

On comparing equations (1) and (2) we get,

$S = x_1x_2'$

$R' = x_1 + x_2' \quad \Rightarrow \quad R = (x_1 + x_2')' = x_1'x_2$

Let us check for SR = 0.

$SR = x_1x_2' \, . \, x_1'x_2 = (x_1x_1') \, . \, (x_2x_2')$

$= 0.0 = 0$

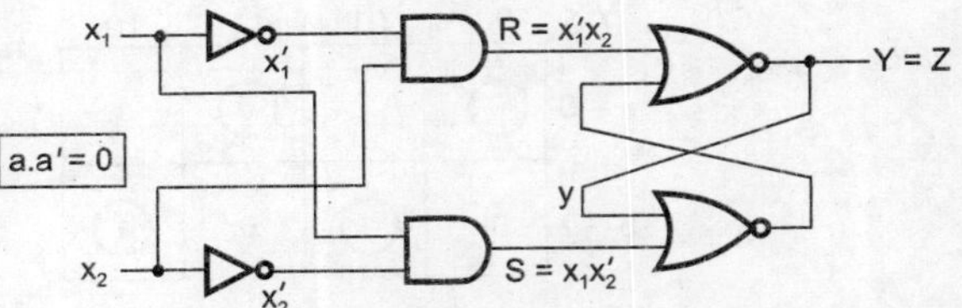

Fig. 1: Logic implementation using NOR based SR latch.

Since, SR = 0, the given equation can be implemented using NOR based SR latch as shown in Fig. 1.

b) Implementation using NAND Latch

The excitation function of the asynchronous sequential circuit is,

$$Y = x_1x_2' + (x_1 + x_2')\,y \qquad .....(3)$$

The excitation function of SR latch using NAND gate is,

$$Y = S' + Ry \qquad .....(4)$$

On comparing equations (3) and (4) we get,

$S' = x_1x_2' \quad \Rightarrow \quad S = (x_1x_2')' = x_1' + x_2$

$R = x_1 + x_2'$

Let us check for S'R' = 0.

$S'R' = x_1x_2'\,(x_1 + x_2)' = x_1x_2' \, . \, x_1'x_2 = (x_1x_1') \, . \, (x_2x_2') = 0.0 = 0$

a.a' = 0

Since, S'R' = 0, the given equation can be implemented using NAND based SR latch as shown in Fig. 2.

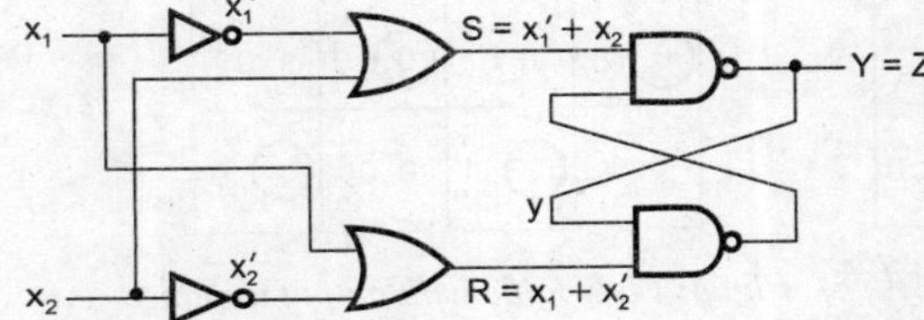

Fig. 2: Logic implementation using NAND based SR latch.

Example 4.16

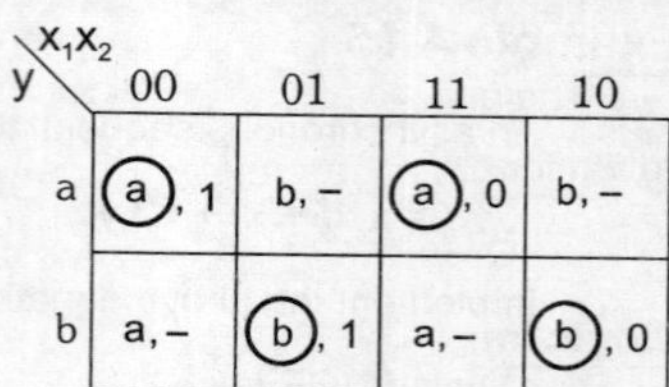

a) Obtain a binary state assignment for the reduced flow table shown in Fig. 1 to avoid critical race condition.

b) Assign output to unstable states and obtain the output equation.

c) Implement the logic circuit using NOR based SR latch.

d) Implement the logic circuit using NAND based SR latch.

y \ x_1x_2	00	01	11	10
a	(a), 1	b, –	(a), 0	b, –
b	a, –	(b), 1	a, –	(b), 0

Fig. 1: Reduced flow table.

Solution

a) Race-Free State Assignment

The given flow table has only two states a and b. Therefore, state transitions can occur either from a to b or b to a. For two states, 1-bit state assignment is sufficient and in 1-bit state assignment there will not be any race condition.

Let us assign binary code 0 to state a and binary code 1 to state b. The transition and flow table with binary code assignment is shown in Fig. 2.

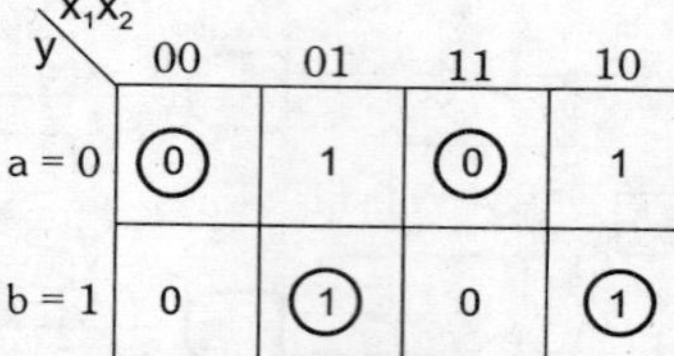

y \ x_1x_2	00	01	11	10
a = 0	(0)	1	(0)	1
b = 1	0	(1)	0	(1)

Fig. a: Transition table.

y \ x_1x_2	00	01	11	10
a = 0	(0), 1	1, –	(0), 0	1, –
b = 1	0, –	(1), 1	0, –	(1), 0

Fig. b: Flow table.

Fig. 2: Transition and flow table with code assignment.

b) To Assign Output and Determine the Output Equation

In flow table, the dashes in the output are assigned values according to the following rules to prevent transient output during a transition.

Rule 1 : Assign 0 to an output of an unstable state which is transient state during a transition between two stable states that have a 0 as output.

Rule 2 : Assign 1 to an output of an unstable state which is transient state during a transition between two stable states that have a 1 as output.

Rule 3 : Assign a don't-care condition to an output of an unstable state which is transient state during a transition between two stable states that have a different values (0 and 1 or 1 and 0) as output.

The state transitions that are used to assign output using the above rules are shown in Fig. 3 and the flow table with output assigned to unstable states is shown in Fig. 4. The output map of Z is shown in Fig. 5.

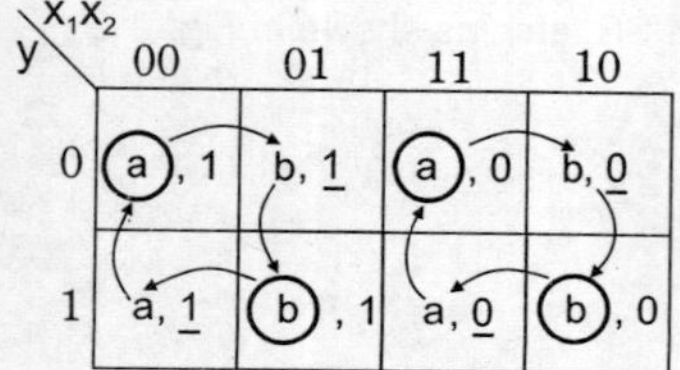

y \ x_1x_2	00	01	11	10
0	(a), 1	b, 1	(a), 0	b, 0
1	a, 1	(b), 1	a, 0	(b), 0

Fig. 3: State transitions used to assign output.

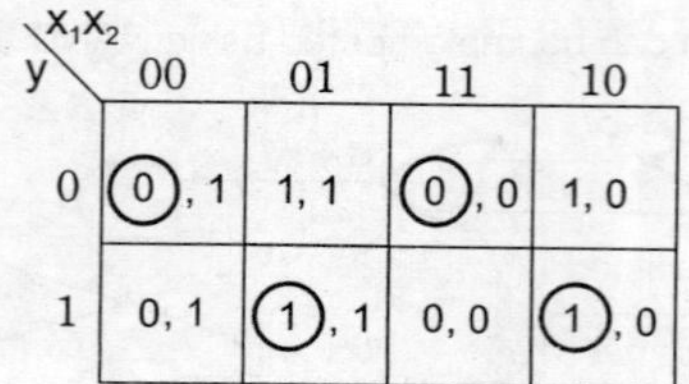

y \ x_1x_2	00	01	11	10
0	(0), 1	1, 1	(0), 0	1, 0
1	0, 1	(1), 1	0, 0	(1), 0

Fig. 4: Flow table with output assigned to unstable states.

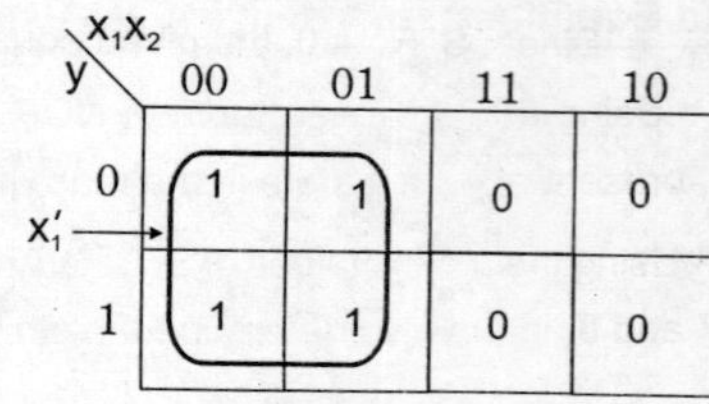

y \ x_1x_2	00	01	11	10
0	1	1	0	0
1	1	1	0	0

Fig. 5: Output map.

Let, Z = output

The output map is constructed as shown in Fig. 8 and from output map we get the following simplified Boolean equation for output.

Output, $Z = x_1'$

c) Logic Implementation using NOR based SR Latch

The binary values inside each square of transition table is the value of next state Y. Hence, using the transition table the map of Y can be drawn as shown in Fig. 6.

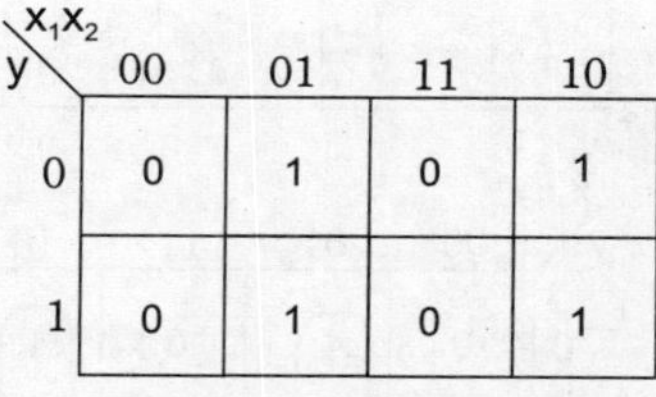

Fig. 6: K-map for Y.

In order to store the one state variable Y, one SR latch is required. Let us choose one NOR based SR latch.

Let, S, R = Inputs for SR latch

To find Equations for Latch Inputs using Map of Y

In this method the equations for S and R that are independent of y can be obtained from map of Y. The map for y = 0 is used to find equation for S by forming prime implicants using adjacent 1's. The map for y = 1 is used to find equation for R′ by forming prime implicants using adjacent 0's. Hence the map of Y is divided into two: One for y = 0 and another for y = 1 as shown in Fig. 7, to form map of S and R′.

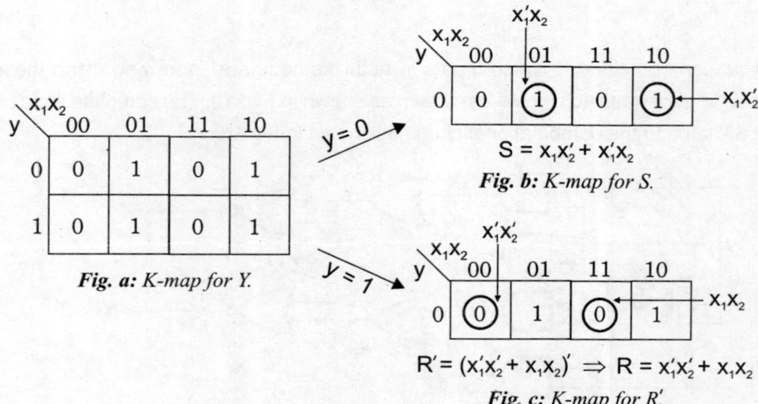

Fig. a: K-map for Y.

Fig. 7: K-maps to find equations for S and R.

Form the maps of S and R′ we get the following equations for SR latch inputs.

$S = x_1x_2' + x_1'x_2$

$R = x_1'x_2' + x_1x_2$

To find Equations for Latch Inputs using Excitation Table of SR Latch

Using the excitation table of NOR based SR latch, the inputs for SR latch can be obtained from the circuit inputs, present and next state in the map of Y and listed in Table 1.

Using the binary values of SR latch inputs listed in Table 1, the K-map for latch inputs are constructed as shown in Figs. 7 and 8, in order to determine the simplified equations for latch inputs.

Table 1: Inputs of SR Latch

y	$x_1x_2 = 00$				$x_1x_2 = 01$				$x_1x_2 = 11$				$x_1x_2 = 10$			
	y	Y	S	R	y	Y	S	R	y	Y	S	R	y	Y	S	R
0	0	0	0	×	0	1	1	0	0	0	0	×	0	1	1	0
1	1	0	0	1	1	1	×	0	1	0	0	1	1	1	×	0

Excitation Table of NOR based SR Latch

Present State y	Next State Y	Latch Inputs S	R
0	0	0	×
0	1	1	0
1	0	0	1
1	1	×	0

y \ x_1x_2	00	01	11	10
0	0	1	0	1
1	0	×	0	×

Groups marked: $x_1'x_2$, x_1x_2'

Fig. 8: K-map for S.

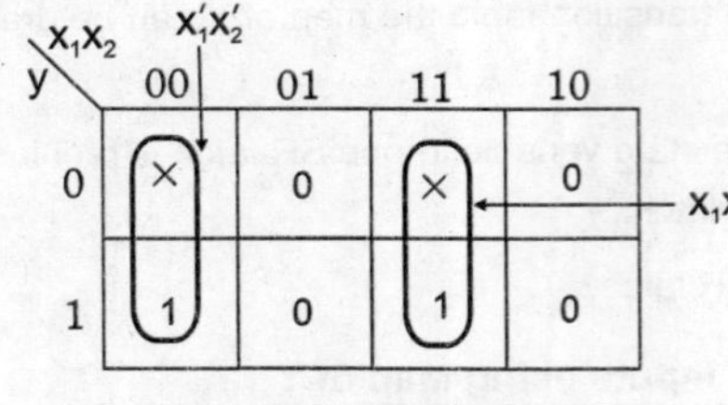

Fig. 9: K-map for R.

From the K-maps shown in Figs. 8 and 9, we get the following simplified equations for SR latch inputs. The output equation is obtained from output map.

$$S = x_1x_2' + x_1'x_2 \quad ; \quad R = x_1'x_2' + x_1x_2 \quad ; \quad Z = x_1'$$

Logic Diagram

The Boolean equations obtained for latch inputs in both methods are same and using these equations the combinational logic circuit to generate latch inputs are drawn as shown in Fig. 10. The complete logic circuit is obtained by adding NOR based SR latch to the combinational circuit as shown in Fig. 10.

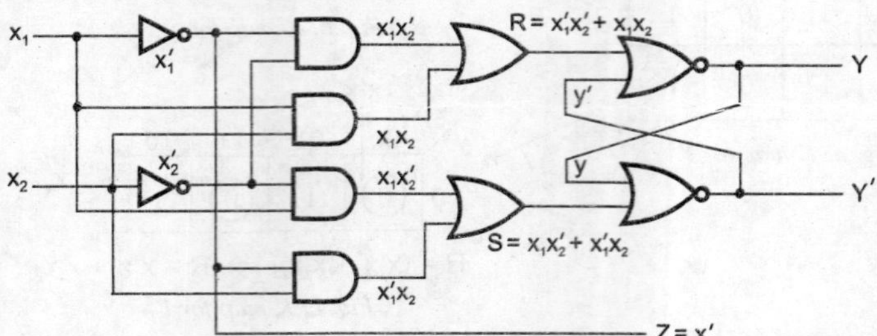

Fig. 10: Logic circuit using NOR based SR latch.

d) Logic Implementation using NAND based SR Latch

The Boolean equations to generate inputs for NAND based SR latches are given by complemented of the equations for NOR based SR latch.

Therefore, the boolean equations to generate inputs for NAND based SR latches are,

$$S = (x_1x_2' + x_1'x_2)' = (x_1x_2')' \cdot (x_1'x_2)' = (x_1' + x_2) \cdot (x_1 + x_2')$$

$$R = (x_1'x_2' + x_1x_2)' = (x_1'x_2')' \cdot (x_1x_2)' = (x_1 + x_2) \cdot (x_1' + x_2')$$

$$Z = x_1'$$

The output equation is obtained from output map. Using the above equations the combinational logic circuit to generate latch inputs are drawn as shown in Fig. 11. The complete logic is obtained by adding NAND based SR latch to the combinational circuit as shown in Fig. 11.

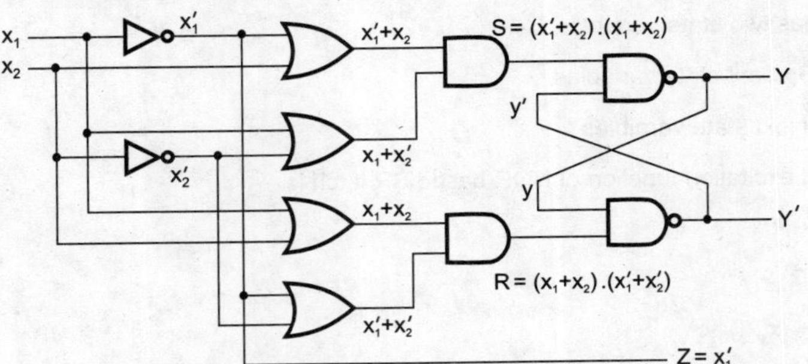

Fig. 11: Logic circuit using NAND based SR latch.

Example 4.17

The Boolean functions for the S and R iinputs of SR latch based asynchronous logic circuit are,

$S_1 = x_1 y_2$ | $S_2 = x_1 x_2$

$R_1 = x_1' x_2'$ | $R_2 = x_2' y_1$

Analyze the circuit and find the stabilitiy and race condition.

Solution

The given Boolean equations are,

$S_1 = x_1 y_2$ | $S_2 = x_1 x_2$

$R_1 = x_1' x_2'$ | $R_2 = x_2' y_1$

Let us check for $S_1 R_1 = 0$ and $S_2 R_2 = 0$

$$S_1 R_1 = x_1 y_2 \,.\, x_1' x_2 = (x_1 x_1') \,.\, y_2 x_2' = 0 \,.\, y_2 x_2' = 0$$ a.a′ = 0

$$S_2 R_2 = x_1 x_2 \,.\, x_2' y_1 = x_1 \,.\, (x_2 x_2') \,.\, y_1 = x_1 . 0 . y = 0$$

Since the given equations satisfy the condition SR = 0, the equations can be implemented by a logic circuit using NOR based SR latch as shown in Fig. 1.

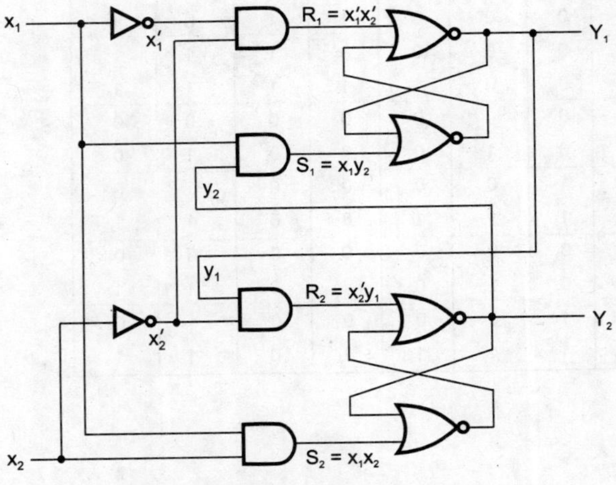

Fig. 1: Circuit with SR latch.

The circuit has two state variables.

Let, $y_1\ y_2$ = Present state variables

Y_1Y_2 = Next state varaibles

The reduced excitation function of NOR based SR latch is,

$$Y = S + R'y$$

$$\therefore\ Y_1 = S_1 + R_1'\,y_1$$

$$= x_1y_2 + (x_1'x_2')'y_1$$

$$= x_1y_2 + (x_1 + x_2)\,y_1$$

$$= x_1y_2 + x_1y_1 + x_2y_1$$

$$Y_2 = S_2 + R_2'y_2$$

$$= x_1x_2 + (x_2'y_1)'\,y_2$$

$$= x_1x_2 + (x_2 + y_1')\,y_2$$

$$= x_1x_2 + x_2y_2 + y_1'y_2$$

Using the above two equations the next state variables are evaluated for all possible choice of present state and inputs and listed in Table 1. Using Table1 the maps for Y_1 and Y_2 are drawn as shown in Figs. 2 and 3, and they are combined to form transition table as shown in Fig. 4.

Table 1: Evaluation of Excitation Variables and Outputs

Present States				Product Terms						Next States	
y_1	y_2	x_1	x_2	x_1y_2	x_1y_1	x_2y_1	x_1x_2	x_2y_2	$y_1'y_2$	Y_1	Y_2
0	0	0	0	0	0	0	0	0	0	0	0
0	0	0	1	0	0	0	0	0	0	0	0
0	0	1	0	0	0	0	0	0	0	0	0
0	0	1	1	0	0	0	1	0	0	0	1
0	1	0	0	0	0	0	0	0	1	0	1
0	1	0	1	0	0	0	0	1	1	0	1
0	1	1	0	1	0	0	0	0	1	1	1
0	1	1	1	1	0	0	1	1	1	1	1
1	0	0	0	0	0	0	0	0	0	0	0
1	0	0	1	0	0	1	0	0	0	1	0
1	0	1	0	0	1	0	0	0	0	1	0
1	0	1	1	0	1	1	1	0	0	1	1
1	1	0	0	0	0	0	0	0	0	0	0
1	1	0	1	0	0	1	0	1	0	1	1
1	1	1	0	1	1	0	0	0	0	1	0
1	1	1	1	1	1	1	1	1	0	1	1

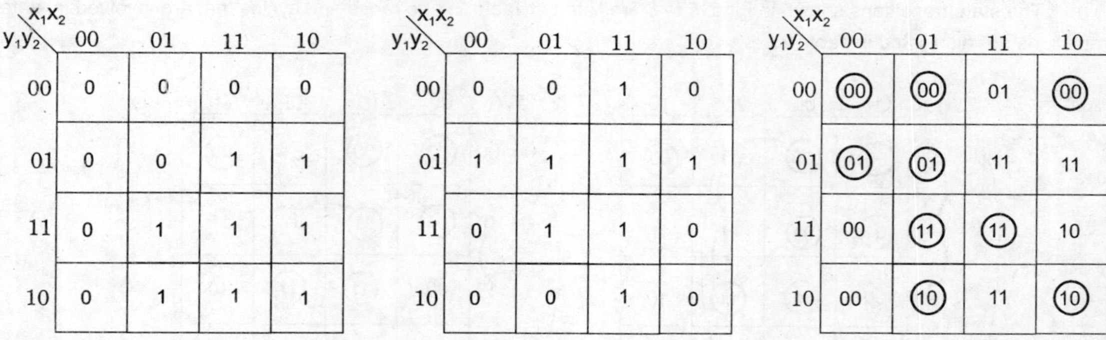

y_1y_2 \ x_1x_2	00	01	11	10
00	0	0	0	0
01	0	0	1	1
11	0	1	1	1
10	0	1	1	1

***Fig. 2:** K-map for Y_1.*

y_1y_2 \ x_1x_2	00	01	11	10
00	0	0	1	0
01	1	1	1	1
11	0	1	1	0
10	0	0	1	0

***Fig. 3:** K-map for Y_2.*

y_1y_2 \ x_1x_2	00	01	11	10
00	(00)	(00)	01	(00)
01	(01)	(01)	11	11
11	00	(11)	(11)	10
10	00	(10)	11	(10)

***Fig. 4:** Transition table.*

Stability Condition

In the transition table of Fig. 4, it can be observed that every column has at least one stable state, so that the circuit remain in a stable state for every possible input and so the circuit is stable.

State Transitions due to Change in Input

State transitions occurs only when there is a change in input. Normally, when the input is stable or when no change in input the circuit remain in one of the stable state corresponding to the input.

Here, there are four possible combinations of inputs. When the circuit remain in a stable state at any one input then state transitions will occur when the input changes to other possible inputs with one bit change.

When input is 00, the circuit remain in either state 00 or 01. The input can change from 00 to 01 or 00 to 10 and the corresponding state transitions will occur as shown in Fig. 5.

When input is 01, the circuit remain in either state 00 or 01. The input can change from 01 to 11 or 01 to 00 and the corresponding state transitions will occur as shown in Fig. 6.

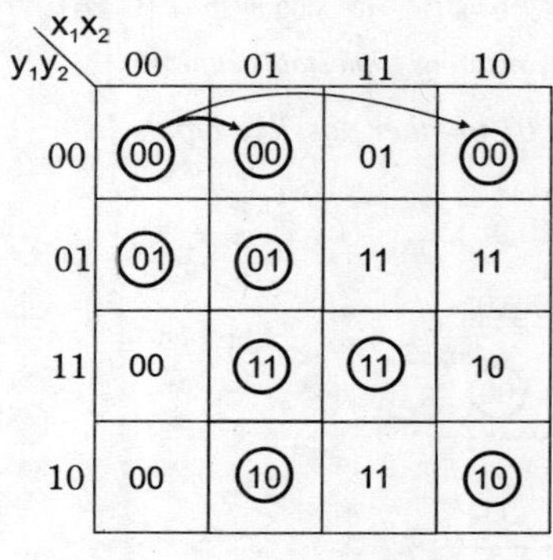

y_1y_2 \ x_1x_2	00	01	11	10
00	(00)	(00)	01	(00)
01	(01)	(01)	11	11
11	00	(11)	(11)	10
10	00	(10)	11	(10)

00 → 00 ; 00 → 00

***Fig.a:** Transitions from stable state 00.*

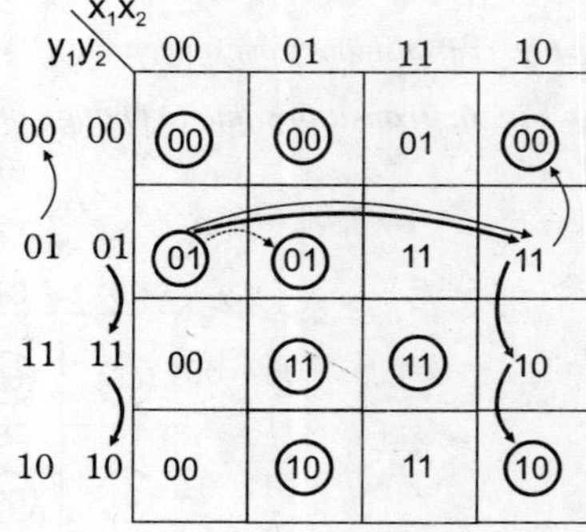

	y_1y_2 \ x_1x_2	00	01	11	10
00	00	(00)	(00)	01	(00)
01	01	(01)	(01)	11	11
11	11	00	(11)	(11)	10
10	10	00	(10)	11	(10)

01 → 01 ; 01 → 11 → 10 ; 01 → 00

***Fig.b:** Transitions from stable state 01.*

***Fig. 5:** Transitions due to change in input from 00 to other possible inputs.*

When input is 11, the circuit remain in state 11. The input can change from 11 to 10 or 11 to 01 and the corresponding state transitions will occur as shown in Fig. 7.

When input is 10, the circuit remain in state 00. The input can change from 10 to 00 or 10 to 11 and the corresponding state transitions will occur as shown in Fig. 8.

The state transitions shown in Figs. 5 to 8 are listed in Table 2. The races and cycles that are involved in state transitions are also listed in Table 2.

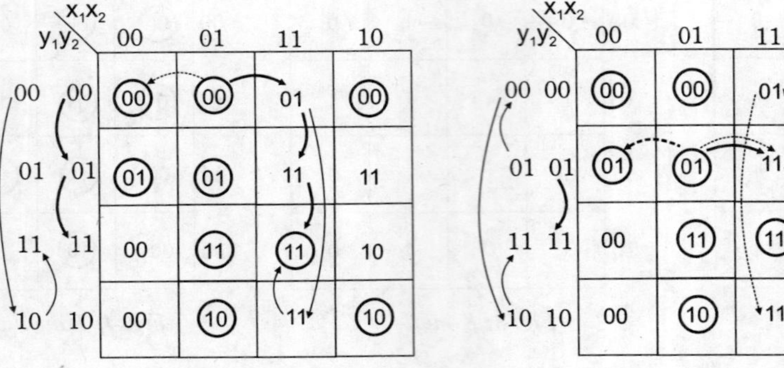

00 → 00 ; 00 → 01 → 11 ; 00 → 10 → 11

Fig. a: Transitions from stable state 00.

01 → 11 ; 01 → 00 → 10 → 11 ; 01 → 01

Fig. b: Transitions from stable state 01.

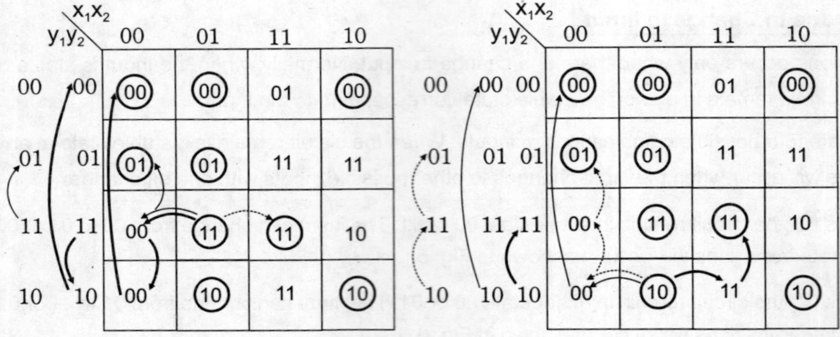

11 → 11 ; 11 → 10 → 00 ; 11 → 01

Fig. c: Transitions from stable state 11.

10 → 11 ; 10 → 00 ; 10 → 11 → 01

Fig. d: Transitions from stable state 10.

***Fig. 6:** Transitions due to change in input from 01 to other possible inputs.*

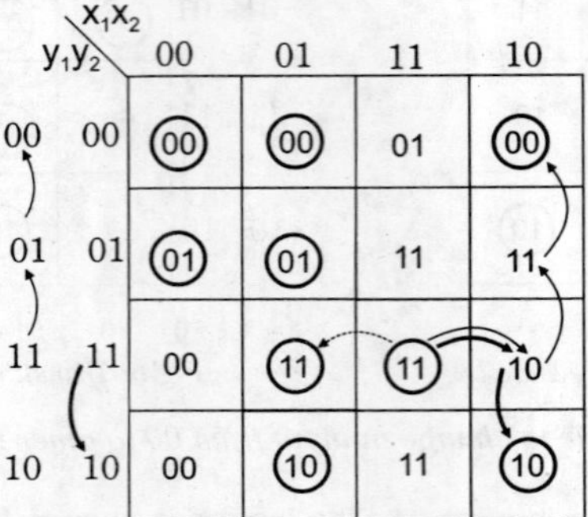

11 → 11 ; 11 → 10 ; 11 → 01 → 00

***Fig. 7:** Transitions due to change in input from 11 to other possible inputs.*

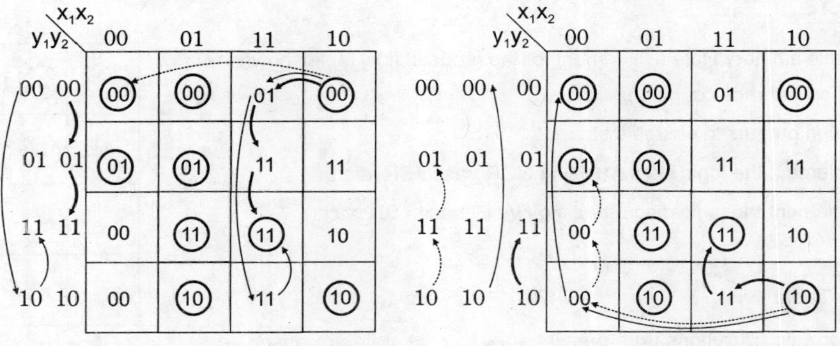

00 → 00 ; 00 → 01 → 11 ; 00 → 10 → 11

Fig. a: *Transitions from stable state 00.*

10 → 11 ; 10 → 00 ; 10 → 11 → 01

Fig. b: *Transitions from stable state 10.*

Fig. 8: *Transitions due to change in input from 10 to other possible inputs.*

Table 2: List of Races and Cycles

Present Input	Present Stable State	Change in Input	State Transitions	Race	Cycles
00	00	00 → 01	00 → 00	-	-
		00 → 10	00 → 00	-	-
	01	00 → 01	01 → 01	-	-
		00 → 10	01 → 11 → 10	Critical	00 → 11 → 10
			01 → 00		-
01	00	01 → 11	00 → 01 → 11	Noncritical	00 → 01 → 11
			00 → 10 → 11		00 → 10 → 11
		01 → 00	00 → 00	-	-
	01	01 → 11	01 → 11	Noncritical	-
			01 → 00 → 10 → 11		01 → 00 → 10 → 11
		01 → 00	01 → 01	-	-
	11	01 → 11	11 → 11	-	-
		01 → 00	11 → 10 → 00	Critical	11 → 10 → 00
			11 → 01		-
	10	01 → 11	10 → 11	-	-
		01 → 00	10 → 00	Critical	-
			10 → 11 → 01		10 → 11 → 01
11	01	11 → 01	11 → 11	-	-
		11 → 10	11 → 10	Critical	-
			11 → 01 → 00		11 → 01 → 00
10	00	10 → 11	00 → 01 → 11	Noncritical	00 → 01 → 11
			00 → 10 → 11		00 → 10 → 11
		10 → 00	00 → 00	-	-
	10	10 → 11	10 → 11	-	-
		10 → 00	10 → 00	Critical	-
			10 → 11 → 01		10 → 11 → 01

Example 4.18

a) Obtain a binary state assignment for the reduced flow table shown in Fig. 1 to avoid critical race condition.

b) Assign outputs to unstable states.

c) Implement the logic diagram using NOR based SR latch.

d) Implement the logic diagram using NAND based SR latch.

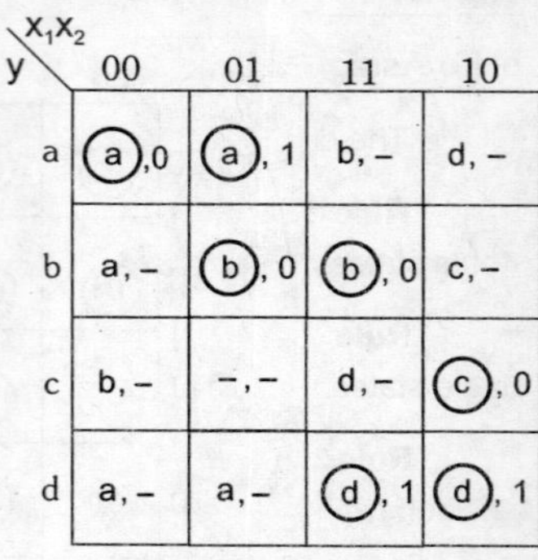

y \ x_1x_2	00	01	11	10
a	(a),0	(a), 1	b, –	d, –
b	a, –	(b), 0	(b), 0	c, –
c	b, –	–, –	d, –	(c), 0
d	a, –	a, –	(d), 1	(d), 1

Fig. 1: Reduced flow table.

Solution

a) Transition Table

The following transitions from present state to next state are observed in transition table in Fig. 2.

From a	From b	From c	From d
a → b	b → a	c → b	d → a
a → d	b → c	c → d	

The above transitions are combined as a single diagram in the transition diagram shown in Fig. 3.

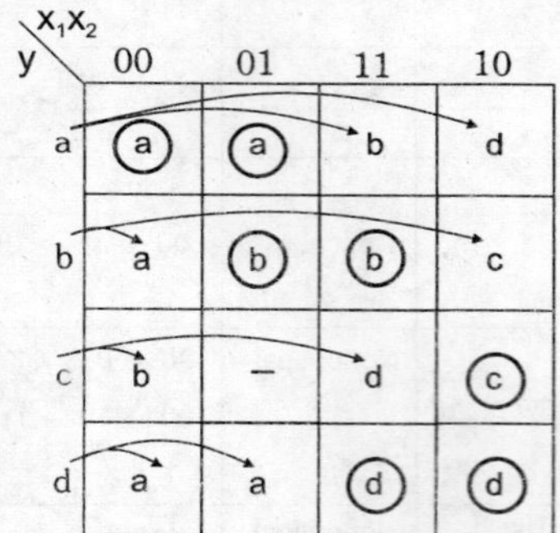

y \ x_1x_2	00	01	11	10
a	(a)	(a)	b	d
b	a	(b)	(b)	c
c	b	–	d	(c)
d	a	a	(d)	(d)

Fig. 2: Transitions from present to next state.

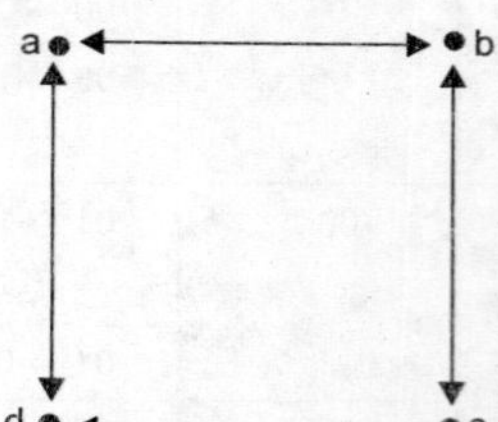

Fig. 3: Transition diagram.

For a critical race-free state assignment the state transitions in Fig. 2 should have adjacent state assignments. Here, it is possible to achieve adjacent state assignments with the following 2-bit binary codes as shown in Fig. 4.

a → 00 | c → 11 | $y = y_1 y_2$

b → 01 | d → 10 |

Let the two bits of binary code be represented by the state variables y_1 and y_2. The transition table with binary code assignment is shown in Fig. 5.

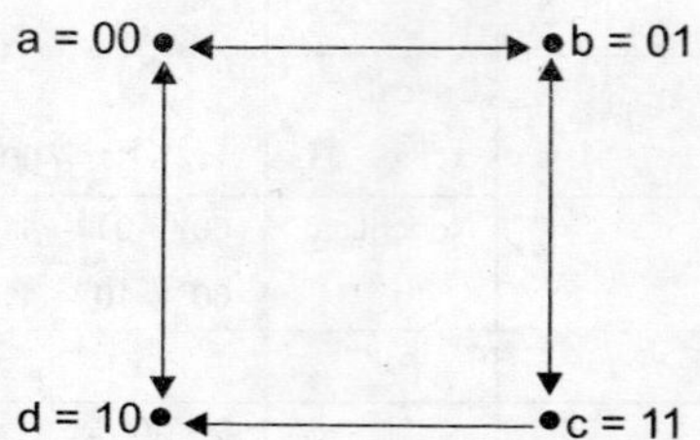

Fig. 4: Transition diagram with binary with assignment.

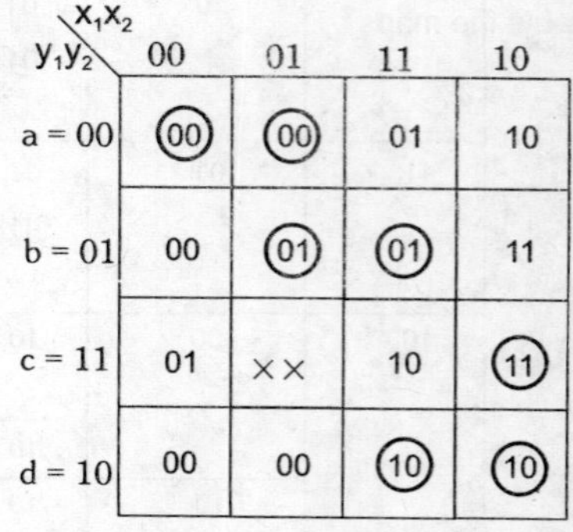

y_1y_2 \ x_1x_2	00	01	11	10
a = 00	(00)	(00)	01	10
b = 01	00	(01)	(01)	11
c = 11	01	××	10	(11)
d = 10	00	00	(10)	(10)

Fig. 5: Transition table.

b) To Assign Output for Unstable States

The dashes in the output are assigned values according to the following rules.

Rule 1 : Assign 0 to an output of an unstable state which is transient state during a transition between two stable states that have a 0 as output.

Rule 2 : Assign 1 to an output of an unstable state which is transient state during a transition between two stable states that have a 1 as output.

Rule 3 : Assign a don't-care condition to an output of an unstable state which is transient state during a transition between two stable states that have a different values (0 and 1 or 1 and 0) as output.

The state transitions that are used to assign output using the above rules are shown in Fig. 6 and the flow table with output assigned to unstable states is shown in Fig. 7.

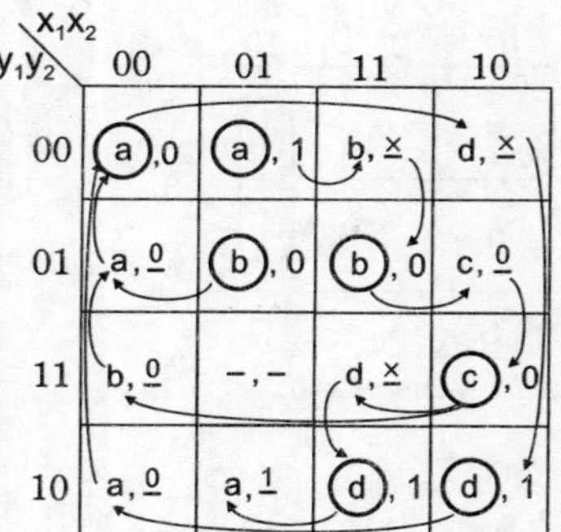

Fig. 6: State transitions used to assign output.

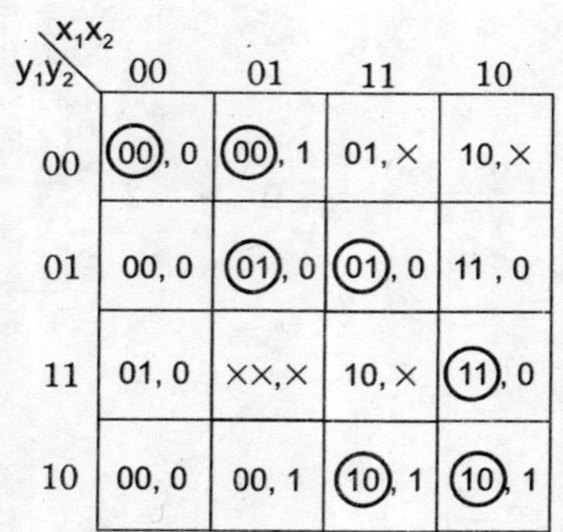

y_1y_2 \ x_1x_2	00	01	11	10
00	(00), 0	(00), 1	01, ×	10, ×
01	00, 0	(01), 0	(01), 0	11 , 0
11	01, 0	××, ×	10, ×	(11), 0
10	00, 0	00, 1	(10), 1	(10), 1

Fig. 7: Flow table with output assigned to unstable states.

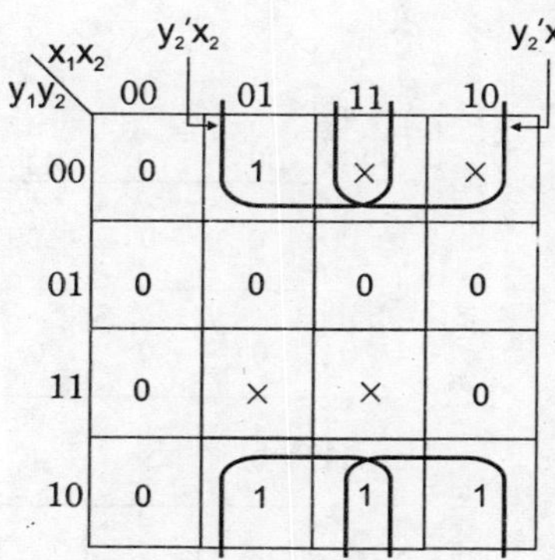

y_1y_2 \ x_1x_2	00	01	11	10
00	0	1	×	×
01	0	0	0	0
11	0	×	×	0
10	0	1	1	1

Fig. 8: Output map.

Let, Z = Output

The output map is constructed as shown in Fig. 8 and from output map we get the following simplified Boolean equation for output.

Output, $Z = y_2'x_2 + y_2'x_1$

c) Logic Implementation using NOR based SR Latch

The binary values inside each square of transition table is the value of next state Y_1Y_2. Hence, using the transition table the maps of Y_1 and Y_2 are drawn as shown in Figs. 9 and 10.

y_1y_2 \ x_1x_2	00	01	11	10
00	0	0	0	1
01	0	0	0	1
11	0	×	1	1
10	0	0	1	1

Fig. 9: K-map for Y_1.

y_1y_2 \ x_1x_2	00	01	11	10
00	0	0	1	0
01	0	1	1	1
11	1	×	0	1
10	0	0	0	0

Fig. 10: K-map for Y_2.

In order to store the two state variables Y_1 and Y_2, two SR latches are required. Let us choose two NOR based SR latches.

Let, S_1, R_1 = Inputs for SR latch-1

S_2, R_2 = Inputs for SR latch-2

To Find Equations for Latch Inputs using Map of Y_1 and Y_2

In latch-1, the input equations for S_1 and R_1 will be independent y_1. Therefore, the map of Y_1 can be split into two maps: One map when y_1 = 0 and other map when y_1 = 1 as shown in Fig. 11.

The map y_1 = 0 is used to find equation for S_1 by forming prime implicants using 1's.

The map y_1 = 1 is used to find equation for R_1' by forming prime implicants using 0's.

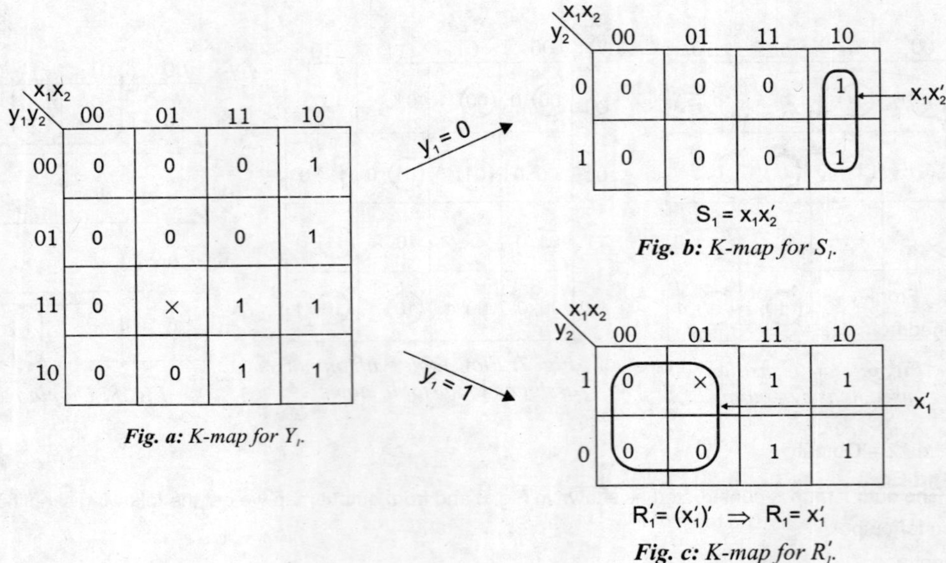

Fig. a: K-map for Y_1.

Fig. b: K-map for S_1.

Fig. c: K-map for R_1'.

Fig. 11: K-maps to find equations for S_1 and R_1.

In latch-2, the input equations for S_2 and R_2 will be independent y_2. Therefore, the map of Y_2 can be split into two maps: One map when y_2 = 0 and other map when y_2 = 1 as shown in Fig. 12.

The map y_2 = 0 is used to find equation for S_2 by forming prime implicants using 1's.

The map y_2 = 1 is used to find equation for R_2' by forming prime implicants using 0's.

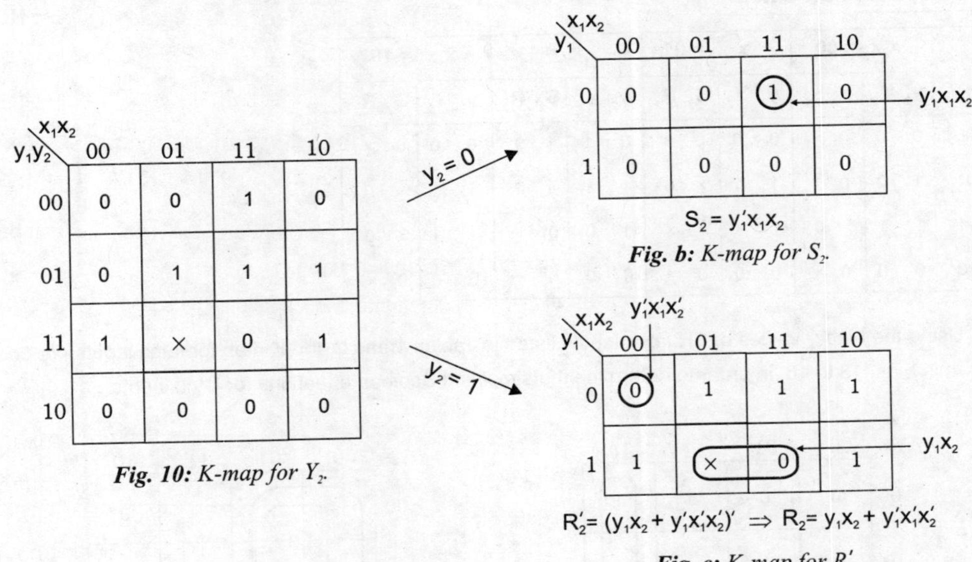

Fig. 10: K-map for Y_2.

Fig. b: K-map for S_2.

Fig. c: K-map for R_2'.

***Fig. 12:** K-maps to find equations for S_2 and R_2.*

From the maps of S_1, R_1', S_2 and R_2' we get the following equations for SR latch inputs. The input equations is obtained from output map.

$S_1 = x_1x_2'$	$R_1 = x_1'$
$S_2 = y_1'x_1x_2$	$R_2 = y_1x_2 + y_1'x_1'x_2'$

To find Equations for Latch Inputs using Excitation Table of SR Latch

Using the excitation table of NOR based SR latch, the inputs for SR latch can be obtained for the circuit inputs, present and next states in the maps of Y_1 and Y_2 and listed in Tables 1 and 2.

Table 1: Inputs of SR Latch-1

	$x_1x_2 = 00$				$x_1x_2 = 01$				$x_1x_2 = 11$				$x_1x_2 = 10$			
y_1y_2	y_1	Y_1	S_1	R_1	y_1	Y_1	S_1	R_1	y_1	Y_1	S_1	R_1	y_1	Y_1	S_1	R_1
00	0	0	0	×	0	0	0	×	0	0	0	×	0	1	1	0
01	0	0	0	×	0	0	0	×	0	0	0	×	0	1	1	0
11	1	0	0	1	1	×	×	×	1	1	×	0	1	1	×	0
10	1	0	0	1	1	0	0	1	1	1	×	0	1	1	×	0

Excitation Table of NOR based SR Latch

Present State y	Next State Y	Latch Inputs S	R
0	0	0	×
0	1	1	0
1	0	0	1
1	1	×	0

Table 2: Inputs of SR Latch-2

	$x_1x_2 = 00$				$x_1x_2 = 01$				$x_1x_2 = 11$				$x_1x_2 = 10$			
y_1y_2	y_2	Y_2	S_2	R_2	y_2	Y_2	S_2	R_2	y_2	Y_2	S_2	R_2	y_2	Y_2	S_2	R_2
00	0	0	0	×	0	0	0	×	0	1	1	0	0	0	0	×
01	1	0	0	1	1	1	×	0	1	1	×	0	1	1	×	0
11	1	1	×	0	1	×	×	×	1	0	0	1	1	1	×	0
10	0	0	0	×	0	0	0	×	0	0	0	×	0	0	0	×

Using the binary values of SR latch inputs listed in Tables 1 and 2, the K-map for latch inputs are constructed as shown in Figs. 13 to 16, in order to determine the simplified Boolean equations for latch inputs.

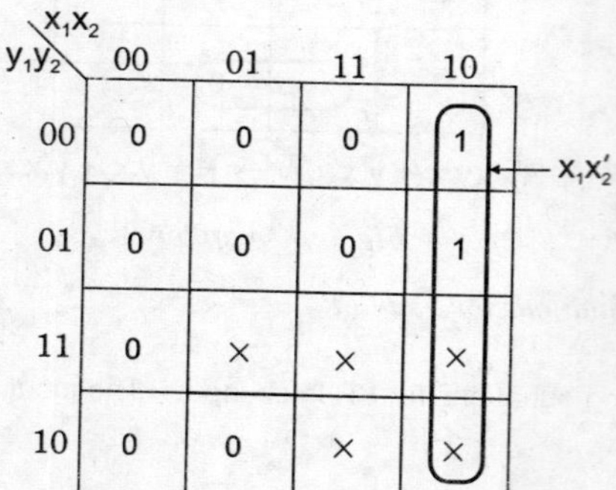

Fig. 13: K-map for S_1.

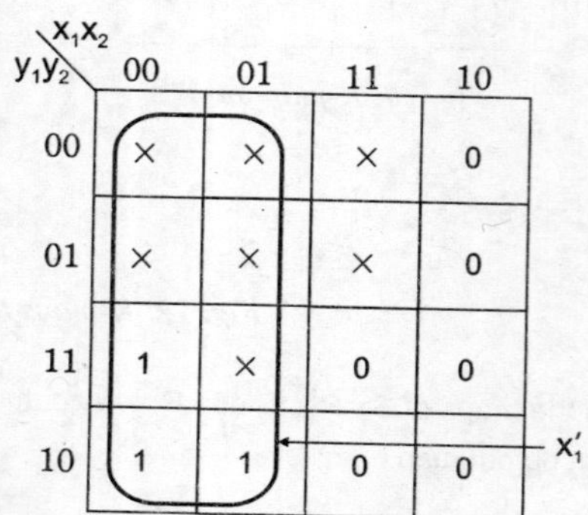

Fig. 14: K-map for R_1.

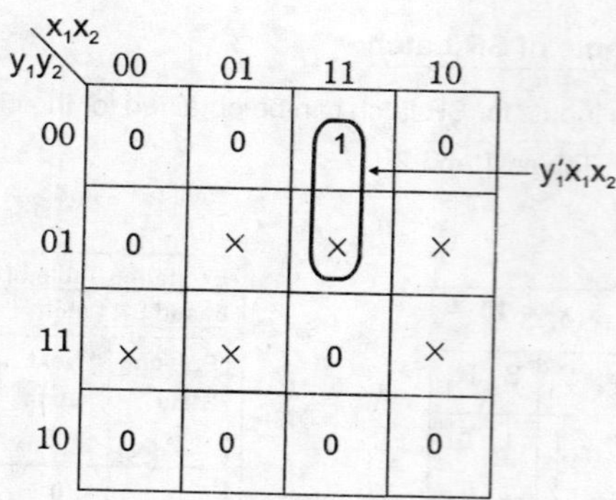

Fig. 15: K-map for S_2.

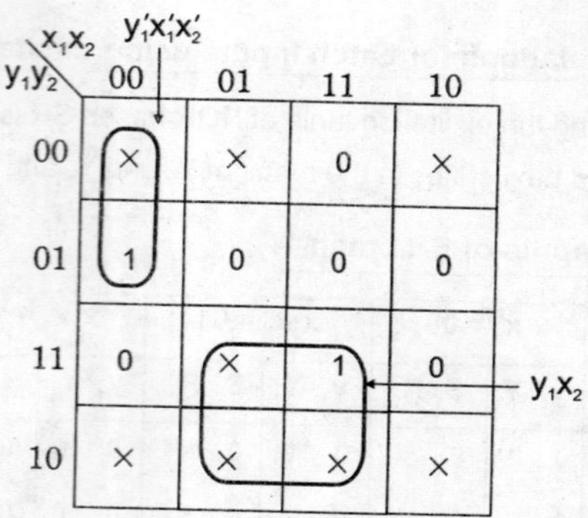

Fig. 16: K-map for R_2.

From the K-maps shown in Figs. 13 to 16, we get the following simplified equations for SR latch inputs. The output equation is obtained from output map.

$S_1 = x_1x_2'$	$S_2 = y_1'x_1x_2$	$Z = y_2'x_2 + y_2'x_1$
$R_1 = x_1'$	$R_2 = y_1'x_1'x_2' + y_1x_2$	

Logic Diagram

The Boolean equations obtained for latch inputs in both methods are same and using these equations the combinational logic circuit to generate the latch inputs are drawn as shown in Fig. 17. The complete logic circuit is obtained by adding NOR based SR latch to the combinational circuit as shown in Fig. 17.

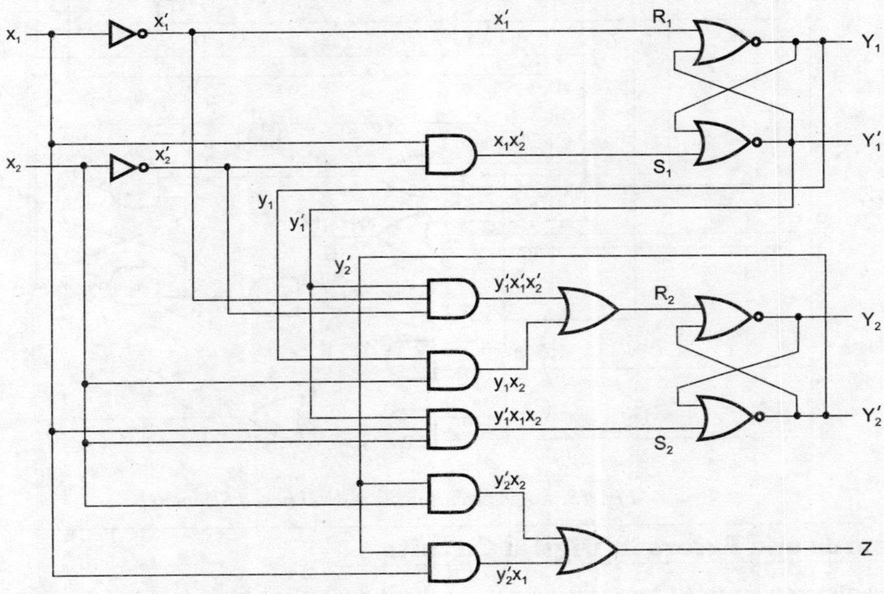

Fig. 17: Logic circuit using NOR based SR latch.

d) Logic Implementation using NAND based SR Latch

The Boolean equations to generate inputs for NAND based SR latches are given by complement of the equations for NOR based SR latch.

Therefore, the Boolean equations to generate inputs for NAND based SR latches are,

$S_1 = (x_1 x_2')'$	$S_2 = (y_1' x_1 x_2)'$	$R_1 = (x_1')'$	$R_2 = (y_1' x_1' x_2' + y_1 x_2)'$	$Z = y_2' x_2 + y_2' x_1$
$= x_1' + x_2$	$= y_1 + x_1' + x_2'$	$= x_1$	$= (y_1' x_1' x_2')' . (y_1 x_2)'$	
			$= (y_1 + x_1 + x_2) . (y_1' + x_2')$	

The output equation is obtained from output map. Using the above equations the combinational logic circuit to generate latch inputs are drawn as shown in Fig. 16. The complete logic circuit is obtained by adding NAND based SR latches to the combinational circuit as shown in Fig. 16.

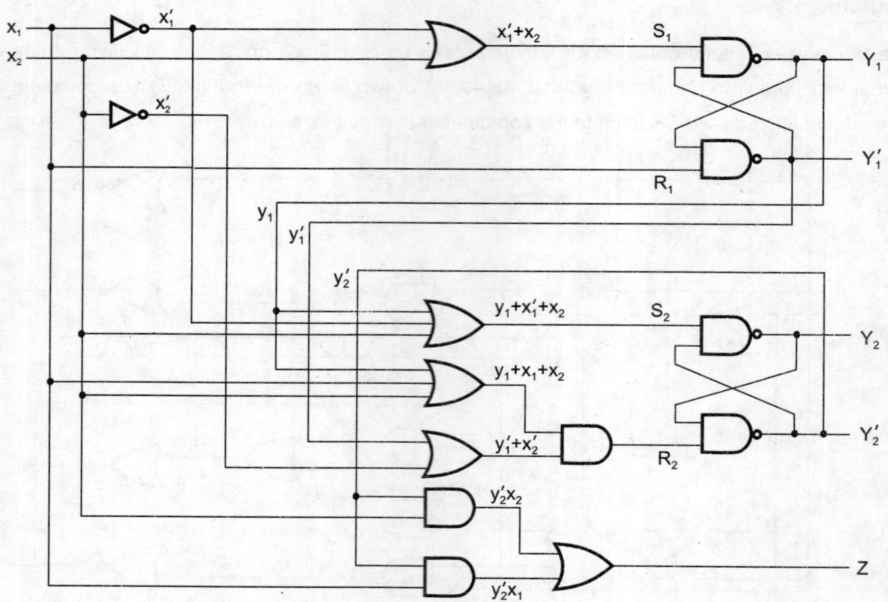

Fig. 18: Logic circuit using NAND based SR latch.

4.6 Hazards and Errors in Digital Circuits

Hazards are unwanted transient outputs in the digital circuits that may arise due to gates and connecting wires with different propagation delay.

The different types of hazards are static-1 hazard, static-0 hazard and dynamic hazard.

Static-1 hazard: When the expected output is 1 for a change in input, if the circuit generates a temporary 0 as shown in Fig. 4.19(a) then the hazard is called static-1 hazard.

Static-0 hazard: When the expected output is 0 for a change in input, if the circuit generates a temporary 1 as shown in Fig. 4.19(b) then the hazard is called static-0 hazard.

Dynamic hazard: When the expected output is either 1 or 0 for a change in input, if the circuit output temporarily changes from 0 to 1 and 1 to 0 three or more times as shown in Fig. 4.19(c), then the hazard is called dynamic hazard.

Fig. a: *Static-1 hazard.* **Fig. b:** *Static-0 hazard.* **Fig. c:** *Dynamic hazard.*

Fig. 4.19: *Types of hazards.*

Hazards may occur in combinational circuits as well as asynchronous sequential circuits. The synchronous sequential circuits will not have hazards because in those circuits the changes are synchronized with a clock.

4.6.1 Hazards in Combinational Circuits

In combinational circuits the output may have one of the following possible values for a change in input.

- The output may change from 1 to 0
- The output may change from 0 to 1
- The output may remain in previous value, either 1 or 0.

When the combinational circuit has a hazard some unwanted temporary output are generated than one of the expected output listed above.

4.6.2 Static-0 and Static-1 Hazards

Some examples of combinational circuits with static-0 and static-1 hazards are shown in Figs. 4.20, 4.22, 4.24 and 4.26. In all these examples the hazards are due to longer propagation delay in inverter or NOT gate. In each example, circuit with and without hazard are drawn and change in signals are shown in a timing diagram. Also, the signal values at various timings are listed in a table.

In the example shown in Fig. 4.20, initially all the inputs are 1 and at time, t_2 the input x_2 changes from 1 to 0. When the inverter or NOT gate has negligible delay the signal propagation through gates 2 and 3 will be same as in the path through gates 1 and 3 and so the output will not have hazard. On the other hand, if the inverter or NOT gate has longer delay then the signal propagation through the two different paths will take different timings which creates a temporary 0 output as shown in Fig. 4.20(b). The signal timings are shown in Fig. 4.21 and values of signals at various timings are listed in Table 4.12.

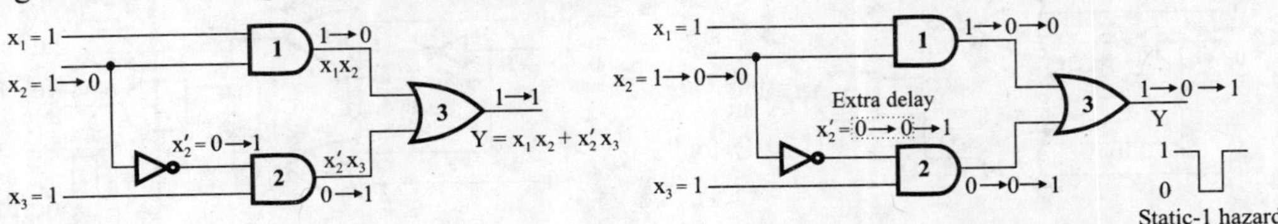

Fig. a: Circuit without hazard. *Fig. b: Circuit with hazard.*

***Fig. 4.20:** Example of combinational circuit with static-1 hazard.*

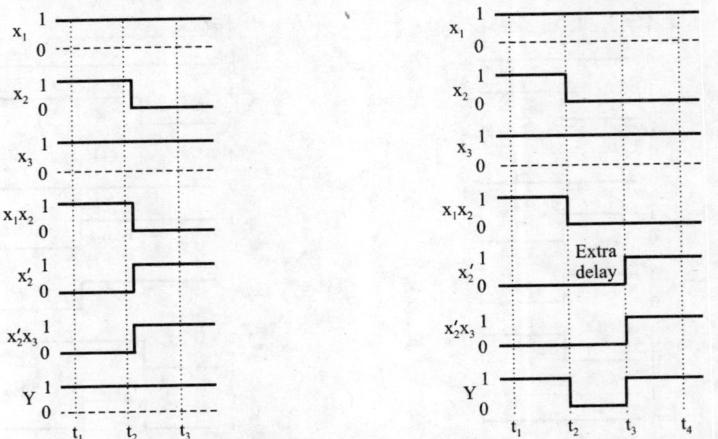

Fig. a: When inverter has negligible delay. *Fig. b: When inverter has longer delay.*

***Fig. 4.21:** Signal timings in the circuit of Fig. 4.20.*

Table 4.12: Signal Values at Various Timings in Circuit of Fig. 4.20

	Signal Values at Various Timings						
	When Inverter has Negligible Delay			When Inverter has Longer Delay			
	t_1	t_2	t_3	t_1	t_2	t_3	t_4
x_1	1	1	1	1	1	1	1
x_2	1	0	0	1	0	0	0
x_3	1	1	1	1	1	1	1
$x_1 x_2$	1	0	0	1	0	0	0
x_2'	0	1	1	0	0	1	1
$x_2' x_3$	0	1	1	0	0	1	1
Y	1	1	1	1	0	1	1

In the example shown in Fig. 4.22, initially all the inputs are 0 and at time, t_2 the input x_2 changes from 0 to 1. When the inverter or NOT gate has negligible delay the signal propagation through gates 1 and 3 will be same as in the path through gates 2 and 3 and so the output will not have hazard. On the other hand, if the inverter or NOT gate has longer delay then the signal propagation through the two different paths will take different timings which creates a temporary 1 output as shown in Fig. 4.22(b). The signal timings are shown in Fig. 4.23 and values of signals at various timings are listed in Table 4.13.

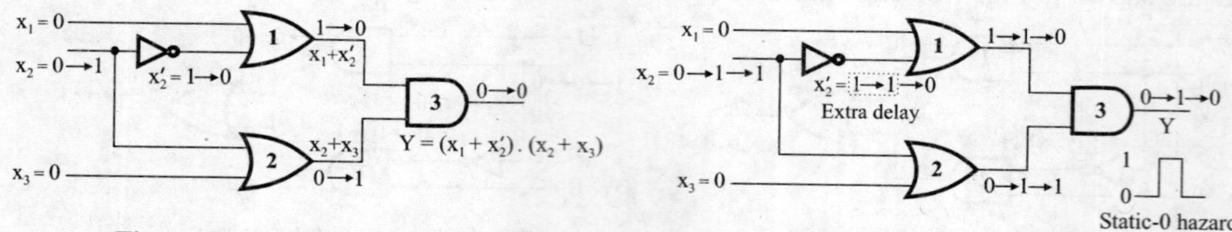

Fig. a: Circuit without hazard. *Fig. b: Circuit with hazard.*

Fig. 4.22: Example of combinational circuit with static-0 hazard.

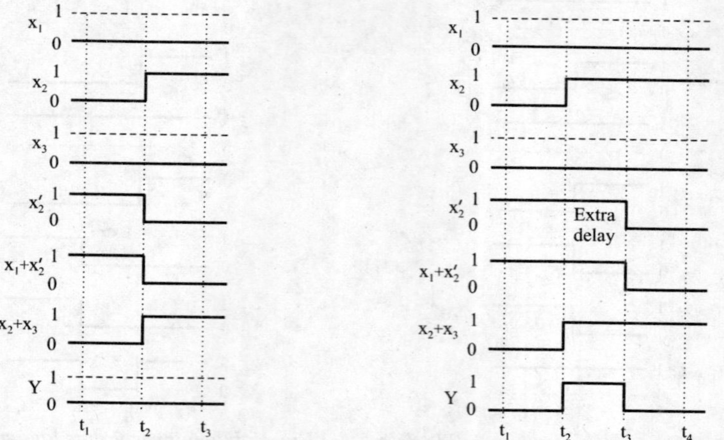

Fig. a: When inverter has negligible delay. *Fig. b: When inverter has longer delay.*

Fig. 4.23: Signal timings in the circuit of Fig. 4.22.

Table 4.13: Signal Values at Various Timings in Circuit of Fig. 4.22

	Signal Values at Various Timings						
	When Inverter has Negligible Delay			When Inverter has Longer Delay			
	t_1	t_2	t_3	t_1	t_2	t_3	t_4
x_1	0	0	0	0	0	0	0
x_2	0	1	1	0	1	1	1
x_3	0	0	0	0	0	0	0
x_2'	1	0	0	1	1	0	0
$x_1 + x_2'$	1	0	0	1	1	0	0
$x_2 + x_3$	0	1	1	0	1	1	1
Y	0	0	0	0	1	0	0

In the example shown in Fig. 4.24, initially all the inputs are 1 and at time, t_2 the input x_2 changes from 1 to 0. When the inverter or NOT gate has negligible delay the signal propagation through gates 2 and 3 will be same as in the path through gates 1 and 3 and so the output will not have hazard. On the other hand, if the inverter or NOT gate has longer delay then the signal propagation through the two different paths will take different timings which creates a temporary 0 output as shown in Fig. 4.24(b). The signal timings are shown in Fig. 4.25 and values of signals at various timings are listed in Table 4.14.

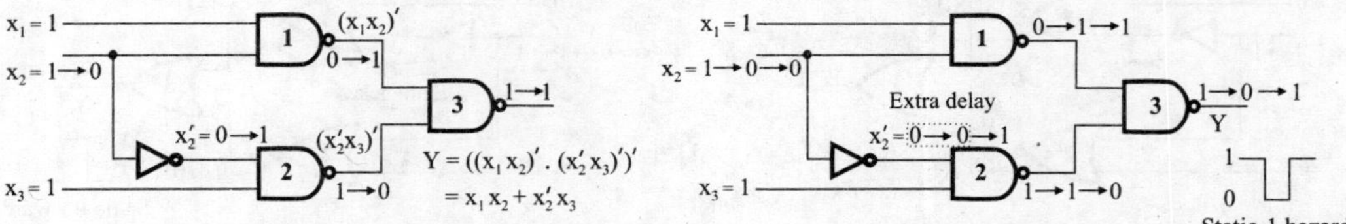

Fig. a: Circuit without hazard. *Fig. b: Circuit with hazard.*

Fig. 4.24: Example of combinational circuit with static-1 hazard.

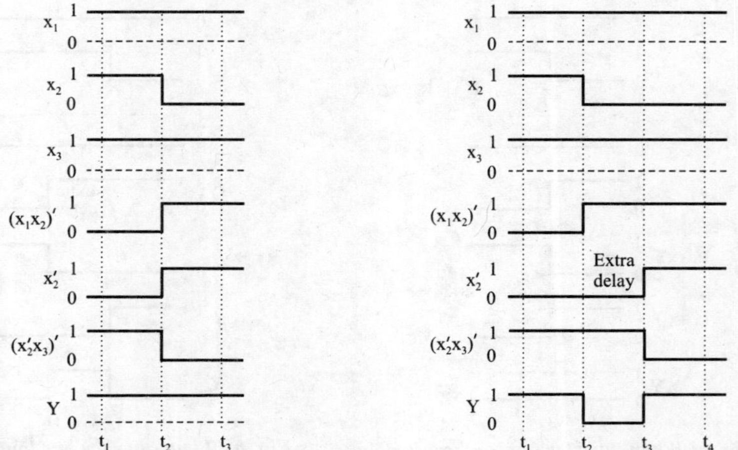

Fig. a: When inverter has negligible delay. *Fig. b: When inverter has longer delay.*

Fig. 4.25: Signal timings in the circuit of Fig. 4.24.

Table 4.14: Signal Values at Various Timings in Circuit of Fig. 4.24

	Signal Values at Various Timings						
	When Inverter has Negligible Delay			When Inverter has Longer Delay			
	t_1	t_2	t_3	t_1	t_2	t_3	t_4
x_1	1	1	1	1	1	1	1
x_2	1	0	0	1	0	0	0
x_3	1	1	1	1	1	1	1
$(x_1 x_2)'$	0	1	1	0	1	1	1
x_2'	0	1	1	0	0	1	1
$(x_2' x_3)'$	1	0	0	1	1	0	0
Y	1	1	1	1	0	1	1

In the example shown in Fig. 4.26, initially all the inputs are 0 and at time, t_2 the input x_2 changes from 0 to 1. When the inverter or NOT gate has negligible delay the signal propagation through gates 1 and 3 will be same as in the path through gates 2 and 3 and so the output will not have hazard. On the other hand, if the inverter or NOT gate has longer delay then the signal propagation through the two different paths will take different timings which creates a temporary 0 output as shown in Fig. 4.26(b). The signal timings are shown in Fig. 4.27 and values of signals at various timings are listed in Table 4.15.

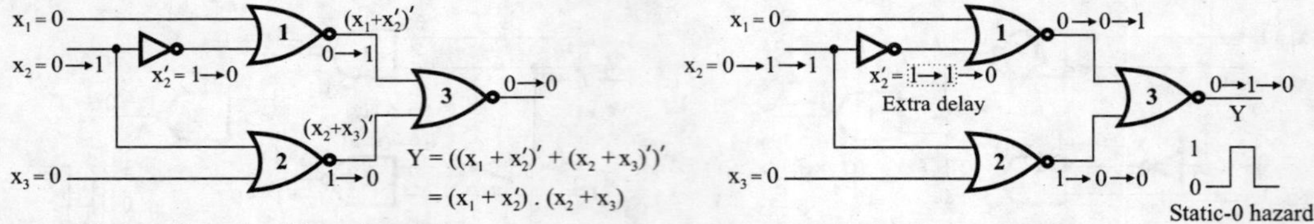

Fig. a: Circuit without hazard. *Fig. b: Circuit with hazard.*

Fig. 4.26: Example of combinational circuit with static-0 hazard.

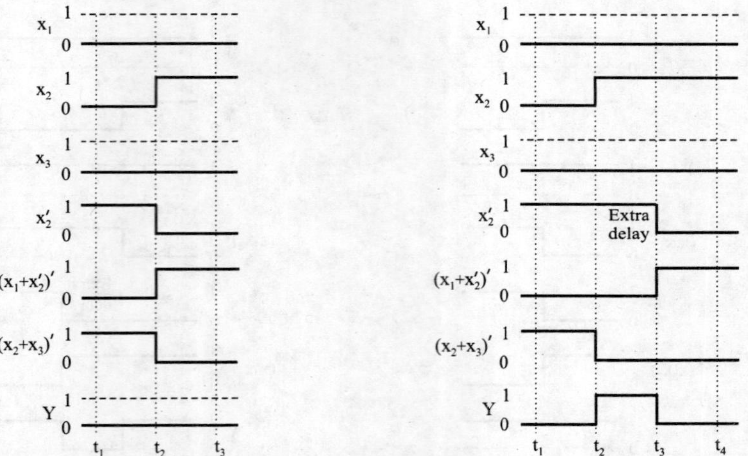

Fig. a: When inverter has negligible delay. *Fig. b: When inverter has longer delay.*

Fig. 4.27: Signal timings in the circuit of Fig. 4.26.

Table 4.15: Signal Values at Various Timings in Circuit of Fig. 4.26

	Signal Values at Various Timings						
	When Inverter has Negligible Delay			When Inverter has Longer Delay			
	t_1	t_2	t_3	t_1	t_2	t_3	t_4
x_1	0	0	0	0	0	0	0
x_2	0	1	1	0	1	1	1
x_3	0	0	0	0	0	0	0
x_2'	1	0	0	1	1	0	0
$(x_1 + x_2')'$	0	1	1	0	0	1	1
$(x_2 + x_3)'$	1	0	0	1	0	0	0
Y	0	0	0	0	1	0	0

4.6.3 Dynamic Hazard

Dynamic hazard occurs only if there are three or more paths between input and output. An example of dynamic hazard with three paths between input and output is shown in Fig. 4.28.

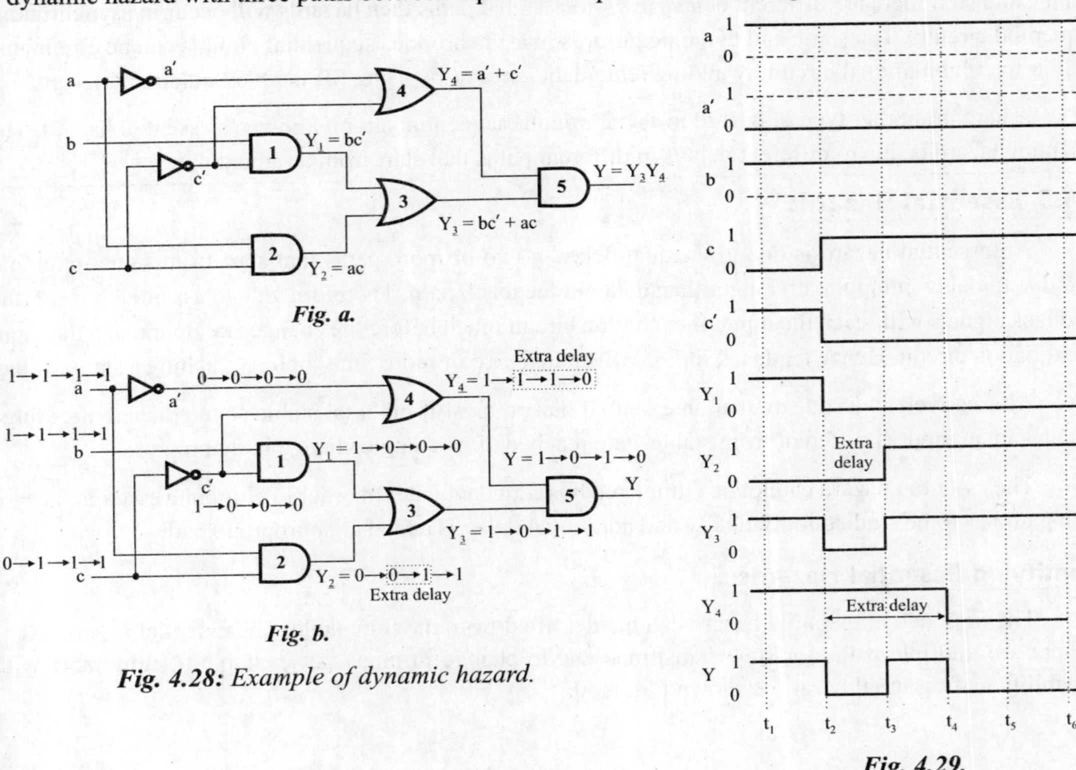

Fig. 4.28: Example of dynamic hazard.

Fig. 4.29.

In the example shown in Fig. 4.28, initially a and b are 1 and c is 0. At time, t_2 the input c changes from 0 to 1. Here, the paths through gates 2 and 4 has delay and the delay in gate 4 is longer than gate 2. Therefore, the change input c creates first change in output, the delay in gate 2 creates second change in output and the delay in gate 4 creates third change in output. The three changes in signals at the output of gates are shown in Fig. 4.28(b). The signal timings are shown in Fig. 4.29 and the values of signals at various timings are listed in Table 4.16.

Table 4.16:

	t_1	t_2	t_3	t_4	t_5	t_6
a	1	1	1	1	1	1
a'	0	0	0	0	0	0
b	1	1	1	1	1	1
c	0	1	1	1	1	1
c'	1	0	0	0	0	0
Y_1	1	0	0	0	0	0
Y_2	0	0	1	1	1	1
Y_3	1	0	1	1	1	1
Y_4	0	0	0	1	1	1
Y	1	0	1	0	0	0

4.6.4 Hazards in Sequential Circuits

In synchronous sequential circuits the changes in circuit conditions are synchronized by a clock and the clock time period is selected long enough to overcome internal delays in signal transitions. Hence hazards will not occur in synchronous sequential circuits.

In asynchronous sequential circuits the internal changes will occur immediately after an input change and so if there are different delays in various signal paths then hazards will occur in asynchronous sequential circuits. The static and dynamic hazards in asynchronous sequential circuits can be eliminated similar to combinational circuit by adding redundant gates or by using SR latch based circuit design.

There is another type of hazard in asynchronous sequential circuit known as essential hazard. The essential hazard is due to different delays in different paths that start from a same input.

4.6.5 Essential Hazards

An essential hazard is due to unequal delay in two or more paths that start from same input and also due to delay in input path longer than delay in feedback path. Therefore, due to essential hazard, the feedback signals will reach the input of excitation circuit much before the changes in input reach the input of excitation circuit. Hence feedback inputs will change two or more times before reaching a stable value.

Alternatively, we can say that an essential hazard exist if the total stable state reached after three changes in an input is different from stable state reached after a single change in that input.

The essential hazard cannot be eliminated by redundant gate. In order to eliminate essential hazard each path has to be studied individually and additional delay is added at appropriate path.

Identifying Essential Hazards

The presence of essential hazard can be identified from the flow table. The essential hazard exists if there are multiple paths for state transitions due to change in input. An example of flow table with possibilities of essential hazard is shown Fig. 4.30.

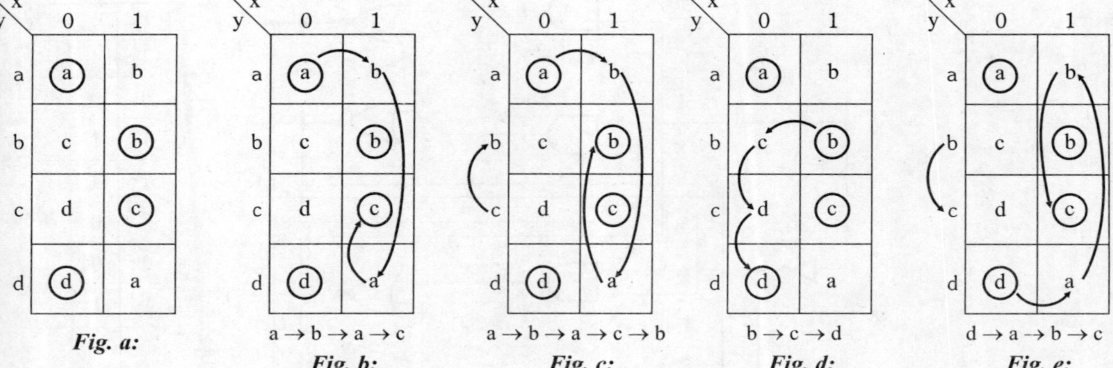

Fig. a: *Fig. b:* *Fig. c:* *Fig. d:* *Fig. e:*

Fig. 4.30: Flow table with essential hazards.

Example of Essential Hazard

Consider the following excitation functions of an asynchronous circuit.

$$Y_1 = xy_2' + (x' + y_2')y_1$$

$$Y_2 = x'y_1 + (x + y_1)y_2$$

Here, y_1, y_2 = Present state variables

Y_1, Y_2 = Next state varaibles

x = Input

The circuit has two state variables Y_1 and Y_2 and so let us choose two NOR based SR latches to store Y_1 and Y_2.

Consider the simplified excitation function of NOR based SR latch, $Y = S + R'y$

Now the excitation function of two NOR based SR latches are,

$$Y_1 = S_1 + R_1'y_1$$

$$Y_2 = S_2 + R_2'y_2$$

On comparing the excitation functions of the circuit and SR latches, we get the following equations for SR latch inputs.

$S_1 = xy_2'$	$S_2 = x'y_1$
$R_1 = (x' + y_2')' = xy_2$	$R_2 = (x + y_1)' = x'y_1'$

Using the above equations, the combinational circuit for latch inputs are constructed as shown in Fig. 4.31. The complete logic circuit is obtained by adding SR latches as shown in Fig. 4.31.

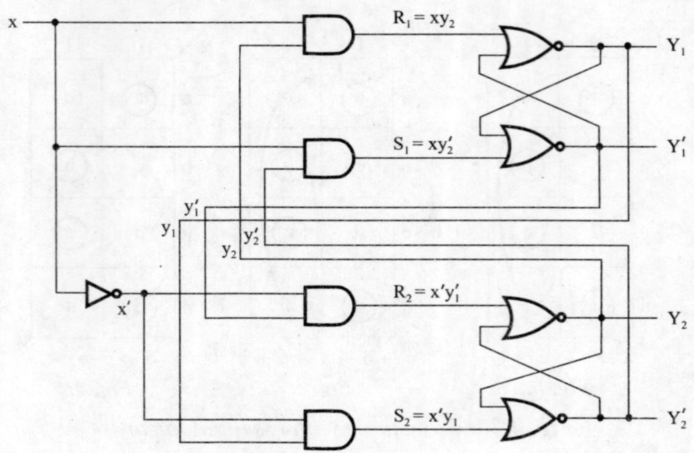

Fig. 4.31: *Example of asynchronous sequential circuit with essential hazard.*

Table 4.17:

y_1	y_2	x	y_1'	y_2'	x'	xy_2'	$x'y_1$	$(x'+y_2')$	$(x+y_1)$	$(x'+y_2')y_1$	$(x+y_1)y_2$	Y_1	Y_2
0	0	0	1	1	1	0	0	1	0	0	0	0	0
0	1	0	1	0	1	0	0	1	0	0	0	0	0
1	0	0	0	1	1	0	1	1	1	1	0	1	1
1	1	0	0	0	1	0	1	1	1	1	1	1	1
0	0	1	1	1	0	1	0	1	1	0	0	1	0
0	1	1	1	0	0	0	0	0	1	0	1	0	1
1	0	1	0	1	0	1	0	1	1	1	0	1	0
1	1	1	0	0	0	0	0	0	1	0	1	0	1

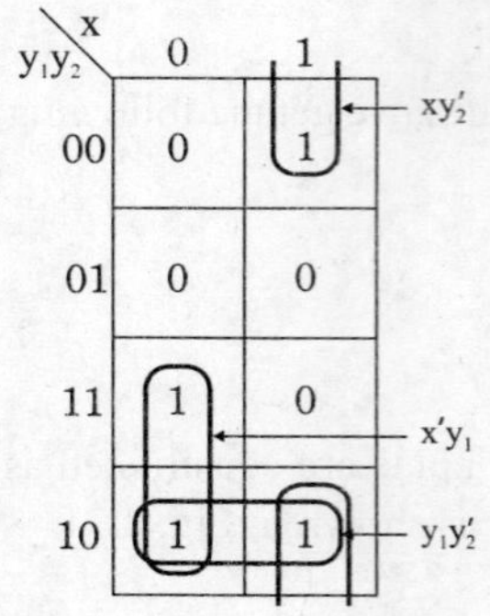

Fig. a: *K-map for Y_1.*

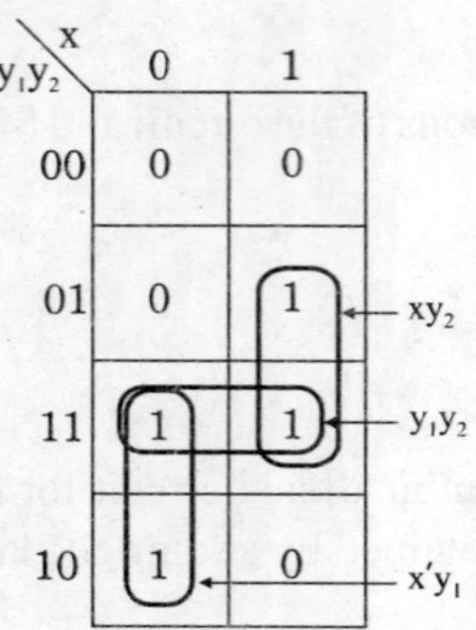

Fig. b: *K-map for Y_2.*

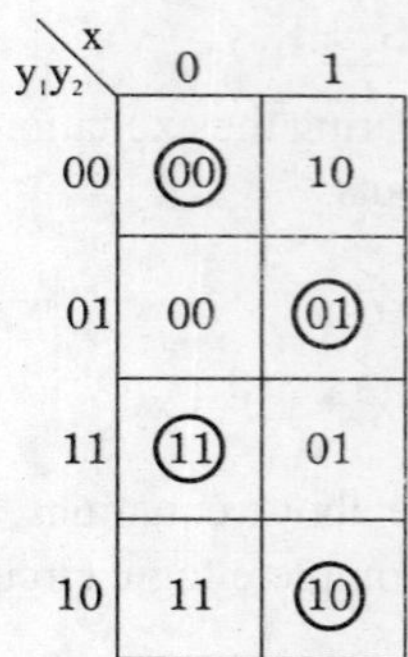

Fig. c: *Transition table.*

Fig. 4.32.

The values of next state Y_1 and Y_2 are evaluated and listed in Table 4.15 and using these values the maps of Y_1 and Y_2 are drawn as shown in Figs. 4.32(a) and (b). The maps of Y_1 and Y_2 are combined to form transition table as shown in Fig. 4.32(c).

Let us mark the product terms of Y_1 and Y_2 in the maps of Y_1 and Y_2. Now, we can observed that all the product terms are overlapped and so there is no possibility of static hazards.

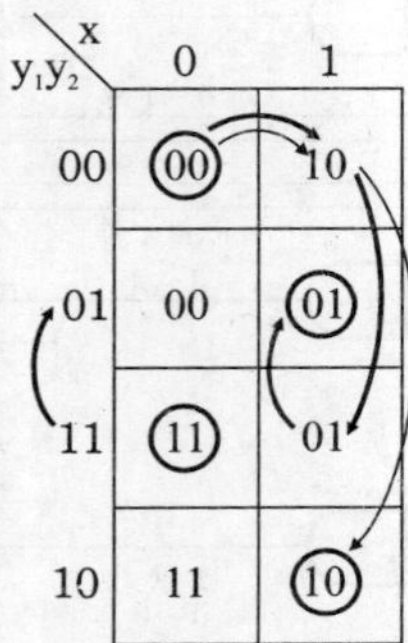

***Fig. 4.33:** State transitions from stable state 00 due to change in input from 0 to 1.*

Let us examine the essential hazard when the input change from 0 to 1 with circuit in stable state 00. If there is no essential hazard the state transition will occur from 00 to 10

Due to essential hazard there is a possibility of state transitions from 00 → 10 → 11 → 01 and the circuit will settle in state 01 instead of 10. The signal changes during these transitions are shown in Fig. 4.34.

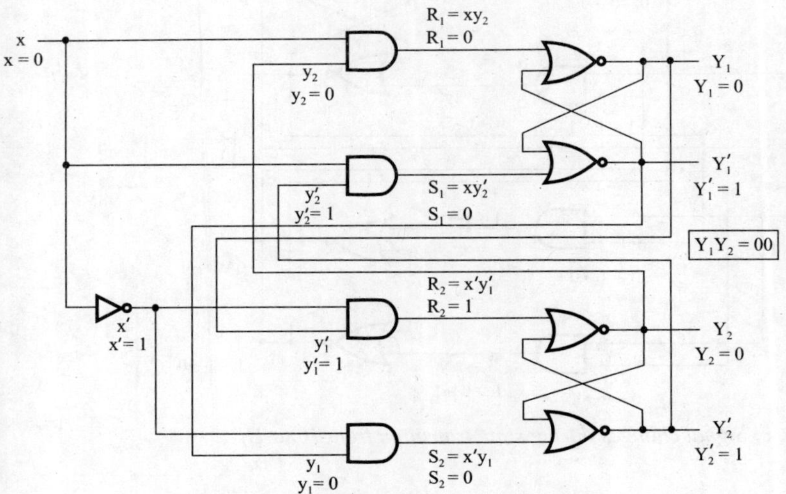

Truth Table of NOR Latch			
Inputs		**Outputs**	
S	**R**	**Q**	**Q'**
0	0	No change	
1	0	1	0
0	1	0	1
1	1	Indeterminate output	

***Fig. a:** Initial value of signals when circuit in stable state 00.*

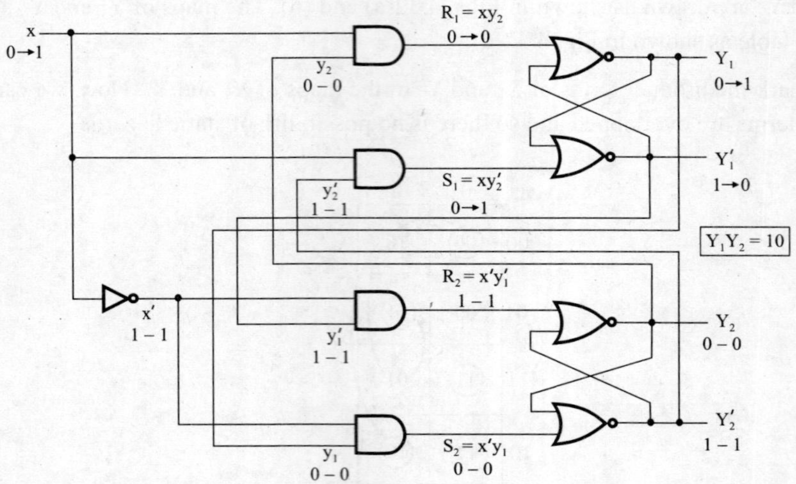

Fig. b: *Signal changes during state transition from 00 to 10.*

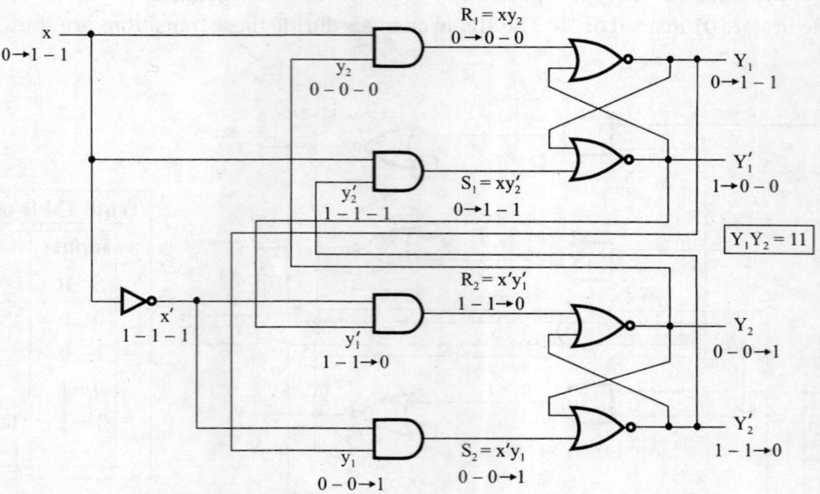

Fig. c: *Signal changes during state transition from 10 to 11.*

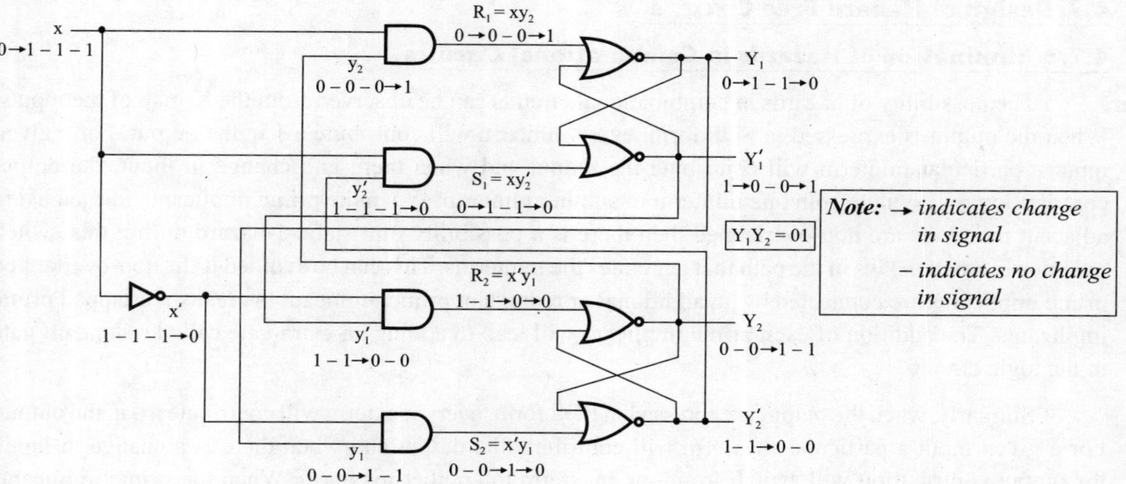

Fig. d: Signal changes during state transition from 11 to 01.

Fig. 4.34: Signal changes during state transitions when the input is changed from 0 to 1.

Table 4.18: Signals during State Transition 00 → 10 → 11 → 01

x	x'	y_1	y_2	y_1'	y_2'	R_1 xy_2	S_1 xy_2'	R_2 $x'y_1'$	S_2 $x'y_1$	Y_1	Y_2	Y_1	Y_2
0	1	0	0	1	1	0	0	1	0	0	0	1 1	1 1
1	1	0	0	1	1	0	1	1	0	1	0		
1	1	1	0	0	1	0	1	0	1	1	1	0 0	0 0
1	0	1	1	0	0	1	0	0	0	0	1	Hazard in Y_1	No hazard in Y_2

No change due to delay

The initial value of signals when the circuit is in stable state 00 with input 0 are shown shown in Fig. 4.34(a). The latch outputs can be determined from the truth table of SR latch.

When the input change from 0 to 1, the inputs to latch-1 S_1 and R_1 changes and so the output of latch-1 will change as shown in Fig. 4.34(b) and so the circuit will go to state 10. The signals at other part of circuit will not change due to propagation delay in inverter and feedback paths in latch-1.

After a small delay, the output of latch-1 propagates to input of latch-2 and so the output latch-2 will change as shown in Fig. 4.34(c) and the circuit will go to state 11.

Let the signal propagation delay in inverter be longer than the combined propagation delay shown in Figs. 4.34(b) and (c). Hence the inverter output will change only after the signal changes shown in Figs. 4.34(b) and (c).

When the output of inverter changes from 1 to 0. The inputs to latch-2 becomes 0 and so no change in its output but this outputs propagates to input of latch-1 as shown in Fig. 4.34(d) and so output of latch-1 changes and the circuit go to state 01. Now the circuit will remain in stable state 01.

The signals during state transition 00 → 10 → 11 → 01 are listed in Table 4.18.

4.7 Design of Hazard Free Circuits

4.7.1 Elimination of Hazards in Combinational Circuits

The possibility of hazards in combinational circuits can be observed from the K-map of the inputs. When the output is expressed in SOP form, every minterm will contribute a 1 in the output. For a given input a particular minterm will contribute the output and when there is a change in input, the output contribution will switch from one minterm to another minterm. When the prime implicants formed using adjacent minterms are non-overlapped then there is a possibility of a static-1 hazard during this switch over due to small delays in the path that generates the minterms. This can be avoided if the non-overlapped prime implicants are connected by an additional or redundant prime implicant to create overlapped prime implicants. The addition of each prime implicant will lead to adding an extra gate called redundant gate in the logic circuit.

Similarly, when the output is expressed in POS form, every maxterm will contribute a 0 in the output. For a given input a particular maxterm will contribute the output and when there is a change in input, the output contribution will switch from one maxterm to another maxterm. When the prime implicants formed using adjacent maxterms are non-overlapped then there is a possibility of a static-0 hazard during this switch over due to small delays in the path that generates the maxterms. This can be avoided if the non-overlapped prime implicants are connected by an additional or redundant prime implicant to create overlapped prime implicants. The addition of each prime implicant will lead to adding an extra gate called redundant gate in the logic circuit.

Consider the combinational circuit shown in Fig. 4.20. The output equation is,

$$Y = x_1x_2 + x_2'x_3$$

The K-map for the above equation is shown in Fig. 4.35. Here, the two prime implicants are non-overlapped prime implicants and so a hazard may occur when the output contribution switch from minterm of a prime implicant to minterm of another prime implicant. The hazard is due to delay in the path that generates minterm. This can be avoided by introducing an additional prime implicant x_1x_3 that makes overlapped prime implicants as shown in Fig. 4.35.

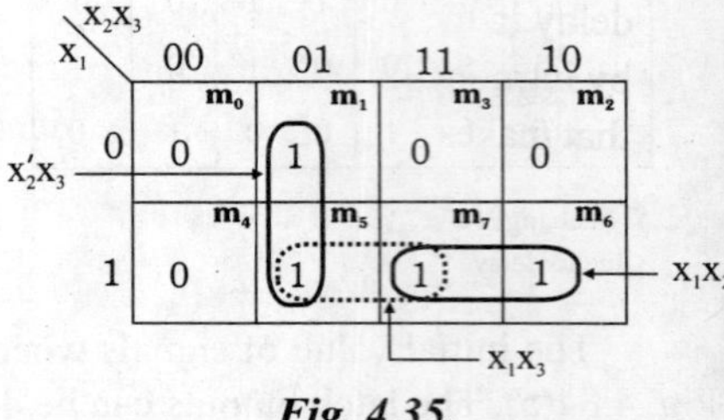

Fig. 4.35.

Now, the output equation is,

$$Y = x_1x_2 + x_2'x_3 \boxed{+ x_1x_3}$$

The additional product term x_1x_3 can be implemented by an additional AND gate as shown in Fig. 4.36.

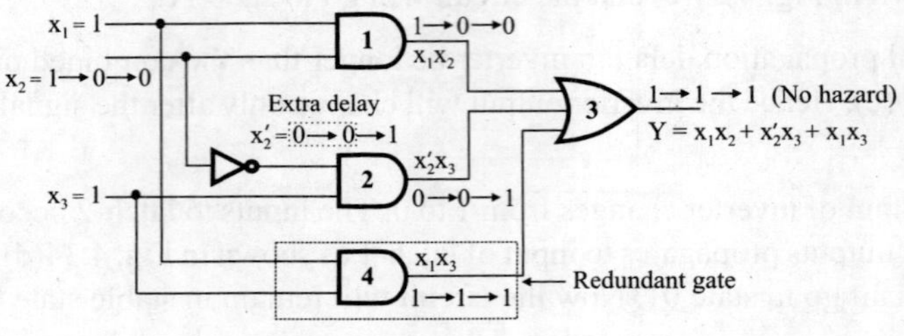

Fig. 4.36.

Alternatively, the additional product term can be implemented by an additional NAND gate as shown in Fig. 4.37.

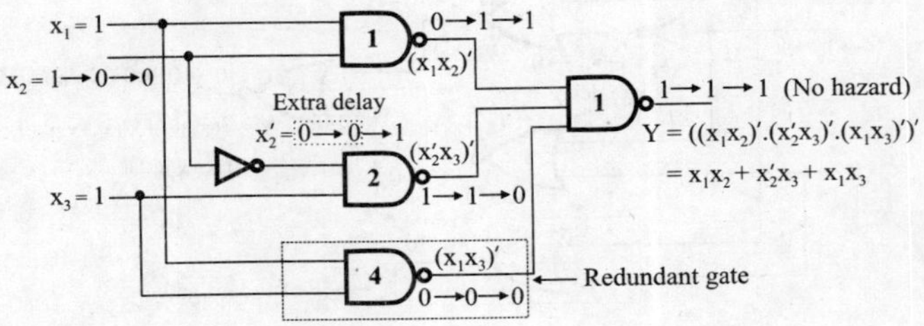

Fig. 4.37.

Consider the combinational circuit shown in Fig. 4.22. The output equation is,

$$Y = (x_1 + x_2') \cdot (x_2 + x_3)$$
$$= (((x_1 + x_2') \cdot (x_2 + x_3))')'$$
$$= ((x_1 + x_2')' \cdot (x_2 + x_3)')'$$
$$= (x_1'x_2 + x_2'x_3')'$$

The K-map for the above equation is shown in Fig. 4.38. Here, the two prime implicants are non-overlapped prime implicants and so a hazard may occur when the output contribution switch from maxterm of a prime implicant to maxterm of another prime implicant. The hazard is due to delay in the path that generates maxterm. This can be avoided by introducing an additional prime implicant $(x_1'x_3')' = (x_1 + x_3)$ that makes overlapped prime implicants as shown in Fig. 4.38.

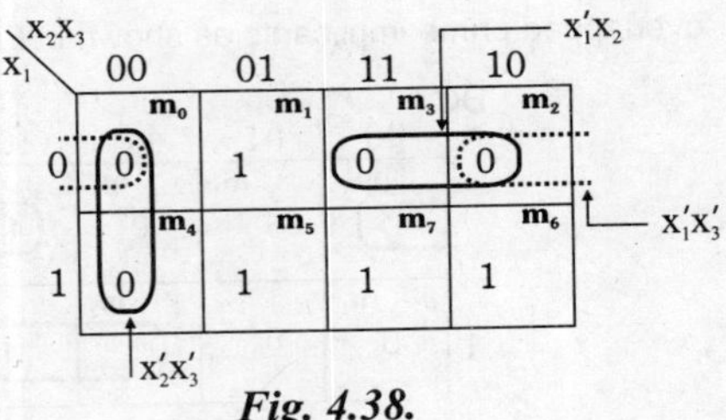

Fig. 4.38.

Now, the output equation is,

$$Y = (x_1'x_2 + x_2'x_3' + x_1'x_3')'$$
$$= (x_1'x_2)' \cdot (x_2'x_3')' \cdot (x_1'x_3')'$$
$$= (x_1 + x_2') \cdot (x_2 + x_3) \cdot \boxed{(x_1 + x_3)}$$

The additional sum term can be implemented by an additional OR gate as shown in Fig. 4.39.

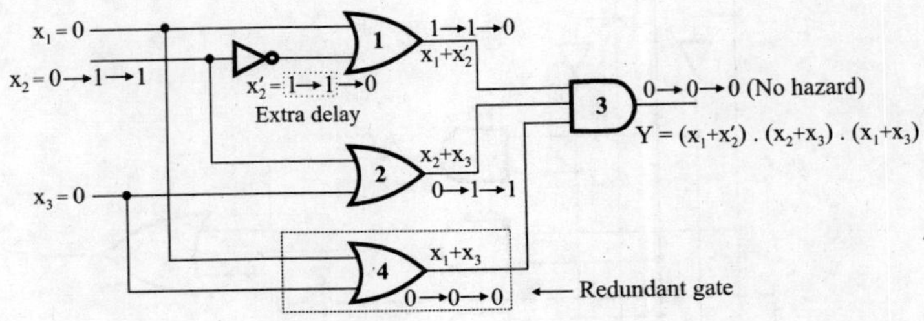

Fig. 4.39.

Alternatively, the additional sum term can be implemented by an additional NOR gate as shown in Fig. 4.40.

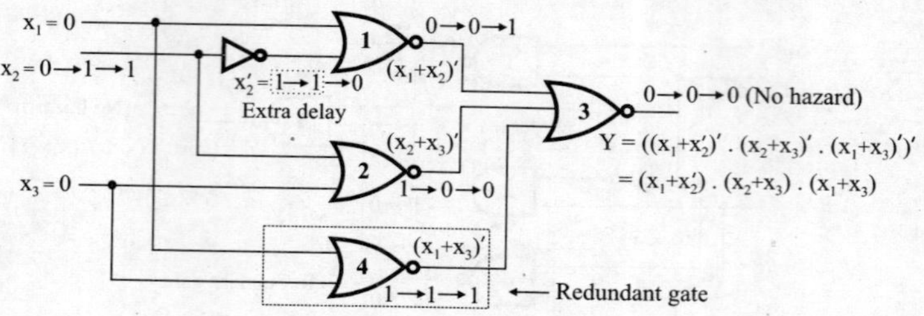

Fig. 4.40.

Example 4.19

Implement the following Boolean function by a logic circuit that has no hazards.

$F(A, B, C) = \Sigma(0, 2, 6, 7)$

Solution

The K-map of the given function is shown in Fig. 1. Here, the prime implicants A′C′ and AB are non-overlapped prime implicants that can create a static-1 hazard. Hence, let us add an additional prime implicant BC′ that creates overlapped prime implicants as shown in Fig. 2.

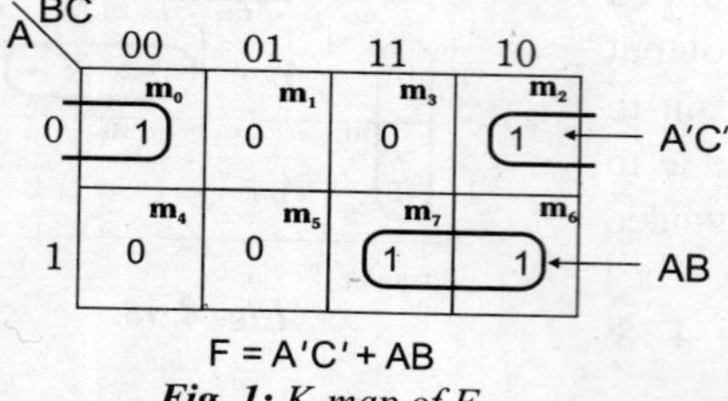

Fig. 1: K-map of F.

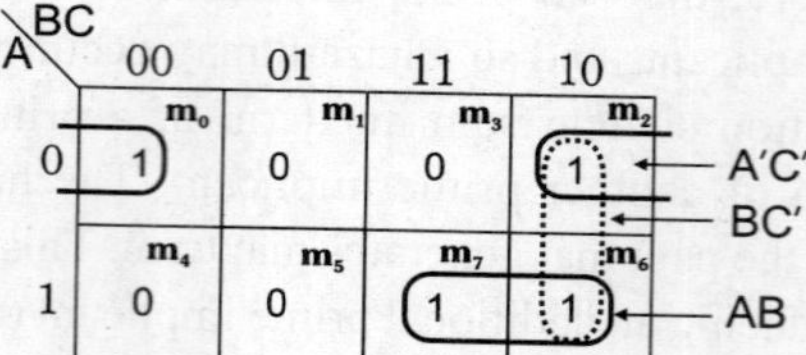

Fig. 2: Overlapping prime implicants to eliminate hazard.

The following Boolean equation is obtained from K-map of Fig. 2.

F = A′C′ + AB + BC′

Using the above Boolean equation the hazard free logic circuit is drawn as shown in Fig. 3. Here, the prime implicant BC′ add an extra redundant gate.

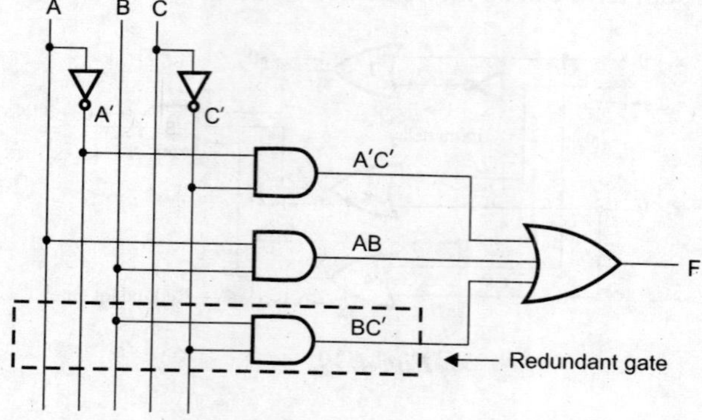

Fig. 3: Hazard free logic circuit.

Example 4.20

Implement the following Boolean function by a logic circuit that has no hazards.

F(A, B, C, D) = Σ(0, 2, 6, 7, 8, 10, 12)

Solution

The K-map of the given function is shown in Fig. 1. Here, the prime implicants B'D', AC'D' and A'BC are non-overlapped prime implicants that can create a static-1 hazard. Hence, let us add an additional prime implicant A'CD' that creates overlapped prime implicant as shown in Fig. 2.

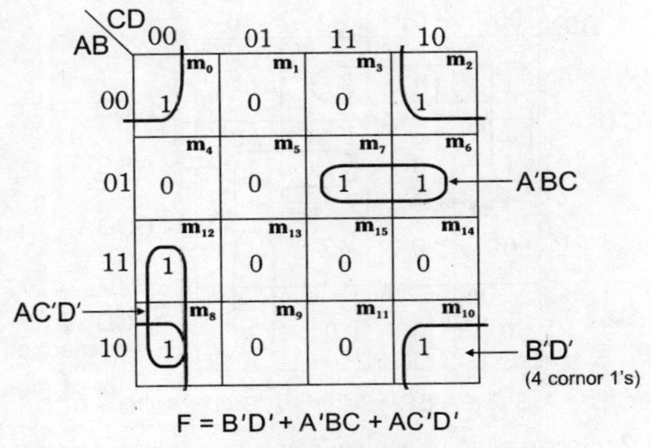

Fig. 1: K-map of F.

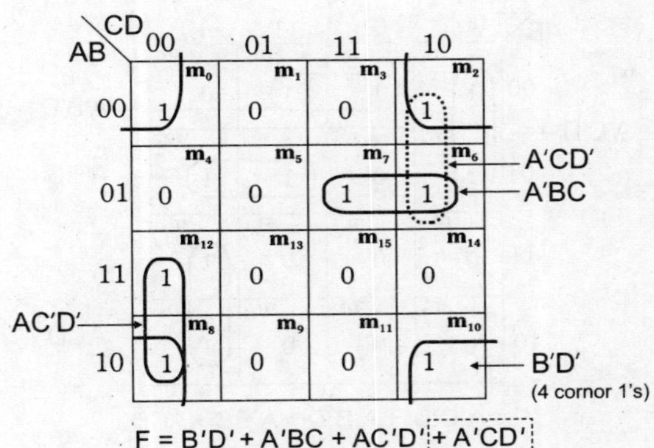

Fig. 2: Overlapping prime implicants to eliminate hazard.

The following Boolean equation is obtained from K-map of Fig. 2.

F = B'D' + A'BC + AC'D' + A'CD'

Using the above Boolean equation the hazard free logic circuit is drawn as shown in Fig. 3. Here, the prime implicant A'CD' add an extra redundant gate.

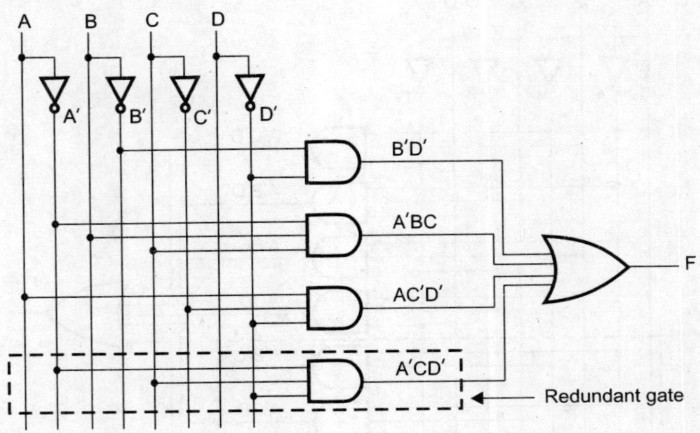

Fig. 3: Hazard free logic circuit.

Example 4.21

For a given Boolean function obtain the hazard free circuit.

$$F = \Sigma m(0, 1, 3, 4, 6, 7, 10, 14)$$

Solution

The K-map of the given function is shown in Fig. 1. Here, the prime implicants are non-overlapped prime implicants. Hence, three additional prime implicants are introduced to make overlapped prime implicants as shown in Fig. 2.

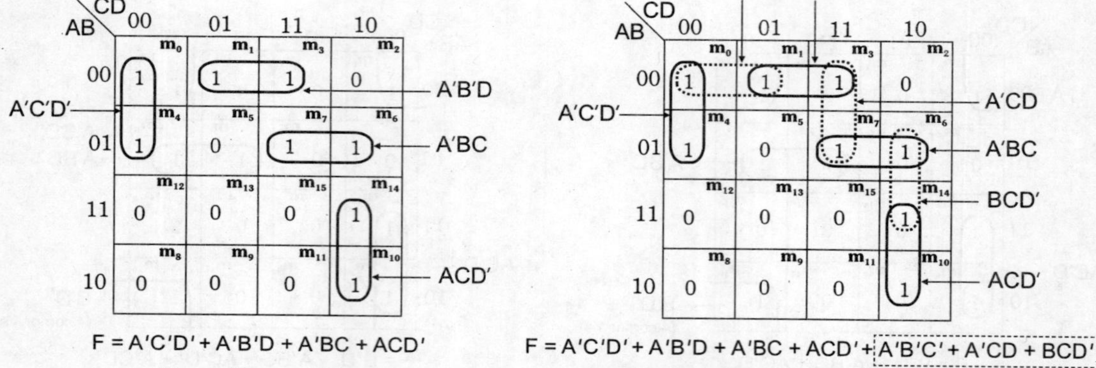

Fig. 1: K-map of F. *Fig. 2: Overlapping prime implicants to eliminate hazard.*

The following Boolean equation is obtained from K-map of Fig. 2.

$$F = A'C'D' + A'B'D + A'BC + ACD' + \boxed{A'B'C' + A'CD + BCD'}$$

Using the above Boolean equation the hazard free logic circuit is drawn as shown in Fig. 3. Hence, the thee extra prime implicants adds three redundant gates in the logic circuit.

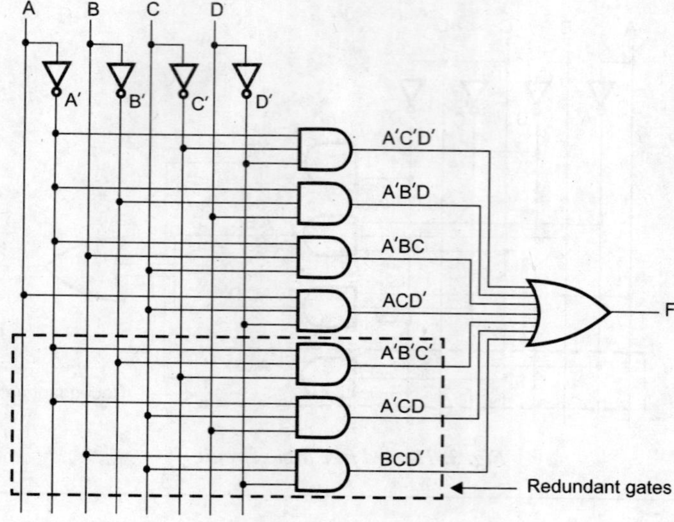

Fig. 3: Hazard free logic circuit.

Example 4.22

Implement the following Boolean function by a logic circuit that has no hazards.

$Y = \prod M(0, 1, 5, 7, 14, 15)$

Solution

The K-map of the given function is shown in Fig. 1. Here, the prime implicants B'D' and A'BC are non-overlapped prime implicants that can create a static-0 hazard. Hence, let us add an additional prime implicant A'CD' that creates overlapped prime implicants as shown in Fig. 2.

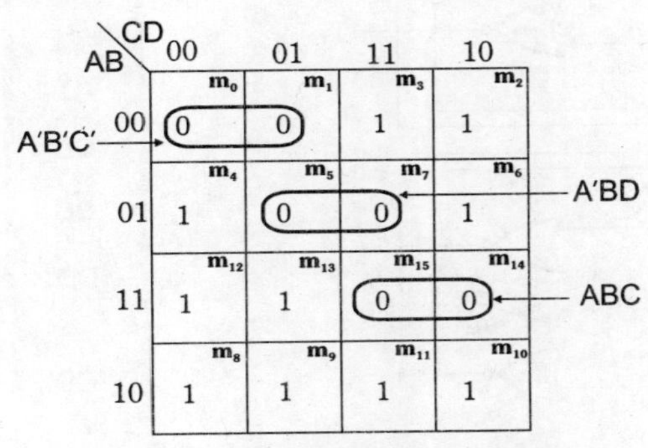

Fig. 1: K-map of F.

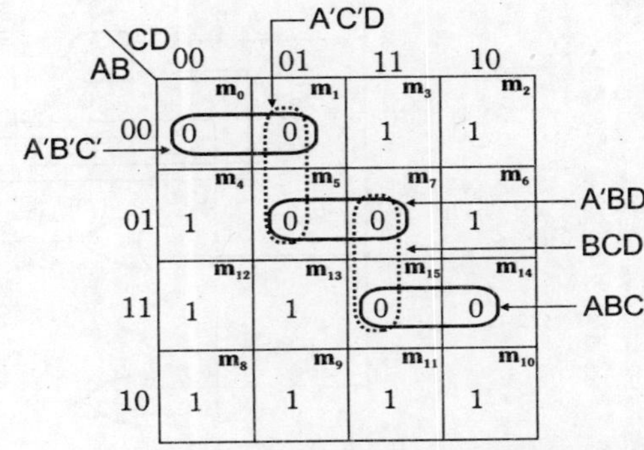

Fig. 2: Overlapping prime implicants to eliminate hazard.

The following Boolean equation is obtained from K-map of Fig. 2.

$$F = (ABC + A'B'C' + A'BD + A'C'D + BCD)'$$
$$= (ABC)' + (A'B'C')' + (A'BD)' + (A'C'D)' + (BCD)'$$
$$= (A' + B' + C') \,.\, (A + B + C) \,.\, (A + B' + D') \,.\, (A + C + D') \,.\, (B' + C' + D')$$

Using the above Boolean equation the hazard free logic circuit is drawn as shown in Fig. 3. Here, the prime implicants (A + C + D') and (B' + C' + D') adds an extra redundant gate each.

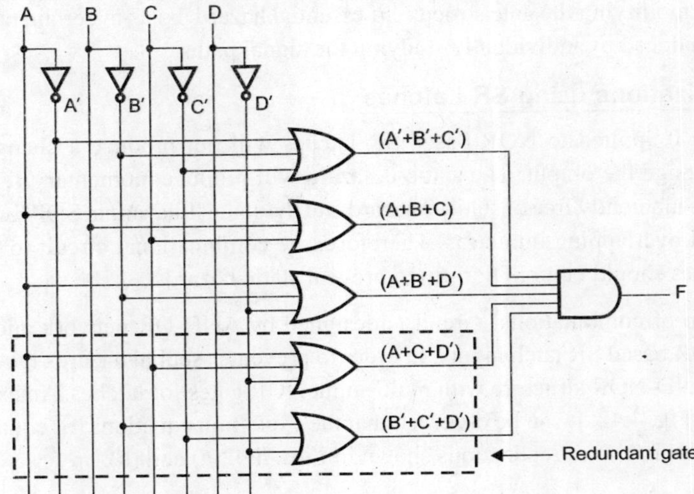

Fig. 3: Hazard free logic circuit.

4.7.2 Elimination of Hazards in Sequential Circuits

(AU, Nov/Dec'18, 13 Marks)

The essential hazard can be eliminated by rearranging the gate structure of the combinational circuit that generates inputs to latches such that the effect of input changes propagates to all latch inputs before propagation of feedback signals (or next state variables) to latch inputs.

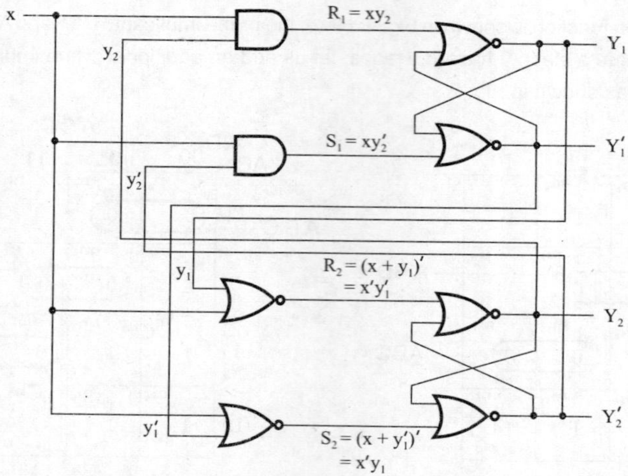

Fig. 4.41: Essential hazard free realization of the circuit shown in Fig. 4.34(a).

The essential hazard in the circuit of Fig. 4.34 can be eliminated by replacing the inverter and AND gates in the circuit of latch-2 by NOR gates as shown in Fig. 4.41.

Here, it is assumed that the gate delay is concentrated at the output. The replacement of NOT-AND combination by NOR gate will ensure that any change input, x will propagate to inputs of latch-2 before latch-1 change state. This ensure that the change in circuit inputs and feedback from latch-1 output arrive at same time to latch-2 inputs and so essential hazard is eliminated.

In general, the essential hazard is minimized or eliminated by reducing the complexity of gate structure. Even after modifying the gate structure if essential hazard exists in certain path then appropriate delays has to be introduced by individually studying the signal paths.

Hazard Free Realizations using SR Latches

A momentary 0 applied to NOR based SR latches will not produce a change in output. But a momentary 1 can change the output. The static-0 hazard will produce momentary 1. The combinational circuit in SOP form is inherently free of static-0 hazard and the static-1 hazard in SOP form can be prevented by adding additional overlapping minterms. Therefore, the combinational circuit to generate inputs for NOR based SR latches should be in SOP form to prevent static hazards.

The SOP form of combinational circuit is designed by AND-OR network and the output of OR gate is applied to NOR based SR latch inputs. In order to prevent essential hazards the AND-OR structure can be replaced by AND-NOR structure with multi-input NOR gates for latches. An example of modified structure is shown in Fig. 4.42. It can be observed that the excitation equations (or equations for Y and Y′) are same in both the logic circuit realizations shown in Figs. 4.42(a) and (b).

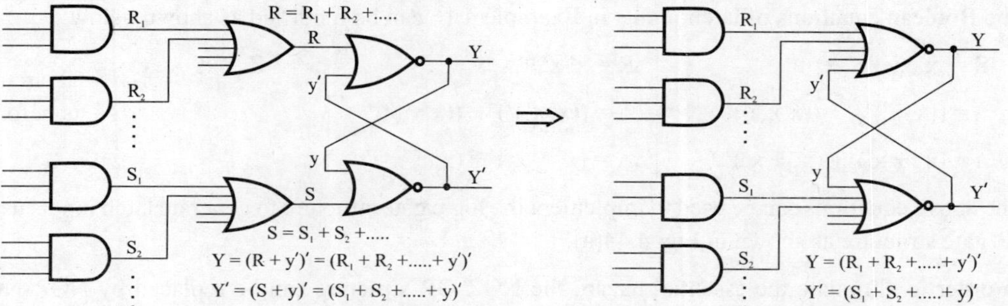

Fig. a: Combinational circuit in SOP form. *Fig. b: Equivalent circuit to eliminate essential hazard.*

Fig. 4.42: Elimination of essential hazard with modified gate structure.

Consider the logic circuit implementation using NOR based SR latch shown in Example 4.16.

The Boolean equations for latch inputs are,

$$S = x_1 x_2' + x_1' x_2 \quad ; \quad R = x_1' x_2' + x_1 x_2 \quad ; \quad Z = x_1'$$

The above equations are implemented using AND-OR gate structure as shown in Fig. 4.43(a). In this implementation the OR gate can be eliminated by using multi-input NOR gates for latch as shown in Fig. 4.43(b). It can be observed that the excitation equations (or equations for Y and Y′) are same in both the logic circuits shown in Figs. 4.43(a) and (b).

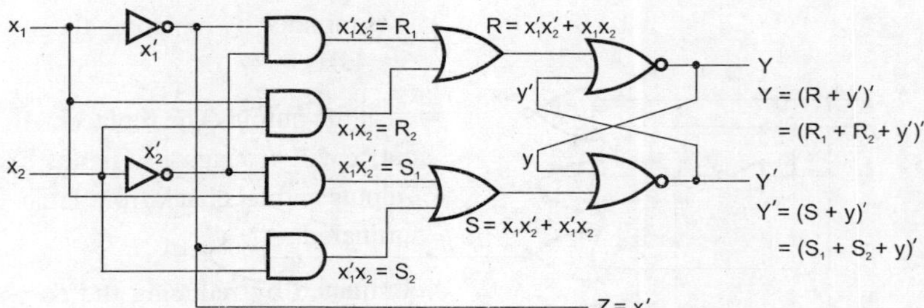

Fig. a: Combinational circuit with AND-OR gate structure.

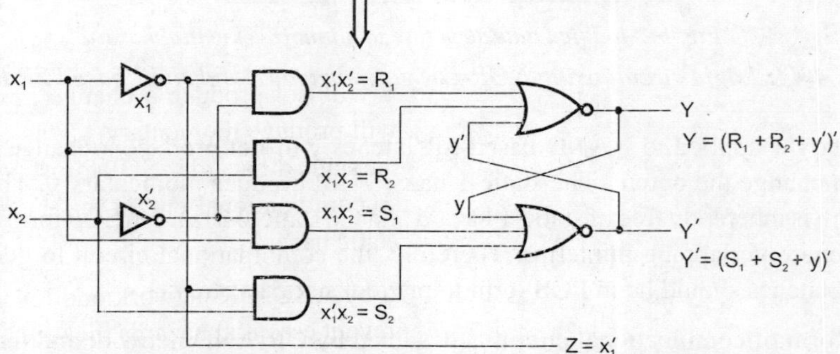

Fig. b: Modified gate structure to minimize essential hazard.

Fig. 4.43: Logic circuit using AND-OR gate structure and NOR based SR latch.

The Boolean equations of latch inputs in Example 4.16 can be modified as shown below:

$S = x_1 x_2' + x_1' x_2$	$R = x_1' x_2' + x_1 x_2$	$Z = x_1'$
$= ((x_1 x_2')')' + ((x_1' x_2)')'$	$= ((x_1' x_2')')' + ((x_1 x_2)')'$	
$= (x_1' + x_2)' + (x_1 + x_2')'$	$= (x_1 + x_2)' + (x_1' + x_2')'$	

The above equations can be used to implement the logic gate structure to generate latch inputs using NOR-OR gate structure as shown in Fig. 4.44(a).

In order to eliminate the essential hazard the NOR-OR structure can be replaced by NOR-NOR structure as shown in Fig. 4.44(b). It can be observed that the excitation equations (or equations for Y and Y') are same in both the logic circuits shown in Figs. 4.44(a) and (b).

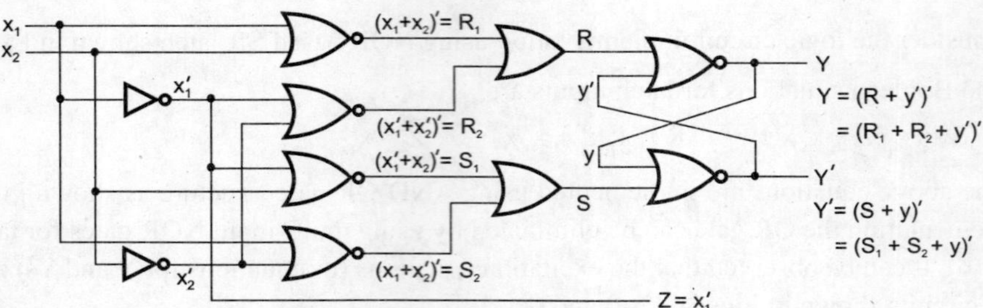

Fig. a: Combinational circuit with NOR-OR gate structure.

⇓

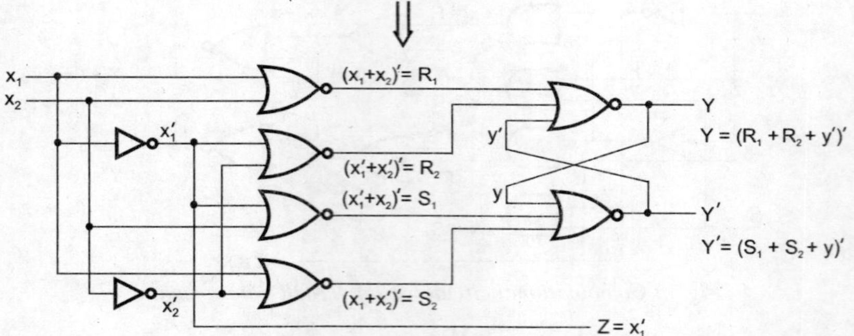

Fig. b: Modified gate structure to minimize essential hazard.

***Fig. 4.44:** Logic circuit using NOR-OR gate structure and NOR based SR latch.*

A momentary 1 applied to NAND based SR latches will not produce a change in output. But a momentary 0 can change the output. The static-1 hazard will produce momentary 0. The combinational circuit in POS form is inherently free of static-1 hazard and the static-0 hazard in SOP form can be prevented by adding additional overlapping minterms. Therefore, the combinational circuit to generate inputs for NAND based SR latches should be in POS form to prevent static hazards.

The POS form of combinational circuit is designed by OR-AND network and the output of AND gate is applied to NAND based SR latch inputs. In order to prevent essential hazards the OR-AND structure can be replaced by OR-NAND structure with multi-input NAND gates for latches. An example of modified structure is shown in Fig. 4.45. It can be observed that the excitation equations (or equations for Y and Y') are same in both the logic circuit realizations shown in Figs. 4.45(a) and (b).

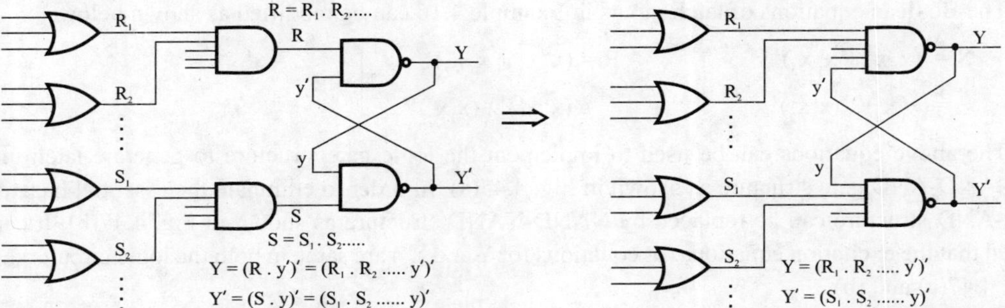

***Fig. a:** Combinational circuit in POS form.* ***Fig. b:** Equivalent circuit to eliminate essential hazard.*

***Fig. 4.45:** Elimination of essential hazard with modified gate structure.*

Consider the logic circuit implementation using NAND based SR latch shown in Example 4.16.

The Boolean equations for latch inputs in Example 4.16 are modified as shown below:

$$S = (x_1' + x_2) . (x_1 + x_2') \quad ; \quad R = (x_1 + x_2)(x_1' + x_2') \quad ; \quad Z = x_1'$$

The above equations are implemented using AND-OR gate structure as shown in Fig. 4.46(a). In this implementation the AND gate can be eliminated by using multi-input NAND gates for latch as shown in Fig. 4.46(b). It can be observed that the excitation equations (or equations for Y and Y′) are same in both the logic circuits shown in Figs. 4.46(a) and (b).

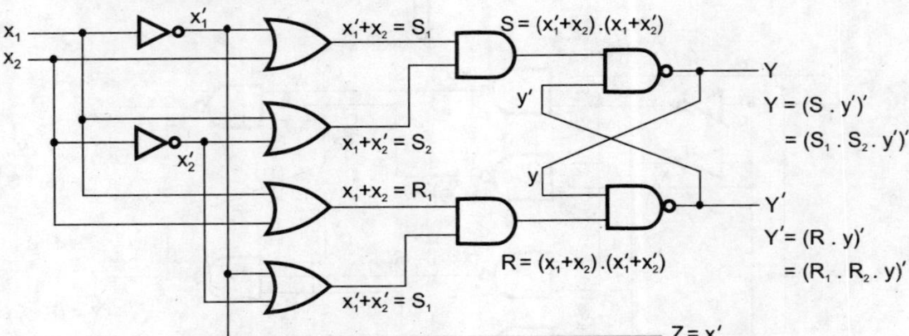

***Fig. a:** Combinational circuit with OR-AND gate structure.*

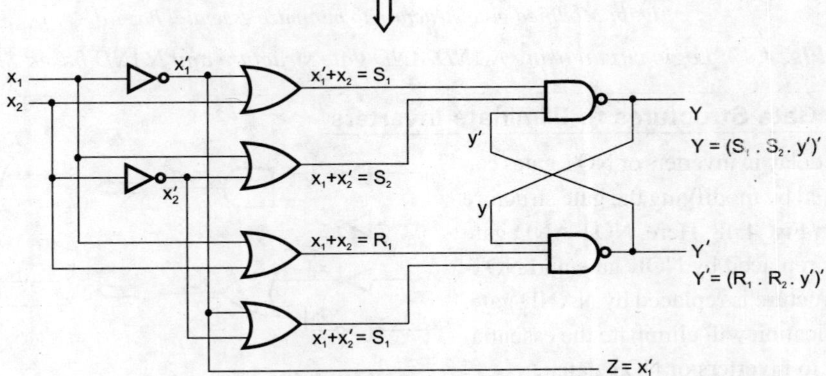

***Fig. b:** Modified gate structure to minimize essential hazard.*

***Fig. 4.46:** Logic circuit using OR-AND gate structure and NAND based SR latch.*

The Boolean equations of latch inputs in Example 4.16 can be modified as shown below:

$$S = (x_1 x_2' + x_1' x_2)' = (x_1 x_2')' . (x_1' x_2)' \quad \Big| \quad R = (x_1' x_2' + x_1 x_2)' = (x_1' x_2')' . (x_1 x_2)' \quad \Big| \quad Z = x_1'$$

The above equations can be used to implement the logic gate structure to generate latch inputs using NAND-AND gate structure as shown in Fig. 4.45(a). In order to eliminate the essential hazard the NAND-AND structure can be replaced by NAND-NAND structure as shown in Fig. 4.47(b). It can be observed that the excitation equations (or equations for Y and Y') are same in both the logic circuits shown in Figs. 4.47(a) and (b).

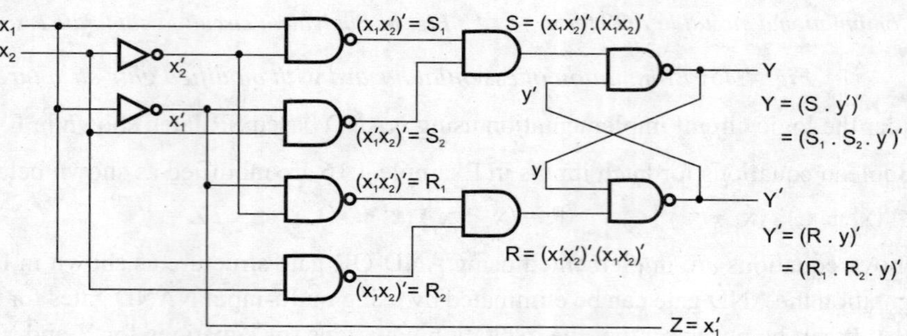

Fig. a: Combinational circuit with NAND-AND gate structure.

⇓

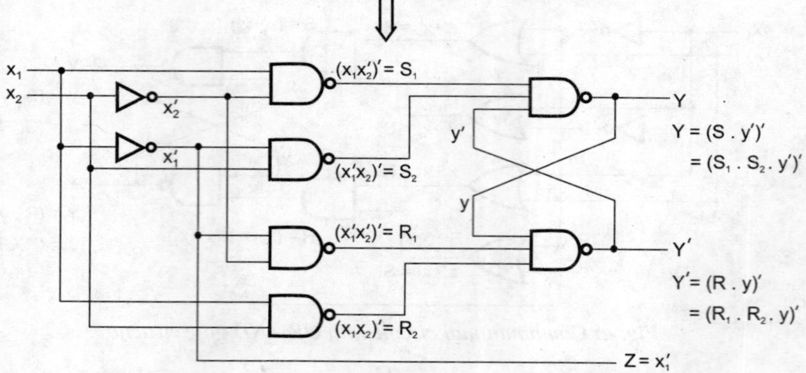

Fig. b: Modified gate structure to minimize essential hazard.

***Fig. 4.47:** Logic circuit using NAND-AND gate structure and NAND based SR latch.*

Modified Gate Structures to Eliminate Inverters

The delay in inverters or NOT gates can be eliminated by modifying the gate structure as shown in Fig. 4.48. Here, NOT-AND gate structure is replaced by NOR gate and NOT-OR gate structure is replaced by NAND gate. This modification will eliminate the essential hazard due to inverters or NOT gate.

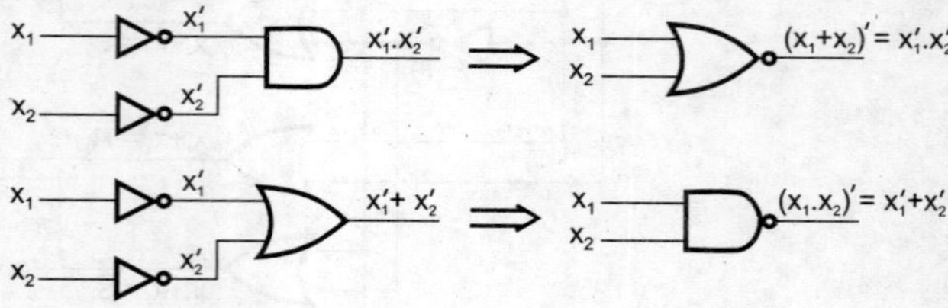

***Fig. 4.48:** Gate structure to replace inverters.*

Example 4.23

Assign output values to the don't-care states in the flow tables of Fig. 1 in such a way to avoid tranisent output pulses or hazard.

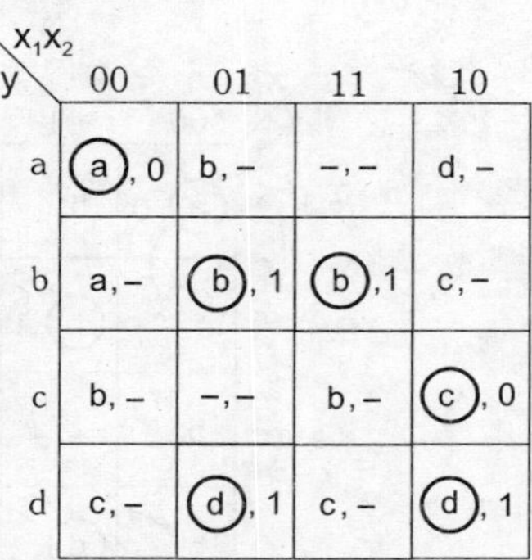

y \ x_1x_2	00	01	11	10
a	(a), 0	b, –	–, –	d, –
b	a, –	(b), 1	(b),1	c, –
c	b, –	–, –	b, –	(c), 0
d	c, –	(d), 1	c, –	(d), 1

Fig. 1: Reduced flow table.

Solution

In the reduced flow table of Fig. 1, the dashes in the output are assigned values according to the following rules.

Rule 1: Assign 0 to an output variable of an unstable state which is transient state between two stable states that have a 0 in the corresponding output variable.

Rule 2: Assign 1 to an output variable of an unstable state which is transient state between two stable states that have a 1 in the corresponding output variable.

Rule 3: Assign a don't-care condition to an output of an unstable state which is transient state during a transition between two stable states that have a different values (0 and 1 or 1 and 0) as output.

The state transitions that are used to assign output using the above rules are shown in Figs. 2 to 5. Here, there are four combinations of inputs. The input is allowed to change only one bit at any one time. Here, the outputs assigned in Fig. 2(b), are retained during next state transitions in Fig. 3(b). Similarly, the outputs assigned in previous transitions are retained during next state transitions in Fig. 3(d) and so on. The state transitions and the assigned outputs are listed in Table 1 and from this table it can be observed that the assigned output does not produce any hazard.

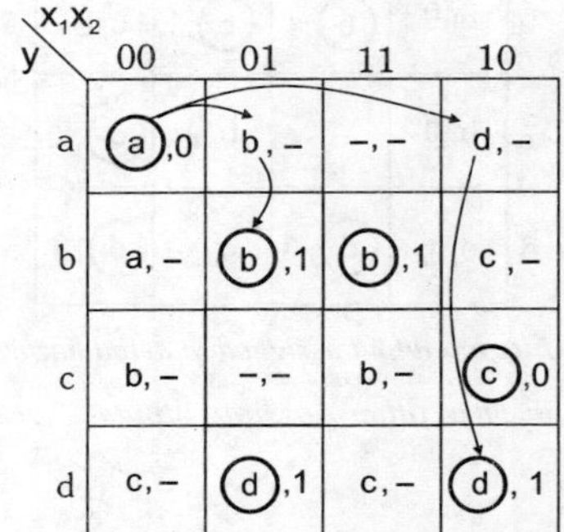

y \ x_1x_2	00	01	11	10
a	(a),0	b, –	–, –	d, –
b	a, –	(b),1	(b),1	c , –
c	b, –	–,–	b,–	(c),0
d	c, –	(d),1	c, –	(d), 1

Fig. a: Transitions from stable state a.

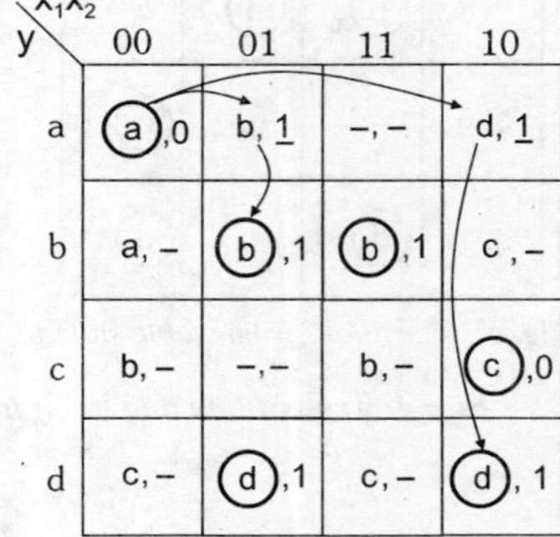

y \ x_1x_2	00	01	11	10
a	(a),0	b, 1	–, –	d, 1
b	a, –	(b),1	(b),1	c , –
c	b, –	–,–	b,–	(c),0
d	c, –	(d),1	c, –	(d), 1

Fig. b: Output assigned to avoid hazard.

Fig. 2: Transitions due to change in input from 00 to other possible inputs.

y \ x_1x_2	00	01	11	10
a	(a), 0	b, 1	–, –	d, 1
b	a, _	(b),1	(b),1	c, –
c	b, –	–, –	b, –	(c), 0
d	c, –	(d), 1	c, –	(d), 1

Fig. a: Transitions from stable state b.

y \ x_1x_2	00	01	11	10
a	(a), 0	b, 1	–, –	d, 1
b	a, ×	(b),1	(b),1	c, –
c	b, –	–, –	b, –	(c), 0
d	c, –	(d), 1	c, –	(d), 1

Fig. b: Output assigned to aviod hazard.

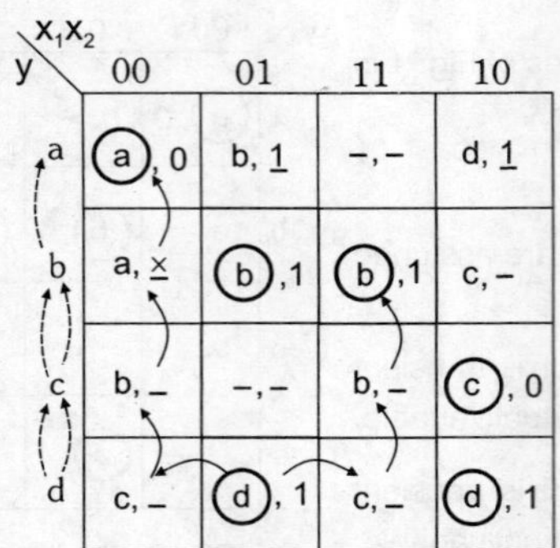

y \ x_1x_2	00	01	11	10
a	(a), 0	b, 1	–, –	d, 1
b	a, ×	(b), 1	(b), 1	c, –
c	b, _	–, –	b, _	(c), 0
d	c, –	(d), 1	c, –	(d), 1

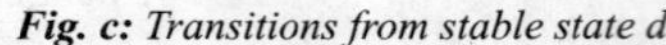

Fig. c: Transitions from stable state d.

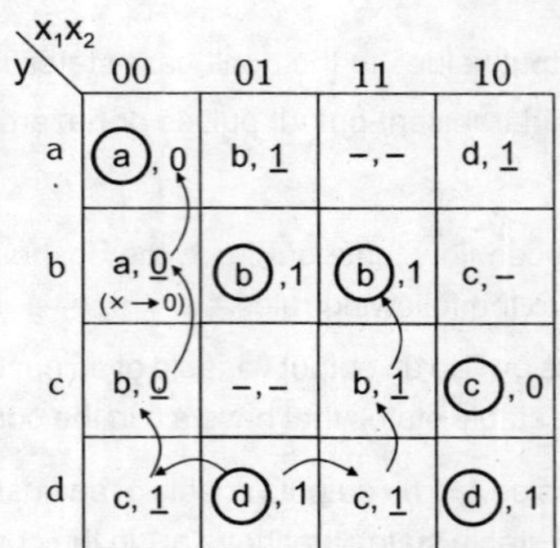

y \ x_1x_2	00	01	11	10
a	(a), 0	b, 1	–, –	d, 1
b	a, 0 (× → 0)	(b), 1	(b), 1	c, –
c	b, 0	–, –	b, 1	(c), 0
d	c, 1	(d), 1	c, 1	(d), 1

Fig. d: Output assigned to avoid hazard.

***Fig. 3:** Transitions due to change in input from 01 to other possible inputs.*

y \ x_1x_2	00	01	11	10
a	(a), 0	b, 1	–, –	d, 1
b	a, 0	(b), 1	(b), 1	c, –
c	b, 0	–, –	b, 1	(c), 0
d	c, 1	(d), 1	c, 1	(d), 1

Fig. a: Transitions from stable state b.

y \ x_1x_2	00	01	11	10
a	(a), 0	b, 1	–, –	d, 1
b	a, 0	(b), 1	(b), 1	c, ×
c	b, 0	–, –	b, 1	(c), 0
d	c, 1	(d), 1	c, 1	(d), 1

Fig. b: Output assigned to aviod hazard.

***Fig. 4:** Transitions due to change in input from 11 to other possible inputs.*

y \ x_1x_2	00	01	11	10
a	(a), 0	b, 1	–, –	d, 1
b	a, 0	(b), 1	(b), 1	c, ×
c	b, 0	–, –	b, 1	(c), 0
d	c, 1	(d), 1	c, 1	(d), 1

Fig. a: Transitions from stable state c.

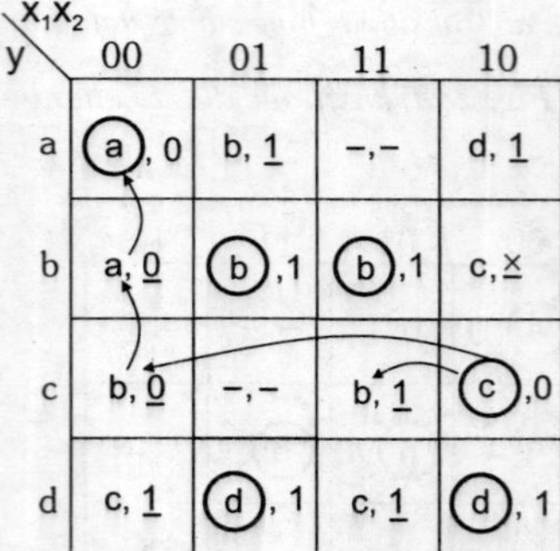

y \ x_1x_2	00	01	11	10
a	(a), 0	b, 1	–, –	d, 1
b	a, 0	(b), 1	(b), 1	c, ×
c	b, 0	–, –	b, 1	(c), 0
d	c, 1	(d), 1	c, 1	(d), 1

Fig. b: Output assigned to avoid hazard.

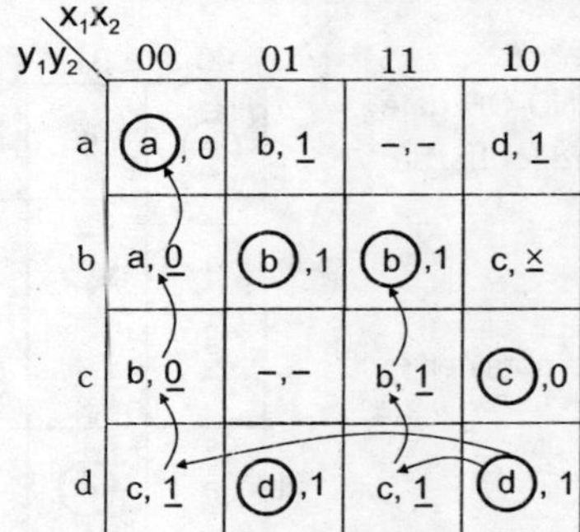

***Fig. c:** Transitions from stable state d.*

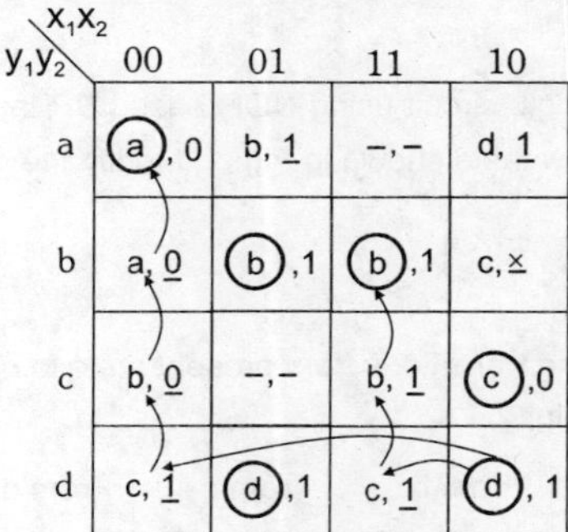

***Fig. d:** Output assigned to avoid hazard.*

***Fig. 5:** Transitions due to change in input from 10 to other possible inputs.*

y_1y_2 \ x_1x_2	00	01	11	10
a	(a), 0	b, 1	–, –	d, 1
b	a, 0	(b), 1	(b), 1	c, ×
c	b, 0	–, –	b, 1	(c), 0
d	c, 1	(d), 1	c, 1	(d), 1

***Fig. 6:** Flow table with output assigned to avoid hazards.*

Table 1:

Present Input	Present Stable State	Change in Input	State Transitions	Output Changes	Hazard
00	a	00 → 01	a → b → b	0 → 1 → 1	No hazard
		00 → 10	a → d → d	0 → 1 → 1	No hazard
01	b	01 → 00	b → a → a	1 → 0 → 0	No hazard
01	d	01 → 00	d → c → b → a → a	1 → 1 → 0 → 0 → 0	No hazard
		01 → 11	d → c → b → b	1 → 1 → 1 → 1	No hazard
11	b	11 → 10	b → c → c	1 → 0 → 0 (× → 0)	No hazard
10	c	10 → 00	c → b → a → a	0 → 0 → 0 → 0	No hazard
		10 → 11	c → b → b	0 → 1 → 1	No hazard
10	d	10 → 00	d → c → b → a → a	1 → 1 → 0 → 0 → 0	No hazard
		10 → 11	d → c → b → b	1 → 1 → 1 → 1	No hazard

Example 4.24

Realize a logic circuit using NOR based SR latches and AND-OR gate structure for the flow table shown in Fig. 1. Modify the gate structure for hazard free realization.

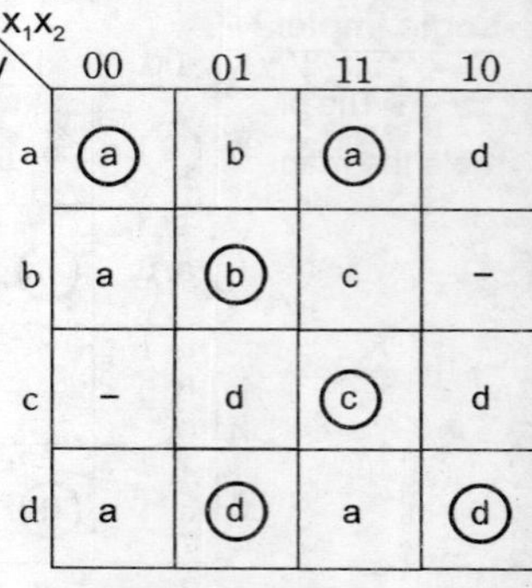

y \ x_1x_2	00	01	11	10
a	(a)	b	(a)	d
b	a	(b)	c	–
c	–	d	(c)	d
d	a	(d)	a	(d)

Fig. 1: Reduced flow table.

Solution

The following transitions from present state to next state are obtained in transition table in Fig. 2.

From a	From b	From c	From d
a → b	b → a	c → d	d → a
a → d	b → c		

The above transitions are combined as a single diagram in the transition diagram shown in Fig. 3. For a critical race free state assignment the state transitions in Fig. 3 should have adjacent state assignments. Here, it is possible to achieve adjacent state assignments with the following 2-bit binary codes as shown in Fig. 4.

The two bits of binary codes are represented by two state variables y_1 and y_2. The transition table with binary code assignment is shown in Fig. 5.

a → 00	c → 11	$y = y_1y_2$
b → 01	d → 10	

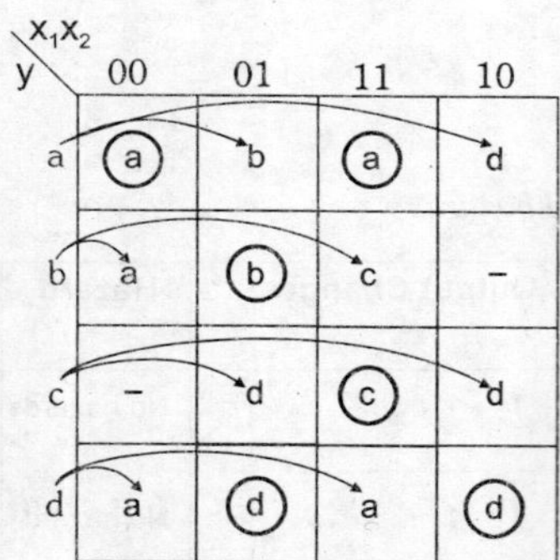

y \ x_1x_2	00	01	11	10
a	(a)	b	(a)	d
b	a	(b)	c	–
c	–	d	(c)	d
d	a	(d)	a	(d)

Fig. 2: Transitions from present state to next state.

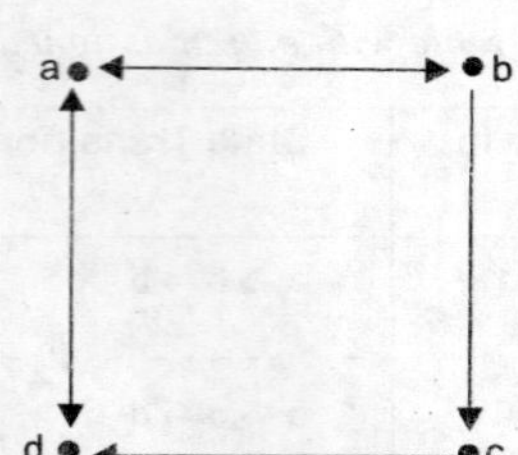

Fig. 3: Transition diagram.

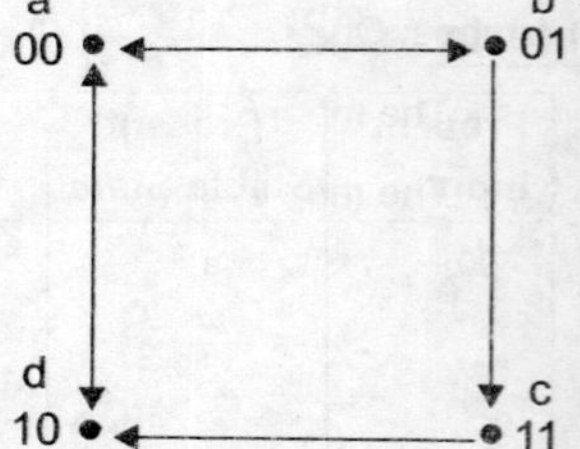

Fig. 4: Transition diagram with binary code assignment.

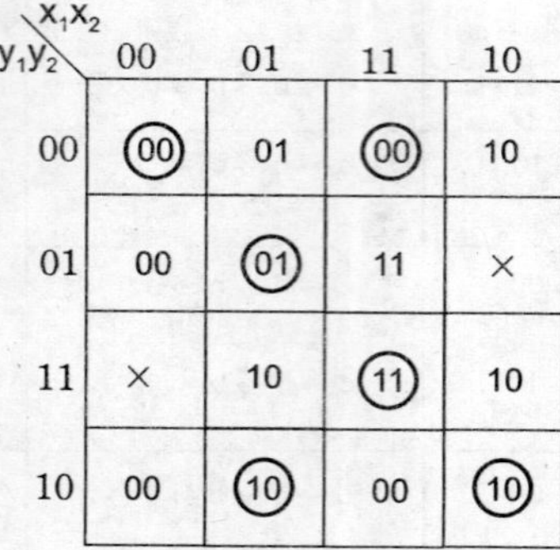

y_1y_2 \ x_1x_2	00	01	11	10
00	(00)	01	(00)	10
01	00	(01)	11	×
11	×	10	(11)	10
10	00	(10)	00	(10)

Fig. 5: Transition table with binary code assignment.

Logic Implementation using NOR based SR Latch

The binary values inside each square of transition table is the value of next state $Y_1 Y_2$. Hence, using the transition table the maps of Y_1 and Y_2 are drawn as shown in Figs. 6 and 7.

y_1y_2 \ x_1x_2	00	01	11	10
00	0	0	0	1
01	0	0	1	×
11	×	1	1	1
10	0	1	0	1

***Fig. 6:** K-map for Y_1.*

y_1y_2 \ x_1x_2	00	01	11	10
00	0	1	0	0
01	0	1	1	×
11	×	0	1	0
10	0	0	0	0

***Fig. 7:** K-map for Y_2.*

In order to store the two state variables Y_1 and Y_1, two SR latches are required. Let us choose two NOR based SR latches.

Let, S_1, R_1 = Inputs for SR latch-1

S_2, R_2 = Inputs for SR latch-2

To Find Equations for Latch Inputs using Maps of Y_1 and Y_2

In latch-1, the input equations for S_1 and R_1 will be independent of y_1. Therefore, the map of Y_1 can be split into two maps: One map when $y_1 = 0$ and other map when $y_1 = 1$ as shown in Fig. 8.

The map $y_1 = 0$ is used to find equation for S_1 by forming prime implicants using 1's.

The map $y_1 = 1$ is used to find equation for R_1' by forming prime implicants using 0's.

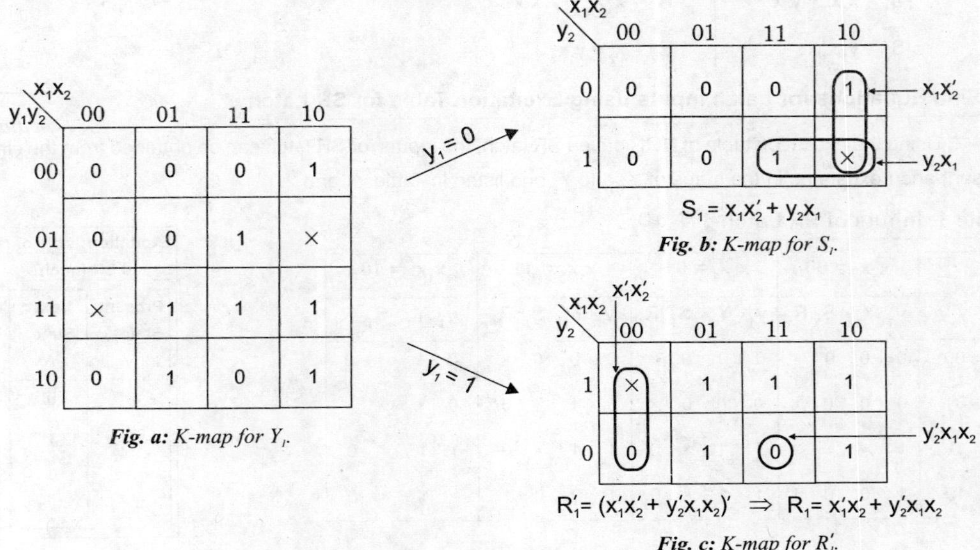

***Fig. 8:** K-maps to find equations for S_1 and R_1.*

In latch-2, the input equations for S_2 and R_2 will be independent of y_2. Therefore, the map of Y_2 can be split into two maps: One map when $y_2 = 0$ and other map when $y_2 = 1$ as shown in Fig. 9.

The map $y_2 = 0$ is used to find equation for S_2 by forming prime implicants using 1's.

The map $y_2 = 1$ is used to find equation for R_2' by forming prime implicants using 0's.

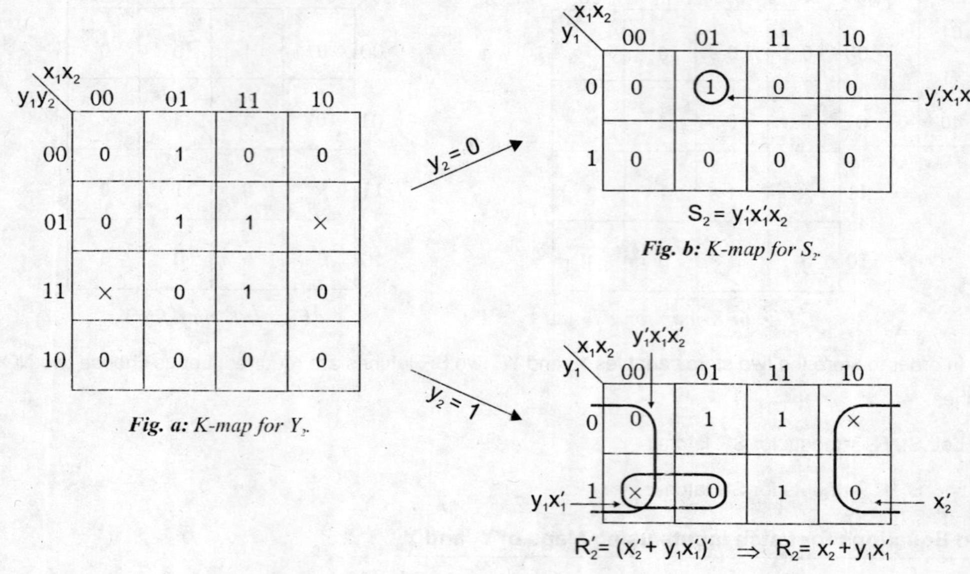

Fig. a: K-map for Y_2.

Fig. b: K-map for S_2.

Fig. c: K-map for R_2'.

Fig. 9: K-maps to find equations for S_2 and R_2.

Form the maps of S_1, R_1', S_2 and R_2' we get the following equations for SR latch inputs.

$S_1 = x_1x_2' + y_2x_1$ | $R_1 = x_1'x_2' + y_2'x_1x_2$

$S_2 = y_1'x_1'x_2$ | $R_2 = x_2' + y_1x_1'$

To Find Equations for Latch Inputs using Excitation Table for SR Latch

Using the excitation table of NOR based SR latch, the inputs for SR latch can be obtained from the circuit inputs, present and next states in the maps of Y_1 and Y_2 and listed in Tables 1 and 2.

Table 1: Inputs of SR Latch-1

	$x_1x_2 = 00$				$x_1x_2 = 01$				$x_1x_2 = 11$				$x_1x_2 = 10$			
y_1y_2	y_1	Y_1	S_1	R_1	y_1	Y_1	S_1	R_1	y_1	Y_1	S_1	R_1	y_1	Y_1	S_1	R_1
00	0	0	0	×	0	0	0	×	0	0	0	×	0	1	1	0
01	0	0	0	×	0	0	0	×	0	1	1	0	0	×	×	×
11	1	×	×	×	1	1	×	0	1	1	×	0	1	1	×	0
10	1	0	0	1	1	1	×	0	1	0	0	1	1	1	×	0

Excitation Table of NOR based SR Latch

Present State y	Next State Y	Inputs for Next state S	R
0	0	0	×
0	1	1	0
1	0	0	1
1	1	×	0

Table 2: Inputs of SR Latch-2

	$x_1x_2 = 00$				$x_1x_2 = 01$				$x_1x_2 = 11$				$x_1x_2 = 10$			
y_1y_2	y_2	Y_2	S_2	R_2	y_2	Y_2	S_2	R_2	y_2	Y_2	S_2	R_2	y_2	Y_2	S_2	R_2
00	0	0	0	×	0	1	1	0	0	0	0	×	0	0	0	×
01	1	0	0	1	1	1	×	0	1	1	×	0	1	×	×	×
11	1	×	×	×	1	0	0	1	1	1	×	0	1	0	0	1
10	0	0	0	×	0	0	0	×	0	0	0	×	0	0	0	×

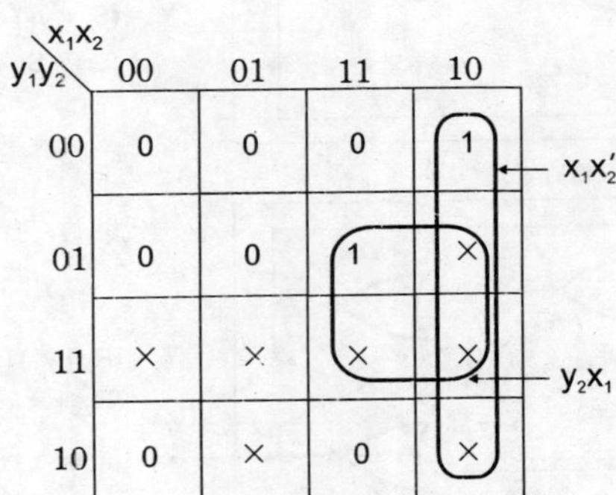

Fig. 10: K-map for S_1.

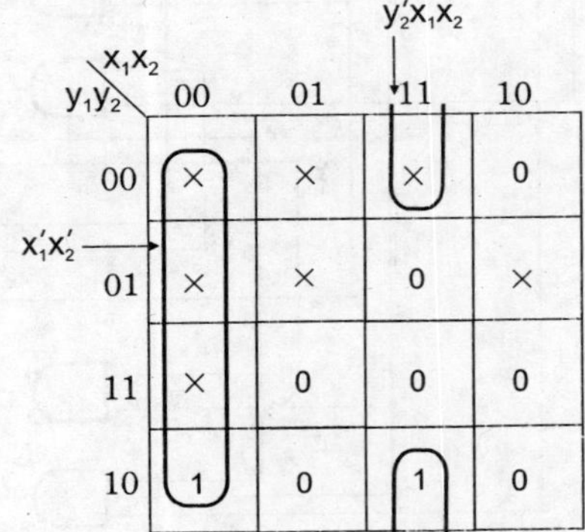

Fig. 11: K-map for R_1.

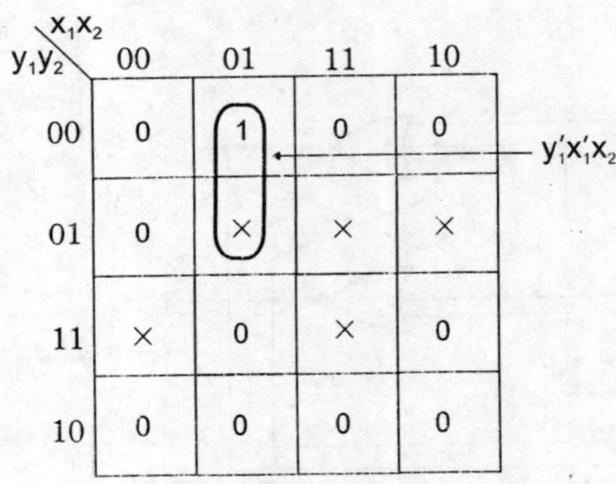

Fig. 12: K-map for S_2.

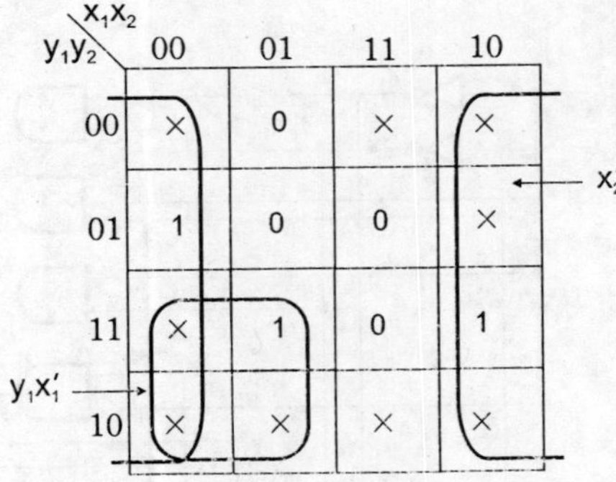

Fig. 13: K-map for R_2.

From Figs. 10 to 13, the maps for S_1, R_1, S_2, R_2 expressions are as shown below:

$S_1 = x_1x_2' + y_2x_1$	$R_1 = x_1'x_2' + y_2'x_1x_2$
$S_2 = y_1'x_1'x_2$	$R_2 = x_2' + y_1x_1'$

Logic Diagram

The Boolean equations obtained for latch inputs in both methods are same and using these equations the combinational logic circuit to generate latch inputs are drawn as shown in Fig. 14(a). The complete logic circuit is obtained by adding NOR based SR latches to the combinational circuit as shown in Fig. 14(a).

In Fig. 14(a), the combinational circuit is implemented using AND-OR gate structure. In order to prevent essential hazards the AND-OR structure can be replaced by AND-NOR structure with multi-input NOR gates for latches as shown in Fig. 14(b).

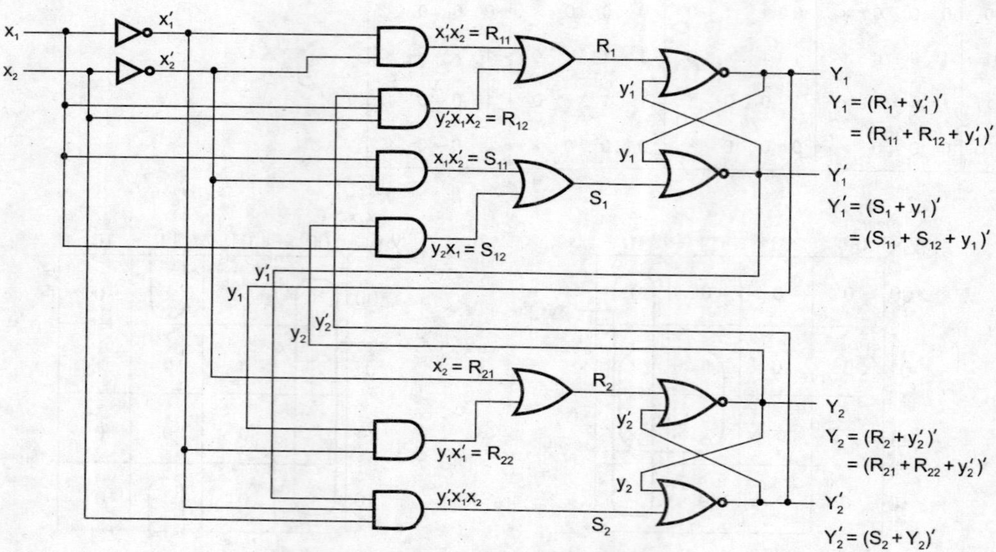

Fig. a: Combinational circuit with AND-OR gate structure.

⇓

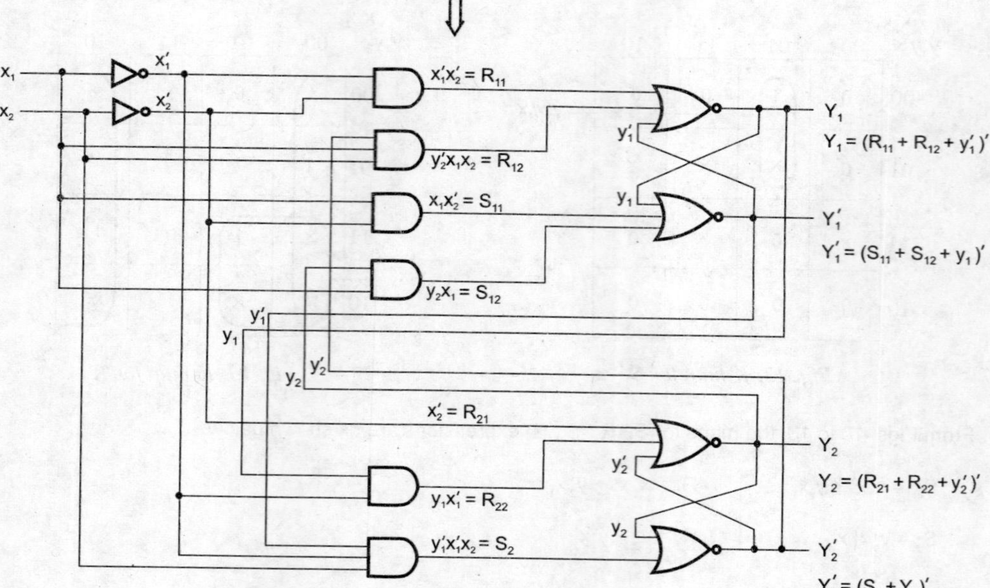

Fig. b: Modified gate structure to minimize essential hazard.

***Fig. 14:** Logic circuit using AND-OR gate structure and NOR based SR latch.*

Modification of Gate Structure to Eliminate Essential Hazard Due to Inverters

In the combinational logic circuit to generate latch inputs in Fig. 14(b), the inverters can be replaced by modified gate structure shown in Fig. 15. This will eliminate the essential hazards due to inveters. The modified complete logic circuit is shown in Fig. 16.

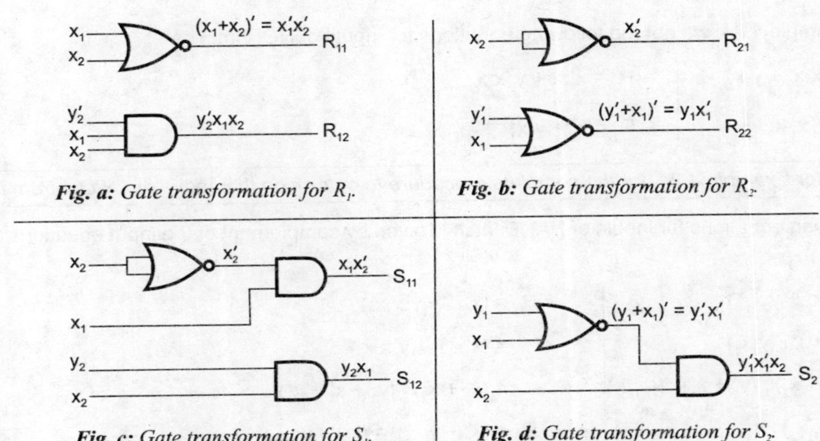

Fig. a: Gate transformation for R_1. *Fig. b: Gate transformation for R_2.*

Fig. c: Gate transformation for S_1. *Fig. d: Gate transformation for S_2.*

***Fig. 15:** Gate transformation to eliminate essential hazards.*

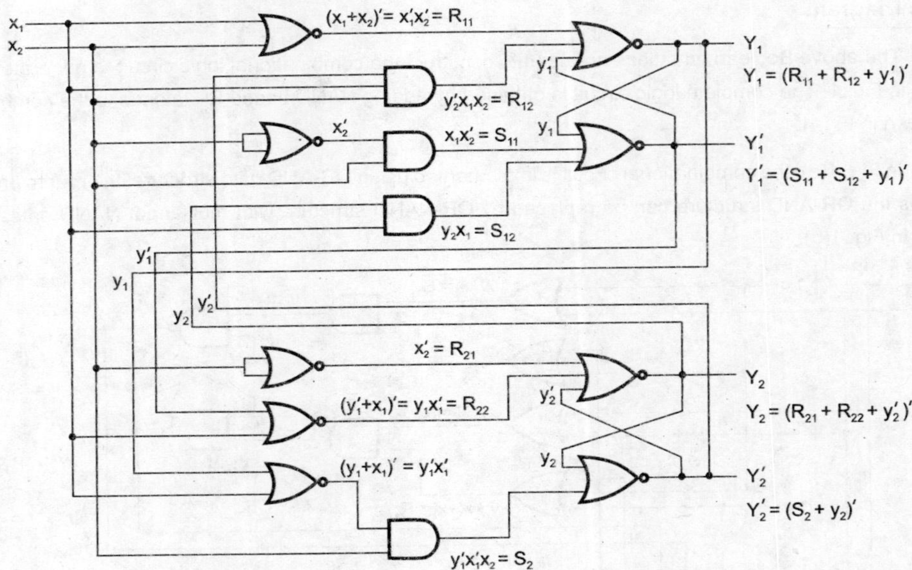

***Fig. 16:** Logic circuit using NOR based SR latch with modified gate structure to eliminate essential hazard due to inverters.*

Example 4.25

Realize a logic circuit using NAND based SR latches and OR-AND gate structure for the flow table of Example 4.24. Modify the gate structure for hazard free realization.

Solution

From Example 4.24, we get the following equations for inputs of NOR latches.

$S_1 = x_1x_2' + y_2x_1$ | $R_1 = x_1'x_2' + y_2'x_1x_2$

$S_2 = y_1'x_1'x_2$ | $R_2 = x_2' + y_1x_1'$

Note: *Refer Example 4.24, for the complete procedure to determine the above equations from the flow table.*

The Boolean equations for inputs of NAND latch is given by complement of the input equations of NOR based SR latch.

$$\therefore S_1 = (x_1x_2' + y_2x_1)'$$
$$= (x_1x_2')' . (y_2x_1)'$$
$$= (x_1' + x_2) . (y_2' + x_1')$$
$$S_2 = (y_1'x_1'x_2)'$$
$$= y_1 + x_1 + x_2'$$

$$R_1 = (x_1'x_2' + y_2'x_1x_2)'$$
$$= (x_1'x_2')' . (y_2'x_1x_2)'$$
$$= (x_1 + x_2) . (y_2 + x_1' + x_2')$$
$$R_2 = (x_2' + y_1x_1')' = (x_2')' . (y_1x_1')'$$
$$= x_2 . (y_1' + x_1)$$

Logic Diagram

The above Boolean equations are used to construct the combinational logic circuit to generate latch inputs as shown in Fig. 1. The complete logic circuit is obtained by adding NAND based SR latches to the combinational circuit as shown in Fig. 1.

In Fig. 1(a), the combinational circuit is implemented using OR-AND gate structure. In order to prevent essential hazards the OR-AND structure can be replaced by OR-NAND structure with multi-input NAND gates for latches as shown in Fig. 1(b).

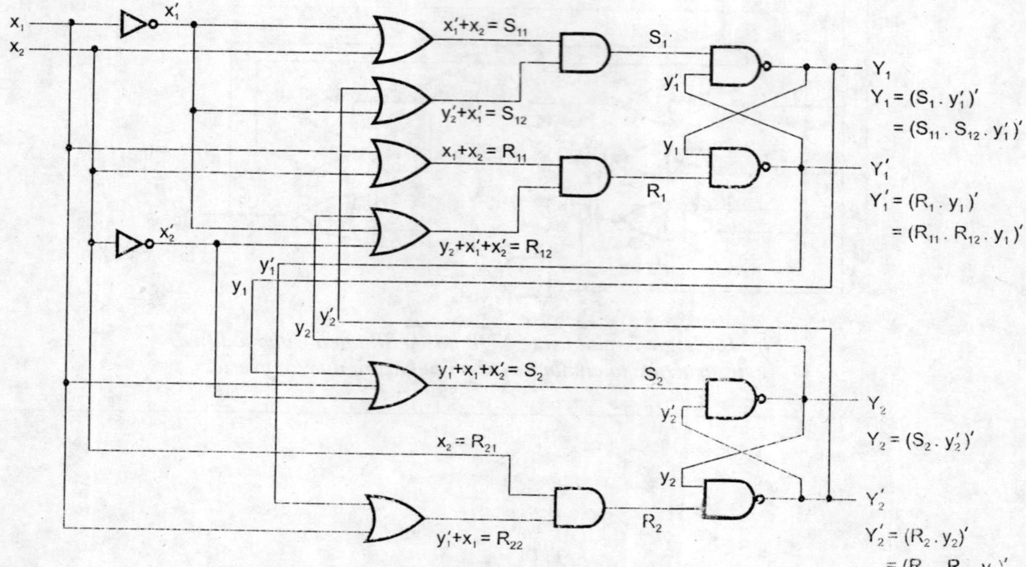

Fig. a: Combinational circuit with OR-AND gate structure.

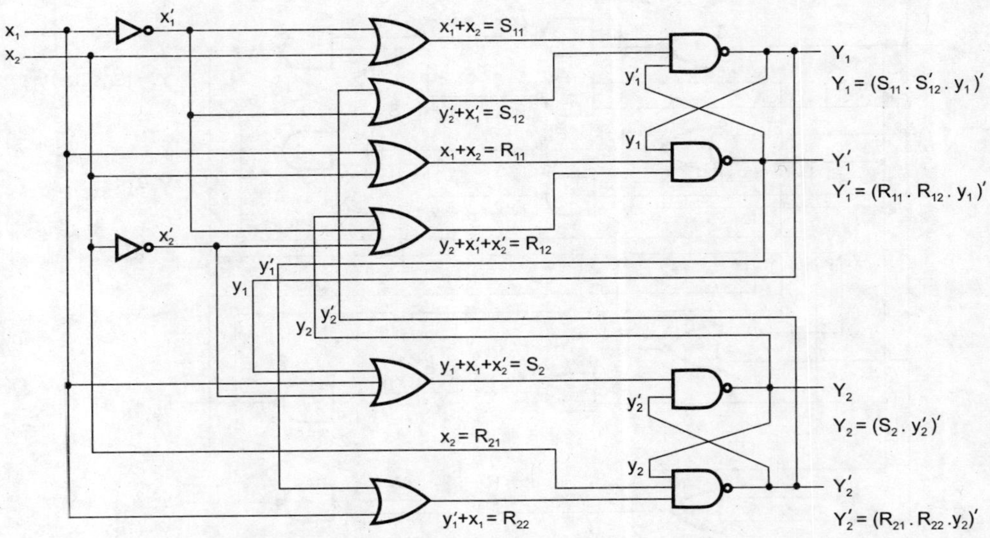

Fig. b: Modified gate structure to minimize essential hazard.

Fig. 1: Logic circuit using OR-AND gate structure and NAND based SR latch.

Modification of Gate Structure to Eliminate Essential Hazard

In the combinational circuit to generate latch input in the circuit of Fig. 1(b), the inverters can be replaced by modified gate structure shown in Fig. 2. This will eliminate the essential hazards due to inveters. The modified complete logic circuit is shown in Fig. 3.

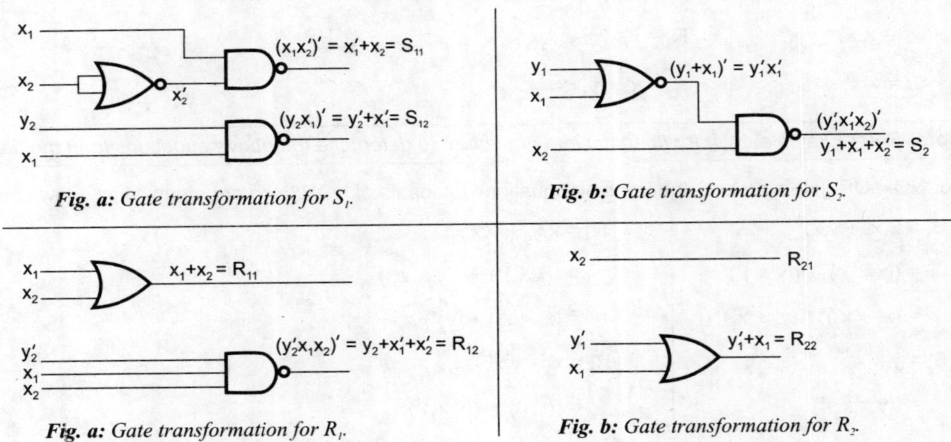

Fig. a: Gate transformation for S_1.

Fig. b: Gate transformation for S_2.

Fig. a: Gate transformation for R_1.

Fig. b: Gate transformation for R_2.

Fig. 2: Gate transformation to eliminate essential hazards.

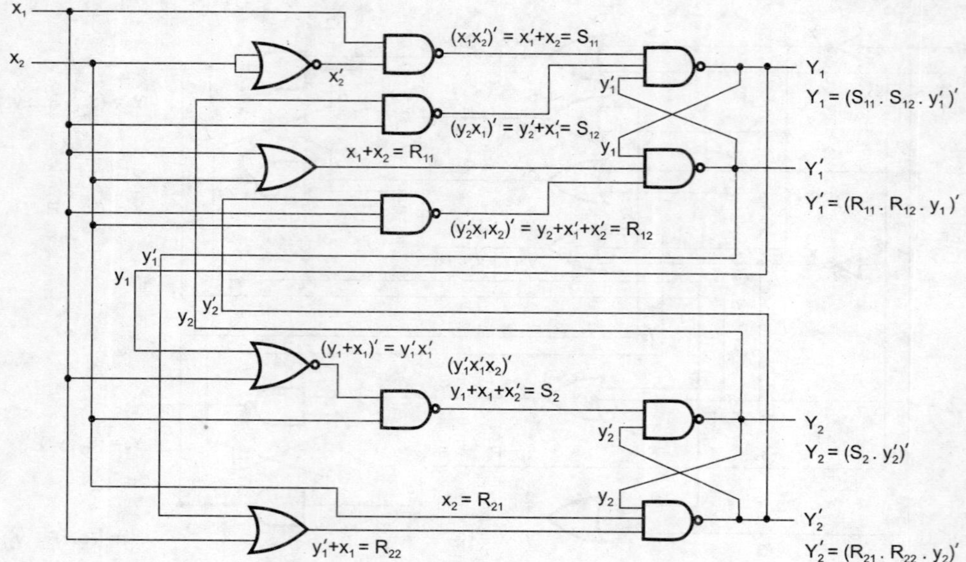

Fig. 3: *Logic circuit using NAND based SR latch with modified gate structure to eliminate essential hazard due to inverters.*

Example 4.26

Realize a logic circuit using NOR based SR latches and NOR-OR gate structure for the flow table of Example 4.24. Modify the gate structure for hazard free realization.

Solution

From Example 4.24, we get the following equations for inputs of NOR latches.

$S_1 = x_1x_2' + y_2x_1$	$R_1 = x_1'x_2' + y_2'x_1x_2$
$S_2 = y_1'x_1'x_2$	$R_2 = x_2' + y_1x_1'$

Note: *Refer Example 4.24, for the complete procedure to determine the above equations from the flow table.*

For NOR-OR gate structure, the Boolean equations for inputs of NOR latch are modified as shown below:

$$\therefore S_1 = x_1x_2' + y_2x_1 = ((x_1x_2')')' + ((y_2x_1)')' = (x_1' + x_2)' + (y_2' + x_1')'$$

$$R_1 = x_1'x_2' + y_2'x_1x_2 = ((x_1'x_2')')' + ((y_2'x_1x_2)) = (x_1 + x_2)' + (y_2 + x_1' + x_2')'$$

$$S_2 = y_1'x_1'x_2 = ((y_1'x_1'x_2)')' = (y_1 + x_1 + x_2')'$$

$$R_2 = x_2' + y_1x_1' = ((x_2')')' + ((y_1x_1')')' = x_2' + (y_1' + x_1)'$$

Logic Diagram

The above Boolean equations are used to construct the combinational logic circuit to generate latch inputs as shown in Fig. 1. The complete logic circuit is obtained by adding NOR based SR latches to the combinational circuit as shown in Fig. 1.

In Fig. 1(a), the combinational logic circuit is implemented using NOR-OR gate structure. In order to prevent essential hazards the NOR-OR structure can be replaced by NOR-NOR structure with multi-input NOR gates for latches as shown in Fig. 1(b).

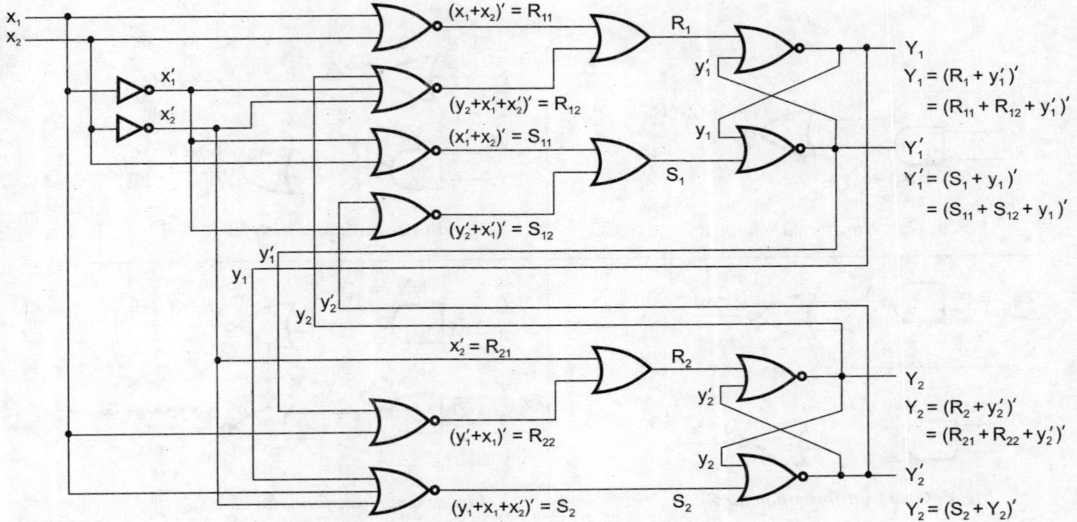

***Fig. a:** Combinational circuit with NOR-OR gate structure.*

⇓

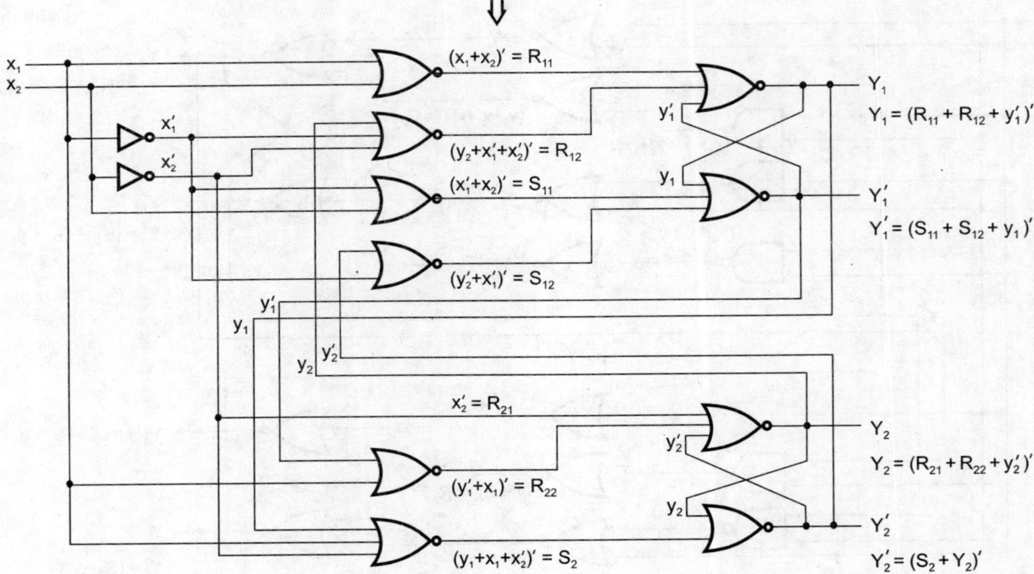

***Fig. b:** Modified gate structure to minimize essential hazard.*

***Fig. 1:** Logic circuit using NOR-OR gate structure and NOR based SR latch.*

Modification of Gate Structure to Eliminate Essential Hazard Due to Inverters

In the combinational logic circuit to generate latch inputs in Fig. 1(b), the inverters can be replaced by modified gate structure shown in Fig. 2. This will eliminate the essential hazards due to inveters. The modified complete logic circuit is shown in Fig. 3.

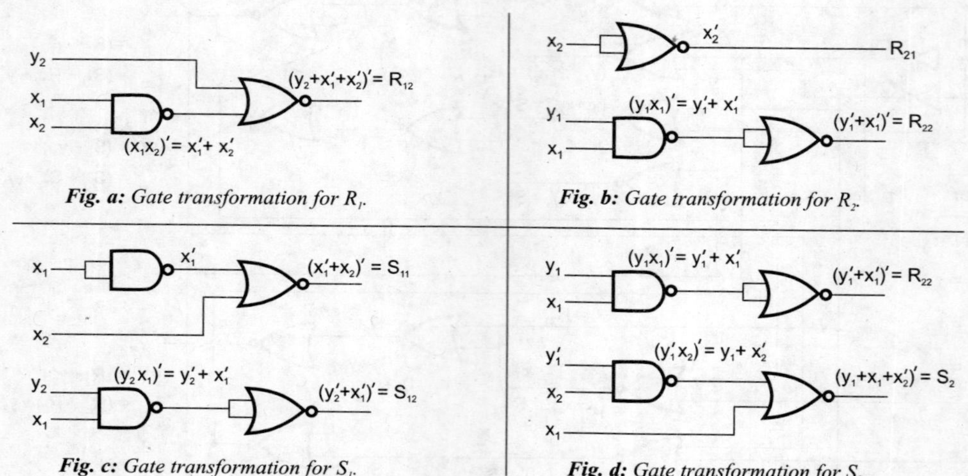

***Fig. a:** Gate transformation for R_1.*

***Fig. b:** Gate transformation for R_2.*

***Fig. c:** Gate transformation for S_1.*

***Fig. d:** Gate transformation for S_2.*

***Fig. 2:** Gate transformation to eliminate essential hazards.*

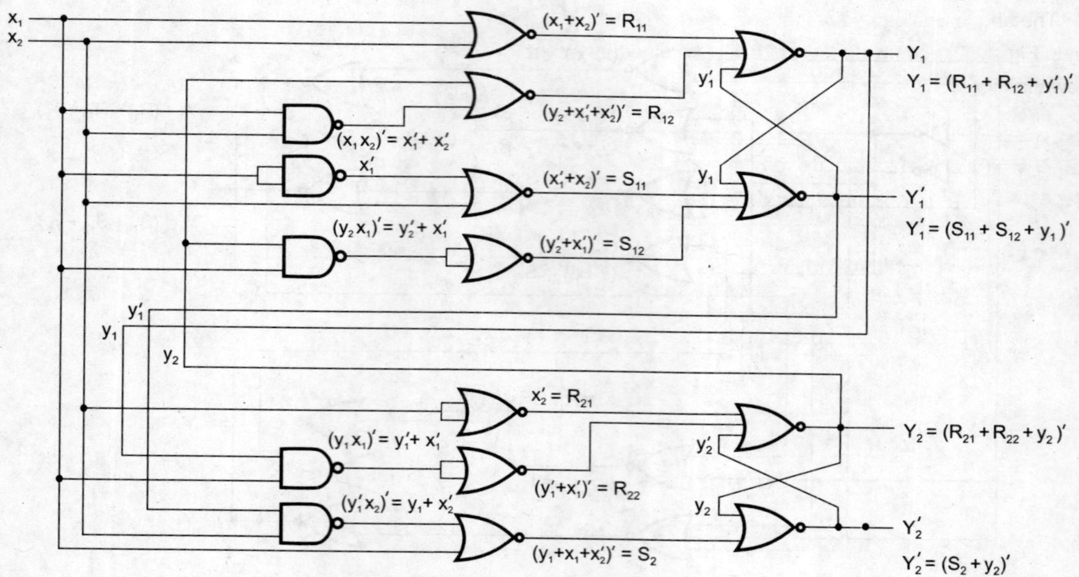

***Fig. 3:** Logic circuit using NOR based SR latch with modified gate structure to eliminate essential hazard.*

Example 4.27

Realize a logic circuit using NAND based SR latches and NAND-AND gate structure for the flow table of Example 4.24. Modify the gate structure for hazard free realization.

Solution

From Example 4.24, we get the following equations for inputs of NOR latches.

$S_1 = x_1x_2' + y_2x_1$	$R_1 = x_1'x_2' + y_2'x_1x_2$
$S_2 = y_1'x_1'x_2$	$R_2 = x_2' + y_1x_1'$

> ***Note:** Refer Example 4.24, for the complete procedure to determine the above equations from the flow table.*

For NAND-AND gate structure, the Boolean equations for inputs of NAND latch are modified as shown below:

$$\therefore S_1 = x_1x_2' + y_2x_1$$
$$= (x_1x_2' + y_2x_1)'$$
$$= (x_1x_2')' . (y_2x_1)'$$
$$S_2 = y_1'x_1'x_2$$
$$= (y_1'x_1'x_2)'$$

$$R_1 = x_1'x_2' + y_2'x_1x_2$$
$$= (x_1'x_2' + y_2'x_1x_2)'$$
$$= (x_1'x_2')'.(y_2'x_1x_2)'$$
$$R_2 = x_2' + y_1x_1'$$
$$= ((x_2' + y_1x_1')'$$
$$= x_2 + (y_1 + x_1')'$$

Logic Diagram

The above Boolean equations are used to construct the combinational logic circuit to generate latch inputs as shown in Fig. 1. The complete logic circuit is obtained by adding NAND based SR latches to the combinational circuit as shown in Fig. 1.

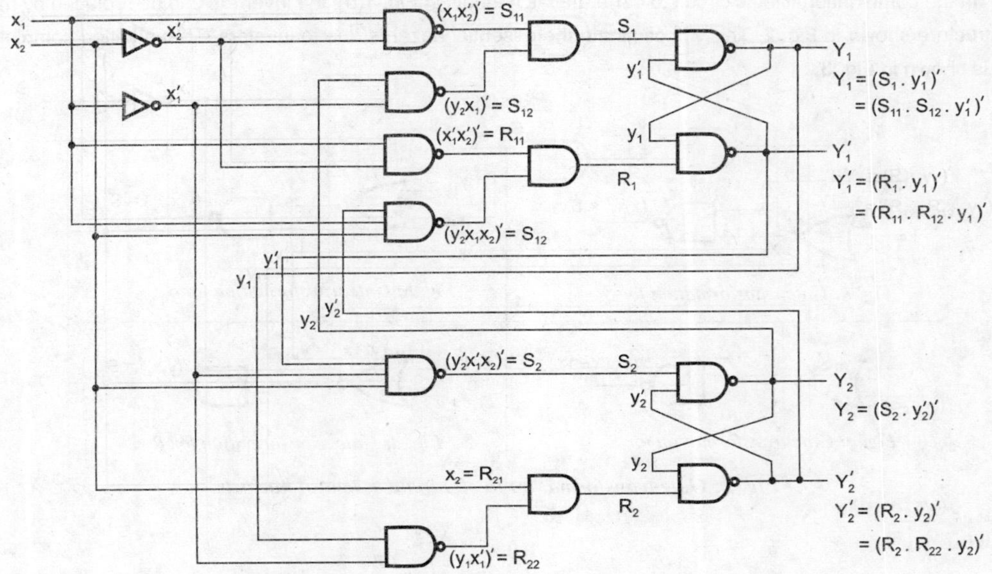

Fig. a: Combinational circuit with NAND-AND gate structure.

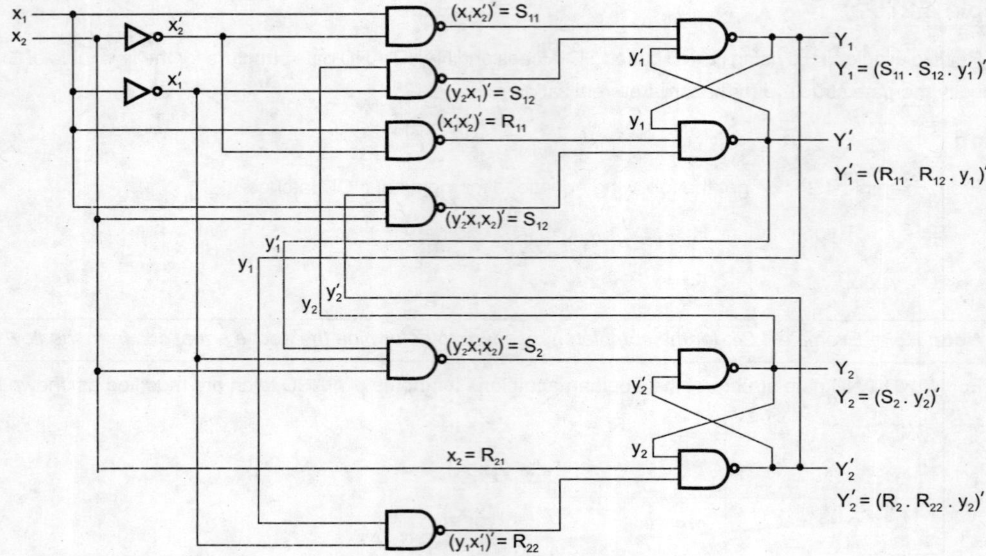

***Fig. b:** Modified gate structure to minimize essential hazard.*

***Fig. 1:** Logic circuit using NAND-AND gate structure and NAND based SR latch.*

In Fig. 1(a), the combinational logic circuit is implemented using NAND-AND gate structure. In order to prevent essential hazards the NAND-AND structure can be replaced by NAND-NAND structure with multi-input NAND gates for latches as shown in Fig. 1(b).

Modification of Gate Structure to Eliminate Essential Hazard Due to Inverters

In the combinational logic circuit to generate latch inputs in Fig. 1(b), the inverters can be replaced by modified gate structure shown in Fig. 2. This will eliminate the essential hazards due to inveters. The modified complete logic circuit is shown in Fig. 3.

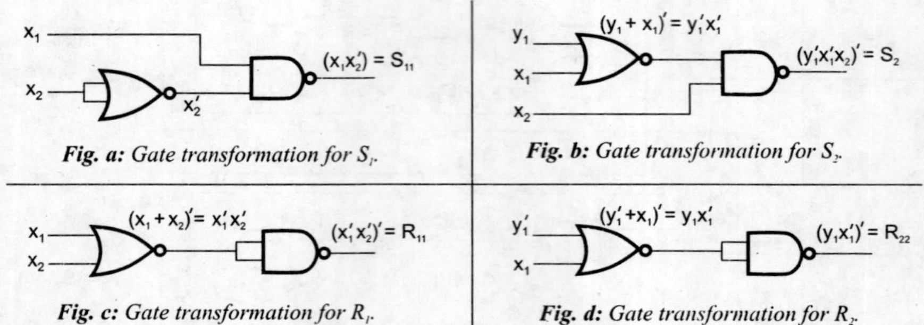

***Fig. a:** Gate transformation for S_1.*

***Fig. b:** Gate transformation for S_2.*

***Fig. c:** Gate transformation for R_1.*

***Fig. d:** Gate transformation for R_2.*

***Fig. 2:** Gate transformation to eliminate essential hazards.*

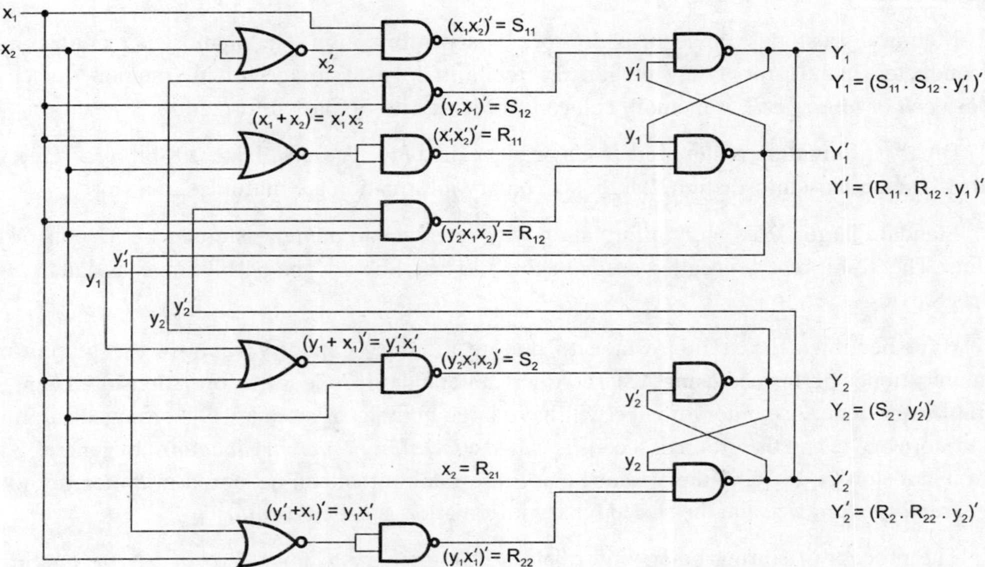

***Fig. 3:** Logic circuit using NAND based SR latch with modified gate structure to eliminate essential hazard.*

4.8 Introduction to Programmable Logic Devices

The digital ICs (Integrated Circuits) can be broadly classified into fixed function ICs and programmable ICs. The fixed function ICs are manufactured to perform a fixed logical function. Examples of fixed function ICs are flip-flops, decoders, multiplexers, counters, etc. The programmable ICs are designed to be programmed by the user to implement a variety of logical functions.

Some of the programmable logical devices are,

- PROM (Programmable ROM)
- PLA (Programmable Logic Array)
- PAL (Programmable Array Logic)
- CPLD (Complex Programmable Logic Device)
- FPGA (Field Programmable Gate Array)

The PROM, PAL and PLA are one-time programmable and consists of programmable logic gate arrays and can be used to implement only combinational logic functions. The CPLD and FPGA devices are reprogrammable and consists of programmable logic gate arrays and flip-flops and so can be used to implement both combinational and sequential logic functions.

The advantages of using PLDs to implement a logical function are,

- PLDs require smaller chip and board area to implement a logic circuit.
- Logical functions can be easily altered or modified without rewiring and changing the board.
- Implementation is much faster when compared to fixed function ICs.

4.8.1 Memory

Memory is a storage device used to store binary information. The memory is manufactured using semiconductors in the form of ICs (Integrated Circuits). The basic storage unit in a memory device is called ***memory cell*** or ***binary cell***. A memory cell can store one bit of binary information.

Binary information is stored in group of bits called ***byte*** or ***word***. Byte is 8-bit binary information. Word is 16-bit binary information. Larger size binary information are multiples of a word.

Standard larger size binary information with reference to memory storage are 32 bits, 64 bits and 128-bits. The 32-bit binary is called ***double word*** and has 4 bytes. The 64-bit binary is called ***long word*** and has 8 bytes.

Byte becomes the standard size to denote memory capacity because the human-machine communication is performed using ASCII (American Standard Code for Information Interchange) which is 8-bit binary code. A memory device will have large number of storage cells arranged in groups of 8 cells to store a byte and the group of 8 cells is called a ***location*** or ***memory location***. In general, a memory location can store m bits and the size of binary information that can be stored in a memory location is called ***memory word size*** and the stored binary information is called ***memory word***.

The process of storing binary information in a memory location is called ***write operation***. The process of reading back the stored information from the memory location is called ***read operation***. The read operation is non-destructive which means that after the read operation the memory content remains same until it is rewritten with new information.

Each memory location has a unique binary number called ***address***. The address is used to access the location for read and write operation. The number of addresses for memory locations depends on size of binary used to generate addresses. We know that with n-bit binary we can generate 2^n different binary codes. Therefore, with n-bit binary we can generate 2^n different binary addresses.

$n = 1,\ 2^1 = 2$	$n = 4,\ 2^4 = 16$	$n = 7,\ 2^7 = 128$
$n = 2,\ 2^2 = 4$	$n = 5,\ 2^5 = 32$	$n = 8,\ 2^8 = 256$
$n = 3,\ 2^3 = 8$	$n = 6\ \ 2^6 = 64$	$n = 9,\ 2^9 = 512$

$n = 10,\ \ 2^{10} = 1024 = 1\ kilo$

$n = 11,\ \ 2^{11} = 2048 = 2 \times 1024 = 2\ kilo$

$n = 12,\ \ 2^{12} = 4096 = 4 \times 1024 = 4\ kilo$

$n = 16,\ \ 2^{16} = 65536 = 64 \times 1024 = 64\ kilo$ and so on.

A memory with 1 *kilo* locations that can store one byte in each location is called 1 *kB (kilo-byte)* memory. This memory will have 8 data lines and 10 address lines. Apart from that it will have read control, write control and chip select/enable signal as shown in Fig. 4.49.

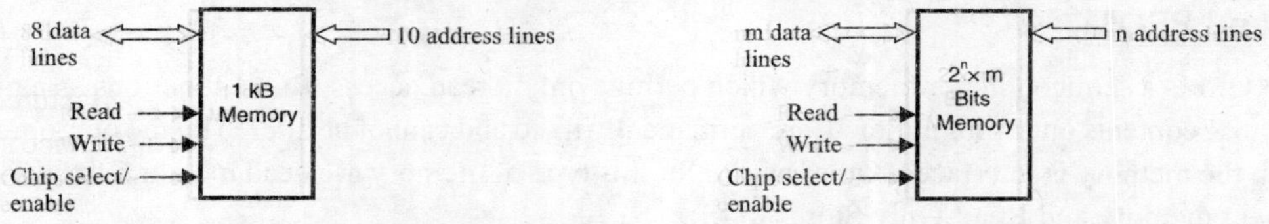

***Fig. 4.49:** Block diagram of 1 kB memory.* ***Fig. 4.50:** Block diagram of $2^n \times m$ bits memory.*

In general, a memory will have m data lines and n address lines as shown in Fig. 4.50. The capacity of memory is $2^n \times m$ bits. Here, m is always multiple of 8, so that memory capacity can be expressed in bytes.

> ***Note:*** $2^n \times m \text{ bits} = \frac{2^n \times m}{8} \text{ bytes} = \frac{2^n \times m}{8 \times 1024} \text{ kilo-bytes.}$

Memory Unit with Capacity 1024 × 16

A memory unit with 1024 × 16 capacity will have 1024 locations and each locations can store 16-bit binary.

$1024 = 2^{10} \Rightarrow$ Memory has 10 address lines

The range of 10-bit address is as follows.

Binary Address			Hexa Address
0 0	0 0 0 0	0 0 0 0	000_H
0 0	0 0 0 0	0 0 0 1	001_H
0 0	0 0 0 0	0 0 1 0	002_H
0 0	0 0 0 0	0 0 1 1	003_H
0 0	0 0 0 0	0 1 0 0	004_H
⋮	⋮	⋮	⋮
1 1	1 1 1 1	1 1 1 1	$3FF_H$

$3FF_H = 1023_{10}$

Here, memory address and data are both in binary. Hexa is short-hand form of binary.

A simplified schematic of memory is shown below:

Memory Address $a_9\, a_8\;\; a_7\, a_6\, a_5\, a_4\;\; a_3\, a_2\, a_1\, a_0$	**Memory Content** $d_{15}\, d_{14}\, d_{13}\, d_{12}\;\; d_{11}\, d_{10}\, d_9\, d_8\;\; d_7\, d_6\, d_5\, d_4\;\; d_3\, d_2\, d_1\, d_0$
0 0 0 0 0 0 0 0 0 0	× × × × × × × × × × × × × × × ×
0 0 0 0 0 0 0 0 0 1	× × × × × × × × × × × × × × × ×
0 0 0 0 0 0 0 0 1 0	× × × × × × × × × × × × × × × ×
0 0 0 0 0 0 0 0 1 1	× × × × × × × × × × × × × × × ×
⋮	⋮
1 1 1 1 1 1 1 1 1 1	× × × × × × × × × × × × × × × ×

> ***Note:*** × × × × × × × × × × × × × × × × *is any 16-bit data or binary content available in memory location.*

ROM and PROM

ROM is a semiconductor memory which permits only a read access. ROM functions as a memory array whose contents once programmed, are permanently fixed and cannot be altered by the microprocessor to which the memory is interfaced. Other names for this type of memory are dead memory, fixed memory, permanent memory and **R**ead-**O**nly **S**tore (ROS).

ROM is a nonvolatile memory, i.e., loss of power or system malfunction does not change the contents of the memory. Also, the ROM memories have the feature of random access for read operation, which means that the access time (or read time) for a given memory location is same as that for all other locations. The process of storing information in ROM is called ***programming***. The technique employed for storing information in the ROM provides a convenient method for classifying ROMs into one of the following three categories:

1. Custom programmed or Mask programmed ROM (ROM).
2. Programmable or Field programmable ROM (PROM).
3. Reprogrammable or Erasable-Programmable ROM (EPROM).

The custom programmed ROMs are programmed by the manufacturer as specified by the user during fabrication and the contents cannot be changed after packaging. The programmable ROMs are one-time programmable by the user.

The reprogrammable ROMs have facilities for programming as well as for erasing its content and reprogramming the memory. The reprogrammable ROMs are erased either by passing electrical current or Ultraviolet light.

The programming of ROMs can be carried out using ROM (EPROM) programmer. Usually the ROM programmer is a digital system interfaced to a **P**ersonal **C**omputer (PC). The information to be programmed are first stored as a file in the PC and converted to the required binary format using a conversion software. Then the information is transferred from the PC to ROM programmer.

Custom or Mask Programmed ROM

ROM devices are programmed during manufacturing as per customers requirements. After manufacturing the content of ROM connot be altered. The ROM device will have n-bit address input and m-bit data output as shown in Fig. 4.51.

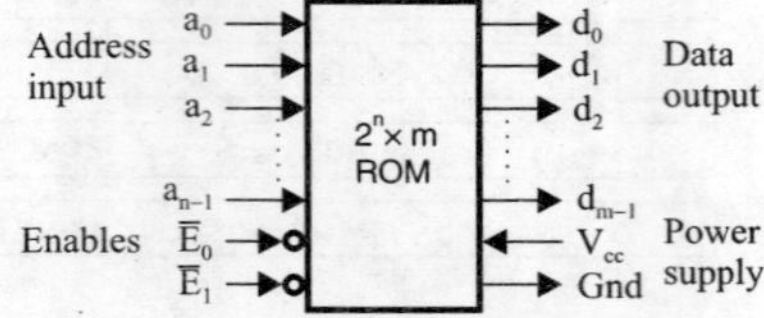

Fig. 4.51: Symbolic representation of ROM.

In mask programmed ROM memory, the memory cell (storage unit) will have a MOS transistor either with open gate or closed gate as shown in Fig. 4.52. The transistors with closed gate represent **1's** and with open gate represent **0's**. Since the configuration is fixed, they permanently store **1's** and **0's**.

A simple structure of a 4 × 4 bit ROM using a single MOS transistor for each shell is shown in Fig. 4.53.

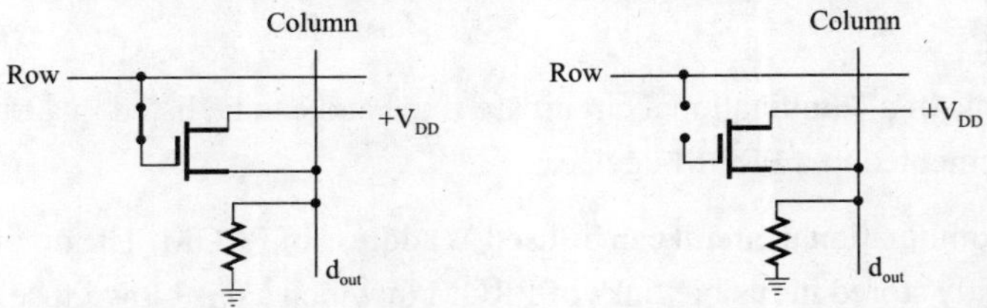

Fig. a: A ROM cell with a 1 stored in it.

Fig. b: A ROM cell with a 0 stored in it.

Fig. 4.52: MOS transistor based ROM cell.

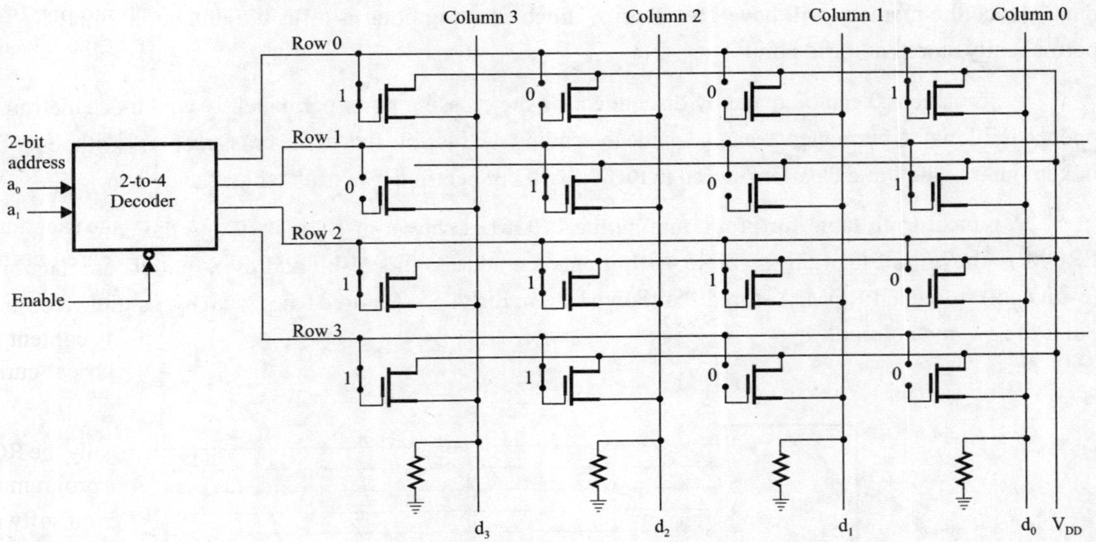

Fig. 4.53: Structure of a 4 × 4 bit ROM.

Programmable or Field Programmable ROM

The programmable ROMs are one-time programmable (OTP) by the user. Hence it is also called **OTP-ROM**. The structure of PROM cells will be same as ROM but instead of permanent open or close a fusible link is provided in each memory cell. The fusible link can be left intact to store a 1 and can be blown open by passing high current to store a 0 and this process is irreversible.

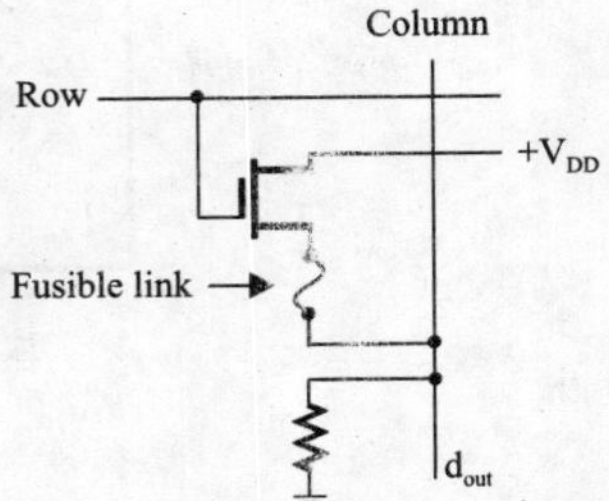

Fig. 4.54: MOS transistor based PROM cell.

Usually, the fusible link is provided in the source connection in order to pass higher current to blow the fuse. A simple MOS transistor based PROM cell is shown in Fig. 4.54. The PROM devices will have n-bit address input and m-bit data output similar to ROM device as shown in Fig. 4.51. The programming of PROM device is carried by fixing the PROM in a programming device and send an address to select a row and then passing current pulses through data lines where ever 0 has to be stored. This process is repeated for all rows one by one.

4.8.2 PROM as PLD

The inputs-outputs of a combinational circuit are fixed and can be listed in a truth table. Any fixed truth table can be implemented in a PROM device.

The inputs of a combinational circuit can be used as address of PROM. The output of combinational circuit can be permanently stored in fusible links of PROM in which a 0 is blown fuse and 1 is intact fuse.

An example of 16 × 8-bit PROM is shown in Fig. 4.55 in which 4 address lines ($16 = 2^4$) and 8 data lines are provided. A decoder (4-to-16 decoder) is employed to decode the 4-bit address to 16 rows. Each row has 8 storage locations to store 8-bit data and 8 column lines are provided to access the 8-bit data. Hence the memory will have 16 × 8 array. Each crossing point is a fusible link and 0's and 1's are permanently stored at these point.

A permanent 0 can be made by disconnecting the crossing point permanently which is called fuse opened or blown. A permanent 1 can be made by leaving the fussible link at the crossing point without open or kept intact. The 1 and 0 pattern stored in PROM will be specified by a truth table as shown in Table 4.19.

A typical truth table for programming the PROM is shown in Table 4.19 and the corresponding PROM is shown in Fig. 4.55. In Table 4.19, inputs are address bits (4-bit address), outputs are data bits (8-bit data) stored in PROM. A stored 1 is shown by x in Fig. 4.55. A stored 0 is shown by a simple crossing point of row and column.

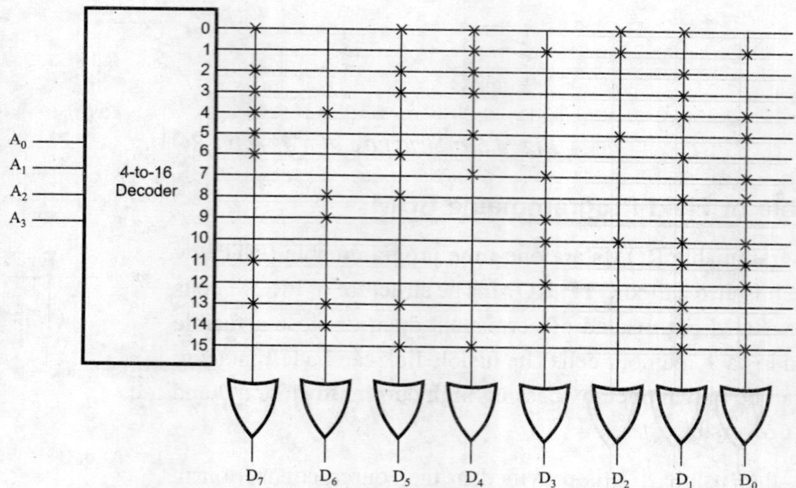

Fig. 4.55: Programming the PROM according to Table 4.19.

Table 4.19: Truth Table Implemented in PROM shown in Fig. 4.49

Inputs				Outputs							
A_3	A_2	A_1	A_0	D_7	D_6	D_5	D_4	D_3	D_2	D_1	D_0
0	0	0	0	1	0	1	1	0	1	1	0
0	0	0	1	0	0	0	1	1	1	0	1
0	0	1	0	1	0	1	1	0	0	1	0
0	0	1	1	1	0	1	1	0	0	1	0
0	1	0	0	1	1	0	0	0	0	1	1
0	1	0	1	1	0	0	1	0	1	0	1
0	1	1	0	1	0	1	0	0	0	1	0
0	1	1	1	0	0	0	1	1	0	0	1
1	0	0	0	0	1	1	0	0	0	1	1
1	0	0	1	0	1	0	0	1	0	0	0
1	0	1	0	0	0	0	0	1	1	1	1
1	0	1	1	1	0	0	0	0	0	1	1
1	1	0	0	0	0	0	0	1	0	0	1
1	1	0	1	1	1	1	0	0	0	1	0
1	1	1	0	0	1	0	0	1	0	1	0
1	1	1	1	0	0	1	1	0	0	1	1

Example 4.28

(AU, Apr/May'19, 6 Marks)

Derive the PROM programming table for the combinational circuit that squares a 4-bit number. Minimize the number of product terms.

Solution

Table 1: Truth Table

Inputs (4-Bit Binary Number) A B C D	Minterm	Decimal Value	Squared Decimal Value	Outputs (Binary Value of 4-bit Squared Number)							
				B_7	B_6	B_5	B_4	B_3	B_2	B_1	B_0
0 0 0 0	m_0	0	0	0	0	0	0	0	0	0	0
0 0 0 1	m_1	1	1	0	0	0	0	0	0	0	1
0 0 1 0	m_2	2	4	0	0	0	0	0	1	0	0
0 0 1 1	m_3	3	9	0	0	0	0	1	0	0	1
0 1 0 0	m_4	4	16	0	0	0	1	0	0	0	0
0 1 0 1	m_5	5	25	0	0	0	1	1	0	0	1
0 1 1 0	m_6	6	36	0	0	1	0	0	1	0	0
0 1 1 1	m_7	7	49	0	0	1	1	0	0	0	1
1 0 0 0	m_8	8	64	0	1	0	0	0	0	0	0
1 0 0 1	m_9	9	81	0	1	0	1	0	0	0	1
1 0 1 0	m_{10}	10	100	0	1	1	0	0	1	0	0
1 0 1 1	m_{11}	11	121	0	1	1	1	1	0	0	1
1 1 0 0	m_{12}	12	144	1	0	0	1	0	0	0	0
1 1 0 1	m_{13}	13	169	1	0	1	0	1	0	0	1
1 1 1 0	m_{14}	14	196	1	1	0	0	0	1	0	0
1 1 1 1	m_{15}	15	225	1	1	1	0	0	0	0	1

The truth table of the combinational circuit that squares a 4-bit binary number is shown in Table 1. Using the truth table the PROM is programmed as shown in Fig. 1. The x indicates a 1 at the crossing points.

A16 × 8 PROM grid is used to implement the truth table. A 4-to-16 decoder is used to select one of the 16 lines of the PROM grid. The decoder will accept the 4-bit binary input and select the corresponding output line to generate the output.

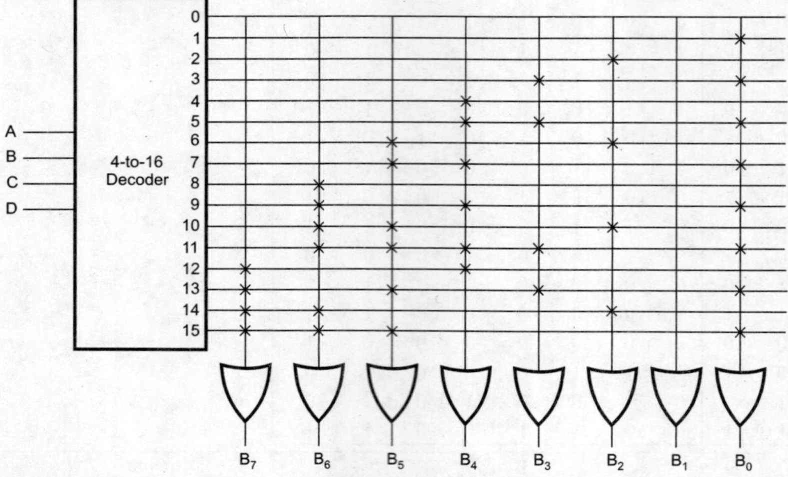

Fig. 1: Programming the PROM according to Table 1.

Minimization of Literals

The Boolean equation of an output of the combinational circuit can be written directly from the truth table by taking the sum of minterms for which the output is 1.

The Boolean equations are,

$B_0 = \Sigma(m_1, m_3, m_5, m_7, m_9, m_{11}, m_{13}, m_{15})$

$B_1 = 0$

$B_2 = \Sigma(m_2, m_6, m_{10}, m_{14})$

$B_3 = \Sigma(m_3, m_5, m_{11}, m_{13})$

$B_4 = \Sigma(m_4, m_5, m_7, m_9, m_{11}, m_{12})$

$B_5 = \Sigma(m_6, m_7, m_{10}, m_{11}, m_{13}, m_{15})$

$B_6 = \Sigma(m_8, m_9, m_{10}, m_{11}, m_{14}, m_{15})$

$B_7 = \Sigma(m_{12}, m_{13}, m_{14}, m_{15})$

The Boolean can be converted to minimum literal form by constructing K-map. The K-map for the outputs are drawn as shown in Figs. 2 to 7. Here $B_1 = 0$ and $B_0 = D$ and so K-maps are not drawn for there two outputs.

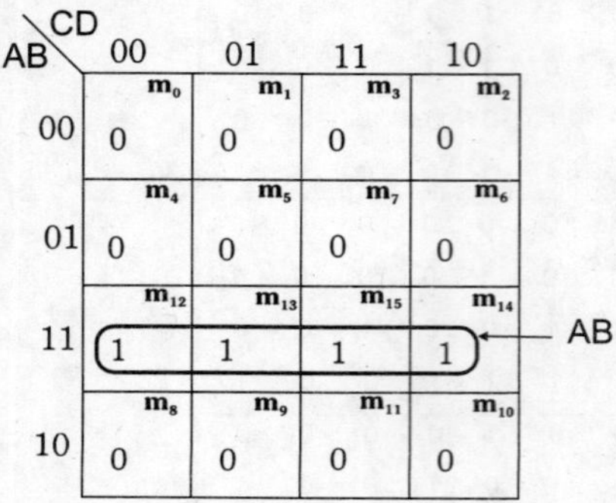

Fig. 2: K-map for B_7.

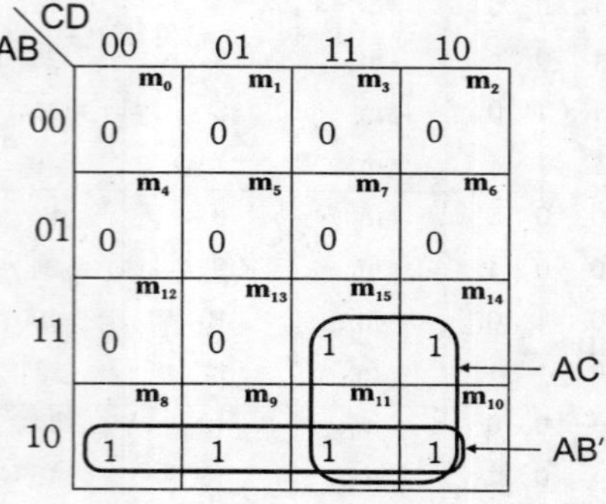

Fig. 3: K-map for B_6.

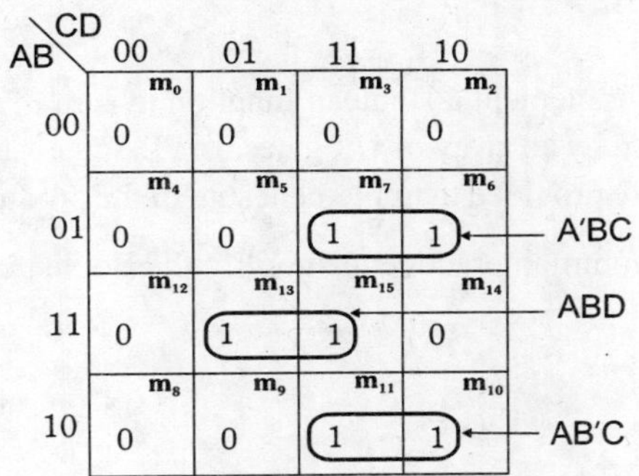

Fig. 4: K-map for B_5.

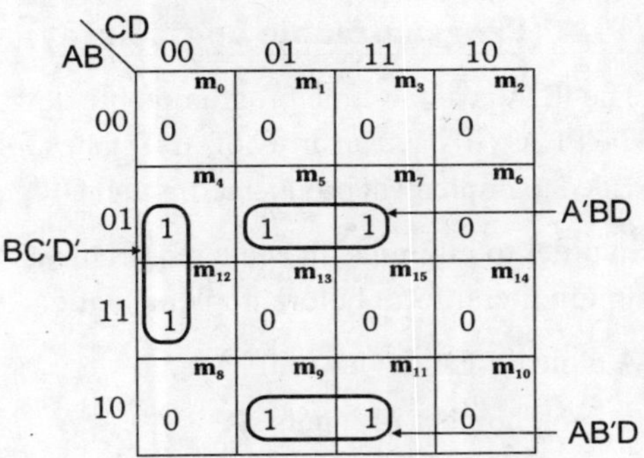

Fig. 5: K-map for B_4.

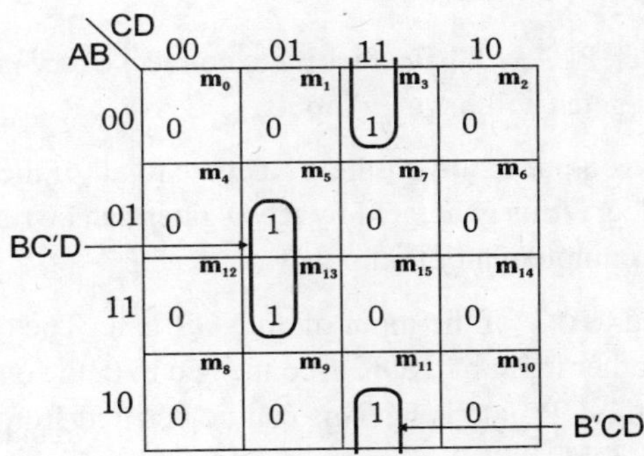

Fig. 6: K-map for B_3.

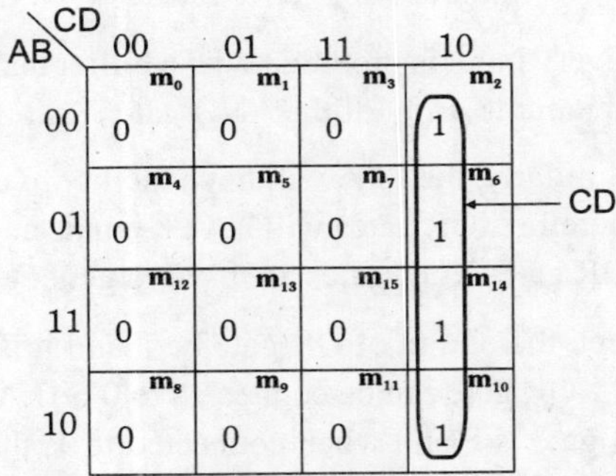

Fig. 7: K-map for B_2.

The following simplified Boolean functions of outputs are obtained from K-maps shown in Figs. 2 to 7.

$B_7 = AB$

$B_6 = AB' + AC$

$B_5 = A'BC + AB'C + ABC$

$B_4 = BC'D' + A'BD + AB'D$

$B_3 = BC'D + B'CD$

$B_2 = CD'$

The output column B_1 is all zero values.

$\therefore B_1 = 0$

The input column D and output column B_0 is same.

$\therefore B_0 = D$

Conclusion: It can be observed that the number of literals in the simplified Boolean equations are lesser than the Boolean equations obtained from truth table.

4.8.3 PLA (Programmable Logic Array)

(AU, Nov/Dec'18, 2 Marks)

The PLA is a one-time programmable device to implement a Boolean function in sum of products form. The PLA will have an array of AND gates followed by an array of OR gates. In PLA, the AND gates are provided to implement product terms and OR gates are provided to implement sum of the product terms.

In order to minimize the gate requirements it is common practice to simplify the Boolean function to minimum literal form before implementation.

A typical PLA device will have,

- n number of inputs
- k number of AND gates
- m number of OR gates
- m number of XOR gates

Every input is provided with a buffer and an inverter. PLA will have facility to connect every input/inverted input to array of the AND gate. Therefore, AND gates will have 2n inputs.

Similarly, the PLA will have facility to connect the output of any of the AND gate to all of the OR gate. Therefore, OR gates will have k inputs and the product terms generated by AND gates can be routed to any OR gate. XOR gates are used to generate normal /complement of output.

Let, the output of OR gate be F and it is connected to one of the input of an XOR gate. The other input of XOR gate can be connected to 0 or 1. When the other input of XOR is connected to 0, the output of XOR gate is F and when connected to 1, the output of XOR gate is F′. This can be verified from the truth table of XOR operation of function, F with 0 and 1 (Table 4.20).

Table 4.20: XOR Operation of function, F with 0 and 1

Input		Output of	Symbolic
Fixed	F	XOR Gate	Representation
0	0	$0 \oplus 0 = 0 = F$	0, F → XOR → F
0	1	$0 \oplus 1 = 1 = F$	
1	0	$1 \oplus 0 = 1 = F'$	1, F → XOR → F′
1	1	$1 \oplus 1 = 0 = F'$	

The structure of PLA with three inputs, five AND gates, three OR gates and three XOR gates is shown in Fig. 4.56. In PLA device the connections to AND gates and OR gates are programmable.

The structure of PLA in Fig. 4.56 can be simplified as shown in Fig. 4.57 by combining the input buffers to a single buffer with two outputs and multiple input lines to gates by a single input line.

A bubble is placed in one of the output of input buffer to indicate that this output will be inverted value of input. A slash with a number in the input line to gate indicates the number of input lines in that gate.

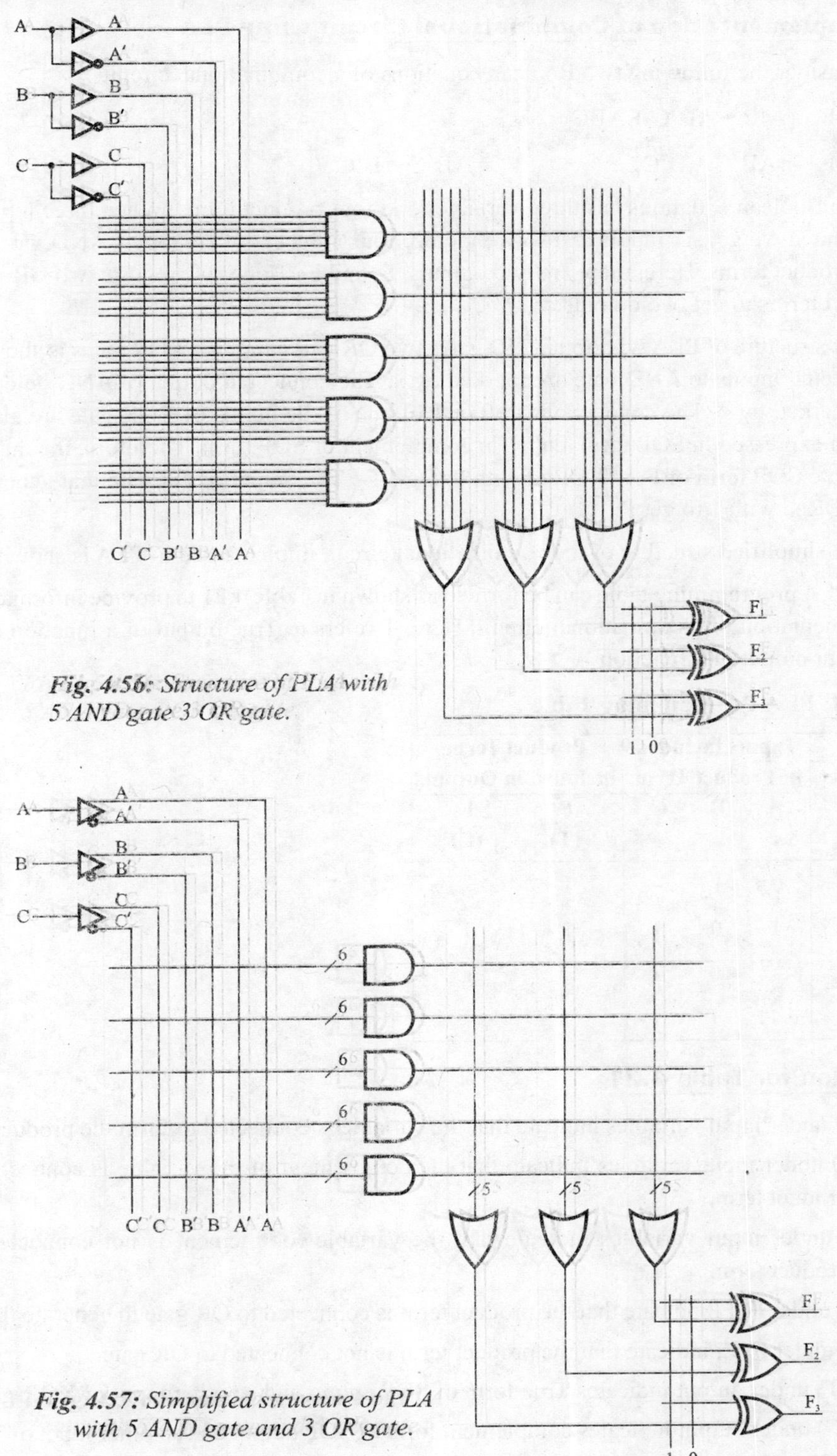

Fig. 4.56: Structure of PLA with 5 AND gate 3 OR gate.

Fig. 4.57: Simplified structure of PLA with 5 AND gate and 3 OR gate.

4.8.4 Implementation of Combinational Circuit using PLA

Consider the following two Boolean equations of a combinational circuit.

$$F_1 = A'B + AB'C + ABC$$

$$F_2 = (AB'C + B'C)' \quad \Rightarrow \quad F_2 = AB'C + B'C$$

The Boolean equations has three variables and four product terms. When three input variables are complemented we get six inputs and so we require four numbers of six inputs AND gates to implement the four product terms. Here, there are two outputs, F_1 and F_2. Hence we require two OR gates to sum up the product terms to get two outputs.

The structure of PLA with four AND gates two OR gates and two XOR gates is shown in Fig. 4.58. The connected inputs to AND gate are marked by x. The connected output of AND gate to input of OR gate are marked by x. The connections to 0 or 1 of one of the inputs of XOR gate are also shown by x. Here, F_1 in expressed in SOP form and F_2 in complement of SOP form. Therefore, the output of OR gate that generate SOP form of F_1 is XORed with 0 to get F_1. The output of OR gate that generates SOP form of F_2 is XORed with 1 to get F_2.

The simplified structure of the combinational circuit implemented in PLA is shown in Fig. 4.59.

A PLA programming table can be formed as shown in Table 4.21 to provide informations regarding the implementation of combinational circuit. Here, T refers to True output of a function and C refers to complement output of a function

Table 4.21: PLA Programming Table

Product Term	Inputs included in Product Term			Product Term included in Outputs	
	A	B	C	F_1 (T)	F_2 (C)
A'B	0	1	-	1	-
AB'C	1	0	1	1	1
B'C	-	0	1	-	1
ABC	1	1	1	1	-

Explanation for Table 4.21

- 1 under input variables indicate that the variable is connected to form the product term
- 0 under input variables indicate that the complement of the variable is connected to form the product term.
- - under input variables indicate that the variable/complement is not connected to form the product term.
- 1 under F_1/F_2 indicate that the product term is connected to OR gate to generate function output.
- - under F_1/F_2 indicate that the product term is not connected to OR gate.
- (T) under output indicates True form of SOP output and so other input of XOR gate is 0.
- (C) under output indicates complement form of SOP output and so other input of XOR gate is 1.

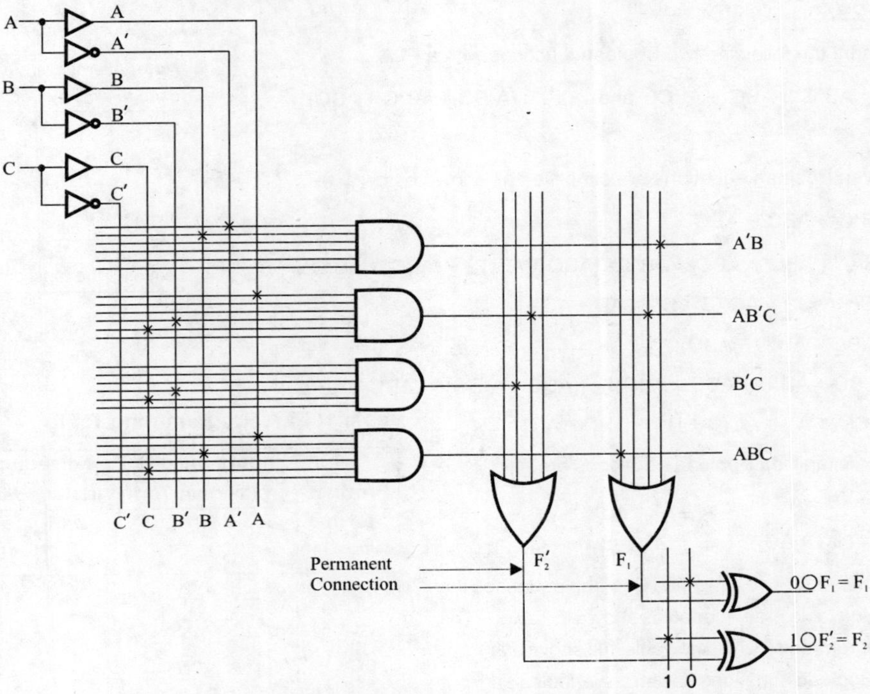

***Fig. 4.58:** Structure of a combinational circuit implemented in PLA.*

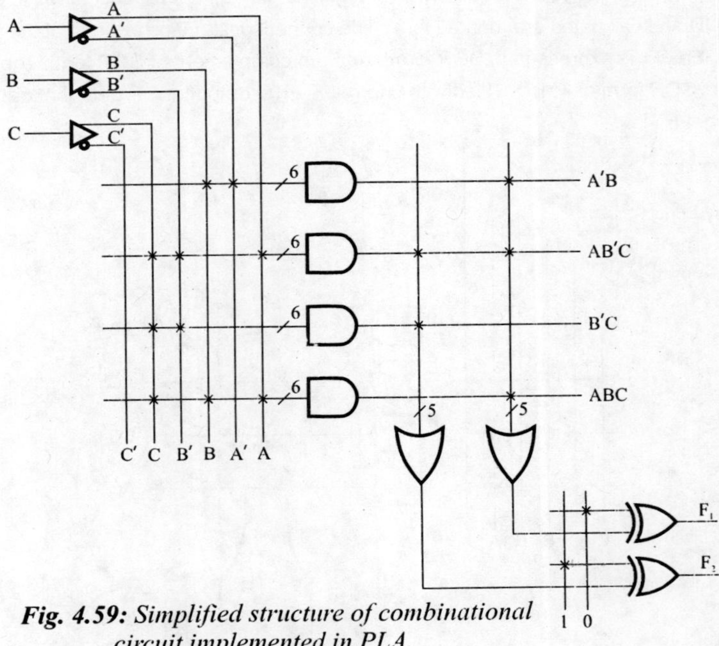

***Fig. 4.59:** Simplified structure of combinational circuit implemented in PLA.*

Example 4.29

Implement the following two Boolean functions with a PLA.

$F_1 = AB'C' + ABC' + ABC$ and $F_2 = (A'BC + AB'C + ABC)'$

> ***Note:*** *Alternatively, K-map can be used to simplify Boolean function.*

Solution

The given Boolean equations are simplified as shown below:

$F_1 = AB'C' + ABC' + ABC$

$= AB'C' + ABC' + ABC' + ABC = (ABC + ABC') + (ABC' + AB'C')$

> Adding an existing minterm will not alter the function value.

$= AB(C + C') + AC'(B + B') = AB + AC'$

> $x + x' = 1$

$F_2 = (A'BC + AB'C + ABC)'$

$= (A'BC + AB'C + ABC + ABC)' = ((ABC + A'BC) + (ABC + AB'C))'$

$= (BC(A + A') + AC(B + B'))' = (BC + AC)'$

Now the simplified Boolean equations are,

$F_1 = AB + AC'$

$F_2 = (BC + AC)' \Rightarrow F_2 = BC + AC$

Table 1: PLA Programming Table

Product Term	Inputs included in Product Term			Product Term included in Outputs	
	A	B	C	F_1 (T)	F_2 (C)
AB	1	1	-	1	-
AC′	1	-	0	1	-
BC	-	1	1	-	1
AC	1	-	1	-	1

Using the above two simplified Boolean equations the PLA programming table is developed as shown in Table 1. The structure of programmed PLA using Table 1 is shown in Fig. 1. The simplified Boolean equations has three variables and four product terms. When three input variables are complemented we get six inputs and so we require four numbers of six inputs AND gates to implement the four product terms. Here, there are two outputs, F_1 and F_2. Hence we require two OR gates to sum up the product terms to get two outputs. The connected inputs to AND and OR gates are marked by ×. The connections to 0 or 1 of one of the inputs of XOR gates are also shown by ×. Here, F_1 is expressed in SOP form and F_2 in complement of SOP form. Therefore, the output of OR gate that generates SOP form of F_1 is XORed with 0 to get F_1. The output of OR gate that generates SOP form of F_2 is XORed with 1 to get F_2.

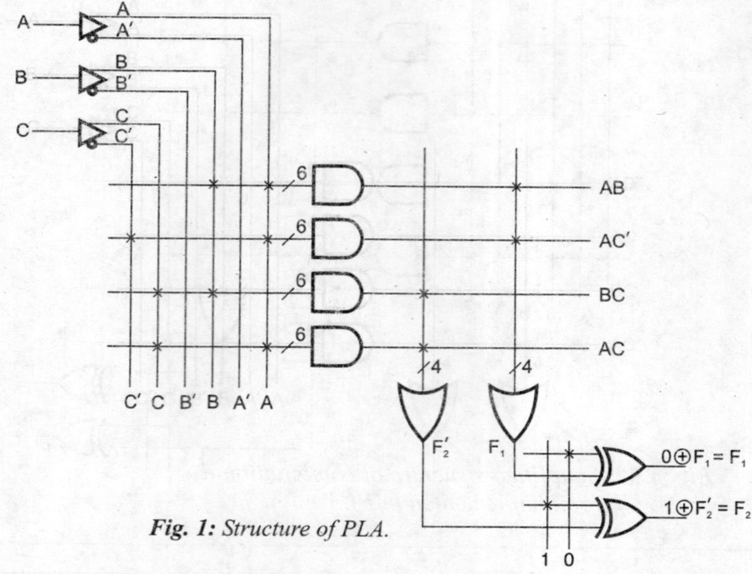

Fig. 1: Structure of PLA.

Example 4.30

Implement the full adder circuit using PLA by deriving the PLA programming table.

Solution

The truth table of full adder is shown in Table 1. The simplified Boolean equations can be obtained by constructing K-maps for s and c_o as shown in Figs. 1 and 2.

Table 1: Truth Table of Full Adder

Inputs			Minterm	Outputs	
a	b	c_i		s	c_o
0	0	0	m_0	0	0
0	0	1	m_1	1	0
0	1	0	m_2	1	0
0	1	1	m_3	0	1
1	0	0	m_4	1	0
1	0	1	m_5	0	1
1	1	0	m_6	0	1
1	1	1	m_7	1	1

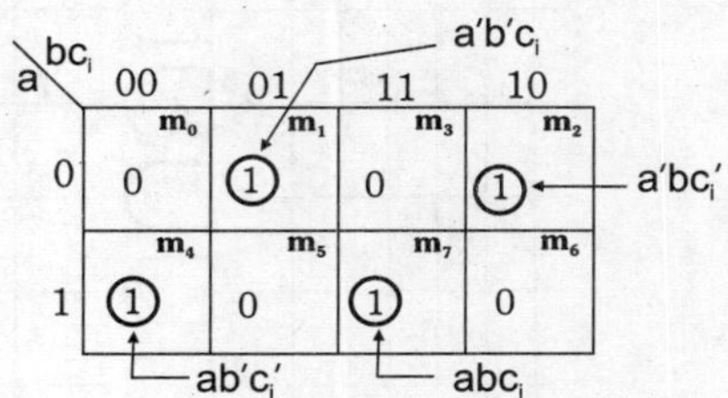

Fig. 1: K-map for s.

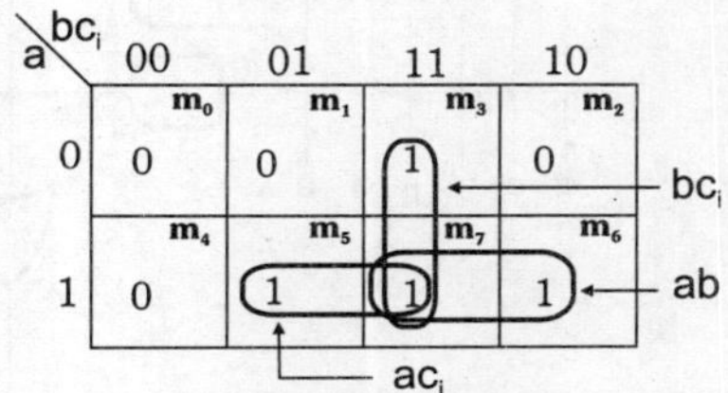

Fig. 2: K-map for c_o.

From the K-map the following Boolean equations are obtained for full adder.

$$s = a'b'c_i + a'bc_i' + ab'c_i' + abc_i$$

$$c_o = ab + ac_i + bc_i$$

Using the above Boolean equations the PLA programming table is developed as shown in Table 2. The structure of full adder implemented in PLA using Table 2 is shown in Fig. 1.

The simplified Boolean equations has three variables and seven product terms. When three input variables are complemented we get six inputs and so we require seven numbers of six input AND gates to implement the seven product terms. Here, there are two outputs, s and c_o. Hence we require two OR gates to sum up the product terms to get two outputs. The connected inputs to AND and OR gates are marked by ×.

Table 2: PLA Programming Table

Product Term	Inputs included in Product Term			Product Term included in Outputs	
	a	b	c_i	s (T)	c_o (T)
$a'b'c_i$	0	0	1	1	-
$a'bc_i'$	0	1	0	1	-
$ab'c_i'$	1	0	0	1	-
abc_i	1	1	1	1	-
ab	1	1	-	-	1
ac_i	1	-	1	-	1
bc_i	-	1	1	-	1

The connections to 0 or 1 of one of the inputs of XOR gates are also shown by ×. Here, s and c_o are expressed in SOP form. Therefore, the outputs of OR gate are XORed with 0 to get s and c_o.

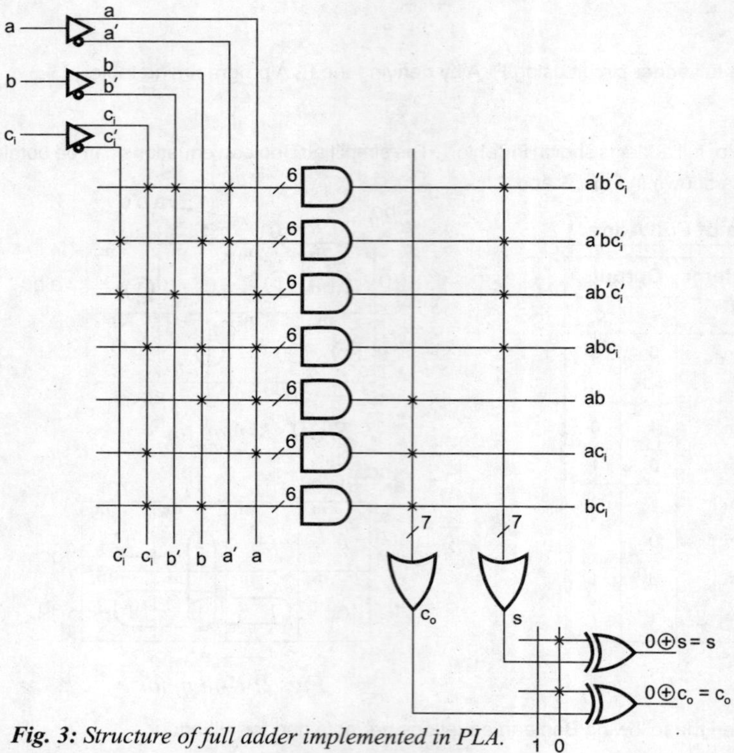

Fig. 3: Structure of full adder implemented in PLA.

4.8.5 PAL (Programmable Array Logic)

PAL is a programmable logic device to implement a Boolean function in sum of products form. PAL will have programmable AND gate array followed by a fixed OR gate. A typical PAL device will have m sets of AND-OR array with each array having k number of AND gates and one OR gate. In PAL device one set of AND-OR array combination is used to independently implement one function.

The only difference between PLA and PAL is that the OR gate array is programmable or not. In PLA device, the OR array is programmable but in PAL device the connections to OR array is fixed. In order to minimize the gate requirements, it is common practice to simplify the Boolean function to minimum literal form before implementation.

In PAL, the AND array is used to generate product terms. The product terms of AND array is routed to fixed OR gate in PAL, so that product terms of one function cannot be directly connected to another function. Whereas in PLA the product terms generated by AND gates can be routed to any OR gate. The PAL will have programmable facility to connect the output of an AND-OR array as input to AND gate of any other AND-OR array. This facility can be used to cascade the AND-OR array so that the cascade combination can be used to implement a larger number of product terms than permitted by a single AND-OR array.

A typical PAL device will have,

- n number of inputs
- k number of AND gates
- m number of OR gates
- m number of XOR gates

Every input is provided with a buffer and an inverter. PAL will have facility to connect every input/ inverted input to any of the AND gate. XOR gates are used to generate True/complement of output.

The output of each OR gate is connected to one of the inputs of an XOR gate. The other input of XOR gate can be connected to 0 or 1. When the other input of XOR is connected to 0 we get True output and when connected to 1 we get complemented output. This can be verified from Table 4.20.

(AU, Apr/May'19, 2 Marks)

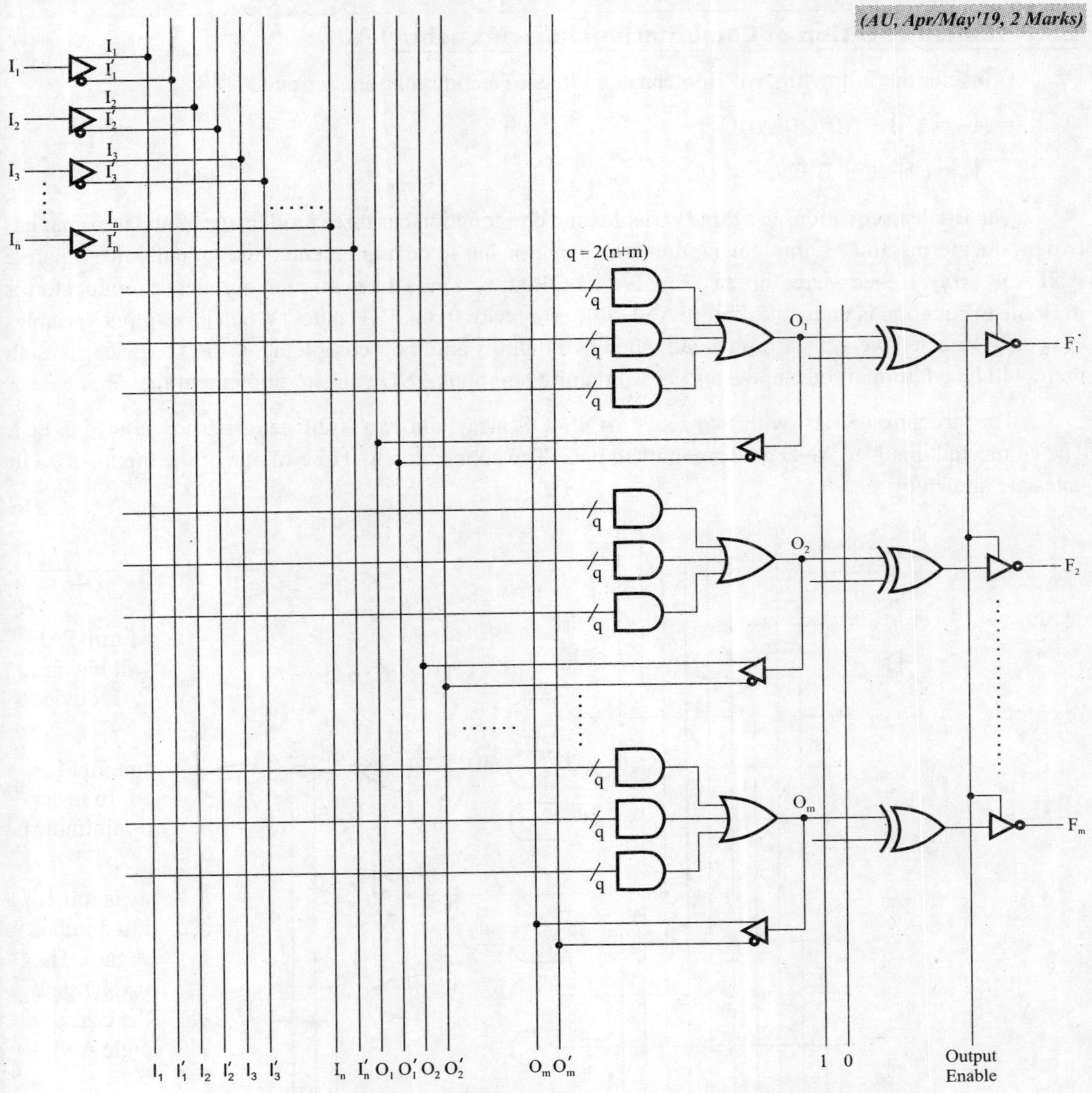

***Fig. 4.60:** Architecture or structure of a PAL device.*

The internal structure or architecture of PAL device with n inputs and m outputs is shown in Fig. 4.60. One set of AND-OR combination has 3 AND gates and so only three product terms can be generated. The three product terms are summed up using an OR gate. When we need a larger number of product terms, the output of an AND-OR array can be connected as input to any AND gate so that cascade connection of AND-OR array will have sum of all the product terms. The number of inputs to AND gate will be 2(n + m).

In some of the PAL devices, the outputs of PAL will have tristate logic with active low output and provided with complement facility using XOR gate similar to PLA.

4.8.6 Implementation of Combinational Circuit using PAL

Consider the following two Boolean equations of a combinational circuit.

$$F_1 = A'B + AB'C + ABC$$

$$F_2 = (AB'C + B'C)'$$

The Boolean equations has three variables and one equation has three product terms and the other has two product terms. In PAL implementation each equation has to be implemented individually in a separate AND-OR array. Hence we require two set of AND-OR array. Since the maximum number of product terms in Boolean equation is three, we require AND-OR array with three AND gates. When three input variables are complemented we get six inputs and when two outputs and their complements are feedback to input there will be additional four inputs and so we require ten input AND gates to implement the circuit.

The structure of PAL with two sets of AND-OR array and two XOR gates is shown in Fig. 4.61. The connected inputs to AND gate are marked by x. The connection to 0 or 1 of one of the inputs of XOR gates are shown by x.

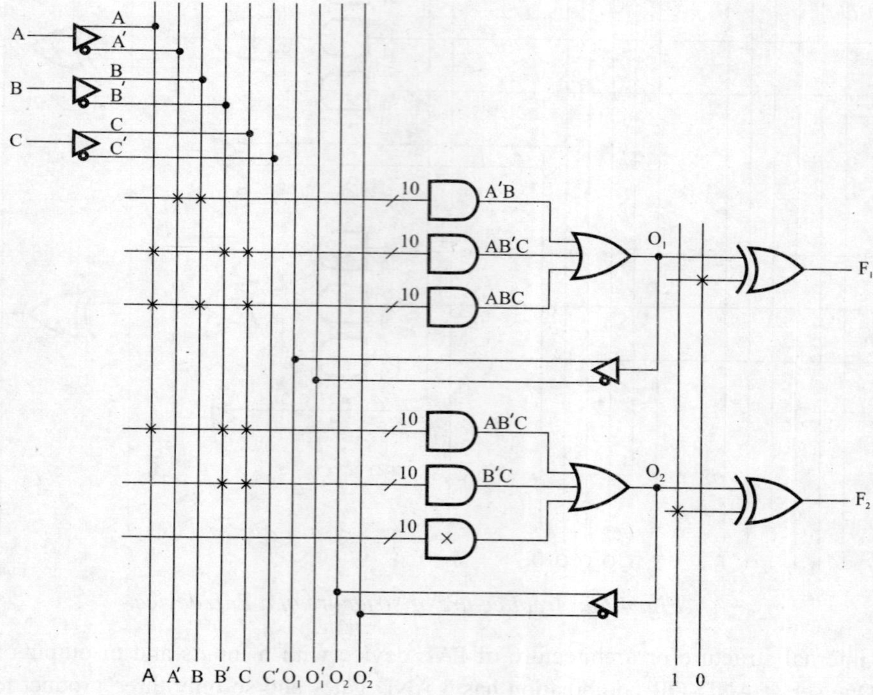

Fig. 4.61: Structure of combinational circuit implemented in PAL.

A PAL programming table can be formed as shown in Table 4.22 to provide informations regarding the implementation of combinational circuit.

Table 4.22: PAL Programming Table

Product Term	Inputs included in Product Term			Product Term included in Outputs
	A	B	C	
A′B	0	1	-	$F_1 = A'B + AB'C + ABC$ (T)
AB′C	1	0	1	
ABC	1	1	1	
AB′C	1	0	1	$F_2 = (AB'C + B'C)'$ (C)
B′C	-	0	1	

Explanation for Table 4.22

- 1 under input variables indicate that the variable is connected to form the product term
- 0 under input variables indicate that the complement of the variable is connected to form the product term.
- - under input variables indicate that the variable/complement is not connected to form the product term.
- (T) under output indicates True form of SOP output and so other input of XOR gate is 0.
- (C) under output indicates complement form of SOP output and so other input of XOR gate is 1.

Example 4.31

Implement the following Boolean equations using PAL.

$F_1 = A'B + AC' + A'BC'$ and $F_2 = AB + AC + BC'$

Solution

The programming table to implement the given Boolean equations is shown in Table 1.

The Boolean equations has three variables and both equations has three product terms. In PAL implementation each equation has to be implemented individually in a separate AND-OR array. Hence we require two set of AND-OR array.

Since the maximum number of product terms in Boolean equation is three, we require AND-OR array with three AND gates. When three input variables are complemented we get six inputs and when two outputs and their complements are feedback to input there will be additional four inputs and so we require ten input AND gates to implement the circuit. The connected inputs to AND gate are marked by ×.

Here, outputs of OR gates are function outputs and so one of the inputs of XOR gates are connected to 0 and the connections are marked by × as shown in Fig. 1.

Table 1: PAL Programming Table

Product Term	Inputs included in Product Term			Product Term included in Outputs
	A	B	C	
A′B	0	1	-	
AC′	1	-	0	F_1 = A′B + AC′ + A′BC′ (T)
A′BC′	0	1	0	
AB	1	1	0	
AC	1	-	1	F_2 = AB + AC + BC′ (T)
BC′	-	1	0	

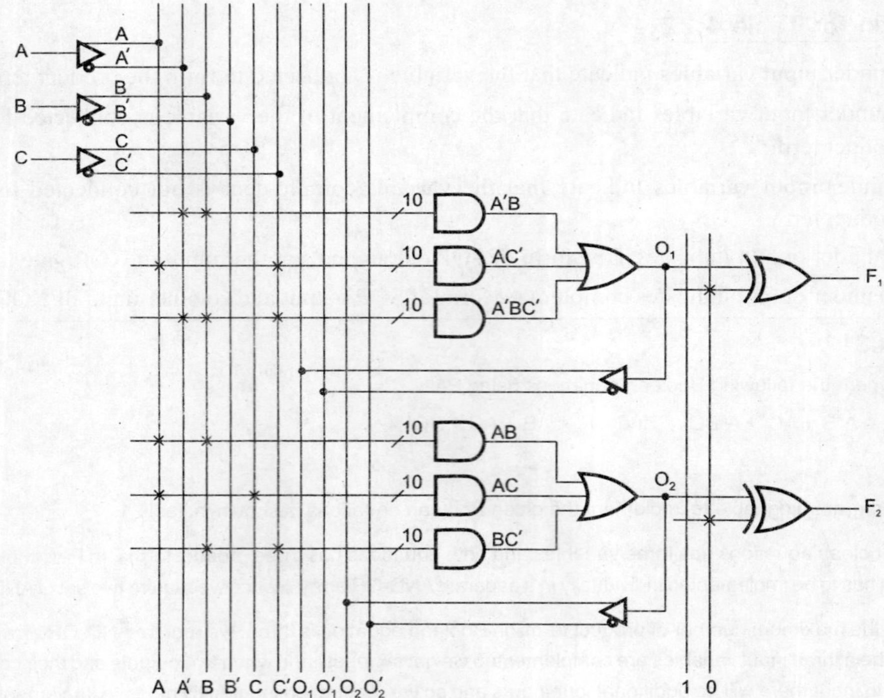

Fig. 1: Structure of PAL.

Example 4.32

(AU, Apr/May'19, 7 Marks)

Implement the following Boolean functions in a PAL device.

$F_1(X, Y, Z) = \sum m(1, 2, 4, 5)$

$F_2(X, Y, Z) = \sum m(0, 1, 3, 4)$

$F_3(X, Y, Z) = \sum m(2, 3, 6, 7)$.

Solution

The truth table of Boolean functions is formed as shown in Table 1. Here, the function output is 1 for the minterms included in that function. The simplified Boolean equations of the functions are obtained by constructing K-maps shown in Figs. 1 to 3.

Table 1: Truth Table

Inputs			Minterm	Outputs		
X	Y	Z		F_1	F_2	F_3
0	0	0	m_0	0	1	0
0	0	1	m_1	1	1	0
0	1	0	m_2	1	0	1
0	1	1	m_3	0	1	1
1	0	0	m_4	1	1	0
1	0	1	m_5	1	0	0
1	1	0	m_6	0	0	1
1	1	1	m_7	0	0	1

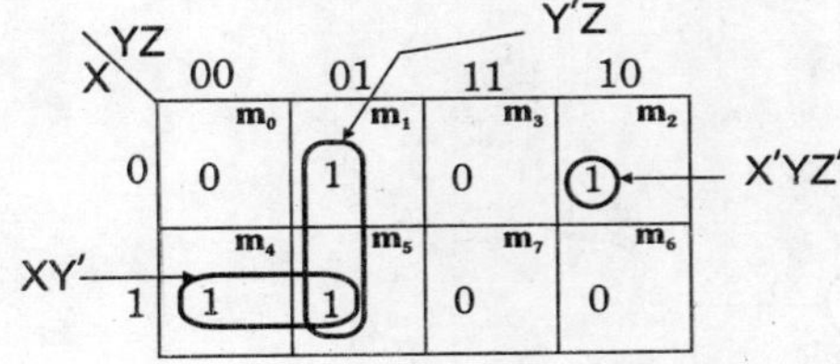

Fig. 1: K-map for F_1.

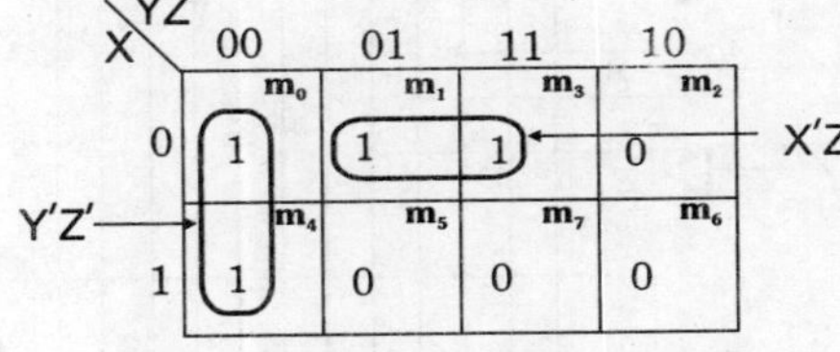

Fig. 2: K-map for F_2.

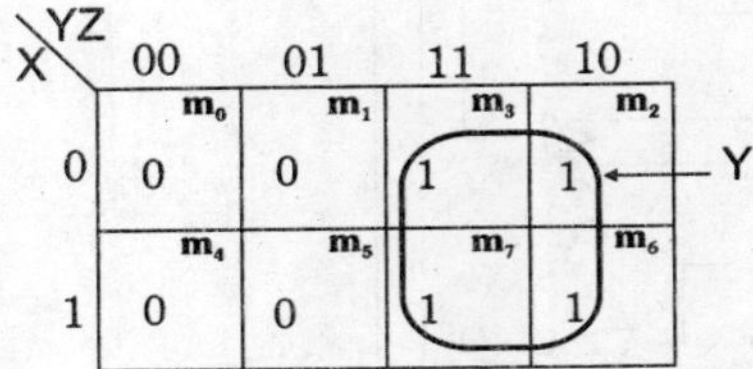

Fig. 3: K-map for F_3.

The following Boolean functions are obtained from K-map of Figs. 1 to 3.

$$F_1 = XY' + Y'Z + X'YZ'$$

$$F_2 = X'Z + Y'Z'$$

$$F_3 = Y$$

Using the above Boolean equations the PAL programming table is developed as shown in Table 2. The structure of programmed PAL using Table 2 is shown in Fig. 4.

The Boolean equations has three variables and first equation has three product terms, second equation has two product terms and the third equation has a single variable product term. In PAL implementation each equation has to be implemented individually in a separate AND-OR array. Hence we require three sets of AND-OR array. Since the maximum number of product terms in Boolean equation is three, we require AND-OR array with three AND gates. When three input variables are complemented we get six inputs and so when three outputs and their complements are feedback to input there will be additional six inputs and so we require twelve input AND gates to implement the circuit. The connected inputs to AND gate are marked by ×. Here, ouptut of OR gates are function outputs and so one of the inputs of XOR gates are connected to 0 and the connections are marked by × as shown in Fig. 4.

Table 2: PAL Programming Table

Product Term	Inputs included in Product Term X	Y	Z	Product Term included in Outputs
XY′	1	0	-	F_1 = XY′ + Y′Z + X′YZ′ (T)
Y′Z	-	0	1	
X′YZ′	0	1	0	
X′Z	0	-	1	F_2 = X′Z + Y′Z′ (T)
Y′Z′	-	-	0	
Y	-	1	-	F_3 = Y (T)

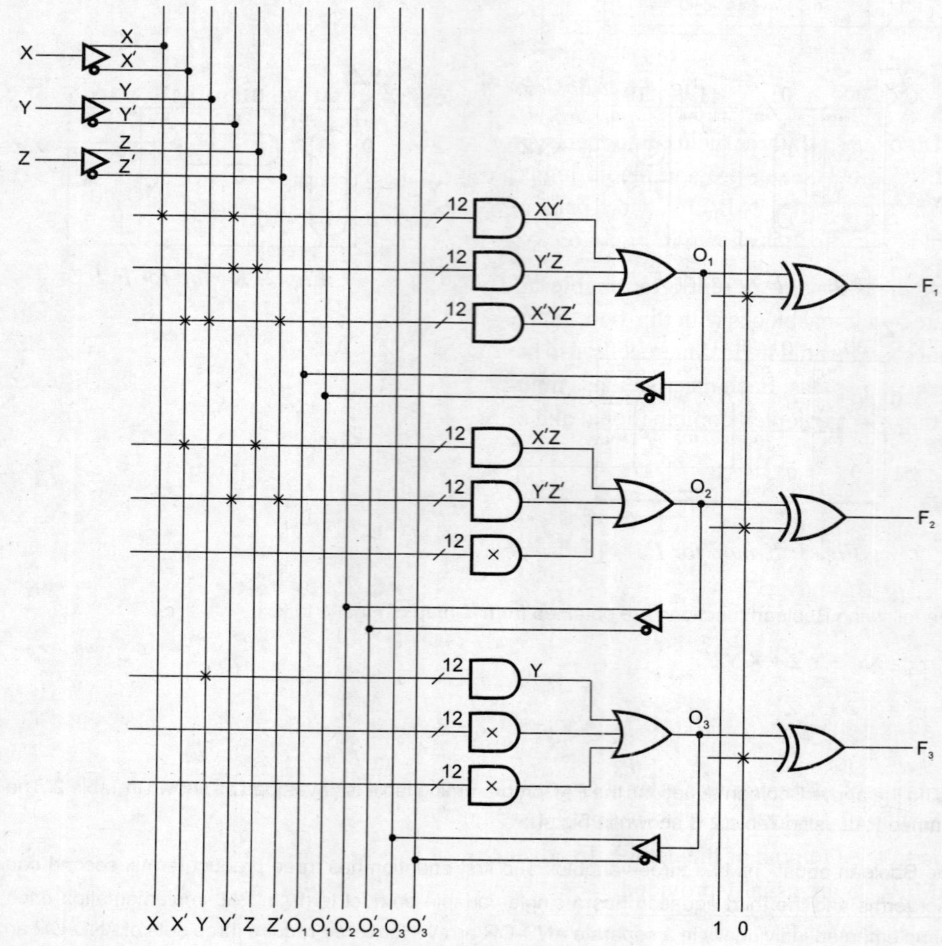

Fig. 4: Structure of PAL.

4.9 Complex Programmable Logic Device (CPLD)

(AU, Apr/May'19, 6 Marks)

A CPLD integrates a number of PLDs in a single device so that large complex combinational and sequential digital systems can be implemented in a CPLD.

The CPLDs of different manufactures differ in architecture and functionality but have some common features. The common basic architecture of a CPLD is shown in Fig. 4.62.

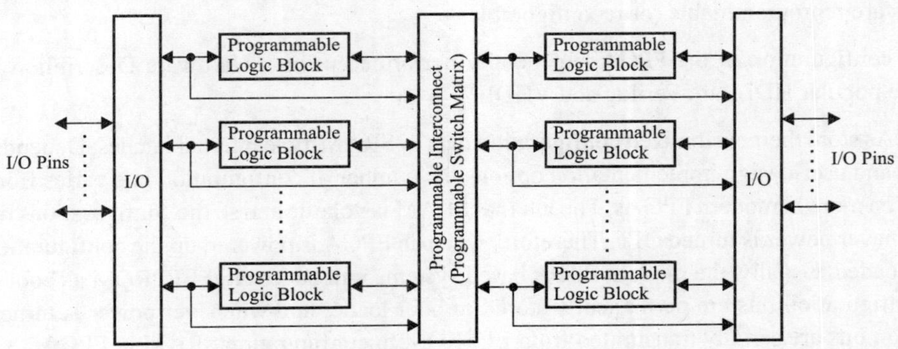

***Fig. 4.62:** Basic architecture of a CPLD.*

A CPLD consists of three main components: programmable logic blocks, programmable interconnect channel and I/O block. Each programmable logic block of a CPLD consists of a fully programmable AND-OR array similar to a PAL/PLA and a bank of macrocells. A simplified block diagram of a typical programmable logic block is shown in Fig. 4.63.

The AND/OR array is reprogrammable and can perform a large number of logic functions. Macrocells are functional blocks with flip-flop, data selector and other control logic circuits that can perform combinatorial or sequential logic. Macrocells also have the added flexibility for true or complement output, along with feedback paths. Each macrocell has multiple configurations and each macrocell can be used in cascade, so that more complex combinational and sequential logic functions can be realized.

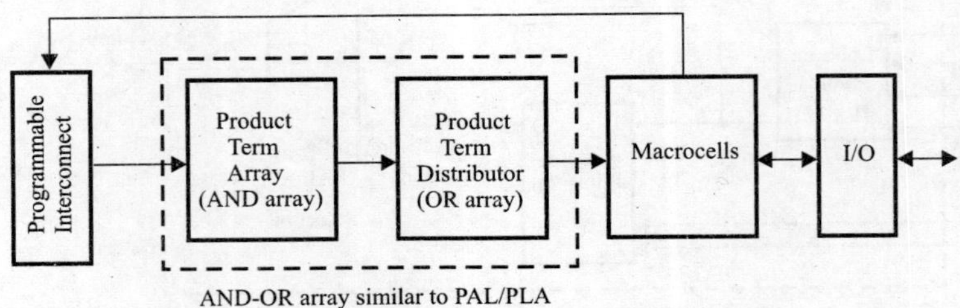

***Fig. 4.63:** Simplified block diagram of a programmable logic block in a CPLD.*

Programmable interconnect channels provide interconnection networks between logic blocks, macrocells and input/output pins. The input/output blocks (I/O blocks) provide the interface between the internal logic to the I/O pins of the device. In advanced version of CPLDs, embedded array blocks with on-chip RAM/ROM are usually provided.

4.10 Field Programmable Gate Arrays (FPGA)

R.H.Freeman invented the first practical FPGA in 1985.

FPGA is a programmable semiconductor logic device consisting of large number of logic gates which can be configured by the user to implement any kind of digital circuit or system. FPGAs are manufactured with thousands to millions of logic gates. The programming of logic gates of FPGA is called configuration and FPGAs are reprogrammable (or reconfigurable).

The configuration of the FPGA is generally performed using a Hardware Description Language (HDL). The popular HDLs are Verilog and VHDL.

FPGAs store their customized configuration data in SRAM-type internal latches. Depending on the device size and user-design implementation options the number of configuration bits varies from 1 *Mb* to 33 *Mb* or even more in modern FPGAs. The internal SRAM is volatile and so the configurations bits will be erased whenever power is turned OFF. Therefore, when an FPGA is powered up the configuration storage must be reloaded. Usually, the modern FPGA based systems will have serial EEPROM as boot loader, so that the configuration bits are permanently stored in boot loader and whenever power is turned ON the configuration bits are serially transmitted from EEPROM to configuration SRAM in FPGA.

Xilinx and Altera are two major manufacturers of FPGA holding nearly 90% of market share. The discussions in this section are with reference to Xilinx FPGAs. Altera FPGAs may also have resources similar to Xilinx FPGAs but there will be differences in architecture and functionality.

4.10.1 Basic Architecture of FPGA

FPGAs are evolved from gate arrays and early versions of FPGA basically have simple Configurable Logic Blocks (CLB) and small number of inputs and outputs as shown in Fig. 4.64. Each CLB will consist of a Look-Up Table (LUT) and a Flip-Flop (FF). The basic idea of FPGA's is to inter-connect small "truth tables" to emulate complex digital circuits. In an FPGA, these tables are called "Lookup Table" (LUT).

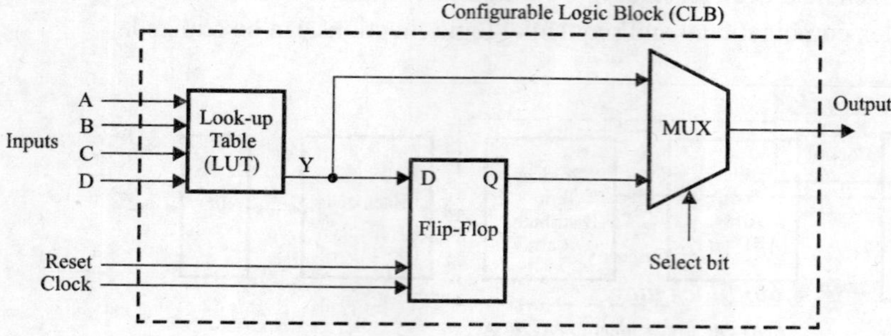

LUT	
Inputs A B C D	Output Y
0 0 0 0	0
0 0 0 1	1
0 0 1 0	1
⋮	⋮
1 1 1 1	0

Fig. 4.64: Simple CLB architecture.

The basic architecture of an FPGA is shown in Fig. 4.65.

The FPGA architecture consists of four major components

- Programmable/Configurable Logic Blocks
- Programmable switch blocks
- Programmable Routing (interconnect lines)
- I/O blocks

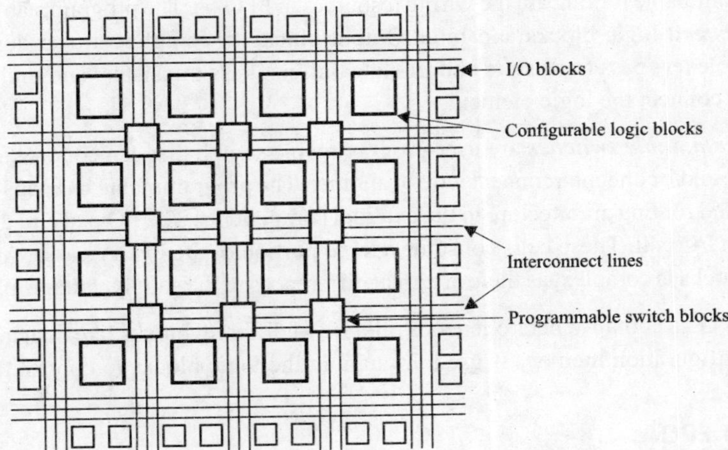

Fig. 4.65: Basic architecture of FPGA.

A configurable logic block (CLB) is the basic logic resource on an FPGA. The configurable logic block (CLB) includes logic and look-up tables (LUTs) that can be configured to implement combinational logic circuits. When CLBs are linked together by routing resources (carry chain), the components in CLBs execute complex logic functions, implement memory functions and synchronize code on the FPGA. The block diagram of a typical CLB in an FPGA is shown in Fig. 4.66.

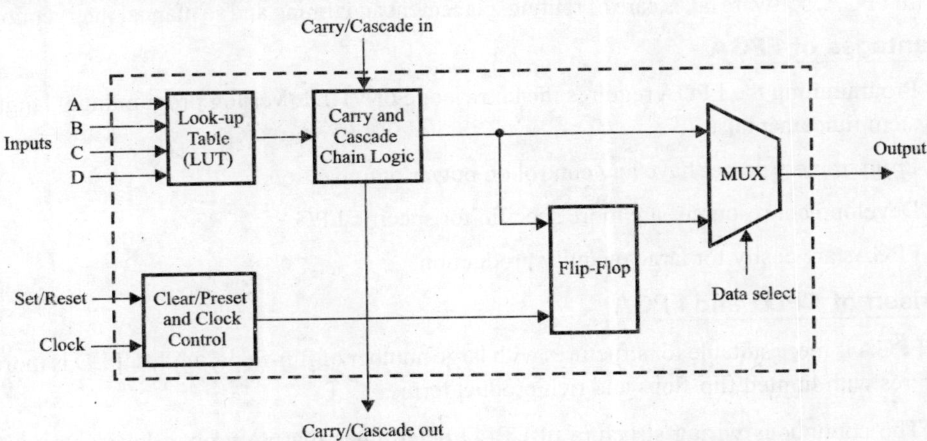

Fig. 4.66: Block diagram of a typical CLB in an FPGA.

The CLB shown in Fig. 4.66 contain look-up-table (LUT), carry logic, flip-flop (FF) and multiplexer. It also has a clear/preset and clock control unit. A LUT can speed up processing by providing the output for a given input without logical processing. A flip-flop is a primitive storage device that can store a single bit of information.

Carry look-ahead logic is implemented in each CLB with a combination of dedicated multiplexer and XOR gates that are used within the carry/cascade chain. The carry/cascade chain logic in an individual CLB is the circuit that directly connects a column of CLBs together. It is possible to cascade a carry chain across multiple CLBs in order to quickly implement addition, subtraction, and multiplication operations on operands that are too large to be processed by a single CLB.

The programmable routing are the wiring resources in FPGAs. The programmable routing establishes a connection between logic blocks and Input/Output blocks to complete a user-defined design unit. It consists of multiplexers pass transistors and tri-state buffers. Pass transistors and multiplexers are used in a logic cluster to connect the logic elements.

The ***programmable switches*** are user-programmable switch that can connect a logic element to an interconnect wire and/or one interconnect wire to another. The programmable I/O pads are used to interface the logic blocks and routing architecture to the external components. The I/O pad and the surrounding logic circuit form as an I/O cell. These cells consume a large portion of the FPGA's area. And the design of I/O programmable blocks is complex, as there are great differences in the supply voltage and reference voltage.

Modern FPGAs contain a heterogeneous mixture of different blocks like dedicated memory blocks, multiplexers. Configuration memory is used throughout the logic blocks to control the specific function of each element.

Advantages of FPGA

- FPGAs can implement faster and parallel processing.
- FPGAs are programmable IC, programmed by software and can be re-programmed or reused any number of times.
- Since FPGA ICs are programmable the solution is available faster to the market.
- FPGA development is cheaper due to non recurring expenses.
- In FPGA, software takes care of routing, placement and timing and so manual intervention is lesser.

Disadvantages of FPGA

- Programming the FPGA requires the knowledge of VHDL/Verilog programming languages and digital system fundamentals.
- Programmers do not have any control on power optimization.
- Development resources are more specific for specific FPGA.
- FPGAs are costly for large quantity production.

Comparison of CPLD and FPGA

- FPGA is more suitable for structures with large number of flip-flops, while CPLD is more suitable for structures with limited flip-flops and rich product terms.
- The continuous wiring structure of CPLD determines that its timing delays are uniform and predictable, while the segmented wiring structure of FPGA determines the unpredictability of its delays.
- CPLD is programmed by modifying the logic function with fixed interconnect circuit, while FPGA is programmed by changing the wiring of internal wires.
- FPGA can be programmed under the logic gate, while CPLD is programmed under the logic block. Hence, CPLDs are faster than FPGAs and have greater time predictability.
- FPGAs are more integrated than CPLDs and have more complex wiring structures and logic implementations.

- CPLD is programmed using internal EEPROM or FASTFLASH technology, no external memory chip is needed. While the FPGA programming information needs to be stored in external memory, and transferred to internal RAM during switch ON.

- CPLD is mainly based on EEPROM or FLASH memory programming, which can be programmed up to 10,000 times, and has the advantage that the programming information is not lost even when the system is powered off.

4.10.2 Xilinx FPGA Family of Devices

Xilinx has released a number of families of FPGA devices with a large number of internal logical devices in each family with enhanced and variable features. XC2000 series is the first family of FPGA devices released by Xilinx in the year 1985 manufactured using 2000 nm (nano-meter) process.

> ***Note:*** *The nano-meter manufacturing process refer to the size of fundamental transistor in the FPGA IC. Xilinx started with 2000 nm process and realized up to 7 nm process in their latest Versal family of FPGAs.*

This is followed by XC3000, XC4000, XC5200, XC6200 and XC8100 series of FPGA devices released over a period of 10 years, from 1985, with enhanced features in every family of devices. These FPGA devices are outdated and no longer manufactured by Xilinx.

The Family of Xilinx FPGA devices released after 1995 are named as Spartan, Artix, Kintex, Virtex and Versal. The evolution of various families of FPGA devices has made improvements in density, performance, power consumption, voltage levels, IO support and other functionality. The various Xilinx FPGAs are listed in Appendix 7. A brief discussion about Spartan-6 family of FPGAs is presented in the following section.

4.10.3 Spartan-6 Family of FPGA

The Spartan-6 family includes thirteen types of FPGA devices with logic cell densities ranging from 3,840 to 1,47,443 logic cells and manufactured using 45 nm low-power copper process technology. The specific features of Spartan-6 family members are summarized in Tables 4.23 and 4.24. The salient features of Spartan-6 family of FPGAs are,

- Dual-register 6-input lookup table (LUT) logic.
- 18 *kB* (2 x 9 *kB*) block RAMs.
- Second generation DSP48 slices.
- SDRAM memory controllers.
- Enhanced mixed-mode clock management blocks.
- Select IO technology.
- Power optimized high-speed serial transceiver blocks.
- PCI Express compatible Endpoint blocks.
- Advanced system-level power management modes.
- Auto-detect configuration options.
- Enhanced IP security with AES and Device DNA protection.

Table 4.23: Spartan-6 FPGA Feature Summary by Device

Device	Logic Cells	Configurable Logic Blocks (CLBs)			DSP48A1 Slices	Block RAM Blocks		Total I/O Banks	Max User I/O
		Slices	Flip-Flops	Max Distributed RAM (Kb)		18Kb	Max (Kb)		
XC6SLX4	3,840	600	4,800	75	8	12	216	4	132
XC6SLX9	9,152	1,430	11,440	90	16	32	576	4	200
XC6SLX16	14,579	2,278	18,224	136	32	32	576	4	232
XC6SLX25	24,051	3,758	30,064	229	38	52	936	4	266
XC6SLX45	43,661	6,822	50,576	401	58	116	2,088	4	358
XC6SLX75	74,637	11,662	93,296	692	132	172	3,096	6	408
XC6SLX100	101,261	15,822	126,576	976	180	268	4,824	6	480
XC6SLX150	147,443	23,038	184,304	1,355	180	268	4,824	6	576
XC6SLX25T	24,051	3,758	30,064	229	38	52	936	4	250
XC6SLX45T	43,661	6,822	54,576	401	58	116	2,088	4	296
XC6SLX75T	74,637	1,662	93,296	692	132	172	3,096	6	348
XC6SLX100T	101,261	15,822	126,576	976	180	268	4,824	6	498
XC6SLX150T	147,443	23,038	184,304	1,355	180	268	4,824	6	540

Table 4.24: Spartan-6 FPGA Logic Resources

Device	Logic Cells	Total Slices	SLICEMs	SLICELs	SLICEXs	Number of 6-Input LUTs	Maximum Distributed RAM (Kb)	Shift Register (Kb)	Number of Flip-Flops
XC6SLX4	3,840	600	300	0	300	2,400	75	38	4,800
XC6SLX9	9,152	1,430	360	355	715	5,720	90	45	11,440
XC6SLX16	14,579	2,278	554	595	1,139	9,112	136	68	18,224
XC6SLX25	24,051	3,758	916	963	1,879	15,032	229	115	30,064
XC6SLX45	43,661	6,822	1,602	1,809	3,411	27,288	401	200	54,576
XC6SLX75	74,637	11,662	2,768	3,063	5,831	46,648	692	346	93,296
XC6SLX100	101,261	15,822	3,904	4,007	7,911	63,288	976	488	126,576
XC6SLX150	147,443	23,038	5,420	6,099	11,519	92,152	1,355	678	184,304
XC6SLX25T	24,051	3,758	916	963	1,879	15,032	229	115	30,064
XC6SLX45T	43,661	6,822	1,602	1,809	3,411	27,288	401	200	54,576
XC6SLX75T	74,637	1,662	2,768	3,063	5,831	46,648	692	346	93,296
XC6SLX100T	101,261	15,822	3,904	4,007	7,911	63,288	976	488	126,576
XC6SLX150T	147,443	23,038	5,420	6,099	11,519	92,152	1,355	678	184,304

4.10.4 Spartan-6 FPGA Architecture

Spartan-6 FPGA architecture is shown in Fig. 4.67 and it consists of following elements.

- Configurable logic blocks (CLB)
- Embedded memories (Block RAM) for distributed data storage.
- Clock management tiles (CMT) with PLLs for driving the FPGA fabric at different clock rates.
- On-chip memory controller blocks (MCB).
- Multiply-accumulate blocks (DSP 48-bit blocks).
- High-speed serial transceivers.
- PCIE endpoint
- I/O banks

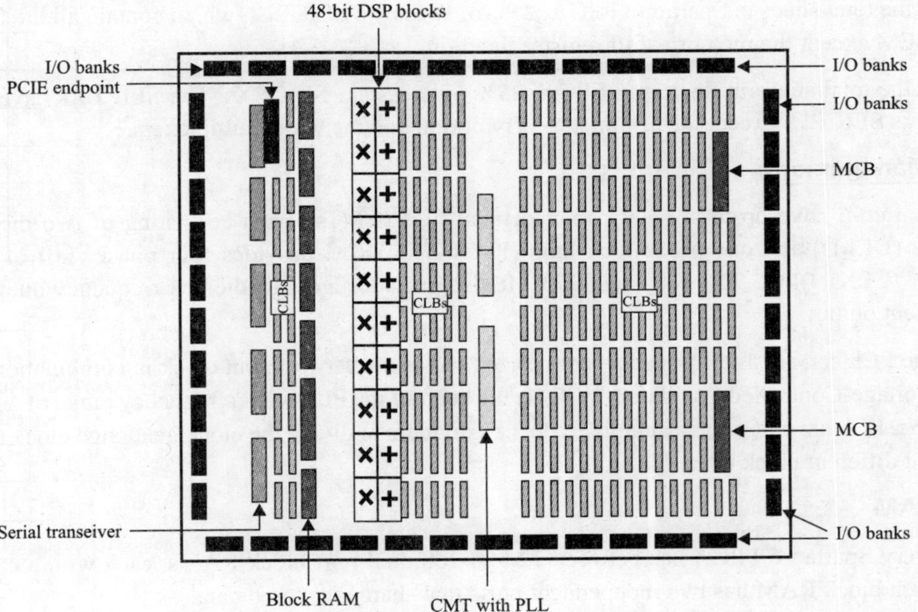

Fig. 4.67: Architecture of Spartan-6 FPGA.

Spartan-6E FPGA contains the following basic resources.

1. Logic resources
   - Slices (grouped into CLB)
   - Memory
   - Multipliers
2. Interconnect resources
   - Programmable interconnect
   - I/O banks
3. Other resources
   - Global clock buffers
   - Boundary scan logic

CLBs, Slices and LUTs

Each configurable logic block (CLB) in Spartan-6 FPGAs consists of two slices, arranged side-by-side as two vertical columns. There are three types of CLB slices in the Spartan-6 architecture: SLICEM, SLICEL and SLICEX. Each slice contains four LUTs, eight flip-flops and other control logic. The LUTs are used to implement combinational and sequential logic functions.

In the total slices in Spartan-6 FPGA, 25% of slices are SLICEMs. Each of the four SLICEM LUTs can be configured as either a 6-input LUT with one output, or as dual 5-input LUTs with identical 5-bit addresses and two independent outputs. These LUTs can also be used as distributed 64-bit or 32-bit RAM, or as a single 32-bit shift register, or as two 16-bit shift registers. Each LUT output can be registered in a flip-flop within the CLB. For arithmetic operations, a high-speed carry chain propagates carry signals in a column of slices.

In the total slices in Spartan-6 FPGA, 25% of slices are SLICELs, which contain all the features of the SLICEM except the memory/shift register function.

In the total slices in Spartan-6 FPGA, 25% of slices are SLICEXs. The SLICEXs have the same structure as SLICELs except the arithmetic carry option and the wide multiplexers.

Clock Management

Spartan-6 have up to six clock management tiles (CMTs), each consisting of two digital clock managers (DCMs) and one phase lock loop (PLL). The DCM provides four phase shift of the input frequency (CLKIN): 0°, 90°, 180°, and 270°. It also provides a doubled clock frequency output and its complement output.

The PLL can act as a frequency synthesizer and as a filter for input clock in combination with the DCMs. Voltage-Controlled Oscillator (VCO) is the heart of the PLL with a frequency range of 400 *MHz* to 1,080 *MHz.* PLL have of programmable frequency dividers to divide the clock generated clock to convert to clock at different clock rates.

Block RAM

Every Spartan-6 FPGA has between 12 and 268 dual-port block RAMs, each with a capacity of 18 *kB*. Each block RAM has two independent ports that share the stored data.

Each memory access is controlled by the clock. All the inputs, data, address, clock enables, and write enables are stored in registers. The data output retains data until the next operation.

The memory data width is programmable and hence, each port can be configured as 16K × 1, 8K × 2, 4K × 4, 2K × 9 (or 8), 1K × 18 (or 16), or 512 x 36 (or 32). Parity bits are included in the x9, x18, and x36 configurations.

Memory Controller Blocks

Spartan-6 devices have dedicated memory controller blocks (MCBs) for interfacing external DRAM and support access rates up to 800 Mb/s. The MCB has dedicated routing to predefined I/Os. If the MCB is not used, these I/Os are used as general purpose I/Os. The MCB can be connected to 4–, 8–, or 16-bit external DRAM. The MCB provides a faster DRAM interface.

The memory controller offers a complete multi-port arbitrated interface to the logic inside the Spartan-6 FPGA. Commands and data can be pushed to and poped from independent built-in FIFOs, using conventional FIFO control signals.

Digital Signal Processing-DSP48 Slice

Spartan-6 FPGAs have many dedicated DSP slices consisting of binary multipliers and accumulators to support DSP computations. Each DSP48 slice consists of a dedicated 18 × 18 bit two's complement multiplier and a 48-bit accumulator.

The DSP48 slice provides extensive pipelining and extension capabilities that enhance speed and efficiency of many applications, such as wide dynamic bus shifters, memory address generators, wide bus multiplexers and memory-mapped I/O register files. The accumulator can also be used as a synchronous up/down counter. The multiplier can perform barrel shifting.

Input/Output Banks

The number of I/O pins varies from 102 to 576, depending on device and package size. Each I/O pin is configurable and can comply with a large number of standards, using up to 3.3V.

With the exception of supply pins and a few dedicated configuration pins, all other package pins have the same I/O capabilities. All user I/Os are bidirectional. All I/O pins are organized in banks, with four banks on the smaller devices and six banks on the larger devices.

Serial Transceiver

Spartan-6 LXT devices have 2 to 8 transceiver circuits. Each transceiver is a combined transmitter and receiver capable of operating at data rates up to 3.2 Gb/s. The transmitter and receiver are independent circuits that use separate PLLs to multiply the reference frequency input by certain programmable numbers between 2 and 25, to become the bit-serial data clock.

The transmitter is fundamentally a parallel-to-serial converter with a conversion ratio of 8, 10, 16, or 20. The transmitter output drives the PC board with a single-channel differential current-mode logic (CML) output signal. The receiver is fundamentally a serial-to-parallel converter, changing the incoming bit serial differential signal into a parallel stream of words, each 8, 10, 16, or 20 bits wide.

Integrated Endpoint Block for PCI Express Designs

The PCI Express standard is a packet-based, point-to-point serial interface standard. The PCI Express Base Specification 1.1 defines bit rate of 2.5 Gb/s per lane, per direction (transmit and receive). The Spartan-6 LXT devices include one integrated Endpoint block for PCI Express technology that is compliant with the PCI Express Base Specification Revision 1.1.

4.11 Summary of Important Concepts

1. The combinational logic circuits with feedback and without clock are called asynchronous sequential circuits.
2. The present and next state variables of asynchronous sequential circuit are called secondary and excitation variables.
3. The circuit will attain a stable condition when excitation (or next state) variables are same as secondary (or present state) variables.
4. In asynchronous sequential circuit, the state table is called transition table.
5. A state is a circuit condition represented by a set of variables called state variables.
6. In a state transition, if the next state is not equal to present state then the corresponding state is called unstable state.

7. All possible combinations of state variables and inputs are called total states of the circuit.
8. The continuous change between two unstable states is also called metastability which refers to an equilibrium condition in an unstable circuit.
9. The flow table lists all state transitions from present state to next state along with output for all possible changes in input.
10. When there is only one stable state in each row of a flow table then it is called a primitive flow table.
11. The diagrammatic way of representing all possible state transitions between unstable states in every row of transition table is called transition diagram.
12. The race condition is classified into critical and noncritical race condition.
13. In a state transition, if the final stable state depends on the order of change in state variables, then the race is called critical race.
14. In a state transition, if the final stable state does not depends on the order of change in state variables, then the race is called noncritical race.
15. When the state transitions in a circuit go through a unique sequence of unstable states then that condition is called a cycle.
16. Races can also be avoided by using cycles, which terminate on a stable state.
17. Two binary codes are said to be adjacent if they differ by only one variable.
18. Assigning adjacent codes to states to avoid critical races is called race-free assignment.
19. In fundamental mode asynchronous sequential circuits the inputs are level triggered inputs.
20. In asynchronous sequential circuits, the state reduction is performed via an implication table or merger table.
21. State pairs that have same next state and output for the same input are equivalent states.
22. State pairs having different next state and output for same input are not equivalent states.
23. If a pair of state having same outputs for the same inputs and go to same equivalent next state then the pair is said to be implied pair.
24. A merger diagram or merger graph is a circle with dots representing the states and every equivalent pair of states is connected by a straight line.
25. Asynchronous circuits with incompletely specified states, the states can be reduced by identifying compatible pairs.
26. Maximum compatible states is a group of compatible states that includes all possible combinations of compatible states.
27. The closed-covering condition refers to set of compatibles that covers all the states and satisfy covering condition.
28. The set of compatibles that satisfy the closed-covering condition form the minimal set of compatibles.
29. Rows of reduced flow table can be formed by taking one element of each of the minimal compatible set.
30. The latches are used as memory elements in asynchronous circuits.
31. SR latch is constructed using two cross coupled NOR gates or two cross coupled NAND gates.
32. Hazards are unwanted transient outputs in the digital circuits that may arise due to gates and connecting wires with different propagation delay.

33. When the expected output is 1 for a change in input, if the circuit generates a temporary 0 then the hazard is called static-1 hazard.
34. When the expected output is 0 for a change in input, if the circuit generates a temporary 1 then the hazard is called static-0 hazard.
35. When the expected output is either 1 or 0 for a change in input, if the circuit output temporarily changes from 0 to 1 and 1 to 0 three or more times, then the hazard is called dynamic hazard.
36. The synchronous sequential circuits will not have hazards because in those circuits the changes are synchronized with a clock.
37. In synchronous sequential circuits the changes in circuit conditions are synchronized by a clock with time period long enough to overcome internal delays in signal transitions.
38. Hazards will occur in asynchronous sequential circuits if there are different delays in various signal paths.
39. The static and dynamic hazards in asynchronous sequential circuits can be eliminated by adding redundant gates or by using SR latch based circuit design.
40. The essential hazard is due to different delays in different paths that start from a same input.
41. An essential hazard is due to unequal delay in two or more paths that start from same input and also due to delay in input path longer than delay in feedback path.
42. The essential hazard can be eliminated by rearranging the gate structure of the combinational circuit that generates inputs to latches.
43. The digital ICs (Integrated Circuits) can be broadly classified into fixed function ICs and programmable ICs.
44. PROM, PAL and PLA are one-time programmable and consists of programmable logic gate arrays and so can be used to implement only combinational logic functions.
45. FPGA devices are reprogrammable and consists of programmable logic gate arrays and flip-flops and so can be used to implement both combinational and sequential logic functions.
46. PLA is a one-time programmable device to implement a Boolean function in sum of products form.
47. PAL device will have m sets of AND-OR array with each array having k number of AND gates and one OR gate.
48. A CPLD integrates a number of PLDs in a single device.
49. A CPLD consists of three main components: programmable logic blocks, programmable inter connect channel and I/O block.
50. Each programmable logic block of a CPLD consists of a fully programmable AND-OR array similar to a PAL/PLA and a bank of macrocells.
51. FPGA is a programmable semiconductor logic device consisting of large number of logic gates.
52. The programming of logic gates of FPGA is called configuration and FPGAs are reprogrammable (or reconfigurable).
53. FPGAs store their customized configuration data in SRAM-type internal latches.
54. The basic idea of FPGA's is to inter-connect small "truth tables" to emulate complex digital circuits.
55. The FPGA architecture consists of four major components: programmable/configurable logic blocks, programmable switch blocks, programmable routing (interconnect lines), I/O blocks.
56. A configurable logic block (CLB) is the basic logic resource on an FPGA.

57. The configurable logic block (CLB) includes logic and look-up tables (LUTs) that can be configured to implement combinational logic circuits.
58. Carry look-ahead logic is implemented in each CLB with a combination of dedicated multiplexer and XOR gates.
59. The programmable routing are the wiring resources in FPGAs.
60. The programmable switches are user-programmable switch that can connect a logic element to an interconnect wire and/or one interconnect wire to another.

4.12 Short-Answer Questions

Q4.1 Define asynchronous sequential circuits.

The combinational logic circuits with feedback and without clock are called asynchronous sequential circuits.

Q4.2 What are the characteristic of asynchronous sequential circuits?

The changes in internal states occur immediately when there is a change in input and hence the response in an asynchronous sequential circuit will be fast. In the asynchronous sequential circuit the change in input may lead to races and unstable condition and therefore, sufficient care must be taken to avoid races and instablity.

Q4.3 List the difference between asynchronous and synchronous sequential circuits.

Synchronous Sequential Circuit	Asynchronous Sequential Circuit
1. Provided with clock	1. Do not have a clock
2. Changes in internal states occur in response to clock.	2. Changes in internal states occur when there is a change in input variables.
3. Memory elements are clocked flip-flops.	3. Memory elements are either latches (unclocked flip-flops) or time-delay elements.

Q4.4 What are secondary and excitation variables.

The present and next state variables of asynchronous sequential circuits are called secondary and excitation variables.

Q4.5 Draw the general model of asynchronous sequential machine (ASM). ***(AU, Nov/Dec'18, 2 Marks)***

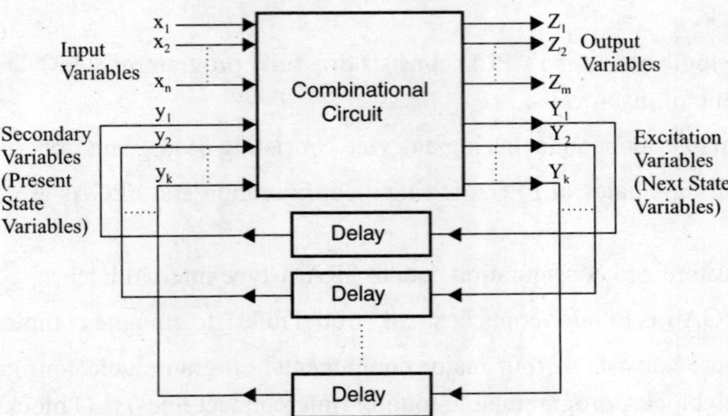

Fig. Q4.5: Block diagram representation of an asynchronous sequential circuit.

Q4.6 Explain stable and unstable states.

The asynchronous sequential circuit will change from one state to another state when there is a change in input. During a state transition when the next state is same as present state then the circuit attains a stable condition and the corresponding state is called a stable stable. On the other hand in a state transition, if the next state is not equal to present state then the corresponding state is called an unstable state.

Q4.7 Define total states.

All possible combinations of state variables and inputs are called total states of the circuit.

Q4.8 Define metastable state. **(AU, Apr/May'19, 2 Marks)**

The continuous change between two unstable states is also called metastability (or metastable state) which refers to an equilibrium condition in an unstable circuit.

Q4.9 What is difference between an internal state and a total state?

When the excitation variables are fedback to input they are called internal states or secondary variables. The excitation variables are called external states.

The internal states along with inputs are called total states.

Q4.10 Define flow table in asynchronous sequential circuit.

The listing of all state transitions from present state to next state along with output for all possible changes in input is called flow table.

Q4.11 What is primitive flow table.

A flow table may have one or more stable states in a row. When there is only one stable state in each row of a flow table then it is called a primitive flow table.

Q4.12 What is cycle?

When the state transitions in a circuit go through a unique sequence of unstable states then that condition is called a cycle.

Q4.13 What is meant by race? (or What is race around condition?)

In an asynchronous sequential circuit, when there are changes in two or more state variables due to change in input, a race condition may exist, in which changes in variables may occur at different timings due to difference in delay in the feedback paths. This leads to unpredictable transition paths during a change in state which in turn settle at different stable state.

Q4.14 Define critical race and non critical race.

The race condition may be classified into critical and noncritical race.

Critical race : In a state transition, if the final stable state depends on the order of change in state variables, then the race is called ***critical race***.

Noncritical race : In a state transition, if the final stable state does not depends on the order of change in state variables, then the race is called ***noncritical race***.

Q4.15 Distinguish between noncritical and critical race.

In noncritical race the final stable state will be same irrespective of the order of changes in state variables and hence final stable state will be same for a change in input.

But in critical race the final stable state depends on the order of changes in state variables and hence the final stable state will be different for a change in input.

Q4.16 *What is meant by race-free condition in sequential circuit?*

During state transitions if the order of changes in internal states does not affect the final stable state then the condition is called race-free condition.

Q4.17 *What is race-free state assignment?*

Assigning adjacent codes to states to avoid critical races is called race-free assignment. Two binary codes are said to be adjacent if they differ only in one variable.

Q4.18 *An asynchronous sequential circuit is described by the excitation function and output function,*

$$Y_1 = x_1x_2 + x_1y_2 + x_2'y_1 \quad ; \quad Y_2 = x_2' + x_1y_1y_2 + x_1y_1 \quad ; \quad Z = x_2' + y_2$$

Draw the logic diagram of the circuit.

Solution:

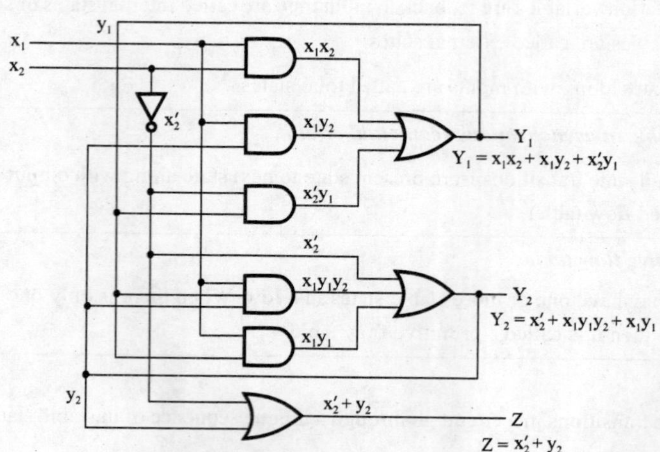

Fig. Q4.18: Logic diagram.

Q4.19 *Write the Boolean equation for the given circuit.*

Solution:

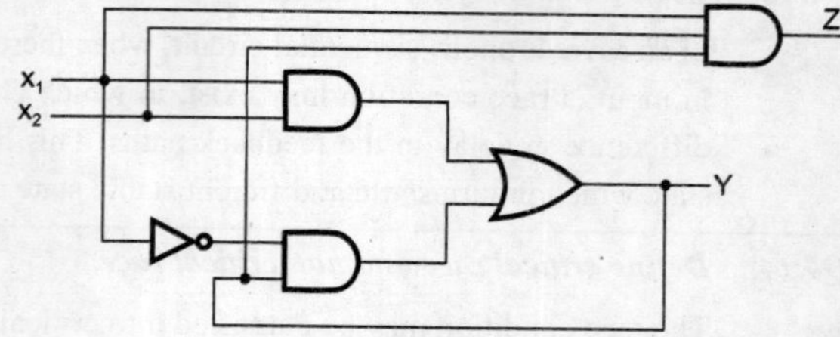

Fig. Q4.19.1: Logic circuit.

From the circuit shown in Fig. Q4.19.2 we get the following Boolean equations for excitation variable and output of the asynchronous circuit.

$$Y = x_1 x_2 + x_1' y$$

$$Z = x_1 x_2 y$$

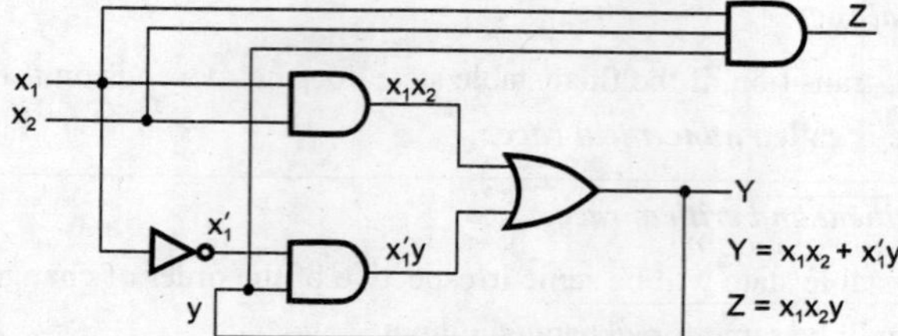

Fig. Q4.19.2: Logic circuit with intermediate signals.

Q4.20 Give an example for critical race.

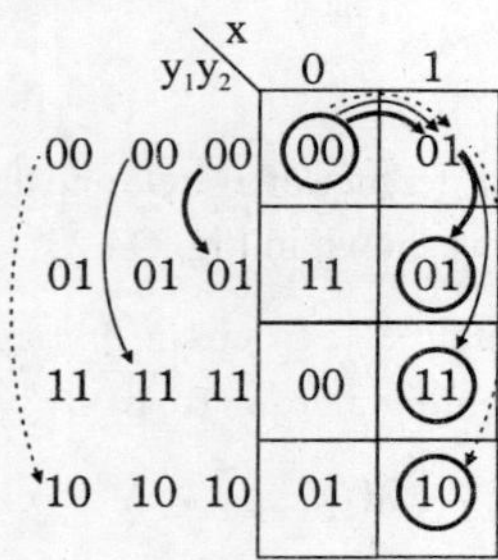

00 → 01 ; 00 → 11 ; 00 → 10

Fig. Q4.20: Critical race during state transition from 00.

Q4.21 Define merger graph and merger table.

Merger graph : Merger graph is a tool used to reduce number of states in the incompletely specified machine. Merger graph is also known as merger diagram.

Merger table : The table in which each cell represents a compatible pair and their implications is called merger table. Merger table is also known as implication table.

Q4.22 Draw the transition diagram for the transition table shown in Fig. Q4.22.1.

y \ x_1x_2	00	01	11	10
a	a	b	c	a
b	a	b	b	c
c	a	c	c	c
d	d	a	c	d

Fig. Q4.22.1: Transition table.

Solution:

The following state transitions from present state to next state are observed in transition table of Fig. Q4.22.2.

From a	From b	From c	From d
a → b	b → a	c → a	d → a
a → c	b → c		d → c

The above transitions are combined as a single diagram as shown in Fig. Q4.22.3 to get transition diagram.

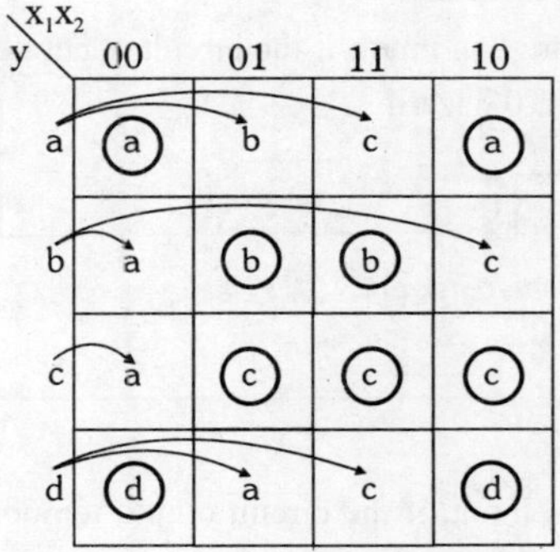

Fig. Q4.22.2: Transition table.

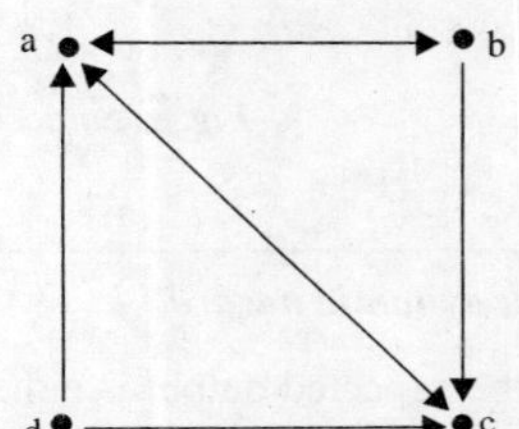

Fig. Q4.22.3: Transition diagram.

Q4.23 ***Draw the the merger diagram for the equivalent state pairs (a, d), (a, f), (b, d), (b, e), (c, d), (d, f), (d, e) and hence find the equivalent state groups.***

Solution:

The states are represented by dots on the circumference of a circle and the equivalent pairs are joined by straignt lines as shown in Fig. Q4.23.

The merger diagram has two triangles representing state groups (a, d, f) and (b, d, e). Also, there is a straight line representing a group (c, d).

∴ Equivalent state groups : (a, d, f), (b, d, e), (c, d)

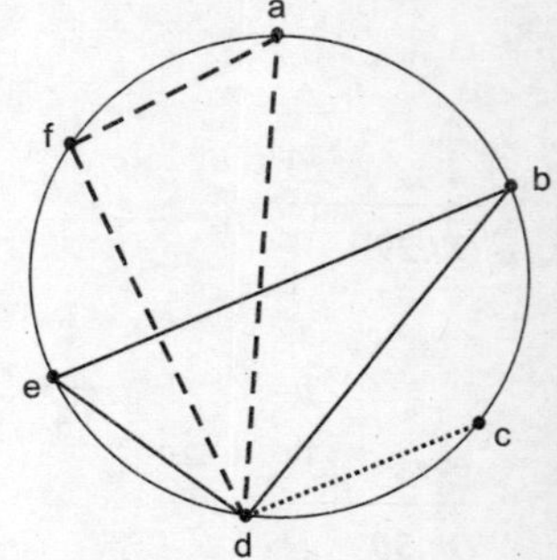

Fig. Q4.23: *Merger diagram.*

Q4.24 ***What is hazard? Give its types.***

Hazards are unwanted transient outputs in the digital circuits that may arise due to gates and connecting wires with different propagation delay.

The different types of hazards are,

1. Static hazard
    1. Static-0 hazard
    2. Static-1 hazard
2. Dynamic hazard
3. Essential hazard

Q4.25 ***Define static hazard. How it can be avoided?***

A single temporary change in output during state transition is called static hazard. This can be eliminated by converting all prime implicants in the K-map of SOP or POS form into overlapped prime implicants.

Q4.26 ***What are static-1 and static-0 hazards?***

Static-1 hazard: When the expected output is 1 for a change in input, if the circuit generates a temporary 0 as shown in Fig. Q4.26(a) then the hazard is called static-1 hazard.

Static-0 hazard: When the expected output is 0 for a change in input, if the circuit generates a temporary 1 as shown in Fig. Q4.26(b) then the hazard is called static-0 hazard.

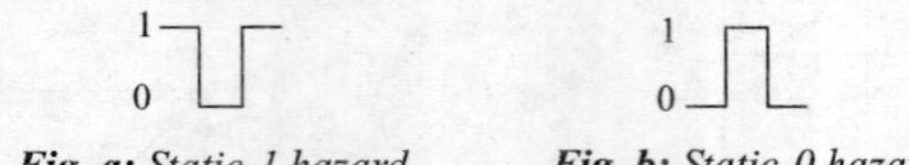

Fig. a: *Static-1 hazard.* ***Fig. b:*** *Static-0 hazard.*

Fig. Q4.26.

Q4.27 ***Explain dynamic hazard.***

When the expected output is either 1 or 0 for a change in input, if the circuit output temporarily changes from 0 to 1 and 1 to 0 three or more times as shown in Fig. Q4.27, then the hazard is called dynamic hazard.

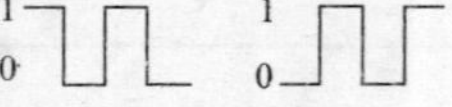

Fig. Q4.27: *Dynamic hazard.*

Q4.28 What is an essential hazard?

An essential hazard is due to unequal delay in two or more paths that start from same input.

Alternatively, we can say that an essential hazard exist if the total stable state reached after three changes in an input is different from stable state reached after a single change in that input.

Q4.29 What is the cause of essential hazard?

The essential hazard is due to delay in changes in input reaching the input of excitation circuit before the feedback output reaching the input of excitation circuit. Hence feedback inputs will change two or more times before reaching a stable value.

Q4.30 How to eliminate essential hazard?

In general, the essential hazard is minimized or eliminated by reducing the complexity of gate structure. Even after modifying the gate structure if essential hazard exists in certain path then appropriate delays has to be introduced by individually studying the signal paths.

Q4.31 Draw the hazard free circuit, $F = \Sigma m(1, 5, 6, 7)$.

Solution:

The K-map of the given function is shown in Fig. Q4.31.1. Here, the prime implicants are non-overlapped prime implicants. Hence, one additional prime implicant is added to create overlapped prime implicants and to prevent hazard as shown in Fig. Q4.31.2.

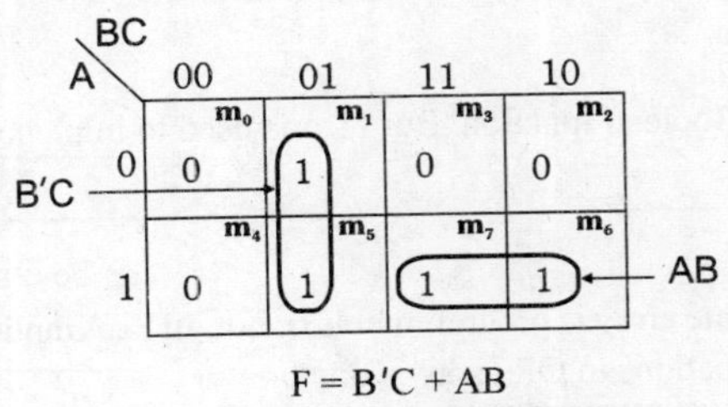

Fig. Q4.31.1: K-map of F.

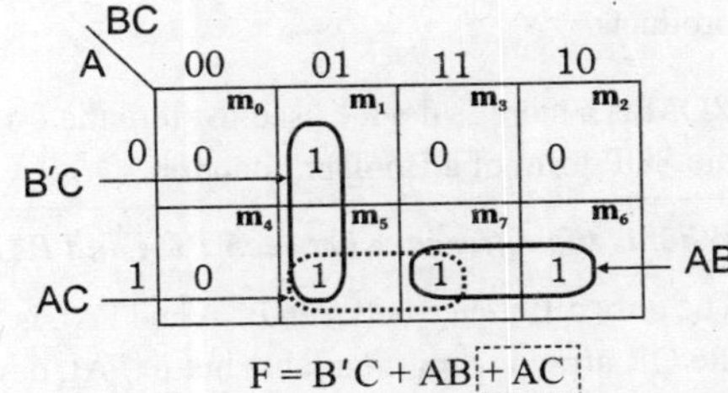

Fig. Q4.31.2: Overlapping minterms to eliminate hazard.

The following Boolean equations are obtained from K-map of Fig. Q4.31.2.

$$F = B'C + AB + AC$$

Using the above Boolean equation the hazard free logic circuit is drawn as shown in Fig. Q4.31.3.

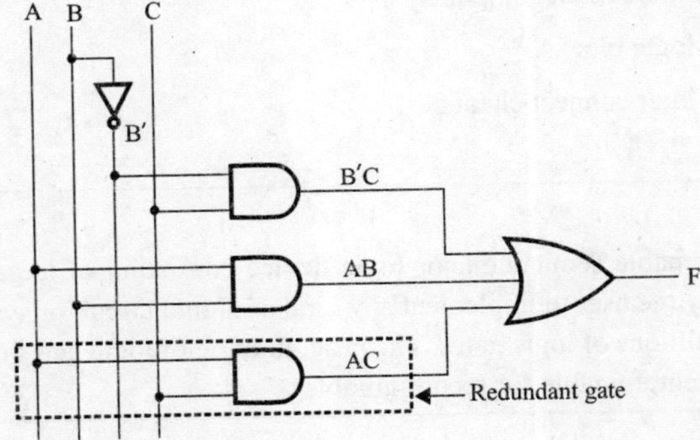

Fig. Q4.31.3: Logic circuit.

Q4.32 What are the types of asynchronous circuits?

There are two types of asynchronous sequential circuits.

1. Fundamental mode circuits. 2. Pulse mode circuits.

Q4.33 What is fundamental mode asynchronous sequential circuit?

In fundamental mode circuits the inputs are level triggered inputs and only one input is allowed to change at any one time when the circuit is in a stable state.

Q4.34 What are steps for design of fundamental mode sequential circuit.

1. Develop a primitive flow table from the word description of asynchronous circuit
2. Minimize the number of states in the flow table.
3. Assign race-free binary codes to states.
4. Construct the transition table.
5. Construct excitation maps and find Boolean equations for a hazard free realization.
6. Check for essential hazards.
7. Draw the logic circuit by implementing methods to eliminate essential hazards.

Q4.35 What is programmable logic array? How it differs from ROM?

The PLA is a device to implement Boolean function in sum of products form. The PLA will have an array of AND gates to implement product terms followed by an array of OR gates to implement the sum of the products.

ROM is a storage device used to store the truth table of a Boolean function. But PLA is used to implement the SOP form of a Boolean function.

Q4.36 What is the difference between PLA and PAL?

The only difference between PLA and PAL is that the OR gate array is programmable or not. In PLA device, the OR array is programmable but in PAL device the connections to OR array is fixed.

Q4.37 What is CPLD?

A CPLD integrates a number of PLDs in a single device so that large complex combinational and sequential digital systems can be implemented in a CPLD.

Q4.38 What are the basic internal components of CPLD?

A CPLD consists of three main components.

- Programmable logic blocks.
- Programmable inter connect channel.
- I/O block.

Q4.39 What is FPGA?

FPGA is a programmable semiconductor logic device consisting of large number of logic gates which can be configured by the user to implement any kind of digital circuit or system. FPGAs are manufactured with thousands to millions of logic gates. The programming of logic gates of FPGA is called configuration and FPGAs are reprogrammable (or reconfigurable).

Q4.40 What are basic internal components of FPGA?

The FPGA architecture consists of four major components.

- Programmable/Configurable Logic Blocks
- Programmable switch blocks
- Programmable Routing (interconnect lines)
- I/O blocks

Q4.41 What is configuration in FPGA?

The programming of logic gates of FPGA is called configuration and FPGAs are reprogrammable (or reconfigurable). The configuration of the FPGA is generally performed using a Hardware Description Language (HDL). The popular HDLs are Verilog and VHDL.

Q4.42 What are the difference between CPLD and FPGA?

- FPGA is more suitable for structures with large number of flip-flops, while CPLD is more suitable for structures with limited flip-flops and rich product terms.
- The continuous wiring structure of CPLD determines that its timing delays are uniform and predictable, while the segmented wiring structure of FPGA determines the unpredictability of its delays.
- CPLD is programmed by modifying the logic function with fixed interconnect circuit, while FPGA is programmed by changing the wiring of internal wires.
- FPGA can be programmed under the logic gate, while CPLD is programmed under the logic block.

4.13 Exercises

I. Fill in the blanks

1. The combinational logic circuits with feedback and without clock are called _____ circuits.
2. The present and next state variables of asynchronous sequential circuit are called ______ and ______ variables.
3. In asynchronous sequential circuit, the state table is called ______ .
4. A state is a circuit condition represented by a set of variables called ______ .
5. The continuous change between two unstable states is also called ______ .
6. When there is only one stable state in each row of a flow table then it is called a ______ .
7. The diagrammetic way of representing all possible state transitions between unstable states in every row of transition table is called ______ .
8. The race condition may be classified into ______ and ______ race.
9. In a state transition, if the final stable state depends on the order of change in state variables, then the race is called ______ .
10. In a state transition, if the final stable state does not depends on the order of change in state variables, then the race is called ______ .
11. When the state transitions in a circuit go through a unique sequence of unstable states then that condition is called a ______ .

12. In asynchronous sequential circuits, the state reduction is performed via an ______ .
13. State pairs that have same next state and output for the same input are ______ states.
14. State pairs having different next state and output for same input are ______ states.
15. The ______ is a group of compatible states that includes all the possible combinations of compatible states.
16. A minimum set of compatibles can be determined from ______ condition.
17. The ______ are used as memory elements in asynchronous circuits
18. ______ occurs only if there are three or more paths between input and output.
19. A ______ integrates a number of PLDs in a single device.
20. FPGA is a programmable semiconductor logic device consisting of large number of ______.
21. FPGAs store their customized configuration data in ______ type internal latches.
22. ______ is the basic logic resource on an FPGA.
23. The ______ are the wiring resources in FPGAs.
24. The ______ are devices that connect a logic element to an interconnect wire and/or one interconnect wire to another.

Answers

1. asynchronous sequential	9. critical race	17. latches
2. secondary, excitation	10. noncritical race	18. dynamic hazard
3. transition table	11. cycle	19. CPLD
4. state variables	12. implication table	20. logic gates
5. metastable	13. equivalent	21. SRAM
6. primitive flow table	14. not equivalent	22. Configurable logic block (CLB)
7. transition diagram	15. maximum compatible	23. programmable routing
8. critical, noncritical	16. closed-covering	24. programmable swithces

II. State whether the following statements are True or False

1. The combinational logic circuits without feedback and with clock are called asynchronous sequential circuits.
2. The combination of output values and state variables are called total states of the circuit.
3. In state table of asynchronous circuit, at least one entry of each row should have same present state and next state.
4. When the state transitions in a circuit go through a unique sequence of stable states then that condition is called a cycle.
5. Races can also be avoided by using cycles, which terminate on a stable state.
6. Two binary codes are said to be adjacent codes if they differ only in one variable.
7. State pairs that have different next state and output for the same input are equivalent states.
8. State pairs having same next state and output for same input are not equivalent states.
9. In implication table each square represent a pair of state.

10. In asynchronous circuits with incompletely specified states, the states can be reduced by identifying compatible pairs.
11. The reduced flow table can be formed by taking one member of each maximum compatible set as a row in reduced flow table.
12. Hazards are unwanted transient inputs in the digital circuits.
13. The synchronous sequential circuits will not have hazards.
14. The static dynamic and hazards can be eliminated by adding redundant gates or by using SR latch based circuit design.
15. The essential hazard is due to different delays in different paths that start from a same input.
16. In fundamental mode asynchronous sequential circuits the inputs are pulse signals.
17. PLA is a one-time programmable device to implement a Boolean function in sum of products form.
18. A CPLD integrates a number of PLDs in a single device.
19. FPGA is a programmable semiconductor logic device consisting of large number of logic gates.
20. FPGAs store their customized configuration data in DRAM-type internal latches.
21. A configurable logic block (CLB) is the basic logic resource on an CPLD.
22. Carry look-ahead logic is implemented in each CLB with a combination of dedicated multiplexer and XOR gates
23. The programmable switches are user-programmable switch that can connect a logic element to an interconnect wire and/or one interconnect wire to another.
24. CPLD is mainly based on SRAM memory programming, which can be programmed up to 10000 times.

Answers

1. False	5. True	9. True	13. True	17. True	21. False
2. False	6. True	10. True	14. True	18. True	22. True
3. True	7. False	11. True	15. True	19. True	23. True
4. False	8. False	12. False	16. False	20. False	24. False

III. Choose the right answer for the following questions

1. *An asynchronous sequential circuit,*

   a) does not use clock pulses

   b) only one input can change at a time

   c) consists of combinational circuit and latches

   d) all the above

2. *In an asynchronous sequential circuit, _______ number of inputs can change at a time.*

   a) 1 **b)** 2

   c) 3 **d)** n

3. ***The total states are formed with***
   a) Inputs and outputs
   b) Outputs and state variables
   c) Inputs and state variables
   d) Internal and external states

4. ***After a change in input the asynchronous circuit attains a stable condition when***
   a) Internal state and output are equal
   b) Internal and external states are equal
   c) Internal state and input are equal
   d) Input and output are equal

5. ***State pairs are said to be equivalent if the states have,***
   a) Same next state and output for same input
   b) Same next data and output for different inputs
   c) Same present and next states for same input
   d) Same present and next states for different inputs

6. ***We can design an asynchronous sequential circuit using***
   a) Truth table
   b) Excitation table
   c) Primitive state table
   d) State table

7. ***Asynchronous sequential circuit can be designed by using***
   a) T flip-flop
   b) JK flip-flop
   c) D flip-flop
   d) all the above

8. ***Races occur due to***
   a) Clock input
   b) Pulse input
   c) Delay in one feedback path
   d) Delay in more than two feedback path

9. ***Static-0 hazard occurs when***
   a) The output becomes '1' for sometime when it should be '0'
   b) The output becomes '0' for sometime when it should be '1'
   c) Output is always 1
   d) Output is always 0

10. ***Static-1 hazard occurs when***
    a) The output becomes '1' for sometime when it should be '0'
    b) The output becomes '0' for sometime when it should be '1'
    c) Output is always 1
    d) Output is always 0

11. ***Static hazard can be avoided by***

a) Nonoverlapped prime implicants

b) Overlapped prime implicants

c) Adding don't-cares to some inputs

d) Taking POS equation

---

12. ***The hazard in which the output changes from 0 to 1, three or more times is,***

a) Static b) Dynamic

c) Essential d) Functional

---

13. ***_______ hazard occurs only in asynchronous sequential circuits with 2 or more feedbacks.***

a) Static b) Dynamic

c) Essential d) Functional

---

14. ***Static hazard can be eliminated by,***

a) Converting POS to SOP equation b) Using redundant gate

c) Using NAND gates only d) Using NOR gates only

---

15. ***Essential hazard can be eliminated by,***

a) Converting POS to SOP equation

b) Converting SOP to POS equation

c) Rearranging gate structure

d) Using NOR or NAND gates only

---

16. ***In the fundamental mode asynchronous sequential circuit, the inputs are,***

a) Level triggered b) Positive edge triggered

c) Negative edge triggered d) Positive or negative edge triggered

---

17. ***In FPGA store their customized configuration data in ______ type internal latches.***

a) SRAM b) DRAM c) RAM d) ROM

---

18. ***A configurable logic block (CLB) is the basic logic resource on an ________.***

a) CPLD b) FPGA c) PLD d) LUT

Answers

1. d	6. c	11. b	16. a
2. a	7. d	12. b	17. a
3. c	8. d	13. c	18. b
4. b	9. a	14. b	
5. a	10. b	15. c	

V. Answer the following questions

1. What is the difference between flow table and transition table?
2. Define transition diagram.
3. What is state diagram? Give an example.
4. What are the significance of state assignment?
5. What is state reduction?
6. How races are avoided in digital circuits?
7. What are hazard free digital circuits?
8. Write a short note on hazards in asynchronous sequnetial circuits.
9. Write detailed notes on hazards in combinational circuits and sequential circuits.
10. Draw hazard free circuit for $F = \Sigma(0, 1, 5, 7)$
11. What is PAL.
12. Why was PAL developed?
13. Draw the basic architecture of CPLD.
14. What are the advantages of FPGA.
15. List the features of spartan-6 family of FPGA.
16. What are the elements of spartan-6 FPGA architecture.

VI. Solve the following problems

E4.1 The Boolean equations for excitation variable and output of an asynchronous circuit are,

$$Y = x_1 x_2 + x_2' y$$

$$Z = x_2 + y'$$

Draw the logic diagram of the circuit. Derive the transition table and output map. Obtain a flow table for the circuit.

E4.2 The Boolean equations for excitation variables and output of an asynchronous circuit are,

$$Y_1 = x'y_1 + xy_2'$$

$$Y_2 = xy_1' + x'y_2$$

$$Z = Y_2$$

Draw the logic diagram of the circuit. Derive the transition table and output map. Obtain a flow table for the circuit.

E4.3 Derive the flow table for the circuit shown in Fig. E4.3.

E4.4 Investigate the transition table of Fig. E4.4, and determine all race conditions and whether they are critical or noncritical. Determine also whether there are any cycles.

E4.5 Investigate the transition table of Fig. E4.5, and determine all race conditions and whether they are critical or noncritical. Determine also whether there are any cycles.

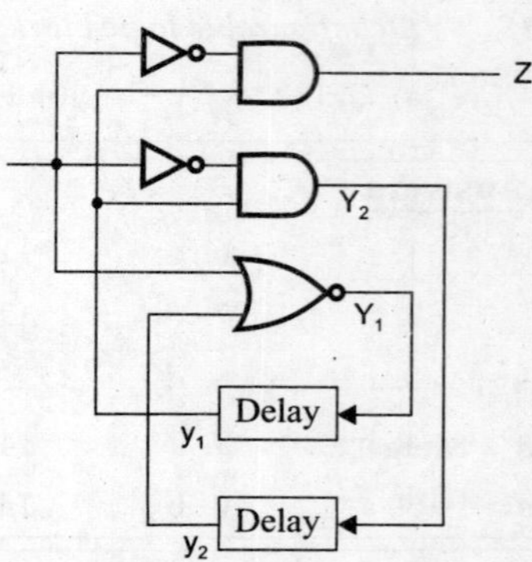

Fig. E4.3: Logic circuit.

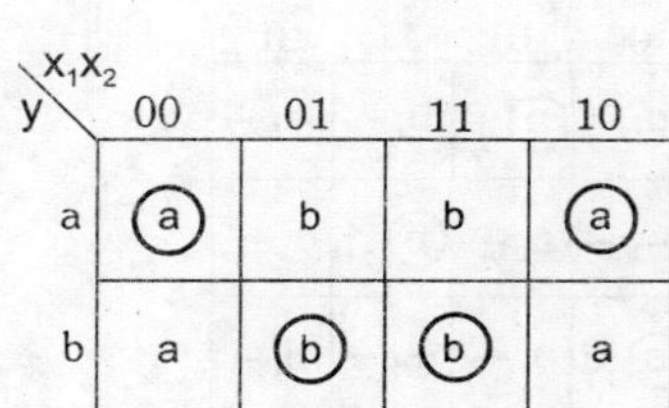

Fig. E4.4: Transition table.

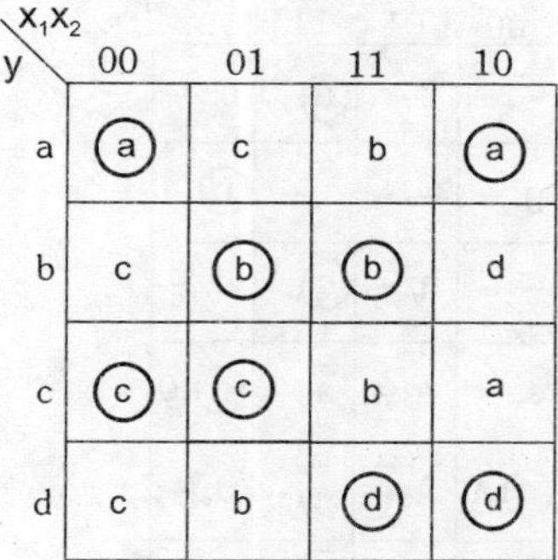

Fig. E4.5: Reduced flow table.

E4.6 Reduce the state table shown in Table E4.6 to a minimum number of states using the implication table method.

Table E4.6: State Table

Present State	Next State x = 0	Next State x = 1	Output x = 0	Output x = 1
a	a	a	1	0
b	a	f	1	1
c	b	e	0	0
d	a	g	0	0
e	d	c	1	1
f	b	e	0	0
g	d	c	1	1

E4.7 Reduce the state table shown in Table E4.7 to a minimum number of states using the implication table method.

Table E4.7: State Table

Present State	Next State x = 0	Next State x = 1	Output x = 0	Output x = 1
a	a	e	0	1
b	a	e	1	1
c	b	f	1	1
d	b	f	1	1
e	c	g	0	0
f	c	g	0	0
g	d	h	0	0
h	d	h	0	0

E4.8 Merge the primitive flow table shown in Fig. E4.8. Find all compatible pairs by means of an implication table. Find the maximal compatibles by means of a merger diagram. Find a minimal set of compatibles that covers all the states and is closed.

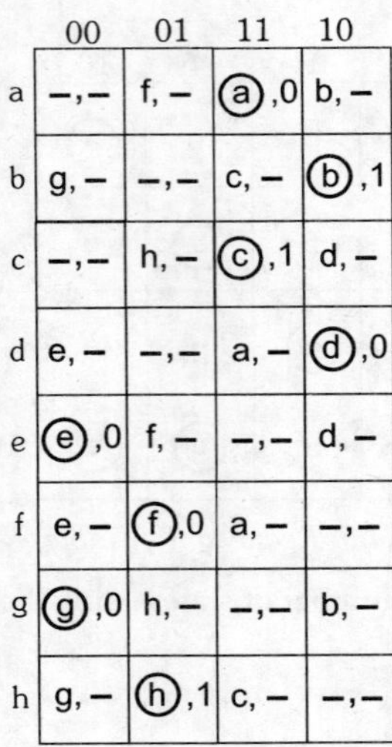

	00	01	11	10
a	–,–	f, –	(a),0	b, –
b	g, –	–,–	c, –	(b),1
c	–,–	h, –	(c),1	d, –
d	e, –	–,–	a, –	(d),0
e	(e),0	f, –	–,–	d, –
f	e, –	(f),0	a, –	–,–
g	(g),0	h, –	–,–	b, –
h	g, –	(h),1	c, –	–,–

Fig. E4.8: *Primitive flow table.*

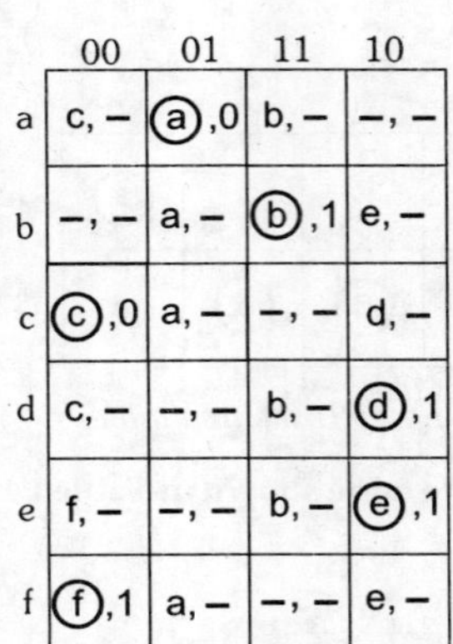

	00	01	11	10
a	c, –	(a),0	b, –	–, –
b	–, –	a, –	(b),1	e, –
c	(c),0	a, –	–, –	d, –
d	c, –	–, –	b, –	(d),1
e	f, –	–, –	b, –	(e),1
f	(f),1	a, –	–, –	e, –

Fig. E4.9: *Primitive flow table.*

E4.9 **Merge the primitive flow table shown in Fig. E4.9. Find all compatible pairs by means of an implication table. Find the maximal compatibles by means of a merger diagram. Find a minimal set of compatibles that covers all the states and is closed.**

E4.10 **Find a critical race-free state assignment for the given transition table shown in Fig. E4.10.**

y \ x_1x_2	00	01	11	10
a	(a)	b	(a)	c
b	a	(b)	(b)	d
c	d	(c)	a	(c)
d	(d)	(d)	e	(d)
e	f	c	(e)	c
f	(f)	b	a	(f)

Fig. E4.10: *Transition table.*

E4.11 **a) Obtain a binary state assignment for the reduced flow table shown in Fig. E4.11 to avoid critical race condition.**

b) Assign output to unstable states and obtain the output equation.

c) Implement the logic circuit using NOR based SR latch.

d) Implement the logic circuit using NAND based SR latch.

E4.12 **Consider the reduced flow table shown in Fig. E4.12.**

a) Obtain the transition diagram, and show that three state variables are needed for a race-free binary state assignment.

b) Obtain the expanded flow table, using the multiple row method of assignment.

y \ x_1x_2	00	01	11	10
a	d, –	(a),0	(a),0	b, –
b	(b),1	(b),1	c, –	(b), 1
c	b, –	(c),1	(c),1	d, –
d	(d), 0	(d), 0	a, –	(d), 0

Fig. E4.11: *Reduced flow table.*

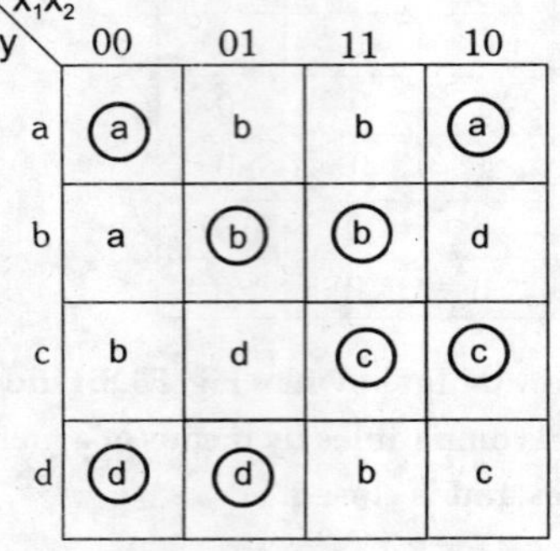

y \ x_1x_2	00	01	11	10
a	(a)	b	b	(a)
b	a	(b)	(b)	d
c	b	d	(c)	(c)
d	(d)	(d)	b	c

Fig. E4.12: *Reduced flow table.*

E4.13 **a) Obtain a binary state assignment for the reduced flow table shown in Fig. E4.13 to avoid critical race condition.**

b) Assign output to unstable states and obtain the output equation.

c) Implement the logic diagram using NOR based SR latch.

d) Implement the logic diagram using NAND based SR latch.

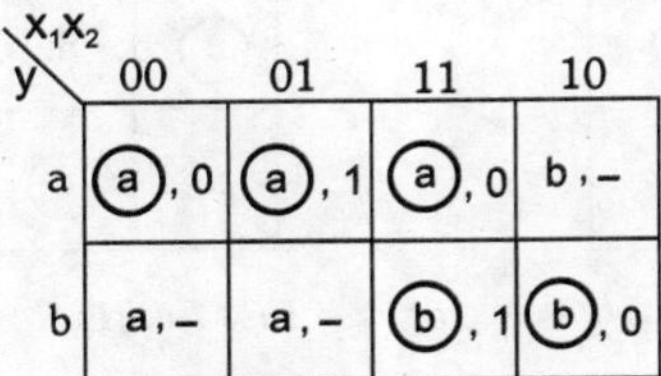

y \ x_1x_2	00	01	11	10
a	(a), 0	(a), 1	(a), 0	b, –
b	a, –	a, –	(b), 1	(b), 0

Fig. E4.13: Reduced flow table.

E4.14 **The Boolean functions for the inputs of an SR latch are,**

$$S = x_1' x_2' x_3 + x_1x_2x_3 \quad ; \quad R = x_1x_2' + x_2x_3'$$

Obtain the circuit diagram using only NAND gates.

E4.15 **Implement the circuit defined in Fig. E4.15 with NOR SR latch. Repeat with NAND SR latch.**

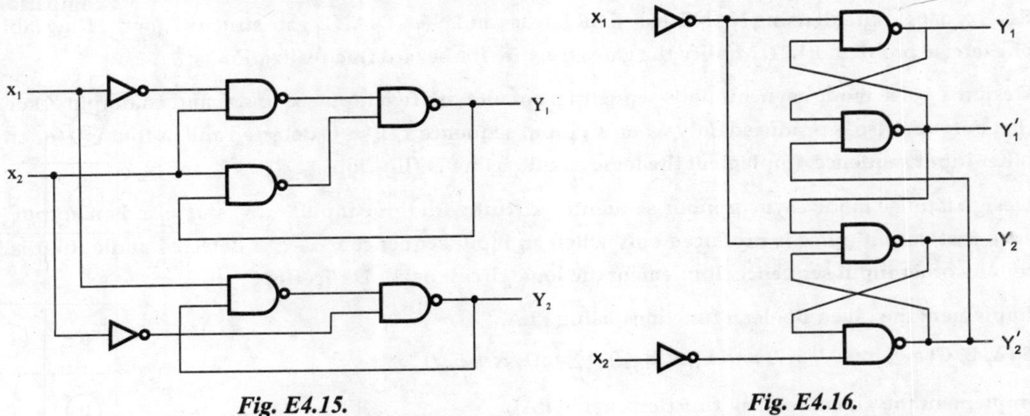

Fig. E4.15. *Fig. E4.16.*

E4.16 **For the asynchronous sequential circuit shown in Fig. E4.16. Derive the Boolean functions for the outputs of the two SR latches Y_1 and Y_2.**

E4.17 **Implement the switching function, $F = \sum m(1, 3, 5, 7, 10, 11, 12, 13)$ by a static hazard free two level AND-OR gate network.**

E4.18 **For a given Boolean function obtain the hazard free circuit.**

$$F = \sum m(0, 1, 3, 4, 6, 7)$$

E4.19 **Implement the following Boolean function by a logic circuit that has no hazards.**

$$Y = \prod M(2, 3, 4, 6)$$

E4.20 **Assign output values to the unstable states in the flow tables of Fig. E4.20 in such a way to avoid tranisent output pulses or hazard.**

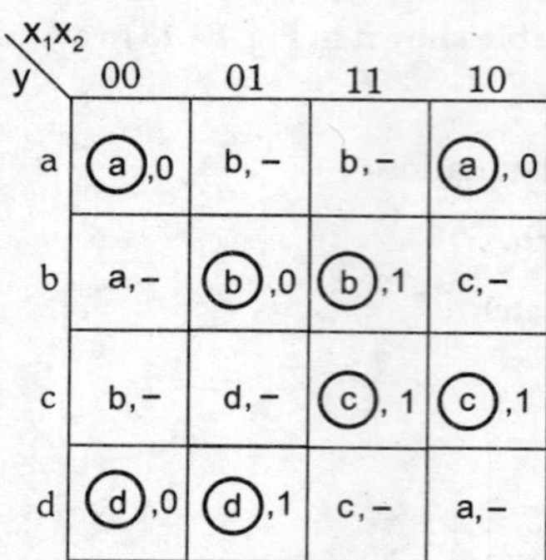

$y \backslash x_1x_2$	00	01	11	10
a	(a),0	b,–	b,–	(a),0
b	a,–	(b),0	(b),1	c,–
c	b,–	d,–	(c),1	(c),1
d	(d),0	(d),1	c,–	a,–

Fig. E4.20: Reduced flow table.

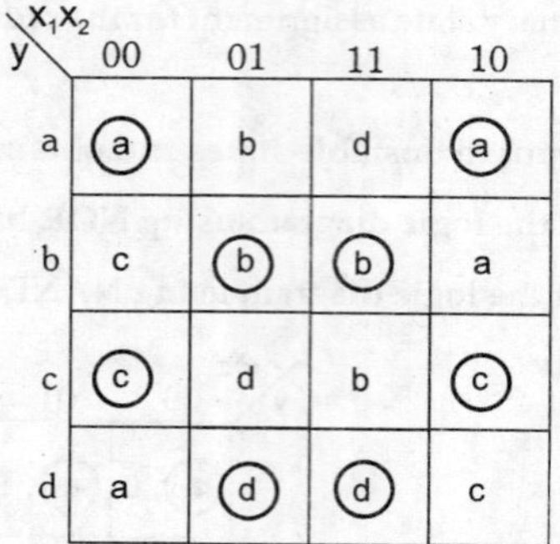

$y \backslash x_1x_2$	00	01	11	10
a	(a)	b	d	(a)
b	c	(b)	(b)	a
c	(c)	d	b	(c)
d	a	(d)	(d)	c

Fig. E4.21: Reduced flow table.

E4.21 **Realize a logic circuit using NOR based SR latches and AND-OR gate structure for the flow table shown in Fig. E4.21. Modify the gate structure for hazard free realization.**

E4.22 **Realize a logic circuit using NOR based SR latches and NOR-OR gate structure for the flow table of exercise problem E4.21. Modify the gate structure for hazard free realization.**

E4.23 **Realize a logic circuit using NAND based SR latches and OR-AND gate structure for the flow table of exercise problem E4.21. Modify the gate structure for hazard free realization.**

E4.24 **Realize a logic circuit using NAND based SR latches and NAND-AND gate structure for the flow table of exercise problem E4.21. Modify the gate structure for hazard free realization.**

E4.25 **Design a pulse mode asynchronous sequential circuit with two inputs x_1 and x_2 and an output Z such that output pulse is produced only when an input sequence x_1-x_1-x_2 is detected and output is 0 for any other input sequence. Implement the logic circuit using D flip-flop.**

E4.26 **Design a pulse mode asynchronous sequential circuit with three inputs x_1, x_2 and x_3 and an output Z such that output pulse is produced only when an input sequence x_1-x_3-x_2 is detected and output is 0 for any other input sequence. Implement the logic circuit using D flip-flop.**

E4.27 **Implement the given Boolean functions using PLA.**

$F_1(A, B, C) = \sum m(0, 2, 6, 7)$ and $F_2(A, B, C) = \sum m(1, 3, 4, 5, 7)$

E4.28 **Implement the given Boolean functions using PAL.**

$F_1(A, B, C) = \sum m(0, 4, 6, 7)$ and $F_2(A, B, C) = \sum m(1, 2, 3, 5, 7)$

E4.29 **Implement the following logical functions using PLA.**

$Y_1 = AB + A'C + ABCD$ $\quad Y_2 = AB'C + ABC' + AC'$ $\quad Y_3 = AB$

E4.30 **Implement the following logical functions using PAL.**

$Y_1 = AB + A'CB' + ABC'$ $\quad Y_2 = AB'C + AB + AC'$ $\quad Y_3 = AB + BC + CA$

E4.31 **Implement the following functions using a single PROM grid.**

$F_1(x, y, z) = \sum m(1, 2, 4, 5)$, $F_2(x, y, z) = \sum m(0, 1, 3, 4)$,

$F_3(x, y, z) = \sum m(2, 3, 6, 7)$, $F_4(x, y, z) = \sum m(1, 4, 5, 7)$

Answers

E4.1

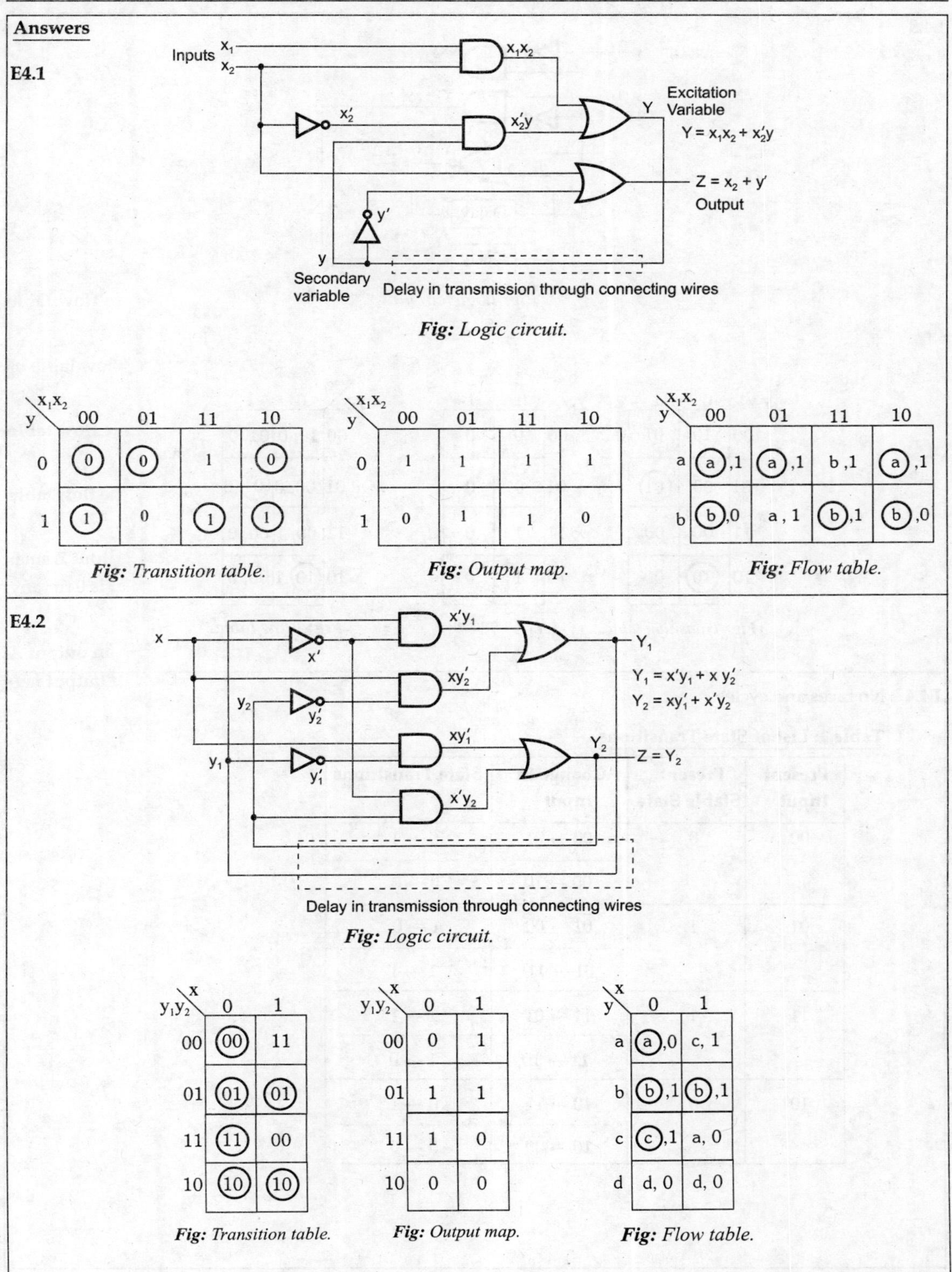

Fig: Logic circuit.

y \ x_1x_2	00	01	11	10
0	(0)	(0)	1	(0)
1	(1)	0	(1)	(1)

Fig: Transition table.

y \ x_1x_2	00	01	11	10
0	1	1	1	1
1	0	1	1	0

Fig: Output map.

y \ x_1x_2	00	01	11	10
a	(a),1	(a),1	b ,1	(a),1
b	(b),0	a , 1	(b),1	(b),0

Fig: Flow table.

E4.2

Fig: Logic circuit.

y_1y_2 \ x	0	1
00	(00)	11
01	(01)	(01)
11	(11)	00
10	(10)	(10)

Fig: Transition table.

y_1y_2 \ x	0	1
00	0	1
01	1	1
11	1	0
10	0	0

Fig: Output map.

y \ x	0	1
a	(a),0	c, 1
b	(b),1	(b),1
c	(c),1	a, 0
d	d, 0	d, 0

Fig: Flow table.

E4.3

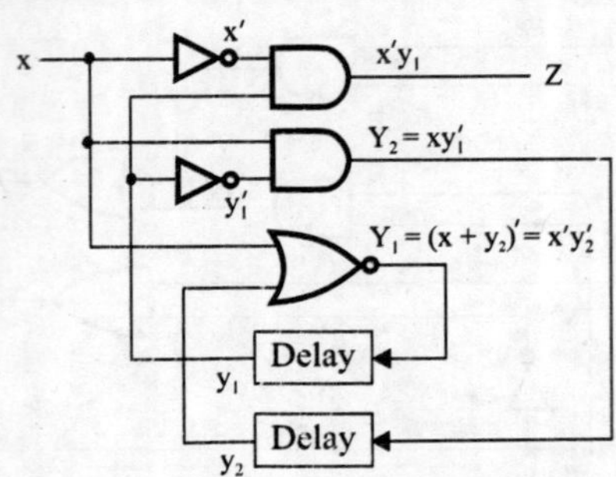

Fig: *Logic circuit.*

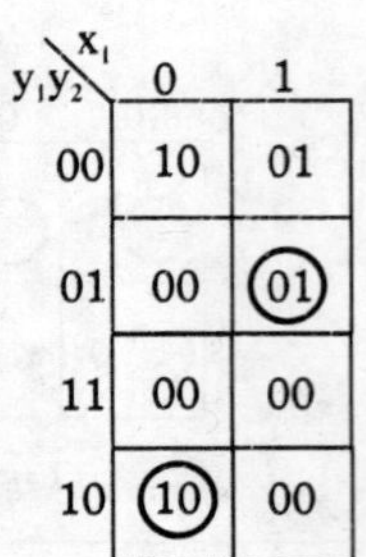

y_1y_2 \ x_1	0	1
00	10	01
01	00	(01)
11	00	00
10	(10)	00

Fig: *Transition table.*

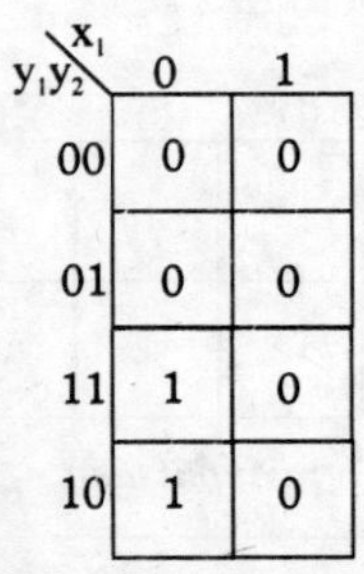

y_1y_2 \ x_1	0	1
00	0	0
01	0	0
11	1	0
10	1	0

Fig: *Output map.*

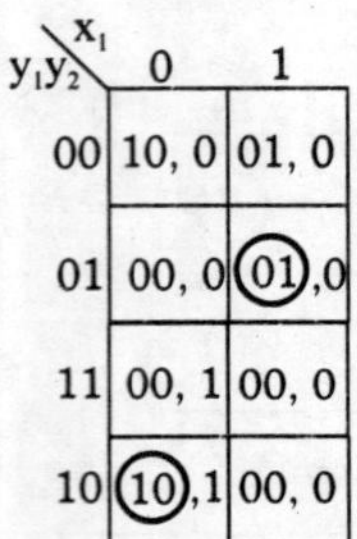

y_1y_2 \ x_1	0	1
00	10, 0	01, 0
01	00, 0	(01),0
11	00, 1	00, 0
10	(10),1	00, 0

Fig: *Flow table.*

E4.4 No races and cycles.

Table 1: List of State Transitions

Present Input	Present Stable State	Change in Input	State Transitions
00	0	00 → 01	0 → 1
		00 → 10	0 → 0
01	1	01 → 00	1 → 0
		01 → 11	1 → 1
11	1	11 → 01	1 → 1
		11 → 10	1 → 0
10	0	10 → 11	0 → 1
		10 → 00	0 → 0

E4.5 Table 1: List of Races and Cycles

Present Input	Present Stable State	Change in Input	State Transitions	Races	Cycles
00	00	00 → 01	00 → 01 00 → 10 → 11	Critical	- 00 → 10 → 11
		00 → 10	00 → 00	-	-
00	11	00 → 01	11 → 10 11 → 01 → 00	Critical	- 11 → 01 → 00
		00 → 10	11 → 11	-	-
01	01	01 → 00	01 → 00 01 → 11	Critical	- -
		01 → 11	01 → 01	-	-
01	11	00 → 00	11 → 11	-	-
		00 → 11	11 → 10 11 → 01	Critical	- -
11	01	11 → 01	01 → 00 01 → 11→ 10	Critical	- 01 → 11→ 10
		11 → 10	01 → 01	-	-
11	10	11 → 01	10 → 11 10 → 00 → 01	Critical	- 10 → 00 → 01
		11 → 10	10 → 10	-	-
10	00	10 → 00	00 → 00	-	-
		10 → 11	00 → 01 00 → 10	Critical	- -
10	10	10 → 00	10 → 11 10 → 00	Critical	- -
		10 → 11	10 → 10	-	-

E4.6

Table: Reduced State Table

Present State	Next State x = 0	Next State x = 1	Output x = 0	Output x = 1
a	a	a	1	0
b	a	c (f → c)	1	1
c	b	e	0	0
d	a	e (g → e)	0	0
e	d	c	1	1

E4.7

Table: Reduced State Table

Present State	Next State x = 0	Next State x = 1	Output x = 0	Output x = 1
a	a	e	0	1
b	a	e	1	1
c	b	f (f → e)	1	1
e	c	e (g → e)	0	0

E4.8 The compatible pairs that satisfy closed-covering condition are, (a, f), (b, g), (c, h), (d, e)

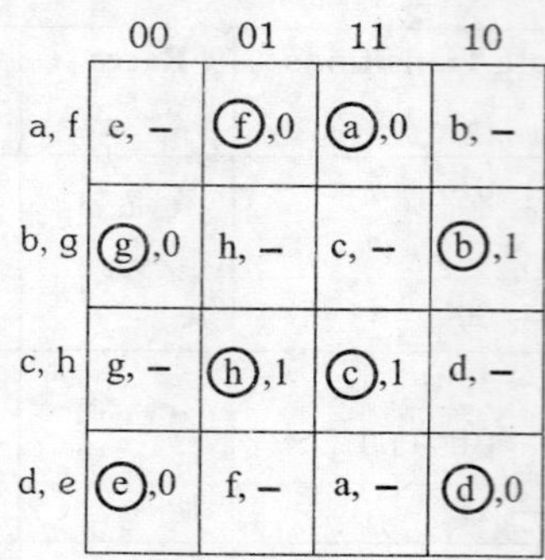

	00	01	11	10
a, f	e, –	(f),0	(a),0	b, –
b, g	(g),0	h, –	c, –	(b),1
c, h	g, –	(h),1	(c),1	d, –
d, e	(e),0	f, –	a, –	(d),0

Fig. a.

	00	01	11	10
a	d, – (e→d)	(a),0 (f→a)	(a),0	b, –
b	(b),0 (g→b)	c, – (h→c)	c, –	(b),1
c	b, – (g→b)	(c),1 (h→c)	(c),1	d, –
d	(d),0 (e→d)	a, – (f→a)	a, –	(d),0

Fig. b.

***Fig:** Reduced flow table.*

E4.9 The compatible pairs that satisfy closed-covering condition are, (a, c, d), (b, e, f)

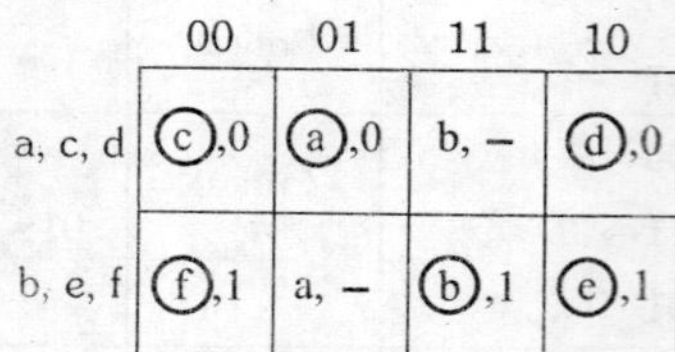

	00	01	11	10
a, c, d	(c),0	(a),0	b, –	(d),0
b, e, f	(f),1	a, –	(b),1	(e),1

Fig. a.

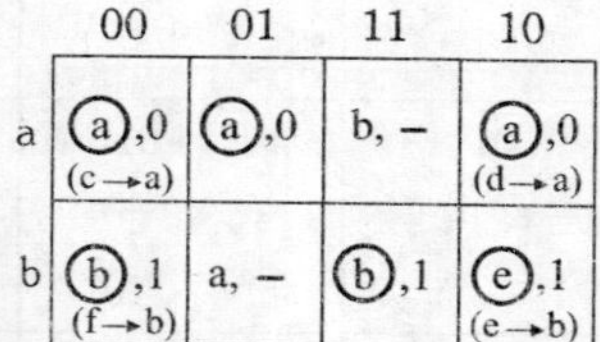

	00	01	11	10
a	(a),0 (c→a)	(a),0	b, –	(a),0 (d→a)
b	(b),1 (f→b)	a, –	(b),1	(e),1 (e→b)

Fig. b.

***Fig:** Reduced flow table.*

E4.10

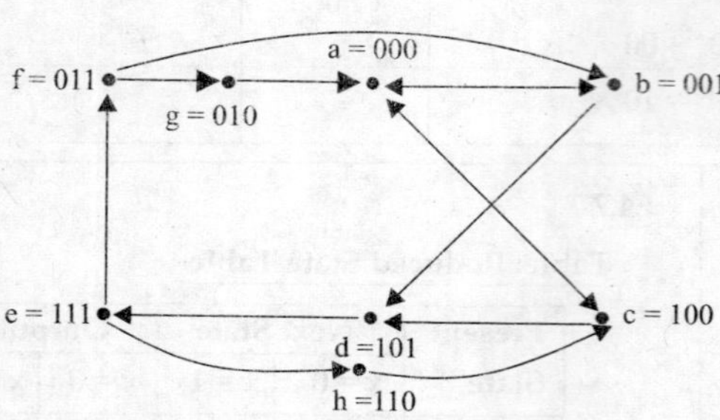

***Fig:** Transition diagram with binary assignment.*

$y_3y_2y_1$ \ x_1x_2	00	01	11	10
000 = a	(a)	b	(a)	c
001 = b	a	(b)	(b)	d
011 = f	(f)	b	g	(f)
010 = g	–	–	a	–
110 = h	–	c	–	c
111 = e	f	h	(e)	h
101 = d	(d)	(d)	e	(d)
100 = c	d	(c)	a	(c)

***Fig:** Modified flow table with race free state assignment.*

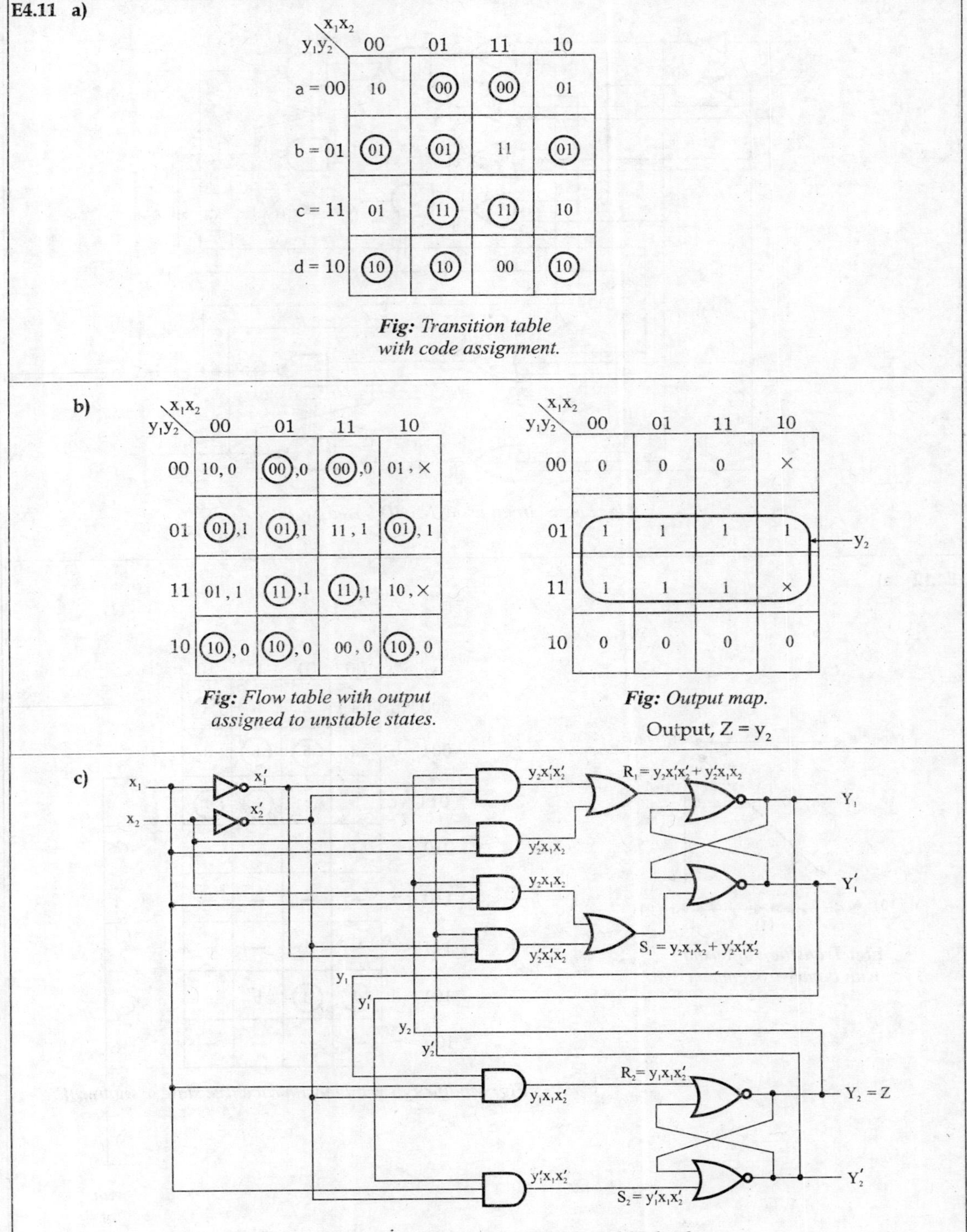

E4.11 a)

y_1y_2 \ x_1x_2	00	01	11	10
a = 00	10	(00)	(00)	01
b = 01	(01)	(01)	11	(01)
c = 11	01	(11)	(11)	10
d = 10	(10)	(10)	00	(10)

Fig: *Transition table with code assignment.*

b)

y_1y_2 \ x_1x_2	00	01	11	10
00	10, 0	(00),0	(00),0	01 , ×
01	(01),1	(01),1	11 , 1	(01), 1
11	01 , 1	(11),1	(11),1	10 , ×
10	(10), 0	(10), 0	00 , 0	(10), 0

Fig: *Flow table with output assigned to unstable states.*

y_1y_2 \ x_1x_2	00	01	11	10
00	0	0	0	×
01	1	1	1	1
11	1	1	1	×
10	0	0	0	0

Fig: *Output map.*

Output, $Z = y_2$

c)

Fig: *Logic circuit using NOR based SR latch.*

d)

x_1 x_1' x_2 x_2'

$y_2' + x_1' + x_2'$

$S_1 = (y_2' + x_1' + x_2').(y_2 + x_1 + x_2)$

$y_2 + x_1 + x_2$

$y_2' + x_1 + x_2$

$y_2 + x_1' + x_2'$

$R_1 = (y_2' + x_1 + x_2).(y_2 + x_1' + x_2')$

Y_1 Y_1'

y_1 y_1' y_2 y_2'

$S_2 = y_1 + x_1' + x_2$

$R_2 = y_1' + x_1' + x_2$

$Y_2 = Z$

Y_2'

***Fig:** Logic circuit using NAND based SR latch.*

E4.12 a)

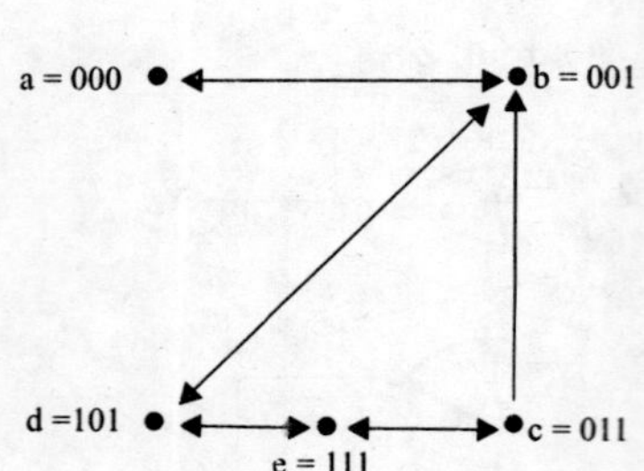

***Fig:** Transition diagram with binary assignment.*

$y_3y_2y_1$ \ x_1x_2	00	01	11	10
000 = a	(a)	b	b	(a)
001 = b	a	(b)	(b)	d
011 = c	b	e	(c)	(c)
010 = –	–	–	–	–
110 = –	–	–	–	–
111 = e	–	d	–	c
101 = d	(d)	(d)	b	e
100 = –	–	–	–	–

***Fig:** Modified flow table with race free state assignment.*

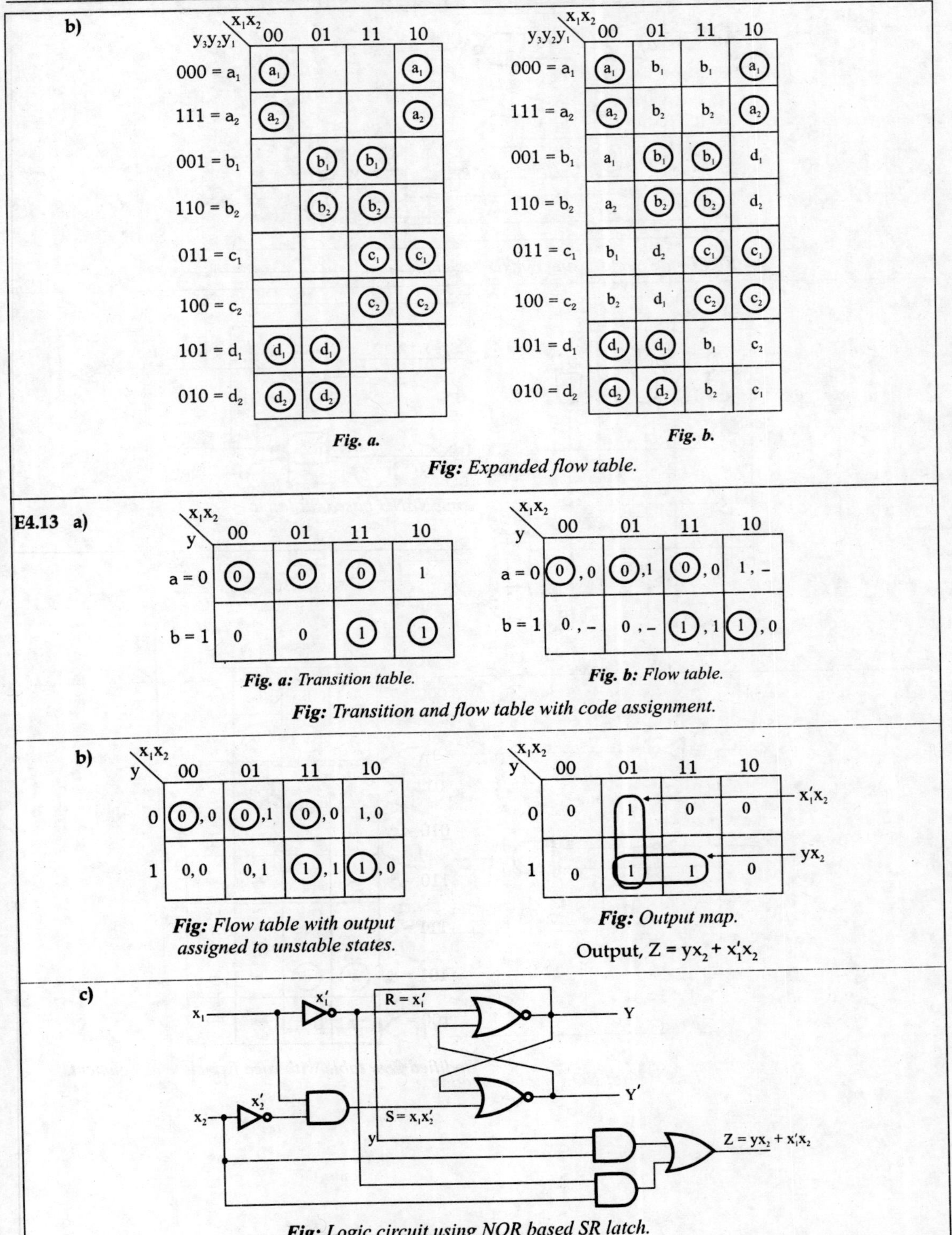

b)

$y_3y_2y_1$ \ x_1x_2	00	01	11	10
$000 = a_1$	(a_1)			(a_1)
$111 = a_2$	(a_2)			(a_2)
$001 = b_1$		(b_1)	(b_1)	
$110 = b_2$		(b_2)	(b_2)	
$011 = c_1$			(c_1)	(c_1)
$100 = c_2$			(c_2)	(c_2)
$101 = d_1$	(d_1)	(d_1)		
$010 = d_2$	(d_2)	(d_2)		

Fig. a.

$y_3y_2y_1$ \ x_1x_2	00	01	11	10
$000 = a_1$	(a_1)	b_1	b_1	(a_1)
$111 = a_2$	(a_2)	b_2	b_2	(a_2)
$001 = b_1$	a_1	(b_1)	(b_1)	d_1
$110 = b_2$	a_2	(b_2)	(b_2)	d_2
$011 = c_1$	b_1	d_2	(c_1)	(c_1)
$100 = c_2$	b_2	d_1	(c_2)	(c_2)
$101 = d_1$	(d_1)	(d_1)	b_1	c_2
$010 = d_2$	(d_2)	(d_2)	b_2	c_1

Fig. b.

***Fig:** Expanded flow table.*

E4.13 a)

y \ x_1x_2	00	01	11	10
a = 0	(0)	(0)	(0)	1
b = 1	0	0	(1)	(1)

***Fig. a:** Transition table.*

y \ x_1x_2	00	01	11	10
a = 0	(0), 0	(0), 1	(0), 0	1, –
b = 1	0, –	0, –	(1), 1	(1), 0

***Fig. b:** Flow table.*

***Fig:** Transition and flow table with code assignment.*

b)

y \ x_1x_2	00	01	11	10
0	(0), 0	(0), 1	(0), 0	1, 0
1	0, 0	0, 1	(1), 1	(1), 0

***Fig:** Flow table with output assigned to unstable states.*

y \ x_1x_2	00	01	11	10
0	0	1	0	0
1	0	1	1	0

***Fig:** Output map.*

Output, $Z = yx_2 + x_1'x_2$

c)

***Fig:** Logic circuit using NOR based SR latch.*

d)

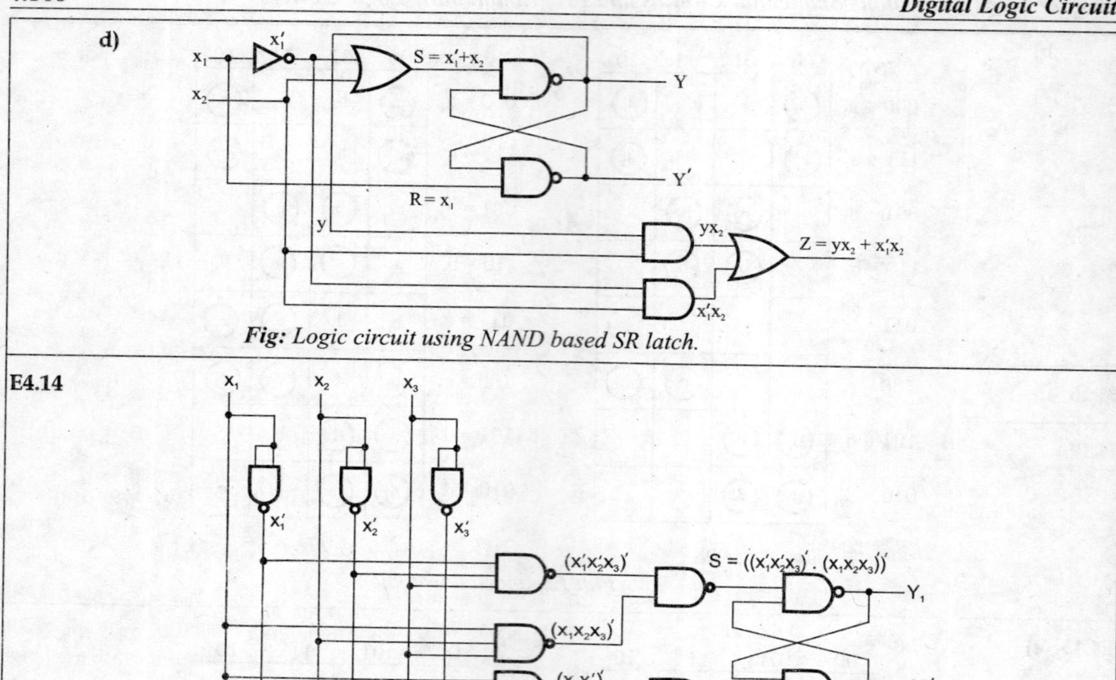

Fig: *Logic circuit using NAND based SR latch.*

E4.14

Fig.

E4.15

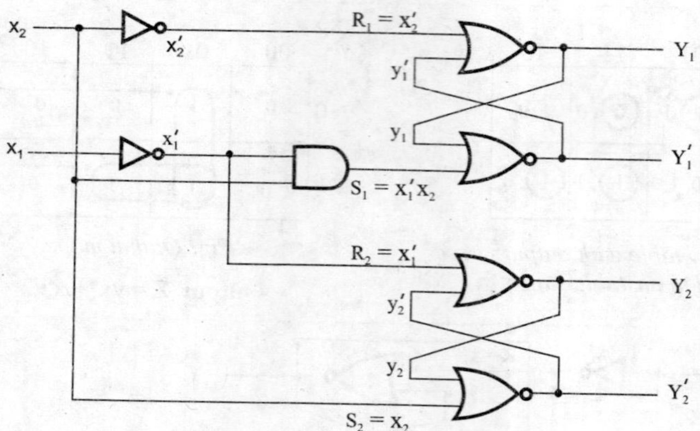

Fig: *NOR implementation*

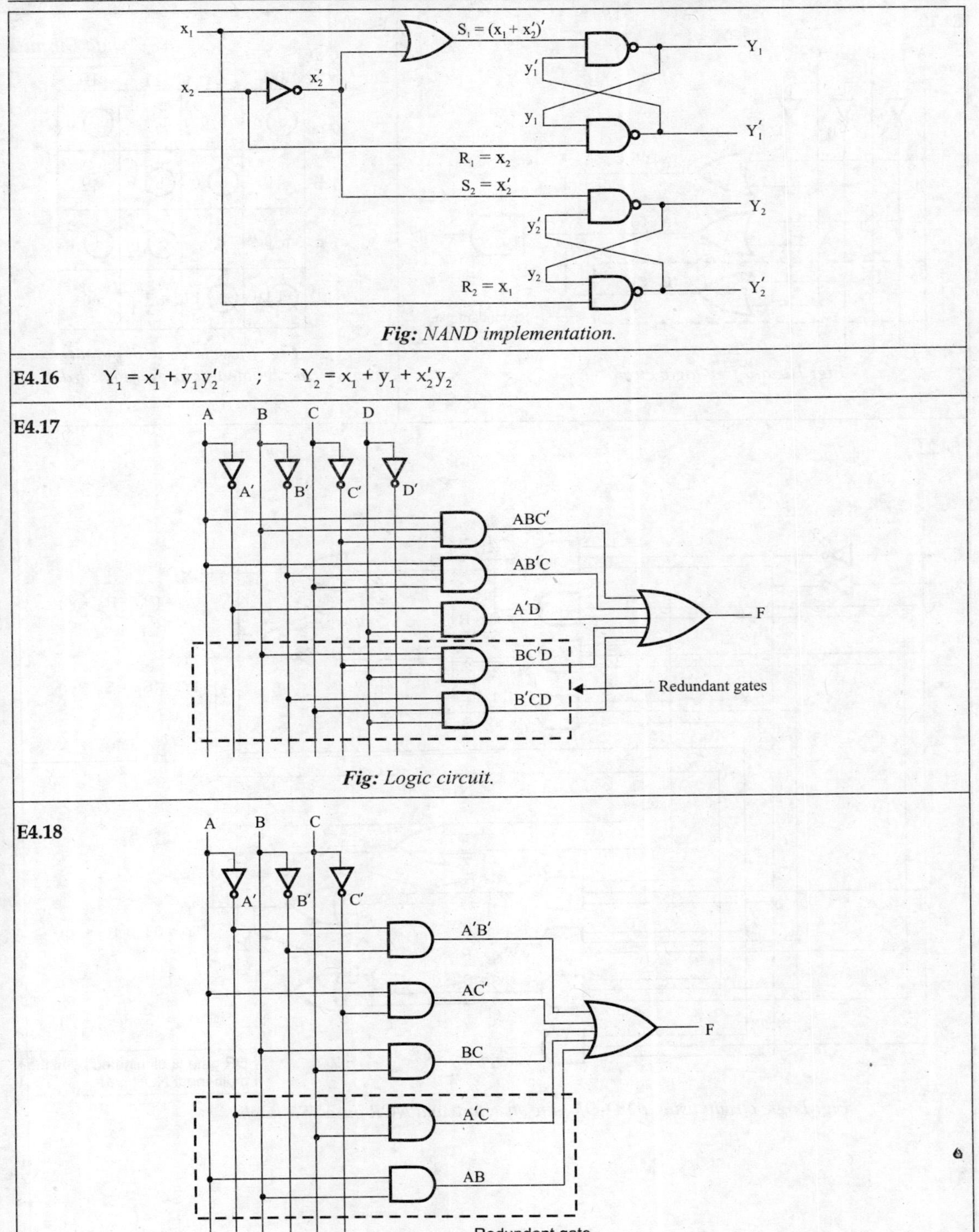

***Fig:** NAND implementation.*

E4.16 $Y_1 = x_1' + y_1 y_2'$; $Y_2 = x_1 + y_1 + x_2' y_2$

E4.17

***Fig:** Logic circuit.*

E4.18

***Fig:** Hazard free logic circuit.*

E4.19

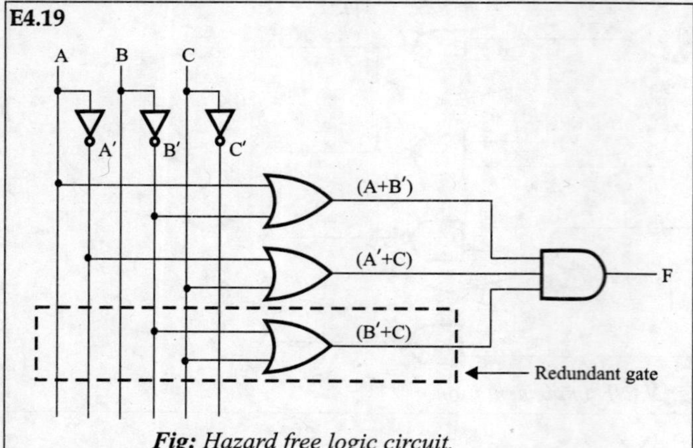

Fig: Hazard free logic circuit.

E4.20

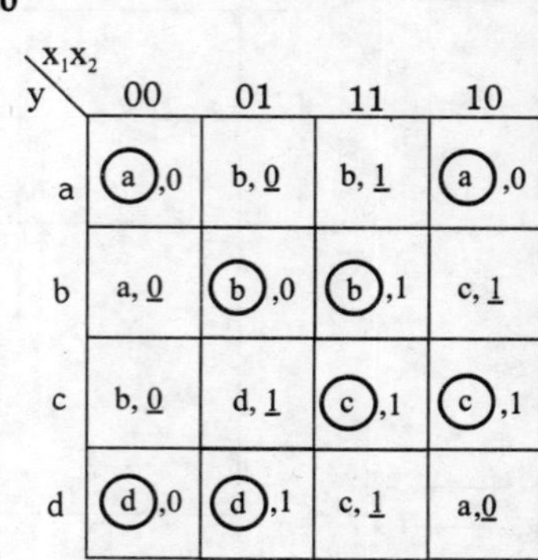

y \ x_1x_2	00	01	11	10
a	(a),0	b, 0	b, 1	(a),0
b	a, 0	(b),0	(b),1	c, 1
c	b, 0	d, 1	(c),1	(c),1
d	(d),0	(d),1	c, 1	a,0

Fig: Flow table with output assignment to prevent hazard.

E4.21

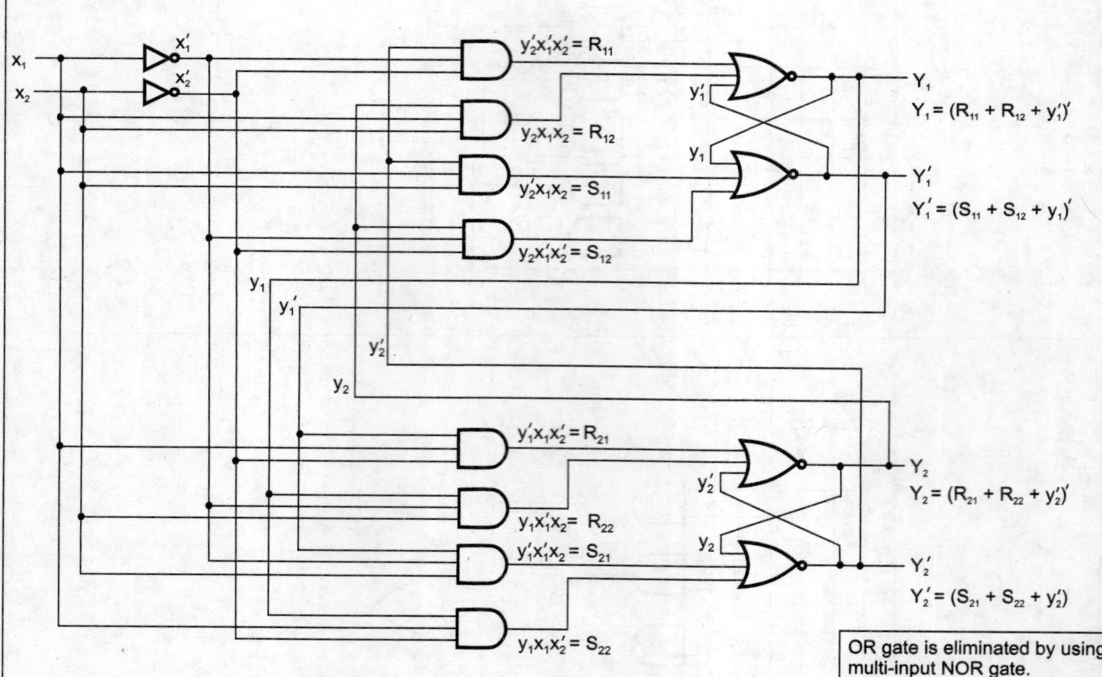

Fig: Logic circuit using AND-OR gate structure and NOR based SR latch.

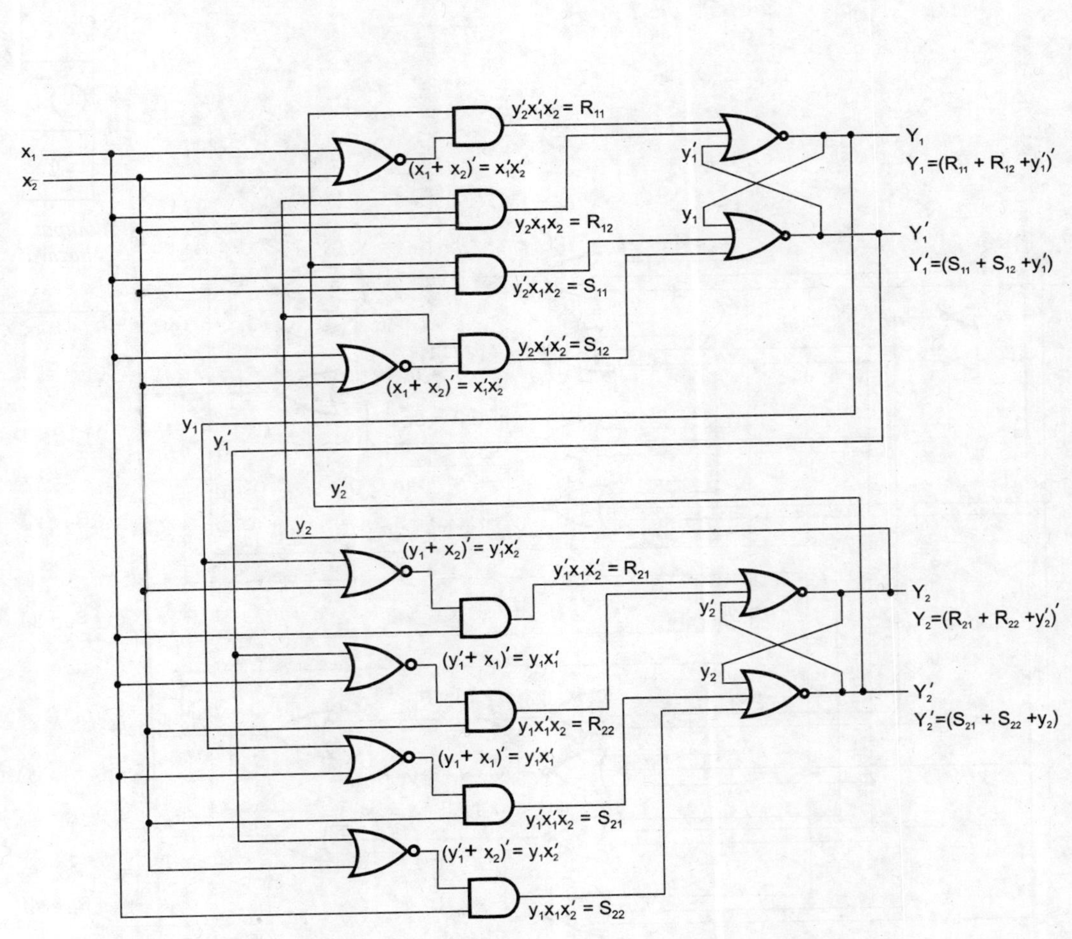

***Fig:** Logic circuit using NOR based SR latch with modified gate structure to eliminate essential hazard due to inverters.*

E4.22

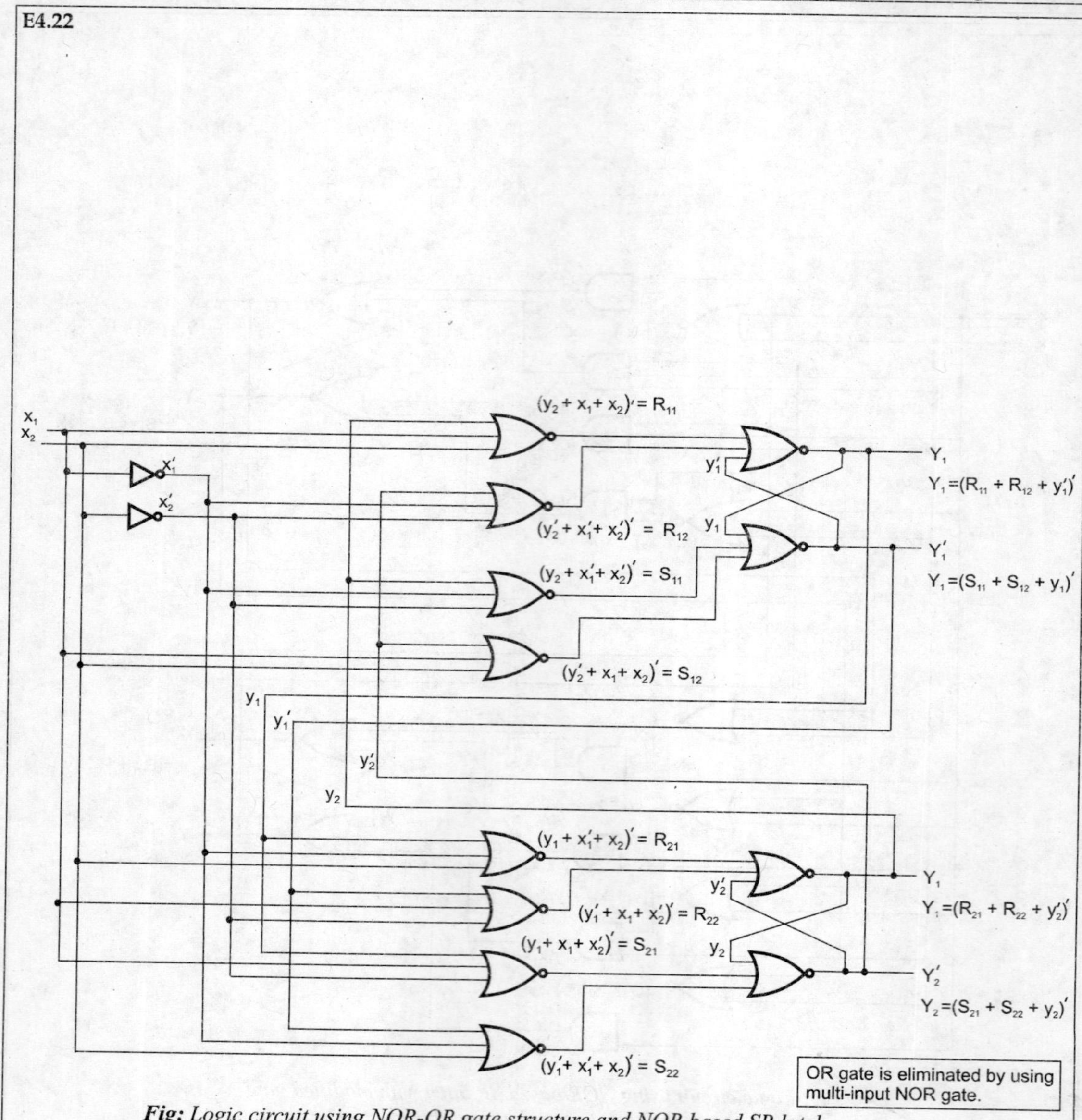

Fig: *Logic circuit using NOR-OR gate structure and NOR based SR latch.*

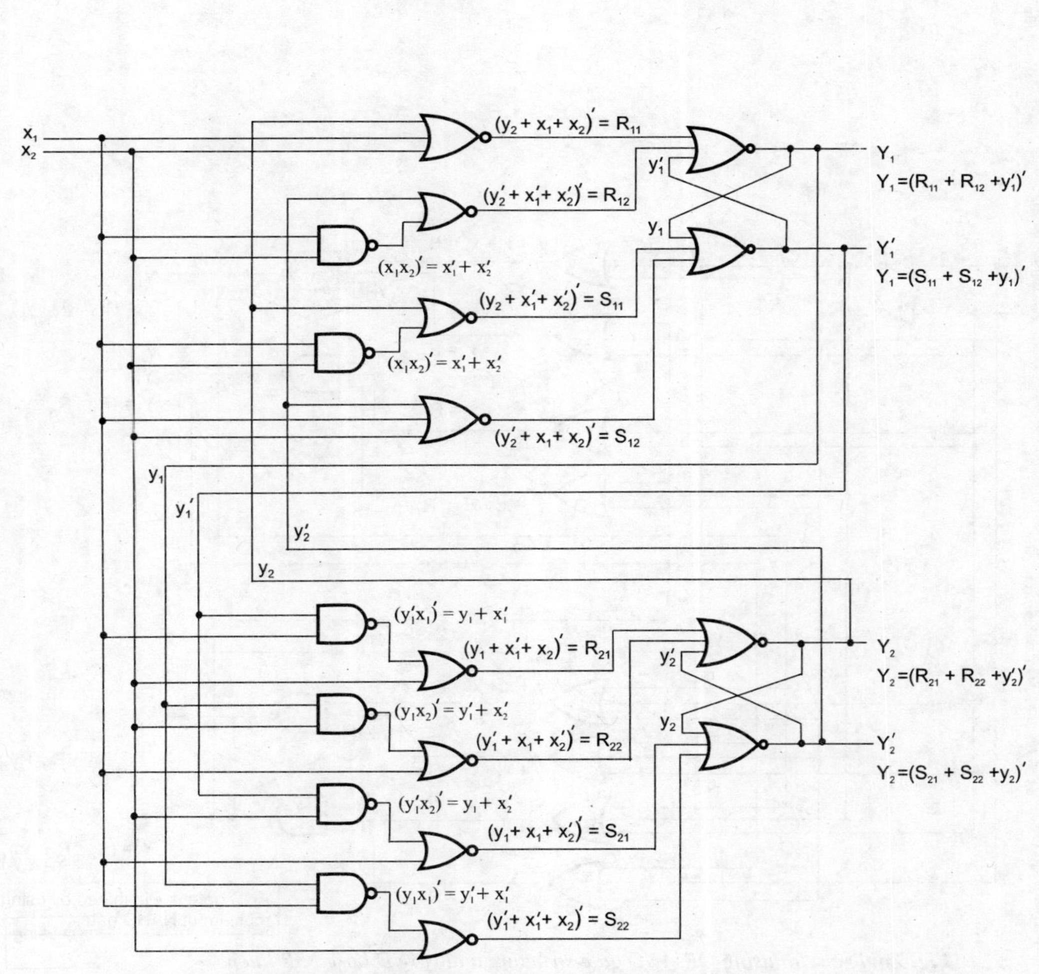

***Fig:** Logic circuit using NOR based SR latch with modified gate structure to eliminate essential hazard due to inverters.*

E4.23

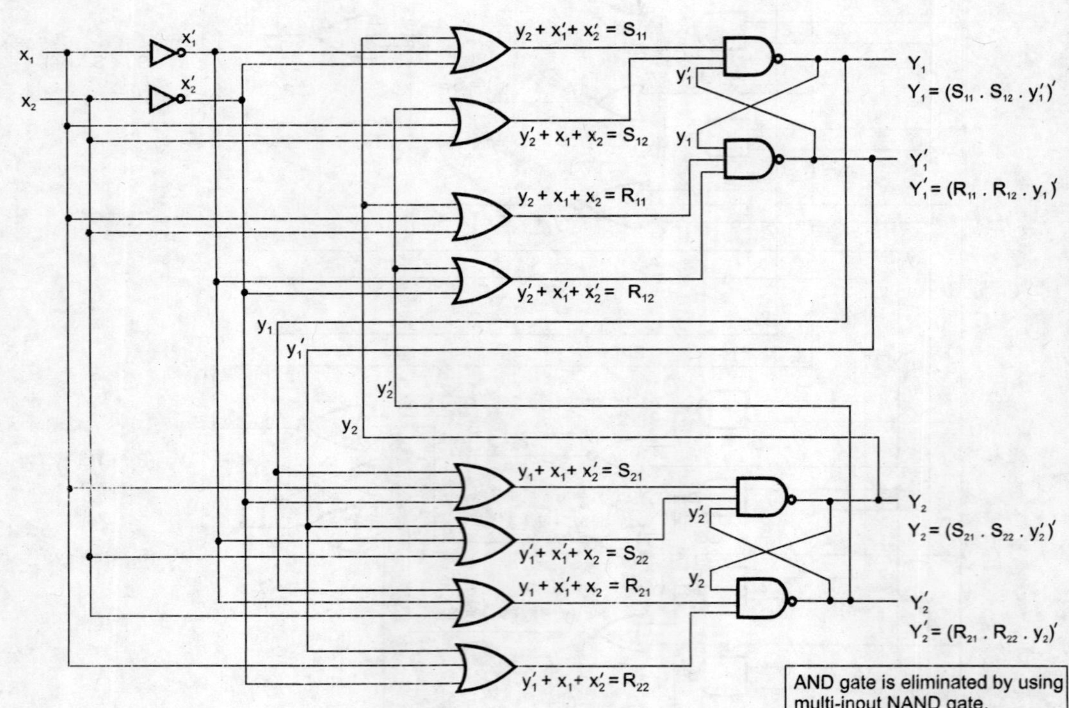

Fig: *Logic circuit using OR-AND gate structure and NAND based SR latch.*

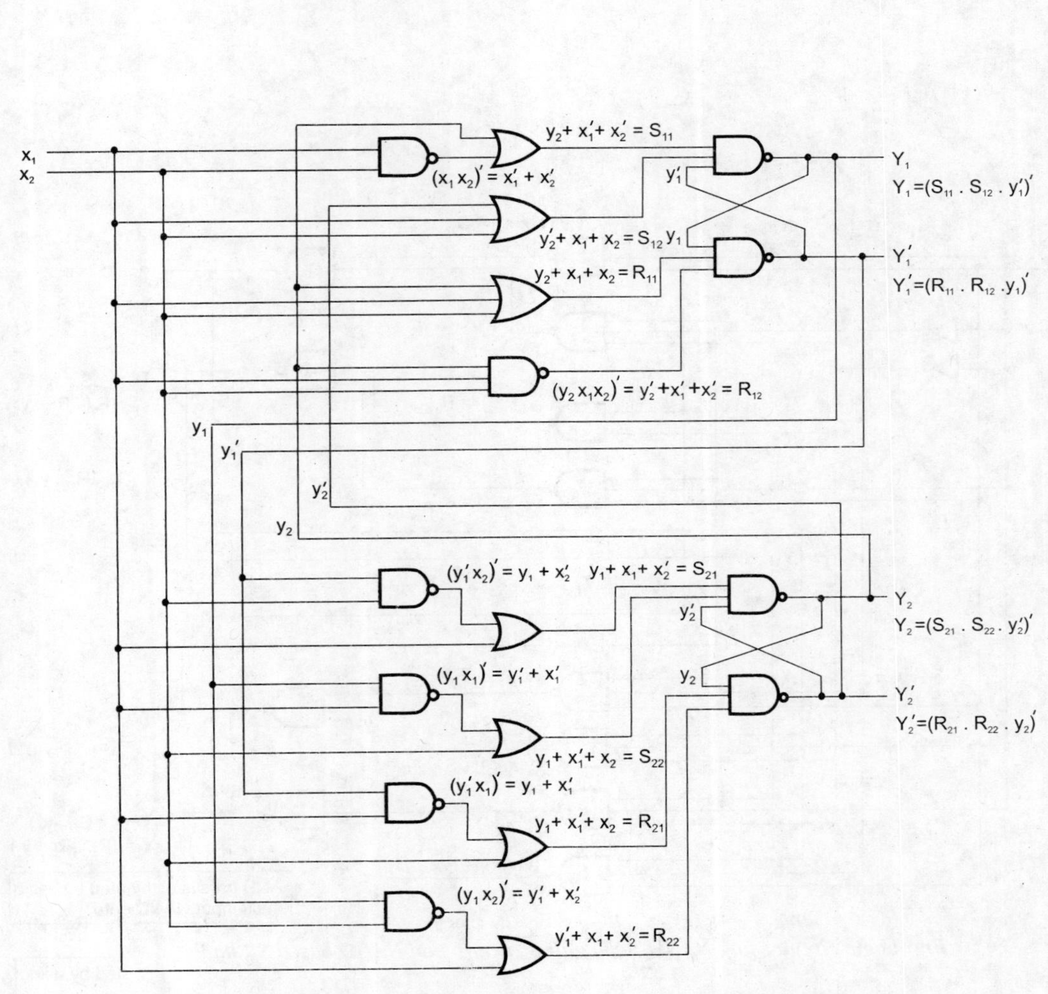

***Fig:** Logic circuit using NAND based SR latch with modified gate structure to eliminate essential hazard due to inverters.*

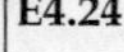

E4.24

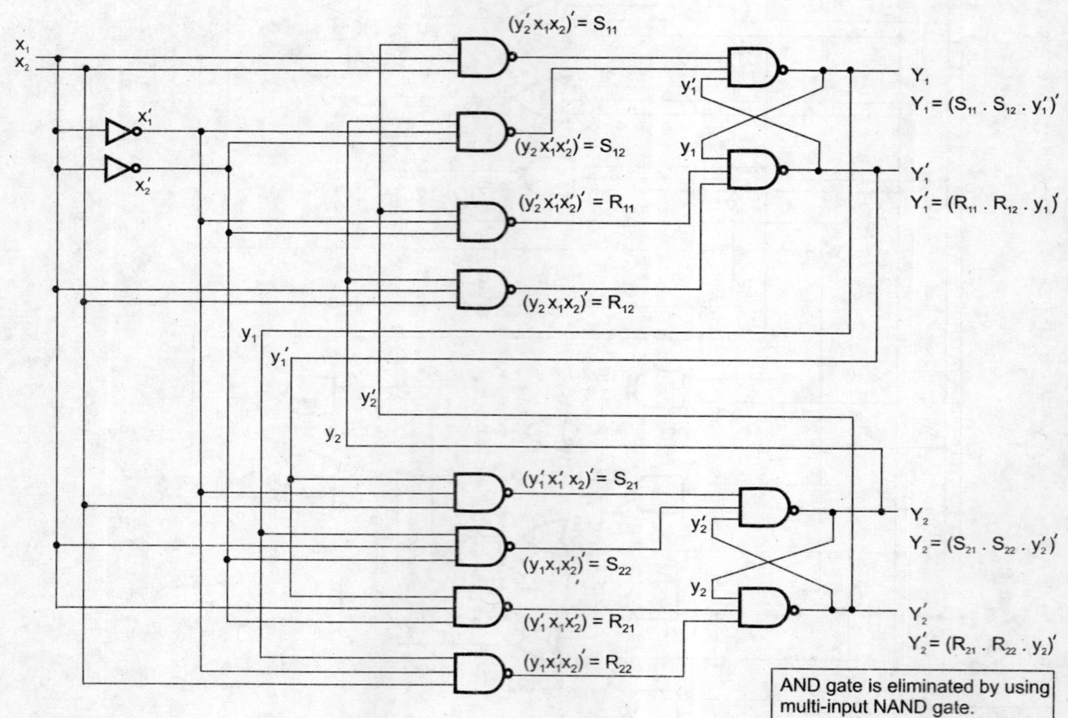

Fig: *Logic circuit using NAND-AND gate structure and NAND based SR latch.*

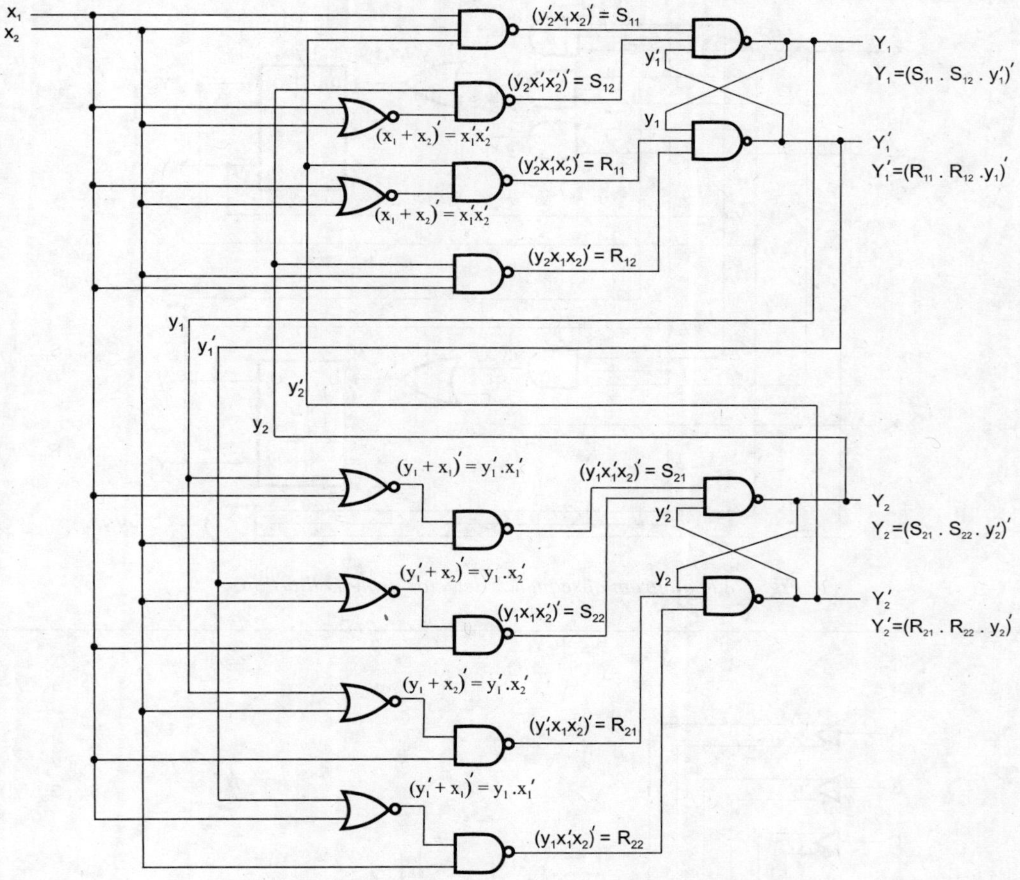

x_1
x_2
$(y_2'x_1x_2)' = S_{11}$
$(y_2x_1'x_2')' = S_{12}$
$(x_1 + x_2)' = x_1'x_2'$
$(y_2'x_1'x_2')' = R_{11}$
$(x_1 + x_2)' = x_1'x_2'$
$(y_2x_1x_2)' = R_{12}$
y_1'
y_1
Y_1
$Y_1 = (S_{11} \cdot S_{12} \cdot y_1')'$
Y_1'
$Y_1' = (R_{11} \cdot R_{12} \cdot y_1)'$
y_1
y_1'
y_2'
y_2
$(y_1 + x_1)' = y_1'.x_1'$
$(y_1'x_1'x_2)' = S_{21}$
y_2'
Y_2
$Y_2 = (S_{21} \cdot S_{22} \cdot y_2')'$
$(y_1' + x_2)' = y_1.x_2'$
$(y_1x_1x_2')' = S_{22}$
y_2
Y_2'
$Y_2' = (R_{21} \cdot R_{22} \cdot y_2)'$
$(y_1 + x_2)' = y_1'.x_2'$
$(y_1'x_1x_2')' = R_{21}$
$(y_1' + x_1)' = y_1.x_1'$
$(y_1x_1'x_2)' = R_{22}$

***Fig:** Logic circuit using NAND based SR latch with modified gate structure to eliminate essential hazard due to inverters.*

E4.25

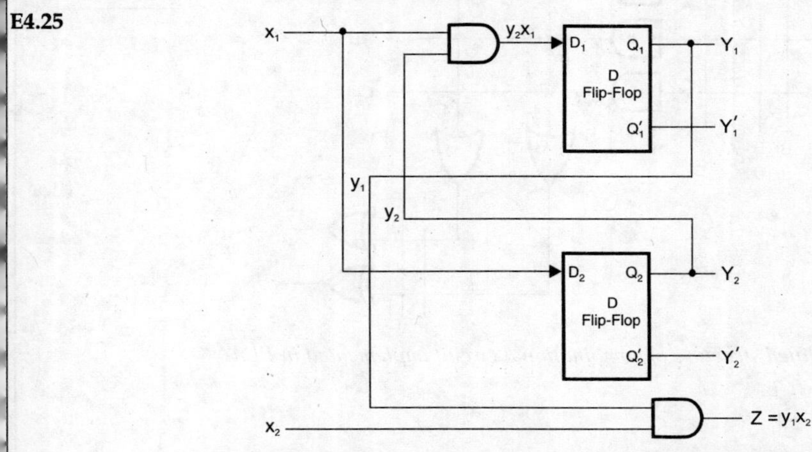

***Fig:** Logic diagram of sequence detector using D flip-flop.*

E4.26

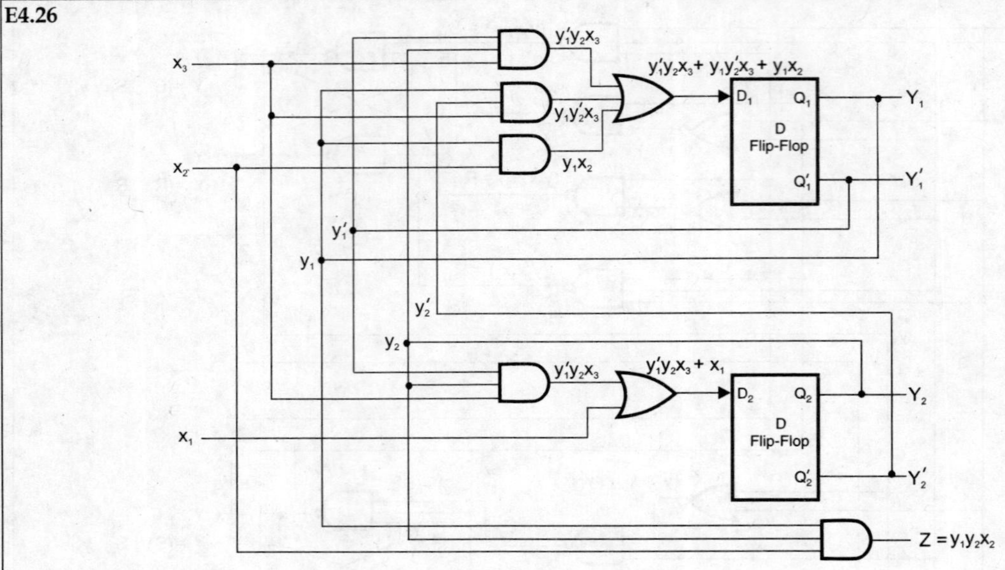

Fig: *Logic diagram of sequence detector using D flip-flop.*

E4.27

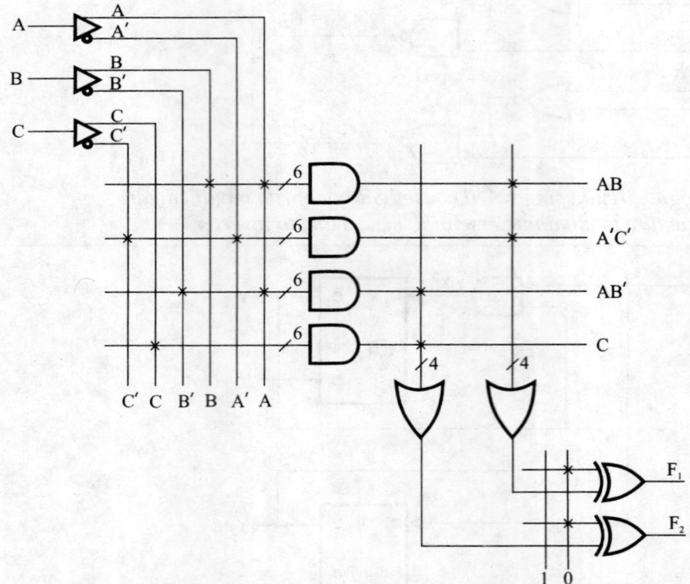

Fig: *Simplified structure of combinational circuit implemented in PLA.*

E4.28

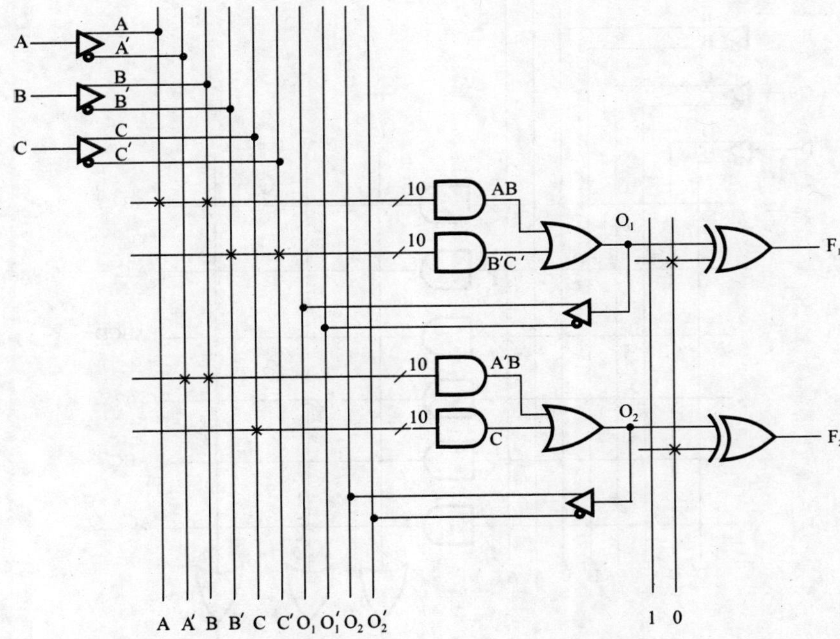

***Fig:** Structure of combinational circuit implemented in PAL.*

E4.29

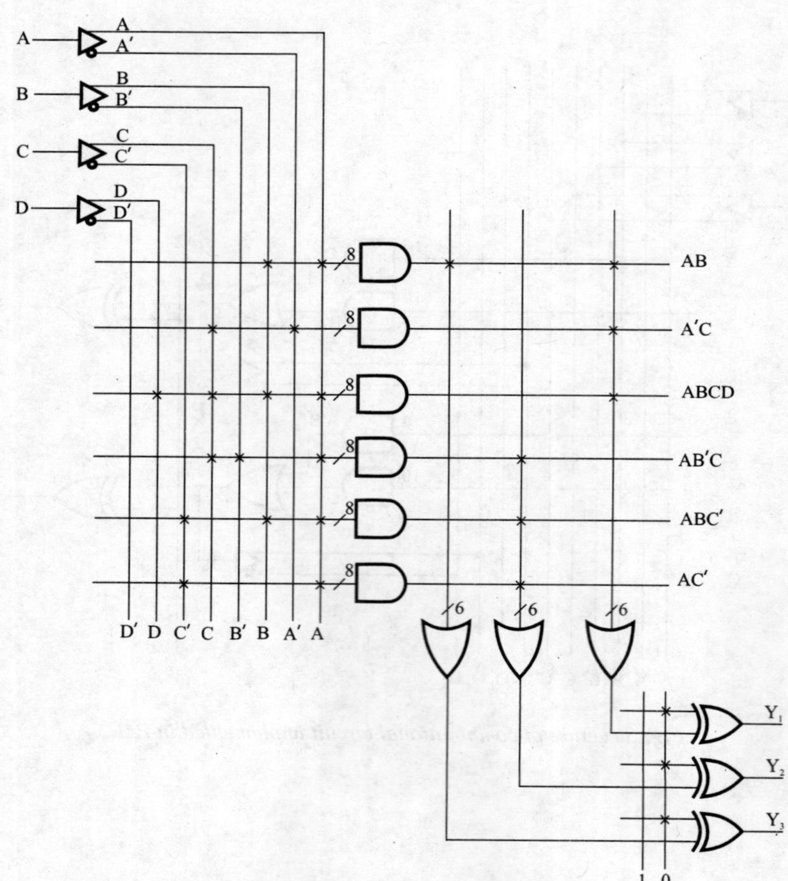

Fig: *Simplified structure of combinational circuit implemented in PLA.*

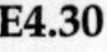

E4.30

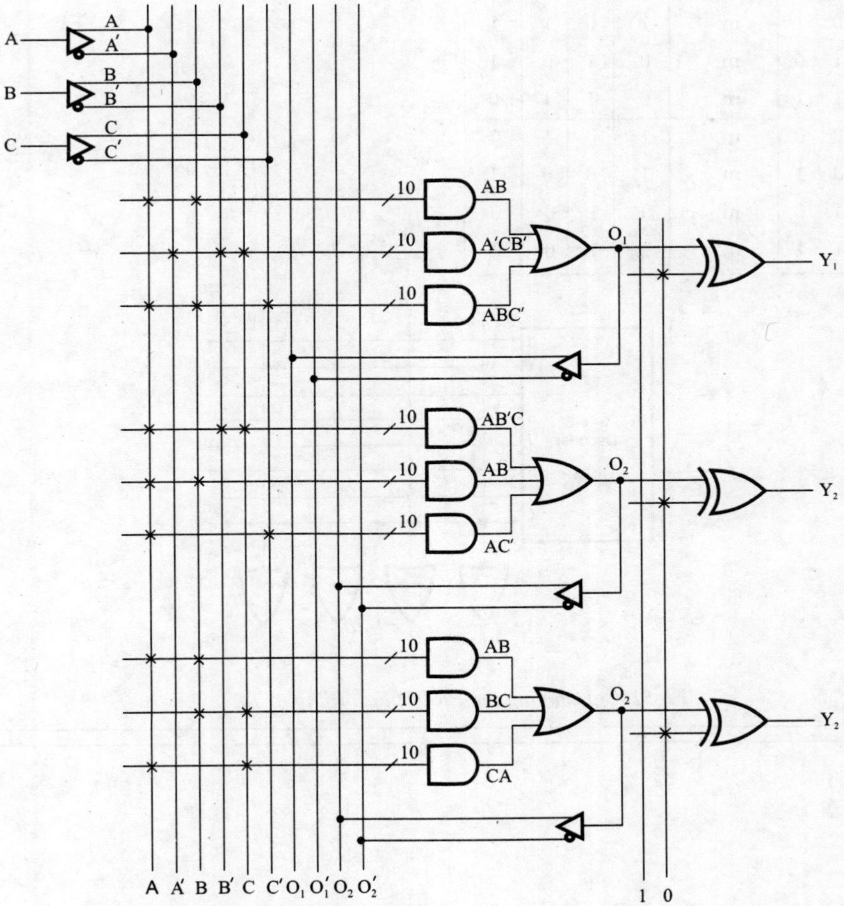

Fig: *Structure of combinational circuit implemented in PAL.*

E4.31

Table 1: Truth Table

Inputs			Minterm	Function Outputs			
x	y	z		F_4	F_3	F_2	F_1
0	0	0	m_0	0	0	1	0
0	0	1	m_1	1	0	1	1
0	1	0	m_2	0	1	0	1
0	1	1	m_3	0	1	1	0
1	0	0	m_4	1	0	1	1
1	0	1	m_5	1	0	0	1
1	1	0	m_6	0	1	0	0
1	1	1	m_7	1	1	0	0

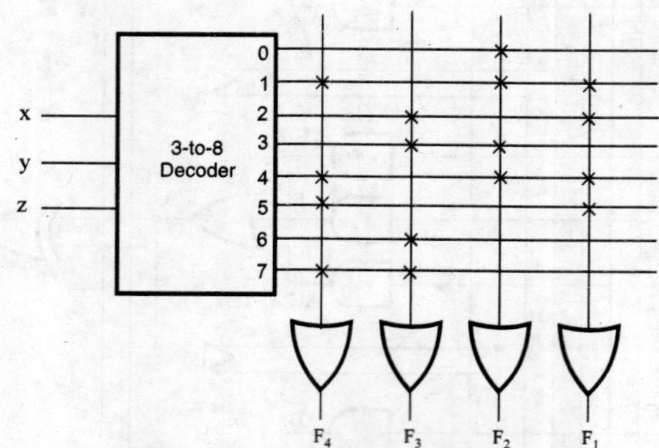

***Fig:** Programming the PROM according to Table 1.*

CHAPTER 5

VHDL

5.1 Introduction to VHDL

VHDL is an acronym for **V**HSIC (Very High Speed Integrated Circuits) **H**ardware **D**escription **L**anguage.

In 1983, VHDL was originally developed by the direction of the U.S. Department of Defense to the ASICs (Application Specific ICs) manufacturers in order to standardize the hardware description. Since 1987, VHDL has been standardized by the Institute of Electrical and Electronics Engineers (IEEE) as **IEEE Standard 1076**. The latest version of VHDL is **IEEE standard 1076-2019**.

VHDL is used to model a digital system at multiple levels of abstraction. An abstraction is a description of a system with a set of minimum details or components required for modelling the system and hiding all other associated details or components of the system. VHDL supports the following four levels of abstraction for digital system design.

- Transistor level
- Gate level
- Register-transfer level (RTL)
- Processor level

The various levels of abstraction will differ in the following aspects of system design.

- Basic building block
- Signal representation
- Time representation
- Behavioral representation
- Physical representation

The basic building blocks are the parts or components used in various levels of abstraction. Signal and time representations refer to expressing signal values and changes in signal over time. Behavioral and physical representations are descriptions concerned with behavioral and physical views.

In transistor level abstraction, basic building blocks are basic electronic components like transistors, resistors, capacitors, etc. In gate level abstraction, basic building blocks are logic gates, flip-flops, multiplexers, etc. In register-transfer level (RTL) abstraction, the basic building blocks are modules developed using registers, adders, multiplexers, etc. Alternatively, the RTL abstraction can be called module-level abstraction. In processor level abstraction, the basic building blocks are processors, memory modules, bus interfaces, etc.

5.2 Elements of VHDL Language

The VHDL language has been developed by integrating the following languages:

- Sequential language
- Concurrent language
- Net-list language
- Timing specifications
- Waveform generation language

Therefore, VHDL language has constructs to express the concurrent or sequential behaviour of a digital system with or without timing, to model the system as an interconnection of components and to generate test waveforms in a single model of the system.

The basic elements of the VHDL language are literals, comments, identifiers, data objects and operators.

5.2.1 Literals

Explicit data values that are assigned to objects or used within expressions are called literals. Literals can be classified into following six general categories.

- Character literals
- String literals
- Bit string literals
- Numeric literals
- Based numeric literals
- Physical literals

Character literals: ***Character literals*** are 1-character ASCII values that are enclosed in single-quotes.

Examples of character literals
'0', 'X', '$'

String literals: ***String literals*** are collections of one or more ASCII characters enclosed in double-quote characters.

Examples of string literals
"my_state", "count", "result"

Bit string literals: ***Bit string literals*** are special forms of string literals that are used to represent binary string in 1's and 0's in binary, octal, or hexadecimal numeric data values. Base specifier B, O and X are used for binary, octal and hexadecimal respectively.

Examples of bit string literals	
B"1111_1101"	-- decimal value 253
O"654"	-- decimal value 428
O"146_231"	-- decimal value 52,377
X"C300"	-- decimal value 49,920

Numeric literals: Numeric literals are decimal numbers. There are two basic forms of numeric literals in VHDL and they are ***integer literals*** and ***real literals***.

Integer literals are signed decimal numbers without fraction. The general range of integers is –2,147,483,647 to +2,147,483,647 (32 bits of precision, including the sign bit).

Real literals are decimal numbers with an integer part and a fraction part separated by a decimal point. For large numbers, scientific notation is also allowed using the character 'e', where the number to the left of the 'e' represents the mantissa of the real number, while the number to the right of the 'e' represents the exponent. The general range of real numbers is – 1.0e38 to +1.0e38.

Examples of numeric literals

```
5.0
- 12.9
1.6e10
1.2e - 20
```

Based numeric literals: *Based numeric literals* are another form of integer or real values, with a base specification such as 2, 8 or 16 and enclose the integer or mantissa with a pair of '# ' characters.

Examples of based numeric literals

```
2#10010001#    --integer value 145
16#FFCC#       --integer value 65,484
2#101.0#E10    --real value 5,120.0
```

Physical literals: *Physical literals* are special types of literals used to represent physical quantities such as time, voltage, current, distance, etc. Physical literals include both an integer numeric part and a unit specification.

Examples of physical literals

```
300 ns    --300 nanoseconds
900 ps    --900 picoseconds
40 ma     --40 mille-amperes
```

5.2.2 Comments

Comments are texts that are added as explanations, wherever necessary. A comment starts with a minimum of two dashes (--) followed by the text to explain the concepts. The comment is provided for explaining the various concepts of VHDL program and hence, anything after two or more dashes will be ignored while compiling the code.

5.2.3 Identifiers

Identifiers are name of objects that are used in VHDL programs. Identifiers are formed using English letters/alphabets, decimal numbers and underscore. Special characters are not allowed in identifiers. Reserved words of VHDL language cannot be used as identifiers. The list of reserved words is listed in Table 5.1. Identifiers are case insensitive and hence uppercase and lower case letters are equivalent.

The identifiers have to satisfy the following rules.

- Identifiers can have only English letters, decimal numbers and underscores.
- First character of an identifier must be an alphabet.

- Last character of an identifier cannot be an underscore.
- Identifiers cannot have two underscore in succession.

Table 5.1: List of Reserved Words

abs	access	after	alias
all	and	architecture	array
asserts	attribute		
begin	block	body	buffer
bus			
case	component	configuration	constant
disconnect	downto		
else	elsif	end	entity
exit			
file	for	function	
generate	generic	guarded	
if	in	inout	is
label	library	linkage	loop
map	mod		
nand	new	next	nor
not	null		
of	on	open	or
others	out		
package	port	procedure	process
range	record	register	rem
report	return		
select	severity	signal	subtype
then	to	transport	type
units	until	use	
variable			
wait	when	while	with
xor			

5.2.4 Predefined Data Types

The VHDL IEEE standard package defines a number of standard data types which are known as predefined data types. The predefined data types in VHDL are,

- Integer
- Boolean
- Bit
- Character
- String
- Bit_vector
- Std_logic
- Std_logic_vector

The predefined data types of VHDL are listed in Table 5.2.

Integer data type is integer decimal number in the range $-(2^{31}-1)$ to $+(2^{31}-1)$.

Boolean data type is true, false.

Bit data type is '0', '1'.

Bit_vector is one dimensional binary array.

The std_logic and std_logic_vector are defined in the std_logic_1164 package of IEEE standard 1164. The std_logic data type has nine possible values listed in Table 5.3.

The std_logic_vector is an array of elements with std_logic data type. A group of n-bit binary signals or an n-bit bus in a logic circuit is represented by std_logic_vector.

Example of std_logic_vector data type

```
signal A: std_logic_vector (7 downto 0);
```

In the above example, signal A is declared as an 8-bit binary array indexed from 7 to 0, with most significant bit as index 7 and least significant bit as index 0. Each bit of signal A can be accessed as A(0), A(1), etc. A part of signal A can be accessed as A(3 downto 0).

Table 5.2: Predefined VHDL Data Types

VHDL Data Type	Value
Bit	'0' or '1'
Boolean	false or true
Integer	$-(2^{31}-1) \leq$ Integer value $\leq (2^{31}-1)$
Positive	$1 \leq$ Integer value $\leq (2^{31}-1)$
Natural	$0 \leq$ Integer value $\leq (2^{31}-1)$
Real	$-1.0e38 \leq$ Floating point value $\leq 1.0e38$
Character	Alphabetical characters (a .... z, A .... Z), digits (0 .... 9), special characters (e.g.,%) each enclosed in single quotes
Time	Integer with units fs, ps, ns, ms, sec, min, hr

Table 5.3: Std_logic Data Types of the IEEE std_logic_1164 Package

Logic Symbols	Description
'U'	Uninitialized
'X'	Strong drive, unknown logic value
'0'	Strong drive, logic 0
'1'	Strong drive, logic 1
'Z'	High impedence
'W'	Weak drive, unknown logic value
'L'	Weak drive, logic 0
'H'	Weak drive, logic 1
'–'	Don't-care

5.2.5 Data Types and Subtypes

The data types and subtypes other than predefined data types that can be declared in VHDL programs are discussed in this section. In VHDL, each data object has data type and associated operators. The data type is declared using a type statement.

Syntax for type declaration

```
type object_name is data_type;
```

Example of type declaration

```
type my_count is ('2', '4', '6', '8');
```

Subtype is a part or subset of a data type that is specified by a condition or constraint.

Syntax for type and subtype declaration

```
type object_name is data_type;
subtype subtype_object_name is object_name data_type;
```

Example of type and subtype

```
type pin_count is range 0 to 40;
subtype address_pin is pin_count range 25 to 40;
```

The data types that are used in VHDL programs can be classified into following four groups.

- Scalar types
- Composite types
- Access types
- File types

5.2.6 Scalar Data Types

The scalar data types can be again classified into following four groups.

- Enumeration
- Integer
- Physical
- Floating point

Enumeration type is associating user defined set of values called enumeration literals to an object. The values of enumeration literals are formed using alphabets and numeric character literals.

Example of enumeration data type

```
type my_state is (S0, S1, S2, S3, S4);
type rain_sevirity is (green, yellow, orange, red);
```

Integer type is associating an integer with values within a specified range to an object.

Example of integer data type

```
type count_int is range 0 to 64;
```

Physical type associates numeric value of a physical quantity such as time, length, mass, voltage, current, etc to an object. Each value of a physical type has a position number that is an integer value. Any value of a physical type is a straight multiple of the primary unit of measurement for that type.

Syntax for physical type

```
type type_name is range left_bound to right_bound
units
    primary_unit_name;
    secondary_unit_name = number primary_unit_name;
    secondary_unit_name = number primary_unit_name;
    . . .
end units type_name;
```

Example of physical data type

```
type length is 0 to 1e3
    units
    mm;
    cm   = 10 mm;
    m    = 1000 mm;
    inch = 25 mm;
    ft   = 12 inch;
end units;
```

Floating point type is associating a real decimal number within a specified range to an object.

Example of floating point data type

```
type probability is range 0.0 to 1.0;
type signal_a_is range – 10.0 to +10.0;
```

5.2.7 Composite Data Types

A composite data type represents a collection of signals or objects. Composite types can be further classified into array types and record types. An array type represents a collection of signals or objects belong to a same data type. A record type represents a collection of signals or objects belong to same or different data types.

Examples of array data type

```
type word is array (15 downto 0) of bit:
type food is array (1 to 10) of character;
```

Example of record type

```
type uart_transmit is record
    rx  : std_logic:
    cts : std_logic;
    tx  : std_logic;
    rts : std_logic;
end record uart_transmit;
```

5.2.8 Access Data Types

Access type is a pointer to an object of some other data type.

Example of access data type

type *fifo_stack* **is array** *(0* **to** *63, 0* **to** *7)* **of bit;**
type *fifo_ptr* **is access** *fifo_stack;*

File Data Types

Data stored as a file in the host environment (the system that simulate and synthesize the code) is called file data type.

Syntax of a file data type declaration

type *file_type_name* **is file of** *data_type;*

Example of file data type

type *input_data* **is file of bit_vector;**
type *employee_names* **is file of string;**

5.2.9 Data Objects (or Objects)

Data object or simply object is an identifier that holds a specific data type, which is declared while creating the object. A data object is created by declaring the class, identifier/name, data type and optional initial value.

The objects in the VHDL language are classified into the following four categories.

- Constants
- Signals
- Variables
- Files

Constant: A constant object is assigned a fixed value and cannot be changed in other part of the code.

Syntax for declaring constant

constant *name_of_constant: data_type := value_of_constant;*

Examples of constants

constant *address_size :* **integer** *:= 16;*
constant *sram :* **bit_vector** *(15* **downto** *0) := X"F0F0";*
constant *message :* **string** *:= "this is a string";*
constant *error_flag :* **boolean** *:= true;*

Signal: A signal object is created for intermediate signals in the design entity apart from input and output signals. Signals are declared in the architecture and outside the process. Initial value is meaning less in signal objects that are assigned a fixed value and cannot be changed in other part of the code.

Syntax for declaring signal
signal *name_of_signal : data_type := initial_value_of_signal;*
Examples of signals
signal *propagation_delay* : **time** := *12ns;*
signal *data_size* : **bit_vector** *(0* **to** *15);*

Variable: An object which takes different values during program execution is called variable. When an object is declared as a variable, the value of the variable can be changed in any part of the code using an assignment statement. Variables are declared inside a process and they are local to a process.

Syntax for declaring variable
variable *name_of_variable : data_type := initial_value_of_variable;*
Examples of variables
variable *propagation_delay:* **time** := *12ns;*
variable *data_size:* **bit_vector** *(0* **to** *15);*

File: A file with a list of input and/or output data can be declared as an object. The file can used to read the stored input data and write the output data while simulating a test bench.

Syntax for declaring a file
type *file_type_name* **is file of** *data_type;* **type** *file_name: file_type_name* **is mode** *"file path";* ***Note:*** *mode is either* **in** *or* **out**.
Examples of file
type *input_data* **is file of bit_vector**; **file** *sel_values: input_data* **is in** *"/usr/home/vhdl_ex";* **type** *employee_names* **is file of** *string;* **file** *sel_employee: employee_names* **is out** *"/usr/home/vhdl_ex"*

5.2.10 Operators

The predefined operators in the VHDL language are classified into the following five categories.

- Logical operators
- Relational operators
- Adding operators
- Multiplying operators
- Miscellaneous operators

The operators of VHDL are listed in Table 5.4 and the precedence of the operators is listed in Table 5.5.

Table 5.4: VHDL Operators

Operator Group	Operator Symbol	Operation / Operator Name
Binary logical	**and**	Logical and operation
	or	Logical or operation
	nand	Logical nand operation
	nor	Logical nor operation
	xor	Logical xor operation
	xnor	Logical xnor operation
Relational	=	Equality operator
	/ =	Inequality operator
	<	Less than
	>	Greater than
	<=	Less than or equal to
	>=	Greater than or equal to
Shift operators	**sll**	Shift-left logical
	srl	Shift-right logical
	sla	Shift-left arithmetic
	sra	Shift-right arithmetic
	rol	Rotate left
	ror	Rotate right
Addition operators	**+**	Addition
	–	Subtraction
Concatenation operator	&	Concatenation
Unary sign operator	**+**	Positive sign
	–	Negative sign
Multiplication operators	*	Multiplication
	/	Division
	mod	Modulus
	rem	Remainder
Miscellaneous operators	**not**	Complement
	abs	Absolute
	**	Exponentiation
Assignment operators	<=	Assign data of one object to another object
	:=	Assign a value to object
	=>	Associate one object to another object

Table 5.5: Precedence of the VHDL Operators

Precedence	Operators
Highest	**, abs, not
	*, /, mod, rem
	+, – (identity of negation)
	&, +, – (addition and subtraction)
	sll, srl, sla, sra, rol, ror
	=, /=, <, <=, >, >=
Lowest	and, or, nand, nor, xor, xnor

5.3 Process and Process Statement

The executable functionality of an entity or a component is defined as a process in behavioral modelling. Hence the process is the key programming structure in behavioral modelling. All components in the model must be defined using one or more processes inside the architecture.

The executable statements of a process are enclosed in a process statement. In a single process model, the process statement includes all executable statements of the model. In a multi-process model each process is defined by a process statement that includes executable statement of the individual process.

Initially, the process will be executed once using the current value of objects in the sensitivity list and the process execution will be repeated only when a change occurs in one or more objects in the sensitivity list. The process label is optional and made to identify a process in a multi-process model. In process without sensitivity list, a wait statement should be provided inside the process statement.

Syntax of process statement

```
process_label : process (sensitivity list)
                    -- declaration statements;
                    begin
                    -- sequential statements;
                    end process (process_label);
```

Example of process statement

```
--program segment showing only a process statement
--a, b, c are declared as inputs and z as output in entity
    process (a, b, c)
        variable m1, m2 : std_logic;
    begin
        m1 :=  a nand b;
        m2 :=  m1 nand c;
        z   <= m1 nand m2;
    end process;
```

Note: All VHDL statement/program examples in this book are simulated and verified in license free Xilinx VHDL tool Vivado 2019.2 web edition which can be directly downloaded from the following link.

https://www.xilinx.com/support/download/index.html/content/xilinx/en/downloadNav/vivado-design-tools/archive.html

***File name:** Vivado HLx 2019.2: All OS installer Single-File Download (TAR/GZIP-26.55GB).*

5.4 Sequential Statements

Sequential statements allow the user to describe signal assignments in an algorithmic fashion. The sequential statements are written inside a process statement. All statements in a process are evaluated sequentially and therefore the order of the statement is important.

The various types of sequential statement available in VHDL are,

- Variable assignment statement
- Signal assignment statement
- Wait statement
- If statement
- Case statement
- Loop statement
- Null statement
- Exit statement
- Next statement

5.4.1 Variable Assignment Statement

The variable objects will take different values whenever the process is executed and the variable assignment statements are used to assign the next value to variable objects.

Syntax of variable assignment statement

```
variable_name := value_expression;
```

Example of variable assignment statement inside a process

```
--program segment showing only a process statement
--a, b, c are declared as inputs and y, z as outputs in entity
process (a, b, c)
    variable m1, m2 : integer range 0 to 10 ;
-- variables are declared inside the process
begin
    m1 := (a + b) – c; -- variable assignment statement
    y   <= – m1
    m2 := m1 * 2;      -- variable assignment statement
    z   <= m2;
end process;
```

5.4.2 Signal Assignment Statement

Signals are assigned values using a signal assignment statement.

Syntax of signal assignment statement

```
signal_name <= value expression;
```

Example of signal assignment statement inside a process

```
--program segment showing only a process statement
--a, b, c are declared as inputs and y as output in entity
signal m1, m2 : std_logic;   -- signals are declared outside the process
    begin
        process (a, b, c)
            begin
                m1 <= a or c;          -- signal assignment statement
                m2 <= a and b;         -- signal assignment statement
                y <= m1 and m2;
        end process;
```

5.4.3 Wait Statement

The wait statement is used to suspend a process without sensitivity list. The wait statement will have a condition and the process will resume only when the condition is true. There are three basic forms of wait statement.

Syntax of a wait statement

```
wait on sensitivity_list;
wait until boolean_expression;
wait for time_expression;
```

Examples of wait statements

```
wait on a, b, c;       --wait until a change occurs in one or more object ( a, b, c)
wait until (a = b);    --wait until object a is equal to object b
wait for 30 ns;        --Resume the process after a delay of 30 ns
```

Example of a process with wait statement

```
--program segment showing only a process statement
--a, b, c, d are declared as inputs and y as output in entity
    process
        variable m1, m2 : std_logic;
    begin
        m1 := a and b;
        m2 := c and d;
        y  <= m1 or m2;
        wait on a, b, c, d;
    end process;
```

5.4.4 If Statement

If statement is used to execute one or more statements when a condition specified in the if statement is true. The general syntax of the if statement shown below can be used to execute a number of statements based on different conditions.

An if statement will have one **if-condition-then** statement, one or more optional **elsif-condition-then** statement and one optional **else** statement. At any one time only one condition will be true and hence the statements followed by the true condition are executed. If no condition is true then the statements followed by else will be executed and if there is no else statement then no statement will be executed

Syntax of if statement when multiple conditions have to be checked

```
if condition_1 then
    sequential_statements ;
elsif condition_2 then
    sequential_statements;
    .
    .
    .
elsif condition_n then
    sequential_statements;
else
    sequential_statements;
end if;
```

Syntax of if statement for only one condition

```
if condition then
    sequential_statements;
else
   sequential_statements;
end if;
```

Syntax of if statement without else statement for only one condition

```
if condition then
    sequential_statements;
end if;
```

Example of if statement inside a process

```
--program segment of 4:1 multiplexer based on if statement
--i0, i1, i2, i3 are declared as inputs and y as output in entity
--s is declared as 2-bit std_logic_vector input in entity
    process (i0, i1, i2, i3)
        begin
            if (s = "00") then
                y <= i0;
            elsif (s = "01") then
                y <= i1;
            elsif (s = "10") then
                y <= i2;
            else
                y <= i3;
            end if;
    end process;
```

5.4.5 Case Statement

A case statement will check the value of an expression or object that can take multiple values. At any onetime only one value will be true and one (or a set of) sequential statement after the true case is executed. The values of the expression or object are denoted as choice_1 to choice_n.

Syntax of case statement

```
case_expression is
    when choice_1 =>
        sequential_statements;
    when choice_2 =>
        sequential_statements;
        .
        .
        .
    when choice_n =>
        sequential_statements;
end case;
```

Example of case statement inside a process

```
--program segment of 4:1 multiplexer based on case statement
--i0, i1, i2, i3 are declared as inputs and y as output in entity
--sel_in is declared as 2-bit std_logic_vector input in entity
    process (i0, i1, i2, i3)
        begin
        case sel_in is
            when "00" =>
                y <= i0;
            when "01" =>
                y <= i1;
            when "10" =>
                y <= i2;
            when others =>
                y <= i3;
        end case;
    end process;
```

5.4.6 Loop Statement

The loop statement is used to execute a sequential statement (or a set of sequential statements) repeatedly with in a process. There are three types of loop statements in VHDL.

- For-loop
- While-loop
- Loop terminated using an exit or next statement

For-loop

Syntax of for-loop statement

```
(loop_label:) for index in loop_range loop
        sequential_statements;
        end loop (loop_label);      -- loop label is optional
```

Example of for-loop

```
sum_int := 0;
for int_num in 1 to 10 loop
    sum_int := sum_int + int_num;
end loop;
```

VHDL Program 5.1

```
--VHDL program using for-loop to find sum of integers
library ieee;
use ieee.std_logic_1164.all;
    entity test_for_loop is
        port (y : out integer range 0 to 60);
    end test_for_loop;
    architecture bm_test_for_loop of test_for_loop is
    begin
        process
            variable sum_int : integer range 0 to 60;
        begin
            sum_int := 0;
            for int_num in 1 to 10 loop
            sum_int := sum_int + int_num;
        end loop;
                y <= sum_int;
            wait for 100ns;
        end process;
    end bm_test_for_loop;
```

While-loop

Syntax of while-loop statement

```
while Boolean_expression loop
    --sequential_statements;
end loop;
```

Example of while-loop

```
sum_int := 0;
int_num := 1;
while int_num <= 10 loop
    sum_int := sum_int + int_num;
    int_num:=int_num+1;
end loop;
```

Note: *The example program given for for-loop can be run with while-loop, by replacing the statements corresponding to for-loop with statements for while-loop.*

Loop terminated using an exit or next statement

Syntax of loop statement

```
loop_label: loop
        --sequential_statements;
        exit when boolean_expression;
        end loop loop_label;           -- loop label is optional
```

Example of simple loop

```
sum_int := 0;
int_num := 1;
L1: loop
    sum_int := sum_int + int_num;
    int_num := int_num + 1;
    exit when int_num>10;
end loop L1;
```

Note: *The example program given for for-loop can be run with simple loop, by replacing the statements corresponding to for-loop with statements for simple loop.*

5.4.7 Exit Statement

The exit statement is a sequential statement that can be used only inside a loop. The exit statement will check the specified condition/Boolean_expression and if the condition / Boolean_expression is true then It stops execution of loop statement immediately and jump out of the loop. Hence it is used to terminate the iterations in the loop statement.

Syntax of exit statement

```
exit when Boolean_expression:
```

Note: *Refer example of simple loop for usage of exit statement.*

5.4.8 Next Statement

The next statement is also a sequential statement that can be used only inside a loop. It is used to skip a part of statements in the iteration. When a next statement is encountered and if the specified condition is true then the execution jump to start of next iteration by skipping the statements after the next statement.

Syntax of next statement

```
next when Boolean_expression:
```

Example of next statement

```
sum_int := 0;
int_num := 0;
while int_num < 10 loop
    int_num := int_num + 1;
    next when int_num = 5;        --integer 5 will not be added in sum
    sum_int := sum_int + int_num;
end loop;
```

5.4.9 Null Statement

The null statement is a sequential statement and used in places where no action is required for a particular condition.

Example of null statement

```
--program segment showing usage of null statement inside a case statement
--a, b, c are declared as inputs and y as output in entity
--opcode is declared as signal in architecture
   case opcode is
      when "000"    => y <= a and b;
      when "000"    => y <= a or b;
      when "000"    => y <= a xor b;
      when others => null;
   end case;
```

5.4.10 Assertion Statement

An assertion statement is used to verify a specified condition and to report if the condition is violated. If the condition is false then the string_expression in the report statement is displayed and the string_expression in the severity statement is used to display the severity clause.

***Note:** During simulation in vivado, the assert statement is included in test bench. The string_expression in report statement is displayed in log file and the string_expression in severity statement is displayed in TCL console.*

Syntax of assertion statement

```
assert condition report "string_expression " severity "string_expression";
```

Example of assertion statement

```
assert (s = '0') report "se1" severity "error";
assert (co = '1') report "ce1" severity "error";
```

VHDL Program 5.2

```
--VHDL program example showing usage of assert statement
--code for half adder
library ieee;
use ieee.std_logic_1164.all;
   entity test_ha is
      port ( a, b  : in std_logic;
             s, co : out std_logic );
   end test_ha;

   architecture dfm_test_ha of test_ha is
      begin
         s   <= a xor b;
         co <= a and b;
   end dfm_test_ha;
```

```
--code for test bench of half adder
library ieee;
use ieee.std_logic_1164.all;
```

```
entity test_tb_ha is
    --port ()
end test_tb_ha;

architecture bm_test_tb_ha of test_tb_ha is
    component test_ha is
        port ( a, b  : in std_logic;
               s, co : out std_logic );
    end component;
    signal a, b, s, co : std_logic;

    begin
        mapping : test_ha port map (a, b, s, co);
    process
        variable e_c : integer := 0;
    begin
```

```
--test1
    a <= '1';
    b <= '0';
wait for 10 ns;

    assert (s = '0') report "se1";          --without severity clause
    assert (co = '1') report "ce1";         --without severity clause

  --assert (s = '0') report "se1" severity "error";
  --assert (co = '1') report "ce1" severity "error";

  if (s /= '1' or co /= '0') then
      e_c := e_c + 1;
  end if;
```

```
--test2
    a <= '1';
    b <= '1';
wait for 10 ns;

    assert (s = '0') report "se2";          --without severity clause
    assert (co = '1') report "ce2";         --without severity clause

  --assert (s = '0') report "se2" severity "error";
  --assert (co = '1') report "ce2" severity "error";

    if (s /= '0' or co /= '1') then
        e_c := e_c + 1;
    end if;
  end process;
end bm_test_tb_ha;
```

5.5 Concurrent Statements

The concurrent statements are used in dataflow architecture. VHDL supports the following concurrent statements.

- Concurrent signal assignment statement
- Conditional signal assignment statement
- Selected signal assignment statement
- Concurrent assertion statement

5.5.1 Concurrent Signal Assignment Statement

Signal assignment statements that appear outside of a process are called concurrent signal assignment statements and they are used to assign a value to an object. The syntax for concurrent signal assignment statement is same as that of signal assignment statement discussed under sequential statements. An architecture body can have any number of concurrent signal assignment statements. Concurrent signal assignment statements are event triggered and hence, they are executed whenever there is an event or change in a signal that appears in the expression in the statement.

5.5.2 Conditional Signal Assignment Statement

Conditional signal assignment statement assigns different values for the target signal based on the specified conditions similar to if statement.

Syntax of conditional signal assignment statement

```
signal_name <= value_exp_1 when condition_1 else
               value_exp_2 when condition_2 else
               .
               .
               value_exp_n
```

Example of conditional assignment statement in a dataflow architecture

```
--program segment of 4:1 multiplexer based on
--conditional assignment statement
--i0, i1, i2, i3 are declared as inputs and y as output in entity
--sel_in is declared as 2-bit std_logic_vector input in entity
   y <= i0 when (sel_in = "00") else
        i1 when (sel_in = "01") else
        i2 when (sel_in = "10") else
        i3;
```

5.5.3 Selected Signal Assignment Statement

Selected signal assignment statement assigns different values for a target signal based on the value of a select_expression similar to case statement.

Syntax of selected signal assignment statement

```
with select_expression select
signal_name  <=  value_exp_1 when choice_1,
                 value_exp_2 when choice_2,
                 .
                 .
                 value_exp_n when choice_n;
```

Example of selected signal assignment statement in a dataflow architecture

```
--program segment of 4:1 multiplexer based on
--selected signal assignment statement
--i0, i1, i2, i3 are declared as inputs and y as output in entity
--sel_in is declared as 2-bit std_logic_vector input in entity
   with sel_in select
      y <= i0 when "00",
           i1 when "01",
           i2 when "10",
           i3 when others;
```

5.5.4 Concurrent Assertion Statement

Assertion statements that appear outside of a process are called concurrent assertion statements and they are used to report the error conditions. The syntax for concurrent assertion statement is same as that of assertion statement discussed under sequential statements.

5.6 VHDL Design Entity (or Model)

In order to describe a design entity (or model), VHDL provides the following five different types of primary constructs called design units.

- Entity declaration
- Architecture body
- Configuration declaration
- Package declaration
- Package body

In the simplest form, a VHDL design entity (or model) includes an entity declaration and at least one architecture body.

The entity declaration describes the input and output signal names. The architecture body contains a set of interconnected components which represents the structure of the entity, or as a set of concurrent or sequential statements that represents the behaviour of the entity. Figure 5.1 shows a simple VHDL design entity or model.

Entity declaration: Describes the input output parameter

Architecture body: Describes the functionality of the design

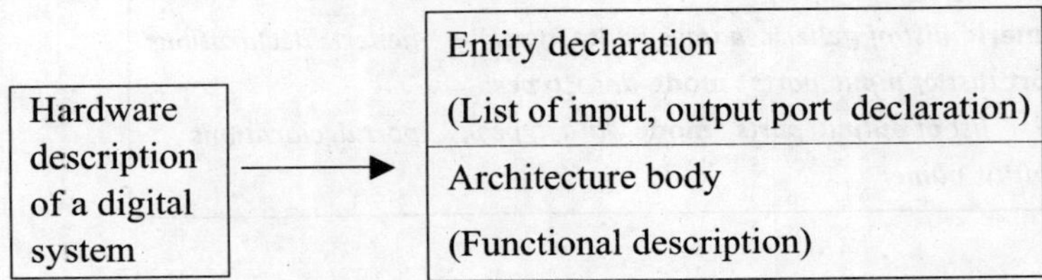

Fig. 5.1: VHDL design entity (or model)

VHDL separates entity declaration and architecture body into two independent design units. ***Configuration*** is the process of binding a component with an entity and architecture. In a simple VHDL program configuration is not needed but configuration will be useful in development and testing of large systems.

A ***package declaration*** is used to store a set of common declarations like components, types, procedures, and functions. These declarations can then be imported into other design units.

A package body is always associated with a package declaration. A ***package body*** is used to store the definitions of functions and procedures that were declared in the corresponding package declaration and also the constant declarations for any constants that appear in the package declaration.

5.6.1 Entity Declaration

Entity declaration specifies the name of the entity and a set of interface ports. Ports are signals through which the entity communicates with the other models in its external environment.

Syntax for entity declaration

```
entity entity_name is
    port ( list of input_port  : mode data_type;
           list of output_port : mode data_type );
end entity_name;
```

Note: **mode** *is in, out, inout, buffer.*

Port

Each input and output signal in an entity is called a port. Each port should have a name, data type and direction (or mode). The mode can be assigned one of the four possible choices: in, out, inout, buffer. The set of ports defined in an entity is called port declaration.

Generic

The generic construct of VHDL is provided to pass information into an entity and a component. Generics are like parameters. They are first declared in entity and component declaration and later assigned a value during component instantiation. In order to use generics, an additional declaration called generic declaration is added in entity.

Syntax for entity declaration with generics

```
entity entity_name is
    generic (list of generic_names : data_types);  --generic declarations
    port (list of input_ports : mode data_types;
          list of output_ports : mode data_types);  --port declarations
end entity_name;
```

5.6.2 Architecture Body

The internal details of a design/model are specified in architecture body. The different types of architecture models are,

- Dataflow model
- Behavioral model
- Structural model
- Combination of the above three.

Dataflow Model

A dataflow model specifies the functionality of a design entity or model without specifying the structure of the design. It shows the flow of information through the design entity

Syntax for architecture body of dataflow model

```
architecture architecture_name of entity_name is
                    --architecture item declarations
begin
                    --concurrent statements;
end architecture_name;
```

Behavioral Model

The behavior of the entity is defined using a set of sequentially executed statements .

A process statement is the main mechanism used to model the behavior of an entity.

Architecture Body

The architecture body consist of a set of interconnected components which represents the structure of the entity, or as a set of concurrent or sequential statements that represents the behavior of the entity

Syntax for architecture body of behavior model

```
architecture architecture_name of entity_name is
            --architecture item declarations
begin
            --concurrent statements ;
                process (sensitivity list)
                begin
                    --statements
                    .
                    .
                    .
                end process
end architecture_name;
```

Structural Model

In structural model, a design entity is designed with the set of components connected by signals.

Components are VHDL modules compiled as separate entity and can be used in any other structural model of digital systems.

Instead of coding a complex design in single VHDL Code, it can be divided into submodules as components and combine them using port map technique in a main module.

The component declaration and component instantiation statements are the primary structures used to describe the design entity.

5.6.3 Component Declaration

A component declaration declares the name of the component and a set of interface ports.

The port interface specifies the mode and the type of ports, similar to an entity declaration.

Syntax of component declaration

```
component component-name
    port (list_of_interface_ports) ;
end component;
```

5.6.4 Component Instantiation

A component instantiation statement is used to define the subcomponent of the entity in which it is appearing.

It connects the signals in the entity with the ports of the subcomponent.

The ***association-list*** associates signals in the entity called ***actuals***, with the ports of a component, called ***locals***.

Syntax for component instantiation

```
component_label: component_name port map (association_list)
```

Syntax for architecture body of Structural model

```
architecture architecture_name of entity_name is
    component component_name1
        port (list_of_interface_ports) ;
    end component;

    component component_name2
        port ( list_of_interface_ports ) ;
    end component;

        begin
            component_label1: component_name1 port map ( association_list1 );
            component_label2: component_name2 port map ( association_list2 );
            end architecture_name;
```

5.6.5 VHDL Programs for Half Adder with Three Different Architecture

The Boolean equations for half adder are, ***(AU, Nov/Dec'18, 13 Marks)***

$$s = a \oplus b$$

$$c_o = a \cdot b$$

Where, a and b are 1-bit input data sum and carry are 1-bit output data. The VHDL codes for dataflow model of half adder are developed using the above equations. (Refer Chapter 2, Section 2.6.1 for design of half adder).

VHDL Program 5.3

VHDL Program for Dataflow Model of Half Adder

```
library ieee;
use ieee.std_logic_1164.all;
    entity half_adder is
        port ( a, b  : in std_logic;
               s, co : out std_logic );
    end half_adder;

    architecture dfm_ha of half_adder is
    begin
        s  <= a xor b;
        co <= a and b;
    end dfm_ha;
```

VHDL Program for Behavioral Model of Half Adder

```
library ieee;
use ieee.std_logic_1164.all;
    entity half_adder is
        port ( a, b  : in std_logic;
               s, co : out std_logic );
    end half_adder;

    architecture bm_ha of half_adder is
    begin
        process ( a, b )
            begin
                s  <= a xor b;
                co <= a and b;
        end process;
    end bm_ha;
```

Program Logic for Structural Model

The VHDL code for structural model of half adder is developed using the logic circuit shown in Fig. 5.1. In structural model, each component involved in the half adder are coded separately and then the components are added in the code for half adder.

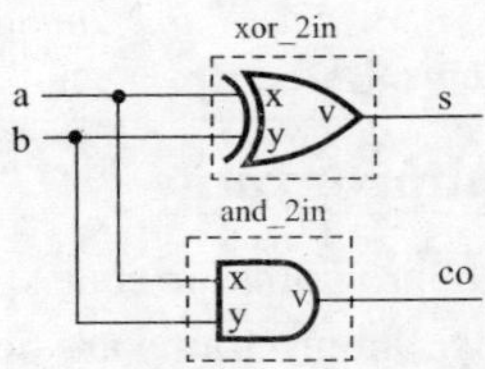

Fig. 5.1: Logic circuit of half adder.

VHDL Program for Structural Model of Half Adder

```
--code for 2-input XOR gate
library ieee;
use ieee.std_logic_1164.all;
    entity xor_2in is
        port ( x, y : in std_logic;
               v    : out std_logic );
    end xor_2in;

    architecture dfm_xor_2in of xor_2in is
    begin
        v <= x xor y;
    end dfm_xor_2in;
```

```
--code for 2-input AND gate
library ieee;
use ieee.std_logic_1164.all;
    entity and_2in is
        port ( x, y : in std_logic;
                v   : out std_logic );
    end and_2in;

    architecture dfm_and_2in of and_2in is
        begin
            v <= x and y;
    end dfm_and_2in;
```

```
--code for half adder
library ieee;
use ieee.std_logic_1164.all;
    entity half_adder is
        port ( a, b  : in std_logic;
                s, co : out std_logic );
    end half_adder;

    architecture sm_ha of half_adder is
        component xor_2in is
            port ( x, y : in std_logic;
                    v   : out std_logic );
        end component;

        component and_2in is
            port ( x, y : in std_logic;
                    v   : out std_logic );
        end component;
            begin
            a1 : xor_2in port map (a, b, s);
            a2 : and_2in port map (a, b, co);
end sm_ha;
```

5.7 Subprograms

Subprograms are codes for repeated operations in a VHDL program, similar to any high level language programs. Subprograms are part of a main program and cannot be compiled separately. Procedures and functions are two types of subprograms supported by VHDL. Procedures and functions are first declared in the declaration part of architecture and called later any number of times in the execution part of architecture.

5.7.1 Procedure

Procedures are subprograms written for repeated operations in a main program. Procedure requires a declaration which includes inputs, outputs and return values. Any number of values can be returned from a procedure.

VHDL Program 5.4

```
--example program for usage of procedure
library ieee;
use ieee.std_logic_1164.all;
use ieee.numeric_std.all;
```

```
entity pro_ex is
   generic (N : integer := 7);
   port ( x, y       : in signed (N – 1 downto 0);
          sum, dif : inout signed (N – 1 downto 0));
end pro_ex;

architecture arch_pro_ex of pro_ex is
signal p : signed (2 * N – 1 downto 0);

--procedure to add and subtract two numbers
procedure sum_dif ( signal a    : in signed (N – 1 downto 0);
                    signal b    : in signed (N – 1 downto 0);
                    signal s, d : out signed (N – 1 downto 0)) is
begin
   s <= a + b;
   d <= a – b;
end sum_dif;                  -- procedure ends, procedure returns two values s and d


begin
   sum_dif ( a => x, b => y, d => dif, s => sum);   -- procedure call
          p <= sum * dif;                            -- product of sum and difference
end arch_pro_ex;
```

5.7.2 Functions

Functions are subprograms written for repeated operations in a main program. Function requires a declaration which includes inputs and a return value. In a function outputs need not be declared and it returns only one value.

The differences between procedure and function are,

- In a function, outputs need not be declared whereas in a procedure, the outputs have to be declared.
- A function will return only one value, whereas a procedure can return multiple values.

VHDL Program 5.5

```
--example program for usage of function
library ieee;
use ieee.std_logic_1164.all;
use ieee.numeric_std.all;
   entity func_ex is
      generic ( N : integer := 7);
      port ( x, y : in signed (N – 1 downto 0);
             sum : inout signed (N – 1 downto 0);
             dif   : inout signed (N – 1 downto 0) );
   end func_ex;

   architecture arch_func_ex of func_ex is
   signal p : signed (2 * N – 1 downto 0);

--function to find sum of two integers
   function sum2int ( signal a : in signed (N – 1 downto 0);
                      signal b : in signed (N – 1 downto 0) );
   return signed is variable s : signed (N –1 downto 0);
```

```
begin
    s := a + b;
    return s;
end sum2int;      -- end of function, function returns the value of s.

--function to find difference of two integers
function dif2int ( signal a : in signed (N – 1 downto 0);
                   signal b : in signed (N – 1 downto 0));
return signed is variable d : signed (N – 1 downto 0);
begin
    d := a – b;
    return d;
end dif2int;                    --end of function, function returns the value of d;
begin
    sum <= sum2int (a => x, b => y);  --function call
    dif  <= dif2int (a => x, b => y);  --function call
    p    <= sum * dif;                --product of sum and difference
end arch_func_ex;
```

5.8 Introduction to Packages

A package provides a convenient method to store and share declarations that are common to many design units. A Package is a VHDL file, contains data types, constants, functions, procedures, etc, that can be shared across many VHDL designs.

A package is created by a package declaration and package body. A package declaration contains a set of declarations that can be shared by many design units. A package body contains the functions and procedures that are declared in the corresponding package.

A package can be compiled or simulated separately and stored in a default library or in any user defined library and called in any other program using package name. The default library in Vivado is "xil_defaultlib". Also, the global include should be activated in Vivado before simulating the package.

Syntax of a package declaration

```
package package_name is
    -- subprogram declarations
    -- subtype declarations
    -- constant declarations
    -- signal declarations
    -- file declarations
    -- alias declarations
    -- component declarations
    -- attribute declarations
    -- attribute specifications
    -- disconnection specifications
    -- use clauses
end package [ package_name ]; -- package name is optional
```

5.8.1 Package Body

A package body primarily contains the behavior of the subprograms and the values of the deferred constants declared in a package declaration. It may contain other declarations as well, as shown by the following syntax of a package body.

```
Syntax of a package declaration
    package body package_name is
        - subprogram bodies
        - subprogram declarations
        - type and subtype declarations
        - file and alias declarations
        - use clauses
    end package body [ package_name ];  -- package name is optional
```

The package body name must be the same as the name of its corresponding package declaration. A package body is not necessary, if its associated package declaration does not have any subprogram or deferred constant declarations.

A package body is used to store local declarations that cannot be access by other design, while a package declaration is used to store public declarations which can be accessed by other design units.

VHDL Program 5.6

```
--example program for package declaration and package body
library ieee;
use ieee.std_logic_1164.all;
use ieee.numeric_std.all;
-- package declaration
    package mypackage3 is
-- declare a constant with value 1
    constant k : signed (3 downto 0) := "0001";
-- procedure to add two integers
    procedure sum_2int ( signal a   : in signed (3 downto 0);
                         signal b   : in signed (3 downto 0);
                         signal sum : out signed (3 downto 0));
-- declare signals for if loop
    signal m : unsigned (2 downto 0):=  (others => '0');
    type state_type is (start_state, continue_state, stop_state);
    end package mypackage3;
-- package body declaration
    package body mypackage3 is
-- procedure to add two integers
    procedure sum_2int ( signal a   : in signed (3 downto 0);
                         signal b   : in signed (3 downto 0);
                         signal sum : out signed (3 downto 0)) is
begin
    sum <= a + b;
end sum_2int;
end package body mypackage3;
```

```
--example program that calls the above package (mypackage3)
library ieee;
library xil_defaultlib;                          --mypackage3 is compiled in xil_defaultlib
use ieee.std_logic_1164.all;
use ieee.numeric_std.all;
use xil_defaultlib.mypackage3.all;               --use clause for mypackage3
   entity mypackage3_call is
      port ( clk                   : std_logic;
             x                     : in unsigned (4 downto 0);
             count                 : out unsigned (2 downto 0);
             num1, num2            : in signed (3 downto 0);
             sum_num1_constant     : out signed (3 downto 0);
             sum_num1_2            : inout signed (3 downto 0) );
   end mypackage3_call;

architecture arch_mypackage3_call of mypackage3_call is
signal current_state : state_type;               --state_type is defined in package
begin
-- add constant k to num1
   sum_num1_constant <= num1 + k;                --k is defined in package
-- add num1 and num2
   sum_2int(num1, num2, sum_num1_2);             --procedure sum_2int is defined in package

--process for start-stop control of a count sequence
   process (clk, current_state, m)
   begin
      if (m <= x) then                           --m is defined in package
         current_state <= continue_state;        --continue_state is defined in package
      else
         current_state <= stop_state;            --stop_state is defined in package
      end if;
   end process;

--process to generate a count sequence
   process (clk,  current_state)
   begin
      if (current_state = continue_state) then
         m <= m + 1;
      else
         m <= (others => '0');
      end if;                                    --if loop end
         count <= m;
      end process;
      end arch_mypackage3_call;
```

5.9 Library

A library is a collection of packages that are compiled and stored in a directory in the host environment (or personal computer) in which the VHDL software tool is installed.

A library is a directory and each package is a file in that directory. The package file is a database containing information about the components in the package.

The two most commonly used libraries are IEEE and WORK. The IEEE library is a standard library standardized by IEEE and has a large number of packages. The WORK library is always available for user defined library packages. In vivado, the default library for user defined library and packages is "xil_defaultlib". The programmer or user can select either WORK library or "xil_defaultlib" library.

The keyword "library" identifies a library in VHDL followed by the logical library name.

Example

```
library IEEE;
library WORK;
library xil_defaultlib;
library mylib1;
```

A library may have large number of packages and all the packages may not be required for a design unit. Normally, the library statement shown above has to be combined with use statements to access the required packages in the library. In general programming practice, the library and use statements are placed in the beginning of the source file.

```
use ieee.std_logic_1164.all;
    --makes all components in std_logic_1164 library from ieee library visible
use ieee.std_logic-arith.all;
    --makes all components in std_logic_arith library from ieee library visible
use ieee.std_logic_unsigned.all;
    --makes all components in std_logic_unsigned library from ieee library visible
use ieee.std_logic_signed.all;
    --makes all components in std_logic_signed library from ieee library visible
use ieee.std_logic_textio.all;
    --makes all components in std_logic_textio library from ieee library visible
use ieee.numeric_bit.all;
    --makes all components in numeric_bit library from ieee library visible
use ieee.numeric_std.all;
    --makes all components in numeric_std library from ieee library visible
```

Design Library

(AU, Apr/May'19, 6 Marks)

Users can create small design units (with a design entity and architecture) and compile and store in a user defined library so that they can be accessed in any other larger design units. A design unit can be compiled or simulated separately and stored in a default library or WORK library or in any user defined library and called in any other program using the design entity name.

The default library in Vivado is "xil_defaultlib". Also, the global include should be activated in Vivado before simulating the design units. In order to use a design entity in another design, the library name in which the design unit is available should be included in design in which the library is called and use clause statement is not required.

VHDL Program 5.7

```
--example of VHDL file compiled in the library named "mylib1"
library ieee;
use ieee.std_logic_1164.all;
    entity not_1in is
        port ( a : in std_logic;
               y : out std_logic );
    end not_1in;
```

```
architecture dfm_not_1in of not_1in is
begin
    y <= not a;
end dfm_not_1in;
```

```
--example of VHDL file that use above library the compiled in the library named "mylib1"
library ieee;
use ieee.std_logic_1164.all;
library mylib1;                          --not_1in is compiled in the user library "mylib1"
    entity top is
        port ( a : in std_logic;
               y : out std_Logic);
    end top;
architecture dfm of top is
begin
    u0 : entity mylib1.not_1in
    port map (a => a, y => y);
end dfm;
```

5.10 Test Bench

A test bench is also a VHDL program that is used to verify the correctness of the VHDL program of the hardware model. A test bench VHDL program has three main purposes:

- Generation of necessary inputs and clock for simulation.
- Apply this inputs and clock to the entity under test and to monitor the output
- Comparing the output of VHDL program model with expected output.

The application of inputs and clock to the entity under test is accomplished automatically by instantiating the entity in the test bench model and then specifying the appropriate interface signals.

Format of a test bench that drives an entity under test

```
entity test_bench is
    --port
end test_bench;

architecture tb_behavior of test_bench is
    component entity_under_test
        port (list_of_ports_their_types_and_modes);
    end component;
    local_signal_declarations;
begin
        generate_waveforms_using_behavioral_constructs;
    apply_to_entity_under_test;
        eut: entity_under_test port map (port_associations);
        monitor_values_and_compare_with_expected_values;
end tb_behavior;
```

VHDL program examples without test bench are presented in Section 5.12 and VHDL programs with test bench along with the output waveform are presented in Section 5.13. All the programs are simulated in Vivado 2019.2, which is a VHDL tool from XILINX.

5.11 RTL Design

In the VLSI design flow, the designer describes the behavior of the circuit using a hardware description language (HDL) such as VHDL or Verilog. These descriptions are then synthesized into a gate-level netlist, which is a representation of the circuit at the level of gates and interconnections. This netlist can be used to create a physical layout of the circuit on a FPGA IC using computer-aided design (CAD) tools.

(AU, Apr/May'19, 7 Marks)

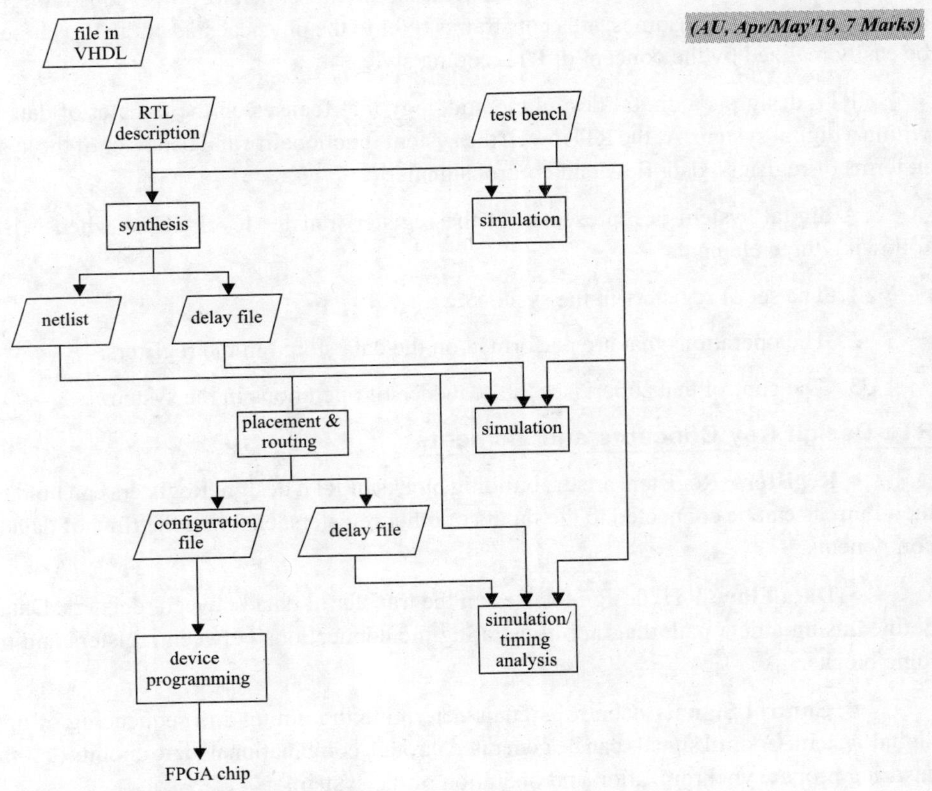

Fig. 5.2: Flowchart for IC design using VHDL.

The various steps in the design of a an IC using VHDL and implementing in FPGA are,

- Develop the design file and testbench in VHDL.
- Use the design file as the circuit description, and perform a simulation to verify the functionality of the design.
- Perform a synthesis and generate a netlist.
- Use the output netlist file of the synthesizer as the circuit description, and perform a simulation of timing analysis.
- Perform placement and routing.
- Update the timing information to the netlist, and perform a simulation and timing analysis to verify the correctness of the placement and routing

- Generate the configuration file and program the FPGA device.
- Verify the working of FPGA device.

A flowchart for IC design using VHDL and implementing in FPGA is shown in Fig. 5.1.

In IC design, it is important to consider various factors such as timing, power and area constraints. Timing constraints refer to the speed at which the circuit must operate, power constraints refer to the amount of power the circuit consumes, area constraints refer to the physical size of the IC. These requirements can be easily realized by the concept of RTL coding style.

RTL design is a digital design methodology that focuses on the transfer of data between registers within a digital system. At the RTL level, the system functionality and behavior of the system are described in terms of registers, data flow and control signals.

A digital system is represented at the register transfer level (RTL) when it is specified by the following three elements:

1. The set of registers in the system.
2. The operations that are performed on the data stored in the registers.
3. The control that supervises the sequence of operations in the system.

RTL Design Key Concepts and Elements

- **Registers:** Registers are the building blocks of RTL design. Registers can hold binary values, and their outputs can be connected to the inputs of other registers, enabling the flow of data between different components.

- **Data Flow:** RTL design focuses on the transfer of data between registers. Data flow is typically defined using a data path diagram, illustrating the connections between registers and the combinational logic blocks.

- **Control Signals:** Control signals determine the timing and sequencing of operations within a digital system. Control signals can be generated through combinational logic circuits or finite state machines, ensuring proper synchronization and operation of the system.

- **Combinational Logic:** Combinational logic circuits perform operations on input data to produce desired output values. Combinational logic circuits include various components such as arithmetic and Boolean operations, multiplexers, decoders, encoders, and logic gates.

The programming structure of RTL design can be broadly classified into data path and control path. The basic structure of RTL design is shown in Fig. 5.3. The data path has routing network, combinational logic and output registers. The control path has next state logic, registers and output logic.

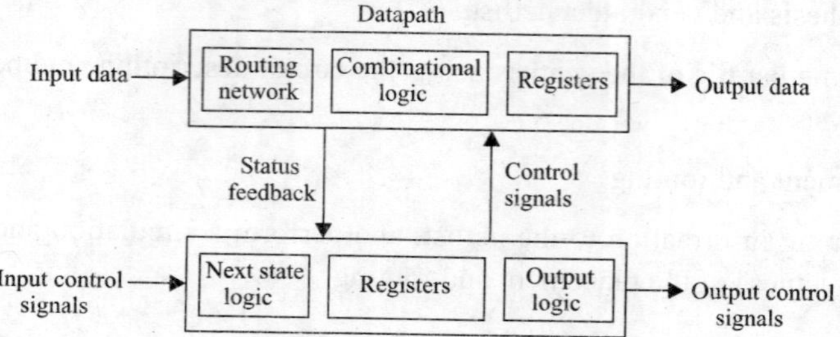

Fig. 5.3: Basic structure of RTL design.

Advantages of RTL Design

- **Performance Optimization:** RTL design allows for performance optimization by carefully managing the data flow, control signals and interconnections between registers.

- **Design Validation:** RTL design provides a structured approach to validate the functionality of a digital system before proceeding to physical implementation.

- **Reusability and Modularity:** Designers can create libraries of RTL blocks, such as adders, multipliers, memory controllers, or communication interfaces. These reusable blocks can be easily integrated into new designs.

RTL Design Methodologies

- **Manual RTL Design:** In manual RTL design, designers explicitly specify the behavior and functionality of the digital system using a hardware description language (HDL) such as Verilog or VHDL. The designer defines the registers, data flow, control signals, and interconnections. Manual RTL design allows for precise control over the system implementation and provides in-depth understanding of the design details. Some examples of manual RTL design are presented in Section 5.14.

- **RTL Synthesis:** RTL synthesis is an automated process that transforms a behavioral description written in VHDL into RTL description using a VLSI software tool. The synthesis tool translates the algorithmic representation into a register-level implementation, optimizing the design for area, power, and timing considerations. The Vivado tool used to synthesize the VHDL codes in this book will follow this automated process of RTL simulation and synthesis. All the example VHDL programs in Sections 5.12, 5.13 and 5.14 are simulated in Vivado 2019.2 as RTL project and outputs are verified.

Table 5.1: Differences between RTL Design and Sequential Logic Design

RTL Design	Sequential Logic Design
1. In RTL design the basic building blocks are registers, multiplexers, adders.	1. In sequential logic design the basic building blocks are the logic gates, flip-flops.
2. RTL design is much closer to the behavioural design of a logic circuit as it models the data flow among different registers, and hence more realistic.	2. Sequential logic design process is more mechanical in nature as compared to RTL design process.
3. RTL modelling allows us to synthesize complex circuits with a large number of states with ease as compared to sequential logic design.	3. The sequential logic design techniques are only applicable to circuits having a small number of states.

5.11.1 Combinational Logic

A combinational circuit, by definition, is a circuit whose output, is a function of current input. It has no internal state and therefore, it is system without memory about the past events (or past inputs).

Combinational RTL design involves the specification of circuits that perform logical operations on input signals to produce output signals, without any internal memory or state. Even though a digital system requires a clock for functioning, it is not mandatory to include clock signals in combinational logic circuit design. But the combinational digital system working will depend on FPGA IC default clock.

Combinational logic circuits can be designed using VHDL concurrent and sequential statements in dataflow or behavioral or structural architectures. Examples of combinational logic circuits are half adder, full adder, multiplexers, demultiplexers, etc. The RTL synthesisable VHDL codes for some combinational logic circuits are presented in Example programs in Section 5.12 and 5.13. All the example programs are simulated in Vivado 2019.2 as RTL project and outputs are verified.

5.11.2 Sequential Circuit

A sequential circuit has an ***internal state***, or ***memory***. Its output is a function of current input as well as the internal state. The internal state essentially "memorizes" the effect of the past input values. Sequential RTL design involves the specification of circuits that perform logical operations on input signals to produce output signals, with internal memory or states.

In addition, the operations of a sequential circuit depend on a clock signal. The "event" attribute of VHDL can be attached to a signal named "clk" to create a clock dependent process in a sequential circuit design. An example VHDL program of D flip-flop using positive edge triggered clock signal is given below:

> ***Note:*** *The attribute is an advanced feature in VHDL. An attribute is a value or function or type or range or signal or a constant that can be associated with certain names in a VHDL description. The name can be an entity name or an architecture name or a label or a signal.*

VHDL Program 5.8

```
--VHDL program for positive edge triggered D flip-flop
library ieee;
use ieee. std_logic_1164.all;
entity pos_dff is
    port ( clk : in std_logic;
           d   : in std_logic;
           q   : out std_logic );
end pos_dff ;

architecture arch of pos_dff is
    begin
        process (clk)
            begin
                if (clk'event and clk = '1') then
                    q <= d;
                end if;
        end process;
end arch;
```

In the above program, the key expression that creates clock sensitive process is,

```
if (clk'event and clk = '1') then
        q <= d;
end if;
```

The 'event term is a VHDL attribute returning true when there is a change in clk signal value (i.e., an event). When the clk = '1' expression is true, it means that the new value of clk is '1' . If statement states that at the rising of the clk signal, q gets the value of d. Since there is no else branch, it means that q keeps its previous value otherwise.

In the above program, the clock can be modified to have negative edge sensitivity by using the following if statement.

```
        if (clk'event and clk = '0') then
                q <= d;
        end if;
```

Sequential logic circuits can be designed using VHDL concurrent and sequential statements in dataflow or behavioral or structural architectures. Examples of sequential logic circuits are flip-flops, counters, shift registers, etc. The RTL synthesisable VHDL codes for some sequential logic circuits are presented in Example programs in Section 5.12 and 5.13. All the example programs are simulated in Vivado 2019.2 as RTL project and outputs are verified.

5.12 VHDL Programs Without Test Bench

VHDL Program 5.9

Simplfy the Boolean function F by means of K-map and write a dataflow and structural HDL to implement the reduced Boolean function.

f(a, b, c, d) = ∑m(0, 6, 7, 9, 11, 13, 15), d(a, b, c, d) = ∑m(2, 5, 8, 10)

To find Boolean Equation using K-map

The given function has four variables and so a 4-variable K-map with 16 squares ($2^4 = 16$) is drawn as shown in Fig. 1. The function is defined as sum of minterms m_0, m_6, m_7, m_9, m_{11}, m_{13} and m_{15} and so a "1" is filled in the corresponding squares in K-map. The function also has don't-care outputs for the minterms m_2, m_5, m_8 and m_{10} and so a "x" is filled in the corresponding squares in K-map and the remaining squares are filled with "0".

For NAND implementation, the Boolean function should be formed in SOP form. In the K-map of Fig. 1, by considering the don't-care condition in m_2 and m_5 as "1", three prime implicants can be formed with four adjacent 1's and the corresponding product terms are ab', bd, a'bc and b'd'. The simplified Boolean function is given by sum of these three product terms.

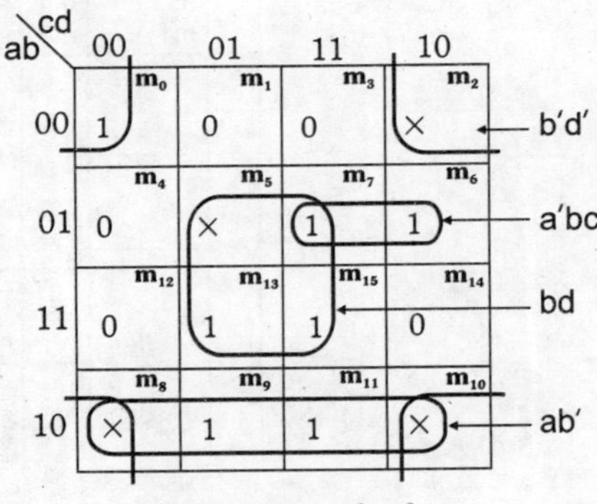

Fig. 1: K-map for f.

∴ f = ab' + bd + a'bc + b'd'

Using the above Boolean function the logic circuit using only NAND gates is drawn as shown in Fig. 2.

Program Logic for Dataflow Model

The Boolean function for K-map is,

f = ab' + bd + a'bc + b'd'

The VHDL code for dataflow model of Boolean function is developed using the above equations.(Refer Chapter 1, Section 1.6 for design of Boolean function).

VHDL Program for Dataflow Model of Boolean Function

```
library ieee;
use ieee.std_logic_1164.all;
   entity boolean_func is
        port ( a, b, c, d : in std_logic;
                  f          : out std_logic );
   end boolean_func;
```

```
architecture dfm_boolean_func of boolean_func is
signal m1, m2, m3 : std_logic;
   begin
      m1 <= not a;
      m2 <= not b;
      m3 <= not d;
      f    <= ((a and m1) or (b and d) or (m1 and b and c) or (m2 and m3));
end dfm_boolean_func;
```

Program Logic for Structural Model

The VHDL code for structural model of Boolean function is developed using the logic circuit shown in Fig. 2. In structural model, each component involved in the logic circuit of Fig. 2 are coded separately and then the components are added in the code for function, f.

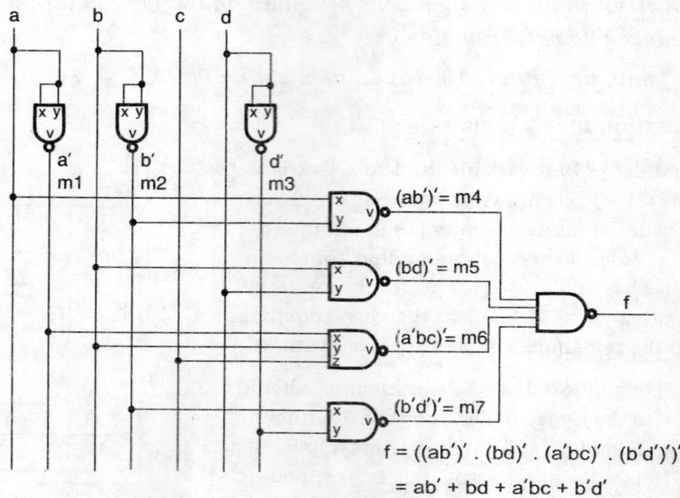

Fig. 2: Logic circuit of function, f using only NAND gates.

VHDL Program for Structural Model of Boolean Function

```
--code for 2-input NAND gate
library ieee;
use ieee.std_logic_1164.all;
   entity nand_2in is
      port ( x, y : in std_logic;
             v    : out std_logic );
   end nand_2in;

   architecture dfm_nand_2in of nand_2in is
      begin
         v <= x nand y;
   end dfm_nand_2in;
```

```
--code for 3-input NAND gate
library ieee;
use ieee.std_logic_1164.all;
```

```
    entity nand_3in is
        port ( x, y, z : in std_logic;
               v       : out std_logic );
    end nand_3in;

    architecture dfm_nand_3in of nand_3in is
        begin
            v <= not (x and (y and z));
    end dfm_nand_3in;
```

```
--code for 4-input NAND gate
library ieee;
use ieee.std_logic_1164.all;
    entity nand_4in is
        port ( w, x, y, z : in std_logic;
               v          : out std_logic );
    end nand_4in;

    architecture dfm_nand_4in of nand_4in is
        begin
            v <= not (w and (x and (y and z)));
    end dfm_nand_4in;
```

```
--code for Boolean function
library ieee;
use ieee.std_logic_1164.all;
    entity boolean_func is
        port ( a, b, c, d : in std_logic;
               f          : out std_logic;
    end boolean_func ;

    architecture sm_boolean_func of boolean_func is
    signal m1, m2, m3, m4, m5, m6, m7 : std_logic;
        component nand_2in is
            port ( x, y : in std_logic;
                   v    : out std_logic );
        end component;

        component nand_3in is
            port ( x, y, z : in std_logic;
                   v       : out std_logic );
        end component;

        component nand_4in is
            port ( w, x, y, z : in std_logic;
                   v          : out std_logic );
        end component;

            begin
                a1 : nand_2in port map (a, a, m1);
                a2 : nand_2in port map (b, b, m2);
                a3 : nand_2in port map (d, d, m3);
                a4 : nand_2in port map (a, m2, m4);
                a5 : nand_2in port map (b, d, m5);
```

```
        a6 : nand_3in port map (m1, b, c, m6);
        a7 : nand_2in port map (m2, m3, m7);
        a8 : nand_4in port map (m4, m5, m6, m7, f);
end sm_boolean_func ;
```

VHDL Program 5.10

Write a VHDL code for behavioral and structural model of a 4*1 multiplexer.

Program Logic for Behavioral Model

The Boolean equations for 4*1 multiplexer are,

$$Y = S0'\,S1'\,I0 + S0'\,S1\,I1 + S0\,S1'\,I2 + S0\,S1\,I3$$

Where, S0 and S1 are 1-bit selection input and Y is 1-bit output data. The VHDL code for behavioral model of 4*1 multiplexer is developed using the above equation. (Refer Chapter 2, Section 2.3 for design of 4*1 multiplexer).

VHDL Program for Behavioral Model of 4*1 Multiplexer using 'if' Statement

```
library ieee;
use ieee.std_logic_1164.all;
   entity mux_4_1 is
      port ( s             : in std_logic_vector (1 downto 0);
             i0, i1, i2, i3 : in std_logic;
             y             : out std_logic);
   end mux_4_1;

   architecture bm_mux_4_1 of mux_4_1 is
      begin
         process (s, i0, i1, i2, i3)
            begin
               if s = "00" then
                  y <= i0;
               elsif s = "01" then
                  y <= i1;
               elsif s = "10" then
                  y <= i2;
               else
                  y <= i3;
               end if;
         end process;
   end bm_mux_4_1;
```

Program Logic for Structural Model

The VHDL code for structural model of 4*1 multiplexer is developed using the logic circuit shown in Fig. 1. In structural model, each component involved in the multiplexer are coded separately and then the components are added in the code for multiplexer.

VHDL Program for Structural Model of 4*1 Multiplexer

```
--code for 1-input NOT gate
library ieee;
use ieee.std_logic_1164.all;
   entity not_1in is
      port ( x : in std_logic;
             v : out std_logic );
   end not_1in;
```

```
    architecture dfm_not_1in of not_1in is
        begin
            v <= not x;
    end dfm_not_1in;
```

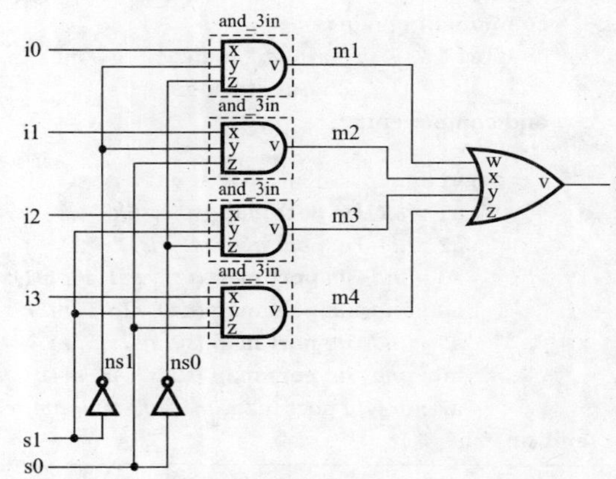

Fig. 1: Logic circuit of 4:1 multiplexer.

```
--code for 3-input AND gate
library ieee;
use ieee.std_logic_1164.all;
    entity and_3in is
        port ( x, y, z : in std_logic;
               v       : out std_logic );
    end and_3in;

    architecture dfm_and_3in of and_3in is
        begin
            v <= x and y and z;
    end dfm_and_3in;
```

```
--code for 4-input OR gate
library ieee;
use ieee.std_logic_1164.all;
    entity or_4in is
        port ( w, x, y, z : in std_logic;
               v          : out std_logic );
    end or_4in;

    architecture dfm_or_4in of or_4in is
        begin
            v <= w or x or y or z;
    end dfm_or_4in;
```

```
--code for 4*1 multiplexer
library ieee;
use ieee.std_logic_1164.all;
    entity mux_4_1 is
        port ( s0, s1         : in std_logic;
               i0, i1, i2, i3 : in std_logic;
               y              : out std_logic );
    end mux_4_1;

    architecture sm_mux_4_1 of mux_4_1 is
    signal ns0, ns1          : std_logic;
    signal m1, m2, m3, m4 : std_logic;
        component not_1in is
            port ( x : in std_logic;
                   v : out std_logic );
        end component;

        component and_3in is
            port ( x, y, z : in std_logic;
                   v       : out std_logic );
        end component;
```

```
        component or_4in is
            port ( w, x, y, z : in std_logic;
                   v          : out std_logic );
        end component;

            begin
                a1 : not_1in port map (s0, ns0);
                a2 : not_1in port map (s1, ns1);
                a3 : and_3in port map (ns0, ns1, i0, m1);
                a4 : and_3in port map (ns0, s1, i1, m2);
                a5 : and_3in port map (s0, ns1, i2, m3);
                a6 : and_3in port map (s0, s1, i3, m4);
                a7 : or_4in port map (m1, m2, m3, m4, y);
    end sm_mux_4_1;
```

VHDL Program 5.11

(AU, Apr/May'19, 6 Marks)

Write a VHDL code for behavioral model of a 8*1 multiplexer.

Program Logic for Behavioral Model

The Boolean equations for 8*1 multiplexer are,

$$Y = I0\,S2'S1'S0' + I0\,S2'S1'S0 + I0\,S2'S1\,S0' + I0\,S2'S1\,S0$$
$$+ I0\,S2\,S1'S0' + I0\,S2\,S1'S0 + I0\,S2\,S1\,S0' + I0\,S2\,S1\,S0$$

Where, S0, S1 and S2 are 1-bit selection input and Y is 1-bit output data. The VHDL code for behavioral model of 8*1 multiplexer is developed using the above equations. (Refer Chapter 2, Section 2.3 for design of 8*1 multiplexer).

VHDL Program for Behavioral of 8*1 Multiplexer

```
library ieee;
use ieee.std_logic_1164.all;
    entity mux_8_1 is
        port ( s                               : in std_logic_vector (2 downto 1);
               i0, i1, i2, i3, i4, i5, i6, i7 : in std_logic;
               y                               : out std_logic );
    end mux_8_1;

    architecture bm_mux_8_1 of mux_8_1 is
        begin
            process (s, i0, i1, i2, i3, i4, i5, i6, i7)
                begin
                    if s = "000" then
                        y <= i0;
                    elsif s = "001" then
                        y <= i1;
                    elsif s = "010" then
                        y <= i2;
                    elsif s = "011" then
                        y <= i3;
                    elsif s = "100" then
                        y <= i4;
                    elsif s = "101" then
                        y <= i5;
```

```
            elsif s = "110" then
                y <= i6;
            else
                y <= i7;
            end if;
        end process;
    end bm_mux_8_1;
```

VHDL Program 5.12

Write a VHDL code for dataflow model of a 1*4 demultiplexer.

Program Logic for Dataflow Model

The Boolean equations for 1*4 demultiplexer are,

Y0 = Ii S1' S0'

Y1 = Ii S1' S0

Y2 = Ii S1 S0'

Y3 = Ii S1 S0

Where, S1 and S0 are 1-bit selection input and Y0, Y1, Y2 and Y3 are 1-bit output data. The VHDL code for dataflow model of 1*4 demultiplexer is developed using the above equations. (Refer Chapter 2, Section 2.4 for design of 1*4 demultiplexer).

VHDL Program for Dataflow Model of 1*4 Demultiplexer

```
library ieee;
use ieee.std_logic_1164.all;
    entity demux_1_4 is
        port ( ii, s0, s1          : in std_logic;
                y0, y1, y2, y3 : out std_logic );
    end demux_1_4;

    architecture dfm_demux_1_4 of demux_1_4 is
    signal m1, m2 : std_logic;
        begin
            m1 <= not s0;
            m2 <= not s1;
            y0  <= ii and (m1 and m2);
            y1  <= ii and (m1 and s1);
            y2  <= ii and (s0 and m2);
            y3  <= ii and (s0 and s1);
    end dfm_demux_1_4;
```

VHDL Program 5.13

Write a VHDL code for structural model of a 1*4 demultiplexer.

Program Logic for Structural Model

The VHDL code for structural model of 1*4 demultiplexer is developed using the logic circuit shown in Fig. 1. In structural model, each component involved in the demultiplexer are coded separately and then the components are added in the code for demultiplexer.

VHDL Program for Structural Model of 1*4 Demultiplexer

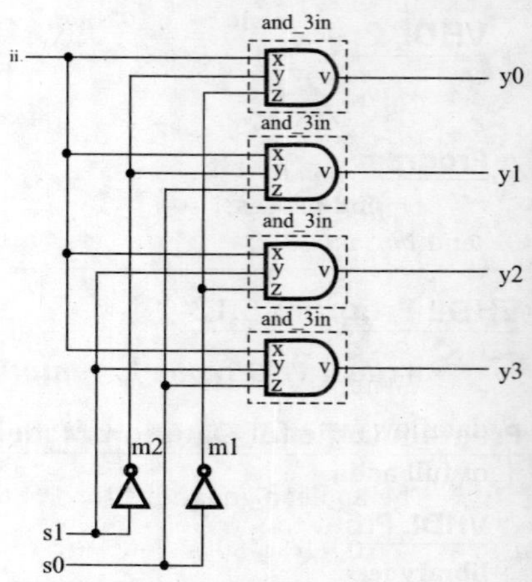

Fig. 1: Logic circuit of 1:4 demux.

```
--code for 1-input NOT gate
library ieee;
use ieee.std_logic_1164.all;
  entity not_1in is
    port ( x : in std_logic;
           v : out std_logic );
  end not_1in;

  architecture dfm_not_1in of not_1in is
    begin
      v <= not x;
  end dfm_not_1in;
```

```
--code for 3-input AND gate
library ieee;
use ieee.std_logic_1164.all;
  entity and_3in is
    port ( x, y, z : in std_logic;
           v       : out std_logic );
  end and_3in;

  architecture dfm_and_3in of and_3in is
    begin
      v <= x and y and z;
  end dfm_and_3in;
```

```
--code for 1*4 demultiplexer
library ieee;
use ieee.std_logic_1164.all;
  entity demux_1_4 is
    port ( ii, s1, s0        : in std_logic;
           y0, y1, y2, y3 : out std_logic );
  end demux_1_4;

  architecture sm_demux_1_4 of demux_1_4 is
  signal m1, m2 : std_logic;
    component not_1in is
      port ( x : in std_logic;
             v : out std_logic );
    end component;

    component and_3in is
      port ( x, y, z : in std_logic;
             v       : out std_logic );
    end component;

      begin
        a1 : not_1in port map (s0, m1);
        a2 : not_1in port map (s1, m2);
        a3 : and_3in port map (ii, m1, m2, y0);
        a4 : and_3in port map (ii, m2, s0, y1);
        a5 : and_3in port map (ii, s1, m1, y2);
        a6 : and_3in port map (ii, s1, s0, y3);
  end sm_demux_1_4;
```

VHDL Program 5.14

(AU, Apr/May'19, 7 Marks)

Write VHDL code for dataflow and structural models of a full adder.

Program Logic for Dataflow Model

The Boolean equations for full adder are,

$$s = a \oplus b \oplus c_i$$

$$c_o = (a\,b) + (b\,c_i) + (c_i\,a)$$

Where, a, b and ci are 1-bit input data, sum (s) and carry (co) are 1-bit output data. The VHDL code for dataflow model of full adder is developed using the above equations. (Refer Chapter 2, Section 2.6.2 for design of full adder).

VHDL Program for Dataflow Model of Full Adder

```
library ieee;
use ieee.std_logic_1164.all;
    entity full_adder is
        port ( a, b, ci  : in std_logic;
               s, co     : out std_logic );
    end full_adder;

    architecture dfm_fa of full_adder is
    signal m1, m2, m3 : std_logic;
        begin
            s    <= a xor b xor ci;
            m1 <= a and b;
            m2 <= b and ci;
            m3 <= ci and a;
            co   <= m1 or m2 or m3;
    end dfm_fa;
```

Program Logic for Structural Model

The VHDL code for structural model of full adder is developed using the logic circuit shown in Fig. 1. In structural model, each component involved in full adder are coded separately and then the components are added in the code for full adder.

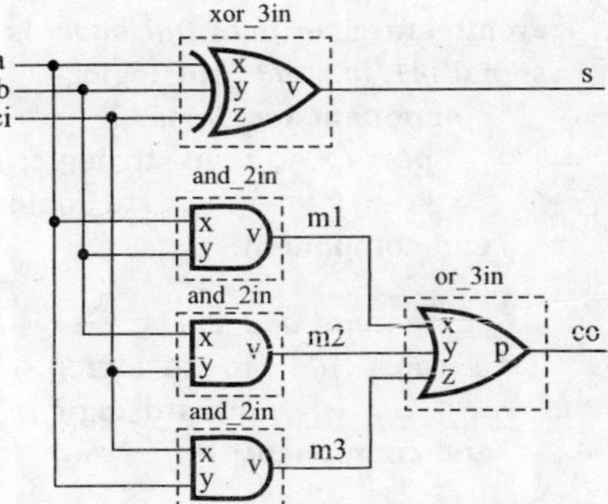

Fig. 1: Logic circuit of full adder.

VHDL Program for Structural Model of Full Adder

```
--code for 3-input XOR gate
library ieee;
use ieee.std_logic_1164.all;
    entity xor_3in is
        port ( x, y, z : in std_logic;
               v         : out std_logic );
    end xor_3in;

    architecture dfm_xor_3in of xor_3in is
        begin
            v <= x xor y xor z;
    end dfm_xor_3in;
```

```
--code for 2-input AND gate
library ieee;
use ieee.std_logic_1164.all;
    entity and_2in is
        port ( x, y : in std_logic;
               v     : out std_logic );
    end and_2in;

    architecture dfm_and_2in of and_2in is
        begin
            v <= x and y;
    end dfm_and_2in;
```

```
--code for 3-input OR gate
library ieee;
use ieee.std_logic_1164.all;
    entity or_3in is
        port ( x, y, z : in std_logic;
               v        : out std_logic );
    end or_3in;

    architecture dfm_or_3in of or_3in is
        begin
            v <= x or y or z;
    end dfm_or_3in;
```

```
--code for full adder
library ieee;
use ieee.std_logic_1164.all;
    entity full_adder is
        port ( a, b, ci  : in std_logic;
               s, co     : out std_logic );
    end full_adder;

    architecture sm_fa of full_adder is
    signal m1, m2, m3 : std_logic;
        component xor_3in is
            port ( x, y, z : in std_logic;
                   v        : out std_logic );
        end component;

        component and_2in is
            port ( x, y : in std_logic;
                   v     : out std_logic );
        end component;

        component or_3in is
            port ( x, y, z : in std_logic;
                   v        : out std_logic );
        end component;
            begin
                a1 : xor_3in port map (a, b, ci, s);
                a2 : and_2in port map (a, b, m1);
                a3 : and_2in port map (b, ci, m2);
                a4 : and_2in port map (ci, a, m3);
                a5 : or_3in port map (m1, m2, m3, co);
    end sm_fa;
```

VHDL Program 5.15

Write VHDL code for dataflow and structural models of a half subtractor.

Program Logic for Dataflow Model

The Boolean equations for half subtractor are,

$$d = a \oplus b = ab' + a'b$$

$$c_o = a'b$$

Where, a and b are 1-bit input data, difference (d) and borrow (co) are 1-bit output data. The VHDL code for dataflow model of half subtractor is developed using the above equations. (Refer Chapter 2, Section 2.6.3 for design of half subtractor).

VHDL Program for Dataflow Model of Half Subtractor

```
library ieee;
use ieee.std_logic_1164.all;
  entity half_subtractor is
     port ( a, b  : in std_logic;
          d, co : out std_logic );
  end half_subtractor;

  architecture dfm_hs of half_subtractor is
  signal m1 : std_logic;
     begin
        d   <= a xor b;
        m1  <= not a;
        co  <= m1 and b;
  end dfm_hs;
```

Program Logic for Structural Model

The VHDL code for structural model of half subtractor is developed using the logic circuit shown in Fig. 1. In structural model, each component involved in half subtractor are coded separately and then the components are added in the code for half subtractor.

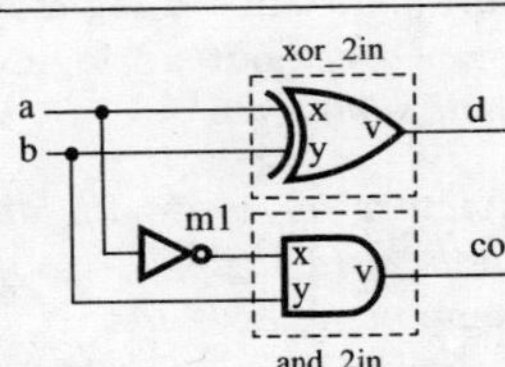

Fig. 1: Logic circuit of half subtractor.

VHDL Program for Structural Model of Half Subtractor

```
--code for 2-input XOR gate
library ieee;
use ieee.std_logic_1164.all;
  entity xor_2in is
     port ( x, y : in std_logic;
          v    : out std_logic );
  end xor_2in;

  architecture dfm_xor_2in of xor_2in is
     begin
        v <= x xor y;
  end dfm_xor_2in;
```

```
--code for 2-input AND gate
library ieee;
use ieee.std_logic_1164.all;
  entity and_2in is
      port ( x, y : in std_logic;
             v    : out std_logic );
  end and_2in;

  architecture dfm_and_2in of and_2in is
      begin
         v <= x and y;
  end dfm_and_2in;
```

```
--code for 1-input NOT gate
library ieee;
use ieee.std_logic_1164.all;
  entity not_1in is
      port ( x : in std_logic;
             v : out std_logic );
  end not_1in;

  architecture dfm_not_1in of not_1in is
      begin
         v <= not x;
  end dfm_not_1in;
```

```
--code for half subtractor
library ieee;
use ieee.std_logic_1164.all;
  entity half_subtractor is
      port ( a, b  : in std_logic;
             d, co : out std_logic );
  end half_subtractor;

  architecture sm_hs of half_subtractor is
  signal m1 : std_logic;
      component xor_2in is
         port ( x, y : in std_logic;
                v    : out std_logic );
      end component;

      component and_2in is
         port ( x, y : in std_logic;
                v    : out std_logic );
      end component;

      component not_1in is
         port ( x : in std_logic;
                v : out std_logic );
      end component;

         begin
            a1 : xor_2in port map (a, b, d);
            a2 : not_1in port map (a, m1);
            a3 : and_2in port map (m1, b, co);
  end sm_hs;
```

VHDL Program 5.16

Write VHDL code for dataflow and structural models of a full subtractor.

Program Logic Dataflow Model

The Boolean equations for full subtractor are,

$$d = a'b'c_i \oplus a'bc_i' \oplus a'bc_i' \oplus abc_i$$

$$c_o = (a'b) + (b\,c_i) + (c_i\,a')$$

Where, a, b and ci are 1-bit input data, difference (d) and borrow (co) are 1-bit output data. The VHDL code for dataflow model of full subtractor is developed using the above equations. (Refer Chapter 2, Section 2.6.4 for design of full subtractor).

VHDL Program for Dataflow Model of Full Subtractor

```
library ieee;
use ieee.std_logic_1164.all;
   entity full_subtractor is
      port ( a, b, ci  : in std_logic;
             d, co     : out std_logic );
   end full_subtractor;

   architecture dfm_fs of full_subtractor is
   signal m1 : std_logic;

      begin
         d    <= a xor b xor ci;
         m1 <= not a;
         co   <= (m1 and b) or (b and ci) or (ci and m1);
   end dfm_fs;
```

Program Logic for Structural Model

The VHDL code for structural model of full subtractor is developed using the logic circuit shown in Fig. 1. In structural model, each component involved in full subtractor are coded separately and then the components are added in the code for full subtractor.

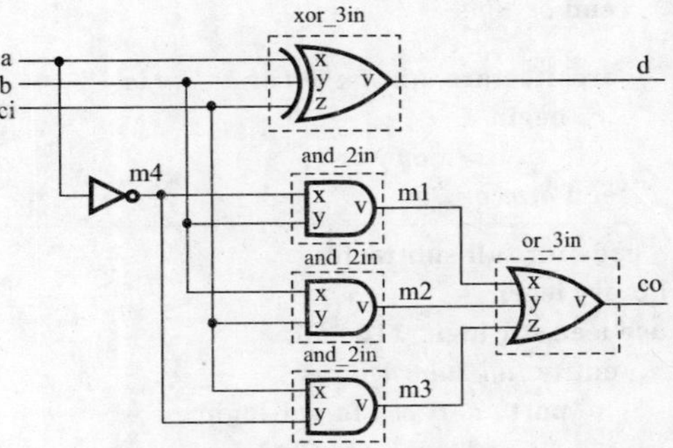

Fig. 1: Logic circuit of full subtractor.

VHDL Program for Structural Model of Full Subtractor

```
--code for 3-input XOR gate
library ieee;
use ieee.std_logic_1164.all;
   entity xor_3in is
      port ( x, y, z : in std_logic;
             v       : out std_logic );
   end xor_3in;

   architecture dfm_xor_3in of xor_3in is
      begin
         v <= x xor y xor z;
   end dfm_xor_3in;
```

```
--code for 1-input NOT gate
library ieee;
use ieee.std_logic_1164.all;
   entity not_1in is
      port ( x : in std_logic;
             v : out std_logic );
   end not_1in;

   architecture dfm_not_1in of not_1in is
      begin
         v <= not x;
   end dfm_not_1in;
```

```
--code for 2-input AND gate
library ieee;
use ieee.std_logic_1164.all;
   entity and_2in is
      port ( x, y : in std_logic;
             v    : out std_logic );
   end and_2in;

   architecture dfm_and_2in of and_2in is
      begin
         v <= x and y;
   end dfm_and_2in;
```

```
-- code for 3-input OR gate
library ieee;
use ieee.std_logic_1164.all;
   entity or_3in is
      port ( x, y, z : in std_logic;
             v       : out std_logic );
   end or_3in;

   architecture dfm_or_3in of or_3in is
      begin
         v <= x or y or z;
   end dfm_or_3in;
```

```
-- code for full subtractor
library ieee;
use ieee.std_logic_1164.all;
   entity full_subtractor is
      port ( a, b, ci  : in std_logic;
             d, co     : out std_logic );
   end full_subtractor;

   architecture sm_fs of full_subtractor is
   signal m1, m2, m3, m4 : std_logic;
      component xor_3in is
         port ( x, y, z : in std_logic;
                v       : out std_logic );
      end component;
```

```
component not_1in is
    port ( x : in std_logic;
           v : out std_logic );
end component;

component and_2in is
    port ( x, y : in std_logic;
           v    : out std_logic );
end component;

component or_3in is
    port ( x, y, z : in std_logic;
           v       : out std_logic );
end component;
    begin
        a1 : xor_3in port map (a, b, ci, d);
        a2 : not_1in port map (a, m4);
        a3 : and_2in port map (b, m4, m1);
        a4 : and_2in port map (b, ci, m2);
        a5 : and_2in port map (ci, m4, m3);
        a6 : or_3in port map (m1, m2, m3, co);
end sm_fs;
```

VHDL Program 5.17

Write VHDL code for structural model of a 4-bit parallel adder.

Program Logic for Structural Model

The Boolean equations for 4-bit parallel adder performs arithmetic sum of two 4-bit binary numbers. A full adder can add two 1-bit binary along with previous carry. In order to add two 4-bit binary numbers, 4 full adders are required. Each ful adder will add one bit of binary numbers. Consider two 4-bit binary numbers a and b shown below:

Binary number 1 : a3, a2, a1, a0 (data-a) ; Binary number 2 : b3, b2, b1, b0 (data-b)

Now, the logic circuit of 4-bit parallel adder will be as shown in Fig. 1. (Refer Chapter 2, Section-2.6.5 for design of 4-bit parallel adder).

The VHDL code for structural model of 4-bit parallel adder is developed using the logic circuit shown in Fig. 1. In structural model, each component involved in 4-bit adder are coded separately and then the components are added in the code for 4-bit adder.

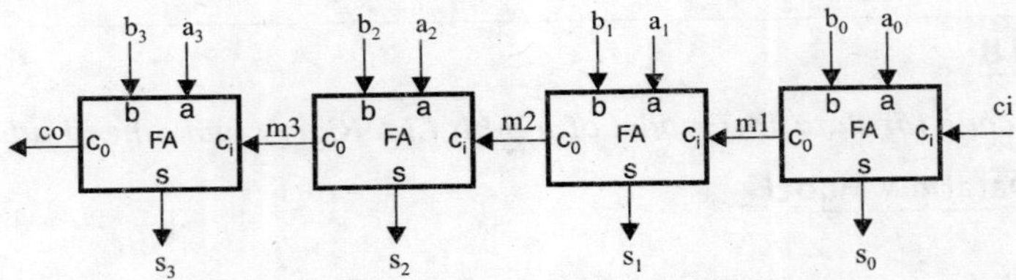

Fig. 1: *4-bit binary adder using full adders.*

VHDL Program for Structural Model of 4-bit Parallel Adder

```
--code for full adder
library ieee;
use ieee.std_logic_1164.all;
   entity full_adder is
       port ( a, b, ci  : in std_logic;
              s, co     : out std_logic );
   end full_adder;

   architecture dfm_fa of full_adder is
   signal m1, m2, m3 : std_logic;
       begin
           s    <= a xor b xor ci;
           m1  <= a and b;
           m2  <= b and ci;
           m3  <= ci and a;
           co   <= m1 or m2 or m3;
   end dfm_fa;
```

```
--code for 4-bit parallel adder
library ieee;
use ieee.std_logic_1164.all;
   entity parallel_adder is
       port ( a, b : in std_logic_vector (3 down to 0);
              ci   : in std_logic;
              s    : out std_logic_vector (3 down to 0);
              co   : out std_logic );
   end parallel_adder;

   architecture sm_parallel_adder of parallel_adder is
   signal m1, m2, m3 : std_logic;
       component full_adder is
           port ( a, b, ci : in std_logic;
                  s, co    : out std_logic );
       end component;

       begin
           fa0 : full_adder port map (a(0), b(0), ci, s(0), m1);
           fa1 : full_adder port map (a(1), b(1), s(1), m1, m2);
           fa2 : full_adder port map (a(2), b(2), s(2), m2, m3);
           fa3 : full_adder port map (a(3), b(3), s(3), m3, co);
   end sm_parallel_adder;
```

VHDL Program 5.18

Write a VHDL code for dataflow model of a 4-bit CLA (Carry look ahead adder).

Program Logic for Dataflow Model

$$s_i = p_i \oplus c_i$$

$$c_{i+1} = g_i + p_i c_i$$

$$p_i = a_i \oplus b_i$$

$$g_i = a_i b_i$$

Where, p_i is carry propagate from i_{th} stage and g_i is carry generate of i_{th} stage. The VHDL code for dataflow model of 4-bit CLA is developed using the above equations. (Refer Chapter 2, Section 2.6.6 for design of 4-bit CLA).

VHDL Program for Dataflow Model of 4-bit CLA

```
library ieee;
use ieee.std_logic_1164.all;
   entity cla_4bit is
      port ( a3, a2, a1, a0     : in std_logic;
             b3, b2, b1, b0     : in std_logic;
             c0                 : in std_logic;
             c4, s3, s2, s1, s0 : out std_logic);
   end cla_4bit;

   architecture dfm_cla_4bit of cla_4bit is
   signal m1, m2, m3    : std_logic;
   signal p0, p1, p2, p3 : std_logic;
   signal g0, g1, g2, g3 : std_logic;

      begin
         p0  <= a0 xor b0;
         p1  <= a1 xor b1;
         p2  <= a2 xor b2;
         p3  <= a3 xor b3;

         g0  <= a0 and b0;
         g1  <= a1 and b1;
         g2  <= a2 and b2;
         g3  <= a3 and b3;

         s0  <= p0 xor c0;
         m1  <= g0 or (p0 and c0);
         s1  <= p1 xor m1;
         m2  <= g1 or (p1 and g0) or (p1 and p0 and c0);
         s2  <= p2 xor m2;
         m3  <= g2 or (p2 and g1) or (p2 and p1 and g0) or (p2 and p1 and p0 and g0) ;
         s3  <= p3 xor m3;
         c4  <= g3 or (p3 and g2) or (p3 and p2 and g1) or (p3 and p2 and p1 and g0) or (p3 and p2 and
                  p1 and p0 and c0) ;
   end dfm_cla_4bit;
```

VHDL Program 5.19

Write a VHDL code for dataflow and structural models of a 4*2 encoder.

Program Logic for Dataflow Model

The Boolean equations for 4*2 encoder are,

$Y1 = I2 + I3$

$Y0 = I1 + I3$

Where, I0, I1, I2 and I3 are 1-bit input data, Y0 and Y1 are 1-bit output data. The VHDL code for dataflow model of 4*2 encoder is developed using the above equations. (Refer Chapter 2, Section 2.7.1 for design of 4*2 encoder).

VHDL Program for Dataflow Model of 4*2 Encoder

```
library ieee;
use ieee.std_logic_1164.all;
    entity encoder_4_2 is
        port ( i0, i1, i2, i3 : in std_logic;
               y0, y1          : out std_logic);
    end encoder_4_2;

    architecture dfm_encoder_4_2 of encoder_4_2 is
        begin
            y1 <= i2 or i3;
            y0 <= i1 or i3;
    end dfm_encoder_4_2;
```

Program Logic for Structural Model

The VHDL code for structural model of 4*2 encoder is developed using the logic circuit shown in Fig. 1. In structural model, each component involved in the 4*2 encoder are coded separately and then the components are added in the code for 4*2 encoder.

Fig. 1: 4-to-2 encoder.

Program for Structural Model of 4*2 Encoder

```
--code for 2-input OR gate
library ieee;
use ieee.std_logic_1164.all;
    entity or_2in is
        port ( x, y : in std_logic;
               v    : out std_logic );
    end or_2in;

    architecture dfm_or_2in of or_2in is
        begin
            v <= x or y;
    end dfm_or_2in;
```

```
--code for 4*2 encoder
library ieee;
use ieee.std_logic_1164.all;
    entity encoder_4_2 is
        port ( i0, i1, i2, i3 : in std_logic;
               y0, y1          : out std_logic );
    end encoder_4_2;

    architecture sm_encoder_4_2 of encoder_4_2 is
        component or_2in is
            port ( x, y : in std_logic;
                   v    : out std_logic );
        end component;

        begin
            a1 : or_2in port map (i2, i3, y1);
            a2 : or_2in port map (i1, i3, y0);
    end sm_encoder_4_2;
```

VHDL Program 5.20

Write a VHDL code for dataflow model of priority encoder.

Program Logic for Dataflow Model

The Boolean equations for priority encoder are,

$$Y1 = I2 + I3 \quad ; \quad Y0 = I1\, I2' + I3 \quad ; \quad v = I0 + I1 + I2 + I3$$

Where, I0, I1, I2 and I3 are 1-bit input data and Y0, Y1 and v are 1-bit output data. The VHDL code for dataflow model of priority encoder is developed using the above equations. (Refer Chapter 2, Section 2.8.2 for design of prioroty encoder).

VHDL Program for Dataflow Model of Priority Encoder

```
library ieee;
use ieee.std_logic_1164.all;
   entity priority_encoder is
      port ( i0, i1, i2, i3 : in std_logic;
            y0, y1, v      : out std_logic );
   end priority_encoder;

   architecture dfm_priority_encoder of priority_encoder is
   signal m1 : std_logic;

      begin
         m1 <= not i2;
         y0  <= i2 or i3;
         y1  <= (i1 and m1) or i3;
         v   <= i0 or i1 or i2 or i3;
   end dfm_priority_encoder;
```

VHDL Program 5.21

Write VHDL code for dataflow, behavioral and structural models of a 2*4 decoder.

Program Logic for Dataflow and Behavioral Model

The Boolean equations for 2*4 decoder are,

$$Y0 = I1'\, I0'$$

$$Y1 = I1'\, I0$$

$$Y2 = I1\, I0'$$

$$Y3 = I1\, I0$$

Where, I1 and I0 are 1-bit input data and Y0, Y1, Y2 and Y3 are 1-bit output data. The VHDL codes for dataflow and behavioral model of 2*4 decoder are developed using the above equations. (Refer Chapter 2, Section 2.9.1 for design of 2*4 decoder).

VHDL Program for Dataflow Model of 2*4 Decoder

```
library ieee;
use ieee.std_logic_1164.all;
   entity decoder_2_4 is
      port ( i1, i0           : in std_logic;
            y0, y1, y2, y3 : out std_logic );
   end decoder_2_4;
```

```
architecture dfm_decoder_2_4 of decoder_2_4 is
signal m1, m2 : std_logic;
    begin
        m1 <= not i1;
        m2 <= not i0;
        y0  <= m1 and m2;
        y1  <= m1 and i0;
        y2  <= i1 and m2;
        y3  <= i1 and i0;
end dfm_ decoder_2_4;
```

VHDL Program for Behavioral Model of 2*4 Decoder using "case" Statement

```
library ieee;
use ieee.std_logic_1164.all;
    entity decoder_2_4 is
    port ( i                : in std_logic_vector (1 downto 0);
        y0, y1, y2, y3 : out std_logic );
    end decoder_2_4;

    architecture bm_decoder_2_4 of decoder_2_4 is
    signal m1, m2 : std_logic;
        begin
            process (i)
                begin
                    m1 <= not i(0);
                    m2 <= not i(1);
                case i is
                    when "00"     => y0 <= m1 and m2;
                    when "01"     => y1 <= m1 and i(1);
                    when "10"     => y2 <= i(0) and m2;
                    when others => y3 <= i(0) and i(1);
                end case;
            end process;
    end bm_decoder_2_4;
```

VHDL Program for Behavioral Model of 2*4 Decoder using "if elsif" Statement

```
library ieee;
use ieee.std_logic_1164.all;
    entity decoder_2_4 is
        port ( i                  : in std_logic_vector (1 downto 0);
            y0, y1, y2, y3  : out std_logic );
    end decoder_2_4;

    architecture bm_decoder_2_4 of decoder_2_4 is
    signal m1, m2 : std_logic;
        begin
            process (i)
                begin
                    m1<= not i(0);
                    m2<= not i(1);
                if i = "00" then
                    y0 <= m1 and m2;
```

```
            elsif i = "01" then
                y1 <= m1 and i(1);
            elsif i = "10" then
                y2 <= i(0) and m2;
            else
                y3 <= i(0) and i(1);
            end if;
        end process;
    end bm_decoder_2_4;
```

Program Logic for Structural Model

The VHDL code for structural model of 2*4 decoder is developed using the logic circuit shown in Fig. 1. In structural model, each component involved in the 2*4 decoder are coded separately and then the components are added in the code for 2*4 decoder.

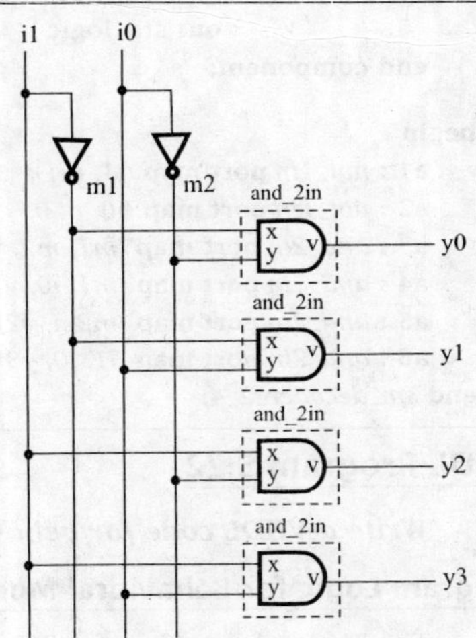

***Fig. 1:** Logic circuit for 2-to-4 decoder.*

Program for Structural Model of 2*4 Decoder

```
--code for 1-input NOT gate
library ieee;
use ieee.std_logic_1164.all;
    entity not_1in is
        port ( x : in std_logic;
               v : out std_logic );
    end not_1in;

    architecture dfm_not_1in of not_1in is
        begin
            v <= not x;
    end dfm_not_1in;
```

```
--code for 2-input AND gate
library ieee;
use ieee.std_logic_1164.all;
    entity and_2in is
        port ( x, y : in std_logic;
               v    : out std_logic );
    end and_2in;

    architecture dfm_and_2in of and_2in is
        begin
            v <= x and y;
    end dfm_and_2in;
```

```
--code for 2*4 decoder
library ieee;
use ieee.std_logic_1164.all;
    entity decoder_2_4 is
        port ( i1, i0           : in std_logic;
               y0, y1, y2, y3 : out std_logic );
    end decoder_2_4;
```

```
architecture sm_decoder_2_4 of decoder_2_4 is
signal m1, m2 : std_logic;

    component not_1in is
        port ( x : in std_logic;
               y : out std_logic );
    end component;

    component and_2in is
        port ( x, y : in std_logic;
               v    : out std_logic );
    end component;

begin
    a1 : not_1in port map (i1, m1);
    a2 : not_1in port map (i0, m2);
    a3 : and_2in port map (m1, m2, y0);
    a4 : and_2in port map (m1, i0, y1);
    a5 : and_2in port map (m2, i, y2);
    a6 : and_2in port map (i1, i0, y3);
end sm_decoder_2_4;
```

VHDL Program 5.22

Write a VHDL code for behavioral model of a 3*8 decoder.

Program Logic for Behavioral Model

The Boolean equations for 3*8 decoder are,

Y0 = I2' I1' I0' ; Y1 = I2' I1' I0 ; Y2 = I2' I1 I0 ; Y3 = I2 I1 I0

Y4 = I2 I1' I0' ; Y5 = I2 I1' I0 ; Y6 = I2 I1 I0' ; Y7 = I2 I1 I0

Where, S0, S1 and S2 are 1-bit selection input and Y is 1-bit output data. The VHDL code for behavioral model of 3*8 decoder are developed using the above equations. (Refer Chapter 2, Section 2.9.7 for design of 3*8 decoder).

VHDL Program for Behavioral Model of 3*8 Decoder

```
library ieee;
use ieee.std_logic_1164.all;
use ieee.std_logic_arith.all;
use ieee.std_logic_unsigned.all;
    entity decoder3_8 is
        port ( s : in bit_vector   (2 downto 0);
               y : out bit_vector (7 downto 0) );
    end decoder3_8;

    architecture bm_decoder3_8 of decoder3_8 is
        begin
            process (s)
                begin
                    if s ="000" then
                        y <= "10000000";
                    elsif s = "001" then
                        y <= "01000000";
```

```
            elsif s = "010" then
                  y <= "00100000";
            elsif s = "011" then
                  y<="00010000";
            elsif s = "100" then
                  y <= "00001000";
            elsif s = "101" then
                  y <= "00000100";
            elsif s = "110" then
                  y <= "00000010";
            else
                  y <= "00000001";
            end if;
      end process;
  end bm_decoder3_8;
```

VHDL Program 5.23

Write a VHDL code for dataflow, behavioral and structural model of a SR flip-flop.

Program Logic for Dataflow and Behavioral Model

The Boolean equations for SR flip-flop are,

$Q = S + R'.Q$

$Qbar = Q'$

Where, S and R are 1-bit input data, Q and Qbar are 1-bit output data. The VHDL code for dataflow model of SR flip-flop is developed using the above equations. The behavioral model of SR flip-flop is developed using the truth table of SR flip-flop. (Refer Chapter 3, Section 3.3.1 for design of SR flip-flop).

Truth Table of SR Flip-Flop

Inputs		Outputs	
S	R	Q	Q'
0	0	No change	
0	1	0	1
1	0	1	0
1	1	Indeterminate	

VHDL Program for Dataflow Model of SR Flip-Flop

```
library ieee;
use ieee.std_logic_1164.all;
  entity sr_ff is
      port ( s, r       : in std_logic;
            q, qbar : inout std_logic);
  end sr_ff;

  architecture dfm_sr_ff of sr_ff is
  signal m1 : std_logic;

      begin
            q       <= s or (not r and m1);
            qbar <= not q;
  end dfm_sr;
```

VHDL Program for Behavioral Model of SR Flip-Flop using 'if' Statement

```
library ieee;
use ieee.std_logic_1164.all;
  entity sr_ff is
      port ( s, r, clk : in std_logic;
            q, qbar : inout std_logic);
  end sr_ff;
```

```
architecture bm_sr_ff of sr_ff is
signal q_temp, qbar_temp : std_logic;

    begin
        process (s, r, clk) begin
            if (clk' event and clk = '1') then
            if (s = '0' and r = '0' ) then
                q_temp      <= q_temp;
                qbar_temp <= qbar_temp;
            elsif (s = '0' and r = '1' ) then
                q_temp      <= '0';
                qbar_temp <= '1';
            elsif (s = '1' and r = '0' ) then
                q_temp      <= '1';
                qbar_temp <= '0';
            elsif (s = '1' and r = '1') then
                q_temp      <='z';
                qbar_temp <= 'z';
            end if;
            end if;
        end process;
                q       <= q_temp;
                qbar <= qbar_temp;
    end bm_sr_ff;
```

Program Logic for Structural Model

The VHDL code for structural model of SR flip-flop is developed using the logic circuit shown in Fig. 1. In structural model, each component involved in the SR flip-flop are coded separately and then the components are added in the code for SR flip-flop.

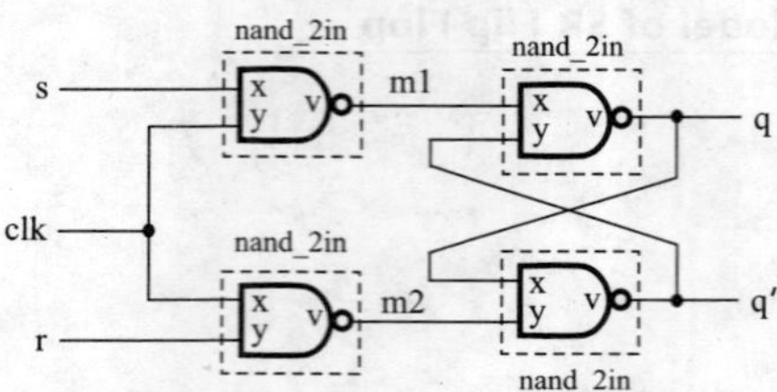

***Fig. 1:** NAND latch based SR flip-flop.*

Program for Structural Model of SR Flip-Flop

```
--code for 2-input NAND gate
library ieee;
use ieee.std_logic_1164.all;
    entity nand_2in is
        port ( x, y : in std_logic;
                 v    : out std_logic);
    end nand_2in;

    architecture dfm_nand_2in of nand_2in is
        begin
            v <= x nand y;
    end dfm_nand_2_in;
```

```
--code for SR flip-flop
library ieee;
use ieee.std_logic_1164.all;
    entity rs_ff is
        port ( r, s, clk : in std_logic;
               q, qbar : inout std_logic );
    end rs_ff;

    architecture sm_ rs_ff of rs_ff is
    signal m1, m2 : std_logic;

        component nand_2in is
            port ( x, y : in std_logic;
                   v    : out std_logic );
        end component;

            begin
                a1 : nand_2in port map (s, clk, m1);
                a2 : nand_2in port map (r, clk, m2);
                a3 : nand_2in port map (m1, qbar, q);
                a4 : nand_2in port map (m2, q, qbar);
    end sm_ rs_ff;
```

VHDL Program 5.24

Write a VHDL code for dataflow, behavioral and structural model of a JK flip-flop.

Program Logic for Dataflow and Behavioral Model

The Boolean equations for JK flip-flop are,

$Q = JQ' + K'Q$

$Qbar = Q'$

Where, J and K are 1-bit input data, Q and Qbar are 1-bit output data. The VHDL code for dataflow model of JK flip-flop is developed using the above equations. The behavioral model of JK flip-flop is developed using the truth table of JK flip-flop. (Refer Chapter 3, Section 3.3.2 for design of JK flip-flop).

Truth Table of JK Flip-Flop

Inputs		Outputs	
J	K	Q	Q'
0	0	No change	
0	1	0	1
1	0	1	0
1	1	Toggle	

VHDL Program for Dataflow Model of JK Flip-Flop

```
library ieee;
use ieee.std_logic_1164.all;
    entity jk_ff is
        port ( j, k, q   : in std_logic;
               q, qbar   : out std_logic );
    end jk_ff;

    architecture dfm_jk_ff of jk_ff is
        begin
            q      <= (j and (not q)) or ((not k) and q);
            qbar <= not q;
    end dfm_jk_ff;
```

VHDL Program for Behavioral Model of JK Flip-Flop

```
library ieee;
use ieee.std_logic_1164.all;
  entity jk_ff is
     port ( j, k        : in std_logic;
            clk, reset : in std_logic;
            q, qbar    : out std_logic);
  end jk_ff;

  architecture bm_jk_ff of jk_ff is
     begin
        qbar <= not q;
     process (clk, reset)
        begin
           if (reset = '1') then
              q     <= 0;
              qbar <= 1;
           elsif (clk' event and clk = '1') then
           if (j = '0' and k = '0') then
              q     <= q;
              qbar <= qbar;
           if (j = '0' and k = '1') then
              q     <= 0;
              qbar <= 1;
           if (j = '1' and k = '0') then
              q     <= 1;
              qbar <= 0;
           else
              q     <= qbar;
              qbar <= not q;
           end if;
           end if;
           end if;
           end if;
     end process;
  end bm_jk_ff;
```

Program Logic for Structural Model

The VHDL code for structural model of JK flip-flop is developed using the logic circuit shown in Fig. 1. In structural model, each component involved in the JK flip-flop are coded separately and then the components are added in the code for JK flip-flop.

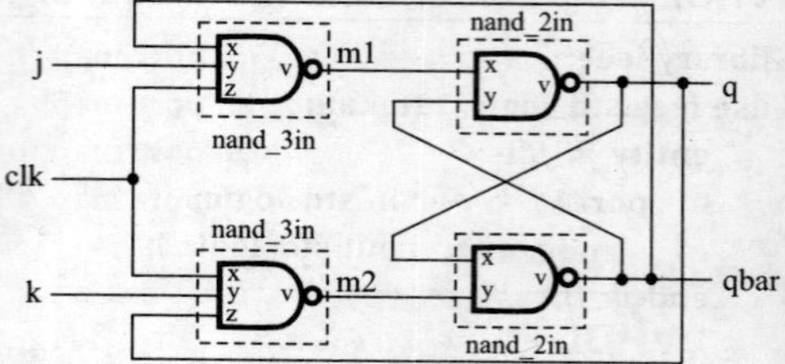

Fig. 1: JK flip-flop.

VHDL Program for Structural Model of JK Flip-Flop

```
--code for 3-input NAND gate
library ieee;
use ieee.std_logic_1164.all;
  entity nand_3in is
     port ( x, y, z : in std_logic;
            v         : out std_logic );
  end nand_3in;
```

```
    architecture dfm_nand_3in of nand_3in is
        begin
            v <= not (x and (y and z);
    end dfm_nand_3in;
```

```
--code for 2-input NAND gate
library ieee;
use ieee.std_logic_1164.all;
    entity nand_2in is
        port ( x, y : in std_logic;
                 v     : out std_logic );
    end nand_2in;

    architecture dfm_nand_2in of nand_2in is
        begin
            v <= x nand y;
    end dfm_nand_2in;
```

```
--code for JK flip-flop
library ieee;
use ieee.std_logic_1164.all;
    entity jk_ff is
        port (j, k, clk : in std_logic;
                 q, qbar : inout std_logic);
    end jk_ff;

    architecture sm_jk_ff of jk_ff is
    signal m1, m2 : std logic;

        component nand_2in is
            port ( x, y : in std_logic;
                     v     : out std_logic);
        end component;

        component nand_3in is
            port ( x, y, z : in std_logic;
                     v          : out std_logic );
        end component;

            begin
                a1 : nand_3in port map (j, clk, qbar, m1);
                a2 : nand_3in port map (k, clk, q, m2);
                a3 : nand_2in port map (m1, qbar, q);
                a4 : nand_2in port map (m2, q, qbar);
    end sm_jk_ff;
```

VHDL Program 5.25

Write a VHDL code for behavioral model of a D flip-flop.

Program Logic for Behavioral Model

The Boolean equations for D flip-flop are,

Q = D

Qbar = Q′

Where, D is 1-bit input data Q and Q bar are 1-bit output data. The VHDL code for behavioral model of D flip-flop is developed using the above equations. (Refer Chapter 3, Section 3.3.4 for design of D flip-flop).

VHDL Program for Behavioral Model of D Flip-Flop

```
library ieee;
use ieee.std_logic_1164.all;
  entity d_ff is
    port ( d, clk, reset : in std_logic;
           q, qbar       : inout std_logic);
  end d_ff;

  architecture bm_d_ff of d_ff is
    begin
      process (d, clk, reset)
        begin
            qbar <= not q;
          if reset = '1' then
            q      <= '0';
            qbar <= not q;
          elsif (clk' event and clk= '1') then
            q      <= d;
            qbar <= not q;
          end if;
      end process;
  end bm_d_ff;
```

VHDL Program 5.26

Write a VHDL code for structural model of a T flip-flop.

Program Logic for Structural Model

The VHDL code for structural model of T flip-flop is developed using the logic circuit shown in Fig. 1. In structural model, each component involved in the T flip-flop are coded separately and then the components are added in the code for T flip-flop.

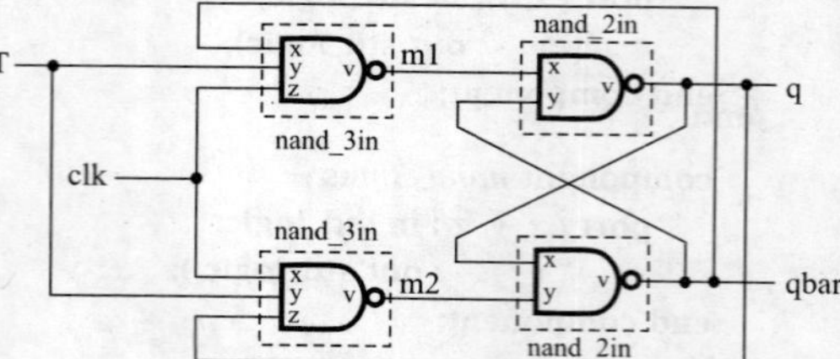

Fig. 1: *T flip-flop.*

VHDL Program for Structural Model of T Flip-Flop

```
--code for 3-input NAND gate
library ieee;
use ieee.std_logic_1164.all;
  entity nand_3in is
    port ( x, y, z : in std_logic;
           v       : out std_logic );
  end nand_3in;

  architecture dfm_nand_3in of nand_3in is
    begin
      v <= not (x and (y and z));
  end dfm_nand_3in;
```

```
--code for 2-input NAND gate
library ieee;
use ieee.std_logic_1164.all;
    entity nand_2in is
        port ( x, y : in std_logic;
               v    : out std_logic );
    end nand_2in;

    architecture dfm_nand_2in of nand_2in is
        begin
            v <= x nand y;
    end dfm_nand_2in;
```

```
--code for T flip-flop
library ieee;
use ieee.std_logic_1164.all;
    entity t_ff is
        port ( t, clk    : in std_logic;
               q, qbar : inout std_logic);
    end t_ff;

    architecture sm_t_ff of t_ff is
    signal m1, m2 : std logic;

        component nand_2in is
            port ( x, y : in std_logic;
                   v    : out std_logic );
        end component;

        component nand_3in is
            port ( x, y, z : in std_logic;
                   v       : out std_logic );
        end component;
            begin
                a1 : nand_3in port map (t, qbar, clk, m1);
                a2 : nand_3in port map (t, q, clk, m2);
                a3 : nand_2in port map (m1, qbar, q);
                a4 : nand_2in port map (m2, q, qbar);
    end sm_t_ff;
```

VHDL Program 5.27

Write a VHDL code for behavioral model of a 2-bit up/down counter.

Program Logic for Behavioral Model

A selection bit, a is required to select up count when a = 1 and down count when a = 0. Initial count is kept 0 and increment by 1 for up count and decrement by 1 for down count for every clock pulse. (Refer Chapter 3, Section 3.8.3 for design of up/down counter).

VHDL Program for Behavioral Model of 2-bit up/down counter

```
library ieee;
use ieee.std_logic_1164.all;
use ieee.std_logic_arith.all;
use ieee.std_logic_unsigned.all;
```

```
entity updown_2bit is
    port ( clk, a : in std_logic;
           dout   : out std_logic_vector (1 downto 0) );
end updown_2bit;

architecture bm_updown_2bit of updown_2bit is
signal m1 : std_logic_vector (1 downto 0) := "00";

    begin
        process (clk)
            begin
                if (clk = '1') then
                if a = '1' then
                    m1   <= m1 + "01";
                    dout <= m1;
                else
                    m1   <= m1 - "01";
                    dout <= m1;
                end if;
                end if;
        end process;
end bm_updown_2bit;
```

VHDL Program 5.28

Translate the following VHDL statement using if else statement.

A <= B1 when C1 = 1' else

B2 when C2 = 1' else

B3 when C3 = 1'

where all the signals are of type bit.

VHDL Program using if else Statement

```
library ieee;
use ieee.std_logic_1164.all;
    entity signals_of_type_bit is
        port ( b1, b2, b3, c1, c2, c3 : in std_logic;
               a                      : out std_logic );
    end signals_of_type_bit;

    architecture bm_signals_of_type_bit of signals_of_type_bit is
        begin
            process (c1, c2, c3)
                begin
                    if c1 = '1' then
                        a <= b1;
                    elsif c2 = '1' then
                        a <= b2;
                    elsif c3 = '1' then
                        a <= b3;
                    end if;
            end process;
    end bm_signals_of_type_bit;
```

5.13 VHDL Programs With Test Bench

VHDL Program 5.29

Write VHDL code for dataflow model of a binary to gray code converter.

Program Logic for Dataflow Model

The following procedure will be useful for bit-by-bit conversion of n-bit binary to n-bit gray code.

Step 1: The most significant bit of binary and gray code are same. Hence retain the most significant bit of binary as most significant bit of gray code.

Step 2: Starting from most significant bit of binary, add two consecutive bits of binary and record the sum as bit of gray code and discard the carry in every addition.

The VHDL code for dataflow model of binary to gray code converter is developed using the above procedure.(Refer Chapter 1, Section 1.3.8 for design of binary gray code).

VHDL Program for Dataflow Model of Binary Gray Code

```
library ieee;
use ieee.std_logic_1164.all;
   entity binary_gray is
      port ( b3, b2, b1, b0 : in std_logic;
             g3, g2, g1, g0 : out std_logic );
   end binary_gray;

   architecture dfm_binary_gray of binary_gray is
      begin
         g3 <= b3;
         g2 <= b3 xor b2;
         g1 <= b2 xor b1;
         g0 <= b1 xor b0;
   end dfm_binary_gray;
```

Test Bench

```
library ieee;
use ieee.std_logic_1164.all;
   entity tb_binary_gray is
      --port ( );
   end tb_binary_gray;

   architecture dfm_tb_binary_gray of tb_binary_gray is
      component binary_gray
         port ( b3, b2, b1, b0 : in std_logic;
                g3, g2, g1, g0 : out std_logic );
      end component;

   signal b3 : std_logic := '0';
   signal b2 : std_logic := '0';
   signal b1 : std_logic := '0';
   signal b0 : std_logic := '0';
   signal g3 : std_logic;
   signal g2 : std_logic;
   signal g1 : std_logic;
   signal g0 : std_logic;
```

```
begin
    uut : binary_gray port map
        ( b3 => b3,
          b2 => b2,
          b1 => b1,
          b0 => b0,
          g3 => g3,
          g2 => g2,
          g1 => g1,
          g0 => g0 );
    stimulus_process : process
        begin
            b3 <= '0';
            b2 <= '1';
            b1 <= '0';
            b0 <= '1';
        wait for 10ns;

            b3 <= '0';
            b2 <= '1';
            b1 <= '1';
            b0 <= '1';
        wait for 10ns;

            b3 <= '1';
            b2 <= '0';
            b1 <= '0';
            b0 <= '1';
        wait for 10ns;

            b3 <= '1';
            b2 <= '0';
            b1 <= '1';
            b0 <= '0';
        wait for 10ns;
    end process;
end dfm_tb_binary_gray;
```

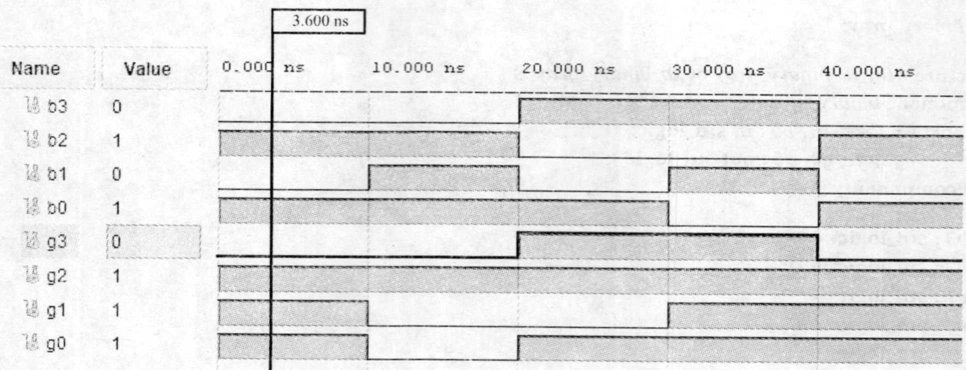

Fig. a:

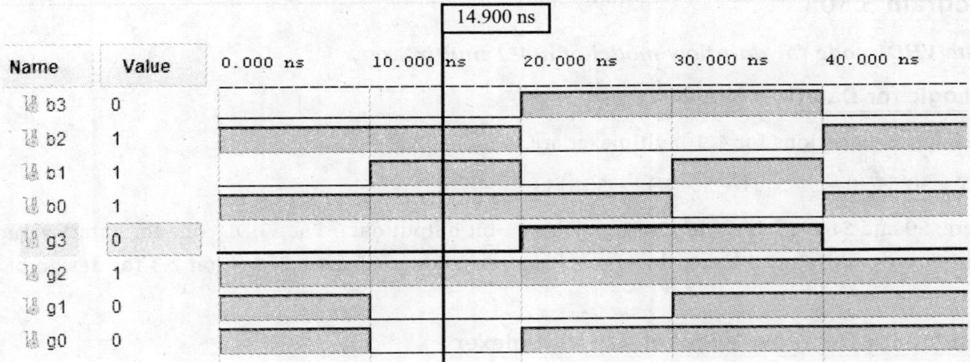

Fig. b:

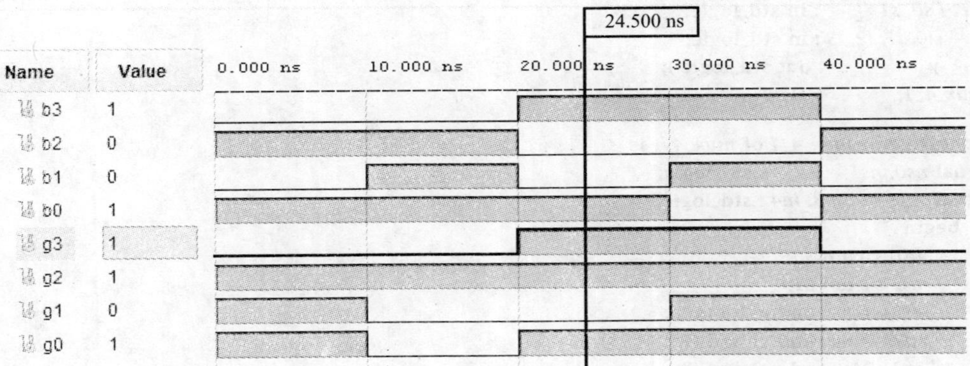

Fig. c:

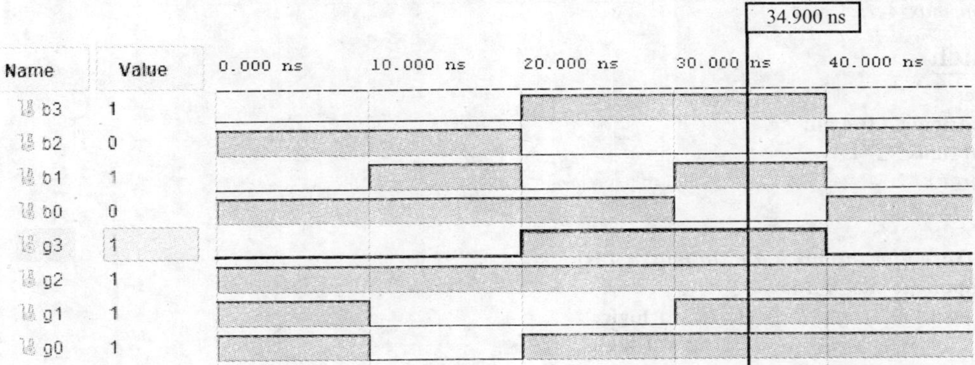

Fig. d:

Fig. 1: *Input-output waveform of binary gray code.*

VHDL Program 5.30

Write VHDL code for dataflow model of a 4*1 multiplexer.

Program Logic for Dataflow Model

The Boolean equations for 4*1 multiplexer are,

$$Y = S0'\,S1'\,I0 + S0'\,S1\,I1 + S0\,S1'\,I2 + S0\,S1\,I3$$

Where, S0 and S1 are 1-bit selection input, Y is 1-bit output data. The VHDL code for dataflow model of 4*1 multiplexer is developed using the above equations. (Refer Chapter 2, Section 2.3 for design of 4*1 multiplexer).

VHDL Program for Dataflow Model of 4*1 Multiplexer

```
library ieee;
use ieee.std_logic_1164.all;
   entity mux_4_1 is
      port ( s0, s1         : in std_logic;
             i0, i1, i2, i3 : in std_logic;
             y              : out std_logic );
   end mux_4_1;

   architecture dfm_mux_4_1 of mux_4_1 is
      signal ns0, ns1          : std_logic;
      signal m1, m2, m3, m4 : std_logic;
         begin
            ns0 <= not s0;
            ns1 <= not s1;
            m1  <= ns0 and ns1 and i0;
            m2  <= ns0 and s1 and i1;
            m3  <= s0 and ns1 and i2;
            m4  <= s0 and s1 and i3;
            y   <= m1 or m2 or m3 or m4;
   end dfm_mux_4_1;
```

Test Bench

```
library ieee;
use ieee.std_logic_1164.all;
   entity tb_mux_4_1 is
      --port ( );
   end tb_mux_4_1;

   architecture dfm_tb_mux_4_1 of tb_mux_4_1 is
      component mux_4_1 is
         port ( i0, i1, i2, i3, s0, s1 : in std_logic;
                y                      : out std_logic);
      end component;

   signal i0 : std_logic := '0';
   signal i1 : std_logic := '0';
   signal i2 : std_logic := '0';
   signal i3 : std_logic := '0';
   signal s0 : std_logic := '0';
   signal s1 : std_logic := '0';
   signal y  : std_logic;
```

```
begin
    uut : mux_4_1 port map
        ( i0 => i0,
          i1 => i1,
          i2 => i2,
          i3 => i3,
          s0 => s0,
          s1 => s1,
          y  => y );
    stimulus_process : process

        begin
            i0 <= '1';
            i1 <= '0';
            i2 <= '0';
            i3 <= '0';
            s0 <= '0';
            s1 <= '0';
        wait for 10ns;

            i0 <= '0';
            i1 <= '0';
            i2 <= '0';
            i3 <= '0';
            s0 <= '0';
            s1 <= '1';
        wait for 10ns;

            i0 <= '0';
            i1 <= '0';
            i2 <= '1';
            i3 <= '0';
            s0 <= '1';
            s1 <= '0';
        wait for 10ns;
            i0 <= '0';
            i1 <= '0';
            i2 <= '0';
            i3 <= '0';
            s0 <= '1';
            s1 <= '1';
        wait for 10ns;
    end process;
end dfm_tb_mux_4_1;
```

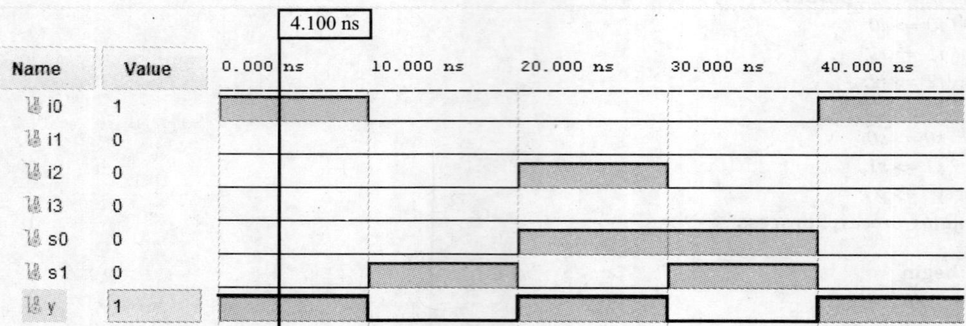

Fig. a:

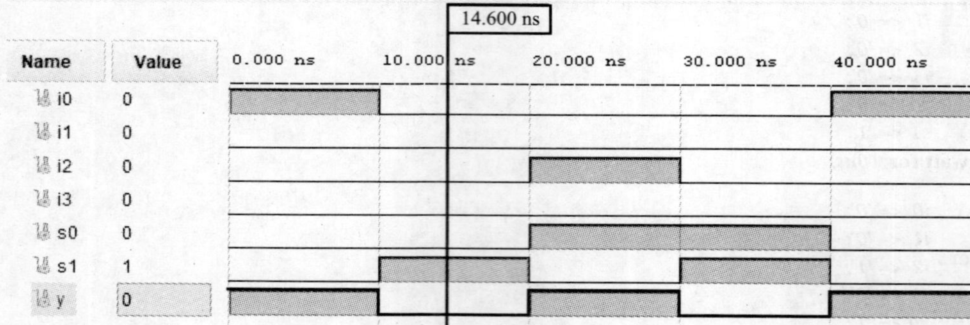

Fig. b:

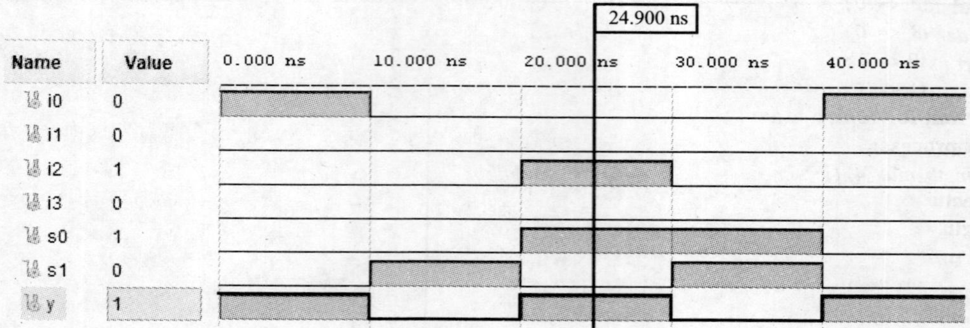

Fig. c:

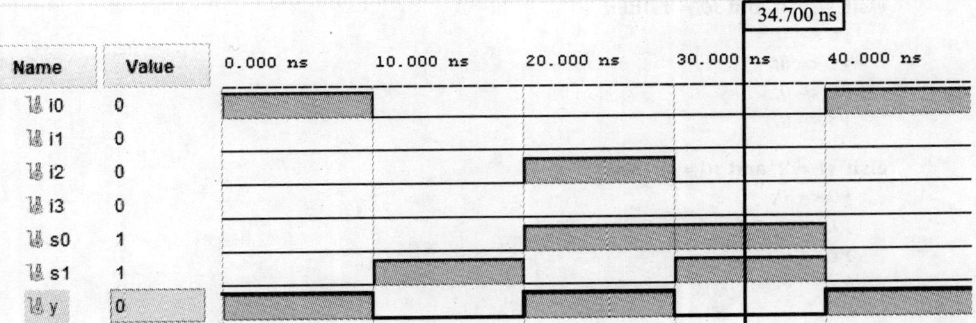

Fig. d:

Fig. 1: *Input-output waveform of 4*1 multiplexer.*

VHDL Program 5.31

Write VHDL code for behavioral model of a 1*4 demultiplexer.

Program Logic for Behavioral Model

The Boolean equations for 1*4 demultiplexer are,

Y0 = Ii S1' S0'

Y1 = Ii S1' S0

Y2 = Ii S1 S0'

Y3 = Ii S1 S0

Truth Table of 1:4 Demux

S_1	S_0	Y_0	Y_1	Y_2	Y_3
0	0	I_i	0	0	0
0	1	0	I_i	0	0
1	0	0	0	I_i	0
1	1	0	0	0	I_i

Where, Ii, S1 and S0 are 1-bit selection input, Y0, Y1, Y2 and Y3 are 1-bit output data. The VHDL code for behavioral model of 1*4 demultiplexer is developed using the above equations. The behavioral models of 1*4 demultiplexer is developed using the truth table of 1*4 demultiplexer. (Refer Chapter 2, Section 2.4 for design of 1*4 demultiplexer).

VHDL Program for Behavioral Model of 1*4 Demultiplexer

```
library ieee;
use ieee.std_logic_1164.all;
    entity demux_1_4 is
        port ( s1, s0           : in std_logic;
                ii              : in std_logic;
            y0, y1, y2, y3 : out std_logic );
    end demux_1_4;

    architecture bm_demux_1_4 of demux_1_4 is
        begin
            process ( s1, s0, ii )
                begin
                    if s1 = '0' and s0 = '0' then
                        y0 <= ii;
                        y1 <= '0';
                        y2 <= '0';
                        y3 <= '0';
```

```
            elsif s1 = '0' and s0 = '1' then
                y0 <= 0;
                y1 <= ii;
                y2 <= 0;
                y3 <= 0;

            elsif s1 = '1' and s0 = '0' then
                y0 <= 0;
                y1 <= 0;
                y2 <= ii;
                y3 <= 0;

            else
                y0 <= 0;
                y1 <= 0;
                y2 <= 0;
                y3 <= ii;
            end if;
        end process;
    end bm_demux_1_4;
```

Test Bench

```
library ieee;
use ieee.std_logic_1164.all;
    entity tb_demux is
        --port ( );
    end tb_demux;

    architecture bm_tb_demux of tb_demux is
        component demux_1_4
            port ( s1, s0         : in std_logic;
                   ii             : in std_logic;
                   y0, y1, y2, y3 : out std_logic );
        end component;

    signal s1, s0         : std_logic;
    signal ii             : std_logic := '0';
    signal y0, y1, y2, y3 : std_logic;
        begin
            uut : demux_1_4 port map
                ( s1  => s1,
                  s0  => s0,
                  ii  => ii,
                  y0 => y0,
                  y1 => y1,
                  y2 => y2,
                  y3 => y3 );
            stimulus_process : process
            begin
                s1 <= '0';
                s0 <= '0';
                ii  <= '1';
            wait for 10ns;
```

```
            s1 <= '0';
            s0 <= '1';
            ii  <= '1';
        wait for 10ns;

            s1 <= '1';
            s0 <= '0';
            ii  <= '1';
        wait for 10ns;

            s1 <= '1';
            s0 <= '1';
            ii  <= '1';
        wait for 10ns;
    end process;
end bm_tb_demux;
```

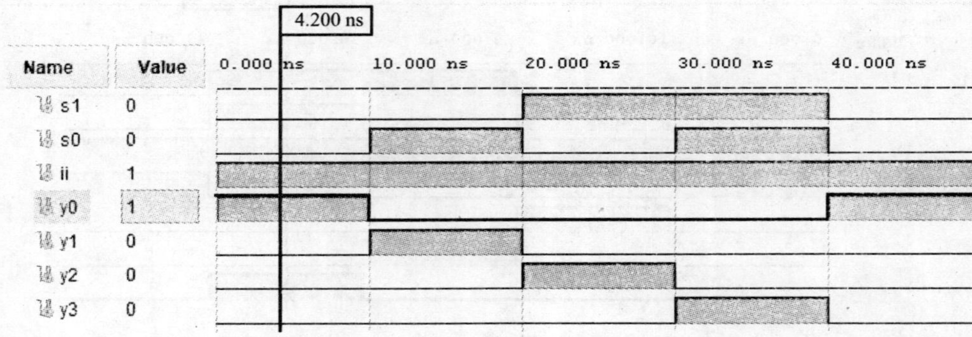

Fig. a:

14.700 ns
Name
Value
0.000 ns
10.000 ns
20.000 ns
30.000 ns
40.000 ns
s1 0
s0 1
ii 1
y0 0
y1 1
y2 0
y3 0

Fig. b:

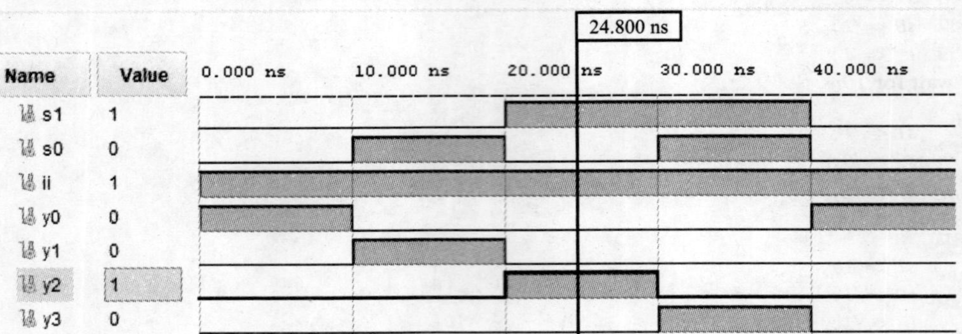

Fig. c:

35.000 ns

Fig. d:

*Fig. 1: Input-output waveform of 1*4 demultiplexer.*

VHDL Program 5.32

Write VHDL code for behavioral model of a half adder.

Program Logic for Behavioral Model

The Boolean equations for half adder are,

$$s = a \oplus b$$

$$c_o = a\,b$$

Where, a and b are 1-bit input data, sum (s) and carry (co) are 1-bit output data. The VHDL code for behavioral model of half adder is developed using the above equations. (Refer Chapter 2, Section 2.6.1 for design of half adder).

VHDL Program for Behavioral Model of Half Adder

```
library ieee;
use ieee.std_logic_1164.all;
   entity half_adder is
      port ( a, b  : in std_logic;
            s, co : out std_logic );
   end half_adder;
```

```
architecture bm_ha of half_adder is
    begin
        process ( a, b )
            begin
                s  <= a xor b;
                co <= a and b;
        end process;
end bm_ha;
```

Test Bench

```
library ieee;
use ieee.std_logic_1164.all;
    entity tb_halfadder is
        --port ( );
    end tb_halfadder;

    architecture bm_tb_halfadder of tb_halfadder is
        component half_adder
            port ( a, b  : in std_logic;
                  s, c0 : out std_logic );
        end component;

    signal a  : std_logic := '0';
    signal b  : std_logic := '0';
    signal s  : std_logic;
    signal c0 : std_logic;

        begin
            uut : half_adder port map
                ( a  => a,
                  b  => b,
                  s  => s,
                  c0 => c0 );
            stimulus_process : process

                begin
                    a <= '0';
                    b <= '0';
                wait for 10ns;

                    a <= '0';
                    b <= '1';
                wait for 10ns;

                    a <=' 1';
                    b <= '0';
                wait for 10ns;

                    a <= '1';
                    b <= '1';
                wait for 10ns;
            end process;
    end bm_tb_halfadder;
```

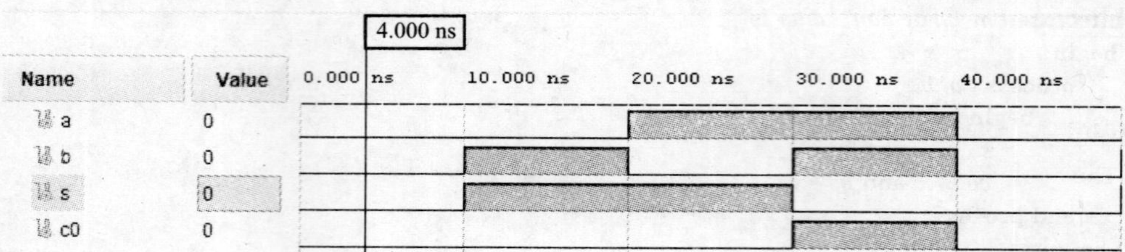

Fig. a: *Inputs: a = 0, b = 0; Outputs: s = 0, c0 = 0.*

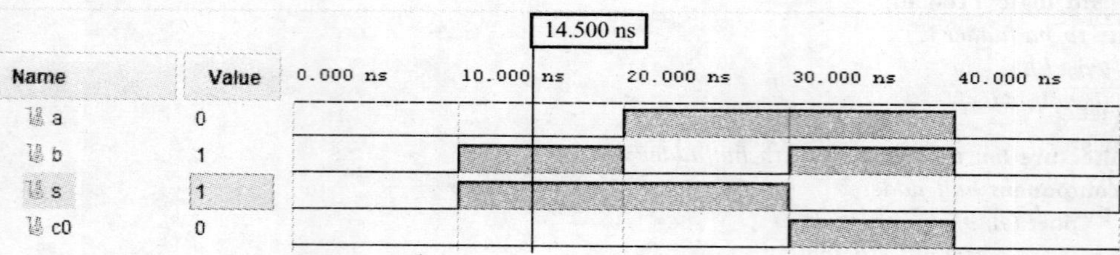

Fig. b: *Inputs: a = 0, b = 1; Outputs: s = 1, c0 = 0.*

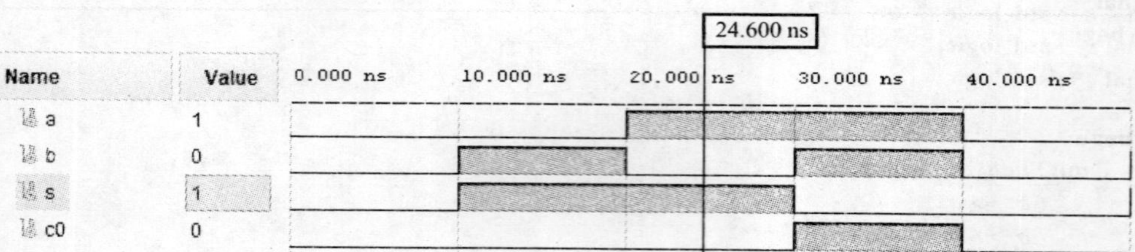

Fig. c: *Inputs: a = 1, b = 0; Outputs: s = 1, c0 = 0.*

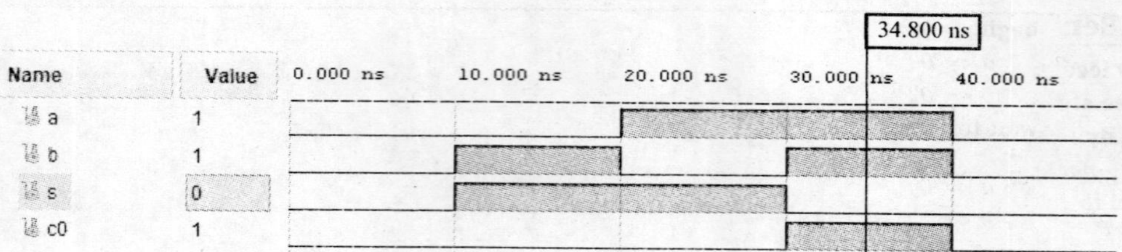

Fig. d: *Inputs: a = 1, b = 1; Outputs: s = 0, c0 = 1.*

Fig. 1: ***Input-output waveform of half adder.***

VHDL Program 5.33

Write VHDL code for behavioral model of a full adder.

Program Logic for Behavioral Model

The Boolean equations for full adder are,

$$s = a \oplus b \oplus c_i$$

$$c_o = (a\,b) + (b\,c_i) + (c_i\,a)$$

Where, a, b and ci are 1-bit input data, sum (s) and carry (co) are 1-bit output data. The VHDL code for behavioral model of full adder is developed using the above equations. (Refer Chapter 2, Section 2.6.2 for design of full adder).

VHDL Program for Behavioral Model of Full Adder

```
library ieee;
use ieee.std_logic_1164.all;
    entity full_adder is
        port ( a, b, ci  :  in std_logic;
               s, co     :  out std_logic );
    end full_adder;

    architecture bm_full_adder of full_adder is
    signal m1, m2, m3 : std_logic := '0';
        begin
            process (a, b, ci)
                begin
                    s   <= a xor b xor ci;
                    m1  <= a and b;
                    m2  <= b and ci;
                    m3  <= ci and a;
                    co  <= m1 or m2 or m3;
            end process;
    end bm_full_adder;
```

Test Bench

```
library ieee;
use ieee.std_logic_1164.all;
    entity tb_ full_adder is
        --port ( );
    end tb_full_adder;

    architecture bm_tb_full_adder of tb_full_adder is
        component full_adder
            port ( a, b, ci  :  in std_logic;
                   s, c0     :  out std_logic );
        end component;

    signal a  : std_logic := '0';
    signal b  : std_logic :=' 0';
    signal ci : std_logic := '0';
    signal s  : std_logic := '0';
    signal c0 : std_logic := '0';
```

```
begin
    uut : full_adder port map
        ( a  => a,
          b  => b,
          ci => ci,
          s  => s,
          c0 => c0 );
    stimulus_process : process

        begin
            a  <= '0';
            b  <= '0';
            ci <= '0';
        wait for 10ns;

            a  <= '0';
            b  <= '0';
            ci <= '1';
        wait for 10ns;

            a  <= '0';
            b  <= '1';
            ci <= '0';
        wait for 10ns;

            a  <= '0';
            b  <= '1';
            ci <= '1';
        wait for 10ns;

            a  <= '1';
            b  <= '0';
            ci <= '0';
        wait for 10ns;

            a  <= '1';
            b  <= '0';
            ci <= '1';
        wait for 10ns;

            a  <= '1';
            b  <= '1';
            ci <= '0';
        wait for 10ns;

            a  <= '1';
            b  <= '1';
            ci <= '1';
        wait for 10ns;
    end process;
end bm_tb_full_adder;
```

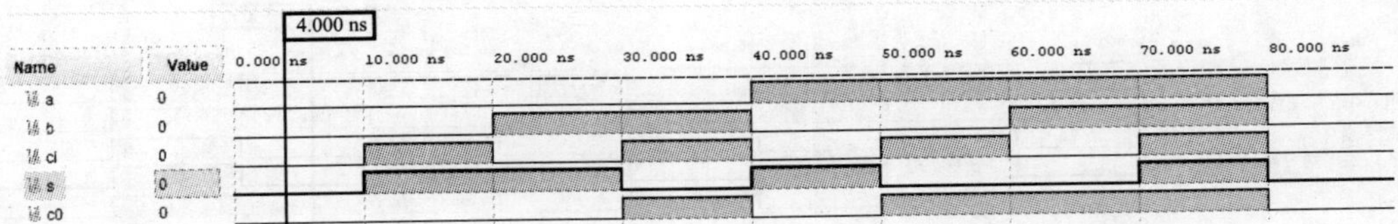

Fig. a: *Inputs: a = 0, b = 0, ci = 0; Outputs: s = 0, c0 = 0.*

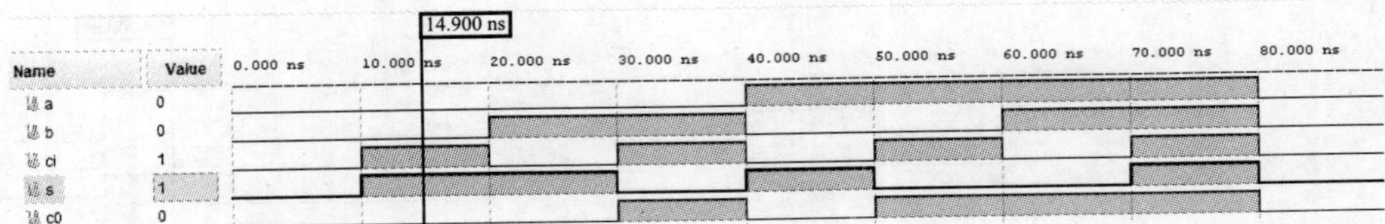

Fig. b: *Inputs: a = 0, b = 0, ci = 1; Outputs: s = 1, c0 = 0.*

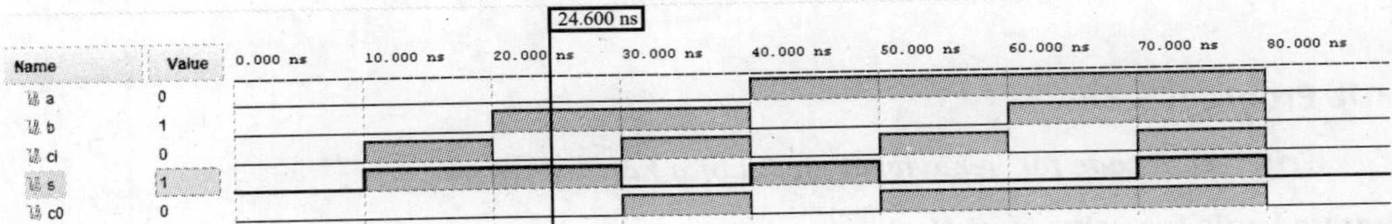

Fig. c: *Inputs: a = 0, b = 1, ci = 0; Outputs: s = 1, c0 = 0.*

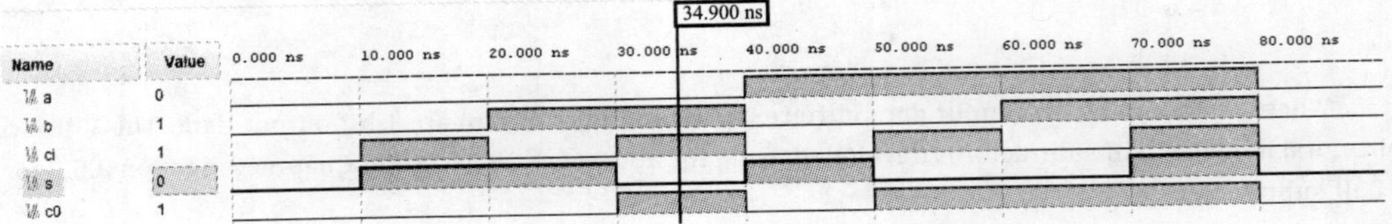

Fig. d: *Inputs: a = 0, b = 1, ci = 1; Outputs: s = 0, c0 = 1.*

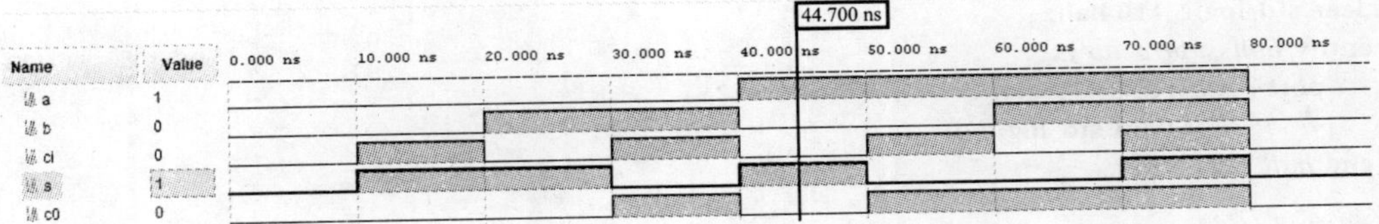

Fig. e: *Inputs: a = 1, b = 0, ci = 0; Outputs: s = 1, c0 = 0.*

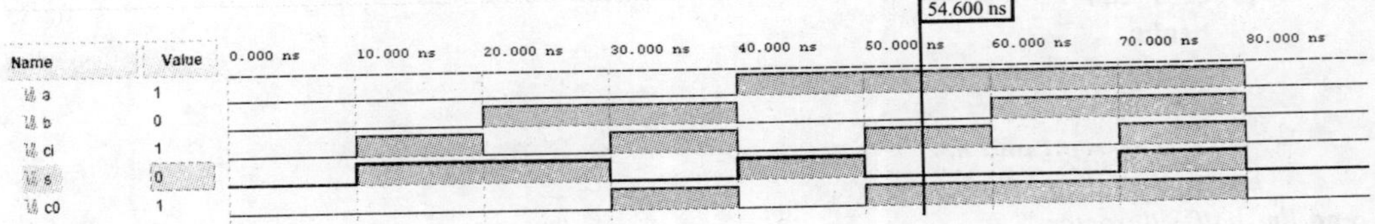

Fig. f: *Inputs: a = 1, b = 0, ci = 1; Outputs: s = 0, c0 = 1.*

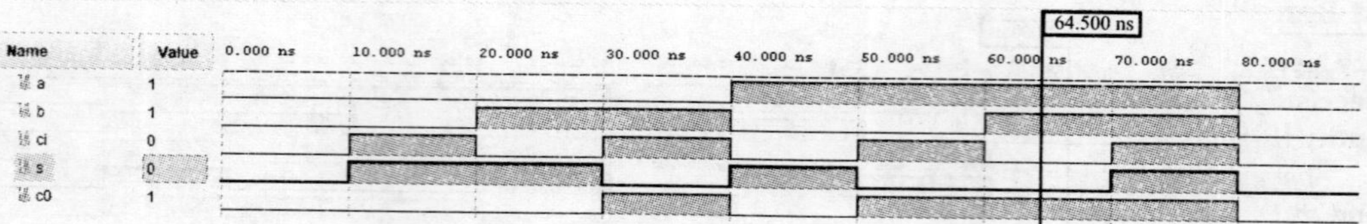

Fig. g: Inputs: a = 1, b = 1, ci = 0; Outputs: s = 0, c0 = 1.

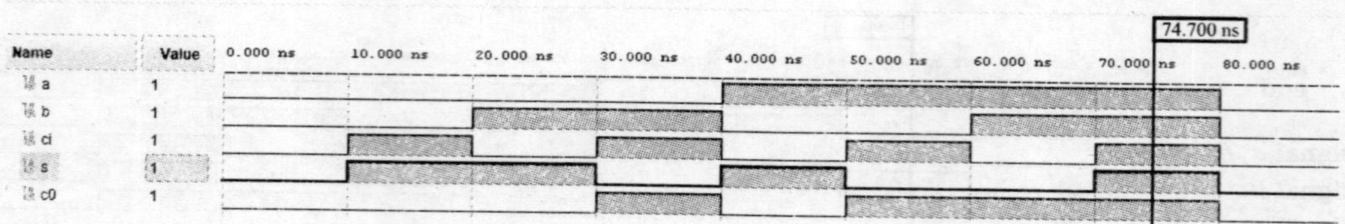

Fig. h: Inputs: a = 1, b = 1, ci = 1; Outputs: s = 1, c0 = 1.

Fig. 1: Input-output waveform of full adder.

VHDL Program 5.34

Write VHDL code for behavioral model of a half subtractor.

Program Logic for Behavioral Model

The Boolean equations for half subtractor are,

$$d = a \oplus b$$

$$c_o = a'b$$

Where, a and b are 1-bit input data, difference (d) and borrow (co) are 1-bit output data. The VHDL code for behavioral model of half subtractor is developed using the above equations. (Refer Chapter 2, Section 2.6.3 for design of half subtractor).

VHDL Program for Behavioral Model of Half Subtractor

```
library ieee;
use ieee.std_logic_1164.all;
   entity half_subtractor is
      port ( a, b  : in std_logic;
             d, co : out std_logic );
   end half_subtractor;

   architecture bm_half_subtractor of half_subtractor is
   signal m1 : std_logic;
      begin
         process ( a, b )
            begin
               d   <= a xor b;
               m1 <= not a;
               co  <= m1 and b;
         end process;
   end bm_half_subtractor;
```

Test Bench

```
library ieee;
use ieee.std_logic_1164.all;
    entity tb_half_subtractor is
        --port ( );
    end tb_half_subtractor;

    architecture bm_tb_half_subtractor of tb_half_subtractor is
        component half_subtractor
            port ( a, b  : in std_logic;
                   d, c0 : out std_logic );
        end component;

    signal a  : std_logic := '0';
    signal b  : std_logic := '0';
    signal d  : std_logic;
    signal c0 : std_logic;

        begin
            uut : half_subtractor port map
                ( a => a,
                  b => b,
                 d  => d,
                 c0 => c0 );
            stimulus_process : process

        begin
            a <= '0';
            b <= '0';
        wait for 20ns;

            a <= '0';
            b <= '1';
        wait for 20ns;

            a <= '1';
            b <= '0';
        wait for 20ns;

            a <= '1';
            b <= '1';
        wait for 20ns;
    end process;
    end bm_half_subtractor;
```

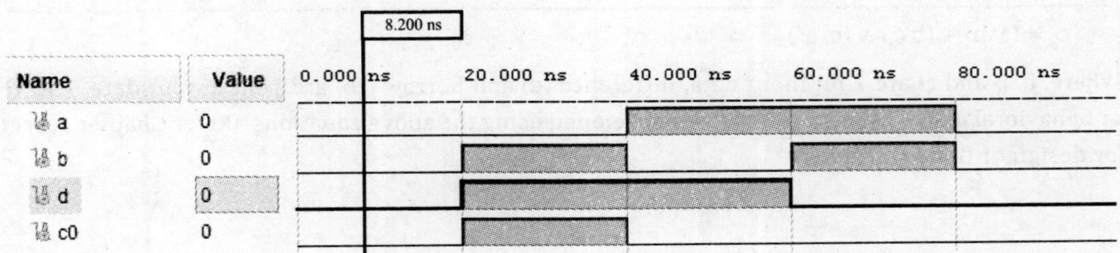

***Fig. a:** Inputs: a = 0, b = 0; Outputs: d = 0, c0 = 0.*

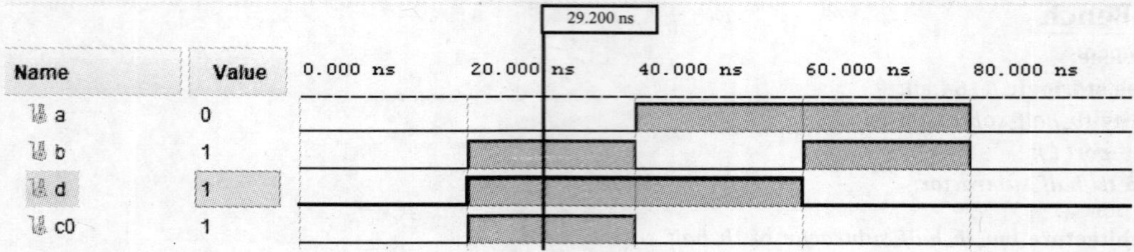

Fig. b: Inputs: a = 0, b = 1; Outputs: d = 1, c0 = 1.

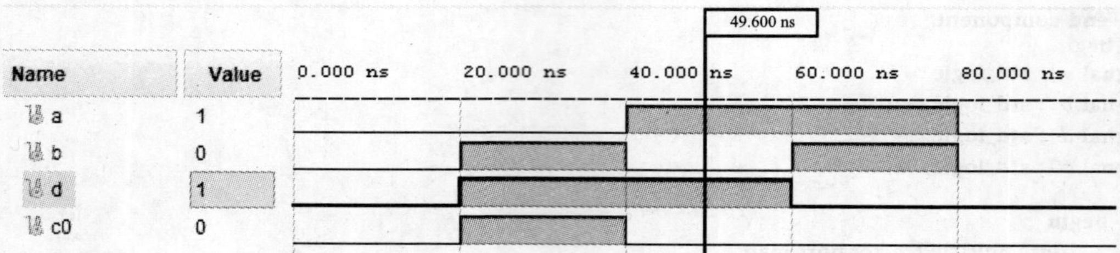

Fig. c: Inputs: a = 1, b = 0; Outputs: d = 1, c0 = 0.

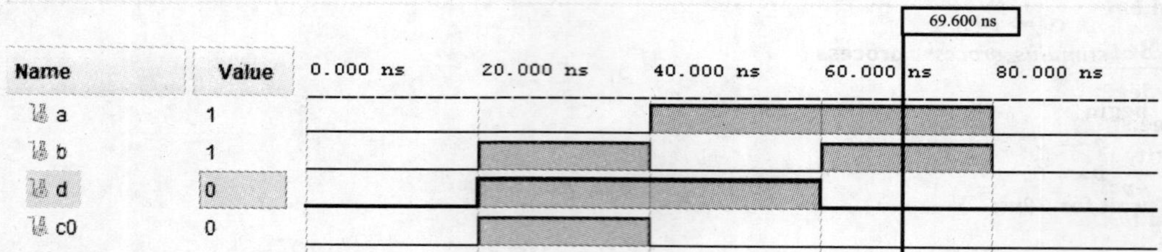

Fig. d: Inputs: a = 1, b = 1; Outputs: d = 0, c0 = 0.

Fig. 1: Input-output waveform of half subtractor.

VHDL Program 5.35

Write VHDL code for behavioral model of a full subtractor.

Program Logic for Behavioral Model

The Boolean equations for full subtractor are,

$$d = a \oplus b \oplus c_i$$

$$c_o = (a'b) + (b\,c_i) + (c_i\,a')$$

Where, a, b and ci are 1-bit input data, difference (d) and borrow (co) are 1-bit output data. The VHDL code for behavioral model of full subtractor is developed using the above equations. (Refer Chapter 2, Section 2.6.4 for design of full subtractor).

VHDL Program for Behavioral Model of Full Subtractor

```
library ieee;
use ieee.std_logic_1164.all;
use ieee.std_logic_arith.all;
use ieee.std_logic_unsigned.all;
    entity full_subtractor is
        port ( a, b, ci : in std_logic;
               d, co    : out std_logic );
    end full_subtractor;

    architecture bm_full_subtractor of full_subtractor is
    signal m1, m2, m3, m4 : std_logic := '0';
        begin
            process (a, b, ci)
                begin
                    d    <= a xor b xor ci;
                    m1 <= not a;
                    m2 <= m1 and b;
                    m3 <= b and ci;
                    m4 <= ci and m1;
                    co  <= m2 or m3 or m4;
            end process;
    end bm_full_subtractor;
```

Test Bench

```
library ieee;
use ieee.std_logic_1164.all;
    entity tb_full_subtractor is
        --port ( );
    end tb_full_subtractor;

    architecture bm_tb_full_subtractor of tb_full_subtractor is
        component full_subtractor
            port ( a, b, ci : in std_logic;
                   d, c0    : out std_logic );
        end component;

    signal a   : std_logic := '0';
    signal b   : std_logic := '0';
    signal ci  : std_logic := '0';
    signal d   : std_logic;
    signal c0 : std_logic;

        begin
            uut : full_subtractor port map
                ( a  => a,
                  b  => b,
                  ci  => ci,
                  d  => d,
                  c0 => c0 );
            stimulus_process : process
```

```
begin
    a <= '0';
    b <= '0';
    ci <= '0';
wait for 20ns;

    a <= '0';
    b <= '0';
    ci <= '1';
wait for 20ns;

    a <= '0';
    b <= '1';
    ci <= '0';
wait for 20ns;

    a <= '0';
    b <= '1';
    ci <= '1';
wait for 20ns;

    a <= '1';
    b <= '0';
    ci <= '0';
wait for 20ns;

    a <= '1';
    b <= '0';
    ci <= '1';
wait for 20ns;

    a <= '1';
    b <= '1';
    ci <= '0';
wait for 20ns;

    a <= '1';
    b <= '1';
    ci <= '1';
wait for 20ns;
  end process;
end bm_tb_full_subtractor;
```

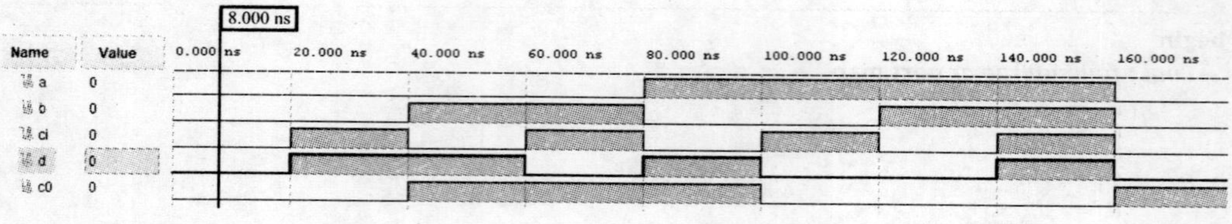

Fig. a: *Inputs: a = 0, b = 0, ci = 0; Outputs: d = 0, c0 = 0.*

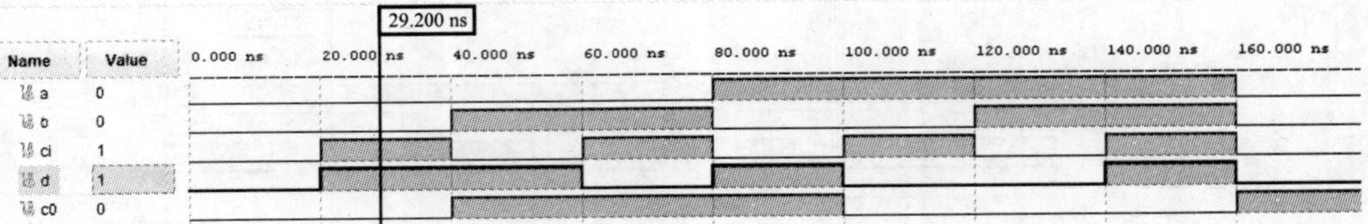

Fig. b: *Inputs: a = 0, b = 0, ci = 1; Outputs: d = 1, c0 = 0.*

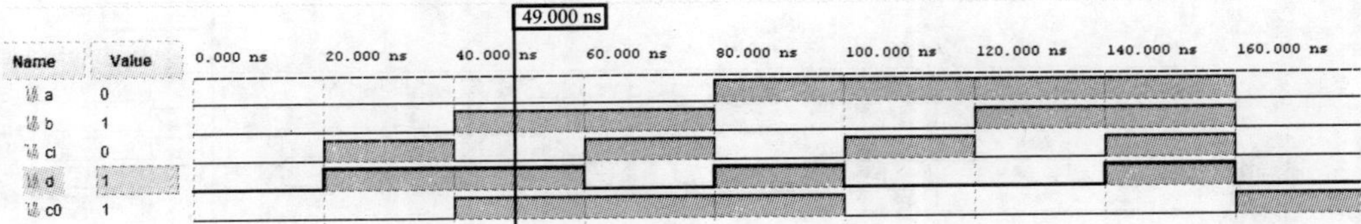

Fig. c: *Inputs: a = 0, b = 1, ci = 0; Outputs: d = 1, c0 = 1.*

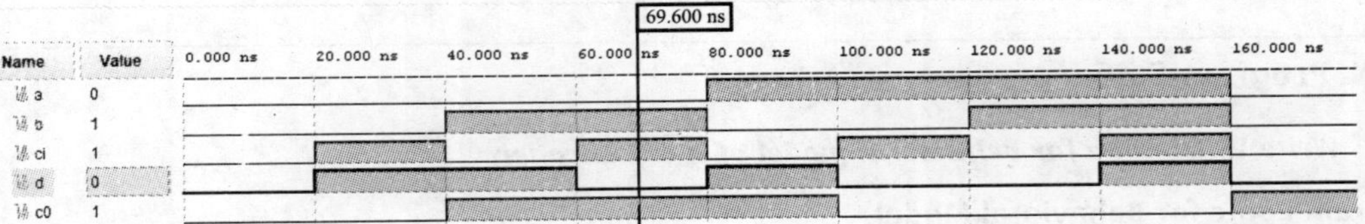

Fig. d: *Inputs: a = 0, b = 1, ci = 1; Outputs: d = 0, c0 = 1.*

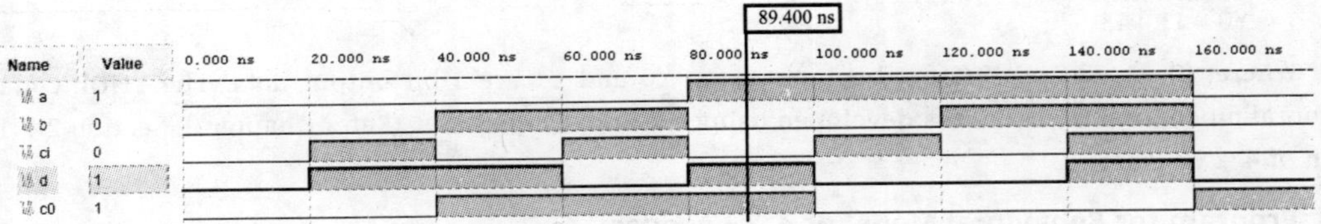

Fig. e: *Inputs: a = 1, b = 0, ci = 0; Outputs: d = 1, c0 = 1.*

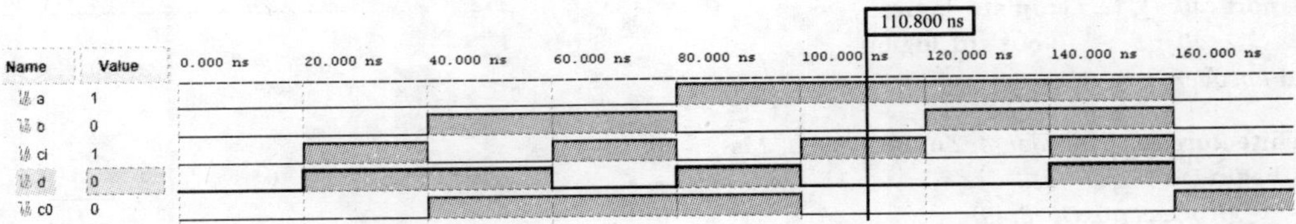

Fig. f: *Inputs: a = 1, b = 0, ci = 1; Outputs: d = 0, c0 = 0.*

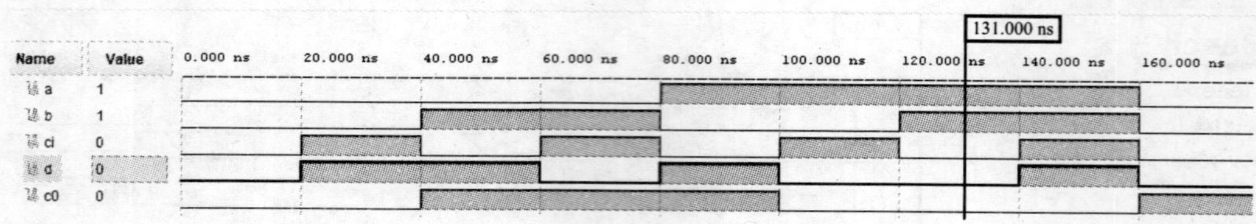

Fig. g: Inputs: a = 1, b = 1, ci = 0; Outputs: d = 0, c0 = 0.

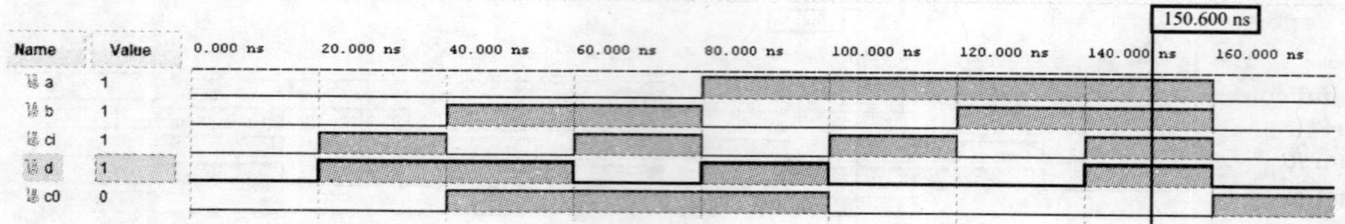

Fig. h: Inputs: a = 1, b = 1, ci = 1; Outputs: d = 1, c0 = 0.

Fig. 1: Input-output waveform of full subtractor.

VHDL Program 5.36

Write VHDL code for behavioral model of a 4*2 encoder.

Program Logic for Behavioral Model

The Boolean equations for 4*2 encoder are,

$$Y1 = I2 + I3$$

$$Y0 = I1 + I3$$

Where, I0, I1, I2 and I3 are 1-bit input data, Y0 and Y1 are 1-bit output data. The VHDL code for behavioral model of 4*2 encoder is developed using the above equations. (Refer Chapter 2, Section 2.7.1 for design of 4*2 encoder).

VHDL Program for Behavioral Model of 4*2 Encoder

```
library ieee;
use ieee.std_logic_1164.all;
   entity encoder_4_2 is
      port ( i0, i1, i2, i3 : in std_logic;
             y0, y1         : out std_logic);
   end encoder_4_2;

   architecture bm_ encoder_4_2 of encoder_4_2 is
      begin
         process (i0, i1, i2, i3)
            begin
               y1 <= i2 or i3;
               y0 <= i1 or i3;
         end process;
   end bm_encoder_4_2;
```

Test Bench

```
library ieee;
use ieee.std_logic_1164.all;
    entity tb_encoder_4_2 is
        --port ( );
    end tb_encoder_4_2;

    architecture bm_tb_encoder_4_2 of tb_encoder_4_2 is
        component encoder_4_2
            port ( i0, i1, i2, i3 : in std_logic;
                   y0, y1         : out std_logic );
        end component;

    signal i0       : std_logic := '0';
    signal i1       : std_logic := '0';
    signal i2       : std_logic := '0';
    signal i3       : std_logic := '0';
    signal y0, y1 : std_logic;
    begin
        uut : encoder_4_2 port map
            ( i0  => i0,
              i1  => i1,
              i2  => i2,
              i3  => i3,
              y0 => y0,
              y1 => y1 );
        stimulus_process : process
        begin
            i0 <= '1';
            i1 <= '0';
            i2 <= '0';
            i3 <= '0';
        wait for 20ns;

            i0 <= '0';
            i1 <= '1';
            i2 <= '0';
            i3 <= '0';
        wait for 20ns;

            i0 <= '0';
            i1 <= '0';
            i2 <= '1';
            i3 <= '0';
        wait for 20ns;

            i0<='0';
            i1<='0';
            i2<='0';
            i3<='1';
        wait for 20ns;
    end process;
    end bm_tb_encoder_4_2;
```

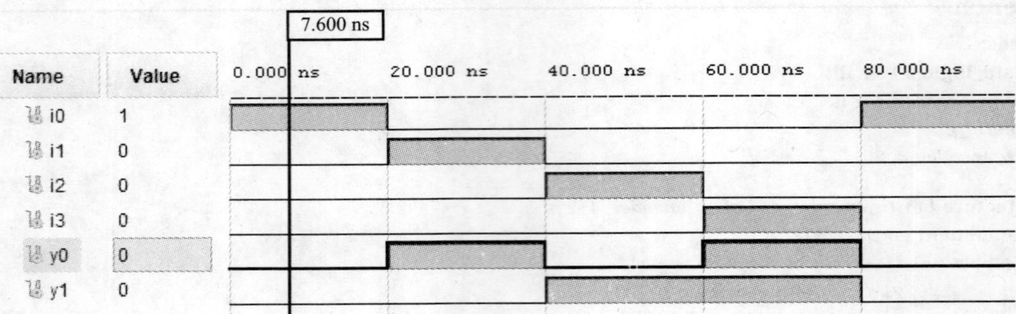

Fig. a: *Inputs: i0 = 1, i1 = i2 = i3 = 0; Outputs: y0, y1 = 0.*

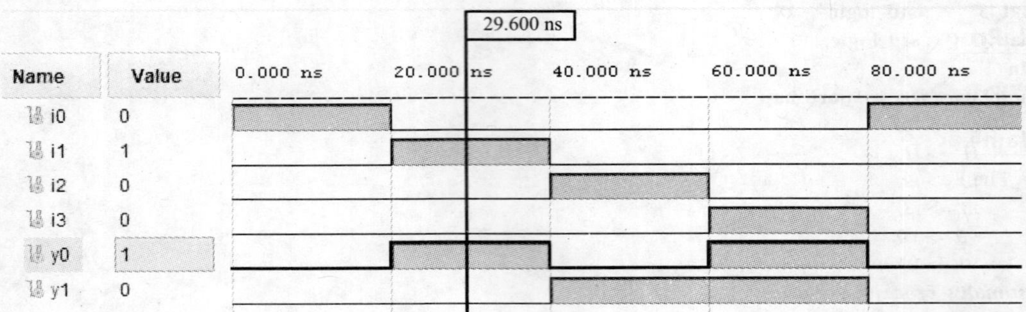

Fig. b: *Inputs: i0 = i2 = i3 = 0, i1 = 1; Outputs: y0 = 1, y1 = 0.*

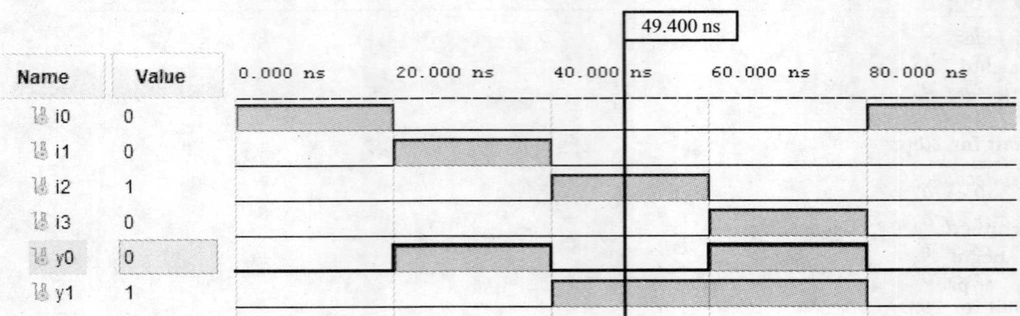

Fig. c: *Inputs: i0 = i1 = i3 = 0, i2 = 1; Outputs: y0 = 0, y1 = 1.*

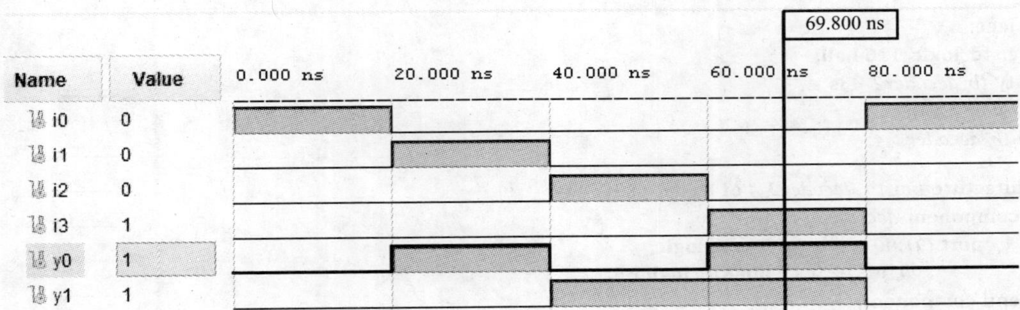

Fig. d: Inputs: i0 = i1 = i2 = 0, i3 = 1; Outputs: y0 = y1 = 1.

***Fig. 1:** Input-output waveform of 4*2 encoder.*

VHDL Program 5.37

Write VHDL code for behavioral model of a 2*4 decoder.

Program Logic for Behavioral Model

The Boolean equations for 2*4 decoder are,

Y0 = I1'I0'

Y1 = I1'I0

Y2 = I1 I0'

Y3 = I1 I0

Where, I1 and I0 are 1-bit input data, Y0, Y1, Y2 and Y3 are 1-bit output data. The VHDL code for behavioral model of 2*4 decoder is developed using the above equations. (Refer Chapter 2, Section 2.9.1 for design of 2*4 decoder).

VHLD Program for Behavioral Model of 2*4 Decoder

```
library ieee;
use ieee.std_logic_1164.all;
   entity decoder2_4 is
      port ( i1, i0          : in std_logic;
             y0, y1, y2, y3  : out std_logic );
   end decoder2_4;

   architecture bm_decoder2_4 of decoder2_4 is
      begin
         process ( i1, i0 )
            begin
               y0 <= not i1 and not i0;
               y1 <= not i1 and i0;
               y2 <= i1 and not i0;
               y3 <= i1 and i0;
         end process;
   end bm_decoder2_4;
```

Test Bench

```
library ieee;
use ieee.std_logic_1164.all;
   entity tb_decoder2_4 is
      --port ( );
   end tb_decoder2_4;

   architecture bm_tb_decoder2_4 of tb_decoder2_4 is
      component decoder2_4
         port ( i1, i0          : in std_logic;
                y0, y1, y2, y3  : out std_logic);
      end component;

   signal i1  : std_logic := '0';
   signal i0  : std_logic := '0';
   signal y0 : std_logic;
   signal y1 : std_logic;
   signal y2 : std_logic;
   signal y3 : std_logic;

   begin
      uut : decoder2_4 port map
         ( i1 => i1,
           i0 => i0,
           y0 => y0,
           y1 => y1,
           y2 => y2,
           y3 => y3 );
      stimulus_process : process
            begin
               i1 <= '0';
               i0 <= '0';
            wait for 20ns;

               i1 <= '0';
               i0 <= '1';
            wait for 20ns;

               i1 <= '1';
               i0 <= '0';
            wait for 20ns;

               i1 <= '1';
               i0 <= '1';
            wait for 20ns;
         end process;
   end bm_tb_decoder2_4;
```

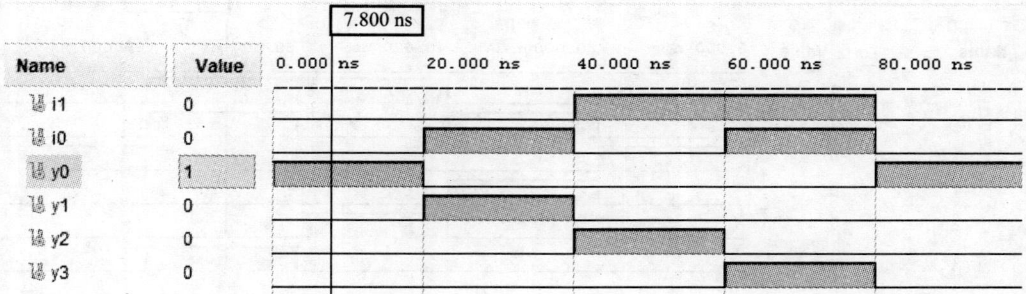

Fig. a: *Inputs: i0 = i1 = 0; Outputs: y0 = 1, y1 = y2 = y3 = 0.*

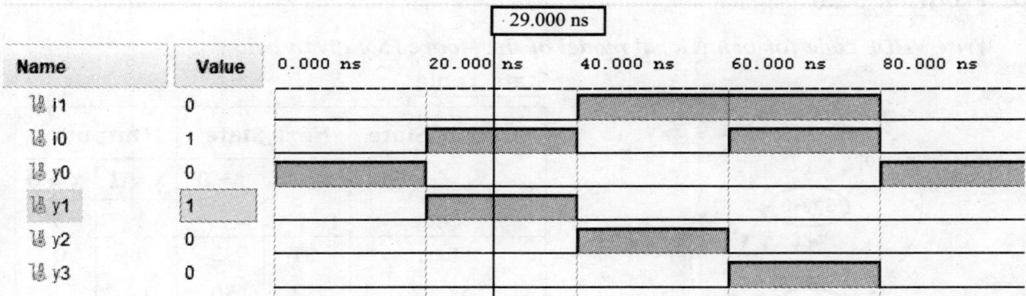

Fig. b: *Inputs: i0 = 1, i1 = 0; Outputs: y0 = y2 = y3 = 0, y1 = 1.*

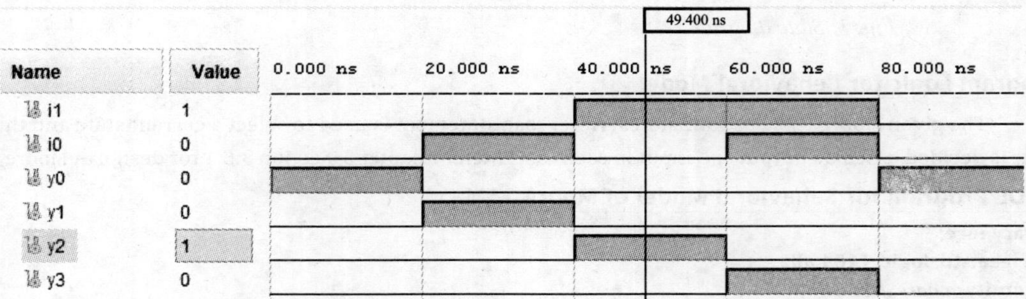

Fig. c: *Inputs: i0 = 0, i1 = 1; Outputs: y0 = y1 = y3 = 0, y2 = 1.*

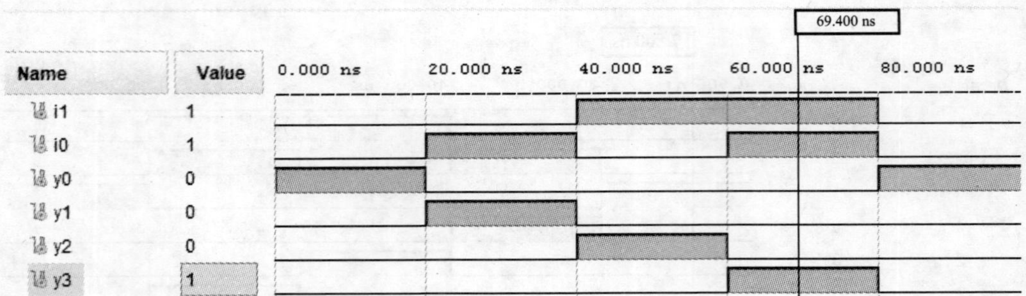

***Fig. d:** Inputs: i1 = i0 = 1; Outputs: y0 = y1 = y2 = 0, y3 = 1.*

***Fig. 1:** Input-output waveform of 2*4 decoder.*

VHDL Program 5.38

Write VHDL code for behavioral model of the Moore FSM given below.

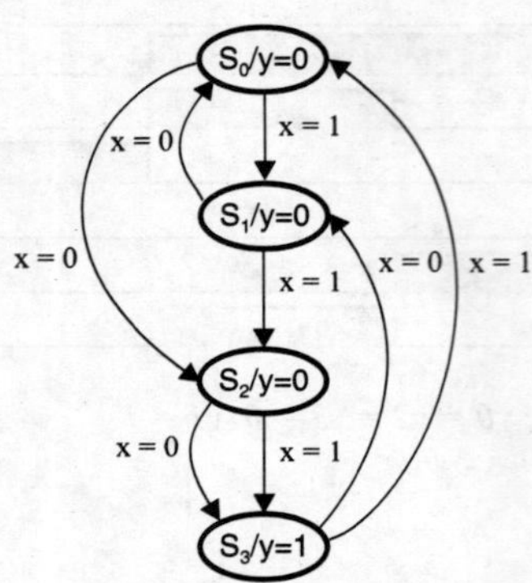

***Fig. 1:** State diagram.*

State Table

Current_State	Next_State		Output	
	x = 1	x= 0	x = 1	x = 0
			y	y
S0	S1	S2	0	0
S1	S2	S0	0	0
S2	S3	S3	0	0
S3	S0	S1	1	1

Program Logic for Behavioral Model

The given Moore FSM has four states. Here, case statement is used to select a current state and the next state is decided based on output and input in each case. (Refer Chapter 3, Section 3.6.1 for design of Moore_fsm).

VHDL Program for Behavioral Model of Moore_FSM

```
library ieee;
use ieee.std_logic_1164.all;
    entity moore_fsm is
        port ( clk : in std_logic;     -- clock signal
               x   : in std_logic;     -- binary sequence input
               y   : out std_logic; -- output of the VHDL sequence detector);
    end moore_fsm;

    architecture bm_moore_fsm of moore_fsm is
    type moore_type is (s0, s1, s2, s3);
    signal current_state, next_state : moore_type;
```

```
begin
    process (clk, current_state, x)
        begin
            case (current_state) is
            when s0 =>
                if (clk = '1' and clk' event) then
                    y <= '0';
                if (x = '1') then
                    next_state <= s1;
            else
                next_state <= s2;
            end if;
            end if;

            when s1 =>
                if (clk = '1' and clk' event) then
                    y <= '0';
                if (x = '1') then
                    next_state <= s2;
            else
                next_state <= s0;
            end if;
            end if;

            when s2 =>
                if (clk = '1' and clk' event) then
                    y <= '0';
                if (x = '1') then
                    next_state <= s3;
            else
                next_state <= s3;
            end if;
            end if;

            when s3 =>
                if (clk = '1' and clk' event) then
                    y <= '1';
                if (x = '1') then
                    next_state <= s0;
            else
                next_state <= s1;
            end if;
            end if;
        end case;
            current_state <= next_state;
    end process;
end bm_moore_fsm;
```

Test Bench

```
library ieee;
use ieee.std_logic_1164.all;
    entity tb_moore_fsm is
        --port ( );
    end tb_moore_fsm;

    architecture bm_tb_moore_fsm of tb_moore_fsm is
        component moore_fsm
            port ( clk : in std_logic;
                   x   : in std_logic;
                   y   : out std_logic );
        end component;

            signal clk : std_logic;
            signal x   : std_logic;
            signal y   : std_logic;
            type moore_type is (s0, s1, s2, s3);
            signal current_state, next_state : moore_type;
            constant clk_period : time := 10 ns;

            begin
                uut : moore_fsm port map
                    ( clk => clk,
                      x   => x,
                      y   => y );
                clock_process : process

                begin
                        clk <= '1';
                    wait for clk_period/2;
                        clk <= '0';
                    wait for clk_period/2;
            end process;
                stimulus_process : process
                    begin
                        x <= '1';
                        current_state <= s0;
                        next_state    <= s1;
                        wait for 5 ns;

                        x <= '0';
                        current_state <= s0;
                        next_state    <= s2;
                        wait for 5 ns;

                        x <= '1';
                        current_state <= s1;
                        next_state    <= s2;
                        wait for 5 ns;

                        x <= '0';
                        current_state <= s1;
```

```
            next_state    <= s0;
            wait for 5 ns;

            x <= '1';
            current_state <= s2;
            next_state    <= s3;
            wait for 5 ns;

            x <= '0';
            current_state <= s2;
            next_state    <= s3;
            wait for 5 ns;

            x <= '1';
            current_state <= s3;
            next_state    <= s0;
            wait for 5 ns;

            x <= '0';
            current_state <= s3;
            next_state    <= s1;
            wait for 5 ns;
      end process;
end bm_tb_moore_fsm;
```

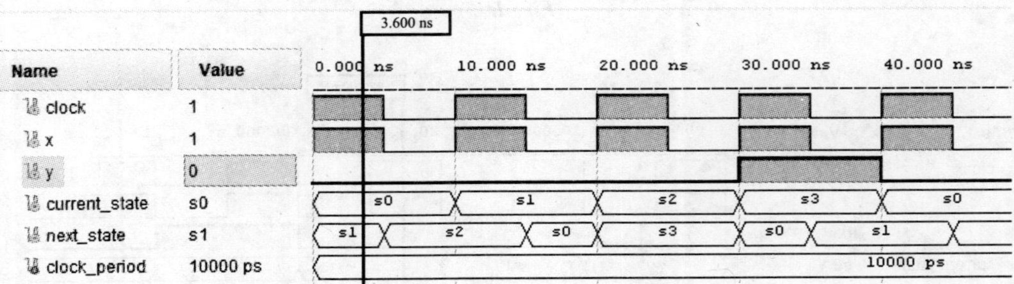

Fig. a:

Fig. b:

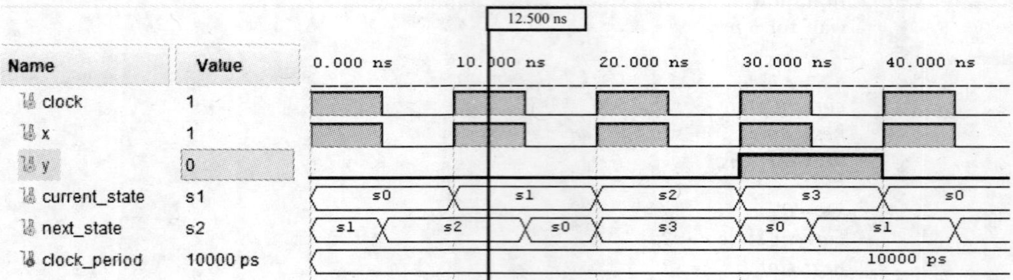

Fig. c:

Fig. d:

Fig. e:

Fig. f:

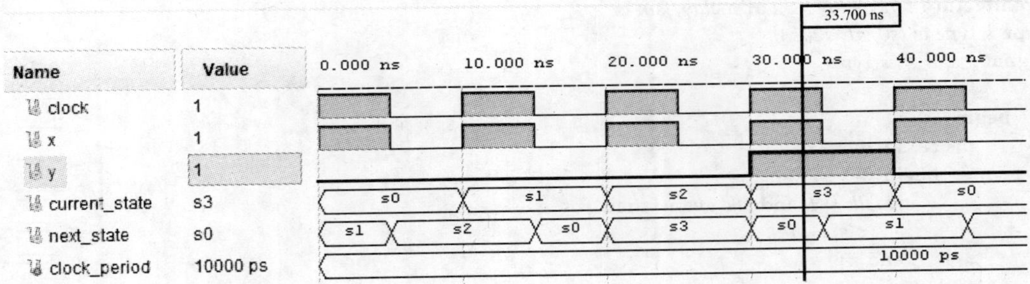

Fig. g:

Fig. h:

Fig. 1: Input-output waveform of Moore_FSM.

VHDL Program 5.39

Write VHDL code for behavioral model of the Mealy FSM given below.

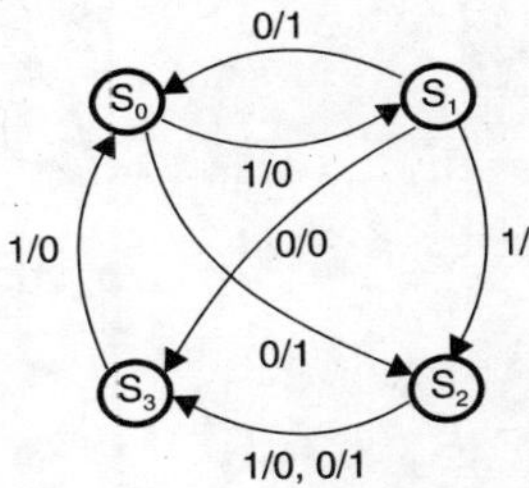

Fig. 1: State diagram.

State Table

Present_State	Next_State		Output	
	x = 1	x= 0	x = 1	x = 0
			y	y
S0	S1	S2	0	1
S1	S2	S0	0	1
S2	S3	S3	0	1
S3	S0	S1	0	1

Program Logic for Behavioral Model

The given Mealy FSM has four states. Here, case statement is used to select a current state and the next state is decided based on output and input in each case. (Refer Chapter 3, Section 3.6.1 for design of Melay_fsm).

VHDL Program for Behavioral Model of Mealy_FSM

```
library ieee;
use ieee.std_logic_1164.all;
   entity melay_fsm is
      port ( x, clk : in bit;
             y      : out bit );
   end melay_fsm;
```

```
architecture bm_melay_fsm of melay_fsm is
type s_type is (s0, s1, s2, s3);
signal c_s, n_s : s_type;

  begin
    process (clk)
      begin
        if (clk = '1' and clk' event) then
          c_s <= n_s;
        end if;
    end process;

    process (x, c_s, n_s)
      begin
        case c_s is
          when s0 => if x = '1' then
            n_s <= s1;
            y   <= '0';
          else
          n_s  <= s2;
          y    <= '1';
        end if;

        when s1 => if x = '1' then
          n_s <= s2;
          y   <= '0';
        else
          n_s <= s0;
          y   <= '1';
        end if;

        when s2=> if x='1' then
          n_s <= s3;
          y   <= '0';
        else
          n_s <= s3;
          y   <= '1';
        end if;

        when s3 => if x = '1' then
          n_s <= s0;
          y   <= '0';
        else
          n_s <= s1;
          y   <= '1';
        end if;
      end case;
    end process;
end bm_melay_fsm;
```

Test Bench

```
library ieee;
use ieee.std_logic_1164.all;
    entity tb_melay_fsm is
        --port ( );
    end tb_melay_fsm;

    architecture bm_tb_melay_fsm of tb_melay_fsm is
        component melay_fsm
            port ( x, clk : in bit;
                   y      : out bit);
        end component;

    type s_type is (s0, s1, s2, s3);
    signal c_s, n_s : s_type;
    signal x   : bit;
    signal clk : bit;
    signal y   : bit;
    constant clk_period : time := 20 ns;
        begin
            uut : melay_fsm port map
                ( x   => x,
                  clk => clk,
                  y   => y );
            clock_process : process
                begin
                    clk <= '1';
                wait for clk_period/2;
                    clk <= '0';
                wait for clk_period/2;
            end process;

            stimulus_process : process
                begin
                    x   <= '1';
                    c_s <= s0;
                    n_s <= s1;
                wait for 10ns;

                    x   <= '0';
                    c_s <= s0;
                    n_s <= s2;
                wait for 10ns;

                    x   <= '1';
                    c_s <= s1;
                    n_s <= s2;
                wait for 10ns;

                    x   <= '0';
                    c_s <= s1;
                    n_s <= s0;
                wait for 10ns;
```

```
            x    <= '1';
            c_s <= s2;
            n_s <= s3;
        wait for 10ns;

            x    <= '0';
            c_s <= s2;
            n_s <= s3;
        wait for 10ns;

            x    <= '1';
            c_s <= s3;
            n_s <= s0;
        wait for 10ns;

            x    <= '0';
            c_s <= s3;
            n_s <= s1;
        wait for 10ns;
    end process;
end bm_tb_melay_fsm;
```

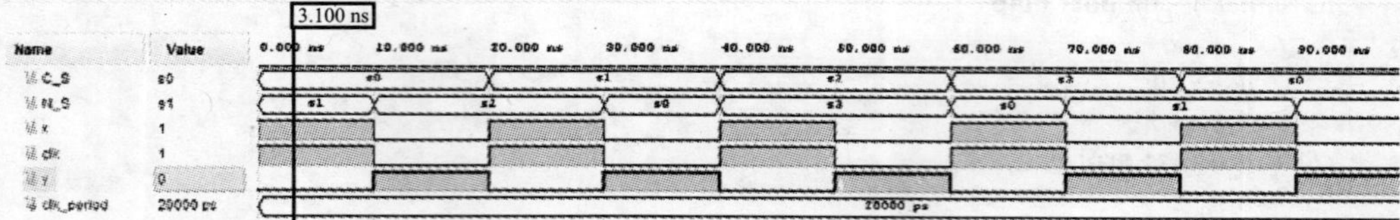

Fig. a:

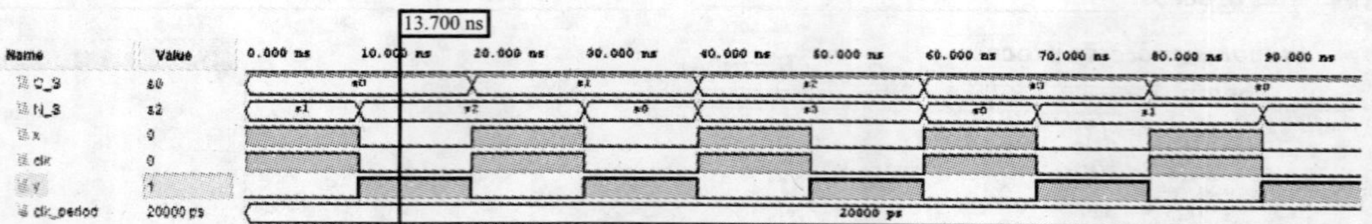

Fig. b:

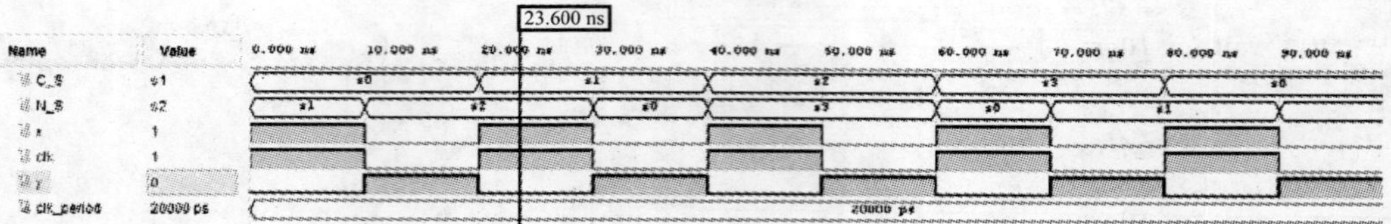

Fig. c:

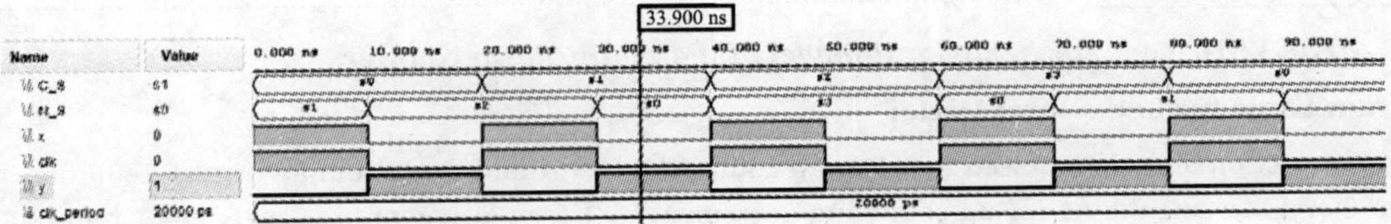

Fig. d:

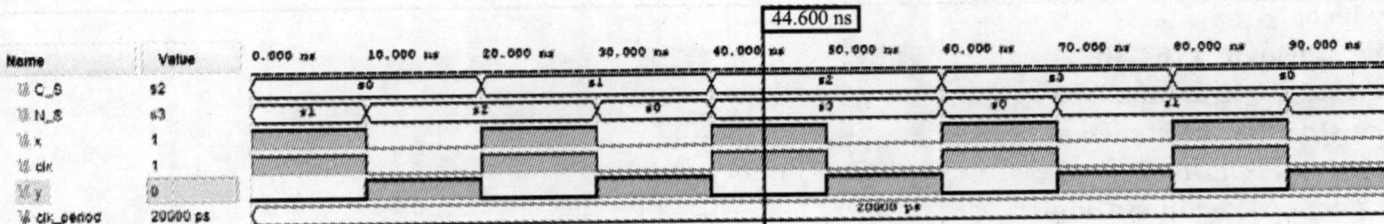

Fig. e:

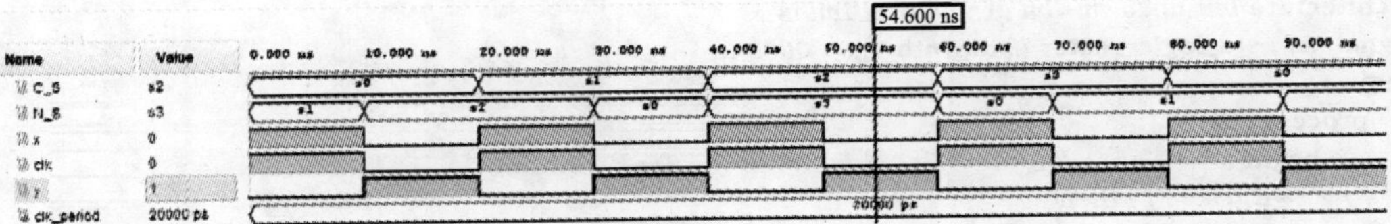

Fig. f:

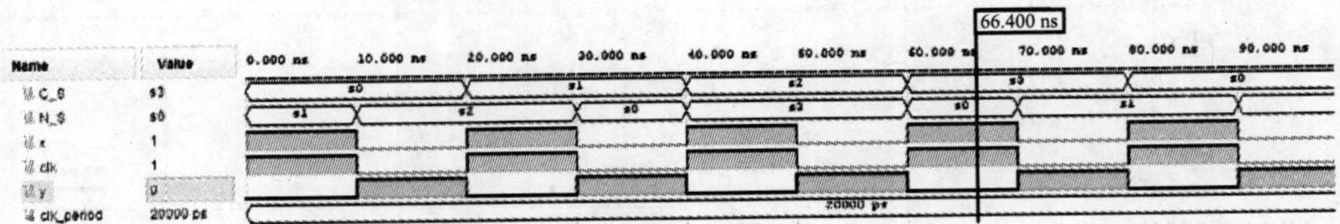

Fig. g:

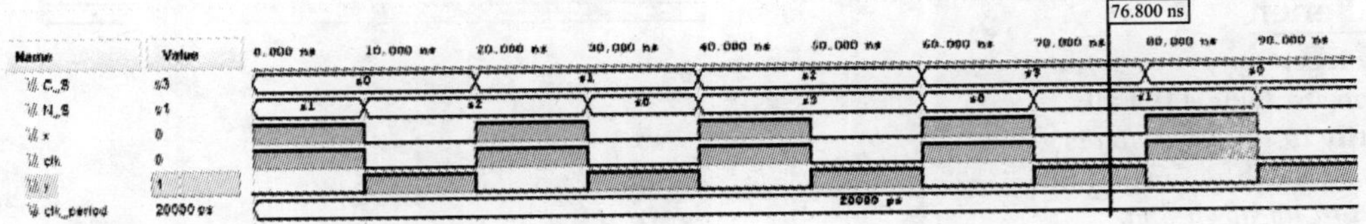

Fig. h:

***Fig. 1:** Input-output waveform of Mealy_FSM.*

VHDL Program 5.40

Write VHDL code for behavioral model of a 4-bit up/down counter.

Program Logic for Behavioral Model

A selection bit, a is required to select up count when a = 1 and down count when a = 0. Initial count is kept 0 and increment by 1 for up count and decrement by 1 for down count for every clock pulse. (Refer Chapter 3, Section 3.8.3 for design of up/down counter).

VHDL Program for Behavioral Model of 4-bit up/down Counter

```
library ieee;
use ieee.std_logic_1164.all;
use ieee.std_logic_arith.all;
use ieee.std_logic_unsigned.all;
    entity updown_4bit is
        port ( clk, a : in std_logic;
                dout  : inout std_logic_vector (3 downto 0) );
    end updown_4bit;

    architecture bm_updown_4bit of updown_4bit is
    signal m1 : std_logic_vector (3 downto 0) := "0000";
    begin
        process (clk)
            begin
                if (clk = '1') then
                if a ='1' then
                    m1   <= m1 + "0001";
                    dout <= m1;
                else
                    m1   <= m1 - "0001";
                    dout <= m1;
                end if;
                end if;
        end process;
    end bm_updown_4bit;
```

Test Bench

```
library ieee;
use ieee.std_logic_1164.all;
    entity tb_up_down_4bit is
        --port ( );
    end tb_up_down_4bit;

    architecture bm_tb_up_down_4bit of tb_up_down_4bit is
        component updown_4bit
            port ( clk, a : in std_logic;
                    dout  : out std_logic_vector (3 downto 0) );
        end component;

    signal clk   : std_logic := '0';
    signal a     : std_logic := '1';  -- :='1' for upcount, :='0' for downcount,
    signal dout : std_logic_vector (3 downto 0);
```

```
  begin
     uut : updown_4bit port map
          ( clk   => clk,
            a     => a,
            dout => dout );
     stimulus_process : process
     begin
          clk <= '1';
     wait for 1ns;

          clk <= '0';
     wait for 1ns;
  end process;
end bm_tb_up_down_4bit;
```

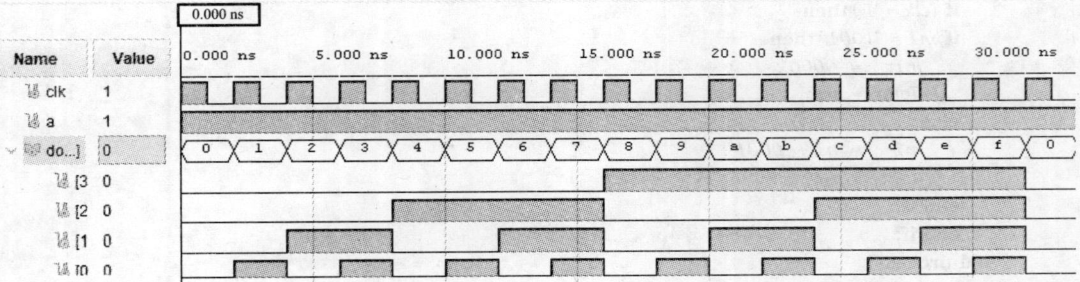

Fig. 1: Input-output waveform of up counter.

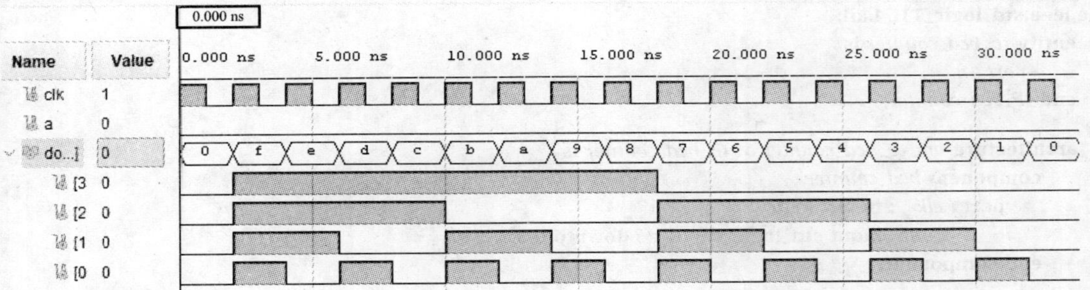

Fig. 2: Input-output waveform of down counter.

VHDL Program 5.41

Write VHDL code for behavioral model of a BCD counter/MOD10 counter.

Program Logic for Behavioral Model

Initially the count value is kept 0. BCD counter is basically an up counter and hence the count value is incremented by 1 for every clock pulse. When the count reaches 1001 the count value is clear to 0.

(Refer Chapter 3,Section 3.8.5 for design of BCD counter/MOD10 counter).

VHDL Program for Behavioral Model of BCD Counter/MOD10 Counter

```
library ieee;
use ieee.std_logic_1164.all;
use ieee.std_logic_arith.all;
use ieee.std_logic_unsigned.all;
    entity bcd_counter is
            port ( clk   : in std_logic;
                   dout : inout std_logic_vector (3 downto 0) );
    end bcd_counter;

    architecture bm_bcd_counter of bcd_counter is
        signal m1 : std_logic_vector (3 downto 0) := "0000";
            begin
                process (clk)
            begin
                if (clk = '1') then
                if m1 = "1001" then
                    m1   <= "0000";
                    dout <= m1;
                else
                    m1   <= m1 + "0001";
                    dout <= m1;
                end if;
                end if;
            end process;
    end bm_bcd_counter;
```

Test Bench

```
library ieee;
use ieee.std_logic_1164.all;
    entity tb_bcd_counter is
        --port ( );
    end tb_bcd_counter;

    architecture bm_tb_bcd_counter of tb_bcd_counter is
        component bcd_counter
            port ( clk   : in std_logic;
                   dout : inout std_logic_vector (3 downto 0) );
        end component;

    signal clk   : std_logic := '0';
    signal dout : std_logic_vector (3 downto 0);
        begin
            uut : bcd_counter port map
                ( clk    => clk,
                  dout => dout);
            stimulus_process : process
        begin
            clk <= '1';
            wait for 10ns;
            clk <= '0';
            wait for 10ns;
```

```
    end process;
end bm_tb_bcd_counter;
```

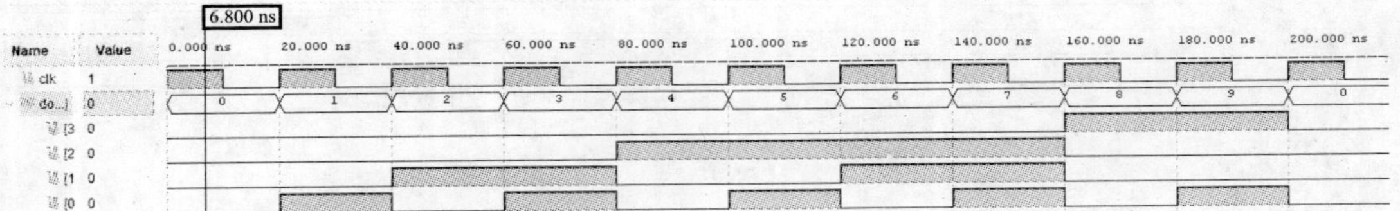

Fig. a:

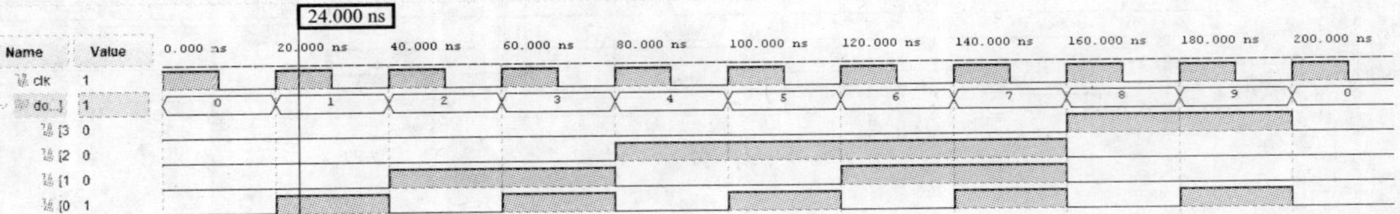

Fig. b:

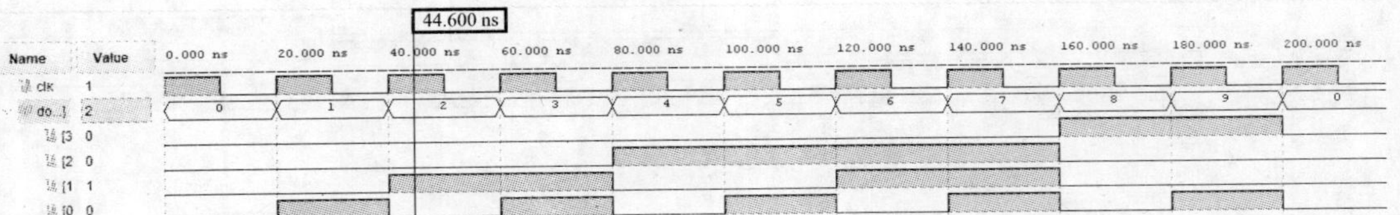

Fig. c:

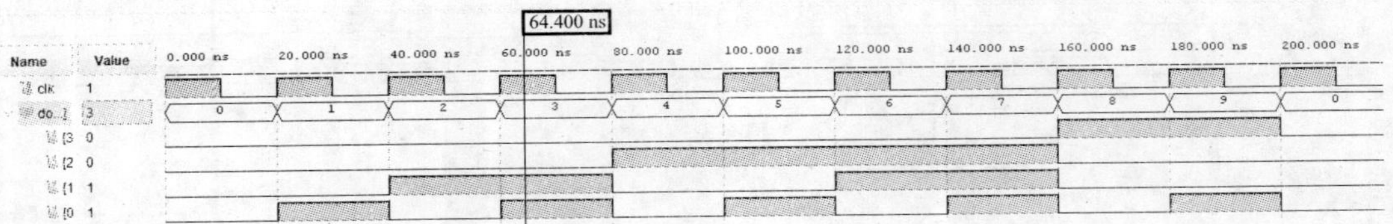

Fig. d:

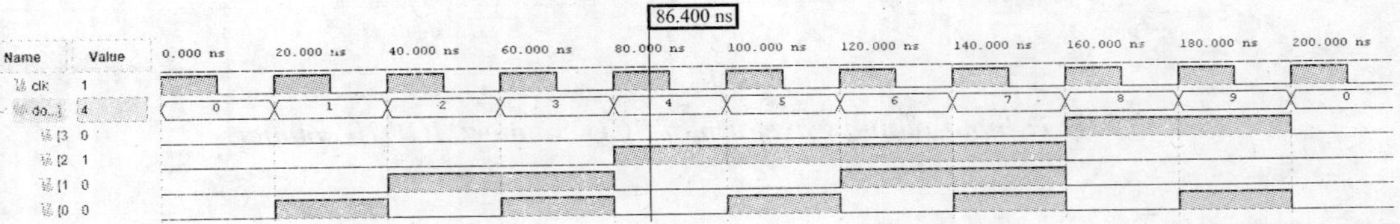

Fig. e:

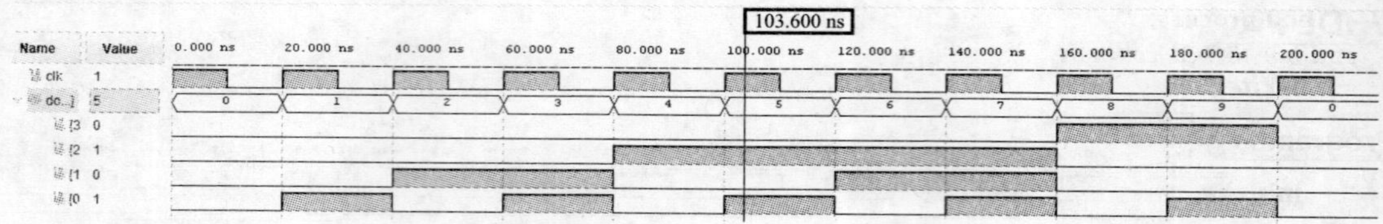

Fig. f:

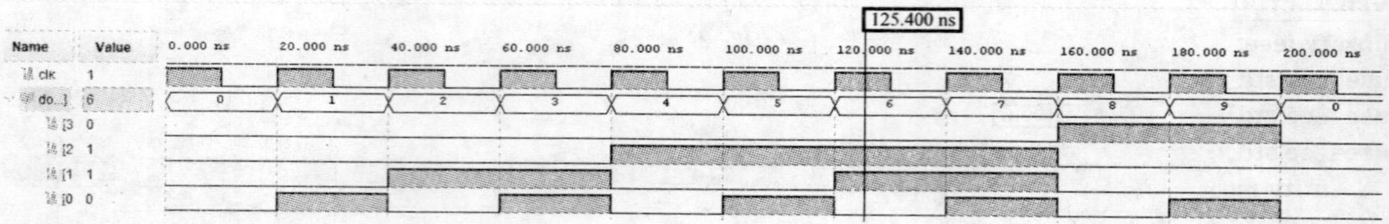

Fig. g:

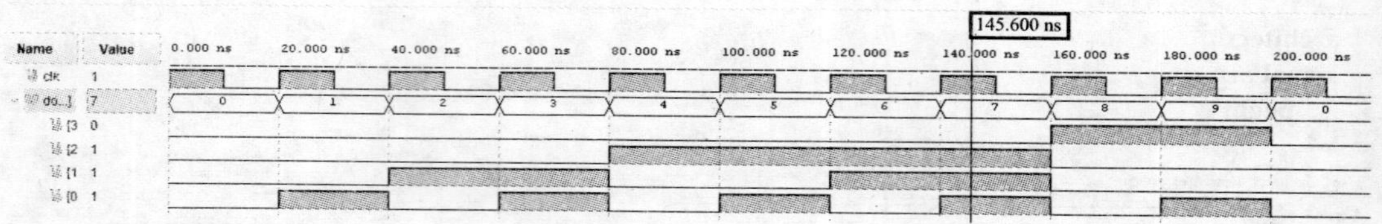

Fig. h:

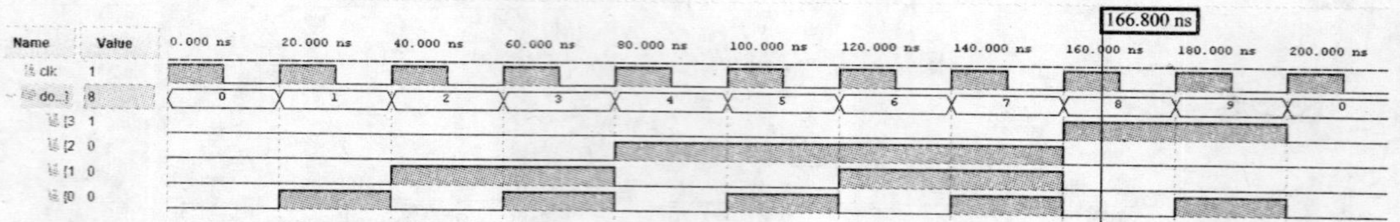

Fig. i:

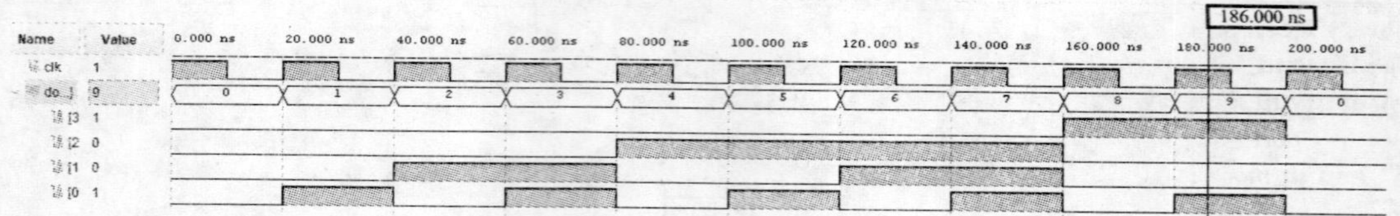

Fig. j:

Fig. 1: *Input-output waveform of BCD counter/MOD10 counter.*

VHDL Program 5.42

Write VHDL code for behavioral model of a MOD6 counter.

Program Logic for Behavioral Model

Initially the count value is kept 0. MOD6 counter is basically an up counter and hence the count value is incremented by 1 for every clock pulse. When the count reaches 101 the count value is clear to 0. (Refer Chapter 3, Section 3.8.6 for design of MOD6 counter).

VHDL Program for Behavioral Model of MOD6 Counter

```
library ieee;
use ieee.std_logic_1164.all;
use ieee.std_logic_arith.all;
use ieee.std_logic_unsigned.all;
    entity mod6_counter is
        port ( clk    : in std_logic;
               dout : inout std_logic_vector (2 downto 0));
    end mod6_counter;

    architecture bm_mod6_counter of mod6_counter is
    signal m1 : std_logic_vector (2 downto 0) := "000";
        begin
            process (clk)
                begin
                    if (clk = '1') then
                    if m1 = "101" then
                        m1   <= "000";
                        dout <= m1;
                     else
                        m1   <= m1 + "001";
                        dout <= m1;
                    end if;
                    end if;
            end process;
    end bm_mod6_counter;
```

Test Bench

```
library ieee;
use ieee.std_logic_1164.all;
    entity tb_mod6_counter is
        --port ( );
    end tb_mod6_counter;

    architecture bm_tb_mod6_counter of tb_mod6_counter is
        component mod6_counter
            port ( clk   : in std_logic;
                   dout : inout std_logic_vector (2 downto 0));
        end component;

    signal clk    : std_logic := '0';
    signal dout : std_logic_vector (2 downto 0);
```

```
begin
    uut : mod6_counter port map
        ( clk  => clk,
          dout => dout );
    stimulus_process : process
        begin
                clk <= '1';
            wait for 10ns;
                clk <= '0';
            wait for 10ns;
    end process;
end bm_tb_mod6_counter;
```

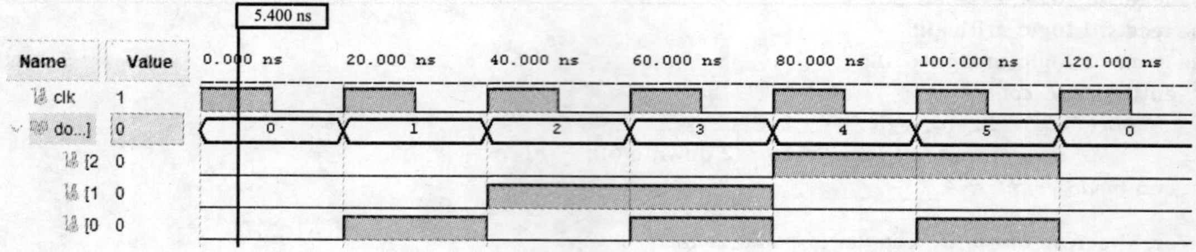

Fig. a:

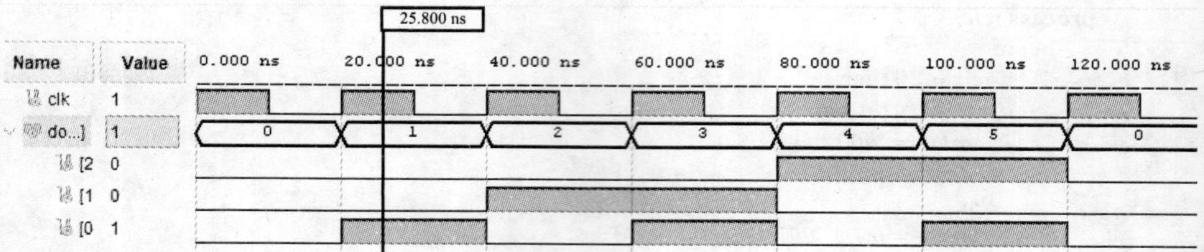

Fig. b:

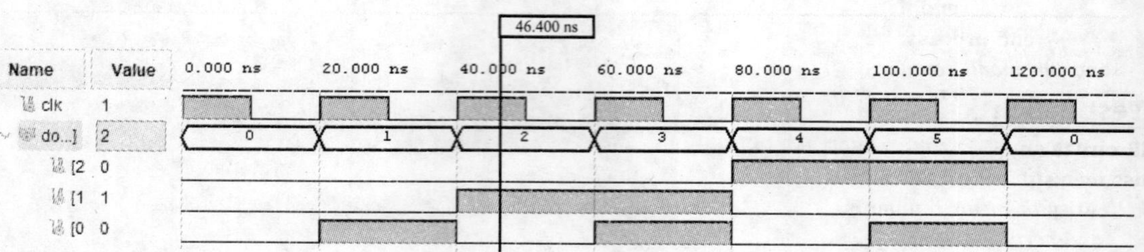

Fig. c:

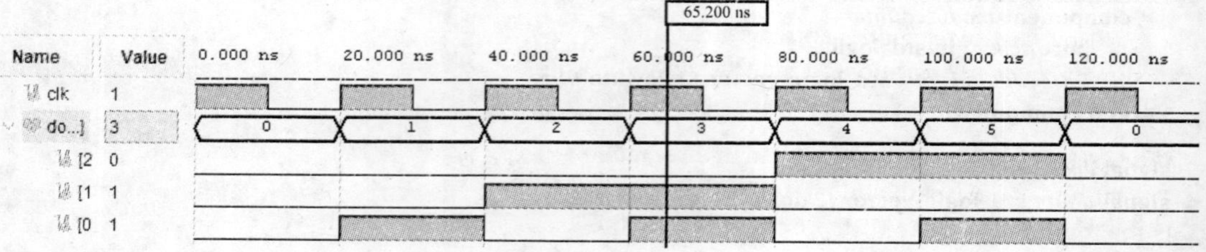

Fig. d:

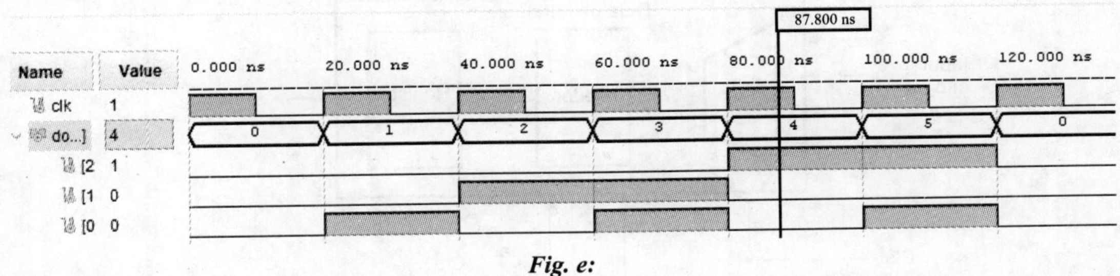

Fig. e:

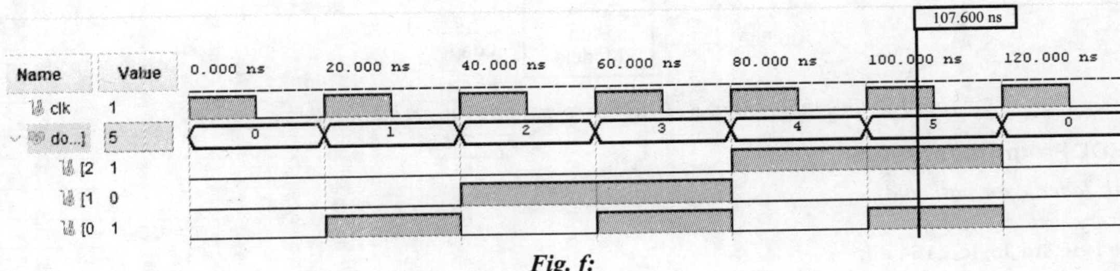

Fig. f:

Fig. 1: Input-output waveform of MOD6 counter.

5.14 VHDL Programs in RTL Methodology

VHDL Program 5.43

Write a VHDL program to perform addition of four 8-bit unsigned integer numbers by using RTL design methodology.

RTL Design Program Logic for Sum of Four Integers

Let a, d, c and d is the four integers.

Now, sum of four integers can be expressed as.

Sum = ((a + b) + c) + d

In RTL implementation, the computation task is breakdown into a sequence of steps, each one involving only a single addition and each operation is executed sequentially.

sum = 0

sum = sum + a

sum = sum + b

sum = sum + c

sum = sum + d

Now the logic components required in data path module are one adder, a register to hold the value of sum in-between operations, a multiplexer to select the input to be added, and a circuit to clear sum at the start of computation. A simple block diagram showing data and control path of RTL design is shown in Fig. 1.

The RTL design is implemented in three modules: rtl_adder_data_path, rtl_adder_control_path , rtl_adder. The rtl_adder_data_path and rtl_adder_control_path are compiled separately and used as in rtl_adder which is the top module. The VHDL programs for the three modules are given ahead:

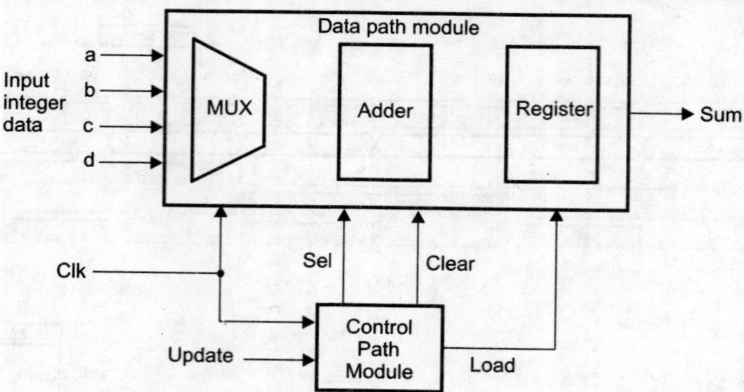

Fig. 1: Block diagram of RTL design for adding four 8-bit intergers.

VHDL Program for Adder

```
--rtl_adder_data_path.vhd
library ieee;
use ieee.std_logic_1164.all;
use ieee.std_logic_arith.all;
   entity rtl_adder_data_path is
      port ( a, b, c, d       : in unsigned (7 downto 0);
             sum              : out unsigned (7 downto 0);
             sel              : in std_logic_vector (1 downto 0);
             load, clear, clk : in std_logic );
   end rtl_adder_data_path;
   architecture arch_rtl_adder_data_path of rtl_adder_data_path is
   signal mux_out, sum_reg, next_sum_reg : unsigned (7 downto 0);
   constant sum_zero
                                         : unsigned (7 downto 0) := "00000000";
   begin

   -- mux to select input data
      with sel select mux_out <=
         a when "00",
         b when "01",
         c when "10",
         d when others;

   -- mux to select register input
      next_sum_reg <=
         sum_reg + mux_out when load = '1' else
         sum_zero when clear         = '1' else
         sum_reg;

   -- process to load sum_reg
   process (clk)
      begin
         if clk'event and clk = '1' then
            sum_reg <= next_sum_reg ;
         end if;
   end process;
```

```
    --output reg_sum
        sum <= sum_reg ;
    end arch_rtl_adder_data_path;
```

```
--rtl_adder_control_path.vhd
library ieee;
use ieee.std_logic_1164.all;
    entity rtl_adder_control_path is
        port ( update      : in std_logic;
               sel         : out std_logic_vector (1 downto 0);
               load, clear : out std_logic;
               clk         : in std_logic );
    end rtl_adder_control_path;

    architecture arch_rtl_adder_control_path of rtl_adder_control_path is
    type states is (clr, add_a, add_b, add_c, add_d, hold);
    signal cs, holdns, ns : states;
    signal tmp            : std_logic_vector (3 downto 0);
    begin
        -- select next state
        with cs select ns <=
            add_a  when clr,
            add_b  when add_a,
            add_c  when add_b,
            add_d  when add_c,
            hold   when add_d,
            holdns when others;

     --process to decide current state (cs)
     process (clk)
         begin
             if clk'event and clk = '1' then
                 cs <= ns;
             end if;
     end process;

     --output control signal to select mux input
         with cs select sel<=
             "00" when add_a,
             "01" when add_b,
             "10" when add_c,
             "11" when others;

     --output control signal to clear register
         holdns <=
             clr when update = '1' else
             hold;
         load  <= '0' when cs = clr or cs = hold else '1';
         clear <= '1' when cs = clr else '0';
     end arch_rtl_adder_control_path;
```

```
--rtl_adder.vhd, top module of VHDL program
--rtl_adder_data_path.vhd and rtl_adder_control_path.vhd are added as components in this top module
library ieee;
use ieee.std_logic_1164.all;
use ieee.std_logic_arith.all;
    entity rtl_adder is
        port ( a, b, c, d   : in unsigned (7 downto 0);
               sum          : out unsigned (7 downto 0);
               update, clk : in std_logic );
    end rtl_adder;

    architecture arch_rtl_adder of rtl_adder is
    signal sel          : std_logic_vector (1 downto 0);
    signal load, clear : std_logic;

      component rtl_adder_data_path
          port ( a, b, c, d           : in unsigned (7 downto 0);
                 sum                  : out unsigned (7 downto 0);
                 sel                  : in std_logic_vector (1 downto 0);
                 load, clear, clk : in std_logic );
      end component;
      component rtl_adder_control_path
          port ( update          : in std_logic;
                 sel             : out std_logic_vector (1 downto 0);
                 load, clear : out std_logic;
                 clk             : in std_logic );
      end component;

      begin
          d1 : rtl_adder_data_path port map (a => a, b => b, c => c, d => d, sum => sum,
                                  sel => sel, load => load, clear => clear, clk => clk);
          c1: rtl_adder_control_path port map (update => update, sel => sel, load => load,
                                  clear => clear, clk => clk );
    end arch_rtl_adder;
```

VHDL Program 5.44

Write a VHDL program to generate a pulse of width equal to width of four input clock pulse using RTL design methodology.

RTL Design Program Logic for Pulse Genereation

A simple block diagram showing data and control path of RTL design is shown in Fig. 1. The data path has a count register that counts the input clock pulses and compares the count with a specified pulse width count. Three input control signals start, stop and reset are used control the pulse generation process. The reset can be used as general system reset. The stop signal is used to stop the process at any desired time. The program is designed to generate an output pulse for a specified width whenever the start pulse goes high.

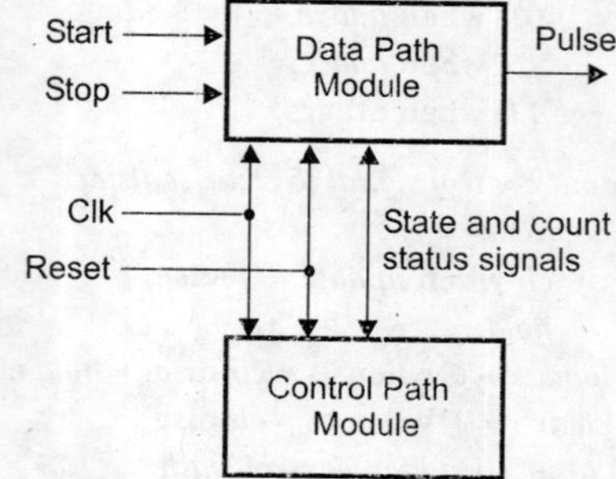

Fig. 1: Block diagram of RTL design for FSMD implementation of one-shot pulse generator.

The RTL design is implemented in three modules: rtl_pulse_gen_data_path, rtl_pulse_gen_control_path, rtl_pulse_gen. The rtl_pulse_gen_data_path and rtl_pulse_gen_control_path are compiled separately and used as in rtl_pulse_gen which is the top module. The VHDL programs for the three modules are given below:

VHDL Program for Pulse Genereation

```
--rtl_pulse_gen_data_path.vhd
library ieee;
use ieee.std_logic_1164.all;
use ieee.std_logic_arith.all;
use ieee.std_logic_unsigned.all;
  entity rtl_pulse_gen_data_path is
    port ( clk, reset  : in std_logic;
          start, stop : in std_logic;
          pulse       : out std_logic );
  end rtl_pulse_gen_data_path;

  architecture arch_rtl_pulse_gen_data_path of rtl_pulse_gen_data_path is
  constant pulse_width : integer := 4;
  type state_type is (idle, delay);
  signal state_reg, state_next   : state_type;
  signal count_reg, count_next : unsigned (3 downto 0);
  begin
  -- next_state logic & datapath functional units/routing
    process (state_reg, start, stop, count_reg)
      begin
        pulse        <= '0';
        count_next <= count_reg;
      case state_reg is
      when idle =>
        if (start = '1') then
          state_next <= delay;
        else
          state_next <= idle;
        end if;
          count_next <= (others => '0');

      when delay =>
        if (stop = '1') then
          state_next <= idle;
        elsif (count_reg = pulse_width – 1) then
          state_next <= idle;
        else
          state_next  <= delay;
          count_next <= count_reg + 1;
        end if;
          pulse <= '1';
        end case;
      end process;
  end arch_rtl_pulse_gen_data_path;
```

```
--rtl_pulse_gen_control_path.vhd
library ieee;
use ieee.std_logic_1164.all;
use ieee.std_logic_arith.all;
use ieee.std_logic_unsigned.all;
   entity rtl_pulse_gen_control_path is
      port ( clk, reset : in std_logic );
   end rtl_pulse_gen_control_path;

   architecture arch_rtl_pulse_gen_control_path of rtl_pulse_gen_control_path is
   type state_type is (idle, delay);
   signal state_reg, state_next   : state_type;
   signal count_reg, count_next : unsigned (3 downto 0);
   begin
      process (clk, reset)
         begin
            if (reset = '1') then
               state_reg  <= idle;
               count_reg <= (others => '0');
            elsif (clk'event and clk = '1') then
               state_reg  <= state_next;
               count_reg <= count_next;
            end if;
      end process;
   end arch_rtl_pulse_gen_control_path;
```

```
--rtl_adder.vhd, top module of VHDL program
--rtl_pulse_gen_data_path.vhd and rtl_pulse_gen_control_path.vhd are added as components in this top module
library ieee;
use ieee.std_logic_1164.all;
use ieee.std_logic_arith.all;
use ieee.std_logic_unsigned.all;
   entity rtl_pulse_gen is
      port ( clk, reset  : in std_logic;
            start, stop : in std_logic;
            pulse       : out std_logic );
   end rtl_pulse_gen;

   architecture arch_rtl_pulse_gen of rtl_pulse_gen is
   constant pulse_width : integer := 4;
   type state_type is (idle, delay);
   signal state_reg, state_next   : state_type;
   signal count_reg, count_next : unsigned (3 downto 0);
      component rtl_pulse_gen_data_path
         port ( clk, reset   : in std_logic;
               start, stop  : in std_logic;
               pulse        : out std_logic );
      end component;

      component rtl_pulse_gen_control_path
         port ( clk, reset : in std_logic );
      end component;
```

```
begin
    d1 : rtl_pulse_gen_data_path port map (clk => clk, reset => reset, start => start, stop => stop,
                                            pulse => pulse);
    c1 : rtl_pulse_gen_control_path port map (clk => clk, reset => reset);
end arch_rtl_pulse_gen;
```

VHDL Program 5.45

Write a VHDL program to design a UART receiver using RTL design methodology.

RTL Design Program Logic for UART receiver

Universal asynchronous receiver and transmitter (UART) is a scheme that sends bytes of data through a serial line. Normally, the serial line is in the '1' state when it is idle and the start of reception is indicated by asserting this line '0', which is considered as start bit. The start bit is followed by eight data bits and ended with a stop bit, which is ' 1'.

UART receiver uses an oversampling scheme to ensure that the data bits are retrieved at the right point. This scheme utilizes a high-frequency sampling signal to estimate the middle point of a data bit and then retrieve data bits at these points. Here, the sampling rate is 16 times the baud rate 1200 and hence, there are 16 sampling pulses for each received bit. The program uses a sampling pulse counter of size 16 bits.

The procedure to retrieve the incoming data bits is follows:

Step-1: When the serial line goes '0' indicating beginning of the start bit, initiate the sampling pulse counter.

Step-2: When the counter reaches 7, clear it to 0 and restart. At this point, the incoming signal reaches about a half of the start bit (or the middle point of the start bit).

Step-3: When the counter reaches 15, clear it to 0 and restart. At this point, the incoming signal progresses for one bit and reaches the middle of the first data bit. The data in the serial line should be retrieved and shifted into a register.

Step-4: Repeat Step 3 seven times to retrieve the remaining seven data bits.

Step-5: Repeat Step 3 one more time but without shifting. The incoming signal should reach the middle of the stop bit at this point, and its value should be '1'.

Here, the system clock is assumed as 1 MHz and the baud rate is 1200 baud. The frequency of the sampling enable pulse should be 16 * 1200, which can be obtained by a mod-52 counter. The syn_pulse signal is the enable pulse whose frequency is 16 times that of the baud rate.

The program uses four states idle, start, data and stop in order to keep track of various activities in the data bit receiving process. Note that the system, remains in the same state unless the syn-pulse signal is activated.

There are two counters namely, syn_pulsecount and syn_bitcount registers. The syn_pulsecount register keeps track of the number of sampling pulses.

The syn_bitcount register keeps track of the number of data bits received in the data state. The retrieved bits are shifted into and reassembled in the data bit register.

The RTL design is implemented in three modules: rtl_uart_rec_data_path, rtl_ uart_rec _control_path, rtl_ uart_rec. The rtl_ uart_rec _data_path and rtl_ uart_rec _control_path are compiled separately and used as in rtl_ uart_rec which is the top module. The VHDL programs for the three modules are given ahead:

VHDL Program for UART Receiver

```
--rtl_uart_rec_data_path.vhd
library ieee;
use ieee.std_logic_1164.all;
use ieee.std_logic_arith.all;
use ieee.std_logic_unsigned.all;
   entity rtl_uart_rec_data_path is
      port ( clk, reset, rec_bit : in std_logic;
            ready               : out std_logic;
            pout                : out std_logic_vector (7 downto 0));
   end rtl_uart_rec_data_path;

   architecture arch_rtl_uart_rec_data_path of rtl_uart_rec_data_path is
   type state_type is (idle, start, data, stop);
   signal syn_pulse                          : std_logic;
   signal state_reg, state_next              : state_type;
   signal syn_pulsecount_reg, pulse_next : unsigned (3 downto 0);
   signal syn_bitcount_reg, bit_next        : unsigned (2 downto 0);
   signal databit_reg, databit_next         : std_logic_vector (7 downto 0);
   begin
      process (state_reg, syn_pulsecount_reg, syn_bitcount_reg, databit_reg, syn_pulse, rec_bit)
         begin
            pulse_next   <= syn_pulsecount_reg;
            bit_next     <= syn_bitcount_reg;
            databit_next <= databit_reg;
            ready        <= '0';
         case state_reg is
            when idle =>
               if (rec_bit ='0') then
                  state_next <= start;
               else
                  state_next <= idle;
               end if;
                  ready <= '1';
            when start =>
               if (syn_pulse = '0') then
                  state_next <= start;
               elsif (syn_pulsecount_reg = 7) then
                  state_next  <= data;
                  pulse_next <= (others => '0');
               else
                  state_next  <= start;
                  pulse_next <= syn_pulsecount_reg + 1;
               end if;
            when data =>
               if (syn_pulse = '0') then
                  state_next <= data;
               elsif (syn_pulsecount_reg = 15) then
                  pulse_next    <= (others => '0');
                  databit_next <= rec_bit & databit_reg (7 downto 1);
```

```
                if (syn_bitcount_reg = 7) then
                    state_next <= stop;
                    bit_next    <= (others => '0');
                else
                    state_next <= data;
                    bit_next    <= syn_bitcount_reg + 1;
                end if;
                else
                    state_next  <= data;
                    pulse_next <= syn_pulsecount_reg + 1;
                end if;

            when stop =>
                if (syn_pulse = '0') then
                    state_next <= stop;
                elsif (syn_pulsecount_reg = 15) then
                    state_next <= idle;
                    pulse_next <= (others => '0');
                else
                    state_next  <= stop;
                    pulse_next <= syn_pulsecount_reg + 1;
                end if;
        end case;
    end process;
        pout <= databit_reg;
  end arch_rtl_uart_rec_data_path;
```

```
--rtl_uart_rec_control_path.vhd
library ieee;
use ieee.std_logic_1164.all;
use ieee.std_logic_arith.all;
use ieee.std_logic_unsigned.all;
    entity rtl_uart_rec_control_path is
        port (clk, reset : in std_logic);
    end rtl_uart_rec_control_path;

    architecture arch_rtl_uart_rec_control_path of rtl_uart_rec_control_path is
    type state_type is (idle, start, data, stop);
    signal syn_pulse                          : std_logic;
    signal clk16_reg, clk16_next              : unsigned (5 downto 0);
    constant dvsr                             : integer := 52;
    signal state_reg, state_next              : state_type;
    signal syn_pulsecount_reg, pulse_next : unsigned (3 downto 0);
    signal syn_bitcount_reg, bit_next         : unsigned (2 downto 0);
    signal databit_reg, databit_next          : std_logic_vector (7 downto 0);
    begin
    -- free running mod-52 counter, independent of fsmd.
        process (clk, reset)
            begin
                if (reset = '1') then
                    clk16_reg <= (others => '0');
```

```
            elsif (clk'event and clk = '1') then
                clk16_reg <= clk16_next;
            end if;

            if (reset = '1') then
                state_reg           <= idle;
                syn_pulsecount_reg <= (others => '0');
                syn_bitcount_reg    <= (others => '0');
                databit_reg         <= (others => '0');
            elsif (clk'event and clk = '1') then
                state_reg           <= state_next;
                syn_pulsecount_reg <= pulse_next;
                syn_bitcount_reg    <= bit_next;
                databit_reg         <= databit_next;
            end if;
        end process;

    --next state/output logic
            clk16_next <= (others => '0') when clk16_reg = (dvsr - 1)
        else clk16_reg + 1;
            syn_pulse <= '1' when clk16_reg = 0 else '0';
    end arch_rtl_uart_rec_control_path;
```

```
--rtl_rtl_uart_rec.vhd top module of VHDL program
--rtl_ uart_rec._data_path.vhd and rtl_ uart_rec. control_path.vhd are added as components in this top module
library ieee;
use ieee.std_logic_1164.all;
use ieee.std_logic_arith.all;
use ieee.std_logic_unsigned.all;
    entity rtl_uart_rec is
        port ( clk, reset, rec_bit : in std_logic;
            ready               : out std_logic;
            pout                : out std_logic_vector (7 downto 0) );
    end rtl_uart_rec;

    architecture arch_rtl_uart_rec of rtl_uart_rec is
        component rtl_uart_rec_data_path
            port ( clk, reset, rec_bit : in std_logic;
                ready               : out std_logic;
                pout                : out std_logic_vector (7 downto 0) );
        end component;

        component rtl_uart_rec_control_path
            port (clk, reset: in std_logic);
        end component;
        begin
            d1 : rtl_uart_rec_data_path port map (clk => clk, reset => reset, rec_bit => rec_bit,
                                                  ready => ready, pout => pout);
            c1 : rtl_uart_rec_control_path port map (clk => clk, reset => reset);
    end arch_rtl_uart_rec;
```

5.15 Summary of Important Concepts

1. VHDL is an acronym for VHSIC (Very High Speed Integrated Circuits) Hardware Description Language.
2. VHDL is used to model a digital system at multiple levels of abstraction.
3. VHDL supports four levels of abstraction for digital system design: Transistor level, gate level, register-transfer level (RTL), processor level.
4. In transistor level abstraction, basic building blocks are basic electronic components like transistors, resistors, capacitors, etc.
5. In gate level abstraction, basic building blocks are logic gates, flip-flops, multiplexers, etc.
6. In register-transfer level (RTL) abstraction, the basic building blocks are modules developed using registers, adders, multiplexers, etc.
7. The RTL abstraction can be called as module-level abstraction.
8. In processor level abstraction, basic building blocks are processors, memory modules, bus interfaces, etc.
9. The basic elements of the VHDL language are literals, comments, identifiers, data objects and operators.
10. Explicit data values that are assigned to objects or used within expressions are called literals.
11. Physical literals are special types of literals used to represent physical quantities such as time, voltage, current, distance, etc.
12. A comment starts with a minimum of two dashes (--) followed by the text to explain the concepts.
13. The VHDL IEEE standard package defines a number of standard data types which are known as predefined data types.
14. Integer data type is integer decimal number in the range $-(2^{31}-1)$ to $+(2^{31}-1)$.
15. Enumeration type is associating user defined set of values called enumeration literals to an object.
16. A composite data type represents a collection of signals or objects.
17. Composite types can be further classified into array types and record types.
18. An array type represents a collection of signals or objects belong to a same data type.
19. A record type represents a collection of signals or objects belong to same or different data types.
20. Data stored as a file in the host environment (the system that simulate and synthesize the code) is called file data type.
21. A constant object is assigned a fixed value and cannot be changed in other part of the code.
22. A file with a list of input and/or output data can be declared as an object.
23. The predefined operators in the VHDL language are classified into five categories: logical operators, relational operators, adding operators, multiplying operators, miscellaneous operators.
24. The concurrent statements are used in dataflow architecture.
25. Signal assignment statements that appear outside of a process are called concurrent signal assignment statements and they are used to assign a value to an object.
26. Assertion statements that appear outside of a process are called concurrent assertion statements and they are used to report the error conditions.
27. Configuration is the process of binding a component with an entity and architecture.
28. A package declaration is used to store a set of common declarations like components, types, procedures, and functions.

29. A package body is always associated with a package declaration.
30. A dataflow model specifies the functionality of a design entity or model without specifying the structure of the design.
31. The behavior of the entity is defined using a set of sequentially executed statements.
32. A process statement is the main mechanism used to model the behavior of an entity.
33. The association-list associates signals in the entity called actuals, with the ports of a component, called locals.
34. A package provides a convenient method to store and share declarations that are common to many design units.
35. A Package is a VHDL file, contains data types, constants, functions, procedures, etc, that can be shared across many VHDL designs.
36. A library is a collection of packages that are compiled and stored in a directory in the host environment (or personal computer) in which the VHDL software tool is installed.
37. The two most commonly used libraries are IEEE and WORK.
38. A test bench is also a VHDL program that is used to verify the correctness of the VHDL program of the hardware model.

5.16 Short-Answer Questions

Q5.1 ***What are components in VHDL?***

Components are reusable VHDL modules compiled as separate entity and can be used in any other structural model of digital systems.

Instead of coding a complex design in single VHDL Code, it can be divided into sub-modules as components and combine them using port map technique in a main module.

Q5.2 ***List the operators in VHDL.***

The predefined operators in the VHDL language are classified into the following five categories.

- Logical operators
- Relational operators
- Adding operators
- Multiplying operators
- Miscellaneous operators

Q5.3 ***Define data objects in VHDL.***

Data object or simply object is an identifier that holds a specific data type, which is declared while creating the object. A data object is created by declaring the class, identifier/name, data type and optional initial value.

Q5.4 ***List the language elements of VHDL.*** ***(AU, Nov/Dec'18, 2 Marks)***

The VHDL language has been developed by integrating the following languages:

- Sequential language
- Concurrent language
- Net-list language
- Timing specifications
- Waveform generation language

Q5.5 ***Write a VHDL program for an 2-input AND gate.***

```
library ieee;
use ieee.std_logic_1164.all;
entity and_2in is
      port ( x, y : in std_logic;
             v    : out std_logic );
end and_2in;
architecture dfm_and_2in of and_2in is
begin
      v <= x and y;
end dfm_and_2in;
```

Q5.6 ***Write a VHDL program for an 2-input NAND gate.***

```
library ieee;
use ieee.std_logic_1164.all;
entity nand_2in is
      port ( x, y : in std_logic;
             v    : out std_logic );
end nand_2in;
architecture dfm_nand_2in of nand_2in is
begin
      v <= x nand y;
end dfm_nand_2in;
```

Q5.7 ***Write a VHDL program for an 2-input OR gate.***

```
library ieee;
use ieee.std_logic_1164.all;
entity or_2in is
      port ( x, y : in std_logic;
             v    : out std_logic );
end or_2in;

architecture dfm_or_2in of or_2in is
begin
      v <= x or y;
end dfm_or_2in;
```

Q5.7 ***Write a VHDL program for an 2-input NOR gate.***

```
library ieee;
use ieee.std_logic_1164.all;
entity nor_2in is
      port ( x, y : in std_logic;
             v    : out std_logic );
end nor_2in;
architecture dfm_nor_2in of nor_2in is
begin
      v <= x nor y;
end dfm_nor_2in;
```

Q5.8 ***Write a VHDL program for an 2-input XOR gate.***

```
library ieee;
use ieee.std_logic_1164.all;
entity xor_2in is
      port ( x, y : in std_logic;
             v    : out std_logic );
end xor_2in;
```

```
architecture dfm_xor_2in of xor_2in is
begin
    v <= x xor y;
end dfm_xor_2in;
```

Q5.9 *Write a VHDL program for an 2-input XNOR gate.*

```
library ieee;
use ieee.std_logic_1164.all;
entity xnor_2in is
    port ( x, y : in std_logic;
           v    : out std_logic );
end xnor_2in;

architecture dfm_xnor_2in of xnor_2in is
begin
    v <= x xnor y;
end dfm_xnor_2in;
```

Q5.10 *Write a VHDL program for an 2-input XNOR gate using behavioral model.* **(AU, Apr/May'19, 2 Marks)**

```
library ieee;
use ieee.std_logic_1164.all;
entity xnor_2in is
    port ( x, y : in std_logic;
           v    : out std_logic );
end xnor_2in;

architecture bm_xnor_2in of xnor_2in is
begin
    process (x, y)
        begin
            v <= x xnor y;
    end process;
end bm_xnor_2in;
```

Q5.11 *Write a VHDL program for an 1-input NOT gate.*

```
library ieee;
use ieee.std_logic_1164.all;
entity not_1in is
    port ( x : in std_logic;
           v : out std_logic );
end not_1in;

architecture dfm_not_1in of not_1in is
begin
    v <= not x;
end dfm_not_1in;
```

Q5.12 *Formulate the syntax of process statement in VHDL.*

Syntax of process statement

```
process_label : process (sensitivity list)
                -- declaration statements;
                begin
                -- sequential statements;
                end process (process_label);
```

Q5.13 Formulate the syntax of selected signal assignment statement in VHDL.

Syntax of selected signal assignment statement

```
with select_expression select
signal_name  <=  value_exp_1 when choice_1,
                 value_exp_2 when choice_2,
                      :
                 value_exp_n when choice_n;
```

Q5.14 Formulate the syntax of conditional signal assignment statement in VHDL.

Syntax of conditional signal assignment statement

```
signal_name <= value_exp_1 when condition_1 else
               value_exp_2 when condition_2 else
                    :
               value_exp_n
```

Q5.16 State the purpose of test bench. ***(AU, Apr/May'19, 2 Marks)***

A test bench is also a VHDL program that is used to verify the correctness of the VHDL program of the hardware model. A test bench VHDL program has three main purposes:

1. Generation of necessary inputs and clock for simulation.
2. Apply this inputs and clock to the entity under test and to monitor the output
3. Comparing the output of VHDL program model with expected output.

5.17 Exercises

I. Fill in the blanks

1. The RTL abstraction can be called ______.
2. Explicit data values that are assigned to objects or used within expressions are called ______.
3. The VHDL IEEE standard package defines a number of standard data types which are known as ______.
4. Integer data type is integer decimal number in the range ______.
5. A composite data type represents a collection of ______ or ______.
6. Composite types can be further classified into ______ and ______.
7. Data stored as a file in the host environment is called ______.
8. The concurrent statements are used in ______.
9. Signal assignment statements that appear outside of a process are called ______ statements.
10. ______ is the process of binding a component with an entity and architecture.
11. A ______ is always associated with a package declaration.
12. A ______ is the main mechanism used to model the behavior of an entity.
13. A ______ is a collection of packages that are compiled and stored in a directory in the host environment (or personal computer) in which the VHDL software tool is installed.
14. The two most commonly used libraries are ______ and ______.
15. A ______ is also a VHDL program that is used to verify the correctness of the VHDL program of the hardware model.

Answers

1. module-level abstraction	6. array types, record types	11. package body
2. literals	7. file data type	12. process statement
3. predefined data types	8. data flow architecture	13. library
4. $-(2^{31}-1)$ to $+(2^{31}-1)$	9. concurrent signal assignment	14. IEEE, WORK
5. signals or objects	10. configuration	15. test bench

II. State whether the following statements are True or False

1. VHDL is used to model a digital system at multiple levels of abstraction.
2. A comment starts with a minimum of two dashes (--) followed by the text to explain the concepts.
3. Character literals are 1-character ASCII values that are enclosed in double-quotes.
4. The general range of integers is – 2,147,483,647 to + 2,147,483,647.
5. The general range of real number is – 1.0e39 to + 1.0e39.
6. An array type represents a collection of signals or objects of same data type.
7. A record type represents a collection of signals or objects belong to different data types.
8. Reserved words of VHDL language can be used as identifiers.
9. String literals are collections of one or more ASCII characters enclosed in double-quote characters.
10. Special characters are allowed in identifiers.
11. An entity can have more than one architecture.
12. A process is the basic unit of execution in VHDL.
13. Functions and procedures are both same.
14. Procedure can return any number of values.

Answers

1. True	6. True	11. True
2. True	7. True	12. True
3. False	8. False	13. False
4. True	9. True	14. True
5. False	10. False	

III. Choose the right answer for the following questions

1. *The general range of integers is ______*

   a) – 2,147,483,645 to + 2,147,483,645

   b) – 2,147,483,646 to + 2,147,483,646

   c) – 2,147,483,647 to + 2,147,483,647

   d) – 2,147,483,648 to + 2,147,483,648

2. *The general range of real number is ______*

   a) – 1.0e37 to + 1.0e37

   b) – 1.0e38 to + 1.0e38

   c) – 1.0e39 to + 1.0e39

   d) – 1.0e40 to + 1.0e40

3. ***Which one of the following is a VHDL relational operator.***

   a) NOT b) NAND

   c) AND d) Equality

4. ***What is the standard form of SLL?***

   a) Shift left logical

   b) Shift line logical

   c) Side left logical

   d) None of the above

5. ***Loop is a ______ statement.***

   a) Concurrent

   b) Sequential

   c) Assignment

   d) Functional

6. ***The addition operators in VHDL are categorized into _____ types?***

   a) Two b) Three

   c) Four d) Five

7. ***What is the expansion of HDL?***

   a) Hardware Describe Language

   b) Hardware Definition Language

   c) Hardware Description Language

   d) None of the above

8. ***Which one of the following is not a reserverd word in VHDL?***

   a) Constant b) Identifier

   c) Variable d) Signal

9. ***signal x : std_logic; In this statement x is***

   a) Variable b) Identifier

   c) Name d) Literal

10. ***Which of the following keyword is not associated with IF statement?***

    a) else b) then

    c) elsif d) when

11. ***What is the correct syntax for using assertion statement?***

    a) **assert** *condition* **report** *"string_expression"* **severity** *"string_expression"*;

    b) *condition* **report** *"string_expression"* **severity** *"string_expression"* **assert**;

    c) *condition* **severity** *"string_expression"* **report** *"string_expression"* **assert** ;

    d) **assert** *condition* **severity** *"string_expression"* **report** *"string_expression "* ;

12. ***Which of the following is not a type of VHDL modeling?***

    a) Behavioral modeling

    b) Dataflow modeling

    c) Structural modeling

    d) Component modeling

13. ***Which of the following is basic unit of behavioral description?***

    a) Structure

    b) Sequence

    c) Process statement

    d) Dataflow

14. ***_______ modeling uses logic gates and basic blocks to describe the functionality of system.***

    a) Behavioral modeling

    b) Dataflow modeling

    c) Structural modeling

    d) Component modeling

Answers

1. c	6. a	11. a
2. b	7. c	12. d
3. d	8. b	13. c
4. a	9. b	14. b
5. b	10. d	

IV. Answer the following questions

1. List the modules of HDL.
2. Explain the various data objects supported by VHDL. Give the necessary examples.
3. Define the following terms relevant to VHDL.

   1. Parameters 2. Keywords 3. Constants 4. Identifiers
4. Explain the difference in program structure of VHDL and any other procedural language. Give an example.
5. Explain various architectural bodies/modeling styles in VHDL with examples.
6. Explain the various data types supported by VHDL. Give the necessary examples.
7. Explain the program structure of VHDL with the help of block diagram.
8. List out the different operators available in VHDL. Explain with example.
9. Explain the different concurrent statements and sequential statements in VHDL?
10. Discuss about the difference between CASE and IF statements in VHDL with examples?
11. List the various advantages and features of VHDL.

12. Describe RTL in HDL with example.

13. How the package declaration is different from entity declaration? Give the syntax for each?

14. Give the syntax for package declaration and package body in VHDL.

V. Write Solution for Following VHDL Programs

E5.1 Design a 4-bit unsigned adder in VHDL.

E5.2 Translate the 4:1 mux if else statement to case statement.

```
if s = "00" then
    y <= i0;
elsif s = "01" then
    y <= i1;
elsif s = "10" then
    y <= i2;
else
    y <= i3;
end if;
```

E5.3 Translate the following code to a case-when statement.

```
with s select
    d <= "0000" when "00",
         "0001" when "01",
         "0010" when "10",
         "0011" when "11",
         "1111" when others;
```

E5.4 Translate the following code to, two with-select-when statements.

```
case state is
    when "00"     => a <= "1111"; b <= "0000";
    when "01"     => a <= "0001"; b <= "0010";
    when "10"     => a <= "1000"; b <= "1001";
    when others => a <= "1100"; b <= "1010";
end case;
```

E5.5 Find the errors in the following code.

```
architecture has_errors of design
begin
    p1 : process
        begin
            if clk'event and clk = '1' then
                q <= a or b and c;
            end if;
        end;
end;
```

E5.6 Write the VHDL code for T flip-flop in behavioral model.

E5.7 Write a VHDL code to demonstrate the use of variables in counter design.

E5.8 Write a VHDL code to perform bit-by-bit logical and of an 8-bit data using variables.

E5.9 Write a VHDL code for D flip-flop with rising-edge clock and asynchronous reset and preset.

E5.10 Write a VHDL code for traffic light sequencer.

E5.11 Write a VHDL code for the following Boolean expressions.

E = A + BC + B'D

F = B'C + BC'D'

E5.12 Write an example VHDL code to demonstrate the use of for-loop statement.

E5.13 Write a VHDL code for logical not gate using variables.

E5.14 Write a VHDL program to generate sum of continuous integer input data.

APPENDIX 1: Pin Configuration of Some Standard IC's

Pin (left)	No.	7404	No.	Pin (right)
a_1	1		14	V_{cc} (+5V)
y_1	2		13	a_6
a_2	3		12	y_6
y_2	4		11	a_5
a_3	5		10	y_5
y_3	6		9	a_4
(0V) GND	7		8	y_4

Pin	Description
a_1 - a_6	Inputs
y_1 - y_6	Outputs
V_{cc}	Supply voltage (+5V)
GND	Ground (0V)

Fig. A1.1: *Inverter gate IC 7404.*

Pin (left)	No.	7408	No.	Pin (right)
a_1	1		14	V_{cc} (+5V)
b_1	2		13	b_4
y_1	3		12	a_4
a_2	4		11	y_4
b_2	5		10	b_3
y_2	6		9	a_3
(0V) GND	7		8	y_3

Pin	Description
a_1 - a_4	Inputs
b_1 - b_4	Inputs
y_1 - y_4	Outputs
V_{cc}	Supply voltage (+5V)
GND	Ground (0V)

Fig. A1.2: *Quad 2-input AND gate IC 7408.*

Pin (left)	No.	7411	No.	Pin (right)
a_1	1		14	V_{cc} (+5V)
b_1	2		13	c_1
a_2	3		12	y_1
b_2	4		11	c_3
c_2	5		10	b_3
y_2	6		9	a_3
(0V) GND	7		8	y_3

Pin	Description
a_1 - a_3	Inputs
b_1 - b_3	Inputs
c_1 - c_3	Inputs
y_1 - y_3	Outputs
V_{cc}	Supply voltage (+5V)
GND	Ground (0V)

Fig. A1.3: *Triple 3-input AND gate IC 7411.*

Note: *The suffix 1, 2, 3, 4, 5 and 6 in pin details refer to device number in the IC's.*

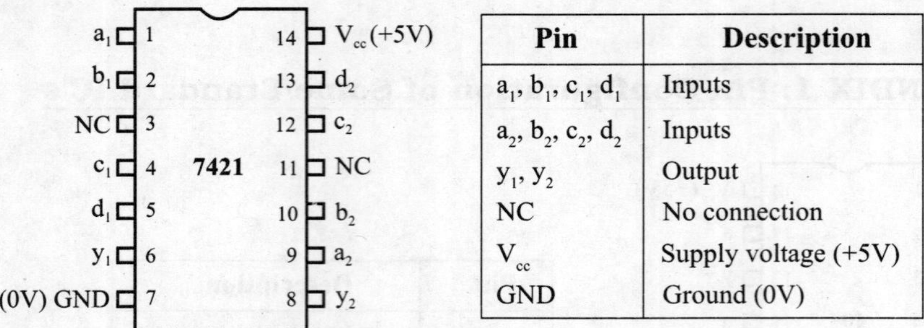

Pin	Description
a_1, b_1, c_1, d_1	Inputs
a_2, b_2, c_2, d_2	Inputs
y_1, y_2	Output
NC	No connection
V_{cc}	Supply voltage (+5V)
GND	Ground (0V)

***Fig. A1.4:** Dual 4-input AND gate IC 7421.*

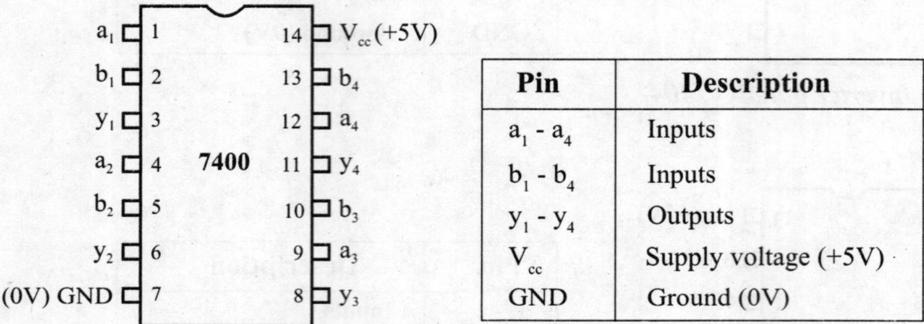

Pin	Description
a_1 - a_4	Inputs
b_1 - b_4	Inputs
y_1 - y_4	Outputs
V_{cc}	Supply voltage (+5V)
GND	Ground (0V)

***Fig. A1.5:** Quad 2-input NAND gate IC 7400.*

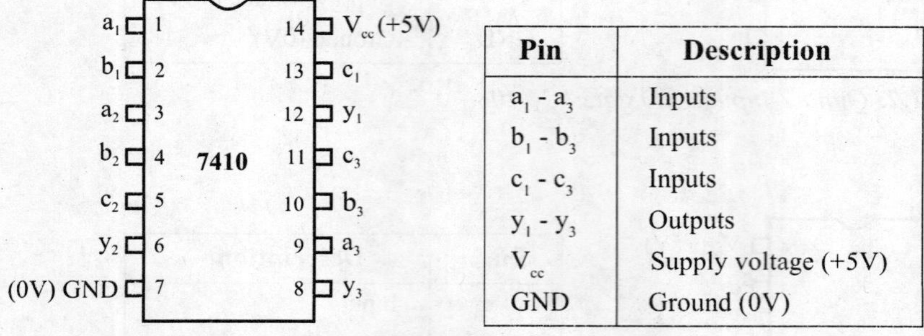

Pin	Description
a_1 - a_3	Inputs
b_1 - b_3	Inputs
c_1 - c_3	Inputs
y_1 - y_3	Outputs
V_{cc}	Supply voltage (+5V)
GND	Ground (0V)

***Fig. A1.6:** Triple 3-input NAND gate IC 7410.*

a1 1 — 14 Vcc (+5V)
b1 2 — 13 d2
NC 3 — 12 c2
c1 4 — 7422 — 11 NC
d1 5 — 10 b2
y1 6 — 9 a2
(0V) GND 7 — 8 y2

Pin	Description
a, b, c, d	Inputs
y_1, y_2	Outputs
NC	No connection
V_{cc}	Supply voltage (+5V)
GND	Ground (0V)

***Fig. A1.7:** Dual 4-input NAND gate IC 7422.*

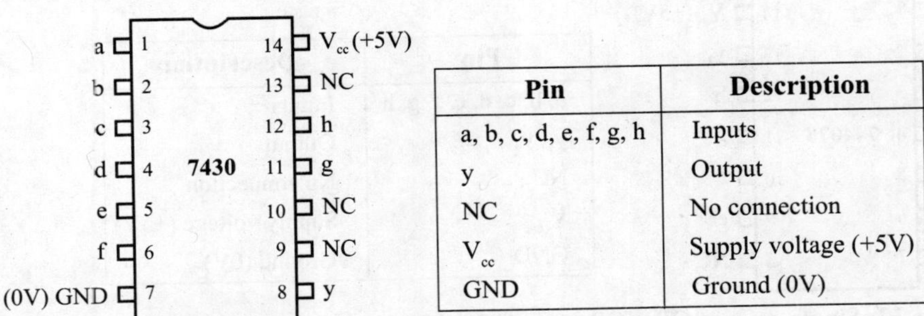

Pin	Description
a, b, c, d, e, f, g, h	Inputs
y	Output
NC	No connection
V_{cc}	Supply voltage (+5V)
GND	Ground (0V)

Fig. A1.8: *Single 8-input NAND gate IC 7430.*

a_1 1, b_1 2, y_1 3, a_2 4, b_2 5, y_2 6, (0V) GND 7 — **7432** — 14 V_{cc} (+5V), 13 b_4, 12 a_4, 11 y_4, 10 b_3, 9 a_3, 8 y_3

Pin	Description
a_1 - a_4	Inputs
b_1 - b_4	Inputs
y_1 - y_4	Outputs
V_{cc}	Supply voltage (+5V)
GND	Ground (0V)

Fig. A1.9: *Quad 2-input OR gate IC 7432.*

a_2 1, b_2 2, a_1 3, b_1 4, c_1 5, y_1 6, (0V) GND 7 — **744075** — 14 V_{cc} (+5V), 13 c_3, 12 b_3, 11 a_3, 10 y_3, 9 y_2, 8 c_2

Pin	Description
a_1 - a_3	Inputs
b_1 - b_3	Inputs
c_1 - c_3	Inputs
y_1 - y_3	Outputs
V_{cc}	Supply voltage (+5V)
GND	Ground (0V)

Fig. A1.10: *Triple 3-input OR gate IC 744075.*

y_1 1, a_1 2, b_1 3, c_1 4, d_1 5, NC 6, (0V) GND 7 — **744072** — 14 V_{cc} (+5V), 13 y_2, 12 a_2, 11 b_2, 10 c_2, 9 d_2, 8 NC

Pin	Description
a_1, b_1, c_1, d_1	Inputs
a_2, b_2, c_2, d_2	Inputs
y_1, y_2	Output
NC	No connection
V_{cc}	Supply voltage (+5V)
GND	Ground (0V)

Fig. A1.11: *Dual 4-input OR gate IC 744072.*

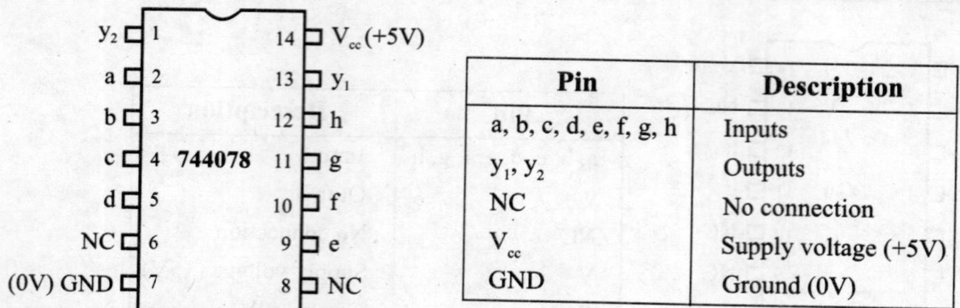

Pin	Description
a, b, c, d, e, f, g, h	Inputs
y_1, y_2	Outputs
NC	No connection
V_{cc}	Supply voltage (+5V)
GND	Ground (0V)

Fig. A1.12: *Single 8-input OR/NOR gate IC 744078.*

y1 1, a1 2, b1 3, y2 4, a2 5, b2 6, (0V) GND 7 — 7402 — 14 Vcc (+5V), 13 y4, 12 b4, 11 a4, 10 y3, 9 b3, 8 a3

Pin	Description
a_1 - a_4	Inputs
b_1 - b_4	Inputs
y_1 - y_4	Outputs
V_{cc}	Supply voltage (+5V)
GND	Ground (0V)

Fig. A1.13: *Quad 2-input NOR gate IC 7402.*

a1 1, b1 2, a2 3, b2 4, c2 5, y2 6, (0V) GND 7 — 7427 — 14 Vcc (+5V), 13 c1, 12 y1, 11 c3, 10 b3, 9 a3, 8 y3

Pin	Description
a_1 - a_3	Inputs
b_1 - b_3	Inputs
c_1 - c_3	Inputs
y_1 - y_3	Outputs
V_{cc}	Supply voltage (+5V)
GND	Ground (0V)

Fig. A1.14: *Triple 3-input NOR gate IC 7427.*

a1 1, b1 2, NC 3, c1 4, d1 5, y1 6, (0V) GND 7 — 7429 — 14 Vcc (+5V), 13 d2, 12 c2, 11 NC, 10 b2, 9 a2, 8 y2

Pin	Description
a_1, b_1, c_1, d_1	Inputs
a_2, b_2, c_2, d_2	Inputs
y_1, y_2	Output
NC	No connection
V_{cc}	Supply voltage (+5V)
GND	Ground (0V)

Fig. A1.15: *Dual 4-input NOR gate IC 7429.*

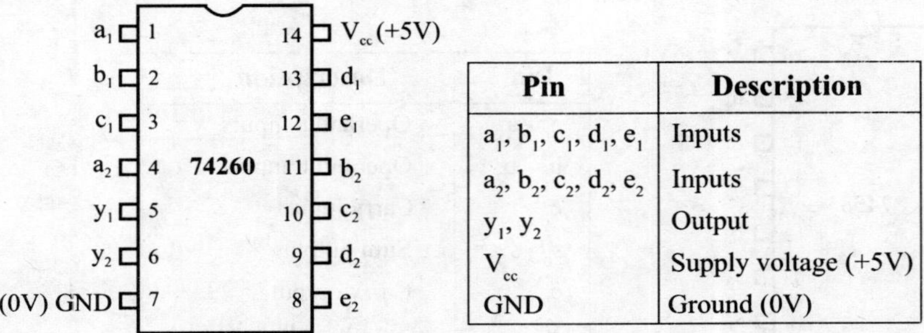

Pin	Description
a_1, b_1, c_1, d_1, e_1	Inputs
a_2, b_2, c_2, d_2, e_2	Inputs
y_1, y_2	Output
V_{cc}	Supply voltage (+5V)
GND	Ground (0V)

***Fig. A1.16:** Dual 5-input NOR gate IC 74260.*

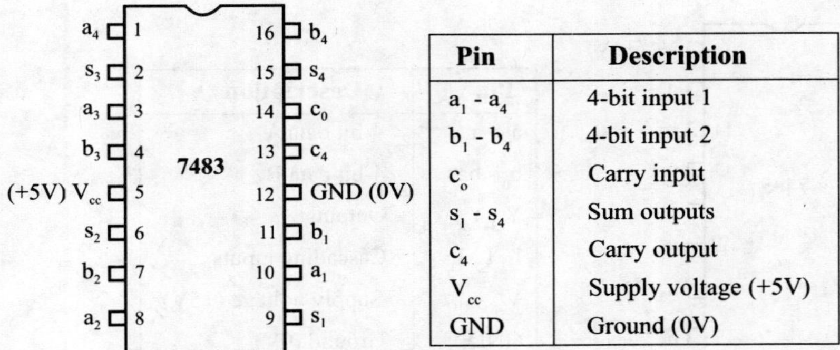

Pin	Description
a_1 - a_4	4-bit input 1
b_1 - b_4	4-bit input 2
c_o	Carry input
s_1 - s_4	Sum outputs
c_4	Carry output
V_{cc}	Supply voltage (+5V)
GND	Ground (0V)

***Fig. A1.17:** 4-bit binary full adder IC 7483.*

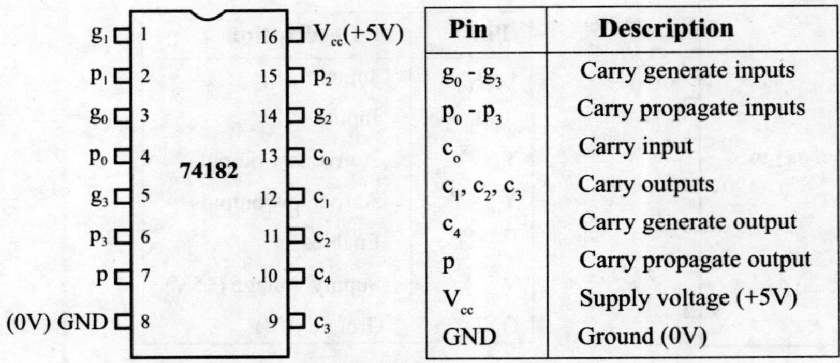

Pin	Description
g_0 - g_3	Carry generate inputs
p_0 - p_3	Carry propagate inputs
c_o	Carry input
c_1, c_2, c_3	Carry outputs
c_4	Carry generate output
p	Carry propagate output
V_{cc}	Supply voltage (+5V)
GND	Ground (0V)

***Fig. A1.18:** Look ahead carry generator IC 74182.*

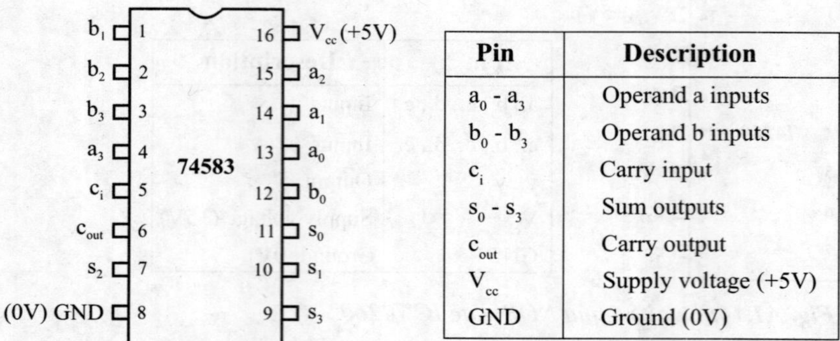

Pin	Description
a_0 - a_3	Operand a inputs
b_0 - b_3	Operand b inputs
c_i	Carry input
s_0 - s_3	Sum outputs
c_{out}	Carry output
V_{cc}	Supply voltage (+5V)
GND	Ground (0V)

***Fig. A1.19:** 4-bit BCD adder IC 74583.*

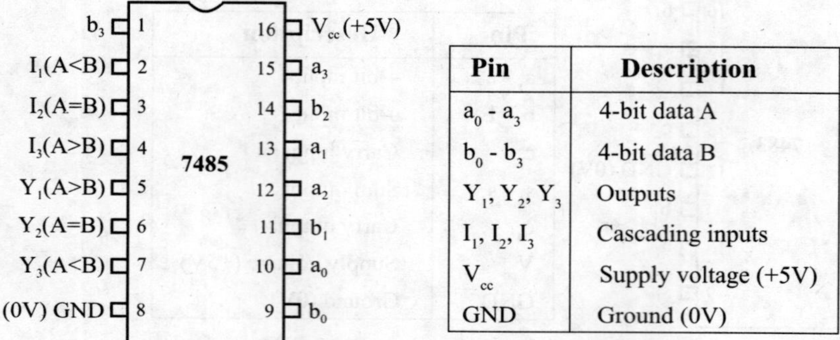

Pin	Description
a_0 - a_3	4-bit data A
b_0 - b_3	4-bit data B
Y_1, Y_2, Y_3	Outputs
I_1, I_2, I_3	Cascading inputs
V_{cc}	Supply voltage (+5V)
GND	Ground (0V)

***Fig. A1.20:** 4-bit magnitude comparator IC 7485.*

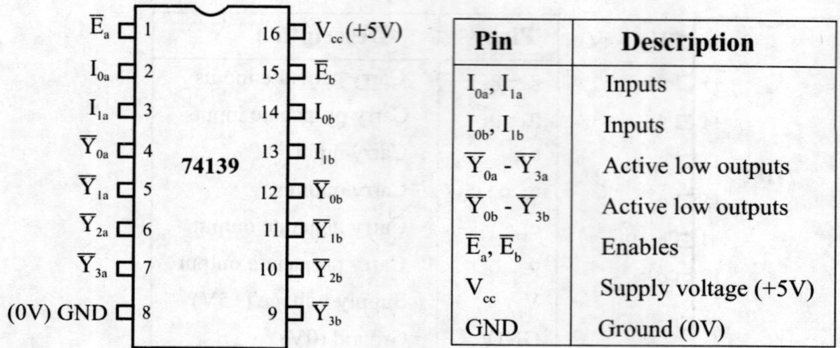

Pin	Description
I_{0a}, I_{1a}	Inputs
I_{0b}, I_{1b}	Inputs
$\overline{Y}_{0a}$ - $\overline{Y}_{3a}$	Active low outputs
$\overline{Y}_{0b}$ - $\overline{Y}_{3b}$	Active low outputs
$\overline{E}_a$, $\overline{E}_b$	Enables
V_{cc}	Supply voltage (+5V)
GND	Ground (0V)

***Fig. A1.21:** Dual 2-to-4 decoder IC 74139.*

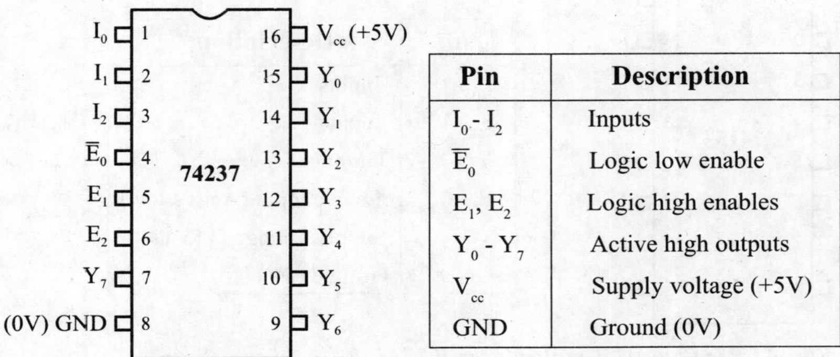

Pin	Description
I_0 - I_2	Inputs
$\overline{E}_0$	Logic low enable
E_1, E_2	Logic high enables
Y_0 - Y_7	Active high outputs
V_{cc}	Supply voltage (+5V)
GND	Ground (0V)

Fig. A1.22: *Logic high 3-to-8 decoder IC 74237.*

74138

Pin	No.	No.	Pin
I_0	1	16	V_{cc} (+5V)
I_1	2	15	$\overline{Y}_0$
I_2	3	14	$\overline{Y}_1$
$\overline{E}_0$	4	13	$\overline{Y}_2$
$\overline{E}_1$	5	12	$\overline{Y}_3$
E_2	6	11	$\overline{Y}_4$
$\overline{Y}_7$	7	10	$\overline{Y}_5$
(0V) GND	8	9	$\overline{Y}_6$

Pin	Description
I_0 - I_2	Inputs
$\overline{E}_0$, $\overline{E}_1$	Logic low enables
E_2	Logic high enable
$\overline{Y}_0$ - $\overline{Y}_7$	Active low outputs
V_{cc}	Supply voltage (+5V)
GND	Ground (0V)

Fig. A1.23: *Logic low 3-to-8 decoder IC 74138.*

74148

Pin	No.	No.	Pin
I_4	1	16	V_{cc} (+5V)
I_5	2	15	EO
I_6	3	14	v
I_7	4	13	I_3
EI	5	12	I_2
Y_2	6	11	I_1
Y_1	7	10	I_0
(0V) GND	8	9	Y_0

Pin	Description
I_0 - I_7	Inputs
Y_0 - Y_2	Outputs
v	Outputs
V_{cc}	Supply voltage (+5V)
GND	Ground (0V)
EI	Input Enable
EO	Output Enable

Fig. A1.24: *8-to-3 priority encoder IC 74148.*

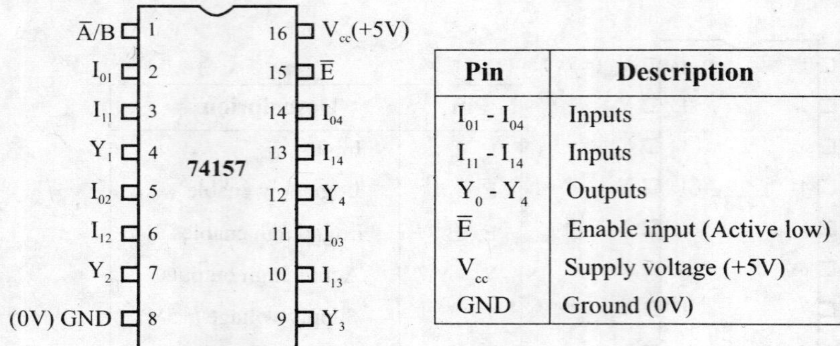

Pin	Description
I_{01} - I_{04}	Inputs
I_{11} - I_{14}	Inputs
Y_0 - Y_4	Outputs
$\overline{E}$	Enable input (Active low)
V_{cc}	Supply voltage (+5V)
GND	Ground (0V)

Fig. A1.25: *Quad 2:1 multiplexer IC 74157.*

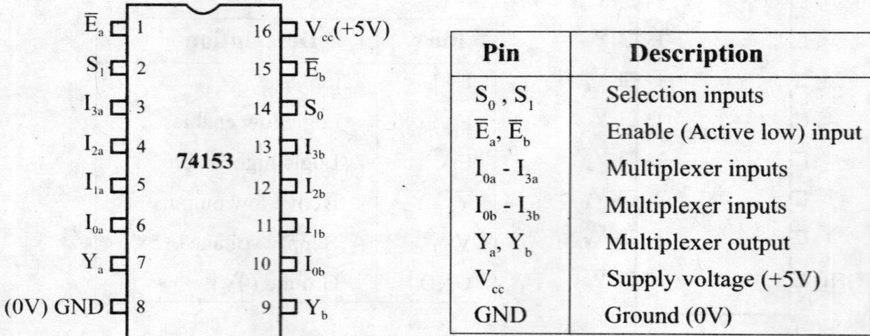

Pin	Description
S_0 , S_1	Selection inputs
$\overline{E}_a$, $\overline{E}_b$	Enable (Active low) input
I_{0a} - I_{3a}	Multiplexer inputs
I_{0b} - I_{3b}	Multiplexer inputs
Y_a, Y_b	Multiplexer output
V_{cc}	Supply voltage (+5V)
GND	Ground (0V)

Fig. A1.26: *Dual 4:1 multiplexer IC 74153.*

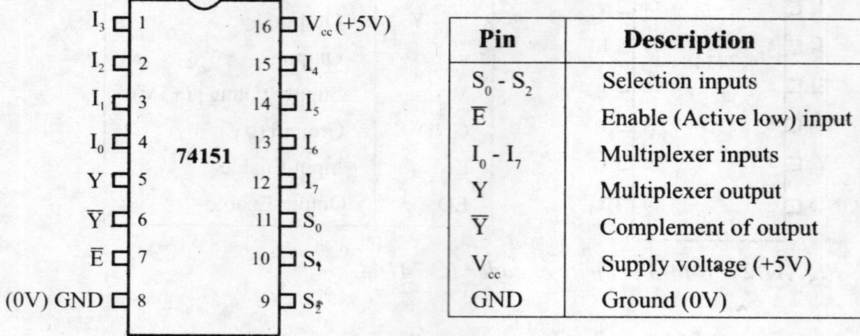

Pin	Description
S_0 - S_2	Selection inputs
$\overline{E}$	Enable (Active low) input
I_0 - I_7	Multiplexer inputs
Y	Multiplexer output
$\overline{Y}$	Complement of output
V_{cc}	Supply voltage (+5V)
GND	Ground (0V)

Fig. A1.27: *8:1 multiplexer IC 74151.*

Pin	Label	Pin	Label
1	I_7	24	V_{cc} (+5V)
2	I_6	23	I_8
3	I_5	22	I_9
4	I_4	21	I_{10}
5	I_3	20	I_{11}
6	I_2	19	I_{12}
7	I_1	18	I_{13}
8	I_0	17	I_{14}
9	$\overline{E}$	16	I_{15}
10	Y	15	S_0
11	S_3	14	S_1
12	(0V) GND	13	S_2

74150

Pin	Description
I_0 - I_{15}	Inputs
$\overline{E}$	Enable input (Active low)
S_0 - S_3	Selection inputs
Y	Output
V_{cc}	Supply voltage (+5V)
GND	Ground (0V)

Fig. A1.28: *16:1 multiplexer IC 74150.*

Pin	Label	Pin	Label
1	$\overline{E}_1$	16	V_{cc} (+5V)
2	S_0	15	$\overline{E}_2$
3	S_1	14	S_2
4	$\overline{Y}_{01}$	13	S_3
5	$\overline{Y}_{11}$	12	$\overline{Y}_{02}$
6	$\overline{Y}_{21}$	11	$\overline{Y}_{12}$
7	$\overline{Y}_{31}$	10	$\overline{Y}_{22}$
8	(0V) GND	9	$\overline{Y}_{32}$

74139

Pin	Description
$\overline{E}_1$, $\overline{E}_2$	Active low enables
S_0, S_1, S_2, S_3	Selection inputs
$\overline{Y}_{01}$ - $\overline{Y}_{31}$	Inverted outputs
$\overline{Y}_{02}$ - $\overline{Y}_{32}$	Inverted outputs
V_{cc}	Supply voltage (+5V)
GND	Ground (0V)

Fig. A1.29: *Dual 1:4 demultiplexer IC 74139.*

Pin	Label	Pin	Label
1	$\overline{R}_1$	16	V_{cc} (+5V)
2	$\overline{S}_{11}$	15	$\overline{S}_{21}$
3	$\overline{S}_{12}$	14	$\overline{R}_4$
4	Q_1	13	Q_4
5	$\overline{R}_2$	12	$\overline{S}_{22}$
6	$\overline{S}_{13}$	11	$\overline{S}_{23}$
7	Q_2	10	$\overline{R}_3$
8	(0V) GND	9	Q_3

74279

Pin	Description
$\overline{S}_{11}$ - $\overline{S}_{12}$	Active low set inputs
$\overline{S}_{21}$ - $\overline{S}_{23}$	Active low set inputs
$\overline{R}_1$ - $\overline{R}_4$	Active low reset inputs
Q_1 - Q_4	Outputs
V_{cc}	Supply voltage (+5V)
GND	Ground (0V)

Fig. A1.30: *Quad SR latch IC 74279.*

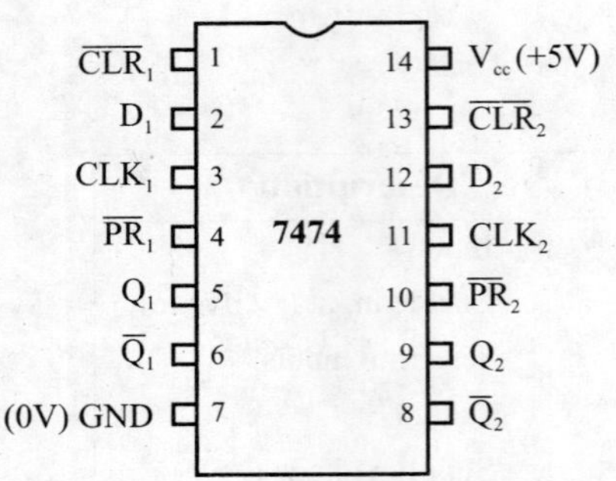

Pin	Description
D_1, D_2	Inputs
$\overline{PR}_1$, $\overline{PR}_2$	Preset inputs
$\overline{CLR}_1$, $\overline{CLR}_2$	Clear inputs
CLK_1, CLK_2	Clock inputs
Q_1, Q_2	Outputs
$\overline{Q}_1$, $\overline{Q}_2$	Complement of outputs
V_{cc}	Supply voltage (+5V)
GND	Ground (0V)

Fig. A1.31: *Dual D flip-flop IC 7474.*

$\overline{Q}_1$ 1
$\overline{Q}_2$ 2
Q_2 3
D_2 4
D_3 5
Q_3 6
$\overline{Q}_3$ 7
(0V) GND 8
74171
16 V_{cc} (+5V)
15 Q_1
14 D_1
13 $\overline{CLR}$
12 CLK
11 D_4
10 Q_4
9 $\overline{Q}_4$

Pin	Description
D_1 - D_4	Inputs
$\overline{CLR}$	Clear input
CLK	Clock input
Q_1 - Q_4	Outputs
$\overline{Q}_1$ - $\overline{Q}_4$	Complement of outputs
V_{cc}	Supply voltage (+5V)
GND	Ground (0V)

Fig. A1.32: *Quad D flip-flop IC 74171.*

CLK_1 1
K_1 2
J_1 3
$\overline{PR}_1$ 4
Q_1 5
$\overline{Q}_1$ 6
$\overline{Q}_2$ 7
(0V) GND 8
74112
16 V_{cc} (+5V)
15 $\overline{CLR}_1$
14 $\overline{CLR}_2$
13 CLK_2
12 K_2
11 J_2
10 $\overline{PR}_2$
9 Q_2

Pin	Description
J_1, K_1, J_2, K_2	Inputs
$\overline{PR}_1$, $\overline{PR}_2$	Preset inputs
$\overline{CLR}_1$, $\overline{CLR}_2$	Clear inputs
CLK_1, CLK_2	Clock inputs
Q_1, Q_2	Outputs
$\overline{Q}_1$, $\overline{Q}_2$	Complement of outputs
V_{cc}	Supply voltage (+5V)
GND	Ground (0V)

Fig. A1.33: *Dual JK flip-flop IC 74112.*

NC 1
$\overline{CLR}$ 2
S_1 3
S_2 4
S_3 5
$\overline{Q}$ 6
(0V) GND 7
7471
14 V_{cc} (+5V)
13 $\overline{PRE}$
12 CLK
11 R_3
10 R_2
9 R_1
8 Q

Pin	Description
S_1 - S_3	Active high set inputs
R_1 - R_3	Active high reset inputs
$\overline{CLR}$	Clear input
CLK	Clock input
Q, $\overline{Q}$	Outputs
V_{cc}	Supply voltage (+5V)
GND	Ground (0V)

Fig. A1.34: *RS master slave flip-flop IC 7471.*

APPENDIX 2: Summary of Various Binary Codes

Table A2.1: BCD Codes

Decimal Digit	8421 BCD	2421 BCD	5421 BCD	8 4 −2 −1 BCD	Excess-3 BCD
0	0000	0000	0000	0000	0011
1	0001	0001	0001	0111	0100
2	0010	0010	0010	0110	0101
3	0011	0011	0011	0101	0110
4	0100	0100	0100	0100	0111
5	0101	1011	1000	1011	1000
6	0110	1100	1001	1010	1001
7	0111	1101	1010	1001	1010
8	1000	1110	1011	1000	1011
9	1001	1111	1100	1111	1100

Table A2.2: Gray Code

Decimal Number	1-bit Gray Code	2-bit Gray Code	3-bit Gray Code	4-bit Gray Code
0	0	00	000	0000
1	1	01	001	0001
2		11	011	0011
3		10	010	0010
4			110	0110
5			111	0111
6			101	0101
7			100	0100
8				1100
9				1101
10				1111
11				1110
12				1010
13				1011
14				1001
15				1000

Table A2.3: ASCII Codes

ASCII in Hex	Symbol	ASCII in Hex	Symbol	ASCII in Hex	Symbol	ASCII in Hex	Symbol
00	NUL	20	SP	40	@	60	`
01	SOH	21	!	41	A	61	a
02	STX	22	“	42	B	62	b
03	ETX	23	#	43	C	63	c
04	EOT	24	$	44	D	64	d
05	ENQ	25	%	45	E	65	e
06	ACK	26	&	46	F	66	f
07	BEL	27	‘	47	G	67	g
08	BS	28	(	48	H	68	h
09	HT	29	)	49	I	69	i
0A	LF	2A	*	4A	J	6A	j
0B	VT	2B	+	4B	K	6B	k
0C	FF	2C	,	4C	L	6C	l
0D	CR	2D	–	4D	M	6D	m
0E	SO	2E	.	4E	N	6E	n
0F	SI	2F	/	4F	O	6F	o
10	DLE	30	0	50	P	70	p
11	DC1	31	1	51	Q	71	q
12	DC2	32	2	52	R	72	r
13	DC3	33	3	53	S	73	s
14	DC4	34	4	54	T	74	t
15	NAK	35	5	55	U	75	u
16	SYN	36	6	56	V	76	v
17	ETB	37	7	57	W	77	w
18	CAN	38	8	58	X	78	x
19	EM	39	9	59	Y	79	y
1A	SUB	3A	:	5A	Z	7A	z
1B	ESC	3B	;	5B	[	7B	{
1C	FS	3C	<	5C	\	7C	|
1D	GS	3D	=	5D	]	7D	}
1E	RS	3E	>	5E	^	7E	~
1F	US	3F	?	5F	_	7F	DEL

APPENDIX 3: Unsigned and Signed Binary Number System

Table A3.1: Decimal Number and its Equivalent Unsigned Binary Number

Decimal Number	Unsigned Binary Number	8-bit Unsigned Binary Number
0	0	0000 0000
1	1	0000 0001
2	10	0000 0010
3	11	0000 0011
4	100	0000 0100
5	101	0000 0101
6	110	0000 0110
7	111	0000 0111
8	1000	0000 1000
9	1001	0000 1001
10	1010	0000 1010
11	1011	0000 1011
12	1100	0000 1100
13	1101	0000 1101
14	1110	0000 1110
15	1111	0000 1111

Table A3.2: Decimal and Sign-Magnitude form of Binary Number

Decimal Number	4-bit Sign-magnitude Binary Number	Standard 8-bit Sign-magnitude Binary Number
+0	0000	0000 0000
+1	0001	0000 0001
+2	0010	0000 0010
+3	0011	0000 0011
+4	0100	0000 0100
+5	0101	0000 0101
+6	0110	0000 0110
+7	0111	0000 0111
–0	1000	1000 0000
–1	1001	1000 0001
–2	1010	1000 0010
–3	1011	1000 0011
–4	1100	1000 0100
–5	1101	1000 0101
–6	1110	1000 0110
–7	1111	1000 0111

Table A3.3: Decimal and One's Complement form of Binary Number

Decimal Number	4-bit One's Complement Binary Number	Sign-extended One's Complement 8-bit Binary Number
+0	0000	0000 0000
+1	0001	0000 0001
+2	0010	0000 0010
+3	0011	0000 0011
+4	0100	0000 0100
+5	0101	0000 0101
+6	0110	0000 0110
+7	0111	0000 0111
–7	1000	1111 1000
–6	1001	1111 1001
–5	1010	1111 1010
–4	1011	1111 1011
–3	1100	1111 1100
–2	1101	1111 1101
–1	1110	1111 1110
–0	1111	1111 1111

Table A3.4: Decimal and Two's Complement form of Binary Number

Decimal Number	4-bit Two's Complement Binary Number	Sign-extended Two's Complement 8-bit Binary Number
+0	0000	0000 0000
+1	0001	0000 0001
+2	0010	0000 0010
+3	0011	0000 0011
+4	0100	0000 0100
+5	0101	0000 0101
+6	0110	0000 0110
+7	0111	0000 0111
–8	1000	1111 1000
–7	1001	1111 1001
–6	1010	1111 1010
–5	1011	1111 1011
–4	1100	1111 1100
–3	1101	1111 1101
–2	1110	1111 1110
–1	1111	1111 1111

APPENDIX 4: Boolean Theorems

Theorem 1: $x + x = x$

$x \cdot x = x$

Theorem 2: $x + 1 = 1$

$x \cdot 0 = 0$

Theorem 3: (Involution) $(x')' = x$

Theorem 4: (Associative) $x + (y + z) = (x + y) + z$

$x \cdot (y \cdot z) = (x \cdot y) \cdot z$

Theorem 5: (DeMorgan's Law or DeMorgan's Theorem) $(x + y)' = x' \cdot y'$

$(x \cdot y)' = x' + y'$

Theorem 6: (Absorption) $x + (x \cdot y) = x$

$x \cdot (x + y) = x$

(y is absorbed)

APPENDIX 5: Two Variable Boolean Function

Table A5.1: Two-Variable Boolean Functions

Functions	x	y	F	Functions	x	y	F
Null operation	0	0	0	NOR operation	0	0	1
$F = F_0 = 0$	0	1	0	$F = F_8 = (x + y)'$	0	1	0
	1	0	0		1	0	0
	1	1	0		1	1	0
AND operation	0	0	0	Exclusive NOR	0	0	1
$F = F_1 = x \cdot y$	0	1	0	(or equivalence)	0	1	0
	1	0	0	$F = F_9 = (x \oplus y)'$	1	0	0
	1	1	1	$= x \cdot y + x' \cdot y'$	1	1	1
Inhibition	0	0	0	Complement of y	0	0	1
$F = F_2 = x \cdot y'$	0	1	0	$F = F_{10} = y'$	0	1	0
(x true, but y false)	1	0	1		1	0	1
	1	1	0		1	1	0
Transfer	0	0	0	Implication	0	0	1
$F = F_3 = x$	0	1	0	$F = F_{11} = x + y'$	0	1	0
(Transfer x)	1	0	1	(If $y = 1, F = x$	1	0	1
	1	1	1	If $y = 0, F = y'$)	1	1	1
Inhibition	0	0	0	Complement of x	0	0	1
$F = F_4 = x' \cdot y$	0	1	1	$F = F_{12} = x'$	0	1	1
(y true, but x false)	1	0	0		1	0	0
	1	1	0		1	1	0
Transfer	0	0	0	Implication	0	0	1
$F = F_5 = y$	0	1	1	$F = F_{13} = x' + y$	0	1	1
(Transfer y)	1	0	0	(If $x = 1, F = y$	1	0	0
	1	1	1	If $x = 0, F = x'$)	1	1	1
XOR operation	0	0	0	NAND operation	0	0	1
$F = F_6 = x \oplus y$	0	1	1	$F = F_{14} = (x \cdot y)'$	0	1	1
$= x \cdot y' + x' \cdot y$	1	0	1		1	0	1
	1	1	0		1	1	0
OR operation	0	0	0	Identity	0	0	1
$F = F_7 = x + y$	0	1	1	$F = F_{15} = 1$	0	1	1
	1	0	1		1	0	1
	1	1	1		1	1	1

APPENDIX 6: Summary of Flip-Flop Tables

SR Flip-Flop

Table A6.1: Truth Table

Inputs		Outputs	
S	R	Q	Q′
0	0	No change	
0	1	0	1
1	0	1	0
1	1	Indeterminate	

Table A6.2: Characteristics Table

Inputs		Present State	Next State
S	R	Q	Q(t+1)
0	0	0	0
0	0	1	1
0	1	0	0
0	1	1	0
1	0	0	1
1	0	1	1
1	1	0	×
1	1	1	×

Table A6.3: Excitation Table

Present State	Next State	Flip-Flop Inputs	
Q(t)	Q(t+1)	S	R
0	0	0	×
0	1	1	0
1	0	0	1
1	1	×	0
0	×	1	1
1	×	1	1

JK Flip-Flop

Table A6.4: Truth Table

Inputs		Outputs		Comment
J	K	Q	Q′	
0	0	No change		-
0	1	0	1	Reset state
1	0	1	0	Set state
1	1	Toggle		Toggle previous output

Table A6.5: Characteristics Table

Inputs		Present State	Next State
J	K	Q	Q(t+1)
0	0	0	0
0	0	1	1
0	1	0	0
0	1	1	0
1	0	0	1
1	0	1	1
1	1	0	1
1	1	1	0

Table A6.6: Excitation Table

Present State	Next State	Flip-Flop Inputs	
Q(t)	Q(t+1)	J	K
0	0	0	×
0	1	1	×
1	0	×	1
1	1	×	0

T Flip-Flop

Table A6.7: Truth Table

T	CLK	Q	Q′
0	↑	No change	
1	↑	Toggles previous output	

Table A6.8: Characteristics Table

Input T	Present State Q	Next State Q(t+1)
0	0	0
0	1	1
1	0	1
1	1	0

Table A6.9: Excitation Table

Present State Q(t)	Next State Q(t + 1)	Flip-Flop Input T
0	0	0
0	1	1
1	0	1
1	1	0

D Flip-Flop

Table A6.10: Truth Table

Input	Outputs	
D	Q	Q′
0	0	1
1	1	0

Table A6.11: Characteristics Table

Input D	Present State Q	Next State Q(t+1)
0	0	0
0	1	0
1	0	1
1	1	1

Table A6.12: Excitation Table

Present State Q(t)	Next State Q(t+1)	Flip-Flop Input D
0	0	0
0	1	1
1	0	0
1	1	1

APPENDIX 7: List of Xillinx FPGAs

Table A7.1: List of Xillinx FPGAs

Generation	Family Name	Year of Launch	Manufacturing Process	Internal Operating Voltage
XC2000	XC2000	1985	2000 nm	5 V
	XC2000L	1993		3.3 V
XC3000	XC3000	1988	1200 nm	5 V
	XC3000A	1993	800 nm	
	XC3000L	1993		3.3 V
	XC3100	1992	800 nm	5 V
	XC3100A	1994	500 nm	
	XC3000L	1995		3.3 V
XC4000	XC4000	1991		5 V
	XC4000A	1991		5 V
	XC4000D	1994		5 V
	XC4000H	1993		5 V
	XC4000E	1995	500 nm	5 V
	XC4000L	1995		3.3 V
	XC4000EX	1996	500 nm	5 V
	XC4000XL	1997	350 nm	3.3 V
	XC4000XLA	1998	350 nm, 250 nm	3.3 V
	XC4000XV	1998	250 nm	2.5 V
	Spartan	1998	500 nm, 350 nm	5 V
	Spartan XL	1998	350 nm, 250 nm	3.3 V
XC5200	XC5200	1994	600 nm	5 V
	XC5200L		500 nm	3.3 V
XC6200		1995	650 nm	5 V
XC8100		1995		5 V or 3.3 V
Virtex	Virtex	1998	220 nm	2.5 V
	Spartan-II	2000		
	Virtex E	1999		
	Virtex EM	2000	180 nm	1.8 V
	Spartan-IIE	2001		
Virtex-II	Virtex-II	2001	150 nm	1.5 V
	Virtex-II Pro	2002	130 nm	1.2 V
	Virtex-II Pro X	2003		
Spartan-3	Spartan-3	2003	90 nm	1.2 V
	Spartan-3E	2004		
	Spartan-3A	2006		
	Spartan-3 AN	2007		
	Spartan-3A DSP			

Continued Table A7.1...

Generation	Family Name	Year of Launch	Manufacturing Process	Internal Operating Voltage
Virtex-4	Virtex-4 LX	2004	90 nm	1.2 V
	Virtex-4 SX			
	Virtex-4 FX			
Virtex-5	Virtex-5 LX	2006	65 nm	1.0 V
	Virtex-5 LXT			
	Virtex-5 SXT			
	Virtex-5 FXT			
	Virtex-5 TXT			
Virtex-6	Virtex-6 LX	2009	40 nm	0.9 V or 1.0 V
	Virtex-6 LXT			
	Virtex-6 SXT			
	Virtex-6 HXT			
	Virtex-6 CXT			
Spartan-6	Spartan-6 LX	2009	45 nm	1.0 V or 1.2 V
	Spartan-6 LXT			
7 Series	Spartan 7	2017	28 nm	0.9 V, 0.95 V or 1.0 V
	Artix-7	2010		
	Kintex-7	2010		
	Virtex-7	2010		
	Virtex-7 3D	2011		
	Virtex-7 HT	2012		
	Zynq-7000	2011		
		2013		
Ultrascale	Kintex Ultrascale	2013	20 nm	0.9 V, 0.85 V or 1.0 V
	Kintex Ultrascale	2013		
	Virtex Ultrascale	2014		
Ultrascale+	Artix Ultrascale+	2021	16 nm	0.72 V, 0.85 V or 0.9 V
	Kintex Ultrascale	2015		
	Virtex Ultrascale+	2016		
	Virtex Ultrascale+ 58G			
	Virtex Ultrascale+ HBM			
	Zynq Ultrascale+ MPSoC 2015			
	Zynq Ultrascale+ RFSoC 2017			
	Alveo	2018		
Versal	Versal Prime	2019	7 nm	0.7 V, 0.8 V or 0.88 V
	Versal AI Core	2019		
	Versal Premium			

APPENDIX 8: Spartan-6 FPGA Feature

Table A8.1: Spartan-6 FPGA Feature Summary by Device

Device	Logic Cells	Configurable Logic Blocks (CLBs)			DSP48A1 Slices	Block RAM Blocks		Total I/O Banks	Max User I/O
		Slices	Flip-Flops	Max Distributed RAM (Kb)		18Kb	Max (Kb)		
XC6SLX4	3,840	600	4,800	75	8	12	216	4	132
XC6SLX9	9,152	1,430	11,440	90	16	32	576	4	200
XC6SLX16	14,579	2,278	18,224	136	32	32	576	4	232
XC6SLX25	24,051	3,758	30,064	229	38	52	936	4	266
XC6SLX45	43,661	6,822	50,576	401	58	116	2,088	4	358
XC6SLX75	74,637	11,662	93,296	692	132	172	3,096	6	408
XC6SLX100	101,261	15,822	126,576	976	180	268	4,824	6	480
XC6SLX150	147,443	23,038	184,304	1,355	180	268	4,824	6	576
XC6SLX25T	24,051	3,758	30,064	229	38	52	936	4	250
XC6SLX45T	43,661	6,822	54,576	401	58	116	2,088	4	296
XC6SLX75T	74,637	1,662	93,296	692	132	172	3,096	6	348
XC6SLX100T	101,261	15,822	126,576	976	180	268	4,824	6	498
XC6SLX150T	147,443	23,038	184,304	1,355	180	268	4,824	6	540

APPENDIX 9: Spartan-6 FPGA Logic Resources

Table A9.1: Spartan-6 FPGA Logic Resources

Device	Logic Cells	Total Slices	SLICEMs	SLICELs	SLICEXs	Number of 6-Input LUTs	Maximum Distributed RAM (Kb)	Shift Register (Kb)	Number of Flip-Flops
XC6SLX4	3,840	600	300	0	300	2,400	75	38	4,800
XC6SLX9	9,152	1,430	360	355	715	5,720	90	45	11,440
XC6SLX16	14,579	2,278	554	595	1,139	9,112	136	68	18,224
XC6SLX25	24,051	3,758	916	963	1,879	15,032	229	115	30,064
XC6SLX45	43,661	6,822	1,602	1,809	3,411	27,288	401	200	54,576
XC6SLX75	74,637	11,662	2,768	3,063	5,831	46,648	692	346	93,296
XC6SLX100	101,261	15,822	3,904	4,007	7,911	63,288	976	488	126,576
XC6SLX150	147,443	23,038	5,420	6,099	11,519	92,152	1,355	678	184,304
XC6SLX25T	24,051	3,758	916	963	1,879	15,032	229	115	30,064
XC6SLX45T	43,661	6,822	1,602	1,809	3,411	27,288	401	200	54,576
XC6SLX75T	74,637	1,662	2,768	3,063	5,831	46,648	692	346	93,296
XC6SLX100T	101,261	15,822	3,904	4,007	7,911	63,288	976	488	126,576
XC6SLX150T	147,443	23,038	5,420	6,099	11,519	92,152	1,355	678	184,304

APPENDIX 10: Reserved Words in VHDL

Table A10.1: Reserved Words in VHDL

abs all	access	after architecture	alias array
			buffer
	component	configuration	constant
[illegible]ct	downto	end	entity
else exit			
file	for	function	
generate	generic	guarded	
if	in	inout	is
label	library	linkage	loop
map	mod		
nand not	new null	next	nor
of others	on out	open	or
package	port	procedure	process
range report	record return	register	rem
select	severity	signal	subtype
then	to	transport	type
units	until	use	
variable			
wait	when	while	with
xor			

APPENDIX 11: Data Types and Operators in VHDL

Table A11.1: Std_logic Data Types of the IEEE std_logic_1164 Package

Logic Symbols	Description
'U'	Uninitialized
'X'	Strong drive, [illegible]
'0'	Strong drive, logic 0
'1'	Strong drive, logic 1
'Z'	High impedence
'W'	Weak drive, unknown logic value
'L'	Weak drive, logic 0
'H'	Weak drive, logic 1
'–'	Don't-care

Table A11.2: Predefined VHDL Data Types

VHDL Data Type	Value
bit	'0' or '1'
boolean	FALSE or TRUE
integer	$-(2^{31}-1) \leq$ INTEGER VALUE $\leq (2^{31}-1)$
positive	$1 \leq$ INTEGER VALUE $\leq (2^{31}-1)$
natural	$0 \leq$ INTEGER VALUE $\leq (2^{31}-1)$
real	– 1.0e38 ≤ FLOATING POINT VALUE ≤ 1.0E38
character	Alphabetical characters (a .... z, A .... Z), digits (0, ... 9), special characters (e.g.,%) each enclosed in single quotes
time	integer with units fs, ps, ns, ms, sec, min or hr

Table A11.3: VHDL Operators

Operator Group	Symbol	Operation/Operator
Binary logical	**and**	Logical and operation
	or	Logical or operation
	nand	Logical nand operation
	nor	Logical nor operation
	xor	Logical xor operation
	xnor	Logical xnor operation
Relational	=	Equality operator
	/=	Inequality operator
	<	Less than
	>	Greater than
	<=	Less than or equal to
	>=	Greater than or equal to
Shift operators	**sll**	Shift-left logical
	srl	Shift-right logical
	sla	Shift-left arithmetic
	sra	Shift-right arithmetic
	rol	Rotate left
	ror	Rotate right
Addition operators	**+**	Addition
	–	Subtraction
Concatenation operator	&	Concatenation
Unary sign operator	**+**	Positive sign
	–	Negative sign
Multiplication operators	*	Multiplication
	/	Division
	mod	Modulus
	rem	Remainder
Miscellaneous operators	**not**	Complement
	abs	Absolute
	**	Exponentiation
Assignment operators	<=	Assign data of one object to another object
	:=	Assign a value to object
	=>	Associate one object to another object

Table A11.4: Precedence of the VHDL Operators

Precedence	Operators
Highest	** abs not
	* / mod rem
	+ − (identity of negation)
	& + − (addition and subtraction)
	sll srl sla sra rol ror
	= /= < <= > >=
Lowest	and or nand nor xor xnor

B.E DEGREE EXAMINATION APRIL/MAY 2019

Third semester

Electrical and Electronics Engineering

EE 8351 - DIGITAL LOGIC CIRCUITS

(Common to Electronics and Instrumentation Engineering/ Instrumentation and Control Engineering)

(Regulation 2017)

Time: Three hours | Maximum: 100 Marks

Answer All questions

PART A - (10 × 2 = 20 Marks)

1. Convert $(101.01)_2$ to decimal number.

   Chapter 1, SA - Q1.3 ***[Page No - 1.180]***

2. Give each one example for error detecting code and error correcting code.

   Chapter 1, Example 1.10 ***[Page No - 1.27]***

3. Determine the exact number of half adders and full adders required for performing the addition of two binary numbers of 5-bits length each.

   Chapter 2, SA - Q2.6 ***[Page No - 2.100]***

4. Find the result of $A + A'D + AC'$.

   Chapter 1, SA - Q1.25 ***[Page No - 1.188]***

5. Write down the characteristic table of JK flip-flop.

   Chapter 3, Sec 3.4.2, Table 3.23 ***[Page No - 3.26]***

6. What is FSM? List its two basic types.

   Chapter 3, SA - Q3.10 ***[Page No - 3.177]***

7. Define metastable state.

   Chapter 4, SA - Q4.8 ***[Page No - 4.132]***

8. Draw the structure of PAL.

   Chapter 4, Sec 4.8.4 ***[Page No - 4.124]***

9. State the purpose of test bench.

   Chapter 5, SA - Q5.16 ***[Page No - 5.125]***

10. Write a VHDL program for an EX-NOR gate using behavioural coding.

    Chapter 5, SA - Q5.10 ***[Page No - 5.124]***

PART - B (5 × 13 = 65 Marks)

11. a) i) Design a 3-input NAND gate circuit using TTL logic. (7)

Chapter 1, Sec 1.12.2 *[Page No - 1.143]*

ii) Explain in detail the generation of Hamming code for 4-bit data. (6)

Chapter 1, Sec 1.2.2 *[Page No - 1.22]*

(OR)

b) i) Design a 2 input NOR gate using CMOS logic. (7)

Chapter 1, Sec 1.14.2 *[Page No - 1.165]*

ii) Explain the operation of RTL inverter circuit with relevant diagrams. (6)

Chapter 1, Sec 1.10.5 *[Page No - 1.136]*

12. a) i) Design a 3 × 8 decoder using 2 × 4 decoders. Draw the truth table. (7)

Chapter 2, Example 2.16, *[Page No - 2.87]*

ii) Design Full adder circuit using logic gates. (6)

Chapter 2, Sec 2.6.2 *[Page No - 2.37]*

(OR)

b) i) Simplify and implement the logic function $F(A, B, C) = \sum m(0, 1, 4, 5, 7)$ using logic gates. (7)

Chapter 1, Example 1.30 *[Page No - 1.95]*

ii) Design a 4 × 2 priority encoder using logic gates. (6)

Chapter 2, Sec 2.8.2 *[Page No - 2.59]*

13. a) i) Design a 2-bit synchronous sequential down counter. (7)

Chapter 3, Sec 3.9.2 *[Page No - 3.113]*

ii) Explain the operation of a 3-bit universal shift register. (6)

Chapter 3, Sec 3.11.5 *[Page No - 3.170]*

(OR)

b) i) Explain Moore and Mealy models with the help of block diagrams. (7)

Chapter 3, Sec 3.6.1 *[Page No - 3.41]*

ii) Draw the state table for the following state diagram. (6)

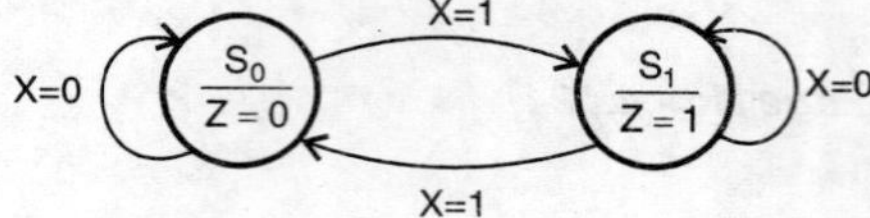

Chapter 3, Example 3.3 *[Page No - 3.46]*

14. a) i) Design a Modulo-6 asynchronous binary up counter. (7)

Chapter 3, Sec 3.8.6 ***[Page No - 3.108]***

ii) Implement the functions $F_1(X, Y, Z) = \sum m(1, 2, 4, 5)$, $F_2(X, Y, Z) = \sum m(0, 1, 3, 4)$ and $F_3(X, Y, Z) = \sum m(2, 3, 6, 7)$ using a single PROM grid. (6)

Chapter 4, Example 4.28 ***[Page No - 4.113]***

(OR)

b) i) Diffentiate PAL and PLA implementations with the help of the same example $F_2(a, b, c) = \sum m(0, 1, 3, 4, 6, 7)$. (7)

Chapter 4, Example 4.32 ***[Page No - 4.126]***

ii) Explain the structure of CPLD with the help of a block diagram. (6)

Chapter 4, Sec 4.9 ***[Page No - 4.133]***

15. a) i) Draw the VLSI design flow chart used for IC design and fabrication. (7)

Chapter 5, Sec 5.11, Fig 5.2 ***[Page No - 5.33]***

ii) Write down a VHDL code for 8:1 multiplexer. (6)

Chapter 5, VHDL Program 5.11 ***[Page No - 5.42]***

(OR)

b) i) Illustrate the two approaches used in VHDL coding with full adder design as your examples. (7)

Chapter 5, VHDL Program 5.14 ***[Page No - 5.45]***

ii) What are components in VHDL? Show step-by-step how a NOR gate component can be created added in the library. (6)

Chapter 5, Sec 5.9 ***[Page No - 5.31]***

PART - C (1 × 15 = 15 Marks)

16. a) Design a synchronous sequential logic circuit that goes through the sequence 0, 2, 4, 6, 8, 10, 12, 14 repeatedly. Use D flip-flops for your design. (15)

Chapter 3, Example 3.20 ***[Page No - 3.141]***

(OR)

b) Simplify the following function and implement it using NAND gates only:
$F(w, x, y, z) = \sum m(1, 3, 5, 7, 9, 11, 13, 15)$, with don't-care states
$d(w, x, y, z) = \sum m(0, 2, 4, 6, 8)$. (15)

Chapter 1, Example 1.39 ***[Page No - 1.110]***

B.E DEGREE EXAMINATION NOVEMBER/DECEMBER 2018

Third semester

Electrical and Electronics Engineering

EE 8351 - DIGITAL LOGIC CIRCUITS

(Common to Electronics and Instrumentation Engineering/ Instrumentation and Control Engineering)

(Regulation 2017)

Time: Three hours Maximum: 100 Marks

Answer All questions

PART A - (10 × 2 = 20 Marks)

1. Draw the DTL based NAND gate.

   Chapter 1, Sec 1.11.1 ***[Page No - 1.136]***

2. Perform subtraction on the following unsigned binary numbers using the 2's complement of the subtrahend **(a)** 11011 – 11001 **(b)** 110100 – 10101

   Chapter 1, SA - Q1.18 ***[Page No - 1.181]***

3. Mention the dependency of output in combinational circuits.

   Chapter 2, SA - Q2.1 ***[Page No - 2.99]***

4. Draw the NAND gate circuit using NOT, AND and OR gates.

   Chapter 1, SA - Q1.23 ***[Page No - 1.187]***

5. Write the role of master clock generator in synchronous circuits.

   Chapter 3, SA - Q3.15 ***[Page No - 3.177]***

6. Comment about a presettable counter and ripple counter.

   Chapter 3, SA - Q3.23 ***[Page No - 3.178]***

7. Draw the block diagram of asynchronous sequential circuit.

   Chapter 4, SA - Q4.5 ***[Page No - 4.131]***

8. Outline about PLA.

   Chapter 4, Sec 4.8.3 ***[Page No - 4.120]***

9. Draw the basic structure of MOS transistor.

   Chapter 1, Sec 1.9.7 ***[Page No - 1.129]***

10. List the languages that are combined together to get VHDL language.

    Chapter 5, SA - Q5.4 ***[Page No - 5.122]***

PART - B (5 × 13 = 65 Marks)

11. a) Assume a 3-input AND gate with output F and a 3-input OR gate with G output. Show the signals of the outputs F and G as functions of the three inputs ABC. Use all 8 possible combinations of inputs ABC. (13)

Chapter 2, Example 2.4 ***[Page No - 2.5]***

(OR)

b) Show that a positive logic NAND gate is a negative logic NOR gate and vice versa. (13)

Chapter 1, Sec 1.4.6 ***[Page No - 1.65]***

12. a) Given the following Boolean function F = A'C + A'B + AB'C + BC (13)

i) Express it in sum of minterms.

ii) Find the minimal sum of products expression.

Chapter 1, Q1.37 ***[Page No - 1.191]***

(OR)

b) Draw the logic diagram of 2-to-4 decoder using NOR gates only. Include an enable input. (13)

Chapter 2, Sec 2.9.2 ***[Page No - 2.67]***

13. a) Explain the operation, state diagram and characteristics of a T flip-flop and master slave JK flip-flop. (13)

Chapter 3, Sec 3.3.3 ***[Page No - 3.19]***

(OR)

b) Describe the design procedure with neat diagram about 4-bit bidirectional shift register with parallel load. (13)

Chapter 3, Sec 3.10 ***[Page No - 3.165]***

14. a) Discuss the operation of SR Latch with NOR and NAND gates analysis. (13)

Chapter 3, Sec 3.2.2 ***[Page No - 3.3]***

Chapter 3, Sec 3.2.3 ***[Page No - 3.5]***

(OR)

b) Illustrate about hazards in sequential circuits and the steps to avoid hazards in it. (13)

Chapter 4, Sec 4.7.2 ***[Page No - 4.90]***

15. a) Explain the structure and working principles of TTL based Totem-pole output configuration. (13)

Chapter 1, Sec 1.12.1 ***[Page No - 1.142]***

(OR)

b) Write a VHDL code to realize a half adder using behavioral modeling and structural modeling. (13)

Chapter 5, VHDL Program 5.2 ***[Page No - 5.24]***

PART - C (1 × 15 = 15 Marks)

16. a) Design a sequential circuit with two D flip-flops A and B and one input x. When x = 0, the state of the circuit remains the same. When x = 1, the circuit goes through the state transitions from 00 to 01 to 11 to 10 back to 00, and repeats. (15)

Chapter 3, Example 3.16 *[Page No - 3.92]*

(OR)

b) Design a combinational circuit with three inputs, x,y and z, and the three outputs, A, B, and C, when the binary input is 0, 1, 2, or 3, the binary output is one greater than the input. When the binary input is 1, 5, 6 or 7 the binary output is one less than the input. (15)

Chapter 2, Example 2.2 *[Page No - 2.4]*

INDEX

F

G

H

I

K

L

M

N